Handbook of Reliability Engineering

Springer
London
Berlin
Heidelberg
New York
Hong Kong
Milan
Paris
Tokyo

http://www.springer.de/phys/

Hoang Pham (Editor)

Handbook of Reliability Engineering

Springer

Hoang Pham, PhD
Rutgers University
Piscataway
New Jersey, USA

British Library Cataloguing in Publication Data
Handbook of reliability engineering
1. Reliability (Engineering)
I. Pham, Hoang
620′.00452
ISBN 1852334533

Library of Congress Cataloging-in-Publication Data
Handbook of reliability engineering/Hoang Pham (ed.).
p.cm.
ISBN 1-85233-453-3 (alk. paper)
1. Reliability (Engineering)–Handbooks, manuals, etc. I. Pham, Hoang.
TA169.H358 2003
620′.00452–dc21 2002030652

ISBN 1-85233-453-3 Springer-Verlag London Berlin Heidelberg
a member of BertelsmannSpringer Science+Business Media GmbH
http://www.springer.co.uk

Typesetting: Sunrise Setting Ltd, Torquay, Devon, UK
Printed and bound in the United States of America
69/3830-543210 Printed on acid-free paper SPIN 10795762

To

Michelle, Hoang Jr., and David

Preface

In today's technological world nearly everyone depends upon the continued functioning of a wide array of complex machinery and equipment for our everyday safety, security, mobility and economic welfare. We expect our electric appliances, lights, hospital monitoring control, next-generation aircraft, nuclear power plants, data exchange systems, and aerospace applications, to function whenever we need them. When they fail, the results can be catastrophic, injury or even loss of life.

As our society grows in complexity, so do the critical reliability challenges and problems that must be solved. The area of reliability engineering currently received a tremendous attention from numerous researchers and practitioners as well.

This **Handbook of Reliability Engineering**, altogether 35 chapters, aims to provide a comprehensive state-of-the-art reference volume that covers both fundamental and theoretical work in the areas of reliability including optimization, multi-state system, life testing, burn-in, software reliability, system redundancy, component reliability, system reliability, combinatorial optimization, network reliability, consecutive-systems, stochastic dependence and aging, change-point modeling, characteristics of life distributions, warranty, maintenance, calibration modeling, step-stress life testing, human reliability, risk assessment, dependability and safety, fault tolerant systems, system performability, and engineering management.

The Handbook consists of five parts. Part I of the Handbook contains five papers, deals with different aspects of *System Reliability and Optimization*.

Chapter 1 by Zuo, Huang and Kuo studies new theoretical concepts and methods for performance evaluation of multi-state k-out-of-n systems. Chapter 2 by Pham describes in details the characteristics of system reliabilities with multiple failure modes. Chapter 3 by Chang and Hwang presents several generalizations of the reliability of consecutive-k-systems by exchanging the role of working and failed components in the consecutive-k-systems. Chapter 4 by Levitin and Lisnianski discusses various reliability optimization problems of multi-state systems with two failure modes using combination of universal generating function technique and genetic algorithm. Chapter 5 by Sung, Cho and Song discusses a variety of different solution and heuristic approaches, such as integer programming, dynamic programming, greedy-type heuristics, and simulated annealing, to solve various combinatorial reliability optimization problems of complex system structures subject to multiple resource and choice constraints.

Part II of the Handbook contains five papers, focuses on the *Statistical Reliability Theory*. Chapter 6 by Finkelstein presents stochastic models for the observed failure rate of systems with periods of operation and repair that form an alternating process. Chapter 7 by Lai and Xie studies a general concept of stochastic dependence including positive dependence and dependence orderings. Chapter 8 by Zhao discusses some statistical reliability change-point models which can be used to model the reliability of both software and hardware systems. Chapter 9 by Lai and Xie discusses the basic concepts for the stochastic univariate and multivariate aging classes including bathtub shape failure rate. Chapter 10 by Park studies characteristics of a new class of NBU-t_0 life distribution and its preservation properties.

Part III of the Handbook contains six papers, focuses on *Software Reliability*. Chapter 11 by Dalal presents software reliability models to quantify the reliability of the software products for early stages as well as test and operational phases. Some further research and directions useful to practitioners and researchers are also discussed. Chapter 12 by Ledoux provides an overview of some aspects of software reliability modeling including black-box modeling, white-box modeling, and Bayesian-calibration modeling. Chapter 13 by Tokuno and Yamada presents software availability models and its availability measures such as interval software reliability and the conditional mean available time.

Chapter 14 by Dohi, Goševa-Popstojanova, Vaidyanathan, Trivedi and Osaki presents the analytical modeling and measurement based approach for evaluating the effectiveness of preventive maintenance in operational software systems and determining the optimal time to perform software rejuvenation. Chapter 15 by Kimura and Yamada discusses nonhomogeneous death process, hidden-Markov process, and continuous state-space models for evaluating and predicting the reliability of software products during the testing phase. Chapter 16 by Pham presents basic concepts and recent studies nonhomogeneous Poisson process software reliability and cost models considering random field environments. Some challenge issues in software reliability are also included.

Part IV contains nine chapters, focuses on *Maintenance Theory and Testing*. Chapter 17 by Murthy and Jack presents overviews of product warranty and maintenance, warranty policies and contracts. Further research topics that link maintenance and warranty are also discussed. Chapter 18 by Pulcini studies stochastic point processes maintenance models with imperfect preventive maintenance, sequence of imperfect and minimal repairs and imperfect repairs interspersed with imperfect preventive maintenance. Chapter 19 Dohi, Kaio and Osaki presents the basic preventive maintenance policies and their extensions in terms of both continuous and discrete-time modeling. Chapter 20 by Nakagawa presents the basic maintenance policies such as age replacement, block replacement, imperfect preventive maintenance, and periodic replacement with minimal repair for multi-component systems.

Chapter 21 by Wang and Pham studies various imperfect maintenance models that minimize the system maintenance cost rate. Chapter 22 by Elsayed presents basic concepts of accelerated life testing and a detailed test plan that can be designed before conducting an accelerated life test. Chapter 23 by Owen and Padgett focuses on the Birnbaum-Saunders distribution and its application in reliability and life testing. Chapter 24 by Tang discusses the two related issues for a step-stress accelerated life testing (SSALT) such as how to design a multiple-steps accelerated life test and how to analyze the data obtained from a SSALT. Chapter 25 by Xiong deals with the statistical models and estimations based on the data from a SSALT to estimate the unknown parameters in the stress-response relationship and the reliability function at the design stress.

Part V contains nine chapters, primarily focuses on *Practices and Emerging Applications.* Chapter 26 by Phillips presents proportional and non-proportional hazard reliability models and its applications in reliability analysis using non-parametric approach. Chapter 27 by Yip, Wang and Chao discusses the capture-recapture methods and the Horvits Thompson estimator to estimate the number of faults in a computer system. Chapter 28 by Billinton and Allan provides overviews and deals with the reliability evaluation methods of electric power systems. Chapter 29 by Dhillon discusses various aspects of human and medical device reliability.

Chapter 30 by Bari presents the basic of probabilistic risk assessment methods that developed and matured within the commercial nuclear power reactor industry. Chapter 31 by Akersten and Klefsjö studies methodologies and tools in dependability and safety management. Chapter 32 by Albeanu and Vladicescu deals with the approaches in software quality assurance and engineering management. Chapter 33 by Teng and Pham presents a generalized software reliability growth model for N-version programming systems which considers the error-introduction rate, the error-removal efficiency, and multi-version coincident failures, based on the non-homogeneous Poisson process. Chapter 34 by Carrasco presents Markovian models for evaluating the dependability and performability of fault tolerant systems.

Chapter 35 by Lee discusses a new random-request availability measure and presents closed-form mathematical expressions for random-request availability which incorporate the random task arrivals, the system state, and the operational requirements of the system.

All the chapters are written by over 45 leading reliability experts in academia and industry. I am deeply indebted and wish to thank all of them for their contributions and cooperation. Thanks are also due to the Springer staff, especially Peter Mitchell, Roger Dobbing and Oliver Jackson, for their editorial work. I hope that practitioners will find this Handbook useful when looking for solutions to practical problems; researchers can use it for quick access to the background, recent research and trends, and most important references regarding certain topics, if not all, in the reliability.

Hoang Pham
Piscataway, New Jersey

Contents

Contributors

Professor Per Anders Akersten
Division of Quality Technology and Statistics
Lulea University of Technology
Sweden

Professor G. Albeanu
The Technical University of Denmark
Denmark

Professor Ronald N. Allan
Department of Electrical Engineering
and Electronics
UMIST, Manchester
United Kingdom

Dr. Robert A. Bari
Energy, Environment and National Security
Brookhaven National Laboratory
USA

Professor Roy Billinton
Department of Electrical Engineering
University of Saskatchewan
Canada

Professor Juan A. Carrasco
Dep. d'Enginyeria Electronica, UPC
Spain

Professor Jen-Chun Chang
Department of Information Management
Ming Hsin Institute of Technology
Taiwan, ROC

Professor Anne Chao
Institute of Statistics
National Tsing Hua University
Taiwan

Dr. Yong Kwon Cho
Technology and Industry Department
Samsung Economic Research Institute
Republic of Korea

Dr. Siddhartha R. Dalal
Information Analysis and
Services Research Department
Applied Research
Telcordia Technologies
USA

Professor B. S. Dhillon
Department of Mechanical Engineering
University of Ottawa
Canada

Professor Tadashi Dohi
Department of Information Engineering
Hiroshima University
Japan

Professor Elsayed A. Elsayed
Department of Industrial Engineering
Rutgers University
USA

Professor Maxim S. Finkelstein
Department of Mathematical Statistics
University of the Orange Free State
Republic of South Africa

Professor Katerina Goševa-Popstojanova
Lane Dept. of Computer Science and
Electrical Engineering
West Virginia University
USA

Dr. Jinsheng Huang
Stantec Consulting
Canada

Professor Frank K. Hwang
Department of Applied Mathematics
National Chao Tung University
Taiwan, ROC

Dr. Nat Jack
School of Computing
University of Abertay Dundee
Scotland

Professor Naoto Kaio
Department of Economic Informatics
Faculty of Economic Sciences
Hiroshima Shudo University
Japan

Professor Mitsuhiro Kimura
Department of Industrial and Systems Engineering
Faculty of Engineering
Hosei University
Japan

Professor Bengt Klefsjö
Division of Quality Technology and Statistics
Lulea University of Technology
Sweden

Professor Way Kuo
Department of Industrial Engineering
Texas A&M University
USA

Professor C. D. Lai
Institute of Information Sciences and Technology
Massey University
New Zealand

Professor James Ledoux
Centre de mathématiques
Institut National des Sciences Appliquees
France

Dr. Gregory Levitin
Reliability Department
The Israel Electric Corporation
Israel

Dr. Anatoly Lisnianski
Reliability Department
The Israel Electric Corporation
Israel

Professor D. N. P. Murthy
Engineering and Operations Management Program
Department of Mechanical Engineering
The University of Queensland
Australia

Professor Toshio Nakagawa
Department of Industrial Engineering
Aichi Institute of Technology
Japan

Professor Shunji Osaki
Department of Information and Telecommunication Engineering
Faculty of Mathematical Sciences and Information Engineering
Nanzan University
Japan

Dr. W. Jason Owen
Mathematics and Computer Science Department
University of Richmond
USA

Professor William J. Padgett
Department of Statistics
University of South Carolina
USA

Professor Dong Ho Park
Department of Statistics
Hallym University
Korea

Professor Hoang Pham
Department of Industrial Engineering
Rutgers University
USA

Dr. Michael J. Phillips
Department of Mathematics and Computer Science
University of Leicester
United Kingdom

Professor Gianpaolo Pulcini
Statistics and Reliability Department
Istituto Motori CNR
Italy

Mr. Sang Hwa Song
Department of Industrial Engineering
Korea Advanced Institute of Science and Technology
Republic of Korea

Professor Chang Sup Sung
Department of Industrial Engineering
Korea Advanced Institute of Science and Technology
Republic of Korea

Professor Loon-Ching Tang
Department of Industrial and Systems Engineering
National University of Singapore
Singapore

Dr. Xiaolin Teng
Department of Industrial Engineering
Rutgers University
USA

Professor Koichi Tokuno
Department of Social Systems Engineering
Tottori University
Japan

Professor Kishor S. Trivedi
Department of Electrical and Computer Engineering
Duke University
USA

Dr. Kalyanaraman Vaidyanathan
Department of Electrical and Computer Engineering
Duke University
USA

Professor Florin Popentiu Vladicescu
The Technical University of Denmark
Denmark

Dr. Yan Wang
School of Mathematics
City West Campus, UniSA
Australia

Professor Min Xie
Department of Industrial and Systems Engineering
National University of Singapore
Singapore

Professor Chengjie Xiong
Division of Biostatistics
Washington University in St. Louis
USA

Professor Shigeru Yamada
Department of Social Systems Engineering
Tottori University
Japan

Professor Paul S. F. Yip
Department of Statistics and Actuarial Science
University of Hong Kong
Hong Kong

Professor Ming Zhao
Department of Technology
University of Gävle
Sweden

Professor Ming J. Zuo
Department of Mechanical Engineering
University of Alberta
Canada

PART I

System Reliability and Optimization

Multi-state k-out-of-n Systems

Ming J. Zuo, Jinsheng Huang and Way Kuo

Chapter 1

1.1 Introduction

In traditional reliability theory, both the system and its components are allowed to take only two possible states: either working or failed. In a multi-state system, both the system and its components are allowed to experience more than two possible states, *e.g.* completely working, partially working or partially failed, and completely failed. A multi-state system reliability model provides more flexibility for modeling of equipment conditions. The terms *binary* and *multi-state* will be used to indicate these two fundamental assumptions in system reliability models.

1.2 Relevant Concepts in Binary Reliability Theory

The following notation will be used:

- x_i: state of component i, $x_i = 1$ if component i is working and zero otherwise;
- $\mathbf{x}$: an n-dimensional vector representing the states of all components, $\mathbf{x} = (x_1, x_2, \ldots, x_n)$;
- $\phi(\mathbf{x})$: state of the system, which is also called the structure function of the system;
- $(j_i, \mathbf{x})$: a vector $\mathbf{x}$ whose ith argument is set equal to j, where $j = 0, 1$ and $i = 1, 2, \ldots, n$.

A component is irrelevant if its state does not affect the state of the system at all. The *structure function* of the system indicates that the state of the system is completely determined by the states of all components. A system of components is said to be *coherent* if: (1) its structure function is non-decreasing in each argument; (2) there are no irrelevant components in the system. These two requirements of a coherent system can be stated as: (1) the improvement of any component does not degrade the system performance; (2) every component in the system makes some non-zero contribution to the system's performance. A mathematical definition of a coherent system is given below.

Definition 1. A binary system with n components is a coherent system if its structure function $\phi(\mathbf{x})$ satisfies:

1. $\phi(\mathbf{x})$ is non-decreasing in each argument x_i, $i = 1, 2, \ldots, n$;
2. there exists a vector $\mathbf{x}$ such that $0 = \phi(0_i, \mathbf{x}) < \phi(1_i, \mathbf{x}) = 1$;
3. $\phi(\mathbf{0}) = 0$ and $\phi(\mathbf{1}) = 1$.

Condition 1 in Definition 1 requires that $\phi(\mathbf{x})$ be a monotonically increasing function of each argument. Condition 2 specifies the so-called *relevancy condition*, which requires that every component has to be relevant to system performance. Condition 3 states that the system fails when all components are failed and system works when all components are working.

A minimal path set is a minimal set of components whose functioning ensures the functioning of the system. A minimal cut set is a minimal set of components whose failure ensures the failure of the system. The following mathematical definitions of minimal path and cut sets are given by Barlow and Proschan [1].

Definition 2. Define $C_0(\mathbf{x}) = i \mid x_i = 0$ and $C_1(\mathbf{x}) = i \mid x_i = 1$. A *path vector* is a vector $\mathbf{x}$ such that $\phi(\mathbf{x}) = 1$. The corresponding *path set* is $C_1(\mathbf{x})$. A *minimal path vector* is a path vector $\mathbf{x}$ such that $\phi(\mathbf{y}) = 0$ for any $\mathbf{y} < \mathbf{x}$. The corresponding *minimal path set* is $C_1(\mathbf{x})$. A *cut vector* is a vector $\mathbf{x}$ such that $\phi(\mathbf{x}) = 0$. The corresponding *cut set* is $C_0(\mathbf{x})$. A *minimal cut vector* is a cut vector $\mathbf{x}$ such that $\phi(\mathbf{y}) = 1$ for any $\mathbf{y} > \mathbf{x}$. The corresponding *minimal cut set* is $C_0(\mathbf{x})$.

The reliability of a system is equal to the probability that at least one of the minimal path sets works. The unreliability of the system is equal to the probability that at least one minimal cut set is failed. For a minimal path set to work, each component in the set must work. For a minimal cut set to fail, all components in the set must fail.

1.3 Binary k-out-of-n Models

A system of n components that *works* (or is "good") if and only if at least k of the n components *work* (or are "good") is called a k-out-of-n:G system. A system of n components that *fails* if and only if at least k of the n components fail is called a k-out-of-n:F system. The term k-out-of-n system is often used to indicate either a G system, an F system, or both. Since the value of n is usually larger than the value of k, redundancy is built into a k-out-of-n system. Both the parallel and the series systems are special cases of the k-out-of-n system. A series system is equivalent to a 1-out-of-n:F system and to an n-out-of-n:G system. A parallel system is equivalent to an n-out-of-n:F system and to a 1-out-of-n:G system.

The k-out-of-n system structure is a very popular type of redundancy in *fault-tolerant* systems. It finds wide applications in both industrial and military systems. Fault-tolerant systems include the multi-display system in a cockpit, the multi-engine system in an airplane, and the multi-pump system in a hydraulic control system. For example, in a V-8 engine of an automobile it may be possible to drive the car if only four cylinders are firing. However, if less than four cylinders fire, then the automobile cannot be driven. Thus, the functioning of the engine may be represented by a 4-out-of-8:G system. It is tolerant of failures of up to four cylinders for minimal functioning of the engine. In a data-processing system with five video displays, a minimum of three displays operable may be sufficient for full data display. In this case, the display subsystem behaves as a 3-out-of-5:G system. In a communications system with three transmitters the average message load may be such that at least two transmitters must be operational at all times or critical messages may be lost. Thus, the transmission subsystem functions as a 2-out-of-3:G system. *Systems with spares* may also be represented by the k-out-of-n system model. In the case of an automobile with four tires, for example, the vehicle is usually equipped with one additional spare tire. Thus, the vehicle can be driven as long as at least four out of five tires are in good condition.

In the following, we will also adopt the following notation:

- n: number of components in the system;
- k: minimum number of components that must work for the k-out-of-n:G system to work;
- p_i: reliability of component i, $i = 1, 2, \ldots, n$, $p_i = \Pr(x_i = 1)$;
- p: reliability of each component when all components are i.i.d.;
- q_i: unreliability of component i, $q_i = 1 - p_i$, $i = 1, 2, \ldots, n$;
- q: unreliability of each component when all components are i.i.d., $q = 1 - p$;
- $R_e(k, n)$: probability that exactly k out of n components are working;
- $R(k, n)$: reliability of a k-out-of-n:G system or probability that at least k out of the n components are working, where $0 \le k \le n$ and both k and n are integers;
- $Q(k, n)$: unreliability of a k-out-of-n:G system or probability that less than k out of the n components are working, where $0 \le k \le n$ and both k and n are integers, $Q(k, n) = 1 - R(k, n)$.

1.3.1 The k-out-of-n:G System with Independently and Identically Distributed Components

The reliability of a k-out-of-n:G system with independently and identically distributed (i.i.d.) components is equal to the probability that the number of working components is greater than or equal to k:

$$R(k, n) = \sum_{i=k}^{n} \binom{n}{i} p^i q^{n-i} \tag{1.1}$$

Other equations that can be used for system reliability evaluation include

$$R(k, n) = p^k \sum_{i=k}^{n} \binom{i-1}{k-1} q^{i-k} \tag{1.2}$$

and

$$R(k, n) = \binom{n-1}{k-1} p^k q^{n-k} + R(k, n+1) \tag{1.3}$$

with the boundary condition:

$$R(k, n) = 0 \quad \text{for } n = k \tag{1.4}$$

1.3.2 Reliability Evaluation Using Minimal Path or Cut Sets

In a k-out-of-n:G system, there are $\binom{n}{k}$ minimal path sets and $\binom{n}{n-k+1}$ minimal cut sets. Each minimal path set contains exactly k different components and each minimal cut set contains exactly $n - k + 1$ components. Thus, all minimal path sets and minimal cut sets are known. To find the reliability of a k-out-of-n:G system, one may choose to evaluate the probability that at least one of the minimal path sets contains all working components or one minus the probability that at least one minimal cut set contains all failed components.

The inclusion–exclusion (IE) method can be used for reliability evaluation of a k-out-of-n:G system. However, it has the disadvantage of involving many canceling terms. Heidtmann [2] and McGrady [3] provide improved versions of the IE method for reliability evaluation of the k-out-of-n:G system. In their improved algorithms, the canceling terms are eliminated. However, these algorithms are still enumerative in nature. For example, the formula provided by Heidtmann [2] is as follows:

$$R(k, n) = \sum_{i=k}^{n} (-1)^{i-k} \binom{i-1}{k-1} \sum_{j_1 < j_2 < \cdots < j_i} \prod_{l=1}^{i} p_{j_l} \tag{1.5}$$

In this equation, for each fixed i value, the inner summation term gives us the probability that i components are working properly regardless of whether the other $n - i$ components are working or not. The total number of terms to be summed together in the inner summation series is equal to $\binom{n}{i}$.

The sum-of-disjoint-product (SDP) method can also be used for reliability evaluation of the k-out-of-n:G systems. Let S_i indicate the ith minimal path of a k-out-of-n:G system ($i = 1, 2, \ldots, m$, where $m = \binom{n}{i}$). The SDP method uses the following equation for system reliability evaluation:

$$R(k, n) = \Pr(S_1) + \Pr(\overline{S_1} S_2) + \Pr(\overline{S_1}\,\overline{S_2} S_3) + \cdots + \Pr(\overline{S_1}\,\overline{S_2} \ldots \overline{S_{m-1}} S_m) \tag{1.6}$$

Like the improved IE method given in Equation 1.5, the SDP method is pretty easy to use for the k-out-of-n:G systems. However, we will see later that there are much more efficient methods than the IE (and its improved version) and the SDP method for the k-out-of-n:G systems.

1.3.3 Recursive Algorithms

Under the assumption that components are s-independent, several efficient recursive algorithms have been developed for system reliability evaluation of the k-out-of-n:G systems. Barlow and Heidtmann [4] and Rushdi [5] independently provide an algorithm with complexity $O(k(n-k+1))$ for system reliability evaluation of the k-out-of-n:G systems. The approaches used to derive the algorithm are the generating function approach (Barlow and Heidtmann) and the symmetric switching function approach (Rushdi). The following equation summarizes the algorithm:

$$R(i,j) = p_j R(i-1, j-1) + q_j R(i, j-1) \quad i \geq 0,\ j \geq 0 \tag{1.7}$$

with boundary conditions

$$R(0,j) = 1 \quad j \geq 0 \tag{1.8}$$

$$R(j+1, j) = 0 \quad j \geq 0 \tag{1.9}$$

Chao and Lin [6] were the first to use the Markov chain technique in analyzing reliability system structures; in their case, it was for the consecutive-k-out-of-n:F system. Subsequently, Chao and Fu [7, 8] standardized this approach of using the Markov chain in the analysis of various system structures and provided a general framework and general results for this technique. The system structures that can be represented by a Markov chain were termed *linearly connected systems* by Fu and Lou [9]. Koutras [10] provides a systematic summary of this technique and calls these systems *Markov chain imbeddable* (MIS) systems. Koutras [10] applied this technique to the k-out-of-n:F system and provided recursive equations for system reliability evaluation of the k-out-of-n:F systems. In the following, we provide the equations for the k-out-of-n:G systems.

$$a_0(t) = q_t a_0(t-1) \quad t \geq 1 \tag{1.10}$$

$$a_j(t) p_t a_{j-1}(t-1) + q_t a_j(t-1) \quad 1 \leq j < k,\ j \leq t \leq n \tag{1.11}$$

$$a_k(t) p_t a_{k-1}(t-1) + a_k(t-1) \quad k \leq t \leq n \tag{1.12}$$

where $a_j(t)$ is the probability that there are exactly j working components in a system with t components for $0 \leq j < k$ and $a_k(t)$ is the probability that there are at least k working components in the t component subsystem. The following boundary conditions are immediate:

$$a_0 = 1 \tag{1.13}$$

$$a_j(0) = 0 \quad j > 0 \tag{1.14}$$

$$a_j(t) = 0 \quad \text{for } t < j \tag{1.15}$$

The reliability of the system, $R(k,n) = a_k(n)$. The computational complexity of the recursive Equations 1.10–1.12 for system reliability of a k-out-of-n:G system is also $O(k(n-k+1))$.

Belfore [11] used the generating function approach as used by Barlow and Heidtmann [4] and applied a fast Fourier transform (FFT) in computation of the products of the generating functions. An algorithm for reliability evaluation of k-out-of-n:G systems results from such a combination that has a computational complexity of $O(n[\log_2(n)]^2)$. This algorithm is not easy to use for manual calculations or when the system size is small. For details of this algorithm, readers are referred to Belfore [11].

1.3.4 Equivalence Between a k-out-of-n:G System and an $(n-k+1)$-out-of-n:F system

Based on the definitions of these two types of systems, a k-out-of-n:G system is equivalent to an $(n-k+1)$-out-of-n:F system. Similarly, a k-out-of-n:F system is equivalent to an $(n-k+1)$-out-of-n:G system. This means that provided the systems have the same set of component reliabilities, the reliability of a k-out-of-n:G system is equal to the reliability of an $(n-k+1)$-out-of-n:F system and the reliability of a k-out-of-n:F system is equal to the reliability of an $(n-k+1)$-out-of-n:G system. As a result, we can use the algorithms that have been covered in the previous section for the k-out-of-n:G systems in reliability evaluation of the k-out-of-n:F systems. The procedure is simple and is outlined below.

Procedure 1. Procedure for using algorithms for the G systems in reliability evaluation of the F systems utilizing the equivalence relationship:

1. given: $k, n, p_1, p_2, \ldots, p_n$ for a k-out-of-n:F system;
2. calculate $k_1 = n - k + 1$;
3. use $k_1, n, p_1, p_2, \ldots, p_n$ to calculate the reliability of a k_1-out-of-n:G system. This reliability is also the reliability of the original k-out-of-n:F system.

1.3.5 The Dual Relationship Between the k-out-of-n G and F Systems

Definition 3. (Barlow and Proschan [1]) Given a structure ϕ, its dual structure ϕ^{D} is given by

$$\phi^{\mathrm{D}}(\mathbf{x}) = (\mathbf{1} - \mathbf{x}) \tag{1.16}$$

where $\mathbf{1} - \mathbf{x} = (1 - x_1, 1 - x_2, \ldots, 1 - x_n)$.

With a simple variable substitution of $\mathbf{y} = \mathbf{1} - \mathbf{x}$ and then writing $\mathbf{y}$ as $\mathbf{x}$, we have the following equation:

$$\phi^{\mathrm{D}}(\mathbf{1} - \mathbf{x}) = 1 - \phi(\mathbf{x}) \tag{1.17}$$

We can interpret Equation 1.17 as follows. Given a primal system with component state vector $\mathbf{x}$ and the system state represented by $\phi(\mathbf{x})$, the state of the dual system is equal to $1 - \phi(\mathbf{x})$ if the component state vector for the dual system can be expressed by $\mathbf{1} - \mathbf{x}$. In the binary system context, each component and the system may only be in two possible states: either working or failed. We say that two components with different states have *opposite states*. For example, if component 1 is in state 1 and component 2 is in state 0, components 1 and 2 have opposite states. Suppose a system (called system 1) has component state vector $\mathbf{x}$ and system state $\phi(\mathbf{x})$. Consider another system (called system 2) having the same number of components as system 1. If each component in system 2 has the opposite state of the corresponding component in system 1 and the state of system 2 becomes the opposite of the state of system 1, then system 1 and system 2 are duals of each other.

Now let us examine the k-out-of-n G and F systems. Suppose that in the k-out-of-n:G system, there are exactly j *working* components and the system is working (in other words, $j \geq k$). Now assume that there are exactly j *failed* components in the k-out-of-n:F system. Since $j \geq k$, the k-out-of-n:F system must be in the failed state. If $j < k$, the k-out-of-n:G system is failed, and at the same time, the k-out-of-n:F system is working. Thus, the k-out-of-n G and F systems are duals of each other. The dual and equivalence relationships between the k-out-of-n G and F systems are summarized below.

1. A k-out-of-n:G system is equivalent to an $(n-k+1)$-out-of-n:F system.
2. A k-out-of-n:F system is equivalent to an $(n-k+1)$-out-of-n:G system.
3. The dual of a k-out-of-n:G system is a k-out-of-n:F system.
4. The dual of a k-out-of-n:G system is an $(n-k+1)$-out-of-n:G system.
5. The dual of a k-out-of-n:F system is a k-out-of-n:G system.
6. The dual of a k-out-of-n:F system is an $(n-k+1)$-out-of-n:F system.

Using the dual relationship, we can summarize the following procedure for reliability evaluation of the dual system if the available algorithms are for the primal system.

Procedure 2. Procedure for using algorithms for the G systems in reliability evaluation of the F systems utilizing the dual relationship:

1. given: $k, n, p_1, p_2, \ldots, p_n$ for a k-out-of-n:F system;
2. calculate $q_i = 1 - p_i$ for $i - 1, 2, \ldots, n$;

3. treat q_i as the reliability of component i in a k-out-of-n:G system and use the algorithms for the G system discussed in the previous section to evaluate the reliability of the G system;
4. subtract the calculated reliability of the G system from one to obtain the reliability of the original k-out-of-n:F system.

Using the dual relationship, we can also obtain algorithms for k-out-of-n:F system reliability evaluation from those developed for the k-out-of-n:G systems. We only need to change reliability measures to unreliability measures and *vice versa*. Take the algorithm developed by Rushdi [5] as an example. The formulas for reliability and unreliability evaluation of a k-out-of-n:G system are given in Equation 1.7 with boundary conditions in Equations 1.8 and 1.9. By changing $R(i, j)$ to $Q(i, j)$, $Q(i, j)$ to $R(i, j)$, p_i to q_i, and q_i to p_i in those equations, we obtain the following equations for unreliability evaluation of a k-out-of-n:F system:

$$Q_{\mathrm{F}}(i, j) = q_j Q_{\mathrm{F}}(i-1, j-1) + p_j Q_{\mathrm{F}}(i, j-1) \quad (1.18)$$

with the following boundary conditions

$$Q_{\mathrm{F}}(0, j) = 1 \quad (1.19)$$

$$Q_{\mathrm{F}}(j+1, j) = 0 \quad (1.20)$$

The subscript "F" is added to indicate that these measures are for the F system to avoid confusion. Similar steps can be applied to other algorithms for the G systems to derive the corresponding algorithms for the F systems.

1.4 Relevant Concepts in Multi-state Reliability Theory

In the multi-state context, the definition domains of the state of the system and its components are expanded from $\{0, 1\}$ to $\{0, 1, \ldots, M\}$ where M is an integer greater than one. Such multi-state system models are *discrete models*. There are also *continuous-state* system reliability models. However, we will focus on discrete state models in this chapter.

We will adopt the following notation:

- x_i: state of component i, $x_i = j$ if component i is in state j, $0 \le j \le M$;
- $\mathbf{x}$: an n-dimensional vector representing the states of all components, $\mathbf{x} = (x_1, x_2, \ldots, x_n)$;
- $\phi(\mathbf{x})$: state of the system, which is also called the structure function of the system, $0 \le \phi(\mathbf{x}) \le M$;
- $(j_i, \mathbf{x})$: a vector $\mathbf{x}$ whose ith argument is set equal to j, where $j = \{0, 1, \ldots, M\}$ and $i = 1, 2, \ldots, n$.

In extending Definition 1 for a binary coherent system to the multi-state context, conditions 1 and 3 can be easily extended, namely, $\phi(\mathbf{x})$ is still a monotonically increasing function in each argument and the state of the system is equal to the state of all components when all components are in the same state. However, there are many different ways to extend condition 2, namely, the relevancy condition. One definition of a coherent system in the multi-state context is given below.

Definition 4. A multi-state system with n components is defined to be a coherent system if its structure function satisfies:

1. $\phi(\mathbf{x})$ is monotonically increasing in each argument;
2. there exists a vector $\mathbf{x}$ such that $\phi(0_i, \mathbf{x}) < \phi(M_i, \mathbf{x})$ for each $i = 1, 2, \ldots, n$;
3. $\phi(\mathbf{j}) = j$ for $j = 0, 1, \ldots, M$.

The reason that there are many different ways to define the relevancy condition (condition 2) is that there are different degrees of relevancy that a component may have on the system in the multi-state context. A component has $M+1$ levels and the system has $M+1$ levels. A level of a component may be relevant to some levels of the system but not others. Every level of a component may be relevant to every level of the system (this is the case in the binary case, since there are only two levels). There are many papers discussing relevancy conditions, for example, see Andrzejezak [12] for a review. However, the focus of this chapter is not on relevancy conditions.

There are two different definitions of minimal path vectors and minimal cut vectors in the multi-state context. One is given in terms of resulting in the system state to be exactly in a certain state while the other is given in terms of resulting in the system state to be in or above a certain state.

Definition 5. (El-Neweihi *et al.* [13]) A vector $\mathbf{x}$ is a connection vector to level j if $\phi(\mathbf{x}) = j$. A vector $\mathbf{x}$ is a lower critical connection vector to level j if $\phi(\mathbf{x}) = j$ and $\phi(\mathbf{y}) < j$ for all $\mathbf{y} < \mathbf{x}$. A vector $\mathbf{x}$ is an upper critical connection vector to level j if $\phi(\mathbf{x}) = j$ and $\phi(\mathbf{y}) > j$ for all $\mathbf{y} > \mathbf{x}$.

A connection vector to level j defined here is a path vector that results in a system state of level j. A lower critical connection vector to level j can be called a minimal path vector to level j. An upper critical connection vector can be called a maximal path vector to level j. These terms refer to path vectors instead of cut vectors. If these vectors are known, one can use them in evaluation of the probability distribution of the system state.

Definition 6. (Natvig [14]) A vector $\mathbf{x}$ is called a minimal path vector to level j if $\phi(\mathbf{x}) \geq j$ and $\phi(\mathbf{y}) < j$ for all $\mathbf{y} < \mathbf{x}$. A vector $\mathbf{x}$ is called a minimal cut vector to level j if $\phi(\mathbf{x}) < j$ and $\phi(\mathbf{y}) \geq j$ for all $\mathbf{y} \geq \mathbf{x}$.

In this definition, both minimal path vectors and minimal cut vectors are defined in terms of whether they result in a system state "equal to or greater than j" or not. Based on the definitions given by El-Neweihi *et al.* [13] and Natvig [14], a minimal path vector to level j may or may not be a connection vector to level j, as it may result in a system state higher than j. A minimal cut vector to level j may or may not be a connection vector to level $j - 1$, as it may result in a system state below $j - 1$. We find that the minimal path vectors and minimal cut vectors defined by Natvig [14] are easier to use than the connection vectors defined by El-Neweihi *et al.* [13].

Definition 7. (Xue [15]) Let $\phi(\mathbf{x})$ be the structure function of a multi-state system. The structure function of its dual system, $\phi^{\mathrm{D}}(\mathbf{x})$, is defined to be

$$\phi^{\mathrm{D}}(\mathbf{x}) = M - \phi(\mathbf{M} - \mathbf{x})$$

where $\mathbf{M} - \mathbf{x} = (M - x_1, M - x_2, \ldots, M - x_n)$.

Based on this definition of duality, the following two results are immediate.

1. $(\phi^{\mathrm{D}}(\mathbf{x}))^{\mathrm{D}} = \phi(\mathbf{x})$ [15].
2. Vector $\mathbf{x}$ is an upper critical connecting vector to level j of ϕ if and only if $\mathbf{M} - \mathbf{x}$ is a lower critical connection vector to level $M - j$ of ϕ^{D}. Vector $\mathbf{x}$ is a lower critical connecting vector to level j of ϕ if and only if $\mathbf{M} - \mathbf{x}$ is an upper critical connection vector to level $M - j$ of ϕ^{D}.

Reliability is the most widely used performance measure of a binary system. It is the probability that the system is in state 1. Once the reliability of a system is given, we also know the probability that the system is in state 0. Thus, the reliability uniquely defines the distribution of a binary system in different states. In a multi-state system, it is often assumed that the *state distribution*, *i.e.* the distribution of each component in different states, is given. The performance of the system is represented by its state distribution. Thus, the most important performance measure in a multi-state system is the state distributions. A state distribution may be given in terms of a probability distribution function, a cumulative distribution function, or a reliability function.

We define the following notation:

- p_{ij}: probability that component i is in state j, $1 \leq i \leq n, 0 \leq j \leq M$;
- p_j: probability that a component is in state j when all components are i.i.d.;
- P_{ij}: probability that component i is in state j or above;
- P_j: probability that a component is in state j or above when all components are i.i.d.;
- $Q_{ij} = 1 - P_{ij}$: the probability that component i is in a state below j;
- $Q_j = 1 - P_j$;
- $R_{sj} = \Pr(\phi(\mathbf{x}) \geq j)$;
- $Q_{sj} = 1 R_{sj}$;
- $r_{sj} = \Pr(\phi(\mathbf{x}) = j)$.

Based on this notation, we have the following facts:

$$P_{i0} = 0 \quad 1 \le i \le n$$

$$\sum_{j=0}^{M} p_{ij} = 1 \quad 1 \le j \le n$$

$$p_{iM} = P_{iM}, \quad p_{ij} = P_{ij} - P_{i,j+1}$$
$$1 \le i \le n, \; 0 \le j \le M-1$$

$$R_{s0} = 1, \quad R_{sM} = r_{sj} = R_{sj} - R_{s(j+1)}$$
$$0 \le j \le M-1$$

1.5 A Simple Multi-state k-out-of-n:G Model

The following notation is adopted:

- $R(k, n; j)$: probability that the k-out-of-n:G system is in state j or above.

The state of a multi-state series system is equal to the state of the worst component in the system [13], *i.e.*

$$\phi(\mathbf{x}) = \min_{1 \le i \le n} x_i$$

The state of a parallel system is equal to the state of the best component in the system [13], *i.e.*

$$\phi(\mathbf{x}) = \max_{1 \le i \le n} x_i$$

These definitions are natural extensions from the binary case to the multi-state case. System performance evaluation for the defined multi-state parallel and series systems is straightforward.

$$R_{sj} = \prod_{i=1}^{n} P_{ij} \quad j = 1, 2, \ldots, M \qquad (1.21)$$

for a series system, and

$$Q_{sj} = \prod_{i=1}^{n} Q_{ij} \quad j = 1, 2, \ldots, M \qquad (1.22)$$

for a parallel system.

A simple extension of the binary k-out-of-n:G system results in the definition of a simple multi-state k-out-of-n:G system. We call it *simple* because there are more general models of the k-out-of-n:G systems, which will be discussed later.

Definition 8. (El-Neweihi *et al.* [13]) A system is a k-out-of-n:G system if its structure function satisfies

$$\phi(\mathbf{x}) = x_{(n-k+1)}$$

where $x_{(1)} \le x_{(2)} \le \cdots \le x_{(n)}$ is an non-decreasing arrangement of $x_1, x_2, \ldots,$ and x_n.

Based on this definition, a k-out-of-n:G system has the following properties.

1. The multi-state series and parallel systems defined above satisfy this definition of the multi-state k-out-of-n:G system.
2. The state of the system is determined by the worst state of the best k components.
3. Each state j for $1 \le j \le M$ has $\binom{n}{k}$ minimal path vectors and $\binom{n}{k-1}$ minimal cut vectors. The numbers of minimal path and minimal cut vectors are the same for each state. We can say that the defined k-out-of-n:G system has the same structure at each system state j for $1 \le j \le M$. In other words, the system is in state j or above if and only if at least k components are in state j or above for each j $(1 \le j \le M)$. Because of this property, Boedigheimer and Kapur [16] actually define the k-out-of-n:G system as one that has $\binom{n}{k}$ minimal path vectors and $\binom{n}{k-1}$ minimal cut vectors to system state j for $1 \le j \le M$.

System performance evaluation for the simple k-out-of-n:G system is straightforward. For any specified system state j, we can use the system reliability evaluation algorithms for a binary k-out-of-n:G system to evaluate the probability that the multi-state system is in state j or above. For example, Equation 1.7 and its boundary condition can be used as follows, for $1 \le j \le M$:

$$R(k, n; j) = P_{nj} R(k-1, n-1; j) + Q_{nj} R(k, n-1; j) \qquad (1.23)$$

$$R(0, n; j) = 1 \qquad (1.24)$$

$$R(n+1, n; j) = 0 \qquad (1.25)$$

1.6 A Generalized Multi-state k-out-of-n:G System Model

The simple multi-state k-out-of-n:G system model limits that the system structure at each system level must be the same. In practical situations, a *multi-state* system may have different structures at different system levels. For example, consider a three-component system with four possible states. The system could be a 1-out-of-3:G structure at level 1; in other words, it requires at least one component to be in state 1 or above for the system to be in state 1 or above. It may have a 2-out-of-3:G structure at level 2; in other words, for the system to be in state 2 or above, at least two components must be in state 2 or above. It may have a 3-out-of-3:G structure at level 3; namely, at least three components have to be in state 3 for the system to be in state 3. Such a k-out-of-n system model is more flexible for modeling real-life systems. Huang *et al.* [17] proposed a definition of the generalized multi-state k-out-of-n:G system and developed reliability evaluation algorithms for the following multi-state k-out-of-n:G system model.

Definition 9. (Huang [17]) $\phi(\mathbf{x}) \geq j$ $(j = 1, 2, \ldots, M)$ if there exists an integer value l $(j \leq l \leq M)$ such that at least k_l components are in state l or above. An n-component system with such a property is called a *multi-state* k-out-of-n:G system.

In this definition, k_j values do not have to be the same for different system states j $(1 \leq j \leq M)$. This means that the structure of the *multi-state* system may be different for different system state levels. Generally speaking, k_j values are not necessarily in a monotone ordering. But the following two special cases of this definition will be particularly considered.

- When $k_1 \leq k_2 \leq \cdots \leq k_M$, the system is called an increasing multi-state k-out-of-n:G system. In this case, for the system to be in a higher state level j or above, a larger number of components must be in state j or above. In other words, there is an *increasing* requirement on the number of components that must be in a certain state or above for the system to be in a higher state level. That is why we call it the increasing multi-state k-out-of-n:G system.
- When $k_1 \geq k_2 \geq \cdots \geq k_M$, the system is called a decreasing multi-state k-out-of-n:G system. In this case, for a higher system state level j, there is a *decreasing* requirement on the number of components that must be in state level j or above.

When k_j is a constant, *i.e.* $k_1 = k_2 = \cdots = k_M = k$, the structure of the system is the same for all system state levels. This reduces to the definition of the simple multi-state k-out-of-n:G system discussed in the previous section. We call such systems constant multi-state k-out-of-n:G systems. All the concepts and results of binary k-out-of-n:G systems can be easily extended to the constant *multi-state* k-out-of-n:G systems. The constant multi-state k-out-of-n:G system is treated as a special case of the increasing multi-state k-out-of-n:G system in our later discussions.

For an increasing multi-state k-out-of-n:G system, *i.e.* $k_1 \leq k_2 \leq \cdots \leq k_M$, Definition 9 can be rephrased as follows:

$$\phi(x) \geq j$$

if and only if at least k_j components are in state j or above. If at least k_j components are in state j or above (these components can be considered "functioning" as far as state level j is concerned), then the system will be in state j or above (the system is considered to be "functioning") for $1 \leq j \leq M$. The only difference between this case of Definition 9 and Definition 8 is that the number of components required to be in state j or above for the system to be in state j or above may change from state to state. Other characteristics of this case of the generalized multi-state k-out-of-n:G system are exactly the same as that defined with Definition 8. Algorithms for binary k-out-of-n:G system reliability evaluation can also be extended for the increasing multi-state k-out-of-n:G system reliability evaluation:

$$R_j(k_j, n) = P_{nj} R_j(k_j - 1, n - 1) + (1 - P_{nj}) R_j(k_j, n - 1) \quad (1.26)$$

where $R_j(b, a)$ is the probability that at least b out of a components are in state j or above. The following boundary conditions are needed for Equation 1.26:

$$R_j(0, a) = 1 \quad \text{for } a \geq 0 \tag{1.27}$$

$$R_j(b, a) = 0 \quad \text{for } b > a > 0 \tag{1.28}$$

When all the components have the same state probability distribution, *i.e.* $p_{ij} = p_j$ for all i, the probability that the system is in j or above, R_{sj}, can be expressed as:

$$R_{sj} = \sum_{k=k_j}^{n} \binom{n}{k} P_j^k (1 - P_j)^{n-k} \tag{1.29}$$

Example 1. Consider an increasing multi-state k-out-of-n:G system with $k_1 = 1$, $k_2 = 2$, $k_3 = 3$, $p_0 = 0.1$, $p_1 = 0.3$, $p_2 = 0.4$, and $p_3 = 0.2$. We can use Equation 1.29 to calculate the probability that the system is at each level.

We have $P_1 = 0.9$, $P_2 = 0.6$, $P_3 = 0.2$.

At level 3, $k_3 = 3$.

$$R_{s3} = P_3^3 = 0.2^3 = 0.008$$

At level 2, $k_2 = 2$.

$$R_{s2} = \binom{3}{2} \times 0.6^2 \times (1 - 0.6) + 0.6^3 = 0.648$$

At level 1, $k_1 = 1$.

$$R_{s1} = \binom{3}{1} \times 0.9 \times (1 - 0.9)^2 + \binom{3}{2} \times 0.9^2 \times (1 - 0.9) + 0.9^3 = 0.999$$

The system probabilities at all levels are as follows:

$$r_{s3} = 0.008$$
$$r_{s2} = R_{s2} - R_{s3} = 0.64$$
$$r_{s1} = R_{s1} - R_{s2} = 0.351$$
$$r_{s0} = 1 - 0.008 - 0.64 - 0.351 = 0.001$$

For a decreasing multi-state k-out-of-n:G system, *i.e.* $k_1 \geq k_2 \geq \cdots \geq k_M$, the wording of its definition is not as simple. The system is in level M if at least k_M components are in level M. The system is in level $M - 1$ or above if at least k_{M-1} components are in level $M - 1$ or above or at least k_M components are in level M. Generally speaking, the system is in level j or above ($1 \leq j \leq M$) if at least k_j components are in level j or above, at least k_{j+1} components are in level $j + 1$ or above, at least k_{j+2} components are in level $j + 2$ or above, ..., or at least k_M components are in level M. The definition of the decreasing multi-state k-out-of-n:G system can also be stated as the following in terms of the system being exactly in a certain state:

> $\phi(\mathbf{x}) = j$ if and only if at least k_j components are in state j or above and at most $k_l - 1$ components are at state l or above for $l = j + 1, j + 2, \ldots, M$ where $j = 1, 2, \ldots, M$.

When all the components have the same state probability distribution, the following equation can be used to calculate the probability that the system is in state j for a decreasing multi-state k-out-of-n:G system:

$$r_{sj} = \sum_{k=k_j}^{n} \binom{n}{k} \left(\sum_{m=0}^{j-1} p_m \right)^{n-k} \times \left(p_j^k + \sum_{l=j+1, k_l>1}^{M} \beta_l(k) \right) \tag{1.30}$$

where $\beta_l(k)$ is the probability that there is at least one and at most $k_l - 1$ components that are in state l, at most $k_u - 1$ components that are in state u for $j \leq u < l$, and the total number of components that are in states between j and l inclusive is k. To calculate $\beta_l(k)$, we can use the following equation:

$$\beta_l(k) = \sum_{i_1=1}^{k_l-1} \binom{k}{i_1} p_l^{i_1} \sum_{i_2=0}^{k_{l-1}-1-i_1} \binom{k - i_1}{i_2} p_{l-1}^{i_2} \times \cdots \times \sum_{i_{l-j}=0}^{k_{j+1}-1-I_{l-j-1}} \binom{k - I_{l-j-1}}{i_{l-j}} p_{j+1}^{i_{l-j}} p_j^{k-I_{l-j}} \tag{1.31}$$

where $I_{l-j-1} = \sum_{m=1}^{l-j-1} i_m$ and $I_l = \sum_{m=1}^{l-j} i_m$.

We can see that even when the components are i.i.d., Equations 1.30 and 1.31 are enumerative in nature. When the components are not necessarily identical, the procedure for evaluation of system state distribution is more complicated. Huang *et al.* [17] provide an enumerative procedure for this purpose. These algorithms need further improvement.

1.7 Properties of Generalized Multi-state k-out-of-n:G Systems

We adopt the following notation:

- x_{ij}: a binary indicator,

$$x_{ij} = \begin{cases} 1, & \text{if } x_i \geq j \\ 0, & \text{otherwise} \end{cases}$$

- $\mathbf{x}^j$: a binary indicator vector,

$$\mathbf{x}^j = (x_{1j}, x_{2j}, \ldots, x_{nj})$$

- $\phi^j(\mathbf{x})$: a binary indicator,

$$\phi^j = \begin{cases} 1, & \text{if } \phi(\mathbf{x}) \geq j \\ 0, & \text{otherwise} \end{cases}$$

- ψ^j: binary structure function for system state j;
- ψ: multi-state structure function of binary variables;
- ϕ^{D}: the dual structure function of ϕ.

In order to develop more efficient algorithms for evaluation of system state distribution, properties of multi-state systems should be investigated.

Definition 10. (Hudson and Kapur [18]) Two-component state vectors $\mathbf{x}$ and $\mathbf{y}$ are said to be equivalent if and only if there exists a j such that $\phi(\mathbf{x}) = \phi(\mathbf{y}) = j$, $j = 0, 1, \ldots, M$. We use notation $\mathbf{x} \leftrightarrow \mathbf{y}$ to indicate that these two vectors are equivalent.

In the generalized k-out-of-n:G system, all permutations of the elements of a component state vector are equivalent to the component state vector, since the positions of the components in the system are not important. For example, consider a k-out-of-n:G system with three components and four possible states. Component state vectors $(1, 2, 3)$ and its permutations, namely $(1, 3, 2)$, $(2, 1, 3)$, $(2, 3, 1)$, $(3, 1, 2)$ and $(3, 2, 1)$, are equivalent to one another.

Definition 11. (Huang and Zuo [19]) A *multi-state* coherent system is called a dominant system if and only if its structure function ϕ satisfies: $\phi(\mathbf{y}) > \phi(\mathbf{x})$ implies either (1) $\mathbf{y} > \mathbf{x}$ or (2) $\mathbf{y} > \mathbf{z}$ and $\mathbf{x} \leftrightarrow \mathbf{z}$.

This dominance condition says that if $\phi(\mathbf{y}) > \phi(\mathbf{x})$, then vector $\mathbf{y}$ must be larger than a vector that is in the same equivalent class as vector $\mathbf{x}$. A vector is larger than another vector if every element of the first vector is at least as large as the corresponding element in the second vector and at least one element in the first vector is larger than the corresponding one in the second vector. For example, vector $(2, 2, 0)$ is larger than $(2, 1, 0)$, but not larger than $(0, 1, 1)$. We use the word "dominant" to indicate that vector $\mathbf{y}$ dominates vector $\mathbf{x}$, even though we may not necessarily have $\mathbf{y} > \mathbf{x}$. If a *multi-state* system does not satisfy Definition 11, we call it a non-dominant system.

The increasing k-out-of-n:G system is a dominant system, whereas a decreasing k-out-of-n:G system is a non-dominant system. Consider a decreasing k-out-of-n:G system with $n = 3$, $M = 3$, $k_1 = 3$, $k_2 = 2$, and $k_3 = 1$. Then, we have $\phi(3, 0, 0) = 3$, $\phi(2, 2, 0) = 2$, and $\phi(1, 1, 1) = 1$. The dominance conditions given in Definition 11 are not satisfied, since $(3, 0, 0) \not> (2, 2, 0)^* \not> (1, 1, 1)^*$ even though $\phi(3, 0, 0) > \phi(2, 2, 0) > \phi(1, 1, 1)$. The asterisk is used to indicate the vector or any one of its permutations.

The concept of a binary image has been used to indicate whether a multi-state system can be analyzed with the algorithms for binary systems. Based on Ansell and Bendell [20], a multi-state system has a binary image if condition $\mathbf{x}^j = \mathbf{y}^j$ implies $\phi^j(\mathbf{x}) = \phi^j(\mathbf{y})$. Huang and Zuo [19] point out that this requirement is too strong. A system may have a binary image even if this condition is

Table 1.1. Structure functions of the multi-state systems in Example 2

	System A			System B			System C		
$\phi(\mathbf{x})$:	0	1	2	0	1	2	0	1	2
$\mathbf{x}$:	(0, 0)	(1, 0) (0, 1) (1, 1) (2, 0) (0, 2) (2, 1) (1, 2)	(2, 2)	(0, 0) (0, 1)	(1, 0) (1, 1) (0, 2) (1, 2)	(2, 0) (2, 1) (2, 2)	(0, 0) (1, 0) (0, 1)	(1, 1)	(2, 0) (0, 2) (2, 1) (1, 2) (2, 2)

not satisfied. The following definition of binary-imaged systems is provided.

Definition 12. (Huang and Zuo [19]) A multi-state system has a binary-image if and only if its structure indicator functions satisfy: $\phi^j(\mathbf{x}) = \phi^j(\mathbf{y})$ implies either (1) $\mathbf{x}^j = \mathbf{y}^j$ or (2) $\mathbf{x}^j = \mathbf{z}^j$ and $\mathbf{z} \leftrightarrow \mathbf{x}$ for each j $(1 \le j \le M)$.

Based on Definitions 11 and 12, we can see that a binary-imaged multi-state system must be a dominant system, whereas a dominant multi-state system may not have a binary image. The increasing k-out-of-n:G system has a binary image. The decreasing multi-state k-out-of-n:G system does not have a binary image. If a multi-state system has a binary image, algorithms for binary systems can be used for evaluation of its system state distribution. These binary algorithms are not efficient for non-dominant systems. The following example illustrates a dominant system with binary image, a dominant system without binary image, and a non-dominant system.

Example 2. Consider three different multi-state systems, each with two components and three possible states. The structure functions of the systems are given in Table 1.1, where System A is a dominant system with binary image, System B is a dominant system without binary image, and System C is a non-dominant system.

The following properties of multi-state systems are identified by Huang and Zuo [19].

1. Let $P_1^j, P_2^j, \ldots, P_r^j$ be the minimal path vectors to level j of a dominant system, then $\phi(P_i^j) = j$ for $i = 1, 2, \ldots, r$.
2. Let $K_1^j, K_2^j, \ldots, K_s^j$ be the minimal cut vectors to level j of a dominant system, then $\phi(K_i^j) = j - 1$ for $i = 1, 2, \ldots, s$.
3. The minimal path vectors to level j of a binary-imaged dominant system are of the form $(j, \ldots, j, 0, \ldots, 0)$ for $j = 1, 2, \ldots, M$ or one of its permutations.

Example 3. A *multi-state* coherent system consists of three i.i.d. components. Both the system and the components are allowed to have four possible states: 0, 1, 2, 3. Assume that $p_0 = 0.1$, $p_1 = 0.3$, $p_2 = 0.4$, and $p_3 = 0.2$. The structures of the system are shown in Figure 1.1.

We say that the system has a parallel structure at level 1 because it is at level 1 or above if and only if at least one of the components is at level 1 or above. The system has a mixed parallel–series structure at level 2, because it is at level 2 or above if and only if at least one of components 1 and 2 is at level 2 or above and component 3 is at level 2 or above. The system has a series structure at level 3, because it is at level 3 if and only if all three components are at level 3. The structures change from strong to weak as the level of the system increases. The minimal path vectors to level 1 are (1, 0, 0), (0, 1, 0), and (0, 0, 1); to level 2 are (2, 0, 2) and (0, 2, 2); and to level 3 is (3, 3, 3). The system is a binary-imaged system. To calculate the probability of the system at each level, we can extend the algorithms for binary reliability

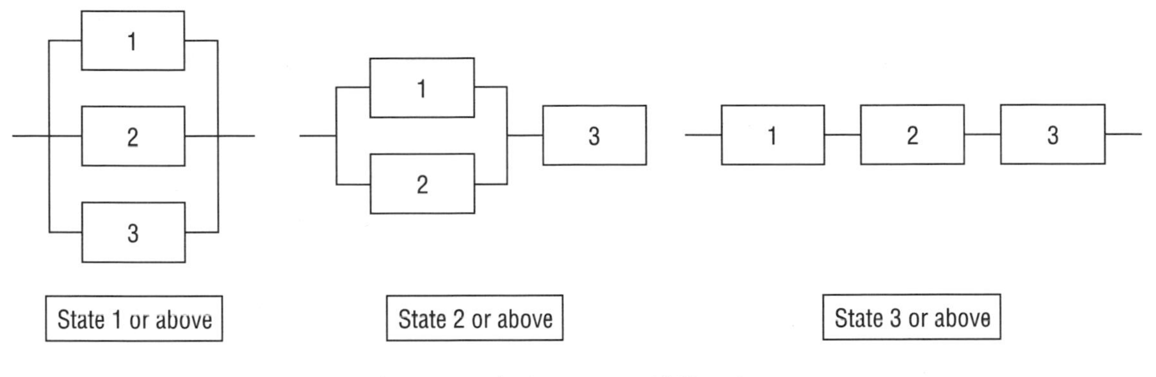

Figure 1.1. A dominant system with binary image

systems:

$$\begin{aligned}
P_1 &= 0.9, \quad P_2 = 0.6, \quad P_3 = 0.2 \\
R_{s3} &= P_{13} \times P_{23} \times P_{33} = 0.008 \\
R_{s2} &= [1 - (1 - P_{12})(1 - P_{22})] \times P_{33} \\
&= 2 \times P_2^2 - P_2^3 = 0.504 \\
R_{s1} &= 1 - (1 - P_{11})(1 - P_{12})(1 - P_{13}) \\
&= 1 - (1 - P_1)^3 = 0.999 \\
r_{s3} &= R_{s3} = 0.008 \\
r_{s2} &= R_{s2} - R_{s3} = 0.496 \\
r_{s1} &= R_{s1} - R_{s2} = 0.495 \\
r_{s0} &= 1 - R_{s1} = 0.001
\end{aligned}$$

1.8 Equivalence and Duality in Generalized Multi-state k-out-of-n Systems

Following the way that a generalized multi-state k-out-of-n:G system is defined, we propose a definition of the generalized multi-state k-out-of-n:F system as follows.

Definition 13. $\phi(\mathbf{x}) < j$ $(j = 1, 2, \ldots, M)$ if at least k_l components are in states below l for all l such that $j \le l \le M$. An n-component system with such a property is called a *multi-state* k-out-of-n:F system.

A generalized multi-state k-out-of-n:F system is labeled *increasing* if $k_1 \le k_2 \le \cdots \le k_M$ and *decreasing* if $k_1 \ge k_2 \ge \cdots \ge k_M$. Similar to an increasing k-out-of-n:G system, the definition of a decreasing k-out-of-n:F system can be stated in a simpler form. A decreasing k-out-of-n:F system is in a state below j if and only if at least k_j components are in states below j for $1 \le j \le M$. For such a system, we do not need to state the requirements for those states above j because they will be automatically satisfied by the requirement on state j. However, this cannot be said of an increasing multi-state k-out-of-n:F system. An increasing multi-state k-out-of-n:F system is in a state below j if and only if all the following conditions are satisfied: at least k_j components are below state j, at least k_{j+1} components are below state $j+1$, at least k_{j+2} components are below state $j+2, \ldots,$ and at least k_M components are below state M. Since the increasing k-out-of-n:G system has a binary image, there is an efficient algorithm for evaluation of its system state distribution, as discussed in a previous section. For the decreasing k-out-of-n:F system, there is a similar efficient algorithm for evaluation of system state distribution

$$Q_j(k_j, n) = (1 - P_j)Q_j(k_j - 1, n - 1) + P_{nj} Q_j(k_j, n - 1) \tag{1.32}$$

where $Q_j(b, a)$ is the probability that at least b out of a components are below state j. $Q_j(k_j, n)$ is the probability that the system is below state j. The following boundary conditions are needed for

Equation 1.32:

$$Q_j(0, a) = 1 \quad \text{for } a \geq 0 \tag{1.33}$$

$$Q_j(b, a) = 0 \quad \text{for } b > a > 0 \tag{1.34}$$

When all the components have the same state probability distribution, *i.e.* $p_{ij} = p_j$ for all i, the probability Q_{sj} that the system is below state j can be expressed as:

$$Q_{sj} = \sum_{k=k_j}^{n} \binom{n}{k} (1 - P_j)^k P_j^{n-k} \tag{1.35}$$

The evaluation of system state distribution for an increasing multi-state k-out-of-n:F system is much more complicated than that for the decreasing case. A similar algorithm to that for a decreasing multi-state k-out-of-n:G system developed by Huang *et al.* [17] can be derived. However, that kind of algorithm is enumerative in nature. More efficient algorithms are needed for the increasing k-out-of-n:F and the decreasing k-out-of-n:G systems.

The definition of a multi-state k-out-of-n:G system with vector $\mathbf{k} = (k_1, k_2, \ldots, k_M)$ given in Definition 9 can be stated in an alternative way as follows. A k-out-of-n:G system is below state j if at least $n - k_l + 1$ components are below state l for all l such that $j \leq l \leq M$. Similarly, the definition of a multi-state k-out-of-n:F system with $\mathbf{k} = (k_1, k_2, \ldots, k_M)$ given in Definition 13 can be stated in an alternative way as follows. A k-out-of-n:F system is in state j or above if at least $n - k_l + 1$ components are in state l or above for at least one l such that $j \leq l \leq M$. An equivalence relationship exists between the generalized multi-state k-out-of-n:F and G systems. An increasing k-out-of-n:F system with vector $\mathbf{k}_\text{F} = (k_1, k_2, \ldots, k_M)$ is equivalent to a decreasing k-out-of-n:G system with vector $\mathbf{k}_\text{G} = (n - k_1 + 1, n - k_2 + 1, \ldots, n - k_M + 1)$. A decreasing k-out-of-n:F system with vector $\mathbf{k}_\text{F} = (k_1, k_2, \ldots, k_M)$ is equivalent to an increasing k-out-of-n:G system with vector $\mathbf{k}_\text{G} = (n - k_1 + 1, n - k_2 + 1, \ldots, n - k_M + 1)$.

Using the definition of duality for multi-state systems given in Definition 7, we can verify that, unlike the binary case, there does not exist any duality relationship between the generalized multi-state k-out-of-n:F and G systems. In a binary k-out-of-n:G system, if vector $\mathbf{x}$ has k components being in state 1, then vector $\mathbf{y} = 1 - \mathbf{x}$ must have k components below state 1. However, in a multi-state k-out-of-n:G system, if vector $\mathbf{x}$ has k components being in state j or above, then vector $\mathbf{y} = \mathbf{M} - \mathbf{x}$ does not necessarily have k components being in states below j.

The issues that should be investigated further on generalized multi-state k-out-of-n systems and general multi-state systems include the following.

1. Should a different definition of duality be given for multi-state systems?
2. Does the dual of a non-dominant system become a dominant system?
3. What techniques can be developed to provide bounds on the probability that the system is in a certain state?
4. How should we evaluate the system state distribution of a dominant system without binary image?
5. How should we evaluate the system state distribution of a non-dominant system?

References

[1] Barlow RE, Proschan F. Statistical theory of reliability and life testing: probability models. New York: Holt, Rinehart and Winston; 1975.

[2] Heidtmann KD. Improved method of inclusion-exclusion applied to k-out-of-n systems. IEEE Trans Reliab 1982;31(1):36–40.

[3] McGrady PW. The availability of a k-out-of-n:G network. IEEE Trans Reliab 1985;34(5):451–2.

[4] Barlow RE, Heidtmann KD. Computing k-out-of-n system reliability. IEEE Trans Reliab 1984;R-33:322–3.

[5] Rushdi AM. Utilization of symmetric switching functions in the computation of k-out-of-n system reliability. Microelectron Reliab 1986;26(5):973–87.

[6] Chao MT, Lin GD. Economical design of large consecutive-k-out-of-n:F systems. IEEE Trans Reliab 1984;R-33(5):411–3.

[7] Chao MT, Fu JC. A limit theorem of certain repairable systems. Ann Inst Stat Math 1989;41:809–18.

[8] Chao MT, Fu JC. The reliability of large series systems under Markov structure. Adv Appl Prob 1991;23:894–908.

[9] Fu JC, Lou WY. On reliabilities of certain large linearly connected engineering systems. Stat Prob Lett 1991;12:291–6.

[10] Koutras MV. On a Markov chain approach for the study of reliability structures. J Appl Prob 1996;33:357–67.

[11] Belfore LA. An $o(n(\log_2(n))^2)$ algorithm for computing the reliability of k-out-of-n:G & k-to-l-out-of-n:G systems. IEEE Trans Reliab 1995;44(1):132–6.

[12] Andrzejczak K. Structure analysis of multi-state coherent systems. Optimization 1992;25:301–16.

[13] El-Neweihi E, Proschan F, Sethuraman J. Multi-state coherent system. J Appl Prob 1978;15:675–88.

[14] Natvig B. Two suggestions of how to define a multi-state coherent system. Appl Prob 1982;14:391–402.

[15] Xue J. On multi-state system analysis. IEEE Trans Reliab 1985;R-34(4):329–37.

[16] Boedigheimer RA, Kapur KC. Customer-driven reliability models for multi-state coherent systems. IEEE Trans Reliab 1994;43(1):46–50.

[17] Huang J, Zuo MJ, Wu YH. Generalized multi-state k-out-of-n:G systems. IEEE Trans Reliab 2002;in press.

[18] Hudson JC, Kapur KC. Reliability analysis for multi-state systems with multi-state components. IIE Trans 1983;15:127–35.

[19] Huang J, Zuo MJ. Dominant multi-state systems. IEEE Trans Reliab 2002;in press.

[20] Ansell JI, Bendell A. On alternative definitions of multi-state coherent systems. Optimization 1987;18(1):119–36.

Reliability of Systems with Multiple Failure Modes

Hoang Pham

Chapter 2

2.1 Introduction

A component is subject to failure in either open or closed modes. Networks of relays, fuse systems for warheads, diode circuits, fluid flow valves, *etc.* are a few examples of such components. Redundancy can be used to enhance the reliability of a system without any change in the reliability of the individual components that form the system. However, in a two-failure mode problem, redundancy may either increase or decrease the system's reliability. For example, a network consisting of n relays in series has the property that an open-circuit failure of any one of the relays would cause an open-mode failure of the system and a closed-mode failure of the system. (The designations "closed mode" and "short mode" both appear in this chapter, and we will use the two terms interchangeably.) On the other hand, if the n relays were arranged in parallel, a closed-mode failure of any one relay would cause a system closed-mode failure, and an open-mode failure of all n relays would cause an open-mode failure of the system. Therefore, adding components in the system may decrease the system reliability. Diodes and transistors also exhibit open-mode and short-mode failure behavior.

For instance, in an electrical system having components connected in series, if a short circuit occurs in one of the components, then the short-circuited component will not operate but will permit flow of current through the remaining components so that they continue to operate. However, an open-circuit failure of any of the components will cause an open-circuit failure of the system. As an example, suppose we have a number of 5 W bulbs that remain operative in satisfactory conditions at voltages ranging between 3 V and 6 V. Obviously, on using the well-known formula in physics, if these bulbs are arranged in a series network to form a two-failure mode system, then the maximum and the minimum number of bulbs at these voltages are $n = 80$ and $k = 40$, respectively, in a situation when the system is operative at 240 V. In this case, any of the bulbs may fail either in closed or in open mode till the system is operative with 40 bulbs. Here, it is clear that, after each failure in closed mode, the rate of failure of a bulb in open mode increases due to the fact that the voltage passing through each bulb increases as the number of bulbs in the series decreases.

System reliability where components have various failure modes is covered in References [1–11]. Barlow *et al.* [1] studied series–parallel and parallel–series systems, where the size of each subsystem was fixed, but the number of subsystems was varied to maximize reliability. Ben-Dov [2] determined a value of k that maximizes the reliability of k-out-of-n systems. Jenney and Sherwin [4] considered systems in which the components are i.i.d. and subject to mutually exclusive open and short failures. Page and Perry [7] discussed the problem of designing the most reliable structure of a given number of i.i.d. components and proposed an alternative algorithm for selecting near-optimal configurations for large systems. Sah and Stiglitz [10] obtained a necessary and sufficient condition for determining a threshold value that maximizes the mean profit of k-out-of-n systems. Pham and Pham [9] further studied the effect of system parameters on the optimal k or n and showed that there does not exist a (k, n) maximizing the mean system profit.

This chapter discusses in detail the aspects of the reliability optimization of systems subject to two types of failure. It is assumed that the system component states are statistically independent and identically distributed, and that no constraints are imposed on the number of components to be used. Reliability optimization of series, parallel, parallel–series, series–parallel, and k-out-of-n systems subject to two types of failure will be discussed next.

In general, the formula for computing the reliability of a system subject to two kinds of failure is [6]:

$$\begin{aligned}\text{System reliability} &\\ &= \Pr\{\text{system works in both modes}\}\\ &= \Pr\{\text{system works in open mode}\}\\ &\quad - \Pr\{\text{system fails in closed mode}\}\\ &\quad + \Pr\{\text{system fails in both modes}\}\end{aligned} \quad (2.1)$$

When the open- and closed-mode failure structures are dual of one another, *i.e.* Pr{system fails in both modes} = 0, then the system reliability given by Equation 2.1 becomes

$$\begin{aligned}\text{System reliability} &\\ &= 1 - \Pr\{\text{system fails in open mode}\}\\ &\quad - \Pr\{\text{system fails in closed mode}\}\end{aligned} \quad (2.2)$$

We adopt the following notation:

- q_o the open-mode failure probability of each component ($p_o = 1 - q_o$)
- q_s the short-mode failure probability of each component ($p_s = 1 - q_s$)
- $\bar{\Phi}$ implies $1 - \Phi$ for any Φ
- $\lfloor x \rfloor$ the largest integer not exceeding x
- $*$ implies an optimal value.

2.2 The Series System

Consider a series system consisting of n components. In this series system, any one component failing in an open mode causes system failure, whereas all components of the system must malfunction in short mode for the system to fail.

The probabilities of system fails in open mode and fails in short mode are

$$F_o(n) = 1 - (1 - q_o)^n$$

and

$$F_s(n) = q_s^n$$

respectively. From Equation 2.2, the system reliability is:

$$R_s(n) = (1 - q_o)^n - q_s^n \quad (2.3)$$

where n is the number of identical and independent components. In a series arrangement, reliability with respect to closed system failure increases with the number of components, whereas reliability with respect to open system failure falls. There exists an optimum number of components, say n^*, that maximizes the system reliability. If we define

$$n_0 = \frac{\log\left(\frac{q_o}{1 - q_s}\right)}{\log\left(\frac{q_s}{1 - q_o}\right)}$$

then the system reliability, $R_s(n^*)$, is maximum for

$$n^* = \begin{cases} \lfloor n_0 \rfloor + 1 & \text{if } n_0 \text{ is not an integer} \\ n_0 \text{ or } n_0 + 1 & \text{if } n_0 \text{ is an integer} \end{cases} \quad (2.4)$$

Example 1. A switch has two failure modes: fail-open and fail-short. The probability of switch open-circuit failure and short-circuit failure are 0.1 and 0.2 respectively. A system consists of n switches wired in series. That is, given $q_o = 0.1$ and $q_s = 0.2$. From Equation 2.4

$$n_0 = \frac{\log\left(\frac{0.1}{1 - 0.2}\right)}{\log\left(\frac{0.2}{1 - 0.1}\right)} = 1.4$$

Thus, $n^* = \lfloor 1.4 \rfloor + 1 = 2$. Therefore, when $n^* = 2$ the system reliability $R_s(n) = 0.77$ is maximized.

2.3 The Parallel System

Consider a parallel system consisting of n components. For a parallel configuration, all the components must fail in open mode or at least one component must malfunction in short mode to cause the system to fail completely.

The system reliability is

$$R_p(n) = (1 - q_s)^n - q_o^n \quad (2.5)$$

where n is the number of components connected in parallel. In this case, $(1 - q_s)^n$ represents the probability that no components fail in short mode, and q_o^n represents the probability that all components fail in open mode. If we define

$$n_0 = \frac{\log\left(\frac{q_s}{1 - q_o}\right)}{\log\left(\frac{q_o}{1 - q_s}\right)} \quad (2.6)$$

then the system reliability $R_p(n^*)$ is maximum for

$$n^* = \begin{cases} \lfloor n_0 \rfloor + 1 & \text{if } n_0 \text{ is not an integer} \\ n_0 \text{ or } n_0 + 1 & \text{if } n_0 \text{ is an integer} \end{cases} \quad (2.7)$$

It is observed that, for any range of q_o and q_s, the optimal number of parallel components that maximizes the system reliability is one, if $q_s > q_o$. For most other practical values of q_o and q_s, the optimal number turns out to be two. In general, the optimal value of parallel components can be easily obtained using Equation 2.6.

2.3.1 Cost Optimization

Suppose that each component costs d dollars and system failure costs c dollars of revenue. We now wish to determine the optimal system size n that minimizes the average system cost given that the costs of system failure in open and short modes are known. Let T_n be a total of the system. The average system cost is given by

$$E[T_n] = dn + c[1 - R_p(n)]$$

where $R_p(n)$ is defined as in Equation 2.5. For given q_o, q_s, c, and d, we can obtain a value of n, say n^*, minimizing the average system cost.

Theorem 1. *Fix q_o, q_s, c, and d. There exists a unique value n^* that minimizes the average system cost, and*

$$n^* = \inf\left\{n < n_1 : (1 - q_o)q_o^n - q_s(1 - q_s)^n < \frac{d}{c}\right\} \quad (2.8)$$

where $n_1 = \lfloor n_0 \rfloor + 1$ and n_0 is given in Equation 2.6.

The proof is straightforward and left for an exercise. It was assumed that the cost of system failure in either open mode or short mode was the same. We are now interested in how the cost of system failure in open mode may be different from that in short mode.

Suppose that each component costs d dollars and system failure in open mode and short mode costs c_1 and c_2 dollars of revenue respectively. Then the average system cost is given by

$$E[T_n] = dn + c_1 q_o^n + c_2[1 - (1 - q_s)^n] \quad (2.9)$$

In other words, the average system cost of system size n is the cost incurred when the system has failed in either open mode or short mode plus the cost of all components in the system. We can determine the optimal value of n, say n^*, which minimizes the average system cost as shown in the following theorem [5].

Theorem 2. *Fix q_o, q_s, c_1, c_2, and d. There exists a unique value n^* that minimizes the average system cost, and*

$$n^* = \begin{cases} 1 & if\, n_a \le 0 \\ n_0 & otherwise \end{cases}$$

where

$$n_0 = \inf\left\{ n \le n_a : h(n) \le \frac{d}{c_2 q_s} \right\}$$

and

$$h(n) = q_o^n \left[\frac{1 - q_o}{q_s} \frac{c_1}{c_2} - \left(\frac{1 - q_s}{q_o} \right)^n \right]$$

$$n_a = \frac{\log \left(\dfrac{1 - q_o}{q_s} \dfrac{c_1}{c_2} \right)}{\log \left(\dfrac{1 - q_s}{q_o} \right)}$$

Example 2. Suppose $d = 10$, $c_1 = 1500$, $c_2 = 300$, $q_s = 0.1$, $q_o = 0.3$. Then

$$\frac{d}{c_2 q_s} = 0.333$$

From Table 2.1, $h(3) = 0.216 < 0.333$; therefore, the optimal value of n is $n^* = 3$. That is, when $n^* = 3$ the average system cost (151.8) is minimized.

Table 2.1. The function $h(n)$ vs n

n	$h(n)$	$R_p(n)$	$E[T_n]$
1	9.6	0.6	490.0
2	2.34	0.72	212.0
3	0.216	0.702	151.8
4	−0.373	0.648	155.3
5	−0.504	0.588	176.5
6	−0.506	0.531	201.7

2.4 The Parallel–Series System

Consider a system of components arranged so that there are m subsystems operating in parallel, each subsystem consisting of n identical components in series. Such an arrangement is called a parallel–series arrangement. The components could be a logic gate, a fluid-flow valve, or an electronic diode, and they are subject to two types of failure: failure in open mode and failure in short mode. Applications of the parallel–series systems can be found in the areas of communication, networks, and nuclear power systems. For example, consider a digital circuit module designed to process the incoming message in a communication system. Suppose that there are, at most, m ways of getting a message through the system, depending on which of the branches with n modules are operable. Such a system is subject to two failure modes: (1) a failure in open circuit of a single component in each subsystem would render the system unresponsive; or (2) a failure in short circuit of all the components in any subsystem would render the entire system unresponsive.

We adopt the following notation:

- m — number of subsystems in a system (or subsystem size)
- n — number of components in each subsystem
- $F_o(m)$ — probability of system failure in open mode
- $F_s(m)$ — probability of system failure in short mode.

The systems are characterized by the following properties.

1. The system consists of m subsystems, each subsystem containing n i.i.d. components.
2. A component is either good, failed open, or failed short. Failed components can never become good, and there are no transitions between the open and short failure modes.
3. The system can be (a) good, (b) failed open (at least one component in each subsystem fails open), or (c) failed short (all the components in any subsystem fail short).
4. The unconditional probabilities of component failure in open and short modes are known and are constrained: $q_o, q_s > 0; q_o + q_s < 1$.

The probabilities of a system failing in open mode and failing in short mode are given by

$$F_o(m) = [1 - (1 - q_o)^n]^m \tag{2.10}$$

and

$$F_s(m) = 1 - (1 - q_s^m)^m \tag{2.11}$$

respectively. The system reliability is

$$R_{ps}(n, m) = (1 - q_s^n)^m - [1 - (1 - q_o)^n]^m \tag{2.12}$$

where m is the number of identical subsystems in parallel and n is the number of identical components in each series subsystem. The term $(1 - q_s^n)^m$ represents the probability that none of the subsystems has failed in closed mode. Similarly, $[1 - (1 - q_o)^n]^m$ represents the probability that all the subsystems have failed in open mode.

An interesting example in Ref. [1] shows that there exists no pair n, m maximizing system reliability, since R_{ps} can be made arbitrarily close to one by appropriate choice of m and n. To see this, let

$$a = \frac{\log q_s - \log(1 - q_o)}{\log q_s + \log(1 - q_o)}$$

$$M_n = q_s^{-n/(1+a)} \qquad m_n = \lfloor M_n \rfloor$$

For given n, take $m = m_n$; then one can rewrite Equation 2.12 as:

$$R_{ps}(n, m_n) = (1 - q_s^n)^{m_n} - [1 - (1 - q_o)^n]^{m_n}$$

A straightforward computation yields

$$\lim_{n\to\infty} R_{ps}(n, m_n) = \lim_{n\to\infty} \{(1 - q_s^n)^{m_n} - [1 - (1 - q_o)^n]^{m_n}\} = 1$$

For fixed n, q_o, and q_s, one can determine the value of m that maximizes R_{ps}, and this is given below [8].

Theorem 3. *Let n, q_o, and q_s be fixed. The maximum value of $R_{ps}(m)$ is attained at $m^* = \lfloor m_0 \rfloor + 1$, where*

$$m_0 = \frac{n(\log p_o - \log q_s)}{\log(1 - q_s^n) - \log(1 - p_o^n)} \tag{2.13}$$

If m_0 is an integer, then m_0 and $m_0 + 1$ both maximize $R_{ps}(m)$.

2.4.1 The Profit Maximization Problem

We now wish to determine the optimal subsystem size m that maximizes the average system profit. We study how the optimal subsystem size m depends on the system parameters. We also show that there does not exist a pair (m, n) maximizing the average system profit.

We adopt the following notation:

$A(m)$	average system profit
β	conditional probability that the system is in open mode
$1 - \beta$	conditional probability that the system is in short mode
c_1, c_3	gain from system success in open, short mode
c_2, c_4	gain from system failure in open, short mode; $c_1 > c_2, c_3 > c_4$.

The average system profit is given by

$$A(m) = \beta\{c_1[1 - F_o(m)] + c_2 F_o(m)\} + (1 - \beta)\{c_3[1 - F_s(m)] + c_4 F_s(m)\} \tag{2.14}$$

Define

$$a = \frac{\beta(c_1 - c_2)}{(1 - \beta)(c_3 - c_4)}$$

and

$$b = \beta c_1 + (1 - \beta)c_4 \tag{2.15}$$

We can rewrite Equation 2.14 as

$$A(m) = (1 - \beta)(c_3 - c_4) \times \{[1 - F_s(m)] - aF_o(m)\} + b \tag{2.16}$$

When the costs of the two kinds of system failure are identical, and the system is in the two modes with equal probability, then the optimization criterion becomes the same as maximizing the system reliability. Here, the following analysis deals with cases that need not satisfy these special restrictions.

For a given value of n, one wishes to find the optimal number of subsystems m (m^*) that maximizes the average system profit. Of course, we would expect the optimal value of m to depend on the values of both q_o and q_s. Define

$$m_0 = \frac{\ln a + n \ln\left(\dfrac{1 - q_o}{q_s}\right)}{\ln\left[\dfrac{1 - q_s^n}{1 - (1 - q_o)^n}\right]} \tag{2.17}$$

Theorem 4. *Fix β, n, q_o, q_s, and c_i for $i = 1, 2, 3, 4$. The maximum value of $A(m)$ is attained at*

$$m^* = \begin{cases} 1 & \text{if } m_0 < 0 \\ \lfloor m_0 \rfloor + 1 & \text{if } m_0 \geq 0 \end{cases} \tag{2.18}$$

If m_0 is a non-negative integer, both m_0 and $m_0 + 1$ maximize $A(m)$.

The proof is straightforward. When m_0 is a non-negative integer, the lower value will provide the more economical optimal configuration for the system. It is of interest to study how the optimal subsystem size m^* depends on the various parameters q_o and q_s.

Theorem 5. *For fixed n, c_1, c_2, c_3, and c_4.*

(a) *If $a \geq 1$, then the optimal subsystem size m^* is an increasing function of q_o.*
(b) *If $a \leq 1$, then the optimal subsystem size m^* is a decreasing function of q_s.*
(c) *The optimal subsystem size m^* is an increasing function of β.*

The proof is left for an exercise. It is worth noting that we cannot find a pair (m, n) maximizing average system-profit $A(m)$. Let

$$x = \frac{\ln q_s - \ln p_o}{\ln q_s + \ln p_o} \qquad M_n = q_s^{-n/l+x} \qquad m_n = \lfloor M_n \rfloor \tag{2.19}$$

For given n, take $m = m_n$. From Equation 2.14, the average system profit can be rewritten as

$$A(m_n) = (1 - \beta)(c_3 - c_4) \times \{[1 - F_s(m_n)] - aF_o(m_n)\} + b \tag{2.20}$$

Theorem 6. *For fixed q_o and q_s*

$$\lim_{n \to \infty} A(m_n) = \beta c_1 + (1 - \beta)c_3 \tag{2.21}$$

The proof is left for an exercise. This result shows that we cannot seek a pair (m, n) maximizing the average system profit $A(m_n)$, since $A(m_n)$ can be made arbitrarily close to $\beta c_1 + (1 - \beta)c_3$.

2.4.2 Optimization Problem

We show how design policies can be chosen when the objective is to minimize the average total system cost given that the costs of system failure in open mode and short mode may not necessarily be the same.

The following notation is adopted:

d	cost of each component
c_1	cost when system failure in open
c_2	cost when system failure in short
$T(m)$	total system cost
$E[T(m)]$	average total system cost.

Suppose that each component costs d dollars, and system failure in open mode and short mode costs c_1 and c_2 dollars of revenue, respectively. The average total system cost is

$$E[T(m)] = dnm + c_1 F_o(m) + c_2 F_s(m) \tag{2.22}$$

In other words, the average system cost is the cost incurred when the system has failed in either the open mode or the short mode plus the cost of all components in the system. Define

$$h(m) = [1 - (p_{\text{o}})^n]^m \left[c_1 p_{\text{o}}^n - c_2 q_{\text{s}}^n \left(\frac{1 - q_{\text{s}}^n}{1 - p_{\text{o}}^n} \right)^m \right] \tag{2.23}$$

$$m_1 = \inf\{m < m_2 : h(m) < dn\}$$

and

$$m_2 = \left\lfloor \frac{\ln \left[\frac{c_1}{c_2} \left(\frac{p_{\text{o}}}{q_{\text{s}}} \right)^n \right]}{\ln \left(\frac{1 - q_{\text{s}}^n}{1 - p_{\text{o}}^n} \right)} \right\rfloor + 1 \tag{2.24}$$

From Equation 2.23, $h(m) > 0$ if and only if

$$c_1 p_{\text{o}}^n > c_2 q_{\text{s}}^n \left(\frac{1 - q_{\text{s}}^n}{1 - p_{\text{o}}^n} \right)^m$$

or equivalently, that $m < m_2$. Thus, the function $h(m)$ is decreasing in m for all $m < m_2$. For fixed n, we determine the optimal value of m, m^*, that minimizes the expected system cost, as shown in the following theorem [11].

Theorem 7. *Fix q_{o}, q_{s}, d, c_1, and c_2. There exists a unique value m^* such that the system minimizes the expected cost, and*

(*a*) *if $m_2 > 0$ then*

$$m^* = \begin{cases} m_1 & \text{if } E[T(m_1)] \le E[T(m_2)] \\ m_2 & \text{if } E[T(m_1)] > E[T(m_2)] \end{cases} \tag{2.25}$$

(*b*) *if $m_2 \le 0$ then $m^* = 1$.*

The proof is straightforward. Since the function $h(m)$ is decreasing in m for $m < m_2$, again the resulting optimization problem in Equation 2.25 is easily solved in practice.

Example 3. Suppose $n = 5, d = 10, c_1 = 500, c_2 = 700$, $q_{\text{s}} = 0.1$, and $q_{\text{o}} = 0.2$. From Equation 2.25, we obtain $m_2 = 26$. Since $m_2 > 0$, we determine the optimal value of m by using Theorem 7(a).

The subsystem size m, $h(m)$, and the expected system cost $E[T(m)]$ are listed in Table 2.2; from this table, we have

$$m_1 = \inf\{m < 26 : h(m) < 50\} = 3$$

Table 2.2. The data for Example 3

m	$h(m)$	$E[T(m)]$
1	110.146	386.17
2	74.051	326.02
3	49.784	301.97
4	33.469	302.19
5	22.499	318.71
6	15.124	346.22
7	10.166	381.10
8	6.832	420.93
9	4.591	464.10
10	3.085	509.50

and

$$E[T(m_1)] = 301.97$$

For $m_2 = 26$, $E[T(m_2)] = 1300.20$. From Theorem 7(a), the optimal value of m required to minimize the expected total system cost is 3, and the expected total system cost corresponding to this value is 301.97.

2.5 The Series–Parallel System

The series–parallel structure is the dual of the parallel–series structure in Section 2.4. We study a system of components arranged so that there are m subsystems operating in series, each subsystem consisting of n identical components in parallel. Such an arrangement is called a series–parallel arrangement. Applications of such systems can be found in the areas of communication, networks, and nuclear power systems. For example, consider a digital communication system consisting of m substations in series. A message is initially sent to substation 1, is then relayed to substation 2, *etc.*, until the message passes through substation m and is received. The message consists of a sequence of 0's and 1's and each digit is sent separately through the series of m substations. Unfortunately, the substations are not perfect and can transmit as output a different digit than that received as input. Such a system is subject to two failure modes: errors in digital transmission occur in such a manner that either (1) a one appears instead of a zero, or (2) a zero appears instead of a one.

Failure in open mode of all the components in any subsystem makes the system unresponsive. Failure in closed (short) mode of a single component in each subsystem also makes the system unresponsive. The probabilities of system failure in open and short mode are given by

$$F_o(m) = 1 - (1 - q_o^n)^m \tag{2.26}$$

and

$$F_s(m) = [1 - (1 - q_s)^n]^m \tag{2.27}$$

respectively. The system reliability is

$$R(m) = (1 - q_o^n)^m - [1 - (1 - q_s)^n]^m \tag{2.28}$$

where m is the number of identical subsystems in series and n is the number of identical components in each parallel subsystem.

Barlow *et al.* [1] show that there exists no pair (m, n) maximizing system reliability. For fixed m, q_o, and q_s, however, one can determine the value of n that maximizes the system reliability.

Theorem 8. *Let n, q_o, and q_s be fixed. The maximum value of $R(m)$ is attained at $m^* = \lfloor m_0 \rfloor + 1$, where*

$$m_0 = \frac{n(\log p_s - \log q_o)}{\log(1 - q_o^n) - \log(1 - p_s^n)} \tag{2.29}$$

If m_0 is an integer, then m_0 and $m_0 + 1$ both maximize $R(m)$.

2.5.1 Maximizing the Average System Profit

The effect of the system parameters on the optimal m is now studied. We also determine the optimal subsystem size that maximizes the average system profit subject to a restricted type I (system failure in open mode) design error.

The following notation is adopted:

β conditional probability (given system failure) that the system is in open mode

$1 - \beta$ conditional probability (given system failure) that the system is in short mode

c_1 gain from system success in open mode

c_2 gain from system failure in open mode ($c_1 > c_2$)

c_3 gain from system success in short mode

c_4 gain from system failure in short mode ($c_3 > c_4$).

The average system-profit, $P(m)$, is given by

$$P(m) = \beta\{c_1[1 - F_o(m)] + c_2 F_o(m)\} + (1 - \beta)\{c_3[1 - F_s(m)] + c_4 F_s(m)\} \tag{2.30}$$

where $F_o(m)$ and $F_s(m)$ are defined as in Equations 2.26 and 2.27 respectively. Let

$$a = \frac{\beta(c_1 - c_2)}{(1 - \beta)(c_3 - c_4)}$$

and

$$b = \beta c_1 + (1 - \beta)c_4$$

We can rewrite Equation 2.30 as

$$P(m) = (1 - \beta)(c_3 - c_4) \times [1 - F_s(m) - aF_o(m)] + b \tag{2.31}$$

For a given value of n, one wishes to find the optimal number of subsystems m, say m^*, that maximizes the average system-profit. We would anticipate that m^* depends on the values of both q_o and q_s. Let

$$m_0 = \frac{n \ln\left(\dfrac{1 - q_s}{q_o}\right) - \ln a}{\ln\left[\dfrac{1 - q_o^n}{1 - (1 - q_s)^n}\right]} \tag{2.32}$$

Theorem 9. *Fix β, n, q_o, q_s, and c_i for $i = 1, 2, 3, 4$. The maximum value of $P(m)$ is attained at*

$$m^* = \begin{cases} 1 & \text{if } m_0 < 0 \\ \lfloor m_0 \rfloor + 1 & \text{if } m_0 \geq 0 \end{cases}$$

If $m_0 \geq 0$ and m_0 is an integer, both m_0 and $m_0 + 1$ maximize $P(m)$.

The proof is straightforward and left as an exercise. When both m_0 and $m_0 + 1$ maximize the average system profit, the lower of the two values costs less. It is of interest to study how m^* depends on the various parameters q_o and q_s.

Theorem 10. *For fixed n, c_i for $i = 1, 2, 3, 4$.*

(*a*) *If $a \geq 1$, then the optimal subsystem size m^* is a decreasing function of q_o.*
(*b*) *If $a \leq 1$, the optimal subsystem size m^* is an increasing function of q_s.*

This theorem states that when q_o increases, it is desirable to reduce m as close to one as is feasible. On the other hand, when q_s increases, the average system-profit increases with the number of subsystems.

2.5.2 Consideration of Type I Design Error

The solution provided by Theorem 9 is optimal in terms of the average system profit. Such an optimal configuration, when adopted, leads to a type I design error (system failure in open mode), which may not be acceptable at the design stage. It should be noted that the more subsystems we add to the system the greater is the chance of system failure by opening ($F_o(m)$, Equation 2.26)); however, we do make the probability of system failure in short mode smaller by placing additional subsystems in series. Therefore, given β, n, q_o, q_s and c_i for $i = 1, 2, \ldots, 4$, we wish to determine the optimal subsystem size m^* in order to maximize the average system profit $P(m)$ in such a way that the probability of system type I design error (*i.e.* the probability of system failure in open mode) is at most α.

Theorem 9 remains unchanged if m^* obtained from Theorem 9 is kept within the tolerableα level, namely $F_o(m) \leq \alpha$. Otherwise, modifications are needed to determine the optimal system size. This is stated in the following result.

Corollary 1. *For given values of β, n, q_o, q_s, and c_i for $i = 1, 2, \ldots, 4$, the optimal value of m, say m^*, that maximizes the average system profit subject to a restricted type I design error α is attained at*

$$m^* = \begin{cases} 1 \quad \text{if } \min\{\lfloor m_o \rfloor, \lfloor m_1 \rfloor\} \\ \min\{\lfloor m_o \rfloor + 1, \lfloor m_1 \rfloor\} \\ \quad \text{otherwise} \end{cases}$$

where $\lfloor m_o \rfloor + 1$ is the solution obtained from Theorem 9 and

$$m_1 = \frac{\ln(1-a)}{\ln(1-q_o^n)}$$

2.6 The k-out-of-n Systems

Consider a model in which a k-out-of-n system is composed of n identical and independent components that can be either good or failed. The components are subject to two types of failure: failure in open mode and failure in closed mode. The system can fail when k or more components fail in closed mode or when $(n - k + 1)$ or more components fail in open mode. Applications of k-out-of-n systems can be found in the areas of target detection, communication, and safety monitoring systems, and, particularly, in the area of human organizations. The following is an example in the area of human organizations. Consider a committee with n members who must decide to accept or reject innovation-oriented projects. The projects are of two types: "good" and "bad". It is assumed that the communication among the members is limited, and each member will make a yes–no decision on each project. A committee member can make two types of error: the error of accepting a bad project and the error of rejecting a good project. The committee will accept a project when k or more members accept it, and will reject a project when $(n - k + 1)$ or more members reject it. Thus, the two types of potential error of the committee are: (1) the acceptance of a bad project (which occurs when k or more members make the error of accepting a bad project); (2) the rejection of a good project (which occurs when $(n - k + 1)$ or more members make the error of rejecting a good project). This section determines the:

- optimal k that minimizes the expected total system cost;
- optimal n that minimizes the expected total system cost;
- optimal k and n that minimizes the expected total system cost.

We also study the effect of the system's parameters on the optimal k or n. The system fails in closed mode if and only if at least k of its n components fail in closed mode, and we obtain

$$F_s(k,n) = \sum_{i=k}^{n} \binom{n}{i} q_s^i p_s^{n-i} = 1 - \sum_{i=0}^{k-1} \binom{n}{i} q_s^i p_s^{n-i} \tag{2.33}$$

The system fails in open mode if and only if at least $n-k+1$ of its n components fail in open mode, that is:

$$F_o(k,n) = \sum_{i=n-k+1}^{n} \binom{n}{i} q_o^i p_o^{n-i} = \sum_{i=0}^{k-1} \binom{n}{i} p_o^i q_o^{n-i} \tag{2.34}$$

Hence, the system reliability is given by

$$\begin{aligned} R(k,n) &= 1 - F_o(k,n) - F_s(k,n) \\ &= \sum_{i=0}^{k-1} \binom{n}{i} q_s^i p_s^{n-i} - \sum_{i=0}^{k-1} \binom{n}{i} p_o^i q_o^{n-i} \end{aligned} \tag{2.35}$$

Let

$$b(k;\, p, n) = \binom{n}{k} p^k (1-p)^{n-k}$$

and

$$b\,\mathrm{inf}(k;\, p, n) = \sum_{i=0}^{k} b(i;\, p, n)$$

We can rewrite Equations 2.33–2.35 as

$$\begin{aligned} F_s(k,n) &= 1 - b\,\mathrm{inf}(k-1;\, q_s, n) \\ F_o(k,n) &= b\,\mathrm{inf}(k-1;\, p_o, n) \\ R(k,n) &= 1 - b\,\mathrm{inf}(k-1;\, q_s, n) \\ &\quad - b\,\mathrm{inf}(k-1;\, p_o, n) \end{aligned}$$

respectively. For a given k, we can find the optimum value of n, say n^*, that maximizes the system reliability.

Theorem 11. *For fixed k, q_o, and q_s, the maximum value of $R(k,n)$ is attained at $n^* = \lfloor n_0 \rfloor$ where*

$$n_0 = k \left[1 + \frac{\log\left(\frac{1-q_o}{q_s}\right)}{\log\left(\frac{1-q_s}{q_o}\right)} \right]$$

If n_0 is an integer, both n_0 and n_0+1 maximize $R(k,n)$.

This result shows that when n_0 is an integer, both n^*-1 and n^* maximize the system reliability $R(k,n)$. In such cases, the lower value will provide the more economical optimal configuration for the system. If $q_o = q_s$, the system reliability $R(k,n)$ is maximized when $n = 2k$ or $2k-1$. In this case, the optimum value of n does not depend on the value of q_o and q_s, and the best choice for a decision voter is a majority voter; this system is also called a majority system [12].

From the above Theorem 11 we understand that the optimal system size n^* depends on the various parameters q_o and q_s. It can be shown the optimal value n^* is an increasing function of q_o and a decreasing function of q_s. Intuitively, these results state that when q_s increases it is desirable to reduce the number of components in the system as close to the value of threshold level k as possible. On the other hand, when q_o increases, the system reliability will be improved if the number of components increases.

For fixed n, q_o, and q_s, it is straightforward to see that the maximum value of $R(k,n)$ is attained at $k^* = \lfloor k_0 \rfloor + 1$, where

$$k_0 = n \frac{\log\left(\frac{q_o}{p_s}\right)}{\log\left(\frac{q_s q_o}{p_s p_o}\right)}$$

If k_0 is an integer, both k_0 and k_0+1 maximize $R(k,n)$.

We now discuss how these two values, k^* and n^*, are related to one another. Define α by

$$\alpha = \frac{\log\left(\frac{q_o}{p_s}\right)}{\log\left(\frac{q_s q_o}{p_s p_o}\right)}$$

then, for a given n, the optimal threshold k is given by $k^* = \lceil n\alpha \rceil$, and for a given k the optimal n is $n^* = \lfloor k/\alpha \rfloor$. For any given q_o and q_s, we can easily show that

$$q_s < \alpha < p_o$$

Therefore, we can obtain the following bounds for the optimal value of the threshold k:

$$nq_s < k^* < np_o$$

This result shows that for given values of q_o and q_s, an upper bound for the optimal threshold k^* is the expected number of components working in open mode, and a lower bound for the optimal threshold k^* is the expected number of components failing in closed mode.

2.6.1 Minimizing the Average System Cost

We adopt the following notation:

d each component cost
c_1 cost when system failure is in open mode
c_2 cost when system failure is in short mode

$$b\inf(k;\, q_s, n) = 1 - b\inf(k-1;\, q_s, n)$$

The average total system cost $E[T(k,n)]$ is

$$E[T(k,n)] = dn + [c_1 F_o(k,n) + c_2 F_s(k,n)] \tag{2.36}$$

In other words, the average total system cost is the cost of all components in the system (dn), plus the average cost of system failure in the open mode ($c_1 F_o(k,n)$) and the average cost of system failure in the short mode ($c_2 F_s(k,n)$).

We now study the problem of how design policies can be chosen when the objective is to minimize the average total system cost when the cost of components, the costs of system failure in the open, and short modes are given. We wish to find the:

- optimal k (k^*) that minimizes the average system cost for a given n;
- optimal n (n^*) that minimizes the average system cost for a given k;
- optimal k and n (k^*, n^*) that minimize the average system cost.

Define

$$k_0 = \frac{\log\left(\frac{c_2}{c_1}\right) + n\log\left(\frac{p_s}{q_o}\right)}{\log\left(\frac{p_o p_s}{q_o q_s}\right)} \tag{2.37}$$

Theorem 12. *Fix n, q_o, q_s, c_1, c_2, and d. The minimum value of $E[T(k,n)]$ is attained at*

$$k^* = \begin{cases} \max\{1, \lfloor k_0 \rfloor + 1\} & \text{if } k_0 < n \\ n & \text{if } k_0 \ge n \end{cases}$$

If k_0 is a positive integer, both k_0 and $k_0 + 1$ minimize $E[T(k,n)]$.

It is of interest to study how the optimal value of k, k^*, depends on the probabilities of component failure in the open mode (q_o) and in the short mode (q_s).

Corollary 2. *Fix n,*

1. *if $c_1 \ge c_2$, then k^* is decreasing in q_o;*
2. *if $c_1 \le c_2$, then k^* is increasing in q_s.*

Intuitively, this result states that if the cost of system failure in the open mode is greater than or equal to the cost of system failure in the short mode, then, as q_o increases, it is desirable to reduce the threshold level k as close to one as is feasible. Similarly, if the cost of system failure in the open mode is less than or equal to the cost of system failure in the short mode, then, as q_s increases, it is desirable to increase k as close to n as is feasible. Define

$$a = \frac{c_1}{c_2}$$

$$n_0 = \left\lfloor \frac{\log a + k\log\left(\frac{p_o P_s}{q_o q_s}\right)}{\log\left(\frac{p_s}{q_o}\right)} - 1 \right\rfloor$$

$$n_1 = \left\lceil \frac{k-1}{1-q_o} - 1 \right\rceil$$

$$f(n) = \left(\frac{p_s}{q_o}\right)^n \frac{(n+1)q_s - (k-1)}{(n+1)p_o - (k-1)}$$

$$B = a\left(\frac{p_o p_s}{q_o q_s}\right)^k \frac{q_o}{p_s}$$

and

$$n_2 = f^{-1}(B) \quad \text{for } k \le n_2 \le n_1$$

Let

$$n_3 = \inf\left\{n \in [n_2,\, n_0] : h(n) < \frac{d}{c_2}\right\}$$

where

$$h(n) = \binom{n}{k-1} p_o^k q_o^{n-k+1} \times \left[a - \left(\frac{q_o q_s}{p_s p_o}\right)^k \left(\frac{p_s}{q_o}\right)^{n+1} \right] \quad (2.38)$$

It is easy to show that the function $h(n)$ is positive for all $k \leq n \leq n_0$, and is increasing in n for $n \in [k, n_2)$ and is decreasing in n for $n \in [n_2, n_0]$. This result shows that the function $h(n)$ is unimodal and achieves a maximum value at $n = n_2$. Since $n_2 \leq n_1$, and when the probability of component failure in the open mode q_o is quite small, then $n_1 \approx k$; so $n_2 \approx k$. On the other hand, for a given arbitrary q_o, one can find a value n_2 between the values of k and n_1 by using a binary search technique.

Theorem 13. *Fix q_o, q_s, k, d, c_1, and c_2. The optimal value of n, say n^*, such that the system minimizes the expected total cost is $n^* = k$ if $n_0 \leq k$. Suppose $n_0 > k$. Then:*

1. *if $h(n_2) < d/c_2$, then $n^* = k$;*
2. *if $h(n_2) \geq d/c_2$ and $h(k) \geq d/c_2$ then $n^* = n_3$;*
3. *if $h(n_2) \geq d/c_2$ and $h(k) < d/c_2$, then*

$$n^* = \begin{cases} k & \text{if } E[T(k,k)] \leq E[T(k,n_3)] \\ n_3 & \text{if } E[T(k,k)] > E[T(k,n_3)] \end{cases} \quad (2.39)$$

Proof. Let $\Delta E[T(n)] = E[T(k, n+1)] - E[T(k, n)]$. From Equation 2.36, we obtain

$$\Delta E[T(n)] = d - c_1 \binom{n}{k-1} p_o^k q_o^{n-k+1} + c_2 \binom{n}{k-1} q_s^k p_s^{n-k+1} \quad (2.40)$$

Substituting $c_1 = ac_2$ into Equation 2.40, and after simplification, we obtain

$$\Delta E[T(n)] = d - c_2 \binom{n}{k-1} p_o^k q_o^{n-k+1} \times \left[a - \left(\frac{q_o q_s}{p_o p_s}\right)^k \left(\frac{p_s}{q_o}\right)^{n+1} \right] = d - c_2 h(n)$$

The system of size $n+1$ is better than the system of size n if, and only if, $h(n) \geq d/c_2$. If $n_0 \leq k$, then $h(n) \leq 0$ for all $n \geq k$, so that $E[T(k,n)]$ is increasing in n for all $n \geq k$. Thus $n^* = k$ minimizes the expected total system cost. Suppose $n_0 > k$. Since the function $h(n)$ is decreasing in n for $n_2 \leq n \leq n_0$, there exists an n such that $h(n) < d/c_2$ on the interval $n_2 \leq n \leq n_0$. Let n_3 denote the smallest such n. Because $h(n)$ is decreasing on the interval $[n_2, n_0]$ where the function $h(n)$ is positive, we have $h(n) \geq d/c_2$ for $n_2 \leq n \leq n_3$ and $h(n) < d/c_2$ for $n > n_3$. Let n^* be an optimal value of n such that $E[T(k,n)]$ is minimized.

(a) If $h(n_2) < d/c_2$, then $n_3 = n_2$ and $h(k) < h(n_2) < d/c_2$, since $h(n)$ is increasing in $[k, n_2)$ and is decreasing in $[n_2, n_0]$. Note that incrementing the system size reduces the expected system cost only when $h(n) \geq d/c_2$. This implies that $n^* = k$ such that $E[T(k,n)]$ is minimized.

(b) Assume $h(n_2) \geq d/c_2$ and $h(k) \geq d/c_2$. Then $h(n) \geq d/c_2$ for $k \leq n < n_2$, since $h(n)$ is increasing in n for $k < n < n_2$. This implies that $E[T(k, n+1)] \leq E[T(k,n)]$ for $k \leq n < n_2$. Since $h(n_2) \geq d/c_2$, then $h(n) \geq d/c_2$ for $n_2 < n < n_3$ and $h(n) < d/c_2$ for $n > n_3$. This shows that $n^* = n_3$ such that $E[T(k,n)]$ is minimized.

(c) Similarly, assume that $h(n_2) \geq d/c_2$ and $h(k) < d/c_2$. Then, either $n = k$ or $n^* = n_3$ is the optimal solution for n. Thus, $n^* = k$ if $E[T(k,k)] \leq E[T(k,n_3)]$; on the other hand, $n^* = n_3$ if $E[T(k,k)] > E[T(k,n_3)]$. □

In practical applications, the probability of component failure in the open mode q_o is often quite small, and so the value of n_1 is close to k. Therefore, the number of computations for finding a value of n_2 is quite small. Hence, the result of the Theorem 13 is easily applied in practice.

In the remaining section, we assume that the two system parameters k and n are unknown. It is of interest to determine the optimum values of (k, n), say (k^*, n^*), that minimize the expected total system cost when the cost of components and the costs of system failures are known. Define

$$\alpha = \frac{\log(p_s/q_o)}{\log(p_o p_s/q_o q_s)} \qquad \beta = \frac{\log(c_2/c_1)}{\log(p_o p_s/q_o q_s)} \quad (2.41)$$

We need the following lemma.

Lemma 1. *For $0 \le m \le n$ and $0 \le p \le 1$:*

$$\sum_{i=0}^{m} \binom{n}{i} p^i (1-p)^{n-i} < \sqrt{\frac{n}{2\pi m(n-m)}}$$

Proof. See [5], Lemma 3.16, for a detailed proof. □

Theorem 14. *Fix q_o, q_s, d, c_1, and c_2. There exists an optimal pair of values (k_n, n), say (k_{n^*}, n^*), such that average total system cost is minimized at (k_{n^*}, n^*), and*

$$k_{n^*} = \lfloor n^* \alpha \rfloor$$

and

$$n^* \le \frac{\frac{(1-q_o-q_s)}{2\pi}\left(\frac{c_1}{d}\right)^2 + 1 + \beta}{\alpha(1-\alpha)} \tag{2.42}$$

Proof. Define $\Delta E[T(n)] = E[T(k_{n+1}, n+1)] - E[T(k_n, n)]$. From Equation 2.36, we obtain

$$\begin{aligned}\Delta E[T(n)] = d &+ c_1[b\inf(k_{n+1}-1;\ p_o, n+1) \\ &- b\inf(k_n - 1;\ p_o, n)] \\ &- c_2[b\inf(k_{n+1}-1;\ q_s, n+1) \\ &- b\inf(k_n - 1;\ q_s, n)]\end{aligned}$$

Let $r = c_2/c_1$, then

$$\begin{aligned}\Delta E[T(n)] &= d - c_1 g(n) \\ g(n) &= r[b\inf(k_{n+1}-1;\ q_s, n+1) \\ &\quad - b\inf(k_n - 1;\ q_s, n)] \\ &\quad - [b\inf(k_{n+1}-1;\ p_o, n+1) \\ &\quad - b\inf(k_n - 1;\ p_o, n)]\end{aligned} \tag{2.43}$$

Case 1. Assume $k_{n+1} = k_n + 1$. We have

$$g(n) = \binom{n}{k_n} p_o^{k_n} q_o^{n-k_n+1} \times \left[r\left(\frac{q_o q_s}{p_o p_s}\right)^{k_n} \left(\frac{p_s}{q_o}\right)^{n+1} - 1\right]$$

Recall that

$$\left(\frac{p_o p_s}{q_o q_s}\right)^{\beta} = r$$

then

$$\left(\frac{q_o q_s}{p_o p_s}\right)^{n\alpha+\beta} \left(\frac{p_s}{q_o}\right)^{n+1} = \frac{1}{r}\frac{p_s}{q_o} \tag{2.44}$$

since $n\alpha + \beta \le k_n \le (n+1)\alpha + \beta$, we obtain

$$\begin{aligned} r\left(\frac{q_o q_s}{p_s p_s}\right)^{k_n} \left(\frac{p_s}{q_o}\right)^{n+1} &\le r\left(\frac{q_o q_s}{p_o p_s}\right)^{n\alpha+\beta} \left(\frac{p_s}{q_o}\right)^{n+1} \\ &= \frac{p_s}{q_o}\end{aligned}$$

Thus

$$\begin{aligned} g(n) &\le \binom{n}{k_n} p_o^{k_n} q_o^{n-k_n+1} \left(\frac{p_s}{q_o} - 1\right) \\ &= \binom{n}{k_n} p_o^{k_n} q_o^{n-k_n} (p_s - q_o)\end{aligned}$$

From Lemma 1, and $n\alpha + \beta \le k_n \le (n+1)\alpha + \beta$, we obtain

$$\begin{aligned} g(n) &\le (p_s - q_o)\left[2\pi n \frac{k_n}{n}\left(1 - \frac{k_n}{n}\right)\right]^{-1/2} \\ &\le (p_s - q_o)\left\{2\pi n\left(\alpha + \frac{\beta}{n}\right) \times \left[1 - \alpha\left(\frac{n+1}{n}\right) - \frac{\beta}{n}\right]\right\}^{-1/2} \\ &\le (1 - q_s - q_s) \\ &\quad \times \{2\pi n[n\alpha(1-\alpha) - \alpha(\alpha+\beta)]\}^{-1/2}\end{aligned} \tag{2.45}$$

Case 2. Similarly, if $k_{n+1} = k_n$ then from Equation 2.43, we have

$$g(n) = \binom{n}{k_n - 1} q_s^{k_n} p_s^{n-k_n+1} \times \left[\left(\frac{p_o p_s}{q_o q_s}\right)^{k_n} \left(\frac{q_o}{p_s}\right)^{n+1} - r\right]$$

since $k_n = \lceil n\alpha + \beta \rceil \le n\alpha + \beta + 1$, and from Equation 2.44 and Lemma 1, we have

$$\begin{aligned} g(n) &\le q_s \binom{n}{k_n - 1} q_s^{k_n-1} p_s^{n-k_n+1} \\ &\quad \times \left[\left(\frac{p_o p_s}{q_o q_s}\right)^{n\alpha+\beta+1} \left(\frac{q_o}{p_s}\right)^{n+1} - r \right] \\ &\le q_s \sqrt{\frac{n}{2\pi(k_n-1)[n-(k_n-1)]}} \\ &\quad \times \left[\left(\frac{p_o p_s}{q_o q_s}\right)^{n\alpha+\beta} \left(\frac{q_o}{p_s}\right)^{n+1} \left(\frac{p_o p_s}{q_o q_s}\right) - r \right] \\ &\le q_s \sqrt{\frac{n}{2\pi(k_n-1)[n-(k_n-1)]}} \\ &\quad \times \left[r\left(\frac{q_o}{p_s}\right)\left(\frac{p_o p_s}{q_o q_s}\right) - r \right] \\ &\le \sqrt{\frac{n}{2\pi(k_n-1)[n-(k_n-1)]}} (1 - q_o - q_s) \end{aligned}$$

Note that $k_{n+1} = k_n$, then $n\alpha - (1 - \alpha - \beta) \le k_n - 1 \le n\alpha + \beta$. After simplifications, we have

$$\begin{aligned} g(n) &\le (1 - q_o - q_s) \\ &\quad \times \left[2\pi n \left(\frac{k_n - 1}{n}\right) \left(1 - \frac{k_n - 1}{n}\right) \right]^{-1/2} \\ &\le (1 - q_o - q_s) \\ &\quad \times \left[2\pi n \left(\alpha - \frac{1 - \alpha - \beta}{n}\right) \right. \\ &\quad \left. \times \left(1 - \alpha - \frac{\beta}{n}\right) \right]^{-1/2} \\ &\le \frac{1 - q_o - q_s}{\sqrt{2\pi[n\alpha(1-\alpha) - (1-\alpha)^2 - (1-\alpha)\beta]}} \end{aligned} \tag{2.46}$$

From the inequalities in Equations 2.45 and 2.46, set

$$(1 - q_s - q_o) \frac{1}{\sqrt{2\pi[n\alpha(1-\alpha) - \alpha(\alpha+\beta)]}} \le \frac{d}{c_1}$$

and

$$\frac{1 - q_o - q_s}{\sqrt{2\pi[n\alpha(1-\alpha) - (1-\alpha)^2 - \alpha(\alpha+\beta)]}} \le \frac{d}{c_1}$$

we obtain

$$\begin{aligned} &(1 - q_o - q_s)^2 \left(\frac{c_1}{d}\right)^2 \frac{1}{2\pi} \\ &\quad \le \min\{n\alpha(1-\alpha) - \alpha(\alpha+\beta), n\alpha(1-\alpha) \\ &\qquad - (1-\alpha)^2 - (1-\alpha)\beta\} \\ &\qquad \Delta E[T(n)] \ge 0 \end{aligned}$$

when

$$n \ge \frac{\frac{(1 - q_o - q_s)^2}{2\pi}\left(\frac{c_1}{d}\right) + 1 + \beta}{\alpha(1-\alpha)}$$

Hence

$$n^* \le \frac{\frac{(1 - q_o - q_s)^2}{2\pi}\left(\frac{c_1}{d}\right) + 1 + \beta}{\alpha(1-\alpha)} \qquad \square$$

The result in Equation 2.42 provides an upper bound for the optimal system size.

2.7 Fault-tolerant Systems

In many critical applications of digital systems, fault tolerance has been an essential architectural attribute for achieving high reliability. It is universally accepted that computers cannot achieve the intended reliability in operating systems, application programs, control programs, or commercial systems, such as in the space shuttle, nuclear power plant control, *etc.*, without employing redundancy. Several techniques can achieve fault tolerance using redundant hardware [12] or software [13]. Typical forms of redundant hardware structures for fault-tolerant systems are of two types: fault masking and standby. Masking redundancy is achieved by implementing the functions so that they are inherently error correcting, *e.g.* triple-modular redundancy (TMR), N-modular redundancy (NMR), and self-purging redundancy. In standby redundancy, spare units are switched into the system when working units break down. Mathur and De Sousa [12] have analyzed, in detail, hardware redundancy in the design of fault-tolerant digital systems. Redundant software structures for fault-tolerant systems based on the acceptance tests have been proposed by Horning *et al.* [13].

This section presents a fault-tolerant architecture to increase the reliability of a special class of digital systems in communication [14]. In this system, a monitor and a switch are associated with each redundant unit. The switches and monitors can fail. The monitors have two failure modes: failure to accept a correct result, and failure to reject an incorrect result. The scheme can be used in communication systems to improve their reliability.

Consider a digital circuit module designed to process the incoming messages in a communication system. This module consists of two units: a converter to process the messages, and a monitor to analyze the messages for their accuracy. For example, the converter could be decoding or unpacking circuitry, whereas the monitor could be checker circuitry [12]. To guarantee a high reliability of operation at the receiver end, n converters are arranged in "parallel". All, except converter n, have a monitor to determine if the output of the converter is correct. If the output of a converter is not correct, the output is cancelled and a switch is changed so that the original input message is sent to the next converter. The architecture of such a system has been proposed by Pham and Upadhyaya [14]. Systems of this kind have useful application in communication and network control systems and in the analysis of fault-tolerant software systems.

We assume that a switch is never connected to the next converter without a signal from the monitor, and the probability that it is connected when a signal arrives is p_s. We next present a general expression for the reliability of the system consisting of n non-identical converters arranged in "parallel". An optimization problem is formulated and solved for the minimum average system cost. Let us define the following notation, events, and assumptions.

The notation is as follows:

p_i^{c}	Pr{converter i works}
p_i^{s}	Pr{switch i is connected to converter $(i+1)$ when a signal arrives}
p_i^{m1}	Pr{monitor i works when converter i works} = Pr{not sending a signal to the switch when converter i works}
p_i^{m2}	Pr{i monitor works when converter i has failed} = Pr{sending a signal to the switch when converter i has failed}
R_{n-k}^{k}	reliability of the remaining system of size $n-k$ given that the first k switches work
R_n	reliability of the system consisting of n converters.

The events are:

C_i^{w}, C_i^{f}	converter i works, fails
M_i^{w}, M_i^{f}	monitor i works, fails
S_i^{w}, S_i^{f}	switch i works, fails
W	system works.

The assumptions are:

1. the system, the switches, and the converters are two-state: good or failed;
2. the module (converter, monitor, or switch) states are mutually statistical independent;
3. the monitors have three states: good, failed in mode 1, failed in mode 2;
4. the modules are not identical.

2.7.1 Reliability Evaluation

The reliability of the system is defined as the probability of obtaining the correctly processed message at the output. To derive a general expression for the reliability of the system, we use an adapted form of the total probability theorem as translated into the language of reliability. Let A denote the event that a system performs as desired. Let X_i and X_j be the event that a component X (*e.g.* converter, monitor, or switch) is good or failed respectively. Then

$$\begin{aligned}
&\Pr\{\text{system works}\}\\
&\quad = \Pr\{\text{system works when unit } X \text{ is good}\}\\
&\qquad \times \Pr\{\text{unit } X \text{ is good}\}\\
&\qquad + \Pr\{\text{system works when unit } X \text{ fails}\}\\
&\qquad \times \Pr\{\text{unit } X \text{ is failed}\}
\end{aligned}$$

The above equation provides a convenient way of calculating the reliability of complex systems. Notice that $R_1 = p_i^c$, and for $n \geq 2$, the reliability of the system can be calculated as follows:

$$\begin{aligned} R_n = &\Pr\{W \mid C_1^w \text{ and } M_1^w\} \Pr\{C_1^w \text{ and } M_1^w\} \\ &+ \Pr\{W \mid C_1^w \text{ and } M_1^f\} \Pr\{C_1^w \text{ and } M_1^f\} \\ &+ \Pr\{W \mid C_1^f \text{ and } M_1^w\} \Pr\{C_1^f \text{ and } M_1^w\} \\ &+ \Pr\{W \mid C_1^f \text{ and } M_1^f\} \Pr\{C_1^f \text{ and } M_1^f\} \end{aligned}$$

In order for the system to operate when the first converter works and the first monitor fails, the first switch must work and the remaining system of size $n-1$ must work:

$$\Pr\{W \mid C_1^w \text{ and } M_1^f\} = p_1^s R_{n-1}^1$$

Similarly:

$$\Pr\{W \mid C_1^f \text{ and } M_1^w\} = p_1^s R_{n-1}^1$$

then

$$R_n p_1^c p_1^{ml} + [p_1^c(1 - p_1^{ml}) + (1 - p_1^c) p_1^{m2}] p_1^s R_{n-1}^1$$

The reliability of the system consisting of n non-identical converters can be easily obtained:

$$R_n = \sum_{i=1}^{n-1} p_i^c p_i^{ml} \pi_{i-1} + \pi_{n-1} p_n^c \quad \text{for } n > 1 \tag{2.47}$$

and

$$R_1 = p_1^c$$

where

$$\pi_k^j = \prod_{i=j}^{k} A_i \quad \text{for } k \geq 1$$

$$\pi_k = \pi_k^1 \quad \text{for all } k, \text{ and } \pi_0 = 1$$

and

$$A_i \equiv [p_i^c(1 - p_i^{ml}) + (1 - p_i^c) p_i^{m2}]$$

for all $i = 1, 2, \ldots, n$. Assume that all the converters, monitors, and switches have the same reliability, that is:

$$p_i^c = p^c, \quad p_i^{ml} = p^{ml}, \quad p_i^{m2} = p^{m2}, \quad p_i^s = p^s$$

for all i, then we obtain a closed form expression for the reliability of system as follows:

$$R_n = \frac{p^c p^{ml}}{1 - A}(1 - A^{n-1}) + p^c A^{n-1} \tag{2.48}$$

where

$$A = [p^c(1 - p^{ml}) + (1 - p^c) p^{m2}] p^s$$

2.7.2 Redundancy Optimization

Assume that the system failure costs d units of revenue, and that each converter, monitor, and switch module costs a, b, and c units respectively. Let T_n be system cost for a system of size n. The average system cost for size n, $E[T_n]$, is the cost incurred when the system has failed, plus the cost of all n converters, $n-1$ monitors, and $n-1$ switches. Therefore:

$$E[T_n] = an + (b + c)(n - 1) + d(1 - R_n)$$

where R_n is given in Equation 2.48. The minimum value of $E[T_n]$ is attained at

$$n^* = \begin{cases} 1 & \text{if } A \leq 1 - p^{ml} \\ \lfloor n_0 \rfloor & \text{otherwise} \end{cases}$$

where

$$n_0 = \frac{\ln(a + b + c) - \ln[dp^c(A + p^{ml} - 1)]}{\ln A} + 1$$

Example 4. [14] Given a system with $p^c = 0.8$, $p^{ml} = 0.90$, $p^{m2} = 0.95$, $p^s = 0.90$, and $a = 2.5$, $b = 2.0$, $c = 1.5$, $d = 1200$. The optimal system size is $n^* = 4$, and the corresponding average cost (81.8) is minimized.

2.8 Weighted Systems with Three Failure Modes

In many applications, ranging from target detection to pattern recognition, including safety-monitoring protection, undersea communication, and human organization systems, a decision has to be made on whether or not to accept the hypothesis based on the given information so that the probability of making a correct decision is maximized. In safety-monitoring protection systems, *e.g.* in a nuclear power plant, where the system state is monitored by a multi-channel sensor system, various core groups of sensors monitor the status of neutron flux density, coolant temperature at the reaction core exit (outlet temperature), coolant temperature at the core entrance (inlet temperature), coolant flow rate, coolant level in pressurizer, on–off status of coolant pumps. Hazard-preventive actions should be performed when an unsafe state is detected by the sensor system. Similarly, in the case of chlorination of a hydrocarbon gas in a gas-lined reactor, the possibility of an exothermic, runway reaction occurs whenever the Cl_2/hydrocarbon gas ratio is too high, in which case a detonation occurs, since a source of ignition is always present. Therefore, there are three unsafe phenomena: a high chlorine flow y_1, a low hydrocarbon gas flow y_2, and a high chlorine-to-gas ratio in the reactor y_3. The chlorine flow must be shut off when an unsafe state is detected by the sensor system. In this application, each channel monitors a different phenomenon and has different failure probabilities in each mode; the outputs of each channel will have different weights in the decision (output). Similarly, in each channel, there are distinct number of sensors and each sensor might have different capabilities, depending upon its physical position. Therefore, each sensor in a particular channel might have different failure probabilities; thereby, each sensor will have different weights on the channel output. This application can be considered as a two-level weighted threshold voting protection systems.

In undersea communication and decision-making systems, the system consists of n electronic sensors each scanning for an underwater enemy target [16]. Some electronic sensors, however, might falsely detect a target when none is approaching. Therefore, it is important to determine a threshold level that maximizes the probability of making a correct decision.

All these applications have the following working principles in common. (1) System units make individual decisions; thereafter, the system as an entity makes a decision based on the information from the system units. (2) The individual decisions of the system units need not be consistent and can even be contradictory; for any system, rules must be made on how to incorporate all information into a final decision. System units and their outputs are, in general, subject to different errors, which in turn affects the reliability of the system decision.

This chapter has detailed the problem of optimizing the reliability of systems with two failure modes. Some interesting results concerning the behavior of the system reliability function have also been discussed. Several cost optimization problems are also presented. This chapter also presents a brief summary of recent studies in reliability analysis of systems with three failure modes [17–19]. Pham [17] studied dynamic redundant system with three failure modes. Each unit is subject to stuck-at-0, stuck-at-1 and stuck-at-x failures. The system outcome is either good or failed. Focusing on the dynamic majority and k-out-of-n systems, Pham derived optimal design policies for maximizing the system reliability. Nordmann and Pham [18] have presented a simple algorithm to evaluate the reliability of weighted dynamic-threshold voting systems, and they recently presented [19] a general analytic method for evaluating the reliability of weighted-threshold voting systems. It is worth considering the reliability of weighted voting systems with time-dependency.

References

[1] Barlow RE, Hunter LC, Proschan F. Optimum redundancy when components are subject to two kinds of failure. J Soc Ind Appl Math 1963;11(1):64–73.

[2] Ben-Dov Y. Optimal reliability design of k-out-of-n systems subject to two kinds of failure. J Opt Res Soc 1980;31:743–8.

[3] Dhillon BS, Rayapati SN. A complex system reliability evaluation method. Reliab Eng 1986;16:163–77.

[4] Jenney BW, Sherwin DJ. Open and short circuit reliability of systems of identical items. IEEE Trans Reliab 1986;R-35:532–8.

[5] Pham H. Optimal designs of systems with competing failure modes. PhD Dissertation, State University of New York, Buffalo, February 1989 (unpublished).

[6] Malon DM. On a common error in open and short circuit reliability computation. IEEE Trans Reliab 1989;38: 275–6.

[7] Page LB, Perry JE. Optimal series–parallel networks of 3-stage devices. IEEE Trans Reliab 1988;37:388–94.

[8] Pham H. Optimal design of systems subject to two kinds of failure. Proceedings Annual Reliability and Maintainability Symposium, 1990. p.149–52.

[9] Pham H, Pham M. Optimal designs of $\{k, n-k+1\}$ out-of-n: F systems (subject to 2 failure modes). IEEE Trans Reliab 1991;40:559–62.

[10] Sah RK, Stiglitz JE. Qualitative properties of profit making k-out-of-n systems subject to two kinds of failures. IEEE Trans Reliab 1988;37:515–20.

[11] Pham H, Malon DM. Optimal designs of systems with competing failure modes. IEEE Trans Reliab 1994;43: 251–4.

[12] Mathur FP, De Sousa PT. Reliability modeling and analysis of general modular redundant systems. IEEE Trans Reliab 1975;24:296–9.

[13] Horning JJ, Lauer HC, Melliar-Smith PM, Randell B. A program structure for error detection and recovery. Lecture Notes in Computer Science, vol. 16. Springer; 1974. p.177–93.

[14] Pham H, Upadhyaya SJ. Reliability analysis of a class of fault-tolerant systems. IEEE Trans Reliab 1989;38:333–7.

[15] Pham H, editor. Fault-tolerant software systems: techniques and applications. Los Alamitos (CA): IEEE Computer Society Press; 1992.

[16] Pham H. Reliability analysis of digital communication systems with imperfect voters. Math Comput Model J 1997;26:103–12.

[17] Pham H. Reliability analysis of dynamic configurations of systems with three failure modes. Reliab Eng Syst Saf 1999;63:13–23.

[18] Nordmann L, Pham H. Weighted voting human-organization systems. IEEE Trans Syst Man Cybernet Pt A 1997;30(1):543–9.

[19] Nordmann L, Pham H. Weighted voting systems. IEEE Trans Reliab 1999;48:42–9.

Reliabilities of Consecutive-k Systems

Jen-Chun Chang and Frank K. Hwang

Chapter 3

3.1 Introduction

In this chapter we introduce the consecutive-k systems, including the original well-known consecutive-k-out-of-n:F system. Two general themes are considered: computing the system reliability and maximizing the system reliability through optimally assigning components to positions in the system.

3.1.1 Background

The consecutive-k-out-of-n:F system is a system of n components arranged in a line such that the system fails if and only if some k consecutive components fail. It was first studied by Kontoleon [1] under the name of r-successive-out-of-n:F system, but Kontoleon only gave an enumerative reliability algorithm for the system. Chiang and Niu [2] motivated the study of the system by some real applications. They proposed the current name "consecutive-k-out-of-n:F system" for the system and gave an efficient reliability algorithm by recursive equations. From then on, the system became more and more popular.

There are many variations and generalizations of the system, such as circular consecutive-k-out-of-n:F systems [3], weighted-consecutive-k-out-of-n:F systems [4, 5], f-or-consecutive-k-out-of-n:F systems [6, 7], f-within-consecutive-k-out-of-n:F systems [8, 9], consecutive-k-out-of-n:F networks [10, 11], consecutive-k-out-of-n:F flow networks [8], and consecutive-$k - r$-out-of-n:DFM systems [12], *etc.* Reliability analysis for these systems has been widely studied in recent years.

There are other variations called consecutive-k:G systems, which are defined by exchanging the role of working and failed components in the consecutive-k systems. For example, the consecutive-k-out-of-n:G system works if and only if some k consecutive components all work. The reliability of the G system is simply the unreliability of the F system computed by switching the component reliabilities and the unreliabilities. However, to maximize the reliability of a G system is not equivalent to minimize the reliability of an F system.

Two basic assumptions for the consecutive-k systems are described as follows.

1. Binary system. The components and the system all have two states: working or failed.
2. IND model. The states of the components are independent. In addition, if the working probabilities of all components are identical, we call it the IID model.

We consider two general problems in this chapter: computing the system reliability and maximizing the system reliability through optimally assigning components to positions in the system.

Two general approaches are used to compute the reliability for the IND probability models: one is the recursive equation approach, and the other is the Markov chain approach. The recursive equation approach was first pursued by Chiang and Niu [2] and Derman *et al.* [3]. The Markov chain approach was first proposed by Griffith and Govindarajula [13], Griffith [14], Chao and Lin [15], and perfected by Fu [16].

The general reliability optimization problem is to allocate $m \geq n$ components with non-identical reliabilities to the n positions in a system, where a position can be allocated one or more components to maximize the system reliability (it is assumed that the components are functionally equivalent). Two different versions of this general problem are:

1. Sequential. Components are allocated one at a time. Once a component is placed into a position, its state is known.
2. Non-adaptive. Components are allocated simultaneously.

Since the general optimization problem is very difficult, in most cases we do not have a globally optimal algorithm. Sometimes heuristics are used. The special case which we study here is $m = n$ under the IND model.

3.1.2 Notation

The following notation is used in this chapter.

$\Pr(e)$	probability of event e
$\Pr(e_1 \mid e_2)$	probability of event e_1 under the condition that event e_2 occurs
$E(X)$	expectation of a random variable X
$E(X \mid e)$	expectation of a random variable X under the condition that event e occurs
$\text{Trace}(\mathbf{M})$	trace of a matrix $\mathbf{M}$
n	number of components in the system
p_i	probability that component i functions
q_i	$1 - p_i$
$L(1, n)$ or L	consecutive-k-out-of-n linear system, where k and all p_i are assumed understood

$C(1, n)$ or C	circular consecutive-k-out-of-n linear system, where k and all p_i are assumed understood
$R_x(1, n)$ or R_x	reliability of the system $x(1, n)$ where $x \in \{L, C\}$, k and all p_i are assumed understood
$\bar{R}_x(1, n)$ or R_x	$1 - R_x(1, n)$, unreliability of the system $x(1, n)$
$R_x(n)$	$R_x(1, n)$ when all $p_i = p$ and all $q_i = q$
$\bar{R}_x(n)$	$1 - R_x(n)$
s_i	working state of component i
$\bar{s}_i$	failure state of component i
$x(i, j)$	x system containing components $i, i+1, \ldots, j$
$R_x(i, j)$	reliability of the subsystem $x(i, j)$
$\bar{R}_x(i, j)$	$1 - R_x(i, j)$, unreliability of the subsystem $x(i, j)$
$S_x(i, j)$	set of all component states including a working $x(i, j)$
$\bar{S}_x(i, j)$	set of all component states including a failed $x(i, j)$
$N_x(d, n, k)$	number of ways that n nodes, including exactly d failed nodes, ordered on a line (if $x = L$) or a cycle (if $x = C$) contain no k consecutive failed nodes.

3.2 Computation of Reliability

We introduce various approaches to compute the reliability of the consecutive-k-out-of-n:F system and its circular version. In addition, we also compare the time complexities of these approaches.

3.2.1 The Recursive Equation Approach

The first solution of the reliability of the linear consecutive-k-out-of-n:F system was given by Chiang and Niu [2] (for the IID model). The solution is in the form of recursive equations. From then on, many other recursive equations have been proposed to improve the efficiency of the reliability computation. We only present the fastest ones.

Shanthikumar [17] and Hwang [18] gave the following recursive equation for the reliability of the linear consecutive-k-out-of-n:F system.

Theorem 1.

$$R_L(1, n) = R_L(1, n-1) - R_L(1, n-k-1)p_{n-k} \prod_{j=1}^{k} q_{n-k+j}$$

This equation holds for any $n \geq k+1$. Since $R_L(1, n)$ can be computed after $R_L(1, 1)$, $R_L(1, 2), \ldots, R_L(1, n-1)$ are all computed in that order, the system reliability $R_L(1, n)$ can be computed in $O(n)$ time.

The solution for the reliability of the circular consecutive-k-out-of-n:F system was first given by Derman *et al.* [3]. Note that the indices are taken modulo n for circular systems.

Theorem 2.

$$R_C(1, n) = \sum_{\substack{1 \leq i \leq k \\ n-k+i \leq j \leq n}}^{n} p_j \left(\prod_{h=j+1}^{n+i-1} q_h \right) \times p_i R_L(i+1, j-1)$$

Hwang [18] observed that there are $O(k^2)$ R_L terms in the right-hand side of the equation, each needing $O(n)$ time to compute. Therefore, Hwang announced that $R_C(1, n)$ can be computed in $O(k^2 n)$ time. Wu and Chen [19] observed that $R_L(i, i), R_L(i, i+1), \ldots, R_L(i, n)$ can be computed together in $O(n)$ time. Hence, they claimed that $R_C(1, n)$ can be computed in $O(kn)$ time.

Wu and Chen [20] also showed a trick of using fictitious components to make the $O(kn)$ time more explicit. Let $L^W(1, n+i)$ be a linear system with $p_1 = p_2 = \cdots = p_i = p_{n+1} = p_{n+2} = \cdots = p_{n+i} = 0$. They gave the following recursive equation.

Theorem 3.

$$R_C(1,n) = R_L(1,n) - \sum_{i=1}^{k-1} \prod_{j=1}^{i} q_j \times [R_{L^W}(1, n+i-1) - R_{L^W}(1, n+i)]$$

Because there are only $O(kn)$ terms of R_L, $R_C(1,n)$ can be computed in $O(kn)$ time.

In addition, Antonopoulou and Papastavridis [21] gave the following recursive equation.

Theorem 4.

$$R_C(1,n) = p_n R_L(1, n-1) + q_n R_C(1, n-1) - \sum_{i=1}^{k} p_{n-k+i-1} \left(\prod_{j=n-k+i}^{n+i-1} q_j \right) \times p_i R_L(i+1, n-k+i-2)$$

Antonopoulou and Papastavridis claimed that this is an $O(kn)$ algorithm, but Wu and Chen [20] found that the computational complexity is $O(kn^2)$ rather than $O(kn)$. Later, Hwang [22] gave a different, more efficient implementation of the same recursive equation, which is an $O(kn)$ algorithm.

3.2.2 The Markov Chain Approach

The Markov chain approach was first introduced into the reliability study by Griffith [14] and Griffith and Govindarajula [13], but they did not give an efficient algorithm. The first efficient algorithm based on the Markov chain approach was given by Chao and Lin [15].

We briefly describe the Markov chain approach as follows. Chao and Lin [15] defined S_v as the aggregate of states of nodes $v-k+1, v-k+2, \ldots, v$. Hence $\{S_v\}$ forms a Markov chain, and S_{v+1} depends only on S_v. Fu [16] lowered the number of states by defining S_v to be in state i, $i \in \{0, 1, \ldots, k\}$ if the last i nodes including node v are all failed. Note that all states (except the failed state k) are working states. The following transition probability matrix for S_v was given by Fu and Hu [23].

$$\Lambda_v = \begin{bmatrix} p_v & q_v & & & \\ p_v & & q_v & & \\ \vdots & & & \ddots & \\ p_v & & & & q_v \\ 0 & & & & 1 \end{bmatrix}$$

Let $\boldsymbol{\pi}_0$ and $\mathbf{U}_0$ denote the $1 \times (k+1)$ row vector $\{1, 0, \ldots, 0\}$ and the $(k+1) \times 1$ column vector of all 1 except the last bit is 0, respectively. The system reliability of the linear consecutive-k-out-of-n:F system can be computed as

$$R_L(1,n) = \boldsymbol{\pi}_0 \left(\prod_{v=1}^{n} \Lambda_v \right) \mathbf{U}_0$$

Fu and Hu did not give any time complexity analysis. Koutras [24] restated this method and claimed that $R_L(1,n)$ can be computed in $O(kn)$ time.

For the circular system, the Markov chain method does not work, since a Markov chain must have a starting point. Hwang and Wright [25] gave a different Markov chain method (called the transfer matrix method) that works well for the circular system. For simplicity, assume k divides n. The n nodes are divided into n/k sets of k each. The ith set consists of nodes $(i-1)k+1, (i-1)k+2, \ldots, ik$. A state of a set is simply a vector of states of its k elements. Let $S_i = \{S_{iu}\}$ be the state space of set i. Then $|S_i| = 2^k$. Let $\mathbf{M}_i$ denote the matrix whose rows are indexed by the elements of S_i, and columns are indexed by the elements of S_{i+1}. Cell $(u, v) = 1$ if $S_{iu} \cup S_{(i+1)v}$ does not contain k consecutive failed nodes, and 0 otherwise. For the circular system, let $S_{n/k+1} = S_1$. The number of working cases out of the 2^n total cases is

$$\text{Trace}\left(\prod_{i=1}^{n/k} \mathbf{M}_i \right)$$

Hwang and Wright [25] described a novel way to compute the reliability by substituting each entry 1 in cell (u, v) of $\mathbf{M}_i$ with

$$\sqrt{\Pr(S_{iu}) \Pr(S_{(i+1)v})}$$

since each S_{iu} appears once in $\mathbf{M}_i$ and once in $\mathbf{M}_{i+1}$. This method can also be applied to linear systems. For a linear system, the corresponding term is then

$$\prod_{i=1}^{n/k-1} \mathbf{M}_i$$

and $M_1(u, v) = 1$ needs to be replaced with $\Pr(S_{1u})\sqrt{\Pr(S_{2v})}$ and $M_{n/k-1}(u, v) = 1$ with $\sqrt{\Pr(S_{n/k-1,v})}\,\Pr(S_{n/k,u})$. In this approach, there are $O(n/k)$ matrix multiplications where each takes $O(2^{3k})$ time. Therefore, the total computational complexity is $O(2^{3k}n/k)$. Hwang and Wright [25] also found a clever way to speed up the multiplication time from $O(2^{3k})$ to $O(k^6)$. For a working state S_{iu}, let l denote the index of the first working node and r ($\geq l$) denote the last. For each S_i, delete the unique failed state and regroup the working states according to (l, r). The size of $|S_i|$ is lowered down to $O(k^2)$, while the ability to determine whether $S_{iu} \cup S_{(i+1)v}$ contains k consecutive failed nodes is preserved. Note that

$$\Pr(S_{i,(l,r)}) = \left(\prod_{j=1}^{l-1} q_{(i-1)k+j}\right) p_{(i-1)k+l}\, p_{(i-1)k+r} \times \left(\prod_{j=1}^{k-1} q_{(i-1)k+r+j}\right)$$

but without $p_{(i-1)k+r}$ if $l = r$. The total computational complexity for the reliability is $O(k^5 n)$.

3.2.3 Asymptotic Analysis

Chao and Lin [15] conjectured, and Fu [26], proved the following result.

Theorem 5. *For any integer $k \geq 1$, if the component reliability $p_n = 1 - \lambda n^{-1/k}$ ($\lambda > 0$) holds for IID model, then*

$$\lim_{n\to\infty} R_L(n) = \exp\{-\lambda^k\}$$

Chao and Fu [27] embedded the consecutive-k-out-of-n line into a finite Markov chain $\{X_0, X_1, \ldots, X_n\}$ with states $\{1, \ldots, k+1\}$ and transition probability matrix (blanks are zeros)

$$\Lambda_n(t) = \begin{pmatrix} p_t & q_t & & & \\ p_t & & q_t & & \\ \vdots & & & \ddots & \\ p_t & & & & q_t \\ 0 & & & & 1 \end{pmatrix}_{(k+1)\times(k+1)}$$

Let $\boldsymbol{\pi}_0 = (\pi_1, \ldots, \pi_{k+1})$ be the initial probability vector. Then

$$R_L(n) = \boldsymbol{\pi}_0 \prod_{t=1}^{n} \Lambda_n(t)\mathbf{U}_0$$

where $\mathbf{U}_0$ is the $(k+1) \times 1$ column vector $(1, \ldots, 1, 0)^{\mathrm{T}}$.

Define

$$b(t, n) = \Pr\{X_t = k+1 \mid X_{t-1} \leq k\}$$

and

$$\lambda_n = \sum_{j=1}^{\infty} \frac{1}{j} \sum_{t=1}^{n} b^j(t, n)$$

Chao and Fu proved Theorem 5 under a more general condition.

Theorem 6. *Suppose $\pi_0 U_0 = 1$ and $\lim_{n\to\infty} \lambda_n = \lambda > 0$. Then*

$$\lim_{n\to\infty} R_L(n, k) = \exp\{-\lambda^k\}$$

3.3 Invariant Consecutive Systems

Given n components with reliabilities $p_1, p_2, \ldots, p_n$, then a permutation of the n reliabilities defines a consecutive-k system and its reliability. An optimal system is one with the largest reliability. An optimal system is called *invariant* if it depends only on the ranks, but not the actual values, of p_i.

3.3.1 Invariant Consecutive-2 Systems

Derman *et al.* [3] proposed two optimization problems for linear consecutive-2 systems and solved one of them. Let the reliabilities of the

n components be $p_1, p_2, \ldots, p_n$. Assume $p_{[i]}$ denotes the ith smallest p_j. In the *sequential assignment problem*, components are assigned one at a time to the system, and the state of the component is determined as soon as it is connected to the system. In the *non-adaptive assignment problem*, the system is constructed all at once.

For the sequential assignment problem, Derman *et al.* [3] proposed the following assignment rule: Assign the least reliable component first. Afterwards, if the last assigned component works, then assign the least reliable component in the remaining pool next; otherwise, if the last assigned component is failed, then assign the most reliable component in the remaining pool next. This rule is called DLR, which is optimal.

Theorem 7. *The assignment rule DLR is optimal.*

For the non-adaptive assignment problem, Derman *et al.* [3] conjectured the optimal sequence is

$$\tilde{L}_n = (p_{[1]}, p_{[n]}, p_{[3]}, p_{[n-2]}, \ldots, p_{[n-3]}, p_{[4]}, p_{[n-1]}, p_{[2]})$$

where $p_{[i]}$ denotes the ith smallest of $p_1, p_2, \ldots, p_n$. This conjecture was extended to the circular system by Hwang [18] as follows:

$$\tilde{C}_n = (p_{[1]}, p_{[n-1]}, p_{[3]}, p_{[n-3]}, \ldots, p_{[n-4]}, p_{[4]}, p_{[n-2]}, p_{[2]}, p_{[n]})$$

Since any consecutive-k-out-of-n linear system can be formulated as a consecutive-k-out-of-n circular system by setting $p_{n+1} = 1$, the line conjecture holds if the cycle conjecture holds. Du and Hwang [28] proved the cycle conjecture, while Malon [29] claimed a proof of the line conjecture (see [30] for comment). Later, Chang *et al.* [30] gave a proof for the cycle conjecture that combines Malon's technique for the line conjecture and Du and Hwang's technique for the cycle conjecture, but is simpler than both. These results can be summarized as follows.

Theorem 8. *$\tilde{L}_n$ is the unique optimal consecutive-2 line.*

Theorem 9. *$\tilde{C}_n$ is the unique optimal consecutive-2 cycle.*

Note that the optimal dynamic consecutive-2 line, the optimal static consecutive-2 line, and cycle are all *invariant*. One would expect that the remaining case, the optimal dynamic consecutive-2 cycle, is also invariant. However, Hwang and Pai [31] recently gave a negative result.

3.3.2 Invariant Consecutive-k Systems

The invariant consecutive-k line problem is completely solved by Malon [29]. The result is quoted as follows.

Theorem 10. *There exist invariant consecutive-k lines if and only if $k \in \{1, 2, n-2, n-1, n\}$. The invariant lines are given below:*

$$\begin{array}{ll}
(\text{any arrangement}) & \text{if } k = 1; \\
(p_{[1]}, p_{[n]}, p_{[3]}, p_{[n-2]}, \ldots, p_{[n-3]}, p_{[4]}, p_{[n-1]}, p_{[2]}) & \text{if } k = 2; \\
(p_{[1]}, p_{[4]}, (\text{any arrangement}), p_{[3]}, p_{[2]}) & \text{if } k = n-2; \\
(p_{[1]}, (\text{any arrangement}), p_{[2]}) & \text{if } k = n-1; \\
(\text{any arrangement}) & \text{if } k = n.
\end{array}$$

Hwang [32] extended Malon's result to the cycle.

Theorem 11. *There exist invariant consecutive-k cycles if and only if $k \in \{1, 2, n-2, n-1, n\}$. For $k \in \{1, n-1, n\}$, any cycle is optimal. For $k \in \{2, n-2\}$, the unique invariant cycle is $\tilde{C}_n$.*

Sometimes there is no universal invariant system, but there are some local invariant assignments. Such knowledge can reduce the number of systems we need to search as candidates of optimal systems. Tong [33] gave the following result for interchanging two adjacent components.

Theorem 12. *Assume that $(n-1)/2 \le k \le n-2$ and $p_1 > p_2$. Let C' be obtained from $C = (p_1, p_2, \ldots, p_n)$ by interchanging p_1 and p_2. Then $R(C') \ge R(C)$ if and only if*

$$\left(\prod_{i=3}^{k+1} q_i\right) p_{k+2} - p_{n-k+1} \left(\prod_{i=n-k+2}^{n} q_i\right) \le 0$$

For $n < 2k$, the following theorem was independently given by Tong [34] and Malon [29]. Kuo *et al.* [35] observed that it holds in general.

Theorem 13. *In an optimal consecutive-k line,* $p_1 \le p_2 \le \cdots \le p_n$.

3.3.3 Invariant Consecutive-k G System

A consecutive-k G system is the counterpart of a consecutive-k (F) system by interchanging the notions of working and failed components. That is, a G system works if and only if some k consecutive components all work. The G system was first suggested by Tong [34]. For reliability computation, formulas for the F systems also work well for the G systems by interchanging p_i and q_i for all i, and interchanging R with $\bar{R}$.

However, the G system brought out new problems in reliability optimization and in component importance. The main reason is that maximizing the reliability of the G system is equivalent to minimizing the reliability of the F system, which has not been studied before. Zuo and Kuo [36] gave the following result.

Theorem 14. *There does not exist an invariant consecutive-k-out-of-n G line for* $2 \le k < n/2$.

This result can be extended to the circular system. That is:

Theorem 15. *There does not exist an invariant consecutive-k-out-of-n G cycle for* $2 \le k < (n-1)/2$.

Zuo and Kuo [36] first observed the following result.

Theorem 16. *In an optimal assignment of a G system,* $p_1 \le p_2 \le \cdots \le p_k$ *and* $p_n \le p_{n-1} \le \cdots \le p_{n-k+1}$.

For the G line, the case of $n \le 2k$ was studied by Kuo *et al.* [35], and the case $n \le 2k+1$ for the G cycle was studied by Zuo and Kuo [36]. But both proofs are incomplete. Recently, Jalali *et al.* [37] proposed the following lemma.

Lemma 1. *In an optimal assignment of a G system,* $p_1 = p_{[1]}$, $p_k = p_{[n-1]}$, $p_{k+1} = p_{[n]}$ *and* $p_n = p_{[2]}$.

With Lemma 1, Jalali *et al.* [37] proved the following theorem for the G line.

Theorem 17. *The unique invariant consecutive-k-out-of-n G line for* $n = 2k$ *is* $\alpha = (p_{[1]}, p_{[3]}, p_{[5]}, \ldots, p_{[6]}, p_{[4]}, p_{[2]})$.

This result can be extended to the $n < 2k$ case.

Theorem 18. *For* $n < 2k$, *the invariant consecutive-k-out-of-n G line is*

$$(p_{[1]}, p_{[3]}, p_{[5]}, \ldots, B, \ldots, p_{[6]}, p_{[4]}, p_{[2]}),$$

where B *is a center block of* $2k - n$ *largest reliabilities in any arrangement.*

Du *et al.* [38] used a completely different method to prove the more general cycle case.

Theorem 19. *For* $n \le 2k+1$, *the unique invariant consecutive-k-out-of-n G cycle is*

$$(p_{[1]}, p_{[3]}, p_{[5]}, \ldots, p_{[6]}, p_{[4]}, p_{[2]}),$$

where the two ends are considered adjacent.

3.4 Component Importance and the Component Replacement Problem

Component importance measures the relative importance of a component, or sometimes the position of a component in the system, with respect to system reliability. The importance index can be used to assist the allocation of redundant components or replacement by giving priority to the more important positions. The most popular importance index is the Birnbaum importance, but very few results have been obtained. Weaker versions have been introduced to obtain more results.

3.4.1 The Birnbaum Importance

The Birnbaum (reliability) importance of component i in system $x \in \{L, C\}$ is defined as

$$I_x(i) = \frac{\partial R_x(p_1, p_2, \ldots, p_n)}{\partial p_i}$$

It is the rate at which the system reliability grows when the reliability of component i grows. When necessary, we use $I_x(i, n)$ to denote $I_x(i)$ with n fixed. Independently, Griffith and Govindarajulu [13] and Papastavridis [39] first studied the Birnbaum importance for consecutive lines. Their results can be quoted as follows.

Theorem 20.

$$I_L(i) = [R_L(1, i-1)R_L(i+1, n) - R_L(1, n)]/q_i$$

and

$$I_C(i) = [R_L(p_{i+1}, p_{i+2}, \ldots, p_{i-1}) - R_C(1, n)]/q_i.$$

It can be shown that $I_L(i)$ observes the same recursive equations as observed by $L(1, n)$. Thus similar to Theorem 1, the following theorem holds.

Theorem 21. *For an IID model*

1. $I_L(i, n) = I_L(i, n-1) - pq^k I_L(i, n-k-1)$ *if* $n - i \geq k+1$
2. $I_L(i, n) = I_L(i-1, n-1) - pq^k I_L(i-k-1, n-k-1)$ *if* $i - 1 \geq k+1$.

The comparison of Birnbaum importance under the IND model is valid only for the underlying set $(p_1, p_2, \ldots, p_n)$. On the other hand, the IID model is the suitable one if the focus is on comparing the positions by neutralizing the differences in p_i. Even for the IID model, Hwang *et al.* [40] showed that the comparison is not independent of p.

The following theorem is given by Chang *et al.* [30].

Theorem 22. *Consider the consecutive-2 line under the IID model. Then*

$$\begin{aligned} I_L(2i) &> I_L(2i-1) \quad \text{for } 2i \leq (n+1)/2 \\ I_L(2i) &< I_L(2i-2) \quad \text{for } 2i \leq (n+1)/2 \\ I_L(2i+1) &> I_L(2i-1) \quad \text{for } 2i+1 \leq (n+1)/2 \end{aligned}$$

For general k, not much is known. Kuo *et al.* [35] first observed the following result.

Theorem 23. *For the consecutive-k line under the IID model*

$$\begin{aligned} &I_L(1) < I_L(2) < \cdots < I_L(k) \quad \text{if } n \geq 2k \\ &I_L(n-k+1) = I_L(n-k+2) \\ &\qquad = \cdots = I_L(k) \quad \text{if } n < 2k \end{aligned}$$

The following theorem was proved (partially) by Zuo [41] and (partially) by Zakaria *et al.* [42].

Theorem 24. *Consider the consecutive-k line under the IID model. Then $I_L(1) \leq I_L(i)$ for all $i \leq n/2$ and $I_L(k) > I_L(k+1)$ for $n > 2k$.*

Chang *et al.* [43,44] gave a method to compare $I_L(i)$ with $I_L(i+1)$, and derived the following results.

Theorem 25.

$$I_L(2k+1) < I_L(2k), \quad I_L(k+1) < I_L(k+2)$$

and

$$I_L(2k-1, 4k-1) < I_L(2k, 4k-1).$$

Recently, Chang *et al.* [45] extended the results in Theorem 25 as follows:

Theorem 26.

(i) $I_L((t-2)k-1, tk-1) < I_L[(t-2)k, tk-1]$ *for* $t \geq 3$
(ii) $I_L(3k+1, 6k+1) < I_L(3k, 6k+1)$.

Hwang [46] defined a new importance measure. In Hwang's definition, component i is said to be more important than component j, written as $H(i) > H(j)$, if for every $d = k, k+1, k+2, \ldots, n$, $|CS_{i,d}|$, the number of d-cutsets containing i is never fewer than $|CS_{j,d}|$. Hwang [46] proved that H more importance implies Birnbaum more importance. He gave the following theorem.

Theorem 27. *$H(i) \geq H(j)$ implies $I_L(i) \geq I_L(j)$ under the IID model for all p.*

Note that one cannot use computation to prove $I_L(i) \geq I_L(j)$ for all p since there is an infinite number of them. But for any finite system, we can verify $H(i) \geq H(j)$ since d is bounded by n. Once $H(i) \geq H(j)$ is verified, then the previously impossible-to-verify relation $I_L(i) \geq I_L(j)$ is also verified. Chang *et al.* [43] also proved that:

Theorem 28. $H(k) \geq H(i)$ *for all* $i \leq (n+1)/2$.

3.4.2 Partial Birnbaum Importance

Chang *et al.* [45] proposed the half-line importance I^{h}, which requires $I^{\mathrm{h}}(i) > I^{\mathrm{h}}(j)$ only for all $p \geq 1/2$ for a comparison. They justified this half-line condition by noting that, in most practical cases, $p \geq 1/2$. For the consecutive-k-out-of-n line, they were able to establish:

Theorem 29.

$$I^{\mathrm{h}}(1) < I^{\mathrm{h}}(2) < \cdots < I^{\mathrm{h}}(k-1) < I^{\mathrm{h}}(k+1) < I^{\mathrm{h}}(i) < I^{\mathrm{h}}(2k) < I^{\mathrm{h}}(k)$$

for all $i > k+1$ *and* $i \neq 2k$.

The Birnbaum importance for the special case $p = 1/2$ is known as the "structure importance" in the literature. Chang *et al.* [30] suggested calling it the "combinatorial importance" so that the term "structure importance" can be reserved for general use (there are other importance indices depending on structure only). Denote the combinatorial importance by $I^C(i)$; Lin *et al.* [47] found an interesting correspondence between $I^C(i)$ and $f_{k,n}$, the Fibonacci numbers of order k, which is defined by

$$f_{k,n} = \begin{cases} 0 & \text{if } 1 \leq n \leq k-1 \\ 1 & \text{if } n = k \\ \sum_{i=1}^{k} f_{k,n-i} & \text{if } n \geq k+1 \end{cases}$$

They proved:

Theorem 30. *For* $p = 1/2$,

$$R_L(n) = (1/2)^n f_{k,n+k+1}.$$

Thus $f_{k,n+k+1}$ can be interpreted as the number of working consecutive-k-out-of-n lines.

Theorem 31.

$$I^{\mathrm{C}}(i) = (1/2)^{n-1}(2 f_{k,i+k} f_{k,n-i+k-1} - f_{k,n+k+1}).$$

Chang and Hwang [48] considered the case that p tends to zero. The importance index, denoted by $I^R(i)$, actually measures the number of minimum pathsets (a subset of components whose collective successes induce a system success) containing component i. Since, in practice, p is not likely to approach zero, $I^R(i)$ is not of interest *per se*. However, it could be a useful tool for the comparison of Birnbaum importance. While it is not easy to establish $I(i) > I(j)$, sometimes it is also difficult to establish the falsity of it. By proving $I^R(i) < I^R(j)$, we automatically establish the above falsity. Further, if we have proved $I^{\mathrm{h}}(i) > I^{\mathrm{h}}(j)$, then proving $I^R(i) \geq I^R(j)$ would add a lot of credibility to the conjecture that $I(i) > I(j)$ since $I^R(i) \geq I^R(j)$ provides evidence from the other end of the p spectrum.

Represent n as $n = qk + r$ with $0 \leq r < k$. Then q is the minimum number of working components for a pathset to exist. Let $\mathrm{ps}_q(k, n)$ denote the number of pathsets with q working components and let $\mathrm{ps}_{i,q}(k, n)$ the number of those containing component i. Chang and Hwang [48] proved:

Theorem 32.

$$\mathrm{ps}_q(k, n) = \binom{q+k-r-1}{k-r-1}$$

Represent i as $i = uk + v$ with $0 < v \leq k$:

Theorem 33.

$$\mathrm{ps}_{i,q}(k, n) = \binom{u+k-v}{k-v}\binom{q-u+v-r-2}{v-r-1}$$

Theorem 34. $I^R(uk+v) \leq I^R[(u+1)k+v]$ *for* $1 \leq v \leq (k+1)/2$ *and* $(u+1)k+v < (n+1)/2$.

$$I^R(uk+1) < I^R(uk) \quad \textit{for } uk+1 \leq (n+1)/2$$

$$I^R(uk+1) \leq I^R(j) \quad \textit{for } uk+1 \leq j \leq (n+1/2)$$

3.4.3 The Optimal Component Replacement

Consider the problem: "When a new extra component is given to replace a component in a linear consecutive-k-out-of-n:F system in order to raise the system reliability, which component should be replaced such that the resulting system reliability is maximized?" When a component is replaced, the change of system reliability is not only dependent on the working probabilities of the removed component and the new component, but also on the working probabilities of all other components. A straightforward algorithm is first re-computing the resulting system reliabilities of all possible replacements and then selecting the best position to replace a component. Even using the most efficient $O(n)$ reliability algorithm for linear consecutive-k-out-of-n:F systems, the computational complexity of the straightforward component replacement algorithm is $O(n^2)$.

Chang *et al.* [49] proposed an $O(n)$-time algorithm for the component replacement problem based on the Birnbaum importance. They first observed the following results.

Lemma 2. *$I_L(i)$ is independent of p_i.*

Let the reliability of the new extra component be p^*. Chang *et al.* [49] derived that:

Theorem 35.

$$\begin{aligned} R_L(p_1, \ldots, p_{i-1}, p^*, p_{i+1}, \ldots, p_n) \\ - R_L(p_1, \ldots, p_{i-1}, p_i, p_{i+1}, \ldots, p_n) \\ = I_L(i)(p^* - p_i) \end{aligned}$$

They provided an algorithm to find the optimal location where the component should be replaced. The algorithm is quoted as follows.

Algorithm 1. (Linear component replacement algorithm)

1. Compute $R_L(1, 1), R_L(1, 2), \ldots, R_L(1, n)$ in $O(n)$ time.
2. Compute $R_L(n, n), R_L(n-1, n), \ldots, R_L(2, n)$ in $O(n)$ time.
3. Compute $I_L(i)$ for $i = 1, 2, \ldots, n$, with the equation given in Theorem 14.
4. Compute $I_L(i)(p^* - p_i)$ for $i = 1, 2, \ldots, n$.
5. Choose the i in step 4 with the largest $I_L(i)(p^* - p_i)$ value. Then replace component i with the new extra component. The reliability of the resulting system is $I_L(i)(p^* - p_i) + R_L(1, n)$.

In Algorithm 1, each of the five steps takes at most $O(n)$ time. Therefore, the total computational complexity is $O(n)$.

Consider the component replacement problem for the circular system. As with the linear case, a straightforward algorithm can be designed as first re-computing the resulting system reliabilities of all possible replacements and then selecting the best position to replace a component. However, even using the most efficient $O(kn)$ reliability algorithm for the circular systems, the computational complexity of the straightforward algorithm is still $O(kn^2)$. Chang *et al.* [49] also proposed a similar algorithm for the circular component replacement problem. The computational complexity is $O(n^2)$.

If the circular consecutive-k-out-of-n:F system contains some components with zero working probabilities such that the whole system reliability is zero, then the computational complexity of the circular component replacement algorithm can be further improved. The following theorem for the special case was given by Chang *et al.* [49].

Theorem 36. *If $R_C(1, n) = 0$ and there is an i in $\{1, 2, \ldots, n\}$ such that $I_C(i) > 0$, then the following three conditions must be satisfied.*

1. *$C(1, n)$ has just one run of at least k consecutive components, where the working probability of each component is 0.*
2. *The run that mentioned in condition 1 contains fewer than $2k$ components.*
3. *If the run that mentioned in condition 1 contains components $1, 2, \ldots, m$, where $k \le m < 2k$, then:*

$$\begin{aligned} &I_C(i) > 0 \\ &\quad \textit{for all } i \in \{m-k+1, m-k+2, \ldots, k\}, \\ &I_C(i) = 0 \\ &\quad \textit{for all } i \notin \{m-k+1, m-k+2, \ldots, k\}. \end{aligned}$$

Based on Theorem 36, the circular component replacement algorithm can be modified as follows.

Algorithm 2. (Modified circular component replacement algorithm for $R_C(1, n) = 0$)

1. Find the largest run of consecutive components consisting of components with 0 working probabilities 0. Then re-index the components such that this run contains components $1, 2, \ldots, m$ $(m \geq k)$.
2. If $m \geq 2k$, then any replacement is optimal. STOP.
3. Compute $R_L(i+1, n+i-1)$ for $i \in \{m-k+1, m-k+2, \ldots, k\}$.
4. Choose the i in step 3 with the largest $R_L(i+1, n+i-1)$ value. Then replace component i with the new extra component. The reliability of the resulting system is $R_L(i+1, n+i-1)p^*$. STOP.

In the modified algorithm, step 1 takes $O(n)$ time, step 2 takes $O(1)$ time, step 3 takes at most $O(kn)$ time, and step 4 takes at most $O(n)$ time. The total computational complexity is thus $O(kn)$.

3.5 The Weighted-consecutive-k-out-of-n System

The weighted-consecutive-k-out-of-n F system was first proposed by Wu and Chen [5]. A weighted-consecutive-k-out-of-n F system consists of n components; each component has its own working probability and a positive integer weight such that the whole system fails if and only if the total weight of some consecutive failed components is at least k. The ordinary consecutive-k-out-of-n F system is a special case with all weights set to 1. Similar to the original consecutive-k-out-of-n F system, this weighted system also has two types: the linear and the circular.

3.5.1 The Linear Weighted-consecutive-k-out-of-n System

For the linear weighted-consecutive-k-out-of-n system, Wu and Chen [5] gave an $O(n)$-time algorithm to compute the set of minimal cutsets. The algorithm scans the system from node 1 and adds the weights one by one of nodes $1, 2, \ldots$ until a sum $K \geq k$ is found, say at node j. Then subtract from K the weights one by one of nodes $1, 2, \ldots$ until a further subtraction would reduce K to below k. Suppose i is the new beginning node. Then $(i, i+1, \ldots, j)$ is the first minimal cutset. Repeating this procedure by starting from node $i+1$, we can find other minimal cutsets.

Once all minimal cutsets are found, the system reliability can be computed easily. Wu and Chen [5] gave the following results, where m is the total number of cutsets, $F(1, n) = 1 - R(1, n)$, $\mathrm{Beg}(i)$ is the index of the first component of ith cutset, and $\mathrm{End}(i)$ is the index of the last component of ith cutset.

Theorem 37. *In a linear weighted-consecutive-k-out-of-n system:*

1. $F(1, i) = 0$, *for* $i = 0, 1, \ldots, \mathrm{End}(1) - 1$;
2. $F(1, i) = F(1, \mathrm{End}(j))$, *for* $i = \mathrm{End}(j), \mathrm{End}(j)+1, \ldots, \mathrm{End}(j+1) - 1$ *and* $j = 1, 2, \ldots, m-1$;
3. $F(1, i) = F(1, \mathrm{End}(m))$, *for* $i = \mathrm{End}(m), \mathrm{End}(m)+1, \ldots, n$;
4. $F(1, \mathrm{End}(j)) - F_{LW}(1, \mathrm{End}(j-1))$

$$= \sum_{i=0}^{\mathrm{Beg}(j)-\mathrm{Beg}(j-1)-1} R[1, \mathrm{Beg}(j-1)+i-1] \times p_{\mathrm{Beg}(j-1)+i} \left(\prod_{t=\mathrm{Beg}(j-1)+i-1}^{\mathrm{End}(j)} q_t \right)$$

for $j = 2, 3, \ldots, m$

Based on the recursive equations given in Theorem 37, the system reliability can be computed in $O(n)$ time.

3.5.2 The Circular Weighted-consecutive-k-out-of-n System

For the circular weighted-consecutive-k-out-of-n system, Wu and Chen [5] also gave an $O(n \cdot \min\{n, k\})$-time reliability algorithm. They consider the system in two cases: $n \geq k$ and $n < k$, and proposed the following equations, where w_i is the weight of component i.

Theorem 38. *In a circular weighted-consecutive-k-out-of-n system where* $n \geq k$

$$R(1, n) = \sum_{s=1}^{k} \sum_{l=n-k+s}^{n} \delta(s, l)$$

where

$$\delta(s, l) = \begin{cases} \left(\prod_{i=1}^{s-1} q_1\right) p_s R(s+1, l-1) p_l \prod_{i=l+1}^{n} q_i \\ \quad \textit{if} \sum_{i=1}^{s-1} w_i + \sum_{j=l+1}^{n} w_j < k \\ 0 \quad \textit{otherwise} \end{cases}$$

Theorem 39. *In a circular weighted-consecutive-k-out-of-n system where* $n < k$

$$R(1, n) = \sum_{s=1}^{n-2} \sum_{l=s+2}^{n} \delta(s, l)$$

where

$$\delta(s, l) = \begin{cases} \left(\prod_{i=1}^{s-1} q_i\right) p_s R(s+1, l-1) p_l \prod_{i=l+1}^{n} q_i \\ \quad \textit{if} \sum_{i=1}^{s-1} w_1 + \sum_{j=l+1}^{n} w_j < k \\ 0 \quad \textit{otherwise} \end{cases}$$

By combining the equations in both cases, Wu and Chen [5] claimed that the reliability of the circular weighted system can be computed in $O(n \cdot \min\{n, k\})$-time.

Though Wu and Chen's algorithm seems to work well, Chang *et al.* [4] found that it is incomplete. In some special circular weighted systems, Wu and Chen's algorithm will result in wrong reliabilities. Chang *et al.* [4] also gave an $O(Tn)$-time reliability algorithm for the circular weighted system, where

$$T = \max\left\{i \,\middle|\, \sum_{j=1}^{i-1} w_j < k, \ 1 \leq i \leq n+1\right\}$$

The basis of Chang *et al.*'s algorithm [4] is an equation that expresses the reliability of the circular weighted system in reliability terms of linear weighted systems. We describe the equation below.

Theorem 40. *In a circular weighted-consecutive-k-out-of-n system*

$$R_C(1, n) = \begin{cases} 1 \quad n < T \\ \sum_{i=1}^{T} \left(\prod_{j=1}^{i-1} q_j\right) p_i R_L(i+1, n+i-1) \\ \quad n \geq T \end{cases}$$

where

$$T = \max\left\{i \,\middle|\, \sum_{j=1}^{i-1} w_j < k, \ 1 \leq i \leq n+1\right\}$$

In order to analyze the computational complexity, Chang *et al.* [4] also gave an upper bound of T.

Theorem 41. $T \leq \min\{n, \lceil (k - w_{\max})/w_{\min} \rceil + 1\}$, *where* $w_{\max}$ *and* $w_{\min}$ *are the maximum and minimum weights of all components, respectively.*

Chang *et al.*'s [4] reliability algorithm is described as follows.

Algorithm 3. (CCH)

1. Compute T.
2. If $T > n$, then $R_C(1, n) = 1$. STOP.
3. Compute $R_L(i+1, n+i-1)$, for $i = 1, 2, \ldots, T$.
4. Compute $R_C(1, n)$ with the equation given in Theorem 40. STOP.

In this algorithm, the bottleneck step is step 3, which costs $O(Tn)$ time. Therefore, the total computational complexity of this reliability algorithm is $O(Tn)$.

3.6 Window Systems

A sequence of k consecutive nodes is called a k-window. In various problems, the definition of a "bad" window varies. In this section, the system failure is defined in terms of windows. It could be that the system fails if it contains a bad window. But we can also define a system to be failed if every window is bad, or, equivalently, the system works if it contains a good window. In general, we can define a system to be failed if it contains b bad windows, called a b-fold-window system.

The window systems have many practical applications. For example, consider a linear flow network consisting of $n+2$ nodes (node 0 to node $n+1$) and directed links from node i to node j ($0 \le i < j \le n+1, j-i \le k$). In this flow network, the source (node 0), the sink (node $n+1$), and all links are infallible, but the intermediate nodes (nodes $1, 2, \ldots, n$) may fail. When an intermediate node fails, no flow can go through it; when it works, the flow capacity is unity. Then, the probability that the maximum flow from the source to the sink is at least f is equal to the probability that the intermediate nodes do not contain a bad k-window, where a bad window is a window that contains at least $k-f+1$ failed nodes. Such a flow network can be realized as a circuit switching wireless telecommunication network, where the communication between nodes has a distance limitation and each intermediate node has an identical bandwidth limitation.

3.6.1 The f-within-consecutive-k-out-of-n System

An f-within-consecutive-k-out-of-n system, or abbreviated as an (f, k, n) system, is a linear or circular system consisting of n components that fails if and only if there exist some k consecutive components (a k-window) containing at least f failed components.

The problem of computing the reliability of an (f, k, n) system can be viewed as the binomial version of the generalized birthday problem studied by Saperstein [50]. The generalized birthday problem is described as follows:

> Given a random assignment of w 1's and $n-w$ 0's, what is the probability that there is no set of k consecutive bits in the arrangement containing f or more 1's?

Naus [51] first studied the case $k=2$. Saperstein considered the binomial version of the generalized birthday problem in which each bit has a probability p to be 0 and a probability q to be 1. By interpreting each 0 as a working component and each 1 as a failed component, this problem is equivalent to the problem of computing the reliability of the (f, k, n) system.

Hwang and Wright [9] proposed an $O(2^{3k}n)$-time reliability algorithm for the (f, k, n) system. Their algorithm is implemented using Griffith's [14] Markov chain approach. Let $w(l)$ denote the number of 1's in a binary k-vector l (that is, the weight of l). They define the state space of the Markov chain $\{Y(t) : t \ge 0\}$ as

$$S = \{l \in \{0, 1\}k : 0 < w(l) < f\} \cup \{s_1\} \cup \{s_N\}$$

Therefore

$$N = |S| = \sum_{i=0}^{f-1} \binom{k}{i} + 1 = O(2^k)$$

And the Markov chain $\{Y(t)\}$ is defined as follows.

1. The k-vector l is encoded from the last k components of $L_{f,k}(1, t)$ where a working component is encoded as 0 and a failed component is encoded as 1. If $t < k$, attach leading 0's to l.
2. $Y(t) = s_1$ if $w(l) = 0$.
3. $Y(t) = l$ if $0 < w(l) < f$.
4. $Y(t) = s_N$ if $w(l) \ge f$.

The transition matrix for $\{Y(t) : t \ge 0\}$ is Λ_t, an $N \times N$ matrix. Each (except the last) row of Λ_t contains two nonzero elements: one is p_t, and the other is q_t. The last row of Λ_t only contains a non-zero element, which is 1, in the last column. With the transition matrix Λ_t, the system reliability can be computed with the following equation:

$$R_{f,k}(1, n) = \pi_0 \left(\prod_{t=1}^{n} \Lambda_t \right) \mathbf{U}^{\mathrm{T}}$$

where

$$\pi_0 = (1, 0, \ldots, 0)_{1 \times N}$$

and

$$\mathbf{U} = (1, \ldots, 1, 0)_{1\times N}$$

Since a multiplication of two $N \times N$ matrices costs $O(N^3)$ time (or less by Strassen's algorithm), the total computational complexity of Hwang and Wright's reliability algorithm is $O(N^3 n) = O(2^{3k} n)$.

Chang [8] proposed another reliability algorithm for the (f, k, n) system, which is more efficient than the $O(2^{3k} n)$ one. Their algorithm is a Markov chain approach, but the Markov chain is derived from an automaton with minimal number of states. The automaton is a mathematical model of a system, with discrete inputs and outputs. The system can be in any one of a finite number of internal configurations or "states". The state of a system summarizes the information concerning past inputs that is needed to determine the behavior of the system on subsequent inputs. In computer science, the theory of automata is a useful design tool for many finite state systems, such as switching circuits, text editors, and lexical analyzers. Chang [8] also employed the sparse matrix data structure to speed up the computation of reliabilities. Their method is described as follows.

Consider an f-within-consecutive-k-out-of-n system. Every system state can be viewed as a binary string where the ith bit is 1 if and only if component i works. Chang [8] defined M as the set of all strings corresponding to working systems. Therefore, $\bar{M} = \{0, 1\}^* - M$ can be expressed as

$$\bar{M} = (0+1)^* \left[\sum_{\substack{x_1, x_2, \ldots, x_k \in \{0,1\} \\ x_1 + x_2 + \cdots + x_k \le k-f}} x_1 x_2 \ldots x_k \right] (0+1)^*$$

Based on automata theory, the minimal state automaton accepting exactly the strings in M has the following set of states Q. They labeled the states with k-bit binary strings.

$$Q = Q_0 \cup Q_1 \cup \cdots \cup Q_f$$

where

$$Q_0 = \left\{ \underbrace{1 \ldots 1}_{k-f} \underbrace{0 \ldots 0}_{f} \right\}$$

$$Q_1 = \left\{ b_1 b_2 \ldots b_k \,\middle|\, b_1 = 1,\ \sum_{i=2}^{k} b_i = k - f \right\}$$

$$Q_2 = \left\{ b_1 b_2 \ldots b_k \,\middle|\, b_1 = b_2 = 1, \ldots \right.$$

$$\left. \ldots \sum_{i=3}^{k} b_i = k - f \right\}$$

$$\ldots$$

$$Q_f = \left\{ b_1 b_2 \ldots b_k \,\middle|\, b_1 = b_2 = \cdots \right.$$

$$\left. = b_f = 1,\ \sum_{i=f+1}^{k} b_i = k - f \right\}$$

$$= \left\{ \underbrace{1 \ldots 1}_{k} \right\}$$

The only state in Q_f is the initial state. The only state in Q_0 is the rejecting state; all other states (including the initial state) are accepting states. The state transition function $\delta : Q \times \{0, 1\} \to Q$ is defined as:

$$\delta(b_1 \ldots b_k, 0) = \begin{cases} \underbrace{1 \ldots 1}_{k-f} \underbrace{0 \ldots 0}_{f} & \text{if } \sum_{i=2}^{k} b_i = k - f \\ b_2 \ldots b_k 0 & \text{otherwise} \end{cases}$$

$$\delta(b_1 \ldots b_k, 1) = \begin{cases} \underbrace{1 \ldots 1}_{k-f} \underbrace{0 \ldots 0}_{f} & \text{if } \sum_{i=1}^{k} b_i = k - f \\ \underbrace{1 \ldots 1}_{t-2} b_t \ldots b_k 1 & \text{otherwise,} \end{cases}$$

where

$$t = \min \left\{ x \,\middle|\, \sum_{i=x}^{k} b_i = k - f - 1 \right\}$$

Thus, in the minimal state automaton

$$\begin{aligned}|Q| &= 1 + \binom{k-1}{k-f} + \binom{k-2}{k-f} + \cdots \\ &\quad + \binom{k-f}{k-f} \\ &= 1 + \binom{k-1}{f-1} + \binom{k-2}{f-2} + \cdots \\ &\quad + \binom{k-f}{f-f} \\ &= 1 + \binom{k}{f-1}\end{aligned}$$

When this minimum state automaton is interpreted as a heterogeneous Markov chain, a reliability algorithm can be derived easily. For convenience, they rewrote the states in Q such that $Q = \{s_1, s_2, \ldots, s_N\}$ where

$$s_1 = \Bigl(\underbrace{1, 1, \ldots, 1}_{k}\Bigr)$$

is the initial state and

$$S_N = \Bigl(\underbrace{1, \ldots, 1}_{k-f}, \underbrace{0, \ldots, 0}_{f}\Bigr)$$

is the only one failed state. When the working probability of component i in the f-within-consecutive-k-out-of-n system is p_i, the transition probability matrix of $\{Y(t)\}$ is

$$\Lambda_t = \begin{bmatrix} m_{1,1} & \cdots & m_{1,N} \\ \vdots & & \vdots \\ m_{N,1} & \cdots & m_{N,N} \end{bmatrix}$$

where

$$m_{i,j} = \begin{cases} p_t & \text{if } i \neq N,\ \delta(s_i, 0) = s_j \\ 1 - p_t & \text{if } i \neq N,\ \delta(s_i, 1) = s_j \\ 0 & \text{if } i = N,\ j \neq N \\ 1 & \text{if } i = N,\ j = N \end{cases}$$

Therefore, the reliability $R_{f,k}(1, n)$ can be computed as follows:

$$R_{f,k}(1, n) = \pi_0 \left(\prod_{t=1}^{n} \Lambda_t \right) \mathbf{U}^{\mathrm{T}}$$

where

$$\pi_0 = (1, 0, \ldots, 0)_{1 \times N}$$

and

$$\mathbf{U} = (1, \ldots, 1, 0)_{1 \times N}.$$

The total computational complexity of Chang *et al.*'s algorithm is $O\bigl(\binom{k}{f-1} n\bigr)$, which is the most efficient one so far.

3.6.2 The 2-within-consecutive-k-out-of-n System

The 2-within-consecutive-k-out-of-n system, abbreviated as the $(2, k, n)$ system, is a special case of the f-within-consecutive-k-out-of-n system. Most results obtained from the consecutive system can be extended to the $(2, k, n)$ system. Let $N_S(d, f, k, n)$ denote the number of the (f, k, n) system which works, $S \in \{L, C\}$, containing exactly d failed components. Naus [51] gave the following result for $f = 2$.

Theorem 42.

$$N_L(d, 2, k, n) = \binom{n - (d-1)(k-1)}{d}$$

Sfaniakakis *et al.* [52] gave the following result for the circular case:

Theorem 43.

$$N_C(d, 2, k, n) = \frac{n}{n - d(k-1)} \binom{n - d(k-1)}{d}$$

Thus, for the IID model and $f = 2$:

$$R_L(n) = \sum_{d=0}^{n} \binom{n - (d-1)(k-1)}{d} q^d p^{n-d}$$

$$R_C(n) = \sum_{d=0}^{n} \frac{n}{n - d(k-1)} \times \binom{n - d(k-1)}{d} q^d p^{n-d}$$

Higashiyama *et al.* [53] gave a recursive equation for the IND model:

Theorem 44. *For $f = 2$ and $n \geq k$:*

$$R_L(1, n) = p_n R_L(1, n-1) + q_n\left(\sum_{i=n-k+1}^{n-1} p_i\right) R_L(1, n-k)$$

They also gave a recursive equation for the IND circular case:

Theorem 45. *For $f = 2$:*

$$R_C(1, n) = \left(\prod_{i=n-k+2}^{n} p_i\right) R_L(1, n-k+1) + \sum_{i=1}^{k-1} q_{n-k+i+1} \times R_L(i+1, n-2k+i+1) \times \left(\prod_{j=n-2k+i+2}^{n-k+i} p_j\right)\left(\prod_{j=n-k+i+2}^{n+i} p_j\right)$$

3.6.3 The b-fold-window System

First consider the b-fold-non-overlapping-consecutive system when a window is a consecutive system and the system fails if and only if there exist b non-overlapping bad windows. Let $R_L^b(1, n)$ be the reliability of such a (linear) system. Papastavridis [54] gave the following result. Based on this result, $R_L^b(1, n)$ can be computed in $O(b^2 n)$ time.

Theorem 46.

$$R_L^b(1, n) = R_L^b(1, n-1) - \sum_{j=1}^{b} p_{n-jk}\left(\prod_{j=1}^{jk} q_{n-jk+i}\right) \times [R_L^{b-j}(1, n-jk-1) - R_L^{b-j+1}(1, n-jk-1)]$$

Alevizos *et al.* [55] gave a similar argument for the circular system. With their equation, $R_C^b(1, n)$ can be computed in $O(b^3 kn)$ time.

Theorem 47.

$$R_C^b(1, n) = p_n R_L^b(1, n) + q_n R_C^b(1, n-1) - \sum_{j=1}^{b}\sum_{i=1}^{jk}\left(\prod_{l=n-jk+i}^{n+i-1} q_l\right) \times p_{n-jk+i-1} p_{n+i} \times [R_L^{b-j}(i+1, n-jk+i-2) - R_L^{b-j+1}(i+1, n-jk+i-2)]$$

For an IID model, Papastavridis proved the following results:

Theorem 48.

$$R_L^b(n) = \sum_{d=0}^{n} N_L^b(d, n, k) p^{n-d} q^d$$

where

$$N_L^b(d, n, k) = \binom{n-d+b}{b} \times \sum_{i \geq 0} (-1)^i \binom{n-d+1}{i} \times \binom{n-k(i+b)}{n-d}$$

Theorem 49.

$$R_C^b(n) = \sum_{d=0}^{n} N_C^b(d, n, k) p^{n-d} q^d$$

where

$$N_C^b(d, n, k) = \binom{n-d+b-1}{b}\binom{n-d}{d} \times \sum_{i \geq 0} (-1)^i \binom{n-d+1}{i} \times \binom{n-k(i+b)}{n-d}$$

Windows are called *isolated* if there exists a working component between any two windows. Chang *et al.* [30] considered the case that the system fails if and only if there are b isolated bad windows. Let $R^{(b)}$ denote the system reliability. Their results are summarized as follows:

Theorem 50.

$$R_L^{(b)}(1, n) = R_L^{(b)}(1, n-1) - \left(\prod_{i=n-k+1}^{n} q_i \right) p_{n-k}$$

Theorem 51.

$$\begin{aligned} R_C^{(b)}(1, n) &= p_n R_L^{(b)}(1, n) + q_n R_C^{(b)}(1, n-1) \\ &- \sum_{i=n-k+1}^{n} \left\{ \left(\prod_{j=i}^{i+k+1} q_j \right) p_{i-1} p_{i+k} \right. \\ &\times [R_L^{(b-1)}(i+k+1, i-2) \\ &\left. - R_L^{(b)}(i+k+1, i-2)] \right\} \end{aligned}$$

Based on these equations, the reliability $R_L^{(b)}(1, n)$ can be computed in $O(bn)$ time, and the reliability $R_C^{(b)}(1, n)$ in $O(bkn)$ time.

3.7 Network Systems

In the literature, a consecutive-k-out-of-n system is a system consisting of n components arranged in a sequence such that the system fails if and only if some k consecutive components all fail. In many practical applications, such as an oil pipeline or a telecommunications system, the objective of the system is to transmit something from the source to the sink. For these cases, the system can be represented by a network with $n+2$ nodes (nodes $0, 1, 2, \ldots, n+1$) and directed links from node i to node j ($0 \leq i < j \leq n+1$, $j - i \leq k$). We refer to node 0 as the source and node $n+1$ as the sink, and assume that they are infallible. Nodes 1 to n correspond to the n components. If some k consecutive nodes all fail, then there exists no path from the source to the sink. On the other hand, if no k consecutive nodes all fail, then the path starting from the source can always move forward to a node until it lands at the sink. Therefore, the system fails if and only if there exists no path from the source to the sink.

The path interpretation of the consecutive-k-out-of-n system brings about the notion of links that connect adjacent pairs of nodes. Therefore, one can study the link failure model in which nodes always work but links can fail. In many practical situations this could be the more appropriate model. For example, nodes could be some physical entities, like telephone switching centers or airports, and the links are routes interconnecting them. When a phone call or a flight is blocked, it is usually due to the non-availability of routes (wires busy or tickets sold out) rather than a breakdown of hardware. However, previous research suggested that the reliability of the link failure system is much harder to compute than that of the node failure system. The network system is more general than the link failure system and the node failure system, since nodes and links can all fail.

3.7.1 The Linear Consecutive-2 Network System

Chen *et al.* [10, 56] first proposed the network model for the consecutive-2-out-of-n line. They assumed that the reliabilities of all nodes are all equal to p. And p_1 (p_2) is the reliability of the link from node i to node $i+1$ ($i+2$). Let $R_1(n)$ denote the probability that node 0 has a working path to node $n+1$. They gave a recursive equation as follows:

$$\begin{aligned} R_1(n) &= p(1 - q_1 q_2) R_1(n-1) \\ &+ p p_2 (1 - p p_1) R_1(n-2) \\ &- p_2 q_1 p_2^2 R_1(n-3) \end{aligned}$$

from which a closed-form solution is obtained.

Theorem 52. *For* $0 < p(1 - q_1 q_2) < 1$ *we have:*

$$R(n) = a\alpha^{n+2} + b\beta^{n+2} + c\gamma^{n+2}$$

where α, β, γ *are the roots of*

$$\begin{aligned} f(x) &= x^3 - p(1 - q_1 q_2) x_2 \\ &- p p_2 (1 - p p_1) x + p_2 q_1 p_2^2 \end{aligned}$$

satisfying

$$\begin{aligned} 1 > \alpha > p(1 - q_1 q_2) > b \geq p p_2 q_1 \\ \geq 0 > -\beta > \gamma > -\alpha > -1 \end{aligned}$$

and

$$a = [pp_2q_1 - \alpha]/[p(\alpha - \beta)(\gamma - \alpha)] > 0$$
$$b = [pp_2q_1 - \beta]/[p(\beta - \gamma)(\alpha - \beta)] < 0$$
$$c = [pp_2q_1 - \gamma]/[p(\gamma - \alpha)(\beta - \gamma)] < 0$$

Newton's method can be employed to approximate the three roots α, β, γ of $f(x)$. Once the roots are known, the reliability $R_1(n)$ can be computed in another $O(\log n)$ time. However, the precision of the final value of $R_1(n)$ is usually worse than that of recursively employing the equation given in Theorem 52, which needs $O(n)$ time.

Chen *et al.* [10] also compared the importance of the three types of component (nodes, 1-links, and 2-link) with respect to the system reliability. Let $r > s > t$ be three given reliabilities. The question is how do we map $\{r, s, t\}$ to $\{p, p_1, p_2\}$. They found:

Theorem 53. *$R(n)$ is maximized by setting $p = r$, $p_1 = s$ and $p_2 = t$.*

3.7.2 The Linear Consecutive-k Network System

Chen *et al.* [10] also claimed that their approach can be extended to compute the reliability of the consecutive-k-out-of-n network, but the details are messy. Let M be a nonempty subset of the first k nodes including node 0. They said one can define:

$$R_M(n) = \Pr\{\text{a node in } M \text{ has a path to the sink}\}$$

Then one can write 2^{k-1} recursive equations, one for each M. The reliability of the consecutive-k-out-of-n:F network is $R_{\{1\}}(n)$.

Chang *et al.* [11] proposed a new Markov chain method and gave a closed-form expression for the reliability of the consecutive-k-out-of-n network system where each of the nodes and links has its own reliability. In their algorithm, they try to embed the consecutive-k-out-of-n network into a Markov chain $\{Y(t) : t \geq k\}$ defined on the state space $S = \{0, 1, 2, \ldots, 2^k - 1\}$. Then, they showed that the reliability of the consecutive-k-out-of-n network is equal to the probability that $Y(n + 1)$ is odd. The Markov chain $\{Y(t) : t \geq k\}$ was defined as follows.

Definition 1. $\{Y(t) : t \geq k\}$ is a Markov chain defined on the state space S, where

$$S = \{0, 1, 2, \ldots, 2^k - 1\}$$

Define $(x_1x_2 \ldots x_m)_2 = \sum_{i=1}^{m} 2^{m-i}x^i$. Initially:

$$\Pr\{Y(k) = 0\} = \Pr\{(d_1d_2 \ldots d_k)_2 = 0\}$$
$$\Pr\{Y(k) = 1\} = \Pr\{(d_1d_2 \ldots d_k)_2 = 1\}$$
$$\Pr\{Y(k) = 2\} = \Pr\{(d_1d_2 \ldots d_k)_2 = 2\}$$
$$\vdots$$
$$\Pr\{Y(k) = 2k - 1\} = \Pr\{(d_1d_2 \ldots d_k)_2 = 2k - 1\}$$

The transition probability matrix of the Markov chain $\{Y(t) : t \geq k\}$ is

$$\Lambda_t = \begin{bmatrix} m_{0,0,t} & m_{0,1,t} & \cdots & m_{0,2^k-1,t} \\ m_{1,0,t} & m_{1,1,t} & \cdots & m_{1,2^k-1,t} \\ \vdots & & & \vdots \\ m_{2^k-1,0,t} & m_{2^k-1,1,t} & \cdots & m_{2^k-1,2^k-1,t} \end{bmatrix}$$

where, for $0 \leq i \leq 2^k - 1, 0 \leq j \leq 2^k - 1, k < t \leq n + 1$

$$m_{i,j,t} = \Pr\{Y(t) = j \mid Y(t-1) = i\}$$
$$= \begin{cases} 1 & \text{if } i = j = 0 \\ \left[1 - \prod\limits_{\substack{x:1\leq x\leq k,\\ (i \bmod 2^x)\geq 2^{x-1}}} q_l(t-x, t)\right] p_n(t) & \\ \quad \text{if } j = (2i + 1) \bmod (2^k) & \\ 1 - \left[1 - \prod\limits_{\substack{x:1\leq x\leq k,\\ (i \bmod 2^x)\geq 2^{x-1}}} q_l(t-x, t)\right] p_n(t) & \\ \quad \text{if } j = (2i) \bmod (2^k) & \\ 0 & \text{otherwise} \end{cases}$$

Theorem 54.

$$R_N(k, n) = \Pr\{Y(n + 1) \textit{ is odd}\}$$
$$= \pi_k \left(\prod_{t=k+1}^{n+1} \Lambda_t \right) \mathbf{U}$$

where

$$\pi_k = [\Pr\{Y(k)=0\}, \Pr\{Y(k)=1\}, \ldots, \Pr\{Y(k)=2^k-1\}]$$
$$\mathbf{U} = [0, 1, 0, 1, \ldots, 0, 1]^{\mathrm{T}}$$

Using a straightforward algorithm to compute the initial probabilities, the computational complexity for $\boldsymbol{\pi}_k$ is at least $\Omega(2^{k(k+3)/2})$. In order to simplify the computation further, Chang *et al.* [11] transformed the network system into an extended network system by adding $k-1$ dummy nodes before node 0. The reliability of the extended consecutive-k-out-of-n:F network was defined as the probability that there is a working path from "node 0" to node $n+1$. Therefore, dummy nodes and their related links cannot contribute to a working path from node 0 to node $n+1$. The reliability of the extended consecutive-k-out-of-n network is identical to that of the original consecutive-k-out-of-n network.

When computing the reliability of the extended consecutive-k-out-of-n:F network, previous definitions must be modified. In Definition 1, $\{Y'(t)\} = \{Y(t) : t \geq 0\}$ can be defined in a similar way except that Λ_t is defined for $0 < t \leq n+1$ and initially

$$\Pr\{Y(0)=0\} = \Pr\{(d_{-k+1}d_{-k+2}\ldots d_0)_2 = 0\}$$
$$\Pr\{Y(0)=1\} = \Pr\{(d_{-k+1}d_{-k+2}\ldots d_0)_2 = 1\}$$
$$\Pr\{Y(0)=2\} = \Pr\{(d_{-k+1}d_{-k+2}\ldots d_0)_2 = 2\}$$
$$\vdots$$
$$\Pr\{Y(0)=2^k-1\} = \Pr\{(d_{-k+1}d_{-k+2}\ldots d_0)_2 = 2^k-1\}$$

Furthermore, in the extended consecutive-k-out-of-n:F network, one can let

$$p_n(i) = 0 \quad \text{for } -(k-1) \leq i < 0$$
$$p_l(i, j) = 0 \quad \text{for } -(k-1) \leq i < 0, \quad 0 < j - i \leq k$$

Again, the extended consecutive-k-out-of-n:F network was embedded into the Markov chain $\{Y\prime(t)\} = \{Y(t) : t \geq 0\}$. Considering the nodes $-k+1, -k+2, \ldots, -1$, it is true that they always fail and there is no working path from node 0 to them. Thus the binary number $(d_{-k+1}d_{-k+2}\ldots d_0)_2$ is always equal to one.

Let

$$\pi_0 = [\Pr\{Y(0)=0\}, \Pr\{Y(0)=1\}, \ldots, \Pr\{Y(0)=2^k-1\}] = [0, 1, 0, 0, \ldots, 0]$$

The system reliability can be computed with

$$R_n(k, n) = \boldsymbol{\pi}_0\left(\prod_{t=1}^{n+1} \Lambda_t\right)\mathbf{U}$$

Chang *et al.*'s [11] reliability algorithm is described as follows.

Algorithm 4.

1. $\pi \leftarrow \pi_0, t \leftarrow 1$.
2. Construct Λ_t in a sparse matrix data structure.
3. $\pi \leftarrow \pi\Lambda_t$.
4. If $t < n+1$ then
   $t \leftarrow t+1$,
   go to Step 2.
5. Return $\boldsymbol{\pi}U$. STOP.

With the efficient sparse matrix data structure, this algorithm can compute $R_N(k, n)$ in $O(2^k n)$ time. This is very efficient, since in most practical situations where n is large and k is small, the time complexity becomes $O(n)$.

3.7.3 The Linear Consecutive-k Flow Network System

The linear consecutive-k flow network system, first proposed by Chang [8], was modified from the network system. The structure of the flow network system is similar to that of the network system. In the flow network, the source (node 0), the sink (node $n+1$), and all links are infallible, but the intermediate nodes (nodes $1, 2, \ldots, n$) may fail. When an intermediate node fails, no flow can go through it; when it works, the flow capacity is unity.

The flow network can be realized as a circuit switching wireless telecommunications network, where the communication between nodes has a distance limitation and each intermediate node has an identical bandwidth limitation. Chang [8] defined the f-flow-reliability, denoted as $R_{FN}(f, k, n)$, as the probability that the maximum flow from the source to the sink is at least f. The maximum flow from the source to the sink is f if and only if there are at most f node-disjoint paths from the source to the sink.

$$R_{FN}(f, k, n) = \Pr\{\text{The maximum flow from node 0 to node } n+1 \text{ is at least } f\}$$

Then, they proved the following lemma:

Lemma 3. *For $1 \le f \le k$, if and only if there is no k consecutive intermediate nodes containing at least $k - f + 1$ failed nodes, the maximum flow from node 0 to node $n + 1$ is at least f.*

Based on Lemma 3, the following equation holds immediately:

Theorem 55.

$$\begin{aligned} R_{FN}(f, k, n) &= \Pr\{\textit{There are no k consecutive intermediate nodes containing at least } k - f + 1 \textit{ failed nodes}\} \\ &= R_{f-k+1,k(n)} \end{aligned}$$

That is, the f-flow-reliability of the consecutive-k-out-of-n:F flow network is equal to the reliability of the $(k - f + 1)$-within-consecutive-k-out-of-n:F system. Thus, by substituting f by $k - f + 1$, the reliability algorithm for the f-within-consecutive-k-out-of-n system can be used to compute the f-flow-reliability of the consecutive-k-out-of-n flow network system. The total computational complexity for $R_{FN}(f, k, n)$ is $O\left(\binom{k}{k-f+1-1}n\right) = O\left(\binom{k}{f}n\right)$.

Furthermore, if the mission is to compute the probability distribution of the maximum flow (from node 0 to node $n + 1$), the following equations can be used:

$$\Pr\{\text{the maximum flow} = 0\} = 1 - R_{FN}(1, k, n)$$

$$\Pr\{\text{the maximum flow} = 1\} = R_{FN}(1, k, n) - R_{FN}(2, k, n)$$

$$\Pr\{\text{the maximum flow} = 2\} = R_{FN}(2, k, n) - R_{FN}(3, k, n)$$

$$\vdots$$

$$\Pr\{\text{the maximum flow} = k-1\} = R_{FN}(k-1, k, n) - R_{FN}(k, k, n)$$

$$\Pr\{\text{the maximum flow} = k\} = R_{FN}(k, k, n)$$

Or equivalently:

$$\Pr\{\text{the maximum flow} = 0\} = 1 - R_k(n)$$

$$\Pr\{\text{the maximum flow} = 1\} = R_k(n) - R_{f,k}(k-1, k, n)$$

$$\Pr\{\text{the maximum flow} = 2\} = R_{f,k}(k-1, k, n) - R_{f,k}(k-2, k, n)$$

$$\vdots$$

$$\Pr\{\text{the maximum flow} = k-1\} = R_{f,k}(2, k, n) - \prod_{i=1}^{n} p_i$$

$$\Pr\{\text{the maximum flow} = k\} = \prod_{i=1}^{n} p_i$$

Another topic that relates to the flow network system is to find the route of the maximum flow. Chang [8] proposed an on-line routing algorithm to route the maximum flow. In executing their routing algorithm, each working intermediate node will dynamically choose a proper flow route through itself in order to maximize the total flow from the source to the sink.

The on-line routing algorithm is described in Algorithm 6. Note that for each i, $1 \le i \le n$, Algorithm 6 is periodically executed on node i. Algorithm 5 is a subroutine called by the main procedure in Algorithm 5.

Algorithm 5. (Algorithm_Mark(i))

```
{  if (Ind(i) ≤ k) then
           Mark(i) ← Ind(i);
   else
   {       for j ← 1 to k do temp(j) ← 0;
           for j ← 1 to k do
           {       (x, y) ← (Ind(i−), Mark(i − j));
                   // Get messages from node i − j.
                   // If node i − j fails (timeout), skip this iteration.
                   if (temp(y) = 0) then temp(y) ← x;
           }
           x ← n;
           for j ← 1 to k do
                   if (temp(j) > 0 and temp(j) < x) then
                   {       x ← temp(j);
                           Mark(i) ← j;
                   }
   }
}
```

Algorithm 6. (Algorithm_Route(i))

```
{       call Algorithm_Mark(i);
if (Ind(i) ≤ k) then
                receive a unity flow from node 0;
        else
        {       for j ← 1 to k do
                        if (Mark(i) = Mark(i − j)) then break;
                receive a unity flow from node i − j;
        }
}
```

The computational complexity of the routing algorithm is $O(k)$, which is optimal, since it is necessary for each node to collect messages from its k predecessor nodes in order to maximize the total flow.

3.8 Conclusion

Owing to a limitation of 10,000 words for this survey, we have to leave out topics so that we can do justice to cover some other more popular topics. Among others, we leave out the following:

The lifetime distribution model. Component i has a lifetime distribution $f_i(t)$; namely, the probability that i is still working at time t. Note that, at a given time t, $f_i(t)$ can be interpreted as p_i, the parameter in the static model we study. Hence, our model studies the availability of the system at any time t. However, there are problems unique to the lifetime model, like mean time to failure, the increasing failure rate property, and estimation of parameters in the lifetime distribution.

The dependence model. The failures among components are dependent. This dependence could be due to batch production or common causes of failure. Typically, Markov chains are used to study this model.

The consecutively connected systems. The reachability parameter k of a node is no longer constant. It is assumed that node i can reach the next k_i nodes. More generally, it is assumed that the nodes are multi-state nodes and node i has probability p_{ij} to be in state (i, j), which allows it to reach the next j nodes.

The multi-failure models. In one model, a component has two failure modes: open-circuit and short-circuit. Consequently, the system also has two failure modes: consecutive k open-circuits and consecutive r short-circuits. In another model, the system fails if either some d components fail, or some k $(<d)$ consecutive components all fail.

The redundancy model. Suppose there are many redundant components, how do we allocate them to the n positions in a consecutive-k system? Though Birnbaum importance does provide some guideline, it has only a figure-of-merit value, and there are very few Birnbaum importance comparisons. More definite answers have been obtained by combinatorial analysis.

The 2-dimension system. Components are arranged in an $m \times n$ rectangle and the rectangle is declared failed if there exists an $r \times s$ subrectangle consisting of all failed components. It is surprisingly difficult to write down efficient recursive equations to compute the reliability of the 2-dimension system.

The graph model. The components are nodes in a graph, and the system fails if and only if there exist two failed adjacent nodes. Computing the reliability is possible only for some special classes of graphs. A theory of the invariant consecutive-2 graph has been developed.

We refer the reader to the recent book by Chang *et al.* [30], which covers the above topics and also gives more details on the topics we covered here.

References

[1] Kontoleon JM. Reliability determination of r-successive-out-of-n:F system. IEEE Trans Reliab 1980;R-29:437.

[2] Chiang DT, Niu SC. Reliability of consecutive-k-out-of-n:F system. IEEE Trans Reliab 1981;R-30:87–9.

[3] Derman C, Lieberman GJ, Ross SM. On the consecutive-k-out-of-n:F system. IEEE Trans Reliab 1982;R-31:57–63.

[4] Chang JC, Chen RJ, Hwang FK. A fast reliability algorithm for the circular consecutive-weighted-k-out-of-n:F system. IEEE Trans Reliab 1998;47:472–4.

[5] Wu JS, Chen RJ. Efficient algorithms for k-out-of-n and consecutive-weighted-k-out-of-n:F system. IEEE Trans Reliab 1994;43:650–5.

[6] Chang GJ, Cui L, Hwang FK. Reliability for (n, f, k) systems. Stat Prob Lett 1999;43:237–42.

[7] Chang JC, Chen RJ, Hwang FK. Faster algorithms to compute reliabilities of the (n, f, k) systems. Preprint 2000.

[8] Chang JC. Reliability algorithms for consecutive-k systems. PhD Thesis, Department of Computer Science and Information Engineering, National Chiao Tung University, December 2000.

[9] Hwang FK, Wright PE. An $O(n \log n)$ algorithm for the generalized birthday problem. Comput Stat Data Anal 1997;23:443–51.

[10] Chen RW, Hwang FK, Li WC. Consecutive-2-out-of-n:F systems with node and link failures. IEEE Trans Reliab 1993;42:497–502.

[11] Chang JC, Chen RJ, Hwang FK. An efficient algorithm for the reliability of consecutive-$k - n$ networks. J Inform Sci Eng 2002; in press.

[12] Koutras MV. Consecutive-k,r-out-of-n:DFM systems. Microelectron Reliab 1997;37:597–603.

[13] Griffith WS, Govindarajulu Z. Consecutive-k-out-of-n failure systems: reliability, availability, component importance and multi state extensions. Am J Math Manag Sci 1985;5:125–60.

[14] Griffith WS. On consecutive k-out-of-n failure systems and their generalizations. In: Basu AP, editor. Reliability and quality control. Amsterdam: Elsevier Science; 1986. p.157–65.

[15] Chao MT, Lin GD. Economical design of large consecutive-k-out-of-n:F systems. IEEE Trans Reliab 1984;R-33:411–3.

[16] Fu JC. Reliability of consecutive-k-out-of-n:F systems with $(k - 1)$-step Markov dependence. IEEE Trans Reliab 1986;R-35:602–6.

[17] Shanthikumar JG. Recursive algorithm to evaluate the reliability of a consecutive-k-out-of-n:F system. IEEE Trans Reliab 1982;R-31:442–3.

[18] Hwang FK. Fast solutions for consecutive-k-out-of-n:F system. IEEE Trans Reliab 1982;R-31:447–8.

[19] Wu JS, Chen RJ. Efficient algorithm for reliability of a circular consecutive-k-out-of-n:F system. IEEE Trans Reliab 1993;42:163–4.

[20] Wu JS, Chen RJ. An $O(kn)$ algorithm for a circular consecutive-k-out-of-n:F system. IEEE Trans Reliab 1992;41:303–5.

[21] Antonopoulou I, Papastavridis S. Fast recursive algorithms to evaluate the reliability of a circular consecutive-k-out-of-n:F system. IEEE Trans Reliab 1987;R-36:83–4.

[22] Hwang FK. An $O(kn)$-time algorithm for computing the reliability of a circular consecutive-k-out-of-n:F system. IEEE Trans Reliab 1993;42:161–2.

[23] Fu JC, Hu B. On reliability of a large consecutive-k-out-of-n:F system with $(k - 1)$-step Markov dependence. IEEE Trans Reliab 1987;R-36:75–7.

[24] Koutras MV. On a Markov approach for the study of reliability structures. J Appl Prob 1996;33:357–67.

[25] Hwang FK, Wright PE. An $O(k^3 \log(n/k))$ algorithm for the consecutive-k-out-of-n:F system. IEEE Trans Reliab 1995;44:128–31.

[26] Fu JC. Reliability of a large consecutive-k-out-of-n:F system. IEEE Trans Reliab 1985;R-34:127–30.

[27] Chao MT, Fu JC. The reliability of a large series system under Markov structure. Adv Appl Prob 1991;23:894–908.

[28] Du DZ, Hwang FK. Optimal consecutive-2-out-of-n:F system. Math Oper Res 1986;11:187–91.

[29] Malon DM. Optimal consecutive-k-out-of-n:F component sequencing. IEEE Trans Reliab 1985;R-34:46–9.

[30] Chang GJ, Cui L, Hwang FK. Reliabilities of consecutive-k systems. Boston: Kluwer, 2000.

[31] Hwang FK, Pai CK. Sequential construction of a consecutive-2 system. Inform Proc Lett 2002; in press.

[32] Hwang FK. Invariant permutations for consecutive-k-out-of-n cycles. IEEE Trans Reliab 1989;R-38:65–7.

[33] Tong YL. Some new results on the reliability of circular consecutive-k-out-of-n:F system. In: Basu AP, editor. Reliability and quality control. Elsevier Science; 1986. p.395–9.

[34] Tong YL. A rearrangement inequality for the longest run, with an application to network reliability. J Appl Prob 1985;22:386–93.

[35] Kuo W, Zhang W, Zuo M. A consecutive-k-out-of-n:G system: the mirror image of consecutive-k-out-of-n:F system. IEEE Trans Reliab 1990;39:244–53.

[36] Zuo M, Kuo W. Design and performance analysis of consecutive-k-out-of-n structure. Nav Res Logist 1990;37:203–30.

[37] Jalali A, Hawkes AG, Cui L, Hwang FK. The optimal consecutive-k-out-of-n:G line for $n \leq 2k$. Preprint 1999.

[38] Du DZ, Hwang FK, Jung Y, Hgo HQ. Optimal consecutive-k-out-of-$(2k + 1)$:G cycle. J Global Opt 2001;19:51–60.

[39] Papastavridis S. The most important component in a consecutive-k-out-of-n:F system. IEEE Trans Reliab 1987;R-36:266–8.

[40] Hwang FK, Cui LR, Chang JC, Lin WD. Comments on "Reliability and component importance of a consecutive-k-out-of-n system by Zuo". Microelectron Reliab 2000;40:1061–3.

[41] Zuo MJ. Reliability and component importance of a consecutive-k-out-of-n system. Microelectron Reliab 1993;33:243–58.

[42] Zakaria RS, David HA, Kuo W. The nonmonotonicity of component importance measures in linear consecutive-k-out-of-n systems. IIE Trans 1992;24:147–54.

[43] Chang GJ, Cui L, Hwang FK. New comparisons in Birnbaum importance for the consecutive-k-out-of-n system. Prob Eng Inform Sci 1999;13:187–92.

[44] Chang GJ, Hwang FK, Cui L. Corrigenda on New comparisons in Birnbaum importance for the consecutive-k-out-of-n system. Prob Eng Inform Sci 2000;14:405.

[45] Chang HW, Chen RJ, Hwang FK. The structural Birnbaum importance of consecutive-k system. J Combin Opt 2002; in press.

[46] Hwang FK. A new index for component importance. Oper Res Lett 2002; in press.

[47] Lin FH, Kuo W, Hwang FK. Structure importance of consecutive-k-out-of-n systems. Oper Res Lett 1990;25:101–7.

[48] Chang HW, Hwang FK. Rare event component importance of the consecutive-k system. Nav Res Logist 2002; in press.

[49] Chang JC, Chen RJ, Hwang FK. The Birnbaum importance and the optimal replacement for the consecutive-k-out-of-n:F system. In: International Computer Symposium: Workshop on Algorithms 1998; p.51–4.

[50] Saperstein B. The generalized birthday problem. J Am Stat Assoc 1972;67:425–8.

[51] Naus J. An extension of the birthday problem. Am Stat 1968;22:27–9.

[52] Sfakianakis M, Kounias S, Hillaris A. Reliability of a consecutive k-out-of-r-from-n:F system. IEEE Trans Reliab 1992;41:442–7.

[53] Higashiyama Y, Ariyoshi H, Kraetzl M. Fast solutions for consecutive-2-out-of-r-from-n:F system. IEICE Trans Fundam Electron Commun Comput Sci 1995;E78A:680–84.

[54] Papastavridis S. The number of failed components in a consecutive-k-out-of-n:F system. IEEE Trans Reliab 1989;38:338–40.

[55] Alevizos PD, Papastavridis SG, Sypsas P. Reliability of cyclic m-consecutive-k-out-of-n:F system, reliability, quality control and risk assessment. In: Proceedings IASTED Conference 1992.

[56] Chen RW, Hwang FK, Li WC. A reversible model for consecutive-2-out-of-n:F systems with node and link failures. Prob Eng Inform Sci 1994;8:189–200.

Multi-state System Reliability Analysis and Optimization (Universal Generating Function and Genetic Algorithm Approach)

G. Levitin and A. Lisnianski

Chapter 4

4.1 Introduction

Modern large-scale technical systems are distinguished by their structural complexity. Many of them can perform their task at several different levels. In such cases, the system failure can lead to decreased ability to perform the given task, but not to complete failure.

In addition, each system element can also perform its task with some different levels. For example, the generating unit in power systems has its nominal generating capacity, which is fully available if there are no failures. Some types of failure can cause complete unit outage, whereas other types of failure can cause a unit to work with reduced capacity. When a system and its components can have an arbitrary finite number of different states (task performance levels) the system is termed a multi-state system (MSS) [1].

The physical characteristics of the performance depend on the physical nature of the system outcome. Therefore, it is important to measure performance rates of system components by their contribution into the entire MSS output performance. In the practical cases, one should deal with various types of MSS corresponding to the physical nature of MSS performance. For example, in some applications the performance measure is defined as productivity or capacity. Examples of such MSSs are continuous materials or energy transmission systems, power generation systems [2, 3], *etc.* The main task of these systems is to provide the desired throughput or transmission capacity for continuous energy, material, or information flow. The data processing speed can also be considered as a performance measure [4,5] and the main task of the system is to complete the task within the desired time. Some other types of MSS were considered by Gnedenko and Ushakov [6].

Much work in the field of reliability analysis was devoted to the binary-state systems, where only the complete failures are considered. The reliability analysis of an MSS is much more complex.

The MSS was introduced in the middle of the 1970s [7–10]. In these studies, the basic concepts of MSS reliability were formulated, the system structure function was defined for a coherent MSS, and its properties were investigated. Griffith [11] generalized the coherence definition and studied three types of coherence. The reliability importance was extended to MSSs by Griffith [11] and Butler [12]. An asymptotic approach to MSS reliability evaluation was developed by Koloworcki [13]. An engineering method for MSS unavailability boundary points estimation based on binary model extension was proposed by Pourret *et al.* [14].

Practical methods of MSS reliability assessment are based on three different approaches [15]: the structure function approach, where Boolean models are extended for the multi-valued case; the stochastic process (mainly Markov) approach; and Monte Carlo simulation. Obviously, the stochastic process method can be applied only to relatively small MSSs, because the number of system states increases drastically with the increase in number of system components. The structure function approach is also extremely time consuming. A Monte Carlo simulation model may be a fairly true representation of the real world, but the main disadvantage of the simulation technique is the time and expense involved in the development and execution of the model [15]. This is an especially important drawback when the optimization problems are solved. In spite of these limitations, the above-mentioned methods are often used by practitioners, for example in the field of power systems reliability analysis [2].

In real-world problems of MSS reliability analysis, the great number of system states that need to be evaluated makes it difficult to use traditional techniques in various optimization problems. On the contrary, the universal generating function (UGF) technique is fast enough to be used in these problems. In addition, this technique allows practitioners to find the entire MSS performance distribution based on the performance distributions of its components. An engineer can find it by using the same procedures for MSSs with different physical natures of performance. In the following sections, the application of the UGF to MSS reliability analysis and optimization is considered.

To solve the wide range of reliability optimization problems, one has to choose an optimization tool that is robust and universal and that imposes minimal requirements as to the knowledge of the structure of the solution space. A genetic algorithm (GA) has all these properties and can be applied for optimizing vectors of binary and real values, as well as for combinatorial optimization. GAs have been proven to be effective optimization tools in reliability engineering. The main areas of GA implementation in this field are redundancy allocation and structure optimization subject to reliability constraints [16, 17], optimal design of reliable network topology [18,19], optimization of reliability analysis procedures [20], fault diagnosis [21], and maintenance optimization [22,23].

4.1.1 Notation

We adopt the following notation herein:

$\Pr\{e\}$	probability of event e
G	random output performance rate of MSS
G_k	output performance rate of MSS in at state k
p_k	probability that a system is at state k
W	random system demand
W_m	mth possible value of system demand
q_m	$\Pr\{W = W_m\}$
$F(G, W)$	function representing the desired relation between MSS performance and demand
A	MSS availability
E_A	stationary probability that MSS meets a variable demand
E_G	expected MSS performance
E_U	expected MSS unsupplied demand
$u_j(z)$	u-function representing performance distribution of individual element
$U(z)$	u-function representing performance distribution of MSS
Ω_ω	composition operator over u-functions
ω	function determining performance rate for a group of elements
$\delta_x(U(z), F, W)$	operator over MSS u-function which determines E_x index, $x \in \{A, G, U\}$
S	MSS sensitivity index
g	nominal performance rate of individual MSS element
a	availability of individual MSS element
h	version of element chosen for MSS
$r(n, h)$	number of elements of version h, included to nth MSS component
$\Re$	MSS configuration.

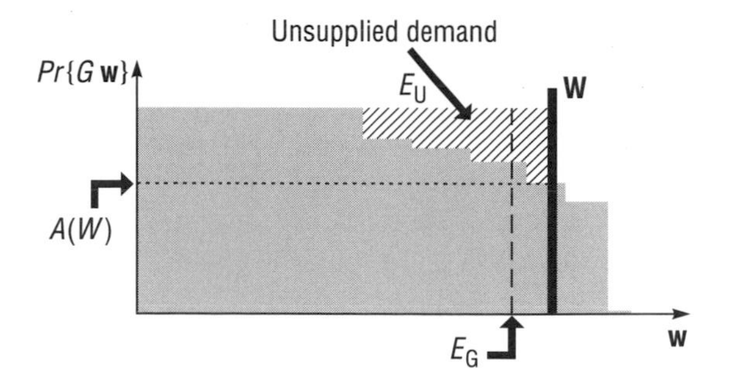

Figure 4.1. MSS reliability indices for failure criterion $F(G, W) = G - W < 0$

4.2 Multi-state System Reliability Measures

Consider a system consisting of n units. We suppose that any system unit i can have k_i states: from complete failure up to perfect functioning. The entire system has K different states as determined by the states of its units. Denote a MSS state at instance t as $Y(t) \in \{1, 2, \ldots, K\}$. The performance rate G_k is associated with each state $k \in \{1, 2, \ldots, K\}$ and the system output performance distribution (OPD) can be defined by two finite vectors $\mathbf{G}$ and $\mathbf{p} = \{p_k(t)\} = \Pr\{G(t) = G_k\}$ $(1 \leq k \leq K)$, where $G(t)$ is the random output performance rate of the MSS and $\Pr\{x\}$ is a probability of event x.

The MSS behavior is characterized by its evolution in the space of states. To characterize numerically this evolution process, one has to determine the MSS reliability indices. These indices can be considered as extensions of the corresponding reliability indices for a binary-state system.

The MSS reliability measures were systematically studied by Aven [15] and Brunelle and Kapur [24]. In this paper, we consider three measures that are most commonly used by engineers, namely MSS availability, MSS expected performance, and MSS expected unsupplied demand (lost throughput) (Figure 4.1).

In order to define the MSS ability to perform its task, we determine a function $F(G, W)$ representing the desired relation between the MSS random performance rate G and some specified performance level (demand) W. The condition $F(G, W) < 0$ is used as the criterion of an MSS failure. Usually $F(G, W) = G - W$, which means that states with system performance rate less than the demand are interpreted as failure states.

MSS availability $A(t)$ is the probability that the MSS will be in the states with performance level satisfying the condition $F(G, W) \geq 0$ at a specified moment $t > 0$, where the MSS initial state at the instance $t = 0$ is j and $F(G_j, W) \geq 0$. For large t, the initial state has practically no influence on the availability. Therefore, the index A is usually used for the steady-state case and is called the stationary availability coefficient, or simply the MSS availability. MSS availability is the function of demand W. It may be defined as

$$A(W) = \sum_{F(G_k, W) \geq 0} p_k \tag{4.1}$$

where p_k is the steady-state probability of MSS state k. The resulting sum is taken only for the states satisfying the condition $F(G, W) \geq 0$.

In practice, the system operation period T is often partitioned into M intervals T_m $(1 \leq m \leq M)$, and each T_m has its own demand level W_m. The following generalization of the availability index [4] is used in these cases:

$$E_{\mathrm{A}}(\mathbf{W}, \mathbf{q}) = \sum_{m=1}^{M} A(W_m) q_m \tag{4.2}$$

where $\mathbf{W}$ is the vector of possible demand levels $\mathbf{W} = \{W_1, \ldots, W_M\}$ and $\mathbf{q} = \{q_1, \ldots, q_M\}$ is the vector of steady-state probabilities of demand levels:

$$q_m = T_m \Big/ \sum_{m=1}^{M} T_m \tag{4.3}$$

For example, in power system reliability analysis, the index $(1 - E_{\mathrm{A}})$ is often used and treated as loss of load probability [2]. The MSS performance in this case is interpreted as power system generating capacity.

The value of MSS expected performance could be determined as

$$E_{\mathrm{G}} = \sum_{k=1}^{K} p_k G_k \tag{4.4}$$

One can note that the expected MSS performance does not depend on demand W. Examples of the E_{G} measure are the average productivity (capacity) [2] or processing speed of the system.

When penalty expenses are proportional to the unsupplied demand, the expected unsupplied demand E_{U} may be used as a measure of system output performance. If in the case of failure $(F(G_k, W) < 0)$ the function $-F(G_k, W)$ expresses the amount of unsupplied demand at state k, the index E_{U} may be presented by the following expression:

$$E_{\mathrm{U}}(\mathbf{W}, \mathbf{q}) = \sum_{m=1}^{M} \sum_{k=1}^{K} p_k q_m \max\{-F(G_k, W_m), 0\} \tag{4.5}$$

Examples of the E_{U} measure are the unsupplied power in power distribution systems and expected output tardiness in information processing systems.

In the following section we consider MSS reliability assessment based on the MSS reliability indices introduced above. The reliability assessment methods presented are based on the UGF technique.

4.3 Multi-state System Reliability Indices Evaluation Based on the Universal Generating Function

Ushakov [25] introduced the UGF and formulated its principles of application [26]. The most systematical description of mathematical aspects of the method can be found in [6], where the method is referred to as a generalized generating sequences approach. A brief overview of the method with respect to its applications for MSS reliability assessment was presented by Levitin *et al.* [4]. The method was first applied to the real system reliability assessment and optimization in [27].

The UGF approach is based on definition of a u-function of discrete random variables and composition operators over u-functions. The u-function of a variable X is defined as a polynomial

$$u(z) = \sum_{k=1}^{K} p_k z^{X_k} \tag{4.6}$$

where the variable X has K possible values and p_k is the probability that X is equal to X_k. The composition operators over u-functions of different random variables are defined in order to determine the probabilistic distribution for some functions of these variables.

The UGF extends the widely known ordinary moment generating function [6]. The essential difference between the ordinary generating function and a UGF is that the latter allows one to evaluate an OPD for a wide range of systems characterized by different topology, different natures of interaction among system elements, and the different physical nature of elements' performance measures. This can be done by introducing different composition operators over UGF (the only composition operator used with an ordinary moment generating function is the product of polynomials). The UGF composition operators will be defined in the following sections.

In our case, the UGF, represented by polynomial $U(z)$ can define an MSS OPD, *i.e.* it represents all the possible states of the system (or element) by relating the probabilities of each state p_k to performance G_k of the MSS in that state in the following form:

$$U(t, z) = \sum_{k=1}^{K} p_k(t) z^{G_k} \tag{4.7}$$

Having an MSS OPD in the form of Equation 4.6, one can obtain the system availability for the arbitrary t and W using the following operator δ_A:

$$\begin{aligned} A(t, F, W) &= \delta_A[U(t, z), F, W] \\ &= \delta_A\left(\sum_{k=1}^{K} p_k(t) z^{G_k}, F, W\right) \\ &= \sum_{k=1}^{K} p_k(t)\alpha[F(G_k, W)] \end{aligned} \tag{4.8}$$

where

$$\alpha(x) = \begin{cases} 1, & x \geq 0 \\ 0, & x < 0 \end{cases} \tag{4.9}$$

A multi-state stationary (steady state) availability was introduced as $\Pr\{G(t) \geq W\}$ after enough time had passed for this probability to become constant. In the steady state, the distribution of state probabilities is

$$p_k = \lim_{t\to\infty} \Pr\{G(t) = G_k\} \quad G(t) \in \{G_1, \ldots, G_K\} \tag{4.10}$$

The MSS stationary availability may be defined according to Equation 4.1 when the demand is constant or according to Equation 4.2 in the case of variable demand. Thus, for the given MSS OPD represented by polynomial $U(z)$, the MSS availability can be calculated as

$$E_A(\mathbf{W}, \mathbf{q}) = \sum_{m=1}^{M} q_m \delta_A(U(z), F, W_m) \tag{4.11}$$

The expected system output performance value during the operating time (Figure 4.1) defined by Equation 4.4 can be obtained for given $U(z)$ using the following δ_G operator:

$$E_G = \delta_G(U(z)) = \delta_G\left(\sum_{k=1}^{K} p_k z^{G_k}\right) = \sum_{k=1}^{K} p_k G_k \tag{4.12}$$

In order to obtain the expected unsupplied demand E_U for the given $U(z)$ and variable demand according to Equation 4.5, the following δ_U operator should be used:

$$E_U(\mathbf{W}, \mathbf{q}) = \sum_{m=1}^{M} q_m \delta_U(U(z), F, W_m) \tag{4.13}$$

where

$$\begin{aligned} E_U(W_m) &= \delta_U(U(z), F, W_m) \\ &= \delta_U\left(\sum_{k=1}^{K} p_k z^{G_k}, F, W_m\right) \\ &= \sum_{k=1}^{K} p_k \max(-F(G_k, W_m), 0) \end{aligned} \tag{4.14}$$

Example 1. Consider, for example, two power system generators with nominal capacity 100 MW as two different MSSs. In the first generator some types of failure require the capacity to be reduced to 60 MW and some types lead to a complete generator outage. In the second generator some types of failure require the capacity to be reduced to 80 MW, some types lead to capacity reduction to 40 MW, and some types lead to a complete generator outage. So, there are three possible relative capacity levels that characterize the performance of the first generator:

$$G_1^1 = 0.0 \quad G_2^1 = \frac{60}{100} = 0.6 \quad G_3^1 = \frac{100}{100} = 1.0$$

and four relative capacity levels that characterize the performance of the second one:

$$G_1^2 = 0.0 \quad G_2^2 = \frac{40}{100} = 0.4$$
$$G_3^2 = \frac{80}{100} = 0.8 \quad G_4^2 = \frac{100}{100} = 1.0$$

The corresponding steady-state probabilities are the following:

$$p_1^1 = 0.1 \quad p_2^1 = 0.6 \quad p_3^1 = 0.3$$

for the first generator and

$$p_1^2 = 0.05 \quad p_2^2 = 0.25 \quad p_3^2 = 0.3 \quad p_4^2 = 0.4$$

for the second generator.

Now we find reliability indices for both MSSs for $W = 0.5$ (required capacity level is 50 MW).

1. The MSS u-functions for the first and the second generator respectively, according to Equation 4.7, are as follows:

$$U_{\mathrm{MSS}}^1(z) = p_1^1 z^{G_1^1} + p_2^1 z^{G_2^1} + p_3^1 z^{G_3^1}$$
$$= 0.1 + 0.6z^{0.6} + 0.3z^{1.0}$$
$$U_{\mathrm{MSS}}^2(z) = p_1^2 z^{G_1^2} + p_2^2 z^{G_2^2} + p_3^2 z^{G_3^2} + p_4^2 z^{G_4^2}$$
$$= 0.05 + 0.25z^{0.4} + 0.3z^{0.8} + 0.4z^{1.0}$$

2. The MSS stationary availability (Equation 4.8) is

$$E_{\mathrm{A}}^1(W) = A^1(0.5)$$
$$= \sum_{G_k^1 - W \geq 0} p_k = 0.6 + 0.3 = 0.9$$
$$E_{\mathrm{A}}^2(W) = A^2(0.5)$$
$$= \sum_{G_k^2 - W \geq 0} p_k = 0.3 + 0.4 = 0.7$$

3. The expected MSS performance (Equation 4.12) is

$$E_{\mathrm{G}}^1 = \sum_{k=1}^{3} p_k^1 G_k^1$$
$$= 0.1 \times 0 + 0.6 \times 0.6 + 0.3 \times 1.0 = 0.66$$

which means 66% of the nominal generating capacity for the first generator, and

$$E_{\mathrm{G}}^2 = \sum_{k=1}^{4} p_k^2 G_k^2$$
$$= 0.05 \times 0 + 0.25 \times 0.4 + 0.3 \times 0.8$$
$$+ 0.4 \times 1.0 = 0.74$$

which means 74% of the nominal generating capacity for the second generator.

4. The expected unsupplied demand (Equation 4.14) is

$$E_{\mathrm{U}}^1(W) = \sum_{G_k - W < 0} p_k(W - G_k)$$
$$= 0.1 \times (0.5 - 0.0) = 0.05$$
$$E_{\mathrm{U}}^2(W) = \sum_{G_k - W < 0} p_k(W - G_k)$$
$$= 0.05 \times (0.5 - 0.0)$$
$$+ 0.25 \times (0.5 - 0.4) = 0.05$$

In this case, E_{U} may be interpreted as expected electric power unsupplied to consumers. The absolute value of this unsupplied demand is 5 MW for both generators. Multiplying this index by the considered system operating time, one can obtain the expected unsupplied energy.

Note that since the reliability indices obtained have different nature, they cannot be used interchangeably. In Example 1, for instance, the first generator performs better than the second one when availability is considered ($E_A^1(0.5) > E_A^2(0.5)$), the second generator performs better than the first one when expected productivity is considered ($E_G^2 > E_G^1$), and both generators have the same unsupplied demand ($E_U^1(0.5) = E_U^2(0.5)$).

4.4 Determination of u-function of Complex Multi-state System Using Composition Operators

Real-world MSSs are often very complex and consist of a large number of elements connected in different ways. To obtain the MSS OPD and the corresponding u-function, we must develop some rules to determine the system u-function based on the individual u-functions of its elements.

In order to obtain the u-function of a subsystem (component) containing a number of elements, composition operators are introduced. These operators determine the subsystem u-function expressed as polynomial $U(z)$ for a group of elements using simple algebraic operations over individual u-functions of elements. All the composition operators for two different elements with random performance rates $g_1 \in \{g_{1i}\}$ $(1 \le i \le I)$ and $g_2 \in \{g_{2j}\}$ $(1 \le j \le J)$ take the form

$$\begin{aligned}
&\Omega_\omega(u_1(z), u_2(z)) \\
&= \Omega_\omega\left[\sum_{i=1}^{I} p_{1i} z^{g_{1i}}, \sum_{j=1}^{J} p_{2j} z^{g_{2j}}\right] \\
&= \sum_{i=1}^{I}\sum_{j=1}^{J} p_{1i} p_{2j} z^{\omega(g_{1i}, g_{2j})} \qquad (4.15)
\end{aligned}$$

where $u_1(z)$, $u_2(z)$ are individual u-functions of elements and $\omega(\cdot)$ is a function that is defined according to the physical nature of the MSS performance and the interactions between MSS elements. The function $\omega(\cdot)$ in composition operators expresses the entire performance of a subsystem consisting of different elements in terms of the individual performance of the elements. The definition of the function $\omega(\cdot)$ strictly depends on the type of connection between the elements in the reliability diagram sense, *i.e.* on the topology of the subsystem structure. It also depends on the physical nature of system performance measure.

For example in an MSS, where performance measure is defined as capacity or productivity (MSS_c), the total capacity of a pair of elements connected in parallel is equal to the sum of the capacities of elements. Therefore, the function $\omega(\cdot)$ in composition operator takes the form

$$\omega(g_1, g_2) = g_1 + g_2 \qquad (4.16)$$

For a pair of elements connected in series, the element with the least capacity becomes the bottleneck of the system. In this case, the function $\omega(\cdot)$ takes the form

$$\omega(g_1, g_2) = \min(g_1, g_2) \qquad (4.17)$$

In MSSs where the performances of elements are characterized by their processing speed (MSS_s) and parallel elements cannot share their work, the task is assumed to be completed by the group of parallel elements when it is completed by at least one of the elements. The entire group processing speed is defined by the maximum element processing speed:

$$\omega(g_1, g_2) = \max(g_1, g_2) \qquad (4.18)$$

If a system contains two elements connected in series, then the total processing time is equal to the sum of processing times t_1 and t_2 of individual elements: $T = t_1 + t_2 = g_1^{-1} + g_2^{-1}$.

Therefore, the total processing speed of the system can be obtained as $T^{-1} = g_1 g_2/(g_1 + g_2)$ and the $\omega(\cdot)$ function for a pair of elements is defined as

$$\omega(g_1, g_2) = g_1 g_2/(g_1 + g_2) \qquad (4.19)$$

Ω operators were determined in [4, 27] for several important types of series–parallel MSS. Some additional composition operators were also derived for bridge structures [5, 28].

One can see that the Ω operators for series and parallel connection satisfy the following conditions:

$$\begin{aligned}&\Omega_\omega\{u_1(z),\ldots,u_k(z),\ u_{k+1}(z),\ldots,u_n(z)\}\\&\quad=\Omega_\omega\{u_1(z),\ldots,u_{k+1}(z),u_k(z),\ldots,u_n(z)\}\\&\Omega_\omega\{u_1(z),\ldots,u_k(z),u_{k+1}(z),\ldots,u_n(z)\}\\&\quad=\Omega_\omega\{\Omega_\omega\{u_1(z),\ldots,u_k(z)\},\\&\qquad\Omega_\omega\{u_{k+1}(z),\ldots,u_n(z)\}\}\end{aligned}\tag{4.20}$$

Therefore, applying the Ω operators in sequence, one can obtain the u-function representing the system performance distribution for an arbitrary number of elements connected in series and in parallel.

Example 2. Consider a system consisting of two elements with total failures connected in parallel. The elements have nominal performances g_1 and g_2 ($g_1 < g_2$) and constant availabilities a_1 and a_2 respectively. In the failure state the elements have performance zero. Even when a system consists of elements with total failures it should be considered as multi-state one when its output performance rate is of interest. Indeed, the u-functions of the individual elements are $(1-a_1)z^0 + a_1z^{g_1}$ and $(1-a_2)z^0 + a_2z^{g_2}$ respectively. The u-function for the entire MSS is

$$\begin{aligned}U_{\mathrm{MSS}}(z) &= \Omega_\omega[u_1(z), u_2(z)]\\&=\Omega_\omega[(1-a_1)z^0 + a_1z^{g_1},\\&\qquad(1-a_2)z^0 + a_2z^{g_2}]\end{aligned}$$

which for $\mathrm{MSS_c}$ takes the form

$$\begin{aligned}U(z) &= (1-a_1)(1-a_2)z^0 + a_1(1-a_2)z^{g_1}\\&\quad+a_2(1-a_1)z^{g_2} + a_1a_2z^{g_1+g_2}\end{aligned}$$

and for $\mathrm{MSS_s}$ takes the form

$$\begin{aligned}U(z) &= (1-a_1)(1-a_2)z^0 + a_1(1-a_2)z^{g_1}\\&\quad+a_2(1-a_1)z^{g_2} + a_1a_2z^{\max(g_1,g_2)}\\&=(1-a_1)(1-a_2)z^0\\&\quad+a_1(1-a_2)z^{g_1} + a_2z^{g_2}\end{aligned}$$

The measures of the system output performance obtained for failure criterion $F(G,W) = G - W < 0$ according to Equations 4.11–4.13 for both types of MSS are presented in Table 4.1.

4.5 Importance and Sensitivity Analysis of Multi-state Systems

Methods for evaluating the relative influence of component availability on the availability of the entire system provide useful information about the importance of these elements. Importance evaluation is a key point in tracing bottlenecks in systems and in the identification of the most important components. It is a useful tool to help the analyst find weaknesses in design and to suggest modifications for system upgrade.

Some various importance measures have been introduced previously. The first importance measure was introduced by Birnbaum [29]. This index characterizes the rate at which the system reliability changes with respect to changes in the reliability of a given element. So, an improvement in the reliability of an element with the highest importance causes the greatest increase in system reliability. Several other measures of elements and minimal cut sets importance in coherent systems were developed by Barlow and Proschan [30] and Vesely [31].

The above importance measures have been defined for coherent systems consisting of binary components. In multi-state systems the failure effect will be essentially different for system elements with different performance levels. Therefore, the performance levels of system elements should be taken into account when their importance is estimated. Some extensions of importance measures for coherent MSSs have been suggested, *e.g.* [11,30], and for non-coherent MSSs [32].

From Equations 4.11 and 4.14 one can see that the entire MSS reliability indices are complex functions of the demand W, which is an additional factor having a strong impact on an element's importance in multi-state systems. Availability of a certain component may be very important for one demand level and less important for another.

For the complex system structure, where each system component can have a large number of possible performance levels and there can be a large number of demand levels M, the importance

Table 4.1. Measures of system performance obtained for MSS

MSS type	E_A	E_U	E_G	W
MSS_c	0	$W - a_1g_1 - a_2g_2$	$a_1g_1 + a_2g_2$	$W > g_1 + g_2$
	a_1a_2	$g_1a_1(a_2 - 1) + g_2a_2(a_1 - 1) + W(1 - a_1a_2)$		$g_2 < W \leq g_1 + g_2$
	a_2	$(1 - a_2)(W - g_1a_1)$		$g_1 < W \leq g_2$
	$a_1 + a_2 - a_1a_2$	$(1 - a_1)(1 - a_2)W$		$0 < W \leq g_1$
MSS_s	0	$W - a_1g_1 - a_2g_2 + a_1a_2g_1$	$a_1(1 - a_2)g_1 + a_2g_2$	$W > g_2$
	a_2	$(1 - a_2)(W - g_1a_1)$		$g_1 < W \leq g_2$
	$a_1 + a_2 - a_1a_2$	$(1 - a_1)(1 - a_2)W$		$0 < W \leq g_1$

evaluation for each component obtained using existing methods requires an unaffordable effort. Such a problem is quite difficult to formalize because of the great number of logical functions for the top-event description when we use the logic methods, and the great number of states when the Markov technique is used.

In this section we demonstrate the method for the Birnbaum importance calculation, based on the UGF technique. The method provides the importance evaluation for complex MSS with different physical nature of performance and also takes into account the demand.

The natural generalization of Birnbaum importance for MSSs consisting of elements with total failure is the rate at which the MSS reliability index changes with respect to changes in the availability of a given element i. For the constant demand W, the element importance can be obtained as

$$\frac{\partial A(W)}{\partial a_i} \tag{4.21}$$

where a_i is the availability of the ith element at the given moment, $A(W)$ is the availability of the entire MSS, which can be obtained for MSSs with a given structure, parameters, and demand using Equation 4.8.

For the variable demand represented by vectors $\mathbf{W}$ and $\mathbf{q}$ the sensitivity of the generalized MSS availability E_A to the availability of the given element i is

$$S_A(i) = \frac{\partial E_A(\mathbf{W}, \mathbf{q})}{\partial a_i} \tag{4.22}$$

where index E_A can be obtained using Equation 4.9.

In the same way, one can obtain the sensitivity of the E_G and E_U indices for an MSS to the availability of the given element i as

$$S_G(i) = \frac{\partial E_G}{\partial a_i} \tag{4.23}$$

and

$$S_U(i) = \left| \frac{\partial E_U(\mathbf{W}, \mathbf{q})}{\partial a_i} \right| \tag{4.24}$$

Since E_U is a decreasing function of a_i for each element i, the absolute value of the derivative is considered to estimate the degree of influence of element reliability on the unsupplied demand.

It can easily be seen that all the suggested measures of system performance (E_A, E_G, E_U) are linear functions of elements' availability. Therefore, the corresponding sensitivities can easily be obtained by calculating the performance measures for two different values of availability.

The sensitivity indices for each MSS element depend strongly on the element's place in the system, its nominal performance level, and system demand.

Example 3. *Analytical example.* Consider the system consisting of two elements with total failures connected in parallel. The availabilities of the elements are a_1 and a_2 and the nominal performance rates are g_1 and g_2 ($g_1 < g_2$). The analytically obtained measures of the system output performance for MSSs of both types are presented in

Table 4.2. Sensitivity indices obtained for MSS

MSS type	$S_A(1)$	$S_A(2)$	$S_U(1)$	$S_U(2)$	$S_G(1)$	$S_G(2)$	W
MSS_c	0	0	$-g_1$	$-g_2$	g_1	g_2	$W > g_1 + g_2$
	a_2	a_1	$(a_2-1)g_1$ $-a_2(W-g_2)$	$(a_1-1)g_2$ $-a_1(W-g_1)$			$g_2 < W \le g_1 + g_2$
	0	1	$(a_2-1)g_1$	$a_1g_1 - W$			$g_1 < W \le g_2$
	$1-a_2$	$1-a_1$	$(a_2-1)W$	$(a_1-1)W$			$0 < W \le g_1$
MSS_s	0	0	$(a_2-1)g_1$	$a_1g_1 - g_2$	$g_1(1-a_2)$	$g_2 - a_1g_1$	$W > g_2$
	0	1	$(a_2-1)g_1$	$a_1g_1 - W$			$g_1 < W \le g_2$
	$1-a_2$	$1-a_1$	$(a_2-1)W$	$(a_1-1)W$			$0 < W \le g_1$

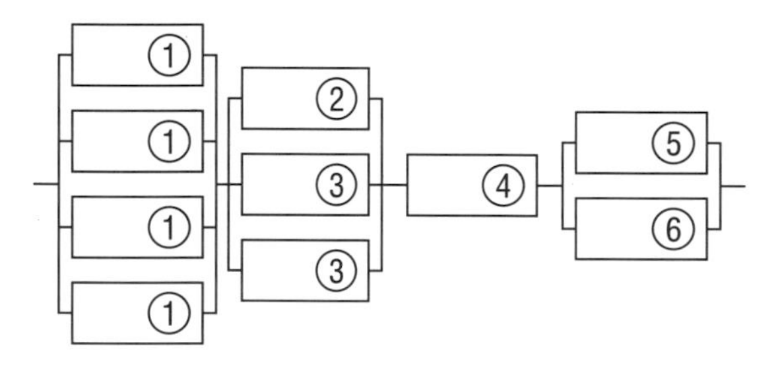

Figure 4.2. Example of series–parallel MSS

Table 4.3. Parameters of system elements

No. of element i	Nominal performance rate g_i	Availability a_i
1	0.40	0.977
2	0.60	0.945
3	0.30	0.918
4	1.30	0.983
5	0.85	0.997
6	0.25	0.967

Table 4.4. Cumulative demand curves

$\mathbf{q}$	0.15	0.25	0.35	0.25
$\mathbf{W}_1$ (ton h^{-1})	1.00	0.90	0.70	0.50
$\mathbf{W}_2$ (kbytes s^{-1})	0.30	0.27	0.21	0.15

the Table 4.1. The sensitivity indices can also be obtained analytically. These indices for MSS_c and MSS_s are presented in Table 4.2. Note that the sensitivity indices are different for MSSs of different types, even in this simplest case.

Numerical example. The series–parallel system presented in Figure 4.2 consists of ten elements of six different types. The nominal performance rate g_i and the availability a_i of each type of element are presented in Table 4.3.

We will consider the example of the system with the given structure and parameters as MSS_c and MSS_s separately. The cumulative demand curves $\mathbf{q}_1$, $\mathbf{W}_1$ for MSS_c and $\mathbf{q}_2$, $\mathbf{W}_2$ for MSS_s are presented in Table 4.4 ($\mathbf{q}_1 = \mathbf{q}_2 = \mathbf{q}$).

The sensitivity indices estimated for both types of system are presented in Table 4.5. These indices depend strongly on the element's place in the system, on the element's nominal performance, and on the system demand.

The dependencies of sensitivity index S_A on the system demand are presented in Figures 4.3 and 4.4. Here, system demand variation is defined as vector $k\mathbf{W}$, where $\mathbf{W}$ is the initial demand vector given for each system in Table 4.4 and k is the demand variation coefficient. As can be seen from Figures 4.3 and 4.4, the $S_A(k)$ are complicated non-monotonic piecewise continuous functions for both types of system. The order of elements according to their S_A indices changes when the demand varies. By using the graphs, all the MSS elements can be ordered according to their importance for any required demand level. On comparing graphs in Figures 4.3

Table 4.5. Sensitivity indices obtained for MSS (numerical example)

Element no.	MSS_C			MSS_S		
	S_A	S_G	S_U	S_A	S_G	S_U
1	0.0230	0.1649	0.0035	0.1224	0.0194	0.0013
2	0.7333	0.4976	0.2064	0.4080	0.0730	0.0193
3	0.1802	0.2027	0.0389	0.1504	0.0292	0.0048
4	0.9150	1.0282	0.7273	0.9338	0.3073	0.2218
5	0.9021	0.7755	0.4788	0.7006	0.1533	0.0709
6	0.3460	0.2022	0.0307	0.1104	0.0243	0.0023

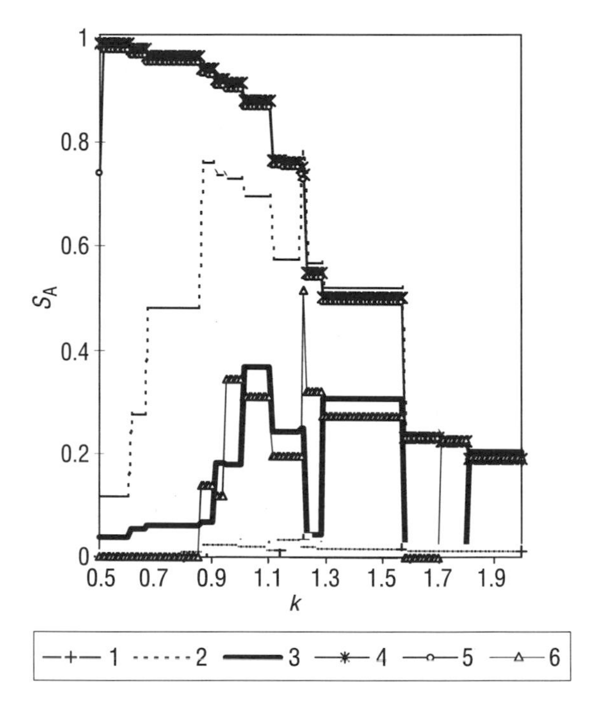

Figure 4.3. Sensitivity S_A as a function of demand for MSS_C

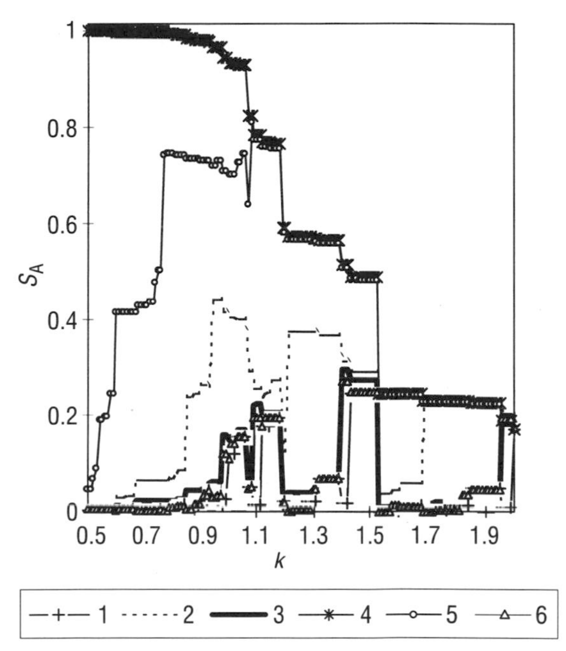

Figure 4.4. Sensitivity S_A as a function of demand for MSS_S

and 4.4, one can notice that the physical nature of MSS performance has a great impact on the relative importance of elements. For example, for $0.6 \leq k \leq 1.0$, $S_A(4)$ and $S_A(5)$ are almost equal for MSS_C and essentially different for MSS_S.

The fourth element is the most important one for both types of system and for all the performance criteria. The order of elements according to their sensitivity indices changes when different system reliability measures are considered. For example, the sixth element is more important than the third one when S_A is considered for MSS_C, but is less important when S_G and S_U are considered.

The UGF approach makes it possible to evaluate the dependencies of the sensitivity indices on the nominal performance rates of elements easily. The sensitivities $S_G(5)$, which are increasing functions of g_5 and decreasing functions of

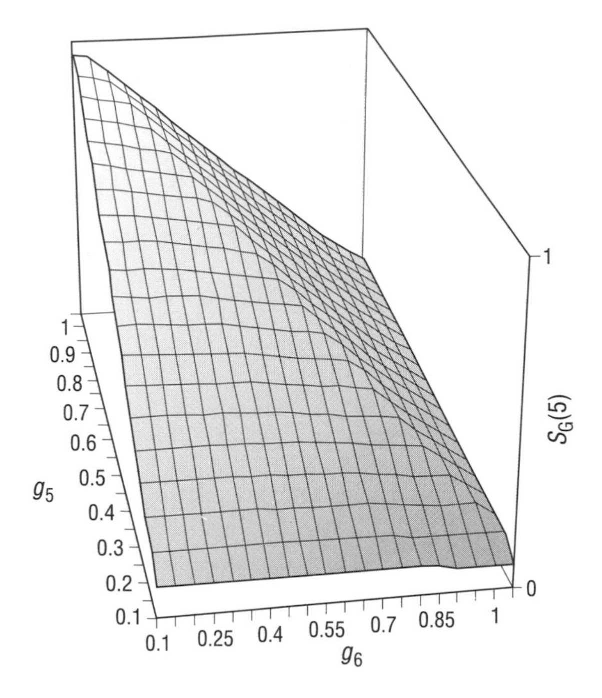

Figure 4.5. Sensitivity S_G as a function of nominal performance rates of elements 5 and 6 for MSS_c

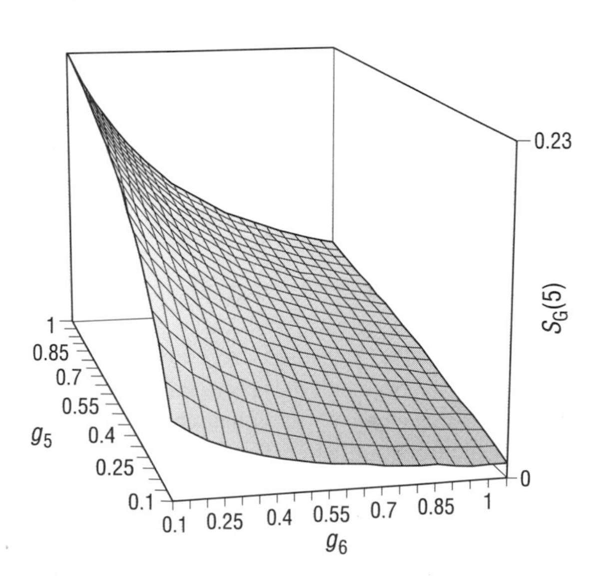

Figure 4.6. Sensitivity S_G as a function of nominal performance rates of elements 5 and 6 for MSS_s

g_6 for both types of system, are presented in Figures 4.5 and 4.6. As elements 5 and 6 share their work, the importance of each element is proportional to its share in the total component performance.

4.6 Multi-state System Structure Optimization Problems

The UGF technique allows system performance distribution, and thereby its reliability indices, to be evaluated based on a fast procedure. The system reliability indices can be obtained as a function of its structure (topology and number of elements), performance rates, and reliability indices of its elements. Therefore, numerous optimization problems can be formulated in which the optimal composition of all or part of the factors influencing the entire MSS reliability has to be found subject to different constraints (*e.g.* system cost). The performance rates and reliability indices of elements composing the system can be supplied by the producer of the elements or obtained from field testing.

In order to provide a required level of system reliability, redundant elements are included. Usually, engineers try to achieve this level with minimal cost. The problem of total investment cost minimization, subject to reliability constraints, is well known as the redundancy optimization problem. The redundancy optimization problem for an MSS, which may consist of elements with different performance rates and reliability, is a problem of system structure optimization.

In order to solve practical problems in which a variety of products exist on the market and analytical dependencies are unavailable for the cost of system components, the reliability engineer should have an optimization methodology in which each product is characterized by its productivity, reliability, price, and/or other parameters. To distinguish among products with different characteristics, the notion of element version is introduced. To find the optimal system structure, one should choose the appropriate versions from a list of available products for each type of equipment, as well as the number of parallel elements of these versions. The objective is to minimize some criterion of the MSS quality (*e.g.* the total MSS cost) subject to the requirement of meeting the demand with the desired level of reliability or to maximize the reliability subject to constraints imposed on the system parameters.

The most general formulation of the MSS structure optimization problem is as follows. Given a system consisting of N components. Each component of type n contains a number of different elements connected in parallel. Different versions and numbers of elements may be chosen for any given system component. Element operation is characterized by its availability and nominal performance rate.

For each component n there are H_n element versions available. A vector of parameters (cost, availability, nominal performance rate, *etc.*) can be specified for each version h of element of type n. The structure of system component n is defined, therefore, by the numbers of parallel elements of each version $r(n, h)$ for $1 \leq h \leq H_n$. The vectors $\mathbf{r}_n = \{r(n, h)\}$, $(1 \leq n \leq N,\ 1 \leq h \leq H_n)$, define the entire system structure. For a given set of vectors $\mathfrak{R} = \{\mathbf{r}_1, \mathbf{r}_2, \ldots, \mathbf{r}_N\}$ the entire system reliability indices $E = (W, \mathfrak{R})$ can be obtained, as well as additional MSS characteristics $C = (W, \mathfrak{R})$, such as cost, weight, *etc.*

Now consider two possible formulations of the problem of system structure optimization:

Formulation 1. Find a system configuration $\mathfrak{R}^*$ that provides a maximal value of system reliability index subject to constraints imposed on other system parameters:

$$\mathfrak{R}^* = \arg\{E(W, \mathfrak{R}) \rightarrow \max \mid C(W, \mathfrak{R}) < C^*\} \tag{4.25}$$

where C^* is the maximal allowable value of the system parameter C.

Formulation 2. Find a system configuration $\mathfrak{R}^*$ that minimizes the system characteristic $C = (W, \mathfrak{R})$ while providing a desired level of system reliability index $E = (W, \mathfrak{R})$

$$\mathfrak{R}^* = \arg\{C(W, \mathfrak{R}) \rightarrow \min \mid E(W, \mathfrak{R}) \geq E^*\} \tag{4.26}$$

where E^* is the minimal allowable level of system reliability index.

4.6.1 Optimization Technique

Equations 4.25 and 4.26 formulate a complicated optimization problem having an enormous number of possible solutions. An exhaustive examination of all these solutions is not realistic, considering reasonable time limitations. As in most optimal allocation problems, the quality of a given solution is the only information available during the search for the optimal solution. Therefore, a heuristic search algorithm is needed that uses only estimates of solution quality and which does not require derivative information to determine the next direction of the search.

The recently developed family of GAs is based on the simple principle of evolutionary search in solution space. GAs have been proven to be effective optimization tools for a large number of applications.

It is recognized that GAs have the theoretical property of global convergence [33]. Despite the fact that their convergence reliability and convergence velocity are contradictory, for most practical, moderately sized combinatorial problems, the proper choice of GA parameters allows solutions close enough to the global optimum to be obtained in a relatively short time.

4.6.1.1 Genetic Algorithm

The basic notions of GAs were originally inspired by biological genetics. GAs operate with "chromosomal" representation of solutions, where crossover, mutation, and selection procedures are applied. Unlike various constructive optimization algorithms that use sophisticated methods to obtain a good singular solution, the GA deals with a set of solutions (population) and tends to manipulate each solution in the simplest manner.

"Chromosomal" representation requires the solution to be coded as a finite length string (in our GA we use strings of integers).

Detailed information on GAs can be found in Goldberg's comprehensive book [34], and recent developments in GA theory and practice can be found in books by Gen and Cheng [17] and Back [33]. The basic structure of the steady-state version of GA referred to as GENITOR [35] is as follows.

First, an initial population of N_S randomly constructed solutions (strings) is generated. Within this population, new solutions are obtained during the genetic cycle by using crossover and mutation operators. The crossover produces a new solution (offspring) from a randomly selected pair of parent solutions, facilitating the inheritance of some basic properties from the parents by the offspring. The probability of selecting the solution as a parent is proportional to the rank of this solution. (All the solutions in the population are ranked in order of their fitness increase.) In this work, we use the so-called two-point (or fragment) crossover operator, which creates the offspring string for the given pair of parent strings by copying string elements belonging to the fragment between two randomly chosen positions from the first parent and by copying the rest of string elements from the second parent. The following example illustrates the crossover procedure (the elements belonging to the fragment are in bold):

Parent string: 7 1 **3 5 5 1 4** 6
Parent string: 1 1 8 6 2 3 5 7
Offspring string: 1 1 **3 5 5 1 4** 7

Each offspring solution undergoes mutation, which results in slight changes to the offspring's structure and maintains a diversity of solutions. This procedure avoids premature convergence to a local optimum and facilitates jumps in the solution space. The positive changes in the solution code, created by the mutation can be later propagated throughout the population via crossovers. In our GA, the mutation procedure swaps elements initially located in two randomly chosen positions on the string. The following example illustrates the mutation procedure (the two randomly chosen positions are in bold):

Solution encoding string before mutation:
1 1 **3** 5 5 1 **4** 7
Solution encoding string after mutation:
1 1 **4** 5 5 1 **3** 7

Each new solution is decoded and its objective function (fitness) values are estimated. These values, which are a measure of quality, are used to compare different solutions. The comparison is accomplished by a selection procedure that determines which solution is better: the newly obtained solution or the worst solution in the population. The better solution joins the population, while the other is discarded. If the population contains equivalent solutions following selection, redundancies are eliminated and the population size decreases as a result.

After new solutions are produced N_{rep} times, new randomly constructed solutions are generated to replenish the shrunken population, and a new genetic cycle begins.

Note that each time the new solution has sufficient fitness to enter the population, it alters the pool of prospective parent solutions and increases the average fitness of the current population. The average fitness increases monotonically (or, in the worst case, does not vary) during each genetic cycle. But in the beginning of a new genetic cycle the average fitness can decrease drastically due to inclusion of poor random solutions into the population. These new solutions are necessary to bring into the population a new "genetic material", which widens the search space and, as well as the mutation operator, prevents premature convergence to the local optimum.

The GA is terminated after N_C genetic cycles. The final population contains the best solution achieved. It also contains different near-optimal solutions, which may be of interest in the decision-making process.

The choice of GA parameters depends on the specific optimization problem. One can find information about parameters choice in [4, 5, 36, 37].

To apply the GA to a specific problem, a solution representation and decoding procedure must be defined.

4.6.1.2 Solution Representation and Decoding Procedure

To provide a possibility of choosing a combination of elements of different versions, our GA deals with L length integer strings, where L is the total number of versions available:

$$L = \sum_{i=1}^{N} H_i \tag{4.27}$$

Each solution is represented by string $\mathbf{S} = \{s_1, s_2, \ldots, s_L\}$, where for each

$$j = \sum_{i=1}^{n-1} H_i + h \tag{4.28}$$

s_j denotes the number of parallel elements of type n and version h: $r(n, h) = s_j$.

For example, for a problem with $N = 3$, $H_1 = 3$, $H_2 = 2$, and $H_3 = 3$, $L = 8$ and string $\{0\ 2\ 1\ 0\ 3\ 4\ 0\ 0\}$ represents a solution in which the first component contains two elements of version 2 and one element of version 3, the second component contains three elements of version 2, and the third component contains four elements of version 1.

Any arbitrary integer string represents a feasible solution, but, in order to reduce the search space, the total number of parallel elements belonging to each component should be limited. To provide this limitation, the solution decoding procedure transforms the string $\mathbf{S}$ in the following way: for each component n for versions from $h = 1$ to $h = H_n$ in sequence define

$$r(n, h) = \min\left\{s_j, D - \sum_{i=1}^{h-1} r(n, i)\right\} \tag{4.29}$$

where D is a specified constant that limits the number of parallel elements and j is the number of elements of string $\mathbf{S}$ defined using Equation 4.28. All the values of the elements of $\mathbf{S}$ are produced by a random generator in the range $D \geq s_j \geq 0$.

For example, for a component n with $H_m = 6$ and $D = 8$ its corresponding substring of $\mathbf{S}$ $\{1\ 3\ 2\ 4\ 2\ 1\}$ will be transformed into substring $\{1\ 3\ 2\ 2\ 0\ 0\}$.

In order to illustrate the MSS structure optimization approach, we present three optimization problems in which MSSs of different types are considered. In the following sections, for each type of MSS we formulate the optimization problem, describe the method of quality evaluation for arbitrary solution generated by the GA, and present an illustrative example.

4.6.2 Structure Optimization of Series–Parallel System with Capacity-based Performance Measure

4.6.2.1 Problem Formulation

A system consisting of N components connected in series is considered (Figure 4.7). Each component of type n contains a number of different elements with total failures connected in parallel.

For each component n there are a number of element versions available in the market. A vector of parameters g_{nh}, a_{nh}, c_{nh} can be specified for each version h of element of type n. This vector contains the nominal capacity, availability, and cost of the element respectively. The element capacity is measured as a percentage of nominal total system capacity.

The entire MSS should provide capacity level not less then W (the failure criterion is $F(G, W) = G - W < 0$).

For given set of vectors $\mathbf{r}_1, \mathbf{r}_2, \ldots, \mathbf{r}_N$ the total cost of the system can be calculated as

$$C = \sum_{n=1}^{N} \sum_{h=1}^{H_i} r(n, h) c_{nh} \tag{4.30}$$

The problem of series–parallel system structure optimization is as follows. Find the minimal cost system configuration $\mathfrak{R}^*$ that provides the required availability level E_{A}^*:

$$\mathfrak{R}^* = \arg\{C(\mathfrak{R}) \to \min \mid E(W, \mathfrak{R}) \geq E_{\mathrm{A}}^*\} \tag{4.31}$$

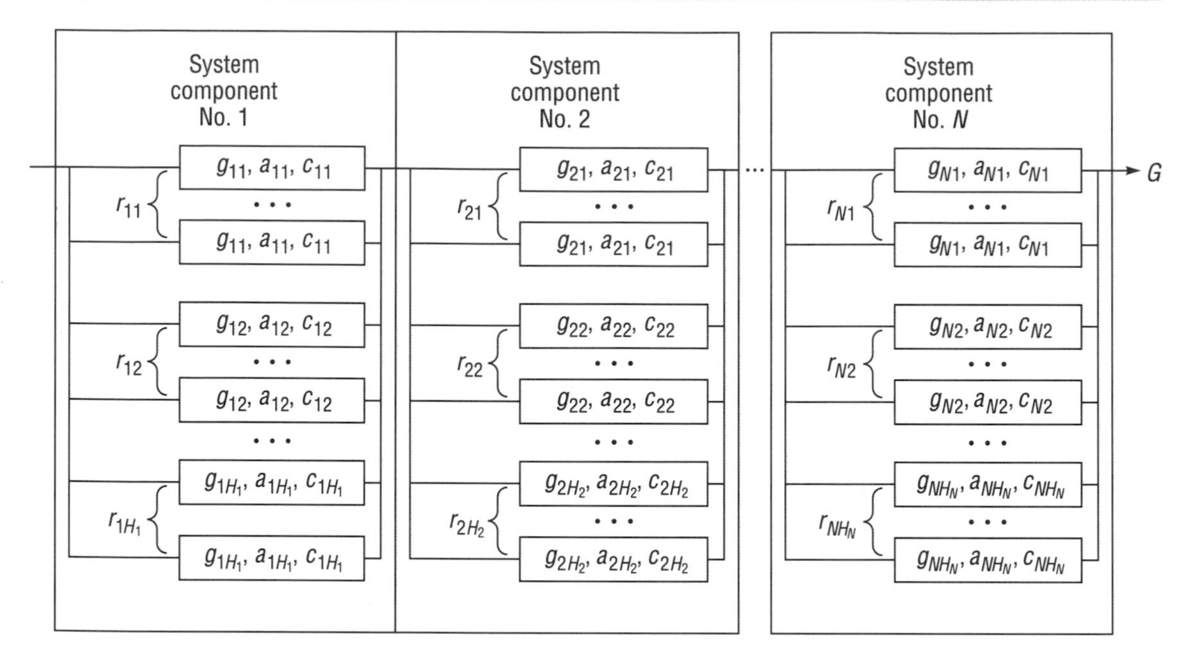

Figure 4.7. Series–parallel MSS structure

4.6.2.2 Solution Quality Evaluation

Using the u-transform technique, the following procedure for E_A index evaluation is used. Since the capacity g_{nh} and availability a_{nh} are given for each element of type n and version h and the number of such elements is determined by the jth element of string **S** (Equation 4.28), one can represent the u-function of the subsystem containing $r(n, h)$ parallel identical elements as

$$u_{nh}(z) = [a_{nh}z^{g_{nh}} + (1 - a_{nh})]^{r(n,h)}$$
$$= \sum_{k=0}^{r(n,h)} \frac{r(n, h)!}{k!(r(n, h) - k)!} \times a_{nh}^k (1 - a_{nh})^{r(n,h)-k} z^{kg_{nh}} \quad (4.32)$$

(This equation is obtained by using the Ω operator (Equation 4.15) with the function in Equation 4.16 corresponding to parallel elements with capacity-based performance.)

To obtain the u-function for the entire component n represented by elements of string **S** with position numbers from $\sum_{i=1}^{n-1} H_i + 1$ to $\sum_{i=1}^{n} H_i$, one can use the same Ω operator over u-functions $u_{nh}(z)$ for $1 \leq h \leq H_n$:

$$U_n(z) = \Omega(u_{n1}(z), \ldots, u_{nH_n}(z))$$
$$= \prod_{j=1}^{H_n} u_{nj}(z) = \sum_{k=1}^{V_n} \mu_k z^{\beta_k} \quad (4.33)$$

where V_n is the total number of different states of the component n, β_k is the output performance rate of the component in state k, and μ_k is the probability of the state k.

In the case of a capacity-based MSS in any combination of states of components connected in series in which at least one component has a performance rate lower than the demand, the output performance of the entire system is also lower than the demand. In this case, there is no need to obtain the u-function of the entire system from u-functions of its components connected in series. Indeed, if considering only that part of the polynomial $U_n(z)$ that provides a capacity exceeding the given demand level W_i, then one has to take into account only elements for which $\beta_k \geq W_i$. Therefore, the following sum should be calculated:

$$p_n(W_i) = \delta_A(U_n(z), F, W_m) = \sum_{\beta_k \geq W_m} \mu_k \quad (4.34)$$

One can calculate the probability of providing capacity that exceeds the level W_i for the entire system containing N components connected in series as

$$A(W_m) = \prod_{n=1}^{N} p_n(W_m) \quad (4.35)$$

and obtain the total E_A index for the variable demand using Equation 4.11.

In order to let the GA look for the solution with minimal total cost and with an E_A that is not less than the required value E_A^*, the solution quality (fitness) is evaluated as follows:

$$\Lambda = \lambda - \Theta(E_A^* - E_A) - \sum_{n=1}^{N} \sum_{j=1}^{H_n} r(n, j) c_{ij} \quad (4.36)$$

where

$$\Theta(x) = \begin{cases} \theta x & x \geq 0 \\ 0 & x < 0 \end{cases} \quad (4.37)$$

and θ is a sufficiently large penalty that is a constant much greater than any possible system cost. For solutions meeting the requirement $E_A > E^*$ the fitness of solution depends only on total system cost.

Example 4. A power station coal transportation system that supplies the boiler consists of five basic components:

1. primary feeder, which loads the coal from the bin to the primary conveyor;
2. primary conveyor, which transports the coal to the stacker–reclaimer;
3. stacker–reclaimer, which lifts the coal up to the burner level;
4. secondary feeder, which loads the secondary conveyor;
5. secondary conveyor, which supplies the burner feeding system of the boiler.

Each element of the system is considered as a unit with total failures. The characteristics of products available in the market for each type of equipment are presented in Table 4.6. This table shows availability a, nominal capacity g (given as a percentage of the nominal boiler capacity), and unit cost c. Table 4.7 contains the data of the piecewise cumulative boiler demand curve.

For this type of problem, we define the minimal cost system configuration that provides the desired reliability level $E_A \geq E_A^*$. The results obtained for different desired values of index E_A^* are presented in Table 4.8. Optimal solutions for a system in which each component can contain only identical parallel elements are given for comparison. System structure is represented by the ordered sequence of strings. Each string has format n: $r_1 * h_1, \ldots, r_i * h_i, \ldots, r_k * h_k$, where n is a number denoting system component, and r_i is the number of elements of version h_i belonging to the corresponding component.

One can see that the algorithm allocating different elements within a component allows for much more effective solutions to be obtained. For example, the cost of the system configuration with different parallel elements obtained for $E^* = 0.975$ is 21% less than the cost of the optimal configuration with identical parallel elements.

The detailed description of the optimization method applied to MSSs with capacity- and processing-speed-based performance measures can be found in [4, 5, 36]. Structure optimization for MSSs with bridge topology is described in [5, 28].

4.6.3 Structure Optimization of Multi-state System with Two Failure Modes

4.6.3.1 Problem Formulation

Systems with two failure modes (STFMs) consist of devices that can fail in either of two different modes. For example, switching systems not only can fail to close when commanded to close but can also fail to open when commanded to open. A typical example of a switching device with two failure modes is a fluid flow valve. It is known that

Table 4.6. Characteristics of the system elements available in the market

Component no.	Description	Version no.	g (%)	a	c (\$$10^6$)
1	Primary feeder	1	120	0.980	0.590
		2	100	0.977	0.535
		3	85	0.982	0.470
		4	85	0.978	0.420
		5	48	0.983	0.400
		6	31	0.920	0.180
		7	26	0.984	0.220
2	Primary conveyor	1	100	0.995	0.205
		2	92	0.996	0.189
		3	53	0.997	0.091
		4	28	0.997	0.056
		5	21	0.998	0.042
3	Stacker–reclaimer	1	100	0.971	7.525
		2	60	0.973	4.720
		3	40	0.971	3.590
		4	20	0.976	2.420
4	Secondary feeder	1	115	0.977	0.180
		2	100	0.978	0.160
		3	91	0.978	0.150
		4	72	0.983	0.121
		5	72	0.981	0.102
		6	72	0.971	0.096
		7	55	0.983	0.071
		8	25	0.982	0.049
		9	25	0.977	0.044
5	Secondary conveyor	1	128	0.984	0.986
		2	100	0.983	0.825
		3	60	0.987	0.490
		4	51	0.981	0.475

Table 4.7. Parameters of the cumulative demand curve

W_m (%)	100	80	50	20
T_m (h)	4203	788	1228	2536

redundancy, introduced to increase the reliability of a system without any change in the reliability of the individual devices that form the system, in the case of an STFM may either increase or decrease entire system reliability [38].

Consider a series–parallel switching system designed to connect or disconnect its input A_1 and output B_N according to a command (Figure 4.8). The system consists of N components connected in series. Each component contains a number of elements connected in parallel. When commanded to close (connect), the system can perform the task when the nodes A_n and B_n $(1 \leq n \leq N)$ are connected in each component by at least one element. When commanded to open (disconnect), the system can perform the task if, in at least one of the component's nodes, A_n and B_n are disconnected, which can occur only if all elements of this component are disconnected. This duality in element and component roles in the two operation modes creates a situation in which redundancy, while increasing system reliability in an open mode, decreases it in a closed mode

Table 4.8. Parameters of the optimal solutions

E_A^*	Identical elements			Different elements		
	E_A	C	Structure	E_A	C	Structure
0.975	0.977	16.450	1 : 2 ∗ 2 2 : 2 ∗ 3 3 : 3 ∗ 2 4 : 3 ∗ 7 5 : 1 ∗ 2	0.976	12.855	1 : 2 ∗ 4, 1 ∗ 6 2 : 6 ∗ 5 3 : 1 ∗ 1, 1 ∗ 4 4 : 3 ∗ 7 5 : 3 ∗ 4
0.980	0.981	16.520	1 : 2 ∗ 2 2 : 6 ∗ 5 3 : 3 ∗ 2 4 : 3 ∗ 7 5 : 1 ∗ 2	0.980	14.770	1 : 2 ∗ 4, 1 ∗ 6 2 : 2 ∗ 3 3 : 1 ∗ 2, 2 ∗ 3 4 : 3 ∗ 7 5 : 2 ∗ 3, 1 ∗ 4
0.990	0.994	17.050	1 : 2 ∗ 2 2 : 2 ∗ 3 3 : 3 ∗ 2 4 : 3 ∗ 7 5 : 3 ∗ 4	0.992	15.870	1 : 2 ∗ 4, 1 ∗ 6 2 : 2 ∗ 3 3 : 2 ∗ 2, 1 ∗ 3 4 : 3 ∗ 7 5 : 3 ∗ 4

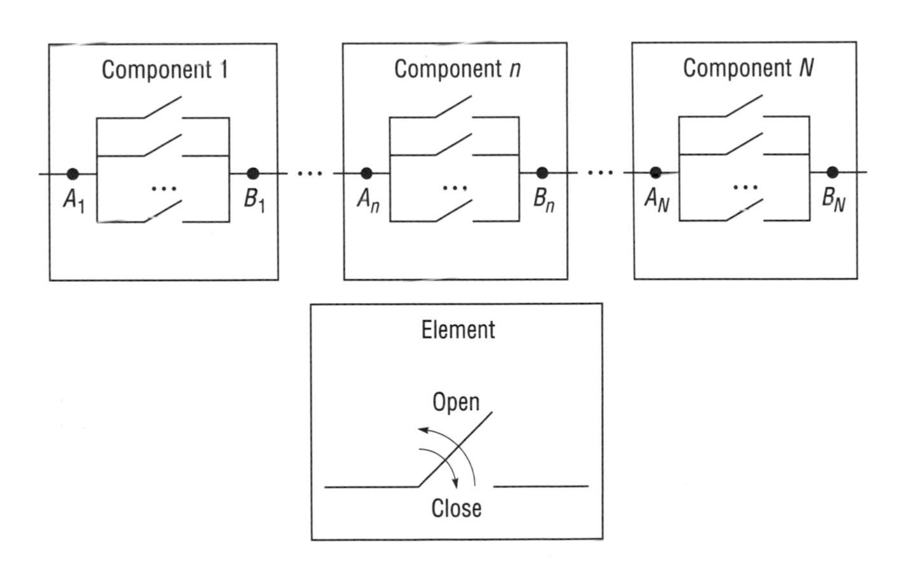

Figure 4.8. Structure of a series–parallel switching system

and *vice versa*. Therefore, the optimal number of redundant elements can be found, and this provides the minimal probability of failures in both states.

In many practical cases, some measures of element (system) performance must be taken into account. For example, fluid-transmitting capacity is an important characteristic of a system containing fluid valves. A system can have different levels of output performance depending on the combination of elements available at any given moment. Therefore, the system should be considered to be MSS.

Since STFMs operate in two modes, two corresponding failure conditions should be defined: $F_o(G_o, W_o) < 0$ in the open mode and $F_c(G_c, W_c) < 0$ in the closed mode, where G_o and G_c are the output performance of the MSS at time t in the open and closed modes respectively, and W_o and W_c are the required system output performances in the open and closed modes respectively. Since the failures in open and closed modes, which have probabilities

$$Q_o = \Pr\{F_o(G_o, W_o) < 0\} = 1 - E_{Ao}(F_o, W_o)$$

and

$$Q_c = \Pr\{F_c(G_c, W_c) < 0\} = 1 - E_{Ac}(F_c, W_c) \quad (4.38)$$

respectively, are mutually exclusive events, the entire system reliability can be defined as

$$1 - Q_o - Q_c = E_{Ao}(F_o, W_o) + E_{Ac}(F_c, W_c) - 1 \quad (4.39)$$

One can also determine the expected system performance in the both modes E_{Go} and E_{Gc}.

Consider two possible formulations of the problem of system structure optimization for a given list of elements' versions characterized by a vector of parameters g_o, g_c, a_o and a_c, where g_x is a nominal element capacity in mode x and a_x is its availability in mode x:

Formulation 3. Find a system configuration $\mathfrak{R}^*$ that provides maximal system availability:

$$\mathfrak{R}^* = \arg\{E_{Ao}(W_o, F_o, \mathfrak{R}) + E_{Ac}(W_c, F_c, \mathfrak{R}) - 1 \to \max\} \quad (4.40)$$

Formulation 4. Find a system configuration $\mathfrak{R}^*$ that provides the maximal proximity of expected system performance to the desired levels for both modes while satisfying the reliability requirements:

$$\mathfrak{R}^* = \arg\{|E_{Go}(\mathfrak{R}) - E^*_{Go}| + |E_{Gc}(\mathfrak{R}) - E^*_{Gc}| \to \min \mid E_{Ao}(\mathfrak{R}) \geq E^*_{Ao}, E_{Ac}(\mathfrak{R}) \geq E^*_{Ac}\} \quad (4.41)$$

4.6.3.2 Solution Quality Evaluation

To determine the u-function of an individual flow-transmitting element with total failures (*e.g.* a fluid flow valve) in the closed mode, note that in the operational state, which has probability a_c, the element should transmit nominal flow f and in the failure state it fails to transmit any flow. Therefore, according to Equation 4.17, the u-function of the element takes the form

$$u_c(z) = a_c z^f + (1 - a_c) z^0 \quad (4.42)$$

In the open mode the element has to prevent flow transmission through the system. If it succeeds in doing this (with probability a_o), the flow is zero; if it fails to do so, the flow is equal to its nominal value in the closed mode f. The u-function of the element in the open mode takes the form

$$u_o(z) = a_o z^0 + (1 - a_o) z^f \quad (4.43)$$

Since the system of flow-transmitting elements is a capacity-based MSS, one has to use operator Ω with the functions in Equations 4.16 and 4.17 over u-functions of the individual elements of Equations 4.42 and 4.43 to determine the u-function of the entire system.

Note that the u-function of a subsystem containing n identical parallel elements can be obtained by applying operator $\Omega(u(z), \ldots, u(z))$ with function ω determined in Equation 4.16 over n u-functions $u(z)$ of an individual element represented by Equations 4.42 or 4.43. The u-function of this subsystem takes the form

$$\sum_{k=0}^{n} \frac{n!}{k!(n-k)!} a_c^k (1 - a_c)^{n-k} z^{kf} \quad (4.44)$$

for the closed mode and

$$\sum_{k=0}^{n} \frac{n!}{k!(n-k)!} a_o^{n-k}(1-a_o)^k z^{kf} \qquad (4.45)$$

for the open mode.

Having u-functions of individual elements and applying corresponding composition operators Ω one obtains the u-functions $U_c(z)$ and $U_o(z)$ of the entire system for both modes. The expected values of flows through the system in open and closed modes E_{Go} and E_{Gc} are obtained using operators $\delta_G(U_o(z))$ and $\delta_G(U_c(z))$ (Equation 4.12).

To determine system reliability one has to define conditions of its successful functioning. For the flow-transmitting system it is natural to require that in its closed mode the amount of flow should exceed some specified value W_c, whereas in the open mode it should not exceed a value W_o. Therefore, the conditions of system success are

$$F_c(G_c, W_c) = G_c - W_c \geq 0$$

and

$$F_o(G_o, W_o) = W_o - G_o \geq 0 \qquad (4.46)$$

Having these conditions one can easily evaluate system availability using operators $\delta_A(U_o(z), F_o, W_o)$ and $\delta_A(U_c(z), F_c, W_c)$ (Equation 4.8).

In order to let the GA look for the solution meeting the requirements of Equation 4.40 or 4.41, the following universal expression of solution quality (fitness) is used:

$$\Lambda = \lambda - |E_{Go} - E^*_{Go}| - |E_{Gc} - E^*_{Gc}| - \theta(\max\{0, E^*_{Ao} - E_{Ao}\} + \max\{0, E^*_{Ac} - E_{Ac}\}) \qquad (4.47)$$

where θ and λ are constants much greater than the maximal possible value of system output performance.

Note that the case $E^*_{Ao} = E^*_{Ac} = 1$ corresponds to the formulation in Equation 4.40. Indeed, since θ is sufficiently large, the value to be minimized in order to maximize Λ is $\theta(E_{Ao} + E_{Ac})$. On the other hand, when $E^*_{Ao} = E^*_{Ac} = 0$, all reliability limitations are removed and expected performance becomes the only factor in determining system structure.

Example 5. Consider the optimization of a fluid-flow transmission system consisting of four components connected in series. Each component can contain a number of elements connected in parallel. The elements in each component should belong to a certain type (*e.g.* each component can operate in a different medium, which causes specific requirements on the valves). There exists a list of available element versions. In this list each version is characterized by element parameters: availability in open and close modes, and transmitting capacity f. The problem is to find the optimal system configuration by choosing elements for each component from the lists presented in Table 4.9.

We want the flow to be not less than $W_c = 10$ in the closed mode and not greater than $W_o = 2$ in the open (disconnected) mode. Three different solutions were obtained for Formulation 3 of

Table 4.9. Parameters of elements available for flow transmission system

Component no.	Element version no.	f	a_c	a_o
1	1	1.8	0.96	0.92
	2	2.2	0.99	0.90
	3	3.0	0.94	0.88
2	1	1.0	0.97	0.93
	2	2.0	0.99	0.91
	3	3.0	0.97	0.88
3	1	3.0	0.95	0.88
	2	4.0	0.93	0.87
4	1	1.5	0.95	0.89
	2	3.0	0.99	0.86
	3	4.5	0.93	0.85
	4	5.0	0.94	0.86

Table 4.10. Solutions obtained for flow transmission system

Component	$E^*_{\text{Ao}} = E^*_{\text{Ac}} = 1$			$E^*_{\text{Ao}} = E^*_{\text{Ac}} = 0.95$		
	$N \geq 13$	$N = 9$	$N = 5$	$N = 13$	$N = 9$	$N = 5$
1	$9*1$	$9*1$	$1*1,\ 1*2,\ 3*3$	$13*1$	$6*1,\ 3*3$	$1*2,\ 4*3$
2	$13*1$	$5*1,\ 4*2$	$2*2,\ 3*3$	$7*1,\ 1*2,\ 5*3$	$3*1,\ 1*2,\ 5*3$	$5*3$
3	$2*1,\ 4*2$	$2*1,\ 4*2$	$2*1,\ 3*2$	$7*2$	$6*2$	$5*2$
4	$11*4$	$8*1,\ 1*2$	$1*1,\ 4*4$	$5*1,\ 4*4$	$6*1,\ 3*4$	$1*1,\ 4*4$
$Q_c = 1 - E_{\text{Ac}}$	0.001	0.002	0.026	0.0001	0.0006	0.034
$Q_o = 1 - E_{\text{Ao}}$	0.007	0.010	0.037	0.049	0.050	0.049
$E_{\text{Ao}} + E_{\text{Ac}} - 1$	0.992	0.988	0.937	0.951	0.949	0.917
E_{Gc}	12.569	12.669	12.026	21.689	18.125	13.078
E_{Go}	0.164	0.151	0.118	0.390	0.288	0.156

the problem, where minimal overall failure probability was achieved ($E^*_{\text{Ao}} = E^*_{\text{Ac}} = 1$). The optimal solutions were obtained under constraints on the maximal number of elements within a single component: $D = 15$, $D = 9$, and $D = 5$. These solutions are presented in Table 4.10, where system structure is represented by strings $r_1 * h_1, \ldots, r_i * h_i, \ldots, r_k * h_k$ for each component, where r_i is the number of elements of version h_i belonging to the corresponding component. Table 4.10 also contains fault probabilities $1 - E_{\text{Ac}}$ and $1 - E_{\text{Ao}}$, availability index $E_{\text{Ao}} + E_{\text{Ac}} - 1$, and expected flows through the system in open and closed mode E_{Go} and E_{Gc} obtained for each solution. Note that when the number of elements was restricted by 15, the optimal solution contains no more than 13 elements in each component. A further increase in the number of elements cannot improve system availability.

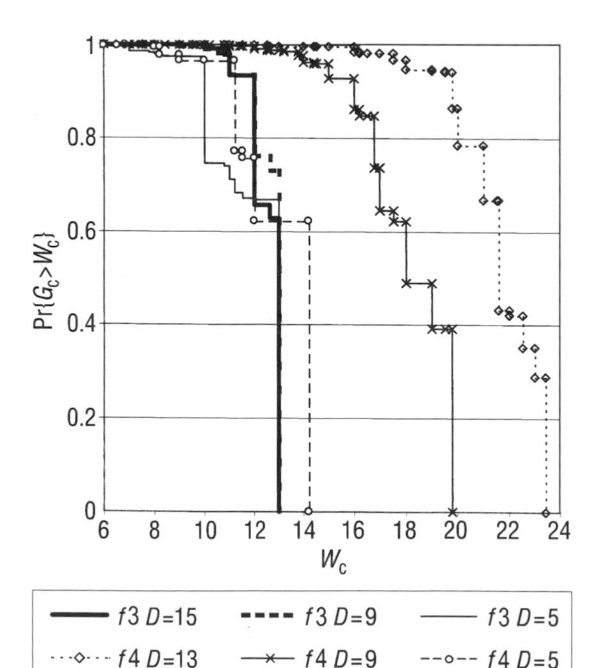

Figure 4.9. Cumulative performance distribution of solutions obtained for a flow-transmission system in closed mode (fk stands for problem formulation k)

The solutions obtained for Formulation 4 of the optimization problem for $E^*_{\text{Ao}} = E^*_{\text{Ac}} = 0.95$ are also presented in Table 4.10. Here, we desired to obtain the flow as great as possible ($E^*_{\text{Gc}} = 25$) in the closed mode and as small as possible ($E^*_{\text{Go}} = 0$) in the open mode, while satisfying conditions $\Pr\{G_c \geq 10\} \geq 0.95$ and $\Pr\{G_o < 2\} \geq 0.95$. The same constraints on the number of elements within a single component were imposed. One can see that this set of solutions provides a much greater difference $E_{\text{Gc}} - E_{\text{Go}}$, while the system availability is reduced when compared with Formulation 3 solutions. The probabilistic distributions of flows through the system

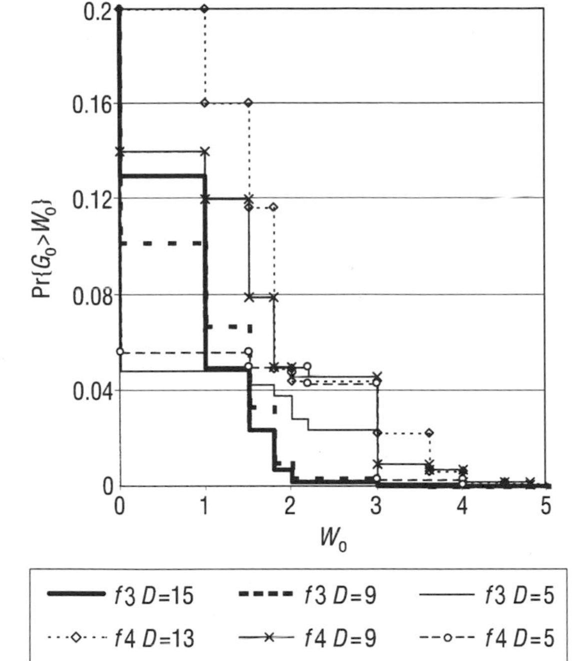

Figure 4.10. Cumulative performance distribution of solutions obtained for a flow-transmission system in open mode (fk stands for problem formulation k)

in closed and open modes for all the solutions obtained are presented in Figures 4.9 and 4.10 in the form of cumulative probabilities $\Pr\{G_c \geq W_c\}$ and $\Pr\{G_o \geq W_o\}$.

Levitin and Lisnianski [39] provide a detailed description of the optimization method applied to STFM with capacity-based (flow valves) and processing-speed-based (electronic switches) performance measures.

4.6.4 Structure Optimization for Multi-state System with Fixed Resource Requirements and Unreliable Sources

4.6.4.1 Problem Formulation

Although the algorithms suggested in previous sections cover a wide range of series–parallel systems, these algorithms are restricted to systems with continuous flows, which are comprised of elements that can process any piece of product (resource) within its capacity (productivity) limits. In this case, the minimal amount of product, which can proceed through the system, is not limited.

In practice, there are technical elements that can work only if the amount of some resources is not lower than specified limits. If this requirement is not met, the element fails to work. An example of such a situation is a control system that stops the controlled process if a decrease in its computational resources does not allow the necessary information to be processed within the required cycle time. Another example is a metalworking machine that cannot perform its task if the flow of coolant supplied is less than required. In both these examples the amount of resources necessary to provide the normal operation of a given composition of the main producing units (controlled processes or machines) is fixed. Any deficit of the resource makes it impossible for all the units from the composition to operate together (in parallel), because no one unit can reduce the amount of resource it consumes. Therefore, any resource deficit leads to the turning off of some producing units.

This section considers systems containing producing elements with fixed resource consumption. The systems consist of a number of resource-generating subsystems (RGSs) that supply different resources to the main producing subsystem (MPS). Each subsystem consists of different elements connected in parallel. Each element of the MPS can perform only by consuming a fixed amount of resources. If, following failures in the RGSs, there are not enough resources to allow all the available producing elements to work, then some of these elements should be turned off. We assume that the choice of the working MPS elements is made in such a way as to maximize the total performance rate of the MPS under given resources constraints.

The problem is to find the minimal cost RGS and MPS structure that provides the desired level of entire system ability to meet the demand.

In spite of the fact that only two-level RGS–MPS hierarchy is considered in this section, the

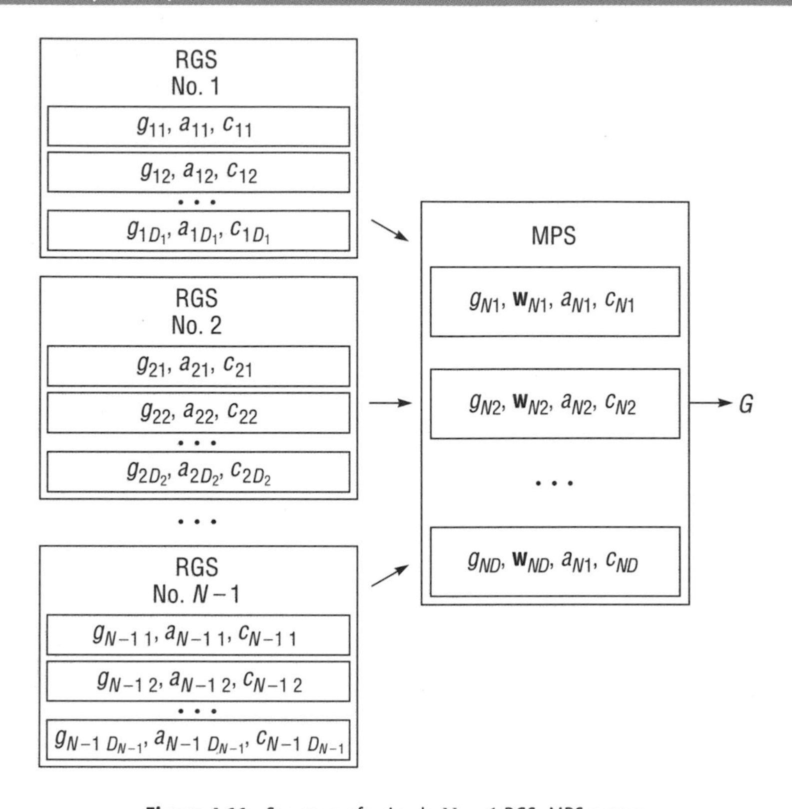

Figure 4.11. Structure of a simple $N-1$ RGS–MPS system

method can easily be expanded to systems with multilevel hierarchy. When solving the problem for multilevel systems, the entire RGS–MPS system (with OPD defined by its structure) may be considered in its turn as one of an RGS for a higher level MPS.

Consider a system consisting of an MPS and $N-1$ different RGSs (Figure 4.11). The MPS can have up to D different elements connected in parallel. Each producing element consumes resources supplied by the RGS and produces a final product. There are H_N versions of producing elements available. Each version h $(1 \le h \le H_N)$ is characterized by its availability a_{Nh}, performance rate (productivity or capacity) g_{Nh}, cost c_{Nh}, and vector of required resources $\mathbf{w}_h = \{w_{nh}\}$ $1 \le n \le N-1$. An MPS element of version h can work only if it receives the amount of each resource defined by vector $\mathbf{w}_h$.

Each resource n $(1 \le n \le N-1)$ is generated by the corresponding RGS, which can contain up to D_n parallel resource-generating elements of different versions. The total number of available versions for the nth RGS is H_n. Each version of an element of RGS supplying the nth resource is characterized by its availability, productivity, and cost.

The problem of system structure optimization is as follows: find the minimal cost system configuration $\mathfrak{R}^*$ that provides the required reliability level E_A^* (Equation 4.31), where for any given system structure $\mathfrak{R}$ the total cost of the system can be calculated using Equation 4.30.

4.6.4.2 Solution Quality Evaluation

Since the considered system contains elements with total failure represented by two-term u-functions and each subsystem consists of a number of parallel elements, one can obtain u-functions $U_n(z)$ for the each RGS n representing the probabilistic distribution of the amount of nth resource, which can be supplied to the MPS as follows:

$$U_n(z) = \prod_{h=1}^{H_n} [(1 - a_{nh})z^0 + a_{nh}z^{g_{nh}}]^{r(n,h)} = \sum_{k=1}^{V_n} \mu_{nk} z^{\beta_{nk}} \quad (4.48)$$

where V_n is the number of different levels of resource n generation, β_{nk} is the available amount of the nth resource at state k, and μ_{nk} is the probability of state k for RGS n.

The same equation can be used in order to obtain the u-function representing the OPD of the MPS $U_N(z)$. In this case, V_N is the number of different possible levels of MPS productivity. The function $U_N(z)$ represents the distribution of system productivity defined only by MPS elements' availability. This distribution corresponds to situations in which there are no limitations on required resources.

4.6.4.3 The Output Performance Distribution of a System Containing Identical Elements in the Main Producing Subsystem

If a producing subsystem contains only identical elements, then the number of elements that can work in parallel when the available amount of nth resource at state k is β_{nk} is $\lfloor \beta_{nk}/w_n \rfloor$, which corresponds to total system productivity $\gamma_{nk} = g\lfloor \beta_{nk}/w_n \rfloor$. In this expression, g and w_n are respectively the productivity of a single element of the MPS and the amount of resource n required for this element, and the function $\lfloor x \rfloor$ rounds a number x down to the nearest integer. Note that γ_{nk} represents the total theoretical productivity, which can be achieved using available resource n by an unlimited number of producing elements. In terms of entire system output, the u-function of the nth RGS can be obtained in the following form:

$$U_n(z) = \sum_{k=1}^{V_n} \mu_{nk} z^{\gamma_{nk}} = \sum_{k=1}^{V_n} \mu_{nk} z^{g\lfloor \beta_{nk}/w_n \rfloor} \quad (4.49)$$

The RGS that provides the work of a minimal number of producing units becomes the bottleneck of the system. Therefore, this RGS defines the total system capacity. To calculate the u-function for all the RGSs $U^{\mathrm{r}}(z)$ one has to apply the Ω operator with the ω function in Equation 4.17. Function $U^{\mathrm{r}}(z)$ represents the entire system OPD for the case of an unlimited number of producing elements.

The entire system productivity is equal to the minimum of total theoretical productivity which can be achieved using available resources and total productivity of available producing elements. To obtain the OPD of the entire system taking into account the availability of the MPS elements, the same Ω operator should be applied:

$$U(z) = \Omega(U_N(z), U^{\mathrm{r}}(z)) = \Omega(U_1(z), U_2(z), \ldots, U_N(z)) \quad (4.50)$$

4.6.4.4 The Output Performance Distribution of a System Containing Different Elements in the Main Producing Subsystem

If an MPS has D different elements, there are 2^D possible states of element availability composition. Each state k may be characterized by set π_k ($1 \le k \le 2^D$) of available elements. The probability of state k can be evaluated as follows:

$$p_k = \prod_{j \in \pi_k} a_{Nj} \prod_{i \notin \pi_k} (1 - a_{Ni}) \quad (4.51)$$

The maximal possible productivity of the MPS and the corresponding resources consumption in state k are $\sum_{j\in\pi_k} g_j$ and $\sum_{j\in\pi_k} w_{nj}$ ($1 \le n < N$) respectively.

The amount of resources generated by an RGS is defined by its OPD. There are not always enough resources to provide the maximal possible productivity of the MPS at state k. In order to define the maximum possible productivity G of the MPS under resource constraints one has to solve the following integer linear programming problem:

$$G(\beta_1, \beta_2, \ldots, \beta_N, \pi_k) = \max \sum_{j \in \pi_k} g_j y_j$$

subject to

$$\sum_{j\in\pi_k} w_{nj} y_j \le \beta_n \quad \text{for } 1 \le n < N,\ \ y_j \in \{0, 1\} \tag{4.52}$$

where β_n is the available amount of nth resource, $y_j = 1$ if element j works (producing g_j units of main product and consuming w_{nj} of each resource $(1 \le n < N)$), and $y_j = 0$ if element j is turned off.

The OPD of the entire system can be defined by evaluating all the possible combinations of available resources generated by the RGS and the states of the MPS. For each combination, a solution of the above formulated optimization problem defines the system productivity. The u-function representing the OPD of the entire system can be defined as follows:

$$U(z) = \sum_{i_1=1}^{V_1}\sum_{i_2=1}^{V_2}\cdots\sum_{i_{N-1}=1}^{V_{N-1}}\left\{\left(\prod_{n=1}^{N-1}\mu_{ni_n}\right) \times \sum_{k=1}^{2^D} p_k z^{G(\beta_{1i_1},\beta_{2i_2},\ldots,\beta_{N-1i_{N-1}},\pi_k)}\right\} \tag{4.53}$$

To evaluate the E_A index for the entire system having its OPD as the probability that the total productivity of the system is not less than a specified demand level W, we can use the function $F = G - W$ and operator δ_A (Equation 4.8).

To obtain the system OPD, its productivity should be determined for each unique combination of available resources and for each unique state of the MPS. From Equation 4.53 one can see that in the general case the total number of integer linear programs to be solved to obtain $U(z)$ is $2^D \prod_{n=1}^{N-1} V_n$. In practice, the number of programs to be solved can be drastically reduced using the following rules.

1. If for the given vector $(\beta_1, \ldots, \beta_n, \ldots, \beta_{N-1})$ and for the given set of MPS elements π_k there exists n for which $\beta_n < \min_{j\in\pi_k} w_{nj}$, then the system productivity $G = 0$.
2. If for each element j from π_k there exists n for which $\beta_n < w_{nj}$, then the system productivity $G = 0$.
3. If there exists element $j \in \pi_k$ for which $\beta_n < w_{nj}$ for some n, this means that in the program (Equation 4.52) y_j must be zeroed. In this case the dimension of the integer program can be reduced by removing all such elements from π_k.
4. If for the given vector $(\beta_1, \ldots, \beta_n, \ldots, \beta_{N-1})$ and for the given set π_k the solution of the integer program (Equation 4.52) determines subset π_k^* of turned on MPS elements ($j \in \pi_k^*$ if $y_j = 1$), the same solution must be optimal for the MPS states characterized by any set $\pi_k' : \pi_k^* \subset \pi_k' \subset \pi_k$. This allows one to avoid solving many integer programs by assigning value of $G(\beta_1, \ldots, \beta_{N-1}, \pi_k)$ to all the $G(\beta_1, \ldots, \beta_{N-1}, \pi_k')$.

It should be noted that, for systems with a large number of elements and/or resources, the required computational effort for solving the redundancy optimization problem could be unaffordable, even when applying the computational complexity reduction technique presented. In this case, the use of fast heuristics for solving integer programs is recommended instead of exact algorithms.

By applying the technique described for obtaining $U(z)$ and using the δ_A operator for determining the E_A index, one can estimate the solution fitness in the GA using Equation 4.36.

Example 6. The main producing component of the system may have up to six parallel producing elements (chemical reactors) working in parallel. To perform their task, producing elements require three different resources:

1. power, generated by the energy supply subsystem (group of converters);
2. computational resource, provided by the control subsystem (group of controllers);
3. cooling water, provided by the water supply subsystem (group of pumps).

Each of these RGSs can have up to five parallel elements. Both producing units and resource-generating units may be chosen from the list

Table 4.11. Parameters of the MPS units available

Version no.	Cost c	Performance rate g	Availability a	Resources required $\mathbf{w}$		
				Resource 1	Resource 2	Resource 3
1	9.9	30.0	0.970	2.8	2.0	1.8
2	8.1	25.0	0.954	0.2	0.5	1.2
3	7.9	25.0	0.960	1.3	2.0	0.1
4	4.2	13.0	0.988	2.0	1.5	0.1
5	4.0	13.0	0.974	0.8	1.0	2.0
6	3.0	10.0	0.991	1.0	0.6	0.7

Table 4.12. Parameters of the RGS units available

Type of resource	Version no.	Cost c	Performance rate g	Availability a
1	1	0.590	1.8	0.980
	2	0.535	1.0	0.977
	3	0.370	0.75	0.982
	4	0.320	0.75	0.978
2	1	0.205	2.00	0.995
	2	0.189	1.70	0.996
	3	0.091	0.70	0.997
3	1	2.125	3.00	0.971
	2	2.720	2.60	0.973
	3	1.590	2.40	0.971
	4	1.420	2.20	0.976

Table 4.13. Parameters of the cumulative demand curve

W_m	65.0	48.0	25.0	8.0
T_m	60	10	10	20

of products available in the market. Each producing unit is characterized by its availability, its productivity, its cost, and the amount of resources required for its work. The characteristics of available producing units are presented in Table 4.11. The resource-generating units are characterized by their availability, their generating capacity (productivity), and their cost. The characteristics of available resource-generating units are presented in Table 4.12. Each element of the system is considered to be a unit with total failures.

The demand for final product varies with time. The demand distribution is presented in Table 4.13 in the form of a cumulative demand curve.

Table 4.14 contains minimal cost solutions for different required levels of system availability E_{A}^{*}. The structure of each subsystem is presented by the list of numbers of versions of the elements included in the subsystem. The actual estimated availability of the system and its total cost are also presented in the table for each solution.

The solutions of the system structure optimization problem when the MPS can contain only identical elements are presented in Table 4.15 for comparison. Note that, when the MPS is composed from elements of different types, the same system availability can be achieved by much lower cost. Indeed, using elements with different availability and capacity (productivity) provides much greater flexibility for optimizing the entire system performance in different states. Therefore, the algorithm for solving the problem for different MPS elements, which requires much greater computational effort, usually yields better solutions then one for identical elements.

A detailed description of the optimization method applied to an MSS with fixed resource requirements has been given by Levitin [40].

4.6.5 Other Problems of Multi-state System Optimization

In this section we present a short description of reliability optimization problems that can be

Table 4.14. Parameters of the optimal solutions for system with different MPS elements

E_A^*	E_A	C	System structure			
			MPS	RGS 1	RGS 2	RGS 3
0.950	0.951	27.790	3, 6, 6, 6, 6	1, 1, 1, 1, 1	1, 1, 2, 3	1, 1
0.970	0.973	30.200	3, 6, 6, 6, 6, 6	1, 1, 1, 1	1, 1, 2, 3	1, 1
0.990	0.992	33.690	3, 3, 6, 6, 6, 6	1, 1, 1, 1	1, 1, 2, 3	4, 4
0.999	0.999	44.613	2, 2, 3, 3, 6, 6	1, 1, 1	1, 2, 2	4, 4, 4

Table 4.15. Parameters of the optimal solutions for system with identical MPS elements

E_A^*	E_A	C	System structure			
			MPS	RGS 1	RGS 2	RGS 3
0.950	0.951	34.752	4, 4, 4, 4, 4, 4	1, 4, 4, 4, 4	2, 2, 2	1, 3, 3, 3, 4
0.970	0.972	35.161	4, 4, 4, 4, 4, 4	1, 1, 1, 4	2, 2, 2, 2	1, 3, 3, 3, 4
0.990	0.991	37.664	2, 2, 2, 2	4, 4	3, 3, 3, 3	4, 4, 4
0.999	0.999	47.248	2, 2, 2, 2, 2	3, 4	2, 2	4, 4, 4, 4

solved using a combination of the UGF technique and GAs.

In practice, the designer often has to include additional elements in the existing system rather than to develop a new one from scratch. It may be necessary, for example, to modernize a system according to new demand levels or according to new reliability requirements. The problem of optimal single-stage MSS expansion to enhance its reliability and/or performance is an important extension of the structure optimization problem. In this case, one has to decide which elements should be added to the existing system and to which component they should be added. Such a problem was considered by Levitin *et al.* [36].

During the MSS life time, the demand and reliability requirements can change. To provide a desired level of MSS performance, management should develop a multistage expansion plan. For the problem of optimal multistage MSS expansion [41], it is important to answer not only the question of what must be included into the system, but also the question of when.

By optimizing the maintenance policy one can achieve the desired level of system reliability (availability) requiring the minimal cost. The UGF technique allows the entire MSS reliability to be obtained as a function of the reliabilities of its elements. Therefore, by having estimations of the influence of different maintenance actions on the elements' reliability, one can evaluate their influence on the entire complex MSS containing elements with different performance rates and reliabilities. An optimal policy of maintenance can be developed that would answer the questions about which elements should be the focus of maintenance activity and what should the intensity of this activity be [42,43].

Since the maintenance activity serves the same role in MSS reliability enhancement as does incorporation of redundancy, the question arises as to what is more effective. In other words, should the designer prefer a structure with more redundant elements and less investment in maintenance or *vice versa*? The optimal compromise should minimize the MSS cost while providing its desired reliability. The joint maintenance and redundancy optimization problem [37] is to find this optimal compromise taking into account differences in reliability and performance rates of elements composing the MSS.

Finally, the most general optimization problem is optimal multistage modernization of an MSS subject to reliability and performance requirements [44]. In order to solve this problem, one should develop a minimal-cost modernization plan that includes maintenance, modernization of elements, and system expansion actions. The objective is to provide the desired reliability level while meeting the increasing demand during the lifetime of the MSS.

References

[1] Aven T, Jensen U. Stochastic models in reliability. New York: Springer-Verlag; 1999.

[2] Billinton R, Allan R. Reliability evaluation of power systems. Pitman; 1990.

[3] Aven T. Availability evaluation of flow networks with varying throughput-demand and deferred repairs. IEEE Trans Reliab 1990;38:499–505.

[4] Levitin G, Lisnianski A, Ben-Haim H, Elmakis D. Redundancy optimization for series–parallel multi-state systems. IEEE Trans Reliab 1998;47(2):165–72.

[5] Lisnianski A, Levitin G, Ben-Haim H. Structure optimization of multi-state system with time redundancy. Reliab Eng Syst Saf 2000;67:103–12.

[6] Gnedenko B, Ushakov I. Probabilistic reliability engineering. New York: John Wiley & Sons; 1995.

[7] Murchland J. Fundamental concepts and relations for reliability analysis of multistate systems. In: Barlow R, Fussell S, Singpurwalla N, editors. Reliability and fault tree analysis, theoretical and applied aspects of system reliability. Philadelphia: SIAM; 1975. p.581–618.

[8] El-Neveihi E, Proschan F, Setharaman J. Multistate coherent systems. J Appl Probab 1978;15:675–88.

[9] Barlow R, Wu A. Coherent systems with multistate components. Math Oper Res 1978;3:275–81.

[10] Ross S. Multivalued state component systems. Ann Probab 1979;7:379–83.

[11] Griffith W. Multistate reliability models. J Appl Probab 1980;17:735–44.

[12] Butler D. A complete importance ranking for components of binary coherent systems with extensions to multistate systems. Nav Res Logist Q 1979;26:556–78.

[13] Koloworcki K. An asymptotic approach to multi-state systems reliability evaluation. In: Limios N, Nikulin M, editors. Recent advances in reliability theory, methodology, practice, inference. Birkhauser; 2000. p.163–80.

[14] Pourret O, Collet J, Bon J-L. Evaluation of the unavailability of a multi-state component system using a binary model. Reliab Eng Syst Saf 1999;64:13–7.

[15] Aven T. On performance measures for multistate monotone systems. Reliab Eng Syst Saf 1993:41:259–66.

[16] Coit D, Smith A. Reliability optimization of series–parallel systems using genetic algorithm, IEEE Trans Reliab 1996;45:254–66.

[17] Gen M, Cheng R. Genetic algorithms and engineering design. New York: John Wiley & Sons; 1997.

[18] Dengiz B, Altiparmak F, Smith A. Efficient optimization of all-terminal reliable networks, using an evolutionary approach. IEEE Trans Reliab 1997;46:18–26.

[19] Cheng S. Topological optimization of a reliable communication network. IEEE Trans Reliab 1998;47:23–31.

[20] Bartlett L, Andrews J. Efficient basic event ordering schemes for fault tree analysis. Qual Reliab Eng Int 1999;15:95–101.

[21] Zhou Y-P, Zhao B-Q, Wu D-X. Application of genetic algorithms to fault diagnosis in nuclear power plants. Reliab Eng Syst Saf 2000;67:2:153–60.

[22] Martorell S, Carlos S, Sanchez A, Serradell V. Constrained optimization of test intervals using a steady-state genetic algorithm. Reliab Eng Syst Saf 2000;67:215–32.

[23] Marseguerra M, Zio E. Optimizing maintenance and repair policies via a combination of genetic algorithms and Monte Carlo simulation. Reliab Eng Syst Saf 2000;68:69–83.

[24] Brunelle R, Kapur K. Review and classification of reliability measures for multistate and continuum models. IIE Trans 1999;31:1171–80.

[25] Ushakov I. A universal generating function. Sov J Comput Syst Sci 1986;24:37–49.

[26] Ushakov I. Reliability analysis of multi-state systems by means of a modified generating function. J Inform Process Cybernet 1988;34:24–9.

[27] Lisnianski A, Levitin G, Ben-Haim H, Elmakis D. Power system structure optimization subject to reliability constraints. Electr Power Syst Res 1996;39:145–52.

[28] Levitin G, Lisnianski A. Survivability maximization for vulnerable multi-state systems with bridge topology. Reliab Eng Syst Saf 2000;70:125–40.

[29] Birnbaum ZW. On the importance of different components in a multicomponent system. In: Krishnaiah PR, editor. Multivariate analysis 2. New York: Academic Press; 1969.

[30] Barlow RE, Proschan F. Importance of system components and fault tree events. Stochast Process Appl 1975;3:153–73.

[31] Vesely W. A time dependent methodology for fault tree evaluation. Nucl Eng Des 1970;13:337–60.

[32] Bossche A. Calculation of critical importance for multi-state components. IEEE Trans Reliab 1987;R-36(2): 247–9.

[33] Back T. Evolutionary algorithms in theory and practice. Evolution strategies. Evolutionary programming. Genetic algorithms. Oxford University Press; 1996.

[34] Goldberg D. Genetic algorithms in search, optimization and machine learning. Reading (MA): Addison Wesley; 1989.

[35] Whitley D. The GENITOR algorithm and selective pressure: why rank-based allocation of reproductive trials is best. In: Schaffer D, editor. Proceedings 3rd International Conference on Genetic Algorithms. Morgan Kaufmann; 1989. p.116–21.

[36] Levitin G, Lisnianski A, Elmakis D. Structure optimization of power system with different redundant elements. Electr Power Syst Res 1997;43:19–27.

[37] Levitin G, Lisnianski A. Joint redundancy and maintenance optimization for multistate series-parallel systems. Reliab Eng Syst Saf 1998;64:33–42.

[38] Barlow R. Engineering reliability. Philadelphia: SIAM; 1998.

[39] Levitin G, Lisnianski A. Structure optimization of multi-state system with two failure modes. Reliab Eng Syst Saf 2001;72:75–89.

[40] Levitin G. Redundancy optimization for multi-state system with fixed resource requirements and unreliable sources. IEEE Trans Reliab 2001;50(1):52–9.

[41] Levitin G. Multistate series–parallel system expansion scheduling subject to availability constraints. IEEE Trans Reliab 2000;49:71–9.

[42] Levitin G, Lisnianski A. Optimization of imperfect preventive maintenance for multi-state systems. Reliab Eng Syst Saf 2000;67:193–203.

[43] Levitin G, Lisnianski A. Optimal replacement scheduling in multi-state series–parallel systems (short communication). Qual Reliab Eng 2000;16:157–62.

[44] Levitin G, Lisnianski A. Optimal multistage modernization of power system subject to reliability and capacity requirements. Electr Power Syst Res 1999;50: 183–90.

Combinatorial Reliability Optimization

C. S. Sung, Y. K. Cho and S. H. Song

Chapter 5

5.1 Introduction

Reliability engineering as a concept appeared in the late 1940s and early 1950s and was applied first to the fields of communication and transportation. Reliability is understood to be a measure of how well a system meets its design objective during a given operation period without repair work. In general, reliability systems are composed of several subsystems (stages), each having more than one component, including operating and redundant units.

Reliability systems have become so complex that successful operation of each subsystem has been a significant issue in automation and productivity management. One important design issue is to design an optimal system having the greatest quality of reliability or to find the best way to increase any given system reliability, which may be subject to various engineering constraints associated with cost, weight, and volume. Two common approaches to the design issue have been postulated as follows:

1. an approach of incorporating more reliable components (units);
2. an approach of incorporating more redundant components.

The first approach is not always feasible, since high cost may be involved or reliability improvement may be technically limited. Therefore, the second approach is more commonly adapted for economical design of systems, for which optimal redundancy is mainly concerned. In particular, the combinatorial reliability optimization problem is defined in terms of redundancy optimization as the problem of determining the optimal number of redundant units of the component associated with each stage subject to a minimum requirement for the associated whole system reliability and also various resource restrictions, such as cost, weight, and volume consumption.

There are two distinct types of redundant unit, called parallel and standby units. In parallel redundant systems, one original operating unit and all redundant units at each stage operate simultaneously. However, in standby redundant systems, any redundant units are not activated unless the operating unit fails. If the operating unit fails, then it will be replaced with one from among the redundant units. Under the assumptions that the replacement time for each failed unit is ignored and the failure probability of each unit is fixed at a real value, a mathematical programming model for the standby redundancy optimization problem can be derived as the same as that for the parallel redundancy optimization problem. The last assumption may be relaxed for the problem to be treated in a stochastic model approach.

The combinatorial reliability optimization issues concerned with redundancy optimization may be classified into:

1. maximization of system effectiveness (reliability) subject to various resource constraints; and
2. minimization of system cost subject to the condition that the associated system effectiveness (reliability) be required to be equal to or greater than a desired level.

The associated system reliability function can have a variety of different forms, depending on the system structures. For example, the redundancy optimization problem commonly considers three types of system structure, including:

1. series structure;
2. mixed series–parallel structure;
3. general structure, which is in neither series nor parallel, but rather in bridge network type.

The series system has only one path between the source and the terminal node, but the mixed series–parallel system can have multiple paths between them. The general-structured system can also have multiple paths, while it is different from the mixed series–parallel system such that at least two subsystems (stages) are related in a non-series and/or non-parallel structure.

The reliability function of the series system is represented by a product form of all the associated stage reliabilities. However, the other two systems cannot have their reliability functions being represented by any product forms of all stage (subsystem) reliabilities. Rather, their reliabilities may be computed by use of a minimal path set or a minimal cut set. This implies that the series-structured system has the simplest reliability function form.

Thereby, the redundancy optimization problem can be formulated in a variety of different mathematical programming expressions depending on the associated system structures and problem objectives, as shown below.

Series system reliability maximization problem (SRP):

$$Z_{\text{SRP}} = \max \prod_{i \in I} [1 - (1 - r_i)^{y_i}]$$

subject to

$$\sum_{i \in I} g_i^m(y_i) \le C^m \quad \forall m \in M \tag{5.1}$$

$$y_i = \text{non-negative integer} \quad \forall i \in I \tag{5.2}$$

where Z_{SRP} represents the optimal system reliability of the problem (SRP), I is the set of stages, r_i is the reliability of the component used at stage i, y_i is the variable representing the number of redundant units of the component allocated at stage i, $g_i^m(y_i)$ is the consumption of resource m for the component and its redundant units in stage i, and C^m is the total available amount of the resource m.

The constraints in Equation 5.1 represent the resource constraints and the constraints in Equation 5.2 define the decision variables to be non-negative integers.

Non-series system reliability maximization problem (NRP):

$$Z_{\text{NRP}} = \max R(R_1, \ldots, R_i, \ldots, R_{|I|})$$

subject to the constraints in Equations 5.1 and 5.2, where Z_{NRP} represents the optimal system reliability of the problem (NRP), and

$R(R_1, \ldots, R_i, \ldots, R_{|I|})$ is the system reliability function, given the reliability of each stage i at $R_i = [1 - (1 - r_i)^{y_i}]$, which is dependent upon the associated system structure.

Series system cost minimization problem (SCP):

$$Z_{\text{SCP}} = \min \sum_{i \in I} g_i(y_i)$$

subject to the constraints in Equations 5.1 and 5.2:

$$\prod_{i \in I} [1 - (1 - r_i)]^{y_i} \geq R_{\min} \tag{5.3}$$

where Z_{SCP} represents the optimal system cost of the problem (SCP), $g_i(y_i)$ is the cost charged for resource consumption y_i at stage i, and $R_{\min}$ is the minimum level of the system reliability requirement. The constraints in Equation 5.3 represent the system reliability constraint.

Non-series system cost minimization problem (NCP):

$$Z_{\text{NCP}} = \min \sum_{i \in I} g_i(y_i)$$

subject to the constraints in Equations 5.1 and 5.2:

$$R(R_1, \ldots, R_i, \ldots, R_{|I|}) \geq R_{\min} \tag{5.4}$$

where Z_{NCP} represents the optimal system cost of the problem (NCP). The constraints in Equation 5.4 represent the system reliability requirement.

The constraints in Equation 5.1 representing resource consumption are given in general function forms, since the resource consumption for improving the associated system reliability does not increase linearly as the number of redundant units of the component at each stage increases. Therefore, the data of the resource consumption rate are important for the problem formulation. However, only few data are practically available, so that the resource constraints are often expressed in linear functions with the variable representing the number of redundant units. Accordingly, the constraints in Equation 5.1 can be specified as

$$\sum_{i \in I} g_i^m(y_i) = \sum_{i \in I} c_i^m y_i \leq C^m$$

where c_i^m denotes the consumption of resource m at stage i.

The proposed combinatorial reliability optimization problems are now found as being expressed in nonlinear integer programming where the associated system reliability functions are nonlinear functions. It is also found that the reliability function of the non-series-structured system is not separable. In other words, the system reliability cannot be in any product form of all stage reliabilities. This implies that it is not easy to evaluate the reliability of the general-structured system. Valiant [1] has proved that the problem of computing reliabilities (reliability objective function values) for general-structured systems is NP-complete. In the meantime, for the mixed series–parallel types of system among them, Satyanarayana and Wood [2] have proposed a linear time algorithm to compute the reliability objective function values, given their associated stage reliabilities.

The complexity analysis by Chern [3] shows that the simplest reliability maximization problem of series systems with only one resource constraint incorporated is NP-complete. Therefore, even though many algorithms have been proposed in the literature, only a few have been proven effective for large-scale nonlinear reliability programming problems, and none of them has been proven to be superior over any of the others. A variety of different solution approaches have been studied for the combinatorial reliability optimization problems; these are listed as follows.

Optimal solution approaches:

- integer programming;
- partial enumeration method;
- branch-and-bound method;
- dynamic programming.

Heuristic approaches:

- greedy-type heuristics.

Continuous relaxation approaches:

- maximum principle;
- geometric programming;
- sequential unconstrained minimization technique (SUMT);

- modified sequential simplex pattern search;
- Lagrangian multipliers and Kuhn–Tucker conditions;
- generalized Lagrangian function;
- generalized reduced gradient (GRG);
- parametric programming.

The integer programming approach has been considered for integer solutions, whereas the associated problems have nonlinear objective functions and/or constraints that may be too difficult to handle in the approach. However, no integer programming techniques guarantee that the optimal solutions can be obtained for large-sized problems in a reasonable time. Similarly, the dynamic programming approach has the curse of dimensionality, whose computation load may increase exponentially with the number of the associated state variables. Moreover, in general, it gets harder to solve problems with more than two constraints.

Heuristic algorithms have been considered to find integer-valued solutions (local optimums) in a reasonable time, but they do not guarantee that their resulting solutions are globally optimal. However, Malon [4] has shown that greedy assembly is optimal whenever the assembly modules have a series structure.

The continuous relaxation approaches including the sequential unconstrained minimization technique, the generalized reduced gradient method, and the generalized Lagrangian function method have also been developed for reliability optimization problems. Along with these methods, rounding-off procedures have often been applied for redundancy allocation problems in situations where none of the relaxation approaches could yield integer solutions. The associated reference works have been summarized in Tables 1 and 3 of Tillman *et al.* [5].

Any problem with a nonlinear reliability objective function can be transformed into a problem with linear objective function, in a situation where the nonlinear objective function is separable. This has been studied in Tillman and Liittschwager [6]. By taking logarithms of the associated system reliability function and considering appropriate substitutions, the following equivalent problem (LSRP) with linear objective function and constraints can be derived:

Problem (LSRP):

$$Z_{\mathrm{LSRP}} = \max \sum_{i\in I}\sum_{k=1}^{\infty} \Delta f_{ij} y_{ij}$$

subject to

$$\sum_{i\in I}\sum_{k=1}^{\infty} \Delta c_{ik}^{m} y_{ik} \le C^{m} \quad \forall m \in M \tag{5.5}$$

$$y_{ik} - y_{i,k-1} \le 0 \quad \forall k \in Z^{+},\ i \in I \tag{5.6}$$

$$y_{i0} = 1 \quad \forall i \in I \tag{5.7}$$

$$y_{ik} = \text{binary integer} \quad \forall k \in Z^{+}, i \in I \tag{5.8}$$

where y_{ik} is the variable representing the kth redundancy at stage i,

$$\begin{aligned}\Delta f_{ik} &= [1-(1-r_i)^k] - [1-(1-r_i)^{k-1}]\\ &= (1-r_i)^{k-1} - (1-r_i)^k,\end{aligned}$$

which is the change in system reliability due to the kth redundancy added at stage i, and $\Delta c_{ik}^{m} = g_i^m(k) - g_i^m(k-1)$, which is the change in consumption of resource m due to the kth redundancy added at stage i.

Likewise, the problem (SCP) can also be transformed into a problem with linear constraints. The problems (NRP) and (NCP), however, cannot be transformed into any linear integer programming problems, since the associated system reliability functions are not separable.

Note that if any problem with a separable reliability function has linear resource constraints, then its transformed problem can be reduced to a knapsack problem.

This chapter is organized as follows. In Section 5.2, the combinatorial reliability optimization problems for series structures are considered. Various solution approaches, including optimal solution methods and heuristics, are introduced. Section 5.3 considers the problems (NRP) for non-series structured systems. Section 5.4 considers a problem with multiple-choice constraints incorporated and the corresponding solution approach is introduced, which has been studied recently. Finally, concluding remarks are presented.

5.2 Combinatorial Reliability Optimization Problems of Series Structure

This section considers combinatorial reliability optimization problems having series structures. Many industrial systems (*e.g.* heavy machines and military weapons), whose repair work is so costly (difficult), are commonly designed in series structures of their subsystems to have multiple components in parallel at each stage (subsystem) so as to maintain (increase) the whole system reliability. For example, each combat tank (system) is composed of three major subsystems (including power train, suspension, and fire control subsystems) in a series structure. In order to increase the whole combat tank reliability, the combat tank is designed to have multiple blowers in parallel in the power train subsystem, multiple torsion bars and support rollers in parallel in the suspension subsystem, and multiple gunner controllers in parallel in the fire control subsystem. Thus, the associated design issue of how many components to be installed multiply in each subsystem for an optimal combat tank system can be handled as a series system combinatorial reliability optimization problem.

The approach of having multiple components in parallel at each subsystem can also be applied to many production systems. For example, the common wafer-fabrication flow line is composed of three major processes, *viz.* diffusion/oxidation, photolithography, and etching, which are handled by the three corresponding facilities called furnace, stepper, and dry etcher respectively. In the fabrication flow line, it is common to have multiple facilities in parallel for each of the processes so as to increase the whole line reliability. Thus, the associated design issue of how many facilities to be installed multiply in each process for an optimal wafer-fabrication flow line can also be handled as a series system combinatorial reliability optimization problem.

5.2.1 Optimal Solution Approaches

The combinatorial reliability problem of a series structure has drawn research attention, as shown in the literature, since 1960s. All the early studies for the associated reliability optimization problems were concerned with finding optimal solutions for them. Their representative solution approaches include the partial enumeration method, the branch-and-bound method, and the dynamic programming algorithm.

5.2.1.1 Partial Enumeration Method

Misra [7] has applied the partial enumeration method to the redundancy optimization problem. The partial enumeration method was first proposed by Lawler and Bell [8], and was developed for a discrete optimization problem with a monotonic objective function and arbitrary constraints. Misra [7] has dealt with the problem as maximizing reliability or optimizing some other objective functions subject to multiple separable constraints (not necessarily being linear functions). He has transformed the problem into conformable forms to the method.

Using the partial enumeration method of Lawler and Bell [8], the following types of problem can be solved:

Minimize $g_0(\bar{v})$
subject to

$$g_1^m(\bar{v}) - g_2^m(\bar{v}) \geq 0 \quad \forall m \in M$$

where $\bar{v} = (v_1, v_2, \ldots, v_{|I|})$ and $v_i = 0$ or 1, $\forall i \in I$.

Note that each of the functions, $g_1^m(\bar{v})$ and $g_2^m(\bar{v})$, must be monotonically non-decreasing with each of its arguments. With some ingenuity, many problems may be put in this form.

Let vector $\bar{v} = (v_1, v_2, \ldots, v_{|I|})$ be a binary vector in the sense that each v_i is either 0 or 1. Define $\bar{v}^1 \leq \bar{v}^2$ iff $v_i^1 \leq v_i^2$ for $\forall i \in I$, which is called the vector partial ordering. The lexicographic or numerical ordering of these vectors can also be obtained by identifying with each $\bar{v}$ the integer value $N(\bar{v}) = v_1 2^{n-1} + v_2 2^{n-2} + \cdots + v_n 2^0$. The numerical ordering is a refinement of the vector partial ordering such that $\bar{v}^1 \leq \bar{v}^2$ implies $N(\bar{v}^1) \leq N(\bar{v}^2)$, whereas $N(\bar{v}^1) \leq N(\bar{v}^2)$ does not imply $\bar{v}^1 \leq \bar{v}^2$.

The partial enumeration method of Lawler and Bell [8] is basically a search method, which starts with $(0, \ldots, 0)$ and examines the $2^{|I|}$ solution vectors in the numerical ordering described above. By the way, the labor of examination can be considerably cut down by the following three rules. As the examination proceeds, one can retain the least costly up-to-date solution. If $\hat{v}$ is the least costly solution having cost $g_0(\hat{v})$ and $\bar{v}$ is the vector to be examined, then the following steps indicate the conditions under which certain vectors may be skipped.

Rule 1. Test if $g_0(\bar{v}) \leq g_0(\hat{v})$. If yes, skip to v^* and repeat the operation; otherwise, proceed to Rule 2. Here, the value v^* is the first vector following $\bar{v}$ in the numerical order that has the property $\bar{v} < v^*$. For any $\bar{v}$, v^* is calculated by subtracting one from the binary number $\bar{v}$, and $v^* - 1$ is calculated by logical OR operation performed between $\bar{v}$ and $\bar{v} - 1$ (*e.g.* let us have $\bar{v} = (1010)$. Then $\bar{v} - 1 = (1001)$, and so $v^* - 1 = \bar{v}\ \text{OR}\ (\bar{v} - 1) = (1011)$). Finally, add one to obtain v^*.

Rule 2. Examine if $g_1^m(v^* - 1) - g_2^m(\bar{v}) \leq 0\ \forall m \in M$. If yes, proceed to Rule 3; otherwise, skip to v^* and go to Rule 1.

Rule 3. If $g_1^m(v^* - 1) - g_2^m(\bar{v}) \geq 0\ \forall m \in M$, then replace $\hat{v}$ by $\bar{v}$ and skip to v^*; otherwise, change $\bar{v}$ to $\bar{v} + 1$. In either case, transfer to Rule 1 for further execution.

The above three rules are the so-called algorithm skipping rules. By following the rules, it is necessary to examine all the binary vectors and continue scanning until a vector having maximum numerical order, $(1, \ldots, 1)$, is found. In a situation where one has skipped to a vector having numerical order higher than $(1, \ldots, 1)$, the procedure is terminated and the resulting least-costly vector provides the optimal solution. For details, refer to Lawler and Bell [8]. For a numerical example, refer to Misra [7].

5.2.1.2 Branch-and-bound Method

A few researchers have considered the branch-and-bound method for the series system reliability maximization problem. They include Ghare and Taylor [9], Sung and Lee [10], and Kuo *et al.* [11]. The branch-and-bound method is usually described by determining bounding and branching procedures. For a general branch-and-bound technique, refer to Lawler and Wood [12]. In Ghare and Taylor [9], the problem (SRP) was transformed into a problem with binary variables, and then it was proved that there is a correspondence between their feasible solutions. Finally, they proposed a branch-and-bound method for the transformed binary problem.

The transformed binary problem is formulated as follows:

$$Z_{\text{SRP}} = \max \sum_{i \in I} \sum_{k=1}^{k=\infty} a_{ik} x_{ik}$$

subject to

$$\sum_{i \in I} \sum_{k=1}^{k=\infty} c_i^m x_{ik} \leq b^m \quad \forall m \in M \tag{5.9}$$

$$x_{ik} = \text{binary integer} \quad \forall k \in Z^+,\ i \in I \tag{5.10}$$

where x_{ik} indicates whether or not the kth redundant unit is used at stage i, $a_{ik} = \ln[1 - (q_i)^{k+1}] - \ln[1 - (q_i)^k]$, $q_i = (1 - r_i)$, c_i^m denotes the consumption of resource m of at stage i, and $b^m = C^m - \sum_{i \in I} c_i^m$.

Ghare and Taylor [9] have developed a branch-and-bound method for the binary problem in which a simple bound of the series system reliability has been derived by using the inequality of the component reliability and cost consumption. In order to develop a bounding procedure for the associated branch-and-bound algorithm, they have considered a single-dimensional knapsack problem as $Z_{\text{SRP}} = \max \sum_{i \in I} \sum_{k=1}^{k=\infty} a_{ik} x_{ik}$, subject to a single constraint $\sum_{i \in I} \sum_{k=1}^{k=\infty} c_i^m x_{ik} \leq b^m$ for a given m, and defined the ratios $\alpha_{ik} = a_{ik}/c_i^m$.

Then, for a feasible solution, the following relation holds:

$$\begin{aligned} Z_{\mathrm{SRP}} &= \max \sum_{i \in I} \sum_{k=1}^{k=\infty} a_{ik} x_{ik} \\ &= \max \sum_{i \in I} \sum_{k=1}^{k=\infty} \alpha_{ik} c_i^m x_{ik} \\ &\le \max_{i,k}(\alpha_{ik}) \sum_{i \in I} \sum_{k=1}^{k=\infty} c_i^m x_{ik} \\ &\le \max_{i,k}(\alpha_{ik}) b^m \end{aligned}$$

Moreover, since

$$\begin{aligned} \exp(a_{ik}) &= [1-(q_i)^{k+1}]/[1-(q_i)^k] \\ &= 1 + (q_i)^k/[1 + q_i + \cdots + (q_i)^{k-1}] \end{aligned}$$

and

$$\begin{aligned} \exp(a_{i,k+1}) &= [1-(q_i)^{k+2}]/[1-(q_i)^{k+1}] \\ &= 1 + (q_i)^{k+1}/[1 + q_i + \cdots + (q_i)^k] \end{aligned}$$

it can be seen that $a_{ik} > a_{ik+1}$, which implies $\alpha_{ik} > \alpha_{ik+1}$, or $\max_{i,k}(\alpha_{ik}) = \max_i(\alpha_{i1})$. It follows that $Z_{\mathrm{SRP}} \le \max_i(\alpha_{i1}) b^m$.

In the problem, there are $|M|$ constraints, one for each resource m, so that for any feasible solution for the problem, it holds that

$$\begin{aligned} Z_{\mathrm{SRP}} &\le \max_i(\alpha_{i1}) b^m \quad \text{for any } m \\ &\le \min_{m \in M} \left[\max_i(\alpha_{i1}) b^m \right] \end{aligned}$$

Consequently, the optimal objective value Z_{SRP} is bounded by the quantity $\min_{m \in M}[\max_i(\alpha_{i1}) b^m]$, which is the upper bound for the problem. For the complete description of the proposed branch-and-bound procedure, refer to Ghare and Taylor [9] and Balas [13]. Their branch-and-bound procedure has been modified by McLeavey [14] in several subfunctions, including algorithm initialization, branching variable selection, depth of searching, and problem formulation with either binary variables or integer variables. Moreover, in McLeavey and McLeavey [15], all those solution methods, in addition to the continuous relaxation method of Luus [16], the greedy-type heuristic method of Aggarwal *et al.* [17], and the branch-and-bound method of Ghare and Talyor [9], have been compared against one another in performance, being measured in terms of CPU time, optimality rate, relative error for the optimum, and coding efforts.

Recently, a new bounding mechanism has been proposed in Sung and Lee [10], based on which an efficient branch-and-bound algorithm has been developed. In developing the algorithm, they did not deal with any transformed binary problem, but dealt with the problem (SRP) directly.

Given any feasible system reliability $\bar{Z}$ to the problem, Sung and Lee [10] have characterized the two properties that the optimal system reliability Z_{SRP} should be equal to or greater than $\bar{Z}$ and that the system reliability can be represented by a product form of all the stage reliabilities. These properties are used to derive the following relation for each stage s as

$$R_s(y_s^*) \ge \frac{Z_{\mathrm{SRP}}}{\prod_{i \ne s} R_i(y_i^{\mathrm{u}})} \ge \frac{\bar{Z}}{\prod_{i \ne s} R_i(y_i^{\mathrm{u}})}$$

where $R_s(y_s^*) = 1 - (1 - r_s)^{y_s^*}$ and, y_s^* and y_s^{u} are the optimal solution and an upper bound of y_s respectively. Accordingly, y_s^* should be less than or equal to y_s^{u}. Therefore, the lower bound y_s^{l} of y_s can be determined as

$$y_s^{\mathrm{l}} = \max \left\{ y_s \,\middle|\, R_s(y_s) \ge \frac{\bar{Z}}{\prod_{i \ne s} R_i(y_i^{\mathrm{u}})} \right\}$$

In addition, an efficient way of determining such upper bounds can be found based on $y_i^{\mathrm{l}}\ \forall i \in I$, K^m, and R_s^m, where

$$K^m = \max_{i \in I} \left\{ \frac{\ln[R_i(y_i^{\mathrm{l}} + 1)] - \ln[R_i(y_i^{\mathrm{l}})]}{g_i^m(y_i^{\mathrm{l}} + 1) - g_i^m(y_i^{\mathrm{l}})} \right\}$$

and

$$\begin{aligned} R_s^m = K^m C^m - \sum_{i \ne s} \{K^m g_i^m(y_i^{\mathrm{l}}) - \ln[R(y_i^{\mathrm{l}})]\} \\ - \ln \bar{Z} \end{aligned}$$

K^m may be interpreted as the maximum marginal contribution to the reliability from unit addition of resource m and R_s^m is just a value computed in association with $\bar{Z}$ and y_i^{l} $(i \neq s)$. Thus, the upper bound of y_s can be determined as

$$y_s^{\mathrm{u}} = \max \left\{ y_s \;\middle|\; y_s \geq y_s^{l};\ K^m g_s^m(y_s) - \ln[R_s(y_s)] \leq R_s^m \text{ or } g_s^m(y_s) \leq C^m - \sum_{i \neq s} g_i^m(y_i),\ \forall m \right\}$$

From the above discussions, the lower and upper bound values of y_s can now be summarized as giving the basis for finding the optimal value y_s^* such that

(a) $y_s^{\mathrm{l}} \leq y_s^* \leq y_s^{\mathrm{u}}, \forall s \in I$;
(b) each y_s^{l} can be determined at a specific y_i^{u} $(i \neq s)$ value given;
(c) each y_s^{u} can be determined at a specific y_i^{l} $(i \neq s)$ value given; and
(d) the larger lower bounds lead to the smaller upper bounds, and *vice versa*.

Based on these relations between the lower and upper bounds, an efficient search procedure for the bounds has been proposed.

Bounding procedure:

Step 1. Set $t = 1$, $R_i(y_{i0}^{\mathrm{u}}) = 1$ (*i.e.* $y_{i0}^{\mathrm{u}} = \infty$), where the index t represents iteration number. Then, find $\bar{Z}$ by using the greedy-type heuristic algorithm in Sung and Lee [10].

Step 2. Calculate y_{it}^{l}, $\forall i \in I$, by using $y_{i,t-1}^{\mathrm{u}}$ $\forall i \in I$.

Step 3. Calculate y_{it}^{u} $\forall i \in I$, by using y_{it}^{l} $\forall i \in I$. If $y_{it}^{\mathrm{u}} = y_{i,t-1}^{\mathrm{u}}$ $\forall i \in I$, then go to Step 4; otherwise, set $t = t + 1$, and then go to Step 2.

Step 4. Set $y_i^{\mathrm{u}} = y_{it}^{\mathrm{u}}$, $y_i^{\mathrm{l}} = y_{it}^{\mathrm{l}}$, and then terminate the procedure.

Sung and Lee [10] have proved that both the upper and lower bounds obtained from the proposed bounding procedure converge to the heuristic solution $\bar{y}_i$ for every i.

Theorem 1. (Sung and Lee [10]) *The heuristic solution* $(\bar{y}_1, \ldots, \bar{y}_{|I|})$ *satisfies the relations:*

$$y_{i1}^{\mathrm{l}} \leq y_{i2}^{\mathrm{l}} \leq \cdots \leq y_i^{\mathrm{l}} \leq \bar{y}_i \leq y_i^{\mathrm{u}} \leq \cdots \leq y_{i1}^{\mathrm{u}} \quad \forall i \in I$$

In Sung and Lee [10], no theoretical comparison between the bound of Ghare and Taylor [9] and that of Sung and Lee [10] has been made, but a relative efficiency comparison with the branch-and-bound algorithm of Ghare and Taylor [9] for ten-stage reliability problems has been made to find that their proposed algorithm is about ten times more efficient. It is seen in the numerical test of the algorithm of Sung and Lee [10] that the smaller-gap-first-order branching procedure is more efficient than the larger-gap-first-order branching procedure. The smaller-gap-first-order branching procedure implies that the stage with the smaller gap between upper and lower bounds is branched earlier. For details, refer to Sung and Lee [10].

Kuo *et al.* [11] have dealt with the problem (SRP) and proposed a method incorporating a Lagrangian relaxation method and a branch-and-bound method. However, the method does not guarantee the optimal solution, since the branch-and-bound method is applied only to search for an integer solution based on a continuous relaxed solution obtained by the Lagrangian multiplier method.

5.2.1.3 Dynamic Programming

Various research studies have shown how dynamic programming could be used to solve a variety of combinatorial reliability optimization problems. As mentioned earlier, the dynamic programming method is based on the principle of optimality. It can yield an exact solution to a problem, but its computational complexity increases exponentially as the number of constraints or the number of variables in the problem increases. For the combinatorial reliability optimization problems, two dynamic programming approaches have been applied: one is a basic dynamic programming approach [18], and the other is a dynamic programming approach using the concept of

Table 5.1. Problem data for the basic dynamic programming algorithm

Stage	1	2	3
r_i	0.8	0.9	0.85
c_i (= cost)	2	4	3
Constraints:	$C \le 9$		

dominance sequence [19, 20]. The first of these approaches is described in this section, but the latter is introduced in Section 5.3, where the computational complexity of each of the two approaches is discussed.

The basic dynamic programming approach is applied to the redundancy optimization problem for a series system, which is to maximize the system reliability subject to only the budget constraint. The recursive formula of the problem in the basic dynamic programming algorithm is formulated as follows:

for stage 1;

$$f_1(d) = \max_{g_1(y_1)\le d,\, y_1\in Z^+} [R_1(y_1)], \quad 1 \le d \le C$$

for stage 2;

$$f_2(d) = \max_{g_2(y_2)\le d,\, y_2\in Z^+} \{R_2(y_2) f_1[d - g_2(y_2)]\}, \quad 1 \le d \le C$$

for stage i;

$$f_i(d) = \max_{g_i(y_i)\le d,\, y_i\in Z^+} \{R_i(y_i) f_{i-1}[d - g_i(y_i)]\}, \quad 1 \le d \le C$$

for the last stage $|I|$;

$$f_{|I|}(C) = \max_{y_{|I|}\in Z^+} \{R_{|I|}(y_{|I|}) \cdot f_{|I|-1}[C - g_{|I|}(y_{|I|})]\}$$

where C represents the amount of budget available, and $R_i(y_i)$ represents the reliability of stage i when the stage i has y_i redundant units.

In order to illustrate the basic dynamic programming method algorithm, a numerical example shall be solved with the data given in Table 5.1. And its results are summarized in Table 5.2.

5.2.2 Heuristic Solution Approach

This section presents a variety of heuristic methods that can give integer solutions directly. As discussed already, the given problems are known to be in the class of NP-hard problems. Therefore, all the solution methods discussed in the previous sections may give the optimal solutions, but their computational time to obtain the solutions may increase exponentially as the problem size (variables) increases. This is why a variety of heuristic methods have been considered as practical approaches to find approximate solutions in a reasonable time.

Research on the heuristic approaches was extensive from the early 1970s to the mid 1980s. Most of these heuristic approaches are categorized into a kind of greedy-type heuristic method. For example, Sharma and Venkateswaran [21] considered a problem in 1971 whose objective was to maximize the associated system reliability subject to nonlinear constraints, which is the same as the problem (SRP). For the problem, they presented a simple algorithm, that handles the objective function being transformed into an easier form, such as the function of the failure probability function transformed as follows:

$$\begin{aligned}
&\max \prod_{i\in I} [1 - (1 - r_i)^{y_i}] \\
&= \min \left\{1 - \prod_{i\in I} [1 - (1 - r_i)^{y_i}]\right\} \\
&\cong \min \left\{1 - \sum_{i\in I} [1 - (1 - r_i)^{y_i}]\right\} \\
&\cong \min \sum_{i\in I} (1 - r_i)^{y_i} \\
&= \min \sum_{i\in I} q_i^{y_i}
\end{aligned}$$

where q_i is the failure probability of the component at stage i.

The algorithm of Sharma and Venkateswaran [21] adds one component to the associated stage in each iteration until no more components can be added to the system due to the associated resource constraints, in which the component-adding stage has the largest failure probability

Table 5.2. Example of the basic dynamic programming algorithm

Stage 1			Stage 2					Stage 3		
d	Y_1	$f_1(d)$	d	y_2	$R_2(y_2)$	$f_1(d-c_2, y_2)$	$f_2(d)$	y_3	$R_3(y_3)$	$f_2(d-c_3, y_3)$
2	1	0.8	4	1	0.9	0	0	1	0.85	0.72[a]
4	2	0.96	6	1	0.9	0.8	0.72	2	0.978	0
6	3	0.992	8	2	0.99	0	0	3	0.997	0
8	4	0.998	—	—	—	—	—	—	—	—

[a] Optimal solution value.

Table 5.3. Problem data for the heuristic algorithm

Stage	1	2	3
q_i	0.1	0.2	0.15
c_i^1 (= cost)	5	7	7
c_i^2 (= weight)	8	7	8
Constraints:	$C^1 \leq 45$ and $C^2 \leq 54$		

among the stages in the system. For the detailed step-by-step procedure, refer to Sharma and Venkateswaran [21].

In order to illustrate the algorithm of Sharma and Venkateswaran [21], a numerical example shall be solved with the data given in Table 5.3.

Starting with a feasible solution (1, 1, 1), add one unit at each iteration of the algorithm as shown in Table 5.4. After four iterations, the final solution (2, 3, 2) is obtained and its objective function value is 0.96.

The algorithm of Sharma and Venkateswaran [21] may not yield a good (near-optimal) solution if only those components of similar reliability but quite different resource consumption rate are available to all the stages. Besides, in certain cases, slack variables (balance of unused resources) may prevent addition of only one component to a particular stage having the lowest reliability, but permit addition of more than one component to another stage having the highest reliability. The latter case may result in a greater net increase in reliability than the former. However, their discussion on component addition is made just based on component-wide reliability, but without considering any effectiveness of resource utilization. In fact, the effectiveness can be represented by the contribution of each component addition to the whole system reliability, so that any component addition at the stage having the lowest reliability will have greater effectiveness in increasing the whole system reliability than any component addition at any other stage. Therefore, Aggarwal *et al.* [17] proposed a heuristic criterion for solving the resource-constrained problem in consideration of resource utilization effectiveness. The heuristic algorithm is based on the concept that a component is added to the stage where its addition produces the largest ratio of increment in reliability to the product of increments in resource usage. Accordingly, the stage selection criterion is defined as

$$F_i(y_i) \equiv \frac{\Delta(q_i^{y_i})}{\prod_{m \in M} \Delta g_i^m(y_i)}$$

where $\Delta(q_i^{y_i}) = q_i^{y_i} - q_i^{y_i+1}$ for all i and $\Delta g_i^m(y_i) = g_i^m(y_i+1) - g_i^m(y_i)$ for all i and m.

In the case of linear constraints, however, all $F_i(y_i)$ can be evaluated by using the following recursive relation:

$$F_i(y_i + 1) = q_i F_i(y_i)$$

For details, refer to Aggarwal *et al.* [17]. The algorithm of Aggarwal *et al.* [17] is similarly illustrated with the data of Table 5.3, as summarized in Table 5.5.

Table 5.4. Illustration of the stepwise iteration of the algorithm of Sharma and Venkateswaran [21]

Iteration	No. of units in each stage			Failure probability of each stage			Cost	Weight
	1	2	3	1	2	3		
0	1	1	1	0.1	0.2[a]	0.15	19	23
1	1	2	1	0.1	0.04	0.15[a]	26	30
2	1	2	2	0.1[a]	0.04	0.0225	33	38
3	2	2	2	0.01	0.04[a]	0.0225	38	46
4	2	3	2	0.01	0.008	0.0225	45	53

[a] The stage to which a redundant unit is to be added.

Table 5.5. Illustration of the stepwise iteration of the algorithm of Aggarwal *et al.* [17]

Iteration	No. of units in each stage			$F_i(y_i)$			Cost	Weight
	1	2	3	1	2	3		
0	1	1	1	0.002 25	0.003 27[a]	0.002 28	19	23
1	1	2	1	0.002 25	0.000 65	0.002 28[a]	26	30
2	1	2	2	0.002 25[a]	0.000 65	0.000 34	33	38
3	2	2	2	0.000 23	0.000 65[a]	0.000 34	38	46
4	2	3	2	—	—	—	45	53

[a] The stage to which a redundant unit is to be added.

For the example given, after four iterations the algorithm gives the same final solution as that of the algorithm of Sharma and Venkateswaran [21].

Some computational experience has revealed that the performance of the algorithm of Aggarwal *et al.* [17] deteriorates with an increase in the number of constraints. The reason for this deterioration may be due to too much emphasis on resources in the product term in the denominator of the selection factor, which is derived as the ratio of increase in system reliability to the product of increments in consumption of various resources when one more redundant component is added to a stage. In order to take care of this problem, Gopal *et al.* [22] proposed a heuristic criterion for selecting the stage to which a redundant unit is to be added. Their criterion is given as the ratio of the relative decrement in failure probability to the largest relative increments in resources. As in Gopal *et al.* [22], the relative increment in resource consumption is defined as

$$\Delta G_i^m(y_i) = \Delta g_i^m(y_i)/\max_{i\in I}[\Delta g_i^m(y_i)] \quad \forall i \text{ and } m$$

where $\Delta g_i^m(y_i)$ represents increment in $g_i^m(y_i)$ due to increasing y_i by unity. A redundant unit is then to be added to the stage where its addition leads to the least value of the selection factor without violating any of the $|M|$ constraints in Equation 5.1.

$$F_i(y_i) = \max_{m\in M}[\Delta G_i^m(y_i)]/\Delta Q_s(y_i|\bar{y})$$

where $\Delta Q_s(y_i|\bar{y})$ represents the decrement in system failure probability due to increasing y_i by unity.

For a system with linear constraints, the function $\Delta G_i^m(y_i)$ is independent of y_i and hence needs to be computed only once. Accordingly, the selection factor $F_i(y_i)$ can be easily computed via the following forms:

(a) $F_i(y_i) \approx F_i(y_i - 1)/(1 - r_i)$ for $y_i > 1$, and
(b) $F_i(1) = \max_{m\in M}[\Delta G_i^m(y_i)]/[r(1 - r_i)]$.

For details, refer to Gopal *et al.* [22]. With the same data in Table 5.3, the algorithm of Gopal *et al.* [22] is also illustrated in Table 5.6. Likewise, after

Table 5.6. Illustration of the stepwise iteration of the algorithm of Gopal *et al.* [22]

Iteration	No. of units in each stage			$F_i(y_i)$			Cost	Weight
	1	2	3	1	2	3		
0	1	1	1	11.11	6.25[a]	7.84	19	23
1	1	2	1	11.11	31.25	7.84[a]	26	30
2	1	2	2	11.11[a]	31.25	52.29	33	38
3	2	2	2	111.11	31.25[a]	52.29	38	46
4	2	3	2	—	—	—	45	53

[a] The stage to which a redundant unit is to be added.

four iterations, the algorithm of Gopal *et al.* [22] gives the final solution, which is the same as those of the algorithms of Sharma and Venkateswaran [21] and Aggarwal *et al.* [17].

Other heuristic algorithms have been proposed by Misra [23] and by Nakagawa and Nakashima [24]. For example, Misra [23] suggested the approach of reducing one r-constraint problem from among many r-problems, each with one constraint, where an r-constraint problem is a reduced problem with only one constraint considered out of the r constraints, and each r-problem is such a reduced problem. In the approach, a desirability factor (*i.e.* the ratio of the percentage increase in the system reliability to the percentage increase of the corresponding cost) was used to determine the stage to which a redundancy is added. However, the approach has dealt with linear constraint problems. As the number of constraints increases, the computation time becomes remarkably larger. Nakagawa and Nakashima [24] presented a method that can solve the reliability maximization problem with nonlinear constraints for a series system. Balancing between the reliability objective function and resource constraints is a basic consideration in the method. A limitation to the approach is that it cannot be used to solve any problem of complex system configuration. For detailed algorithms and numerical examples, refer to Misra [23] and Nakagawa and Nakashima [24].

Kuo *et al.* [25] have presented a note on some of the heuristic algorithms of Sharma and Venkateswaran [21], Aggarwal *et al.* [17], Misra [23], and Nakagawa and Nakashima [24]. Nakagawa and Miyazaki [26] have compared three heuristic algorithms, each against the other, including those of Sharma and Venkateswaran [21], Gopal *et al.* [22], and Nakagawa and Nakashima [24], in terms of CPU time, optimality rate, and relative error.

5.3 Combinatorial Reliability Optimization Problems of a Non-series Structure

This section considers problems having non-series structures including mixed series–parallel structures and general structures.

5.3.1 Mixed Series–Parallel System Optimization Problems

As mentioned earlier, a mixed series–parallel structure is defined as one in which the relation between any two stages is in series or parallel. For such a structure, the system reliability function is not expressed as a product form of the reliabilities of all stages, but as a multinomial form of the reliabilities of multiple paths. Therefore, the system reliability depends on its system structure, and so is very complicated to compute. However, it is known that the system reliability of a mixed series–parallel structure can be figured out on the order of a polynomial function of the number of stages, so that the complexity of computing the system reliability does not increase exponentially as the number of stages in the system increases.

One often sees mixed series–parallel structures in international telephone systems. For example, each international telephone system between any two countries is composed of two major subsystems. One of them is a wire system, and the other one is a wireless system. In the whole system, all calls from one country to another country are collected together at an international gateway; some of the calls are transmitted through the wireless system from an earth station of one country to the corresponding earth station of the other country via a communications satellite; the rest of the calls are transmitted through the wire system (being composed of terminals and optical cables) from the international gateway to the corresponding international gateway of the other country. For these international communications, both the wire and the wireless systems are equipped with multiple modules in parallel, each being composed of various chipsets so as to increase the whole system reliability. Thus, the associated design issue of how many modules to be installed multiply in each subsystem for an optimal international telephone system can be handled as a mixed series–parallel system optimization problem.

Only a few researches have considered the reliability optimization issue for such mixed series–parallel systems. For example, Burton and Howard [27] have dealt with the problem of maximizing system reliability subject to one resource constraint such as budget restriction. Their algorithm is based on a basic dynamic programming. Recently, Cho [28] has proposed a dominance-sequence-based dynamic programming algorithm for the problem with binary variables and analyzed its computational complexity, and then compared it with the algorithm of Burton and Howard [27].

We now introduce the dominance-sequence-based dynamic programming algorithm of Cho [28]. The problem, dealt with by Cho [28], is one with a mixed series–parallel system structure. For developing the algorithm, the following assumptions are made. First, the resource consumption for the entire system is represented by a discrete linear function. Second, the resource coefficients are all integer valued. The proposed algorithm is composed of two phases. The first phase is to construct a reduction-order graph, which is defined as a directed graph having the precedence relation between nodes in the proposed mixed series–parallel system. Referring to Satyanarayana and Wood [2], the mixed series–parallel system can be reduced to a one-node network in the complexity order of $O(|E|)$, where $|E|$ denotes the cardinality of the set of edges in the given mixed series–parallel system. Two examples of the reduction-order graphs are depicted in Figure 5.1. The order of reducing the given graph is not unique, but the structure of the reduction-order graph remains the same as that of the given graph, which is represented as a tree. The number of end nodes of the graph depends on the given system structure. The second phase is concerned with a process of reducing two stages (including series-stage reduction and parallel-stage reduction) into one as follows.

Series-stage reduction. Suppose that stages i and k are merged together in series relation and to generate a new stage s. Then, it can be processed as $r_{ij}r_{kl} \to r_{s,j\times l}$ and $c_{ij}^m + c_{kl}^m \to c_{s,j\times l}^m$ for all m. Accordingly, the variable $y_{s,j\times l}$ will be included in the stage s with its reliability and consumption of resource m at $r_{s,j\times l}$ and $c_{s,j\times l}^m$, respectively.

Parallel-stage reduction. Suppose that stages i and k are also merged together in parallel relation and to generate a new stage s. Then, it can be processed as $r_{ij} + r_{kl} - r_{ij}r_{kl} \to r_{s,j\times l}$ and $c_{ij}^m + c_{kl}^m \to c_{s,j\times l}^m$. Accordingly, the variable $x_{s,j\times l}$ will be newly included in the stage s with its reliability and consumption of resource m at $r_{s,j\times l}$ and $c_{s,j\times l}^m$, respectively.

The above discussions on stage reduction are now put together to formulate the dominance-sequence-based dynamic programming procedure:

Step 0. (Ordering) Apply the algorithm of Satyanarayana and Wood [2] to the proposed mixed series–parallel system and find the

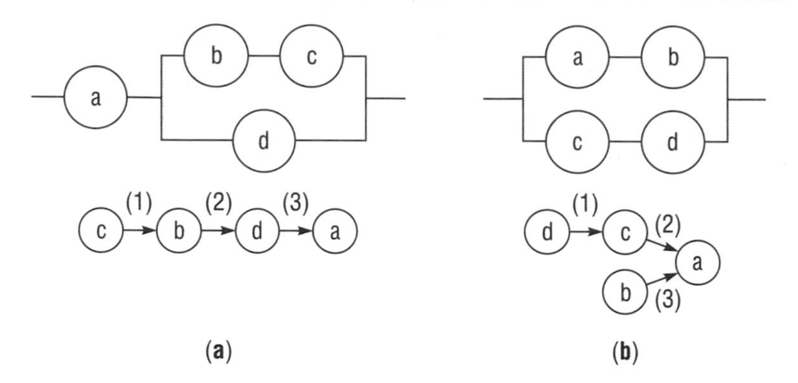

Figure 5.1. Two examples of the reduction-order graph

reduction-order graph of the system. Go to Step 1.

Step 1. (Initialization) Arrange all stages according to the reduction-order graph, and set $\sigma = \emptyset$ and $\bar{\sigma} = \{(1), (2), \ldots, (|I|)\}$, which represents the set of the reduced stages and non-reduced ones respectively. Go to Step 2.

Step 2. (Reduction) If any two selected stages are in series relation, then perform the series-stage reduction procedure for the two stages. Otherwise, perform the parallel-stage reduction procedure. Go to Step 3.

Step 3. (Termination) As $\bar{\sigma}$ becomes a null set, terminate the procedure. Otherwise, go to Step 2.

To illustrate the dominance-sequence-based dynamic programming procedure, a numerical example is solved. The problem data are listed in Table 5.7. And the structure of Figure 5.1(a) is used as representing the example.

At the first iteration, the series-stage reduction procedure is applied to merge stages b and c into stage 3. The variables of stage 3 are generated in Table 5.8.

At the second iteration, the parallel-stage reduction procedure is applied to merge stages 3 and d into stage 2. At the last iteration, the series-stage reduction procedure is applied to merge stages 2 and a into stage 1, whose variables are given in Table 5.9.

The first and second variables are not feasible because the weight consumptions are too great. Thus, the optimal system reliability is found at 0.9761, and its cost and weight consumptions are found at 28 and 15 respectively.

Now, the computational complexity of the dominance-sequence-based dynamic programming algorithm is analyzed and compared with the basic dynamic programming algorithm of Burton and Howard [27]. For the one-dimensional problem, the following properties are satisfied.

Proposition 1. (Cho [28]) *If the reduction-order graph of the given system is a tree with only one end node, then the computational complexity of the dominance-sequence-based dynamic programming algorithm of Cho [28] is in the order of* $O(|I|JC)$,*where* $|I|$ *and* $|J_i|$ *denote the cardinalities of* I *and* J_i *respectively;* J_i *represents the set of variables used at stage* i, *and* $J = \max_{i\in I}\{|J_i|\}$.

Proposition 2. (Cho [28]) *If the reduction-order graph of the given system is a tree with more than one end node, then the computational complexity of the dominance-sequence-based dynamic programming algorithm is in the order of* $O(|I|C^2)$.

Proposition 3. (Cho [28]) *The computational complexity of the basic dynamic programming algorithm of Burton and Howard [27] is in the order of* $O(|I|C^2)$ *for any mixed series–parallel network.*

Table 5.7. Problem data (reliability, cost, weight) for the dominance-sequence-based dynamic programming algorithm[a]

Component type	Stage a	Stage b	Stage c	Stage d
1	0.92, 7, 5	0.90, 3, 9	0.92, 5, 11	0.80, 3, 3
2		0.98, 8, 3	0.98, 11, 4	0.90, 5, 6

[a] Available cost and weight are 30 and 17, respectively.

Table 5.8. Variables generated at the first iteration of the algorithm of Cho [28]

j	$r_{3j}\,(=r_{bj}\times r_{cj})$	$c_{3j}\,(=c_{bj}\times c_{cj})$	$w_{3j}\,(=w_{bj}\times w_{cj})$
1	$0.828\,(=0.9\times 0.92)$[a]	$8\,(=3+5)$	$20\,(=9+11)$
2	0.882	14	13
3	0.9016	13	14
4	0.9604	19	7

[a] The variable is fathomed because its weight consumption is not feasible (*i.e.* $20>17$).

Table 5.9. Variables generated at the third iteration of the algorithm of Cho [28]

j	$r_{1j}\,(=r_{2j}\times r_{aj})$	$c_{1j}\,(=c_{2j}\times c_{aj})$	$w_{1j}\,(=w_{2j}\times w_{aj})$
1	$0.9569\,(=0.9764\times 0.98)$[a]	$21\,(=17+4)$	$18\,(=16+2)$
2	0.9607[a]	20	19
3	0.9723	26	12
4	0.9761	28	15

[a] The variable is fathomed due to resource violation.

As seen in the computational complexity analysis of the one-dimensional case, the dominance-sequence-based dynamic programming algorithm may depend on the system structure, whereas the basic dynamic programming algorithm of Burton and Howard [27] does not. Therefore, the dominance-sequence-based dynamic programming algorithm may require a reduced computational complexity for the systems of a tree structure with one end node. For the detailed proof, refer to Burton and Howard [27] and Cho [28].

For multi-dimensional problems, the computational complexity is characterized below.

Proposition 4. (Cho [28]) *The computational complexity of the dominance-sequence-based dynamic programming algorithm of Cho [28] is in the order of* $O(|M|J^{|I|})$ *for any mixed series–parallel network.*

Proposition 5. (Cho [28]) *The computational complexity of the basic dynamic programming algorithm of Burton and Howard [27] is in the order of* $O(|I|C^{2|M|})$ *for any mixed series–parallel network.*

In general, the number of constraints is smaller than the number of stages, so that the computational complexity of the proposed algorithm is larger than that of Burton and Howard [27], even in the situation where $J \ll C^2$. There exist a number of variables having not the same resource consumption rate but the same reliability, so that the number of the variables generated is not bounded by the value C at each iteration of the algorithm. For the detailed proof, refer to Burton and Howard [27] and Cho [28].

5.3.2 General System Optimization Problems

One often sees general-structure systems in the area of telecommunications. Telecommunication systems are composed of many switches, which are interconnected with one another to form complex mesh types of network, where each switch is equipped with multiple modules in parallel, each being composed of many chipsets to process traffic transmission operations so as to increase the whole switch reliability. Thus, the associated design issue of how many modules to be installed multiply in each switch for an optimal telecommunication system can be handled as a general system reliability optimization problem.

For the reliability optimization problems of complex system structures, no efficient optimal solution method has yet been derived, and only a few heuristic methods [29, 30] have been proposed. The heuristic methods, however, have been applied to small-sized problems, because, as stated earlier, the computational complexity of the system reliability of complex structures may increase exponentially as the number of stages in the system increases.

Aggarwal [29] proposed a simple heuristic algorithm for a reliability maximization problem (NRP) to select a stage having the largest ratio of the relative increment in reliability to the increment in resource usage and to add a redundant unit at the stage for increasing redundancy. That is, a redundant unit is added to the stage where its addition has the largest value of the selection factor $F_i(y_i)$ defined as follows:

$$F_i(y_i) = \frac{\Delta Q_s(y_i)}{\prod_{m \in M} g_i^m(y_i)}$$

where $\Delta Q_s(y_i)$ represents increment in reliability when a redundant unit is added to stage i having y_i redundant units such that

$$\Delta Q_s(y_i) = Q_s(Q_1, \ldots, Q_i, \ldots, Q_{|I|}) - Q_s(\hat{Q}_1, \ldots, \hat{Q}_i, \ldots, \hat{Q}_{|I|})$$

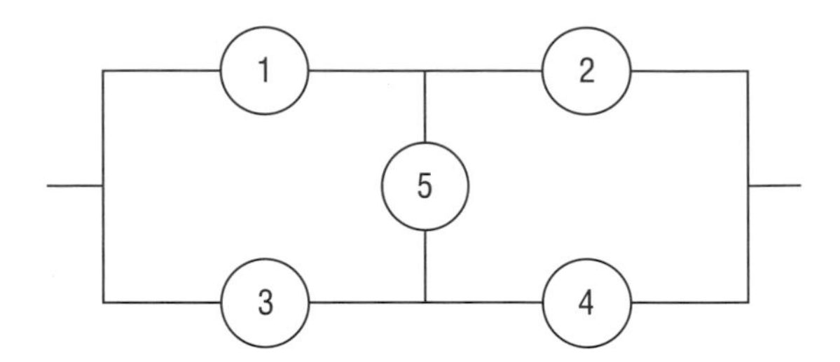

Figure 5.2. A bridge system

where $Q_s(Q_1, \ldots, Q_{|I|})$ and $Q_i = (1 - r_i)^{y_i}$ represent the failure probabilities of the system and stage i, respectively, and $Q_j = \hat{Q}_j \ \forall j \neq i$, and $\hat{Q}_i = (1 - r_i)^{y_i + 1}$. For the detailed step-by-step procedure, refer to Aggarwal [29].

In order to illustrate the algorithm of Aggarwal [29] for the bridge system structure given in Figure 5.2, a numerical example is solved with the data of Table 5.10; the results are presented in Table 5.11.

Shi [30] has also proposed a greedy-type heuristic method for a reliability maximization problem (NRP), which is composed of three phases. The first phase is to choose the minimal path with the highest sensitivity factor a_l from all minimal paths of the system as

$$a_l = \frac{\prod_{i \in l} R_i(y_i)}{\sum_{i \in P_l} \sum_{m \in M} [g_i^m(y_i)/|M|C^m]} \quad \text{for } l \in L$$

which represents the ratio of the minimal path reliability to the percentage of consumed resources by the minimal path. The second phase is to find within the chosen minimal path the stage having the highest selection factor b_i where

$$b_i = \frac{R - i(y_i)}{\sum_{m \in M} [g_i^m(y_i)/|M|C^m]} \quad \text{for } i \in P_l$$

Table 5.10. Problem data for the bridge system problem

Stage	1	2	3	4	5
r_i	0.7	0.85	0.75	0.8	0.9
c_i (= cost)	3	4	2	3	2
Constraints: $C \leq 19$					

Table 5.11. Illustration of the stepwise iteration of the algorithm of Aggarwal [29]

Iteration	Stage					$F_i(y_i)$					$\sum c_i$	Q_s (%)
	1	2	3	4	5	1	2	3	4	5		
1	1	1	1	1	1	1.77	1.34	2.76[a]	1.07	0.27	14	10.9
2	1	1	2	1	1	0.53	0.62	0.69	1.15[a]	0.24	16	5.4
3	1	1	2	2	1	—	—	—	—	—	19	2.6

[a] The stage to which a redundant unit is to be added.

Table 5.12. Illustration of the stepwise iteration of the algorithm of Shi [30]

Stage					a_l					b_i				
1	2	3	4	5	$\sum c_i$	P_1	P_2	P_3	P_4	1	2	3	4	5
1	1	1	1	1	14	1.62	2.28[a]	1.20	1.36	—	—	7.1[b]	5.1	—
1	1	2	1	1	16	1.62	2.04	1.20	1.36	—	—	4.5	5.1[b]	—
1	1	2	2	1	19	—	—	—	—	—	—	—	—	—

[a] The minimal path set having the largest sensitivity factor.
[b] The stage to which a redundant unit is to be added.

The second phase checks feasibility and, if feasible, allocates a component to the selected stage. For the detailed step-by-step procedure, refer to Shi [30].

In order to illustrate the algorithm of Shi [30], a numerical example is solved with the data of Table 5.10, and Table 5.12 shows the stepwise results to obtain the optimal solution, which also shows the minimal path sets $P_1 = \{1, 2\}$, $P_2 = \{3, 4\}$, $P_3 = \{1, 4, 5\}$, $P_4 = \{2, 3, 5\}$.

As seen in the above illustration, the optimal solution obtained from the algorithm of Shi [30] is the same as that of the algorithm of Aggarwal [29].

Recently, Ravi *et al.* [31] proposed a simulated-annealing technique for the reliability optimization problem of a general-system structure having the objective of maximizing the system reliability subject to multiple resource constraints.

5.4 Combinatorial Reliability Optimization Problems with Multiple-choice Constraints

In considering a redundant system, the important practical issue is concerned with handling a variety of different design alternatives in choosing each component type. That is, more than one component type may have to be considered for choosing each redundant unit for a subsystem. This combinatorial situation of choosing each component type makes the associated reliability optimization model incorporate a new constraint, which is called a multiple-choice constraint. For example, a guided-weapons system, such as an anti-aircraft gun, has three major subsystems, including the target detection/tracking subsystem, the control console subsystem, and the fault detection subsystem. A failure of one of these subsystems implies the associated whole system failure. Moreover, each subsystem can have many components. For instance, the target detection/tracking subsystem may have three important components, including radar, TV camera and forward-looking infra-red equipment. These components are normally assembled in a parallel-redundant structure (rather than a standby one) to maintain (promote) the subsystem reliability in situations where no on-site repair work is possible. For such subsystems, a variety of different component types may be available to choose from for each component.

It is often desired at each stage to consider the practical design issue of handling a variety of combinations of different component types, along with various design methods that are concerned with parallel redundancy of two ordinary (similar) units, standby redundancy of main and slave units, and 2-out-of-3 redundancy of ordinary units. For the design issue, the associated reliability optimization problem is required to select only one design combination from among such a variety of different design alternatives.

The reliability optimization problem, called Problem (NP), is expressed in the following binary nonlinear integer programming:

$$Z_{\mathrm{NP}} = \max \prod_{i \in I} \left(\sum_{j \in J_i} r_{ij} x_{ij} \right)$$

subject to

$$\sum_{i \in I} \sum_{j \in J_i} c_{ij}^m x_{ij} \le C^m \quad \forall m \in M \tag{5.11}$$

$$\sum_{j \in J_i} x_{ij} = 1 \quad \forall i \in I \tag{5.12}$$

$$x_{ij} = \{0, 1\} \quad \forall j \in J_i,\ i \in I \tag{5.13}$$

where x_{ij} indicates whether or not the jth design alternative is used at stage i; r_{ij} and c_{ij}^m represent the reliability and its consumption of resource m of the design alternative x_{ij}, respectively. The constraints in Equations 5.11 and 5.12 represent the resource consumption constraints and the multiple-choice constraints respectively, and the constraint in Equation 5.13, defines the decision variables.

5.4.1 One-dimensional Problems

The one-dimensional case of Problem (NP) can have the corresponding constraints in Equation 5.11, newly expressed as

$$\sum_{i \in I} \sum_{j \in J_i} c_{ij} x_{ij} \le C$$

The problem has been investigated in Sung and Cho [32], who proposed a branch-and-bound solution procedure for the problem, along with a reduction procedure of its solution space to improve the solution search efficiency. For the branch-and-bound solution procedure, the lower and the upper bounds for the optimal reliability at each stage are derived as in Theorems 2 and 3, given a feasible system reliability.

Theorem 2. (Sung and Cho [32]) *Let Z^l be a feasible system reliability. Then, R_s^l, where $R_s^l = Z^l / \prod_{i \in I \setminus \{s\}} r_{i,|J_i|}$, serves as a lower bound of the reliability at stage s in the optimal solution. The corresponding lower bound index of x_{sj} for $j \in J_s$ is determined as $j_s^l = \min\{j \mid r_{sj} \ge R_s^l,\ j \in J_s\}$, so that $x_{sj} = 0, \forall j < j_s^l$.*

Theorem 3. (Sung and Cho [32]) *At stage s, $b_s = \sum_{i \in I/\{s\}} \min\{c_{ij} \mid r_{ij} \ge R_i^l,\ j \in J_i\}$ is the minimum amount of budget allocated to the system except stage s. Given the total budget C, $r_{s,s^{\mathrm{u}}}$, where $s^{\mathrm{u}} = \max\{j \mid c_{sj} \le C - b_s,\ j \in J_s\}$, is an upper bound of the reliability at stage s in the optimal solution, and s^{u} is an upper bound index at stage s, so that $x_{sj} = 0, \forall j > s^{\mathrm{u}}$.*

For the detailed proofs of Theorems 2 and 3, refer to Sung and Cho [32]. The computational complexity of one iteration of the full bounding process for finding the stagewise lower and upper bounds (by Theorems 2 and 3 respectively), starting from stage 1 through stage $|I|$, are both on the order of $O[\max[|I|^2, |I|J)]$, where $J = \max_{i \in I} |J_i|$. Thus, the computational complexity concerned with the number of all the iterations of the full bounding process required for the solution space reduction is on the order of $O(|I|J)$. Therewith, the total bounding process for the solution space reduction requires a computational complexity order of $O[\max[|I|^3 J, |I|^2 J^2)]$. Table 5.13 shows how efficient Theorems 2 and 3 are in reducing the solution space.

Sung and Cho [32] derived several different bounds for the branch-and-bound procedure. They include the bound derived in a continuous relaxation approach, the bound derived in a Lagrangian relaxation approach, and the bound derived by use of the results of Theorems 2 and 3. Among them, the Lagrangian relaxation approach has been considered to find the upper bound value

Table 5.13. Problem data ($C = 24$) and stepwise results of the solution space reduction procedure [31]

j	Stage 1		Stage 2		Stage 3		Stage 4	
	R_{1j}	c_{1j}	r_{2j}	c_{2j}	r_{3j}	c_{3j}	r_{4j}	c_{4j}
1	0.9	2[a]	0.85	3[a]	0.8	2[a]	0.75	3[a]
2	0.99	4	0.9775	6	0.96	4	0.938	6
3	0.999	6	0.9966	9	0.99	6	0.984	9
4	$1-0.1^4$	8	0.9995	12[b]	0.998	8	0.996	12[b]
5	$1-0.1^5$	10[b]	0.9999	15[b]	0.9997	10[b]	0.999	15[b]
6	$1-0.1^6$	12[b]			0.9999	12[b]		
7	$1-0.1^7$	14[b]			$1-0.2^7$	14[b]		
8	$1-0.1^8$	18[b]			$1-0.2^8$	16[b]		
	$J_1 = \{2, 3, 4\}$		$J_2 = \{2, 3\}$		$J_3 = \{2, 3, 4\}$		$J_4 = \{2, 3\}$	

[a] Variables discarded by Theorem 2.
[b] Variables discarded by Theorem 3.

of the given objective function. In the continuous relaxation approach, the Lagrangian dual problem of the associated continuous relaxed problem with the constraints in Equation 5.11 dualized has been found as satisfying the integrality property. This implies that the optimal objective value of the Lagrangian dual problem is equal to that obtained by the continuous relaxation approach.

In order to derive the associated solution bounds, some notation is introduced. Let $\sigma_p = \{(1), (2), \ldots, (p)\}$ denote the set of stages that were branched already from level 0 down to level p in the branch-and-bound tree, and $\bar{\sigma}_p = \{(p+1), \ldots, (|I|)\}$ denote the complement of σ_p. For an upper bound derivation, let $N(j_{(1)}, j_{(2)}, \ldots, j_{(p)})$ denote a (branched) node in level p of the branch-and-bound tree. The node represents a partial solution in which the variables of the stages in σ_p have the values being determined already as $x_{(1),j_{(1)}} = 1, \ldots, x_{(p),j_{(p)}} = 1$, but those of the stages in $\bar{\sigma}_p$ are not yet determined. Then, any one of the following three equations can be used as a bound at the branching node $N(j_{(1)}, j_{(2)}, \ldots, j_{(p)})$:

(b.1) $$B_1(j_{(1)}, j_{(2)}, \ldots, j_{(p)}) = \prod_{i\in\sigma_p} r_{i,j_i} \prod_{i\in\bar{\sigma}_p} r_{i,j_i^{\mathrm{u}}}$$

(b.2) $$B_2(j_{(1)}, j_{(2)}, \ldots, j_{(p)}) = \prod_{i\in\sigma_p} r_{i,j_i} \prod_{i\in\bar{\sigma}_p} r_{i,\bar{j}_i}$$

(b.3) $$B_3(j_{(1)}, j_{(2)}, \ldots, j_{(p)}) = \prod_{i\in\sigma_p} r_{i,j_i}\, \mathrm{e}^{Z^*_{\mathrm{CLNP}}(j_{(1)}, j_{(2)}, \ldots, j_{(p)})}$$

where j_i^{u}, $\forall i \in \bar{\sigma}_p$, represents the upper bound derived by using the results of Theorems 2 and 3,

$$\bar{j}_i = \max\{j \mid c_{ij} \le C - \bar{c}_i,\ j \in J_i\},$$
$$\bar{c}_i = \sum_{k\in\sigma_p} c_{k,j_k} + \sum_{k\in\bar{\sigma}_p/\{i\}} \min_{j\in J_k}\{c_{kj}\},$$

and $Z^*_{\mathrm{CLNP}}(j_{(1)}, j_{(2)}, \ldots, j_{(p)})$ denotes the optimal solution value for the partial LP problem (corresponding to the remaining stages from $(p+1)$ through $(|I|)$) at the branching node $N(j_{(1)}, j_{(2)}, \ldots, j_{(p)})$ to be solved by relaxing the variables in $\bar{\sigma}_p$ to become continuous ones.

Bound (b.1) can be obtained from the proposed problem, based on the associated reduced solution space (Theorems 2 and 3), and bound (b.2) can be derived by applying the results of Theorem 3 to node $N(j_{(1)}, j_{(2)}, \ldots, j_{(p)})$. It is easily seen that $B_1(j_{(1)}, j_{(2)}, \ldots, j_{(p)}) \ge B_2(j_{(1)}, j_{(2)}, \ldots, j_{(p)})$.

For bounds (b.2) and (b.3), it has not been proved that the value of bound (b.3) is always less than or equal to that of bound (b.2). However, it is proved that the upper bound value of one of the integer-valued stages in bound (b.3) is less than that of the stage in bound (b.2) under the condition that some stages have integer solution values in the optimal solution for the continuous relaxed problem. This implies that the value of bound (b.3) is probably less than that of bound (b.2). This conclusion is also revealed in the numerical tests. For the detailed derivation, refer to Sung and Cho [32].

Based on the above discussion, Sung and Cho [32] derived a branch-and-bound procedure in the depth-first-search principle.

Branching strategy. At each step of branching, select a node with the largest upper bound value (not fathomed) for the next branching based on the depth-first-search principle. Then, for a such selected node $N(j_{(1)}, j_{(2)}, \ldots, j_{(p)})$, branch all of its immediately-succeeding nodes $N(j_{(1)}, \ldots, j_{(p)}, j'), \forall j' \in J_{(p+1)}$.

Fathoming strategy. For each currently branched node, compute the bound $B(j_{(1)}, \ldots, j_{(p)})$. If the bound satisfies the following condition (a), then fathom the node:

$$\text{condition (a)} \quad B(j_{(1)}, \ldots, j_{(p)}) \leq Z_{\text{opt}}$$

where Z_{opt} represents the current best feasible system reliability obtained from the branch-and-bound procedure. Moreover, any node satisfying the following Proposition 6 needs to be fathomed from any further consideration in the solution search.

Proposition 6. (Sung and Cho [32]) *Each branch node $N(j_{(1)}, \ldots, j_{(p)}, j^*)$ for $j^* \in J_{(p+1)}$ and $j^* > \bar{j}_{(p+1)}$ needs to be removed form the solution search tree.*

The variables satisfying the inequality $j* > \bar{j}_{(p+1)}$ will lead to an infeasible leaf solution in the associated branch-and-bound tree, so that any branching node with such variables can be eliminated from any further consideration in finding the optimal solution, where a leaf solution corresponds to a node in the last (bottom) level of the branch-and-bound tree. Thus, any currently branched node $N(j_{(1)}, \ldots, j_{(p)}, j^*), j^* \in J_{(p+1)}$, satisfying the following condition (b), needs to be removed from the solution search tree:

condition (b)

$$\{j^* \mid c_{(p+1),j^*} > C - \bar{c}_{(p+1)}, \ j^* \in J_{(p+1)}\}$$

Backtracking strategy. If all the currently examined nodes need to be fathomed or any currently examined node is a leaf node, then the associated immediately preceding (parent) node will be backtracked. This backtracking operation continues until any node to be branched is found.

The above-discussed properties are now put together to formulate the branch-and-bound procedure:

Step 0. (Variable Reduction) Reduce the solution space by using Theorems 2 and 3. Go to Step 1.

Step 1. (Initialization) Arrange all stages in the smallest-gap-first ordered sequence. Go to Step 2.

Step 2. (Branching) According to the branching strategy, select the next branching node and implement the branching process at the selected node. Go to Step 3.

Step 3. (Fathoming) For each currently branched node, check if it satisfies the fathoming condition (b). If it does, then it needs to be fathomed; otherwise, compute the upper bound associated with the node, and then use it to decide whether or not the node needs to be fathomed ,according to the fathoming condition (a). If all the currently branched nodes need to be fathomed, then go to Step 5. Otherwise, go to Step 2. If the currently examined (not fathomed) node is a leaf node, then go to Step 4. Otherwise, go to Step 2.

Step 4. (Updating) Whenever a solution (leaf node) is newly found, compare it with the current best solution. Update the current best solution by replacing it with the newly obtained solution and fathom every other

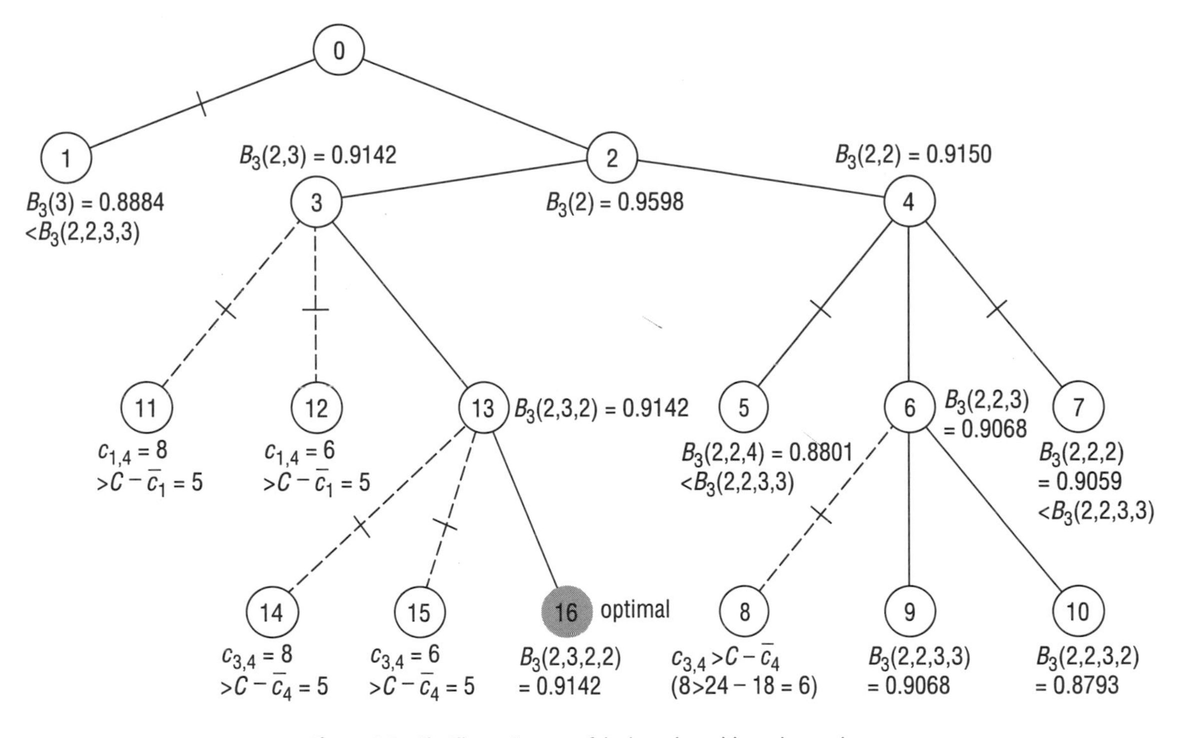

Figure 5.3. The illustrative tree of the branch-and-bound procedure

node that has its bound less than or equal to the reliability of the newly updated best solution. Go to Step 5.

Step 5. (Backtracking and Terminating) Backtrack according to the backtracking strategy. If there are no more branching nodes to be examined, then terminate the search process and the current best solution is optimal. Otherwise, go to Step 2.

The smaller problem resulting from Table 5.13 is then solved by employing the branch-and-bound procedure with bound (b.2). This is depicted in Figure 5.3 for illustration, where the number in each circle denotes the node number. As seen from Figure 5.3, the stage ordering in the branch-and-bound process can be decided as sequencing the associated tree levels in the smallest-gap-first order as 2–4–1–3, since $|J_2| = 2$, $|J_4| = 2$, $|J_1| = 3$, and $|J_3| = 3$. For details, refer to Sung and Cho [32].

5.4.2 Multi-dimensional Problems

The extension of the problem of Section 5.4.1 to a problem with multiple resource constraints, called a multi-dimensional problem, has been handled by Chern and Jan [33] and by Sung and Cho [34], who have proposed branch-and-bound methods for the multi-dimensional problem. Sung and Cho [34] derived solution bounds based on some solution properties and compared them with those of Chern and Jan [33].

For a cost minimization problem with multiple resource constraints, Sasaki *et al.* [35] presented a partial enumeration method referring to the algorithm of Lawler and Bell [8], in which the objective is to minimize the system cost subject to a target system reliability.

Table 5.14. Illustration of the solution space reduction by Theorems 4 and 5 [33]

j	Stage 1				Stage 2				Stage 3			
	x_{1j}	r_{1j}	c^1_{1j}	c^2_{1j}	x_{2j}	r_{2j}	c^1_{2j}	c^2_{2j}	x_{3j}	r_{3j}	c^1_{3j}	c^2_{3j}
1		0.99	4	2		0.8[a]	3	3		0.92[a]	5	6
2		0.9999	8	4		0.9[a]	3	9		0.98	11	4
3		$1 - 10^{-6}$	12	6		0.96[a]	6	6		0.9936	10	12
4		$1 - 10^{-8}$ [a]	16	8		0.98	8	3		0.9984	16	10
5		$1 - 10^{-10}$ [a]	20	10		0.992	9	9		0.9996[a]	22	8
6						0.996	11	6				
7						0.9992	14	9				
8						0.9996	16	6				
9						0.9999[a]	19	9				
	$J_1 = \{1, 2, 3\}$				$J_2 = \{3, 4, \ldots, 8\}$				$J_3 = \{2, 3, 4\}$			

[a] Variables discarded by the reduction properties.

Sung and Cho [34] have proved that the bounding properties of Theorems 2 and 3 can be easily extended to the multi-dimensional problems. Accordingly, Theorems 4 and 5 can be used for reducing the solution space by eliminating the unnecessary variables in order to find the optimal solution.

Theorem 4. (Sung and Cho [34]) *Define*

$$\tilde{R}_s = \bar{Z} / \prod_{i \in I \setminus \{s\}} r_{i,|J_i|}$$

for any stage s, where $\bar{Z}$ is a feasible system reliability of the multi-dimensional problem. Then, the reliability R_s^* of stage s in the optimal solution to the problem satisfies the relation $R_s^* \geq \tilde{R}_s$. Moreover, the set of variables $\{x_{sj},\ j \in J_s : r_{sj} < \tilde{R}_s\}$ take on values of zero in the optimal solution.

Theorem 5. (Sung and Cho [34]) *Define*

$$\bar{b}_s^m = \sum_{i \in I \setminus \{s\}} \min_{j \in J} \{c_{ij}^m : r_{ij} \geq \tilde{R}_i\}$$

and

$$s^{\mathrm{u}} = \arg\max_{j \in J_s} \{x_{sj} : c_{sj}^m \leq C^m - \bar{b}_s^m,\ \forall m \in M\}$$

Then the reliability R_s^* of stage s in the optimal solution to the problem satisfies the relation $R_s^* \leq r_{s,s^{\mathrm{u}}}$. Moreover, the set of variables $\{x_{sj},\ j \in J_i : r_{sj} > r_{s,s^{\mathrm{u}}}\}$ take on the values of zero in the optimal solution.

Table 5.14 shows how efficient Theorems 4 and 5 are in reducing the solution space.

Sung and Cho [34] derived a bounding strategy by using the results of Theorem 5. For an upper bound derivation, let $N(j_{(1)}, j_{(2)}, \ldots, j_{(p)})$ denote a node in level p of the associated branch-and-bound tree. The node represents a partial solution in which the variables of the stages in σ_p have the values being determined already as $x_{(1),j_{(1)}} = 1, \ldots, x_{(p),j_{(p)}} = 1$, but those of the stages in $\bar{\sigma}_p$ not determined yet. Thus, an upper bound $B(j_{(1)}, j_{(2)}, \ldots, j_{(p)})$ on the objective function at the node $N(j_{(1)}, j_{(2)}, \ldots, j_{(p)})$ can be computed as

$$B(j_{(1)}, j_{(2)}, \ldots, j_{(p)}) = \prod_{i \in \sigma_p} r_{i,j_i} \prod_{i \in \bar{\sigma}_p} r_{i,\bar{j}_i}$$

where

$$\bar{j}_i = \arg\max_{j \in J_i} \{x_{ij} : c_{ij}^m \leq C^m - \bar{c}_i^m,\ \forall m \in M\}$$

and

$$\bar{c}_i^m = \sum_{k \in \sigma_p} c_{k,j_k}^m + \sum_{k \in \bar{\sigma}_p \setminus \{i\}} \min_{j \in J_k} \{c_{kj}^m\}$$

Note that the upper bound is less than or equal to that suggested by Chern and Jan [33], since

$$B(j_{(1)}, j_{(2)}, \ldots, j_{(p)}) = \prod_{i \in \sigma_p} r_{i,j_i} \prod_{i \in \bar{\sigma}_p} r_{i,\bar{j}_i}$$
$$\leq \prod_{i \in \sigma_p} r_{i,j_i} \prod_{i \in \bar{\sigma}_p} \max_{j \in J_i} \{r_{ij}\}$$

For the problem, Sung and Cho [34] proposed a branch-and-bound procedure based on the depth-first-search principle, as done for the one-dimensional case, whereas Chern and Jan [33] suggested a branch-and-bound procedure in a binary tree, referred to as a tree of Murphee [36]. It has been shown in numerical tests that the branch-and-bound procedure of Sung and Cho [34] finds the optimal solution much quicker than that of Chern and Jan [33]. For the detailed illustration, refer to Chern and Jan [33] and Sung and Cho [34].

5.5 Summary

This chapter considers various combinatorial reliability optimization problems with multiple resource constraints or multiple-choice constraints incorporated for a variety of different system structures, including series systems, mixed series–parallel systems, and general systems.

Each of the problems is known as an NP-hard problem. This provided us with the motivation to develop various solution approaches, including optimal solution approaches, continuous relaxation approaches, and heuristic approaches, based on various problem-dependent solution properties. In particular, owing to the combinatorial nature of the problems, this chapter focuses on some combinatorial solution approaches, including the branch-and-bound method, the partial enumeration method, the dynamic programming method, and the greedy-type heuristic method, which can give integer optimal or heuristic solutions.

These days, metaheuristic approaches, including the simulated annealing method, the tabu search method, the neural network approach, and genetic algorithms, have been getting popular for combinatorial optimization. However, not many applications of those approaches have been made to the combinatorial reliability optimization problems yet. Thus, those approaches are not discussed here.

References

[1] Valiant LG. The complexity of enumeration and reliability problems. SIAM J Comput 1979;8:410–21.

[2] Satyanarayana A, Wood RK. A linear-time algorithm for computing K-terminal reliability in series–parallel networks. SIAM J Comput 1985;14:818–32.

[3] Chern CS. On the computational complexity of reliability redundancy allocation in a series system. Oper Res Lett 1992;11:309–15.

[4] Malon DM. When is greedy module assembly optimal? Nav Res Logist 1990;37:847–54.

[5] Tillman IA, Hwang CL, Kuo W. Optimization techniques for system reliability with redundancy: a review. IEEE Trans Reliab 1977;R-36:148–55.

[6] Tillman FA, Liittschwager JM. Integer programming formulation of constrained reliability problems. Manage Sci 1967;13:887–99.

[7] Misra KB. A method of solving redundancy optimization problems. IEEE Trans Reliab 1971;R-20:117–20.

[8] Lawler EL, Bell MD. A method of solving discrete optimization problems. Oper Res 1966;14:1098–112.

[9] Ghare PM, Talyer RE. Optimal redundancy for reliability in series systems. Oper Res 1969;17:838–47.

[10] Sung CS, Lee HK. A branch-and-bound approach for spare unit allocation in a series system. Eur J Oper Res 1994;75:217–32.

[11] Kuo E, Lin HH, Xu Z, Zhang W. Reliability optimization with the Lagrange-multiplier and branch-and-bound technique. IEEE Trans Reliab 1987;R-36:624–30.

[12] Lawler EL, Wood DE. Branch-and-bound method: a survey. Oper Res 1966;14:699–719.

[13] Balas E. A note on the branch-and-bound principle. Oper Res 1968;16:442–5.

[14] McLeavey DW. Numerical investigation of optimal parallel redundancy in series systems. Oper Res 1974;22:1110–7.

[15] McLeavey DW, McLeavey JA. Optimization of system reliability by branch-and-bound. IEEE Trans Reliab 1976;R-25:327–9.

[16] Luus R. Optimization of system reliability by a new nonlinear programming procedure. IEEE Trans Reliab 1975;R-24:14–6.

[17] Aggarwal AK, Gupta JS, Misra KB. A new heuristic criterion for solving a redundancy optimization problem. IEEE Trans Reliab 1975;R-24:86–7.

[18] Bellman R, Dreyfus S. Dynamic programming and the reliability of multicomponent devices. Oper Res 1958;6:200–6.

[19] Kettelle JD. Least-cost allocation of reliability investment. Oper Res 1962;10:249–65.

[20] Woodhouse CF. Optimal redundancy allocation by dynamic programming. IEEE Trans Reliab 1972;R-21:60–2.

[21] Sharma J, Venkateswaran KV. A direct method for maximizing the system reliability. IEEE Trans Reliab 1971;R-20:256–9.

[22] Gopal D, Aggarwal KK, Gupta JS. An improved algorithm for reliability optimization. IEEE Trans Reliab 1978;R-27:325–8.

[23] Misra KB. A simple approach for constrained redundancy optimization problem. IEEE Trans Reliab 1972;R-21:30–4.

[24] Nakagawa Y, Nakashima K. A heuristic method for determining optimal reliability allocation. IEEE Trans Reliab 1977;R-26:156–61.

[25] Kuo W, Hwang CL, Tillman FA. A note on heuristic methods in optimal system reliability. IEEE Trans Reliab 1978;R-27:320-324.

[26] Nakagawa Y, Miyazaki S. An experimental comparison of the heuristic methods for solving reliability optimization problems. IEEE Trans Reliab 1981;R-30:181–4.

[27] Burton RM, Howard GT. Optimal system reliability for a mixed series and parallel structure. J Math Anal Appl 1969;28:370–82.

[28] Cho YK. Redundancy optimization for a class of reliability systems with multiple-choice resource constraints. Unpublished PhD Thesis, Department of Industrial Engineering, KAIST, Korea, 2000; p.66–75.

[29] Aggarwal KK. Redundancy optimization in general systems. IEEE Trans Reliab 1976;R-25: 330–2.

[30] Shi DH. A new heuristic algorithm for constrained redundancy-optimization in complex system. IEEE Trans Reliab 1987;R-36:621–3.

[31] Ravi V, Murty BSN, Reddy PJ. Nonequilibrium simulated annealing-algorithm applied to reliability optimization of complex systems. IEEE Trans Reliab 1997;46:233–9.

[32] Sung CS, Cho YK. Reliability optimization of a series system with multiple-choice and budget constraints. Eur J Oper Res 2000;127:159–71.

[33] Chern CS, Jan RH. Reliability optimization problems with multiple constraints. IEEE Trans Reliab 1986;R-35:431–6.

[34] Sung CS, Cho YK. Branch-and-bound redundancy optimization for a series system with multiple-choice constraints. IEEE Trans Reliab 1999;48:108–17.

[35] Sasaki G, Okada T, Shingai S. A new technique to optimize system reliability. IEEE Trans Reliab 1983;R-32:175–82.

[36] Murphee EL, Fenves S. A technique for generating interpretive translators for problem oriented languages. BIT 1970:310–23.

PART II

Statistical Reliability Theory

10 Class of NBU-t_0 Life Distribution

10.1 Introduction
10.2 Characterization of NBU-t_0 Class
10.3 Estimation of NBU-t_0 Life Distribution
10.4 Tests for NBU-t_0 Life Distribution

Modeling the Observed Failure Rate

M. S. Finkelstein

Chapter 6

6.1 Introduction

The notion of failure rate is crucial in reliability and survival analysis. However, obtaining the failure rate in many practical situations is often not so simple, as the structure of the system to be considered, for instance, can be rather complex, or the process of the failure development cannot be described in a simple way. In these cases a "proper model" can help a lot in deriving reliability characteristics. In this chapter we consider several models that can be effectively used for deriving and analyzing the corresponding failure rate, and eventually the survival function. The general approach developed eventually boils down to constructing an equivalent failure rate for different settings (survival in the plane, multiple availability on demand, mixtures of distributions). It can be used for other applications as well.

Denote by $T \geq 0$ a lifetime random variable and assume that the corresponding cumulative distribution function (cdf) $F(t)$ is absolutely continuous. Then the following exponential formula exists:

$$F(t) = 1 - \exp\left[-\int_0^t \lambda(u)\, du\right] \tag{6.1}$$

where $\lambda(t)$, $t \geq 0$, is the failure rate. In many instances a conventional statistical analysis of the overall random variable T presents certain difficulties, as the corresponding data can be scarce (*e.g.* the failure times of the highly reliable systems). On the other hand, information on the structure of the system (object) or on the failure process can often be available. This information can be used for modeling $\lambda(t)$, to be called the observed (equivalent) failure rate, and eventually, for estimation of $F(t)$. The other important problem that can be approached in this way is the analysis of the shape of the failure rate, which is crucial for the study of the aging properties of the cumulative distribution functions under consideration.

The simplest reliability example of this kind is a system of two independent components in parallel with exponentially distributed lifetimes. In this case, the system's lifetime cumulative distribution function is

$$F(t) = 1 - \exp(-\lambda_1 t) - \exp(-\lambda_2 t) + \exp[-(\lambda_1 + \lambda_2)t]$$

where λ_1 (λ_2) is a failure rate of the first (second) component. It can be easily seen that the following relation defines the system's observed (equivalent) failure rate:

$$\lambda(t) = \{\lambda_1 \exp(-\lambda_1 t) + \lambda_2 \exp(-\lambda_2 t) - (\lambda_1 + \lambda_2) \exp[-(\lambda_1 + \lambda_2)t]\} \times \{\exp(-\lambda_1 t) + \exp(-\lambda_2 t) - \exp[-(\lambda_1 + \lambda_2)t]\}^{-1}$$

A simple analysis of this relation shows that the observed failure rate $\lambda(t)$ ($\lambda(0) = 0$) is monotonically increasing in $[0, t_0)$ and monotonically decreasing in $[t_0, \infty)$, asymptotically approaching λ_1 from above as $t \to \infty$ ($\lambda_1 < \lambda_2$). The maximum t_0 is uniquely obtained from the equation:

$$\lambda_2^2 \exp(-\lambda_1 t) + \lambda_1^2 \exp(-\lambda_2 t) = (\lambda_1 - \lambda_2)^2 \quad \lambda_1 \neq \lambda_2$$

It is also evident that $\lambda(t) < \lambda_2$, because the reliability of the considered system cannot be worse than the reliability of a component.

Thus, a given structure of the system creates a possibility of obtaining and analyzing $\lambda(t)$ analytically. There are situations, however, when deriving the observed (equivalent) failure rate is not so straightforward. In this chapter, several applications of this kind are considered where the observed failure rate can be effectively constructed via the specific models describing the process of "failure development".

Sections 6.2 and 6.3 are devoted to two different applications of a method of the per demand failure rate [1]. The first one presents a probabilistic description for the survival in the plane, when a small normally or tangentially oriented interval is moving along a fixed route in the plane, crossing points of initial Poisson random processes. Each crossing leads to a termination of the process with a given probability, and the probability of passing the route without termination is derived. The second one deals with the, so-called, multiple availability, when the system should be available at each instant of demand, and the probability of this event is of interest. The demands occur in accordance with the Poisson process. Some weaker criteria of failure are also considered when it is not necessary that all demands should be serviced in a given interval of time. Though these applications are different in nature, mathematical approaches for obtaining characteristics of interest are similar.

Finally, in Section 6.4, a mixture of distributions is studied. The analysis of the observed failure rate, which in this case is a mixture failure rate, is performed for the given governing and mixing distributions. It appears that the shape of the observed failure rate can dramatically differ from the shape of the governing distribution. This fact is rather surprising, and should be taken into account in practical applications.

6.2 Survival in the Plane

6.2.1 One-dimensional Case

A model of survival in the plane is described in this section, which is based on the following simple reasoning used in the one-dimensional case. Consider a system subject to stochastic point influences (shocks). Each shock can lead with a given probability to a fatal failure of a system, resulting in a termination of the process, and this will be called an "accident". The probability of performance without accidents in the time interval $(0, t]$ is of interest. It is natural to describe the situation in terms of stochastic point processes. Let $\{N(t); t > 0\}$ be a point process of shocks, occurring at times $0 < t_1 < t_2 < \cdots$, where $N(t)$ is the corresponding counting measure in $(0, t]$.

Denote by $h(t)$ the rate function of a point process. For orderly processes, assuming the limits exist [2]:

$$h(t) = \lim_{\Delta t \to 0} \frac{\Pr\{N(t, t+\Delta t) = 1\}}{\Delta t} = \lim_{\Delta t \to 0} \frac{E[N(t, t+\Delta t)]}{\Delta t}$$

Thus, $h(t)\,\mathrm{d}t$ can be interpreted as an approximate probability of a shock occurrence in $(t, t+\mathrm{d}t]$.

Assume now that a shock, which occurs in $(t, t+\mathrm{d}t]$ independently of the previous shocks, leads to an accident with probability $\theta(t)$, and does not cause any changes in the system with the complementary probability $1-\theta(t)$. Denote by T_{a} a random time to an accident and by $F_{\mathrm{a}}(t) = \Pr\{T_{\mathrm{a}} \le t\}$ the corresponding cumulative distribution function. If $F_{\mathrm{a}}(t)$ is absolutely continuous, then similar to Equation 6.1

$$P(t) = 1 - F_{\mathrm{a}}(t) = \exp\left[-\int_0^t \lambda_{\mathrm{a}}(x)\,\mathrm{d}x\right] \quad (6.2)$$

where $\lambda_{\mathrm{a}}(t)$ is a failure (accident) rate, corresponding to $F_{\mathrm{a}}(t)$, and $P(t)$ is the survival function: the probability of performance without accidents in $(0, t]$. Assume that $\{N(t); t > 0\}$ is a non-homogeneous Poisson process, $\theta(x)h(x)$ is integrable in $[0, t)$ $\forall t \in (0, \infty)$ and

$$\int_0^\infty \theta(x)h(x)\,\mathrm{d}x = \infty$$

Then the following important relation takes place [3,4]:

$$\lambda_{\mathrm{a}}(t) = \theta(t)h(t) \quad (6.3)$$

For the time-independent case $\theta(t) \equiv \theta$, this approach has been widely used in the literature. In [1], for instance, it was called the method of the per demand failure rate. This name comes from the following simple reasoning. Let $h(t)$ be the rate of the non-homogeneous Poisson process of demands and θ be the probability that a single demand is not serviced (it is serviced with probability $1-\theta$). Then the probability that all demands in $[0, t)$ are serviced is given by

$$1 - F_{\mathrm{a}}(t) = \sum_0^\infty (1-\theta)^k \exp\left[-\int_0^t h(u)\,\mathrm{d}u\right] \times \frac{\left[\int_0^t h(u)\,\mathrm{d}u\right]^k}{k!} = \exp\left[-\int_0^t \theta h(u)\,\mathrm{d}u\right]$$

Therefore, $\theta h(t)$ can be called the per demand failure rate.

As in Section 6.1, the failure rate in Equation 6.2 is called the observed failure rate, and the relation in Equation 6.3 gives its concrete expression for the setting under consideration. In other words, the modeling of the observed failure rate is performed via the relation in Equation 6.3. This concept will be exploited throughout the current chapter. Considering the Poisson point processes of shocks in the plane (which will be interpreted as obstacles) in the next section will lead to obtaining the corresponding observed failure rate "along the fixed curve". An important application of this model results in deriving the probability of a safe passage for a ship in a field of obstacles with fixed (shallows) or moving (other ships) coordinates [4].

6.2.2 Fixed Obstacles

Denote by $\{N(B)\}$ a non-homogeneous Poisson point process in the plane, where $N(B)$ is the corresponding random measure: *i.e.* the random number of points in $B \subset \Re^2$, where B belongs to the Borel σ-algebra in $\Re^2$. We shall consider points as prospective point influences on the system (shallows for the ship, for instance). Similar to the one-dimensional definition (Equation 6.1), the rate $h_{\mathrm{f}}(\xi)$ can be formally defined as [2]

$$h_{\mathrm{f}}(\xi) = \lim_{S(\delta(\xi)) \to 0} \frac{E[N(\delta(\xi))]}{S(\delta(\xi))} \quad (6.4)$$

where $B = \delta(\xi)$ is the neighborhood of ξ with area $S(\delta(\xi))$ and the diameter tending to zero, and the subscript "f" stands for "fixed".

Assume for simplicity that $h_f(\xi)$ is a continuous function of ξ in an arbitrary closed circle in $\Re^2$. Let R_{ξ_1,ξ_2} be a fixed continuous curve to be called a route, connecting ξ_1 and ξ_2-two distinct points in the plane. A point (a ship in the application) is moving in one direction along the route. Every time it "crosses the point" of the process $\{N(B)\}$ an accident can happen with a given probability. Let r be the distance from ξ_1 to the current point of the route (coordinate) and $h_f(r)$ denote the rate of the process in $(r, r + dr]$. Thus, a one-dimensional parameterization is considered. For defining the corresponding Poisson measure, the dimensions of objects under consideration should be taken into account.

Let $(\gamma_n^+(r), \gamma_n^-(r))$ be a small interval of length $\gamma_n(r) = \gamma_n^+(r) + \gamma_n^-(r)$ in a normal to R_{ξ_1,ξ_2} in the point with coordinate r, where the upper indexes denote the corresponding direction ($\gamma_n^+(r)$ is on one side of R_{ξ_1,ξ_2}, and $\gamma_n^-(r)$ is on the other). Let $\bar{R}$ be the length of $R_{\xi_1,\xi_2} : \bar{R} \equiv |R_{\xi_1,\xi_2}|$ and assume that

$$\bar{R} \gg \gamma_n(r), \quad \forall r \in [0, R]$$

The interval $(\gamma_n^+(r), \gamma_n^-(r))$ is moving along R_{ξ_1,ξ_2}, crossing points of a random field. For the application, it is reasonable to assume the following model for the symmetrical ($\gamma_n^+(r) = \gamma_n^-(r)$) equivalent interval: $\gamma_n(r) = 2\delta_s + 2\delta_o(r)$, where $2\delta_s$ and $2\delta_o(r)$ are the diameters of a ship and of an obstacle respectively. It is assumed for simplicity that all obstacles have the same diameter. There can be other models as well. Using the definition in Equation 6.4 for the specific domain B and the r-parameterization, the corresponding *equivalent rate* of occurrence of points, $h_{ef}(r)$ can be obtained:

$$h_{ef}(r) = \lim_{\Delta r \to 0} \frac{E[N(B(r, \Delta r, \gamma_n(r)))]}{\Delta r} \tag{6.5}$$

where $N(B(r, \Delta r, \gamma_n(r)))$ is the random number of points crossed by the interval $\gamma_n(r)$, moving from r to $r + \Delta r$.

When $\Delta r \to 0$ and $\gamma_n(r)$ is sufficiently small [4]:

$$\begin{aligned} &E[N(B(r, \Delta r, \gamma_n(r)))] \\ &\quad = \int_{B(r,\Delta r,\gamma_n(r))} h_f(\xi)\, dS(\delta(\xi)) \\ &\quad \overset{\Delta r \to 0}{=} \gamma_n(r) h_f(r)\, dr[1 + o(1)] \end{aligned}$$

which leads to the following relation for the equivalent rate of the corresponding one-dimensional non-homogeneous Poisson process:

$$h_{ef}(r) = \gamma_n(r) h_f(r)[1 + o(1)] \tag{6.6}$$

It was also assumed, while obtaining Equation 6.6, that the radius of curvature of the route $R_c(r)$ is sufficiently large compared with $\gamma_n(r)$:

$$\gamma_n(r) \ll R_c(r)$$

uniformly in $r \in [0, \bar{R}]$. This means that the domain covered by the interval $(\gamma_n^+(r), \gamma_n^-(r))$ while it moves from r to $r + \Delta r$ along the route is asymptotically rectangular with an area $\gamma_n(r)\Delta r$, $\Delta r \to 0$. The generalization allowing small values of radius of curvature can be considered. In this case, some points, for instance, can be intersected by the moving interval twice.

Hence, the r-parameterization along the fixed route reduces the problem to the one-dimensional setting of Section 6.2.1. As in the one-dimensional case, assume that crossing a point with a coordinate r leads to an accident with probability $\theta_f(r)$ and to the "survival" with a complementary probability $\bar{\theta}_f(r) = 1 - \theta_f(r)$. Denote by R a random distance from the initial point of the route ξ_1 till a point on the route where an accident had occurred. Similar to Equations 6.2 and 6.3, the probability of passing the route R_{ξ_1,ξ_2} without accidents can be derived in a following way:

$$\begin{aligned} \Pr\{R > \bar{R}\} &\equiv P(\bar{R}) = 1 - F_{af}(\bar{R}) \\ &= \exp\left[-\int_0^{\bar{R}} \theta_f(r) h_{ef}(r)\, dr\right] \\ &\equiv \exp\left[-\int_0^{\bar{R}} \lambda_{af}(r)\, dr\right] \end{aligned} \tag{6.7}$$

where

$$\lambda_{af}(r) \equiv \theta_f(r) h_{ef}(r) \tag{6.8}$$

is the corresponding observed failure rate written in the same way as the per demand failure rate (Equation 6.3) of the previous section.

6.2.3 Failure Rate Process

Assume that $\lambda_{af}(r)$ in Equation 6.8 is now a stochastic process defined, for instance, by an unobserved covariate stochastic process $Y = Y_r$, $r \geq 0$ [5]. By definition, the observed failure rate cannot be stochastic. Therefore, it should be obtained via the corresponding conditioning on realizations of Y. Denote the defined failure (hazard) rate process by $\lambda_{af}(Y, r)$. It is well known, *e.g.* [6], that in this case the following equation holds:

$$P(\bar{R}) = E\left[\exp\left(-\int_0^{\bar{R}} \lambda_{af}(Y, r)\,dr\right)\right] \qquad (6.9)$$

where conditioning is performed with respect to the stochastic process Y. Equation 6.9 can be written via the conditional failure (hazard) rate process [5] as

$$P(\bar{R}) = \exp\left\{-\int_0^{\bar{R}} E[\lambda_{af}(Y, r) \mid R > r]\,dr\right\}$$
$$= \exp\left[-\int_0^{\bar{R}} \bar{\lambda}_{af}(r)\,dr\right] \qquad (6.10)$$

where $\bar{\lambda}_{af}(r)$ now denotes the corresponding observed hazard rate. Thus, the relation of the observed and conditional hazard rates is

$$\bar{\lambda}_{af}(r) = E[\lambda_{af}(Y, r \mid R > r)] \qquad (6.11)$$

As follows from Equation 6.10, Equation 6.11 can constitute a reasonable tool for obtaining $P(\bar{R})$, but the corresponding explicit derivations can be performed only in some of the simplest specific cases. On the other hand, it can help to analyze some important properties.

Consider an important specific case. Let probability $\theta_f(r)$ be indexed by a parameter $Y : \theta_f(Y, r)$, where Y is interpreted as a non-negative random variable with support in $[0, \infty)$ and the probability density function $\pi(y)$. In the sea safety application this randomization can be due to the unknown characteristics of a navigation (or (and) a collision avoidance) onboard system, for instance. There can be other interpretations as well. Thus the specific case of when Y in Equations 6.9 and 6.10 is a random variable is considered. The observed failure rate $\bar{\lambda}_{af}(r)$ reduces in this case to the corresponding mixture failure rate:

$$\bar{\lambda}_{af}(r) = \int_0^{\infty} \lambda_{af}(y, r)\pi(y \mid r)\,dy \qquad (6.12)$$

where $p(y \mid r)$ is the conditional probability density function of Y given that $R > r$. As in [7], it can be defined in the following way:

$$\pi(y \mid r) = \frac{\pi(y)P(y, r)}{\int_0^{\infty} P(y, r)\pi(y)\,dy} \qquad (6.13)$$

where $P(y, r)$ is defined as in Equation 6.7, and $\lambda_{af}(r)$ is substituted by $\theta_f(y, r)h_{ef}(r)$ for the fixed y.

The relations in Equations 6.12 and 6.13 constitute a convenient tool for analyzing the shape of the observed failure rate $\bar{\lambda}_{af}(r)$. This topic will be considered in a more detailed way in Section 6.3. It is only worth noting now that the shape of $\bar{\lambda}_{af}(r)$ can differ dramatically from the shape of the conditional failure rate $\lambda_{af}(y, r)$, and this fact should be taken into consideration in applications. Assume, for instance, a specific multiplicative form of parameterization:

$$\theta_f(Y, r)h_{ef}(r) = Y\theta_f(r)h_{ef}(r)$$

It is well known that if $\theta_f(r)h_f(r)$ is constant, then the observed failure rate is decreasing [8]. But it turns out that even if $\theta_f(r)h_{ef}(r)$ is sharply increasing, the observed failure rate $\bar{\lambda}_{af}(r)$ can still decrease, at least for sufficiently large r! Thus, the random parameter changes the aging properties of the governing distribution function (Section 6.3).

A similar randomization can be performed in the second multiplier in the right-hand side of Equation 6.8: $h_e \approx \gamma_n(r)h_f(r)$, where the length of the moving interval $\gamma_n(r)$ can be considered random as well as the rate of fixed obstacles $h_f(\xi)$. In the latter case the situation can be described in terms of doubly stochastic Poisson processes [2].

For the "highly reliable systems", when, for instance, $h_f \to 0$ uniformly in $r \in [0, \bar{R}]$, one can easily obtain from Equations 6.9 and 6.11 the following obvious "unconditional" asymptotic approximate relations:

$$P(\bar{R}) = \left\{1 - E\left[\int_0^{\bar{R}} \lambda_{af}(Y, r)\,dr\right]\right\}[1 + o(1)]$$
$$= \left\{1 - \int_0^{\bar{R}} E[\lambda_{af}(Y, r)\,dr]\right\}[1 + o(1)]$$
$$\bar{\lambda}_{af}(r) = E[\lambda_{af}(Y, r \mid R > r)]$$
$$= E[\lambda_{af}(Y, r)][1 + o(1)]$$

By applying Jensen's inequality to the right-hand side of Equation 6.9 a simple lower bound for $P(\bar{R})$ can be also derived:

$$P(\bar{R}) \geq \exp\left\{E\left[-\int_0^{\bar{R}} \lambda_{af}(Y, r)\,dr\right]\right\}$$

6.2.4 Moving Obstacles

Consider a random process of continuous curves in the plane to be called paths. We shall keep in mind an application, that the ship routes on a chart represent paths and the rate of the stochastic processes, to be defined later, represents the intensity of navigation in a given sea area. The specific case of stationary random lines in the plane is called a *stationary line process.*

It is convenient to characterize a line in the plane by its (ρ, ψ) coordinates, where ρ is the perpendicular distance from the line to a fixed origin, and ψ is the angle between this perpendicular line and a fixed reference direction. A random process of undirected lines can be defined as a point process on the cylinder $\Re_+ \times S$, where $\Re_+ = (0, \infty)$ and S denote both the circle group and its representations as $(0, 2\pi]$ [9]. Thus, each point on the cylinder is equivalent to the line in $\Re^2$ and for the finite case the point process (and associated stationary line process) can be described. Let V be a fixed line in $\Re^2$ with coordinates (ρ_V, α) and let N_V be the point process on V generated by its intersections with the stationary line process. Then N_V is a stationary point process on V with rate h_V given by [9]:

$$h_V = h \int_S |\cos(\psi - \alpha)| P(d\psi) \qquad (6.14)$$

where h is the constant rate of the stationary line process and $P(d\psi)$ is the probability that an arbitrary line has orientation ψ. If the line process is isotropic, then the relation in Equation 6.14 reduces to:

$$h_V = 2h/\pi$$

The rate h is induced by the random measure defined by the total length of lines inside any closed bounded convex set in $\Re^2$. One cannot define the corresponding measure as the number of lines intersecting the above-mentioned set, because in this case it will not be additive.

Assume that the line process is the homogeneous Poisson process in the sense that the point process N_V generated by its intersections with an arbitrary V is a Poisson point process. Consider, now, a stationary temporal Poisson line process in the plane. Similar to N_V, the Poisson point process $\{N_V(t); t > 0\}$ of its intersections with V in time can be defined. The constant rate of this process, $h_V(1)$, as usual, defines the probability of intersection (by a line from a temporal line process) of an interval of a unit length in V and in a unit interval of time given these units are substantially small. As previously, Equation 6.17 results in $h_V(1) = 2h(1)/\pi$ for the isotropic case.

Let V_{ξ_1,ξ_2} be a finite line route, connecting ξ_1 and ξ_2 in $\Re^2$, and r, as in the previous section, is the distance from ξ_1 to the current point of V_{ξ_1,ξ_2}. Then $h_V(1)\,dr\,dt$ can be interpreted as the probability of intersecting V_{ξ_1,ξ_2} by the temporal line process in

$$(r, r + dr) \times (t, t + dt) \quad \forall r \in (0, \bar{R}),\ t > 0$$

A point (a ship) starts moving along V_{ξ_1,ξ_2} at ξ_1, $t = 0$ with a given speed. We assume that an accident happens with a given probability when "it intersects" the line from the (initial) temporal line process. Note that intersection in the previous section was time independent. A regularization procedure, involving dimensions (of a ship, in particular) can be performed in the following way:

an attraction interval

$$(r - \gamma_{ta}^{-}, r + \gamma_{ta}^{+}) \subset V_{\xi_1,\xi_2} \quad \gamma_{ta}^{+}, \gamma_{ta}^{-} \geq 0$$
$$\gamma_{ta}(r) = \gamma_{ta}^{+}(r) + \gamma_{ta}^{-}(r) \ll \bar{R}$$

where the subscript "ta" stands for tangential, is considered. The attraction interval (which can be defined by the ship's dimensions), "attached to the point r" is moving along the route. The coordinate of the point is changing in time in accordance with the following equation:

$$r(t) = \int_0^t v(s)\, ds \quad t \leq t_{\bar{R}}$$

where $t_{\bar{R}}$ is the total time on the route.

Similar to Equations 6.5 and 6.6, the *equivalent rate of intersections* $h_{em}(r)$ can be derived. Assume for simplicity that the speed $v(t) = v_0$ and the length of the attraction interval γ_{ta} are constants. Then

$$E[N_V((r, r + \Delta r), \Delta t)] = h_V(1)\Delta r \Delta t \quad (6.15)$$

Thus, the equivalent rate is also a constant:

$$h_{em} = \Delta t h_V(1) = \frac{\gamma_{ta}}{v_0} h_V(1) \quad (6.16)$$

where $\Delta t = \gamma_{ta}/v_0$ is the time needed for the moving attraction interval to pass the interval $(r, r + \Delta r)$ as $\Delta r \to 0$. In other words, the number of intersections of $(r, r + dr)$ by the lines from the temporal Poisson line process is counted only during this interval of time. Therefore, the temporal component of the process is "eliminated" and its r-parameterization is performed.

As assumed earlier, the intersection can lead to an accident. Let the corresponding probability of an accident θ_m also be a constant. Then, using the results of Sections 6.2.1 and 6.2.2, the probability of moving along the route V_{ξ_1,ξ_2} without accidents can be easily obtained in a simple form:

$$P(\bar{R}) = \exp(-\theta_m h_{em} \bar{R}) \quad (6.17)$$

The nonlinear generalization, which is rather straightforward, can be performed in the following way. Let the line route V_{ξ_1,ξ_2} turn into the continuous curve R_{ξ_1,ξ_2} and lines of the stochastic line process also turn into continuous curves. As previously, assume that the radius of curvature $R_c(r)$ is bounded from the left by a positive, sufficiently large constant (the same for the curves of the stochastic process). The idea of this generalization is based on the fact that for a sufficiently narrow domain (containing R_{ξ_1,ξ_2}) the paths can be considered as lines and sufficiently small portions of the route can also be considered as approximately linear.

Alternatively it can be assumed from the beginning (and this seems to be quite reasonable for our application) that the point process generated by intersections of R_{ξ_1,ξ_2} with temporal-planar paths is a non-homogeneous Poisson process (homogeneous in time, for simplicity) with the rate $h_R(r, 1)$. Similar to Equation 6.16, an r-dependent equivalent rate can be obtained asymptotically as $\gamma_{ta} \to 0$:

$$h_{em}(r) = \frac{\gamma_{ta}}{v_0} h_R(r, 1)[1 + o(1)] \quad (6.18)$$

It is important that, once again, everything is written in r-parameterization. Marginal cases for Equation 6.18 are quite obvious. When $v_0 \to 0$ the equivalent rate tends to infinity (time of exposure of a ship to the temporal path process tends to infinity); when $v_0 \to \infty$ the rate tends to zero. The relation in Equation 6.17 can be now written in the r-dependent way:

$$P(\bar{R}) = \exp\left[-\int_0^{\bar{R}} \theta_m(r) h_{em}(r)\, dr\right] \quad (6.19)$$

Similar to the previous section, the corresponding randomization approaches can also be applied to the equivalent failure rate $\theta_m(r)h_{em}(r)$. Finally, assuming an independence of two causes of accidents, the probability of survival on the route is given by the following equation, which combines Equations 6.7 and 6.19:

$$P(\bar{R}) = \exp\left\{-\int_0^{\bar{R}} [\theta_f(r)h_{ef}(r) + \theta_m(r)h_{em}(r)]\, dr\right\} \quad (6.20)$$

Other types of external point influence can also be considered, which under certain assumptions will result in additional terms in the integrand in the right-hand side of Equation 6.20. Thus, the method of the per demand failure rate, as in the one-dimensional case, combined with the r-parameterization procedure, resulted in the obtaining of simple relations for the observed failure rate and eventually in deriving the probability of survival in the plane.

6.3 Multiple Availability

6.3.1 Statement of the Problem

The main objective of this section is to show how the method of the per demand failure rate works as a fast repair approximation to obtain some reliability characteristics of repairable systems that generalize the conventional availability. At the same time, exact solutions for the characteristics under consideration or the corresponding integral equations will be also derived.

Consider a repairable system, which begins operating at $t = 0$. A system is repaired upon failure, and then the process repeats. Assume that all time intervals (operation and repair) are s-independent, the periods of operation are exponential, independent, identically distributed (i.i.d.) random variables, and the periods of repair are also exponential, i.i.d. random variables. Thus, a specific alternating renewal process describes the process of system's functioning.

One of the main reliability characteristics of repairable systems is availability—the probability of being in the operating state at an arbitrary instant of time t. This probability can be easily obtained [8] for a system that is functioning and repaired under the above assumptions:

$$A(t) = \frac{\mu}{\mu + \lambda} + \frac{\lambda}{\mu + \lambda} \exp[-(\lambda + \mu)t] \quad (6.21)$$

where λ and μ are parameters of exponential distributions of times to failure and to repair respectively.

It is quite natural to generalize the notion of availability on a setting when a system should be available at a number of time instants (demands).

Define a breakdown (it was called "an accident" in the specific setting of Section 6.2) as an event when a system is in the state of repair at the moment of a demand occurrence. Define multiple availability as the probability of the following event: the system is in the operating state at each moment of demand in $[0, t)$.

Let, at first, the instants of demands are fixed: $0 < t_1 < t_2 < \cdots < t_k < t$. Then, owing to the Markovian properties of the model, multiple availability $V(t)$ can be obtained for this case as:

$$V(t) = A(t_1)A(t_2 - t_1) \cdots A(t_k - t_{k-1}) \quad (6.22)$$

Assume now that a point process of demands is the homogeneous Poisson process with the rate λ_d and it is s-independent of the process of the system functioning and repair. The straightforward integration of Equation 6.22 for t_i, $i = 1, 2, \ldots, k$ from the homogeneous Poisson process and summation for all possible k from 1 to ∞ gives the following general formula [10]:

$$V(t) = \sum_{k=1}^{\infty} \int\int \cdots \int A_{\lambda,\mu}(t_1, t_2, \ldots, t_k)\lambda_d^k \, dt_1 \, dt_2 \cdots dt_k \exp(-\lambda_d t) + \exp(-\lambda_d t) \quad (6.23)$$

where the limits of integration for each t_i, $i = 1, 2, \ldots$ are 0 and t respectively. The term $\exp(-\lambda_d t)$ stands for the probability of absence of demands in $[0, t)$. Of course, computations of $V(t)$ using the relation in Equation 6.23 can be only approximate.

There are many applications, where multiple availability can be considered as an important probabilistic characteristic for repairable systems. Performance of various information-measuring systems, for instance, which operate continuously and deliver information only on users' demand at some instants of time, can be adequately described by multiple availability [11].

The criterion of a breakdown in the above definition of multiple availability is very strong: a system should be available at each instant of demand. Thus, it is reasonable to consider the weaker versions of this criterion, when:

1. not all demands need to be serviced;
2. if a demand occurs, while a system is in the state of repair, but the time from the start of repair (or to the completion of repair) is sufficiently small, then this event may not qualify as a breakdown.

These new assumptions are quite natural. The first describes the fact that the quality of a system performance can still be considered as acceptable if the number of demands when a system is in the state of repair is less than k, $k \geq 2$. The second can be applied, for instance, to information-measuring systems when the quality of information, measured before the system last failure, is still acceptable after $\tilde{\tau}$ units of repair time (information is slowly aging). Two methods are used for deriving expressions for multiple availability for these criteria of a breakdown:

1. the method of integral equations to be solved via the corresponding Laplace transforms;
2. an approximate method of the per demand failure rate.

The method of integral equations leads to the exact explicit solution for multiple availability only for the simplest criterion of a breakdown and exponential distributions of times to failure and repair. In other cases the solution can be obtained via the numerical inversion of Laplace transforms.

On the other hand, simple relations for multiple availability can be obtained via the method of the per demand failure rate on a basis of the fast repair assumption: the mean time of repair is much less than the mean time of a system time to failure and the mean time between the consecutive demands. The fast repair approximation also makes it possible to derive approximate solutions for the problems that cannot be solved by the first method, *e.g.* for the case of arbitrary time to failure and time to repair cumulative distribution functions.

6.3.2 Ordinary Multiple Availability

Multiple availability $V(t)$ defined in Section 6.3.1 will be called ordinary multiple availability as distinct from the availability measures of subsequent sections. It is easy to obtain the following integral equation for $V(t)$ [11]

$$V(t) = \exp(-\lambda_d t) + \int_0^t \lambda_d \exp(-\lambda_d x) A(x) V(t-x)\,dx \tag{6.24}$$

where the first term defines the probability of the absence of demands in $[0, t)$, and the integrand is a joint probability of three events:

- occurrence of the first demand in $[x, x + dx)$
- the system is in the operation state at the time instant x
- the system is in the operation state at each moment of demand in $[x, t)$ (probability: $V(t - x)$).

Equation 6.4 can be easily solved via the Laplace transform. By applying this transformation to both sides of Equation 6.24 and using the Laplace transform of $A(t)$, then, after elementary transformations, the following expression for the Laplace transform of $V(t)$ can be obtained:

$$\tilde{V}(s) = \frac{s + \lambda_d + \lambda + \mu}{s^2 + s(\lambda_d + \lambda + \mu) + \lambda_d \lambda} \tag{6.25}$$

Finally, performing the inverse transform:

$$V(t) = B_1 \exp(s_1 t) + B_2 \exp(s_2 t) \tag{6.26}$$

where the roots of the denominator in Equation 6.5 are

$$s_{1,2} = \frac{-(\lambda_d + \lambda + \mu) \pm \sqrt{(\lambda_d + \lambda + \mu)^2 - 4\lambda_d \lambda}}{2}$$

and $B_1 = s_2/(s_2 - s_1)$ and $B_2 = -s_1/(s_2 - s_1)$. It is worth noting that, unlike the relation in Equation 6.23, a simple explicit expression is obtained in Equation 6.26.

The following assumptions are used while deriving approximations for $V(t)$ and the corresponding errors of approximation:

$$\frac{1}{\lambda} \gg \frac{1}{\mu} \quad (\lambda \ll \mu) \tag{6.27}$$

$$\frac{1}{\lambda_d} \gg \frac{1}{\mu} \quad (\lambda_d \ll \mu) \tag{6.28}$$

$$t \gg \frac{1}{\mu} \tag{6.29}$$

The assumption in Equation 6.27 implies that the mean time to failure far exceeds the mean time of repair, and this is obviously the case in many practical situations. The mean time of repair usually does not exceed several hours, whereas it can be thousands of hours for λ^{-1}. It is usually referred to as a condition of the fast repair [12]. This assumption is important for the method of the current section, because, as it will be shown, it is used for approximating availability (Equation 6.11) by its stationary characteristic in the interval between two consecutive demands and for expansions in series as well. This means that the interval mentioned should be sufficiently large and this is stated by the assumption in Equation 6.28. There exist, of course, systems with a very high intensity of demands (a database system, where queries arrive every few seconds, for instance), but here we assume a sufficiently low intensity of demands, which is relevant for a number of applications. An onboard navigational system will be considered in this section as the corresponding example. The assumption in Equation 6.29 means that the mission time t should be much larger than the mean time of repair, and this is quite natural for the fast repair approximation.

With the help of the assumptions in Equations 6.28 and 6.29 [11], an approximate relation for $V(t)$ can be obtained via the method of the per demand failure rate, Equations 6.2 and 6.3, in the following way:

$$\begin{aligned} V(t) &\equiv \exp\left[-\int_0^t \lambda_b(u)\,du\right] \\ &\approx \exp[-(1-A)\lambda_d t] \\ &= \exp\left(-\frac{\lambda\lambda_d t}{\lambda+\mu}\right) \end{aligned} \qquad (6.30)$$

where $\lambda_b(t)$ is the observed failure rate to be called now the breakdown rate (notation $\lambda_a(t)$ was used for the similar characteristic).

The above assumptions are necessary for obtaining Equation 6.30, as the time-dependent availability $A(t)$ is substituted by its stationary characteristic $A(\infty) \equiv A = \mu/\lambda + \mu$. Thus

$$\lambda_b(t) = \frac{\lambda\lambda_d}{\lambda+\mu}$$

and

$$\theta(t) = (1-A) = \frac{\lambda}{\lambda+\mu}, \qquad h(t) = \lambda_d$$

6.3.3 Accuracy of a Fast Repair Approximation

What is the source of an error in the approximate relation in Equation 6.30? The error results from the fact that the system is actually available on demand with probability $A(t_d)$, where t_d is the time since the previous demand. The assumptions in Equations 6.28 and 6.29 imply that $A(t_d) \approx A$, and this leads to the approximation in Equation 6.30, whereas the exact solution is defined by Equation 6.26. When $t \gg \mu^{-1}$ (assumption in Equation 6.29), the second term in the right-hand side of Equation 6.26 is negligible. Then an error δ is defined as

$$\delta = \frac{s_2}{s_2 - s_1}\exp(s_1 t) - \exp\left(-\frac{\lambda\lambda_d t}{\lambda+\mu}\right)$$

Using Equation 6.27 for expanding s_1 and s_2 in series:

$$\begin{aligned} \delta &= \left[1 + \frac{\lambda\lambda_d}{(\lambda+\lambda_d+\mu)^2}\right] \\ &\quad \times \left(1 - s_1 t + \frac{s_1 t^2}{2}\right)[1+o(1)] \\ &\quad - \left[1 - \frac{\lambda\lambda_d t}{\lambda+\mu} + \frac{(\lambda\lambda_d)^2 t^2}{2(\lambda+\mu)^2}\right][1+o(1)] \\ &= \frac{\lambda\lambda_d}{(\lambda+\lambda_d+\mu)^2}\left[1 + \lambda_d t\left(1 - \frac{\lambda\lambda_d t}{\lambda+\mu}\right)\right] \\ &\quad \times [1+o(1)] \\ &= \frac{\lambda\lambda_d}{\mu^2}(1+\lambda_d t)[1+o(1)] \end{aligned}$$

Notation $o(1)$ is understood asymptotically: $o(1) \to 0$ as $\lambda/\mu \to 0$.

On the other hand, a more accurate result can easily be obtained if, instead of $1 - A$, "an averaged probability" $1 - \tilde{A}$ is used:

$$1 - \tilde{A} \equiv \int_0^\infty [1 - A(t_d)]\lambda_d \exp(-\lambda_d t_d)\, dt_d = \frac{\lambda}{\lambda + \lambda_d + \mu}$$

where Equation 6.1 was taken into account. Hence

$$V(t) \approx \exp[-(1 - \tilde{A})\lambda_d t] = \exp\left(-\frac{\lambda\lambda_d t}{\lambda + \lambda_d + \mu}\right) \tag{6.31}$$

In the same way, as above, it can be shown that

$$\tilde{\delta} \equiv V(t) - \exp\left(-\frac{\lambda\lambda_d t}{\lambda + \lambda_d + \mu}\right) = \frac{\lambda\lambda_d}{\mu^2}\left(1 - \frac{\lambda\lambda_d t}{\lambda + \lambda_d + \mu}\right)[1 + o(1)]$$

The error in Equation 6.31 results from the infinite upper bound of integration in $1 - \tilde{A}$, whereas in realizations it is bounded. Thus, $\exp[-(\lambda\lambda_d t)/(\lambda + \lambda_d + \mu)]$ is a more accurate approximation than $\exp[-(\lambda\lambda_d t)/(\lambda + \mu)]$ as $\lambda_d t$ is not necessarily small.

Consider the following practical example [11].

Example 1. A repairable onboard navigation system is operating in accordance with the above assumptions. To obtain the required high precision of navigational data, corrections from the external source are performed with rate λ_d, s-independent of the process of the system functioning and repair. The duration of the correction is negligible. Corrections can be performed only while the system is operating; otherwise it is a breakdown. These conditions describe multiple availability; thus $V(t)$ can be assessed by Equations 6.30 and 6.31. Let:

$$\lambda = 10^{-2}\,\text{h}^{-1} \quad \mu = 2\,\text{h}^{-1}$$
$$\lambda_d \approx 0.021\,\text{h}^{-1} \quad t = 10^3\,\text{h}^{-1} \tag{6.32}$$

This implies that the mean time between failures is 100 h, the mean time of repair is 0.5 h, and the mean time between successive corrections is approximately 48 h. The data in Equation 6.32 characterize the performance of the real navigation system. On substituting the concrete values of parameters in the exact relation (Equation 6.26):

$$V(t) = 0.901\,812\,8$$

whereas the approximate relation (Equation 6.31) gives:

$$V(t) \approx 0.901\,769\,3$$

Thus, the "real error" is $\tilde{\delta} = 0.000\,043\,5$, and

$$\frac{\lambda\lambda_d}{\mu^2}\left(1 - \frac{\lambda\lambda_d t}{\lambda + \lambda_d + \mu}\right) = 0.000\,047\,1$$

The general case of arbitrary distributions of time to failure $F(t)$ with mean $\bar{T}$ and time to repair $G(t)$ with mean $\bar{\tau}$ under assumptions similar to Equations 6.27–6.29 can also be considered. This possibility follows from the fact that the stationary value of availability A depends only on the mean times to failure and repair. Equation 6.30, for instance, will turn to

$$V(t) \equiv \exp\left[-\int_0^t \lambda_b(u)\, du\right] \approx \exp[-(1 - A)\lambda_d t] = \exp\left(-\frac{\bar{\tau}\lambda_d t}{\bar{T} + \bar{\tau}}\right) \tag{6.33}$$

Furthermore, as follows from [11], the method of the per demand failure rate can be generalized to the non-homogeneous Poisson process of demands. In this case, $\lambda_d t$ in Equation 6.33 should be substituted by $\int_0^t \lambda_d(u)\, du$.

It is difficult to estimate an error of approximation for the case of arbitrary $F(t)$ and $G(t)$ directly. Therefore, strictly speaking, Equation 6.33 can be considered only like a heuristic result. Similar to Equation 6.31, a more accurate solution $\exp[-(1 - \tilde{A})\lambda_d t]$ can be formally defined, but the necessary integration of the availability function can be performed explicitly only for the special choices of $F(t)$ and $G(t)$, like the exponential or the Erlangian, for instance.

6.3.4 Two Non-serviced Demands in a Row

The breakdown in this case is defined as an event when a system is in the repair state at two consecutive moments of demand. Multiple availability $V_1(t)$ is now defined as the probability of a system functioning without breakdowns in $[0, \infty)$.

As was stated before, this setting can be quite typical for some information-processing systems. If, for instance, a correction of the navigation system in the example above had failed (the system was unavailable), we can still wait for the next correction, and consider the situation as normal, if this second correction was performed properly.

Similar to Equation 6.24, the following integral equation for $V_1(t)$ can be obtained:

$$\begin{aligned} V_1(t) &= \exp(-\lambda_{\rm d}t) \\ &\quad + \int_0^t \lambda_{\rm d}\exp(-\lambda_{\rm d}x)A(x)V_1(t-x)\,{\rm d}x \\ &\quad + \int_0^t \Bigg\{\lambda_{\rm d}\exp(-\lambda_{\rm d}x)[1-A(x)] \\ &\quad \times \int_0^{t-x} \lambda_{\rm d}\exp(-\lambda_{\rm d}y)A^*(y) \\ &\quad \times V_1(t-x-y)\,{\rm d}y\Bigg\}\,{\rm d}x \end{aligned} \tag{6.34}$$

where $A^*(t)$ is the availability of the system at the time instant t, if at $t = 0$ the system was in the state of repair:

$$A^*(t) = \frac{\mu}{\mu+\lambda} - \frac{\mu}{\mu+\lambda}\exp[-(\lambda+\mu)t]$$

The first two terms in the right-hand side of Equation 6.34 have the same meaning as in the right-hand side of Equation 6.24, and the third term defines the joint probability of the following events:

- occurrence of the first demand in $[x, x + {\rm d}x)$;
- the system is in the repair state at x (probability: $1 - A(x)$);
- occurrence of the second demand in $[x + y, x + y + {\rm d}y)$, given that the first demand had occurred in $[x, x + {\rm d}x)$;
- the system is in the operation state at $x + y$, given that it was in the repair state on the previous demand (probability: $A^*(y)$);
- the system operates without breakdowns in $[x + y, t)$ (probability: $V_1(t - x - y)$).

Equation 6.34 can also be easily solved via the Laplace transform. After elementary transformations

$$\begin{aligned} \tilde{V}_1(s) &= \{(s+\lambda_{\rm d})(s+\lambda_{\rm d}+\lambda+\mu)^2\} \\ &\quad \times \{s(s+\lambda_{\rm d})(s+\lambda_{\rm d}+\lambda+\mu)^2 \\ &\quad + s\lambda_{\rm d}\lambda(s+2\lambda_{\rm d}+\lambda+\mu) \\ &\quad + \lambda_{\rm d}^2\lambda(\lambda_{\rm d}+\lambda)\}^{-1} \\ &= \frac{P_3(s)}{P_4(s)} \end{aligned} \tag{6.35}$$

where $P_3(s)$ $(P_4(s))$ denotes the polynomial of the third (fourth) order in the numerator (denominator).

The inverse transformation results in

$$V_1(t) = \sum_1^4 \frac{P_3(s_i)}{P_4'(s_i)}\exp(s_i t) \tag{6.36}$$

where $P_4'(s)$ defines the derivative of $P_4(s)$ and $s_{\rm i}$, $i = 1, 2, 3, 4$, are the roots of the denominator in Equation 6.35:

$$P_4(s) = \sum_0^4 b_k s^{4-k} = 0 \tag{6.37}$$

where b_k are defined as:

$$\begin{aligned} b_0 &= 1 \\ b_1 &= 2\lambda + 2\mu + 3\lambda_{\rm d} \\ b_2 &= (\lambda_{\rm d}+\lambda+\mu)(3\lambda_{\rm d}+\lambda+\mu) + \lambda_{\rm d}\lambda \\ b_3 &= \lambda_{\rm d}[(\lambda_{\rm d}+\lambda+\mu)^2 \\ &\quad + \lambda(\lambda_{\rm d}+\lambda+\mu) + \lambda(\lambda_{\rm d}+\lambda)] \\ b_4 &= \lambda_{\rm d}^2\lambda(\lambda_{\rm d}+\lambda) \end{aligned}$$

Equation 6.36 defines the exact solution of the problem. It can be easily obtained numerically by solving Equation 6.37 and substituting the corresponding roots in Equation 6.36. As in the previous section, a simple approximate formula can also be used in practice. Let the assumptions in Equations 6.27 and 6.28 hold. All b_k, $k = 1, 2, 3, 4$, in Equation 6.37 are positive, and this means that there are no positive roots of this equation. Consider the smallest in absolute value root s_1. Owing to the assumption in Equation 6.27, and the definitions of $P_3(s)$ and $P_4(s)$ in Equation 6.35, it can be seen that

$$s_1 \approx -\frac{b_4}{b_3} \approx -\frac{\lambda\lambda_d(\lambda+\lambda_d)}{\mu^2}, \qquad \frac{P_3(s_1)}{P_4'(s_1)} \approx 1$$

and that the absolute values of other roots far exceed $|s_1|$. Thus, Equation 6.36 can be written as the fast repair exponential approximation:

$$V_1(t) \approx \exp\left[-\frac{\lambda\lambda_d(\lambda+\lambda_d)}{\mu^2}t\right] \tag{6.38}$$

It is difficult to assess the approximation error of this approach directly, as it was done in the previous section, because the root s_1 is also defined approximately.

On the other hand, the method of the per demand failure rate can be used for obtaining $V_1(t)$. Similar to Equation 6.10:

$$\begin{aligned} V_1(t) &\equiv \exp\left[-\int_0^t \lambda_b(u)\,du\right] \\ &\approx \exp[-A(1-A)^2\lambda_d t] \\ &= \exp\left[-\frac{\mu\lambda^2\lambda_d t}{(\lambda+\mu)^3}\right] \end{aligned} \tag{6.39}$$

Indeed, the breakdown can happen in $[t, t+dt)$ if a demand occurs in this interval (probability: $\lambda_d\,dt$) and the system is unavailable at this moment of time and at the moment of the previous demand, while it was available before. Owing to the assumptions in Equations 6.28 and 6.29, this probability is approximately $[\mu\lambda/(\lambda+\mu)]^3$, as

$$A \equiv A(\infty) = A^*(\infty) = \frac{\mu}{\lambda+\mu}$$

Similar to Equation 6.31, a more accurate approximation for $V_1(t)$ can be obtained if we substitute in Equation 6.39 more accurate values of unavailability at the moments of current $(1-\tilde{A}^*)$ and previous $(1-\tilde{A})$ demands. Hence

$$\begin{aligned} V_1(t) &\approx \exp[-A(1-\tilde{A})(1-\tilde{A}^*)\lambda_d t] \\ &= \exp\left\{-\frac{\mu\lambda\lambda_d[\mu(\lambda+\lambda_d)+\lambda\lambda_d+\lambda^2]t}{(\lambda+\mu)^2(\lambda+\lambda_d+\mu)^2}\right\} \end{aligned} \tag{6.40}$$

Actually, this is also an obvious modification of the method of the per demand failure rate. The analysis of accuracy of this and subsequent models is much more cumbersome than the one for ordinary multiple availability. Therefore, it is not included in this chapter. The fast repair assumption (Equation 6.27) leads to an interesting result:

$$\frac{\mu\lambda\lambda_d[\mu(\lambda+\lambda_d)+\lambda\lambda_d+\lambda^2]}{(\lambda+\mu)^2(\lambda+\lambda_d+\mu)^2} = \frac{\lambda\lambda_d(\lambda+\lambda_d)}{\mu^2}[1+o(1)]$$

Thus, Equation 6.40 and the approximate formula in Equation 6.38, which was derived via the Laplace transform, are asymptotically equivalent! It is worth noting that a more crude relation (Equation 6.39) does not always lead to Equation 6.38, as it is not necessarily so that $\lambda_d \ll \lambda$.

Taking into account the reasoning that was used while deriving Equation 6.30, the approximation in Equation 6.39 can be generalized to arbitrary $F(t)$ and $G(t)$:

$$\begin{aligned} V_1(t) &\approx \exp[-A(1-A)^2\lambda_d t] \\ &= \exp\left[-\frac{\lambda_d\bar{T}\bar{\tau}^2 t}{(\bar{T}+\bar{\tau})^3}\right] \end{aligned} \tag{6.41}$$

6.3.5 Not More than N Non-serviced Demands

The breakdown in this case is defined as an event, when more than $N \geq 1$ demands are non-serviced in $[0, t)$. Multiple availability $V_{2,N}(t)$ is defined as the probability of a system functioning without breakdowns in $[0, t)$.

Let $N = 1$. Similar to Equations 6.24 and 6.34:

$$\begin{aligned} V_{2,1}(t) &= \exp(-\lambda_d t) \\ &\quad + \int_0^t \lambda_d \exp(-\lambda_d x) A(x) V_{2,1}(t-x)\,dx \\ &\quad + \int_0^t \lambda_d \exp(-\lambda_d x) \\ &\quad \times [1-A(x)]V^*(t-x)\,dx \end{aligned} \tag{6.42}$$

where $V^*(t)$ is an ordinary multiple availability for the system that is at $t = 0$ in the state of repair. Similar to Equation 6.26:

$$V^*(t) = B_1^* \exp(s_1 t) + B_2^* \exp(s_2 t)$$

where s_1 and s_2 are the same as in Equation 6.26 and

$$B_1^* = \frac{s_2 + \lambda_d}{s_2 - s_1} \quad B_2^* = -\frac{s_1 + \lambda_d}{s_2 - s_1}$$

The first term in the right-hand side of Equation 6.42 determines the probability of absence of demands in $[0, t)$. The second term defines the probability that the system is operable on the first demand (renewal point) and is functioning without breakdowns in $[x, t)$. The third term determines the probability of the system being in the state of repair on the first demand and in the operable state on all subsequent demands.

Applying the Laplace transform to both sides of Equation 6.42 and solving the algebraic equation with respect to $\tilde{V}_{2,1}(s)$:

$$\begin{aligned}\tilde{V}_{2,1}(s) = \{&(s + \lambda_d + \lambda + \mu)\\ &\times [s^2 + s(\lambda_d + \lambda + \mu) + \lambda_d\lambda]\\ &+ \lambda_d\lambda(s + \lambda + \mu)\}\\ &\times \{[s^2 + s(\lambda_d + \lambda + \mu) + \lambda_d\lambda]^2\}^{-1}\end{aligned} \tag{6.43}$$

The roots of the denominator s_1 and s_2 are the same as in Equation 6.25, but two-folded. A rather cumbersome exact solution can be obtained by inverting Equation 6.43. The complexity will increase with increase in N. The cumbersome recurrent system of integral equations for obtaining $V_{2,\mathrm{N}}(t)$ can be derived and solved in terms of the Laplace transform (to be inverted numerically). Hence, the fast repair approximation of the previous sections can also be very helpful in this case.

Consider a point process of breakdowns of our system in the sense of Section 6.3.2 (each point of unavailability on demand forms this point process). It is clear that this is the renewal process with $1 - V(t)$ as the cumulative distribution function of the first cycle and $1 - V^*(t)$ as cumulative distribution functions of the subsequent cycles. On the other hand, as follows from Equation 6.31, the point process can be approximated by the Poisson process with rate $(1 - \tilde{A})\lambda_d$. This leads to the following approximate result for arbitrary N:

$$\begin{aligned}V_{2,N}(t) \approx \exp&\left(-\frac{\lambda\lambda_d}{\lambda + \lambda_d + \mu}t\right)\\ &\times \sum_{n=0}^{N} \frac{1}{n!}\left(\frac{\lambda\lambda_d}{\lambda + \lambda_d + \mu}t\right)^n\\ &N = 1, 2, \ldots\end{aligned} \tag{6.44}$$

Thus, a rather complicated problem had been immediately solved via the Poisson approximation, based on the per demand failure rate $\lambda\lambda_d(\lambda + \lambda_d + \mu)^{-1}$. When $N = 0$, we arrive at the case of ordinary multiple availability: $V_{2,0}(t) \equiv V(t)$. As previously:

$$\begin{aligned}V_{2,N}(t) \approx \exp&\left(\frac{\lambda_d\bar{\tau}}{\bar{T} + \bar{\tau}}t\right)\\ &\times \sum_{n=0}^{N} \frac{1}{n!}\left(\frac{\lambda_d\bar{\tau}}{\bar{T} + \bar{\tau}}t\right)^n \quad N = 1, 2, \ldots\end{aligned}$$

for the case of general distributions of failure and repair times.

6.3.6 Time Redundancy

The breakdown in this case is defined as an event, when at the moment of a demand occurrence the system has been in the repair state for a time exceeding $\tilde{\tau} > 0$. As previously, multiple availability $V_3(t)$ is defined as the probability of a system functioning without breakdowns in $[0, t)$.

The definition of a breakdown for this case means that, if the system is "not too long" in the repair state at the moment of a demand occurrence, then this situation is not considered as a breakdown. Various types of sensor in information-measuring systems, for instance, can possess this feature of certain inertia. It is clear that if $\tilde{\tau} = 0$, then $V_3(t) = V(t)$.

The following notation will be used:

$V_3^*(t)$ — multiple availability in $[0, t)$ in accordance with a given definition of a breakdown (the system is in the repair state at $t = 0$)

$\Phi(x, \tilde{\tau})$ probability that the system has been in the repair state at the moment of a demand occurrence x for a time not exceeding $\tilde{\tau}$ (the system is in the operable state at $t = 0$)

$\Phi^*(x, \tilde{\tau})$ probability that the system has been in the repair-state at the moment of a demand occurrence x for a time not exceeding $\tilde{\tau}$ (the system is in the state of repair at $t = 0$).

Generalizing Equation 6.24, the following integral equation can be obtained:

$$V_3(t) = \exp(-\lambda_d t) + \int_0^t \lambda_d \exp(-\lambda_d x) A(x) V_3(t-x)\,dx + \int_0^t \lambda_d \exp(-\lambda_d x)\Phi(x, \tilde{\tau}) V_3^*(t-x)\,dx \tag{6.45}$$

where

$$V_3^*(t) = \exp(-\lambda_d t) + \int_0^t \lambda_d \exp(-\lambda_d x) A^*(x) V_3(t-x)\,dx + \int_0^t \lambda_d \exp(-\lambda_d x)\Phi^*(x, \tilde{\tau}) V_3^*(t-x)\,dx$$

and

$$\Phi(x, \tilde{\tau}) = 1 - A(x)\Phi^*(x, \tilde{\tau}) = 1 - A^*(x)$$

for $x \le \tilde{\tau}$, and

$$\Phi(x, \tilde{\tau}) = \int_0^{\tilde{\tau}} A(x-y) \times \left\{ \int_0^y \lambda \exp(-\lambda z) \times \exp[-\mu(y-z)]\,dz \right\} dy$$

$$\Phi^*(x, \tilde{\tau}) = \int_0^{\tilde{\tau}} A^*(x-y) \times \left\{ \int_0^y \lambda \exp(-\lambda z) \times \exp[-\mu(y-z)]\,dz \right\} dy$$

for $x \ge \tilde{\tau}$.

The integrand in the third term in the right-hand side of Equation 6.45, for instance, defines the joint probability of the following events: a demand in $[x, x + dx)$; the system is in the repair-state for less than $\tilde{\tau}$ at this moment of time; the system is operating without breakdowns in $[x, t - x)$ (the system was in the repair-state at $t = 0$). By solving these equations via the Laplace transform, the cumbersome exact formulas for $V_3(t)$ can also be obtained.

To obtain a simple approximate formula by means of the method of the per demand failure rate, as in the previous section, consider the Poisson process with rate $(1 - \tilde{A})\lambda_d$, which, due to Equation 6.31, approximates the point process of the non-serviced demands (each point of unavailability on demand forms this point process). Multiplying the rate of this initial process $(1 - \tilde{A})\lambda_d$ by the probability of a breakdown on a demand $\exp(-\mu\tilde{\tau})$, the corresponding breakdown rate can be obtained. Eventually:

$$\begin{aligned} V_3(t) &\equiv \exp\left[-\int_0^t \lambda_b(u)\,du\right] \\ &\approx \exp[-\lambda_d(1-\tilde{A})\exp(-\mu\tilde{\tau})t] \\ &= \exp\left[-\frac{\lambda\lambda_d}{\lambda + \lambda_d + \mu}\exp(-\mu\tilde{\tau})t\right] \end{aligned} \tag{6.46}$$

The definition of the breakdown will now be changed slightly. It is defined as an event when the system is in the repair state at the moment of a demand occurrence and remains in this state for at least time $\tilde{\tau}$ after the occurrence of the demand. The corresponding multiple availability is denoted in this case by $V_4(t)$.

This setting describes the following situation: a demand cannot be serviced at an instant of its occurrence (the system is in the repair state), but, if the remaining time of repair is less than $\tilde{\tau}$ (probability $1 - \exp(-\mu\tilde{\tau})$), it can be still serviced after the completion of the repair operation. The integral equation for this case can be easily written:

$$V_4(t) = \exp(-\lambda_d t) + \int_0^t \lambda_d \exp(-\lambda_d x) A(x) V_4(t-x)\,dx$$

$$+\int_0^t \lambda_d \exp(-\lambda_d x)[1 - A(x)]$$
$$\times \int_0^{\tilde{\tau}} \mu \exp(-\mu y) V_4(t - x - y)\, dy\, dx$$

Though this equation seems to be similar to the previous one, it cannot be solved via the Laplace transform, as $V_4(t - x - y)$ is dependent on y. Thus, numerical methods for solving integral equations should be used and the importance of the approximate approach for this setting increases.

It can be easily seen that the method of the per demand failure rate can also be applied in this case, which results in the same approximate relation (Equation 6.46) for $V_4(t)$.

6.4 Modeling the Mixture Failure Rate

6.4.1 Definitions and Conditional Characteristics

In most settings involving lifetimes, the population of lifetimes is not homogeneous. That is, all of the items in the population do not have exactly the same distribution; there is usually some percentage of the lifetimes that is different from the majority. For example, for electronic components, most of the population might be exponential, with long lives, while a certain percentage often has an exponential distribution with short lives and a certain percentage can be characterized by lives of intermediate duration. Thus, a mixing procedure arises naturally when pooling from heterogeneous populations. The observed failure rate, which is actually the mixture failure rate in this case, obtained by usual statistical methods, does not bear any information on this mixing procedure. To say more, as was stated in Section 6.1, obtaining the observed failure rate in this way can often present certain difficulties, because the corresponding data can be scarce. On the other hand, it is clear that the proper modeling can add some valuable information for estimating the mixture failure rate and for analyzing and comparing its shape with the shape of the failure rate of the governing cumulative distribution function. Therefore, the analysis of the shape of the observed failure rate will be the main topic of this section.

Mixtures of decreasing failure rate (DFR) distributions are always DFR distributions [8]. It turns out that, very often, mixtures of increasing failure rate (IFR) distributions can decrease, at least in some intervals of time. This means that the operation of mixing can change the corresponding pattern of aging. This fact is rather surprising, and should be taken into account in various applications owing to its practical importance.

As previously, consider a lifetime random variable $T \geq 0$ with a cumulative distribution function $F(t)$. Assume, similar to Section 6.2.3, that $F(t)$ is indexed by a parameter θ, so that $P(T \leq t \mid \theta) = F(t, \theta)$ and that the probability density function $f(t, \theta)$ exists. Then the corresponding failure rate $\lambda(t, \theta)$ can be defined in the usual way as $f(t, \theta)/\bar{F}(t, \theta)$. Let θ be interpreted as a non-negative random variable with a support in $[0, \infty)$ and a probability density function $\pi(\theta)$. For the sake of convenience, the same notation is used for the random variable and its realization. A mixture cumulative distribution function is defined by

$$F_m(t) = \int_0^\infty F(t, \theta)\pi(\theta)\, d\theta \tag{6.47}$$

and the mixture (observed) failure rate in accordance with this definition is

$$\lambda_m(t) = \frac{\int_0^\infty f(t, \theta)\pi(\theta)\, d\theta}{\int_0^\infty \bar{F}(t, \theta)\pi(\theta)\, d\theta} \tag{6.48}$$

Similar to Equations 6.12 and 6.13, and using the conditional probability density function $\pi(\theta \mid t)$ of θ given $T \geq t$:

$$\pi(\theta \mid t) = \frac{\pi(\theta)\bar{F}(t, \theta)}{\int_0^\infty \bar{F}(t, \theta)\pi(\theta)\, d\theta} \tag{6.49}$$

the mixture failure rate $\lambda_m(t)$ can be written as:

$$\lambda_m(t) = \int_0^\infty \lambda(t, \theta)\pi(\theta \mid t)\, d\theta \tag{6.50}$$

Denote by $E[\theta \mid t]$ the conditional expectation of θ given $T \geq t$:

$$E[\theta \mid t] = \int_0^\infty \theta \pi(\theta \mid t)\, d\theta$$

An important characteristic for further consideration is $E'[\theta \mid t]$, the derivative with respect to t:

$$E'[\theta \mid t] = \int_0^\infty \theta \pi'(\theta \mid t)\, d\theta$$

where

$$\pi'(\theta \mid t) = \lambda_m(t)\pi(\theta \mid t) - \frac{f(t, \theta)\pi(\theta)}{\int_0^\infty \bar{F}(t, \theta)\pi(\theta)\, d\theta}$$

Therefore, the following relation holds [13]:

$$E'[\theta \mid t] = \lambda_m(t)E[\theta \mid t] - \frac{\int_0^\infty \theta f(t, \theta)\pi(\theta)\, d\theta}{\int_0^\infty \bar{F}(t, \theta)\pi(\theta)\, d\theta} \tag{6.51}$$

In the next two sections the specific models of mixing will be considered.

6.4.2 Additive Model

Suppose that

$$\lambda(t, \theta) = \alpha(t) + \theta \tag{6.52}$$

where $\alpha(t)$ is a deterministic continuous increasing function ($\alpha(t) \geq 0, t \geq 0$) to be specified later. Then, noting that $f(t, \theta) = \lambda(t, \theta)\bar{F}(t, \theta)$, and applying the definition given in Equation 6.48 for this concrete model:

$$\begin{aligned}\lambda_m(t) &= \alpha(t) + \frac{\int_0^\infty \theta \bar{F}(t, \theta)\pi(\theta)\, d\theta}{\int_0^\infty \bar{F}(t, \theta)\pi(\theta)\, d\theta} \\ &= \alpha(t) + E[\theta \mid t]\end{aligned} \tag{6.53}$$

Using Equations 6.51 and 6.53, the specific form of $E'[\theta \mid t]$ can be obtained:

$$\begin{aligned}E'[\theta \mid t] &= [E[\theta \mid t]]^2 - \int_0^\infty \theta^2 \pi(\theta \mid t)\, d\theta \\ &= -\operatorname{Var}(\theta \mid t) < 0\end{aligned} \tag{6.54}$$

Therefore, the conditional expectation of θ for the additive model is a decreasing function of $t \in [0, \infty)$.

Upon differentiating Equation 6.53, and using Equation 6.54, the following result on the shape of the mixture failure rate can be obtained [7]:

Let $\alpha(t)$ be an increasing (non-decreasing) convex function in $[0, \infty)$. Assume that $\operatorname{Var}(\theta \mid t)$ is decreasing in $t \in [0, \infty)$.

If

$$\operatorname{Var}(\theta \mid 0) > \alpha'(0)$$

then $\lambda_m(t)$ decreases in $[0, c)$ and increases in $[c, \infty)$, where c can be uniquely defined by the equation: $\operatorname{Var}(\theta \mid t)$ *(bathtub shape).*

If

$$\operatorname{Var}(\theta \mid 0) \leq \alpha'(0)$$

then $\lambda_m(t)$ is increasing in $[0, \infty)$.

6.4.3 Multiplicative Model

Suppose that

$$\lambda(t, \theta) = \theta\alpha(t) \tag{6.55}$$

where $\alpha(t)$ is some deterministic increasing (non-decreasing), at least for sufficiently large t, continuous function ($\alpha(t) \geq 0, t \geq 0$). The model in Equation 6.55 is usually called a proportional hazards (PH) model, whereas the previous one is called an additive hazards model. Applying Equation 6.50:

$$\lambda_m(t) = \int_0^\infty \lambda(t, \theta)\pi(\theta \mid t)\, d\theta = \alpha(t)E[\theta \mid t] \tag{6.56}$$

After differentiating

$$\lambda'_m(t) = \alpha'(t)E[\theta \mid t] + \alpha(t)E'[\theta \mid t]$$

It follows immediately from this equation that when $\alpha(0) = 0$ the failure rate $\lambda_m(t)$ increases in the neighborhood of $t = 0$. Further behavior of this function depends on the other parameters involved. If $\lambda'_m(t) \geq 0$, then the mixture failure rate is increasing (non-decreasing). Thus, the mixture will have an IFR if and only if $\forall t \in [0, \infty)$:

$$\frac{\alpha'(t)}{\alpha(t)} \geq -\frac{E'[\theta \mid t]}{E[\theta \mid t]} \tag{6.57}$$

Similar to Equation 6.54, it can be shown that [13]:

$$E'[\theta \mid t] = -\alpha(t)\,\mathrm{Var}(\theta \mid t) < 0 \qquad (6.58)$$

which means that the conditional expectation of θ for the model (Equation 6.55) is a decreasing function of $t \in [0, \infty)$. Combining this property with Equations 6.55 and 6.56 for the specific case $\alpha(t) = \mathrm{const}$, the well-known result that the mixture of exponentials has a DFR can be easily obtained. Thus, the foregoing can be considered as a new proof of this fact.

With the help of Equation 6.58, the inequality in Equation 6.57 can be written as

$$\frac{\alpha'(t)}{\alpha^2(t)} \geq \frac{\mathrm{Var}(\theta \mid t)}{E[\theta \mid t]} \qquad (6.59)$$

Thus, the first two conditional moments and the function $\alpha(t)$ are responsible for the IFR (DFR) properties of the mixture distribution. These characteristics can be obtained for each specific case.

The following result is intuitively evident, as it is known already that $E[\theta \mid t]$ is monotonically decreasing:

$$E[\theta \mid t] = \int_0^{\infty} \theta\pi(\theta \mid t)\,\mathrm{d}\theta \to 0 \quad t \to \infty \qquad (6.60)$$

It is clear from the definition of $\pi(\theta \mid t)$ and the fact that the weaker populations are dying out earlier [14], that $E[\theta \mid t]$ should converge to zero. The weaker population "should be distinct" from the stronger one as $t \to \infty$. This means that the distance between different realizations of $\lambda(t, \theta)$ should be bounded from below by some constant $d(\theta_1, \theta_2)$:

$$\lambda(t, \theta_2) - \lambda(t, \theta_1) \geq d(\theta_1, \theta_2) > 0$$
$$\theta_2 > \theta_1, \ \theta_1, \theta_2 \in [0, \infty)$$

uniformly in $t \in (c, \infty)$, where c is a sufficiently large constant. This is obviously the case for our model (Equation 6.55), as $\alpha(t)$ is an increasing function for large t. Therefore, as $t \to \infty$, the proportion of populations with larger failure rates $\lambda(t, \theta)$ (larger values of θ) is decreasing. More precisely: for each $\varepsilon > 0$, the proportion of populations with $\theta \geq l$ is asymptotically decreasing to zero no matter how small ε is. This means that $E[\theta \mid t]$ tends to zero as $t \to \infty$ [15]. Alternatively, asymptotic convergence (Equation 6.60) can be proved [13] more strictly based on the fact that

$$\pi(\theta \mid t) \to \delta(0) \quad \text{as } t \to \infty$$

where $\delta(t)$ denotes the Dirac delta function.

From Equation 6.55:

$$\lim_{t\to\infty} \lambda_{\mathrm{m}}(t) = \lim_{t\to\infty} \alpha(t) \int_a^b \theta\pi(\theta \mid t)\,\mathrm{d}\theta = \lim_{t\to\infty} \alpha(t) E[\theta \mid t] \qquad (6.61)$$

Specifically, for the mixture of exponentials when $\alpha(t) = \mathrm{const}$, it follows from Equations 6.60 and 6.61 that the mixture failure rate $\lambda_{\mathrm{m}}(t)$ asymptotically decreases to zero as $t \to \infty$. The same conclusion can be made for the case when $F(t, \theta)$ is a gamma cumulative distribution function. Indeed, it is well known that the failure rate of the gamma distribution asymptotically converges to a constant.

Asymptotic behavior of $\lambda_{\mathrm{m}}(t)$ in Equation 6.61 for a general case depends on the pattern of $\alpha(t)$ increasing ($E[\theta \mid t]$ decreasing). Hence, by using the L'Hospital rule ($\alpha(t) \to \infty$), the following result can be obtained [13]:

The mixture failure rate asymptotically tends to zero:

$$\lim_{t\to\infty} \lambda_{\mathrm{m}}(t) = 0 \qquad (6.62)$$

if and only if

$$\lim_{t\to\infty} -\frac{\alpha^2(t) E'[\theta \mid t]}{\alpha'(t)} = \lim_{t\to\infty} \frac{\alpha^3(t)\,\mathrm{Var}(\theta \mid t)}{\alpha'(t)} = 0 \qquad (6.63)$$

This convergence will be monotonic if, in addition to Equation 6.63:

$$\frac{\alpha'(t)}{\alpha^2(t)} < \frac{\mathrm{Var}(\theta \mid t)}{E[\theta \mid t]} \qquad (6.64)$$

Hence, conditional characteristics $E[\theta \mid t]$ and $\mathrm{Var}[\theta \mid t]$ are responsible for asymptotic and monotonic properties of the mixture failure rate. The failure rate of the mixture can decrease, if $E[\theta \mid t]$ decreases more sharply than $\alpha(t)$ increases (see Equations 6.58 and 6.61). As follows from Equation 6.63, the variance should also be sufficiently small for convergence to zero. Simplifying the situation, one can say that the mixture failure rate tends to increase but, at the same time, another process of "sliding down", due to the effect of dying out of the weakest populations, takes place. The combination of these effects can eventually result in convergence to zero, which is usually the case for the commonly used lifetime distribution functions.

Results, obtained for the multiplicative model are valid, in a simplified way, for the additive model. Indeed:

$$\lambda_{\mathrm{m}}(t) = \int_a^b [\alpha(t) + \theta]\pi(\theta \mid t)\,\mathrm{d}\theta$$
$$= \alpha(t) + \int_a^b \theta\pi(\theta \mid t)\,\mathrm{d}\theta = \alpha(t) + E[\theta \mid t]$$

Similar to Equation 6.60, it can be shown that $E[\theta \mid t] \to 0$ as $t \to \infty$. Therefore, $\lambda_{\mathrm{m}}(t)$ as $t \to \infty$ is asymptotically converging to the increasing function $\alpha(t)$:

$$\lambda_{\mathrm{m}}(t) - [\alpha(t) + a] \to 0$$

6.4.4 Some Examples

Three simple examples analyzing the shape of the mixture failure rate for the multiplicative model will be considered in this section [13]. The corresponding results will be obtained by using the "direct" definition (Equation 6.48).

Example 2. Let $F(t)$ be an exponential distribution with parameter λ, and $\pi(\theta)$ is the exponential probability density function with parameter ϑ. Thus, the corresponding failure rate is constant: $\lambda(t, \theta) = \theta\lambda$. Using Equation 6.48:

$$\lambda_{\mathrm{m}}(t) = \frac{\lambda}{\lambda t + \vartheta} \to \frac{1}{t}[1 + o(1)] \to 0 \quad \text{as } t \to \infty$$

It can be easily shown by formal calculations that the same asymptotic result is valid when $F(t)$ is a gamma distribution. This follows from general considerations, because the failure rate of a gamma distribution tends to a constant as $t \to \infty$. The conditional expectation in this case is defined by

$$E[\theta \mid t] = \frac{1}{\lambda t + \vartheta}$$

Example 3. Consider the specific type of the Weibull distribution with linearly IFR $\lambda(t, \theta) = 2\theta t$ and assume that $\pi(\theta)$ is a gamma probability density function:

$$\pi(\theta) = \frac{\vartheta^{\beta}\theta^{\beta-1}\exp(-\theta\vartheta)}{\Gamma(\beta)} \qquad \beta, \vartheta > 0,\ \theta \geq 0$$

Via direct integration, the mixture failure is defined by

$$\lambda_{\mathrm{m}}(t) = \frac{2\beta t}{\vartheta + t^2}$$

It is equal to zero at $t = 0$ and tends to zero as $t \to \infty$ with a single maximum at $t = \sqrt{\theta}$. Hence, the mixture of IFR distributions has a decreasing (tending to zero!) failure rate for sufficiently large t, and this is rather surprising. Furthermore, the same result asymptotically holds for arbitrary Weibull distributions with IFR.

It is surprising that the mixture failure rate is converging to zero in both examples in a similar pattern defined by t^{-1} because in the Example 2 we are mixing constant failure rates and in this example it is IFRs. The conditional expectation in this case is:

$$E[\theta \mid t] = \frac{2\beta}{\vartheta + t^2}$$

Example 4. Consider the truncated extreme value distribution defined in the following way:

$$F(t, k, \theta) = 1 - \exp\{-\theta k[\exp(t) - 1]\} \quad t \geq 0$$
$$\lambda(t, \theta) = \theta\alpha(t) = \theta k \exp(t)$$

where $k > 0$ is a constant. Assume that $\pi(\theta)$ is an exponential probability density function with parameter ϑ. Then:

$$\int_0^\infty f(t,k,\theta)\pi(\theta)\,\mathrm{d}\theta = \int_0^\infty \theta k \exp(t)\exp\{-\theta k[\exp(t)-1]\} \times \vartheta \exp(-\theta\vartheta)\,\mathrm{d}\theta = \frac{\vartheta k\,\mathrm{e}^t}{\omega^2}$$

$$\omega = k\exp(t) - k + \vartheta$$

$$\int_0^\infty \bar{F}(t,k,\theta)\pi(\theta)\,\mathrm{d}\theta = \vartheta \int_0^\infty \exp(-\theta)\,\mathrm{d}\theta = \frac{\vartheta}{\omega}$$

Eventually, using the definition in Equation 6.48:

$$\lambda_{\mathrm{m}}(t) = \frac{k\exp(t)}{\omega} = \frac{k\exp(t)}{k\exp(t)-k+\vartheta} = 1 + \frac{k-\vartheta}{k\exp(t)-k+\vartheta} \tag{6.65}$$

For analyzing the shape of $\lambda_{\mathrm{m}}(t)$ it is convenient to write Equation 6.65 in the following way:

$$\lambda_{\mathrm{m}}(t) = \frac{1}{1+(d/z)} \tag{6.66}$$

where $z = k\exp(t)$ and $d = -k + \vartheta$.

It follows from Equations 6.65 and 6.66 that $\lambda_{\mathrm{m}}(0) = k/\vartheta$. When $d < 0$ $(k > \vartheta)$, the mixture failure rate is monotonically decreasing, asymptotically converging to unity. For $d > 0$ $(k < \vartheta)$ it is monotonically increasing, asymptotically converging to unity. When $d = 0$ $(k = \vartheta)$, the mixture failure rate is equal to unity: $\lambda_{\mathrm{m}}(t) \equiv 1$, $\forall t[0,\infty)$. The conditional expectation is defined as

$$E[\theta \mid t] = \frac{1}{k\exp(t)-k+\vartheta} = \frac{1}{z+d}$$

This result is even more surprising. The initial failure rate is increasing extremely sharply, and still for some values of parameters the mixture failure rate is decreasing! On the other hand, for $k < \vartheta$ it is increasing but has no resemblance at all with the pattern of $\lambda(t,\theta)$ increasing. The most amazing fact is that this kind of mixing can result in a constant failure rate for $k = \vartheta$!

The mixing rule resulting in Equation 6.66 has a very interesting interpretation in survival analysis. Exponential failure rate (Gompertz law) is used to model human mortality, but recent demographic studies show that the long-term pattern of adult mortality does not follow the exponential shape of the corresponding failure rate. This means that after the age of 80–85 years a substantial deceleration in the observed failure rate is encountered. What is the reason for this deceleration? The model of Example 4 can explain this. Indeed, for various reasons, the population is not homogeneous. This means that parameter θ is random, and the corresponding mixing leads to a result similar to Equation 6.66. If the Weibull cumulative distribution function is chosen as a model for $F_{\mathrm{m}}(t,\theta)$ (and this model is also used in demographic studies), then similar to Example 3, the failure rate can eventually decrease to zero, but this effect can start at ages far beyond the current human life span.

6.4.5 Inverse Problem

The following important inverse problem [15] arises while modeling the mixture failure rate:

Given the mixture failure rate $\lambda_{\mathrm{m}}(t)$ and the mixing distribution $\pi(\theta)$, obtain the failure rate $\lambda(t)$ of the governing distribution $F(t)$.

It will be shown that this problem can be explicitly solved for additive and proportional failure rate models (Equations 6.52 and 6.55 respectively). In both of these cases the baseline failure rate $\alpha(t)$ should be obtained.

The conditional expectation $E[\theta \mid t]$ in Equation 6.53 for the additive model (Equation 6.52) can be simplified to:

$$E[\theta \mid t] = \frac{\int_0^\infty \theta\exp(-\theta t)\pi(\theta)\,\mathrm{d}\theta}{\int_0^\infty \exp(-\theta t)\pi(\theta)\,\mathrm{d}\theta} \tag{6.67}$$

It is important that $E[\theta \mid t]$ does not depend on $\alpha(t)$. We want the mixture failure rate $\lambda_{\mathrm{m}}(t)$ to be equal to some arbitrary continuous function $g(t)$, such that $\int_0^\infty g(t)\,\mathrm{d}t = \infty$, $g(t) > 0$.

Thus, the corresponding equation for solving the inverse problem is $\lambda_m(t) = g(t)$. It follows from Equation 6.53 that

$$\alpha(t) = g(t) - E[\theta \mid t] \tag{6.68}$$

and this is, actually, the trivial solution of the inverse problem for the additive model. It is clear that for the given $\pi(\theta)$ the additional condition should be imposed on $g(t)$:

$$g(t) \geq E[\theta \mid t] \quad \forall t \geq 0$$

It is worth noting that $\alpha(t) \to g(t)$ as $t \to \infty$.

If, for instance, $\pi(\theta)$ is an exponential probability density function with parameter ϑ, then it follows from Equation 6.67 that $E[\theta \mid t] = (t + \vartheta)^{-1}$, and

$$\alpha(t) = g(t) - \frac{1}{t + \vartheta}$$

Assume that $g(t)$ is a constant: $g(t) = c$. This means that mixing will result in a constant failure rate!

Thus, an arbitrary shape of $\lambda_m(t)$ can be constructed. For instance, let us consider the modeling of the bathtub shape [7], which is popular in many applications (decreasing in $[0, t_0)$ and increasing in $[t_0, \infty)$ for some $t_0 > 0$). This operation can be performed via mixing IFR distributions. It follows from the result of Section 6.4.2 that, if $\alpha(t)$ is an increasing convex function in $[0, \infty)$ $\alpha'(0) < \vartheta^{-2}$, then $g(t)$ have a bathtub shape.

The analysis is slightly more complicated for the multiplicative model (Equation 6.65) [15]. As for the additive model, the equation for obtaining $\alpha(t)$ is $\lambda_m(t) = g(t)$. It follows from Equations 6.55 and 6.56 that

$$\beta'(t)\frac{\int_0^\infty \theta \exp[-\theta\beta(t)]\pi(\theta)\,d\theta}{\int_0^\infty \exp[-\theta\beta(t)]\pi(\theta)\,d\theta} = g(t) \tag{6.69}$$

where $\beta(t) = \int_0^t \alpha(u)\,du$. Thus, unlike the first model, the conditional expectation $E[\theta \mid t]$ depends now on $\alpha(t)$.

Denote by $\pi^*(t)$ the Laplace transform of $\pi(\theta)$ for $t \geq 0$:

$$\int_0^\infty \exp(-\theta t)\pi(\theta)\,d\theta \equiv \pi^*(t)$$

Taking into account that $\beta(t)$ is monotonically increasing:

$$\int_0^\infty \exp[-\theta\beta(t)]\pi(\theta)\,d\theta \equiv \pi^*[\beta(t)] \equiv S(t) \tag{6.70}$$

Differentiating both sides of Equation 6.70:

$$S'(t) = -\beta'(t)\int_0^\infty \theta \exp[-\theta\beta(t)]\pi(\theta)\,d\theta \tag{6.71}$$

Combining Equation 6.69 with Equations 6.70 and 6.71, we can obtain now a simple expected result [15], which also follows from the definition of the mixture failure rate. It is very convenient that, in this specific case, the solution of the inverse problem can be written via the Laplace transform as

$$\pi^*[\beta(t)] = S(t) = \exp\left[-\int_0^t g(u)\,du\right] \tag{6.72}$$

$$\Rightarrow \beta(t) = (\pi^*)^{-1}\left\{\exp\left[-\int_0^t g(u)\,du\right]\right\} \tag{6.73}$$

Thus, obtaining $\pi^*(t)$, substituting $\beta(t)$ instead of t, and solving Equation 6.72 with respect to $\beta(t) = \int_0^t \alpha(u)\,du$ leads to the solution of the inverse problem for this case: obtaining the governing distribution, which, due to the model in Equation 6.55, is defined by $\alpha(t)$. Equation 6.72 always has a unique solution, as $\beta(t)$ is a monotonically increasing function and $\pi^*(t)$ $(\pi^*(\beta(t)))$ is a survival function (monotonically decreasing from unity at $t = 0$ $(\beta(t) = 0)$ to zero as $t \to \infty$ $(\beta(t) \to \infty)$). This fact is actually stated by the inverse transform (Equation 6.73). The solution can be easily obtained explicitly for the mixing distributions with a "nice" Laplace transform $\pi^*(t)$.

Example 5. Let $\pi(t) = \vartheta \exp(-\vartheta\theta)$. Then

$$\pi^*(t) = \frac{\vartheta}{t + \vartheta}$$

$$\pi^*[\beta(t)] = \frac{\vartheta}{\beta(t) + \vartheta}$$

and

$$\pi^*[\beta(t)] = \exp\left[-\int_0^t g(u)\,\mathrm{d}u\right]$$
$$\Rightarrow \beta(t) = \vartheta \exp\left[\int_0^t g(u)\,\mathrm{d}u\right] - \vartheta$$
$$\alpha(t) = \vartheta g(t) \exp\left[\int_0^t g(u)\,\mathrm{d}u\right]$$

For the gamma mixing distribution:

$$\pi(\theta) = \exp(-\vartheta\theta)\frac{\theta^{n-1}\vartheta^n}{\Gamma(n)} \qquad n > 0$$

$$\pi^*(t) = \int_0^\infty \exp(-\theta t)\frac{\exp(-\vartheta\theta)\theta^{n-1}}{\Gamma(n)}\vartheta^n\,\mathrm{d}\theta = \frac{\vartheta^n}{(t+\vartheta)^n}$$

Thus:

$$\pi^*[\beta(t)] = \frac{\vartheta^n}{[\beta(t)+\vartheta]^n}$$
$$\beta(t) = \vartheta \exp\left[\frac{1}{n}\int_0^t g(u)\,\mathrm{d}u\right] - \vartheta$$

Finally:

$$\alpha(t) = \frac{\vartheta}{n} g(t) \exp\left[\frac{1}{n}\int_0^t g(u)\,\mathrm{d}u\right] \qquad (6.74)$$

Example 6. Assume for the setting of the previous example the concrete shape of the mixture failure rate: $\lambda_{\mathrm{m}}(t) = t$ (Weibull $F_{\mathrm{m}}(t)$ with linear failure rate). It follows from Equation 6.74 that

$$\alpha(t) = \frac{\vartheta t}{n}\exp\left(\frac{t^2}{2n}\right)$$

And for obtaining the baseline $\alpha(t)$ in this case, the mixture failure rate t should be just multiplied by $(\vartheta/n)\exp(t^2/2n)$.

The corresponding equation for the general model of mixing Equations 6.47–6.50 can be written as

$$\int_0^\infty \exp\left[-\int_0^t \lambda(u,\theta)\,\mathrm{d}u\right]\pi(\theta)\,\mathrm{d}\theta = \exp\left[-\int_0^t g(u)\,\mathrm{d}u\right] \qquad (6.75)$$

where the mixture failure rate $\lambda(t,\theta)$ should be defined by the concrete model of mixing. Specifically, this model can be additive, multiplicative, or the mixing rule, which is specified by the accelerated life model [16]:

$$\lambda(t,\theta) = \theta\alpha(\theta t)$$

where, as in the multiplicative model, $\alpha(t)$ denotes a baseline failure rate. Equation 6.75 will turn, in this case, to:

$$\int_0^\infty \exp\left[-\int_0^{t\theta} \alpha(u)\,\mathrm{d}u\right]\pi(\theta)\,\mathrm{d}\theta = \exp\left[-\int_0^t g(u)\,\mathrm{d}u\right] \qquad (6.76)$$

In spite of a similarity with Equation 6.72, Equation 6.76 cannot be solved for this case in an explicit form.

References

[1] Thompson Jr WA. Point process models with applications to safety and reliability. London: Chapman and Hall; 1988.

[2] Cox DR, Isham V. Point processes. London: Chapman and Hall; 1980.

[3] Block HW, Savits TH, Borges W. Age dependent minimal repair. J Appl Prob 1985;22:370–86.

[4] Finkelstein MS. A point process stochastic model of safety at sea. Reliab Eng Syst Saf 1998;60; 227–33.

[5] Yashin AI, Manton KG. Effects of unobserved and partially observed covariate processes on system failure: a review of models and estimation strategies. Stat Sci 1997;12:20–34.

[6] Kebir Y. On hazard rate processes. Nav Res Logist 1991;38:865–76.

[7] Lynn NJ, Singpurwalla ND. Comment: "Burn-in" makes us feel good. Statist Sci 1997;12:13–9.

[8] Barlow RE, Proschan F. Statistical theory of reliability: probability models. New York: Holt, Rinehart & Winston; 1975.

[9] Daley DJ, Vere-Jones D. An introduction to the theory of point processes. New York: Springer-Verlag; 1988.

[10] Lee WK, Higgins JJ, Tillman FA. Stochastic models for mission effectiveness. IEEE Trans Reliab 1990;39: 321–4.

[11] Finkelstein MS. Multiple availability on stochastic demand. IEEE Trans. Reliab 1999;39:19–24.

[12] Ushakov IA, Harrison RA. Handbook of reliability engineering. New York: John Wiley & Sons; 1994.

[13] Finkelstein MS, Esaulova V. Why the mixture failure rate decreases. Reliab Eng Syst Saf 2001;71:173–7.

[14] Block HW, Mi J, Savits TH. Burn-in and mixed populations. J Appl Prob 1993; 30:692–02.

[15] Finkelstein MS, Esaulova V. On inverse problem in mixture failure rate modeling. Appl Stochast Model Bus Ind 2001;17;N2.

[16] Block HW, Joe H. Tail behavior of the failure rate functions of mixtures. Lifetime Data Anal 1999;3: 269–88.

Concepts of Stochastic Dependence in Reliability Analysis

C. D. Lai and M. Xie

Chapter 7

7.1 Introduction

The concept of dependence permeates the world. There are many examples of interdependence in the medicines, economic structures, and reliability engineering, to name just a few. A typical example in engineering is that all output from a piece of equipment will depend on the input in a broader sense, which includes material, equipment, environment, and others. Moreover, the dependence is not deterministic but of a stochastic nature. In this chapter, we limit the scope of our discussion to the dependence notions that are relevant to reliability analysis.

In the reliability literature, it is usually assumed that the component lifetimes are independent. However, components in the same system are used in the same environment or share the same load, and hence the failure of one component affects the others. We also have the case of so-called common cause failure and components might fail at the same time. The dependence is usually difficult to describe, even for very similar components. From light bulbs in an overhead projector to engines in an aeroplane, we have dependence, and it is essential to study the effect of dependence for better reliability design and analysis. There are many notions of bivariate and multivariate dependence. Several of these concepts were motivated from applications in reliability.

We may have seen abbreviations like PQD, SI, LTD, RTI, *etc.* in recent years. As one probably would expect, they refer to the form of positive dependence between two or more variables. We shall try to explain them and their interrelationships in this chapter. Positive dependence means that large values of Y tend to accompany large values of X, and *vice versa* for small values. Discussion of the concepts of dependence involves refining this basic idea, by means of definitions and deductions.

In this chapter, we focus our attention on a relatively weaker notion of dependence, namely the positive quadrant dependence between two variables X and Y. We think that this easily verified form of positive dependence is more relevant in the subject area under discussion. Also, as might be expected, the notions of dependence are simpler and their relationships are more readily exposed in the bivariate case than the multivariate ones.

Hutchinson and Lai [1] devoted a chapter to reviewing concepts of dependence for a bivariate distribution. Recently, Joe [2] gave a comprehensive treatment of the subject on multivariate dependence. Thus, our goal here is not to provide another review; instead, we shall focus our attention on the positive dependence, in particular the positive quadrant dependence. For simplicity, we confine ourselves mainly to the bivariate case, although most of our discussion can be generalized to the multivariate situations. An important aspect of this chapter is the availability of several examples that are employed to illustrate the concepts of positive dependence.

The present chapter endeavors to develop a fundamental dependence idea with the following structure:

- positive dependence, a general concept;
- important conditions describing positive dependence (we state some definitions, and examine their relative stringency and their interrelationships);
- positive quadrant dependence—conditions and applications;
- examples of positive quadrant dependence;
- positive dependence orderings.

Some terms that we will come across herein are listed below:

PQD (NQD)	Positive quadrant dependent (negative quadrant dependent)
SI (alias PRD)	Stochastically increasing (positively regression dependent)
LTD (RTI)	Left-tail decreasing (right-tail increasing)
TP_2	Totally positive of order 2
PQDE	Positive quadrant dependent in expectation
RCSI	Right corner set increasing
$\bar{F} = 1 - F$	$\bar{F}$ survival function, F cumulative distribution function.

7.2 Important Conditions Describing Positive Dependence

Concepts of stochastic dependence for a bivariate distribution play an important part in statistics. For each concept, it is often convenient to refer to the bivariate distributions that satisfy it as a family, a group. In this chapter, we are mainly concerned with positive dependence. Although negative dependence concepts do exist, they are often obtained by negative analogues of positive dependence via reversing the inequity signs.

Various notions of dependence are motivated from applications in statistical reliability (*e.g.* see Barlow and Proschan [3]). The baseline, or starting point, of the reliability analysis of systems is independence of the lifetimes of the components. As noted by many authors, it is often more realistic to assume some form of positive dependence among components.

Around about 1970, several studies discussed different notions of positive dependence between two random variables, and derived some interrelationships among them, *e.g.* Lehmann [4], Esary *et al.* [5], Esary and Proschan [6], Harris [7], and Brindley and Thompson [8], among others. Yanagimoto [9] unified some of these notions by introducing a family of concepts of positive dependence. Some further notions of positive dependence were introduced by Shaked [10–12].

For concepts of multivariate dependence, see Block and Ting [13] and a more recent text by Joe [2].

7.2.1 Six Basic Conditions

Jogdeo [14] lists four of the following basic conditions describing positive dependence; these are in increasing order of stringency (see also Jogdeo [15]).

1. Positive correlation, $\mathrm{cov}(X, Y) \geq 0$.
2. For every pair of increasing functions a and b, defined on the real line R, $\mathrm{cov}[a(X), b(Y)] \geq 0$. Lehmann [4] showed that this condition is equivalent to

$$\Pr(X \geq x, Y \geq y) \geq \Pr(X \geq x)\Pr(Y \geq y) \tag{7.1}$$

   Or equivalently, $\Pr(X \leq x, Y \leq y) \geq \Pr(X \leq x)\Pr(Y \leq y)$.
   We say that (X, Y) shows positive quadrant dependence if and only if these inequalities hold. Several families of PQD distributions will be introduced in Section 7.4.
3. Esary *et al.* [5] introduced the following condition, termed "association": for every pair of functions a and b, defined on R^2, that are increasing in each of the arguments (separately)

$$\mathrm{cov}[a(X, Y), b(X, Y)] \geq 0 \tag{7.2}$$

   We note, in passing, that a direct verification of this dependence concept is difficult in general. However, it is often easier to verify one of the alternative positive dependence notions that imply association.
4. Y is right-tail increasing in X if $\Pr(Y > y \mid X > x)$ is increasing in x for all y (written as $\mathrm{RTI}(Y \mid X)$). Similarly, Y is left-tail decreasing if $\Pr(Y \leq y \mid X \leq x)$ is decreasing in x for all y (written as $\mathrm{LTD}(Y \mid X)$).
5. Y is said to be stochastically increasing in x for all y (written as $\mathrm{SI}(Y \mid X)$) if, for every y, $\Pr(Y > y \mid X = x)$ is increasing in x. Similarly, we say that X is stochastically increasing in y for all x (written as $\mathrm{SI}(X \mid Y)$) if for every x, $\Pr(X > x \mid Y = y)$ is increasing in y.
   Note that $\mathrm{SI}(Y \mid X)$ is often simply denoted by SI. Some authors refer to this relationship as Y being positively regression dependent on X (abbreviated to PRD) and similarly X being positively regression dependent on Y.
6. Let X and Y have a joint probability density function $f(x, y)$. Then f is said to be totally positive of order two (TP_2) if for all $x_1 < x_2$, $y_1 < y_2$:

$$f(x_1, y_1)f(x_2, y_2) \geq f(x_1, y_2)f(x_2, y_1) \tag{7.3}$$

   Note that if f is TP_2, then F and $\bar{F}$ (survival function) are also TP_2, *i.e.* $F(x_1, y_1)F(x_2, y_2) \geq F(x_1, y_2)F(x_2, y_1)$ and $\bar{F}(x_1, y_1)\bar{F}(x_2, y_2) \geq \bar{F}(x_1, y_2)\bar{F}(x_2, y_1)$. It is easy to see that either F TP_2 or $\bar{F}$ TP_2 implies that F is PQD.

7.2.2 The Relative Stringency of the Conditions

It is well known that these concepts are interrelated, *e.g.* see Joe [2]. The six conditions we listed above can be arranged in an increasing order of stringency, *i.e.* (6) $\Rightarrow$ (5) $\Rightarrow$ (4) $\Rightarrow$ (3) $\Rightarrow$ (2) $\Rightarrow$ (1). More precisely:

$$\mathrm{TP}_2 \Rightarrow \mathrm{SI} \Rightarrow \mathrm{RTI} \Rightarrow \text{Association} \Rightarrow \mathrm{PQD} \Rightarrow \mathrm{cov}(X, Y) \geq 0$$

$$\mathrm{TP}_2 \Rightarrow \mathrm{SI} \Rightarrow \mathrm{LTD} \Rightarrow \text{Association} \Rightarrow \mathrm{PQD} \Rightarrow \mathrm{cov}(X, Y) \geq 0$$

Some of the proofs for the chain of implications are not straightforward, whereas some others are obvious (*e.g.* see Barlow and Proschan [3], p.143–4). For example, it is easy to see that PQD implies positive correlation by applying Hoeffding's lemma, which states:

$$\operatorname{cov}(X, Y) = \int_{-\infty}^{\infty}\int_{-\infty}^{\infty} [F(x, y) - F_X(x)F_Y(y)]\,dx\,dy \quad (7.4)$$

This identity is often useful in many areas of statistics.

7.2.3 Positive Quadrant Dependent in Expectation

We now introduce a slightly less stringent dependence notion that would include the PQD distributions. For any real number x, let Y_x be the random variable with distribution function $\Pr(Y \leq y \mid X > x)$. It is easy to verify that the inequality in the conditional distribution $\Pr(Y \leq y \mid X > x) \leq \Pr(Y \leq y)$ implies an inequality in expectation $E(Y_x) \geq E(Y)$ if Y is a non-negative random variable.

We say that Y is PQDE on X if the last inequality involving expectation holds. Similarly, we say that there is negative quadrant dependent in expectation if $E(Y_x) \leq E(Y)$.

It is easy to show that PQD $\Rightarrow$ PQDE by observing PQD is equivalent to $\Pr(Y > y \mid X > x) \geq \Pr(Y > y)$, which in turn implies $E(Y_x) \geq E(Y)$ (assuming $Y \geq 0$).

Next, we have

$$\begin{aligned}\operatorname{cov}(X, Y) &= \iint [\bar{F}(x, y) - \bar{F}_X(x)\bar{F}_Y(y)]\,dx\,dy \\ &= \int \bar{F}_X(x)\left\{\int [\Pr(Y > y \mid X > x) - \bar{F}_Y(y)]\,dy\right\}dx \\ &= \int \bar{F}_X(x)[E(Y_x) - E(Y)]\,dx\end{aligned}$$

which is greater than zero if X and Y are PQDE. Thus, PQDE implies that $\operatorname{cov}(X, Y) \geq 0$. In other words, PQDE lies between PQD and positive correlation. There are many bivariate random variables being PQDE, because all the PQD distributions with $Y \geq 0$ are also PQDE.

7.2.4 Associated Random Variables

Recall in Section 7.2.1, we say that two random variables X and Y are associated if $\operatorname{cov}[a(X, Y), b(X, Y)] \geq 0$. Obviously, this expression can be represented alternatively by

$$E[a(X, Y)b(X, Y)] \geq E[a(X, Y)]E[b(X, Y)] \quad (7.5)$$

where the inequality holds for all real functions a and b that are increasing in each component and are such that the expectations in Equation 7.5 exist.

It appears from Equation 7.5 that it is almost impossible to check this condition for association directly. What one normally does is to verify one of the dependence conditions in the higher hierarchy that implies association.

Barlow and Proschan [3], p.29, considered some practical reliability situations for which the components lifetimes are not independent, but rather are associated:

1. minimal path structures of a coherent system having components in common;
2. components subject to the same set of stresses;
3. structures in which components share the same load, so that the failure of one component results in increased load on each of the remaining components.

We note that in each case the random variables of interest tend to act similarly. In fact, all the positive dependence concepts share this characteristic.

An important application of the concept of association is to provide probability bounds for system reliability. Many such bounds are presented in Barlow and Proschan [3], in section 3 of chapter 2 and section 7 of chapter 4.

7.2.5 Positively Correlated Distributions

Positive correlation is the weakest notion of dependence between two random variables X and Y. We note that it is easy to construct a positively correlated bivariate distribution. For example, such a distribution may be obtained by simply applying a well-known trivariate reduction technique described as follows.

Set $X = X_1 + X_3$, $Y = X_2 + X_3$, with X_i $(i = 1, 2, 3)$ being mutually independent, then the correlation coefficient of X and Y is

$$\rho = \text{var}\, X_3/[\text{var}(X_1 + X_3)\, \text{var}(X_2 + X_3)]^{1/2} > 0$$

For example, let $X_i \sim \text{Poisson}(\lambda_i)$, $i = 1, 2, 3$. Then $X \sim \text{Poisson}(\lambda_1 + \lambda_3)$, $Y \sim \text{Poisson}(\lambda_2 + \lambda_3)$, with $\rho = \lambda_3/[(\lambda_1 + \lambda_3)(\lambda_2 + \lambda_3)]^{1/2} > 0$. X and Y constructed in this manner are also PQD; see example 1(ii) in Lehmann [4], p.1139.

7.2.6 Summary of Interrelationships

Among the positive dependence concepts we have introduced so far, TP_2 is the strongest. A slightly weaker notion introduced by Harris [7] is called the right corner set increasing (RCSI), meaning $\Pr(X > x_1, Y > y_1 \mid X > x_2, Y > y_2)$ is increasing in x_2 and y_2 for all x_1 and y_1. Shaked [10] showed that $TP_2 \Rightarrow$ RCSI. By choosing $x_1 = -\infty$ and $y_2 = -\infty$, we see that RCSI $\Rightarrow$ RTI.

We may summarize the chain of relations in the following (in which Y is conditional on X whenever there is a conditioning):

$$\begin{array}{ccccccccc} \text{RCSI} & \Rightarrow & \text{RTI} & \Rightarrow & \text{ASSOCIATION} & \Rightarrow & \text{PQD} & \Rightarrow & \text{cov} \geq 0 \\ \Uparrow & & \Uparrow & & \Uparrow & & & & \\ TP_2 & \Rightarrow & \text{SI} & \Rightarrow & \text{LTD} & & & & \end{array}$$

There are other chains of relationships between various concepts of dependence.

7.3 Positive Quadrant Dependent Concept

There are many notions of bivariate dependence known in the literature. The notion of positive quadrant dependence appears to be more straightforward and easier to verify than other notions. The rest of the chapter mainly focuses on this dependence concept. The definition of PQD, which was first given by Lehmann [4], is now reproduced below.

Definition 1. Random variables X and Y are PQD if the following inequality holds:

$$\Pr(X > x, Y > y) \geq \Pr(X > x)\Pr(Y > y) \quad (7.6)$$

for all x and y. Equation 7.6 is equivalent to $\Pr(X \leq x, Y \leq y) \geq \Pr(X \leq x)\Pr(Y \leq y)$.

The reason why Equation 7.6 constitutes a positive dependence concept is that X and Y here are more likely to be large or small together compared with the independent case.

If the inequality in Equation 7.6 is reversed, then X and Y are negatively quadrant dependent.

PQD is shown to be a stronger notion of dependence than the positive (Pearson) correlation but weaker than the "association", which is a key concept of positive dependence in Barlow and Proschan [3], originally introduced by Esary *et al.* [5].

Consider a system of two components that are arranged in series. By assuming that the two components are independent, when they are in fact PQD, we will underestimate the system reliability. For systems in parallel, on the other hand, assuming independence when components are in fact PQD, will lead to overestimation of system reliability. This is because the other component will fail earlier knowing that the first has failed. This, from a practical point of view, reduces the effectiveness of adding parallel redundancy. Thus a proper knowledge of the extent of dependence among the components in a system will enable us to obtain a more accurate estimate of the reliability characteristic in question.

7.3.1 Constructions of Positive Quadrant Dependent Bivariate Distributions

Let $F(x, y)$ denote the distribution function of (X, Y) having continuous marginal cumulative distribution functions $F_X(x)$ and $F_Y(y)$ with marginal probability distribution functions $f_X = F'_X$ and $f_Y = F'_Y$ respectively. For a PQD bivariate distribution, the joint distribution function may be written as

$$F(x, y) = F_X(x)F_Y(y) + w(x, y)$$

satisfying the following conditions:

$$w(x, y) \geq 0 \tag{7.7}$$

$$w(x, \infty) \to 0, \quad w(\infty, y) \to 0,$$
$$w(x, -\infty) = 0, \quad w(-\infty, y) = 0 \tag{7.8}$$

$$\frac{\partial^2 w(x, y)}{\partial x \partial y} + f_X(x) f_Y(y) \geq 0 \tag{7.9}$$

Note that if both $X \geq 0$ and $Y \geq 0$, then the condition in Equation 7.8 may be replaced by $w(x, \infty) \to 0$, $w(\infty, y) \to 0$, $w(x, 0) = 0$, $w(0, y) = 0$.

Lai and Xie [16] used these conditions to construct a family of PQD distributions with uniform marginals.

7.3.2 Applications of Positive Quadrant Dependence Concept to Reliability

The notion of association is used to establish probability bounds on reliability systems, *e.g.* see chapter 3 of Barlow and Proschan [3]. Given a coherent system of n components with minimal path sets P_i $(i = 1, 2, \ldots, p)$ and minimal cut sets K_j $(j = 1, 2, \ldots, k)$. Let T_i denote the lifetime of the ith component and thus $p_i = P(T_i > t)$ is its survival probability at time t. It has been shown, *e.g.* see [3], p.35–8, that if components are independent, the

$$\prod_{j=1}^{k} \coprod_{i \in K_j} p_i \leq \text{System Reliability} \leq \coprod_{j=1}^{p} \prod_{i \in P_j} p_i \tag{7.10}$$

If, in addition, components are associated, then we have an alternative set of bounds

$$\max_{1 \leq r \leq p} \prod_{i \in P_j} p_i \leq \text{System Reliability} \leq \min_{1 \leq s \leq k} \coprod_{i \in K_j} p_i \tag{7.11}$$

As independence implies association, the bounds given in Equation 7.11 are also applicable for a system with independent components, although one cannot conclude that Equation 7.11 is tighter than Equation 7.10 or *vice versa*. A closer examination on the derivations leading to Equation 7.11 would readily reveal that the reliability bounds presented here remain valid if we assume only the positive quadrant dependence of the components.

One can find details related to bounds on reliability of a coherent system with associated components in the text by Barlow and Proschan [3].

To our disappointment, we have not been able to find more examples of real applications of positive dependence in reliability contexts.

We note in passing that the concept of positive quadrant dependence is widely used in statistics, for example:

- partial sums [17];
- order statistics [5];
- analysis of variance [18];
- contingency tables [19].

7.3.3 Effect of Positive Dependence on the Mean Lifetime of a Parallel System

Parallel redundancy is a common device to increase system reliability by enhancing the mean time to failure. Xie and Lai [20] studied the effectiveness of adding parallel redundancy to a single component in a complex system consisting of several independent components. It is shown that adding parallel redundancy to components with increasing failure rate (IFR) is less effective than that for a component with decreasing failure rate (DFR). Motivated to enhance reliability, Mi [21] considered the question of which component should be "bolstered" or "improved" in order to stochastically maximize the lifetime of a parallel system, series system, or, in general, a k-out-of n system.

Xie and Lai [20] assumed that the two parallel components were independent. This assumption is rarely valid in practice. Indeed, in reliability analysis, the component lifetimes are usually dependent. Two components in a reliability structure may share the same load or may be subject to the same set of stresses. This will cause the two lifetime random variables to be related to each other, or to be positively dependent.

When the components are dependent, the problem of effectiveness via adding a component may be different from the case of independent components. In particular, we are interested in investigating how the degree of the correlation affects the increase in the mean lifetime. A general approach may not be possible, and at present we can only consider typical cases when the two components are either PQD or NQD.

Kotz *et al.* [22] studied the increase in the mean lifetime by means of parallel redundancy for several cases when the two components are either PQD or NQD. The findings strongly suggest that the effect of dependence among the components should be taken into consideration when designing reliability systems. The above observation is valid whenever the two components are PQD and the result is stated in the following.

Let X and Y be identically distributed, not necessarily independent, PQD non-negative random variables. Kotz *et al.* [22] showed that

$$E(T) < E(X) + E(Y) - \int_0^\infty \bar{F}^2(t)\,\mathrm{d}t \tag{7.12}$$

where $F(t)$ is the common distribution function and $\bar{F}(t) = 1 - F(t)$. Hence, for increasing failure rate average components that are PQD, we have

$$\begin{aligned} E(T) &< E(X) + E(Y) - \int_0^\infty \bar{F}^2(t)\,\mathrm{d}t \\ &= 2\mu - \frac{\mu}{2} + \frac{\mu\,\mathrm{e}^{-2}}{2} \\ &= \frac{\mu}{2}(3 + \mathrm{e}^{-2}) \end{aligned} \tag{7.13}$$

Similarly, if a pair of non-negative identically distributed random variables X and Y is NQD, then the inequality in Equation 7.6 is reversed, *i.e.*

$$E(T) > E(X) + E(Y) - \int_0^\infty \bar{F}^2(t)\,\mathrm{d}t \tag{7.14}$$

If, in addition, both X and Y are DFR components, then

$$\begin{aligned} E(T) &> E(X) + E(Y) - \int_0^\infty \bar{F}^2(t)\,\mathrm{d}t \\ &\geq \mu\left(\frac{3}{2} - \frac{\mathrm{e}^{-2}}{2}\right) \end{aligned}$$

7.3.4 Inequality Without Any Aging Assumption

If the aging property of the marginal distributions is unknown but the median m of the common marginal is given, then Equation 7.14 reduces to

$$E(T) < E(X) + E(Y) - \int_0^m \bar{F}(t)\,\mathrm{d}t + \frac{1}{2}m$$

7.4 Families of Bivariate Distributions that are Positive Quadrant Dependent

Since the PQD concept is important in reliability applications, it is imperative for a reliability practitioner to know what kinds of PQD bivariate distribution are available for reliability modeling. In this section, we list several well-known PQD distributions, some of which were originally derived from a reliability perspective. Most of these PQD bivariate distributions can be found, for example, in Hutchinson and Lai [1].

7.4.1 Positive Quadrant Dependent Bivariate Distributions with Simple Structures

The distributions whose PQD property can be established easily are now given below.

Example 1. The Farlie–Gumbel–Morgenstern bivariate distribution [23]:

$$F(x, y) = F_X(x)F_Y(y) \times \{1 + \alpha[1 - F_X(x)][1 - F_Y(y)]\} \tag{7.15}$$

For convenience, the above family may simply be denoted by FGM. This general system of bivariate distributions is widely studied in the literature. It is easy to verify that X and Y are PQD if $\alpha > 0$.

Consider a special case of the FGM system where both marginals are exponential. The joint distribution function is then of the form (*e.g.* see Johnson and Kotz [24], p.262–3):

$$F(x, y) = (1 - e^{-\lambda_1 x})(1 - e^{-\lambda_2 y})(1 + \alpha\, e^{-\lambda_1 x - \lambda_2 y})$$

Clearly

$$\begin{aligned} w(x, y) &= F(x, y) - F_X(x)F_Y(y) \\ &= \alpha\, e^{-\lambda_1 x - \lambda_2 y}(1 - e^{-\lambda_1 x})(1 - e^{-\lambda_2 x}) \\ &\qquad 0 < \alpha \leq 1 \end{aligned}$$

satisfies the conditions in Equations 7.7–7.9, and hence X and Y are PQD.

Mukerjee and Sasmal [25] have worked out the properties of a system of two exponential components having the FGM distribution. The properties are such things as the densities, means, moment-generating functions, and tail probabilities of $\min(X, Y)$, $\max(X, Y)$, and $X + Y$, these being of relevance to series, parallel, and standby systems respectively.

Lingappaiah [26] was also concerned with properties of the FGM distribution relevant to the reliability context, but with gamma marginals.

Building upon a paper by Philips [27], Kotz and Johnson [28] considered a model in which components 1 and 2 were subject to "revealed" and "unrevealed" faults respectively, with (X, Y) having an FGM distribution, where X is the time between unrevealed faults and Y is the time from an unrevealed fault to a revealed fault.

Example 2. The bivariate exponential distribution

$$F(x, y) = 1 - e^{-x} - e^{-y} + (e^x + e^y - 1)^{-1}$$

This distribution is not well known, and we could not confirm its source. However, both marginals are exponential, which is used widely in a reliability context. This bivariate distribution function can be rewritten as

$$\begin{aligned} F(x, y) &= 1 - e^{-x} - e^{-y} + e^{-(x+y)} \\ &\quad + (e^x + e^y - 1)^{-1} - e^{-(x+y)} \\ &= F_X(x)F_Y(y) \\ &\quad + (e^x + e^y - 1)^{-1} - e^{-(x+y)} \end{aligned}$$

Now

$$\begin{aligned} &(e^x + e^y - 1)^{-1} - e^{-(x+y)} \\ &= \frac{(e^x - 1)(e^y - 1)}{(e^x + e^y - 1)\, e^{(x+y)}} \\ &= \frac{(1 - e^{-x})(1 - e^{-y})}{(e^x + e^y - 1)} \\ &\geq 0 \end{aligned}$$

and therefore F is PQD.

Example 3. The bivariate Pareto distribution

$$\bar{F}(x, y) = (1 + ax + by)^{-\lambda} \quad a, b, \lambda > 0$$

(*e.g.* see Mardia, [29], p.91). Consider a system of two independent exponential components that share a common environment factor η that can be described by a gamma distribution. Lindley and Singpurwalla [30] showed that the resulting joint distribution has a bivariate Pareto distribution. It is very easy to verify that this joint distribution is PQD. For a generalization to multivariate components, see Nayak [31].

Example 4. The Durling–Pareto distribution

$$\bar{F}(x, y) = (1 + x + y + kxy)^{-a} \quad a > 0,\ 0 \leq k \leq a + 1 \tag{7.16}$$

Obviously, it is a generalization of Example 3 above.

Consider a system of two dependent exponential components having a bivariate Gumbel distribution

$$F(x, y) = 1 - e^{-x} - e^{-y} + e^{-x-y-\theta xy}$$
$$x, y \geq 0, \ 0 \leq \theta \leq 1$$

and sharing a common environment that has a gamma distribution. Sankaran and Nair [32] have shown that the resulting bivariate distribution is specified by Equation 7.16. It follows from Equation 7.16 that

$$\begin{aligned}\bar{F}(x, y) &- \bar{F}_X(x)\bar{F}_Y(y)\\ &= \frac{1}{(1+x+y+kxy)^a} - \frac{1}{[(1+x)(1+y)]^a}\\ &\quad 0 \leq k \leq (a+1)\\ &= \frac{1}{(1+x+y+kxy)^a} - \frac{1}{(1+x+y+xy)^a}\\ &\geq 0 \quad 0 \leq k \leq 1\end{aligned}$$

Hence, F is PQD if $0 \leq k \leq 1$.

7.4.2 Positive Quadrant Dependent Bivariate Distributions with More Complicated Structures

Example 5. Marshall and Olkin's bivariate exponential distribution [33]

$$\begin{aligned}P(X > x, Y > y)&\\ &= \exp[-\lambda_1 x - \lambda_2 y - \lambda_{12}\max(x, y)] \quad \lambda \geq 0\end{aligned} \tag{7.17}$$

This has become a widely used bivariate exponential distribution over the last three decades. The Marshall and Olkin bivariate exponential distribution was derived from a reliability context.

Suppose we have a two-component system subjected to shocks that are always fatal. These shocks are assumed to be governed by three independent Poisson processes with parameters λ_1, λ_2, and λ_{12}, according to whether the shock applies to component 1 only, component 2 only, or to both components. Then the joint survival function is given by Equation 7.17.

Barlow and Proschan [3], p.129, show that X and Y are PQD.

Example 6. Bivariate distribution of Block and of Basu [34]

$$\begin{aligned}\bar{F}(x, y) &= \frac{2+\theta}{2}\exp[-x - y - \theta\max(x, y)]\\ &\quad - \frac{\theta}{2}\exp[-(2+\theta)\max(x, y)]\\ &\quad \theta, x, y \geq 0\end{aligned}$$

This was constructed to modify Marshall and Olkin's bivariate exponential, which has a singular part. It is, in fact, a reparameterization of a special case of Freund's [35] bivariate exponential distribution. The marginal is

$$\bar{F}_X(x) = \frac{1+\theta}{2}\exp[-(1+\theta)x] - \frac{\theta}{2}\exp[(1+\theta)x]$$

and a similar expression for $\bar{F}_Y(y)$. It is easy to show that this distribution is PQD.

Example 7. Kibble's bivariate gamma distribution. The joint density function is

$$\begin{aligned}f_\rho(x, y; \alpha) &= f_X(x) f_Y(y)\\ &\quad \times \exp[-\rho(x+y)/(1-\rho)]\frac{\Gamma(\alpha)}{1-\rho}\\ &\quad \times (xy\rho)^{-(\alpha-1)/2} I_{\alpha-1}\left(\frac{2\sqrt{xy\rho}}{1-\rho}\right)\\ &\quad 0 \leq \rho < 1\end{aligned}$$

with, f_X, f_Y being the marginal gamma probability density function with shape parameter α. Here, $I_\alpha(\cdot)$ is the modified Bessel function of the first kind and the αth order.

Lai and Moore [36] show that the distribution function is given by

$$\begin{aligned}F(x, y; \rho) &= F_X(x) F_X(y)\\ &\quad + \alpha \int_0^\rho f_t(x, y; \alpha+1)\, dt\\ &\geq F_X(x) F_Y(y)\end{aligned}$$

because $u(x, y) = \int_0^\rho f_t(x, y; \alpha+1)\, dt$ is obviously positive.

For the special case of when $\alpha = 1$, Kibble's gamma becomes the well-known Moran–Downton bivariate exponential distribution.

Downton [37] presented a construction from a reliability perspective. He assumed that the two components C_1 and C_2 receive shocks occurring in independent Poisson streams at rates λ_1 and λ_2 respectively, and that the numbers N_1 and N_2 shocks needed to cause failure of C_1 and C_2 respectively have a bivariate geometric distribution.

For applications of Kibble's bivariate gamma, see, for example, Hutchinson and Lai [1].

Example 8. The bivariate exponential distribution of Sarmanov. Sarmanov [38] introduced a family of bivariate densities of the form

$$f(x, y) = f_X(x) f_Y(y)[1 + \omega\phi_1(x)\phi_2(y)] \quad (7.18)$$

where

$$\int_{-\infty}^{\infty} \phi_1(x) f_X(x)\, dx = 0$$

$$\int_{-\infty}^{\infty} \phi_2(y) f_Y(y)\, dy = 0$$

and ω satisfies the condition that $1 + \omega\phi_1(x)\phi_2(y) \geq 0$ for all x and y.

Lee [39] discussed the properties of the Sarmanov family; in particular, she derived the bivariate exponential distribution given below:

$$f(x, y) = \lambda_2\, e^{-(x+y)} \left[1 + \omega\left(e^{-x} + \frac{\lambda_1}{1+\lambda_1}\right) \times \left(e^{-y} + \frac{\lambda_2}{1+\lambda_2}\right)\right] \quad (7.19)$$

where

$$\frac{-(1+\lambda_1)(1+\lambda_2)}{\lambda_1\lambda_2} \leq \omega \leq \frac{(1+\lambda_1)(1+\lambda_2)}{\max(\lambda_1, \lambda_2)}$$

(Here, $\phi_1(x) = e^{-x} - [\lambda_1/(1+\lambda_1)]$ and $\phi_2(y) = e^{-y} - [\lambda_2/(1+\lambda_2)]$.)

It is easy to see that, for $\omega > 0$:

$$F(x, y) = (1 - e^{-\lambda x})(1 - e^{-\lambda y}) + \omega\left(\frac{\lambda}{1+\lambda}\right)^2 \times [e^{-\lambda x} - e^{-(\lambda+1)x}][e^{-\lambda y} - e^{-(\lambda+1)y}] \geq F_X(x)F_Y(y)$$

whence X and Y are shown to be PQD if

$$0 \leq \omega \leq \frac{(1+\lambda_1)(1+\lambda_2)}{\max(\lambda_1, \lambda_2)}$$

Example 9. The bivariate normal distribution has a density function given by

$$f(x, y) = (2\pi\sqrt{1-\rho^2})^{-1} \times \exp\left[-\frac{1}{2(1-\rho^2)}(x^2 - 2\rho xy + y^2)\right] \quad -1 < \rho < 1$$

X and Y are PQD for $0 \leq \rho < 1$ and NQD for $-1 < \rho \leq 0$. This result follows straightaway from the following lemma:

Lemma 1. *Let (X_1, Y_1) and (X_2, Y_2) be two standard bivariate normal distributions, with correlation coefficients ρ_1 and ρ_2 respectively. If $\rho_1 \geq \rho_2$, then*

$$\Pr(X_1 > x, Y_1 > y) \geq \Pr(X_2 > x, Y_2 > y)$$

The above is known as the Slepian inequality [40, p.805].

By letting $\rho_2 = 0$ (thus $\rho_1 \geq 0$), we establish that X and Y are PQD. On the other hand, letting $\rho_1 = 0$ (thus $\rho_2 \leq 0$), X and Y are then NQD.

7.4.3 Positive Quadrant Dependent Bivariate Uniform Distributions

A copula $C(u, v)$ is simply the uniform representation of a bivariate distribution. Hence a copula is just a bivariate uniform distribution. For a formal definition of a copula, see, for example, Nelsen [41]. By a simple marginal transformation, a copula becomes a bivariate distribution with specified marginals. There are many examples of copulas that are PQD, such as that given in Example 10.

Example 10. The Ali–Mikhail–Haq family

$$C(u, v) = \frac{uv}{1 - \theta(1-u)(1-v)} \quad \theta \in [0, 1]$$

It is clear that the copula is PQD. In fact, Bairamov *et al.* [42] have shown that it is a copula that corresponds to the Durling–Pareto distribution given in Example 4.

Nelsen [41], p.152, has pointed out that if X and Y are PQD, then their copula C is also PQD. Nelsen's book provides a comprehensive treatment on copulas and a number of examples of PQD copulas can be found therein.

7.4.3.1 Generalized Farlie–Gumbel–Morgenstern Family of Copulas

The so-called bivariate FGM distribution given in Example 1 was originally introduced by Morgenstern [43] for Cauchy marginals. Gumbel [44] investigated the same structure for exponential marginals.

It is easy to show that the FGM copula is given by

$$C_\alpha(u,v) = uv[1+\alpha(1-u)(1-v)] \qquad 0 \le u, v \le 1, \quad -1 \le \alpha \le 1 \tag{7.20}$$

It is clear that the FGM copula is PQD for $0 \le \alpha \le 1$.

It was Farlie [23] who extended the construction by Morgenstern and Gumbel to

$$C_\alpha(u,v) = uv[1+\alpha A(u)B(v)] \quad 0 \le u, v \le 1 \tag{7.21}$$

where $A(u) \to 0$ and $B(v) \to 0$ as $u, v \to 1$, $A(u)$ and $B(v)$ satisfy certain regularity conditions ensuring that C is a copula. Here, the admissible range of α depends on the functions A and B.

If $A(u) = B(v) = 1-u$, we then have the classical one-parameter FGM family Equation 7.20.

Huang and Kotz [45] consider the two types:

(i) $A(u) = (1-u)^p, \quad B(v) = (1-v)^p,$

$$p > 1, \quad -1 \le \alpha \le \left(\frac{p+1}{p-1}\right)^{p-1}$$

(ii) $A(u) = (1-u^p), \quad B(v) = (1-v^p),$

$$p > 0, \quad -(\max\{1,p\})^{-2} \le \alpha \le p^{-1}$$

We note that copula (ii) was investigated earlier by Woodworth [46].

Bairamov and Kotz [47] introduce further generalizations such that:

(iii) $A(u) = (1-u)^p, \quad B(v) = (1-v)^q,$

$$p > 1,\ q > 1\ (p \ne q),$$
$$-\min\left\{1, \left(\frac{1+p}{p-1}\right)^{p-1} \left(\frac{1+q}{q-1}\right)^{q-1}\right\} \le \alpha \le \min\left\{\left(\frac{1+p}{p-1}\right)^{p-1}, \left(\frac{1+q}{q-1}\right)^{q-1}\right\}$$

(iv) $A(u) = (1-u^n)^p, \quad B(v) = (1-v^n)^q,$

$$p \ge 1;\ n \ge 1,$$
$$-\min\left\{\frac{1}{n^2}\left[\frac{1+np}{n(p-1)}\right]^{2(p-1)}, 1\right\} \le \alpha \le \frac{1}{n}\left[\frac{1+np}{n(p-1)}\right]^{p-1}$$

Recently, Bairamov *et al.* [48] considered a more general model:

(v) $A(u) = (1-u^{p_1})^{q_1}, \quad B(v) = (1-v^{p_2})^{q_2},$

$$p_1, p_2 \ge 1;\ q_1, q_2 > 1,$$
$$-\min\left\{1, \frac{1}{p_1 p_2}\left[\frac{1+p_1q_1}{p_1(q_1-1)}\right]^{q_1-1} \times \left[\frac{1+p_2q_2}{p_2(q_2-1)}\right]^{q_2-1}\right\}$$
$$\le \alpha$$
$$\le \min\left\{\frac{1}{p_1}\left[\frac{1+p_1q_1}{p_1(q_1-1)}\right]^{q_1-1}, \frac{1}{p_2}\left[\frac{1+p_2q_2}{p_2(q_2-1)}\right]^{q_2-1}\right\}$$

Motivated by a desire to construct PQDs, Lai and Xie [16] derived a new family of FGM copulas that possess the PQD property with:

(vi) $A(u) = u^{b-1}(1-u)^a,$

$$B(v) = v^{b-1}(1-v)^a, \quad a, b \ge 1;\ 0 \le \alpha \le 1$$

so that

$$C_\alpha(u,v) = uv + \alpha u^b v^b (1-u)^a (1-v)^a \qquad a, b \ge 1, \quad 0 \le \alpha \le 1 \tag{7.22}$$

Table 7.1. Range of dependence parameter α for some positive quadrant dependent FGM copulas

Copula type	α range for which copula is PQD
(i)	$0 \le \alpha \le \left(\frac{p+1}{p-1}\right)^{p-1}$
(ii)	$0 \le \alpha \le p^{-1}$
(iii)	$0 \le \alpha \le \min\left\{\left(\frac{1+p}{p-1}\right)^{p-1}, \left(\frac{1+q}{q-1}\right)^{q-1}\right\}, p > 1, q > 1$
(iv)	$0 \le \alpha \le \frac{1}{n}\left[\frac{1+np}{n(p-1)}\right]^{p-1}$
(v)	$0 \le \alpha \le \min\left\{\frac{1}{p_1}\left[\frac{1+p_1q_1}{p_1(q_1-1)}\right]^{q_1-1}, \frac{1}{p_2}\left[\frac{1+p_2q_2}{p_2(q_2-1)}\right]^{q_2-1}\right\}$
(vi)	$0 \le \alpha \le \frac{1}{B^+(a,b)B^-(a,b)}$, B^+, B^- are some functions of a and b

Bairamov and Kotz [49] have shown that the range of α in Equation 7.22 can be extended and they also provide the ranges of α for which the copulas (i)–(v) are PQD. These are summarized in Table 7.1.

In concluding this section, we note that Joe [2], p.19, considered the concepts of PQD and the concordance ordering (more PQD) that are discussed in Section 7.6 as being basic to the parametric families of copulas in determining whether a multivariate parameter is a dependence parameter.

7.5 Some Related Issues on Positive Dependence

7.5.1 Examples of Bivariate Positive Dependence Stronger than Positive Quadrant Dependent Condition

So far, we have presented only the families of bivariate distributions that are PQD, which is a weaker notion of the positive dependence discussed in this chapter. We now introduce some bivariate distributions that also satisfy more stringent conditions.

(i) The bivariate normal density is TP_2 if and only if their correlation coefficient $0 \le \rho < 1$ (*e.g.* see Barlow and Proschan [3], p.149). Abdel-Hameed and Sampson [50] have shown that the bivariate density of the absolute normal distribution is TP_2.

(ii) X and Y of Marshall and Olkin's bivariate distribution are associated owing to having a variable in common in the construction procedure. In fact, Barlow and Proschan [3], p.132, showed that Y is stochastically increasing in X (SI), which, in turn, implies association.

(iii) FGM bivariate exponential distribution: Rödel [51] showed that, for an FGM distribution, X and Y are SI (alias positively regression dependent) if $\alpha > 0$. The following is a direct and easy proof for the case with exponential marginals such that $\alpha > 0$:

$$\begin{aligned} P(Y \le y \mid X = x) &= [1 - \alpha(2\,e^{-x} - 1)(1 - e^{-y})] \\ &\quad + \alpha(2\,e^{-x} - 1)(1 - e^{-2y}) \\ &= (1 - e^{-y}) \\ &\quad + \alpha(2\,e^{-x} - 1)(e^{-y} - e^{-2y}) \end{aligned}$$

Thus

$$P(Y > y \mid X = x) = e^{-y} - \alpha(2\,e^{-x} - 1)(e^{-y} - e^{-2y}) \quad (7.23)$$

which is clearly increasing in x, from which we conclude that X and Y are positively regression dependent if $\alpha > 0$.

(iv) Rödel [51] showed that Kibble's bivariate gamma distribution given in Section 7.4.2 is also SI (alias PRD), which is a stronger concept of positive dependence than PQD.

(v) Sarmanov bivariate exponential. The conditional distribution that corresponds to Equation 7.19 is

$$P(Y \le y \mid X = x) = F_Y(y) + \omega\phi_1(x) \int_{-\infty}^{y} \phi_2(z) f_Y(z)\,\mathrm{d}z$$

$$\phi_i(x) = \mathrm{e}^{-x} - \frac{\lambda_i}{1+\lambda_i} \quad i = 1, 2$$

It follows that

$$P(Y > y \mid X = x) = \mathrm{e}^{-\lambda_2 y} - \omega \left(\mathrm{e}^{-x} - \frac{\lambda_1}{1+\lambda_1} \right) \times \int_{-\infty}^{y} \phi_2(z) f_Y(z)\,\mathrm{d}z$$

is increasing in x because

$$\int_{-\infty}^{y} \phi_2(z) f_Y(z)\,\mathrm{d}z \ge 0$$

and thus Y is SI increasing in x if

$$0 \le \omega \le \frac{(1+\lambda_1)(1+\lambda_2)}{\max(\lambda_1, \lambda_2)}$$

Further, it follows from theorem 3 of Lee [39] that (X, Y) is TP_2 since $\omega\phi'(x)\phi'(y) \ge 0$ for $\omega \ge 0$.

(vi) Durling–Pareto distribution: Lai *et al.* [52] showed that X and Y are right tail increasing if $k \le 1$ and right tail decreasing if $k \ge 1$. From the chains of relationships in Section 7.2.6, it is known that right tail increasing implies association. Thus X and Y are associated if $k \le 1$.

(vii) The bivariate exponential of Example 3:

$$F(x, y) = 1 - \mathrm{e}^{-x} - \mathrm{e}^{-y} + (\mathrm{e}^{x} + \mathrm{e}^{y} - 1)^{-1}$$

It can easily be shown that

$$P(Y \le y \mid X = x) = 1 + \frac{1}{(\mathrm{e}^{x} + \mathrm{e}^{y} - 1)^2}$$

and hence

$$P(Y > y \mid X = x) = \frac{-1}{(\mathrm{e}^{x} + \mathrm{e}^{y} - 1)^2}$$

which is increasing in x, so Y is SI in X.

7.5.2 Examples of Negative Quadrant Dependence

Although the main theme of this chapter is on positive dependence, it is a common knowledge that negative dependence does exist in various reliability situations. Several bivariate distributions discussed in Section 7.4, namely the bivariate normal, FGM family, Durling–Pareto distribution, and bivariate exponential distribution of Sarmanov are NQD when the ranges of the dependence parameter are appropriately specified. The two variables of the following example can only be negatively dependent.

Example 11. Gumbel's bivariate exponential distribution

$$F(x, y) = 1 - \mathrm{e}^{-x} - \mathrm{e}^{-y} + \mathrm{e}^{-(x+y+\theta xy)} \quad 0 \le \theta \le 1$$

$$F(x, y) - F_X(x)F_Y(y) = \mathrm{e}^{-(x+y+\theta xy)} - \mathrm{e}^{-(x+y)} \le 0 \quad 0 \le \theta \le 1$$

showing that F is NQD. It is well known, *e.g.* see Kotz *et al.* [53], p.351, that $-0.403\,65 \le \operatorname{corr}(X, Y) \le 0$.

Lehmann [4] presented the following situations for which negative quadrant dependence occurs:

- Consider the rankings of n objects by m persons. Let X and Y denote the rank sum for the ith and the jth object respectively. Then X and Y are NQD.
- Consider a sequence of n multinomial trials with s possible outcomes. Let X and Y denote the number of trials resulting in outcome i and j respectively. Then X and Y are NQD.

7.6 Positive Dependence Orderings

Consider two bivariate distributions having the same pair of marginals F_X and F_Y; and we assume that they are both positively dependent. Naturally, we would like to know which of the two bivariate distributions is more positively dependent. In other words, we wish to order the two given bivariate distributions by the extent of their positive dependence between the two marginal variables with higher in ordering meaning more positively dependent. In this section, the concept of positive dependence ordering is introduced. The following definition is found in Kimeldorf and Sampson [54].

Definition 2. A relation $\ll$ on a family of all bivariate distributions is a positive dependence ordering (PDO) if it satisfies the following ten conditions:

(P0) $F \ll G \Rightarrow F(x, \infty) = G(x, \infty)$ and $F(\infty, y) = G(\infty, y)$;
(P1) $F \ll G \Rightarrow F(x, y) \leq G(x, y)$ for all x, y;
(P2) $F \ll G$ and $G \ll H \Rightarrow F \ll H$;
(P3) $F \ll F$;
(P4) $F \ll G$ and $G \ll F \Rightarrow F = G$;
(P5) $F^- \ll F \ll F^+$, where $F^+(x, y) = \max[F(x, \infty), F(\infty, y)]$ and $F^-(x, y) = \min[F(x, \infty), F(\infty, y) - 1, 0]$;
(P6) $(X, Y) \ll (U, V) \Rightarrow (a(X), Y) \ll (a(U), V)$ where the $(X, Y) \ll (U, V)$ means the relation $\ll$ holds between the corresponding bivariate distributions;
(P7) $(X, Y) \ll (U, V) \Rightarrow (-U, V) \ll (-X, Y)$;
(P8) $(X, Y) \ll (U, V) \Rightarrow (Y, X) \ll (V, U)$;
(P9) $F_n \ll G_n$, $F_n \to F$ in distribution, $G_n \to G$ in distribution $\Rightarrow F \ll G$, where F_n, F, G_n, and G all have the same pair of marginals.

Tchen [55] defined a bivariate distribution G to be more PQD than a bivariate distribution F having the same pair of marginals if $G(x, y) \geq F(x, y)$ for all $x, y \in R^2$. It was shown that PQD partial ordering is a PDO.

Note that more PQD is also known as "more concordant" in the dependence concepts literature.

Example 12. Generalized FGM copula. Lai and Xie [16] constructed a new family of PQD bivariate distributions that is a generalization of the FGM copula:

$$C_\theta(u, v) = uv + \theta u^b v^b (1 - u)^a (1 - v)^a \qquad a, b \geq 1, \quad 0 \leq \theta \leq 1 \tag{7.24}$$

Let the dependence ordering be defined through the ordering of θ. It is clear from Equation 7.24 that, when $\theta < \theta'$, then $C_\theta(u, v) \ll C_{\theta'}(u, v)$.

Example 13. Bivariate normal with positive correlation coefficient ρ. The Slepian inequality in Section 7.4.2 says

$$\Pr(X_1 > x, Y_1 > y) \geq \Pr(X_2 > x, Y_2 > y)$$

if $\rho_1 \geq \rho_2$. Thus a more PQD ordering can be defined in terms of the positive correlation coefficient ρ.

Example 14. $C_\theta(u, v) = uv/[1 - \theta(1 - u)(1 - v)]$, $\theta \in [0, 1]$ (the Ali–Mikhail–Haq family). It is easy to see that $C_\theta \gg C_{\theta'}$ if $\theta > \theta'$, *i.e.* C_θ is more PQD than $C_{\theta'}$.

There are several other types of PDO in the literature and we recommend the reader to consult the text by Shaked and Shantikumar [56], who gave a comprehensive treatment on stochastic orderings.

7.7 Concluding Remarks

Concepts of stochastic dependence are widely applicable in statistics. Given that some of these concepts have arisen from reliability contexts, it seems rather unfortunate that not many reliability practitioners have caught onto this important subject. This observation is transparent, since the assumption of independence is still prevailing in many reliability analyses. Among the dependence concepts, the correlation is a widely used concept in applications. Association is advocated and studied in Barlow and Proschan [3]. On the other hand, PQD is a weaker condition and also seems to be easier to verify. On reflection, this phenomenon may be due in part to the fact that many of the proposed dependence models are often not readily applicable. One would hope that, in the near future, more applied probabilists and reliability engineers would get together to forge a partnership to bridge the gap between the theory and applications of stochastic dependence.

References

[1] Hutchinson TP, Lai CD. Continuous bivariate distributions, emphasising applications. Adelaide, Australia: Rumsby Scientific Publishing; 1990.

[2] Joe H. Multivariate models and dependence concepts. London: Chapman and Hall; 1997.

[3] Barlow RE, Proschan F. Statistical theory of reliability and life testing: probability models. Silver Spring (MD): To Begin With; 1981.

[4] Lehmann EL. Some concepts of dependence. Ann Math Stat 1966;37:1137–53.

[5] Esary JD, Proschan F, Walkup DW. Association of random variables, with applications. Ann Math Stat 1967;38:1466–74.

[6] Esary JD, Proschan F. Relationships among some bivariate dependence. Ann Math Stat 1972;43:651–5.

[7] Harris R. A multivariate definition for increasing hazard rate distribution. Ann Math Stat 1970;41:713–7

[8] Brindley EC, Thompson WA. Dependence and aging aspects of multivariate survival. J Am Stat Assoc 1972;67:822–30.

[9] Yanagimoto T. Families of positive random variables. Ann Inst Stat Math 1972;26:559–73.

[10] Shaked M. A concept of positive dependence for exchangeable random variables. Ann Stat 1977;5:505–15.

[11] Shaked M. Some concepts of positive dependence for bivariate interchangeable distributions. Ann Inst Stat Math 1979;31:67–84.

[12] Shaked M. A general theory of some positive dependence notions. J Multivar Anal 1982;12 : 199–218.

[13] Block HW, Ting, ML. Some concepts of multivariate dependence. Commun Stat A: Theor Methods 1981;10:749–62.

[14] Jogdeo K. Dependence concepts and probability inequalities. In: Patil GP, Kotz S, Ord JK, editors. A modern course on distributions in scientific work - models and structures, vol. 1. Dordrecht: Reidel; 1975. p.271–9.

[15] Jogdeo K. Dependence concepts of. In: Encyclopedia of statistical sciences, vol. 2. New York: Wiley; 1982. p.324–34.

[16] Lai CD, Xie M. A new family of positive dependence bivariate distributions. Stat Probab Lett 2000;46:359–64.

[17] Robbins H. A remark on the joint distribution of cumulative sums. Ann Math Stat 1954;25:614–6.

[18] Kimball AW. On dependent tests of significance in analysis of variance. Ann Math Stat 1951;22:600–2.

[19] Douglas R, Fienberg SE, Lee MLT, Sampson AR, Whitaker LR. Positive dependence concepts for ordinal contingency tables. In: IMS lecture notes monograph series: topics in statistical dependence, vol. 16. Hayward (CA): Institute of Mathematical Statistics; 1990. p.189–202.

[20] Xie M, Lai, CD. On the increase of the expected lifetime by parallel redundancy. Asia–Pac J Oper Res 1996;13:171–9.

[21] Mi J. Bolstering components for maximizing system lifetime. Nav Res Logist 1998;45:497–509.

[22] Kotz S, Lai CD, Xie, M. The expected lifetime when adding redundancy in systems with dependent components. IIE Trans 2003;in press.

[23] Farlie DJG. The performance of some correlation coefficients for a general bivariate distribution. Biometrika 1960;47:307–23.

[24] Johnson NL, Kotz S. Distributions in statistics: continuous multivariate distributions. New York: Wiley; 1972.

[25] Mukerjee SP, Sasmal BC. Life distributions of coherent dependent systems. Calcutta Stat Assoc Bull 1977;26:39–52.

[26] Lingappaiah GS. Bivariate gamma distribution as a life test model. Aplik Mat 1983;29:182–8.

[27] Philips MJ. A preventive maintenance plan for a system subject to revealed and unrevealed faults. Reliab Eng 1981;2:221–31.

[28] Kotz S, Johnson NL. Some replacement-times distributions in two-component systems. Reliab Eng 1984;7:151–7.

[29] Mardia KV. Families of bivariate distributions. London: Griffin; 1970.

[30] Lindley DV, Singpurwalla ND. Multivariate distributions for the life lengths of components of a system sharing a common environment. J Appl Probab 1986;23:418–31.

[31] Nayak TK. Multivariate Lomax distribution: properties and usefulness in reliability theory. J Appl Probab 1987;24:170–7.

[32] Sankaran PG, Nair NU. A bivariate Pareto model and its applications to reliability. Nav Res Logist 1993;40:1013–20.

[33] Marshall AW, Olkin I. A multivariate exponential distribution. J Am Stat Assoc 1967;62:30–44.

[34] Block HW, Basu AP. A continuous bivariate exponential distribution. J Am Stat Assoc 1976;64:1031–7.

[35] Freund J. A bivariate extension of the exponential distribution. J Am Stat Assoc 1961;56:971–7.

[36] Lai CD, Moore T. Probability integrals of a bivariate gamma distribution. J Stat Comput Simul 1984;19:205–13.

[37] Downton F. Bivariate exponential distributions in reliability theory. J R Stat Soc Ser B 1970;32:408–17.

[38] Sarmanov OV. Generalized normal correlation and two-dimensional Frechet classes. Dokl Sov Math 1966;168:596–9.

[39] Lee MLT. Properties and applications of the Sarmanov family of bivariate distributions. Commun Stat A: Theor Methods 1996;25:1207–22.

[40] Gupta SS. Probability integrals of multivariate normal and multivariate *t*. Ann Math Stat 1963;34:792–828.

[41] Nelsen RB. An introduction copulas. Lecture notes in statistics, vol. 139,: New York: Springer-Verlag; 1999.

[42] Bairamov I, Lai CD, Xie M. Bivariate Lomax distribution and generalized Ali–Mikhail–Haq distribution. Unpublished results.

[43] Morgenstern D. Einfache Beispiele zweidimensionaler Verteilungen. Mitteilungsbl Math Stat 1956;8:234–5.

[44] Gumbel EJ. Bivariate exponential distributions. J Am Stat Assoc 1960;55:698–707.

[45] Huang JS, Kotz S. Modifications of the Farlie–Gumbel–Morgenstern distributions. A tough hill to climb. Metrika 1999;49:135–45.

[46] Woodworth GG. On the asymptotic theory of tests of independence based on bivariate layer ranks. Technical Report No 75, Department of Statistics, University of Minnesota. See also Abstr Ann Math Stat 1966;36:1609.

[47] Bairamov I, Kotz S. Dependence structure and symmetry of Huang–Kotz FGM distributions and their extensions. Metrika 2002;in press.

[48] Bairamov I, Kotz S, Bekci M. New generalized Farlie–Gumbel-Morgenstern distributions and concomitants of order statistics. J Appl Stat 2001;28:521–36.

[49] Bairamov I, Kotz S. On a new family of positive quadrant dependent bivariate distributions. GWU/IRRA/TR No 2000/05. The George Washington University, 2001.

[50] Abdel-Hameed M, Sampson AR. Positive dependence of the bivariate and trivariate absolute normal t, χ^2 and F distributions. Ann Stat 1978;6:1360–8.

[51] Rödel E. A necessary condition for positive dependence. Statistics 1987;18 :351–9.

[52] Lai CD, Xie M, Bairamov I. Dependence and ageing properties of bivariate Lomax distribution. In: Hayakawa Y, Irony T, Xie M, editors. A volume in honor of Professor R. E. Barlow on his 70th birthday. Singapore: WSP; 2001. p.243–55.

[53] Kotz S, Balakrishnan N, Johnson NL. Continuous multivariate distributions, vol. 1: models and applications. New York: Wiley; 2000.

[54] Kimeldorf G, Sampson AR. Positive dependence orderings. Ann Inst Stat Math 1987;39:113–28.

[55] Tchen A. Inequalities for distributions with given marginals. Ann Probab 1980;8:814–27.

[56] Shaked M, Shantikumar JG, editors. Stochastic orders and their applications. New York: Academic Press; 1994.

Statistical Reliability Change-point Estimation Models

Ming Zhao

Chapter 8

8.1 Introduction

The classical change-point problem arises from the observation of a sequence of random variables $X_1, X_2, \ldots, X_n$, such that $X_1, X_2, \ldots, X_\tau$ have a common distribution F while $X_{\tau+1}, X_{\tau+2}, \ldots, X_n$ have the distribution G with $F \neq G$. The index τ, called the change point, is usually unknown and has to be estimated from the data.

The change-point problem has been widely discussed in the literature. Hinkley [1] used the maximum likelihood (ML) method to estimate the change-point τ in the situations where F and G can be arbitrary known distributions and belong to the same parametric family. The non-parametric estimation of the change point has been discussed by Carlstein [2]. Joseph and Wolfson [3] generalized the change-point problem by studying multipath change points where several independent sequences are considered simultaneously and each sequence has one change point. The bootstrap and empirical Bayes methods are also suggested for estimating the change points. If the sequence $X_1, X_2, \ldots, X_n$ is the observation of arrival times at which n events occur in a Poisson process, which is widely applied in reliability analysis, this is a Poisson process model with one change point. Raftery and Akiman [4] proposed using Bayesian analysis for the Poisson process model with a change point. The parametric estimation in the Poisson process change-point model has been given by Leonard [5].

In reliability analysis, change-point models can be very useful. In software reliability modeling, the initial number of faults contained in a program is always unknown and its estimation is a great concern. One has to execute the program in a specific environment and improve its reliability by detecting and correcting the faults. Many software reliability models assume that, during the fault detection process, each failure caused by a fault occurs independently and randomly in time according to the same distribution, *e.g.* see Musa *et al.* [6]. The failure distribution can be affected by many different factors, such as the running environment, testing strategy, and the resource. Generally, the running environment may not be homogeneous and can be changed with the human learning process. Hence, the change-point models are of interest in modeling the fault detection process. The change point can occur when the testing strategy and the resource allocation are changed. Also, with increasing knowledge of the program, the testing facilities and other random factors can be the causes of change points. Zhao [7] has discussed different software reliability change-point models and the method of parametric estimation. The non-homogeneous Poisson process (NHPP) model with change points has been studied by Chang [8] and the parameters in NHPP change-point model are estimated by the weighted least-squares method.

Change-point problems can also arise in hardware reliability and survival analysis, for example, when reliability tests are conducted in a random testing environment. Nevertheless, the problem may not be as complex as with the software reliability problem. This is because the sample size in a lifetime test, which is comparable to the initial number of faults in software that needs to be estimated, is usually known in hardware reliability analysis. Unlike the common change-point models, the observed data in hardware reliability or survival analysis are often grouped and dependent. For example, if we are investigating a new treatment against some disease, the observed data may be censored because some patients may be out of the trials and new patients may come into the trials. There is a need to develop new models for a reliability change-point problem.

In this chapter, we address the change-point problems in reliability analysis. A change-point model, which is applicable for both software and hardware reliability systems, is considered in a general form. Nevertheless, more focus is given to software reliability change-point models, since the hardware reliability change-point models can be analyzed as a special case where the sample size is assumed to be known. When it is necessary, however, some discussion is given on the differences between hardware and software reliability models. The ML method is considered to estimate the change-point and model parameters. A numerical example is also provided to show the application.

8.2 Assumptions in Reliability Change-point Models

Let F and G be two different lifetime distributions with density functions $f(t)$ and $g(t)$, $X_1, X_2, \ldots, X_\tau, X_{\tau+1}, X_{\tau+2}, \ldots, X_n$ be the inter-failure times of the sequential failures in a lifetime testing. Before we consider a number of specific reliability change-point models, we need to make some assumptions.

Assumption 1. *There are a finite number of items N under reliability test; the parameter N may be unknown.*

In hardware reliability analysis, the parameter N, which is usually assumed to be known, is the total number of units under the test. However, where software reliability analysis is concerned, the parameter N, the initial number of faults in the software, is assumed to be unknown.

Assumption 2. *At the beginning, all of the items have the same lifetime distribution F. After τ failures are observed, the remaining $(N-\tau)$ items have the distribution G. The change point τ is assumed unknown.*

In software reliability, this assumption means that all faults in software have the same detection rate, but it is changed when a number of faults have been detected. The most common reason is that the test team has learned a lot from the testing process, so that the detection process is more efficient. On the other hand, it could be that the fault detection becomes increasingly difficult due to the complexity of the problem. In both situations, a change-point model is an alternative to apply.

Assumption 3. *The sequence $\{X_1, X_2, \ldots, X_\tau\}$ is statistically independent of the sequence $\{X_{\tau+1}, X_{\tau+2}, \ldots, X_n\}$.*

This assumption may not be realistic in some cases. We use it only for the sake of model simplicity. However, it is easy to modify this assumption. Note that we do not assume the independence between the variables within sequence $\{X_1, X_2, \ldots, X_\tau\}$ or $\{X_{\tau+1}, X_{\tau+2}, \ldots, X_n\}$, because the inter-failure times of failures in lifetime testing are usually dependent.

8.3 Some Specific Change-point Models

To consider the reliability change-point model, we further assume that a lifetime test is performed according to the Type-II censoring plan, in which the number of failures n is determined in advance.

Denote by $T_1, T_2, \ldots, T_n$ the arrival times of sequential failures in the lifetime test. Then we have

$$\begin{aligned} T_1 &= X_1 \\ T_2 &= X_2 \\ &\vdots \\ T_n &= X_1 + X_2 + \cdots + X_n \end{aligned} \tag{8.1}$$

According to Assumptions 1–3, the failure times $T_1, T_2, \ldots, T_\tau$ are the first τ order statistics of a sample with size N from parent distribution F, $T_{\tau+1}, T_{\tau+2}, \ldots, T_n$ are the first $(n-\tau)$-order statistics of a sample with size $(N-\tau)$ from parent distribution G.

8.3.1 Jelinski–Moranda De-eutrophication Model with a Change Point

8.3.1.1 Model Review

One of the earliest software reliability models is the de-eutrophication model developed by Jelinski and Moranda [9]. It was derived based on the following assumptions.

Assumption 4. *There are N initial faults in the program.*

Assumption 5. *A detected fault is removed instantaneously and no new fault is introduced.*

Assumption 6. *Each failure caused by a fault occurs independently and randomly in time according to an exponential distribution.*

Based on the model assumptions, we can determine that the inter-failure times $X_i = T_i - T_{i-1}$, $i = 1, 2, \ldots n$, are independent exponentially distributed random variables with failure rate $\lambda(N - i + 1)$, where λ is the initial fault detection rate. One can see that each fault is discovered with a failure rate reduced by the proportionality constant λ. This means that the impact of each fault removal is the same and the faults are of equal size; see [10, 11] for details.

The Jelinski–Moranda de-eutrophication model is the simplest and most cited software reliability model. It is also the most criticized model, because Assumptions 5 and 6 imply that all faults in a program have the same size and each removal of the detected faults reduces the failure intensity by the same amount. However, the model remains central to the topic of software reliability. Many other models are derived based on assumptions similar to that of the de-eutrophication model.

8.3.1.2 Model with One Change Point

We assume that F and G are exponential distributions with failure rate parameters λ_1 and λ_2 respectively. From Assumptions 1–3, it is easy to show that the inter-failure times $X_1, X_2, \ldots, X_n$ are independently exponentially distributed. Specifically, $X_i = T_i - T_{i-1}, i = 1, 2, \ldots, \tau$, are exponentially distributed with parameter $\lambda_1(N - i + 1)$ and $X_j = T_j - T_{j-1}$, $j = \tau + 1, \tau + 2, \ldots, n$, are exponentially distributed with parameter $\lambda_2(N - \tau - j + 1)$.

Note that the original Jelinski–Moranda de-eutrophication model implies that all faults are of the same size; the debugging process is perfect. However, an imperfect debugging assumption is more realistic; see [10] for details. By considering the change point in this model, this implies that there are at least two groups of faults with different size. If more change points are considered in the model, one can think that the debugging could be imperfect and the faults are of different size. The proposed model is expected to be closer to the reality.

8.3.2 Weibull Change-point Model

A Weibull change-point model appears when F and G are Weibull distribution functions with parameters (λ_1, β_1) and (λ_2, β_2) respectively. That is

$$F(t) = 1 - \exp(-\lambda_1 t^{\beta_1}) \tag{8.2}$$

$$G(t) = 1 - \exp(-\lambda_2 t^{\beta_2}) \tag{8.3}$$

In this case, the sequence $\{X_1, X_2, \ldots, X_\tau\}$ is not independent. The Weibull model without change points has been used by Wagoner [12] to describe the fault detection process. In particular, when the shape parameter $\beta = 2$, the Weibull model is the model proposed by Schick and Wolverton [13]. In application, one can use the simplified model in which the shape parameters β_1 and β_2 are assumed to be equal.

8.3.3 Littlewood Model with One Change Point

Assume that F and G are Pareto distribution functions given by

$$F(t) = 1 - (1 + t/\lambda_1)^{\beta_1} \tag{8.4}$$

$$G(t) = 1 - (1 + t/\lambda_2)^{\beta_2} \tag{8.5}$$

Then we have a Littlewood model with one change point. The model without a change point was given by Littlewood [14]. It is derived using Assumptions 4–6, but the failure rates associated with each fault are assumed to be identical and independent random variables with gamma distributions. Under the Bayesian framework, the failure distribution is shown to be a Pareto distribution; see [14] for details. This model tries to account for the possibility that the software program could become less reliable than before.

8.4 Maximum Likelihood Estimation

In previous sections we have presented some reliability change-point models. In order to consider the parametric estimation, we assume that the distributions belong to parametric families $\{F(t \mid \theta_1), \theta_1 \in \Theta_1\}$ and $\{G(t \mid \theta_2), \theta_2 \in \Theta_2\}$. Because $T_1, T_2, \ldots, T_\tau$ are the first τ-order statistics of a sample with size N from parent distribution F, $T_{\tau+1}, T_{\tau+2}, \ldots, T_n$ are the first $(n-\tau)$-order statistics of a sample with size $(N-\tau)$ from parent distribution G, then the log-likelihood without the constant term is given by

$$\begin{aligned} L(\tau, N, \theta_1, \theta_2 | T_1, T_2, \ldots, T_n) \\ = \sum_{i=1}^{n}(N - i + 1) + \sum_{i=1}^{\tau} f(T_i \mid \theta_1) \\ + \sum_{i=\tau+1}^{n} g(T_i \mid \theta_2) \\ + (N-\tau)\log[1 - F(T_\tau \mid \theta_1)] \\ + (N-n)\log[1 - G(T_n \mid \theta_2)] \end{aligned} \tag{8.6}$$

Note that where hardware reliability analysis is concerned, the simple size N is known, and the likelihood function is then given by

$$\begin{aligned} L(\tau, \theta_1, \theta_2 \mid T_1, T_2, \ldots, T_n) \\ = \sum_{i=1}^{\tau} f(T_i \mid \theta_1) + \sum_{i=\tau+1}^{n} g(T_i \mid \theta_2) \\ + (N-\tau)\log[1 - F(T_\tau \mid \theta_1)] \\ + (N-n)\log[1 - G(T_n \mid \theta_2)] \end{aligned}$$

In the following, we consider only the case where the sample size N is unknown because it is more general.

The ML estimator (MLE) of the change point is the value $\hat{\tau}$ that together with $(\hat{N}, \hat{\theta}_1, \hat{\theta}_2)$ maximizes the function in Equation 8.6. Unfortunately, there is no closed form for $\hat{\tau}$. However, it can be obtained by calculating the log-likelihood for each possible value of τ, $1 \le \tau \le (n-1)$, and selecting as $\hat{\tau}$ the value that maximizes the log-likelihood function.

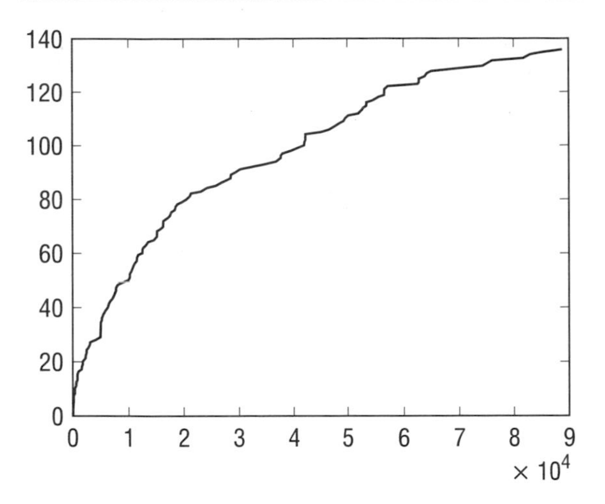

Figure 8.1. The plot of cumulative number of failures against their occurrence times in the SYS1 data set

In principle, the likelihood function (Equation 8.6) is valid for the change-point models discussed in the previous section. Here, we only consider the estimation for the Jelinski–Moranda de-eutrophication model with one change-point.

For this model, the MLEs of parameters$(N, \lambda_1, \lambda_2)$, given the change-point τ in terms of the inter-failure times, are determined by the following equations:

$$\hat{\lambda}_1 = \frac{\tau}{\sum_{i=1}^{\tau} (\hat{N} - i + 1)x_i} \qquad (8.7)$$

$$\hat{\lambda}_2 = \frac{(n - \tau)}{\sum_{i=\tau+1}^{n} (\hat{N} - i + 1)x_i} \qquad (8.8)$$

$$\sum_{i=1}^{n} \frac{1}{(\hat{N} - i + 1)} = \hat{\lambda}_1 \sum_{i=1}^{\tau} x_i + \hat{\lambda}_2 \sum_{i=\tau+1}^{n} x_i \qquad (8.9)$$

In order to find out the estimation, the numerical method has to be applied. In a regular situation, one can expect that the MLEs exist. Therefore, we can see that the likelihood function is only dependent on τ, since the other parameters can be determined from Equations 8.7–8.9 if τ is given. Consequently, we can compare the values of the likelihood function taken at $\tau = 1, 2, \ldots, n-1$ respectively. The value of τ that makes the likelihood maximum is therefore the MLE of τ.

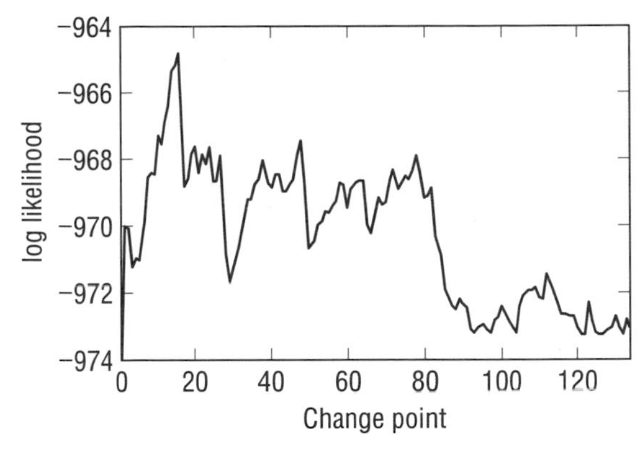

Figure 8.2. The values of the log-likelihood function against the change points

On the distribution of $\hat{\tau}$, it is not easy to have a general conclusion. The asymptotic distribution of $\hat{\tau}$ may be obtained using the method in [1]. It is also possible to use the bootstrap approach to estimate the distribution of $\hat{\tau}$ following the method in [3]. We do not discuss these topics further here, but we should point out that there exists a serious problem when the parameter N is unknown. In some cases, the MLE of N does not exist; *e.g.* see [15–17], where the Jelinski–Moranda model and other models are considered. The conditions for the existence of the MLE of N have been given. Consequently, when the MLE of N does not exist for some data set, the MLE of τ does not exist either.

8.5 Application

In this section, we use a real data set in software reliability engineering to the Jelinski–Moranda de-eruphication model. The data set we used is called SYS1, from Musa [18], and has frequently been analyzed in the literature. Figure 8.1 shows the plot of the cumulative failure numbers against their occurrence times in seconds. There are 136 failures in total in the data set.

The log-likelihood for the SYS1 data set is plotted in Figure 8.2 for possible values of the change-point τ. We can see that the log-likelihood takes a maximum value near to 20. Actually, the MLEs of these parameters in the model with one change point are

$$\hat{\tau} = 16, \quad \hat{N} = 145, \quad \hat{\lambda}_1 = 1.1086 \times 10^{-4}, \quad \hat{\lambda}_2 = 2.9925 \times 10^{-5}$$

When no change points are considered, the MLEs of parameters N and λ are

$$\hat{N} = 142, \quad \hat{\lambda} = 3.4967 \times 10^{-5}$$

We can see that the estimates for the number of faults N by these two models are slightly different.

In order to see the differences between these two models we can check out the prediction ability, which is a very important criterion for evaluating the models, since the main objective in software reliability is to predict the behavior of future failures as precisely as possible; *e.g.* see [19, 20] for more discussion on the prediction ability.

The u-plot and prequential likelihood techniques [21, 22] are used to evaluate the prediction ability in this case. In general, the u-plot of a good predicting system should be close to the diagonal line with unit slope, and the maximum vertical distance between the u-plot and the diagonal line, which is called the Kolmogrov distance, is a measurement of the closeness. On the other hand, it can be shown that model A is favored of model B if the prequential likelihood ratio $\mathrm{PL}^{\mathrm{A}}/\mathrm{PL}^{\mathrm{B}}$ is consistently increasing as the predicting process continues.

Figure 8.3 shows the u-plots of the change-point model and the original model starting after 35 failures. The Kolmogrov distance for the change-point model is equal to 0.088 and for the original model the distance is 0.1874.

Figure 8.4 is the plot of the log prequential likelihood ratio between the change-point model and the Jelinski–Moranda model. The increasing trend is strong, so the change-point model makes the prediction better.

The analysis above has shown that the fitness and the prediction have been greatly improved by considering the change point in the Jelinski–Moranda model for this specific data set.

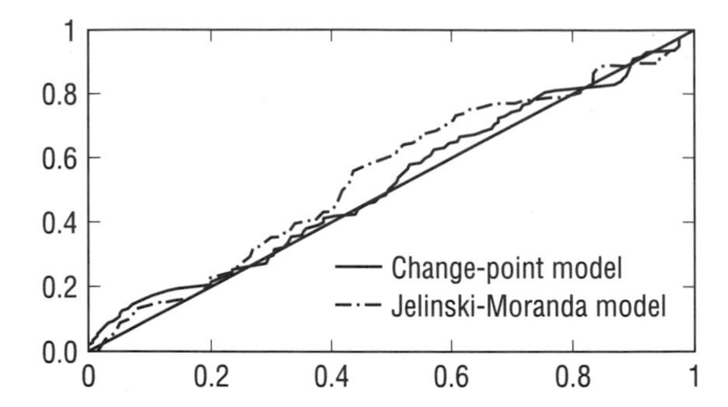

Figure 8.3. A comparison of the u-plots between the change-point model and the Jelinski–Moranda model

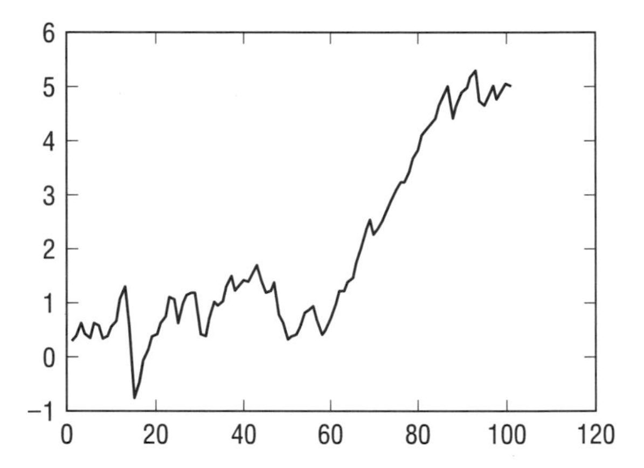

Figure 8.4. The log-prequential likelihood ratios between the change-point model and the Jelinski–Moranda model

8.6 Summary

We have presented some change-point models that can be used in reliability analysis. The main differences between the models discussed and the classical change-point model are the dependent sample and the unknown sample size, which is particular to software reliability analysis. When the sample size is assumed to be known, the model discussed is the one used for hardware reliability analysis. By using a classical software reliability model, we have shown how a change-point model can be developed and applied for real failure data. The improvement in modeling and prediction by introducing the change point has been demonstrated for one data set. The MLEs of the change point are yet to be solved, because in some software reliability models the ML estimation of the number of faults can be infinity. In such cases, another estimation procedure has to be developed. An extension of the change-point problems in reliability analysis could be made by considering multipath change points.

References

[1] Hinkley DV. Inference about the change-point in a sequence of random variables. Biometrika 1970;57:1–16.

[2] Carlstein E. Nonparametric change-point estimation. Ann Stat 1988;16:188–97.

[3] Joseph L, Wolfson DB. Estimation in multi-path change-point problems. Commun Stat Theor Methods 1992;21:897–913.

[4] Raftery AE, Akiman VE. Bayesian analysis of a Poisson process with a change-point. Biometrika 1986;73:85–9.

[5] Leonard T. Density estimation, stochastic processes and prior information (with discussion). J R Stat Soc B 1978;40:113–46.

[6] Musa JD, Iannino A, Okumoto K. Software reliability: measurement, prediction, application. New York: McGraw-Hill; 1987.

[7] Zhao M. Statistical reliability change-point estimation models. Commun Stat Theor Methods 1993;22:757–68.

[8] Chang YP. Estimation of parameters for nonhomogeneous Poisson process: software reliability with change-point model. Commun Stat Simul C 2001;30(3):623–35.

[9] Jelinski Z, Moranda PB. Software reliability research. In: Freiberger W, editor. Statistical computer performance evaluation. New York: Academic Press; 1972. p.465–97.

[10] Pham H. Software reliability. Singapore: Springer-Verlag; 2000.

[11] Xie M. Software reliability modelling. Singapore: World Scientific; 1991.

[12] Wagoner WL. The final report on a software reliability measurement study. In: Report TOR-007(4112)-1. Aerospace Corporation, 1973.

[13] Schick GJ, Wolverton RW. Assessment of software reliability. In: Proceedings, Operations Research. Wurzburg Wein: Physica-Verlag; 1973. p.395–422.

[14] Littlewood B. Stochastic reliability growth: a model for fault-removal in computer-programs and hardware-design. IEEE Trans Reliab 1981;30:312–20.

[15] Joe H, Reid, N. Estimating the number of faults in a system. J Am Stat Assoc 1985;80:222–6.

[16] Wright DE, Hazelhurst CE. Estimation and prediction for a simple software reliability model. Statistician 1988;37:319–25.

[17] Huang XZ. The limit conditions of some time between failure models of software reliability. Microelectron Reliab 1990;30:481–5.

[18] Musa, JD. Software reliability data. New York: RADC; 1979.

[19] Musa JD. Software reliability engineered testing. UK: McGraw-Hill; 1998.

[20] Zhao M. Nonhomogeneous Poisson processes and their application in software reliability. PhD Dissertation, Linköping University, 1994.

[21] Littlewood B. Modelling growth in software reliability. In: Rook P, editor. Software reliability handbook. UK: Elsevier Applied Science; 1991. p.111–36.

[22] Dawid AP. Statistical theory: the prequential approach. J R Stat Soc A 1984;147:278–92.

Concepts and Applications of Stochastic Aging in Reliability

C. D. Lai and M. Xie

Chapter 9

9.1 Introduction

The notion of aging plays an important role in reliability and maintenance theory. "No aging" means that the age of a component has no effect on the distribution of residual lifetime of the component. "Positive aging" describes the situation where residual lifetime tends to decrease, in some probabilistic sense, with increasing age of the component. This situation is common in reliability engineering, as systems or components tend to become worse with time due to increased wear and tear. On the other hand, "negative aging" has an opposite effect on the residual lifetime. Although this is less common, when a system undergoes regular testing and improvement, there are cases for which we have reliability growth phenomena.

Concepts of aging describe how a component or system improves or deteriorates with age. Many classes of life distributions are categorized or defined in the literature according to their aging properties. The classifications are usually based on the behavior of the conditional survival function, the failure rate function, or the mean residual life of the component concerned. Within a given class of life distributions, closure properties under certain reliability operations are often investigated. In several cases, for example, the increasing failure rate average, arise naturally as a result of damages being accumulated from random shocks. Moreover, there are a number of chains of implications among these aging classes. The exponential distribution is the simplest life distribution that is characterized by a constant failure rate or no aging. It is also nearly always a member of every known aging class, and thus several tests have been proposed for testing a life distribution being exponential against an alternative that it belongs to a specified aging class. Some of the life classes have been derived more recently and, as far as we know, no test statistics have been proposed. On the other hand, there are several tests available for some classes.

By "life distributions" we mean those for which negative values do not occur, *i.e.* $F(x)=0$ for $x<0$. The non-negative variate X is thought of as the time to failure (or death) of an electrical or mechanical component (or organism), but other interpretations may be possible-an inter-event time is normally necessarily positive.

In this chapter, we focus on classes of life distributions based on notions of aging; increasing failure rate (IFR) is perhaps the best known, but we shall meet many others also, and study their interrelationships whenever possible. The major parts of the chapter are devoted to: (i) the naming and interpretation of univariate reliability classes, and showing how these notions may be extended to the bivariate case; (ii) presenting the interrelationships between different classes; and (iii) summarizing various statistical tests on aging in three tables for various univariate and bivariate aging classes. An expository paper by Bergman [1] and chapter 18 of Hutchinson and Lai [2] are helpful for (i) and (ii); details on (iii) can be found in Lai [3].

From the definitions of the life distribution classes, results may be derived concerning such things as properties of systems (based upon properties of components), bounds for survival functions, moment inequalities, and algorithms for use in maintenance policies [4].

Most readers will know that statistical theory applied to distributions of lifetime lengths plays an important part in both the reliability engineering literature and the biometrics literature. We may also note a third applications area: Heckman and Singer [5] review econometric work on duration variables (*e.g.* lengths of periods of unemployment, or time intervals between purchases of a certain good), much of which, they say, has borrowed freely and often uncritically from reliability theory and biostatistics. We believe the topics under discussion are indeed important, and hope that this brief review will provide a clear picture of the recent developments in this area.

The chapter considers the following aspects:

- univariate reliability classes;
- interrelationship between the classes;
- bathtub-shaped life distributions;
- life classes based on mean residual lifetime;
- partial ordering of life distributions;
- bivariate reliability classes;
- tests of univariate and bivariate aging.

In Section 9.2 we begin with giving a review of some basic aging notions and their interrelationships, together with some key references. Section 9.3 discusses the properties of these basic aging classes and Section 9.4 is devoted to the bathtub-shaped life distributions that are important in reliability applications. Section 9.5 characterizes life classes based on their mean residual lifetimes, and Section 9.6 tidies up the aging classifications with the inclusion of some further, but less well known, classes. Section 9.7 provides an introduction to partial ordering, through which the strength of the aging property of the two life distributions within the same class is compared. Univariate aging concepts are extended in Section 9.8, in which bivariate aging classes are introduced. Section 9.9 considers the subject of tests of aging, with the exponential distribution being taken as the null hypothesis against various alternatives. We omit details of these test procedures, as the subject matter may have less appeal to engineers than to statisticians. Finally, in Section 9.10, we tidy up the loose ends on stochastic aging and the section ends with some remarks concerning future research directions that may bridge the theory and applications.

The following notation is adopted herein:

F cumulative distribution function
f density function given by F'

$\bar{F}$ reliability function given $1 - F$ (also known as the survival probability)
$\bar{X}$ sample mean
ψ indicator function
H_0 null hypothesis
H_1 alternative hypothesis
ϕ non-negative function
$r(t)$ hazard rate (or failure rate) function
$m(t)$ mean residual lifetime.

A list of abbreviations used in the text is given Table 9.1.

9.2 Basic Concepts for Univariate Reliability Classes

9.2.1 Some Acronyms and the Notions of Aging

The concepts of increasing and decreasing failure rates for univariate distributions have been found very useful in reliability theory. The classes of distributions having these aging properties are designated the IFR and DFR distributions respectively, and they have been studied extensively. Other classes, such as "increasing failure rate on average" (IFRA), "new better than used" (NBU), "new better than used in expectation" (NBUE), and "decreasing mean residual life" (DMRL), have also been of much interest. For fuller accounts of these classes see, for example, Bryson and Siddiqui [6], Barlow and Proschan [7], Klefsjö [8], and Hollander and Proschan [4].

A class that slides between NBU and NBUE, known as "new better than used in convex ordering" (NBUC), has also attracted some interest recently.

The notion of "harmonically new better than used in expectation" (HNBUE) was introduced by Rolski [9] and studied by Klefsjö [10, 11]. Further generalizations along this line were given by Basu and Ebrahimi [12]. A class of distributions denoted by $\mathfrak{L}$ has an aging property that is based on the Laplace transform, and was put forward by Klefsjö [8]. Deshpande *et al.* [13] used stochastic dominance comparisons to describe positive aging and suggested several new positive aging criteria based on these ideas (see their paper for details). Two further classes, NBUFR ("new better than used in failure rate") and NBUFRA ("new better than used in failure rate average") require the absolute continuity of the distribution function, and have been discussed by Loh [14, 15], Deshpande *et al.* [13], Kochar and Wiens [16], and Abouammoh and Ahmed [17].

Rather than $F(x)$, we often think of $\bar{F}(x) = \Pr(X > x) = 1 - F(x)$, which is known as the survival distribution or reliability function. The expected value of X is denoted by μ. The function

$$\bar{F}(t \mid x) = \bar{F}(x+t)/\bar{F}(x) \qquad (9.1)$$

represents the survival function of a unit of age x, *i.e.* the conditional probability that a unit of age x will survive for an additional t units of time. The expected value of the remaining (residual) life, at age x, is $m(x) = E(X - x \mid X > x)$, which may be shown to be $\int_0^\infty \bar{F}(t \mid x)\, dt$.

When $F'(x) = f(x)$ exists, we can define the failure rate (or hazard rate, or force of mortality) as

$$r(x) = f(x)/\bar{F}(x) \qquad (9.2)$$

for x such that $\bar{F}(x) > 0$. It follows that, if $r(x)$ exists:

$$-\log \bar{F}(t) = \int_0^t r(x)\, dx \qquad (9.3)$$

which represents the cumulative failure rate. We are now ready to define several reliability classes.

9.2.2 Definitions of Reliability Classes

Most of the reliability classes are defined in terms of the failure rate $r(t)$, conditional survival function $\bar{F}(t \mid x)$, or the mean residual life $m(t)$. All these three functions provide probabilistic information on the residual lifetime, and hence aging classes may be formed according to the behavior of the aging effect on a component. The ten reliability classes mentioned above are defined as follows.

Definition 1. F is said to be IFR if $\bar{F}(t \mid x)$ is decreasing in $0 \leq x < \infty$ for each $t \geq 0$.

Table 9.1. List of abbreviations used in the text

Abbreviation	Aging class
BFR	Bathtub-shaped failure rate
DMRL (IMRL)	Decreasing mean residual life (Increasing mean residual life)
HNBUE	Harmonically new better than used in expectation
IFR (DFR)	Increasing failure rate (Decreasing failure rate)
IFRA (DFRA)	Increasing failure rate average (Decreasing failure rate average)
$\mathfrak{L}$-class	Laplace class of distributions
NBU (NWU)	New better than used (New worse than used)
NBUE (NWUE)	New better than used in expectation (New better than used in expectation)
NBUC	New better than used in convex ordering
NBUFR	New better than used in failure rate
NBUFRA	New better than used in failure rate average
NBWUE (NWBUE)	New better then worse than used in expectation (New worse then better than used in expectation)

When the density exists, this is equivalent to $r(x) = f(x)/\bar{F}(x)$ being increasing in $x \geq 0$ [7].

Definition 2. F is said to be IFRA if $-(1/t) \times \log \bar{F}(t)$ is increasing in t. (That is, if $-\log \bar{F}$ is a star-shaped function; for this notion, see Dykstra [18].)

As $-\log \bar{F}$ given in Equation 9.3 represents the cumulative failure rate, the name given to this class is appropriate. Block and Savits [19] showed that IFRA is equivalent to $E^{\alpha}[h(X)] \leq E[h^{\alpha}(X/\alpha)]$ for all continuous non-negative increasing functions h and all such that $0 < \alpha < 1$. This class is perhaps the most important in reliability analysis. It is the smallest closed class containing an exponential distribution under the formation of coherent systems. It has been shown that a device subject to shocks governed by a Poisson process, which fails when the accumulated damage exceeds a fixed threshold, has an IFRA distribution [20].

Definition 3. F is said to be DMRL if the mean remaining life function $\int_0^{\infty} \bar{F}(x \mid t)\, dt$ is decreasing in x, *i.e.* the older the device is, the smaller is its mean residual life [6].

Definition 4. F is said to be NBU if $\bar{F}(t \mid x) \leq \bar{F}(t)$ for $x \geq 0, t \geq 0$.

This means that a device of any particular age has a stochastically smaller remaining lifetime than does a new device [7].

Definition 5. F is said to be NBUE if $\int_0^{\infty} \bar{F}(t \mid x)\, dt \leq \mu$ for $x \geq 0$.

This means that a device of any particular age has a smaller mean remaining lifetime than does a new device [7].

Definition 6. F is said to be HNBUE if $\int_x^{\infty} \bar{F}(t \mid x)\, dt \leq \mu \exp(-x/\mu)$ for $x \geq 0$. There is an alternative definition in terms of the mean residual life [9].

Definition 7. F is said to be an $\mathfrak{L}$-distribution if $\int_0^{\infty} e^{-st} \bar{F}(t)\, dt \geq \mu/(1+s)$ for $s \geq 0$.

The expression $\mu/(1+s)$ can be written as $\int \exp(-sx)\bar{G}(x)\, dx$, where $\bar{G}(x) = \exp(-x/\mu)$. This means that the inequality is one between the Laplace transforms of $\bar{F}$ and of an exponential survival function with the same mean as F [8].

Definition 8. F is said to be NBUFR if $r(x) > r(0)$ for $x > 0$ [13].

Definition 9. F is said to be NBUFRA if $r(0) \leq (1/x) \int_0^{\infty} r(t)\, dt = -\log[\bar{F}(x)]/x$ [14].

Using Laplace transforms, Block and Savits [21] established necessary and sufficient conditions for the IFR, IFRA, DMRL, NBU, and NBUE properties to hold.

Definition 10. F is NBUC if $\int_x^{\infty} \bar{F}(t \mid x)\, dt \leq \int_x^{\infty} \bar{F}(t)\, dt$ [22].

Hendi *et al.* [23] have shown that the new better than used in convex ordering is closed under the formation of parallel systems with independent and identically distributed components. Li *et al.* [24] presented a lower bound of the reliability function for this class based upon the mean and the variance. Cao and Wang [22] have proved that

$$\text{NBU} \Rightarrow \text{NBUC} \Rightarrow \text{NBUE} \Rightarrow \text{HNBUE}$$

9.2.3 Interrelationships

The following chain of implications exists among the aging classes [13, 16]:

$$\begin{array}{ccccccccc}
\text{IFR} & \Rightarrow & \text{IFRA} & \Rightarrow & \text{NBU} & \Rightarrow & \text{NBUFR} & \Rightarrow & \text{NBUFRA}\\
\Downarrow & & & & \Downarrow & & & & \Downarrow\\
\Downarrow & & & & \text{NBUC} & & & & \Downarrow\\
\Downarrow & & & & \Downarrow & & & & \Downarrow\\
\text{DMRL} & & \Rightarrow & & \text{NBUE} & \Rightarrow & \text{HNBUE} & \Rightarrow & \mathfrak{L}
\end{array}$$

In Definitions 1–10, if we reverse the inequalities and interchange "increasing" and "decreasing", we obtain the classes DFR, DFRA, NWU, IMRL, NWUE, HNWUE, $\bar{\mathfrak{L}}$, NWUFR, NWUFRA, and NWUC. They satisfy the same chain of implications.

9.3 Properties of the Basic Concepts

9.3.1 Properties of Increasing and Decreasing Failure Rates

Patel [25] gives the following properties of the IFR and DFR concepts.

1. If X_1 and X_2 are both IFR, so is $X_1 + X_2$; but the DFR property is not so preserved.
2. A mixture of DFR distributions is also DFR; but this is not necessarily true of IFR distributions.
3. Parallel systems of identical IFR units are IFR.
4. Series systems of (not necessarily identical) IFR units are IFR.
5. Order statistics from an IFR distribution have IFR distributions, but this is not true for spacings from an IFR distribution; order statistics from a DFR distribution do not necessarily have a DFR distribution, but spacings from a DFR distribution are DFR.
6. The probability density function (p.d.f.) of an IFR distribution need not be unimodal.
7. The p.d.f. of a DFR distribution is a decreasing function.

9.3.2 Property of Increasing Failure Rate on Average

This aging notion is fully investigated in the book by Barlow and Proschan [7]; in particular, IFRA is closed under the formation of coherent systems, as well as under convolution. The IFRA closure theorem is pivotal to many of the results given in Barlow and Proschan [7].

9.3.3 Properties of NBU, NBUC, and NBUE

The properties of NBU, NWU, NBUE, and NWUE are also well documented in the book by Barlow and Proschan [7]. Chen [26] showed that the distributions of these classes may be characterized through certain properties of the corresponding renewal functions.

Li *et al.* [24] gave a lower bound of the reliability of the NBUC class based on the first two moments. Among other things, they also proved that the NBUC class is preserved under the formation of parallel systems. The last property was also proved earlier by Hendi *et al.* [23].

9.4 Distributions with Bathtub-shaped Failure Rates

A failure rate function falls into one of three categories: (a) monotonic failure rates, where $r(t)$ is either increasing or decreasing; (b) bathtub failure rates, where $r(t)$ has a bathtub or a U shape; and (c) generalized bathtub failure rates, where

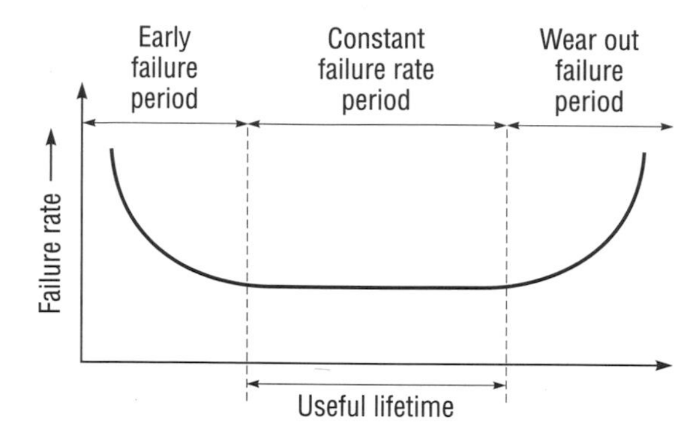

Figure 9.1. A bathtub failure rate $r(t)$

$r(t)$ is a polynomial, or has a roller-coaster shape or some generalization.

The main theme of this section is to give an overview on the class of bathtub-shaped (BFR) life distributions having non-monotonic failure rate functions. For a recent review, see Lai *et al.* [27].

There are several definitions of bathtub-shaped life distributions. Below is the one that gives rise to curves having definite "bathtub shapes".

Definition 11. (Mi [28]) A distribution F is a bathtub-shaped life distribution if there exists $0 \leq t \leq t_0$ such that:

1. $r(t)$ is strictly decreasing, if $0 \leq t \leq t_1$;
2. $r(t)$ is a constant if $t_1 \leq t \leq t_2$;
3. strictly increasing if $t \geq t_2$.

The above definition may be rewritten as

$$r(t) = \begin{cases} r_1(t) & \text{for } t \leq t_1 \\ \lambda & \text{for } t_1 \leq t \leq t_2 \\ r_2(t) & \text{for } t \geq t_2 \end{cases} \tag{9.4}$$

where $r_1(t)$ is strictly decreasing in $[0, t_1]$ and $r_2(t)$ is strictly increasing for $t \geq t_2$. We do not know of many parametric distributions that possess this property, except for some piecewise continuous distributions. One example is the sectional Weibull in Jiang and Murthy [29] involving three Weibull distributions.

A typical bathtub-shaped failure rate function is shown in Figure 9.1. Several comments are in order.

In Definition 11, Mi [28] called the points t_1 and t_2 the change points of $r(t)$. If $t_1 = t_2 = 0$, then a BFR becomes an IFR; and if $t_1 = t_2 \to \infty$, then $r(t)$ is strictly decreasing, so becoming a DFR. In general, if $t_1 = t_2$, then the interval for which $r(t)$ is a constant degenerates to a single point. In other words, the strict monotonic failure rate distributions IFR and DFR may be treated as the special cases of BFR in this definition. Park [30] also used the same definition. The reader will find more information on bathtub-shaped life distributions in Lai *et al.* [27].

The class of lifetime distributions having a bathtub-shaped failure rate function is very important because the lifetimes of electronic, electromechanical, and mechanical products are often modeled with this feature. In survival analysis, the lifetime of human beings exhibits this pattern. Kao [31], Lieberman [32], Glaser [33], and Lawless [34] give many examples of bathtub-shaped life distributions. For more recent examples, see Chen [35] and Lai *et al.* [27]. For applications of the bathtub-shaped life distributions, see Lai *et al.* [27].

9.5 Life Classes Characterized by the Mean Residual Lifetime

Recall, the mean residual lifetime is defined as $m(t) = E(X - x \mid X > x)$ which is equivalent to $\int_0^\infty \bar{F}(t \mid x)\, dt$. Recall also that F is said to be DMRL if the mean remaining life function $\int_0^\infty \bar{F}(x \mid t)\, dt$ is decreasing in x. That is, the older the device is, the smaller is its mean residual life and hence $m(t)$ is monotonic. However, in many real life situations, the mean residual lifetime is non-monotonic, and thus there arise several aging notions defined in terms of the non-monotonic behavior of $m(t)$.

Definition 12. (Guess *et al.* [36]) A life distribution with a finite first moment is called an increasing then decreasing mean residual life (IDMRL) if there exists a turning point $\tau \geq 0$ such that

$$m(s) \begin{cases} \leq m(t) & \text{for } 0 \leq s \leq t < \tau \\ \geq m(t) & \text{for } \tau \leq s \leq t \end{cases} \tag{9.5}$$

The dual class of "decreasing initially, then increasing mean residual life" (DIMRL) is obtained by reversing the above inequality (Equation 9.5).

Gupta and Akman [37] showed that the non-monotonic behavior is related to the non-monotonic failure rates. Their results are summarized below.

Case (A): $r(t)$ is BFR, then:

1. $m(t)$ is decreasing if $r(0) \leq 1/\mu$;
2. IDMRL if $r(0) \geq 1/\mu$.

Case (B) $r(t)$ has an upside-down bathtub shape, then:

1. $m(t)$ is increasing if $r(0) \geq 1/\mu$;
2. $m(t)$ is bathtub shaped if $r(0) \leq 1/\mu$.

A similar result is also given by Mi [28], who stated "if the failure rate has a bathtub-shape, then the associated MRL has an upside-down bathtub-shape".

Mitra and Basu [38] defined another aging notion based on non-monotonic MRL:

Definition 13. A life distribution F having support on $[0, \infty)$ (and finite mean μ) is said to be "new worse then better than used in expectation" (NWBUE) ("new better then worse than used in expectation" (NBWUE)) if there exists a point $\tau \geq 0$ such that

$$m(t) \begin{cases} \geq (\leq) m(0) & \text{for } t < \tau \\ \leq (\geq) m(0) & \text{for } t \geq \tau \end{cases}$$

They showed that non-monotonic aging classes such as BFR and NWBUE arise from the shock model under consideration.

For applications of concepts based on the mean residual life, see Guess and Proschan [39]. See Newby [40] for some simple explanations and examples of analyses in terms of mean residual life, hazard, and hazard rate average.

9.6 Some Further Classes of Aging

In addition to those lifetime classes defined above, there are a number of other aging classes that have been investigated over the years. Without giving details, we just present their acronyms and references in Table 9.2.

Other chains of relationships are given in Joe and Proschan [46]:

NBU $\Leftrightarrow$ NBUP-α for all $0 < \alpha < 1 \Rightarrow$ NBUE
IFR $\Leftrightarrow$ DPRL-α for all $0 < \alpha < 1$
DPRL $\Rightarrow$ DPRL-$\alpha \Rightarrow$ NBUP-α for any $0 < \alpha < 1$

There are other chains of relationships, but we will not elaborate on them here.

9.7 Partial Ordering of Life Distributions

A distribution F is said to be more IFR than G if the ratio of hazard

$$\frac{r_F[F^{-1}(t)]}{r_G[G^{-1}(t)]}$$

is non-decreasing in t. (This definition has assumed that the failure rates exist.) Orderings of each of the other reliability classes may also be defined.

There is a chain of implications connecting the orderings of the classes, that is similar but not identical to that which connects the classes (Section 9.2.3) [16].

Let X and Y be two random variables with distribution functions G and H respectively. We say that X is stochastically larger than Y if $G(x) \leq H(x)$, for all x. In economic theory, this is known as the first-order stochastic dominance. Higher orders of stochastic dominance are defined in Deshpande *et al.* [13].

Table 9.2. Table of further aging classes

IFR(2)	Increasing failure rate (of second order)	[13,41]
NBU(2)	New better than used (of second order)	[13,41,42]
HNBUE(3)	Harmonic new better than used (of third order)	[13]
DMRLHA	Decreasing mean residual life in harmonic average	[13]
SIFR	Stochastically increasing failure rate	[43]
SNBU	Stochastically new better than used	[43]
NBU-t_0	New better than used of age t_0	[4]
BMRL-t_0	Better mean residual life at t_0 (*i.e.* the mean residual life declines during the time 0 to t_0, and thereafter is no longer greater than what it was at t_0)	[44]
DVRL	Decreasing variance of residual life	[45]
DPRL-α	Decreasing 100α percentile residual life	[46]
NBUP-α	New better than used with respect to the 100α percentile	[46]

9.7.1 Relative Aging

Recently, Sengupta and Deshpande [47] studied three types of relative aging of two life distributions. The first of these relative aging concepts is the partial ordering originally proposed by Kalashnikov and Rachev [48], which is defined as follows:

Definition 14. Let F and G be the distribution functions of the random variables X and Y respectively. X is said to be *aging* faster than Y if the random variable $Z = \Lambda_G(X)$ has an IFR distribution. Here $\Lambda_G = -\log \bar{G}$ and $\bar{G}(t) = 1 - G(t)$.

If the failure rates $r_F(t)$ and $r_G(t)$ both exist with $r_F(t) = f(t)/[1 - F(t)]$ and $r_G(t) = g(t)/[1 - G(t)]$, then the above definition is equivalent to $r_F(t)/r_G(t)$ being an increasing function of t.

Several well-known partial orderings are given in Table 9.3 (assuming both failure rates of both distributions do exist).

Lai and Xie [51] established some results on relative aging of two parallel structures. It is observed that the relative aging property may be used to allocate resources and can be used for failure identification when two components (systems) have the same mean. In particular, if "X ages faster than Y" and they have the same mean, then $\mathrm{Var}(X) \le \mathrm{Var}(Y)$. Several examples are given. In particular, it is shown that when two Weibull distributions have equal means, the one that ages faster has a smaller variance.

9.7.2 Applications of Partial Orderings

Most of the stochastic aging definitions involve failure rate in some way. The partial orderings discussed above have important applications in reliability.

Design engineers are well aware that a system where active spare allocation is made at the component level has a lifetime stochastically larger than the corresponding system where active spare allocation is made at the system level. Boland and El-Neweihi [52] investigated this principle in hazard rate ordering and demonstrated that it does not hold in general. However, they discovered that for a 2-out-of-n system with independent and identical components and spares, active spare allocation at the component level is superior to active spare allocation at the system level. They conjectured that such a principle holds in general for a k-out-of-n system. Singh and Singh [50] have proved that for a k-out-of-n system where components and spares have independent and identical life distributions, active spare allocation at the component level is superior to active spare allocation at the system level in likelihood ratio ordering. This is stronger than hazard rate ordering and thus establishes the conjecture of Boland and El-Neweihi [52].

Table 9.3. Partial orderings on aging

Partial ordering	Condition	Reference
$X \leq_{st} Y$ ($Y > X$ in stochastic order)	$\bar{F}(t) \leq \bar{G}(t)$, for all t	[7] p.110
$X \leq_{hr} Y$ ($Y > X$ in hazard order)	$r_F(t) \geq r_G(t)$, for all t	[49]
$X \leq_c Y$ (F is convex with respect to G)	$G^{-1}F(t)$ is convex in t	[7] p.106
$X \leq_* Y$ (F is star shaped with respect to G)	$(1/t)G^{-1}F(t)$ is $\uparrow$ in t	[7] p.106
$X \prec_c Y$ (X is aging faster than Y)	$r_F(t)/r_G(t)$ is $\uparrow$ in t	[47]
$X \prec_* Y$ (X is aging faster than Y on average)	$Z = \Lambda_G(X)$ has an IFRA	[47]
$X \prec_{lr} Y$ (X is smaller than Y in likelihood ratio)	$f(t)/g(t)$ is $\uparrow$ in t	[50]

Boland and El-Neweihi [52] has also established that the active spare allocation at the component level is better than the active spare allocation at the system level in hazard rate ordering for a series system when the components are matching although the components may not be identically distributed. Boland [49] gave an example to show that the hazard rate comparison is what people really mean when they compare the performance of two products. For more on the hazard rate and other stochastic orders, the readers should consult Shaked and Shanthikumar [53].

9.8 Bivariate Reliability Classes

In dealing with multicomponent systems, one wants to extend the whole univariate aging properties to bivariate and multivariate distributions. Consider, for example, the IFR concept in the bivariate case. Let us use the notation $\bar{F}(x_1, x_2)$ to mean the probability that item 1 survives longer than time x_1 and item 2 survives longer than time x_2. Then a possible definition of bivariate IFR is that $\bar{F}(x_1+t, x_2+t)/\bar{F}(x_1, x_2)$ decreases in x_1 and x_2 for all t.

Bivariate as well as multivariate versions of IFR, IFRA, NBU, NBUE, DMRL, HNBUE and of their duals have been defined and their properties have been developed by several authors. In each case, however, several definitions are possible, because of different requirements imposed by various authors. For example, Buchanan and Singpurwalla [54] used one set of requirements whereas Esary and Marshall [55] used a different set. For a bibliography of available results, see Block and Savits [56], Basu *et al.* [57], and Hutchinson and Lai [2]. Table 9.4 gives the relevant references for different bivariate aging concepts.

Table 9.4. Bivariate versions of univariate aging concepts

Ageing	Reference
Bivariate IFR	[58–67]
Bivariate IFRA	[55, 68–73]
Bivariate NBU	[69, 74–76]
Bivariate NBUE	[56]
Bivariate DMRL	[54, 77, 78]
Bivariate HNBUE	[57, 79, 80]

9.9 Tests of Stochastic Aging

Tests of stochastic aging play an important part in reliability analysis. Usually, the null hypothesis is that the lifetime distribution follows an exponential distribution, *i.e.* there is no aging. Let T have the exponential distribution function $F(t) = 1 - \mathrm{e}^{-\lambda t}$ for $t \geq 0$. A property of the exponential distribution that makes it especially important in reliability theory and application is that the remaining life of a used exponential component is independent of its initial age (*i.e.* it has the memoryless property). In other words:

$$\Pr\{T > t + x \mid T > t\} = \mathrm{e}^{-\lambda x} \quad x \geq 0 \tag{9.6}$$

(independent of t). It is easy to verify that the exponential distribution has a constant failure rate, *i.e.* $r(t) = \lambda$, and is the only life distribution with this property.

The exponential distribution is characterized by having a constant failure rate, and thus stands at the boundary between the IFR and DFR classes. For a review of goodness-of-fit tests of the exponential distribution, see Ascher [81].

9.9.1 A General Sketch of Tests

In developing tests for different classes of life distributions, it is almost without exception that the exponential distribution is used as a strawman to be knocked down. As the exponential distribution is always a boundary member of an aging class C (in the univariate case), a usual format for testing is

H_0 (null hypothesis):
  F is exponential versus
H_1 (alternative):
  $F \in C$ but F is not exponential.

If one is merely interested to test whether F is exponential, then no knowledge of the alternative is required; in such a situation, a two-tailed test is appropriate. Recently, Ascher [81] discussed and compared a wide selection of tests for exponentiality. Power computations, using simulations, were done for each procedure. In short, he found:

1. certain tests performed well for alternative distributions with non-monotonic hazard (failure) rates, whereas others fared well for monotonic hazard rates;
2. of all the procedures compared, the score test presented in Cox and Oakes [82] appears to be the best if one does not have a particular alternative in mind.

In what follows, our discussion will focus on tests with alternatives being specified. In other words, our aim is not to test exponentiality, but rather to test whether a life distribution belongs to a specific aging class. A general outline of these procedures is as follows.

1. Find an appropriate measure of deviation from the null hypothesis of exponentiality toward the alternative (under H_1). The measure of deviation is usually based on some characteristic property of the aging class we wish to test against.
2. Based on this measure, some statistics, such as the U-statistic, are proposed. A large value of this statistic often (but not always) indicates the distribution belongs to class C.
3. Large-sample properties, such as asymptotic normality and consistency, are proved.
4. Pitman's asymptotic relative efficiencies are usually calculated for the following families of distributions (with scale factor of unity, $\theta \geq 0$, $x \geq 0$):
   (a) the Weibull distribution $\bar{F}_1(x) = \exp(-x^{\theta})$;
   (b) the linear failure rate distribution:
   $$\bar{F}_2(x) = \exp(-x - \theta x^2/2)$$
   (c) the Makeham distribution:
   $$\bar{F}_3(x) = \exp\{-[x + \theta(x + \mathrm{e}^{-x} - 1)]\}$$
   (d) the Gamma distribution with density
   $$f(x) = [x^{\theta-1}/\Gamma(\theta)]\,\mathrm{e}^{-x}.$$

The above distributions are well known except for the Makeham distribution, which is widely used in life insurance and mortality studies (see [7], p.229). All these distributions are IFR (for an appropriate restriction on θ); hence they all belong to a wider class. Moreover, all these reduce to the exponential distribution with an appropriate value of θ.

Table 9.5. Tests on univariate aging

Test name	Basic statistics	Aging alternatives	Key references
TTT plot	$\tau(X_i)=\sum_{j=1}^{i}(n-j+1)(X_j-X_{j-1}),\quad j=i,\ldots n$	IFR, IFRA,	[57]
	$U_i=\tau(X_i)/\tau(X_n), T=\sum_{j=1}^{n}a_jU_j$ Each test has a different set of a_i values	NBUE DMRL HNBUE	[84]
U_n	$U_n=[4/n(n-1)(n-2)(n-3)]\sum\psi[\min(X_{\gamma_1}X_{\gamma_2}),(X_{\gamma_3}+X_{\gamma_4})/2]$ sum over $\Omega=\{\gamma_i\neq\gamma_j\}\cap\{\gamma_1<\gamma_2\}\cap\{\gamma_3<\gamma_4\}$	IFR	[85–87]
J_b	$J_b=\frac{n}{(n-1)}\sum\psi(X_i,bX_j),\quad 0<b<1$	IFRA	[88]
Q_n	$Q_n=\sum_{i=1}^{n}J\left(\frac{i}{n+i}\right)\Big/n\bar{X},\quad J(u)=2(1-u)[1-\log(1-u)]-1$	IFRA	[89]
Link	$\frac{2}{n(n-1)}\sum_{i<j}X_i/X_j$	IFRA	[90]
V^*	$V^*=V/\bar{X}, V=n^{-4}\sum_{i=1}^{n}C_{i,n}X_i$ $C_{i,n}=\frac{4}{3}i^3-4ni^2+3n^2i-\frac{1}{2}n^3+\frac{1}{2}n^2-\frac{1}{2}i^2+\frac{1}{6}i$	DMRL	[91, 92]
J_n or J	$J_n=2[n(n-1)(n-2)]^{-1}\sum\psi(X_{\gamma_1},X_{\gamma_2}+X_{\gamma_3}),\quad 1\leq\gamma\leq n,$ $\gamma_1\neq\gamma_2,\gamma_1\neq\gamma_3,\gamma_2<\gamma_3$	NBU	[93]
δ	$\delta_n=n^{-2}\sum_{i}^{n}\sum_{j}^{n}\phi(S_{ij}/n),\phi$ increasing; $S_{ij}=\sum_{k=1}^{n}\psi(X_k,X_i+X_j)$	NBU	[94, 95]
Generalized J	J test together with a two-stage test procedure	NBU	[96]
S	$S=U-J$, see above for J test	NBU	[97]
L_n	$L_n(\psi,m)=n^{-1}\sum_{i=1}^{n}\phi(n^{-1}S_i^{(m)}), S_i^{(m)}=\sum_{j=1}^{n}\psi(X_j,mX)$	NBU	[98]
K^*	$K^*=K/\bar{X},K=\frac{1}{2n^2}\sum_{i=1}^{n}\left(\frac{3n}{2}-2i+\frac{1}{2}\right)X_i$	NBUE	[91]
CV	$S/\bar{X},S^2=\sum_{i=1}^{n}(X_i-\bar{X})^2/n$	NBUE, NWUE	[10]
T_n	$T_n=\frac{1}{n}\sum_{i=1}^{n}\exp(-X_i/\bar{X})$	HNBUE	[101]
E_n	$E_n=\frac{1}{n}\sum_{i=1}^{n-1}\log[1-\tau(X_i)],\tau(X_i)=\sum_{j=1}^{i}(n-j+1)(X_j-X_{j-1})$	HNBUE	[102]

Table 9.5. *Continued.*

Test name	Basic statistics	Aging alternatives	Key references
T	$T = [n(n-1)]^{-1} \sum \psi(X_{\alpha_1}, X_{\alpha_2} + t_0) - (2n)^{-1} \sum_{i=1}^{n} \psi(X_i, t_0)$	NBU-t_0	[103]
T_k	$T_k = T_{1k} - T_{2k}$, $T_{1k} = \sum_{i=j=n} \psi(X_i, X_j + kt_0) \Big/ \frac{2}{n(n-1)}$ $T_{2k} = \frac{1}{2} \sum \psi_k(X_{i_1}, \dots, X_{i_k}) \Big/ \binom{n}{k}$ $\psi_k(a_1, \dots, a_k) = \begin{cases} 1 & \text{if } \min a_i > t_0 \\ 0 & \text{otherwise} \end{cases}$	NBU-t_0	[104]
$W_{1:n}$	$W_{1n} = \frac{1}{4} \sum_{i=0}^{n-1} B_1 \frac{i}{n}(X_{i+1} - X_i)$, $B_1(t) = \begin{cases} -\bar{t}^2(\bar{t}^2 - 1) & 0 \le t < \alpha \\ \bar{t}^2(\bar{\alpha}^{-2} - 1)[2(\bar{\alpha}^{-2} + 1)\bar{t}^2 - 1] & \alpha < t < 1 \end{cases}$ $\quad \bar{t} = 1 - t$ $\bar{\alpha} = 1 - \alpha$	DPRL-α	[105, 106]
$W_{2:n}$	$W_{2:n} = \frac{1}{2} F_n^{-1}(\alpha) - \frac{1}{2} \sum_{i=1}^{n} [B_2[(i-1)/n) - B_2(i/n)]$ $B_2(t) = \begin{cases} -\frac{1}{2}[(1-t)^2 - 1] & 1 \le t < \alpha \\ \frac{1}{2}[(1-\alpha)^{-2} - 1](1-t)^2 & \alpha < t \le 1 \end{cases}$	NBUP-α	[105, 106]
TTT	$T = \sum_{j=1}^{n} a_j U_j$, $\quad U_i = \tau(X_i)/\tau(X_n)$	BFR	[107]
Stochastic order	Residual life $X_t = \{X - t \mid X > t\}$ with cumulative distribution function $\bar{F}_X(t + x)/\bar{F}_X(t), x \ge 0$. For $0 \le s < t$		[108]
	F IFR if $X_t \le_{\text{st}} X_s$; $\quad F$ DFR if $X_s \le_{\text{st}} X_t$;	IFR, DFR	
	F NBU if $X_t \le_{\text{st}} X$; $\quad F$ NWU if $X \le_{\text{st}} X_t$.	NBU, NWU	
	Test statistics based on testing stochastic order		
Right censored	$\hat{F}_n(x) = 1 - \prod_{\{i: Z_{(i)} \le x\}} \left(\frac{n-i}{n-i+1} \right)^{\delta_{(i)}}$ (Kaplan–Meier estimator [111]),	IFRA	[109]
	$\delta_i = 0$ or 1 depending on whether object i is censored on the right or not, $Z_i = \min(X_i, Y_i)$, $\quad Y_i$ is random time to the right censorship	NBU NBUE NBU-t_0 DPRL-α NBUP-α	[110] [112] [113] [105] [105]

9.9.2 Summary of Tests of Aging in Univariate Case

We shall now present several statistical procedures for testing exponentiality of a life distribution against different alternatives.

Let ψ be the indicator function defined as

$$\psi(a,b) = \begin{cases} 1 & \text{if } a > b \\ 0 & \text{otherwise} \end{cases}$$

ϕ denotes an increasing function (*i.e.* $\phi(x) \leq \phi(y)$, for $x < y$), and $X_1 < X_2 < \cdots < X_n$ are the order statistics.

Table 9.5 gives an overview of the tests available for various aging classes. We refer our readers to Lai [3] for details.

9.9.3 Summary of Tests of Bivariate Aging

Let X and Y denote the lifetimes of two components having a joint distribution function $F(x, y)$. The joint survival function is given by $\bar{F}(x, y) = \Pr(X > x, Y > y)$.

In testing bivariate aging properties, there are two problems facing us:

- Which bivariate exponential distribution is the null distribution?
- Which version of bivariate aging property in a given class are we dealing with? (There are several versions of bivariate IFR, bivariate NBU, *etc.*)

Generally, the bivariate exponential distribution of Marshall and Olkin [61] (denoted by BVE) is used as the null distribution. This joint distribution has the survival function given by

$$\bar{F}(x, y) = \exp[-\lambda_1 x - \lambda_2 y - \lambda_{12} \max(x, y)] \qquad x, y \geq 0 \tag{9.7}$$

$\lambda_1, \lambda_2 > 0, \lambda_{12} \geq 0$.

BVE is chosen presumably because:

- it was derived from a reliability context;
- it has the property of bivariate lack of memory:

$$\bar{F}(x+t, y+t) = \bar{F}(x, y)\bar{F}(t, t) \tag{9.8}$$

Table 9.6. Tests of bivariate aging

Bivariate version	Key references on bivariate tests
IFR	[114, 115]
IFRA	[116]
DMRL	[117]
NBU	[118, 119]
NBUE	[117]
HNBUE	[120]

Most of the test statistics listed in Table 9.6 are bivariate generalizations of the univariate tests discussed in Section 9.9.2.

It is our impression that tests for bivariate and multivariate stochastic aging are more difficult than for the univariate case; however, they would probably offer a greater scope for applications. We anticipate that more work will be done in this area.

9.10 Concluding Remarks on Aging

It is clear from the material presented in this chapter that research on aging properties (univariate, bivariate, and multivariate) is currently being pursued vigorously. Several surveys are given, *e.g.* see Block and Savits [56, 121, 122], Hutchinson and Lai [2], and Lai [3].

The simple aging concepts such as IFR, IFRA, NBU, NBUE, DMRL, *etc.*, have been shown to be very useful in reliability-related decision making, such as in replacement and maintenance studies. For an introduction on these applications, see Bergman [1] or Barlow and Proschan [7].

It is our impression that the more advanced univariate aging concepts exist in something of a mathematical limbo, and have not yet been applied by the mainstream of reliability practitioners. Also, it seems that, at present, many of the multivariate aging definitions given above lack clear physical interpretations, though it is true that some can be deduced from shock models. The details of these derivations can be found in Marshall and Shaked [73], Ghosh and Ebrahimi [123], and Savits [124].

It is our belief that practical concern is very much with multicomponent systems—in which case, appreciation and clarification of the bivariate concepts (and hence of the multivariate concepts) could lead to greater application of this whole body of work.

References

[1] Bergman B. On reliability theory and its applications. Scand J Stat 1985;12:1-30 (discussion: 30-41).

[2] Hutchinson TP, Lai CD. Continuous bivariate distributions, emphasising applications. Adelaide: Rumsby Scientific Publishing; 1990.

[3] Lai CD. Tests of univariate and bivariate stochastic ageing. IEEE Trans Reliab 1994;R43:233-41.

[4] Hollander M, Proschan F. Nonparametric concepts and methods in reliability. In: Krishnaiah PR, Sen PK, editors. Handbook of statistics: nonparametric methods, vol. 4. Amsterdam: North Holland; 1984. p.613-55.

[5] Heckman JJ, Singer B. Economics analysis of longitudinal data. In: Griliches Z, Intriligator MD, editors. Handbook of econometrics, vol. 3. Amsterdam: North-Holland; 1986. p.1689-763.

[6] Bryson MC, Siddiqui MM. Some criteria for aging. J Am Stat Assoc 1969;64:1472-83.

[7] Barlow RE, Proschan F. Statistical theory of reliability and life testing. Silver Spring: To Begin With; 1981.

[8] Klefsjö B. A useful ageing property based on the Laplace transform. J Appl Probab 1983;20:615-26.

[9] Rolski T. Mean residual life. Bull Int Stat Inst 1975;46:266-70.

[10] Klefsjö B. HNBUE survival under shock models. Scand J Stat 1981;8:39-47.

[11] Klefsjö B. HNUBE and HNWUE classes of life distributions. Nav Res Logist Q 1982;29:615-26.

[12] Basu, AP, Ebrahimi N. On k-order harmonic new better than used in expectation distributions. Ann Inst Stat Math 1984;36:87-100.

[13] Deshpande JV, Kochar SC, Singh H. Aspects of positive ageing. J Appl Probab 1986;23:748-58.

[14] Loh WY. A new generalisation of NBU distributions. IEEE Trans Reliab 1984;R-33:419-22.

[15] Loh WY. Bounds on ARE's for restricted classes of distributions defined via tail orderings. Ann Stat 1984;12:685-701.

[16] Kochar SC, Wiens DD. Partial orderings of life distributions with respect to their ageing properties. Nav Res Logist 1987;34:823-9.

[17] Abouammoh AM, Ahmed AN. The new better than used failure rate class of life distributions. Adv Appl Probab 1988;20:237-40.

[18] Dykstra RL. Ordering, starshaped. In: Encyclopedia of statistical sciences, vol. 6. New York: Wiley; 1985. p.499-501.

[19] Block HW, Savits TH. The IFRA closure problem. Ann Probab 1976;4:1030-2.

[20] Esary JD, Marshall AW, Proschan F. Shock models and wear processes. Ann Probab 1973;1:627-47.

[21] Block HW, Savits TH. Laplace transforms for classes of life distributions. Ann Probab 1980;8:465-74.

[22] Cao J, Wang Y. The NBUC and NWUC classes of life distributions. J Appl Probab 1991;28:473-9.

[23] Hendi MI, Mashhour AF, Montassser MA. Closure of the NBUC class under formation of parallel systems. J Appl Probab 1993;30:975-8.

[24] Li X, Li Z, Jing B. Some results about the NBUC class of life distributions. Stat Probab Lett 2000;46:229-37.

[25] Patel JK. Hazard rate and other classifications of distributions. In: Encyclopedia in statistical sciences, vol. 3. New York: Wiley; 1983. p.590-4.

[26] Chen Y. Classes of life distributions and renewal counting process. J Appl Probab 1994;31:1110-5.

[27] Lai CD, Xie M, Murthy DNP. Bathtub-shaped failure life distributions. In: Balakrishnan N, Rao CR, editors. Handbook of statistics, vol. 20. Amsterdam: Elsevier Science; 2001. p.69-104.

[28] Mi J. Bathtub failure rate and upside-down bathtub mean residual life. IEEE Trans Reliab 1995;R-44:388-91.

[29] Jiang R, Murthy DNP. Parametric study of competing risk model involving two Weibull distributions. Int J Reliab Qual Saf Eng 1997;4:17-34.

[30] Park KS. Effect of burn-in on mean residual life. IEEE Trans Reliab 1985;R-34:522-3.

[31] Kao JHK. A graphical estimation of mixed Weibull parameters in life testing of electronic tubes. Technometrics 1959;1:389-407.

[32] Lieberman GJ. The status and impact of reliability methodology. Nav Res Logist Q 1969;14:17-35.

[33] Glaser RE. Bathtub and related failure rate characterizations. J Am Stat Assoc 1980;75:667-72.

[34] Lawless JF. Statistical models and methods for life time data. New York: John Wiley; 1982.

[35] Chen Z. A new two-parameter lifetime distribution with bathtub shape or increasing failure rate function. Stat Probab Lett 2000;49:155-61.

[36] Guess F, Hollander M, Proschan F. Testing exponentiality versus a trend change in mean residual life. Ann Stat 1986;14:1388-98.

[37] Gupta RC, Akman HO. Mean residual life function for certain types of non-monotonic ageing. Commun Stat Stochast Models 1995;11:219-25.

[38] Mitra M, Basu SK. Shock models leading to non-monotonic aging classes of life distributions. J Stat Plan Infer 1996;55:131-8.

[39] Guess F, Proschan F. Mean residual life: theory and applications. In: Krishnaiah PR, Rao CR, editors. Handbook of statistics, vol. 7. Amsterdam: Elsevier Science; 1988. p.215-24.

[40] Newby M. Applications of concepts of ageing in reliability data analysis. Reliab Eng 1986;14:291–308.

[41] Franco M, Ruiz JM, Ruiz MC. On closure of the IFR(2) and NBU(2) classes. J Appl Probab 2001;38:235–41.

[42] Li XH, Kochar SC. Some new results involving the NBU(2) class of life distributions. J Appl Probab 2001;38:242–7.

[43] Singh H, Deshpande JV. On some new ageing properties. Scand J Stat 1985;12:213–20.

[44] Kulaseker KB, Park HD. The class of better mean residual life at age t_0. Microelectron Reliab 1987;27:725–35.

[45] Launer RL. Inequalities for NBUE and NWUE life distributions. Oper Res 1984;32:660–7.

[46] Joe H, Proschan F. Percentile residual life functions. Oper Res 1984;32:668–78

[47] Sengupta D, Deshpande JV. Some results on relative ageing of two life distributions. J Appl Probab 1994;31:991–1003.

[48] Kalashnikov VV, Rachev ST. Characterization of queuing models and their stability. In: Prohorov YuK, Statulevicius VA, Sazonov VV, Grigelionis B, editors. Probability theory and mathematical statistics, vol. 2. Amsterdam: VNU Science Press; 1986. p.37–53.

[49] Boland PJ. A reliability comparison of basic systems using hazard rate functions. Appl Stochast Models Data Anal 1998;13:377–84.

[50] Singh H, Singh RS. On allocation of spares at component level versus system level. J Appl Probab 1997;34:283–7.

[51] Lai CD, Xie M. Relative ageing for two parallel systems and related problems. Math Comput Model 2002;in press.

[52] Boland PJ, El-Neweihi E. Component redundancy versus system redundancy in the hazard rate ordering. IEEE Trans Reliab 1995;R-8:614–9.

[53] Shaked M, Shanthikumar JG. Stochastic orders and their applications. San Diego: Academic Press; 1994.

[54] Buchanan WB, Singpurwalla ND. Some stochastic characterizations of multivariate survival. In: Tsokos CP, Shimi IN, editors. The theory and applications of reliability. New York: Academic Press; 1977. p.329–48.

[55] Esary JD, Marshall AW. Multivariate distributions with increasing hazard average. Ann Probab 1979;7:359–70.

[56] Block HW, Savits TH. Multivariate classes in reliability theory. Math Oper Res 1981;6:453–61.

[57] Basu AP, Ebrahimi N, Klefsjö B. Multivariate harmonic new better than used in expectation distributions. Scand J Stat 1983;10:19–25.

[58] Harris R. A multivariate definition for increasing hazard rate distribution. Ann Math Stat 1970;41:713–7.

[59] Brindley EC, Thompson WA. Dependence and ageing aspects of multivariate survival. J Am Stat Assoc 1972;67:822–30.

[60] Marshall AW. Multivariate distributions with monotonic hazard rate. In: Barlow RE, Fussel JR, Singpurwalla ND, editors. Reliability and fault tree analysis—theoretical and applied aspects of system reliability and safety assessment. Philadelphia: Society for Industrial and Applied Mathematics; 1975. p.259–84.

[61] Marshall AW, Olkin I. A multivariate exponential distribution. J Am Stat Assoc 1967;62:291–302.

[62] Basu AP. Bivariate failure rate. J Am Stat Assoc 1971;66:103–4.

[63] Block HW. Multivariate reliability classes. In: Krishnaiah PR, editor. Applications of statistics. Amsterdam: North-Holland; 1977. p.79–88.

[64] Johnson NL, Kotz S. A vector multivariate hazard rate. J Multivar Anal 1975;5:53–66.

[65] Johnson NL, Kotz S. A vector valued multivariate hazard rate. Bull Int Stat Inst 1973;45(Book 1): 570–4.

[66] Savits TH. A multivariate IFR class. J Appl Probab 1985:22:197–204.

[67] Shanbhag DN, Kotz S. Some new approaches to multivariate probability distributions. J Multivar Anal 1987;22:189–211

[68] Block HW, Savits TH. Multivariate increasing failure rate distributions. Ann Probab 1980;8:730–801.

[69] Buchanan WB, Singpurwalla ND. Some stochastic characterizations of multivariate survival. In: Tsokos CP, Shimi IN, editors. The theory and applications of reliability, with emphasis on Bayesian and nonparametric methods, vol. 1. New York: Academic Press; 1977. p.329–48.

[70] Block HW, Savits TH. The class of MIFRA lifetime and its relation to other classes. Nav Res Logist Q 1982;29:55–61.

[71] Shaked M, Shantikumar JG. IFRA processes. In: Basu AP, editor. Reliability and quality control. Amsterdam: North-Holland; 1986. p.345–52.

[72] Muhkerjee SP, Chatterjee A. A new MIFRA class of life distributions. Calcutta Stat Assoc Bull 1988;37:67–80.

[73] Marshall AW, Shaked M. Multivariate shock models for distribution with increasing failure rate average. Ann Probab 1979;7:343–58.

[74] Marshall AW, Shaked M. A class of multivariate new better than used distributions. Ann Probab 1982;10:259–64.

[75] Marshall AW, Shaked M. Multivariate new better than used distributions. Math Oper Res 1986;11:110–6.

[76] Marshall AW, Shaked M. Multivariate new better than used distributions: a survey. Scand J Stat 1986;13:277–90.

[77] Arnold BC, Zahedi H. On multivariate mean remaining life functions. J Multivar Anal 1988:25:1–9.

[78] Zahedi H. Some new classes of multivariate survival distribution functions. J Stat Plan Infer 1985;11:171–88.

[79] Klefsjö B. On some classes of bivariate life distributions. Statistical Research Report 1980-9, Department of Mathematical Statistics, University of Umea, 1980.

[80] Basu AP, Ebrahimi N. HNBUE and HNWUE distributions—a survey. In: Basu AP, editor. Reliability and quality control. Amsterdam: North-Holland; 1986. p.33–46.

[81] Ascher S. A survey of tests for exponentiality. Commun Stat Theor Methods 1990;19:1811–25.

[82] Cox DR, Oakes D. Analysis of survival data. London: Chapman and Hall; 1984.

[83] Klefsjö B. Some tests against ageing based on the total time on test transform. Commun Stat Theor Methods 1983;12:907–27.

[84] Klefsjö B. Testing exponentiality against HNBUE. Scand J Stat 1983;10:67–75.

[85] Ahmad IA. A nonparametric test for the monotonicity of a failure rate function. Commun Stat 1975;4:967–74.

[86] Ahmad IA. Corrections and amendments. Commun Stat Theor Methods 1976;5:15.

[87] Hoeffiding W. A class of statistics with asymptotically normal distributions. Ann Math Stat 1948;19:293–325.

[88] Deshpande JV. A class of tests for exponentiality against increasing failure rate average alternatives. Biometrika 1983;70:514–8.

[89] Kochar SC. Testing exponentiality against monotone failure rate average. Commun Stat Theor Methods 1985;14:381–92.

[90] Link WA. Testing for exponentiality against monotone failure rate average alternatives. Commun Stat Theor Methods 1989;18:3009–17.

[91] Hollander M, Proschan F. Tests for the mean residual life. Biometrika 1975;62:585–93.

[92] Hollander M, Proschan F. Amendments and corrections. Biometrika 1980;67:259.

[93] Hollander M, Proschan F. Testing whether new is better than used. Ann Math Stat 1972;43:1136–46.

[94] Koul HL. A test for new is better than used. Commun Stat Theor Methods 1977;6:563–73.

[95] Koul HL. A class of testing new is better than used. Can J Stat 1978;6:249–71.

[96] Alam MS, Basu AP. Two-stage testing whether new is better than used. Seq Anal 1990;9:283–96.

[97] Deshpande JV, Kochar SC. A linear combination of two U-statistics for testing new better than used. Commun Stat Theor Methods 1983;12:153–9.

[98] Kumazawa Y. A class of test statistics for testing whether new better than used. Commun Stat Theor Methods 1983;12:311–21.

[99] Hollander M, Proschan F. Testing whether new is better than used. Ann Math Stat 1972;43:1136–46.

[100] De Souza Borges W, Proschan F. A simple test for new better than used in expectation. Commun Stat Theor Methods 1984;13:3217–23.

[101] Singh H, Kochar SC. A test for exponentiality against HNBUE alternative. Commun Stat Theor Methods 1986;15:2295–2304.

[102] Kochar SC, Deshpande JV. On exponential scores statistics for testing against positive ageing. Stat Probab Lett 1985;3:71–3.

[103] Hollander M, Park HD, Proschan F. A class of life distributions for ageing. J Am Stat Assoc 1986;81:91–5.

[104] Ebrahimi N, Habibullah M. Testing whether survival distribution is new better than used of specific age. Biometrika 1990;77:212–5.

[105] Joe H, Proschan F. Tests for properties of the percentile residual life function. Commun Stat Theor Methods 1983;12:1087–119.

[106] Joe H, Proschan F. Percentile residual life functions. Oper Res 1984;32:668–78.

[107] Xie M. Some total time on test quantiles useful for testing constant against bathtub-shaped failure rate distributions. Scand J Stat 1988;16:137–44.

[108] Belzunce F, Candel J, Ruiz JM. Testing the stochastic order and the IFR, DFR, NBU, NWU ageing classes. IEEE Trans Reliab 1998;R47:285–96.

[109] Wells MT, Tiware RC. A class of tests for testing an increasing failure rate average distributions with randomly right-censored data. IEEE Trans Reliab 1991;R 40:152–6.

[110] Chen YY, Hollander M, Langberg NA. Testing whether new is better than used with randomly censored data. Ann Stat 1983;11:267–74.

[111] Kaplan EL, Meier P. Nonparametric estimation from incomplete observations. J Am Stat Assoc 1958;53:457–81.

[112] Koul HL, Susarla V. Testing for new better than used in expectation with incomplete data. J Am Stat Assoc 1980;75:952–6.

[113] Hollander M, Park HD, Proschan F. Testing whether new is better than used of a specified age, with randomly censored data. Can J Stat 1985;13:45–52.

[114] Sen K, Jain MB. A test for bivariate exponentiality against BIFR alternative. Commun Stat Theor Methods 1991;20:3139–45.

[115] Bandyopadhyay D, Basu AP. A class of tests for exponentiality against bivariate increasing failure rate alternatives. J Stat Plan Infer 1991;29:337–49.

[116] Basu AP Habibullah M. A test for bivariate exponentiality against BIFRA alternative. Calcutta Stat Assoc Bull 1987;36:79–84.

[117] Sen K, Jain MB. Tests for bivariate mean residual life. Commun Stat Theor Methods 1991;20:2549–58.

[118] Basu AP, Ebrahimi N. Testing whether survival function is bivariate new better than used. Commun Stat 1984;13:1839–49.

[119] Sen K, Jain MB. A new test for bivariate distributions: exponential vs new-better-than-used alternative. Commun Stat Theor Methods 1991;20:881–7.

[120] Sen K, Jain MB. A test for bivariate exponentiality against BHNBUE alternative. Commun Stat Theor Methods 1990;19:1827–35.

[121] Block HW, Savits TH. Multivariate distributions in reliability theory and life testing. In: Taillie C, Patel GP, Baldessari BA, editors. Statistical distributions in scientific work, vol. 5. Dordrecht, Boston (MA): Reidel; 1981. p.271–88.

[122] Block HW, Savits, TH. Multivariate nonparametric classes in reliability. In: Krishnaiah PR, Rao CR, editors. Handbook of statistics, vol. 7. Amsterdam: North-Holland; 1988. p.121–9.

[123] Ghosh M, Ebrahimi N. Shock models leading multivariate NBU and NBUE distributions. In: Sen PK, editor. Contributions to statistics: essays in honour of Norman Lloyd Johnson. Amsterdam: North-Holland; 1983. p.175–84.

[124] Savits TH. Some multivariate distributions derived from a non-fatal shock model. J Appl Probab 1988;25:383–90.

Class of NBU-t_0 Life Distribution

Dong Ho Park

Chapter 10

10.1 Introduction

The notion of aging plays an important role in reliability theory, and several classes of life distributions (*i.e.* distributions for which $F(t) = 0$ for $t < 0$) have been proposed and discussed to categorize different aspects of aging. Among those, the classes of life distributions that have been shown to be fundamental in the study of maintenance policies are the new better than used (NBU) and the new better than used in expectation (NBUE) classes. Marshall and Proschan [1] and Esary *et al.* [2] discuss the maintenance policy of a system when the underlying life distributions belong to either the NBU class or the NBUE class.

Definition 1. A life distribution F is an NBU distribution if $\bar{F}(x+y) \le \bar{F}(x)\bar{F}(y)$ for all $x, y \ge 0$, where $\bar{F} \equiv 1 - F$ denotes the survival function. The dual concept of a new worse than used (NWU) distribution is defined by reversing the inequality (*i.e.* F is NWU if $\bar{F}(x+y) \ge \bar{F}(x)\bar{F}(y)$ for all $x, y \ge 0$).

Definition 2. A life distribution F is an NBUE distribution if $\int_0^\infty \bar{F}(x)\,\mathrm{d}x < \infty$ and $\varepsilon_F(0) \ge \varepsilon_F(t)$ for all $t \ge 0$, where $\varepsilon_F(t) = \int_t^\infty \bar{F}(x)\,\mathrm{d}x / \bar{F}(t)$ is the mean residual life at age t. It is new worse than used in expectation (NWUE) if $\varepsilon_F(0) \le \varepsilon_F(t)$ for all $t \ge 0$.

The NBU property may be interpreted as stating that a used item of any age has a stochastically smaller residual life length than does a new item, and the NBUE property implies that the mean residual life length at any age $t \ge 0$ is less than or equal to the mean life length of a new item, *i.e.* the mean residual life length at age zero. The boundary members of the NBU and NBUE classes are the exponential distributions, for which residual life lengths do not improve or deteriorate with age.

Other well-known classes of life distributions that have been categorized according to monotonicity properties of the failure rate, the average failure rate, and the mean residual life length include the following:

Definition 3. A life distribution F is an increasing failure rate (IFR) distribution if $F(0)=0$ and $\bar{F}(x+t)/\bar{F}(t)$ is decreasing in t for $0<t<F^{-1}(1)$ and $x>0$. Here $F^{-1}(1)$ is defined as $\sup_t\{t : F(t)<1\}$. It is a decreasing failure rate (DFR) distribution if $\bar{F}(x+t)/\bar{F}(t)$ is increasing in t for $0<t<F^{-1}(1)$. If F has a density f, then F is IFR (DFR) if the failure rate $r(t)=f(t)/\bar{F}(t)$ is increasing (decreasing) for $0<t<F^{-1}(1)$.

Definition 4. A life distribution F is an increasing failure rate average (IFRA) distribution if $F(0)=0$ and $-(1/t)\log\bar{F}(t)$ is increasing in t for $0<t<F^{-1}(1)$. It is a decreasing failure rate average (DFRA) distribution if $-(1/t)\log\bar{F}(t)$ is decreasing in t for $0<t<F^{-1}(1)$. If F has a density f, then F is IFRA (DFRA) if the average failure rate $(1/t)\int_0^t r(u)\,du$ is increasing (decreasing) for $0<t<F^{-1}(1)$.

Definition 5. A life distribution F is a decreasing mean residual life (DMRL) distribution if $\int_0^\infty \bar{F}(t)\,dt<\infty$ and $\varepsilon_F(s)\geq\varepsilon_F(t)$ for all $0\leq s\leq t$. It is an increasing mean residual life (IMRL) distribution if $\varepsilon_F(0)\leq\varepsilon_F(t)$ for all $0\leq s\leq t$.

It is well known [3] that the following implications among these classes of life distributions hold if F has a finite mean:

$$\begin{array}{lllllll} \text{IFR} & \Rightarrow & \text{IFRA} & \Rightarrow & \text{NBU} & \Rightarrow & \text{NBUE} \\ \text{IFR} & \Rightarrow & \text{DMRL} & \Rightarrow & \text{NBUE} & & \end{array}$$

Similar relationships exist for the dual classes, DFR, DFRA, IMRL, NWU, and NWUE. It is also known that the exponential distributions are the only distributions that are both IFR and DFR, both IFRA and DFRA, both NBU and NWU, both DMRL and IMRL, and both NBUE and NWUE.

We introduce a new class of life distributions, which is obtained by relaxing the conditions for the NBU (NWU) class somewhat.

Definition 6. Let $t_0\geq 0$. A life distribution F is new better than used at t_0 (NBU-t_0) if

$$\bar{F}(x+t_0)\leq\bar{F}(x)\bar{F}(t_0) \quad \text{for all } x\geq 0 \tag{10.1}$$

The dual notion of new worse than used at t_0 (NBU-t_0) is defined analogously by reversing the first inequality in Equation 10.1.

It is obvious that the NBU-t_0 class is related to, but contains and is much larger than, the NBU class. The NBU-t_0 property states that a used item of age t_0 has a stochastically smaller residual life length than does a new item, whereas the NBU property states that a used item of any age has a stochastically smaller residual life length than does a new item. Section 10.2 characterizes the NBU-t_0 and NWU-t_0 classes and considers their preservation properties under several reliability operations. The boundary members of the NBU-t_0 and NWU-t_0 classes are also discussed. In Section 10.3, the estimation of the NBU-t_0 (NWU-t_0) life distribution is described. Section 10.4 presents various nonparametric tests of exponentiality against the NBU-t_0 (NWU-t_0) alternatives based on complete and incomplete data.

10.2 Characterization of NBU-t_0 Class

10.2.1 Boundary Members of NBU-t_0 and NWU-t_0

It is well known that the only life distributions that belong to both IFR and DFR, IFRA and DFRA, NBU and NWU, DMRL and IMRL, and NBUE and NWUE are the class of exponential distributions. However, not only the class of exponential distributions, but also some other life distributions belong to the boundary members of NBU-t_0 and NWU-t_0.

Let C denote the class of boundary members of NBU-t_0 and NWU-t_0. That is, for $t_0>0$

$$C_0=\{F : \bar{F}(x+t_0)=\bar{F}(x)\bar{F}(t_0) \text{ for all } x\geq 0\} \tag{10.2}$$

Using theorem 2 of Marsaglia and Tubilla [4], we may easily verify that the following distributions F_1, F_2, and F_3 are in C_0:

$$\begin{aligned} \bar{F}_1(x) &= \exp(-\lambda x) \quad \lambda>0,\ x\geq 0 \\ \bar{F}_2(x) &= \bar{G}(x) \quad 0\leq x<\infty \end{aligned} \tag{10.3}$$

where $\bar{G}(x)$ is a survival function for which $\bar{G}(0) = 1$ and $\bar{G}(t_0) = 0$.

$$\begin{aligned}
\bar{F}_3 &= \bar{G}(x) && \text{for } 0 \le x < t_0 \\
&= \bar{G}(t_0)\bar{G}(x - t_0) && \text{for } t_0 \le x < 2t_0 \\
&= \bar{G}^2(t_0)\bar{G}(x - 2t_0) && \text{for } 2t_0 \le x < 3t_0 \\
&\vdots \\
&= \bar{G}^n(t_0)\bar{G}(x - nt_0) && \text{for } nt_0 \le x < (n+1)t_0 \\
&\vdots
\end{aligned}$$

where $\bar{G}(x)$ is a survival function defined for $x \ge 0$. Note that if G has a density function on $[0, t_0]$, then the failure rate of F_3 is periodic with period t_0.

Since the NBU-t_0 class contains the NBU class, it is of interest to find what kinds of life distributions belong to the NBU-t_0 class, but not to the NBU class. We consider

$$C_a = \{F : \bar{F}(x + t_0) \le \bar{F}(x)\bar{F}(t_0) \text{ for all } x \ge 0, \text{ and equality holds for some } x \ge 0\} \quad (10.4)$$

Then C_a is the class of NBU-t_0 life distributions excluding the boundary members of NBU-t_0 and NWU-t_0, defined in Equation 10.2. Let C^* be the class of life distributions that are not NBU, but are in C_a. Theorem 1 gives a method of constructing some distribution functions in C^*.

Given a survival function $\bar{H}$, let $\bar{H}_t(x) \equiv \bar{H}(t + x)/\bar{H}(t)$ be the conditional survival function. Recall that for $x \ge 0$:

$$\bar{H}(x) = \exp\left[-\int_0^x r_H(u)\,du\right] \quad (10.5)$$

$$\bar{H}_t(x) = \exp\left[-\int_t^{t+x} r_H(u)\,du\right] \quad (10.6)$$

when H has a failure rate function r_H.

Theorem 1. *Let G be NBU with failure rate function $r_G(x) > 0$ for $0 \le x < \infty$ and let F have a failure rate function r_F satisfying:*

$$r_F(x) \le r_G(x) \quad \textit{for } 0 \le x \le t_0 \quad (10.7)$$

$$r_F(x) = r_G(x) \quad \textit{for } t_0 \le x \le \infty \quad (10.8)$$

and

$$r_F(x) \textit{ is strictly decreasing for } 0 \le t \le t_1 \quad (10.9)$$

where $0 < t_1 < t_0$. Then F is NBU-t_0 but not NBU.

Proof. F is not NBU by Equation 10.9. To show that F is NBU-t_0, note that for $x \ge 0$, $\bar{F}_{t_0}(x) = \bar{G}_{t_0}(x)$ by Equations 10.6 and 10.8. Also, $\bar{G}_{t_0}(x) \le \bar{G}(x)$ since G is NBU, and $\bar{G}(x) \le \bar{F}(x)$ by Equations 10.7 and 10.8. Thus, we conclude that, for $x \ge 0$, $\bar{F}_{t_0}(x) \le \bar{F}(x)$; *i.e.* F is NBU-t_0. □

Example 1. As an example of Theorem 1, let $r_G(x) = 1$ for $0 \le x < \infty$, and let $r_F(x) = 1 - (\theta/t_0)x$ for $0 \le x < t_0$ and $0 < \theta \le 1$, and $r_F(x) = 1$ for $t_0 \le x < \infty$. We do not let θ exceed unity, since we want to ensure that $r_F(x)$ remains positive as $x \to t_0$. Then r_F satisfies Equations 10.7–10.9 and thus F is in C^*. (See Figure 10.1.)

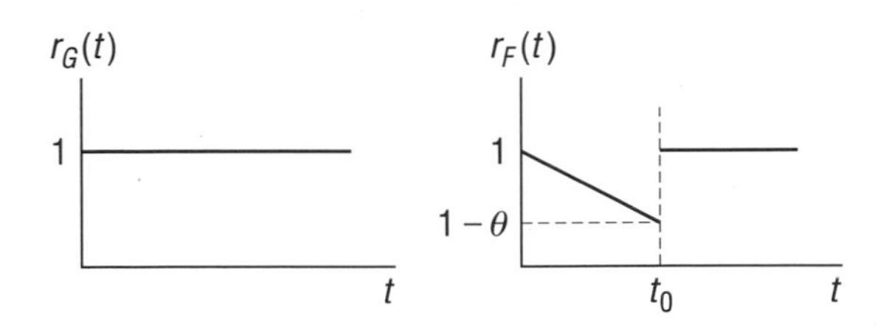

Figure 10.1. Failure rates for F and G of Example 1

Using Equation 10.3 and the failure rates defined in Example 1, if we extend the range of θ to include $\theta = 0$, we can construct

$$\begin{aligned}
\bar{F}(x; \theta) &= \exp\{-[x - \theta(2t_0)^{-1}x^2]\} \quad 0 \le x < t_0 \\
&= \exp\{-[x - \theta(2)^{-1}t_0]\} \quad x \ge t_0
\end{aligned}$$

for $0 \le \theta \le 1$. For $\theta = 0$, F is reduced to the exponential distribution which is both NBU and NBU-t_0. For $0 < \theta \le 1$, F is NBU-t_0, but not NBU. Another possible NBU-t_0 life distribution that is not NBU can be constructed as in the following example.

Example 2. Consider a distribution function F with the following failure rate $r_F(t)$ shown in

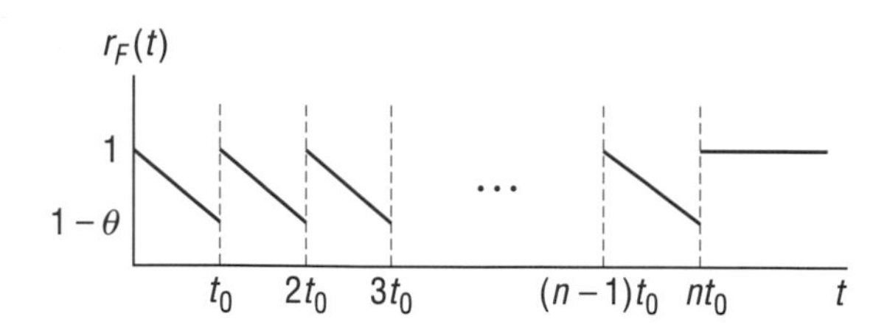

Figure 10.2. Failure rate for F of Example 2

Figure 10.2. Then for $n \geq 1$

$$\begin{aligned}
\bar{F}(x) &= \exp[-(\theta/2t_0)x^2] \\
&\qquad \text{for } 0 \leq x < t_0 \\
&= \exp\{-[-\theta t_0 + (1+\theta)x - (\theta/2t_0)x^2]\} \\
&\qquad \text{for } t_0 \leq x < 2t_0 \\
&= \exp\{-[-3\theta t_0 + (1+2\theta)x - (\theta/2t_0)x^2]\} \\
&\qquad \text{for } 2t_0 \leq x < 3t_0 \\
&\ \vdots \\
&= \exp(-\{-[n(n-1)\theta/2]t_0 \\
&\quad + [1+(n-1)\theta]x - (\theta/2t_0)x^2\}) \\
&\qquad \text{for } (n-1)t_0 \leq x < nt_0 \\
&= \exp\{-[-(n/2)\theta t_0 + x]\} \\
&\qquad \text{for } x \geq nt_0
\end{aligned}$$

To see that F is in C_a, we note that, for $x + t_0 \leq nt_0$

$$\bar{F}(x+t_0) = \bar{F}(t_0)\bar{F}(x+t_0-t_0) = \bar{F}(t_0)\bar{F}(x)$$

and for $x + t_0 > nt_0$

$$\begin{aligned}
\bar{F}(x+t_0) &= \exp\{-[-(n/2)\theta t_0 + t_0 + x]\} \\
&= \bar{F}(t_0)\exp(-\{-[(n-1)/2]\theta t_0 + x\})
\end{aligned}$$

Case 1. $(n-1)t_0 \leq x < nt_0$:

$$\begin{aligned}
\bar{F}(x) &= \exp(-\{-[n(n-1)/2]\theta t_0 \\
&\quad + [1+(n-1)\theta]x - (\theta/2t_0)x^2\}) \\
&= \exp\{-[-(n-1)/2]\theta t_0 + x\} \\
&\quad \times \exp\{(\theta/2t_0)[x-(n-1)t_0]^2\} \\
&\geq \exp\{-[-(n-1)/2]\theta t_0 + x\}
\end{aligned}$$

Thus $\bar{F}(x+t_0) \leq \bar{F}(x)\bar{F}(t_0)$.

Case 2. $x \geq nt_0$:

$$\begin{aligned}
\bar{F}(x) &= \exp\{-[-(n/2)\theta t_0 + x]\} \\
&\geq \exp\{-[-(n-1)/2]\theta t_0 + x\}
\end{aligned}$$

Thus $\bar{F}(x+t_0) \leq \bar{F}(x)\bar{F}(t_0)$. In both cases $\bar{F}(x+t_0) \leq \bar{F}(x)\bar{F}(t_0)$ and so F is in C_a. It is obvious that F is not an NBU distribution.

10.2.2 Preservation of NBU-t_0 and NWU-t_0 Properties under Reliability Operations

Table 10.1 summarizes the preservation properties of several known classes of life distributions under reliability operations. With the exception of the DMRL class, each class representing adverse aging is closed under convolution of distributions. On the other hand, none of these classes is closed under mixture of distributions. For the classes representing beneficial aging, the reverse is true: Each is closed under mixture of distributions (in the NWU and NWUE cases, the distributions being mixed are non-crossing). On the other hand, none of these beneficial aging life distribution classes is closed under convolution of distributions.

We now consider preservation properties of the NBU-t_0 and NWU-t_0 classes under those reliability operations specified in Table 10.1.

Theorem 2. *The NBU-t_0 class is preserved under the formation of coherent system.*

Proof. The proof is exactly analogous to that of the corresponding result for the NBU class. In the proof of theorem 5.1 of Barlow and Proschan [5], p.182–3, simply replace s by x and t by t_0. □

Theorem 3. *The NWU-t_0 class is (a) preserved under mixture of non-crossing distributions and (b) not preserved under arbitrary mixtures.*

Proof. (a) In the proof of the corresponding result for the NWU class (Barlow and Proschan [5], p.186, theorem 5.7), substitute x for s and t_0

Table 10.1. Preservation properties of classes of life distributions

Class of life distribution	Reliability operation	Formation of coherent structure	Convolution of life distributions	Mixture of life distributions	Mixture of non-crossing life distributions
Adverse aging	IFR	Not closed	Closed	Not closed	Not closed
	IFRA	Closed	Closed	Not closed	Not closed
	NBU	Closed	Closed	Not closed	Not closed
	NBUE	Not closed	Closed	Not closed	Not closed
	DMRL	Not closed	Not closed	Not closed	Not closed
	HNBUE	Not closed	Closed	Not closed	Not closed
Beneficial aging	DFR	Not closed	Not closed	Closed	Closed
	DFRA	Not closed	Not closed	Closed	Closed
	NWU	Not closed	Not closed	Not closed	Closed
	NWUE	Not closed	Not closed	Not closed	Closed
	IMRL	Not closed	Not closed	Closed	Closed
	HNWUE	Not closed	Not closed	Closed	Not closed

for t. (b) The example given in 5.9, Barlow and Proschan [5], p.197, may be used. □

Both NBU-t_0 and NWU-t_0 classes are not closed under other reliability operations and such properties are exhibited by providing the following counter examples in those reliability operations.

Example 3. The NBU-t_0 class is not preserved under convolution. Let F be the distribution that places mass $\frac{1}{2}-\varepsilon$ at the point Δ_1, mass $\frac{1}{2}-\varepsilon$ at the point $\frac{5}{8}-\Delta_2$, and mass 2ε at the point $1+\frac{1}{2}\Delta_1$, where Δ_1, Δ_2, and ε are "small" positive numbers. Let $t_0=1$. Then it is obvious that F is NBU-t_0, since a new item with life distribution F survives at least until time Δ_1 with probability of unity, whereas an item of age t_0 has residual life $\frac{1}{2}\Delta_1$ with probability unity.

Next, consider $F^{(2)}(t_0+x)=P[X_1+X_2>t_0+x]$, where X_1 and X_2 are independent, identically distributed (i.i.d.) $\sim F$, $t_0=1$ (as above), and $x=\frac{1}{4}$. Then $F^{(2)}(\frac{5}{4})=(\frac{1}{2}+\varepsilon)^2$, since $X_1+X_2>\frac{5}{4}$ if and only if $X_1\geq\frac{5}{8}-\Delta_2$ and $X_2\geq\frac{5}{8}-\Delta_2$. Similarly, $F^{(2)}(t_0)=F^{(2)}(1)=P[X_1+X_2>1]=(2\varepsilon)^2+2(2\varepsilon)$, since $X_1+X_2>1$ if and only if: (a) $X_1>\frac{5}{8}-\Delta_2$ and $X_2>\frac{5}{8}-\Delta_2$; (b) $X_1=1+\frac{1}{2}\Delta_1$ and X_2 is any value; or (c) $X_2=1+\frac{1}{2}\Delta_2$ and X_1 is any value.

Finally, $F^{(2)}(\frac{1}{4})=1-(\frac{1}{2}-\varepsilon)^2$, since $X_1+X_2>\frac{1}{4}$ except when $X_1=\Delta_1$ and $X_2=\Delta_1$. It follows that, for $t_0=1$ and $x=\frac{1}{4}$, we have

$$\begin{aligned} F^{(2)}(t_0+x) &- F^{(2)}(t_0)F^{(2)}(x) \\ &= (\tfrac{1}{2}+\varepsilon)^2-\{[(\tfrac{1}{2}+\varepsilon)^2+2(2\varepsilon)] \\ &\quad\times[1-(\tfrac{1}{2}-\varepsilon)^2]\} \\ &= \tfrac{1}{4}-(\tfrac{1}{4}\tfrac{3}{4})+o(\varepsilon) \end{aligned}$$

which is greater than zero for sufficiently small ε. Thus $F^{(2)}$ is not NBU-t_0.

Example 4. The NWU-t_0 class is not preserved under convolution. The exponential distribution $F(x)=1-\mathrm{e}^{-x}$ is NWU-t_0. The convolution $F^{(2)}$ of F with itself is the gamma distribution of order two: $F^{(2)}(x)=1-(1+x)\,\mathrm{e}^{-x}$, with strictly increasing failure rate. Thus $F^{(2)}$ is not NWU-t_0.

Example 5. The NWU-t_0 class is not preserved under the formation of coherent systems. This may be shown using the same example as is used for the analogous result for NWU systems in Barlow and Proschan [5], p.183.

Example 6. The NBU-t_0 class is not preserved under mixtures. The following example shows that a mixture of NBU-t_0 distributions need not be NBU-t_0. Let $\bar{F}_\alpha(x)=\mathrm{e}^{-\alpha x}$ and $\bar{G}(x)=\int_0^\infty \bar{F}_\alpha(x)\,\mathrm{e}^{-\alpha}\,\mathrm{d}\alpha=(x+1)^{-1}$. Then the density

Table 10.2.

Class of life distribution	Formation of coherent structure	Convolution of life distributions	Mixture of life distributions	Mixture of non-crossing life distributions
NBU-t_0	Closed	Not closed	Not closed	Not closed
NWU-t_0	Not closed	Not closed	Not closed	Closed

function is $g(x) = (x+1)^{-2}$ and the failure rate function is $r_g(x) = (x+1)^{-1}$, which is strictly decreasing in $x \geq 0$. Thus G is not NBU-t_0.

In summary, we have Table 10.2.

10.3 Estimation of NBU-t_0 Life Distribution

In this section we describe the known estimators of the survival function when the underlying life distribution belongs to the class of NBU-t_0 distributions. Reneau and Samaniego [6] proposed an estimator of the survival function that is known to be a member of the NBU-t_0 class and studied many properties, including the consistency of the estimator and convergence rates. Confidence procedures based on the estimator were also discussed. The Reneau and Samaniego (RS) estimator is shown to be negatively biased in the sense that the expectation of the estimator is bounded above by its survival function. Chang [7] provides sufficient conditions for the weak convergence for the RS estimator to hold. Chang and Rao [8] modified the RS estimator to propose a new estimator, which improved the optimality properties such as mean-squared error and goodness of fit. They also developed an estimator that is positively biased.

Let $X_1, \ldots, X_n$ be a random sample from the survival function $\bar{F}$, which is known to be a member of the NBU-t_0 class. Denote the empirical survival function by $\bar{F}_n$. It is well known that $\bar{F}_n$ is an unbiased estimator of $\bar{F}$ and almost surely converges to $\bar{F}$ at an optimal rate (see Serfling [9], theorem 2.1.4B). However, $\bar{F}_n$ is, in general, not a member of the NBU-t_0 class.

10.3.1 Reneau–Samaniego Estimator

The first non-parametric estimator of $\bar{F}$ when $\bar{F}$ belongs to the NBU-t_0 class was proposed by Reneau and Samaniego [6]. Let $t_0 > 0$ be fixed. To find an NBU-t_0 estimator of $\bar{F}$, the condition in Equation 10.1 can be rewritten as

$$\bar{F}(x) \leq \bar{F}(x - t_0)\bar{F}(t_0) \quad \text{for all } x \geq t_0 \qquad (10.10)$$

The procedure to construct an NBU-t_0 estimator is as follows. For $x \in [0, t_0]$, we let $\hat{\bar{F}}(x) = \bar{F}_n(x)$. Since $\hat{\bar{F}}(x) \leq \hat{\bar{F}}(x - t_0)\hat{\bar{F}}(t_0) = \bar{F}_n(x - t_0)\bar{F}_n(t_0)$ for $t_0 \in [t_0, 2t_0]$, we let $\hat{\bar{F}}(x) = \bar{F}_n(x)$ if $\bar{F}_n(x) \leq \bar{F}_n(x - t_0)\bar{F}_n(t_0)$. Otherwise, we define $\hat{\bar{F}}(x) = \bar{F}_n(x - t_0)\bar{F}_n(t_0)$. The procedure continues in this fashion for $x \in (kt_0, (k+h)t_0]$, where $k \geq 2$ is an integer, and consequently we may express the estimator by the following recursion:

$$\begin{aligned} \hat{\bar{F}}(x) &= \bar{F}_n(x) \quad \text{for } 0 \leq x \leq t_0 \\ &= \min\{\bar{F}_n(x), \hat{\bar{F}}(x - t_0)\hat{\bar{F}}(t_0)\} \\ &\qquad \text{for } x > t_0 \end{aligned} \qquad (10.11)$$

Theorem 4. *The estimator, given in Equation 10.11, uniquely defines an NBU-t_0 survival function that may be written as a function of $\bar{F}_n$ as follows:*

$$\hat{\bar{F}}(x) = \min_{0 \leq k \leq [x/t_0]} \{\bar{F}_n^k(t_0)\bar{F}_n(x - kt_0)\}$$

where $[u]$ denotes the greatest integer less than or equal to u.

Proof. See Reneau and Samaniego [6]. □

Theorem 4 assures that the estimator of Equation 10.11 is indeed in the NBU-t_0 class. It is obvious by the method of construction of the estimator that $\hat{\bar{F}}(x) = \bar{F}_n(x)$ for $0 \leq x \leq t_0$ and

$\hat{\bar{F}}(x) \le \bar{F}_n(x)$ for $x > t_0$. Thus it is shown that $\hat{\bar{F}}(x)$ is the largest NBU-t_0 survival function that is everywhere less than or equal to $\bar{F}_n$ for all $x \ge 0$. As a result, the RS estimator is a negatively biased estimator and the expectation of $\hat{\bar{F}}(x)$ is bounded above by $\bar{F}(x)$ for all $x \ge 0$.

The following results show that the estimator of Equation 10.11 is uniformly consistent and converges uniformly to $\bar{F}$ with probability of one. For the proofs, see Reneau and Samaniego [6].

Lemma 1. *Define*

$$D_n \equiv \sup_{x \ge 0} |\bar{F}_n(x) - \bar{F}(x)|$$

$$T(x) = \bar{F}^k(t_0)\bar{F}(x - kt_0)$$

and

$$T_n(x) = \bar{F}_n^k(t_0)\bar{F}_n(x - kt_0) \text{ for } 0 \le k \le [x/t_0]$$

Then

$$|T_n(x) - T(x)| \le D_n\{k[\bar{F}(t_0) + D_n]^{k-1} + 1\}$$

Theorem 5. *If $\bar{F}$ is in the NBU-t_0 class, then $\hat{\bar{F}}$ converges uniformly to $\bar{F}$ with probability of one.*

It is well known that the empirical survival function $\bar{F}_n$ is mean squared, consistent with rate $O(n^{-1})$ and almost surely uniformly consistent with rate $O[n^{-1/2}(\log\log n)^{1/2}]$. The following results show that the RS estimator converges to $\bar{F}$ at these same rates when $\bar{F}$ is in the NBU-t_0 class.

Theorem 6. $E|\hat{\bar{F}}(x) - \bar{F}(x)|^2 = O(n^{-1})$.

Theorem 7.

$$\sup_{x \ge 0} |\hat{\bar{F}}(x) - \bar{F}(x)| = O[n^{-1/2}(\log\log n)^{1/2}]$$

almost surely.

Reneau and Samaniego [6] also show that, for fixed $x \ge 0$, $\hat{\bar{F}}(x)$ has the same asymptotic distribution as $\bar{F}_n(x)$, which is

$$\hat{\bar{F}}(x) \to N\left(\bar{F}(x), \frac{1}{n}\bar{F}(x)[1 - \bar{F}(x)]\right)$$

as $n \to \infty$ and, based on the asymptotic distribution of $\hat{\bar{F}}(x)$, the large sample confidence procedures are discussed using the bootstrap method.

Reneau and Samaniego [6] presented a simulation study showing that the mean-squared errors of the RS estimator are smaller than those of the empirical survival function in all cases. Such a simulation study implies that the RS estimator, given in Equation 10.11, estimates the NBU-t_0 survival function more precisely than does the empirical survival function. For the simulation study, Reneau and Samaniego [6] used three NBU-t_0 distributions: a linear failure rate distribution, a Weibull distribution, and

$$\begin{aligned} \bar{F} &= \exp[-x + (\theta/2t_0)x^2] && \text{for } 0 \le x < t_0 \\ &= \exp[-x + (\theta/2t_0)] && \text{for } x \ge t_0 \end{aligned} \tag{10.12}$$

which is in the NBU-t_0 class, but not in the NBU class, Such a distribution was originally introduced by Hollander *et al.* [10]. For all three distributions, t_0 is taken to be equal to one.

Although Reneau and Samaniego [6] proved that $\hat{\bar{F}}(x)$ is the largest NBU-t_0 survival function such that $\hat{\bar{F}}(x) \le \bar{F}(x)$ for every $x \ge 0$ and $\hat{\bar{F}}$ strongly uniformly converges to $\bar{F}$ at an optimal rate, the weak convergence of the stochastic process $W_n = \sqrt{n}(\hat{\bar{F}} - \bar{F})$ was not solved. Chang [7] discussed such a weak convergence of W_n. For this purpose, Chang [7] defined two new subclasses of survival functions as follows.

A survival function is said to be New Strictly Better than Used of age t_0 if $\bar{F}(x + t_0) < \bar{F}(x)\bar{F}(t_0)$ for all $x > 0$. A survival function is said to be New is the Same as Used of age t_0 if $\bar{F}(x + t_0) = \bar{F}(x)\bar{F}(t_0)$ for all $x \ge 0$.

Chang [7] showed that the weak convergence of W_n does not hold in general and established sufficient conditions for the weak convergence to hold. Three important cases are as follows. (1) If the underlying survival function $\bar{F}$ is from the subclass of New Strictly Better than Used of age t_0 (*e.g.* gamma and Weibull distributions), then W_n converges weakly with the same limiting distribution as that of the empirical process. (2) If $\bar{F}$ is from the subclass of New is the Same as Used of age t_0 (*e.g.* exponential distributions), then the weak convergence of W_n holds, and the finite-dimensional limiting distribution is estimable.

W_n does not converge weakly to a Brownian bridge, however. (3) If $\bar{F}$ satisfies neither (1) nor (2), then the weak convergence of W_n fails to hold. For more detailed discussions on the weak convergence of W_n, refer to Chang [7].

10.3.2 Chang–Rao Estimator

10.3.2.1 Positively Biased Estimator

Chang and Rao [8] modified the RS estimator to propose several estimators of an NBU-t_0 survival function, which is itself in an NBU-t_0 class. Derivation of an estimator of an NBU-t_0 survival function is motivated by rewriting the NBU-t_0 condition of Equation 10.1 as

$$\bar{F}(x) \geq \frac{\bar{F}(x+t_0)}{\bar{F}(t_0)} \quad \text{for all } x \geq 0$$

Define $\bar{\tilde{F}}_0 = \bar{F}_n$ and for $i = 1, 2, \ldots,$ we let

$$\begin{aligned}\bar{\tilde{F}}_i(x) &= \bar{F}_n(x) \quad \text{if } \bar{F}_n(t_0) = 0 \\ &= \max\{\bar{\tilde{F}}_{i-1}(x), [\bar{\tilde{F}}_{i-1}(x+t_0)]/\bar{\tilde{F}}_{i-1}(t_0)\} \\ &\quad \text{if } \bar{F}_n(t_0) > 0\end{aligned}$$

Then, it is obvious that the functions $\bar{\tilde{F}}_i$ satisfies the inequalities

$$\bar{F}_n = \bar{\tilde{F}}_0 \leq \cdots \leq \bar{\tilde{F}}_i \leq \bar{\tilde{F}}_{i+1} \leq \cdots \leq 1$$

Define

$$\bar{\tilde{F}} = \lim_{i\to\infty} \bar{\tilde{F}}_i \tag{10.13}$$

Then, $\bar{\tilde{F}}$ is a well-defined function, and Chang and Rao [8] refer to $\bar{\tilde{F}}$ as a positively based NBU-t_0 estimator. The following results show that $\bar{\tilde{F}}$ is in the NBU-t_0 class and is a positively biased estimator of $\bar{F}$.

Theorem 8. *The estimator $\bar{\tilde{F}}$, defined in Equation 10.13, is an NBU-t_0 survival function such that $E[\bar{\tilde{F}}(x) - \bar{F}(x)] \geq 0$ for all $x \geq 0$. The empirical survival function $\bar{F}_n$ is a member of an NBU-t_0 class if and only if $\bar{\tilde{F}} = \bar{F}_n$.*

Proof. See Chang and Rao [8]. □

Under the assumption that $\bar{F}$ is an NBU-t_0 survival function with compact support, the consistency of $\bar{\tilde{F}}$ and the convergence rate are stated in Theorem 9. For the proof, refer to Chang and Rao [8].

Theorem 9. *Suppose that $\bar{F}$ is an NBU-t_0 survival function such that $T = \sup\{x : \bar{F}(x) > 0\} < \infty$. Then, $\bar{\tilde{F}}$ converges uniformly to $\bar{F}$ with probability of one.*

Theorem 10. *Under the same conditions as in Theorem 9, $\sup_{x\geq 0} |\bar{\tilde{F}}(x) - \bar{F}(x)| = O[n^{-1/2}(\log\log n)^{1/2}]$ almost surely.*

10.3.2.2 Geometric Mean Estimator

Let $0 \leq \alpha \leq 1$ and define the weighted geometric average of the RS estimator and positively biased estimator, *i.e.*

$$\bar{\hat{F}}_\alpha = [\bar{\hat{F}}]^\alpha [\bar{\tilde{F}}]^{1-\alpha} \tag{10.14}$$

For $\alpha = 1$ and $\alpha = 0$, $\bar{\hat{F}}_\alpha$ is reduced to the RS estimator and the positively biased estimator respectively. Consider the class

$$\mathcal{N} = \{\bar{\hat{F}}_\alpha : 0 \leq \alpha \leq 1\}$$

Then, each member of $\mathcal{N}$ is a survival function and is in an NBU-t_0 class since

$$\begin{aligned}&\bar{\hat{F}}_\alpha(x+t_0) \\ &= [\bar{\hat{F}}(x+t_0)]^\alpha [\bar{\tilde{F}}(x+t_0)]^{1-\alpha} \\ &\leq [\bar{\hat{F}}(x)\bar{\hat{F}}(t_0)]^\alpha [\bar{\tilde{F}}(x)\bar{\tilde{F}}(t_0)]^{1-\alpha} \\ &= [\bar{\hat{F}}(x)]^\alpha [\bar{\tilde{F}}(x)]^{1-\alpha} [\bar{\hat{F}}(t_0)]^\alpha [\bar{\tilde{F}}(t_0)]^{1-\alpha} \\ &= \bar{\hat{F}}_\alpha(x)\bar{\hat{F}}_\alpha(t_0)\end{aligned} \tag{10.15}$$

which shows that $\bar{\hat{F}}_\alpha$ satisfies the NBU-t_0 condition of Equation 10.1. The inequality of Equation 10.15 holds since $\bar{\hat{F}}$ and $\bar{\tilde{F}}$ are the members of the NBU-t_0 class. For the estimator $\bar{\hat{F}}_\alpha$ of Equation 10.14 to be practical, the value of α must be estimated, denoted by $\hat{\alpha}$, so that $\bar{\hat{F}}_{\hat{\alpha}}$ can be used as an estimator of $\bar{F}$.

Chang and Rao [8] recommend the estimator $\hat{\bar{F}}_{\hat{\alpha}}$ where $\hat{\alpha}$ satisfies the minimax criterion

$$|\hat{\bar{F}}_{\hat{\alpha}} - \bar{F}_n| = \inf_{0\leq\alpha\leq 1} \sup_{x\geq 0} |\hat{\bar{F}}_{\alpha}(x) - \bar{F}_n(x)| \quad (10.16)$$

where $\bar{F}_n$ is the empirical survival function. The minimizing value $\hat{\alpha}$ can be found by any standard grid search program. The estimator $\hat{\bar{F}}_{\hat{\alpha}}$ is referred to as the geometric mean (GM) estimator. Theorems 11 and 12 prove that $\hat{\bar{F}}_{\hat{\alpha}}$ is a strongly consistent NBU-t_0 estimator of $\bar{F}$ and converges to $\bar{F}$ at an optimal rate. The proofs are given in Chang and Rao [8].

Theorem 11. *The estimator $\hat{\bar{F}}_{\hat{\alpha}}$ converges uniformly to $\bar{F}$ with probability of one.*

Theorem 12. *The estimator $\hat{\bar{F}}_{\hat{\alpha}}$ converges to $\bar{F}$ at an optimal rate, i.e.* $\sup_{x\geq 0} |\hat{\bar{F}}_{\hat{\alpha}}(x) - \bar{F}(x)| = O[n^{-1/2}(\log\log n)^{1/2}]$ *almost surely.*

Note that neither Theorem 11 nor Theorem 12 requires the assumption that $\bar{F}$ possesses a compact support.

A simulation study is conducted to compare the RS estimator and the GM estimator. For this purpose, the simulated values of the RS and GM estimators are used to calculate four measures of performance of the two estimation procedures. They are bias, mean-squared error, sup-norm and L-2 norm. As an NBU-t_0 life distribution for the simulation study, the distribution given in Equation 10.12 is used. In summary, the simulation study shows that, in addition to having a smaller mean-squared error at all time points of practical interest, the GM estimator provides an improved overall estimate of the underlying NBU-t_0 survival function, as measured in terms of sup-norm and L-2 norm.

10.4 Tests for NBU-t_0 Life Distribution

Since the class of NBU-t_0 life distributions was first introduced in Hollander *et al.* [10], several nonparametric tests dealing with the NBU-t_0 class have been proposed in the literature. The tests of exponentiality versus (non-exponential) NBU-t_0 alternatives on the basis of complete data were proposed by Hollander *et al.* [10], Ebrahimi and Habibullah [11], and Ahmad [12]. Hollander *et al.* [13] also discussed the NBU-t_0 test when the data are censored.

10.4.1 Tests for NBU-t_0 Alternatives Using Complete Data

Let $X_1, \ldots, X_n$ be a random sample from a continuous life distribution F. Based on these complete data, the problem of our interest is to test

$$H_0 : F \text{ is in } C_0 \text{ versus } H_a : F \text{ is in } C_a$$

for $t_0 > 0$ being fixed, where C_0 and C_a are defined in Equations 10.3 and 10.5 respectively. The null hypothesis asserts that a new item is as good as a used item of age t_0, whereas the alternative H_a states that a new item has stochastically greater residual life than does a used item of age t_0.

Examples of situations where it is reasonable to use the NBU-t_0 (NWU-t_0) tests are as follows.

(i) From experience, cancer specialists believe that a patient newly diagnosed as having a certain type of cancer has a distinctly smaller chance of survival than does a patient who has survived 5 years ($= t_0$) following initial diagnosis. (In fact, such survivors are often designated as "cured".) The cancer specialists may wish to test their beliefs. Censorship may occur at the time of data analysis because some patients may still be alive.

(ii) A manufacturer believes that a certain component exhibits "infant mortality", *e.g.* has a decreasing failure rate over an interval $[0, t_0]$. This belief stems from experience accumulated for similar components. He wishes to determine whether a used component of age t_0 has stochastically greater residual life length than does a new component. If so, he will test over the interval $[0, t_0]$ a certain percentage of his output, and then sell the surviving components of age t_0 at higher prices to purchasers who must have high-reliability components (*e.g.* a spacecraft assembler). He wishes to test such a hypothesis to reject or accept his *a priori* belief.

10.4.1.1 Hollander–Park–Proschan Test

The first non-parametric test of H_0 versus H_a was introduced by Hollander *et al.* [10]. Their test statistic (termed HPP here) is motivated by considering the following parameter as a measure of deviation. Let

$$\begin{aligned}T(F) &\equiv \int_0^\infty [\bar{F}(x+t_0) - \bar{F}(x)\bar{F}(t_0)]\,\mathrm{d}F(x)\\ &= \int_0^\infty \bar{F}(x+t_0)\,\mathrm{d}F(x) - \frac{1}{2}\bar{F}(t_0)\\ &\equiv T_1(F) - T_2(F)\end{aligned}$$

Observe that under H_0, $T_1(F) = T_2(F)$, and that under H_a, $T_1(F) \le T_2(F)$, and thus $T(F) \le 0$. In fact, $T(F)$ is strictly less than zero under H_a if F is continuous. $T(F)$ gives a measure of the deviation of F from its specification under H_0, and, roughly speaking, the more negative $T(F)$ is, the greater is the evidence in favor of H_a. It is reasonable to replace F of $T(F)$ by the empirical distribution function F_n of $X_1, \ldots, X_n$ and reject H_0 in favor of H_a if $T(F_n) = T_1(F_n) - T_2(F_n)$ is too small. Instead of using this test statistic, the HPP test uses the asymptotically equivalent U-statistic T_n given in Equation 10.17. Hoeffding's U-statistic theory [14] yields asymptotic normality of T_n and the HPP test is based on this large sample approximation.

Let

$$h_1(x_1, x_2) = \tfrac{1}{2}[\psi(x_1, x_2 + t_0) + \psi(x_2, x_1 + t_0)]$$

and

$$h_2(x_1) = \tfrac{1}{2}\psi(x_1, t_0)$$

be the kernels of degrees 2 and 1 corresponding to $T_1(F)$ and $T_2(F)$ respectively, where $\psi(a, b) = 1$ if $a > b$, and $\psi(a, b) = 0$ if $a \le b$. Let

$$\begin{aligned}T_n &= [n(n-1)]^{-1}{\sum}' \psi(x_{\alpha_1}, x_{\alpha_2} + t_0)\\ &\quad - [2n]^{-1}\sum_{i=1}^{n}\psi(x_i, t_0)\end{aligned} \tag{10.17}$$

where $\sum'$ is the sum taken over all $n(n-1)$ sets of two integers (α_1, α_2) such that $1 \le \alpha_i \le n$, $i = 1, 2$, and $\alpha_1 \ne \alpha_2$. To apply Hoeffding's U-statistic theory, we let

$$\begin{aligned}\xi_1^{[1]} &= E[h_1(x_1, x_2)h_1(x_1, x_3)] - [T_1(F)]^2\\ \xi_2^{[1]} &= E[h_1^2(x_1, x_2)] - [T_1(F)]^2\\ \xi_1^{[2]} &= E[h_2^2(x_1)] - [T_2(F)]^2\end{aligned}$$

and

$$\xi^{[1,2]} = E[h_1(x_1, x_2) - T_1(F)][h_2(x_1) - T_2(F)]$$

Then

$$\begin{aligned}\mathrm{var}(T_n) &= \binom{n}{2}^{-1}\sum_{k=1}^{2}\binom{2}{k}\binom{n-2}{2-k}\xi_k^{[1]}\\ &\quad + n^{-1}\xi_1^{[2]} - (4/n)\xi^{[1,2]}\end{aligned}$$

and

$$\sigma^2 = \lim_{n\to\infty} n\,\mathrm{var}(T_n) = 4\xi_1^{[1]} + \xi_1^{[2]} - 4\xi^{[1,2]} \tag{10.18}$$

It follows from Hoeffding's U-statistic theory that if F is such that $\sigma^2 > 0$, where σ^2 is given in Equation 10.18, then the limiting distribution of $n^{1/2}[T_n - T(F)]$ is normal with mean zero and variance σ^2. Straightforward calculations yield, under H_0

$$\begin{aligned}\mathrm{var}_0(T_n) &= (n+1)[n(n-1)]^{-1}\\ &\quad \times [(1/12)\bar{F}(t_0) + (1/12)\bar{F}^2(t_0)\\ &\quad - (1/6)\bar{F}^3(t_0)]\end{aligned}$$

and

$$\sigma_0^2 = (1/12)\bar{F}(t_0) + (1/12)\bar{F}^2(t_0) - (1/6)\bar{F}^3(t_0)$$

Thus, under H_0, the limiting distribution of $n^{1/2}T_n$ is normal with mean zero and variance σ_0^2. Since the null variance σ_0^2 of $n^{1/2}T_n$ does depend on the unspecified distribution F, σ_0^2 must be estimated from the data. The consistent estimator of σ_0^2 can be obtained by replacing F by its empirical distribution F_n as follows:

$$\sigma_n^2 = (1/12)\bar{F}_n(t_0) + (1/12)\bar{F}_n^2(t_0) - (1/6)\bar{F}_n^3(t_0) \tag{10.19}$$

where $\bar{F}_n = 1 - F_n$. Using the asymptotic normality of T_n and Slutsky's theorem, it can be shown, under H_0, that $n^{1/2}T_n\sigma_n^{-1}$ is asymptotically $N(0, 1)$. Thus the approximate α-level test of

Table 10.3. Asymptotic relative efficiencies

t_0	F					
	$E_{F_1}(J,T)$	$E_{F_1}(S,T)$	$E_{F_2}(J,T)$	$E_{F_2}(S,T)$	$E_{F_3}(J,T)$	$E_{F_3}(S,T)$
0.2	6.569	14.598	3.554	4.443	0.579	0.433
0.6	2.156	4.79	1.695	2.118	0.253	0.531
1.0	1.342	2.983	1.493	1.866	0	0.145
1.4	1.047	2.328	1.607	2.009	0	0
1.8	0.933	2.074	1.929	2.411	0	0
2.0	0.913	2.030	2.172	2.715	0	0

H_0 versus H_a is to reject H_0 in favor of H_a if

$$n^{1/2}T_n\sigma_n^{-1} \leq -z_\alpha \tag{10.20}$$

where z_α is the upper α-percentile point of a standard normal distribution. The test of H_0 versus H_a defined by Equation 10.20 is referred to as the HPP test. Analogously, the approximate α-level test of H_0 against the alternative that a new item has stochastically less residual life length than does a used item of age t_0 is to reject H_0 if $n^{1/2}T_n\sigma_n^{-1} \geq z_\alpha$.

Theorem 13. *If F is continuous, then the HPP test is consistent against C_a, where C_a is the class of non-exponential NBU-t_0 life distributions defined in Equation 10.4.*

Proof. See Hollander *et al.* [10]. □

Note that

$$P\{n^{1/2}T_n\sigma_n^{-1} \leq -z_\alpha\} \\ = P\{n^{1/2}[T_n - T(F)] \leq -z_\alpha\sigma_n - n^{1/2}T(F)\}$$

and for $F \in C_a$, $T(F) < 0$ and that $\sigma_n \xrightarrow{p} \sigma_0 < \infty$. Thus, under H_a, $P\{n^{1/2}T_n\sigma_n^{-1} \leq -z_\alpha\} \geq \alpha$ for sufficiently large n, which shows that the HPP test is asymptotically unbiased against C_a.

To evaluate the performance of the HPP test, the Pitman asymptotic relative efficiencies are computed for the following three distributions.

1. Linear failure rate distribution

$$\bar{F}_1(x;\theta) = \exp\{-[x + (\theta/2)x^2]\} \\ \theta \geq 0, x \geq 0$$

2. Makeham distribution

$$\bar{F}_2(x;\theta) = \exp[-(x + \{\theta[x + \exp(-x) - 1]\})] \\ \theta \geq 0, \; x \geq 0$$

3.

$$\begin{aligned}\bar{F}_3(x;\theta) &= \exp[-\{x - \theta(2t_0)^{-1}x^2\}] \\ &\qquad 0 \leq \theta \leq 1, \; 0 \leq x < t_0 \\ &= \exp[-\{x - \theta(2)^{-1}t_0\}] \\ &\qquad 0 \leq \theta \leq 1, \; x \geq t_0\end{aligned}$$

For $\theta = 0$, F_1, F_2, and F_3 reduce to the exponential distribution. For $\theta > 0$, F_1 and F_2 are NBU. For $0 < \theta \leq 1$, as shown in Example 1, F_3 is in C_a but is not NBU.

Since no other tests of H_0 versus H_a have been proposed, the HPP test is compared with Hollander and Proschan's NBU-test [15] (referred to as the J test) and Bickel and Doksom's NBU-test [16] (referred to as the S test). Both the J test and S test are testing the null hypothesis H_0' : F is exponential, versus the alternative H_a : F is NBU and not exponential. The asymptotic relative efficiency calculations are summarized in Table 10.3, where T refers to the HPP test. The values of zeros for $E_{F_3}(J,T)$ and $E_{F_3}(S,T)$ indicate that the J test and T test are not consistent against F_3 when t_0 exceeds a certain value. More details are given in Hollander *et al.* [10].

The HPP test based on the statistic T_n is not intended to be a competitor of the J or S test. The latter tests are designed for smaller classes of alternatives, whereas the HPP test is designed for the relatively large class of alternatives C_a. As Table 10.3 indicates, when F is NBU-t_0 and is not a member of the smaller classes (such as the NBU class), the HPP test will often be preferred to other tests.

10.4.1.2 Ebrahimi–Habibullah Test

To develop a class of tests for testing H_0 versus H_a, Ebrahimi and Habibullah [11] consider the parameter

$$\begin{aligned}\Delta_k(F) &= \int_0^\infty [\bar{F}(x+kt_0) - \bar{F}(x)\{\bar{F}(t_0)\}^k]\,\mathrm{d}F(x)\\ &= \Delta_{1k}(F) - \Delta_{2k}(F)\end{aligned}$$

where

$$\begin{aligned}\Delta_{1k}(F) &= \int_0^\infty \bar{F}(x+kt_0)\,\mathrm{d}F(x)\\ \Delta_{2k}(F) &= \frac{1}{2}\{\bar{F}(t_0)\}^k\end{aligned}$$

$\Delta_k(F)$ is motivated by the fact that the NBU-t_0 property implies the weaker property $\bar{F}(x+kt_0) \le \bar{F}(x)[\bar{F}(t_0)]^k$ for all $x \ge 0$ and $k = 1, 2, \ldots$. They term such a property as new better than used of order kt_0. Under H_0, $\Delta_k(F) = 0$ and under H_a, $\Delta_k(F) < 0$. Thus, $\Delta_k(F)$ can be taken as an overall measure of deviation from H_0 towards H_a. Ebrahimi and Habibullah [11] utilized the U-statistic to develop a class of test statistics. Define

$$T_k = T_{1k} - T_{2k}$$

where

$$T_{1k} = \left[1\Big/\binom{n}{2}\right] \sum_{1\le i\le j\le n} \psi(X_i, X_j + kt_0)$$

and

$$T_{2k} = \left\{1\Big/\left[2\binom{n}{k}\right]\right\} \sum \psi_k(X_{i_1}, \ldots, X_{i_k})$$

and the sum is over all combinations of k integers $(i_1, \ldots, i_k)$ chosen out of $(1, \ldots, n)$. Here $\psi_k(a_1, \ldots, a_k) = 1$ if $\min_{1\le i\le k} a_i > t_0$, or zero otherwise.

Note that T_{1k} and T_{2k} are the U-statistics corresponding to the parameters $\Delta_{1k}(F)$ and $\Delta_{2k}(F)$ with kernels ψ and ψ_k of degrees 2 and k respectively. Thus, T_k is a U-statistic corresponding to $\Delta_k(F)$ and works as a test statistic for H_0 versus H_a. It follows that $E(T_k) \le 0$ or $E(T_k) \ge 0$ according to whether F is new better than used of age t_0 or new worse than used of age t_0 respectively. If $k = 1$, T_k reduces to the test statistic of Hollander *et al.* [10]. The asymptotic normality of T_k is established by applying Hoeffding's U-statistic theory [14].

Theorem 14. *If F is such that*

$$\sigma^2 = \lim_{n\to\infty} n \operatorname{var}(T_k) > 0$$

then the asymptotic distribution of $n^{1/2}[T_k - \Delta_k(F)]$ is normal with mean zero and variance σ^2.

Corollary 1. *Under H_0, $n^{1/2}T_k$ is asymptotically normal with mean zero and variance*

$$\begin{aligned}\sigma_0^2 &= \tfrac{1}{3}[\bar{F}(t_0)]^k + (k - \tfrac{2}{3} - \tfrac{1}{4}k^2)[\bar{F}(t_0)]^{2k}\\ &\quad + \tfrac{1}{3}[\bar{F}(t_0)]^{3k} + (\tfrac{1}{4}k^2 - \tfrac{1}{2}k)[\bar{F}(t_0)]^{2k-1}\\ &\quad - \tfrac{1}{2}k[\bar{F}(t_0)]^{2k+1}\end{aligned}$$

Approximate α-level tests of H_0 against H_a are obtained by replacing $\bar{F}(t_0)$ by its consistent estimator $\bar{F}_n(t_0) = n^{-1}\sum \psi(X_i, t_0)$ in σ_0^2. It can be easily shown that if F is continuous, then T_k is consistent against H_a. The Pitman asymptotic relative efficiencies of T_k for a given $k \ge 2$ relative to the T test of Hollander *et al.* [10] for the following distributions are evaluated:

$$\begin{aligned}\bar{F}_1(x) &= \exp\{-x + \theta[x + \exp(-x) - 1]\}\\ &\qquad \theta \ge 0,\ x \ge 0\\ \bar{F}_2(x) &= \left[\exp\left(-x + \frac{\theta x^3}{2t_0}\right)\right] I(0 \le x < t_0)\\ &\quad + \left[\exp\left(-x + \frac{\theta t_0}{2}\right)\right] I(x \ge t_0)\\ &\qquad 0 \le \theta \le \frac{2}{3}\end{aligned}$$

For $\theta = 0$, $\bar{F}_1$ and $\bar{F}_2$ reduce to the exponential distribution and thus satisfy H_0. For $\theta > 0$, $\bar{F}_1$ is both NBU and NBU-t_0 and $\bar{F}_2$ is NBU-t_0, but not NBU.

Table 10.4. Pitman asymptotic relative efficiency of T_k with respect to T for $\bar{F}_1$ and $\bar{F}_2$

t_0	(k_1, k_2)	$(E_{F_1}(T_{k_1}, T), E_{F_1}(T_{k_2}, T))$	(k_1, k_2)	$(E_{F_2}(T_{k_1}, T), E_{F_2}(T_{k_2}, T))$
0.2	(2, 16)	(1.154, 3.187)	(2, 5)	(1.138, 1.205)
0.6	(2, 4)	(1.221, 1.634)	(2, 3)	(1.665, 1.779)
1.0	(2, 3)	(1.0, 1.0)	(2, 2)	(2.638, 2.638)
1.4	(1, 1)	(1.0, 1.0)	(2, 2)	(4.241, 4.241)
1.8	(1, 1)	(1.0, 1.0)	(2, 2)	(6.681, 6.681)
2.0	(1, 1)	(1.0, 1.0)	(2, 2)	(8.271, 8.271)

Table 10.4 gives the numerical values of $E_{F_1}(T_k, T)$ and $E_{F_2}(T_k, T)$ for certain specified values of t_0. In both cases, two values of k, namely k_1 and k_2, are reported, where k_1 is the smallest value for which $E(T_k, T) > 1$ and k_2 is the integer that gives optimum Pitman asymptotic relative efficiency. Table 10.4 shows that although the value of k varies, there is always a $k > 1$ for which T_k performs better than T for small t_0. Secondly, if the alternative is NBU-t_0, but not NBU, then for all t_0, there is a $k > 1$ for which T_k performs better than T.

10.4.1.3 Ahmad Test

Ahmad [12] considered the problem of testing H_0 versus H_a when the point t_0 is unknown, but which can be estimated from the data. He proposed the tests when $t_0 = \mu$, the mean of F, and also when $t_0 = \xi_p$, the pth percentile of F. In particular, when $t_0 = \xi_p$ the proposed test is shown to be distribution free, whereas both the HPP test and the Ebrahimi–Habibullah test are not distribution free.

Ahmad [12] utilized the fact that $\bar{F}$ is NBU-t_0 if and only if, for all integers $k \geq 1$, $\bar{F}(x + kt_0) \leq \bar{F}(x)[\bar{F}(t_0)]^k$ and considered the following parameter as the measure of departure of F from H_0. Let

$$\gamma_k(F) = \int_0^\infty [\bar{F}(x)\bar{F}^k(t_0) - \bar{F}(x + kt_0)]\,\mathrm{d}F(x)$$
$$= \frac{1}{2}\bar{F}^k(t_0) - \int_0^\infty \bar{F}(x + kt_0)\,\mathrm{d}F(x) \tag{10.21}$$

When $t_0 = \xi_p$, $\bar{F}(t_0) = 1 - p$ and thus Equation 10.21 reduces to

$$\gamma_k(F) = \frac{1}{2}(1-p)^k - \int_0^\infty \bar{F}(x + k\xi_p)\,\mathrm{d}F(x)$$

In the usual way, ξ_p is estimated by $\hat{\xi}_p = X_{([np])}$, where $X_{(r)}$ denotes the rth-order statistic in the sample. Thus, $\gamma_k(F)$ can be estimated by

$$\begin{aligned}\hat{\gamma}_k(F_n) &= \frac{1}{2}(1-p)^k \\ &\quad - \int_0^\infty \bar{F}_n(x + kX_{([np])})\,\mathrm{d}F_n(x) \\ &= \frac{1}{2}(1-p)^k \\ &\quad - n^{-2}\sum_{i=1}^n\sum_{j=1}^n I(X_i > X_j + kX_{([np])})\end{aligned}$$

where F_n ($\bar{F}_n$) is the empirical distribution (survival) function.

A U-statistic equivalent to $\hat{\gamma}_k(F_n)$ is given by

$$\begin{aligned}\hat{T}_k &= \frac{1}{2}(1-p)^k - \binom{n}{2}^{-1} \\ &\quad \times \sum_{1 \leq i \leq j \leq n} I(X_i > X_j + kX_{([np])})\end{aligned}$$

The statistic $\hat{\gamma}_k$ is used to test H_0 versus H_1 : $\bar{F}$ is NBU-ξ_p for $0 \leq p \leq 1$. The asymptotic distribution of $\hat{\gamma}_k$ is stated in the following theorem.

Theorem 15. *As $n \to \infty$, $n^{1/2}[\hat{\gamma}_k - \gamma_k(F)]$ is asymptotically normal with mean zero and variance $\sigma^2 > 0$, provided that $f = F'$ exists and is bounded. Under H_0, the variance reduces to $\sigma_0^2 =$*

Table 10.5. Efficacies of $\hat{\gamma}_k$ for F_1 and F_2 (the first entry is for F_1 and the second is for F_2)

k	$P=0.05$	$P=0.25$	$P=0.50$	$P=0.75$	$P=0.95$
1	0.0919	0.1318	0.1942	0.2618	0.2319
	14.4689	0.1605	0.0022	0.0002	
3	0.1093	0.2160	0.2741	0.1484	0.0065
	1.0397	0.0002	0.0002		
5	0.1266	0.2646	0.1986	0.0288	0.0001
	0.2346	0.0002			
7	0.1436	0.2750	0.1045	0.0037	
	0.0721	0.0002			

$\frac{1}{3}(1-p)^k[1-(1-p)^k]^2$. *Here,* $\sigma^2 = V[\bar{F}(X_1 + k\xi_p) + F(X_1 - k\xi_p)]$.

Proof. See Ahmad [12]. □

Although there are no other tests in the literature to compare with the Ahmad test for the testing problem of H_0 versus H_1, the Pitman asymptotic efficacies for two members of the NBU-t_0 class have been calculated. Two NBU-t_0 life distributions and their calculated efficacies are given as follows.

(i) Makehem distribution:

$$\bar{F}_1(x) = \exp(-\{x + \theta[x + \exp(-x) - 1]\}) \quad \text{for } \theta \geq 0,\ x \geq 0$$

$$\text{eff}(F_1) = \frac{[2 + 3\log(1-p)^k - 2(1-p)^k]^2(1-p)^k}{12[1-(1-p)^k]^2}$$

(ii) Linear failure rate distribution:

$$\bar{F}_2(x) = \exp[-(x + \theta x^2/2)]\}) \quad \text{for } \theta \geq 0,\ x \geq 0$$

$$\text{eff}(F_2) = \frac{3(1-p)^k[\log(1-p)^k]^2[1+\log(1-p)^k]^2}{64[1-(1-p)^k]^2}$$

Table 10.5 offers some values of the above efficacies for some choices of k and p. Unreported entries in the table are zero to four decimal places. Note that, for F_1, the efficacy is increasing in k. It also increases in p initially, and then decreases for large p. For F_2, however, the efficacy is decreasing in both k and p.

When $t_0 = \mu$, $\gamma_k(F)$ of Equation 10.21 becomes

$$\gamma_k(F) = \frac{1}{2}\bar{F}^k(\mu) - \int_0^\infty \bar{F}(x + k\mu)\,\mathrm{d}F(x)$$

Replacing μ by $\bar{X}$, and F ($\bar{F}$) by its corresponding empirical distribution F_n ($\bar{F}_n$), $\gamma_k(F)$ can be estimated by

$$\hat{\gamma}_k(F) = \frac{1}{2}\bar{F}_n^k(\bar{X}) - \int_0^\infty \bar{F}_n(x + k\bar{X})\,\mathrm{d}F_n(x)$$

The equivalent U-statistic estimator is

$$\hat{T}_k = \frac{1}{2}\hat{T}_{1k} - \hat{T}_{2k}$$

where

$$\hat{T}_{1k} = \binom{n}{k}^{-1} \sum_{1 \leq i < \cdots < i_k \leq n} I\{\min(X_{i_1}, \ldots, X_{i_k}) > \bar{X}\}$$

$$\hat{T}_{2k} = \binom{n}{2}^{-1} \sum_{1 \leq i < j \leq n} I(X_i > X_j + k\bar{X})$$

Theorem 16. *As* $n \to \infty$, $n^{1/2}[\hat{T}_k - \gamma_k(F)]$ *is asymptotically normal with mean zero and variance* σ^2, *provided that* $f = F'$ *exists and is bounded. Under* H_0, *the variance reduces to*

$$\sigma_0^2 = \frac{1}{3}[\bar{F}(\mu)]^k + \left(k - \frac{2}{3} - \frac{k^2}{4}\right)[\bar{F}(\mu)]^{2k} + \frac{1}{3}[\bar{F}(u)]^{3k} + \left(\frac{k^2}{4} - \frac{k}{2}\right)[\bar{F}(\mu)]^{2k-1} - \frac{k}{2}[\bar{F}(\mu)]^{2k+1}$$

Here, $\sigma^2 = \lim_{n\to\infty} nV(\hat{T}_k) > 0$.

Proof. See Ahmad [12]. □

10.4.2 Tests for NBU-t_0 Alternatives Using Incomplete Data

In many practical situations the data are incomplete due to random censoring. For example, in a clinical study, patients may withdraw from the study or they may not yet have experienced the end-point event before the completion of a study. Thus, instead of observing a complete sample $X_1, \ldots, X_n$ from the life distribution F, we are able to observe only pairs (Z_i, δ_i), $i = 1, \ldots, n$, where

$$Z_i = \min(X_i, Y_i)$$

and

$$\delta_i = \begin{cases} 1 & \text{if } Z_i = X_i \\ 0 & \text{if } Z_i = Y_i \end{cases}$$

Here, we assume that $Y_1, \ldots, Y_n$ are i.i.d. according to the continuous censoring distribution H and the X and Y values are mutually independent. Observe that $\delta_i = 1$ means that the ith observation is uncensored, whereas $\delta_i = 0$ means that the ith observation is censored on the right by Y_i. The hypothesis of H_0 against H_a of Section 10.4.1 is tested on the basis of the incomplete observations $(Z_i, \delta_1), \ldots, (Z_n, \delta_n)$. Several consistent estimators of F based on the incomplete observations have been developed by Kaplan and Meier [17], Susarla and Van Ryzin [18], and Kitchin *et al.* (1980), among others. To develop the NBU-t_0 test using a randomly censored model, the Kaplan and Meier [17] estimator (KME) is utilized.

Under the assumption of F being continuous, the KME is given by:

$$\hat{\bar{F}}_n(t) = \prod_{\{i:Z_{(i)}\leq t\}} [(n-i)(n-i+1)^{-1}]^{\delta_{(i)}}$$

$$t \in [0, Z_{(n)})$$

where $Z_{(0)} \equiv 0 < Z_{(1)} < \cdots < Z_{(n)}$ denote the ordered Z values and $\delta_{(i)}$ is the δ corresponding to $Z_{(i)}$. Here, we treat $Z_{(n)}$ as an uncensored observation, whether it is uncensored or censored. When censored observations are tied with uncensored observations, our convention for ordering the Z values is to treat the uncensored observations as preceding the censored observations. Large sample properties of $\hat{\bar{F}}_n$ have been studied by Efron [20], Breslow and Crowley [21], Peterson [22], and Gill [23]. In the sequel, $\hat{\bar{F}}_n$ will denote the KME and $\hat{F}_n = 1 - \hat{\bar{F}}_n$. Hollander *et al.* [13] form the test statistic by replacing F of $T(F)$ defined in Section 10.4.1.1, by $\hat{F}_n$, the KME of F, *i.e.*

$$T(\hat{F}_n) = \int_0^\infty \hat{\bar{F}}_n(x + t_0)\, \mathrm{d}\hat{F}_n(x) - \frac{1}{2}\hat{\bar{F}}_n(t_0)$$

For computational purpose, we may write:

$$\begin{aligned} T_n^c &\equiv T(\hat{F}_n) \\ &= \Bigg(\sum_{i=1}^{n} \prod_{\{k:Z_{(k)}\leq Z_{(i)}+t_0\}} [(n-k)(n-k+1)^{-1}]^{\delta_{(k)}} \\ &\quad \times \left\{ \prod_{r=1}^{i} [(n-r)(n-r+1)^{-1}]^{\delta_{(r)}} \right\} \\ &\quad \times \{1 - [(n-i)(n-i+1)^{-1}]^{\delta_{(i)}}\} \Bigg) \\ &\quad - \frac{1}{2} \prod_{\{k:Z_{(k)}\leq t_0\}} [(n-k)(n-k+1)^{-1}]^{\delta_{(k)}} \end{aligned}$$

Asymptotic normality of $n^{1/2}[T_n^c - T(F)]$ can be established under the following assumptions:

Assumption 1. *The support of both F and H is* $[0, \infty)$.

Assumption 2. $\sup\{[\bar{F}(x)]^{1-\varepsilon}[\bar{H}(x)]^{-1},\ x \in [0, \infty)\} < \infty$, *for some* $\varepsilon \geq 0$.

Assumption 2 restricts the amount of censoring. For example, in the proportional hazard model where $\bar{H}(t) = [\bar{F}(t)]^\beta$, Assumption 2 implies that $\beta < 1$, which means that the expected proportion of censored observations, $\beta/(\beta + 1)$, must be less than 0.5.

Let $\bar{K}(t) = \bar{F}(t)\bar{H}(t)$, $t \in [0, \infty)$, and let $\{\phi(t), t \in (0, \infty)\}$ be a Gaussian process with mean zero and covariance kernel given by

$$E\phi(t)\phi(s) = \bar{F}(t)\bar{F}(s)\int_0^s [\bar{K}(z)\bar{F}(z)]^{-1}\,\mathrm{d}F(z)$$
$$0 \le s < t < \infty$$

Theorem 17. *Assume that Assumptions 1 and 2 hold. Then $n^{1/2}[T_n^c - T(F)]$ converges in distribution to a normal random variable with mean zero and variance σ_c^2, where*

$$\sigma_c^2 = \iint E\left[\phi(x+t_0) - \phi(x-t_0) - \frac{1}{2}\phi(t_0)\right] \times \left[\phi(u+t_0) - \phi(u-t_0) - \frac{1}{2}\phi(t_0)\right]\mathrm{d}F(x)\,\mathrm{d}F(u)$$

Proof. See Hollander *et al.* [13]. □

The null asymptotic mean of $n^{1/2}T_n^c$ is zero, independent of the distributions F and H. However, the null asymptotic variance of $n^{1/2}T_n^c$ depends on both F and H, and thus it must be estimated from the incomplete observations $(Z_i, \delta_1), \ldots, (Z_n, \delta_n)$. Under H_0, it can be shown that the null asymptotic variance of $n^{1/2}T_n^c$ is

$$\begin{aligned}\sigma_{c0}^2 = {} & \frac{1}{4}\bar{F}^2(t_0)\int_0^\infty \bar{F}^3(z)[\bar{K}(z+t_0)]^{-1}\,\mathrm{d}F(z) \\ & + \frac{1}{4}\bar{F}^2(t_0)\int_0^\infty \bar{F}^3(z)[\bar{K}(z)]^{-1}\,\mathrm{d}F(z) \\ & - \frac{1}{2}\bar{F}^4(t_0)\int_0^\infty \bar{F}^3(z)[\bar{K}(z+t_0)]^{-1}\,\mathrm{d}F(z)\end{aligned} \tag{10.22}$$

If there is no censoring, *i.e.* if $\bar{K}(z) = \bar{F}(z)$ for $z \in [0,\infty)$, then σ_{c0}^2 reduces to $\frac{1}{12}\bar{F}(t_0) + \frac{1}{12}\bar{F}^2(t_0) - \frac{1}{6}\bar{F}^3(t_0)$, agreeing with the null asymptotic variance σ_0^2 of $n^{1/2}T_n$ for the complete data case.

By a change of variable in the first and third terms of Equation 10.22, σ_{c0}^2 can be simplified to

$$\begin{aligned}\sigma_{c0}^2 = {} & \frac{1}{4}[\bar{F}(t_0)]^{-2}\int_{t_0}^\infty \bar{F}^3(u)[\bar{K}(u)]^{-1}\,\mathrm{d}F(u) \\ & + \frac{1}{4}\bar{F}^2(t_0)\int_0^\infty \bar{F}^3(u)[\bar{K}(u)]^{-1}\,\mathrm{d}F(u) \\ & - \frac{1}{2}\int_{t_0}^\infty \bar{F}^3(u)[\bar{K}(u)]^{-1}\,\mathrm{d}F(u)\end{aligned} \tag{10.23}$$

Let $Z_{(1)} \le \cdots \le Z_{(n)}$ denote the ordered Z values and let K_n denote the empirical distribution function of the Z values. Then, $nK_n(t) =$ (number of Z values $\le t$). Since $T_n^c = 0$ when $Z_{(n)} \le t_0$, we assume that our sample is such that $Z_{(n)} > t_0$. Replacing $\bar{F}$ by $\hat{\bar{F}}_n$, $\bar{K}$ by $\bar{K}_n$, and ∞ by $Z_{(n)}$ in Equation 10.23 yields the estimate $\hat{\sigma}_{cn}^2$ defined by Equation 10.24.

$$\begin{aligned}\hat{\sigma}_{cn}^2 = {} & \left\{\frac{1}{4}[\hat{\bar{F}}_n(t_0)]^{-2} - \frac{1}{2}\right\} \times \sum_{\{i:t_0 \le W_{(i)} \le Z_{(n)}\}} \hat{\bar{F}}^3(W_{(i)})[\bar{K}_n(W_{(i)})]^{-1} \\ & \times [\hat{F}_n(W_{(i)}) - \hat{F}_n(W_{(i-1)})] + \frac{1}{4}\hat{\bar{F}}_n^2(t_0) \\ & \times \sum_{\{i:1\le i \le \tau(n)\}} \hat{\bar{F}}_n^3(W_{(i)})[\bar{K}_n(W_{(i)})]^{-1} \\ & \times [\hat{F}_n(W_{(i)}) - \hat{F}_n(W_{(i-1)})]\end{aligned} \tag{10.24}$$

where $W_{(0)} \equiv 0 < W_{(1)} < W_{(2)} < \cdots < W_{(\tau(n))}$ are the ordered observed failure times, and $\tau(n) = \sum_{i=1}^n \delta_i$ is the total number of failures among the n observations. In Equation 10.24, when $W_{(i)} = Z_{(n)}$ we redefine $\bar{K}_n(Z_{(n)})$, changing it from zero to $1/n$ so that Equation 10.24 is well defined.

The approximate α-level test of H_0 versus H_a, which rejects H_0 in favor of H_a if $n^{1/2}T_n^c\hat{\sigma}_{cn}^{-1} \le -z_\alpha$ and accepts H_0 otherwise, is called the NBU-t_0 test for incomplete data. The approximate α-level test of H_0 versus the alternative that a new item has a stochastically smaller residual life length than does a used item of age t_0 is called the NWU-t_0 test for incomplete data. The NWU-t_0 test rejects H_0 if $n^{1/2}T_n^c\hat{\sigma}_{cn}^{-1} \ge z_\alpha$ and accepts H_0 otherwise.

References

[1] Marshall AH, Proschan F. Classes of distributions applicable in replacement, with renewal theory implications. In: Proceedings of the 6th Berkeley Symposium on Mathematical Statistics and Probability, vol. I. University of California Press; 1972. p.395–415.

[2] Esary JD, Marshall AH, Proschan F. Shock models and wear processes. Ann Prob 1973;1:627–49.

[3] Bryson MC, Siddiqui MM. Some criteria for aging. J Am Stat Assoc 1969;64:1472–83.

[4] Marsaglia G, Tubilla A. A note on the "lack of memory" property of the exponential distribution. Ann. Prob. 1975;3:353–4.

[5] Barlow RE, Proschan F. Statistical theory of reliability and life testing probability models. Silver Spring: To Begin With; 1981.

[6] Reneau DM, Samaniego FJ. Estimating the survival curve when new is better than used of a specified age. J Am Stat Assoc 1990;85:123–31.

[7] Chang MN. On weak convergence of an estimator of the survival function when new is better than used of a specified age. J Am Stat Assoc 1991;86:173–8.

[8] Chang MN, Rao PV. On the estimation of a survival function when new better than used of a specified age. J Nonparametric Stat 1992;2:45–8.

[9] Serfling RJ. Approximation theorems of mathematical statistics. New York: Wiley; 1980.

[10] Hollander M, Park DH, Proschan F. A class of life distributions for aging. J. Am Stat Assoc 1986;81:91–5.

[11] Ebrahimi N, Habibullah M. Testing whether a survival distribution is new better than used of a specified age. Biometrika 1990;77:212–5.

[12] Ahmad IA. Testing whether a survival distribution is new better than used of an unknown specified age. Biometrika 1998;85:451–6.

[13] Hollander M, Park DH, Proschan F. Testing whether new is better than used of a specified age, with randomly censored data. Can J Stat 1985;13:45–52.

[14] Hoeffding WA. A class of statistics with asymptotically normal distribution. Ann Math Stat 1948;19:293–325.

[15] Hollander M, Prochan F. Testing whether new is better than used. Ann Math Stat 1972;43:1136–46.

[16] Bickel PJ, Doksum KA. Tests for monotone failure rate based on normalized spacings. Ann Math Stat 1969;40:1216–35.

[17] Kaplan EL, Meier P. Nonparametric estimation from incomplete observations. J Am Stat Assoc 1958;53:457–81.

[18] Susarla V, Van Ryzin J. Nonparametric Bayesian estimation of survival curves from incomplete observations. J Am Stat Assoc 1976;71:897–902.

[19] Kitchin J, Langberg NA, Proschan F. A new method for estimating life distributions from incomplete data. Florida State University Department of Statistics Technical Report No. 548, 1980.

[20] Efron B. The two sample problem with censored data. In: Proceedings of the 5th Berkley Symposium, vol. 4, 1967. p.831–53.

[21] Breslow N, Crowley J. A large sample study of the life table and product limit estimates under random censorship. Ann Stat 1974;2:437–53.

[22] Peterson AV. Expressing the Kaplan–Meier estimator as a function of empirical subsurvival function. J Am Stat Assoc 1977;72:854–8.

[23] Gill RD. Large sample behaviour of the product limit estimator on the whole line [Preprint]. Amsterdam: Mathematisch Centrun; 1981.

PART III

Software Reliability

16 Recent Studies in Software Reliability Engineering

Software Reliability Models: A Selective Survey and New Directions

Siddhartha R. Dalal

Chapter 11

11.1 Introduction

Software development, design, and testing have become very intricate with the advent of modern highly distributed systems, networks, middleware, and interdependent applications. The demand for complex software systems has increased more rapidly than the ability to design, implement, test, and maintain them, and the reliability of software systems has become a major concern for our modern society. Within the last decade of the 20th century and the first few years of the 21st century, many reported system outages or machine crashes were traced back to computer software failures. Consequently, recent literature is replete with horror stories due to software problems.

Even discounting the costly "Y2K" problem as a design failure, a problem that occupied tens of thousands of programmers in 1998–99 with the costs running to tens of billions of dollars, there have been many other critical failures. Software failures have impaired several high-visibility programs in space, telecommunications, defense and health industries. The Mars Climate Orbiter crashed in 1999. The Mars Climate Orbiter Mission Failure Investigation Board [1] concluded that "The 'root cause' of the loss of the spacecraft was the failed translation of English units into metric units in a segment of ground-based, navigation-related mission software, . . . ". Besides the costs involved, it set back the space program by more than a year. Current versions of the Osprey aircraft, developed at a cost of billions of dollars, are not deployed because of software-induced field failures. In the health industry [2], the Therac-25 radiation therapy machine was hit by software errors in its sophisticated control systems and claimed several patients' lives in 1985 and 1986. Even in the telecommunications industry, known for its five nines reliability, the nationwide long-distance network of a major carrier suffered an embarrassing network outage on 15 January 1990, due to a software problem. In 1991, a series of local network outages occurred in a number of US cities due to software problems in central office switches [3].

Software reliability is defined as the probability of failure-free software operations for a specified period of time in a specified environment [4]. The software reliability field discusses ways of quantifying it and using it for improvement and control of the software development process. Software reliability is operationally measured by the number of field failures, or failures seen in development, along with a variety of ancillary information. The ancillary information includes the time at which the failure was found, in which part of the software it was found, the state of software at that time, the nature of the failure, *etc.*

Most quality improvement efforts are triggered by lack of software reliability. Thus, software companies recognize the need for systematic approaches to measuring and assuring *software reliability*, and devote a major share of project development resources to this. Almost a third of the total development budget is typically spent on testing, with the expectation of measuring and improving software reliability. A number of standards have emerged in the area of developing reliable software consistently and efficiently. ISO 9000-3 [5] is the weakest amongst the recognized standards, in that it specifies measurement of field failures as the only required quality metric: "... at a minimum, some metrics should be used which represent reported field failures and/or defects form the customer's viewpoint. ... The supplier of software products should collect and act on quantitative measures of the quality of these software products". The Software Engineering Institute has proposed an elaborate standard called the Software Capability Maturity Model (CMM) [6] that scores software development organizations on multiple criteria and gives a numeric grade from one to five. A similar approach is taken by the SPICE standards, which are prevalent in Europe [7].

Formally, software reliability engineering is the field that quantifies the operational behavior of software-based systems with respect to user requirements concerning reliability. It includes data collection on reliability, statistical estimation and prediction, metrics and attributes of product architecture, design, software development, and the operational environment. Besides its use for operational decisions like deployment, it includes guiding software architecture, development, testing, *etc.* Indeed, much of the testing process is driven by software reliability concerns, and most applications of software reliability models are to improve the effectiveness of testing. Many current software reliability engineering techniques and practices are detailed by Musa *et al.* [8] and Lyu [9]. However, in this chapter we take a narrower view and just look at models that are used in software reliability—their efficacy and adequacy—without going into details of the interplay between testing and software reliability models.

Though prevalent software reliability models have their genesis in hardware reliability models, there are clearly a number of differences between hardware and software reliability models. Failures in hardware are typically based on the age of hardware and the stress of the operational environment, whereas failures in software are due to incorrect requirements, design, coding, or the inability to interoperate with other related software. Software failures typically manifest when the software is operated in an environment for which it was not designed or tested. Typically, except for the infant mortality factor in hardware, hardware reliability decreases with age, whereas software reliability increases with age (due to fault fixes).

In this chapter we focus on software reliability models and measurements. A software reliability model specifies the general form of the dependence of the failure process on the principal factors that affect it: fault introduction, fault removal, and the operational environment. During the test phase, the failure rate of a software system is generally decreasing due to the discovery and correction of software faults. With careful record-keeping procedures in place, it is possible to use statistical methods to analyze the historical record. The purpose of these analyses is twofold: (1) to predict the additional time needed to achieve a specified reliability objective; (2) to predict the expected reliability when testing is finished.

Implicit in this discussion is the concept of "time". For some purposes this may be calendar time, assuming that testing proceeds roughly uniformly; another possibility is to use computer execution time, or some other measure of testing effort. Another implicit assumption is that the software system being tested remains fixed throughout (except for the removal of faults as they are found). This assumption is frequently violated.

Software reliability measurement includes two types of model: *static* and *dynamic* reliability estimation, used typically in the earlier and later stages of development respectively. These will be discussed in the following two sections. One of the main weaknesses of many of the models is that they do not take into account ancillary information, like churn in system during testing. Such a model is described in Section 11.4. A key use of the reliability models is in the area of when to stop testing. An economic formulation is discussed in Section 11.5. Additional challenges and conclusions are stated in Section 11.6.

11.2 Static Models

One purpose of reliability models is to perform reliability prediction in an early stage of software development. This activity determines future software reliability based upon available software metrics and measures. Particularly when field failure data are not available (*e.g.* software is in the design or coding stage), the metrics obtained from the software development process and the characteristics of the resulting product can be used to estimate the reliability of the software upon testing or delivery. We discuss two prediction models: the phase-based model by Gaffney and Davis [10] and a predictive development life cycle model from Telcordia Technologies by Dalal and Ho [11].

11.2.1 Phase-based Model: Gaffney and Davis

Gaffney and Davis [10] proposed the phase-based model, which divides the software development cycle into different phases (*e.g.* requirement review, design, implementation, unit test, software integration, systems test, operation, *etc.*) and assumes that code size estimates are available during the early phases of development. Further, it assumes that faults found in different phases follow a Raleigh density function when normalized by the lines of code. Their model makes use of the fault statistics obtained during the early development phases (*e.g.* requirements review, design, implementation, and unit test) to predict the expected fault densities during a later phase (*e.g.* system test, acceptance test and operation).

The key idea is to divide the stage of development along a continuous time (*i.e.* $t =$ "0–1" means requirements analysis, and so on), and overlay the Raleigh density function with a scale parameter. The scale parameter, known as the *fault discovery phase constant*, is estimated by equating the area under the curve between earlier phases with observed error rates normalized by the lines of code. This method gives an estimate of the fault density for any later phase. The model also estimates the number of faults in a given phase by multiplying the fault density estimate by the number of lines of code.

This model is clearly motivated by the corresponding model used in hardware reliability, and the predictions are hardwired in the model based on one parameter. In spite of this criticism, this model is one of the first to leverage information available in earlier development life cycle phases.

11.2.2 Predictive Development Life Cycle Model: Dalal and Ho

In this model the development life cycle is divided into the same phases as in Section 11.2.1. However, unlike in Section 11.2.1, it does not postulate a fixed relationship (*i.e.* Raleigh distribution) between the numbers of faults discovered during different phases. Instead, it leverages past releases of similar products to determine the relationships. The relationships are not postulated beforehand, but are determined from data using only a few releases per product. Similarity is measured by

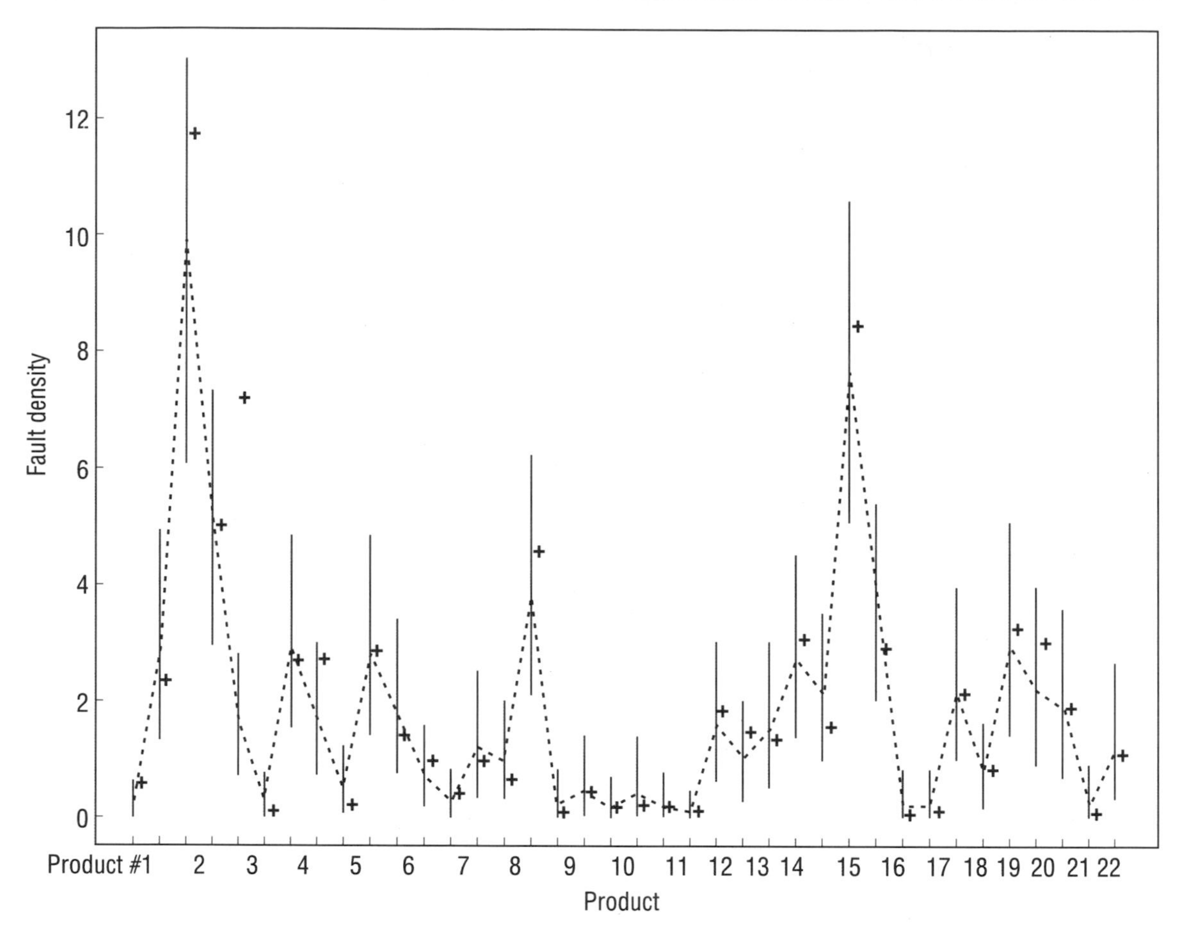

Figure 11.1. 22 Products and their releases versus Observed (+) and Predicted Fault Density connected by dash lines. Solid vertical lines are 90% predictive intervals for Fault Density

using an empirical hierarchical Bayes framework. The number of releases used as data is kept minimal and, typically, only the most recent one or two releases are used for prediction. This is critical, since there are often major modifications to the software development process over time, and these modifications change the interphase relationships between faults. The lack of data is made up for by using as many products as possible that were being developed in a software organization at around the same time. In that sense it is similar to meta analysis [12], where a lack of longitudinal data is overcome by using cross-sectional data.

Conceptually, the basic assumptions behind this model are as follows:

Assumption 1. *Defect rates from different products in the same product life cycle phase are samples from a statistical universe of products coming from that development organization.*

Assumption 2. *Different releases from a given product are samples from a statistical universe of releases for that product.*

Assumption 1 reflects the fact that the products developed within the same organization at the same life cycle maturity are more or less homogeneous. Defect density for a given product is related to covariates like lines of code and number of faults in previous life cycle phases by use of a regression model with the coefficients of the regression model assumed to have a normal prior distribution. The homogeneity assumption is minimally restrictive, since the Bayesian estimates we obtain depend increasingly on the data as more data become available. Based on the detailed model described by Dalal and Ho [11], one obtains predictive distributions of the fault density per life cycle phase conditionally on observing some of the previous product life cycle phases. Figure 11.1 shows the power of prediction of this method. On the horizontal axis we have 22 products, each with either one or two releases (some were new products and had no previous release). On the vertical axis we plot the predicted system's test fault density per million lines of code based on all fault information available prior to the system's test phase, along with the corresponding posterior confidence intervals. A dashed line connects the predicted fault density, and "+" indicates the observed fault density. Except for product number 4, all observed values are quite close to the predicted value.

11.3 Dynamic Models: Reliability Growth Models for Testing and Operational Use

Software reliability estimation determines the current software reliability by applying statistical inference techniques to failure data obtained during system test or during system operation. Since reliability tends to improve over time during the software testing and operation periods because of removal of faults, the models are also called reliability growth models. They model the underlying failure process of the software, and use the observed failure history as a guideline, in order to estimate the residual number of faults in the software and the test time required to detect them. This can be used to make release and deployment decisions. Most current software reliability models fall into this category. Details of these models can be found in Lyu [9], Musa *et al.* [8], Singpurwalla and Wilson [13], and Gokhale *et al.* [14].

11.3.1 A General Class of Models

Now we describe a general class of models. In binomial models the total number of faults is some number N; the number found by time t has a binomial distribution with mean $\mu(t) = NF(t)$, where $F(t)$ is the probability of a particular fault being found by time t. Thus, the number of faults found in any interval of time (including the interval (t, ∞)) is also binomial. $F(t)$ could be any arbitrary cumulative distribution function. Then, a general class of reliability models is obtained by appropriate parameterization of $\mu(t)$ and$N$.

Letting N be Poisson (with some mean ν) gives the related Poisson model; now, the number of faults found in any interval is Poisson, and for disjoint intervals these numbers are independent. Denoting the derivative of F by F', the hazard rate at time t is $F'(t)/[1 - F(t)]$. These models are Markovian but not strongly Markovian, except when F is exponential; minor variations of this case were studied by Jelinski and Moranda [15], Shooman [16], Schneidewind [17], Musa [18], Moranda [19], and Goel and Okomoto [20]. Schick and Wolverton [21] and Crow [22] made F a Weibull distribution; Yamada *et al.* [23] made F a Gamma distribution; and Littlewood's model [24] is equivalent to assuming F to be Pareto. Musa and Okumoto [25] assumed the hazard rate to be an inverse linear function of time; for this "logarithmic Poisson" model the total number of failures is infinite.

The success of a model is often judged by how well it fits an estimated reliability curve $\mu(t)$ to the observed "number of faults versus time" function. On general grounds, having a good fit may be an overfit and may have little to do with how useful the model is in predicting future faults in the present system, or future experience with another system, unless we can establish statistical relationships between the measurable attributes of the system and the estimated parameters of the fitted models.

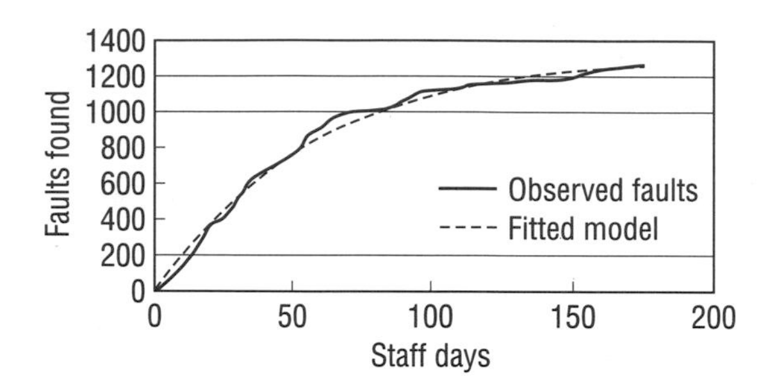

Figure 11.2. The observed versus the fitted model

Let us examine the real example plotted in Figure 11.2 from testing a large software system at a telecommunications research company. The system had been developed over years, and new releases were created and tested by the same development and testing groups respectively. In Figure 11.2, the elapsed testing time in staff days t is plotted against the cumulative number of faults found for one of the releases. It is not clear whether there is some "total number" of bugs to be found, or whether the number found will continue to increase indefinitely. However, from data such as that in Figure 11.2, an estimation of the tail of a distribution with a reasonable degree of precision is not possible. We also fit a special case of the general reliability growth model described above corresponding to N being Poisson and F being exponential.

11.3.2 Assumptions Underlying the Reliability Growth Models

Different sets of assumptions can lead to equivalent models. For example, the assumption that, for each fault, the time-to-detection is a random variable with a Pareto distribution, these random variables being independent, is equivalent to assuming that each fault has an exponential lifetime, with these lifetimes being independent, and the rates for the different faults being distributed according to a Gamma distribution (this is Littlewood's model [24]). A single experience cannot distinguish between a model that assumes a fixed but unknown number of faults and a model that assumes this number is random. Little is known about how well the various models can be distinguished.

Most of the published models are based on common underlying assumptions. These commonly include the following.

1. The system being tested remains essentially unchanged throughout testing, except for the removal of faults as they are found. Some models allow for the possibility that faults are not corrected perfectly. Miller [26] assumed that if faults are not removed as they are found, then each fault causes failures according to a stationary Poisson process; these processes are independent of one another and may have different rates. By specifying the rates, many of the models mentioned in Section 11.3.1 can be obtained. Gokhale *et al.* [27] dealt with the case of imperfect debugging.
2. Removing a fault does not affect the chance that a different fault will be found.
3. "Time" is measured in such a way that testing effort is constant. Musa [18] reported that execution time (processor time) is the most successful way to measure time. Others prefer testing effort measured in staff hours [28].
4. The model is Markovian, *i.e.* at each time, the future evolution of the testing process depends only on the present state (the current time, the number of faults found and remaining, and the overall parameters of the model) and not on details of the past history of the testing process. In some models a stronger property holds, namely that the future depends only on the current state and the parameters, and not on the current time. We call this the "strong Markov" property.
5. All faults are of equal importance (contribute equally to the failure rate). Some extensions have been discussed by Dalal and Mallows [29] in the context of when to stop testing.
6. At the start of testing, there is some finite total number of faults, which may be fixed (known or unknown) or random; if random, its distribution may be known or of known form

with unknown parameters. Alternatively, the "number of faults" is not assumed finite, so that, if testing continues indefinitely, an ever-increasing number of faults will be found.
7. Between failures, the hazard rate follows a known functional form; this is often taken to be simply a constant.

11.3.3 Caution in Using Reliability Growth Models

Here we would also like to offer some caution to the readers regarding the usage of software reliability models.

In fitting any model to a given data set, one must first bear in mind a given model's assumptions. For example, if a model assumes that a fixed number of software faults will be removed within a limited period of time, but in the observed process the number of faults is not fixed (*e.g.* new faults are inserted due to imperfect fault removal, or new code is added), then one should use another model that does not make this assumption.

A second model limitation and implementation issue concerns future predictions. If the software is being operated in a manner different than the way it is tested (*e.g.* new capabilities are being exercised that were not tested before), the failure history of the past will not reflect these changes, and poor predictions may result. Development of operational profiles, as proposed by Musa *et al.* [30], is very important if one wants to predict future reliability accurately in the user's environment.

Another issue relates to the software development environment. Most reliability growth models are primarily applicable from testing onward: the software is assumed to have matured to the point that extensive changes are not being made. These models cannot have a credible performance when the software is changing and churn of software code is observed during testing. In this case, the techniques described in Section 11.4 should be used to handle the dynamic testing situation.

11.4 Reliability Growth Modeling with Covariates

We have so far discussed a number of different kinds of reliability model of varying degrees of plausibility, including phase-based models depending upon a Raleigh curve, growth models like the Goel–Okumoto model, *etc.* The growth models take as input either failure time or failure count data, and fit a stochastic process model to reflect reliability growth. The differences between the models lie principally in assumptions made on the underlying stochastic process generating the data.

However, most existing models assume that no explanatory variables are available. This assumption is assuredly simplistic, when the models are used to model a testing process, for all but small systems involving short development and life cycles. For large systems (*e.g.* greater than 100 KNCSL, *i.e.* thousands of non-commentary source lines) there are variables, other than time, that are very relevant. For example, it is typically assumed that the number of faults (found and unfound) in a system under test remains stable during testing. This implies that the code remains frozen during testing. However, this is rarely the case for large systems, since aggressive delivery cycles force the final phases of development to overlap with the initial stages of system test. Thus, the size of code and, consequently, the number of faults in a large system can vary widely during testing. If these changes in code size are not considered as a *covariate*, one is, at best, likely to have an increase in variability and a loss in predictive performance; at worst, a poor fitting model with unstable parameter estimates is likely. We briefly describe a general approach proposed by Dalal and McIntosh [28] for incorporating covariates along with a case study dealing with reliability modeling during product testing when code is changing.

Example 1. Consider a new release of a large telecommunications system with approximately 7 million NCSL and 300 KNCNCSL (*i.e.* thousands of lines of non-commentary new or changed

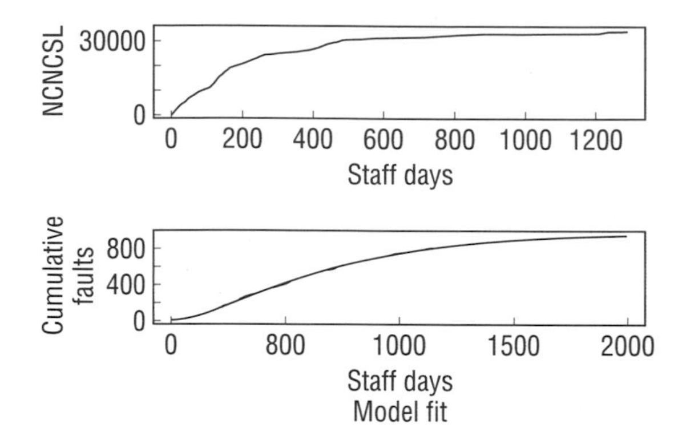

Figure 11.3. Plots of module size (NCNCSL) versus staff time (days) for a large telecommunications software system (top). Observed and fitted cumulative faults versus staff time (bottom). The dotted line (barely visible) represents the fitted model, the solid line represents the observed data, and the dashed line is the extrapolation of the fitted model

source lines). For a faster delivery cycle, the source code used for system test was updated every night throughout the test period. At the end of each of 198 calendar days in the test cycle, the number of faults found, NCNCSL, and the staff time spent on testing were collected. Figure 11.3 portrays the growth of the system in terms of NCNCSL and of faults against staff time. The corresponding numerical data are provided in Dalal and McIntosh [28].

Assume that the testing process is observed at time t_i, $i = 0, \ldots, h$, and at any given time the amount of time it takes to find a specific bug is exponential with rate m. At time t_i, the total number of faults remaining in the system is Poisson with mean l_{i+1}, and NCNCSL is increased by an amount C_i. This change adds a Poisson number of faults with mean proportional to C, say qC_i. These assumptions lead to the mass balance equation, namely that the expected number of faults in the system at t_i (after possible modification) is the expected number of faults in the system at t_{i-1} adjusted by the expected number found in the interval (t_{i-1}, t_i) plus the faults introduced by the changes made at t_i:

$$l_{i+1} = l_i \, e^{-m(t_i - t_{i-1})} + qC_i$$

for $i = 1, \ldots, h$. Note that q represents the number of new faults entering the system per additional NCNCSL, and l_1 represents the number of faults in the code at the start of system test. Both of these parameters make it possible to differentiate between the new code added in the current release and the older code. For the example, the estimated parameters are $q = 0.025, m = 0.002$, and $l_1 = 41$. The fitted and the observed data are plotted against staff time in Figure 11.3 (bottom). The fit is evidently very good. Of course, assessing the model on independent or new data is required for proper validation.

Now we examine the efficacy of creating a statistical model. The estimate of q in the example is highly significant, both statistically and practically, showing the need for incorporating changes in NCNCSL as a covariate. Its numerical value implies that for every additional 10 000 NCNCSL added to the system, 25 faults are being added as well. For these data, the predicted number of faults at the end of the test period is Poisson distributed with mean 145. Dividing this quantity by the total NCNCSL, gives 4.2 per 10 000 NCNCSL as an estimated field fault density. These estimates of the incoming and outgoing quality are valuable in judging the efficacy of system testing and for deciding where resources should be allocated to improve the quality. Here, for example, system testing was effective, in that it removed 21 of every 25 faults. However, it raises another issue: 25 faults per 10 000 NCNCSL entering system test may be too high, and a plan ought to be considered to improve the incoming quality.

None of the above conclusions could have been made without using a statistical model. These conclusions are valuable for controlling and improving the process. Further, for this analysis it was essential to have a covariate other than time.

11.5 When to Stop Testing Software

Dynamic reliability growth models can be used to make decisions about when to stop testing.

Software testing is a necessary but expensive process, consuming one-third to one-half the cost of a typical development project. Testing a large software system costs thousands of dollars per day. Overzealous testing can lead to a product that is overpriced and late to market, whereas fixing a fault in a released system is usually an order of magnitude more expensive than fixing the fault in the testing laboratory. Thus, the question of how much to test is an important economic one. We discuss an economic formulation of the "when to stop testing" issue as proposed by Dalal and Mallows [29, 31]. Other formulations have been proposed by Dalal and Mallows [32] and by Singpurwalla [33].

Like many other reliability models, the Dalal and Mallows stochastic model assumes that there are N (unknown) faults in the software, and the times to find faults are observable and are i.i.d. exponential with rate m. N is Poisson(l), and that l is Gamma (a, b). Further, their economic model defines the cost of testing at time t to be $ft - cK(t)$, where $K(t)$ is the number of faults observed to time t and f is the cost of operating the testing laboratory per unit time. The constant c is the net cost of fixing a fault after rather than before release. Under somewhat more general assumptions, Dalal and Mallows [29] found the exact optimal stopping rule. The structure of the exact rule, which is based on stochastic dynamic programming, is rather complex. However, for large N, which is necessarily the case for large systems, the optimal stopping rule is: stop as soon as $f(e^{mt} - 1)/(mc) \geq K(t)$. Besides the economic guarantee, this rule gives a guarantee on the number of remaining faults, namely that this number has a Poisson distribution with mean $f/(mc)$. Thus, instead of determining the ratio f/c from economic considerations, we can choose it so that there are probabilistic guarantees on the number of remaining faults. Some practitioners may find that this probabilistic guarantee on the number of remaining faults is more relevant in their application. (See Dalal and Mallows [32] for a more detailed discussion.) Finally, by using reasoning similar to that used in deriving equation (4.5) of Dalal and Mallows [29], it can be shown that the current estimate of the additional time required for testing Δt, is given by

$$\frac{1}{m} \log \left[\frac{cmK(t)}{f(e^{mt} - 1)} \right]$$

For applications of this we refer the reader to Dalal and Mallows [31].

11.6 Challenges and Conclusions

Software reliability modeling and measurement have drawn quite a bit of attention recently in various industries due to concerns about the quality of software. Many reliability models have been proposed, many success stories have been reported, several conferences and forums have been formed, and much project experience has been shared.

In spite of this, there are many challenges in getting widespread use of software reliability models. Part of the challenge is that testing and other activities are not as compartmentalized as assumed in the models. As discussed in Section 11.4, code churn constantly takes place during testing, and, except for the Dalal and McIntosh model described in Section 11.4, there is very little work in that area. Further, for decision-making purposes during testing and deployment phases one would like to have a quick estimate of the system reliability. Waiting to collect a substantial amount of data before being able to fit a model is not feasible in many settings. Leveraging information from the early phases of the development life cycle to come up with a quick reliable model would ameliorate this difficulty. An extension of the approach described in Section 11.2.2 seems to be needed. Looking to the future, it would also be worthwhile incorporating the architecture of the system to come up with preliminary estimates; for a survey of architecture-based reliability models, see Goševa-Popstojanova and Trivedi [34]. Finally, to a great extent, the reliability growth models as used in the field do not leverage information about the test cases used. One of the reasons is that each test case is considered to be unique. However, as we are moving in the model-based testing area [35], test case creation is supported by an underlying meta model of the use cases and constraints imposed by users. Creating new reliability models leveraging these meta models would be important.

In conclusion, in this chapter we have described key software reliability models for early stages, as well as for the test and operational phases, and have given some examples of their uses. We have also proposed some new research directions useful to practitioners, which will lead to wider use of software reliability models.

References

[1] Mars Climate Orbiter Mishap Investigation Board Phase I Report, 1999, NASA, ftp://ftp.hq.nasa.gov/pub/pao/reports/1999/MCO_report.pdf

[2] Lee L. The day the phones stopped: how people get hurt when computers go wrong. New York: Donald I. Fine, Inc.; 1992.

[3] Dalal SR, Horgan JR, Kettenring JR. Reliable software and communication: software quality, reliability, and safety. IEEE J Spec Areas Commun 1993;12:33–9.

[4] Institute of Electrical and Electronics Engineers. ANSI/IEEE standard glossary of software engineering terminology, IEEE Std. 729-1991.

[5] ISO 9000-3. Quality management and quality assurance standard—part 3: guidelines for the application of ISO 9001 to the development, supply and maintenance of software. Switzerland: ISO; 1991.

[6] Paulk M, Curtis W, Chrissis M, Weber C. Capability maturity model for software, version 1.1, CMU/SEI-93-TR-24. Carnegie Mellon University, Software Engineering Institute, 1993.

[7] Emam K, Jean-Normand D, Melo W. SPICE: the theory and practice of software process improvement and capability determination. IEEE Computer Society Press; 1997.

[8] Musa JD, Iannino A, Okumoto K. Software reliability—measurement, prediction, application. New York: McGraw-Hill; 1987.

[9] Lyu MR, editor. Handbook of software reliability engineering. New York: McGraw-Hill; 1996.

[10] Gaffney JD, Davis CF. An approach to estimating software errors and availability. SPC-TR-88-007, version 1.0, 1988. [Also in Proceedings of the 11th Minnowbrook Workshop on Software Reliability.]

[11] Dalal SR, and Ho YY. Predicting later phase faults knowing early stage data using hierarchical Bayes models. Technical Report, Telcordia Technologies, 2000.

[12] Thomas D, Cook T, Cooper H, Cordray D, Hartmann H, Hedges L, Light R, Louis T, Mosteller F. Meta-analysis for explanation: a casebook. New York: Russell Sage Foundation; 1992.

[13] Singpurwalla ND, Wilson SP.Software reliability modeling. Int Stat Rev 1994;62(3):289–317.

[14] Gokhale S, Marinos P, Trivedi K. Important milestones in software reliability modeling. In: Proceedings of Software Engineering and Knowledge Engineering (SEKE 96), 1996. p.345–52.

[15] Jelinski Z, Moranda PB. Software reliability research. In: Statistical computer performance evaluation. New York: Academic Press; 1972. p.465–84.

[16] Shooman ML. Probabilistic models for software reliability prediction. In: Statistical computer performance evaluation. New York: Academic Press; 1972. p.485–502.

[17] Schneidewind NF. Analysis of error processes in computer software. Sigplan Note 1975;10(6):337–46.

[18] Musa JD. A theory of software reliability and its application. IEEE Trans Software Eng 1975;SE-1(3):312–27.

[19] Moranda PB. Predictions of software reliability during debugging. In: Proceedings of the Annual Reliability and Maintainability Symposium, Washington, DC, 1975. p.327–32.

[20] Goel AL, Okumoto K. Time-dependent error-detection rate model for software and other performance measures. IEEE Trans Reliab 1979;R-28(3):206–11.

[21] Schick GJ, Wolverton RW. Assessment of software reliability. In: Proceedings, Operations Research. Wurzburg-Wien: Physica-Verlag; 1973. p.395–422.

[22] Crow LH. Reliability analysis for complex repairable systems. In: Proschan F, Serfling RJ, editors. Reliability and biometry. Philadelphia: SIAM; 1974. p.379–410.

[23] Yamada S, Ohba M, Osaki S. S-shaped reliability growth modeling for software error detection. IEEE Trans Reliab 1983;R-32(5):475–8.

[24] Littlewood B. Stochastic reliability growth: a model for fault-removal in computer programs and hardware designs. IEEE Trans Reliab 1981;R-30(4):313–20.

[25] Musa JD, Okumoto K. A logarithmic Poisson execution time model for software reliability measurement. In: Proceedings Seventh International Conference on Software Engineering, Orlando (FL), 1984. p.230–8.

[26] Miller D. Exponential order statistic models of software reliability growth. IEEE Trans Software Eng 1986;SE-12(1):12–24.

[27] Gokhale S, Lyu M, Trivedi K. Software reliability analysis incorporating debugging activities. In: Proceedings of International Symposium on Software Reliability Engineering (ISSRE 98), 1998. p.202–11.

[28] Dalal SR, McIntosh AM. When to stop testing for large software systems with changing code. IEEE Trans Software Eng 1994;20:318–23.

[29] Dalal SR, Mallows CL. When should one stop software testing? J Am Stat Assoc 1988;83:872–9.

[30] Musa JD, Fuoco G. Irving N, Kropfl D, Juhlin B. The operational profile. In: Lyu MR, editor. Handbook of software reliability engineering. New York: McGraw-Hill; 1996. p.167–218.

[31] Dalal SR, Mallows CL. Some graphical aids for deciding when to stop testing software. Software Quality & Productivity [special issue]. IEEE J Spec Areas Commun 1990;8:169–75.

[32] Dalal SR, Mallows CL. Buying with exact confidence. Ann Appl Probab 1992;2:752–65.

[33] Singpurwalla ND. Determining an optimal time interval for testing and debugging software. IEEE Trans Software Eng 1991;17(4):313–9.

[34] Goševa-Popstojanova K, Trivedi K. Architecture-based software reliability. In: Proceedings of ASSM 2000 International Conference on Applied Stochastic System Modeling, March 2000.

[35] Dalal SR, Jain A, Karunanithi N, Leaton J, Lott C, Patton G. Model-based testing in practice. In: International Conference in Software Engineering—ICSE '99, 1999.

Chapter 12

Software Reliability Modeling

James Ledoux

12.1 Introduction

This chapter proposes an overview of some aspects of software reliability (SR) engineering. Most systems are now driven by software. Thus, it is well recognized that assessing the reliability of software applications is a major issue in reliability engineering, particularly in terms of cost. But predicting software reliability is not easy. Perhaps the major difficulty is that we are concerned primarily with design faults, which is a very different situation from that tackled by conventional hardware theory. A *fault* (or bug) refers to a manifestation in the code of a mistake made by the programmer or designer with respect to the specification of the software. Activation of a fault by an input value leads to an incorrect output. Detection of such an event corresponds to an occurrence of a software *failure*. Input values may be considered as arriving to the software randomly. So although software failure may not be generated stochastically, it may be detected in such a manner. Therefore, this justifies the use of stochastic models of the underlying random process that governs the software failures. In Section 12.2 we briefly recall the basic concepts of stochastic modeling for reliability. Two approaches are used in SR modeling. The most prevalent is the so-called *black-box* approach, in which only the interactions of the software with the environment are considered. Following Gaudoin [1] and Singpurwalla and Wilson [2], in Section 12.3 we use the *self-exciting point processes* as a basic tool to model the failure process. This enables an overview of most of the published SR models. A second approach, called the *white-box* approach, incorporates information on the structure of the software in the models. This is presented in Section 12.4. Section 12.5 proposes basic techniques for calibrating black-box models. The final section tries to give an account of the current practices in SR modeling and points out some challenging issues for future research.

Note that this chapter does not aspire to cover the whole topic of SR engineering. In particular, we do not discuss fault prevention, fault removal, or fault tolerance, which are three methods to achieve reliable software. We focus here on methods to forecast failure times. For a more complete view, we refer the reader to Musa *et al.* [3], and the two software reliability handbooks [4, 5]. We have used the two recent books by Singpurwalla and Wilson [2] and by Pham [6] to prepare this chapter. We also recommend reading the short paper by Everett *et al.* [7], which describes, in particular, the available software reliability toolkits (see also Ramani *et al.* [8]). Finally, the references of the chapter give a good account of those journals that propose research and tutorial papers on SR.

12.2 Basic Concepts of Stochastic Modeling

The *reliability* of software, as defined by Laprie [9], is a measure of the continuous delivery of the correct service by the software under a specified environment. This is a measure of the time to failure.

12.2.1 Metrics with Regard to the First Failure

The metrics of the first time to failure of a system, as defined by Barlow and Proschan [10] and Ascher and Feingold [11], are now recalled. The first failure time is a random variable T with distribution function

$$F(t) = \mathbb{P}\{T \le t\} \quad t \in \mathbb{R}$$

If F has a probability density function (p.d.f.) f, then we define the *hazard rate* of the random variable T by

$$r(t) = \frac{f(t)}{R(t)} \quad t \ge 0$$

with $R(t) = 1 - F(t) = \mathbb{P}\{T > t\}$. We will also use the term *failure rate*. Function $R(t)$ is called the *survivor function* of the random variable T. The hazard rate function is interpreted to be

$$\begin{aligned} r(t)\,\mathrm{d}t &\approx \mathbb{P}\{t < T \le t + \mathrm{d}t \mid T > t\} \\ &\approx \mathbb{P}\{\text{a failure occurs in }]t, t + \mathrm{d}t] \\ &\qquad \text{given that no failure occurred} \\ &\qquad \text{up to time } t\} \end{aligned}$$

Thus, the phenomenon of reliability growth ("wear out") may be represented by a decreasing (increasing) hazard rate.

When F is continuous, the hazard rate function characterizes the probability distribution of T through the *exponentiation formula*

$$R(t) = \exp\left(-\int_0^t r(s)\,\mathrm{d}s\right)$$

Finally, the mean time to failure (MTTF) is the expectation $E[T]$ of the waiting time of the first failure. Note that $E[T]$ is also $\int_0^{+\infty} R(s)\,\mathrm{d}s$.

A basic model for the non-negative random variable T is the Weibull distribution with parameters $\lambda, \beta > 0$:

$$\begin{aligned} f(t) &= \lambda\beta t^{\beta-1}\exp(-\lambda t^{\beta})\mathbf{1}_{]0,+\infty[}(t) \\ R(t) &= \exp(-\lambda t^{\beta}) \\ r(t) &= \lambda\beta t^{\beta-1} \end{aligned}$$

$$\text{MTTF} = \frac{1}{\lambda^{1/\beta}} \int_0^{+\infty} u^{1/\beta}\exp(-u)\,\mathrm{d}u$$

Note that the hazard rate is increasing for $\beta > 1$, decreasing for $\beta < 1$, and constant for $\beta = 1$. For $\beta = 1$ we obtain the exponential model with parameter λ.

12.2.2 Stochastic Process of Times of Failure

The failure process can be thought of as a *point process, i.e.* a sequence of random variables $(T_i)_{i \geq 0}$ where T_i is the ith failure time of the software (with $T_0 = 0$). An equivalent point of view is to define the sequence of random variables $X_i = T_i - T_{i-1}$ for $i \geq 1$. X_i is ith *inter-failure time*. We define the *counting process* $N(\cdot)$ associated with a point process by

$$N(t) = \sum_{i \geq 0} \mathbf{1}_{]0,t]}(T_i) \quad (N(0) = 0)$$

$N(t)$ is the number of observed failures up to time t. A point process will refer to any of (T_i), (X_i), or $N(\cdot)$. The standard metrics associated with a counting process are [11]:

- the mean value function: $M(t) = E[N(t)]$
- the *rate of occurrence of failures* at time t:

$$\text{ROCOF}(t) = \frac{\mathrm{d}M}{\mathrm{d}t}(t)$$

In such a context, we define the (conditional) *reliability function* at time $t \geq 0$ by

$$R_t(s) = \mathbb{P}\{N(t+s) - N(t) = 0 \mid N(t), T_1, \ldots, T_{N(t)}\} \quad s \geq 0$$

This is a measure of the continuous delivery of correct service during the mission interval $]t, t+s]$. At time $t = T_i$, this function is nothing else but the conditional survivor function of random variable $X_{i+1} = T_{i+1} - T_i$ given $T_1, \ldots, T_i$. This will be denoted by $R_i(s)$. We also define the (conditional) mean time to failure at time t, MTTF(t), by

$$\text{MTTF}(t) = \int_0^{+\infty} R_t(s)\, \mathrm{d}s$$

The mean time to failure at $t = T_i$ will also be denoted by MTTF$_i$ and is $E[X_{i+1} \mid T_1, \ldots, T_i]$.

During the operational life of the software, repairs are carried out when it fails to perform correctly. In such a case, the time to repair, the time to reboot the system, and other factors affect the dependability of a product. Thus, we may define the software availability as a measure of the delivery correct service with respect to the alternation correct and incorrect service. Availability is highly dependent on the maintenance policies of the software. We do not go into further details on dependability in the operational phase. Indeed, we focus here on the reliability attribute of the software as most of the literature on software reliability modeling does. We refer the reader to Chapter 2 of Lyu [4] for an account of dependability during the operational phase.

12.3 Black-box Software Reliability Models

In this section, only *dynamic models* will be discussed. That is, we are only concerned with models that consider the failure process as a stochastic process. In other words, time is an essential component of the description of the models. On the other hand, *static models* are essentially capture–recapture models. For a good account of static models, we refer to Chapter 5 of Xie [12] and to Pham [6]. A recent evaluation of capture–recapture models in the software engineering context is that of Briand *et al.* [13]. Our overview of dynamic models closely follows Chapter 2 of Gaudoin [1], Singpurwalla and coworkers [2, 14], and Snyder and Miller [15]. We assume throughout this section that any corrective action is instantaneous and each detected fault is removed.

A basic way to represent time evolution in confidence in software is as follows. At instant zero, the first failure occurs at time t_1 according a random variable $X_1 = T_1$ with hazard rate r_1. Given time $T_1 = t_1$, we observe a second failure at time t_2 at rate r_2. Function r_2 is the hazard rate of the inter-failure random variable $X_2 = T_2 - T_1$

given $T_1 = t_1$. The choice of r_2 is based on the fact that one fault was detected at time t_1. At time t_2, a third failure occurs at t_3 with failure rate r_3. Function r_3 is the hazard rate of the random variable $X_3 = T_3 - T_2$ given $T_1 = t_1$, $T_2 = t_2$ and is selected according to the "past" of the failure process at time t_2: two observed failures at times t_1 and t_2. And so on. It is expected that, due to a fault removal activity, confidence in the software's ability to deliver a proper service will be improved during its life cycle. Therefore, a basic model in SR has to capture the phenomenon of *reliability growth*. Reliability growth will basically follow from a sequence of inequalities of the following form

$$r_{i+1}(t - t_i) \leq r_i(t_i) \quad \text{on } t \geq t_i \tag{12.1}$$

and/or from selection of decreasing hazard rates $r_i(\cdot)$. We illustrate this "modeling process" on the celebrated Jelinski–Moranda model (JM) [16]. We assume *a priori* that software includes only a finite number N of faults. The first hazard rate is $r_1(t; \phi, N) = \phi N$, where ϕ is some non-negative parameter. From time $T_1 = t_1$, a second failure occurs with the constant failure rate $r_2(t; \phi, N) = \phi(N - 1), \ldots$. In a more formal setting, the two parameters N and ϕ will be encompassed in what we call a *background history* $\mathcal{F}_0$, which is any background information that we may have about the software. Then "the failure rate" of the software is represented by the function

$$\forall t \geq 0,$$

$$r_C(t; \mathcal{F}_0) = \sum_{i=1}^{+\infty} r_i(t - T_{i-1}; \mathcal{F}_0)\mathbf{1}_{[T_{i-1}, T_i[}(t) \tag{12.2}$$

which is called the *concatenated failure rate function* by Singpurwalla and Wilson [2]. An appealing graphical display of a path of this stochastic function is given in Figure 12.1 for the JM model. We can rewrite Equation 12.2 as

$$\lambda(t; \mathcal{F}_0, N(t), T_1, \ldots, T_{N(t)}) = \phi[N - N(t)] \tag{12.3}$$

Function $\lambda(\cdot)$ will be called the *stochastic intensity* of the point process $N(\cdot)$. We see that the stochastic intensity for the JM model is proportional to

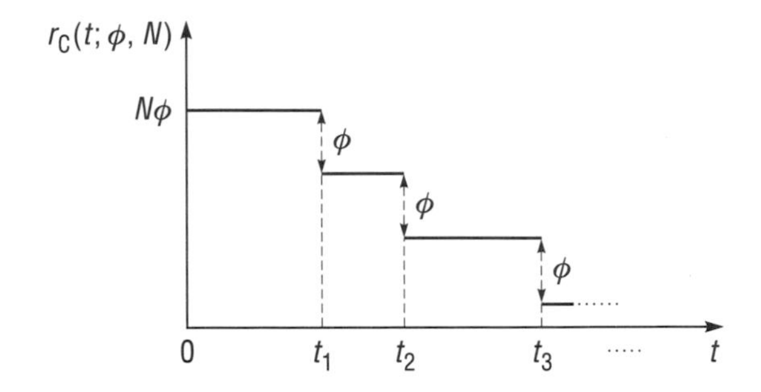

Figure 12.1. Concatenated failure rate function for (JM)

the residual number of bugs at any time t and each detection of failure results in a failure rate whose value decreases by an amount ϕ. This suggests that no new fault is inserted during a corrective action and any bug contributes in the same manner to the "failure rate" of the software.

To go further, we replace our intuitive presentation in a stochastic modeling framework. Justification for what follows is that almost all published software reliability models can be interpreted in the foregoing framework. Specifically, this allows a complete overview of the stochastic properties of the panoply of available models without referring to their original presentations.

12.3.1 Self-exciting Point Processes

A slightly more formal presentation of the previous construction of the point process (T_i) would be that we have to specify all the conditional distributions

$$\mathcal{L}(X_i \mid \mathcal{F}_0, T_{i-1}, \ldots T_1), \quad i \geq 1$$

The sequence of conditioning $\mathcal{F}_0$, $\{\mathcal{F}_0, T_1\}$, $\{\mathcal{F}_0, T_1, T_2\}, \ldots$ should be thought of as the natural or internal history on the point process at times $0, T_1, T_2, \ldots$ respectively. Thus, function r_i is the hazard rate of the conditional distribution $\mathcal{L}(X_i \mid \mathcal{F}_0, T_{i-1}, \ldots, T_1)$, *i.e.* when it has a p.d.f.

$$r_i(t; \mathcal{F}_0, T_{i-1}, \ldots, T_1) = \frac{f_{X_i \mid \mathcal{F}_0, T_{i-1}, \ldots, T_1}(t)}{R_{X_i \mid \mathcal{F}_0, T_{i-1}, \ldots, T_1}(t)} \tag{12.4}$$

This leads to the following expression of the stochastic intensity:

$$\lambda(t;\mathcal{F}_0,N(t),T_1,\ldots,T_{N(t)}) = \sum_{i=1}^{+\infty}\frac{f_{X_i|\mathcal{F}_0,T_{i-1},\ldots,T_1}(t-T_{i-1})}{R_{X_i|\mathcal{F}_0,T_{i-1},\ldots,T_1}(t-T_{i-1})}\mathbf{1}_{[T_{i-1},T_i[}(t) \quad (12.5)$$

If we turn back to the JM model, we have $\mathcal{F}_0 = \{\phi, N\}$ and

$$f_{X_i|\mathcal{F}_0,T_{i-1},\ldots,T_1}(t) = \phi[N-(i-1)]\exp\{-\phi[N-(i-1)]t\} \times \mathbf{1}_{[0,+\infty[}(t)$$

Continuing in this way, this should lead to the martingale approach for analyzing a point process, which essentially adheres to the concepts of compensator and stochastic intensity with respect to the internal history of the counting process. In particular, the left continuous version of the stochastic intensity defined in Equation 12.5 may be thought of as the usual predictable intensity of a point process in the martingale point of view (*e.g.* see Theorem 11 of [17] and references cited therein). Van Pul [18] gives a good account of what can be done using the so-called *dynamic approach* of a point process. We do not go into further details here. We prefer to embrace the engineering point of view developed by Snyder and Miller [15].

It is clear from Equation 12.5 that the stochastic intensity is excited by the history of the point process itself. Such a stochastic process is usually called a *self-exciting point process* (SEPP). What follows is from Gaudoin [1], Singpurwalla and Wilson [2], and Snyder and Miller [15].

1. $\mathcal{H}_t = \{N(t), T_1, \ldots, T_{N(t)}\}$ will denote the internal history of $N(\cdot)$ up to time t.
2. $N(\cdot)$ is said to be conditionally orderly if, for any $Q_t \subseteq \mathcal{H}_t$, we have

$$\mathbb{P}\{N(t+\mathrm{d}t)-N(t)\geq 2 \mid \mathcal{F}_0, Q_t\} = \mathbb{P}\{N(t+\mathrm{d}t)-N(t)=1 \mid \mathcal{F}_0, Q_t\}O(\mathrm{d}t)$$

where $O(\mathrm{d}t)$ is some real-valued function such that $\lim_{\mathrm{d}t\to 0} O(\mathrm{d}t)=0$.

With $Q_t=\emptyset$ and Equation 12.6 we get $\mathbb{P}\{N(t+\mathrm{d}t)-N(t)\geq 2 \mid \mathcal{F}_0\} = o(\mathrm{d}t)$, where $o(\mathrm{d}t)$ is some real-valued function such that $\lim_{\mathrm{d}t\to 0} o(\mathrm{d}t)/\mathrm{d}t = 0$. This is the usual orderliness or regular property of a point process [11]. Conditional orderliness is to be interpreted as saying that, given Q_t and $\mathcal{F}_0$, as $\mathrm{d}t$ decreases to zero, the probability of at least two failures occurring in a time interval of length $\mathrm{d}t$ tends to zero at a rate higher than the probability that exactly one failure in the same interval does.

Definition 1. A point process $N(\cdot)$ is called an SEPP if:

1. $N(\cdot)$ is conditionally orderly;
2. there exists a non-negative function $\lambda(\cdot;\mathcal{F}_0,\mathcal{H}_t)$ such that

$$\mathbb{P}\{N(t+\mathrm{d}t)-N(t)=1 \mid \mathcal{F}_0,\mathcal{H}_t\} = \lambda(t;\mathcal{F}_0,\mathcal{H}_t)\,\mathrm{d}t + o(\mathrm{d}t) \quad (12.6)$$

and $E[\lambda(t;\mathcal{F}_0,\mathcal{H}_t)] < +\infty$ for any $t>0$;
3. $\mathbb{P}\{N(0)=0 \mid \mathcal{F}_0\}=1$.

Function λ is called the stochastic intensity of the SEPP.

We must think of $\lambda(t;\mathcal{F}_0,\mathcal{H}_t)$ as a function of $\mathcal{F}_0$, t, and $N(t), T_1, \ldots, T_{N(t)}$. The degree to which $\lambda(t)$ depends on $\mathcal{H}_t$ is formalized in the notion of *memory*. An SEPP is of m-memory, if:

- for $m=0$: $\lambda(\cdot)$ depends on $\mathcal{H}_t$ only through $N(t)$, the number of observed failures at time t;
- for $m=1$: $\lambda(\cdot)$ depends on $\mathcal{H}_t$ only through $N(t)$ and $T_{N(t)}$;
- for $m\geq 2$: $\lambda(\cdot)$ depends on $\mathcal{H}_t$ only through $N(t), T_{N(t)}, \ldots, T_{N(t)-m+1}$;
- for $m=-\infty$: $\lambda(\cdot)$ is independent of the history $\mathcal{H}_t$ of the point process. We also say that the stochastic intensity has no memory.

When stochastic intensity depends only on a background history $\mathcal{F}_0$, then we get a *doubly stochastic Poisson process* (DSPP). Thus, the class of SEPP also encompasses the family of Poisson processes. If intensity is a non-random constant λ, we have the homogeneous Poisson process (HPP).

An SEPP with no memory and a deterministic intensity function $\lambda(\cdot)$ is a *non-homogeneous Poisson process* (NHPP). In particular, if we turn back to the concatenated failure rate function (see Equation 12.2), then selecting an NHPP model corresponds to selecting some continuous deterministic function as r_C. We see that, given $T_i = t_i$, the hazard rate of X_{i+1} is $r_{i+1}(\cdot) = r_C(\cdot + t_i) = \lambda(\cdot + t_i)$. We retrieve a well-known fact for an NHPP: the (conditional) hazard rate between the ith and the $i+1$th failure times and the intensity function $\lambda(\cdot)$ only differs through the initial time of observation of the two functions. We now list the properties of an SEPP [15] that are of some value for analyzing the main characteristics of models. We will omit writing about the dependence in $\mathcal{F}_0$.

12.3.1.1 Counting Statistics for a Self-exciting Point Process

The probability distribution of random variable $N(t)$ is strongly related to the conditional expectation

$$\hat{\lambda}(t; N(t)) = E[\lambda(t; \mathcal{H}_t) \mid N(t)]$$

This function is called the *count-conditional intensity*. For an SEPP, $\hat{\lambda}(\cdot; N(t))$ satisfies

$$\begin{aligned}\hat{\lambda}(t; N(t)) &= \lim_{dt \to 0} \frac{\mathbb{P}\{N(t+dt) - N(t) = 1 \mid N(t)\}}{dt}\\ &= \lim_{dt \to 0} \frac{\mathbb{P}\{N(t+dt) - N(t) \geq 1 \mid N(t)\}}{dt}.\end{aligned}$$

Then we can obtain the following explicit representation for $\mathbb{P}\{N(t) = n\}$ with $n \geq 1$:

$$\begin{aligned}\mathbb{P}\{N(t) = n\} = &\int_{0<t_1<\cdots<t_n<t} \prod_{i=1}^{n} \hat{\lambda}(t_i; i-1)\\ &\times \exp\left[-\sum_{i=0}^{n} \int_{t_i}^{t_{i+1}} \hat{\lambda}(u; i)\, du\right] dt_1 \ldots dt_n\end{aligned} \tag{12.7}$$

with $t_0 = 0$ and $t_{n+1} = t$.

ROCOF(t), defined as the derivate of $M(t)$, is then

$$\text{ROCOF}(t) = E[\hat{\lambda}(t; N(t))] = E[\lambda(t; \mathcal{H}_t)]$$

We see that the notion of ROCOF(t) and stochastic intensity coincide only if intensity is a deterministic function of time, *i.e.* the point process is an NHPP.

12.3.1.2 Likelihood Function for a Self-exciting Point Process

Assume that we observe a fixed number i of failures. Then the likelihood function is

$$\begin{aligned}&f_{T_1,\ldots,T_i}(t_1, \ldots, t_i)\\ &\quad = \lambda(t_1; 0) \prod_{k=2}^{i} \lambda(t_k; k-1, t_1, \ldots, t_{k-1})\\ &\quad\quad \times \exp\left[-\int_0^{t_1} \lambda(s; 0)\, ds\right.\\ &\quad\quad \left. - \sum_{k=2}^{i} \int_{t_{k-1}}^{t_k} \lambda(s; k-1, t_1, \ldots, t_{k-1})\, ds\right]\end{aligned} \tag{12.8}$$

If we observe the failure process up to time t, the joint distribution of $N(t), T_1, \ldots, T_{N(t)}$ is given by Theorem 6.2.2 in [15].

12.3.1.3 Reliability and Mean Time to Failure Functions

$$\begin{aligned}&R_t(s)\\ &\quad = \exp\left[-\int_t^{t+s} \lambda(u; N(t), T_1, \ldots, T_{N(t)})\, du\right]\end{aligned}$$

In particular, at instant T_i we obtain

$$R_i(s) = \begin{cases} \exp\left[-\int_0^s \lambda(u; 0)\, du\right] & \text{if } i = 0\\ \exp\left[-\int_{T_i}^{T_i+s} \lambda(u; i, T_1, \ldots, T_i)\, du\right] & \text{if } i \geq 1.\end{cases}$$

The following characterization of 0-memory SEPP is intuitively clear from the definition of the stochastic intensity of an SEPP.

Theorem 1. *$N(\cdot)$ is an SEPP with 0-memory is equivalent to $N(\cdot)$ is a Markov process.*

This result explains why a very large part of SR models may be developed in a Markov framework (*e.g.* see Xie [12], and Musa *et al.* [3] chapter 10). In particular, all NHPP models are of Markov type. In fact, it can be shown [15] that an SEPP with m-memory corresponds to a process (T_i) that is an m-order Markov chain, *i.e.* $\mathcal{L}(T_{i+1} \mid T_i, \ldots, T_1) = \mathcal{L}(T_{i+1} \mid T_i, \ldots, T_{i-m+1})$ for $i \geq m$.

A final relevant result is concerned with 1-memory SEPP.

Theorem 2. *A 1-memory SEPP with a stochastic intensity satisfying*

$$\lambda(t; N(t), T_{N(t)}) = f(N(t), t - T_{N(t)})$$

for some real-valued function f, is characterized by a sequence of independent inter-failure durations (X_i). In this case, the density probability function of X_i is

$$f_{X_i}(x_i) = f(i-1; x_i) \exp\left[-\int_0^{x_i} f(i-1; u)\, du\right] \tag{12.9}$$

Such an SEPP was called a generalized renewal process [19] because T_i is the sum of independent but not identically distributed random variables. Moreover, it is an usual renewal process when function f does not depend on $N(t)$.

To close this presentation of self-exciting processes, we point out that we only use a "constructive" point of view. Our purpose, here, is not to discuss the existence of point processes with a fixed concatenated failure rate function or stochastic intensity. However, we emphasize that the orderliness condition and the existence of the limit in Equation 12.6 in Definition 1 are enough to specify the point process (*e.g.* see Cox and Isham [20]). It is also shown by Chen and Singpurwalla [14] (Theorem 4.1) that, under the conditional orderliness condition, the concatenated failure rate function well-defines an SEPP with respect to the operational Definition 1. Moreover, an easily checked criterion for conditional orderliness is given by Chen and Singpurwalla [14] (Theorem 4.2). In particular, if the hazard rates in Equation 12.4 are locally bounded then conditional orderliness holds.

12.3.2 Classification of Software Reliability Models

We obtain from the concept of memory for an SEPP a classification of the existing models. It is appealing to define a model with a high memory. But, as usual, the pay-off is the complexity in the statistical inference and the amount of data to be collected.

12.3.2.1 0-Memory Self-exciting Point Process

The first type of 0-memory SEPP is when $\lambda(t; \mathcal{H}_t, \mathcal{F}_0) = f(N(t), \mathcal{F}_0)$. We obtain the major common properties of this first class of models from Section 12.3.1.

- $N(\cdot)$ is a Markov process (a pure birth Markov process).
- (X_i) are independent (given $\mathcal{F}_0$). From Equation 12.9, random variable X_i has an exponential distribution with parameter $f(i-1, \mathcal{F}_0)$. This easily gives a likelihood function given inter-failure durations (X_i).
- $T_i = \sum_{k=1}^{i} X_k$ is a hypoexponential distributed random variable as the sum of i independent and exponentially distributed random variables [21].
- $R_t(s) = \exp[-f(N(t), \mathcal{F}_0)s]$. The reliability function only depends on the current number of failures $N(t)$. We have $\text{MTTF}(t) = 1/f(N(t), \mathcal{F}_0)$.

Example 1. (JM) The Jelinski–Moranda model was introduced in Section 12.3. This model has to be considered as a benchmark model, since all authors designing a new model emphasize that their model includes the JM model as a particular case. The stochastic intensity is given in Equation 12.3 (see Figure 12.1 for a path). Besides the properties common to the class of SEPP considered in this subsection, we have the following additional assumptions: the software includes a finite N of bugs in the program and no new fault is inserted during debugging. We also noted that each fault has the same contribution to the unreliability of the software.

These assumptions are generally considered as questionable. We derive from Equation 12.7 that the distribution of random variable $N(t)$ is binomial with parameters N and $1 - \exp(-\phi t)$. The main reliability metrics are:

$$\mathrm{ROCOF}(t) = N\phi \exp(-\phi t)$$
$$\mathrm{MTTF}_i = \frac{1}{N - i\phi}$$
$$R_i(s) = \exp(-(N - i\phi)s) \qquad (12.10)$$

We also mention the geometric model of Moranda [22], where the stochastic intensity is $\lambda(t; \mathcal{H}_t, \lambda, c) = \lambda c^{N(t)}$ where $\lambda \geq 0$ and $c \in]0, 1[$.

A second class of 0-memory SEPP is when the stochastic intensity is actually a function of time t, $N(t)$ (and $\mathcal{F}_0$). In fact, we only have in this category of models SEPP with $\lambda(t; \mathcal{H}_t, \mathcal{F}_0) = [N - N(t)]\varphi(t)$ for some deterministic function $\varphi(\cdot)$ of time t. Note that $\varphi(t) = \phi$ gives (JM).

- $N(\cdot)$ is a Markov process.
- Let us denote $\int_0^t \varphi(s)\, ds$ by $\Phi(t)$. We deduce from Equation 12.7 that $N(t)$ is a binomially distributed random variable with parameters N and $p(t) = 1 - \exp[-\Phi(t)]$. This leads to the term binomial-type model in [23]. It follows that $E[N(t)] = Np(t)$ and $\mathrm{ROCOF}(t) = N\varphi(t) \exp[-\Phi(t)]$.
- $R_t(s) = \exp\{-[N - N(t)][\Phi(s + t) - \Phi(t)]\}$.
- We get the likelihood function given failure times (T_i) from Equation 12.8

$$f_{T_1,\ldots,T_i}(t_1, \ldots, t_i) = \frac{N!}{(N-i)!} \exp[-(N - i)\Phi(t_i)] \times \prod_{j=1}^{i} \varphi(t_j) \exp[-\Phi(t_j)]$$

- The p.d.f. of random variable T_i is $f_{T_i}(t_i) = i\binom{N}{i} \exp[-\Phi(t_i)]\varphi(t_i)\{\exp[-\Phi(t_i)] - 1\}^{i-1}$.

Littlewood's model [24] is an instance of a binomial model with $\varphi(t) = \alpha/(\beta + t)$ and $\alpha, \beta > 0$.

12.3.2.2 Non-homogeneous Poisson Process Model: $\lambda(t; \mathcal{H}_t, \mathcal{F}_0) = f(t; \mathcal{F}_0)$ and is Deterministic

A large number of models are of the NHPP type. Refer to Osaki and Yamada [25] and chapter 5 of Pham [6] for a complete list of such models. The very appealing properties of the counting process explain the wide use of this family of point processes in SR modeling.

- $N(\cdot)$ is a Markov process.
- We deduce from Definition 1 that, for any t, $N(t)$ is a Poisson distributed random variable with parameter $\Lambda(t) = \int_0^t f(s; \mathcal{F}_0)\, ds$. In fact, $N(\cdot)$ has independent (but non-stationary) increments, *i.e.* $N(t_1), N(t_2) - N(t_1), \ldots, N(t_i) - N(t_{i-1})$ are independent random variables for any $(t_1, \ldots, t_i)$.
- The mean value function $M(t)$ is $\Lambda(t)$. Then $\mathrm{ROCOF}(t) = f(t; \mathcal{F}_0)$. Such a model is called a finite NHPP model if $M(+\infty) < +\infty$ and an infinite one when $M(+\infty) = +\infty$. Indeed, if $M(+\infty) < +\infty$ then we only consider a finite number of failure times with probability of one.
- $R_t(s) = \exp\{-[\Lambda(t + s) - \Lambda(t)]\}$.
- Likelihood function given failure times (T_i) is from Equation 12.8

$$f_{T_1,\ldots,T_i}(t_1, \ldots, t_i) = \exp[-\Lambda(t_i)] \prod_{k=1}^{i} f(t_k, \mathcal{F}_0) \qquad (12.11)$$

Example 2. (GO) This model is characterized by the following mean value function and intensity function: $\Lambda(t) = M[1 - \exp(-\phi t)]$ and $\lambda(t; \phi, M) = M\phi \exp(-\phi t)$ for $t \geq 0$. Parameters ϕ and M are the failure rate per fault and the finite expected (initial) number of faults contained in the software respectively. We see that, at any time t, the intensity function is proportional to the expected remaining number of faults $\lambda(t; \phi, M) = \phi[M - \Lambda(t)]$. Thus, (GO) is essentially an NHPP version of (JM).

Example 3. (MO) For the Musa–Okumoto model [3], the mean value function and intensity function are $\Lambda(t) = \ln(\lambda\theta t + 1)/\theta$ and $\lambda(t; \theta, \lambda) = \lambda/(\lambda\theta t + 1)$ respectively. λ is the initial value of the intensity function and θ is called the failure intensity decay parameter. It is easily seen that $\lambda(t; \theta, \lambda) = \lambda \exp[-\theta\Lambda(t)]$. Thus, the intensity function decreases exponentially with the expected number of failures and shows that (MO) may be understood as an NHPP version of Moranda's geometric model.

As noted by Littlewood [26], using an NHPP model may appear inappropriate to describe reliability growth of software. Indeed, this is debugging that modifies the reliability. Thus, the true intensity function probably changes in a discontinuous manner during corrective actions. However, Miller [27] showed that the binomial-type model of Section 12.3.2.1 may be transformed in an NHPP variant assuming that the initial number of faults is a Poisson distributed random variable with expectation N. For instance, (JM) is transformed into (GO). Moreover, Miller showed that the binomial-type model and its NHPP variant are indistinguishable from a single realization of the failure process. But these two models differ as a prediction model because the estimates of parameters are different.

12.3.2.3 1-Memory Self-exciting Point Process with $\lambda(t; \mathcal{H}_t, \mathcal{F}_0) = f(N(t), t - T_{N(t)}, \mathcal{F}_0)$

$N(\cdot)$ is not Markovian. But the inter-failure durations (X_i) are independent random variables given $\mathcal{F}_0$ (see Theorem 2) and the p.d.f. of random variable X_i is given in Equation 12.9.

Example 4. (LV) The stochastic intensity of the Littlewood–Verrall model [28] is

$$\lambda(t; \mathcal{H}_t, \alpha, \psi(\cdot)) = \frac{\alpha}{\psi[N(t) + 1] + t - T_{N(t)}} \tag{12.12}$$

for some non-negative function $\psi(\cdot)$. We briefly recall the Bayesian rationale underlying the definition of this model. Uncertainty about the debugging operation is represented by a sequence

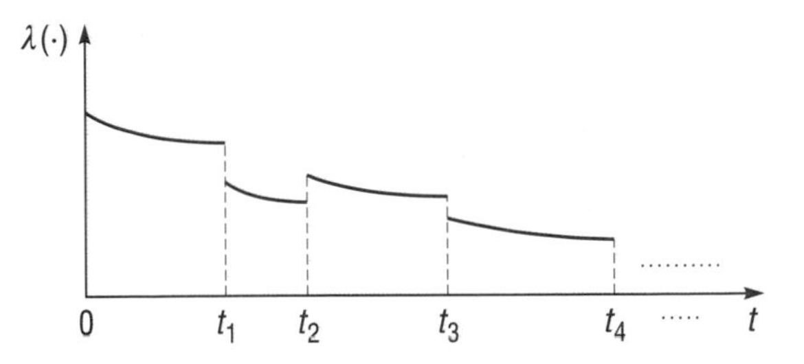

Figure 12.2. A path of the stochastic intensity for (LV)

of stochastically decreasing failure rates $(\Lambda_i)_{i\geq 1}$, *i.e.*

$$\Lambda_j \leq_{\text{st}} \Lambda_{j-1}$$
$$i.e.\ (\forall t \in \mathbb{R} : \mathbb{P}\{\Lambda_j \leq t\} \geq \mathbb{P}\{\Lambda_{j-1} \leq t\}) \tag{12.13}$$

Thus, using stochastic order allows a decay of reliability that takes place when faults are inserted. The prior distribution of random variable Λ_i is Gamma with parameters α and $\psi(i)$. It can be shown that the inequality of Equation 12.13 holds when $\psi(\cdot)$ is a monotonic increasing function of i. Given $\Lambda_i = \lambda_i$, random variable X_i has an exponential distribution with parameter λ_i. The unconditional distribution of random variable X_i is a Pareto distribution with p.d.f.

$$f(x_i; \alpha, \psi(i)) = \frac{\alpha\psi(i)^\alpha}{[x_i + \psi(i)]^{\alpha+1}}$$

Thus, the "true" hazard rate of X_i is $r_i(t; \alpha, \psi(i)) = \alpha/[t + \psi(i)]$. Mazzuchi and Soyer [29] estimated parameters α and ψ using a Bayesian method. The corresponding model is called a hierarchical Bayesian model [2] and is also a 1-memory SEPP. If $\psi(\cdot)$ is linear in i, we have

$$R_i(t) = \left[\frac{\psi(i+1)}{t + \psi(i+1)}\right]^\alpha, \quad \text{MTTF}_i = \frac{\psi(i+1)}{\alpha - 1}$$

We also mention the Schick–Wolverton's model [30], where $\lambda(t; \mathcal{H}_t, \Phi, N) = \phi[N - N(t)](t - T_{N(t)})$ and $\phi > 0$, N are the same parameters as for (JM).

12.3.2.4 $m \geq 2$-Memory

Instances of such models are rare. For $m = 2$, we have the time-series models of Singpurwalla and

Soyer [31, 32]. For $m > 2$, we have the adaptive concatenated failure rate model of Al-Mutairi *et al.* [33] and the Weibull models of Pham and Pham [34]. We refer the reader to the original contributions for details.

12.4 White-box Modeling

Most work on SR assessment adopts the black-box view of the system, in which only the interactions with the environment are considered. The *white-box* (or structural) point of view is an alternative approach in which the structure of the system is explicitly taken into account. This is advocated for instance by Cheung [35] and Littlewood [36]. Specifically, the structure-based approach allows one to analyze the sensitivity of the reliability of the system with respect to the reliability of its components. Up to recently, only a few papers proposed structure-based SR models. Representative samples can be found for discrete time [35,37,38], and continuous time [19,36,39,40]. An up-to-date review on the architecture-based approach is given by Goseva-Popstojanova and Trivedi [41]. We will present the main features of the basic Littlewood model that are common to most previous cited works. The discrete time counterpart is Cheung's model (see Ledoux and Rubino [42]).

In a first step, Littlewood defines an execution model of the software. The basic entity is the standard software engineering concept of *module*, *e.g.* see Cheung [35]. The software structure is then represented by the *call graph* of the set $\mathcal{M}$ of the modules. These modules interact by execution control transfer and, at each instant, control lies in one and only one of the modules, which is called the active module. From such a view of the system, we build up a continuous time stochastic process $(X_t)_{t\geq 0}$ that indicates the active module at each time t. $(X_t)_{t\geq 0}$ is assumed to be a homogeneous Markov process on the set $\mathcal{M}$.

In a second step, Littlewood describes the failure processes associated with execution actions. Failure may happen during a control transfer between two modules or during an execution period of any module. During a sojourn of the execution process in the module i, failures are part of a Poisson process having parameter λ_i. When control is transferred from module $i \in \mathcal{M}$ to module $j \in \mathcal{M}$, a failure may happen with probability $\mu(i, j)$. Given a sequence of executed modules, the failure processes associated with each state are independent. Also, the interface failure events are independent of each other and of the failure processes occurring when a module is active.

The architecture of the software is combined with the failure behavior of the modules and that of interfaces into a single model which can then be analyzed. This method is referred to as the "composite-method" according to the classification of Markov models in the white-box approach proposed by Goseva-Popstojanova and Trivedi [41]. Basically, we are still interested in the counting process $N(\cdot)$.

Another interesting point process is obtained by assuming that the probability of a secondary failure during a control transfer is zero. Thus, assuming that $\mu(i, j) = 0$ for all $i, j \in \mathcal{M}$ in the previous context, we get a Poisson process whose parameter is modulated by the Markov process $(X_t)_{t\geq 0}$. This is also called a Markov modulated Poisson process (MMPP). It is well known (*e.g.* [17]) that the stochastic intensity of an MMPP with respect to the history $\mathcal{H}_t \vee \mathcal{F}_0$, where $\mathcal{F}_0 = \sigma(X_s, s \geq 0)$, is λ_{X_t}. Thus, $N(\cdot)$ is an instance of a DSPP.

Asymptotic analysis and transient assessment of distribution of random variable $N(t)$ may be carried out in observing that the bivariate process $(X_t, N(t))_{t\geq 0}$ is a jump Markov process with state space $\mathcal{M} \times \mathbb{N}$. Computation of all standard reliability metrics may be performed as in [42,43]. We refer the reader to these papers for details and for calibration of the models.

It can be argued that models of the Littlewood type are inappropriate to capture reliability growth of the software. In fact, Laprie *et al.* [19] have developed a method to incorporate such a phenomenon; this can be used, for instance, in the Littlewood model to take into account reliability growth of modules in the assessment of the overall reliability (see Ledoux [43]).

In the original papers about Littlewood's model for modular software, it is claimed that if failure parameters decrease to zero then $N(\cdot)$ is asymptotically an HPP with parameter

$$\lambda = \sum_{i \in \mathcal{M}} \pi(i) \left[\sum_{j \in \mathcal{M}, j \neq i} Q(i, j)\mu(i, j) + \lambda_i \right]$$

where Q is the irreducible generator of the Markov process $(X_t)_{t \geq 0}$ and π its stationary distribution. $\pi(i)$ is to be interpreted as the proportion of time the software passed in module i (over a long time period). We just discuss the case of an MMPP. Convergence to zero of the failure parameters λ_i $(i \in \mathcal{M})$ may be achieved in multiplying each of them by a positive scalar ε, and in considering that ε decreases to zero. Hence, the new stochastic intensity of the MMPP is $\varepsilon \lambda_{X_t}$. As ε tends to zero, it is easily seen from an MMPP with a modulating two-states Markov process $(X_t)_{t \geq 0}$ that, for the first failure time T, probability $\mathbb{P}\{T > t\}$ converges to unity. Therefore, we cannot obtain an exponential approximation to the distribution of random variable T as ε tends to zero. Thus, we cannot expect to derive a Poisson approximation to the distribution of $N(\cdot)$. In fact, the correct statement is: if failure parameters are much smaller than the switching rates between modules, $N(\cdot)$ is approximately an HPP with parameter λ. For an MMPP, a proof is given by Kabanov *et al.* [44] using martingale theory. Moreover, the rate of convergence in total variation of finite dimensional-distributions of $N(\cdot)$ to those of the HPP is shown to be in ε. This last fact is important, because the user has no information on the quality of the Poissonian approximation given in [36]. However, there is no rule to decide *a priori* if the approximation is optimistic or pessimistic (see Ledoux [43]). A similar approach to Kabanov *et al.* [44] is used by Gravereaux and Ledoux [45] to derive the Poisson approximation and rate of convergence for a more general point process than MMPP, including the complete Littlewood's counting process [46], the counting model of Ledoux [43]. Such asymptotic results give a "hierarchical-method" [41] for reliability prediction: we solve the architectural model and superimpose the failure behavior of the modules and that of the interfaces on to the solution to predict reliability.

12.5 Calibration of Model

Suppose that we have selected one of the black-box models of Section 12.3. We obtain reliability metrics that depend on the unknown parameters of the model. Thus, we hate to estimate these metrics from the failure data. We briefly review standard methods to get point estimates in Sections 12.5.1 and 12.5.2.

One major goal of the SR modeling is to predict the future value of metrics from the gathered failure data. It is clear from Section 12.3 that a central problem in SR is in selecting a model, because of the huge number of models available. Criteria to compare SR models are listed by Musa *et al.* [3]. The authors propose quality of assumptions, applicability, simplicity, and predictive validity. To assess the predictive validity of models, we need methods that are not just based on a goodness-of-fit approach. Various techniques may be used: u-plot, prequential likelihood, *etc.* We do not discuss this issue here. Good accounts of predictive validation methods are given by Abdel-Ghaly *et al.* [47], Brocklehurst *et al.* [48], Lyu [4] (chapter 4), Musa *et al.* [3], Ascher and Feingold [11], where comparisons between models are also carried out. Note also that the predictive quality of an SR model may be drastically improved by using preprocessing of data. In particular, statistical tests have been designed to capture trends in data. Thus, reliability trend analysis allows the use of SR models that are adapted to reliability growth, stable reliability, and reliability decrease. We refer the reader to Kanoun [49], and Lyu [4] (chapter 10) and references cited therein for details. Parameters will be denoted by θ (it can be multivariate).

12.5.1 Frequentist Procedures

Parameter θ is considered as taking an unknown but fixed value. Two basic methods to estimate the value of θ are the method of maximum likelihood (ML) and the least squares method. That is, we have to optimize with respect to θ an objective function that depends on θ and collected failure data to get a *point estimate*. Another standard procedure, *interval estimation*, gives an interval of values as estimate of θ. We only present point-estimation by the ML method on (JM) and NHPP models. Others models may be analyzed in a similar way. ML estimations possess several appealing properties that make the procedure widely used. Usually, two of these properties are the consistency and the asymptotic normality of obtaining a *confidence interval* for the point estimate. Another one is that the ML estimates of $f(\theta)$ (for a one-to-one function f) is simply $f(\hat{\theta})$, where $\hat{\theta}$ is the ML estimate of θ. We refer the reader to chapter 12 of Musa *et al.* [3] for a complete view of the frequentist inference procedures in an SR modeling context.

Example 5. (JM) Parameters are ϕ and N. Assume that failure data are given by observed values $\underline{x} = (x_1, \ldots, x_i)$ of random variables $X_1, \ldots, X_i$. If (ϕ, N) are the true values of parameters, then the likelihood to observe $\underline{x}$ is defined as

$$L(\phi, N; \underline{x}) = f_{X_1,\ldots,X_i}(\underline{x}; \phi, N) \qquad (12.14)$$

where $f_{X_1,\ldots,X_i}(\cdot; \phi, N)$ is the p.d.f. of the joint distribution of $X_1, \ldots, X_i$. The estimate $(\hat{\phi}, \hat{N})$ of (ϕ, N) will be the value of (ϕ, N) that maximizes the likelihood to observe data $\underline{x}$: $(\hat{\phi}, \hat{N}) = \arg\max_{\phi,N} L(\phi, N; \underline{x})$. Maximizing the likelihood function is equivalent to maximizing the log-likelihood function $\ln L(\phi, N; \underline{x})$. From the independence of random variables (X_i) and Equation 12.14, we obtain that

$$\ln L(\phi, N; \underline{x}) = \ln \prod_{k=1}^{i} \phi[N - (k-1)] \times \exp\{-\phi[N - (k-1)]x_k\}$$

Estimates $(\hat{\phi}, \hat{N})$ are solutions of $\frac{\partial}{\partial\phi} \ln L(\phi, N; \underline{x}) = \frac{\partial}{\partial N} \ln L(\phi, N; \underline{x}) = 0$:

$$\hat{\phi} = \frac{i}{\hat{N}\sum_{k=1}^{i} x_k - \sum_{k=1}^{i}(k-1)x_k}$$

$$\sum_{k=1}^{i} \frac{1}{\hat{N} - (k-1)} = \frac{i}{\hat{N} - \left(\sum_{k=1}^{i} x_k\right)^{-1} \sum_{k=1}^{i}(k-1)x_k}$$

The second equation here may be solved by numerical techniques and then the solution is put into the first equation to get $\hat{\phi}$. These estimates are plugged into Equation 12.10 to get ML estimates of reliability metrics.

Example 6. (NHPP models) Assume that failure data are $\underline{t} = (t_1, \ldots, t_i)$ the observed failure times. The likelihood function is given by Equation 12.11. Thus, for the (MO) model, ML estimates of parameters β_0 and β_1 (where $\beta_0 = 1/\theta$ and $\beta_1 = \lambda\theta$) are the solution of [3]:

$$\hat{\beta}_0 = \frac{i}{\ln(1 + \hat{\beta}_1 t_i)}$$

$$\frac{1}{\hat{\beta}_1} \sum_{k=1}^{i} \frac{1}{1 + \hat{\beta}_1 t_k} = \frac{i t_i}{(1 + \hat{\beta}_1 t_i)\ln(1 + \hat{\beta}_1 t_i)}$$

Assume now that failure data are the cumulative numbers of failures $n_1, \ldots, n_i$ at some instants $d_1, \ldots, d_i$. The likelihood function is from the independence and Poisson distribution of the increments of the counting process $N(\cdot)$

$$L(\theta; \underline{t}) = \prod_{k=1}^{i} \frac{[\Lambda_\theta(d_k) - \Lambda_\theta(d_{k-1})]^{n_k - n_{k-1}}}{(n_k - n_{k-1})!} \times \exp\{-[\Lambda_\theta(d_k) - \Lambda_\theta(d_{k-1})]\} \quad (d_0 = n_0 = 0)$$

where $\Lambda_\theta(\cdot)$ is the mean value function of $N(\cdot)$ given that θ is the true value of the parameter. For the (GO) model, parameters are M and ϕ and their

ML estimates are

$$\hat{M} = \frac{n_i}{1 - \exp(-\phi d_i)}$$

$$\frac{d_i \exp(-\phi d_i) n_i}{1 - \exp(-\phi d_i)} = \sum_{k=1}^{i} \{(n_k - n_{k-1})[d_k \exp(-\phi d_k) - d_{k-1} \exp(-\phi d_{k-1})]\} \times \{\exp(-\phi d_{k-1}) - \exp(-\phi d_k)\}^{-1}$$

The second equation here is solved by numerical techniques and the solution is incorporated into the first equation to get $\hat{M}$. Estimation of reliability metrics are obtained as for (JM).

12.5.2 Bayesian Procedure

An alternative to frequentist procedures is to use Bayesian statistics. Parameter θ is considered as a value of a random variable Θ. A *prior distribution* of Θ has to be selected. This distribution represents the *a priori* knowledge on the parameter and is assumed to have a p.d.f. $\pi_\Theta(\cdot)$. Now, from the failure data $\underline{x}$, we have to update our knowledge on Θ. If $L(\theta; \underline{x})$ is the likelihood function of the data, then Bayes theorem is used as an updating formula: the p.d.f. of the *posterior distribution* of Θ given data $\underline{x}$ is

$$f_{\Theta|\underline{x}}(\theta) = \frac{L(\theta; \underline{x})\pi_\Theta(\theta)}{\int_{\mathbb{R}} L(\theta; \underline{x})\pi_\Theta(\theta)\, \mathrm{d}\theta}$$

Now, we find an estimate $\hat{\theta}$ of θ by minimizing the so-called *posterior expected loss*

$$\int_{\mathbb{R}} l(\hat{\theta}, \theta) f_{\Theta|\underline{x}}(\theta)\, \mathrm{d}\theta$$

where $l(\cdot, \cdot)$ is the loss function. With a quadratic loss function $l(\hat{\theta}, \theta) = (\hat{\theta} - \theta)^2$, it is well known that the minimum is obtained by the conditional expectation

$$\hat{\theta} = E[\Theta \mid \underline{x}] = \int_{\mathbb{R}} \theta f_{\Theta|\underline{x}}(\theta)\, \mathrm{d}\theta$$

It is just the mean of the posterior distribution of Θ. Note that all SR models involve two or more parameters, so that the previous integral must be considered as multidimensional. Therefore, computation of such integrals is by numerical techniques. For a decade now, progress has been made on such methods: *e.g.* Monte Carlo Markov chain (MCMC) methods (*e.g.* see Gilks *et al.* [50]).

Note that the prior distribution can also be parametric. In general, Gamma or Beta distributions are used. Hence, additional parameters need to be estimated. This may be carried out by using the ML method. For instance, this is the case for function $\psi(\cdot)$ in the (LV) model [28]. However, such estimates may be derived in the Bayesian framework, and we obtain a hierarchical Bayesian analysis of the model. Many of the models of Section 12.3 have been analyzed from a Bayesian point of view using MCMC methods like Gibbs sampling, or the data augmentation method (see Kuo and Yang [51] for (JM) and (LV); see Kuo and Yang [52] for the NHPP model; see Kuo *et al.* [53] for S-shaped models and references cited therein).

12.6 Current Issues

12.6.1 Black-box Modeling

We list some issues that are not new, but which cover well-documented limitations of popular black-box models. Most of them have been addressed recently, and practical validation is needed. We will see that the prevalent approach is to use the NHPP modeling framework. Indeed, an easy way to incorporate the various factors affecting the reliability of software in an NHPP model is to select a suitable parameterization of the intensity function (or ROCOF). Any alternative model combining most of these factors will be of value.

12.6.1.1 Imperfect Debugging

The problem of imperfect debugging may be naturally addressed in a Bayesian framework. Reliability growth is captured through a deterministically non-increasing sequence of failure rates $(r_i(\cdot))$ (see Equation 12.1). In a Bayesian framework, parameters of $r_i(\cdot)$ are considered as random.

Hence, we can deal with stochastically decreasing sequence of random variables $(r_i)_i$ (see Equation 12.13), which allows one to take into account the uncertainty on the effect of a corrective action. An instance of this approach is given by the (LV) model (see also Mazzuchi and Soyer [29], Pham and Pham [34]).

Note that the binomial class of models can incorporate a defective correction of a detected bug. Indeed, assume that each fault detected has a probability p to be removed from the software. The hazard rate after $(i-1)$ repairs is $\phi[N - p(i-1)]$ (see Shanthikumar [23]). But the problem of eventual introduction of new faults is not addressed. Kremer [54] solved the case of a single insertion using a non-homogeneous birth–death Markov model. This has been extended to multiple introductions by Gokhale *et al.* [55]. Shanthikumar and Sumita [56] proposed a multivariate model where multiple removals and insertions of faults are allowed at each repair. This model involved complex computational procedures and is not considered in literature. A recent advance in addressing the problem of eventual insertion of new faults is concerned with finite NHPP models. It consists in generalizing the basic proportionality between the intensity function and the expected number of remaining faults at time t of the (GO) model (see Example 2) in

$$\lambda(t) = \phi(t)[n(t) - \Lambda(t)]$$

where $\phi(t)$ represents a time-dependent detection rate of a fault; $n(t)$ is the number of faults in the software at time t, including those already detected and removed and those inserted during the debugging process ($\phi(t) = \phi$ and $n(t) = M$ in (GO)). Making $\phi(\cdot)$ time dependent allows one to represent the phenomenon of the learning process that is closely related to the changes in the efficiency of testing. This function can monotonically increase during the testing period. Select a non-decreasing S-shaped curve, as $\phi(\cdot)$ gives an usual S-shaped NHPP model. Further details are given by Pham [6] (chapter 5 and references cited therein).

Another basic way to consider insertion of new faults is to use a marked point process (MPP) (*e.g.* see chapter 4 in Snyder and Miller [15]). We have the point process of failure detection times $T_1 < T_2 < \cdots$, and with each date T_i we associate a mark M_i that represents the cumulative number of faults removed and inserted during the debugging phase. We retrieve a usual point process if all marks are unity. For such a model, we are interested in the mark-accumulator process $\sum_{i=0}^{N(t)} M_i$ ($M_0 = 0$). A basic instance of an MPP is the compound Poisson process where $N(\cdot)$ is an HPP and M_i values are i.i.d. and independent of $N(\cdot)$. This may also be used to model clustering of failures (see Sahinoglu [57]). Such MPPs are not an SEPP of Section 12.3.1 because the orderliness condition fails. This framework was used by van Pul [18] to extend models of the (JM) type to incorporate the possibility of inserting new faults during repair.

12.6.1.2 Early Prediction of Software Reliability

A major limitation of the SR models of Section 12.3 for the software engineering community is that they provide no help for managing in the earlier phase of development (or testing) of a product. Indeed, calibration of these black-box models requires a relatively large set of failure data. This is rarely encountered in the earlier life-cycle of software. In some sense, we are now concerned with the general topic of software quality assessment (which includes dependability concepts) with no failure data. Thus, we have to develop statistical models of quality that are not directly related to the knowledge of a part of the failure process. In such a case, the model must be based on *a priori* information on the product: judgment of experts, quality of the development process, similar existing products, software complexity, *etc.* Incorporating subjective information leads naturally to Bayesian statistics. We refer the reader to chapters 5 and 6 of Singpurwalla and Wilson [2] for discussion in this context.

Since we are mainly interested in reliability assessment, we restrict ourselves to more and less recent issues relying quality control to the software reliability.

A widespread idea is that the complexity of software is an influencing factor of the reliability attributes. Much work has been devoted to quantifying the software complexity through *software metrics* (*e.g.* see Fenton [58]). Typically, we compute Halstead and McCabe metrics, which are program size and control flow measures respectively. It is worthy of note that most software complexity metrics are strongly related to the concept of structure of software code. Thus, including a complexity factor in SR may be thought of as a first attempt to take into account the architecture of the software in reliability assessment. We turn back to this issue in Section 12.6.2. Now, how do we include complexity attributes in earlier reliability analysis? Most of the recent research focuses on the identification of software modules that are likely fault-prone from data of various complexity metrics. In fact, we are faced with a typical problem of data analysis that explains why literature on this subject is mainly concerned with procedures of multivariate analysis: linear and nonlinear regression methods, classification methods, techniques of discriminant analysis. We refer the reader to Lyu [4] (chapter 12) and Khoshgoftaar and coworkers [59, 60] and references cited therein for details.

Other empirical evidence suggests that the higher the test coverage, then the higher the reliability of the software would be. Thus, a model that incorporates information on functional testing as soon as it is available is of value. This issue is addressed in an NHPP model proposed by Gokhale and Trivedi [61]. It consists in defining an appropriate parameterization of a finite NHPP model which relates software reliability to the measurements that can be obtained from the code during functional testing. Let a be the expected number of faults that would be detected given infinite time testing. The intensity function $\lambda(\cdot)$ is assumed to be proportional to the expected number of remaining failures: $\lambda(t) = [a - \Lambda(t)]\phi(t)$, where $\phi(t)$ is the hazard rate per fault. Finally, the time-dependent function $\phi(t)$ is of the form

$$\phi(t) = \frac{\mathrm{d}c(t)/\mathrm{d}t}{1 - c(t)}$$

where $c(t)$ is the coverage function. That is, the ratio of the number of potential fault sites covered by time t divided by the total number of potential fault sites under consideration during testing. Function $c(t)$ is assumed to be continuous and monotonic as a function of testing time. Specific forms of function $c(\cdot)$ allow the retrieving of some well-known finite failure models: the exponential function $c(t) = 1 - \exp(-\phi t)$ corresponds to the (GO) model; the Weibull coverage function $c(t) = 1 - \exp(-\phi t^{\gamma})$ corresponds to the generalized (GO) model [62]; the S-shaped coverage function corresponds to S-shaped models [25]; *etc.* Gokhale and Trivedi [61] propose the use of a log-logistic function; see that reference for details. Such a parameterization leads to estimate a and the parameters of function $c(\cdot)$. The model may be calibrated according to the different phase of the software life-cycle. Here, in the early phase of testing, an approach is to estimate a from software metrics (using procedures of multivariate analysis) and measure coverage during the functional testing using a coverage measurement tool (*e.g.* see Lyu [4], chapter 13). Thus, we get early prediction of reliability (see Xie *et al.* [63] for an alternative using information from testing phases of similar past projects).

12.6.1.3 Environmental Factors

Most SR models in Section 12.3 ignore the factors affecting software reliability. In some sense, previous issues discussed in this section can be considered as an attempt to capture some environmental factors. Imperfect debugging is related to the fact that new faults may be inserted during a repair. The complexity attributes of software are strongly correlated to its fault-proneness. Empirical investigations show that the development process, testing procedure, programmer skill, human factors, the operational profile and many others factors affect the reliability of a product (*e.g.* see Pasquini *et al.* [64], Lyu [4] (chapter 13),

Özekici and Sofer [65], Zhang and Pham [66], and references cited therein). A major issue is to incorporate all these attributes into a single model. At the present time, investigations focus on the functional relationship between the hazard rate $r_i(\cdot)$ of the software and quantitative measures of the various factors. In this context, a well-known model is the so-called Cox proportional hazard model (PHM), where $r_i(\cdot)$ is assumed to be an exponential function of the environmental factors:

$$r_i(t) = r(t) \exp\left(\sum_{j=1}^{n} \beta_j z_j(i)\right) \qquad (12.15)$$

where $z_j(\cdot)$, called explanatory variables or *covariates*, are the measures of the factors, and β_j are the regression coefficients. $r(\cdot)$ is a baseline hazard rate that gives the hazard rate when all covariates are set to zero. Therefore, given $\underline{z} = (z_1, \ldots, z_n)$, the reliability function R_i is

$$R_i(t \mid \underline{z}) = R(t) \exp\left(\sum_{j=1}^{n} \beta_j z_j(i)\right) \qquad (12.16)$$

with $R(t) = \exp(-\int_0^t r(s)\,\mathrm{d}s)$. Equation 12.15 expresses the effect of accelerating or decelerating the time to failure given $\underline{z}$. Note that covariates may be time-dependent, random. In this last case the reliability function will be the expectation of the function in Equation 12.16. The baseline hazard rate may be any of the hazard rates used in Section 12.3. The family of parameters can be estimated using ML. Note that one of the reasons for the popularity of PHM is that the unknown β_j may be estimated by the partial likelihood approach without putting a parametric structure on the baseline hazard rate. We refer the reader to Kalbfleisch and Prentice [67], Cox and Oakes [68], Ascher and Feingold [11], and Andersen *et al.* [69] for a general discussion on Cox regression models. Applications of PHM to software reliability modeling are given by Xie [12] (chapter 7), Saglietti [70], Wright [71], and references cited therein. Recently, Pham [72] derived an enhanced proportional hazard (JM) model.

A general way to represent the influence of environmental factors on reliability is to assume that the stochastic intensity of the counting process $N(\cdot)$ is a function of some m stochastic processes $E_1(t), \ldots, E_m(t)$ or covariates

$$\lambda(t; \mathcal{H}_t, \mathcal{F}_0) = f(t, E_1(t), \ldots, E_m(t), T_1, \ldots, T_{N(t)}, N(t))$$

where $\mathcal{H}_t$ is the past up to time t of the point process and $\mathcal{F}_0$ encompasses the specification of the paths of all covariates. Thus, function $\lambda(t; \mathcal{H}_t, \mathcal{F}_0)$ may be thought of as the stochastic intensity of an SEPP driven or modulated by the multivariate environmental process $(E_1(t), \ldots, E_m(t))$. The DSPP of Section 12.3.1 is a basic instance of such models and has been widely used in communication engineering and in reliability. Castillo and Siewiorek [73] proposed a DSPP with a cyclo-stationary stochastic intensity to represent the effect of the workload (measure of system usage) on failure process. That is, the intensity is a stochastic process assumed to have periodic mean and autocorrelation function. A classic form for intensity of an DSPP is $\lambda(E_t)$, where (E_t) is a finite Markov process. This is the MMPP discussed in Section 12.4, where (E_t) represented the control flow structure of the software. In the same spirit of system in a random environment, Özekici and Sofer [65] use a point process whose stochastic intensity is $[N - N(t)]\lambda(E_t)$, where $N - N(t)$ is the remaining number of faults at time t and E_t is the operation performed by the system at t. (E_t) is also assumed to be a finite Markov process. Transient analysis of $N(\cdot)$ may be carried out, as in Ledoux and Rubino [42], from the Markov property of the bivariate process $(N - N(t), E_t)$; see Özekici and Sofer [65] for details. Note that both point processes are instances of SEPP with respective stochastic intensities $E[\lambda(E_t) \mid \mathcal{H}_t]$ and $[N - N(t)]E[\lambda(E_t) \mid \mathcal{H}_t]$ ($\mathcal{H}_t$ is the past of the counting process up to time t). The practical purpose of such models has to be addressed. In particular, further investigations are needed to estimate parameters (see Koch and Spreij [74]).

12.6.1.4 Conclusion

There are other issues that exist which are of value in SR engineering. In the black-box modeling framework, we can think about alternatives to the approaches reported in Section 12.3. The problem of assessing SR growth may be thought of as a problem of statistical analysis of data. Therefore, prediction techniques developed in this area of research can be used. For instance, some authors have considered neural networks (NNs). The main interest for such an SR model is to be non-parametric. Thus, we rejoin discussion on the statistical issues raised in Section 12.6.3. NNs may also be used as a classification tool. For instance, identifying fault-prone modules may be performed with a NN classifier. See Lyu [4] (chapter 17) and references cited therein for an account of the NN approach. Empirical comparison of the predictive performance of NN models and recalibrated standard models (as defined in [75]) is given by Sitte [76]. NNs are found to be a good alternative to the standard models.

Computing dependability metrics is not an end in itself in software engineering. A major question is the time to release the software. In particular, we have to decide when to stop testing. Optimal testing time is a problem of decision making under uncertainty. A good account of Bayesian decision theory for solving such a problem is given by Singpurwalla and Wilson [2] (chapter 6). In general, software release policies are based on reliability requirement and cost factors. We do not go into further details here. See Xie [12] (chapter 8) for a survey up to 1990s, and Pham and Zhang [77, 78] for more recent contributions to these topics.

12.6.2 White-box Modeling

A challenging issue in SR modeling is to define models taking into account information about the architecture of the software. To go further, software interacts with hardware to make a system. In order to derive a model for a system made up of software and hardware, the approach to take is the white-box approach (see Kanoun [49] and Laprie and Kanoun [79] for an account on this topic). We focus on the software product here. Many reasons lead to advocating a structure-based approach in SR modeling:

- advancement and widespread use of object-oriented systems designs. Reuse of components;
- software is developed in a heterogeneous fashion using components-based software development;
- early prediction methods of reliability have to take into account information on the testing and the reliability of the components of the software;
- early failure data are prior to the integration phase and thus concern testing part of the software, not the whole product;
- addressing problem of reliability allocation, resource allocation for modular software;
- analyze sensitivity of the reliability of the software to the reliability of its components.

As noted in Section 12.4, the structure-based approach has been largely ignored. The foundations of the Markovian models presented in Section 12.4 are not new. Some limitations of Littlewood's model have been recently addressed [43], in particular to obtain availability measures. Asymptotic considerations [45] show that such a reliability model tends to be of the Poisson type (homogeneous or not depending on stationarity or not of the failure parameters) when the product has achieved a good level of reliability. It is important to point out that no experience with such kinds of models is reported in the literature. Maybe this is related to questionable assumptions in the modeling, such as the Markov exchanges of control between modules.

An alternative way to represent the interactions between components of software is to use one of the available modeling tools which are based on stochastic Petri nets, SAN networks, *etc.* But whereas many of them offer a high degree of flexibility in the representation of the behavior of the software, the computation of various metrics is very often performed using automatic generation of a Markov chain. So these approaches are subject to the traditional limitation of Markov modeling: the failure rate of the components is not time dependent; the generated state-space is intractable from the computational point of view; *etc.*

To overcome the limitations of an analytic approach, a widespread method in performance analysis of a system is discrete-event simulation. This point of view was initiated by Lyu [4] (chapter 16) for software dependability assessment. The idea is to represent the behavior of each component as a non-homogeneous Markov process whose dynamic evolution only depends on a hazard rate function. At any time t, this hazard rate function depends on the number of failures observed from the component up to time t, as well as on the execution time experienced by the component up to time t. Then a rate-based simulation technique may be used to obtain a possible realization of such a Markovian arrivals process. The overall hazard rate of the software is actually a function of the number of failures observed from each component up to time t and of the amount of execution time experienced by each component. See Lyu [4] (chapter 16) and Lyu *et al.* [80] for details. In some sense, the approach is to simulate the failure process from the stochastic intensity of the counting process of failures.

For a long time now, the theory of a "coherent system" has allowed one to analyze a system made up of n components through the so-called *structure function*. If x_i $(i = 1, \ldots, n)$ denotes the state of component i ($x_i = 1$ if component i is up and zero otherwise), then the state of the system is obtained from computation of the structure function

$$\Phi(x_1, \ldots, x_n) = \begin{cases} 1 & \text{if system is up} \\ 0 & \text{if system is down} \end{cases}$$

Function Φ describes the functional relationship between the state of the system and the state of its components. Many textbooks on reliability review methods for computing reliability from the complex function Φ assuming that the state of each component is a Bernoulli random variable (*e.g.* see Aven and Jensen [17] (chapter 2)). An instance of representation of a 2-module software by a structure function taking into account control flow and data flow is discussed by Singpurwalla and Wilson [2] (chapter 7). This "coherent system" approach is widely used to analyze the reliability of communication networks. However, it is well known that exact computation of reliability is then an NP-hard problem. Thus, only structure functions of a few dozens of components can be analyzed exactly. Large systems have to be assessed by Monte Carlo simulation techniques (*e.g.* see Ball *et al.* [81] and Rubino [82]). Moreover, in the case of fault-tolerant software, we are faced with a highly reliable system that involves sophisticated simulation procedures to overcome the limitations of standard ones. In such a context, an alternative consists in using binary decision diagrams (see Lyu [4] (chapter 15) and Limnios and Rauzy [83]). We point out that structure function is mainly a functional representation of the system. Thus, many issues discussed in the context of black-box modeling also have to be addressed. For instance, how do you incorporate environmental factors identified by Pham and Pham [34]?

As we can see, a great deal of research is needed to obtain a white-box model that offers the advantages motivating development of such an approach. Opening the "black box" to get accurate models is a hard task. Many aspects have to be addressed: the definition of what is the structure or architecture of software; what kind of data can be expected for future calibration of models, and so on. A first study of the potential sources of SR data available during development is given by Smidts and Sova [84]. This would help the creation of some benchmark data sets, which will allow validation of white-box models. What is clear is that actual progress in white-box modeling can only be achieved from an active interaction between the statistics and software engineering communities. All this surely explains why the prevalent approach in SR is the black-box approach.

12.6.3 Statistical Issues

A delicate issue in SR is the statistical properties of the estimators used to calibrate models. The main drawbacks of the ML method are well documented in the literature. Finding ML estimators requires solving equations that may not always have a solution, or which may give an inappropriate solution. For instance, Littlewood and Verrall [85] give a criterion for $\hat{N} = \infty$ and $\hat{\phi} = 0$ (with finite non-zero $\hat{\lambda} = \hat{N}\hat{\phi}$) to be the unique solution of ML equations for the (JM) model. The problem of solving ML equations in SR modeling has also been addressed [86–89]. Another well-known drawback is that such ML estimators are usually unstable with small data sets. This situation is basic in SR. Moreover, note that certain models, like (JM), assume that software contains a finite number of faults. So, using the standard asymptotic properties of ML estimators may be questionable. Such asymptotic results are well known in the case of an i.i.d. sample. In SR, however, samples are not i.i.d. That explains why recent investigations on asymptotic normality and consistency of ML estimators for standard SR models use the framework of martingale theory, which allows dependence in data. Note that overcoming a conceptually finite (expected) number of faults needs unusual concept of asymptotic properties. A detailed discussion is given by van Pul [18] (and references cited therein) and by Zhao and Xie [88] for NHPP models. Such studies are important, because these asymptotic properties are the foundations of interval estimation (a standard alternative to point estimate), of the derivation of confidence interval for parameters, of studies on asymptotic variance of estimators, *etc.* All these topics must be addressed in detail to improve the predictive quality of SR models.

It is clear that any model works well with failure data that correspond to the basic assumptions of the model. But, given data, a large number of models are inappropriate. A natural way to overcome too stringent assumptions, in particular of the distributional type, is to use non-parametric models. However, the parametric approach remains highly prevalent in SR modeling. A major attempt to gap this fill was undertaken by Miller and Sofer [90], where a completely monotonic ROCOF is estimated by regression techniques (see also Brocklehurst and Littlewood [91]). A recent study by Littlewood and co-workers [92] uses non-parametric estimates for the distribution of inter-failure times (X_i). This is based on kernel methods for p.d.f. estimation (see Silverman [93]). The conclusion of the authors is that the results are not very impressive but that more investigation is needed, in particular using various kernel functions. We can think, for instance, of wavelets [94]. A similar discussion may be undertaken with regard to the Bayesian approach of SR modeling. Specifically, most Bayesian inference for NHPP assumes a parametric model for ROCOF and proceeds with prior assumption on the unknown parameters. In such a context, an instance of a non-parametric Bayesian approach has recently been used Kuo and Ghosh [95]. Conceptually, the non-parametric approach is promising, but it is computationally intensive in general and is not easy to comprehend.

These developments may be viewed as preliminary studies using statistical methods based on the so-called dynamic approach of counting processes, as reported for instance in the book by Andersen *et al.* [69] (the bibliography gives a large account for research in this area). The pioneering work of Aalen was on the multiplicative intensity model, which, roughly speaking, writes the stochastic intensity associated with a counting process as

$$\lambda(t; \mathcal{H}_t, \sigma(Y_s, s \leq t)) = \lambda(t)\, Y(t)$$

where $\lambda(\cdot)$ is a non-negative deterministic function, whereas $Y(\cdot)$ is a non-negative observable stochastic process whose value at any time t is known just before t ($Y(\cdot)$ is a predictable process). Non-parametric estimation for such a model is discussed by Andersen *et al.* [69], chapter 4. A Cox-type model may be obtained in choosing $Y(t) = \exp\left(\sum_j \beta_j Z_j(t)\right)$ with stochastic processes as covariates Z_j (see Slud [96]). We can also consider additive intensity models when the multiplicative form involves multivariate functions $\lambda(\cdot)$ and $Y(\cdot)$ (*e.g.* see Pijnenburg [97]). Conceptually, this dynamic approach is appealing because it is well supported by a lot of theoretic results. It is worth noting that martingale theory may be of some value for analyzing static models as capture–recapture models (see Yip and coworkers [98, 99] for details). Thus, the applicability of the dynamic point of view on point processes in the small data set context of software engineering is clearly a direction of further investigations (*e.g.* see van Pul [18] for such an account). Moreover, if it is shown that gain in predictive validity is high with respect to standard approaches, then a user-oriented "transfer of technology" must follow. That is, friendly tools for using such statistical material must be developed.

References

[1] Gaudoin O. Statistical tools for software reliability evaluation (in French). PhD thesis, Université Joseph Fourier–Grenoble I, 1990.

[2] Singpurwalla ND, Wilson SP. Statistical methods in software engineering: reliability and risk. Springer; 1999.

[3] Musa JD, Iannino A, Okumoto K. Software reliability: measurement, prediction, application. Computer Science Series. McGraw-Hill International Editions; 1987.

[4] Lyu MR, editor. Handbook of software reliability engineering. McGraw-Hill; 1996.

[5] Software reliability estimation and prediction handbook. American Institute of Aeronautics and Astronautics; 1992.

[6] Pham H. Software reliability. Springer; 2000.

[7] Everett W, Keene S, Nikora A. Applying software reliability engineering in the 1990s. IEEE Trans Reliab 1998;47:372–8.

[8] Ramani S, Gokhale SS, Trivedi KS. Software reliability estimation and prediction tool. Perform Eval 2000;39:37–60.

[9] Laprie J-C. Dependability: basic concepts and terminology. Springer; 1992.

[10] Barlow RE, Proschan F. Statistical theory of reliability and life testing. New York: Holt, Rinehart and Winston; 1975.

[11] Ascher H, Feingold H. Repairable systems reliability. Lecture Notes in Statistics, vol. 7. New York: Marcel Dekker; 1984.

[12] Xie M. Software reliability modeling. UK: World Scientific Publishing; 1991.

[13] Briand LC, El Emam K, Freimut BG. A comprehensive evaluation of capture-recapture models for estimating software defect content. IEEE Trans Software Eng 2000;26:518–40.

[14] Chen Y, Singpurwalla ND. Unification of software reliability models by self-exciting point processes. Adv Appl Probab 1997;29:337–52.

[15] Snyder DL, Miller MI. Random point processes in time and space. Springer; 1991.

[16] Jelinski Z, Moranda PB. Software reliability research. In: Freiberger W, editor. Statistical methods for the evaluation of computer system performance. Academic Press; 1972. p.465–84.

[17] Aven T, Jensen U. Stochastic models in reliability. Applications of Mathematics, vol. 41. Springer; 1999.

[18] Van Pul MC. A general introduction to software reliability. CWI Q 1994;7:203–44.

[19] Laprie J-C, Kanoun K, Béounes C, Kaâniche M. The KAT (knowledge-action-transformation) approach to the modeling and evaluation of reliability and availability growth. IEEE Trans Software Eng 1991;17:370–82.

[20] Cox DR, Isham V. Point processes. Chapman and Hall; 1980.

[21] Trivedi KS. Probability and statistics with reliability, queuing and computer science applications. John Wiley & Sons; 2001.

[22] Moranda PB. Predictions of software reliability during debugging. In: Annual Reliability and Maintainability Symposium 1975; p.327–32.

[23] Shanthikumar JG. Software reliability models: a review. Microelectron Reliab 1983;23:903–43.

[24] Littlewood B. Stochastic reliability-growth: a model for fault-removal in computer programs and hardware designs. IEEE Trans Reliab 1981;30:313–20.

[25] Osaki S, Yamada S. Reliability growth models for hardware and software systems based on nonhomogeneous Poisson processes: a survey. Microelectron Reliab 1983;23:91–112.

[26] Littlewood B. Forecasting software reliability. In: Bittanti S, editor. Software reliability modeling and identification. Lecture Notes in Computer Science 341. Springer; 1988. p.141–209.

[27] Miller DR. Exponential order statistic models for software reliability growth. IEEE Trans Software Eng 1986;12:12–24.

[28] Littlewood B, Verrall JL. A Bayesian reliability growth model for computer software. Appl Stat 1973;22:332–46.

[29] Mazzzuchi TA, Soyer R. A Bayes empirical-Bayes model for software reliability. IEEE Trans Reliab 1988;37:248–54.

[30] Schick GJ, Wolverton RW. Assessment of software reliability. In: Operation Research. Physica-Verlag; 1973. p.395–422.

[31] Singpurwalla ND, Soyer R. Assessing (software) reliability growth using a random coefficient autoregressive process and its ramification. IEEE Trans Software Eng 1985;11:1456–64.

[32] Singpurwalla ND, Soyer R. Nonhomogeneous autoregressive processes for tracking (software) reliability growth, and their Bayesian analysis. J R Stat Soc Ser B 1992;54:145–56.

[33] Al-Mutairi D, Chen Y, Singpurwalla ND. An adaptative concatenated failure rate model for software reliability. J Am Stat Assoc 1998; 93:1150–63.

[34] Pham L, Pham H. Software reliability models with time-dependent hazard function based on Bayesian approach. IEEE Trans Syst Man Cyber Part A 2000;30:25–35.

[35] Cheung RC. A user-oriented software reliability model. IEEE Trans Software Eng 1980;6:118–25.

[36] Littlewood B. Software reliability model for modular program structure. IEEE Trans Reliab 1979;28:241–6.

[37] Siegrist K. Reliability of systems with Markov transfer of control. IEEE Trans Software Eng 1988;14:1049–53.

[38] Kaâniche M, Kanoun K. The discrete time hyperexponential model for software reliability growth evaluation. In: International Symposium on Software Reliability (ISSRE), 1992; p.64–75.

[39] Littlewood B. A reliability model for systems with Markov structure. Appl Stat 1975;24:172–7.

[40] Kubat P. Assessing reliability of modular software. Oper Res Lett 1989;8:35–41.

[41] Goseva-Popstojanova K, Trivedi KS. Architecture-based approach to reliability assessment of software systems. Perform Eval 2001;45:179–204.

[42] Ledoux J, Rubino G. Simple formulae for counting processes in reliability models. Adv Appl Probab 1997;29:1018–38.

[43] Ledoux J. Availability modeling of modular software. IEEE Trans Reliab 1999;48:159–68.

[44] Kabanov YM, Liptser RS, Shiryayev AN. Weak and strong convergence of the distributions of counting processes. Theor Probab Appl 1983;28:303–36.

[45] Gravereaux JB, Ledoux J. Poisson approximation for some point processes in reliability. Technical report, Institut National des Sciences Appliquées, Rennes, France, 2001.

[46] Ledoux J. Littlewood reliability model for modular software and Poisson approximation. In: Mathematical Methods for Reliability, June, 2002; p.367–370.

[47] Abdel-Ghaly AA, Chan PY, Littlewood B. Evaluation of competing software reliability predictions. IEEE Trans Software Eng 1986;12:950–67.

[48] Brocklehurst S, Kanoun K, Laprie J-C, Littlewood B, Metge S, Mellor P, *et al.* Analyses of software failure data. Technical report No. 91173, Laboratoire d'Analyse et d'Architecture des Systèmes, Toulouse, France, May 1991.

[49] Kanoun K. Software dependability growth: characterization, modeling, evaluation (in French). Technical report 89.320, LAAS, Doctor ès Sciences thesis, Polytechnic National Institute, Toulouse, 1989.

[50] Gilks WR, Richardson S, Spiegelhalter DJ, editors. Markov Chain Monte Carlo in practice. Chapman and Hall; 1996.

[51] Kuo L, Yang TY. Bayesian computation of software reliability. J Comput Graph Stat 1995;4:65–82.

[52] Kuo L, Yang TY. Bayesian computation for nonhomogeneous Poisson processes in software reliability. J Am Stat Assoc 1996;91:763–73.

[53] Kuo L, Lee JC, Choi K, Yang TY. Bayes inference for s-shaped software-reliability growth models. IEEE Trans Reliab 1997;46:76–80.

[54] Kremer W. Birth–death and bug counting. IEEE Trans Reliab 1983;32:37–47.

[55] Gokhale SS, Philip T, Marinos PN. A non-homogeneous Markov software reliability model with imperfect repair. In: International Performance and Dependability Symposium, 1996; p.262–70.

[56] Shanthikumar JG, Sumita U. A software reliability model with multiple-error introduction and removal. IEEE Trans Reliab 1986;35:459–62.

[57] Sahinoglu H. Compound-Poisson software reliability model. IEEE Trans Software Eng 1992;18:624–30.

[58] Fenton NE, Pfleeger SL. Software metrics: a rigorous and practical approach, 2nd edn. International Thomson Computer Press; 1996.

[59] Khoshgoftaar TM, Allen EB, Wendell DJ, Hudepohl JP. Classification-tree models of software-quality over multiple releases. IEEE Trans Reliab 2000;49:4–11.

[60] Khoshgoftaar TM, Allen EB. A practical classification-rule for software-quality models. IEEE Trans Reliab 2000;49:209–16.

[61] Gokhale SS, Trivedi KS. A time/structure based software reliability model. Ann Software Eng 1999;8:85–121.

[62] Goel AL. Software reliability models: assumptions, limitations, and applicability. IEEE Trans Software Eng 1985;11:1411–23.

[63] Xie M, Hong GY, Wohlin C. A practical method for the estimation of software reliability growth in the early stage of testing. In: International Symposium on Software Reliability (ISSRE), 1997; p.116–23.

[64] Pasquini A, Crespo AN, Matrella P. Sensitivity of reliability-growth models to operational profile errors *vs.* testing accuracy. IEEE Trans Reliab 1996;45:531–40.

[65] Özekici S, Soyer R. Reliability of software with an operational profile. Technical report, The George Washington University, Department of Management Science, 2000.

[66] Zhang X, Pham H. An analysis of factors affecting software reliability. J. Syst Software 2000; 50:43–56.

[67] Kalbfleisch JD, Prentice RL. The statistical analysis of failure time data. Wiley; 1980.

[68] Cox CR, Oakes D. Analysis of survival data. London: Chapman and Hall; 1984.

[69] Andersen PK, Borgan O, Gill RD, Keiding N. Statistical models on counting processes. Springer Series in Statistics. Springer; 1993.

[70] Saglietti F. Systematic software testing strategies as explanatory variables of proportional hazards. In SAFECOMP'91, 1991; p.163–7.

[71] Wright D. Incorporating explanatory variables in software reliability models. In Second year report of PDCS, vol. 1. Esprit BRA Project 3092, May 1991.

[72] Pham H. Software reliability. In: Webster JG, editor. Wiley Encyclopedia of Electrical and Electronics Engineering. Wiley; 1999. p.565–78.

[73] Castillo X, Siewiorek DP. A workload dependent software reliability prediction model. In: 12th International Symposium on Fault-Tolerant Computing, 1982; p.279–86.

[74] Koch G, Spreij P. Software reliability as an application of martingale & filtering theory. IEEE Trans Reliab 1983;32:342–5.

[75] Brocklehurst S, Chan PY, Littlewood B, Snell J. Recalibrating software reliability models. IEEE Trans Software Eng 1990;16:458–70.

[76] Sitte R. Comparison of software-reliability-growth predictions: neural networks *vs* parametric-recalibration. IEEE Trans Reliab 1999;49:285–91.

[77] Pham H, Zhang X. A software cost model with warranty and risk costs. IEEE Trans Comput 1999;48:71–5.

[78] Pham H, Zhang X. Software release policies with gain in reliability justifying the costs. Ann Software Eng 1999;8:147–66.

[79] Laprie J-C, Kanoun K. X-ware reliability and availability modeling. IEEE Trans Software Eng 1992;18:130–47.

[80] Lyu MR, Gokhale SS, Trivedi KS. Reliability simulation of component-based systems. In: International Symposium on Software Reliability (ISSRE), 1998; p.192–201.

[81] Ball MO, Colbourn C, Provan JS. Network models. In: Monma C, Ball MO, Magnanti T, Nemhauser G, editors. Handbook of operations research and management science. Elsevier; 1995. p.673–762.

[82] Rubino G. Network reliability evaluation. In: Bagchi K, Walrand J, editors. State-of-the-art in performance modeling and simulation. Gordon and Breach; 1998.

[83] Limnios N, Rauzy A, editors. Special issue on binary decision diagrams and reliability. Eur J Automat 1996;30(8).

[84] Smidts C, Sova D. An architectural model for software reliability quantification: sources of data. Reliab Eng Syst Saf 1999;64:279–90.

[85] Littlewood B, Verrall JL. Likelihood function of a debugging model for computer software reliability. IEEE Trans Reliab 1981;30:145–8.

[86] Huang XZ. The limit condition of some time between failure models of software reliability. Microelectron Reliab 1990;30:481–5.

[87] Hossain SA, Dahiya RC. Estimating the parameters of a non-homogeneous Poisson-process model for software reliability. IEEE Trans Reliab 1993;42:604–12.

[88] Zhao M, Xie M. On maximum likelihood estimation for a general non-homogeneous Poisson process. Scand J Stat 1996;23:597–607.

[89] Knafl G, Morgan J. Solving ML equations for 2-parameter Poisson-process models for ungrouped software-failure data. IEEE Trans Reliab 1996;45:43–53.

[90] Miller DR, Sofer A. A nonparametric software-reliability growth model. IEEE Trans Reliab 1991;40:329–37.

[91] Brocklehurst S, Littlewood B. New ways to get accurate reliability measures. IEEE Software 1992;9:34–42.

[92] Barghout M, Littlewood B, Abdel-Ghaly AA. A non-parametric order statistics software reliability model. J Test Verif Reliab 1998;8:113–32.

[93] Silverman BW. Density estimation for statistics and data analysis. Chapman and Hall; 1986.

[94] Antoniadis A, Oppenheim G, editors. Wavelets and statistics. Lecture Notes in Statistics, vol. 103. Springer; 1995.

[95] Kuo L, Ghosh SK. Bayesian nonparametric inference for nonhomogeneous Poisson processes. Technical report 9718, University of Connecticut, Storrs, 1997.

[96] Slud EV. Some applications of counting process models with partially observed covariates. Telecommun Syst 1997;7:95–104.

[97] Pijnenburg M. Additive hazard models in repairable systems reliability. Reliab Eng Syst Saf 1991;31:369–90.

[98] Lloyd CJ, Yip PSF, Chan Sun K. Estimating the number of errors in a system using a martingale approach. IEEE Trans Reliab 1999;48:369–76.

[99] Yip PSF, Xi L, Fong DYT, Hayakawa Y. Sensitivity-analysis and estimating the number-of-faults in removing debugging. IEEE Trans Reliab 1999;48:300–5.

Software Availability Theory and Its Applications

Koichi Tokuno and Shigeru Yamada

Chapter 13

13.1 Introduction

Many methodologies for software reliability measurement and assessment have been discussed for the last few decades. A mathematical software reliability model is often called a software reliability growth model (SRGM). Several books and papers have surveyed software reliability modeling [1–8]. Most existing SRGMs have described stochastic behaviors of only software fault-detection or software failure-occurrence phenomena during the testing phase of the software development process and the operation phase. A software failure is defined as an unacceptable departure from program operation caused by a fault remaining in the software system. These can provide several measures of software reliability defined as the attribute that the software-intensive systems can perform without software failures for a given time period, under the specified environment; for example, the mean time between software failures (MTBSF), the expected number of faults in the system, and so on. These measures are developer-oriented and utilized for measuring and assessing the degree of achievement of software reliability, deciding the time to software release for operational use, and estimating the maintenance cost for faults undetected during the testing phase. Therefore, the traditional SRGMs have evaluated "the reliability for developers" or "the inherent reliability".

These days, the performance and quality of software systems become evaluated from customers' viewpoints. For example, customers take interest in the information on possible utilization, not the number of faults remaining in the system. One of the customer-oriented software quality attributes is software availability [9–11]; this is the attribute that the software-intensive systems are available at a given time point, under the specified environment. Software availability is also called "the reliability for customers" or "the reliability with maintainability".

When we measure and assess software availability, we need to consider explicitly not only the up time (the software failure time) but also the down time to restore the system and to describe the stochastic behavior of the system alternating between the up and down states.

This chapter surveys the current research on the stochastic modeling for software availability measurement techniques. In particular, we focus on the software availability modeling with Markov processes [12] to describe the time-dependent behaviors of the systems. The organization of this chapter is as follows. Section 13.2 gives the basic ideas of software availability modeling and the several software availability measures. Based on the basic model, Section 13.3 discusses two modified models reflecting the software failure-occurrence phenomenon and the restoration scenario peculiar to the user-operational phase since software availability is a metric applied in the user operation phase. Section 13.4 also refers to the applied models and considers computation performance and combining a hardware and a software subsystem.

13.2 Basic Model and Software Availability Measures

Beforehand, we state the difference in the descriptions of the failure/restoration characteristics between hardware and software systems. The causes of hardware failures are the deterioration or wearing-out of component parts due to secular changes, and the renewal of hardware systems by replacement of failing parts. Therefore, the failure and the restoration characteristics of the hardware systems are often described without consideration of the cumulative numbers of failures or restorations. On the other hand, the causes of software failures are faults latent in the systems introduced in the development process, and the restoration actions include the debugging activities for the detected faults. So the debugging activities decrease the faults and improve software reliability, unless secondary faults are introduced. Accordingly, the failure and the restoration characteristics of software systems are often described with relation to the numbers of software failures and debugging activities. The above remarks have to be reflected in software availability modeling.

The following lists the general assumptions for software availability modeling.

Assumption 1. *A software system is unavailable and starts to be restored as soon as a software failure occurs, and the system cannot operate until the restoration action is complete.*

Assumption 2. *A restoration action implies debugging activity, which is performed perfectly with probability a $(0 < a \leq 1)$ and imperfectly with probability b $(= 1 - a)$. We call a the perfect debugging rate. One fault is corrected and removed from the software system when the debugging activity is perfect.*

Assumption 3. *The software failure-occurrence time (up time) and the restoration time (down time) follow exponential distributions with means $1/\lambda_n$ and $1/\mu_n$, respectively, where n denotes the cumulative number of corrected faults. λ_n and μ_n are functions of n.*

Assumption 4. *The probability that two or more software failures occur simultaneously is negligible.*

Consider the stochastic behavior of the system alternating between up and down states with a Markov process. Let $\{X(t), t \geq 0\}$ be the Markov process representing the state of the system at the time point t and its state space W_n and R_n $(n = 0, 1, 2, \ldots)$; then denote

W_n the system is operating
R_n the system is inoperable due to the restoration action.

In the actual debugging environment, it often happens that some fault corrections are inexact and that secondary faults are introduced. This is the so-called imperfect debugging environment. The assumption of imperfect debugging has given rise to much controversy in software reliability/availability modeling [1, 13]. In this chapter the following refers to the treatment of perfect debugging: the purpose of the debugging activity is to improve software quality/reliability. Therefore, we assume that a debugging activity is perfect when it contributes to the improvement of software reliability. That is, a perfect debugging activity means that the hazard rate mentioned later decreases and that one fault is corrected and removed from the system. We also assume that the increase of the hazard rate due to the introduction of new faults is negligible [14].

From Assumption 2, when the restoration action has been complete in $\{X(t) = R_n\}$,

$$X(t) = \begin{cases} W_n & \text{(with probability } b) \\ W_{n+1} & \text{(with probability } a) \end{cases} \quad (13.1)$$

Next, we refer to the transition probability between the states, *i.e.* the descriptions of the failure/restoration characteristics. For describing the software failure characteristic, several classical SRGMs can be applied. For example, Okumoto and Goel [15] and Kim *et al.* [16] have applied the model of Jelinski and Moranda [17], *i.e.* they have described the hazard rate λ_n as

$$\lambda_n = \phi(N - n)$$
$$(n = 0, 1, 2, \ldots, N; N > 0, \phi > 0) \quad (13.2)$$

where N and ϕ are respectively the initial fault content prior to the testing and the hazard rate per fault remaining in the system. Then $X(t)$ forms a finite-state Markov process. Tokuno and Yamada [18–20] have applied the model of Moranda [21], *i.e.* they have given λ_n as

$$\lambda_n = Dk^n \quad (n = 0, 1, 2, \ldots; D > 0, 0 < k < 1) \quad (13.3)$$

where D and k are the initial hazard rate and the decreasing ratio of the hazard rate respectively. Then $X(t)$ forms an infinite-state Markov process. Equation 13.2 assumes that any faults have the same impact on software reliability growth. On the other hand, Equation 13.3 means that the perfect debugging activities for the faults detected earlier have a higher impact on software reliability growth than those for the later faults [2, 22].

We also mention the restoration rate μ_n. There are many cases where the difficulties of fault isolations and debugging continue to rise and the restoration time tends to be longer as the fault correction progresses [23]. In order that such situations are reflected in the modeling, the same forms as Equations 13.2 and 13.3 that are the decreasing functions of n are often applied to μ_n. When we apply $\mu_n \equiv Er^n$ $(E > 0, 0 < r \leq 1)$, E and r denote the initial restoration rate and the decreasing ratio of the restoration rate respectively.

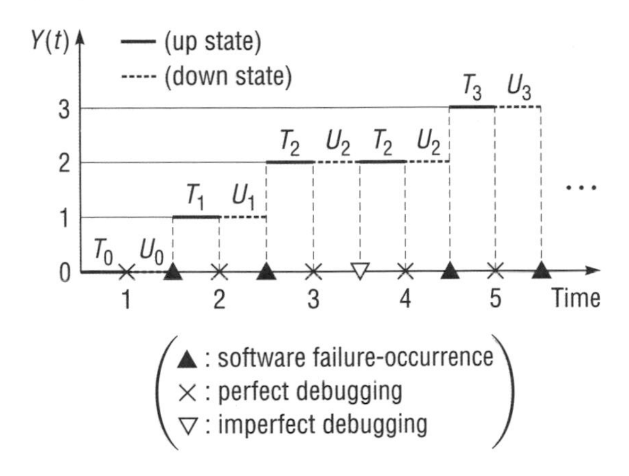

Figure 13.1. A sample realization of $Y(t)$

Let $Y(t)$ be the random variable representing the cumulative number of faults corrected up to the time t. Figure 13.1 illustrates the sample behavior of $Y(t)$, where T_n and U_n $(n = 0, 1, 2, \ldots)$ denote the random variables representing the sojourn times in states W_n and R_n respectively. It is noted that the cumulative number of corrected faults is not always coincident with that of software failures or restoration actions. Furthermore, Figure 13.2 illustrates the sample state transition diagram of $X(t)$.

We can obtain the state occupancy probabilities that the system is in the states W_n and R_n at time point t as

$$P_{W_n}(t) \equiv \Pr\{X(t) = W_n\} = \frac{g_{n+1}(t)}{a\lambda_n} + \frac{g'_{n+1}(t)}{a\lambda_n \mu_n} \quad (n = 0, 1, 2, \ldots) \quad (13.4)$$

$$P_{R_n}(t) \equiv \Pr\{X(t) = R_n\} = \frac{g_{n+1}(t)}{a\mu_n} \quad (n = 0, 1, 2, \ldots) \quad (13.5)$$

respectively, where $g_n(t)$ is the probability density function of the random variable S_n representing the first passage time to the state W_n, and $g'_n(t) \equiv \mathrm{d}g_n(t)/\mathrm{d}t . g_n(t)$ and $g'_n(t)$ can be obtained analytically.

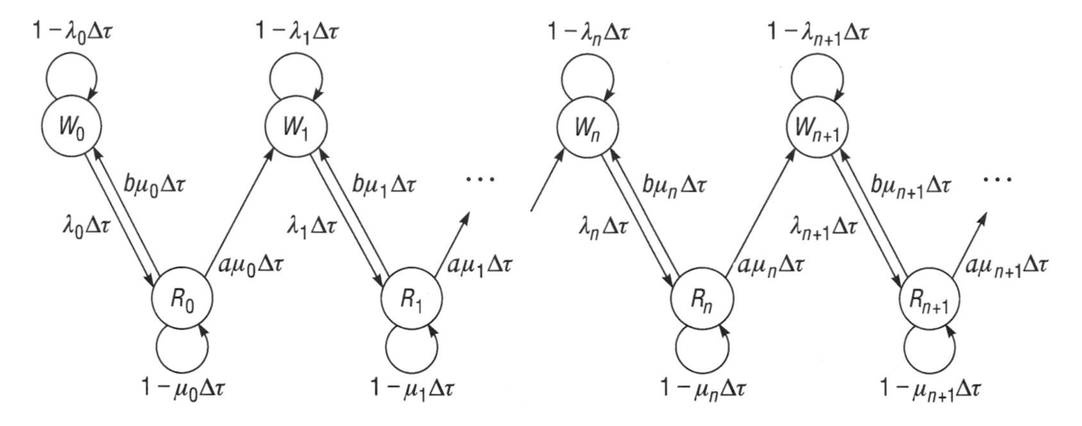

Figure 13.2. State transition diagram for software availability modeling

The following equation holds for the arbitrary time t:

$$\sum_{n=0}^{\infty}[P_{W_n}(t) + P_{R_n}(t)] = 1 \qquad (13.6)$$

The *instantaneous software availability* is defined as

$$A(t) \equiv \sum_{n=0}^{\infty} P_{W_n}(t) \qquad (13.7)$$

which represents the probability that the software system is operating at the time point t. Furthermore, the *average software availability* over $(0, t]$ is defined as

$$A_{\text{av}}(t) \equiv \frac{1}{t}\int_0^t A(x)\,\mathrm{d}x \qquad (13.8)$$

which represents the expected proportion of the system's operating time to the time interval $(0, t]$. Using Equations 13.4 and 13.5, we can express Equations 13.7 and 13.8 as

$$A(t) = \sum_{n=0}^{\infty}\left[\frac{g_{n+1}(t)}{a\lambda_n} + \frac{g'_{n+1}(t)}{a\lambda_n\mu_n}\right] = 1 - \sum_{n=0}^{\infty}\frac{g_{n+1}(t)}{a\mu_n} \qquad (13.9)$$

$$A_{\text{av}}(t) = \frac{1}{t}\sum_{n=0}^{\infty}\left[\frac{G_{n+1}(t)}{a\lambda_n} + \frac{g_{n+1}(t)}{a\lambda_n\mu_n}\right] = 1 - \frac{1}{t}\sum_{n=0}^{\infty}\frac{G_{n+1}(t)}{a\mu_n} \qquad (13.10)$$

respectively, where $G_n(t)$ is the distribution function of S_n.

The proportion of the up and down times is called the maintenance factor. Given that $\lambda_n \equiv Dk^n$ and $\mu_n \equiv Er^n$, the maintenance factor ρ_n is expressed by

$$\rho_n \equiv \frac{\mathrm{E}[U_n]}{\mathrm{E}[T_n]} = \beta v^n \quad (\beta \equiv D/E,\ v \equiv k/r) \qquad (13.11)$$

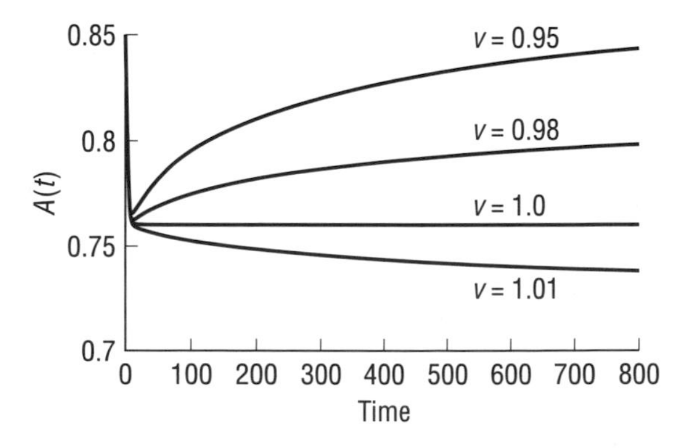

Figure 13.3. Dependence of v on $A(t)$ ($\lambda_n \equiv Dk^n$, $\mu_n \equiv Er^n$; $a = 0.9, D = 0.1, E = 0.3, r = 0.8$)

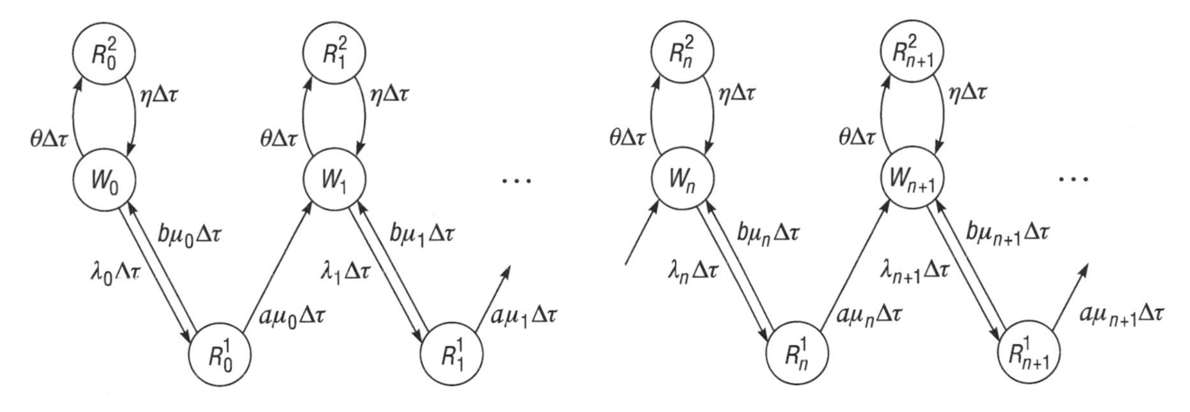

Figure 13.4. State transition diagram for software availability modeling with two types of failure

where β and v are the initial maintenance factor and the availability improvement parameter respectively [24]. In general, the maintenance factor of the hardware system is assumed to be constant and has no bearing on the number of failures, whereas that of the software system depends on n. Figure 13.3 shows the dependence of v on $A(t)$ in Equation 13.9. This figure tells us that we can judge whether the software availability increases or decreases with the value of v, *i.e.* $A(t)$ increases and decreases in the cases of $v < 1$ and $v > 1$ respectively, and it is constant in the case of $v = 1$. We must consider not only the software reliability growth process but also the upward tendency of the difficulty in debugging in software availability measurement and assessment.

13.3 Modified Models

Since software availability is a quality characteristic to be considered in the user operation, we need to continue to adapt the basic model discussed above to the actual operation environment.

13.3.1 Model with Two Types of Failure

In this section, we assume that the following two types of software failure exist during the operation phase [25, 26]:

F1: software failures caused by the faults that could not be detected/corrected during the testing phase

F2: software failures caused by the faults introduced by deviating from the expected operational use.

The state space of the process $\{X(t), t \geq 0\}$ in this section is defined as follows:

W_n the system is operating

R_n^1 the system is inoperable due to F1 and restored

R_n^2 the system is inoperable due to F2 and restored.

The failure and the restoration characteristics of F1 are assumed to be described by the forms dependent on the number of corrected faults, and those of F2 are assumed to be constant. Then, Figure 13.4 illustrates the sample state transition diagram of $X(t)$, where θ and η are the hazard rate and the restoration rate of F2 respectively.

We can express the instantaneous software availability and the average software availability as

$$A(t) = \sum_{n=0}^{\infty} \left[\frac{g_{n+1}(t)}{a\lambda_n} + \frac{g'_{n+1}(t)}{a\lambda_n\mu_n} \right] \tag{13.12}$$

$$A_{\text{av}}(t) = \frac{1}{t} \sum_{n=0}^{\infty} \left[\frac{G_{n+1}(t)}{a\lambda_n} + \frac{g_{n+1}(t)}{a\lambda_n\mu_n} \right] \tag{13.13}$$

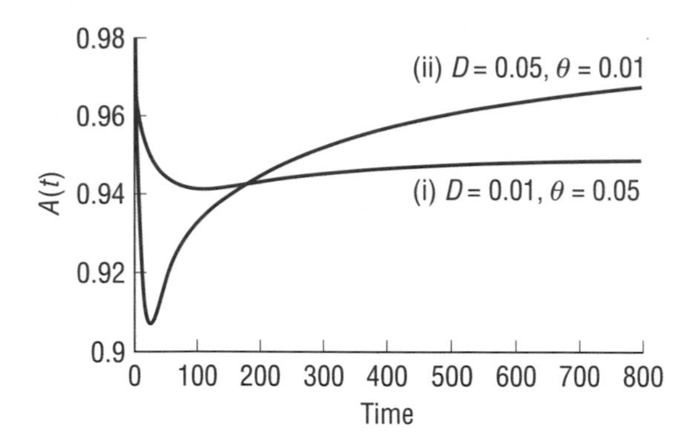

Figure 13.5. Dependence of the ratio of hazard rates of F1 and F2 on $A(t)$ ($\lambda_n \equiv Dk^n$, $\mu_n \equiv Er^n$; $a = 0.9$, $\eta = 1.0$, $k = 0.8$, $E = 0.5$, $r = 0.9$)

respectively. The analytical solutions of $G_n(t)$, $g_n(t)$, and $g'_n(t)$ can also be obtained.

Figure 13.5 shows the dependence of the occurrence proportion of F1 and F2 on Equation 13.12. The hazard rates of (i) and (ii) in Figure 13.5 for the first software failure without the distinction between F1 and F2, $\alpha_0 = D + \theta$, are the same value, *i.e.* both (i) and (ii) are $\alpha_0 = 0.06$. $A(t)$ of (ii) is smaller than (i) in the early stage of the operation phase. However, $A(t)$ of (ii) increases more than (i) with time, since (ii), whose D is larger than θ, means that the system has more room for reliability growth even in the operation phase than (i), whose D is smaller than θ.

13.3.2 Model with Two Types of Restoration

In Section 13.3.1, the basic model was modified from the viewpoint of the software failure characteristics. Tokuno and Yamada [27] have paid attention to the restoration actions for operational use. Cases often exist where the system is restored without debugging activities corresponding to software failures occurring in the operation phase, since protracting an inoperable time may greatly affect the customers. This is a different policy from the testing phase. This section considers two kinds of restoration action during the operation phase: one involves debugging and the other does not. Furthermore, we assume that it is probabilistic whether or not a debugging activity is performed and that the restoration time without debugging has no bearing on the number of corrected faults.

The state space of $\{X(t),\ t \geq 0\}$ in this section is defined as follows:

- W_n the system is operating
- R_n^1 the system is inoperable and restored with the debugging activity
- R_n^2 the system is inoperable and restored without the debugging activity.

Figure 13.6 illustrates the sample state transition diagram of $X(t)$, where p ($0 < p < 1$) denotes the probability that a restoration action with debugging is performed and $q = 1 - p$, and η denotes the restoration rate without debugging.

We can express the instantaneous software availability and the average software availability as

$$A(t) = \sum_{n=0}^{\infty} \left[\frac{g_{n+1}(t)}{pa\lambda_n} + \frac{g'_{n+1}(t)}{pa\lambda_n\mu_n} \right] \tag{13.14}$$

$$A_{\text{av}}(t) = \frac{1}{t} \sum_{n=0}^{\infty} \left[\frac{G_{n+1}(t)}{pa\lambda_n} + \frac{g_{n+1}(t)}{pa\lambda_n\mu_n} \right] \tag{13.15}$$

respectively. The analytical solutions of $G_n(t)$, $g_n(t)$, and $g'_n(t)$ can also be obtained.

Figure 13.7 shows the dependence of p in Equation 13.14. $A(t)$ is lower in the early stage of the operation phase, but it increases more with time as the value of p increases. The larger p means that the system developer intends to improve software reliability even during the operation phase. However, the time of the restoration with debugging tends to be longer than one without debugging. Therefore, the value of p depends on the policy of the developer and is an important factor in software availability assessment.

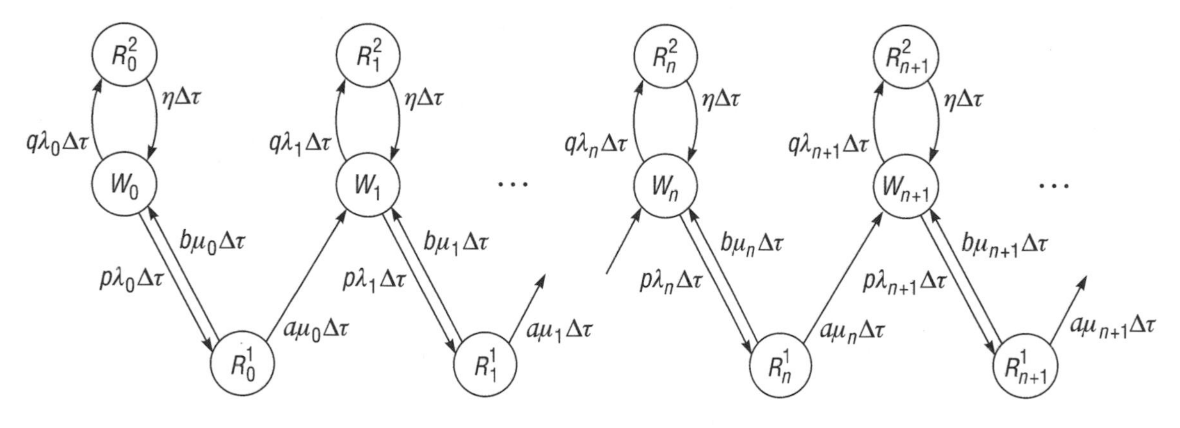

Figure 13.6. State transition diagram for software availability modeling with two types of restoration

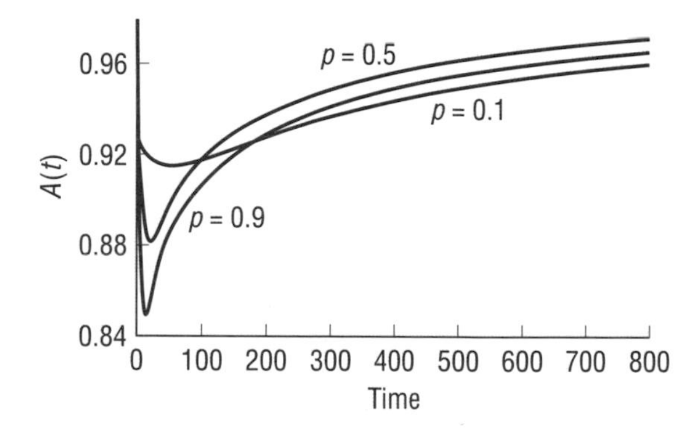

Figure 13.7. Dependence of p on $A(t)$ ($\lambda_n \equiv Dk^n$, $\mu_n \equiv Er^n$; $a = 0.9, D = 0.1, k = 0.8, E = 0.5, r = 0.9, \eta = 1.0$)

13.4 Applied Models

13.4.1 Model with Computation Performance

In Section 13.3, the basic model was modified by detailing the inoperable states. This section discusses a new software availability measure, noting the operable states. Tokuno and Yamada [28] have provided a software availability measure with computation performance by introducing the concept of the computation capacity [29]. The computation capacity is defined as the computation amount that the system is able to process per unit time.

The state space of $\{X(t), t \geq 0\}$ in this section is defined as follows:

- W_n the system is operating with full performance in accordance with the specification
- L_n the system is operating but its performance degenerates
- R_n the system is inoperable and restored.

Suppose that the sojourn times of the states W_n and L_n are distributed exponentially with the rates θ and η respectively. Then, Figure 13.8 illustrates the sample state transition diagram of $X(t)$.

Let C (>0) and $C\delta$ ($0 < \delta < 1$) denote the computation capacities when the system is in the states W_n and L_n respectively. Then the *computation software availability* is given by

$$A_c(t) \equiv C \sum_{n=0}^{\infty} [\Pr\{X(t) = W_n\} + \delta \Pr\{X(t) = L_n\}] \tag{13.16}$$

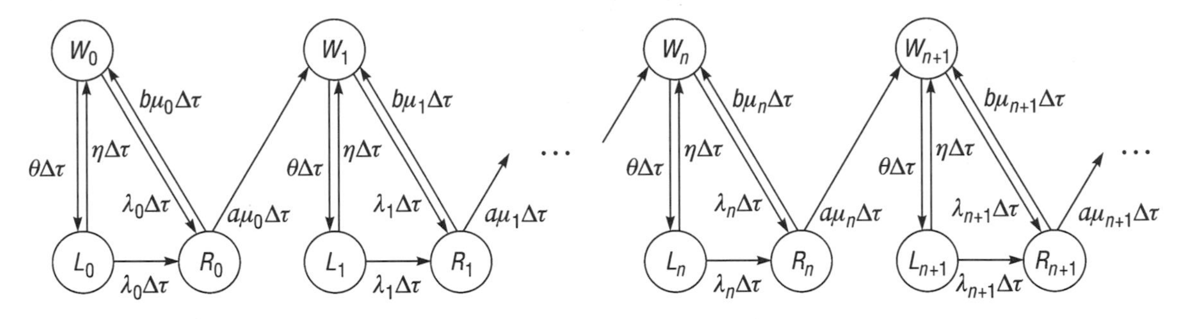

Figure 13.8. State transition diagram for software availability modeling with computation performance

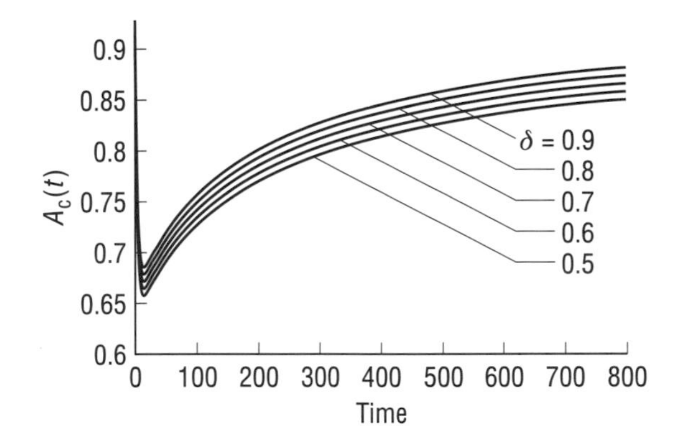

Figure 13.9. Dependence of δ on $A_c(t)$ ($\lambda_n \equiv Dk^n$, $\mu_n \equiv Er^n$; $C = 1.0$, $a = 0.9$, $D = 0.1$, $k = 0.8$, $E = 0.2$, $r = 0.9, \theta = 0.1, \eta = 1.0$)

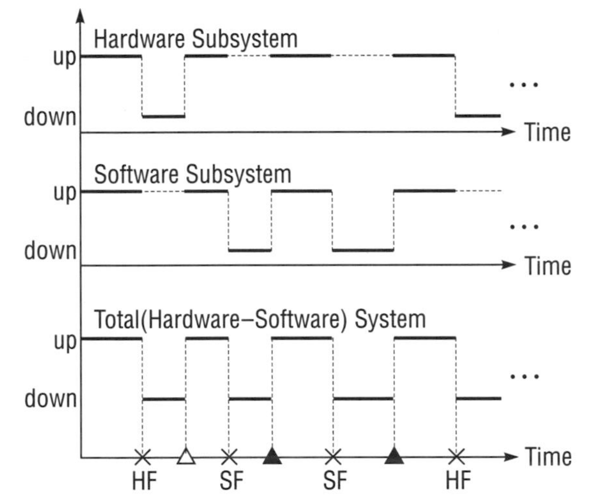

Figure 13.10. Sample behavior of a hardware–software system

which represents the expected value of computation capacity at the time point t. Figure 13.9 shows a numerical example of Equation 13.16.

13.4.2 Model for Hardware–Software System

Recently, the design of computer systems has begun to attach importance to hardware/software co-design (or simply co-design) [30]; this is the concept that the hardware and the software systems should be designed synchronously, not separately, with due consideration of each other. Co-design is not a new concept, but it has received much attention since computer systems have grown in size and complexity and both the hardware and the software systems have to be designed in order to bring out the mutual maximum performances. The concept of co-design is also important in system quality/performance measurement and assessment. Goel and Soenjoto [31] and Tokuno and Yamada [32] have discussed the availability models for a system consisting of one hardware and one software subsystem. Figure 13.10 shows a sample behavior of a hardware–software system.

The state space of $\{X(t), t \geq 0\}$ in this section is defined as follows:

W_n the system is operating
R_n^{S} the system is inoperable due to a software failure
R_n^{H} the system is inoperable due to a hardware failure.

Suppose that the failure and the restoration characteristics of the software subsystem depend on the number of corrected faults and that those of the hardware subsystem have no bearing on the number of failures. Then the instantaneous and the average system availabilities can be obtained in the same forms as with the model with two types of software failure discussed in Section 13.3.1 by replacing the states R_n^{S} and R_n^{H} with the states R_n^1 and R_n^2 respectively.

13.5 Concluding Remarks

In this chapter, we have surveyed the Markovian models for software availability measurement and assessment. Most of the results in this chapter have been obtained as closed forms. Thus we can evaluate performance measures more easily than the simulation models. Then the model parameters have to be estimated based on the actual data. However, it is too difficult to collect the failure data in the operation phase, especially the software restoration times. Under the present situation, we cannot help but decide the parameters experientially from similar systems developed before. We need to provide collection procedures for the field data.

In addition, *software safety* has now begun to draw attention as one of the user-oriented quality characteristics. This is defined as the characteristic that the software systems do not induce any hazards, whether or not the systems maintain their intended functions [33], and is distinct from software reliability. It is of urgent necessity to establish software safety evaluation methodologies. We also add that studies on software safety measurement and assessment [34, 35] are currently being pursued.

References

[1] Goel AL. Software reliability models: assumptions, limitations, and applicability. IEEE Trans Software Eng 1985;SE-11:1411–23.

[2] Lyu MR, editor. Handbook of software reliability engineering. Los Alamitos (CA): IEEE Computer Society Press, 1996.

[3] Malaiya YK, Srimani PK, editors. Software reliability models: theoretical developments, evaluation and applications. Los Alamitos (CA): IEEE Computer Society Press; 1991.

[4] Musa JD. Software reliability engineering. New York: McGraw-Hill; 1999.

[5] Pham H. Software reliability. Singapore: Springer-Verlag; 2000.

[6] Xie M. Software reliability modelling. Singapore: World Scientific; 1991.

[7] Yamada S. Software reliability models: fundamentals and applications (in Japanese). Tokyo: JUSE Press; 1994.

[8] Yamada S. Software reliability models. In: Osaki S, editor. Stochastic models in reliability and maintenance. Berlin: Springer-Verlag; 2002. p.253–80.

[9] Laprie J-C, Kanoun K, Béounes C, Kaâniche M. The KAT (knowledge–action–transformation) approach to the modeling and evaluation of reliability and availability growth. IEEE Trans Software Eng 1991;17: 370–82.

[10] Laprie J-C, Kanoun K. X-ware reliability and availability modeling. IEEE Trans Software Eng 1992;18: 130–47.

[11] Tokuno K, Yamada S. User-oriented software reliability assessment technology (in Japanese). Bull Jpn Soc Ind Appl Math 2000;10:186–97.

[12] Ross SM. Stochastic processes, second edition. New York: John Wiley & Sons; 1996.

[13] Ohba M, Chou X. Does imperfect debugging affect software reliability growth?. In: Proceedings of 11th IEEE International Conference on Software Engineering 1989;p.237–44.

[14] Tokuno K, Yamada S. An imperfect debugging model with two types of hazard rates for software reliability measurement and assessment. Math Comput Modell 2000;31:343–52.

[15] Okumoto K, Goel AL. Availability and other performance measures for system under imperfect maintenance. In: Proceedings of COMPSAC'78, 1978;p.66–71.

[16] Kim JH, Kim YH, Park CJ. A modified Markov model for the estimation of computer software performance. Oper Res Lett 1982;1:253–57.

[17] Jelinski Z, Moranda PB. Software reliability research. In: Freiberger W, editor. Statistical computer performance evaluation. New York: Academic Press, 1972. p.465–84.

[18] Tokuno K, Yamada S. A Markovian software availability measurement with a geometrically decreasing failure-occurrence rate. IEICE Trans Fundam 1995;E78-A:737–41.

[19] Tokuno K, Yamada S. Markovian software availability modeling for performance evaluation. In: Christer AH, Osaki S, Thomas LC, editors. Stochastic modelling in innovative manufacturing: proceedings. Berlin: Springer-Verlag; 1997. p.246–56.

[20] Tokuno K, Yamada S. Software availability model with a decreasing fault-correction rate (in Japanese). J Reliab Eng Assoc Jpn 1997;19:3–12.

[21] Moranda PB. Event-altered rate models for general reliability analysis. IEEE Trans Reliab 1979;R-28:376–81.

[22] Yamada S, Tokuno K, Osaki S. Software reliability measurement in imperfect debugging environment and its application. Reliab Eng Syst Saf 1993;40:139–47.

[23] Nakagawa Y, Takenaka I. Error complexity model for software reliability estimation (in Japanese). Trans IEICE D-I 1991;J74-D-I:379–86.

[24] Tokuno K, Yamada S. Markovian software availability measurement based on the number of restoration actions. IEICE Trans Fundam 2000;E83-A:835–41.

[25] Tokuno K, Yamada S. A Markovian software availability model for operational use (in Japanese). J Jpn Soc Software Sci Technol 1998;15:17–24.

[26] Tokuno K, Yamada S. Markovian availability measurement with two types of software failures during the operation phase. Int J Reliab Qual Saf Eng 1999;6:43–56.

[27] Tokuno K, Yamada S. Operational software availability measurement with two kinds of restoration actions. J Qual Mainten Eng 1998;4:273–83.

[28] Tokuno K, Yamada S. Markovian software availability modeling with degenerated performance. In: Lydersen S, Hansen GK, Sandtorv HA, editors. Proceedings of the European Conference on Safety and Reliability, vol. 1. Rotterdam: AA Balkema, 1998;1:425–31.

[29] Beaudry MD. Performance-related reliability measures for computing systems. IEEE Trans Comput 1978;C-27:540–7.

[30] De Micheli G. A survey of problems and methods for computer-aided hardware/software co-design. J Inform Process Soc Jpn 1995;36:605–13.

[31] Goel AL, Soenjoto J. Models for hardware–software system operational-performance evaluation. IEEE Trans Reliab 1981;R-30:232–9.

[32] Tokuno K, Yamada S. Markovian availability modeling for software-intensive systems. Int J Qual Reliab Manage 2000;17:200–12.

[33] Leveson NG. Safeware: system safety and computers. New York: Addison-Wesley; 1995.

[34] Tokuno K, Yamada S. Stochastic software safety/ reliability measurement and its application. Ann Software Eng 1999;8:123–45.

[35] Tokuno K, Yamada S. Markovian reliability modeling for software safety/availability measurement. In: Pham H, editor. Recent advances in reliability and quality engineering. Singapore: World Scientific; 2001. p.181–201.

Software Rejuvenation: Modeling and Applications

Tadashi Dohi, Katerina Goševa-Popstojanova, Kalyanaraman Vaidyanathan, Kishor S. Trivedi and Shunji Osaki

14.1 Introduction

Since system failures due to software faults are more frequent than failures caused by hardware faults, there is a tremendous need for improvement of software availability and reliability. Present-day applications impose stringent requirements in terms of cumulative downtime and failure-free operation of software, since, in many cases, the consequences of software failure can lead to huge economic losses or risk to human life. However, these requirements are very difficult to design for and guarantee, particularly in applications of non-trivial complexity.

Recently, the phenomenon of *software aging* [1–4] in which error conditions actually accrue with time and/or load, has been observed. In systems with high reliability/availability requirements, software aging can cause outages resulting in high costs. Huang and coworkers [2,5,6] and Jalote *et al.* [7] report this phenomenon in telecommunications billing applications, where over time the application experiences a crash or a hang failure. Avritzer and Weyuker [8] and Levendel [9] discuss aging in telecommunication switching software where the effect manifests as gradual performance degradation. Software aging has also been observed in widely used software like Netscape and xrn. Perhaps the most vivid example of aging in safety critical systems is the Patriot's software [10], where the accumulated errors led to a failure that resulted in loss of human life.

Resource leaking and other problems causing software to age are due to the software faults whose fixing is not always possible because, for example, the application developer may not have access to the source code. Furthermore, it is almost impossible to fully test and verify if a piece of software is fault free. Testing software becomes harder if it is complex, and more so if testing and debugging cycle times are reduced due to smaller release time requirements. Common experience suggests that most software failures are transient in nature [11]. Since transient failures will disappear if the operation is retried in a slightly different context, it is difficult to characterize their root origin. Therefore, the residual faults have to be tolerated in the operational phase. The usual strategies to deal with failures in the operational phase are reactive in nature; they consist of action taken after the failure.

A complementary approach to handling transient software failures, called *software rejuvenation*, was proposed by Huang *et al.* [2]. Software rejuvenation is a preventive and proactive (as opposite to being reactive) solution that is particularly useful for counteracting the phenomenon of software aging. It involves stopping the running software occasionally, cleaning its internal state and restarting it. *Garbage collection*, *flushing operating system kernel tables* and *reinitializing internal data structures* are some examples of what cleaning the internal state of software might involve [12]. An extreme, but well-known example of rejuvenation is a *hardware reboot*.

Apart from being used in an *ad hoc* manner by almost all computer users, rejuvenation has been used to avoid unplanned outages in high availability systems, such as telecommunication systems [2, 8], where the cost of downtime is extremely high. Among typical applications of mission critical systems, periodic software and system rejuvenation have been implemented for long-life deep-space missions [13–15].

Recently, the use of rejuvenation was extended to cluster systems [16]. Using the node failover mechanisms in a high availability cluster, one can maintain operation (though possibly at a degraded level) while rejuvenating one node at a time. The first commercial version of this kind of a software rejuvenation agent for IBM cluster servers has been implemented with collaboration with Duke University researchers [16, 17].

Although the faults in the software still remain, performing rejuvenation periodically removes or minimizes potential error conditions due to these faults, thus preventing failures that might have unacceptable consequences. Rejuvenation has the same motivation and advantages/disadvantages as preventive maintenance policies in hardware systems. Rejuvenation typically involves an overhead, but, on the other hand, it prevents more severe failures from occurring. The application will, of course, be unavailable during rejuvenation, but since this is a scheduled downtime the cost is expected to be much lower than the cost of an unscheduled downtime caused by failure. Hence, an important issue is to determine the optimal schedule to perform software rejuvenation in terms of availability and cost.

In this chapter, we present an overview of the approaches for analyzing software aging and software rejuvenation. In the following section, we describe the model-based approach to determine the optimal rejuvenation times, introducing the basic software rejuvenation model, and examples from transaction-based software systems and cluster systems. In this approach, Markov/semi-Markov processes, queuing processes and stochastic Petri nets are applied to describe the system behavior. Measures such as availability and cost are formulated, and algorithms to calculate the optimal software rejuvenation times are developed. Next, we describe the measurement-based approach for detection and validation of the existence of software aging. For quantifying the effect of aging in UNIX operating system resources, we perform empirical studies on a purely time-based approach and a time and workload-based approach. Here, we periodically monitor and collect data on the attributes responsible for determining the health of the executing software and estimate resource exhaustion times. Finally, the chapter is concluded with some remarks.

14.2 Modeling-based Estimation

The model-based approach is aimed at evaluating the effectiveness of software rejuvenation and determining the optimal schedules to perform rejuvenation. Huang *et al.* [2] used a continuous time Markov chain to model software rejuvenation. They considered the two-step failure model where the application goes from the initial robust (clean) state to a failure probable (degraded) state from which two actions are possible: rejuvenation or transition to failure state. Both rejuvenation and recovery from failure return the software system to the robust state.

This model was recently generalized using semi-Markov reward processes [18–20]. The optimal software rejuvenation schedules are analytically derived under the steady-state availability and the expected cost per unit time in the steady state. Further, non-parametric statistical algorithms to estimate the optimal software rejuvenation schedules are also developed. Thus, this approach does not depend on the form of the failure time distribution function. To deal with deterministic interval between successive rejuvenations, Garg *et al.* [21] used a Markov regenerative stochastic Petri net model.

The fine-grained software rejuvenation model presented by Bobbio and Sereno [22] takes a different approach to characterize the effect of software aging. It assumes that the degradation process consists of a sequence of additive random shocks; the system is considered out of service as soon as the appropriate parameter reaches an assigned threshold level. Garg *et al.* [23] analyzed the effects of checkpointing and rejuvenation used together on the expected completion time of a software program. A fluid stochastic Petri-net-based model that captures the behavior of aging software systems which employ rejuvenation, restoration, and checkpointing was also proposed [24]. The use of preventive on-board maintenance that includes periodic software and system rejuvenation has also been proposed and analyzed [13–15].

Garg *et al.* [25] include arrival and queuing of jobs in the system and compute load and time-dependent rejuvenation policy. The above models consider the effect of aging as crash/hang failure, referred to as hard failures, which result in unavailability of the software. However, the aging of the software system can manifest as soft failures, *i.e.* performance degradation. Pfennig *et al.* [26] modeled performance degradation by the gradual decrease of the service rate. Both effects of aging, hard failures that result in an unavailability and soft failures that result in performance degradation, are considered in the model of a transaction-based software system presented by Garg *et al.* [27]. This model was generalized by considering multiple servers [28] and threshold policies [29].

Recently, stochastic models of time-based and prediction-based rejuvenation as applied to cluster systems were developed [16]. These models capture a multitude of cluster system characteristics, failure behavior, and performability measures.

14.2.1 Examples in Telecommunication Billing Applications

Consider the basic software rejuvenation model proposed by Huang *et al.* [2]. Suppose that the stochastic behavior of the system can be described by a simple continuous-time Markov chain (CTMC) with the following four states.

State 0: highly robust state
(normal operation state).

State 1: failure probable state.

State 2: failure state.

State 3: software rejuvenation state.

Figure 14.1 depicts the state transition diagram of the CTMC. Let Z be the random time interval when the highly robust state changes to the failure probable state, having the exponential distribution $\Pr\{Z \le t\} = F_0(t) = 1 - \exp(-t/\mu_0)$ $(\mu_0 > 0)$. Just after the state becomes the failure probable state, a system failure may occur with a positive probability. Without loss of generality, we assume that the random variable Z is observable during the system operation.

Define the failure time X (from State 1) and the repair time Y, having the exponential distributions $\Pr\{X \le t\} = F_f(t) = 1 - \exp(-t/\lambda_f)$ and $\Pr\{Y \le t\} = F_a(t) = 1 - \exp(-t/\mu_a)$ $(\lambda_f > 0,$ $\mu_a > 0)$. If the system failure occurs before triggering a software rejuvenation, then the repair is started immediately at that time and is completed

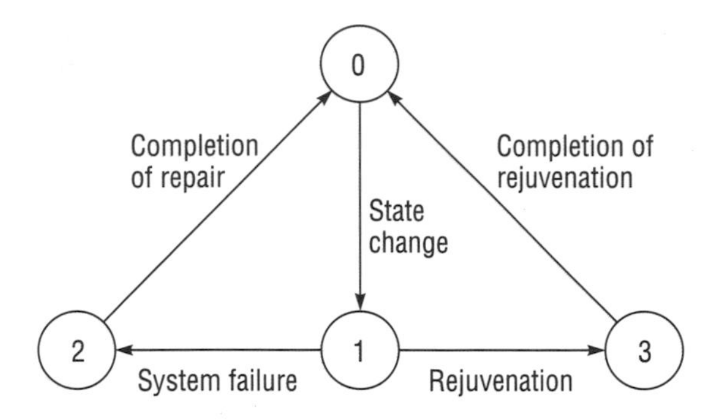

Figure 14.1. State transition diagram of CTMC

after the random time Y elapses. Otherwise, the software rejuvenation is started. Note that the software rejuvenation cycle is measured from the time instant just after the system enters State 1. Define the distribution functions of the time to invoke the software rejuvenation and of the time to complete software rejuvenation by $F_{\mathrm{r}}(t) = 1 - \exp(-t/\mu_{\mathrm{r}})$ and $F_c(t) = 1 - \exp(-t/\mu_c)$ ($\mu_c > 0$, $\mu_{\mathrm{r}} > 0$) respectively. Huang *et al.* [2] analyzed the above CTMC and calculated the expected system down time and the expected cost per unit time in the steady state. However, it should be noted that the rejuvenation schedule for the CTMC model is not feasible, since the preventive maintenance time to rejuvenate the system is an exponentially distributed random variable.

It is not difficult to introduce the periodic rejuvenation schedule and to extend the CTMC model to the general one. Dohi *et al.* [18–20] developed semi-Markov models with the periodic rejuvenation and general transition distribution functions. More specifically, let Z be the random variable having the common distribution function $\Pr\{Z \le t\} = F_0(t)$ with finite mean μ_0 (>0). Also, let X and Y be the random variables having the common distribution functions $\Pr\{X \le t\} = F_f(t)$ and $\Pr\{Y \le t\} = F_a(t)$ with finite means λ_f (>0) and μ_a (>0) respectively. Denote the distribution function of the time to invoke the software rejuvenation and the distribution of the time to complete software rejuvenation by $F_{\mathrm{r}}(t)$ and $F_c(t)$ (with mean μ_c (>0)) respectively. After completing the repair or the rejuvenation, the software system becomes as good as new, and the software age is initiated at the beginning of the next highly robust state. Consequently, we define the time interval from the beginning of the system operation to the next one as one cycle, and the same cycle is repeated again and again.

If we consider the time to software rejuvenation as a constant t_0, then it follows that

$$F_{\mathrm{r}}(t) = U(t - t_0) = \begin{cases} 1 & \text{if } t \ge t_0 \\ 0 & \text{otherwise} \end{cases} \tag{14.1}$$

We call t_0 (≥ 0) the *software rejuvenation interval* in this chapter and $U(\cdot)$ is the unit step function.

The underlying stochastic process is a semi-Markov process with four regeneration states. If the sojourn times in all states are exponentially distributed, of course, this model is the CTMC in Huang *et al.* [2]. From the familiar renewal argument [30], we formulate the steady-state system availability as

$$\begin{aligned} A(t_0) &= \Pr\{\text{software system is operative in the steady state}\} \\ &= \frac{\mu_0 + \int_0^{t_0} \bar{F}_f(t)\,\mathrm{d}t}{\mu_0 + \mu_a F_f(t_0) + \mu_c \bar{F}_f(t_0) + \int_0^{t_0} \bar{F}_f(t)\,\mathrm{d}t} \\ &= S(t_0)/T(t_0) \end{aligned} \tag{14.2}$$

where in general $\bar{\phi}(\cdot) = 1 - \phi(\cdot)$. The problem is to derive the optimal software rejuvenation interval t_0^* that maximizes the system availability in the steady state $A(t_0)$.

We make the following assumption.

Assumption 1. $\mu_a > \mu_c$.

Assumption 1 means that the mean time to repair is strictly larger than the mean time to complete the software rejuvenation. This assumption is quite reasonable and intuitive. The following result gives the optimal software rejuvenation schedule for the semi-Markov model.

Theorem 1.

(1) *Suppose that the failure time distribution is strictly increasing failure rate (IFR) under Assumption 1. Define the following nonlinear function:*

$$q(t_0) = T(t_0) - \{(\mu_a - \mu_c)r_f(t_0) + 1\}S(t_0) \tag{14.3}$$

where $r_f(t) = (\mathrm{d}F_f(t)/\mathrm{d}t)/\bar{F}_f(t)$ *is the failure rate.*

(i) *If* $q(0) > 0$ *and* $q(\infty) < 0$*, then there exists a finite and unique optimal software rejuvenation schedule* t_0^* $(0 < t_0^* < \infty)$ *satisfying* $q(t_0^*) = 0$*, and the maximum system availability is*

$$A(t_0^*) = \frac{1}{(\mu_a - \mu_c)r_f(t_0^*) + 1} \tag{14.4}$$

(ii) *If* $q(0) \leq 0$*, then the optimal software rejuvenation schedule is* $t_0^* = 0$*, i.e. it is optimal to start the rejuvenation just after entering the failure probable state, and the maximum system availability is* $A(0) = \mu_0/(\mu_0 + \mu_c)$.

(iii) *If* $q(\infty) \geq 0$*, then the optimal rejuvenation schedule is* $t_0^* \to \infty$*, i.e. it is optimal not to carry out the rejuvenation, and the maximum system availability is* $A(\infty) = (\mu_0 + \lambda_f)/(\mu_0 + \mu_a + \lambda_f)$.

(2) *Suppose that the failure time distribution is decreasing failure rate (DFR) under Assumption 1. Then, the system availability* $A(t_0)$ *is a convex function of* t_0*, and the optimal rejuvenation schedule is* $t_0^* = 0$ *or* $t_0^* \to \infty$.

Proof. Differentiating $A(t_0)$ with respect to t_0 and setting it equal to zero implies $q(t_0) = 0$. Further differentiation yields

$$\frac{\mathrm{d}q(t_0)}{\mathrm{d}t_0} = -(\mu_a - \mu_c)S(t_0)\frac{\mathrm{d}r_f(t_0)}{\mathrm{d}t_0} \tag{14.5}$$

If $r_f(t_0)$ is strictly increasing, then the function $q_1(t_0)$ is strictly decreasing and the system availability $A(t_0)$ is strictly concave in t_0 under Assumption 1. Further, if $q(0) > 0$ and $q(\infty) < 0$, then there exists a unique optimal software rejuvenation schedule t_0^* $(0 < t_0^* < \infty)$ satisfying $q(t_0^*) = 0$. If $q(0) \leq 0$ or $q(\infty) \geq 0$, then the system availability is monotonically increasing or decreasing in t_0, and the optimal policy becomes $t_0^* = 0$ or $t_0^* \to \infty$. On the other hand, if $r_f(t_0)$ is a decreasing function of t_0, then $A(t_0)$ is a convex function of t_0, and the optimal software rejuvenation schedule is $t_0^* = 0$ or $t_0^* \to \infty$. The proof is thus completed. □

It is easy to check that the result above implies the result in Huang *et al.* [2], although they used the system downtime and its associated cost as criteria of optimality. As is clear from Theorem 1, when the failure time obeys the exponential distribution, the optimal software rejuvenation schedule becomes $t_0^* = 0$ or $t_0^* \to \infty$. This means that the rejuvenation should be performed as soon as the software enters the failure probable state ($t_0 = 0$) or should not be performed at all ($t_0 \to \infty$). Therefore, the determination of the optimal rejuvenation schedule based on the system availability is never motivated in such a situation. Since for a software system that ages it is more realistic to assume that failure time distribution is strictly IFR, our general setting is plausible and the result satisfies our intuition.

Dohi *et al.* [19] developed a non-parametric algorithm to estimate the optimal software rejuvenation time, when the failure time data are obtained but the underlying distribution function is unknown. Before developing the statistical estimation algorithms for the optimal software rejuvenation schedules, we translate the underlying problems $\max_{0 \leq t_0 < \infty} A(t_0)$ to a graphical one. Following Barlow and Campo [31], define the scaled total time on test (TTT) transform of the failure time distribution:

$$\phi(p) = (1/\lambda_f) \int_0^{F_f^{-1}(p)} \bar{F}_f(t)\,\mathrm{d}t \tag{14.6}$$

where

$$F_f^{-1}(p) = \inf\{t_0;\ F_f(t_0) \geq p\} \quad (0 \leq p \leq 1) \tag{14.7}$$

It is well known [31] that $F_f(t)$ is IFR (DFR) if and only if $\phi(p)$ is concave (convex) on $p \in [0, 1]$. After a few algebraic manipulations, we have the following result.

Theorem 2. *Obtaining the optimal software rejuvenation schedule $t_0{}^*$ maximizing the system availability $A(t_0)$ is equivalent to obtaining p^* $(0 \leq p^* \leq 1)$ such as*

$$\max_{0\leq p\leq 1} \frac{\phi(p)+\alpha}{p+\beta} \tag{14.8}$$

where $\alpha = \lambda_f + \mu_0$ and $\beta = \mu_c(\mu_a - \mu_c)$.

The proof is omitted for brevity. From Theorem 2 it follows that the optimal software rejuvenation schedule $t_0{}^* = F_f^{-1}(p^*)$ is determined by calculating the optimal point $p^*(0 \leq p^* \leq 1)$ maximizing the tangent slope from the point $(-\beta, -\alpha) \in (-\infty, 0) \times (-\infty, 0)$ to the curve $(p, \phi(p)) \in [0, 1] \times [0, 1]$.

Suppose that the optimal software rejuvenation schedule needs to be estimated from an ordered sample of observations $0 = x_0 \leq x_1 \leq x_2 \leq \cdots \leq x_n$ of the failure times from an absolutely continuous distribution F_f, which is unknown. Then the scaled TTT statistics based on this sample are defined by $\phi_{nj} = \psi_j/\psi_n$, where

$$\psi_j = \sum_{k=1}^{j}(n-k+1)(x_k - x_{k-1}) \quad (j = 1, 2, \ldots, n;\ \psi_0 = 0) \tag{14.9}$$

The empirical distribution function $F_n(x)$ corresponding to the sample data x_j $(j = 0, 1, 2, \ldots, n)$ is

$$F_n(x) = \begin{cases} j/n & \text{for } x_j \leq x < x_{j+1} \\ 1 & \text{for } x_n \leq x \end{cases} \tag{14.10}$$

From this, we obtain the polygon by plotting the points $(F_n(x), \phi_{nj})$ $(j = 0, 1, 2, \ldots, n)$ and connecting them by line segments known as the *scaled TTT plot*. In other words, the scaled TTT plot can be regarded as a numerical counterpart of the scaled TTT transform.

The following result gives a non-parametric statistical estimation algorithm for the optimal software rejuvenation schedule.

Theorem 3.

(i) *Suppose that the optimal software rejuvenation schedule is to be estimated from n ordered samples $0 = x_0 \leq x_1 \leq x_2 \leq \cdots \leq x_n$ of the failure times from an absolutely continuous distribution F_f, which is unknown. Then, a non-parametric estimator of the optimal software rejuvenation schedule $\hat{t}_0^*$ that maximizes $A(t_0)$ is given by x_{j^*}, where*

$$j^* = \left\{ j \;\middle|\; \max_{0\leq j\leq n} \frac{\phi_{nj}+\alpha_i}{j/n+\beta_i} \right\} \tag{14.11}$$

and the theoretical mean λ_f is replaced by the sample mean $\sum_{k=1}^{n} x_k/n$.

(ii) *The estimator given in (i) is strongly consistent,* i.e. *x_{j^*} converges to the optimal solution $t_0{}^*$ uniformly with probability one as $n \to \infty$, if a unique optimal software rejuvenation schedule exists.*

It is straightforward to prove the above result in (i) from Theorem 2. The uniform convergence property in (ii) is due to the Glivenko–Cantelli lemma (*e.g.* see [31]) and the strong law of large numbers. The graphical procedure proposed here has an educational value for better understanding of the optimization problem and it is convenient for performing sensitivity analysis of the optimal software rejuvenation policy when different values are assigned to the model parameters. Finally, it enables us to estimate the optimal schedule without specifying the failure time distribution. Although some typical theoretical distribution functions, such as the Weibull distribution and the gamma distribution, are often assumed in the reliability analysis, our non-parametric estimation algorithm can generate the optimal software rejuvenation schedule using the on-line knowledge about the observed failure times.

Figure 14.2 shows the estimation result of the optimal software rejuvenation schedule for the semi-Markov model, where the failure time data is generated by the Weibull distribution with shape parameter $\beta = 4.0$ and scale parameter $\theta = 0.9$. For 100 failure data points, the estimates of the optimal rejuvenation schedule and the maximum system availability are $\hat{t}_0^* = 0.565\,92$ and $A(\hat{t}_0^*) = 0.987\,813$ respectively. Actually, we examined through a simulation study in [18–20] that the non-parametric estimator of the optimal software

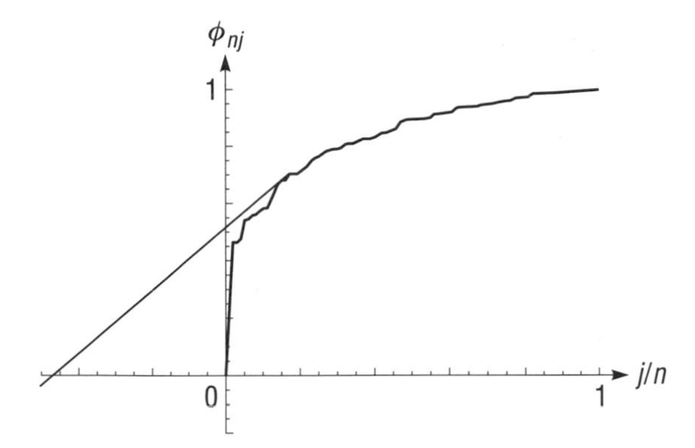

Figure 14.2. Estimation of the optimal software rejuvenation schedule on the two-dimensional graph: $\theta = 0.9$, $\beta = 4.0$, $\mu_0 = 2.0, \mu_a = 0.04, \mu_c = 0.03$

rejuvenation time derived using Theorem 3 has a nice convergence property.

14.2.2 Examples in a Transaction-based Software System

Next, consider the rejuvenation scheme for a single server software system. Garg *et al.* [27] analyzed a queuing system with preventive maintenance as a mathematical model for a transaction-based software system. Suppose that the transactions arrive at the system in accordance with the homogeneous Poisson process with rate λ (>0). The transactions are processed with service rate $\mu(\cdot)$ (>0) based on the first come, first serve (FCFS) discipline. Transactions that arrive during the service operation of the other transaction are accumulated in the buffer whose maximum capacity is K (>1), where without a loss of generality the buffer is empty at time $t = 0$. If a transaction arrives at the system when K transactions have already been accumulated in the buffer, it will be rejected from the system automatically. That is to say, the software system under consideration can continue processing if there exists at least one transaction in the buffer. When the buffer is empty, the system will wait for the arrival of an additional transaction, keeping the system *idle*.

The service rate $\mu(\cdot)$ is assumed to be a function of the operation time t, the number of transactions in the buffer, and/or other variables. For example, let $\{X_t; t \geq 0\}$ be the number of transactions accumulated in the buffer at time t. Then, the service rate may be given by $\mu(t)$, $\mu(X_t)$ or $\mu(t, X_t)$. In this model, the software system is assumed to be unreliable, where the failure rate $\rho(\cdot)$ also depends on the operation time t, the number of transactions in the buffer and/or other variables, *i.e.* it can be described as $\rho(t)$, $\rho(X_t)$ or $\rho(t, X_t)$. The recovery (repair) operation starts immediately when the system fails, where the recovery time is the random variable Y_r with $E[Y_r] = \gamma_r$ (>0). A newly arriving transaction is rejected if it arrives at the system during the recovery operation period. Intuitively, software rejuvenation is beneficially motivated if the software failure rate $\rho(\cdot)$ increases with the passage of time [2, 27]. Thus, it is assumed that the system failure time distribution is IFR, *i.e.* the software failure rate $\rho(\cdot)$ is an increasing function of time t. Let Y_R, the preventive maintenance time for the software rejuvenation, be the random variable with $E[Y_R] = \gamma_R$ (>0). Suppose that the software system operation is started at time $t = 0$ with an empty buffer. The system is regarded as good as new after both the completion of software rejuvenation and the completion of recovery.

Garg *et al.* [27] consider the following software rejuvenation schemes based on the cumulative operation time.

Policy I: rejuvenate the system when the cumulative operation time becomes T (>0). Then, all transactions in the buffer, if any, will be lost.

Policy II: rejuvenate the system at the beginning of the first idle period following the cumulative operation time T (>0).

Figure 14.3 depicts the possible behavior of software system under the two kinds of rejuvenation policies.

Consider three dependability measures: steady-state availability, loss probability of transactions, and mean response time of transactions. We will

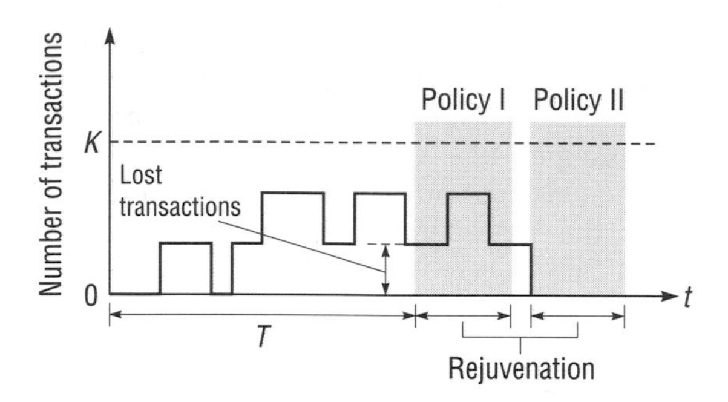

Figure 14.3. Possible realization of transaction-based software system under two policies

apply the well-known theory of the Markov regenerative process (MRGP) to derive the above measures. First, consider a discrete-time Markov chain (DTMC) with three states.

State A: system operation.

State B: system is recovering from failure.

State C: system is undergoing software rejuvenation.

Figure 14.4 is the state transition diagram of the DTMC. Let P_{AB} and P_{AC} denote the transition probabilities from State A to B and from State A to C respectively. Taking account of $P_{\mathrm{AC}} = 1 - P_{\mathrm{AB}}$, the transition probability matrix is given by

$$\mathbf{P} = \begin{bmatrix} 0 & P_{\mathrm{AB}} & P_{\mathrm{AC}} \\ 1 & 0 & 0 \\ 1 & 0 & 0 \end{bmatrix}$$

The steady-state probabilities for the DTMC can be easily found as

$$\pi_{\mathrm{A}} = \tfrac{1}{2} \tag{14.12}$$

$$\pi_{\mathrm{B}} = \frac{P_{\mathrm{AB}}}{2} \tag{14.13}$$

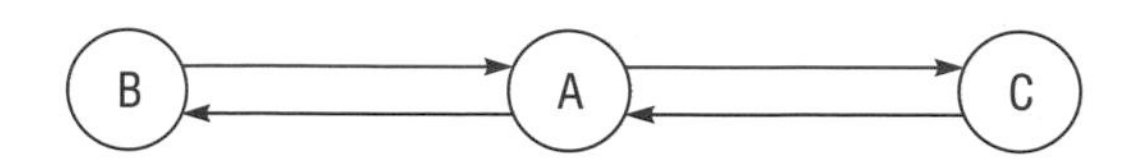

Figure 14.4. State transition diagram of DTMC

and

$$\pi_{\mathrm{C}} = \frac{P_{\mathrm{AC}}}{2}. \tag{14.14}$$

Next, define the following random variables:

U is the sojourn time in State A in the steady state

U_n is the sojourn time in State A when n $(= 0, \ldots, K)$ transactions are accumulated in the buffer, so that

$$U = \sum_{n=0}^{K} U_n, \quad \text{almost surely} \tag{14.15}$$

N_{l} is the number of transactions lost at the transition from State A to State B or C.

Theorem 4.

(i) The steady-state system availability is

$$A_{\mathrm{ss}} = \frac{\pi_{\mathrm{A}} E[U]}{\pi_{\mathrm{A}} E[U] + \pi_{\mathrm{B}} \gamma_{\mathrm{r}} + \pi_{\mathrm{C}} \gamma_{\mathrm{R}}} = \frac{E[U]}{E[U] + P_{\mathrm{AB}} \gamma_{\mathrm{r}} + P_{\mathrm{AC}} \gamma_{\mathrm{R}}} \tag{14.16}$$

(ii) The loss probability of transactions is

$$P_{\mathrm{loss}} = \frac{\lambda(P_{\mathrm{AB}} \gamma_{\mathrm{r}} + P_{\mathrm{AC}} \gamma_{\mathrm{R}} + E[U_K]) + E[N_{\mathrm{l}}]}{\lambda(E[U] + P_{\mathrm{AB}} \gamma_{\mathrm{r}} + P_{\mathrm{AC}} \gamma_{\mathrm{R}})} \tag{14.17}$$

(iii) Let E and W_{s} be the expected number of transactions processed by the system and the expected total processing time for transactions served by the system respectively, where $E = \lambda(E[U] - E[U_K])$. Then an upper bound of the mean response time on transactions $T_{\mathrm{res}} = W_{\mathrm{s}}/(E - E[N_{\mathrm{l}}])$ is given by

$$T_{\mathrm{res}} < \frac{W}{E - E[N_{\mathrm{l}}]} \tag{14.18}$$

In this way, we can formulate implicitly the steady-state availability, the loss probability of transactions, and the upper bound of mean response time on transactions, based on the simple DTMC. Next, we wish to derive the optimal rejuvenation interval T^* under Policy I and Policy II. To do this, we represent the steady-state availability, the loss probability of transactions, and the upper bound of the mean response

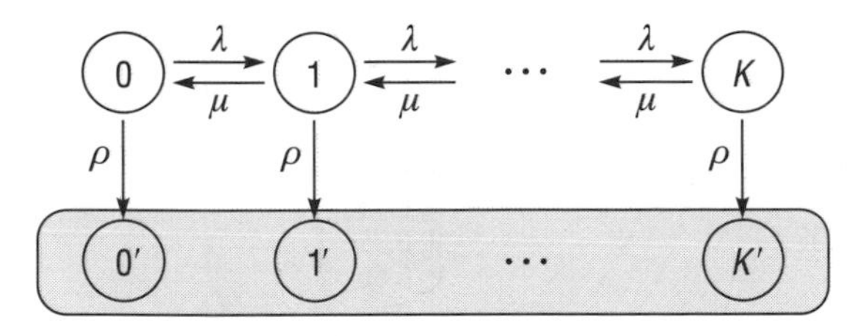

Figure 14.5. State transition diagram of Policy I

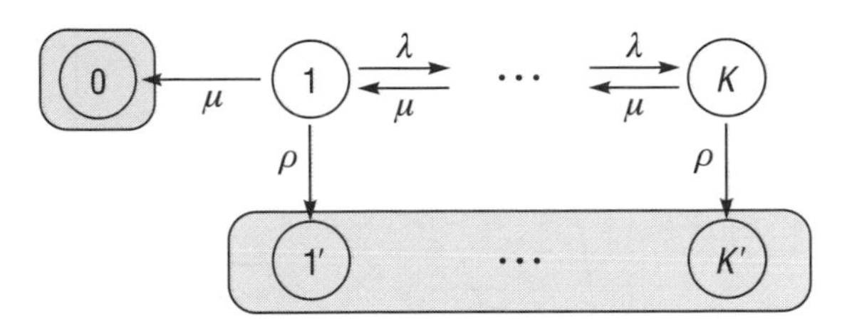

Figure 14.6. State transition diagram of Policy II

time as the functions of T, *i.e.* $A_{ss}(T)$, $P_{loss}(T)$ and $T_{res}(T)$. Since these functions include the parameters P_{AB}, $E[U_n]$ and $E[N_l]$ depending on T, we need to calculate P_{AB}, $E[U_n]$ and $E[N_l]$ numerically. Define the following probabilities:

$p_n(t)$ is the probability that n transactions remain in the buffer at time t

$p_{n'}(t)$ is the probability that the system fails up to time t and that n transactions remain in the buffer at the failure time

where $n = 0, \ldots, K$.

First, we analyze a single server queuing system under Policy I. Consider a non-homogeneous continuous time Markov chain (NHCTMC) with $2K$ states, where:

State $0, \ldots, K$**:** $0, \ldots, K$ denotes the number of transactions in the buffer;

State $0', \ldots, K'$**:** the system fails with $0, \ldots, K$ transactions in the buffer (absorbing states).

Figure 14.5 illustrates the state transition diagram for the NHCTMC with Policy I. Applying the well-known state-space method to the NHCTMC, we can formulate the difference-differential equations (Kolmogorov's forward equations) on $p_n(t)$ and $p_{n'}(t)$ as follows:

$$\frac{\mathrm{d}p_0(t)}{\mathrm{d}t} = \mu(\cdot)p_1(t) - \{\lambda + \rho(\cdot)\}p_0(t) \quad (14.19)$$

$$\frac{\mathrm{d}p_n(t)}{\mathrm{d}t} = \mu(\cdot)p_{n+1}(t) + \lambda p_{n-1}(t) - \{\mu(\cdot) + \lambda + \rho(\cdot)\}p_n(t) \quad n = 1, \ldots, K-1 \quad (14.20)$$

$$\frac{\mathrm{d}p_K(t)}{\mathrm{d}t} = \lambda p_{K-1}(t) - \{\mu(\cdot) + \rho(\cdot)\}p_K(t) \quad (14.21)$$

$$\frac{\mathrm{d}p_{n'}(t)}{\mathrm{d}t} = \rho(\cdot)p_n(t) \quad n = 0, \ldots, K \quad (14.22)$$

By solving the difference-differential equations numerically, we can obtain

$$P_{AB} = \sum_{n=0}^{K} p_{n'}(T) \quad (14.23)$$

$$E[U_n] = \int_0^T p_n(t)\,\mathrm{d}t \quad n = 1, \ldots, K \quad (14.24)$$

$$E[N_l] = \sum_{n=0}^{K} n\{p_n(T) + p_{n'}(T)\} \quad (14.25)$$

and the dependability measures in Equations 14.16, 14.17 and 14.18 based on Policy I.

Figure 14.6 illustrates the state transition diagram for the NHCTMC with Policy II. In a fashion similar to Policy I, we derive the difference-differential equations on $p_n(t)$ and $p_{n'}(t)$ for Policy II. Since the system starts the software rejuvenation when the buffer becomes empty for the first time, the state transition diagrams of Policies I and II are equivalent in $t \leq T$. In the case of $t > T$, we can derive the difference-differential equations under Policy II as follows.

$$\frac{\mathrm{d}p_0(t)}{\mathrm{d}t} = \mu(\cdot)p_1(t) \quad (14.26)$$

$$\frac{\mathrm{d}p_1(t)}{\mathrm{d}t} = \mu(\cdot)p_2(t) - \{\mu(\cdot) + \lambda + \rho(\cdot)\}p_1(t) \quad (14.27)$$

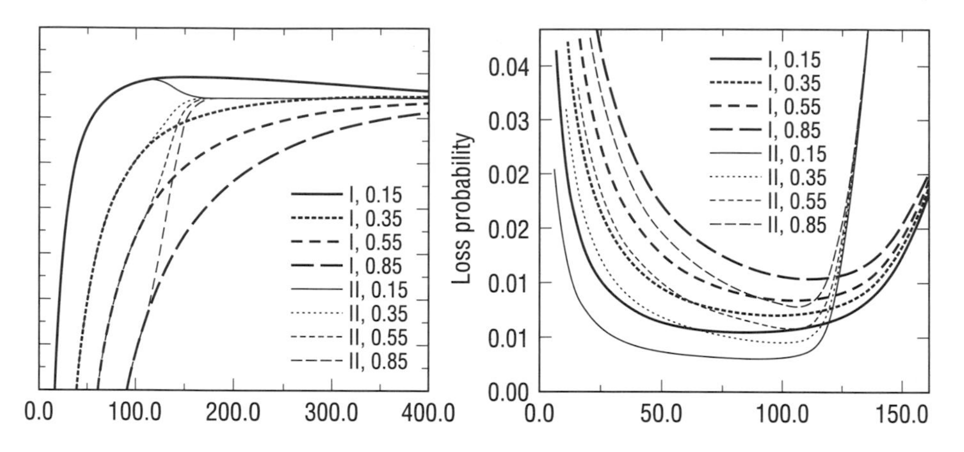

Figure 14.7. A_{SS} and P_{loss} for both policies plotted against T

$$\frac{\mathrm{d}p_n(t)}{\mathrm{d}t} = \mu(\cdot)p_{n+1}(t) + \lambda p_{n-1}(t) - \{\mu(\cdot) + \lambda + \rho(\cdot)\}p_n(t) \quad n = 2, \ldots, K-1 \tag{14.28}$$

$$\frac{\mathrm{d}p_K(t)}{\mathrm{d}t} = \lambda p_{K-1}(t) - \{\mu(\cdot) + \rho(\cdot)\}p_K(t) \tag{14.29}$$

$$\frac{\mathrm{d}p_{n'}(t)}{\mathrm{d}t} = \rho(\cdot)p_n(t) \quad n = 1, \ldots, K. \tag{14.30}$$

Using $p_n^{(i)}(t)$ and $p_{n'}(t)$ in the difference-differential equations, P_{AB}, $E[U_n]$ and $E[N_{\mathrm{l}}]$ are given by

$$P_{\mathrm{AB}} = \sum_{n=0}^{K} p_{n'}(\infty) \tag{14.31}$$

$$E[U_0] = \int_0^T p_0(t)\,\mathrm{d}t \tag{14.32}$$

$$E[U_n] = \int_0^\infty p_n(t)\,\mathrm{d}t \quad n = 1, \ldots, K \tag{14.33}$$

$$E[N_{\mathrm{l}}] = \sum_{n=0}^{K} n p_{n'}(\infty). \tag{14.34}$$

The difference-differential equations derived here can be solved with the standard numerical solution methods, such as the Runge–Kutta method and Adam's method. However, it is not always easy to carry out the numerical calculation, since the computation time depends on the size of the underlying NHCTMC, *i.e.* the number of states.

We illustrate the usefulness of the model presented in determining the optimum value of T on the example adopted from Garg *et al.* [27]. The failure rate and the service rate are assumed to be functions of real time, where $\rho(t)$ is defined to be the hazard function of Weibull distribution, and $\mu(t)$ is defined to be a monotone non-increasing function that approximates the service degradation. Figure 14.7 shows A_{ss} and P_{loss} for both policies plotted against T for different values of the mean time to perform rejuvenation γ_{R}. Under both policies, it can be seen that, for any particular value of T, the higher the value of γ_{R}, the lower is the availability and the higher is the corresponding loss probability. It can also be observed that the value of T which minimizes probability of loss is much lower than that which maximizes availability. In fact, the probability of loss becomes very high at values of T that maximize availability. For any specific value of γ_{R}, Policy II results in a lower minimum in loss probability than that achieved under Policy I. Therefore, if the objective is to minimize long run probability of loss, such as in the case of telecommunication switching software, Policy II always fares better than Policy I.

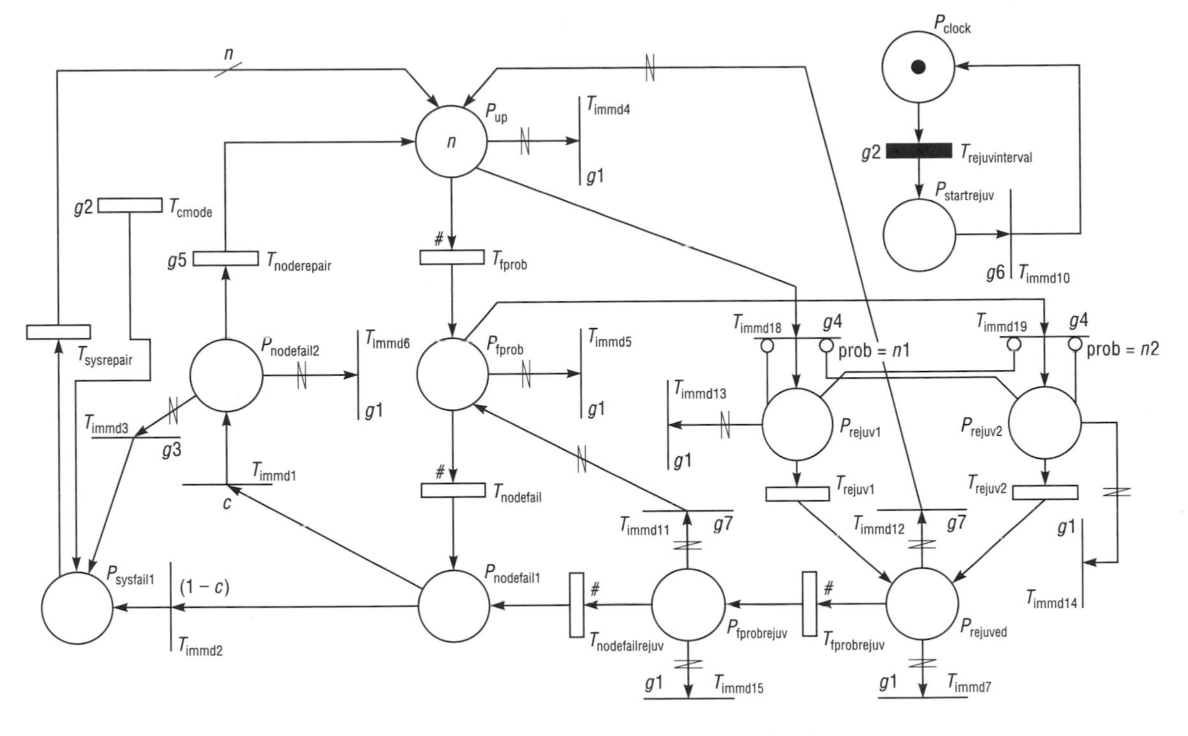

Figure 14.8. SRN model of a cluster system employing time-based rejuvenation

14.2.3 Examples in a Cluster System

In this section we discuss software rejuvenation as applied to cluster systems [16]. This is a novel application, which significantly improves cluster system availability and productivity. Cluster computing [32] has been increasingly popular in the last decade, and is being widely used in various kinds of applications, many of which are not tolerant to system failures and the resulting loss of service. By coupling software rejuvenation with clustering, significant increases in cluster system availability and performance can be achieved. Taking advantage of node failovers in a cluster, one can still maintain operation (though possibly at a degraded level) by rejuvenating one node at a time. The main assumptions here are that a node rejuvenation takes less time to perform, is less disruptive, and is less expensive than recovery from an unplanned node failure. Simple time-based rejuvenation policies, in which the nodes are rejuvenated at regular intervals, can be implemented easily. The cluster system availability and service level can be further enhanced by taking a more proactive approach to detect and predict an impending outage of a specific server in order to initiate planned failover in a more orderly fashion. This approach not only improves the end user's perception of service provided by the system, but also gives the system administrator additional time to work around any system capacity issues that may arise.

The stochastic reward net (SRN) model of a cluster system employing simple time-based rejuvenation is shown in Figure 14.8. The cluster consists of n nodes that are initially in a "robust" working state P_{up}. The aging process is modeled as a two-stage hypo-exponential distribution (increasing failure rate) with transitions T_{fprob} and $T_{noderepair}$. Place P_{fprob} represents a "failure-probable" state in which the nodes are still operational. The nodes then can eventually transit

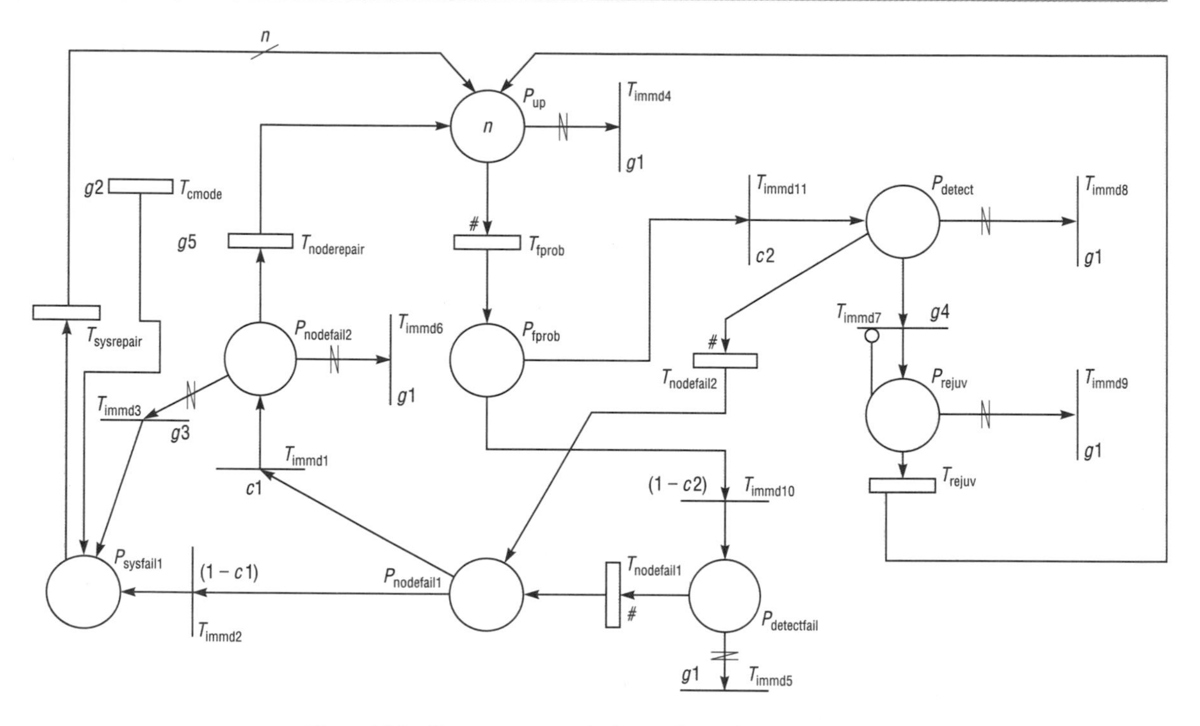

Figure 14.9. Cluster system employing prediction-based rejuvenation

to the fail state, $P_{\text{nodefail1}}$. A node can be repaired through the transition $T_{\text{noderepair}}$, with a coverage c. In addition to individual node failures, there is also a common-mode failure (transition T_{cmode}). The system is also considered down when there are a ($a \leq n$) individual node failures. The system is repaired through the transition $T_{\text{sysrepair}}$.

In the simple time-based policy, rejuvenation is done successively for all the operational nodes in the cluster, at the end of each deterministic interval. The transition $T_{\text{rejuvinterval}}$ fires every d time units, depositing a token in place $P_{\text{startrejuv}}$. Only one node can be rejuvenated at any time (at places P_{rejuv1} or P_{rejuv2}). Weight functions are assigned such that the probability of selecting a token from P_{up} or P_{fprob} is directly proportional to the number of tokens in each. After a node has been rejuvenated, it goes back to the "robust" working state, represented by place P_{rejuved}. This is a duplicate place for P_{up} in order to distinguish the nodes that are waiting to be rejuvenated from the nodes that have already been rejuvenated. A node, after rejuvenation, is then allowed to fail with the same rates as before rejuvenation, even when another node is being rejuvenated. Duplicate places for P_{up} and P_{fprob} are needed to capture this. Node repair is disabled during rejuvenation. Rejuvenation is complete when the sum of nodes in places P_{rejuved}, $P_{\text{fprobrejuv}}$ and $P_{\text{nodefail2}}$ is equal to the total number of nodes n. In this case, the immediate transition T_{immd10} fires, putting back all the rejuvenated nodes in places P_{up} and P_{fprob}. Rejuvenation stops when there are $a - 1$ tokens in place $P_{\text{nodefail2}}$, to prevent a system failure. The clock resets itself when rejuvenation is complete and is disabled when the system is undergoing repair. Guard functions ($g1$ through $g7$) are assigned to express complex enabling conditions textually.

In prediction-based rejuvenation (Figure 14.9), rejuvenation is attempted only when a node transits into the "failure probable" state. In practice,

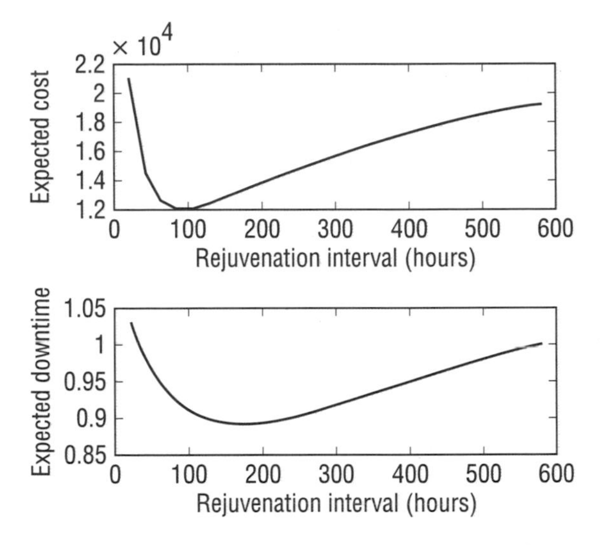

Figure 14.10. Time-based rejuvenation for 8/1 configuration

this degraded state could be predicted in advance by means of analyses of some observable system parameters [33]. In the case of a successful prediction, assuming that no other node is being rejuvenated at that time, the newly detected node can be rejuvenated. A node is allowed to fail even while waiting for rejuvenation.

For the analyses, the following values are assumed. The mean times spent in places P_{up} and P_{fprob} are 240 h and 720 h respectively. The mean times to repair a node, to rejuvenate a node, and to repair the system are 30 min, 10 min and 4 h respectively. In this analysis, the common-mode failure is disabled and node failure coverage is assumed to be perfect. All the models were solved using the Stochastic Petri Net Package (SPNP) tool. The measures computed were expected unavailability and the expected cost incurred over a fixed time interval. It is assumed that the cost incurred due to node rejuvenation is much less than the cost of a node or system failure, since rejuvenation can be done at predetermined or scheduled times. In our analysis, we fix the value for $cost_{nodefail}$ at \$5000/h, the $cost_{rejuv}$ at \$250/h. The value of $cost_{sysfail}$ is computed as the number of nodes n times $cost_{nodefail}$.

Figure 14.10 shows the plots for an 8/1 configuration (eight nodes including one spare) system employing simple time-based rejuvenation. The upper and lower plots show the expected cost incurred and the expected downtime (in hours) respectively in a given time interval, versus rejuvenation interval (time between successive rejuvenation) in hours. If the rejuvenation interval is close to zero, the system is always rejuvenating and thus incurs high cost and downtime. As the rejuvenation interval increases, both expected unavailability and cost incurred decrease and reach an optimum value. If the rejuvenation interval goes beyond the optimal value, the system failure has more influence on these measures than rejuvenation. The analysis was repeated for 2/1, 8/2, 16/1 and 16/2 configurations. For time-based rejuvenation, the optimal rejuvenation interval was 100 h for the one-spare clusters, and approximately 1 h for the two-spare clusters. In our analysis of predictive rejuvenation, we assumed 90% prediction coverage. For systems that have one spare, time-based rejuvenation can reduce downtime by 26% relative to no rejuvenation. Predictive rejuvenation does somewhat better, reducing downtime by 62% relative to no rejuvenation. However, when the system can tolerate more than one failure at a time, downtime is reduced by 98% to 95% via time-based rejuvenation, compared with a mere 85% for predictive rejuvenation.

14.3 Measurement-based Estimation

In this section we describe the measurement-based approach for detection and validation of the existence of software aging. The basic idea is to monitor periodically and collect data on the attributes responsible for determining the health of the executing software, in this case the UNIX operating system. The SNMP-based distributed resource monitoring tool discussed by Garg *et al.* [33] was used to collect operating system resource usage and system activity data from nine heterogeneous UNIX workstations connected by an Ethernet LAN at the Duke Department of Electrical and Computer Engineering. A central monitoring station runs the manager program which sends *get* requests periodically to each of the agent programs running on the monitored workstations. The agent programs, in turn, obtain data for the manager from their respective machines by executing various standard UNIX utility programs like *pstat*, *iostat* and *vmstat*. For quantifying the effect of aging in operating system resources, the metric *Estimated time to exhaustion* is proposed. The earlier work [33] used a purely time-based approach to estimate resource exhaustion times, whereas the work presented by Vaidyanathan and Trivedi [34] takes into account the current system workload as well.

14.3.1 Time-based Estimation

Data were collected from the machines at intervals of 15 min for about 53 days. Time-ordered values for each monitored object are obtained, constituting a time series for that object. The objective is to detect aging or a long-term trend (increasing or decreasing) in the values. Only results for the data collected from the machine Rossby are discussed here.

First, the trends in operating system resource usage and system activity are detected using *smoothing* of observed data by *robust locally weighted regression*, proposed by Cleveland [35]. This technique is used to get the global trend between outages by removing the local variations. Then, the slope of the trend is estimated in order to do prediction. Figure 14.11 shows the smoothed data superimposed on the original data points from the time series of objects for Rossby. The amount of *real memory free* (plot 1) shows an overall decrease, whereas *file table size* (plot 2) shows an increase. Plots of some other resources not discussed here also showed an increase or decrease. This corroborates the hypothesis of aging with respect to various objects.

The seasonal Kendall test [36] was applied to each of these time series to detect the presence of any global trends at a significance level α of 0.05.

Table 14.1. Seasonal Kendall test

Resource name	Rossby
Real memory free	−13.668
File table size	38.001
Process table size	40.540
Used swap space	15.280
No. of disk data blocks	48.840
No. of queues	39.645

The associated statistics are listed in Table 14.1. With $Z_\alpha = 1.96$, all values in the table are such that the null hypothesis H_0 that no trend exists is rejected.

Given that a global trend is present and that its slope is calculated for a particular resource, the time at which the resource will be exhausted because of aging only is estimated. Table 14.2 refers to several objects on Rossby and lists an estimate of the slope (change per day) of the trend obtained by applying Sen's slope estimate for data with seasons [33]. The values for real memory and swap space are in kilobytes. A negative slope, as in the case of *real memory*, indicates a decreasing trend, whereas a positive slope, as in the case of *file table size*, is indicative of an increasing trend. Given the slope estimate, the table lists the estimated time to failure of the machine due to aging only with respect to this particular resource. The calculation of the time to exhaustion is done by using the standard linear approximation $y = mx + c$.

A comparative effect of aging on different system resources can be obtained from the above estimates. Overall, it was found that *file table size* and *process table size* are not as important as *used swap space* and *real memory free*, since they have a very small slope and high estimated times to failure due to exhaustion. Based on such comparisons, we can identify important resources to monitor and manage in order to deal with aging-related software failures. For example, the resource *used swap space* has the highest slope and *real memory free* has the second highest slope. However, *real memory free* has a lower time to exhaustion than *used swap space*.

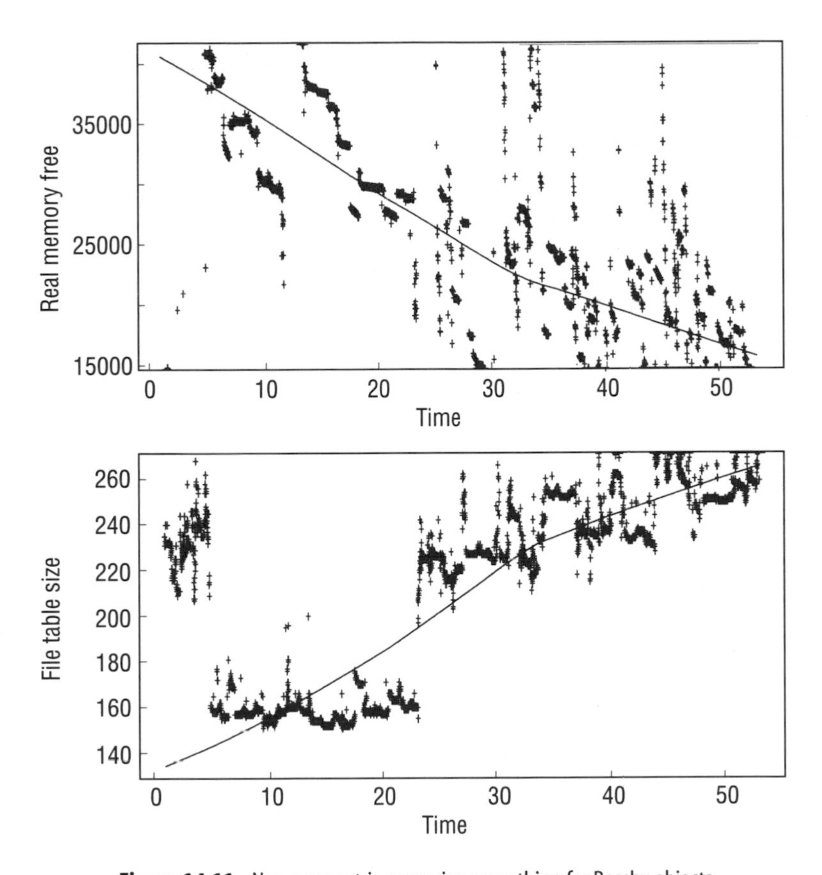

Figure 14.11. Non-parametric regression smoothing for Rossby objects

Table 14.2. Estimated slope and time to exhaustion for Rossby objects

Resource name	Initial value	Max value	Sen's slope estimation	Estimated time to exhaustion
Real memory free	40 814.17	84 980	−252.00	161.96
File table size	220	7110	1.33	5167.50
Process table size	57	2058	0.43	4602.30
Used swap space	39 372	312 724	267.08	1023.50

Table 14.3. Statistics for the workload clusters

No.	Cluster center				% of pts
	cupConSw	sysCall	pgOut	pgIn	
1	48 405.16	94 194.66	5.16	677.83	0.98
2	54 184.56	122 229.68	5.39	81.41	0.98
3	34 059.61	193 927.00	0.02	136.73	0.93
4	20 479.21	45 811.71	0.53	243.40	1.89
5	21 361.38	37 027.41	0.26	12.64	7.17
6	15 734.65	54 056.27	0.27	14.45	6.55
7	37 825.76	40 912.18	0.91	12.21	11.77
8	11 013.22	38 682.46	0.03	10.43	42.87
9	67 290.83	37 246.76	7.58	19.88	4.93
10	10 003.94	32 067.20	0.01	9.61	21.23
11	197 934.42	67 822.48	415.71	184.38	0.93

14.3.2 Time and Workload-based Estimation

The method discussed in Section 14.3.1 assumes that accumulated use of a resource over a time period depends only on the elapsed time. However, it is intuitive that the rate at which a resource is consumed is dependent on the current workload. In this subsection, we discuss a measurement-based model to estimate the rate of exhaustion of operating system resources as a function of both time and the system workload presented by Vaidyanathan and Trivedi [34]. The SNMP-based distributed resource monitoring tool described previously was used for collecting operating system resource usage and system activity parameters (at 10 min intervals) for over 3 months. The *longest* stretch of sample points in which no reboots or failures occurred were used for building the model. A semi-Markov reward model was constructed using the data. First, different workload states were identified using statistical cluster analysis and a state-space model was constructed. Corresponding to each resource, a reward function based on the rate of resource exhaustion in the different states was then defined. Finally, the model is solved to obtain trends and time to exhaustion for the resources.

The following variables were chosen to characterize the system workload: *cpuContextSwitch*, *sysCall*, *pageIn*, and *pageOut*. Hartigan's k-means clustering algorithm [37] was used for partitioning the data points into clusters based on workload. The statistics for the 11 workload clusters obtained are shown in Table 14.3. Clusters whose centroids were relatively close to each other, and those with a small percentage of data points in them, were merged to simplify computations. The resulting clusters are $W_1 = \{1, 2, 3\}$, $W_2 = \{4, 5\}$, $W_3 = \{6\}$, $W_4 = \{7\}$, $W_5 = \{8\}$, $W_6 = \{9\}$, $W_7 = \{10\}$, and $W_8 = \{11\}$.

Transition probabilities from one state to another were computed from data, resulting in transition probability matrix $\mathbf{P}$ of the embedded DTMC shown below:

$$\mathbf{P} = \begin{bmatrix} 0.00 & 0.16 & 0.22 & 0.13 & 0.26 & 0.03 & 0.17 & 0.03 \\ 0.07 & 0.00 & 0.14 & 0.14 & 0.32 & 0.03 & 0.31 & 0.00 \\ 0.12 & 0.26 & 0.00 & 0.10 & 0.43 & 0.00 & 0.11 & 0.02 \\ 0.15 & 0.36 & 0.06 & 0.00 & 0.10 & 0.22 & 0.09 & 0.03 \\ 0.03 & 0.07 & 0.04 & 0.01 & 0.00 & 0.00 & 0.85 & 0.00 \\ 0.07 & 0.16 & 0.02 & 0.54 & 0.12 & 0.00 & 0.02 & 0.07 \\ 0.02 & 0.05 & 0.00 & 0.00 & 0.92 & 0.00 & 0.00 & 0.00 \\ 0.31 & 0.08 & 0.15 & 0.23 & 0.08 & 0.15 & 0.00 & 0.00 \end{bmatrix}$$

Table 14.4. Sojourn time distributions

State	Sojourn time distribution $F(t)$
W_1	$1 - 1.602\,919\,\mathrm{e}^{-0.9t} + 0.602\,9185\,\mathrm{e}^{-2.392\,739t}$
W_2	$1 - 0.9995\,\mathrm{e}^{-0.44\,9902t} - 0.0005\,\mathrm{e}^{-0.007\,110\,07t}$
W_3	$1 - 0.9952\,\mathrm{e}^{-0.327\,4977t} - 0.0048\,\mathrm{e}^{-0.017\,5027t}$
W_4	$1 - 0.841\,362\,\mathrm{e}^{-0.327\,5372t} - 0.158\,638\,\mathrm{e}^{-0.038\,254\,29t}$
W_5	$1 - 1.425\,856\,\mathrm{e}^{-0.56t} + 0.425\,8555\,\mathrm{e}^{-1.875t}$
W_6	$1 - 0.806\,94\,\mathrm{e}^{-0.550\,9307t} - 0.193\,06\,\mathrm{e}^{-0.037\,057\,56t}$
W_7	$1 - 2.865\,33\,\mathrm{e}^{-1.302t} + 1.865\,33\,\mathrm{e}^{-2t}$
W_8	$1 - 0.9883\,\mathrm{e}^{-265\,5196t} - 0.0117\,\mathrm{e}^{-0.027\,101\,47t}$

Table 14.5. Slop estimates (in Kb/10 min)

State	usedSwapSpace		realMemoryFree	
	Slope estimate	95% confidence interval	Slope estimate	95% confidence interval
W_1	119.3	5.5 to 222.4	−133.7	−137.7 to −133.3
W_2	0.57	0.40 to 0.71	−1.47	−1.78 to −1.09
W_3	0.76	0.73 to 0.80	−1.43	−2.50 to −0.62
W_4	0.57	0.00 to 0.69	−1.23	−1.67 to −0.80
W_5	0.78	0.75 to 0.80	0.00	−5.65 to −6.00
W_6	0.81	0.64 to 1.00	−1.14	−1.40 to −0.88
W_7	0.00	0.00 to 0.00	0.00	0.00 to 0.00
W_8	91.8	72.4 to 111.0	91.7	−369.9 to 475.2

The sojourn time distribution for each of the workload states was fitted to either two-stage hyper-exponential or two-stage hypo-exponential distribution functions. The fitted distributions, shown in Table 14.4, were tested using the Kolmogorov–Smirnov test at a significance level of 0.01.

Two resources, *usedSwapSpace* and *realMemoryFree*, are considered for the analysis, since the previous time-based analysis suggested that they are critical resources. For each resource, the reward function is defined as the rate of corresponding resource exhaustion in different states. The true slope (rate of increase/decrease) of a resource at every workload state is estimated by using Sen's non-parametric method [34]. Table 14.5 shows the slopes with 95% confidence intervals.

It was observed that slopes in a given workload state for a particular resource during different visits to that state are almost the same. Further, the slopes across different workload states are different and, generally, the higher the system activity, then the higher is the resource utilization. This validates the assumption that resource usage *does* depend on the system workload and the rates of exhaustion vary with workload changes. It can also be observed from Table 14.5 that the slopes for *usedSwapSpace* in all the workload states are non-negative, and the slopes for *realMemoryFree* are non-positive in all the workload states except in one. It follows that *usedSwapSpace* increases, whereas *realMemoryFree* decreases, over time, which validates the software aging phenomenon.

The semi-Markov reward model was solved using the SHARPE [38]. The slope for the workload-based estimation is computed as the expected reward rate in steady state from the model. As in the case of time-based estimation, the times to resource exhaustion are computed using the linear formula $y = mx + c$. Table 14.6 gives the estimates for the slope and time to exhaustion for *usedSwapSpace* and *realMemoryFree*. It can be

Table 14.6. Estimates for slope (in KB/10 min) and time to exhaustion (in days)

Estimation method	usedSwapSpace		realMemoryFree	
	Slope estimate	Estimated time to exhaustion	Slope estimate	Estimated time to exhaustion
Time based	0.787	2276.46	−2.806	60.81
Workload based	4.647	453.62	−3.386	50.39

seen that workload-based estimations gave a lower time to resource exhaustion than those computed using time-based estimations. Since the machine failures due to resource exhaustion were observed much before the times to resource exhaustion estimated by the time-based method, it follows that the workload-based approach results in better estimations.

14.4 Conclusion and Future Work

In this chapter we have motivated the need for pursuing preventive maintenance in operational software systems on a scientific basis rather than on the current *ad hoc* practice. Thus, an important research issue is to determine the optimal times to perform the preventive maintenance. In this regard, we discuss two possible approaches: analytical modeling, and a measurement-based approach.

In the first part, we discuss analytical models for evaluating the effectiveness of preventive maintenance in operational software systems that experience aging. The aim of the analytical modeling approach is to determine the optimal times to perform rejuvenation. The latter part of the chapter deals with a measurement-based approach for detection and validation of the existence of software aging. Although the SNMP-based distributed resource monitoring tool is specific to UNIX, the methodology can also be used for detection and estimation of aging in other software systems.

In future, a unified approach, by taking account of both analytical modeling and measurement-based rejuvenations, should be established. Then, the probability distribution function of the software failure occurrence time has to be modeled in a consistent way. In other words, the stochastic models to describe the physical failure phenomena in operational software systems should be developed under a random usage environment.

Acknowledgments

This material is based upon the support by the Department of Defense, Alcatel, Telcordia, and Aprisma Management Technologies via a CACC Duke core project. The first author is also supported in part by the Telecommunication Advancement Foundation, Tokyo, Japan.

References

[1] Adams E. Optimizing preventive service of the software products. IBM J Res Dev 1984;28:2–14.

[2] Huang Y, Kintala C, Kolettis N, Fulton ND. Software rejuvenation: analysis, module and applications. In: Proceedings of 25th International Symposium on Fault Tolerant Computer Systems. Los Alamitos: IEEE CS Press; 1995. p.381–90.

[3] Lin F. Re-engineering option analysis for managing software rejuvenation. Inform Software Technol 1993;35:462–467.

[4] Parnas DL. Software aging. In: Proceedings of 16th International Conference on Software Engineering. New York: ACM Press; 1994. p.279–87.

[5] Huang Y, Kintala C. Software fault tolerance in the application layer. In: Lyu MR, editor. Software fault tolerance. Chichester: John Wiley & Sons; 1995. p.231–48

[6] Huang Y, Jalote P, Kintala C. Two techniques for transient software error recovery. In: Banatre M, Lee PA, editors. Hardware and software architectures for fault tolerance: experience and perspectives. Springer Lecture Note in Computer Science, vol. 774. Berlin: Springer-Verlag; 1995. p.159–70.

[7] Jalote P, Huang Y, Kintala C. A framework for understanding and handling transient software failures. In: Proceedings of 2nd ISSAT International Conference on Reliability and Quality in Design, 1995; p.231–7.

[8] Avritzer A, Weyuker EJ. Monitoring smoothly degrading systems for increased dependability. Empirical Software Eng J 1997;2:59–77.

[9] Levendel Y. The cost effectiveness of telecommunication service. In: Lyu MR, editor. Software fault tolerance. Chichester: John Wiley & Sons; 1995. p.279–314.

[10] Marshall E. Fatal error: how Patriot overlooked a Scud. Science 1992;255:1347.

[11] Gray J, Siewiorek DP. High-availability computer systems. IEEE Comput 1991;9:39–48.

[12] Trivedi KS, Vaidyanathan K, Goševa-Postojanova K. Modeling and analysis of software aging and rejuvenation. In: Proceedings of 33rd Annual Simulation Symposium. Los Alamitos: IEEE CS Press; 2000. p.270–9.

[13] Tai AT, Chau SN, Alkalaj L, Hecht H. On-board preventive maintenance: analysis of effectiveness and optimal duty period. In: Proceedings of 3rd International Workshop on Object Oriented Real-Time Dependable Systems. Los Alamitos: IEEE CS Press; 1997. p.40–7.

[14] Tai AT, Alkalaj L, Chau SN. On-board preventive maintenance for long-life deep-space missions: a model-based evaluation. In: Proceedings of 3rd IEEE International Computer Performance & Dependability Symposium. Los Alamitos: IEEE CS Press; 1998. p.196–205.

[15] Tai AT, Alkalaj L, Chau SN. On-board preventive maintenance: a design-oriented analytic study for long-life applications. Perform Eval 1999;35:215–32.

[16] Vaidyanathan K, Harper RE, Hunter SW, Trivedi KS. Analysis and implementation of software rejuvenation in cluster systems. In: Proceedings of Joint International Conference on Measurement and Modeling of Computer Systems, ACM SIGMETRICS 2001/Peformance 2001. New York: ACM; 2001. p.62–71.

[17] IBM Netfinity director software rejuvenation–white paper. URL: http://www.pc.ibm.com/us/techlink/wtpapers/.

[18] Dohi T, Goševa-Popstojanova K, Trivedi KS. Analysis of software cost models with rejuvenation. In: Proceedings of 5th IEEE International Symposium High Assurance Systems Engineering. Los Alamitos: IEEE CS Press; 2000. p.25–34.

[19] Dohi T, Goševa-Popstojanova K, Trivedi KS. Statistical non-parametric algorithms to estimate the optimal software rejuvenation schedule. In: Proceedings of 2000 Pacific Rim International Symposium Dependable Computing. Los Alamitos: IEEE CS Press; 2000. p.77–84.

[20] Dohi T, Goševa-Popstojanova K, Trivedi KS. Estimating software rejuvenation schedule in high assurance systems. Comput J 2001;44:473–85.

[21] Garg G, Puliafito A, Trivedi KS. Analysis of software rejuvenation using Markov regenerative stochastic Petri net. In: Proceedings of 6th International Symposium on Software Reliability Engineering. Los Alamitos: IEEE CS Press; 1995. p.180–7.

[22] Bobbio A, Sereno M. Fine grained software rejuvenation models. In: Proceedings of 3rd IEEE International Computer Performance & Dependability Symposium. Los Alamitos: IEEE CS Press; 1998. p.4–12.

[23] Garg S, Huang Y, Kintala C, Trivedi KS. Minimizing completion time of a program by checkpointing and rejuvenation. In: Proceedings of 1996 ACM SIGMETRICS Conference. New York: ACM; 1996. p.252–61.

[24] Bobbio A, Garg S, Gribaudo M, Horvaáth A, Sereno M, Telek M. Modeling software systems with rejuvenation, restoration and checkpointing through fluid stochastic Petri nets. In: Proceedings of 8th International Workshop on Petri Nets and Performance Models. Los Alamitos: IEEE CS Press; 1999. p.82–91.

[25] Garg S, Huang Y, Kintala C, Trivedi KS. Time and load based software rejuvenation: policy, evaluation and optimality. In: Proceedings of 1st Fault-Tolerant Symposium, 1995. p.22–5.

[26] Pfening A, Garg S, Puliafito A, Telek M, Trivedi KS. Optimal rejuvenation for tolerating soft failures. Perform Eval 1996;27–28:491–506.

[27] Garg S, Puliafito A, Telek M, Trivedi KS. Analysis of preventive maintenance in transactions based software systems. IEEE Trans Comput 1998;47:96–107.

[28] Okamura H, Fujimoto A, Dohi T, Osaki S, Trivedi KS. The optimal preventive maintenance policy for a software system with multi server station. In: Proceedings of 6th ISSAT International Conference on Reliability and Quality in Design, 2000; p.275–9.

[29] Okamura H, Miyahara S, Dohi T, Osaki S. Performance evaluation of workload-based software rejuvenation scheme. IEICE Trans Inform Syst D 2001;E84-D: 1368-75.

[30] Ross SM. Applied probability models and optimization applications. San Francisco (CA): Holden-Day; 1970.

[31] Barlow RE, Campo R. Total time on test processes and applications to failure data. In: Barlow RE, Fussell J, Singpurwalla ND, editors. Reliability and fault tree analysis. Philadelphia: SIAM; 1975. p.451–81.

[32] Pfister G. In search of clusters: the coming battle in lowly parallel computing. Englewood Cliffs (NJ): Prentice-Hall; 1998.

[33] Garg S, van Moorsel A, Vaidyanathan K, Trivedi KS. A methodology for detection and estimation of software aging. In: Proceedings of 9th International Symposium on Software Reliability Engineering. Los Alamitos: IEEE CS Press; 1998. p.282–92

[34] Vaidyanathan K, Trivedi KS. A measurement-based model for estimation of resource exhaustion in operational software systems. In: Proceedings of 10th IEEE International Symposium on Software Reliability Engineering. Los Alamitos: IEEE CS Press; 1999. p.84–93.

[35] Cleveland WS. Robust locally weighted regression and smoothing scatterplots. J Am Stat Assoc 1979;74:829–36.

[36] Gilbert RO. Statistical methods for environmental pollution monitoring. New York (NY):Van Nostrand Reinhold; 1987.

[37] Hartigan JA. Clustering algorithms. New York (NY):John Wiley & Sons; 1975.

[38] Sahner RA, Trivedi KS, Puliafito A. Performance and reliability analysis of computer systems–an example-based approach using the SHARPE software package. Norwell (MA): Kluwer Academic Publishers; 1996.

Software Reliability Management: Techniques and Applications

Mitsuhiro Kimura and Shigeru Yamada

Chapter 15

15.1 Introduction

In this chapter we discuss three new stochastic models for assessing software reliability and the degree of software testing progress, and a software reliability management tool.

Precise software reliability assessment is necessary to measure and predict the reliability and performance of a developed software product. Also, it has become one of the urgent issues in software engineering to develop quality software products and increase the productivity of their development processes [1–5]. In order to solve such software quality management issues, a number of software reliability assessment models have been developed by many researchers over the last two decades [2–5]. Among these models, those that have been built based on some stochastic processes have recently become to be known gradually by software development managers and practitioners.

Section 15.2 considers a model that describes the degree of software testing progress by focusing on the consumption process of test cases during the software testing phase. Measuring the software testing progress may also play a significant role in software development management, because one of the most important factors is traceability in the software development process. To evaluate the degree of the testing progress, we construct a generalized death process model that is able to analyze the data sets of testing time and the number of test cases consumed during the testing phase. The model predicts the mean number of the consumed test cases by extrapolation and estimates the mean completion time of the testing phase. We show several numerical illustrations using the actual data.

In Section 15.3, we propose a new approach to estimating the software reliability assessment model considering the imperfect debugging environment by analyzing the traditional software reliability data sets. Generally, stochastic models for software reliability assessment need several assumptions, and some of them are too strict to describe the actual phenomena. In the research area of software reliability modeling, it is known that the assumption of perfect debugging is very useful in simplifying the models, but many software quality/reliability researchers and practitioners are critical in terms of assuming perfect debugging in a software testing environment. Considering such a background, several imperfect debugging software reliability assessment models have been developed [6–12]. However, it is found that the really practical methods for estimating the imperfect debugging probability have not been proposed yet. We show that the hidden-Markov estimation method [13] enables one to estimate the imperfect debugging probability by only using the traditional data sets.

In Section 15.4, we consider a state dependency in the software reliability growth process. In the literature, almost all of the software reliability assessment models based on stochastic processes are continuous time and discrete state space models. In particular, the models based on a non-homogeneous Poisson process (NHPP) are widely known as software reliability growth models and have actually been implemented in several computer-aided software engineering (CASE) tools [4, 5]. These NHPP models can be easily applied to the observed data sets and expanded to describe several important factors which have an effect on software debugging processes in the actual software testing phase. However, the NHPP models essentially have a linear relation on the state dependency; this is a doubtful assumption with regard to actual software reliability growth phenomena. That is, these models have no choice but to use a mean value function for describing a nonlinear relation between the cumulative number of detected faults and the fault detection rate. Thus we present a continuous state space model to introduce directly the nonlinearity of the state dependency of the software reliability growth process.

Finally, in Section 15.5, we introduce a prototype of a software reliability management tool that includes the models described in this chapter. This tool has been implemented using JAVA and object-oriented analysis.

15.2 Death Process Model for Software Testing Management

In this section we focus on two phenomena that are observable during the software testing phase, which is the last phase of a large-scale software development process. The first is the software failure-occurrence or fault-detection process, and the second is the consumption process of test cases which are provided previously and consumed to examine the system test of the software system implemented. To evaluate the quality of the software development process or the software product itself, we describe a non-homogeneous death process [14] with a state-dependent parameter α $(0 \leq \alpha \leq 1)$.

After the model description, we derive several measures for quantitative assessment. We show that this model includes Shanthikumar's binomial reliability model [15] as a special case. The estimation method of maximum likelihood for the unknown parameters is also shown. Finally, we illustrate several numerical examples for software testing-progress evaluation by applying the actual data sets to our model, and we discuss the applicability of the model.

15.2.1 Model Description

We first consider a death process [16, 17] $\{X(t),\ t \geq 0\}$, which has the initial population K $(K \geq 1)$ and satisfies the following assumptions:

- $\Pr[X(t+\Delta t) = x-1 \mid X(t) = x] = (x\alpha + 1 - \alpha)\phi(t)\Delta t + o(\Delta t)$
- $\Pr[X(t) - X(t+\Delta t) \geq 2] = o(\Delta t)$
- $\Pr[X(0) = K] = 1$
- $\{X(t),\ t \geq 0\}$ has independent increments.

The term *population* has two meanings in this section. The first one is the number of software faults that are latent in the software system, if our new model describes the software fault-detection or failure-occurrence process. The second one means the number of test cases for the testing phase, if the model gives a description of the consumption process of the test cases.

In the above assumptions, the positive function $\phi(t)$ is called an intensity function, and α $(0 \leq \alpha \leq 1)$ is a constant parameter and affects the state-dependent property of the process $\{X(t),\ t \geq 0\}$. That is, the state transition probability $\Pr[X(t+\Delta t) = x-1 \mid X(t) = x]$ does not depend on the state x at time t but the intensity function $\phi(t)$, if $\alpha = 0$. On the other hand, when $\alpha = 1$, the process depends on residual population x and $\phi(t)$ at time t. This model describes a generalized non-homogeneous death process. We consider that the residual population at time t, $X(t)$, represents the number of remaining software faults or remaining test cases at the testing time t.

Under the assumptions listed above, we can obtain the probability that the cumulative number of deaths is equal to n at time t, $\Pr[X(t) = K - n] = P_n(t \mid \alpha)$, as follows:

$$P_n(t \mid \alpha) = \frac{\prod_{i=0}^{n-1}[(K-i)\alpha + 1 - \alpha]}{n!} \times [\mathrm{e}^{-\alpha G(t)}]^{K-n-1} \times \left[\frac{1-\mathrm{e}^{-\alpha G(t)}}{\alpha}\right]^n \mathrm{e}^{-G(t)} \quad (n = 0, 1, 2, \ldots, K-1) \tag{15.1}$$

$$P_K(t \mid \alpha) = 1 - \sum_{n=0}^{K-1} P_n(t \mid \alpha)$$

where

$$G(t) = \int_0^t \phi(x)\,\mathrm{d}x \tag{15.2}$$

and we define

$$\prod_{i=0}^{-1}[(K-i)\alpha + 1 - \alpha] = 1 \tag{15.3}$$

In the above, the term *death* means the elimination of a software fault or the consumption of a testing case during the testing phase.

In Equation (15.1), $\lim_{\alpha\to 0} P_n(t \mid \alpha)$ yields

$$P_n(t \mid 0) = \frac{G(t)^n}{n!}\mathrm{e}^{-G(t)} \quad (n = 0, 1, 2, \ldots, K-1) \tag{15.4}$$

$$P_K(t \mid 0) = \sum_{i=K}^{\infty} \frac{G(t)^i}{i!}\mathrm{e}^{-G(t)}$$

since

$$\lim_{\alpha\to 0} \frac{1-\mathrm{e}^{-\alpha G(t)}}{\alpha} = G(t) \tag{15.5}$$

Similarly, by letting $\alpha \to 1$ we obtain

$$P_n(t \mid 1) = \binom{K}{n}\mathrm{e}^{-G(t)(K-n)}(1-\mathrm{e}^{-G(t)})^n \quad (n = 0, 1, \ldots, K) \tag{15.6}$$

Equation 15.6 is a binomial reliability model by Shanthikumar [15].

Based on the above, we can derive several measures.

15.2.1.1 Mean Number of Remaining Software Faults/Testing Cases

The mean number of remaining software faults/testing cases at time t, $E[X(t)]$, is derived as

$$\begin{aligned} E[X(t)] &= K - \mathrm{e}^{-G(t)} \sum_{n=1}^{K} \frac{\Gamma(K+1/\alpha)}{\Gamma(K+1/\alpha-n)(n-1)!} \\ &\times [\mathrm{e}^{-\alpha G(t)}]^{K-n-1}[1-\mathrm{e}^{-\alpha G(t)}]^n \end{aligned} \quad (15.7)$$

if $0 < \alpha \leq 1$. To obtain Equation 15.7, we use the following equation:

$$\frac{\prod_{i=0}^{n-1}[(K-i)\alpha+1-\alpha]}{\alpha^n} = \frac{\Gamma(K+1/\alpha)}{\Gamma(K+1/\alpha-n)} \quad (15.8)$$

where $\Gamma(a)$ represents a gamma function defined as

$$\Gamma(a) = \int_0^\infty \mathrm{e}^{-x} x^{a-1}\,\mathrm{d}x \quad (15.9)$$

Also, Equation 15.7 can be simplified if $\alpha = 1$ as

$$E[X(t)]_{\alpha=1} = K\mathrm{e}^{-G(t)} \quad (15.10)$$

In the case of $\alpha = 0$, $E[X(t)]$ yields

$$\begin{aligned} E[X(t)]_{\alpha=0} &= KA(K, G(t)) \\ &\quad - G(t)A(K-1, G(t)) \end{aligned} \quad (15.11)$$

where $A(K, G(t))$ represents the incomplete gamma ratio which is defined as follows [14]:

$$A(K, G(t)) = \Gamma(K, G(t))/\Gamma(K, 0) \quad (15.12)$$

$$\Gamma(a, b) = \int_b^\infty \mathrm{e}^{-x} x^{a-1}\,\mathrm{d}x \quad (15.13)$$

15.2.1.2 Mean Time to Extinction

Let $B(t)$ be the cumulative distribution function (CDF) of the event that all of K are disappeared at time t. From Equation 15.1, $B(t)$ is represented as

$$B(t) = P_K(t \mid \alpha) = \sum_{n=K}^{\infty} P_n(t \mid \alpha) \quad (15.14)$$

Therefore, the mean time to extinction $E[T]$ is

$$E[T] = \int_0^\infty t\,\mathrm{d}B(t) \quad (15.15)$$

If the intensity function $\phi(t)$ is assumed by

$$\phi(t) = \lambda\beta t^{\beta-1} \quad (15.16)$$

two special cases of $E[T]$ are obtained as follows:

$$E[T]_{\alpha=0} = \frac{\Gamma(K+1/\beta)}{\Gamma(K)} \left(\frac{1}{\lambda}\right)^{1/\beta} \quad (15.17)$$

$$\begin{aligned} E[T]_{\alpha=1} &= \sum_{r=1}^{K} \left[\binom{K}{r}(-1)^{r+1}\left(\frac{1}{r}\right)^{1/\beta}\right] \\ &\times \left(\frac{1}{\beta}\right)\Gamma\left(\frac{1}{\beta}\right)\left(\frac{1}{\lambda}\right)^{1/\beta} \end{aligned} \quad (15.18)$$

15.2.2 Estimation Method of Unknown Parameters

We show the estimation method of unknown parameters included in the model except for K. We assume that a data set consists of time t_j and cumulative number of deaths y_j, *i.e.* (t_j, y_j) $(j = 1, 2, \ldots, m)$ and $t_1 < t_2 < \cdots < t_m$, $y_1 < y_2 < \cdots < y_m$, and the initial population K.

15.2.2.1 Case of $0 < \alpha \leq 1$

By using the data, the likelihood function l can be constructed as

$$\begin{aligned} l &= \Pr[X(t_1) = K - y_1, X(t_2) \\ &= K - y_2, \ldots, X(t_m) = K - y_m] \\ &= \prod_{j=1}^{m} \frac{\Gamma(K+1/\alpha-y_{j-1})}{(y_j - y_{j-1})!\Gamma(K+1/\alpha-y_j)} \\ &\times (\mathrm{e}^{-\alpha G(t_{j-1})} - \mathrm{e}^{-\alpha G(t_j)})^{y_j-y_{j-1}} \\ &\times (1 - \mathrm{e}^{-\alpha G(t_{j-1})} + \mathrm{e}^{-\alpha G(t_j)})^{K+(1/\alpha)-y_j-1} \end{aligned} \quad (15.19)$$

where $t_0 \equiv 0$, $y_0 \equiv 0$. From the above equation, the log-likelihood function L is represented as

$$\begin{aligned} L &\equiv \log l \\ &= \sum_{j=1}^{m} \log\left[\Gamma\left(K + \frac{1}{\alpha} - y_{j-1}\right)\right] \\ &\quad - \sum_{j=1}^{m} \log[(y_j - y_{j-1})!] \end{aligned}$$

$$-\sum_{j=1}^{m}\log\left[\Gamma\left(K+\frac{1}{\alpha}-y_j\right)\right]$$
$$+\sum_{j=1}^{m}(y_j-y_{j-1})\log[\mathrm{e}^{-\alpha G(t_{j-1})}-\mathrm{e}^{-\alpha G(t_j)}]$$
$$+\sum_{j=1}^{m}\left(K+\frac{1}{\alpha}-y_j-1\right)$$
$$\times\log[(1-\mathrm{e}^{-\alpha G(t_{j-1})}+\mathrm{e}^{-\alpha G(t_j)})] \quad (15.20)$$

If the intensity function is given by Equation 15.16, the simultaneous likelihood equations are shown as

$$\frac{\partial L(\alpha,\lambda,\beta)}{\partial\alpha}=0 \quad \frac{\partial L(\alpha,\lambda,\beta)}{\partial\lambda}=0$$
$$\frac{\partial L(\alpha,\lambda,\beta)}{\partial\beta}=0 \quad (15.21)$$

We can solve these equations numerically to obtain the maximum likelihood estimates.

15.2.2.2 Case of $\alpha = 0$

In this case, the likelihood function l and the log-likelihood L are rewritten respectively as

$$l\equiv\prod_{j=1}^{m}\frac{\int_{t_{j-1}}^{t_j}\phi(x)\,\mathrm{d}x^{y_j-y_{j-1}}}{(y_j-y_{j-1})!}$$
$$\times\exp\left[-\int_{t_{j-1}}^{t_j}\phi(x)\,\mathrm{d}x\right] \quad (15.22)$$

$$L=\log l=\sum_{j=1}^{m}(y_j-y_{j-1})\log[G(t_j)-G(t_{j-1})]$$
$$-\sum_{j=1}^{m}\log(y_j-y_{j-1})!-G(t_m) \quad (15.23)$$

Substituting Equation 15.16 into Equation 15.23, we have

$$L(\lambda,\beta)=y_m\log\lambda+\sum_{j=1}^{m}(y_j-y_{j-1})$$
$$\times\log(t_j^{\beta}-t_{j-1}^{\beta})$$
$$-\sum_{j=1}^{m}\log(y_j-y_{j-1})!-\lambda t_m^{\beta} \quad (15.24)$$

The simultaneous likelihood equations are obtained as follows:

$$\frac{\partial L(\lambda,\beta)}{\partial\lambda}=\frac{y_m}{\lambda}-t_m^{\beta}=0 \quad (15.25)$$
$$\frac{\partial L(\lambda,\beta)}{\partial\beta}=\sum_{j=1}^{m}(y_j-y_{j-1})$$
$$\times\frac{t_j^{\beta}\log t_j-t_{j-1}^{\beta}\log t_{j-1}}{t_j^{\beta}-t_{j-1}^{\beta}}$$
$$-y_m\log t_m=0 \quad (15.26)$$

By solving Equations 15.25 and 15.26 numerically, we obtain the maximum likelihood estimates for λ and β.

15.2.3 Software Testing Progress Evaluation

Up to the present, a lot of stochastic models have been proposed for software reliability/quality assessment. Basically, such models analyze the software fault-detection or failure-occurrence data which can be observed in the software testing phase. Therefore, we first consider the applicability of our model to the software fault-detection process.

The meaning of the parameters included in Equation 15.1 is as follows:

- K is the total number of software faults latent in the software system
- $\phi(t)$ is the fault detection rate at testing time t per fault
- α is the degree of the relation on the fault-detection rate affected by the number of remaining faults.

However, it is difficult for this model to analyze the software fault-detection data, because the estimation for the unknown parameter K is difficult unless $\alpha = 1$. We need to use the other estimation method for K.

We now focus on the analysis of the consumption process of test cases. The software development managers not only want to know the software reliability attained during the testing, but also the testing progress. In the testing phase of software development, many test cases provided previously are examined as to whether each of them satisfies the specification of the software product or not. When the software completes the task that is defined by each test case without causing any software failure, the test case is consumed. On the other hand, if a software failure occurs, the software faults that cause the software failure are detected and removed. In this situation, $\{X(t),\ t \geq 0\}$ can be treated as the consumption process of test cases. The meaning of the parameters is as follows:

- K is the total number of test cases provided
- $\phi(t)$ is the consumption rate of test case at time t per test case
- α is the degree of the relation on the consumption rate affected by the number of remaining test cases.

15.2.4 Numerical Illustrations

We show several numerical examples for software testing-progress evaluation. The data sets were actually obtained in the actual testing processes, and consist of testing time t_i, the cumulative number of consumed test cases y_i, and the total number of test cases K. The data sets are:

DS-2.1: (t_i, y_i) $(i = 1, 2, \ldots, 10)$, $K = 5964$. t_i is measured in days. This data appeared in Shibata [18].

DS-2.2: (t_i, y_i) $(i = 1, 2, \ldots, 19)$, $K = 11\,855$. t_i is measured in days. This data set was observed from a module testing phase and the modules were implemented by high-level programming language [14].

First, we estimate the three unknown parameters λ, β and α using DS-2.1. The results are $\hat{\lambda} = 1.764$, $\hat{\beta} = 1.563$, and $\hat{\alpha} = 1.000$. We illustrate the mean number of consumed test cases and the actual data in Figure 15.1: the small dots represent the realizations of the number of consumed test cases actually observed.

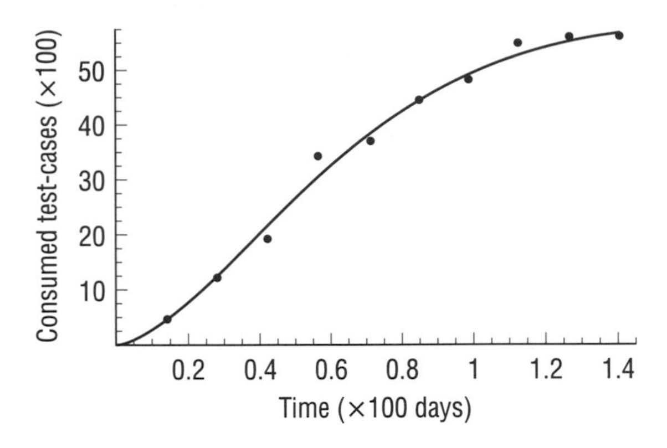

Figure 15.1. Estimated mean number of consumed test cases (DS-2.1; $\hat{\lambda} = 1.764$, $\hat{\beta} = 1.563$, $\hat{\alpha} = 1.000$)

Next, we discuss the predictability of our model in terms of the mean number of consumed test cases by using DS-2.2. That is, we estimate the model parameters by using the first half of DS-2.2 (*i.e.* $t_1 \rightarrow t_{10}$). The estimates are obtained as $\hat{\lambda} = 4.826$, $\hat{\beta} = 3.104$, and $\hat{\alpha} = 0.815$. Figure 15.2 shows the estimation result for DS-2.2. The right-hand side of the plot is the extrapolated prediction of the test case consumption. These data sets seem to fit our model.

Hence the software development managers may estimate the software testing-progress quantitatively by using our model.

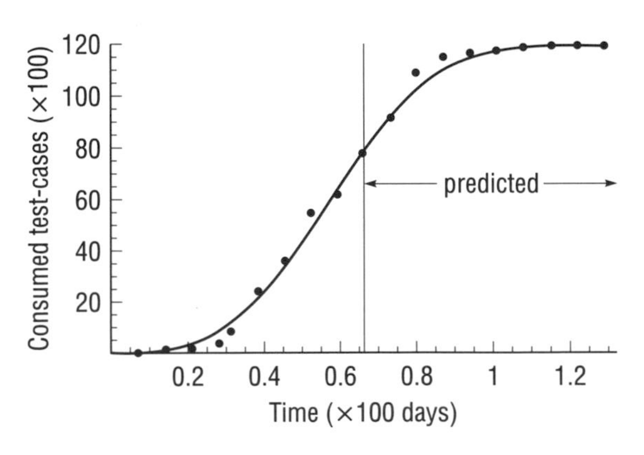

Figure 15.2. Estimated mean number of consumed test cases (DS-2.2; $\hat{\lambda} = 4.826$, $\hat{\beta} = 3.104$, $\hat{\alpha} = 0.815$)

15.2.5 Concluding Remarks

In this section, we formulated a non-homogeneous death process model with a state-dependent parameter, and derived the mean residual population at time t and the mean time to extinction. The estimation method of the unknown parameters included in the model has also been presented based on the method of maximum likelihood. We have considered the applicability of the model to two software quality assessment issues. As a result, it is found that our model is not suitable for assessing software reliability, since the parameter K, which represents the total number of software faults latent in the software system, is difficult to estimate. This difficulty remains as a future problem to be solved. However, our model is useful for such problems when K is known. Thus we have shown numerical examples of testing-progress evaluation by using the model and the actual data.

15.3 Estimation Method of Imperfect Debugging Probability

We study a simple software reliability assessment model taking an imperfect debugging environment into consideration in this section. The model was originally proposed by Goel and Okumoto [12]. We focus on developing a practical method for estimating the imperfect debugging software reliability assessment model.

In Section 15.3.1 we describe the modeling and the several assumptions required for it. We show the Baum–Welch re-estimation procedure, which is commonly applied in hidden-Markov modeling [13, 19], in Section 15.3.2. Section 15.3.3 presents several numerical illustrations based on the actual software fault-detection data set collected. We discuss some computational difficulties of the estimation procedure. Finally, in Section 15.3.4, we summarize with some remarks about the results obtained, and express several issues remaining for future study.

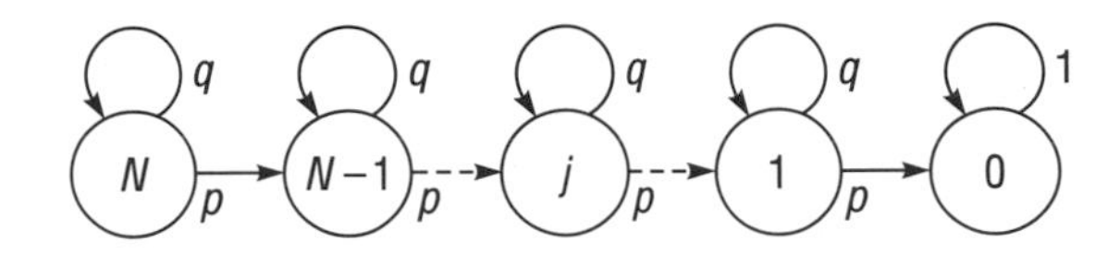

Figure 15.3. State transition diagram of the imperfect debugging model

15.3.1 Hidden-Markov modeling for software reliability growth phenomenon

In this section, we assume the following software testing and debugging environment.

- Software is tested by using test cases that are previously provided based on their specification requirements.
- One software failure is caused by one software fault.
- In the debugging process, a detected software fault is fixed correctly or done correctly but the debugging activity introduces one new fault into the other program path in terms of the test case.

In the usual software testing phase, software is inspected and debugged until a test case can be processed successfully. Consequently, when a test case is performed correctly by the software system, the program path of the test case must have no fault. In other words, if a software debugging worker fails to fix a fault, a new fault might be introduced into the other program path, and it may be detected by the other (later) test case, or not be detected at all.

Based on these assumptions, Goel and Okumoto [12] proposed a simple software reliability assessment model considering imperfect debugging environments. This model represents the behavior of the number of faults latent in the software system as a Markov model. The state transition diagram is illustrated in Figure 15.3. In this figure, the parameters p and q represent perfect and imperfect debugging probabilities, respectively $(p+q=1)$. The sojourn time distribution function of state j is defined as

$$F_j(t) = 1 - \exp[-\lambda j t] \quad (j = 0, 1, \ldots, N) \tag{15.27}$$

where j represents the number of remaining faults, λ is a constant parameter that corresponds to the software failure occurrence rate, and N is the total number of software faults. This simple distribution function was first proposed by Jelinski and Moranda [20]. Goel and Okumoto [12] discussed the method of statistical inference by the methods of maximum likelihood and Bayesian estimation. However, the former method, especially, needs data that includes information on whether each software failure is caused by a fault that is originally latent in the software system or by an introduced one. Generally, during the actual testing phase in the software development, it seems very difficult for testing managers to collect such data. We need to find another estimation method for the imperfect debugging probability.

In this section, we consider this imperfect debugging model as a hidden-Markov model. A hidden-Markov model is a kind of Markov model; however, it has hidden states in its state space. That is, the model includes the states that are unobservable from the sequence of results of the process considered. The advantage of this modeling technique is that there is a proposed method for estimating the unknown parameters included in the hidden states. Hidden-Markov modeling has been mainly used in the research area of speech recognition [13, 19]. We apply this method to estimate the imperfect debugging probability q and other unknown parameters.

Now we slightly generalize the sojourn time distribution function of state j as

$$F_j(t) = 1 - \exp[-\lambda_j t] \quad (j = 0, 1, \ldots, N) \tag{15.28}$$

This $F_j(t)$ corresponds to Equation 15.27 if $\lambda_j = \lambda j$.

We adopt the following notation:

- N is the number of software faults latent in the software system at the beginning of the testing phase of software development
- p is the perfect debugging probability ($0 < p \leq 1$)
- q is the imperfect debugging rate ($q = 1 - p$)
- λ_j is the hazard rate of software failure occurrence at state j ($\lambda_j > 0$)
- j is the state variable representing the number of software faults latent in the software system ($j = 0, 1, \ldots, N$)
- $F_j(t)$ is the sojourn time distribution in state j, which represents the CDF of a time interval measured from the $(N - j)$th to the $(N - j + 1)$th fault-detection (or failure-occurrence) time.

In our model, the state transition diagram is the same as that of Goel and Okumoto (see Figure 15.3). In the figure, each state represents the number of faults latent in the software during the testing phase. From the viewpoint of hidden-Markov modeling, the hidden states in this model are the states themselves. In other words, we cannot find how the sequence of the transitions occurred, based on the observed software fault-detection data.

15.3.2 Estimation Method of Unknown Parameters

We use the Baum–Welch re-estimation formulas to estimate the unknown parameters q, λ_j, and N, with fault-detection time data (i, x_i) $(i = 1, 2, \ldots, T)$ where x_i means the time interval between the $(i - 1)$th and ith fault-detection time $(x_0 \equiv 0)$.

By using this estimation procedure, we can re-estimate the values of q and λ_j $(j = N^+, N^+ - 1, \ldots, 1, 0)$, where N^+ represents the number of initial faults content which is virtually given for the procedure. We can also obtain the re-estimated value of π_j $(j = N^+, N^+ - 1, \ldots, 1, 0)$ through this procedure, where π_j is the initial probability of state j $\left(\sum_{j=0}^{N^+} \pi_j = 1\right)$. The re-estimated values are given as follows:

$$\bar{q} = \frac{\sum_{t=1}^{T-1} \xi_{00}(t)}{\sum_{t=1}^{T-1} \gamma_0(t)} \tag{15.29}$$

$$\bar{\lambda}_j = \frac{\sum_{t=1}^{T} \alpha_j[t]\beta_j[t]}{\sum_{t=1}^{T} \alpha_j[t]\beta_j[t]x_t} \tag{15.30}$$

$$\bar{\pi}_j = \gamma_j(1) \tag{15.31}$$

where T represents the length of the observed data, and

$$a_{ij} = \begin{cases} 1 & (i = 0,\ j = 0) \\ q & (i = j) \\ p & (i - j = 1) \\ 0 & (\text{otherwise}) \end{cases} \tag{15.32}$$

$$b_j(k) = \frac{\mathrm{d}F_j(k)}{\mathrm{d}k} = \lambda_j \exp[-\lambda_j k]$$
$$(j = N^+, N^+ - 1, \ldots, 1, 0)$$

$$\alpha_j(t) = \begin{cases} \pi_j b_j(x_1) \\ \quad (t = 1;\ j = N^+, N^+ - 1, \ldots, 1, 0) \\ \sum_{i=0}^{N^+} \alpha_i(t-1) a_{ij} b_j(x_t) \\ \quad (t = 2, 3, \ldots, T; \\ \quad j = N^+, N^+ - 1, \ldots, 1, 0) \end{cases} \tag{15.33}$$

$$\beta_i(t) = \begin{cases} 1 \quad (t = T;\ i = N^+, N^+ - 1, \ldots, 1, 0) \\ \sum_{j=0}^{N^+} a_{ij} b_j(x_{t+1}) \beta_j(t+1) \\ \quad (t = T - 1, T - 2, \ldots, 1; \\ \quad i = N^+, N^+ - 1, \ldots, 1, 0) \end{cases} \tag{15.34}$$

$$\gamma_i(t) = \frac{\alpha_i(t)\beta_i(t)}{\sum_{i=0}^{N^+} \alpha_i(T)} \tag{15.35}$$

$$\xi_{ij}(t) = \frac{\alpha_i(t) a_{ij} b_j(x_{t+1}) \beta_j(t+1)}{\sum_{i=0}^{N^+} \alpha_i(T)} \tag{15.36}$$

The initial values for the parameters q, λ_j, and $\pi = \{\pi_0, \pi_1, \ldots, \pi_{N^+}\}$ should be given previously. After iterating these re-estimation procedures, we can obtain the estimated values as the convergence ones. Also, we can estimate the optimal state sequence by using $\gamma_i(t)$ in Equation 15.35. The most-likely state, s_i, is given as

$$s_i = \underset{0 \le j \le N^+}{\operatorname{argmax}}[\gamma_j(i)] \quad (1 \le i \le T) \tag{15.37}$$

The estimated value of the parameter N, which is denoted $\hat{N}$, is given by $\hat{N} = s_1$ from Equation 15.37.

Table 15.1. Estimation results for $\hat{\lambda}$ and $\hat{N}$

$\tilde{q}$ (%)	$\hat{\lambda}$	$\hat{N}$	SSE
0.0	0.006 95	31	7171.9
1.0	0.006 88	31	7171.7
2.0	0.006 80	31	7172.4
3.0	0.006 74	31	7173.5
4.0	0.007 13	30	6872.0
5.0	0.007 06	30	7213.6
6.0	0.006 98	30	7182.7
7.0	0.007 40	29	7217.1
8.0	0.007 32	29	6872.7
9.0	0.007 25	29	7248.9
10.0	0.007 17	29	7218.6

15.3.3 Numerical Illustrations

We now try to estimate the unknown parameters included in our model described in the previous section. We analyze a data set that was cited by Goel and Okumoto [21]. This data set (denoted DS-3.1) forms (i, x_i) $(i = 1, 2, \ldots, 26$, *i.e.* $T = 26)$. However, we have actually found some computational difficulty in terms of the re-estimation procedure. This arises from the fact that there are many unknown parameters, namely λ_j $(j = N^+, N^+ - 1, \ldots, 1, 0)$. Hence, we have reduced our model to Goel and Okumoto's one, *i.e.* $\lambda_j = \lambda \cdot j$ (for all j).

We give the initial values for π by taking the possible number of the initial fault content from the data set into consideration, and λ. For computational convenience, we give the values of the imperfect debugging rate q as $\tilde{q} = 0, 1, 2, \ldots, 10$ (%), and we estimate the other parameters for each q.

The estimation results are shown in Table 15.1. In this table, we also calculate the sum of the squared error (SSE), which is defined as

$$\text{SSE} = \sum_{i=1}^{26} (x_i - E[X_i])^2 \tag{15.38}$$

where $E[X_i]$ represents the estimated mean time between the $(i-1)$th and the ith fault-detection times, *i.e.* $E[X_i] = 1/(\hat{\lambda} s_i)$ $(i = 1, 2, \ldots, 26)$.

From the viewpoint of minimizing the SSE, the best model for this data set is $\hat{N} = 30$, $\tilde{q} =$

Table 15.2. Results of data analysis for DS-3.1

i	x_i	$\hat{E}[x_i]$	s_i
1	9	4.68	30
2	12	4.84	29
3	11	5.01	28
4	4	5.19	27
5	7	5.39	26
6	2	5.61	25
7	5	5.84	24
8	8	6.10	23
9	5	6.38	22
10	7	6.68	21
11	1	7.01	20
12	6	7.38	19
13	1	7.79	18
14	9	8.25	17
15	4	8.77	16
16	1	9.35	15
17	3	10.0	14
18	3	10.8	13
19	6	11.7	12
20	1	12.8	11
21	11	14.0	10
22	33	15.6	9
23	7	17.5	8
24	91	20.0	7
25	2	20.0	7[a]
26	1	23.4	6

[a]The occurrence of imperfect debugging.

4.0 (%), $\hat{\lambda} = 0.007\,13$. This model fits better than the model by Jelinski and Moranda [20] (their model corresponds to the case of $\tilde{q} = 0.0$ (%) in Table 15.1). By using Equation 15.37, we additionally obtain the most-likely sequence of the (hidden) states. The result is shown in Table 15.2 with the analyzed data set. From Table 15.2, we find that the estimated number of residual faults is five. Figure 15.4 illustrates the cumulative testing-time versus the number of times of fault-detection based on the results of Table 15.2. We have used the actual data from Table 15.2, and plotted them with the estimated mean cumulative fault-detection time for the first 26 faults. Since an additional five data points are also available from Goel and Okumoto [21], we have additionally plotted them in Figure 15.4 with the prediction curve, which is extrapolated by using

$$E[X_m] = \frac{1}{\hat{\lambda}(1-\tilde{q})(\hat{N}-m+1+I)} \quad (m = 27, 28, \ldots, 31) \qquad (15.39)$$

$$I = 26 - (\hat{N} - s_{26} + 1) \qquad (15.40)$$

where I denotes the estimated number of times of imperfect debugging during the testing phase.

15.3.4 Concluding Remarks

This section considered the method of estimating the imperfect debugging probability which appears in several software reliability assessment models. We have applied hidden-Markov modeling and the Baum–Welch re-estimation procedure to the estimation problems of the unknown model parameters. We have found that there are still some computational difficulties for obtaining the estimates if the model is complicated in terms of the model structure. However, we could estimate the imperfect debugging probability by only using the fault-detection time data.

It seems that hidden-Markov modeling is competent for describing more complicated models for the software fault-detection process. One area of future study is to solve the computational problems appearing in the re-estimation procedure, and to find some rational techniques for giving the initial values.

15.4 Continuous State Space Model for Large-scale Software

One of our interests here is to introduce the nonlinear state dependency of software debugging processes directly into the software reliability assessment model, not as a mean behavior like the NHPP models. By focusing on software debugging speed, we first show the importance of introducing the nonlinear behavior of the debugging speed into the software reliability modeling. The modeling needs the mathematical theory of stochastic differential equations of the Itô type [22–24], since a counting process

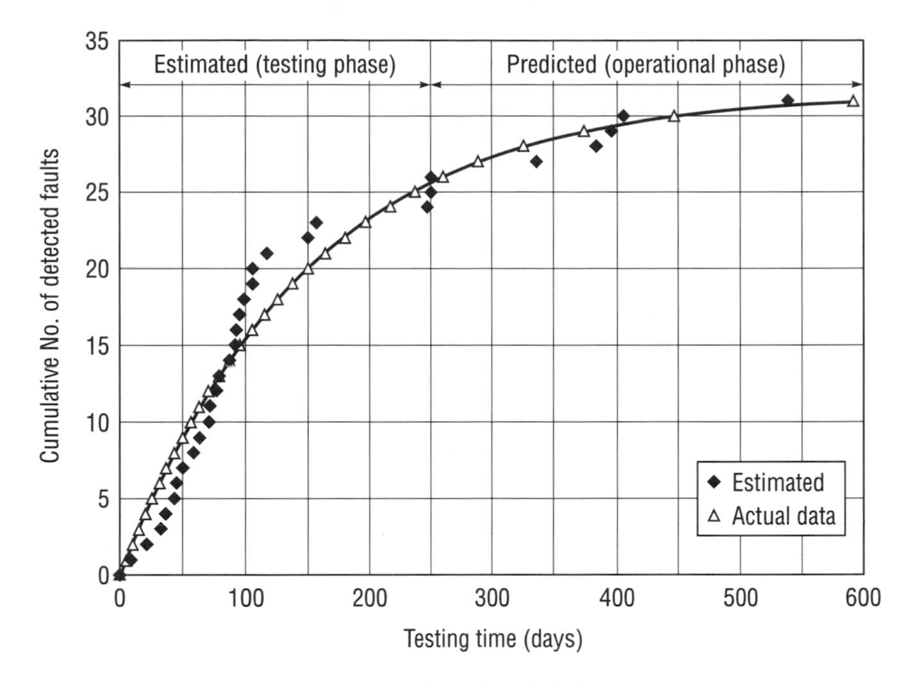

Figure 15.4. Estimated cumulative fault-detection time

(including an NHPP) cannot directly describe the nonlinearity unless it is modeled as a mean behavior. We derive several software reliability assessment measures based on our new model, and also show the estimation method of unknown parameters included in the model.

15.4.1 Model Description

Let $M(t)$ be the number of software faults remaining in the software system at testing time t ($t \geq 0$). We consider a stochastic process $\{M(t),\ t \geq 0\}$, and assume $M(t)$ takes on a continuous real value. In past studies, $\{M(t),\ t \geq 0\}$ has usually been modeled as a counting process for software reliability assessment modeling [2–5, 20, 21, 25]. However, we can suppose that $M(t)$ is continuous when the size of the software system in the testing phase is sufficiently large [26, 27]. The process $\{M(t),\ t \geq 0\}$ may start from a fixed value and gradually decrease with some fluctuation as the testing phase goes on. Thus, we assume that $M(t)$ holds the basic equation as follows:

$$\frac{\mathrm{d}M(t)}{\mathrm{d}t} = -b(t)g(M(t)) \tag{15.41}$$

where $b(t)$ (> 0) represents a fault-detection rate per time, and $M(0)$ ($= m_0$ (const.)) is the number of inherent faults at the beginning of the testing phase. $g(x)$ represents a nonlinear function, which has the following required conditions:

- $g(x)$ is non-negative and has Lipschitz continuity;
- $g(0) = 0$.

Equation 15.41 means that the fault-detection rate at testing time t, $\mathrm{d}M(t)/\mathrm{d}t$, is defined as a function of the number of faults remaining at testing time t, $M(t)$. $\mathrm{d}M(t)/\mathrm{d}t$ decreases gradually as the testing phase goes on. This supposition has often been made in software reliability growth modeling.

In this section, we suppose that $b(t)$ in Equation 15.41 has an irregular fluctuation, *i.e.* we expand Equation 15.41 to the following stochastic differential equation [27]:

$$\frac{\mathrm{d}M(t)}{\mathrm{d}t} = -\{b(t) + \xi(t)\}g(M(t)) \quad (15.42)$$

where $\xi(t)$ represents a noise function that denotes the irregular fluctuation. Further, $\xi(t)$ is defined as

$$\xi(t) = \sigma\gamma(t) \quad (15.43)$$

where $\gamma(t)$ is a standard Gaussian white noise and σ is a positive constant that corresponds to the magnitude of the irregular fluctuation. Hence we rewrite Equation 15.42 as

$$\frac{\mathrm{d}M(t)}{\mathrm{d}t} = -\{b(t) + \sigma\gamma(t)\}g(M(t)) \quad (15.44)$$

By using the Wong–Zakai transformation [28, 29], we obtain the following stochastic differential equation of the Itô type.

$$\begin{aligned}\mathrm{d}M(t) = \{&-b(t)g(M(t)) \\ &+ \tfrac{1}{2}\sigma^2 g(M(t))g'(M(t))\}\,\mathrm{d}t \\ &- \sigma g(M(t))\,\mathrm{d}W(t)\end{aligned} \quad (15.45)$$

where the so-called growth condition is assumed to be satisfied with respect to the function $g(x)$.

In Equation 15.45, $W(t)$ represents a one-dimensional Wiener process, which is formally defined as an integration of the white noise $\gamma(t)$ with respect to testing time t. The Wiener process $\{W(t),\ t \geq 0\}$ is a Gaussian process, and has the following properties:

- $\Pr[W(0) = 0] = 1$;
- $E[W(t)] = 0$;
- $E[W(t)W(\tau)] = \min[t, \tau]$.

Under these assumptions and conditions, we derive a transition probability distribution function of the solution process by solving the Fokker–Planck equation [22] based on Equation 15.45.

The transition probability distribution function $P(m, t \mid m_0)$ is denoted by:

$$P(m, t \mid m_0) = \Pr[M(t) \leq m \mid M(0) = m_0] \quad (15.46)$$

and its density is represented as

$$p(m, t \mid m_0) = \frac{\partial}{\partial m} P(m, t \mid m_0) \quad (15.47)$$

The transition probability density function $p(m, t \mid m_0)$ holds for the following Fokker–Planck equation.

$$\begin{aligned}\frac{\partial p(m,t)}{\partial t} &= b(t)\frac{\partial}{\partial m}\{g(m)p(m,t)\} \\ &\quad - \frac{1}{2}\sigma^2\frac{\partial}{\partial m}\{g(m)g'(m)p(m,t)\} \\ &\quad + \frac{1}{2}\sigma^2\frac{\partial^2}{\partial m^2}\{g(m)^2 p(m,t)\}\end{aligned} \quad (15.48)$$

By solving Equation 15.48 with the following initial condition:

$$\lim_{t\to 0} p(m, t) = \delta(m - m_0) \quad (15.49)$$

we obtain the transition probability density function of the process $M(t)$ as follows:

$$p(m, t \mid m_0) = \frac{1}{g(m)\sigma\sqrt{2\pi t}} \times \exp\left\{ -\frac{\left[\int_{m_0}^{m} \frac{\mathrm{d}m'}{g(m')} + \int_0^t b(t')\,\mathrm{d}t'\right]^2}{2\sigma^2 t} \right\} \quad (15.50)$$

Moreover, the transition probability distribution function of $M(t)$ is obtained as:

$$\begin{aligned}P(m, t \mid m_0) &= \Pr[M(t) \leq m \mid M(0) = m_0] \\ &= \int_{-\infty}^{m} p(m', t \mid m_0)\,\mathrm{d}m' \\ &= \Phi\left(\frac{\int_{m_0}^{m} \frac{\mathrm{d}m'}{g(m')} + \int_0^t b(t')\,\mathrm{d}t'}{\sigma\sqrt{t}} \right)\end{aligned} \quad (15.51)$$

where the function $\Phi(z)$ denotes the standard normal distribution function defined as

$$\Phi(z) = \frac{1}{\sqrt{2\pi}} \int_{-\infty}^{z} \exp\left[-\frac{s^2}{2}\right] \mathrm{d}s \quad (15.52)$$

Let $N(t)$ be the total number of detected faults up to testing time t. Since the condition $M(t) + N(t) = m_0$ holds with probability one, we have the transition probability distribution function of the

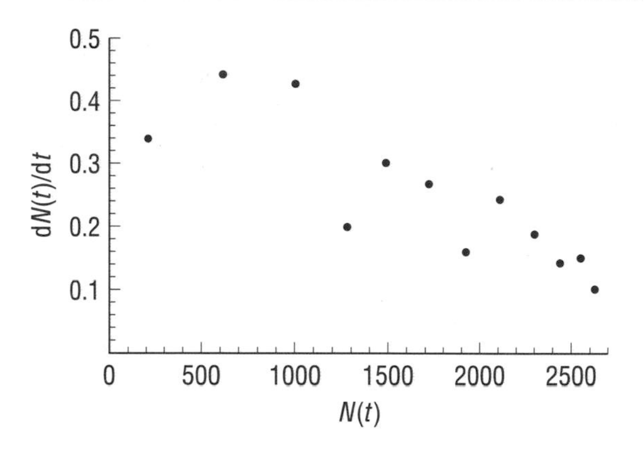

Figure 15.5. Software debugging speed $\mathrm{d}N(t)/\mathrm{d}t$ for DS-4.1

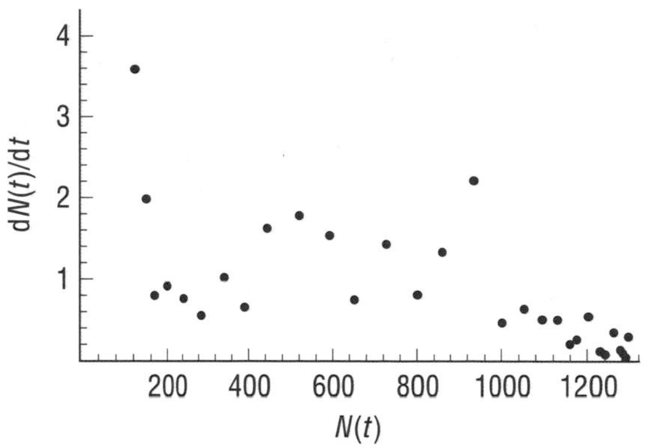

Figure 15.6. Software debugging speed $\mathrm{d}N(t)/\mathrm{d}t$ for DS-4.2

process $N(t)$ as follows:

$$\Pr[N(t) \leq n \mid N(0) = 0,\ M(0) = m_0] = \Phi\left(\frac{\int_0^n \frac{\mathrm{d}n'}{g(m_0 - n')} - \int_0^t b(t')\,\mathrm{d}t'}{\sigma\sqrt{t}}\right) \quad (n < m_0) \tag{15.53}$$

15.4.2 Nonlinear Characteristics of Software Debugging Speed

In order to apply the model constructed in the previous section to the software fault-detection data, we should determine the function $g(x)$ previously. First, we investigate two data sets (denoted DS-4.1 and DS-4.2) that were collected in the actual testing phase of software development to illustrate the behavior of their debugging speed. Each data set forms (t_j, n_j) $(j = 1, 2, \ldots, K; 0 < t_1 < t_2 < \cdots < t_K)$, where t_j represents testing time and n_j is the total number of detected faults up to the testing time t_j [30]. By evaluating numerically, we obtain the debugging speed $\mathrm{d}N(t)/\mathrm{d}t$ from these data sets. Figures 15.5 and 15.6 illustrate the behavior of debugging speed versus the cumulative number of detected faults $N(t)$ for the two data sets. Figure 15.5 shows an approximate linear relation between the debugging speed and the degree of fault-detection. Therefore, we can give $g(x)$ as:

$$g(x) = x \tag{15.54}$$

for such a data set. On the other hand, Figure 15.6 illustrates the existence of a nonlinear relation between them. Thus, we expand Equation 15.54 and assume the function as follows:

$$g(x) = \begin{cases} x^r & (0 \leq x \leq x_\mathrm{c},\ r > 1) \\ r{x_\mathrm{c}}^{r-1}x + (1 - r){x_\mathrm{c}}^r & \\ \quad (x > x_\mathrm{c},\ r > 1) & \end{cases} \tag{15.55}$$

which is a power function that is partially revised in $x > x_\mathrm{c}$ so as to satisfy the growth condition. In Equation 15.55, r denotes a shape parameter and x_c is a positive constant that is needed to ensure the existence of the solution process $M(t)$ in Equation 15.45. The parameter x_c can be given arbitrarily. We assume $x_\mathrm{c} = m_0$ in this model.

15.4.3 Estimation Method of Unknown Parameters

We discuss the estimation method for unknown parameters m_0, b, and σ by applying the maximum likelihood method. The shape parameter r is estimated by the numerical analysis between $\mathrm{d}N(t)/\mathrm{d}t$ and $N(t)$, which was mentioned in the previous section, because r is very sensitive. We assume again that we have a data set of the form (t_j, n_j) $(j = 1, 2, \ldots, K; 0 < t_1 < t_2 < \cdots < t_K)$.

As a basic model, we suppose that the instantaneous fault-detection rate $b(t)$ takes on a constant value, *i.e.*

$$b(t) = b \text{ (const.)} \tag{15.56}$$

Let us denote the joint probability distribution function of the process $N(t)$ as

$$\begin{aligned} P(t_1, n_1; t_2, n_2; \ldots ; t_K, n_K) \\ \equiv \Pr[N(t_1) \le n_1, \ldots, N(t_K) \le n_K \\ \mid N(0) = 0, M(0) = m_0] \end{aligned} \tag{15.57}$$

and denote its density as

$$\begin{aligned} p(t_1, n_1; t_2, n_2; \ldots ; t_K, n_K) \\ \equiv \frac{\partial^K P(t_1, n_1; t_2, n_2; \ldots ; t_K, n_K)}{\partial n_1 \partial n_2 \cdots \partial n_K} \end{aligned} \tag{15.58}$$

Therefore, we obtain the likelihood function l for the data set as follows:

$$l = p(t_1, n_1; t_2, n_2; \ldots ; t_K, n_K) \tag{15.59}$$

since $N(t)$ takes on a continuous value. For convenience in mathematical manipulations, we use the following logarithmic likelihood function L:

$$L = \log l \tag{15.60}$$

The maximum-likelihood estimators m_0^*, b^*, and σ^* are the values making L in Equation 15.60 maximize, which can be obtained as the solutions of the simultaneous equations as follows:

$$\frac{\partial L}{\partial m_0} = 0 \quad \frac{\partial L}{\partial b} = 0 \quad \frac{\partial L}{\partial \sigma} = 0 \tag{15.61}$$

By using Bayes' formula, the likelihood function l can be rewritten as

$$\begin{aligned} l = p(t_1, n_1; t_2, n_2; \ldots ; t_K, n_K \mid t_1, n_1; t_0, n_0) \\ \cdot\, p(t_1, n_1 \mid t_0, n_0) \end{aligned} \tag{15.62}$$

where $p(\cdot|t_0, n_0)$ is the conditional probability density under the condition of $N(t_0) = n_0$, and we suppose that $t_0 = 0$ and $n_0 = 0$. By iterating this transformation under the Markov property, we finally obtain the likelihood function l as follows:

$$l = \prod_{j=1}^{K} p(t_j, n_j \mid t_{j-1}, n_{j-1}) \tag{15.63}$$

The transition probability of $N(t_j)$ under the condition $N(t_{j-1}) = n_{j-1}$ can be obtained as

$$\begin{aligned} \Pr[N(t_j) \le n_j \mid N(t_{j-1}) = n_{j-1}] \\ = \Phi \left[\frac{\int_{n_{j-1}}^{n_j} \frac{\mathrm{d}n'}{g(m_0 - n')} - b(t_j - t_{j-1})}{\sigma \sqrt{t_j - t_{j-1}}} \right] \end{aligned} \tag{15.64}$$

Partially differentiating Equation 15.64 with respect to n_j yields

$$\begin{aligned} p(t_j, n_j \mid t_{j-1}, n_{j-1}) \\ = \frac{1}{g(m_0 - n_j)\sigma\sqrt{2\pi(t_j - t_{j-1})}} \\ \times \exp\left\{ - \frac{\left[\int_{n_{j-1}}^{n_j} \frac{\mathrm{d}n'}{g(m_0 - n')} - b(t_j - t_{j-1})\right]^2}{2\sigma^2(t_j - t_{j-1})} \right\} \end{aligned} \tag{15.65}$$

By substituting Equation 15.65 into 15.63 and taking the logarithm of the equation, we obtain the logarithmic likelihood function as follows:

$$\begin{aligned} L &= \log l \\ &= -K \log\sigma - \frac{K}{2}\log 2\pi - \sum_{j=1}^{K} \log g(m_0 - n_j) \\ &\quad - \frac{1}{2}\sum_{j=1}^{K} \log(t_j - t_{j-1}) \\ &\quad - \frac{1}{2\sigma^2}\sum_{j=1}^{K} \left[\int_{n_{j-1}}^{n_j} \frac{\mathrm{d}n'}{g(m_0 - n')} \right. \\ &\quad \left. - b(t_j - t_{j-1}) \right]^2 \Big/ (t_j - t_{j-1}) \end{aligned} \tag{15.66}$$

Substituting Equation 15.55 into Equation 15.66 yields

$$\begin{aligned} L &= \log l \\ &= -K \log\sigma - \frac{K}{2}\log 2\pi - r\sum_{j=1}^{K} \log(m_0 - n_j) \\ &\quad - \frac{1}{2}\sum_{j=1}^{K} \log(t_j - t_{j-1}) \end{aligned}$$

$$- \frac{1}{2\sigma^2} \sum_{j=1}^{K} \left\{ \frac{1}{r-1} [(m_0 - n_j)^{1-r} - (m_0 - n_{j-1})^{1-r}] - b(t_j - t_{j-1}) \right\}^2 \Big/ (t_j - t_{j-1}) \quad (15.67)$$

We obtain the simultaneous likelihood equations by applying Equation 15.67 to Equation 15.61 as follows:

$$\frac{\partial L}{\partial m_0} = -r \sum_{j=1}^{K} \frac{1}{m_0 - n_j} - \frac{1}{\sigma^2} \sum_{j=1}^{K} \frac{S_{Ij} \partial S_{Ij} / \partial m_0}{t_j - t_{j-1}} = 0 \quad (15.68)$$

$$\frac{\partial L}{\partial b} = -\frac{1}{\sigma^2} \times \sum_{j=1}^{K} \left\{ \frac{[(m_0 - n_j)^{1-r} - (m_0 - n_{j-1})^{1-r}]}{r-1} - b(t_j - t_{j-1}) \right\} = 0 \quad (15.69)$$

$$\frac{\partial L}{\partial \sigma} = -\frac{K}{\sigma} + \frac{1}{\sigma^3} \sum_{j=1}^{K} \frac{S_{Ij}^2}{t_j - t_{j-1}} = 0 \quad (15.70)$$

where

$$S_{Ij} = \frac{1}{r-1} [(m_0 - n_j)^{1-r} - (m_0 - n_{j-1})^{1-r}] - b(t_j - t_{j-1}) \quad (15.71)$$

$$\frac{\partial S_{Ij}}{\partial m_0} = -(m_0 - n_j)^{-r} + (m_0 - n_{j-1})^{-r} \quad (15.72)$$

Since the variables b and σ can be eliminated in Equations 15.68–15.70, we can estimate these three parameters by solving only one nonlinear equation numerically.

15.4.4 Software Reliability Assessment Measures

We derive several quantitative measures to assess software reliability in the testing and operational phase.

15.4.4.1 Expected Number of Remaining Faults and Its Variance

In order to assess software reliability quantitatively, information on the current number of remaining faults in the software system is useful to estimate the situation of the progress on the software testing phase.

The expected number of remaining faults can be evaluated by

$$E[M(t)] = \int_0^{\infty} m \, \mathrm{d}\Pr[M(t) \le m \mid M(0) = m_0] \quad (15.73)$$

The variance of $M(t)$ is also obtained as

$$\mathrm{Var}[M(t)] = E[M(t)^2] - E[M(t)]^2 \quad (15.74)$$

If we assume $g(x) = x$ as a special case, we have

$$E[M(t)] = m_0 \exp[-(b - \tfrac{1}{2}\sigma^2)t] \quad (15.75)$$

$$\mathrm{Var}[M(t)] = m_0^2 \exp[-(2b - \sigma^2)t]\{\exp[\sigma^2 t] - 1\} \quad (15.76)$$

15.4.4.2 Cumulative and Instantaneous Mean Time Between Failures

In the fault-detection process $\{N(t),\ t \ge 0\}$, average fault-detection time-interval per fault up to time t is denoted by $t/N(t)$. Hence, the cumulative mean time between failure (MTBF), $\mathrm{MTBF_C}(t)$, is approximately given by

$$\mathrm{MTBF_C}(t) = E\left[\frac{t}{N(t)}\right] = E\left[\frac{t}{m_0 - M(t)}\right] \approx \frac{t}{m_0 - E[M(t)]} = \frac{t}{m_0\{1 - \exp[-(b - \frac{1}{2}\sigma^2)t]\}} \quad (15.77)$$

when $g(x) = x$ is assumed.

The instantaneous MTBF, $\mathrm{MTBF_I}(t)$, can be evaluated by

$$\mathrm{MTBF_I}(t) = E\left[\frac{1}{\mathrm{d}N(t)/\mathrm{d}t}\right] \quad (15.78)$$

We calculate $\mathrm{MTBF_I}(t)$ approximately by

$$\mathrm{MTBF_I}(t) \approx \frac{1}{E[\mathrm{d}N(t)/\mathrm{d}t]} \quad (15.79)$$

Since the Wiener process has the independent increment property, $W(t)$ and $dW(t)$ are statistically independent and $E[dW(t)] = 0$, we have

$$\text{MTBF}_{\text{I}}(t) = \frac{1}{E[-dM(t)/dt]} = \frac{1}{m_0(b - \frac{1}{2}\sigma^2)\exp[-(b - \frac{1}{2}\sigma^2)t]} \tag{15.80}$$

if $g(x) = x$.

15.4.5 Concluding Remarks

In this section, a new model for software reliability assessment has been proposed by using nonlinear stochastic differential equations of the Itô type. By introducing the debugging speed function, we can obtain a more precise estimation in terms of software reliability assessment. We have derived several measures for software reliability assessment based on this nonlinear stochastic model. The method of maximum-likelihood estimation for the model parameters has also been shown.

In any future study we must evaluate the performance of our model in terms of software reliability assessment and prediction by using data actually collected in the testing phase of software development.

15.5 Development of a Software Reliability Management Tool

Recently, numerous software tools that can assist the software development process have been constructed (*e.g.* see [4, 5]). The tools are constructed as software packages to analyze software faults data that are observed in the testing phase, and to evaluate the quality/reliability of the software product. In the testing phase of the software development process, the testing manager can perform the reliability evaluation easily by using such tools without knowing the details of the process of the faults data analysis. Thus, this section aims at constructing a tool for software management that provides quantitative estimation/prediction of the software reliability and testing management.

We introduce the notion of object-oriented analysis for implementation of the software management tool. The methodology of object-oriented analysis has had great success in the fields of programming language, simulation, the graphical user interface (GUI), and construction of databases in software development. A general idea of object-oriented methodology is developed as a technique that can easily construct and maintain a complex software system. This tool has been implemented using JAVA, which is one of a number of object-oriented languages. Hence we first show the definition of the specification requirements of the tool. Then we discuss the implementation methodology of the software testing management tool using JAVA. An example of software quality/reliability evaluation is presented for a case where the tool is applied to actual observed data. Finally, we discuss the future problems of this tool.

15.5.1 Definition of the Specification Requirement

The definition of the specification requirements for this tool is as follows.

1. It should perform as a software reliability management tool, which can analyze the fault data and evaluate the software quality/reliability and the degree of testing progress quantitatively. The results should be displayed as a window system on a PC.
2. For the data analysis required for quality/reliability and testing-progress control, seven stochastic models should be employed for software reliability analysis and one for testing progress evaluation. For software reliability assessment, the models are:
   (i) exponential software reliability growth (SRGM) model;
   (ii) delayed S-shaped SRGM;
   (iii) exponential-S-shaped SRGM [31];

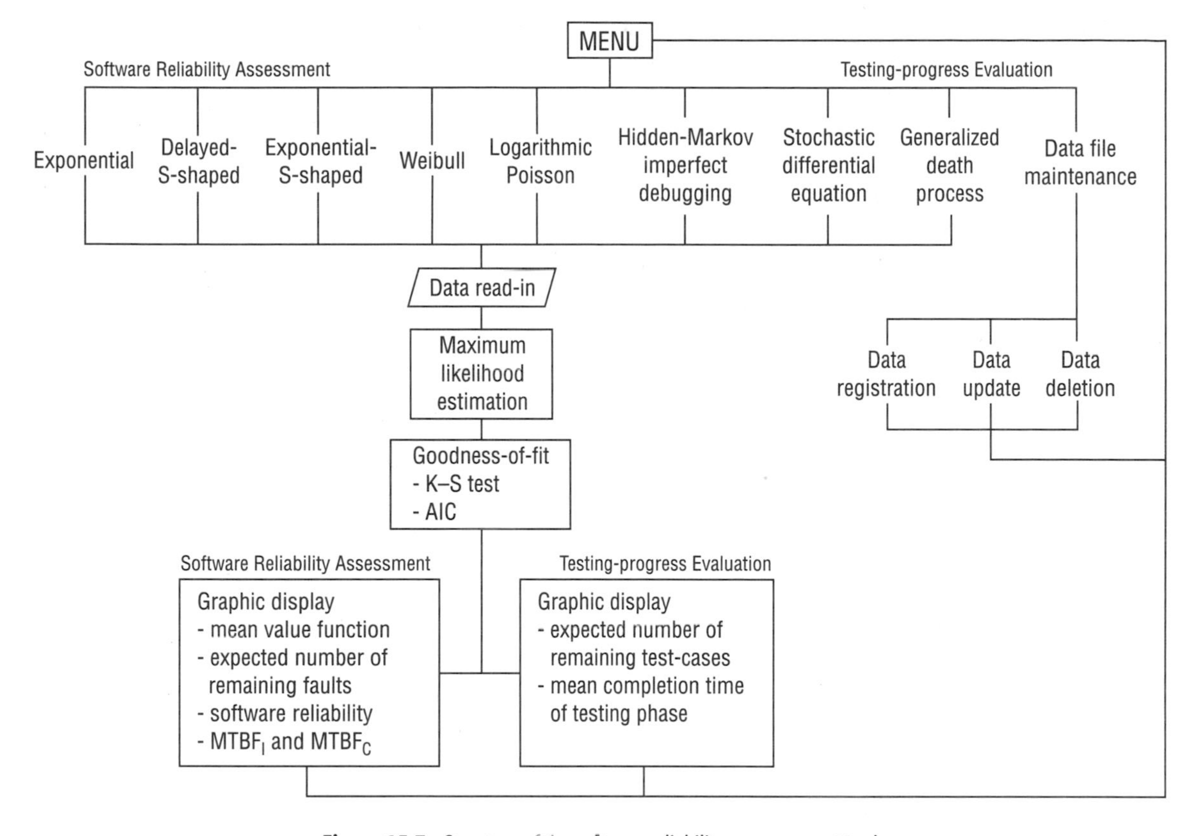

Figure 15.7. Structure of the software reliability management tool

(iv) logarithmic Poisson execution time model [32];
(v) Weibull process model [2, 33];
(vi) hidden-Markov imperfect debugging model;
(vii) nonlinear stochastic differential equation model.

In the above models, the first five models belong to NHPP models. The model for testing progress evaluation is that of the generalized death process model.

The tool additionally examines the goodness-of-fit test to the observed fault data.

3. This tool should be operated by clicking of mouse buttons and typing at the keyboard to input the data through a GUI system.
4. JAVA, which is an object-oriented language, should be used to implement the program. This tool is developed as a stand-alone application on the Windows95 operating system.

15.5.2 Object-oriented Design

This tool should be composed of several processes, such as fault data analysis, estimation of unknown parameters, goodness-of-fit test for estimated models, graphical representation of fault data, and their estimation results (see Figure 15.7). As to searching for the numerical solution of the simultaneous likelihood equations that are obtained from the software reliability assessment models based on NHPP and stochastic differential equation models, the bisection method is adopted. The Baum–Welch re-estimation method is implemented for the imperfect debugging model explained in Section 15.3.

The goodness-of-fit test of each model estimated is examined by the K–S test and the Akaike information criterion (AIC) [34] for the software reliability assessment models and the testing-progress evaluation model. To represent the results of quality/reliability evaluation, the estimated mean value function, the expected number of remaining faults, the software reliability, and the MTBF are displayed graphically for the software reliability assessment models, except for hidden-Markov imperfect debugging model. The imperfect debugging model graphically gives the estimation result of imperfect debugging probability and the expected number of actual detected faults.

To perform the testing-progress evaluation, the generalized death process model described in Section 15.2 predicts the expected number of consumed test cases at testing time t, and the mean completion time of the testing phase.

The object-oriented methodology for implementation of a software program is defined as the general idea that enables us to implement the program code abstractly, and the program logic is performed by passing messages among the objects. This is different from the traditional methodology of structured programming. Thus, it combines the advantages of information hiding and modularization. Moreover, it is comparatively easy for programmers to construct application software program with a GUI by using an object-oriented language, *e.g.* see Arnold *et al.* [35].

This tool is implemented using the JAVA language and the source code of the program conforms to the object-oriented programming methodology by using message communication among the various classes. In the class construction, we implement the source program as each class performs each function. For example, the MOML class is constructed to perform the function of estimating the unknown parameters of the models using the method of maximum likelihood. To protect from improper access among classes, one should avoid using static variables as much as possible.

The proposed tool has been implemented on an IBM-PC/AT compatible personal computer with Windows95 operating system and JAVA interpreter version 1.01 for MS-DOS. We use the integrated JAVA development environment package software "WinCafe", which has a project manager, editor, class browser, GUI debugger, and several utilities.

15.5.3 Examples of Reliability Estimation and Discussion

In this section we discuss a software testing-management tool that can evaluate quantitatively the quality/reliability and the degree of testing progress. The tool analyzes the fault data observed in the testing phase of software development and displays the accompanying various information in visual forms. Using this tool, the users can quantitatively and easily obtain the achievement level of the software reliability, the degree of testing progress, and the testing stability.

By analyzing the actual fault-detection data set, denoted DS-4.1 in Section 15.4, we performed a reliability analysis using this tool and show examples of the quality/reliability assessment in Figure 15.8. This figure shows the estimation results for the expected number of detected faults by using the nonlinear stochastic differential equation model.

In the following we evaluate our tool and discuss several problems. The first point concerns the expandability of our tool. JAVA is simple and compactly developed and we can easily code and debug the source code. Moreover, since JAVA, like the C++ language, is an object-oriented language, we can easily use object-oriented techniques and their attendant advantages, such as modularization and reuse of existing code. We have implemented several classes to perform various functions, such as producing a new data file, estimating the unknown parameters, displaying of results graphically, and

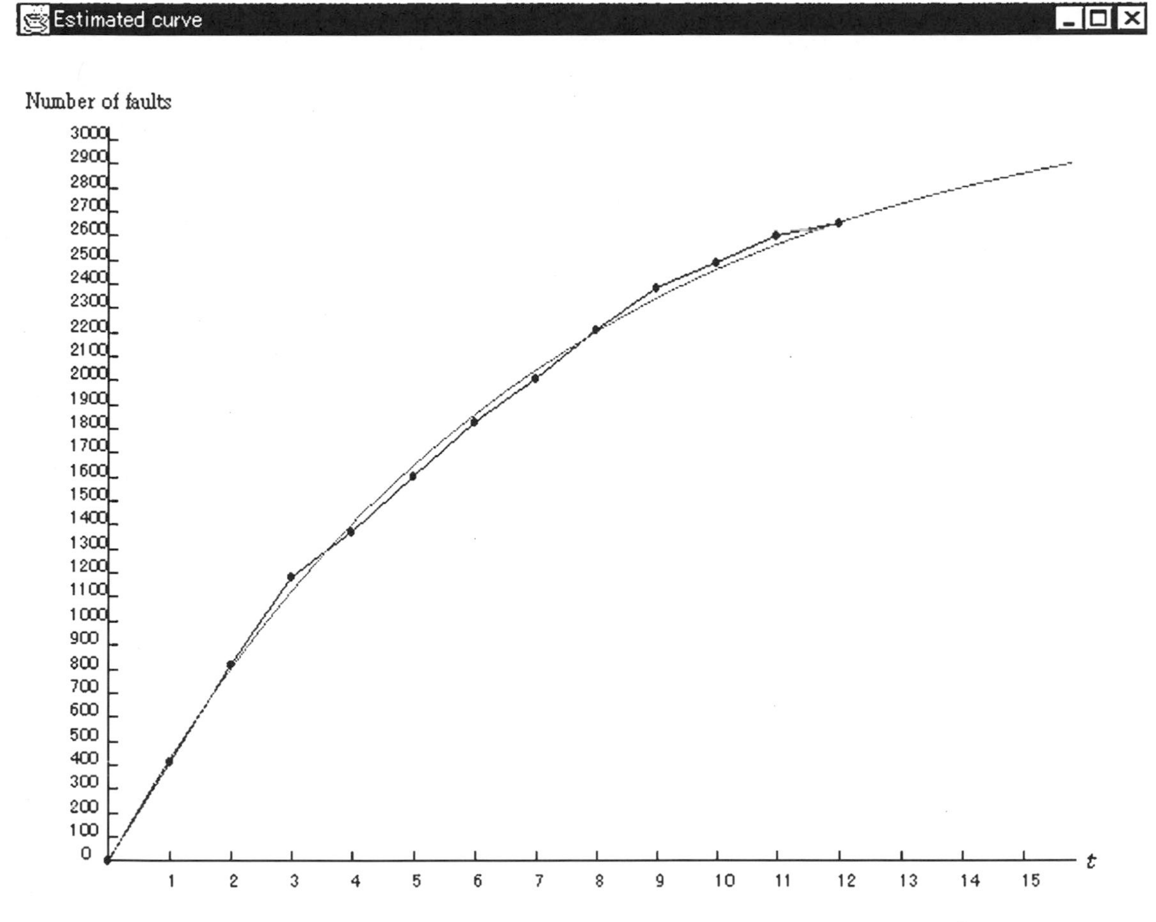

Figure 15.8. Expected cumulative number of detected faults (by SDE model with DS-4.1; $\hat{m}_0 = 3569.8$, $\hat{b} = 0.000\,131$, $\hat{\sigma} = 0.008\,00$, $\hat{r} = 1.0$)

several functions. If new software reliability assessment techniques are developed in the field of software reliability research, our tool can be easily reconstructed with new software components to perform the new assessment techniques. So we can quickly and flexibly cope with extending the function of our tool. The second point is the portability of the program. JAVA is a platform-free programming language, and so does not depend on the specific computer architecture. This tool can be implemented and perform on any kind of computer hardware. The final point is the processing speed. The JAVA compiler makes use of bytecode, which cannot be directly executed by the CPU. This is the greatest difference from a traditional programming language. But we do not need to pay attention to the problem of the processing speed of JAVA bytecode, since the processing speed of our tool is sufficient.

In conclusion, this tool is new and has the advantages that the results of data analysis are represented simply and visually by GUI. The tool is also easily expandable, it is machine portable, and is maintainable because of the use of JAVA.

References

[1] Pressman R. Software engineering: a practitioner's approach. Singapore: McGraw-Hill Higher Education; 2001.

[2] Musa JD, Iannino A, Okumoto K. Software reliability: measurement, prediction, application. New York: McGraw-Hill; 1987.

[3] Bittanti S. Software reliability modeling and identification. Berlin: Springer-Verlag; 1988.

[4] Lyu M. Handbook of software reliability engineering. Los Alamitos: IEEE Computer Society Press; 1995.

[5] Pham H. Software reliability. Singapore: Springer-Verlag; 2000.

[6] Okumoto K, Goel AL. Availability and other performance measurement for system under imperfect debugging. In: Proceedings of COMPSAC'78, 1978; p.66–71.

[7] Shanthikumar JG. A state- and time-dependent error occurrence-rate software reliability model with imperfect debugging. In: Proceedings of the National Computer Conference, 1981; p.311–5.

[8] Kramer W. Birth–death and bug counting. IEEE Trans Reliab. 1983;R-32:37–47.

[9] Sumita U, Shanthikumar JG. A software reliability model with multiple-error introduction & removal. IEEE Trans Reliab. 1986;R-35:459–62.

[10] Ohba M, Chou X. Does imperfect debugging affect software reliability growth? In: Proceedings of the IEEE International Conference on Software Engineering, 1989; p.237–44.

[11] Goel AL, Okumoto K. An imperfect debugging model for reliability and other quantitative measures of software systems. Technical Report 78-1, Department of IE&OR, Syracuse University, 1978.

[12] Goel AL, Okumoto K. Classical and Bayesian inference for the software imperfect debugging model. Technical Report 78-2, Department of IE&OR, Syracuse University, 1978.

[13] Rabinar LR, Juang BH. An introduction to hidden Markov models. IEEE ASSP Mag 1986;3:4–16.

[14] Kimura M, Yamada S. A software testing-progress evaluation model based on a digestion process of test cases. Int J Reliab, Qual Saf Eng 1997;4:229–39.

[15] Shanthikumar JG. A general software reliability model for performance prediction. Microelectron Reliab 1981;21:671–82.

[16] Ross SM. Stochastic processes. New York: John Wiley & Sons; 1996.

[17] Karlin S, Taylor HM. A first course in stochastic processes. San Diego: Academic Press; 1975.

[18] Shibata K. Project planning and phase management of software product (in Japanese). Inf Process Soc Jpn 1980;21:1035–42.

[19] Rabinar LR. A tutorial on hidden Markov models and selected applications in speech recognition. Proc IEEE 1989;7:257–86.

[20] Jelinski Z, Moranda PB. Software reliability research. In: Freiberger W, editor. Statistical computer performance evaluation. New York: Academic Press; 1972. p.465–484.

[21] Goel AL, Okumoto K. Time-dependent and error-detection rate model for software reliability and other performance measures. IEEE Trans Reliab 1979;R-28:206–11.

[22] Arnold L. Stochastic differential equations: theory and applications. New York: John-Wiley & Sons; 1974.

[23] Protter P. Stochastic integration and differential equations: a new approach. Berlin Heidelberg New York: Springer-Verlag; 1995.

[24] Karlin S, Taylor HM. A second course in stochastic processes. San Diego: Academic Press; 1981.

[25] Goel AL, Okumoto K. A Markovian model for reliability and other performance measures of software system. In: Proceedings of the National Computer Conference, 1979; p.769–74.

[26] Yamada S, Kimura M, Tanaka H, Osaki S. Software reliability measurement and assessment with stochastic differential equations. IEICE Trans Fundam 1994;E77-A:109–16.

[27] Kimura M, Yamada S, Tanaka H. Software reliability assessment modeling based on nonlinear stochastic differential equations of Itô type. In: Proceedings of the ISSAT International Conference on Reliability and Quality in Design, 2000; p.160–3.

[28] Wong E, Zakai M. On the relation between ordinary and stochastic differential equations. Int J Eng Sci 1965;3:213–29.

[29] Wong E. Stochastic processes in information and systems. New York: McGraw-Hill; 1971.

[30] Yamada S, Ohtera H, Narihisa H. Software reliability growth models with testing-effort. IEEE Trans Reliab 1986;R-35:19–23.

[31] Kimura M, Yamada S, Osaki S. Software reliability assessment for an exponential-S-shaped reliability growth phenomenon. J Comput Math Appl 1992;24:71–8.

[32] Musa JD, Okumoto K. A logarithmic Poisson execution time model for software reliability measurement. In: Proceedings of the 7th International Conference on Software Engineering, 1984; p.230–8.

[33] Crow LH. On tracking reliability growth. In: Proceedings of the 1975 Annual Reliability and Maintainability Symposium; p.438–43.

[34] Akaike H. A new look at the statistical model identification. IEEE Trans Autom Control 1974;AC-19:716–23.

[35] Arnold K, Gosling J, Holmes D. The Java programming language. Boston Tokyo: Addison-Wesley; 2000.

Recent Studies in Software Reliability Engineering

Hoang Pham

Chapter 16

16.1 Introduction

Computers are being used in diverse areas for various applications, *e.g.* air traffic control, nuclear reactors, aircraft, real-time military weapons, industrial process control, automotive mechanical and safety control, and hospital patient monitoring systems. New computer and communication technologies are obviously transforming our daily life. They are the basis of many of the changes in our telecommunications systems and also a new wave of automation on the farm, in manufacturing, hospital, transportation, and in the office. Computer information products and services have become a major and still rapidly growing component of our global economy.

The organization of this chapter is divided into five sections. Section 16.1 presents the basic concepts of software reliability. Section 16.2 presents some existing non-homogeneous Poisson process (NHPP) software reliability models and their application to illustrate the results. Section 16.3 presents generalized software reliability models with environmental factors. Section 16.4 discusses a risk–cost model. Finally, Section 16.5 presents recent studies on software reliability and cost models with considerations of random field environments. Future research directions in software reliability engineering and challenge issues are also discussed.

16.1.1 Software Reliability Concepts

Research activities in software reliability engineering have been conducted over the past 30 years, and many models have been developed for the estimation of software reliability [1, 2]. Software reliability is a measure of how closely user requirements are met by a software system in actual operation. Most existing models [2–16] for quantifying software reliability are purely based upon observation of failures during the system test of the software product. Some companies are indeed putting these models to use in their

Table 16.1. The real-time control system data for time domain approach

Fault	TBF	Cumulative TBF	Fault	TBF	Cumulative TBF	Fault	TBF	Cumulative TBF	Fault	TBF	Cumulative TBF
1	3	3	35	227	5324	69	529	15806	103	108	42296
2	30	33	36	65	5389	70	379	16185	104	0	42296
3	113	146	37	176	5565	71	44	16229	105	3110	45406
4	81	227	38	58	5623	72	129	16358	106	1247	46653
5	115	342	39	457	6080	73	810	17168	107	943	47596
6	9	351	40	300	6380	74	290	17458	108	700	48296
7	2	353	41	97	6477	75	300	17758	109	875	49171
8	91	444	42	263	6740	76	529	18287	110	245	49416
9	112	556	43	452	7192	77	281	18568	111	729	50145
10	15	571	44	255	7447	78	160	18728	112	1897	52042
11	138	709	45	197	7644	79	828	19556	113	447	52489
12	50	759	46	193	7837	80	1011	20567	114	386	52875
13	77	836	47	6	7843	81	445	21012	115	446	53321
14	24	860	48	79	7922	82	296	21308	116	122	53443
15	108	968	49	816	8738	83	1755	23063	117	990	54433
16	88	1056	50	1351	10089	84	1064	24127	118	948	55381
17	670	1726	51	148	10237	85	1783	25910	119	1082	56463
18	120	1846	52	21	10258	86	860	26770	120	22	56485
19	26	1872	53	233	10491	87	983	27753	121	75	56560
20	114	1986	54	134	10625	88	707	28460	122	482	57042
21	325	2311	55	357	10982	89	33	28493	123	5509	62551
22	55	2366	56	193	11175	90	868	29361	124	100	62651
23	242	2608	57	236	11411	91	724	30085	125	10	62661
24	68	2676	58	31	11442	92	2323	32408	126	1071	63732
25	422	3098	59	369	11811	93	2930	35338	127	371	64103
26	180	3278	60	748	12559	94	1461	36799	128	790	64893
27	10	3288	61	0	12559	95	843	37642	129	6150	71043
28	1146	4434	62	232	12791	96	12	37654	130	3321	74364
29	600	5034	63	330	13121	97	261	37915	131	1045	75409
30	15	5049	64	365	13486	98	1800	39715	132	648	76057
31	36	5085	65	1222	14708	99	865	40580	133	5485	81542
32	4	5089	66	543	15251	100	1435	42015	134	1160	82702
33	0	5089	67	10	15261	101	30	42045	135	1864	84566
34	8	5097	68	16	15277	102	143	42188	136	4116	88682

software product development. For example, the Hewlett-Packard company has been using an existing reliability model to estimate the failure intensity expected for firmware, software embedded in the hardware, in two terminals, known as the HP2393A and HP2394A, to determine when to release the firmware. The results of the reliability modeling enabled it to test the modules more efficiently and so contributed to the terminals' success by reducing development cycle cost while maintaining high rcliability. AT&T software developers also used a software reliability model to predict the quality of software system T. The model predicted consistently and the results were within 10% of predictions. AT&T Bell Laboratories also predicted requested rate for field maintenance to its 5ESS telephone switching system differed from the company's actual experience by only 5 to 13%. Such accuracies could help to make warranties of software performance practical [17].

Most existing models, however, require considerable numbers of failure data in order to obtain an accurate reliability prediction. There are two common types of failure data: time-domain data and interval-domain data. Some existing software reliability models can handle both types of data. The time-domain data are characterized by recording the individual times at which the failure occurred. For example, in the real-time control system data of Lyu [18], there are a total of 136 faults reported and the times between failures (TBF) in seconds are listed in Table 16.1. The interval-domain data are characterized by counting the number of failures occurring during a fixed period. The time-domain data always provides better accuracy in the parameter estimates with current existing software reliability models, but this involves more data collection effort than the interval domain approach.

Information concerning the development of the software product, the method of failure detection, environmental factors, field environments, *etc.*, however, are ignored in almost all the existing models. In order to develop a useful software reliability model and to make sound judgments when using the model, one needs to have an in-depth understanding of how software is produced; how errors are introduced, how software is tested, how errors occur, the types of errors, and environmental factors can help us in justifying the reasonableness of the assumptions, the usefulness of the model, and the applicability of the model under given user environments [19]. These models, in other words, would be valuable to software developers, users, and practitioners if they are capable of using information about the software development process, incorporating the environmental factors, and are able to give greater confidence in estimates based on small numbers of failure data.

Within the non-Bayes framework, the parameters of the assumed distribution are thought to be unknown but fixed and are estimated from past inter-failure times utilizing the maximum likelihood estimation (MLE) technique. A time interval for the next failure is then obtained from the fitted model. However, the TBF models based on this technique tend to give results that are grossly optimistic [20] due to the use of MLE. Various Bayes predictive models [21, 22] have been proposed, hopefully to overcome such problems, since if the prior accurately reflects the actual failure process, then model prediction on software reliability can be improved while there is a reduction in total testing time or sample size requirements. Pham and Pham [14] developed Bayesian software reliability models with a stochastically decreasing hazard rate. Within any given failure time interval, the hazard rate is a function of both total testing time and the number of failures encountered. To improve the predictive performance of these particular models, Pham and Pham [16] recently presented a modified Bayesian model introducing pseudo-failures whenever there is a period when the failure-free execution estimate equals the interval-percentile of a predictive distribution.

In addition to those already defined, the following acronyms will be used throughout the chapter:

AIC	the Akaike information criterion [23]
EF	environmental factors
LSE	least squared estimate
MVF	mean value function
SRGM	software reliability growth model
SSE	sum of squared errors
SRE	software reliability engineering.

The following notation is also used:

$a(t)$	time-dependent fault content function: total number of faults in the software, including the initial and introduced faults
$b(t)$	time-dependent fault detection-rate function per fault per unit time
$\lambda(t)$	failure intensity function: faults per unit time
$m(t)$	expected number of error detected by time t (MVF)
$N(t)$	random variable representing the cumulative number of errors detected by time t
S_j	actual time at which the jth error is detected
$R(x/t)$	software reliability function, *i.e.* the conditional probability of no failure occurring during $(t, t+x)$ given that the last failure occurred at time t
$\hat{}$	estimates using MLE method
y_k	the number of actual failures observed at time t_k

$\hat{m}(t_k)$ estimated cumulative number of failures at time t_k obtained from the fitted MVFs, $k = 1, 2, \ldots, n$.

16.1.2 Software Life Cycle

As software becomes an increasingly important part of many different types of systems that perform complex and critical functions in many applications, such as military defense, nuclear reactors, *etc.*, the risk and impacts of software-caused failures have increased dramatically. There is now general agreement on the need to increase software reliability by eliminating errors made during software development.

Software is a collection of instructions or statements in a computer language. It is also called a computer program, or simply a program. A software program is designed to perform specified functions. Upon execution of a program, an input state is translated into an output state. An input state can be defined as a combination of input variables or a typical transaction to the program. When the actual output deviates from the expected output, a failure occurs. The definition of failure, however, differs from application to application, and should be clearly defined in specifications. For instance, a response time of 30 s could be a serious failure for an air traffic control system, but acceptable for an airline reservation system.

A software life cycle consists of five successive phases: analysis, design, coding, testing, and operation [2]. The *analysis phase* is the most important phase; it is the first step in the software development process and the foundation of building a successful software product. The purpose of the analysis phase is to define the requirements and provide specifications for the subsequence phases and activities. The *design phase* is concerned with how to build the system to behave as described. There are two parts to design: system architecture design and detailed design. The system architecture design includes, for example, system structure and the system architecture document. System structure design is the process of partitioning a software system into smaller parts. Before decomposing the system, we need to do further specification analysis, which is to examine the details of performance requirements, security requirements, assumptions and constraints, and the needs for hardware and software.

The *coding phase* involves translating the design into code in a programming language. Coding can be decomposed into the following activities: identify reusable modules, code editing, code inspection, and final test plan. The final test plan should provide details on what needs to be tested, testing strategies and methods, testing schedules and all necessary resources, and be ready at the coding phase. The *Testing phase* is the verification and validation activity for the software product. Verification and validation are the two ways to check if the design satisfies the user requirements. In other words, verification checks if the product, which is under construction, meets the requirements definition, and validation checks if the product's functions are what the customer wants. The objectives of the testing phase are to: (1) affirm the quality of the product by finding and eliminating faults in the program; (2) demonstrate the presence of all specified functionality in the product; and (3) estimate the operational reliability of the software. During this phase, system integration of the software components and system acceptance tests are performed against the requirements. The *operation phase* is the final phase in the software life cycle. The operation phase usually contains activities such as installation, training, support, and maintenance. It involves the transfer of responsibility for the maintenance of the software from the developer to the user by installing the software product and it becomes the user's responsibility to establish a program to control and manage the software.

16.2 Software Reliability Modeling

Many NHPP software reliability growth models [2, 4, 6–11, 14, 24] have been developed over the past 25 years to assess the reliability of software. Software reliability models based on the NHPP have been quite successful tools in practical software reliability engineering [2]. In this section, we only discuss software reliability models based on NHPP. These models consider the debugging process as a counting process characterized by its MVF. Software reliability can be estimated once the MVF is determined. Model parameters are usually estimated using either the MLE or LSE.

16.2.1 A Generalized Non-homogeneous Poisson Process Model

Many existing NHPP models assume that failure intensity is proportional to the residual fault content. A general class of NHPP SRGMs can be obtained by solving the following differential equation [7]:

$$\frac{\mathrm{d}m(t)}{\mathrm{d}t} = b(t)[(a(t) - m(t))] \tag{16.1}$$

The general solution of the above differential equation is given by

$$m(t) = \mathrm{e}^{-B(t)}\left[m_0 + \int_{t_0}^{t} a(\tau)b(\tau)\,\mathrm{e}^{B(\tau)}\,\mathrm{d}\tau\right] \tag{16.2}$$

where $B(t) = \int_{t_0}^{t} b(\tau)\,\mathrm{d}\tau$ and $m(t_0) = m_0$ is the marginal condition of Equation 16.1 with t_0 representing the starting time of the debugging process. The reliability function based on the NHPP is given by:

$$R(x/t) = \mathrm{e}^{-[m(t+x)-m(t)]} \tag{16.3}$$

Many existing NHPP models can be considered as a special case of Equation 16.2. An increasing function $a(t)$ implies an increasing total number of faults (note that this includes those already detected and removed and those inserted during the debugging process) and reflects imperfect debugging. An increasing $b(t)$ implies an increasing fault detection rate, which could be attributed either to a learning curve phenomenon, or to software process fluctuations, or to a combination of both. Different $a(t)$ and $b(t)$ functions also reflect different assumptions of the software testing processes. A summary of the most NHPP existing models is presented in Table 16.2.

16.2.2 Application 1: The Real-time Control System

Let us perform the reliability analysis using the real-time control system data given in Table 16.1. The first 122 data points are used for the goodness of fit test and the remaining data (14 points) are used for the predictive power test. The results for fit and prediction are listed in Table 16.3.

Although software reliability models based on the NHPP have been quite successful tools in practical software reliability engineering [2], there is a need to fully validate their validity with respect to other applications, such as in communications, manufacturing, medical monitoring, and defense systems.

16.3 Generalized Models with Environmental Factors

We adopt the following notation:

$\tilde{z}$	vector of environmental factors
$\tilde{\beta}$	coefficient vector of environmental factors
$\Phi(\tilde{\beta}\tilde{z})$	function of environmental factors
$\lambda_0(t)$	failure intensity rate function without environmental factors
$\lambda(t, \tilde{z})$	failure intensity rate function with environmental factors
$m_0(t)$	MVF without environmental factors
$m(t, \tilde{z})$	MVF with environmental factors
$R(x/t, \tilde{z})$	reliability function with environmental factors.

The proportional hazard model (PHM), which was first proposed by Cox [25], has been successfully utilized to incorporate environmental factors in survival data analysis, in the medical field and in the hardware system reliability area. The basic assumption for the PHM is that the hazard rates of any two items associated with the settings

Table 16.2. Summary of the MVFs [8]

Model name	Model type	MVF ($m(t)$)	Comments
Goel–Okumoto (G–O)	Concave	$m(t) = a(1 - e^{-bt})$ $a(t) = a$ $b(t) = b$	Also called the exponential model
Delayed S-shaped	S-shaped	$m(t) = a[1 - (1 + bt)\, e^{-bt}]$	Modification of G–O model to make it S-shaped
Inflection S-shaped SRGM	S-shaped	$m(t) = \dfrac{a(1 - e^{-bt})}{1 + \beta\, e^{-bt}}$ $a(t) = a$ $b(t) = \dfrac{b}{1 + \beta\, e^{-bt}}$	Solves a technical condition with the G–O model. Becomes the same as G–O if $\beta = 0$
Yamada exponential	S-shaped	$m(t) = a\{1 - e^{-r\alpha[1-\exp(-\beta t)]}\}$ $a(t) = a$ $b(t) = r\alpha\beta\, e^{-\beta t}$	Attempt to account for testing effort
Yamada Rayleigh	S-shaped	$m(t) = a\{1 - e^{-r\alpha[1-\exp(-\beta t^2/2)]}\}$ $a(t) = a$ $b(t) = r\alpha\beta t\, e^{-\beta t^2/2}$	Attempt to account for testing effort
Yamada exponential imperfect debugging model (Y-ExpI)	Concave	$m(t) = \dfrac{ab}{\alpha + b}(e^{\alpha t} - e^{-bt})$ $a(t) = a\, e^{\alpha t}$ $b(t) = b$	Assume exponential fault content function and constant fault detection rate
Yamada linear imperfect debugging model (Y-LinI)	Concave	$m(t) = a(1 - e^{-bt})\left(1 - \dfrac{\alpha}{b}\right) + \alpha a t$ $a(t) = a(1 + \alpha t)$ $b(t) = b$	Assume constant introduction rate α and the fault detection rate
Pham–Nordmann–Zhang (P–N–Z) model	S-shaped and concave	$m(t) = \dfrac{a(1 - e^{-bt})(1 - \alpha/b) + \alpha a t}{1 + \beta\, e^{-bt}}$ $a(t) = a(1 + \alpha t)$ $b(t) = \dfrac{b}{1 + \beta\, e^{-bt}}$	Assume introduction rate is a linear function of testing time, and the fault detection rate function is non-decreasing with an inflexion S-shaped model
Pham–Zhang (P–Z) model	S-shaped and concave	$m(t) = \dfrac{1}{1 + \beta\, e^{-bt}}\left[(c + a)(1 - e^{-bt}) - \dfrac{a}{b - \alpha}(e^{-\alpha t} - e^{-bt})\right]$ $a(t) = c + a(1 - e^{-\alpha t})$ $b(t) = \dfrac{b}{1 + \beta\, e^{-bt}}$	Assume introduction rate is exponential function of the testing time, and the fault detection rate is non-decreasing with an inflexion S-shaped model

Table 16.3. Parameter estimation and model comparison

Model name	SSE (fit)	SSE (Predict)	AIC	MLEs
G–O model	7615.1	704.82	426.05	$\hat{a} = 125$ $\hat{b} = 0.00006$
Delayed S-shaped	51729.23	257.67	546	$\hat{a} = 140$ $\hat{b} = 0.00007$
Inflexion S-shaped	15878.6	203.23	436.8	$\hat{a} = 135.5$ $\hat{b} = 0.00007$ $\hat{\beta} = 1.2$
Yamada exponential	6571.55	332.99	421.18	$\hat{a} = 130$ $\hat{\alpha} = 10.5$ $\hat{\beta} = 5.4 \times 10^{-6}$
Yamada Rayleigh	51759.23	258.45	548	$\hat{a} = 130$ $\hat{\alpha} = 5 \times 10^{-10}$ $\hat{\beta} = 6.035$
Y-Expl model	5719.2	327.99	450	$\hat{a} = 120$ $\hat{b} = 0.00006$ $\hat{\alpha} = 1 \times 10^{-5}$
Y-Linl model	6819.83	482.7	416	$\hat{a} = 120.3$ $\hat{b} = 0.00005$ $\hat{\alpha} = 3 \times 10^{-5}$
P–N–Z model	5755.93	106.81	415	$\hat{a} = 121$ $\hat{b} = 0.00005$ $\hat{\alpha} = 2.5 \times 10^{-6}$ $\hat{\beta} = 0.002$
P–Z model	14233.88	85.36	416	$\hat{a} = 20$ $\hat{b} = 0.00007$ $\hat{\alpha} = 1.0 \times 10^{-5}$ $\hat{\beta} = 1.922$ $\hat{c} = 125$

of environmental factors, say z_1 and z_2, will be proportional to each other. The environmental factors are also known as covariates in the PHM. When the PHM is applied to the NHPP it becomes the proportional intensity model (PIM). A general fault intensity rate function incorporating the environmental factors based on PIM can be constructed using the following assumptions.

(a) The new fault intensity rate function consists of two components: the fault intensity rate functions without environmental factors, $\lambda_0(t)$, and the environmental factor function $\Phi(\tilde{\beta}\tilde{z})$.
(b) The fault intensity rate function $\lambda_0(t)$ and the function of the environmental factors are independent. The function $\lambda_0(t)$ is also called the baseline intensity function.

Assume that the fault intensity function $\lambda(t, \tilde{z})$ is given in the following form:

$$\lambda(t, \tilde{z}) = \lambda_0(t) \cdot \Phi(\tilde{\beta}\tilde{z})$$

Typically, $\Phi(\tilde{\beta}\tilde{z})$ takes an exponential form, such as:

$$\Phi(\tilde{\beta}\tilde{z}) = \exp(\beta_0 + \beta_1 z_1 + \beta_2 z_2 + \dots)$$

The MVF with environmental factors can then be easily obtained:

$$\begin{aligned} m(t, \tilde{z}) &= \int_0^t \lambda_0(s)\Phi(\tilde{\beta}\tilde{z})\, \mathrm{d}s \\ &= \Phi(\tilde{\beta}\tilde{z}) \int_0^t \lambda_0(s)\, \mathrm{d}s \\ &= \Phi(\tilde{\beta}\tilde{z}) m_0(t) \end{aligned}$$

Therefore, the reliability function with environmental factors is given by [9]:

$$\begin{aligned} R(x/t, \tilde{z}) &= \mathrm{e}^{-[m(t+x,\tilde{z}) - m(t,\tilde{z})]} \\ &= \mathrm{e}^{-[\Phi(\tilde{\beta}\tilde{z}) m_0(t+x,\tilde{z}) - \Phi(\tilde{\beta}\tilde{z}) m_0(t,\tilde{z})]} \\ &= (\exp\{-[m_0(t+x) - m_0(t)]\})^{\Phi(\tilde{\beta}\tilde{z})} \\ &= [R_0(x/t)]^{\Phi(\tilde{\beta}\tilde{z})} \end{aligned}$$

16.3.1 Parameters Estimation

The MLE is a widely used method to estimate unknown parameters in the models and will be used to estimate the model parameters in Section 16.3. Since the environmental factors are considered here, the parameters that need to be estimated include not only those in baseline intensity rate function $\lambda_0(t)$, but also the coefficients β_i in the link function of the environmental factors introduced. For example, we have m unknown parameters in function $\lambda_0(t)$ and we introduced k environmental factors into the model, $\beta_1, \beta_2, \dots, \beta_k$; thus we have $(m + k)$ unknown parameters to estimate. The maximum likelihood function for this model can be expressed as follows:

$$\begin{aligned} &L(\tilde{\theta}, \tilde{\beta}, t, z) \\ &= P\Big\{ \prod_{j=1}^{n} [m(0, z_j) = 0, m(t_{1,j}, z_j) = y_{1,j}, \\ &\qquad m(t_{2,j}, z_j) = y_{2,j}, \dots, \\ &\qquad m(t_{n,j}, z_j) = y_{n,j}] \Big\} \\ &= \prod_{j=1}^{n} \prod_{i=1}^{k_j} \frac{[m(t_{i,j}, z_j) - m(t_{i-1,j}, z_j)]^{(y_{i,j} - y_{i-1,j})}}{(y_{i,j} - y_{i-1,j})!} \\ &\quad \times \mathrm{e}^{-[m(t_{i,j}, z_j) - m(t_{i-1,j}, z_j)]} \end{aligned}$$

where n is the number of total failure data groups, k_j is the number of faults in group j, $(j = 1, 2, \dots, n)$, z_j is the vector variable of the environmental factors in data group j and $m(t_{i,j}, z_j)$ is the mean value function incorporating the environmental factors.

The logarithm likelihood function is given by

$$\begin{aligned} &\ln[L(\tilde{\theta}, \tilde{\beta}, t, \tilde{z})] \\ &= \sum_{j=1}^{n} \sum_{i=1}^{k_j} \{(y_{i,j} - y_{i-1,j}) \\ &\quad \times \ln[m(t_{i,j}, z_j) - m(t_{i-1,j}, z_j)] \\ &\quad - \ln[y_{i,j} - y_{i-1,j})!] \\ &\quad - [m(t_{i,j}, z_j) - m(t_{i-1,j}, z_j)]\} \end{aligned}$$

A series of differential equations can be constructed by taking the derivatives of the log likelihood function with respect to each parameter, and set them equal to zero. The estimates of unknown parameters can be obtained by solving these differential equations.

A widely used method, which is known as the partial likelihood estimate method, can be used to facilitate the parameter estimate process. The partial likelihood method estimates the coefficients of covariates, the β_1 values, separately from the parameters in the baseline intensity function. The likelihood function of the partial likelihood method is given by [25]:

$$L(\beta) = \prod_i \frac{\exp(\beta z_i)}{\left[\sum_{m \in R} \exp(\beta z_m)\right]^{d_i}}$$

where d_i represents the tie failure times.

16.3.2 Application 2: The Real-time Monitor Systems

In this section we illustrate the software reliability model with environmental factors based on the PIM method using the software failure data collected from real-time monitor systems [26]. The software consists of about 200 modules and each module has, on average, 1000 lines of a high-level language like FORTRAN. A total of 481 software faults were detected during the 111 days testing period. Both the information of the testing team size and the software failure data are recorded.

The only environmental factor available in this application is the testing team size. Team size is one of the most useful measures in the software development process. It has a close relationship with the testing effort, testing efficiency, and the development management issues. From the correlation analysis of the 32 environmental factors [27], team size is the only factor correlated to the program complexity, which is the number one significant factor according to our environmental factor study. Intuitively, the more complex the software, the larger the team that is required. Since the testing team size ranges from one to eight, we first categorize the factor of team size into two levels. Let z_1 denote the factor of team size as follows:

$$z_1 = \begin{cases} 0 & \text{team size ranges from 1 to 4} \\ 1 & \text{team size ranges from 5 to 8} \end{cases}$$

After carefully examining the failure data, we find that, after day 61, the software becomes stable and the failures occur with a much slower frequency. Therefore, we use the first 61 data points for testing the goodness-of-fit and estimating the parameters, and use the remaining 50 data points (from day 62 to day 111) as real data for examining the predictive power of software reliability models.

In this application, we use the P–Z model listed in Table 16.2 as the baseline mean value function, *i.e.*,

$$m_0(t) = \frac{1}{1+\beta\,\mathrm{e}^{-bt}}\left[(c+a)(1-\mathrm{e}^{-bt}) - \frac{ab}{b-\alpha}(\mathrm{e}^{-\alpha t}-\mathrm{e}^{-bt})\right]$$

and the corresponding baseline intensity function is:

$$\lambda_0(t) = \frac{1}{1+\beta\,\mathrm{e}^{-bt}}\left[(c+a)(1-\mathrm{e}^{-bt}) - \frac{ab}{b-\alpha}(\mathrm{e}^{-\alpha t}-\mathrm{e}^{-bt})\right] + \frac{-\beta b\,\mathrm{e}^{-bt}}{(1+\beta\,\mathrm{e}^{-bt})^2}\left[(c+a)(1-\mathrm{e}^{-bt}) - \frac{ab(\mathrm{e}^{-\alpha t}-\mathrm{e}^{-bt})}{b-\alpha}\right]$$

Therefore, the intensity function with environmental factor is given by [28]:

$$\begin{aligned}\lambda(t) &= \lambda_0(t)\,\mathrm{e}^{\beta_1 z_1} \\ &= \left\{\frac{1}{1+\beta\,\mathrm{e}^{-bt}}\left[(c+a)(1-\mathrm{e}^{-bt}) - \frac{ab}{b-\alpha}(\mathrm{e}^{-\alpha t}-\mathrm{e}^{-bt})\right] + \frac{-\beta b\,\mathrm{e}^{-bt}}{(1+\beta\,\mathrm{e}^{-bt})^2}\left[(c+a)(1-\mathrm{e}^{-bt}) - \frac{ab(\mathrm{e}^{-\alpha t}-\mathrm{e}^{-bt})}{b-\alpha}\right]\right\}\mathrm{e}^{\beta_1 z_1}\end{aligned}$$

The estimate of β_1 using the partial likelihood estimate method is $\hat{\beta}_1 = 0.0246$, which indicates that this factor is significant to consider. The estimates of parameters in the baseline intensity function are given as follows: $\hat{a} = 40.0$, $\hat{b} = 0.09$, $\hat{\beta} = 8.0$, $\hat{\alpha} = 0.015$, $\hat{c} = 450$.

The results of several existing NHPP models are also listed in Table 16.4.

The results show that incorporating the factor of team size into the P–Z model explains the fault detection better and thus enhances the predictive power of this model. Further research is needed to incorporate application complexity, test effectiveness, test suite diversity, test coverage, code reused, and real application operational environments into the NHPP software reliability models and into the software reliability model in general. Future software reliability models must account for these important situations.

Table 16.4. Model evaluation

Model name	MVF ($m(t)$)	SSE (Prediction)	AIC
G–O Model	$m(t) = a(1 - \mathrm{e}^{-bt})$ $a(t) = a$ $b(t) = b$	1052528	978.14
Delayed S-shaped	$m(t) = a[1 - (1 + bt)\,\mathrm{e}^{-bt}]$ $a(t) = a$ $b(t) = \dfrac{b^2 t}{1 + bt}$	83929.3	983.90
Inflexion S-shaped	$m(t) = \dfrac{a(1 - \mathrm{e}^{-bt})}{1 + \beta\,\mathrm{e}^{-bt}}$ $a(t) = a$ $b(t) = \dfrac{b}{1 + \beta\,\mathrm{e}^{-bt}}$	1051714.7	980.14
Yamada exponential	$m(t) = a\{1 - \mathrm{e}^{-r\alpha[1 - \exp(-\beta t)]}\}$ $a(t) = a$ $b(t) = r\alpha\beta\,\mathrm{e}^{-\beta t}$	1085650.8	979.88
Yamada Rayleigh	$m(t) = a\{1 - \mathrm{e}^{-r\alpha[1 - \exp(-\beta t^2/2)]}\}$ $a(t) = a$ $b(t) = r\alpha\beta t\,\mathrm{e}^{-\beta t^2/2}$	86472.3	967.92
Y-Expl model	$m(t) = \dfrac{ab}{\alpha + b}(\mathrm{e}^{\alpha t} - \mathrm{e}^{-bt})$ $a(t) = a\,\mathrm{e}^{\alpha t}$ $b(t) = b$	791941	981.44
Y-Linl model	$m(t) = a(1 - \mathrm{e}^{-bt})\left(1 - \dfrac{\alpha}{b}\right) + \alpha a t$ $a(t) = a(1 + \alpha t)$ $b(t) = b$	238324	984.62
P–N–Z model	$m(t) = \dfrac{a}{1 + \beta\,\mathrm{e}^{-bt}}\left[(1 - \mathrm{e}^{-bt})\left(1 - \dfrac{\alpha}{b}\right) + \alpha t\right]$ $a(t) = a(1 + \alpha t)$ $b(t) = \dfrac{b}{1 + \beta\,\mathrm{e}^{-bt}}$	94112.2	965.37
P–Z model	$m(t) = \dfrac{1}{1 + \beta\,\mathrm{e}^{-bt}}\left[(c + a)(1 - \mathrm{e}^{-bt}) - \dfrac{ab}{b - \alpha}(\mathrm{e}^{-\alpha t} - \mathrm{e}^{-bt})\right]$ $a(t) = c + a(1 - \mathrm{e}^{-\alpha t})$ $b(t) = \dfrac{b}{1 + \beta\,\mathrm{e}^{-bt}}$	86180.8	960.68
Environmental factor model	$m(t) = \dfrac{1}{1 + \beta\,\mathrm{e}^{-bt}}\left[(c + a)(1 - \mathrm{e}^{-bt}) - \dfrac{ab}{b - \alpha}(\mathrm{e}^{-\alpha t} - \mathrm{e}^{-bt})\right]\mathrm{e}^{\beta_1 z_1}$ $a(t) = c + a(1 - \mathrm{e}^{-\alpha t})$ $c(t) = 1 - \dfrac{1 + \beta}{\mathrm{e}^{bt} + \beta}$	560.82	890.68

16.4 Cost Modeling

The quality of the software system usually depends on how much time testing takes and what testing methodologies are used. On the one hand, the longer time people spend in testing, the more errors can be removed, which leads to more reliable software; however, the testing cost of the software will also increase. On the other hand, if the testing time is too short, the cost of the software could be reduced, but the customers may take a higher risk of buying unreliable software [29–31]. This will also increase the cost during the operational phase, since it is much more expensive to fix an error during the operational phase than the testing phase. Therefore, it is important to determine when to stop testing and release the software. In this section, we present a recent generalized cost model and also some other existing cost models.

16.4.1 Generalized Risk–Cost Models

In order to improve the reliability of software products, testing serves as the main tool to remove faults in software products. However, efforts to increase reliability will require an exponential increase in cost, especially after reaching a certain level of software refinement. Therefore, it is important to determine when to stop testing based on the reliability and cost assessment. Several software cost models and optimal release policies have been studied in the past two decades. Okumoto and Goel [32] discussed a simple cost model addressing a linear developing cost during the testing and operational periods. Ohtera and Yamada [33] also discussed the optimum software-release time problem with a fault-detection phenomenon during operation. They introduced two evaluation criteria for the problem: software reliability and mean time between failures. Leung [34] discussed the optimal software release time with consideration of a given cost budget. Dalal and McIntosh [35] studied the stop-testing problem for large software systems with changing code using graphical methods. They reported the details of a real-time trial of a large software system that had a substantial amount of code added during testing. Yang and Chao [36] proposed two criteria for making decisions on when to stop testing. According to them, software products are released to the market when (1) the reliability has reached a given threshold, and (2) the gain in reliability cannot justify the testing cost.

Pham [6] developed a cost model with an imperfect debugging and random life cycle as well as a penalty cost to determine the optimal release policies for a software system. Hou *et al.* [37] discussed optimal release times for software systems with scheduled delivery time based on the hyper-geometric software reliability growth model. The cost model included the penalty cost incurred by the manufacturer for the delay in software release. Recently, Pham and Zhang [38] developed the expected total net gain in reliability of the software development process (as the economical net gain in software reliability that exceeds the expected total cost of the software development), which is used to determine the optimal software release time that maximizes the expected total gain of the software system.

Pham and Zhang [29] also recently developed a generalized cost model addressing the fault removal cost, warranty cost, and software risk cost due to software failures for the first time. The following cost model calculates the expected total cost:

$$\begin{aligned} E(T) = {} & C_0 + C_1 T^{\alpha} + C_2 m(T)\mu_y \\ & + C_3 \mu_w [m(T + T_w) - m(T)] \\ & + C_R [1 - R(x \mid T)] \end{aligned}$$

where C_0 is the set-up cost for software testing, C_1 is the software test cost per unit time, C_2 is the cost of removing each fault per unit time during testing, C_3 is the cost to remove an fault detected during the warranty period, C_R is the loss due to software failure, $E(T)$ is the expected total cost of a software system at time T, μ_y is the expected time to remove a fault during the testing period, and μ_w is the expected time to remove a fault during the warranty period.

The details on how to obtain the optimal software release policies that minimize the expected total cost can be obtained in Pham and Zhang [29]. The benefits of using the above cost models are that they provide:

1. an assurance that the software has achieved safety goals;
2. a means of rationalizing when to stop testing the software.

In addition, with this type of information, a software manager can determine whether more testing is warranted or whether the software is sufficiently tested to allow its release or unrestricted use Pham and Zhang [29].

16.5 Recent Studies with Considerations of Random Field Environments

In this section, a generalized software reliability study under a random field environment will be discussed with consideration of not only the time to remove faults during in-house testing, the cost of removing faults during beta testing, and the risk cost due to software failure, but also the benefits from reliable executions of the software during the beta testing and final operation. Once a software product is released, it can be used in many different locations, applications, tasks and industries, *etc.* The field environments for the software product are quite different from place to place. Therefore, the random effects of the end-user environment will affect the software reliability in an unpredictable way.

We adopt the following notation:

$R(x \mid T)$ software reliability function; this is defined as the probability that a software failure does not occur in time interval $[t, t+x]$, where $t \geq 0$, and $x > 0$

$G(\eta)$ cumulative distribution function of random environmental factor

γ shape parameter of field environmental factor (gamma density variable)

θ scale parameter of field environmental factor (gamma density variable)

$N(T)$ counting process, which counts the number of software failures discovered by time T

$m(T)$ expected number of software failures by time T, $m(T) = E[N(T)]$

$m_1(T)$ expected number of software failures during in-house testing by time T

$m_2(T)$ expected number of software failures in beta testing and final field operation by time T

$m_F(t \mid \eta)$ expected number of software failures in field by time t

C_0 set-up cost for software testing

C_1 software in-house testing per unit time

C_2 cost of removing a fault per unit time during in-house testing

C_3 cost of removing a fault per unit time during beta testing

C_4 penalty cost due to software failure

C_5 benefits if software does not fail during beta testing

C_6 benefits if software does not fail in field operation

μ_y expected time to remove a fault during in-house testing phase

μ_w expected time to remove a fault during beta testing phase

a number of initial software faults at the beginning of testing

a_F number of initial software faults at the beginning of the field operations

t_0 time to stop testing and release the software for field operations

b fault detection rate per fault

T_w time length of the beta testing

x time length that the software is going to be used

p probability that a fault is successfully removed from the software

β probability that a fault is introduced into the software during debugging, and $\beta \ll p$.

16.5.1 A Reliability Model

Teng and Pham [39] recently propose an NHPP software reliability model with consideration of random field environments. This is a unified software reliability model that covers both testing and operation phases in the software life cycle. The model also allows one to remove faults if a software failure occurs in the field and can be used to describe the common practice of so-called 'beta testing' in the software industry. During beta testing, software faults will still be removed from the software after failures occur, but beta testing is conducted in an environment that is the same as (or close to) the end-user environment, and it is commonly quite different from the in-house testing environment.

In contrast to most existing NHPP models, the model assumes that the field environment only affects the unit failure detection rate b by multiplying by a random factor η. The mean value function $m_F(t \mid \eta)$ in the field can be obtained as [39]:

$$m_F(t \mid \eta) = \frac{a_F}{p-\beta}\{1 - \exp[-\eta b(p-\beta)t]\} \tag{16.4}$$

where η is the random field environmental factor.

The random factor η captures the random field environment effects on software reliability. If η is modeled as a gamma distributed variable with probability density function $G(\gamma, \theta)$, the MVF of this NHPP model becomes [39]:

$$m(T) = \begin{cases} \dfrac{a}{p-\beta}[1 - e^{-b(p-\beta)T}] & T \le t_0 \\ \dfrac{a}{p-\beta}\left\{1 - e^{-b(p-\beta)t_0} \times \left[\dfrac{\theta}{\theta + b(p-\beta)(T-t_0)}\right]^{\gamma}\right\} & T \ge t_0 \end{cases} \tag{16.5}$$

Generally, the software reliability prediction is used after the software is released for field operations, *i.e.* $T \ge t_0$. Therefore, the reliability of the software in the field is

$$R(x \mid T) = \exp\left(a\, e^{-b(p-\beta)t_0} \times \left\{\left[\frac{\theta}{\theta + b(p-\beta)(T+x-t_0)}\right]^{\gamma} - \left[\frac{\theta}{\theta + b(p-\beta)(T-t_0)}\right]^{\gamma}\right\}\right)$$

We now need to determine the time when to stop testing and release the software. In other words, we only need to know the software reliability immediately after the software is released. In this case, $T = t_0$; therefore

$$R(x \mid T) = \exp\left(-a\, e^{-b(p-\beta)T} \times \left\{1 - \left[\frac{\theta}{\theta + b(p-\beta)x}\right]^{\gamma}\right\}\right)$$

16.5.2 A Cost Model

The quality of the software depends on how much time testing takes and what testing methodologies are used. The longer time people spend in testing, the more faults can be removed, which leads to more reliable software; however, the testing cost of the software will also increase. If the testing time is too short, the cost of the development cost is reduced, but the risk cost in the operation phase will increase. Therefore, it is important to determine when to stop testing and release the software [29].

Figure 16.1 shows the entire software development life cycle considered in this software cost model: in-house testing, beta testing, and operation; beta testing and operation are conducted in the field environment, which is commonly quite different from the environment where the in-house testing is conducted.

Let us consider the following.

1. There is a set-up cost at the beginning of the in-house testing, and we assume it is a constant.
2. The cost to do testing is a linear function of in-house testing time.

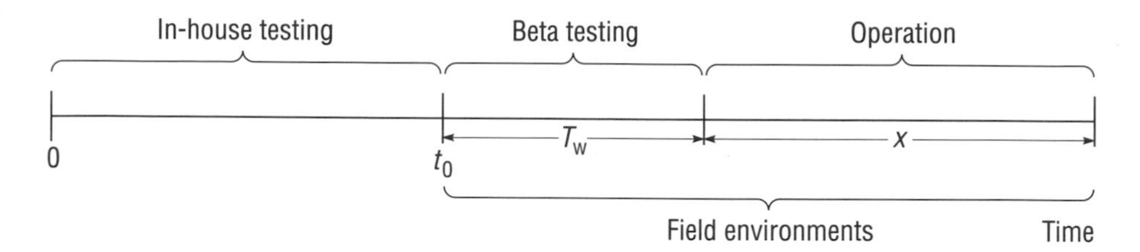

Figure 16.1. Software gain model

3. The cost to remove faults during the in-house testing period is proportional to the total time of removing all faults detected during this period.
4. The cost to remove faults during the beta testing period is proportional to the total time of removing all faults detected in the time interval $[T, T + T_w]$.
5. It takes time to remove faults and it is assumed that the time to remove each fault follows a truncated exponential distribution.
6. There is a penalty cost due to the software failure after formally releasing the software, *i.e.* after the beta testing.
7. Software companies receive the economic profits from reliable executions of their software during beta testing.
8. Software companies receive the economic profits from reliable executions of their software during final operation.

From point 5, the expected time to remove each fault during in-house testing is given by [29]:

$$\mu_y = \frac{1 - (\lambda_y T_0 + 1)\, e^{-\lambda_y T_0}}{\lambda_y (1 - e^{-\lambda_y T_0})}$$

where λ_y is a constant parameter associated with a truncated exponential density function for the time to remove a fault during in-house testing, and T_0 is the maximum time to remove any fault during in-house testing.

Similarly, the expected time to remove each fault during beta testing is given by:

$$\mu_w = \frac{1 - (\lambda_w T_1 + 1)\, e^{-\lambda_w T_1}}{\lambda_w (1 - e^{-\lambda_w T_1})}$$

where λ_w is a constant parameter associated with a truncated exponential density function for the time to remove a fault during beta testing, and T_1 is the maximum time to remove any fault during beta testing.

The expected net gain of the software development process $E(T)$ is defined as the economical net gain in software reliability that exceeds the expected total cost of the software development [38]. That is

$$\begin{aligned} E(T) &= \text{Expected Gain in Reliability} \\ &\quad - (\text{Total Development Cost} + \text{Risk Cost}) \end{aligned}$$

Based on the above assumptions, the expected software system cost consists of the following.

1. Total development cost $E_C(T)$, including:
   (a) A constant set-up cost C_0.
   (b) Cost to do in-house testing $E_1(T)$. We assume it is a linear function of time to do in-house testing T; then

   $$E_1(T) = C_1 T$$

   In some previous literature the testing cost has been assumed to be proportional to the testing time. Here, we keep using this assumption.
   (c) Fault removal cost during in-house testing period $E_2(T)$. The expected total time to remove all faults during in-house testing period is

   $$E\left[\sum_{i=1}^{N(T)} Y_i\right] = E[N(T)]E[Y_i] = m(T)\mu_y$$

   Hence, the expected cost to remove all faults detected by time T, $E_2(T)$, can be expressed as

$$E_2(T) = C_2 E\left[\sum_{i=1}^{N(T)} Y_i\right] = C_2 m(T)\mu_y$$

(d) Fault removal cost during beta testing period $E_3(T)$. The expected total time to remove all faults detected during the beta testing period $[T, T + T_w]$ is given by

$$E\left[\sum_{i=N(T)}^{N(T+T_w)} W_i\right] = E[N(T+T_w) - N(T)]E[W_i] = (m(T+T_w) - m(T))\mu_w$$

Hence, the expected total time to remove all faults detected by time T, $E_3(T)$, is:

$$E_3(T) = C_3 E\left[\sum_{i=N(T)}^{N(T+T_w)} W_i\right] = C_3\mu_w(m(T+T_w) - m(T))$$

Therefore, the total software development cost is

$$E_C(T) = C_0 + E_1(T) + E_2(T) + E_3(T)$$

2. Penalty cost due to software failures after releasing the software $E_4(T)$ is given by

$$E_4(T) = C_4(1 - R(x \mid T + T_w))$$

3. The expected gain $E_p(T)$, including:

(a) Benefits gained during beta testing due to reliable execution of the software, $E_5(T)$ is

$$E_5(T) = C_5 R(T_w \mid T)$$

(b) Benefits gained during field operation due to reliable execution of the software, $E_6(T)$, is

$$E_6(T) = C_6 R(x \mid T + T_w)$$

Therefore, the expected net gain of the software development process $E(T)$ can be expressed as:

$$\begin{aligned}E(T) &= E_p(T) - (E_C(T) + E_4(T))\\ &= (E_5(T) + E_6(T))\\ &\quad - (C_0 + E_1(T) + E_2(T)\\ &\quad + E_3(T) + E_4(T))\\ &= C_5 R(T_w \mid T) + C_6 R(x \mid T + T_w)\\ &\quad - C_0 - C_1 T - C_2\mu_y m(T)\\ &\quad - C_3\mu_w(m(T+T_w) - m(T))\\ &\quad - C_4(1 - R(x \mid T + T_w))\end{aligned}$$

or, equivalently, as

$$\begin{aligned}E(T) &= C_5 R(T_w \mid T)\\ &\quad + (C_4 + C_6)R(x \mid T + T_w)\\ &\quad - (C_0 + C_4) - C_1 T - C_2 m_1(T)\mu_y\\ &\quad - C_3\mu_w(m_2(T+T_w) - m_2(T))\end{aligned}$$

where

$$m_1(T) = \frac{a}{p-\beta}[1 - e^{-b(p-\beta)T}]$$

$$m_2(T) = \frac{a}{p-\beta}\left\{1 - e^{-b(p-\beta)t_0} \times \left[\frac{\theta}{\theta + b(p-\beta)(T-t_0)}\right]^{\gamma}\right\}$$

The optimal software release time T^*, that maximizes the expected net gain of software development process can be obtained as follows.

Theorem 1. *Given C_0, C_1, C_2, C_3, C_4, C_5, C_6, x, μ_y, μ_w, T_w, the optimal value of T, say T^*, that maximizes the expected net gain of the software development process $E(T)$, is as follows:*

1. *If $u(0) \le C$ and:*
   (a) *if $y(0) \le 0$, then $T^* = 0$;*
   (b) *if $y(\infty) > 0$, then $T^* = \infty$;*
   (c) *if $y(0) > 0$, $y(T) \ge 0$ for $T \in (0, T']$ and $y(T) < 0$ for $T \in (T', \infty]$, then $T^* = T'$ where $T' = y^{-1}(0)$.*
2. *If $u(\infty) > C$ and:*
   (a) *if $y(0) \ge 0$, then $T^* = \infty$;*
   (b) *if $y(\infty) < 0$, then $T^* = 0$;*

(c) *if* $y(0) < 0$, $y(T) \le 0$ *for* $T \in (0, T'']$ *and* $y(T) > 0$ *for* $T \in (T'', \infty]$, *then*

$$T^* = \begin{cases} \infty & \text{if } E(0) < E(\infty) \\ 0 & \text{if } E(0) \ge E(\infty) \end{cases}$$

where $T'' = y^{-1}(0)$.

3. *If* $u(0) > C$, $u(T) \ge C$ *for* $T \in (0, T^0]$ *and* $u(T) < C$ *for* $T \in (T^0, \infty]$, *where* $T^0 = u^{-1}(C)$, *then:*

   (a) *if* $y(0) < 0$, *then*
   if $y(T^0) \le 0$, *then* $T^* = 0$
   if $y(T^0) > 0$, *then*

$$T^* = \begin{cases} 0 & \text{if } E(0) \ge E(T_b) \\ T_b & \text{if } E(0) < E(T_b) \end{cases}$$

   where $T_b = y^{-1}(0)$ *and* $T_b \ge T^0$;

   (b) *if* $y(0) \ge 0$, *then* $T^* = T_c$, *where* $T_c = y^{-1}(0)$ *where*

$$\begin{aligned} y(T) = &-C_1 - C_2\mu_y ab\,\mathrm{e}^{-b(p-\beta)T} \\ &+ \mu_w C_3 ab\,\mathrm{e}^{-b(p-\beta)T} \\ &\times \left\{1 - \left[\frac{\theta}{\theta + b(p-\beta)T_w}\right]^\gamma\right\} \\ &+ C_5 ab\,\mathrm{e}^{-b(p-\beta)T} R(T_w \mid T) \\ &\times \left\{1 - \left[\frac{\theta}{\theta + b(p-\beta)T_w}\right]^\gamma\right\} \\ &+ (C_4 + C_6)ab\,\mathrm{e}^{-b(p-\beta)T} \\ &\times R(x \mid T+T_w)\left\{\left[\frac{\theta}{\theta+b(p-\beta)T_w}\right]^\gamma \right. \\ &\left. - \left[\frac{\theta}{\theta + b(p-\beta)(T_w+x)}\right]^\gamma\right\} \end{aligned}$$

$$\begin{aligned} u(T) = &-(C_4+C_6)(p-\beta)ab^2 R(x \mid T+T_w) \\ &\times \left\{\left[\frac{\theta}{\theta + b(p-\beta)T_w}\right]^\gamma \right. \\ &\left. - \left[\frac{\theta}{\theta + b(p-\beta)(T_w+x)}\right]^\gamma\right\} \\ &\times \left(1 - \frac{a}{p-\beta}\mathrm{e}^{-b(p-\beta)T}\right. \\ &\times \left\{\left[\frac{\theta}{\theta + b(p-\beta)T_w}\right]^\gamma \right. \\ &\left.\left. - \left[\frac{\theta}{\theta + b(p-\beta)(T_w+x)}\right]^\gamma\right\}\right) \\ &- C_5(p-\beta)ab^2 R(T_w \mid T) \\ &\times \left\{1 - \left[\frac{\theta}{\theta + b(p-\beta)T_w}\right]^\gamma\right\} \\ &\times \left(1 - \frac{a}{p-\beta}\mathrm{e}^{-b(p-\beta)T}\right. \\ &\left. \times \left\{1 - \left[\frac{\theta}{\theta + b(p-\beta)T_w}\right]^\gamma\right\}\right) \end{aligned}$$

$$\begin{aligned} C = &\mu_w C_3 (p-\beta)ab^2 \\ &\times \left\{1 - \left[\frac{\theta}{\theta + b(p-\beta)T_w}\right]^\gamma\right\} \\ &- C_2\mu_y(p-\beta)ab^2 \end{aligned}$$

This model can help developers to determine when to stop testing the software and release it to beta-testing users and to end-users.

Further research interest is needed to determine the following.

1. How should resources be allocated from the beginning of the software lifecycle to ensure the on-time and efficient delivery of a software product?
2. What information during the system test does a software engineer need to determine when to release the software for a given reliability/risk level?
3. What is the risk cost due to the software failures after release?
4. How should marketing efforts—including advertising, public relations, trade-show participation, direct mail, and related initiatives—be allocated to support the release of a new software product effectively?

16.6 Further Reading

There are several survey papers and books on software reliability research and software cost published in the **last five years**, that can be read at an introductory/intermediate stage. Interested readers are referred to the following articles by Pham [9] and Whittaker and Voas [40], Voas [41], and the *Handbook of Software Reliability Engineering* by Lyu [18], (McGraw-Hill and IEEE CS Press, 1996); The books *Software Reliability* by Pham [2] (Springer-Verlag, 2000); *Software-Reliability-Engineered Testing Practice* by J. Musa, (John Wiley & Sons, 1998), are excellent textbooks and references for students, researchers, and practitioners.

This list is by no means exhaustive, but I believe it will help readers get started learning about the subjects.

Acknowledgment

This research was supported in part by the US National Science Foundation under INT-0107755.

References

[1] Wood A. Predicting software reliability. IEEE Comput 1996;11:69–77.

[2] Pham H. Software reliability. Springer-Verlag; 2000.

[3] Jelinski Z, Moranda PB. Software reliability research. In: Freiberger W, editor. Statistical computer performance evaluation. New York: Academic Press; 1972.

[4] Goel L, Okumoto K. Time-dependent error-detection rate model for software and other performance measures. IEEE Trans Reliab 1979;28:206–11.

[5] Pham H. Software reliability assessment: imperfect debugging and multiple failure types in software development. EG&G-RAAM-10737, Idaho National Engineering Laboratory, 1993.

[6] Pham H. A software cost model with imperfect debugging, random life cycle and penalty cost. Int J Syst Sci 1996;27(5):455–63.

[7] Pham H, Zhang X. An NHPP software reliability models and its comparison. Int J Reliab Qual Saf Eng 1997;4(3):269–82.

[8] Pham H, Nordmann L, Zhang X. A general imperfect software debugging model with s-shaped fault detection rate. IEEE Trans Reliab 1999;48(2):169–75.

[9] Pham H. Software reliability. In: Webster JG, editor. Wiley encyclopedia of electrical and electronics engineering, vol. 19. Wiley & Sons; 1999. p.565–78.

[10] Ohba M. Software reliability analysis models. IBM J Res Dev 1984;28:428–43.

[11] Yamada S, Ohba M, Osaki S. S-shaped reliability growth modeling for software error detection. IEEE Trans Reliab 1983;12:475–84.

[12] Yamada S, Osaki S. Software reliability growth modeling: models and applications. IEEE Trans Software Eng 1985;11:1431–7.

[13] Yamada S, Tokuno K, Osaki S. Imperfect debugging models with fault introduction rate for software reliability assessment. Int J Syst Sci 1992;23(12).

[14] Pham L, Pham H. Software reliability models with time-dependent hazard function based on Bayesian approach. IEEE Trans Syst Man Cybernet Part A: Syst Hum 2000;30(1).

[15] Pham H, Wang H. A quasi renewal process for software reliability and testing costs. IEEE Trans Syst Man Cybernet 2001;31(6):623–31.

[16] Pham L, Pham H. A Bayesian predictive software reliability model with pseudo-failures. IEEE Trans Syst Man Cybernet Part A: Syst Hum 2001;31(3):233–8.

[17] Ehrlich W, Prasanna B, Stampfel J, Wu J. Determining the cost of a stop-testing decision. IEEE Software 1993;10:33–42.

[18] Lyu M., editor. Software reliability engineering handbook. McGraw-Hill; 1996.

[19] Pham H. Software reliability and testing. IEEE Computer Society Press; 1995.

[20] Littlewood B, Softer A. A Bayesian modification to the Jelinski-Moranda software reliability growth model. Software Eng J 1989;30–41.

[21] Mazzuchi TA, Soyer R. A Bayes empirical-Bayes model for software reliability. IEEE Trans Reliab 1988;37:248–54.

[22] Csenki A. Bayes predictive analysis of a fundamental software reliability model. IEEE Trans Reliab 1990;39:142–8.

[23] Akaike H. A new look at statistical model identification. IEEE Trans Autom Control 1974;19:716–23.

[24] Hossain SA, Ram CD. Estimating the parameters of a non-homogeneous Poisson process model for software reliability. IEEE Trans Reliab 1993;42(4):604–12.

[25] Cox DR. Regression models and life-tables (with discussion). J R Stat Soc B 1972;34:187–220.

[26] Tohma Y, Yamano H, Ohba M, Jacoby R. The estimation of parameters of the hypergeometric distribution and its application to the software reliability growth model. IEEE Trans Software Eng 1991;17(5): 483–9.

[27] Zhang X, Pham H. An analysis of factors affecting software reliability. J Syst Software 2000;50:43–56.

[28] Zhang X. Software reliability and cost models with environmental factors. PhD Dissertation, Rutgers University, 1999, unpublished.

[29] Pham H, Zhang X. A software cost model with warranty and risk costs. IEEE Trans Comput 1999;48(1):71–5.

[30] Zhang X, Pham H. A software cost model with error removal times and risk costs. Int J Syst Sci, 1998;29(4):435–42.

[31] Zhang X, Pham H. A software cost model with warranty and risk costs. IIE Trans Qual Reliab Eng 1999;30(12):1135–42.

[32] Okumoto K, Goel AL. Optimum release time for software systems based on reliability and cost criteria. J Syst Software 1980;1:315–8.

[33] Ohtera H, Yamada S. Optimal allocation and control problems for software-testing resources. IEEE Trans Reliab 1990;39:171–6.

[34] Leung YW. Optimal software release time with a given cost budget. J Syst Software 1992;17:233–42.

[35] Dalal SR, McIntosh AA. When to stop testing for large software systems with changing code. IEEE Trans Software Eng 1994;20(4):318–23.

[36] Yang MC, Chao A. Reliability-estimation and stopping rules for software testing based on repeated appearances of bugs. IEEE Trans Reliab 1995;22(2):315–26.

[37] Hou RH, Kuo SY, Chang YP. Optimal release times for software systems with scheduled delivery time based on the HGDM. IEEE Trans Comput 1997;46(2):216–21.

[38] Pham H, Zhang X. Software release policies with gain in reliability justifying the costs. Ann Software Eng 1999;8:147–66.

[39] Teng X, Pham H. A NHPP software reliability model under random environment. Quality and Reliability Engineering Center Report, Rutgers University, 2002.

[40] Whittaker JA, Voas J. Toward a more reliable theory of software reliability. IEEE Comput 2000;(December):36–42.

[41] Voas J. Fault tolerance. IEEE Software 2001;(July/August):54–7.

PART IV

Maintenance Theory and Testing

Chapter 17

Warranty and Maintenance

D. N. P. Murthy and N. Jack

17.1 Introduction

All products are unreliable in the sense that they fail. A failure can occur early in an item's life due to manufacturing defects or late in its life due to degradation that is dependent on age and usage. Most products are sold with a warranty that offers protection to buyers against early failures over the warranty period. The warranty period offered has been getting progressively longer. For example, the warranty period for cars was 3 months in the early 1930s; this changed to 1 year in the 1960s, and currently it varies from 3 to 5 years. With extended warranties, items are covered for a significant part of their useful lives, and this implies that failures due to degradation can occur within the warranty period.

Offering a warranty implies additional costs (called "warranty servicing" costs) to the manufacturer. This warranty servicing cost is the cost of repairing item failures (through corrective maintenance (CM)) over the warranty period. For short warranty periods, the manufacturer can minimize the expected warranty servicing cost through optimal CM decision making. For long warranty periods, the degradation of an item can be controlled through preventive maintenance (PM), and this reduces the likelihood of failures. Optimal PM actions need to be viewed from a life cycle perspective (for the buyer and manufacturer), and this raises several new issues.

The literature that links warranty and maintenance is significant but small in comparison with the literature on these two topics. In this paper we review the literature dealing with warranty and maintenance and suggest areas for future research.

The outline of the paper is as follows. In Sections 17.2 and 17.3 we give brief overviews of product warranty and maintenance respectively, in order to set the background for later discussions. Following this, we review warranty and CM in Section 17.4, and warranty and PM in Section 17.5. Section 17.6 covers extended warranties and service contracts, and in Section 17.7 we state our conclusions and give a brief discussion of future research topics.

17.2 Product Warranties: An Overview

17.2.1 Role and Concept

A warranty is a contract between buyer and manufacturer associated with the sale of a product. Its purpose is basically to establish liability in the event of a premature failure of an item sold, where "failure" is meant as the inability of the item to perform its intended function. The contract specifies the promised performance and, if this is not met, the means for the buyer to be compensated. The contract also specifies the buyer's responsibilities with regards to due care and operation of the item purchased.

Warranties serve different purposes for buyers and manufacturers. For buyers, warranties serve a dual role—protectional (assurance against unsatisfactory product performance) and informational (better warranty terms indicating a more reliable product). For manufacturers, warranties also serve a dual role—promotional (to communicate information regarding product quality and to differentiate from their competitors' products) and protectional (against excessive claims from unreasonable buyers).

17.2.2 Product Categories

Products can be either new or used, and new products can be further divided into the following three categories.

1. Consumer durables, such as household appliances, computers, and automobiles bought by individual households as a single item.
2. Industrial and commercial products bought by businesses for the provision of services (*e.g.* equipment used in a restaurant, aircraft used by airlines, or copy machines used in an office). These are bought either individually (*e.g.* a single X-ray machine bought by a hospital) or as a batch, or lot, of K ($K > 1$) items (*e.g.* tires bought by a car manufacturer). Here we differentiate between "standard" off-the-shelf products and "specialized" products that are custom built to buyer's specifications.
3. Government acquisitions, such as a new fleet of defense equipment (*e.g.* missiles, ships, *etc.*). These are usually "custom built" and are products involving new and evolving technologies. They are usually characterized by a high degree of uncertainty in the product development process.

Used products are, in general, sold individually and can be consumer durables, industrial, or commercial products.

17.2.3 Warranty Policies

Both new and used products are sold with many different types of warranty. Blischke and Murthy [1] developed a taxonomy for new product warranty and Murthy and Chatophadaya [2] developed a similar taxonomy for used products. In this section we briefly discuss the salient features of the main categories of policies for new products and then briefly touch on policies for used items.

17.2.3.1 Warranties Policies for Standard Products Sold Individually

We first consider the case of new standard products where items are sold individually. The first important characteristic of a warranty is the form of compensation to the customer on failure of an item. The most common forms for non-repairable products are (1) a lump-sum rebate (*e.g.* "money-back guarantee"), (2) a free replacement of an item identical to the failed item, (3) a replacement provided at reduced cost to the buyer, and (4) some combination of the preceding terms. Warranties of Type (2) are called Free Replacement Warranties (FRWs). For warranties of Type (3), the amount of reduction is usually a function of the amount of service received by the buyer up to the time of failure, with decreasing discount as time of service increases. The discount is a percentage of the purchase price, which can change one or more times during the warranty period, or it may be a continuous function of the time remaining in the warranty period. These are called pro rata warranties (PRWs). The most common combination warranty is one that provides for free replacements up to a specified time and a replacement at pro-rated cost during the remainder of the warranty period. This is called a combination FRW/PRW. For a repairable product under an FRW policy, the failed item is repaired at no cost to the buyer.

Warranties can be further divided into different sub-groups based on dimensionality (one-dimensional warranties involve only age or usage; two-dimensional warranties involve both age and usage) and whether the warranty is renewing or not. In a renewing warranty, the repaired or replacement item comes with a new warranty identical to the initial warranty.

For used products, warranty coverage may also be limited in many ways. For example, certain types of failure or certain parts may be specifically excluded from coverage. Coverage may include parts and labor or parts only, or parts and labor for a portion of the warranty period and parts only thereafter. The variations are almost endless.

17.2.3.2 Warranty Policies for Standard Products Sold in Lots

Under this type of warranty, an entire batch of items is guaranteed to provide a specified total amount of service, without specifying a guarantee on any individual item. For example, rather than guaranteeing that each item in a batch of K items will operate without failure for W hours, the batch as a whole is guaranteed to provide at least KW hours of service. If, after the last item in the batch has failed, the total service time is less than KW hours, items are provided as specified in the warranty (*e.g.* free of charge or at pro rata cost) until such time as the total of KW hours is achieved.

17.2.3.3 Warranty Policies for Specialized Products

In the procurement of complex military and industrial equipment, warranties play a very different and important role of providing incentives to the seller to increase the reliability of the items after they are put into service. This is accomplished by requiring that the contractor maintain the items in the field and make design changes as failures are observed and analyzed. The incentive is an increased fee paid to the contractor if it can be demonstrated that the reliability of the item has, in fact, been increased. Warranties of this type are called reliability improvement warranties (RIWs).

17.2.3.4 Extended Warranties

The warranty that is an integral part of a product sale is called the base warranty. It is offered by the manufacturer at no additional cost and is factored into the sale price. An extended warranty provides additional coverage over the base warranty and is obtained by the buyer through paying a premium. Extended warranties are optional warranties that are not tied to the sale process and can either be offered by the manufacturer or by a third party (*e.g.* several credit card companies offer extended warranties for products bought using their credit cards, and some large merchants offer extended warranties). The terms of the extended warranty can be either identical to the base warranty or may differ from it in the sense that they might include cost limits, deductibles, exclusions, *etc.*

Extended warranties are similar to service contracts, where an external agent agrees to maintain a product for a specified time period under a contract with the owner of the product. The terms of the contract can vary and can include CM and/or PM actions.

17.2.3.5 Warranties for Used Products

Some of the warranty policies for second-hand products are similar to those for new products, whilst others are different. Warranties for second-hand products can involve additional features, such cost limits, exclusions, and so on. The terms (*e.g.* duration and features) can vary from item to item and can depend on the condition of the item involved. They are also influenced by the negotiating skills of the buyer. Murthy and Chattopadhyay [2] proposed a taxonomy for one-dimensional warranty policies for used items sold individually. These include options such as cost limits, deductibles, cost sharing, money-back guarantees, *etc.*

17.2.4 Issues in Product Warranty

Warranties have been analyzed by researchers from many different disciplines. The various issues dealt with are given in the following list.

1. Historical: origin and use of the notion.
2. Legal: court action, dispute resolution, product liability.
3. Legislative: Magnusson–Moss Act in the USA, warranty requirements in government acquisition (particularly military) in different countries, EU legislation.
4. Economic: market equilibrium, social welfare.
5. Behavioral: buyer reaction, influence on purchase decision, perceived role of warranty, claims behavior.
6. Consumerist: product information, consumer protection, and consumer awareness.
7. Engineering: design, manufacturing, quality control, testing.
8. Statistics: data acquisition and analysis, stochastic modeling.
9. Operations research: cost modeling, optimization.
10. Accounting: tracking of costs, time of accrual, taxation implications.
11. Marketing: assessment of consumer attitudes, assessment of the marketplace, use of warranty as a marketing tool, warranty and sales.
12. Management: integration of many of the previous items, determination of warranty policy, warranty servicing.
13. Society: public policy issues.

Blischke and Murthy [3] examined several of these issues in depth. Here, we briefly discuss two issues of relevance to this chapter.

17.2.4.1 Warranty Cost Analysis

We first look at warranty cost from the manufacturer's perspective. There are a number of approaches to the costing of warranties, and costs are clearly different for buyer and seller. The following are some of the methods for calculating costs that might be considered.

1. Cost per item sold: this per unit cost may be calculated as the total cost of warranty, as determined by general principles of accounting, divided by number of items sold.
2. Cost per unit of time.

These costs are random variables, since claims under warranty and the cost to rectify each claim are uncertain. The selection of an appropriate cost basis depends on the product, the context and perspective. The type of customer—individual, corporation, or government—is also important, as are many other factors.

From the buyer's perspective, the time interval of interest is from the instant an item is purchased to the instant when it is disposed or replaced. This includes the warranty period and the post-warranty period. The cost of rectification over the warranty period depends on the type of warranty. It can vary from no cost (in the case of an FRW) to cost sharing (in the case of a PRW). The cost of rectification during the post-warranty period is borne completely by the buyer. As such, the variable of interest to the buyer is the cost of maintaining an item over its useful life. Hence, the following cost estimates are of interest.

1. Cost per item averaging over all items purchased plus those obtained free or at reduced price under warranty.

2. Life cycle cost of ownership of an item with or without warranty, including purchase price, operating and maintenance cost, *etc.*, and finally the cost of disposal.
3. Life cycle cost of an item and its replacements, whether purchased at full price or replaced under warranty, over a fixed time horizon.

The warranty costs depend on the nature of the maintenance actions (corrective and/or preventive) used and we discuss this further later in the chapter.

17.2.4.2 Warranty Servicing

Warranty servicing costs can be minimized by using optimal servicing strategies. In the case where only CM actions are used, two possible strategies are as follows.

1. Replace versus repair. The manufacturer has the option of either repairing or replacing a failed item by a new one.
2. Cost repair limit strategy. Here, an estimate of the cost to repair a failed item is made and, by comparing it with some specified limit, the failed item is either repaired or replaced by a new one.

17.2.5 Review of Warranty Literature

Review papers on warranties include a three-part paper in the *European Journal of Operational Research*: Blischke and Murthy [1] deal with concepts and taxonomy, Murthy and Blischke [4] deal with a framework for the study of warranties, and Murthy and Blischke [5] deal with warranty cost analysis. Papers by Blischke [6] and Chukova *et al.* [7] deal mainly with warranty cost analysis. Two recent review papers are by Thomas and Rao [8] and Murthy and Djamaludin [9], with the latter dealing with warranty from a broader perspective.

Over the last 6 years, four books have appeared on the subject. Blischke and Murthy [10] deal with cost analysis of over 40 different warranty policies for new products. Blischke and Murthy [3] provide a collection of review papers dealing with warranty from many different perspectives. Sahin and Polotoglu [11] deal with the cost analysis of some basic one-dimensional warranty policies. Brennan [12] deals with warranty administration in the context of defense products.

Finally, Djamaludin *et al.* [13] list over 1500 papers on warranties, dividing these into different categories. This list does not include papers that have appeared in the law journals.

17.3 Maintenance: An Overview

As indicated earlier, maintenance can be defined as actions (i) to control the deterioration process leading to failure of a system, and (ii) to restore the system to its operational state through corrective actions after a failure. The former is called PM and the latter CM.

CM actions are unscheduled actions intended to restore a system from a failed state into a working state. These involve either repair or replacement of failed components. In contrast, PM actions are scheduled actions carried out either to reduce the likelihood of a failure or prolong the life of the system.

17.3.1 Corrective Maintenance

In the case of a repairable product, the behavior of an item after a repair depends on the type of repair carried out. Various types of repair action can be defined.

- *Good as new repair.* Here, the failure time distribution of repaired items is identical to that of a new item, and we model successive failures using an ordinary renewal process. In real life this type of repair would seldom occur.
- *Minimal repair.* A failed item is returned to operation with the same effective age as it possessed immediately prior to failure. Failures then occur according to a non-homogeneous Poisson process with an intensity function having the same form as the hazard rate of the time to first failure distribution. This type of

rectification model is appropriate when item failure is caused by one of many components failing and the failed component being replaced by a new one (see Murthy [14] and Nakagawa and Kowada [15]).

- *Different from new repair (I).* Sometimes when an item fails, not only the failed components are replaced but also others that have deteriorated sufficiently. These major overhauls result in $F_1(x)$, the failure time distribution function of all repaired items, being different from $F(x)$, the failure time distribution function of a new item. The mean time to failure of a repaired item is assumed to be smaller than that of a new item. In this case, successive failures are modeled by a modified renewal process.
- *Different from new repair (II).* In some instances, the failure distribution of a repaired item depends on the number of times the item has been repaired. This situation can be modeled by assuming that the distribution function after the jth repair ($j \geq 1$) is $F_j(x)$ with the mean time to failure μ_j decreasing as j increases.

17.3.2 Preventive Maintenance

PM actions can be divided into the following categories.

- *Clock-based maintenance.* PM actions are carried out at set times. An example of this is the "block replacement" policy.
- *Age-based maintenance.* PM actions are based on the age of the component. An example of this is the "age replacement" policy.
- *Usage-based maintenance.* PM actions are based on usage of the product. This is appropriate for items such as tires, components of an aircraft, and so forth.
- *Condition-based maintenance.* PM actions are based on the condition of the component being maintained. This involves monitoring of one or more variables characterizing the wear process (*e.g.* crack growth in a mechanical component). It is often difficult to measure the variable of interest directly, and in this case some other variable may be used to obtain estimates of the variable of interest. For example, the wear of bearings can be measured by dismantling the crankcase of an engine. However, measuring the vibration, noise, or temperature of the bearing case provides information about wear, since there is a strong correlation between these variables and bearing wear.
- *Opportunity-based maintenance.* This is applicable for multi-component systems, where maintenance actions (PM or CM) for a component provide an opportunity for carrying out PM actions on one or more of the remaining components of the system.
- *Design-out maintenance.* This involves carrying out modifications through redesigning the component. As a result, the new component has better reliability characteristics.

In general, PM is carried out at discrete time instants. In cases where the PM actions are carried out fairly frequently they can be treated as occurring continuously over time. Many different types of model formulations have been proposed to study the effect of PM on the degradation and failure occurrence of items and to derive optimal PM strategies.

17.3.3 Review of Maintenance Literature

Several review papers on maintenance have appeared over the last 30 years. These include McCall [16], Pierskalla and Voelker [17], Sherif and Smith [18], Monahan [19], Jardine and Buzacott [20], Thomas [21], Gits [22], Valdez-Flores and Feldman [23], Pintelon and Gelders [24], Dekker [25], and Scarf [26]. Cho and Parlar [27] and Dekker *et al.* [28] deal with the maintenance of multi-component systems and Pham and Wang [29] review models with imperfect maintenance. These review papers contain references to the large number of papers and books dealing with maintenance.

17.4 Warranty and Corrective Maintenance

The bulk of the literature on warranty and corrective maintenance deals with warranty servicing costs under different CM actions. We first review the literature linking warranties and CM for new products, then we consider the literature for used items.

Although most warranted items are complex multi-component systems, the 'black-box' approach has often been used to model time to first failure. The items are viewed as single entities characterized by two states—working and failed—and $F(x)$, the distribution function for time to first failure, is usually selected, either on an intuitive basis, or from the analysis of failure data. Subsequent failures are modeled by an appropriate stochastic point process formulation depending on the type of rectification action used. If the times to complete rectification actions are very small in comparison with the times between failures, they are ignored in the modeling.

Most of the literature on warranty servicing (for one- and two-dimensional warranties) is summarized by Blischke and Murthy [3, 10]. We confine our discussion to one-dimensional warranties, for standard products, sold individually, and focus on issues relating to maintenance actions and optimal decision making.

Models where items are subjected to different from new repair (I) include those of Biedenweg [30], and Nguyen and Murthy [31, 32]. Biedenweg [30] shows that the optimal strategy is to replace with a new item at any failure occurring up to a certain time measured from the initial purchase and then repair all other failures that occur during the remainder of the warranty period. This technique of splitting the warranty period into distinct intervals for replacement and repair is also used by Nguyen and Murthy [31, 32], where any item failures occurring during the second part of the warranty period are rectified using a stock of used items [31]. Nguyen and Murthy [32] extend Biedenweg's [30] model by adding a third interval, where failed items are either replaced or repaired and a new warranty is given at each failure.

The first warranty servicing model, involving minimal repair and assuming constant repair and replacement costs, is that of Nguyen [33]. As in Biedenweg [30], the warranty period is split into a replacement interval followed by a repair interval. Under this strategy a failed item is always replaced by a new one in the first interval, irrespective of its age at failure. Thus, if the failure occurs close to the beginning of the warranty then the item will be replaced at a higher cost than that of a repair and yet there will be very little reduction in its effective age. This is the major limitation of this model, and makes the strategy clearly sub-optimal.

Using the same assumptions as Nguyen [33], Jack and Van der Duyn Schouten [34] investigated the structure of the manufacturer's optimal servicing strategy using a dynamic programming model. They showed that the repair–replacement decision on failure should be made by comparing the item's current age with a time-dependent control limit function $h(t)$. The item is replaced on failure at time t if and only if its age is greater than $h(t)$. A new servicing strategy proposed by Jack and Murthy [35] involves splitting the warranty period into distinct intervals for carrying out repairs and replacements with no need to monitor the item's age. In intervals near the beginning and end of the warranty period the failed items are always repaired, whereas in the intermediate interval at most one failure replacement is carried out.

Murthy and Nguyen [36] examined the optimal cost limit repair strategy where, at each failure during the warranty period, the item is inspected and an estimate of the repair cost determined. If this estimate is less than a specified limit then the failed item is minimally repaired, otherwise a replacement is provided at no cost to the buyer.

For used items, Murthy and Chattopadhyay [2] deal with both FRW and PRW policies with no cost limits. Chattopadhyay and Murthy [37] deal with the cost analysis of limit on total cost (LTC) policies. Chattopadhyay and Murthy [38] deal with the following three different policies—specific parts exclusion (SPE) policy; limit on individual cost (LIC) policy; and limit on individual and total cost (LITC) policy—and discuss their cost analysis.

17.5 Warranty and Preventive Maintenance

PM actions are carried out either to reduce the likelihood of a failure or to prolong the life of an item. PM can be perfect (restoring the item to "good-as-new") or imperfect (restoring the item to a condition that is between as "good-as new" and as "bad-as-old").

PM over the warranty period has an impact on the warranty servicing cost. It is worthwhile for the manufacturer to carry out this maintenance only if the reduction in the warranty cost exceeds the cost of PM. From a buyer's perspective, a myopic buyer might decide not to invest in any PM over the warranty period, as item failures over this period are rectified by the manufacturer at no cost to the buyer. Investing in maintenance is viewed as an additional unnecessary cost. However, from a life cycle perspective the total life cycle cost to the buyer is influenced by maintenance actions during the warranty period and the post-warranty period. This implies that the buyer needs to evaluate the cost under different scenarios for PM actions.

This raises several interesting questions. These include the following:

1. Should PM be used during the warranty period?
2. If so, what should be the optimal maintenance effort? Should the buyer or the manufacturer pay for this, or should it be shared?
3. What level of maintenance should the buyer use during the post-warranty period?

PM actions are normally scheduled and carried out at discrete time instants. When the PM is carried out frequently and the time between the two successive maintenance actions is small, then one can treat the maintenance effort as being continuous over time. This leads to two different ways (discrete and continuous) of modeling maintenance effort.

Another complicating factor is the information aspect. This relates to issues such as the state of item, the type of distribution function appropriate for modeling failures, the parameters of the distribution function, *etc.* The two extreme situations are complete information and no information, but often the information available to the manufacturer and the buyer lies somewhere between these two extremes and can vary. This raises several interesting issues, such as the adverse selection and moral hazard problems. Quality variations (with all items not being statistically similar) add yet another dimension to the complexity.

As such, the effective study of PM for products sold under warranty requires a framework that incorporates the factors discussed above. The number of factors to be considered and the nature of their characterization result in many different model formulations linking PM and warranty. We now present a chronological review of the models that have been developed involving warranty and PM.

Ritchken and Fuh [39] discuss a preventive replacement policy for a non-repairable item after the expiry of a PRW. Any item failure occurring within the warranty period results in a replacement by a new item with the cost shared between the producer and the buyer. After the warranty period finishes, the item in use is either preventively replaced by the buyer after a period T (measured from the end of the warranty period) or replaced on failure, whichever occurs first. A new warranty is issued with the replacement item and the optimal T^* is found by minimizing the buyer's asymptotic expected cost per unit time.

Chun and Lee [40] consider a repairable item with an increasing failure rate that is subjected to periodic imperfect PM actions both during the warranty period and after the warranty expires. They assume that each PM action reduces the item's age by a fixed amount and all failures between PM actions are minimally repaired. During the warranty period, the manufacturer pays all the repair costs and a proportion of each PM cost, with the proportion depending on when the action is carried out. After the warranty expires, the buyer pays for the cost of all repairs and PM. The optimal period between PM actions is obtained by minimizing the buyer's asymptotic expected cost per unit time over an infinite horizon. An example is given for an item with a Weibull failure distribution.

Chun [41] dealt with a similar problem to Chun and Lee [40], but with the focus instead on the manufacturer's periodic PM strategy over the warranty period. The optimal number of PM actions N^* is obtained by minimizing the expected cost of repairs and PMs over this finite horizon.

Jack and Dagpunar [42] considered the model studied by Chun [41] and showed that, when the product has an increasing failure rate, a strict periodic policy for PM actions is not the optimal strategy. They showed that, for a warranty of length W and a fixed amount of age reduction at each PM, the optimal strategy is to perform N PMs at intervals x apart, followed by a final interval at the end of the warranty of length $W - Nx$, where only minimal repairs are carried out. Performing PMs with this frequency means that the item is effectively being restored to as good-as-new condition.

Dagpunar and Jack [43] assumed that the amount of age reduction is under the control of the manufacturer and the cost of each PM action depends on the operating age of the item and on the effective age reduction resulting from the action. In this model, the optimal strategy can result in the product not being restored to as good as new at each PM. The optimal number of PM actions N^*, optimal operating age to perform a PM s^*, and optimal age reduction x^* are obtained by minimizing the manufacturer's expected warranty servicing cost.

Sahin and Polatoglu [44] discussed two types of preventive replacement policy for the buyer of a repairable item following the expiry of a warranty. Failures over the warranty period are minimally repaired at no cost to the buyer. In the first model, the item is replaced by a new item at a fixed time T after the warranty ends. Failures before T are minimally repaired, with the buyer paying all repair costs. In the second model, the replacement is postponed until the first failure after T. They considered both stationary and non-stationary strategies in order to minimize the long-run average cost to the buyer. The non-stationary strategies depend on the information regarding item age and number of previous failures that might be available to the buyer at the end of the warranty period. Sahin and Polatoglu [45] dealt with a model that examined PM policies with uncertainty in product quality.

Monga and Zuo [46] presented a model for the reliability-based design of a series–parallel system considering burn-in, warranty, and maintenance. They minimized the expected system life cycle cost and used genetic algorithms to determine the optimal values of system design, burn-in period, PM intervals, and replacement time. The manufacturer pays the costs of rectifying failures under warranty and the buyer pays post-warranty costs.

Finally, Jung *et al.* [47] determined the optimal number and period for periodic PMs following the expiry of a warranty by minimizing the buyer's asymptotic expected cost per unit time. Both renewing PRWs and renewing FRWs are considered. The item is assumed to have a monotonically increasing failure rate and the PM actions slow down the degradation.

17.6 Extended Warranties and Service Contracts

The literature on extended warranties deals mainly with the servicing cost to the provider of these extended warranties. This is calculated using models similar to those for the cost analysis of base warranties with only CM actions.

Padmanabhan and Rao [48] and Padmanabhan [49] examined extended warranties with heterogeneous customers with different attitudes to risk and captured through a utility function. Patankar and Mitra [50] considered the case where items are sold with PRW where the customer is given the option of renewing the initial warranty by paying a premium should the product not fail during the initial warranty period.

Mitra and Patankar [51] dealt with the model where the product is sold with a rebate policy and the buyer has the option to extend the warranty should the product not fail during the initial warranty period. Padmanabhan [52] discussed alternative theories and the design of extended warranty policies.

Service contracts also involve maintenance actions. Murthy and Ashgarizadeh [53, 54] dealt with service contracts involving only CM. The authors are unaware of any service contract models that deal with PM or optimal decision making with regard to maintenance actions.

17.7 Conclusions and Topics for Future Research

In this final section we again stress the importance of maintenance modeling in a warranty context, we emphasize the need for model validation, and we then outline some further research topics that link maintenance and warranty.

Post-sale service by a manufacturer is an important element in the sale of a new product, but it can result in substantial additional costs. These warranty servicing costs, which can vary between 0.5 and 7% of a product's sale price, have a significant impact on the competitive behavior of manufacturers. However, manufacturers can reduce the servicing costs by adopting proper CM and PM strategies, and these are found by using appropriate maintenance models.

Models for determining optimal maintenance strategies in a warranty context were reviewed in Sections 17.4 and 17.5. The strategies discussed, whilst originating from more general maintenance models, have often had to be adapted to suit the special finite time horizon nature of warranty problems. However, all warranty maintenance models will only provide useful information to manufacturers provided they can be validated, and this requires the collection of accurate product failure data.

We have seen that some maintenance and warranty modeling has already been done, but there is still scope for new research, and we now suggest some topics worthy of investigation.

1. For complex products (such as locomotives, aircraft, *etc.*) the (corrective and preventive) maintenance needs for different components vary. Any realistic modeling requires grouping the components into different categories based on the maintenance needs. This implies modeling and analysis at the component level rather than the product level (see Chukova and Dimitrov [55]).
2. The literature linking warranty and PM deals primarily with age-based or clock-based maintenance. Opportunity-based maintenance also offers potential for reducing the overall warranty servicing costs. The study of optimal opportunistic maintenance policies in the context of warranty servicing is an area for new research.
3. Our discussion has been confined to one-dimensional warranties. Optimal maintenance strategies for two-dimensional warranties have received very little attention. Iskandar and Murthy [56] deal with a simple model, and there is considerable scope for more research on this topic.
4. The optimal (corrective and preventive) maintenance strategies discussed in the literature assume that the model structure and model parameters are known. In real life, this is often not true. In this case, the optimal decisions must be based on the information available, and this changes over time as more failure data are obtained. This implies that the modeling must be done using a Bayesian framework. Mazzuchi and Soyer [57] and Percy and Kobbacy [58] dealt with this issue in the context of maintenance, and a topic for research is to apply these ideas in the context of warranties and extended warranties.
5. The issue of risk becomes important in the context of service contracts. When the attitude

to risk varies across the population and there is asymmetry in the information available to different parties, several new issues (such as moral hazard, adverse selection) need to be incorporated into the model (see Murthy and Padmanabhan [59]).

References

[1] Blischke WR, Murthy DNP. Product warranty management—I. A taxonomy for warranty policies. Eur J Oper Res 1991;62:127–48.

[2] Murthy DNP, Chattopadhyay G. Warranties for second-hand products. In: Proceedings of the FAIM Conference, Tilburg, Netherlands, June, 1999.

[3] Blischke WR, Murthy DNP. Product warranty handbook. New York: Marcel Dekker; 1996.

[4] Murthy DNP, Blischke WR. Product warranty management—II: an integrated framework for study. Eur J Oper Res 1991;62:261–80.

[5] Murthy DNP, Blischke WR. Product warranty management—III: a review of mathematical models. Eur J Oper Res 1991;62:1–34.

[6] Blischke WR. Mathematical models for warranty cost analysis. Math Comput Model 1990;13:1–16.

[7] Chukova SS, Dimitrov BN, Rykov VV. Warranty analysis. J Sov Math 1993;67:3486–508.

[8] Thomas M, Rao S. Warranty economic decision models: a review and suggested directions for future research. Oper Res 2000;47:807–20.

[9] Murthy DNP, Djamaludin I. Product warranties—a review. Int J Prod Econ 2002;in press.

[10] Blischke WR, Murthy DNP. Warranty cost analysis. New York: Marcel Dekker; 1994.

[11] Sahin I, Polatoglu H. Quality, warranty and preventive maintenance. Boston: Kluwer Academic Publishers; 1998.

[12] Brennan JR. Warranties: planning, analysis and implementation. New York: McGraw-Hill; 1994.

[13] Djamaludin I, Murthy DNP, Blischke WR. Bibliography on warranties. In: Blischke WR, Murthy DNP, editors. Product warranty handbook. New York: Marcel Dekker; 1996.

[14] Murthy DNP. A note on minimal repair. IEEE Trans Reliab 1991;R-40:245–6.

[15] Nakagawa T, Kowada M. Analysis of a system with minimal repair and its application to replacement policy. Eur J Oper Res 1983;12:176–82.

[16] McCall JJ. Maintenance policies for stochastically failing equipment: a survey. Manage Sci 1965;11:493–524.

[17] Pierskalla WP, Voelker JA. A survey of maintenance models: the control and surveillance of deteriorating systems. Nav Res Logist Q 1976;23:353–88.

[18] Sherif YS, Smith ML. Optimal maintenance models for systems subject to failure—a review. Nav Res Logist Q 1976;23:47–74.

[19] Monahan GE. A survey of partially observable Markov decision processes: theory, models and algorithms. Manage Sci 1982;28:1–16.

[20] Jardine AKS, Buzacott JA. Equipment reliability and maintenance. Eur J Oper Res 1985;19:285–96.

[21] Thomas LC. A survey of maintenance and replacement models for maintainability and reliability of multi-item systems. Reliab Eng 1986;16:297–309.

[22] Gits CW. Design of maintenance concepts. Int J Prod Econ 1992;24:217–26.

[23] Valdez-Flores C, Feldman RM. A survey of preventive maintenance models for stochastically deteriorating single-unit systems. Nav Res Logist Q 1989;36:419–46.

[24] Pintelon LM, Gelders LF. Maintenance management decision making. Eur J Oper Res 1992;58:301–17.

[25] Dekker R. Applications of maintenance optimization models: a review and analysis. Reliab Eng Syst Saf 1996;51:229–40.

[26] Scarf PA. On the application of mathematical models in maintenance. Eur J Oper Res 1997;99:493–506.

[27] Cho D, Parlar M. A survey of maintenance models for multi-unit systems. Eur J Oper Res 1991;51:1–23.

[28] Dekker R, Wildeman RE, Van der Duyn Schouten FA. A review of multi-component maintenance models with economic dependence. Math Methods Oper Res 1997;45:411–35.

[29] Pham H, Wang H. Imperfect maintenance. Eur J Oper Res 1996; 94:425–38.

[30] Biedenweg FM. Warranty analysis: consumer value vs manufacturers cost. Unpublished PhD thesis, Stanford University, USA, 1981.

[31] Nguyen DG, Murthy DNP. An optimal policy for servicing warranty. J Oper Res Soc 1986;37:1081–8.

[32] Nguyen DG, Murthy DNP. Optimal replace-repair strategy for servicing items sold with warranty. Eur J Oper Res 1989;39:206–12.

[33] Nguyen DG. Studies in warranty policies and product reliability. Unpublished PhD thesis, The University of Queensland, Australia, 1984.

[34] Jack N, Van der Duyn Schouten FA. Optimal repair-replace strategies for a warranted product. Int J Prod Econ 2000;67:95–100.

[35] Jack N, Murthy DNP. Servicing strategies for items sold with warranty. J Oper Res 2001;52:1284–8.

[36] Murthy DNP, Nguyen DG. An optimal repair cost limit policy for servicing warranty. Math Comput Model 1988;11:595–9.

[37] Chattopadhyay GN, Murthy DNP. Warranty cost analysis for second hand products. J Math Comput Model 2000;31:81–8.

[38] Chattopadhyay GN, Murthy DNP. Warranty cost analysis for second-hand products sold with cost sharing policies. Int Trans Oper Res 2001;8:47–68.

[39] Ritchken PH, Fuh D. Optimal replacement policies for irreparable warrantied items. IEEE Trans Reliab 1986;R-35:621–3.

[40] Chun YH, Lee CS. Optimal replacement policy for a warrantied system with imperfect preventive maintenance operations. Microelectron Reliab 1992;32:839–43.

[41] Chun YH. Optimal number of periodic preventive maintenance operations under warranty. Reliab Eng Syst Saf 1992;37:223–5.

[42] Jack N, Dagpunar JS. An optimal imperfect maintenance policy over a warranty period. Microelectron Reliab 1994;34:529–34.

[43] Dagpunar JS, Jack N. Preventive maintenance strategy for equipment under warranty. Microelectron Reliab 1994;34:1089–93.

[44] Sahin I, Polatoglu H. Maintenance strategies following the expiration of warranty. IEEE Trans Reliab 1996;45:220–8.

[45] Sahin I, Polatoglu H. Manufacturing quality, reliability and preventive maintenance. Prod Oper Manage 1996;5:132–47.

[46] Monga A, Zuo MJ. Optimal system design considering maintenance and warranty. Comput Oper Res 1998;9:691–705.

[47] Jung GM., Lee CH, Park DH. Periodic preventive maintenance policies following the expiration of warranty. Asia-Pac J Oper Res 2000;17:17–26.

[48] Padmanabhan V, Rao RC. Warranty policy and extended service contracts: theory and an application to automobiles. Market Sci 1993;12:230–47.

[49] Padmanabhan V. Usage heterogeneity and extended warranty. J Econ Manage Strat 1995;4:33–53.

[50] Patankar JG, Mitra A. A multicriteria model for a renewable warranty program. J Eng Val Cost Anal 1999; 2:171–85.

[51] Mitra A, Patankar JG. Market share and warranty costs for renewable warranty programs. Int J Prod Econ 1997;50:155–68.

[52] Padmanabhan V. Extended warranties. In: Blischke WR, Murthy DNP, editors. Product warranty handbook. New York: Marcel Dekker; 1996. p.439–51.

[53] Murthy DNP, Ashgarizadeh E. A stochastic model for service contracts. Int J Reliab Qual Saf Eng 1998;5:29–45.

[54] Murthy DNP, Ashgarizadeh E. Optimal decision making in a maintenance service operation. Eur J Oper Res 1999;116:259–73.

[55] Chukova SS, Dimitrov BV. Warranty analysis for complex systems. In: Blischke WR, Murthy DNP, editors. Product warranty handbook. New York: Marcel Dekker; 1996. p.543–84.

[56] Iskandar BP, Murthy DNP. Repair-replace strategies for two-dimensional warranty policies. In: Proceedings of the Third Australia-Japan Workshop on Stochastic Models, Christchurch, September, 1999; p.206–13.

[57] Mazzuchi TA, Soyer R. A Bayesian perspective on some replacement strategies. IEEE Trans Reliab 1996;R-37:248–54.

[58] Percy DF, Kobbacy KAH. Preventive maintenance modelling—a Bayesian perspective. J Qual Maint Eng 1996;2:44–50.

[59] Murthy DNP, Padmanabhan V. A dynamic model of product warranty with consumer moral hazard. Research paper no. 1263. Graduate School of Business, Stanford University, Stanford, CA, 1993.

Mechanical Reliability and Maintenance Models

Gianpaolo Pulcini

18.1 Introduction

This chapter deals with the reliability modeling of repairable mechanical equipment and illustrates some useful models, and related statistical procedures, for the statistical analysis of failure data of repairable mechanical equipment. Repairable equipment is an item that may fail to perform at least one of its required functions many times during its lifetime. After each failure, it is not replaced but is restored to satisfactory performance by repairing or by substituting the failed part (a complete definition of repairable equipment can be found in Ascher and Feingold [1] (p.8)).

Many products are designed to be repaired and put back into service. However, the distinction between repairable and non-repairable units sometimes depends on economic considerations rather than equipment design, since it may be uneconomical, unsafe, or unwanted to repair the failed item. For example, a toaster is repaired numerous times in a poor society (and hence it is repairable equipment), whereas it is often discarded when it fails and is replaced in a rich society (therefore, it is a non-repairable item).

Apart from possible improvement phenomena in an early phase of the operating time, mechanical repairable equipment is subjected, unlike electronic equipment, to degradation phenomena with operating time, so that the failures become increasingly frequent with time. This makes the definition of reliability models for mechanical equipment more complicated, both due to the presence of multiple failure mechanisms and to the complex maintenance policy that is often carried out. In fact, the failure pattern (*i.e.* the sequence of times between successive failures) of mechanical repairable equipment strictly depends not only on the failure mechanisms and physical structure of the equipment, but also on the type of repair that is performed at failure (corrective maintenance). Moreover, the failure pattern also depends on the effect of preventive maintenance actions that may be carried out in the attempt to prevent equipment degradation.

Hence, an appropriate modeling of the failure pattern of mechanical repairable equipment cannot but take into account the type of maintenance policy being employed. On the other hand, corrective and preventive maintenance actions are generally classified just in terms of their effect on the operating conditions of the equipment (*e.g.* see Pham and Wang [2]). In particular:

1. if the maintenance action restores the equipment to the as same as new condition, the maintenance is defined as a *perfect maintenance*;
2. if the maintenance action brings the equipment to the condition it was in just before the maintenance (same as old), the maintenance is a *minimal maintenance*;
3. if the maintenance action significantly improves the equipment condition, even without bringing the equipment to a seemingly new condition (better than old), the maintenance is an *imperfect maintenance*;
4. if the maintenance action brings the equipment to a worse condition than it was in just before the maintenance (worse than old), the maintenance is a *worse maintenance*.

Of course, the classification of a given maintenance action depends not only on the action that is performed (repair or substitution of the failed part, major overhaul of the equipment, and so on), but also on the physical structure and complexity of the equipment. For example, the substitution of a failed part of complex equipment does not generally produce an improvement to the equipment conditions, and hence can be classified as a minimal repair. On the contrary, if the equipment is not complex, then the same substitution can produce a noticeable improvement and could be classified as an imperfect repair.

In Section 18.2, the basic concepts of stochastic point processes are illustrated. Sections 18.3–18.6 deal with stochastic models used to describe the failure pattern of repairable equipment subject to perfect, minimal, imperfect or worse maintenance, as well as to some complex maintenance policies. The final section illustrates models that describe the failure pattern of equipment undergoing development programs that are carried out to improve reliability through design modifications (*reliability growth*). Of course, in this context, the observed failure pattern is also strongly influenced by design changes.

Owing to obvious limitations of space, this discussion cannot be exhaustive, and simply intends to illustrate those models that have been mainly referred to in reliability analysis.

18.2 Stochastic Point Processes

The more commonly used way to model the reliability of repairable equipment is through stochastic point processes. A stochastic point process is a probability model to describe a physical phenomenon that is characterized by highly localized events that occur randomly in the continuum [1] (p.17). In reliability modeling, the events are the equipment failures and the continuum is the operating time (or any other measure of the actual use of the equipment).

Suppose that the failures of repairable equipment, observed over time $t \geq 0$, occur at times $t_1 < t_2 < \cdots$. Time t_i $(i = 1, 2, \ldots)$, which is measured from zero, is called the *arrival time* to the ith failure. If repair times are negligible, or are measured on a different time scale, then the *interarrival times* $x_i = t_i - t_{i-1}$ (where $t_0 = 0$ and $i = 1, 2, \ldots$) represent the operating times between the $(i-1)$th and ith failures, and constitute the failure pattern of the equipment.

Let $N(s, t)$ denote the (integer valued) random variable counting the number of failures that occur in the time interval (s, t). It includes both the number of failures occurring in (s, t) and the times at which they occur. Such a point process can be specified mathematically in several ways, *e.g.* via the joint distributions of the counts of failures in arbitrary time intervals. One convenient way [3] is via its *complete* (or conditional) *intensity function* (CIF) defined as [4] (p.9):

$$\lambda(t; H_t) = \lim_{\Delta t \to 0} \frac{\Pr\{N(t, t + \Delta t) \geq 1 \mid H_t\}}{\Delta t} \tag{18.1}$$

where $H_t = \{N(0, s) : 0 \leq s \leq t\}$ represents the entire *history* of the process through time t. For small Δt values, the product $\lambda(t; H_t)\Delta t$ is approximately equal to the conditional probability that at least one failure (not necessarily the first) occurs in the time interval $(t, t + \Delta t)$, given the history up to t [3].

The expectation of the number $N(t) = N(0, t)$ of failures up to t, say $M(t) = E\{N(t)\}$, is a non-decreasing right continuous function. Its derivative $\mu(t) = M'(t)$, if it exists, is called the *instantaneous rate of occurrence of failures* (ROCOF) and represents the time rate of change of the expected number of failures [1] (p.19). If the probability of a failure at any specified point is zero, then the expected number of failures $M(t)$ is continuous everywhere and the process is said to be *regular* [5] (p.15–16).

If simultaneous failures cannot occur, so that the t_i are distinct, then the point process is said to be *orderly* [1] (p.23). Under orderliness conditions, the ROCOF $\mu(t)$ numerically equals the (*unconditional*) intensity function $\lambda(t)$, if they exist, which is defined as

$$\lambda(t) = \lim_{\Delta t \to 0} \frac{\Pr\{N(t, t + \Delta t) \geq 1\}}{\Delta t} \tag{18.2}$$

The intensity $\lambda(t)$ can be viewed as the expectation of the CIF (Equation 18.1) over the entire space of possible histories. Of course, if the CIF is independent of H_t then the two intensity functions coincide. Also, for small Δt, the product $\lambda(t)\Delta t$ is approximately equal to the (unconditional) probability that at least one failure occurs in the time interval $(t, t + \Delta t)$ [6]. Note that, if the process is not orderly, then the ROCOF is greater than the intensity function $\lambda(t)$ [4] (p.26) and does not specify the process uniquely [3].

The assumption of no simultaneous failures is usually reasonable in most applications and allows useful distribution functions to be derived from the CIF.

Under orderliness conditions, the CIF can be related to the conditional distribution $f_i(t; H_t)$ of the arrival time t_i of the ith failure. Suppose that the $(i-1)$th failure has occurred at t_{i-1}; then, for $t > t_{i-1}$:

$$\lambda(t; H_t) = \frac{f_i(t; H_t)}{1 - F_i(t; H_t)} \qquad t \geq t_{i-1} \tag{18.3}$$

where H_t is the observed history that consists of $(i-1)$ failures. Hence, the conditional density function $f_i(t; H_t)$ of t_i is

$$f_i(t; H_t) = \lambda(t; H_t) \exp\left[-\int_{t_{i-1}}^{t} \lambda(z; H_z)\, \mathrm{d}z\right] \qquad t \geq t_{i-1} \tag{18.4}$$

From Equation 18.4, the conditional density function $f_i(x; H_t)$ of the ith interarrival time

$x_i = t_i - t_{i-1}$ is

$$f_i(x; H_t) = \lambda(t_{i-1} + x; H_{t_{i-1}+x}) \times \exp\left[- \int_0^x \lambda(t_{i-1} + z; H_{t_{i-1}+z}) \, dz \right] \quad x \geq 0 \tag{18.5}$$

The equipment *reliability* $R(t, t + w)$, defined as the probability that the equipment operates without failing over the time interval $(t, t + w)$, generally depends on the entire history of the process, and hence can be expressed as [1] (p.24–25)

$$R(t, t + w) \equiv \Pr\{N(t, t + w) = 0 \mid H_t\} \tag{18.6}$$

For any orderly point process, the equipment reliability is

$$R(t, t + w) = \exp\left[- \int_t^{t+w} \lambda(x; H_x) \, dx \right] \tag{18.7}$$

The *forward waiting time* w_t is the time to the next failure measured from an arbitrary time t: $w_t = t_{N(t)+1} - t$, where $t_{N(t)+1}$ is the time of the next failure. The cumulative distribution function of w_t is related to the equipment reliability (Equation 18.6) by [1] (p.23–25):

$$\begin{aligned} K_t(w) &= \Pr\{w_t \leq w \mid H_t\} \\ &= \Pr\{N(t, t + w) \geq 1 \mid H_t\} \\ &= 1 - R(t, t + w) \end{aligned} \tag{18.8}$$

which, under orderliness conditions, becomes:

$$K_t(w) = 1 - \exp\left[- \int_0^w \lambda(t + x; H_{t+x}) \, dx \right] \tag{18.9}$$

Of course, when the forward waiting time w_t is measured from a failure time, say $t_{N(t)}$, then w_t coincides with the interarrival time $x_{N(t)+1} = t_{N(t)+1} - t_{N(t)}$.

For any orderly process, the likelihood relative to a time-truncated sample $t_1 < t_2 < \cdots < t_n \leq T$ is expressible in the form [7]

$$L(\boldsymbol{\theta}) = \prod_{i=1}^{n} \lambda(t_i; H_{t_i}) \exp\left[- \int_0^T \lambda(t; H_t) \, dt \right] \tag{18.10}$$

where $\boldsymbol{\theta}$ is the vector of unknown parameters of the CIF. For failure-truncated sampling, $T \equiv t_n$.

Finally, the integrated CIF over the time between two successive failures, say $e_i = \int_{t_{i-1}}^{t_i} \lambda(t; H_t) \, dt$, under orderliness conditions, is distributed as a standard exponential random variable. This property allows one to have a useful graphical tool in order to check if the assumed stochastic model describes adequately an observed data set [3]. In fact, if the assumed model is satisfactory, then the estimate of the generalized residuals e_i should look roughly like a standard exponential sample.

Since the CIF characterizes the failure pattern of the repairable equipment, its form depends both on the failure mechanisms acting on the equipment during its operating life and on the maintenance policy that is performed. We assume that the CIF can be discontinuous with operating time and discontinuities can occur only at the (corrective or preventive) maintenance epochs, so that the CIF is continuous in any time interval between two successive maintenance epochs.

In the following, some point processes that satisfy the reasonable orderliness conditions and meet practical applications in the reliability analysis of mechanical equipment are shown depending on the type of maintenance policy being employed.

18.3 Perfect Maintenance

Perfect maintenance is a maintenance action that brings the equipment to a like-new condition. In other words, each time the equipment is perfectly maintained, it is as though we are starting over with a new piece of equipment. Then, if a perfect maintenance is performed at time τ, the CIF at any time $t \geq \tau$ depends only on the history of the process from τ up to t.

If the equipment is perfectly repaired at each failure, then its failure pattern can be described by a *renewal process* (same as new process). The CIF of a renewal process depends on the operating time t and on the history H_t only through the difference $t - t_{N(t)}$, where $t_{N(t)}$ is the time of the most recent failure prior to t:

$$\lambda(t;\, H_t) = h(t - t_{N(t)}) \tag{18.11}$$

The CIF (Equation 18.11) implies that the times x_i $(i = 1, 2, \ldots)$ between successive failures are independent and identically distributed random variables with hazard rate $r(x)$ numerically equal to $h(x)$, then the failure pattern shows no trend with time (times between failures do not tend to become smaller or larger with operating time).

As pointed out by many authors (*e.g.* see Thompson [5] (p.49) and Ascher [8]) a renewal process is generally inappropriate to describe the failure pattern of repairable equipment. In fact, it is implausible that each repair actually brings the equipment to a same as new condition, and thus the equipment experiences deterioration (times between failures tend to become smaller with time) that cannot be modeled by a renewal process. Likewise, the renewal process cannot be used to model the failure pattern of equipment that experiences reliability improvement during the earlier period of its operating life.

A special form of renewal process arises when the CIF (Equation 18.11) is constant with the operating time t: $\lambda(t;\, H_t) = \rho$, so that the interarrival times x_i are independent and identically exponentially distributed random variables. Such a process is also a same as old model, since each repair restores the equipment to the condition it was in just before the repair (each repair is both *perfect* and *minimal*). For convenience of exposition, such a process will be discussed later, in Section 18.4.1.

18.4 Minimal Repair

The minimal repair is a corrective maintenance action that brings the repaired equipment to the conditions it was in just before the failure occurrence. Thus, if a minimal repair is carried out at time t_i, then

$$\lambda(t_i^+;\, H_{t_i^+}) = \lambda(t_i^-;\, H_{t_i^-}) \tag{18.12}$$

and the equipment reliability is unchanged by failure and repair (*lack of memory* property). If the equipment is minimally repaired at each failure, and preventive maintenance is not carried out, then the CIF is continuous everywhere and does not depend on the history of the process but only on the operating time t:

$$\lambda(t;\, H_t) = \rho(t) \tag{18.13}$$

In this case, $\rho(t)$ is both the unconditional intensity function and the ROCOF $\mu(t)$. The assumption of minimal repair implies that the failure process is a *Poisson process* [6,9], *i.e.* a point process with independent Poisson distributed increments. Thus:

$$\begin{aligned}&\Pr\{N(t, t+\Delta) = k\}\\ &\quad = \frac{[M(t, t+\Delta)]^k}{k!}\exp[-M(t, t+\Delta)]\\ &\quad\quad k = 0, 1, 2, \ldots\end{aligned} \tag{18.14}$$

where $M(t, t+\Delta)$ is the expected number of failures in the time interval $(t, t+\Delta)$:

$$M(t, t+\Delta) = E\{N(t, t+\Delta)\} = \int_t^{t+\Delta} \mu(z)\,\mathrm{d}z \tag{18.15}$$

For a Poisson process, the interarrival times $x_i = t_i - t_{i-1}$ generally are neither identically nor independently distributed random variables. In fact, the conditional density function of x_i depends on the occurrence time t_{i-1} of the most recent failure:

$$\begin{aligned}f_i(x;\, t_{i-1}) &= \rho(t_{i-1} + x)\\ &\quad\times \exp\left[-\int_0^x \rho(t_{i-1} + z)\,\mathrm{d}z\right]\\ &\quad x \geq 0\end{aligned} \tag{18.16}$$

Then, unlike the renewal process, only the first interarrival time x_1 and, then, the arrival time t_1 of the first failure have hazard rate numerically equal to $\rho(t)$. Also, the generalized residual $e_i = \int_{t_{i-1}}^{t_i} \lambda(t;\, H_t)\,\mathrm{d}t$ equals the expected number of failures in the time interval (t_{i-1}, t_i), *i.e.* for a Poisson process: $e_i = M(t_{i-1}, t_i)$.

The form of Equation 18.13 enables one to model, in principle, the failure pattern of any repairable equipment subjected to minimal repair, both when a complex behavior is observed in failure data and when a simpler behavior arises. Several testing procedures have been developed for testing the absence of trend with operating time against the alternative of a simple behavior in failure data (such as monotonic or bathtub-type trend).

One of the earliest tests is the Laplace test [10] (p.47), which is based, under a failure-truncated sampling, on the statistic [1] (p.79)

$$\mathrm{LA} = \frac{\sum_{i=1}^{n-1} t_i - (n-1)t_n/2}{t_n\sqrt{(n-1)/12}} \tag{18.17}$$

where t_i $(i = 1, \ldots, n)$ are the occurrence times of the observed failures. Under the null hypothesis of constant intensity, LA is approximately distributed as a standard normal random variable. Then, a large positive (large negative) value of LA provides evidence of a monotonically increasing (decreasing) trend with time, *i.e.* of deteriorating (improving) equipment. The Laplace test is the uniformly most powerful unbiased (UMPU) test when the failure data come from a Poisson process with log-linear intensity (see Section 18.4.2.2).

The test procedure, which generally performs well in testing the null hypothesis of constant intensity against a broad class of Poisson processes with monotonically increasing intensity [11], is based, for failure-truncated samples, on the statistic

$$Z = 2\sum_{i=1}^{n-1} \ln(t_n/t_i) \tag{18.18}$$

Under the null hypothesis of constant intensity, Z is exactly distributed as a χ^2 random variable with $2(n-1)$ degrees of freedom [12]. Small (large) values of Z provide evidence of a monotonically increasing (decreasing) trend. The Z test is UMPU when the failure data actually come from a Poisson process with power-law intensity (see Section 18.4.2.1).

Forms that little different for LA and Z arise when the failure data come from a time-truncated sampling, *i.e.* when the process is observed until a prefixed time T [11]:

$$\mathrm{LA} = \frac{\sum_{i=1}^{n} t_i - nT/2}{T\sqrt{n/12}}$$

and

$$Z = 2\sum_{i=1}^{n} \ln(T/t_i) \tag{18.19}$$

Under a time-truncated sampling, LA is still approximately normally distributed and Z is χ^2 distributed with $2n$ degrees of freedom.

Both LA and Z can erroneously provide evidence of trend when data arise from the more general renewal process, rather than from a Poisson process with constant intensity. Then, if a time trend is to be tested against the null hypothesis of a general renewal process, one should use the Lewis and Robinson test, based on the statistic [13]

$$\mathrm{LR} = \mathrm{LA}\left[\frac{\bar{x}^2}{\sum_{i=1}^{n}(x_i - \bar{x})^2/(n-1)}\right]^{1/2} \tag{18.20}$$

where x_i $(i = 1, \ldots, n)$ are the times between successive failures and $\bar{x} = \sum_{i=1}^{n} x_i/n = t_n/n$. Under the null hypothesis of a general renewal process, LR is approximately distributed as a standard normal random variable, and then large positive (large negative) values of LR provide evidence of deteriorating (improving) equipment.

The above test procedures are often not effective in detecting deviation from no trend when there is a non-monotonic trend, *i.e.* when the process has a bathtub or inverted bathtub intensity. For testing non-monotonic trends, Vaurio [14] proposed three different test statistics under a time-truncated sampling. In particular, the statistic

$$V_1 = \frac{\sum_{i=1}^{n} |t_i - T/2| - nT/4}{T\sqrt{n/48}} \tag{18.21}$$

under the null hypothesis of constant intensity, is approximately distributed as a standard normal random variable. Then, large positive (large negative) values of V_1 indicate a bathtub-type (inverted bathtub-type) behavior in failure data.

A very useful way to investigate the presence and nature of trend in the failure pattern is by plotting the cumulative number i of failures occurred until the observed failure time t_i $(i = 1, 2, \ldots, n)$ against times t_i. If the plot is approximately linear, then no trend is present in the observed data. A concave (convex) plot is representative of an increasing (decreasing) trend, whereas an inverted S-shaped plot suggests the presence of a non-monotonic trend with bathtub-type intensity.

18.4.1 No Trend with Operating Time

If the failure pattern of minimally repaired equipment does not show any trend with time, *i.e.* times between failures do not tend to become smaller or larger with time, then it can be described by a *homogeneous Poisson process* (HPP) [4]. The HPP is a Poisson process with constant intensity:

$$\lambda(t;\, H_t) = \rho \tag{18.22}$$

so that the interarrival times x_i are independent and identically exponentially distributed random variables with mean $1/\rho$. Such a Poisson process is called *homogeneous* since it has stationary increments; the $\Pr\{N(t, t+\Delta) = k\}$ does not depend on time t, only on Δ:

$$\Pr\{N(t, t+\Delta) = k\} = \frac{(\rho\Delta)^k}{k!}\exp(-\rho\Delta) \quad k = 0, 1, 2, \ldots \tag{18.23}$$

The forward waiting time w_t is exponentially distributed with mean $1/\rho$, too, and the failure times t_k are gamma distributed with scale parameter $1/\rho$ and shape parameter k. The expected number of failures increases linearly with time: $M(t) = \rho t$. Hence, if a failure data set follows the HPP, then the plot of the cumulative number of failures i $(i = 1, 2, \ldots)$ versus the observed failure times t_i should be reasonably close to a straight line with slope ρ.

Suppose that a process is observed from $t = 0$ until a prefixed time T, and that n failures occur at times $t_1 < t_2 < \ldots < t_n \leq T$. The likelihood relative to this data set is

$$L(\rho) = \rho^n \exp(-\rho T) \tag{18.24}$$

and the maximum likelihood (ML) estimate of the intensity ρ depends only on the number of observed failures and not on the failure times: $\hat{\rho} = n/T$.

When data consist of interarrival times, then statistical methods developed for the exponential distribution can be used [15]. Methods based on counts of failures in fixed time intervals are discussed by Cox and Lewis [10] (p.29–36) and Bayesian procedures for predicting the number of failures in a future time interval are given in Beiser and Rigdon [16].

Despite the restrictive hypotheses for it, the HPP has been successfully used to describe the failure pattern of repairable mechanical equipment over portions of operating life, such as the so-called *useful life* [17], *i.e.* the phase that follows the initial life period of early failures and precedes the degradation period. Likewise, when defective parts and/or assembling defects are screened out completely by the manufacturer so that early failures do not occur, the HPP can model adequately the entire initial period of the operating life (which generally includes the warranty period) until degradation phenomena appear. Of course, the HPP cannot describe reliability degradation or improvement, and then caution has to be used in making predictions if the HPP is used to model the failure pattern both in the observation period and in future intervals.

Application examples of the HPP are:

- air-conditioning equipment in Boeing 720 aircraft, first given by Proschan [18] and later analyzed by Lawless and Thiagarajah [3] and Cox and Lewis [10]
- the hydraulic systems of some load–haul–dump machines deployed at Kiruna mine [19]
- modified hydraulic excavators supplied to cement plants, coal mines, and iron ore mines, and observed during the warranty period [20].

18.4.2 Monotonic Trend with Operating Time

In many cases, mechanical equipment subject to minimal repairs experiences reliability improvement or deterioration with the operating time, *i.e.* there is a tendency toward less or more frequent failures [8]. In such situations, the failure pattern can be described by a Poisson process whose intensity function (Equation 18.13) monotonically decreases or increases with t. A Poisson process with a non-constant intensity is called *non-homogeneous*, since it does not have stationary increments.

18.4.2.1 The Power Law Process

The best known and most frequently used non-homogeneous Poisson process (NHPP) is the *power law process* (PLP), whose intensity (Equation 18.13) is

$$\rho(t) = \frac{\beta}{\alpha}\left(\frac{t}{\alpha}\right)^{\beta-1} \qquad \alpha, \beta > 0 \tag{18.25}$$

The intensity function (Equation 18.25) increases (decreases) with t when the power-law parameter $\beta > 1$ ($\beta < 1$), and thus the PLP can describe both deterioration and reliability improvement. If $\beta = 1$, the PLP reduces to the HPP. The expected number of failures $M(t) = (t/\alpha)^\beta$ is linear with t in a log–log scale: $\ln[M(t)] = \beta \ln(t) - \beta \ln(\alpha)$. Hence, if a failure data set follows the PLP, then the plot of the cumulative number of failures i ($i = 1, 2, \ldots$) versus the observed failure times t_i should be reasonably close to a straight line with slope β in a coordinate system with both logarithmic scales.

Some objections are raised to the PLP when it describes improving equipment. In fact, in such a situation ($\beta < 1$), the PLP intensity (Equation 18.14) is unrealistically infinite at $t = 0$ and tends to zero as t approaches infinity. Nevertheless, the PLP enjoys large popularity in the analysis of failure data of repairable equipment, partly due to the simplicity of statistical inference procedures. In fact, given the first n successive failures times $t_1 < t_2 < \cdots < t_n$ in a failure-truncated sampling, the likelihood under the hypothesis of a PLP is surprisingly simple:

$$L(\alpha, \beta) = (\beta/\alpha)^n \left[\prod_{i=1}^{n} (t_i/\alpha)^{\beta-1}\right] \exp[-(t_n/\alpha)^\beta] \tag{18.26}$$

so that the closed form of the ML estimators of parameters β and α can be obtained:

$$\hat{\beta} = \frac{n}{\sum_{i=1}^{n} \ln(t_n/t_i)} \qquad \hat{\alpha} = \frac{t_n}{n^{1/\hat{\beta}}} \tag{18.27}$$

These estimators are joint sufficient statistics, and some useful pivotal quantities can be defined. Again, in most cases, confidence intervals can be obtained using existing tables.

Results on ML estimation of PLP parameters, both for time-truncated and failure-truncated data, can be found in Crow [12], Finkelstein [21], Lee and Lee [22], Bain and Engelhardt [23], and Calabria *et al.* [24]. Testing procedures for comparing the scale parameters α or the power-law parameters β of several independent PLPs were given by Crow [12], Lee [25], and Calabria *et al.* [26]. Also, sequential testing procedures for the power-law parameter of one or more PLPs can be found in Bain and Engelhardt [27]. When only count data from several identical PLPs are available, estimation procedures based on parametric bootstrap are given in Calabria *et al.* [28].

Goodness-of-fit tests for checking the hypothesis that the observed failure data came from a PLP are discussed in Crow [12], Rigdon and Basu [29] (who also give a complete review both of PLP properties and of inferential and testing procedures), Park and Kim [30], and Baker [31]. Also, methods for obtaining ML prediction limits for future failure times of a PLP can be found in Lee and Lee [22], Engelhardt and Bain [32], and Calabria *et al.* [33].

A complete discussion on ML inferential, prediction and testing procedures, both for time- and failure-truncated sampling, can be found in Bain and Engelhardt [34] (p.415–449). In addition, a useful FORTRAN program is illustrated by Rigdon *et al.* [35]; this allows ML estimates of PLP parameters and intensity function to be computed, as well as goodness-of-fit tests to be performed, both for single and multiple equipment.

Inferential procedures on the PLP parameters and functions thereof, based on the Bayesian approach, can be found in Guida *et al.* [36], Bar-Lev *et al.* [37], and Campodónico and Singpurwalla [38]. Bayesian prediction procedures for future failure times are given in Calabria *et al.* [39] and Bar-Lev *et al.* [37]. Point and interval predictions for the number of failures in some given future interval can be found in Bar-Lev *et al.* [37], Campodónico and Singpurwalla [38], and Beiser and Rigdon [40]. These Bayes procedures allow the unification of the two sampling protocols (failure- and time-truncated) under a single analysis and can provide more accurate estimate and prediction when prior information or technical knowledge on failure mechanism is available.

Finally, graphical methods for analyzing failure data of several independent PLPs can be found in Lilius [41], Härtler [42], and Nelson [43] on the basis of non-parametric estimates of the intensity function or of the mean cumulative number of failures.

A very large number of numerical applications show the adequacy of the PLP in describing the failure pattern of mechanical equipment experiencing reliability improvement or degradation. Some examples follow:

- a 115 kV transmission circuit in New Mexico given by Martz [44] and analyzed by Rigdon and Basu [29];
- several machines (among which are diesel engines and loading cranes) [45];
- several automobiles (1973 AMC Ambassador) [46];
- the hydraulic systems of some load–haul–dump machines deployed at Kiruna mine [19];
- an airplane air-conditioning equipment [18], denoted as Plane 6 by Cox and Lewis [10] and analyzed by Park and Kim [30] and Ascher and Hansen [47];
- American railroad tracks [38];
- bus engines of the Kowloon Motor Bus Company in Hong Kong [48].

18.4.2.2 The Log–Linear Process

A second well accepted form of the NHPP is the log-linear process (LLP), whose intensity (Equation 18.13) has the form

$$\rho(t) = \exp(\alpha + \beta t) \qquad -\infty < \alpha, \beta < \infty \tag{18.28}$$

This process was firstly proposed by Cox and Lewis [10] (p.45) and can describe monotonic trends in the failure data: reliability improvement when $\beta < 0$, and deterioration when $\beta > 0$. When $\beta = 0$, then the LLP reduces to the HPP. Unlike the PLP, the LLP intensity (Equation 18.28) is finite at $t = 0$ when a reliability improvement is described, and it is greater than zero when deterioration is described. Since the increasing LLP intensity takes non-negligible values at the beginning of the observed period, the LLP is able to describe the occurrence of a small number of failures at the beginning of the operating time, even of equipment that on the whole deteriorates. In addition, the LLP with $\beta > 0$ can adequately model repairable equipment with extremely rapid deterioration, since the intensity (Equation 18.28) is increasing at an exponential rate with time.

The expected number of failures for an LLP is $M(t) = \exp(\alpha)[\exp(\beta t) - 1]/\beta$ and, under a failure-truncated sampling, the likelihood results in

$$L(\alpha, \beta) = \exp\left\{n\alpha + \beta \sum_{i=1}^{n} t_i - \exp(\alpha)[\exp(\beta t_n) - 1]/\beta\right\} \tag{18.29}$$

Since the observed data t_i $(i = 1, 2, \ldots, n)$ enter the likelihood only through $(t_n, \sum_{i=1}^{n} t_i)$, these are sufficient statistics. The ML estimates of LLP parameters are not in closed form and can be found from

$$\sum_{i=1}^{n} t_i + \frac{n}{\hat\beta} - \frac{n t_n}{1 - \exp(-\hat\beta t_n)} = 0$$
$$\hat\alpha = \ln\left[\frac{n\hat\beta}{\exp(\hat\beta t_n) - 1}\right] \tag{18.30}$$

Inferential and testing procedures can be found in Ascher and Feingold [9], Cox and Lewis [10] (p.45) and Lawless [15] (p.497–499). A goodness-of-fit test based on the Cramér–von Mises statistic was described by Ascher and Feingold [9], whereas Lawless [15] used the ML estimates of the generalized residuals

$$\begin{aligned}\hat{e}_i &= \hat{M}(t_{i-1}, t_i)\\ &= \exp(\hat{\alpha})[\exp(\hat{\beta}t_i) - \exp(\hat{\beta}t_{i-1})]/\hat{\beta}\\ &\quad i = 1, \ldots, n \end{aligned} \tag{18.31}$$

in order to assess the adequacy of the LLP. Testing procedures for comparing the improvement or deterioration rate of two types of equipment operating in two different environmental conditions were given by Aven [49]. Bayesian inferential and prediction procedures, based on prior information on the initial intensity $\rho_0 = \exp(\alpha)$ and on the aging rate β, have recently been proposed by Huang and Bier [50].

Since classical inferential procedures for the LLP are not as easy as for the PLP, the LLP has not received the large attention devoted to the PLP. Nevertheless, the LLP has often been used to analyze the failure data of repairable mechanical equipment:

- the air-conditioning equipment [18] in Boeing 720 airplanes, denoted as Planes 2 and 6 by Cox and Lewis [10] and analyzed also by Lawless and coworkers [3, 15] and Aven [49];
- main propulsion diesel engines installed on several submarines [9];
- a number of hydraulic excavator systems [20];
- a turbine-driven pump in a pressurized water reactor auxiliary feedwater system [50].

18.4.2.3 Bounded Intensity Processes

Both the PLP and the LLP intensity functions used to model deterioration tend to infinity as the operating time increases. However, the repeated application of repairs or substitutions of failed parts in mechanical equipment subject to reliability deterioration sometimes produces a finite bound for the increasing intensity function. In such a situation, the equipment, therefore, becomes composed of parts with a randomized mix of ages, so that its intensity settles down to a constant value (*bounded increasing intensity*). This result is known as Drenick's theorem [1] and can be observed when the operating time of the equipment is very large.

A simple NHPP with bounded intensity function was suggested by Engelhardt and Bain [51]:

$$\rho(t) = \frac{1}{\theta}[1 - (1 + t/\theta)^{-1}] \quad \theta > 0 \tag{18.32}$$

However, this one-parameter model does not seem adequate in describing actual failure data because of its analytical simplicity.

A two-parameter NHPP (called the *bounded intensity process* (BIP)) has recently been proposed by Pulcini [52]. Its intensity has the form

$$\rho(t) = \alpha[1 - \exp(-t/\beta)] \quad \alpha, \beta > 0 \tag{18.33}$$

This increasing intensity is equal to zero at $t = 0$ and approaches the asymptote of α as t tends to infinity. The parameter β is a measure of the initial increasing rate of the intensity (Equation 18.33): the smaller β is, the faster the intensity increases initially until it approaches α. As β tends to zero, the intensity (Equation 18.33) approaches α, and then the BIP includes as a limiting form the HPP. The expected number of failures up to a given time t is

$$M(t) = \alpha\{t - \beta[1 - \exp(-t/\beta)]\} \tag{18.34}$$

and the likelihood in a time-truncated sampling is

$$\begin{aligned}L(\alpha, \beta) = \alpha^n \prod_{i=1}^{n} [1 - \exp(-t_i/\beta)]\\ \times \exp(-\alpha\{T - \beta[1 - \exp(-T/\beta)]\})\end{aligned} \tag{18.35}$$

The ML estimates $\hat{\alpha}$ and $\hat{\beta}$ of the BIP parameters are given by

$$\begin{aligned}\hat{\alpha} &= \frac{n}{T - \hat{\beta}[1 - \exp(-T/\hat{\beta})]}\\ \frac{n[1 - \exp(-T/\hat{\beta}) - (T/\hat{\beta})\exp(-T/\hat{\beta})]}{(T/\hat{\beta}) - [1 - \exp(-T/\hat{\beta})]} &= \sum_{i=1}^{n} \frac{(t_i/\hat{\beta})\exp(-t_i/\hat{\beta})}{1 - \exp(-t_i/\hat{\beta})}\end{aligned} \tag{18.36}$$

For the case of failure-truncated sampling, $T \equiv t_n$. Approximate confidence intervals can be obtained on the basis of the asymptotic distribution of $\hat{\alpha}$

and $\hat{\beta}$, and a testing procedure for time trend against the HPP can be performed on the basis of the likelihood ratio statistic. The BIP has been used in analyzing the failure data (in days) of automobile #4 [46] and the failure data of the hydraulic system of the load–haul–dump LHD17 machine [19].

18.4.3 Bathtub-type Intensity

In many cases, mechanical deteriorating equipment is also subject to early failures, because of the presence of defective parts or assembling defects that were not screened out through burn-in techniques by the manufacturer. Thus, the failure pattern has a bathtub-type behavior: at first the interarrival times increase with operating time, and then they decrease. A bathtub-type intensity is more typical of large and complex equipment with many failure modes. As stated by Nelson [43], a bathtub intensity is typical for 10 to 20% of products.

The presence of a bathtub intensity in an observed data set can be detected both by a graphical technique (the plot of the cumulative number of failures i against the failure time t_i should be inverted S-shaped) and by performing the testing procedures of Vaurio [14].

An NHPP suitable to describe bathtub behavior is the three-parameter process with intensity

$$\rho(t) = \lambda\gamma t^{\gamma-1}\exp(\beta t) \qquad \lambda, \gamma > 0;\ -\infty < \beta < \infty \qquad (18.37)$$

This process, for which the PLP and the LLP are special cases, was proposed by Lee [53] in order to assess the adequacy of the simpler processes to analyze a given data set. Ascher and Feingold [1] (p.85) and Lawless [15] (p.506) showed that, for $\lambda > 0$ and $\gamma < 1$, the intensity given by Equation 18.37 has a bathtub-type behavior and so can model the failure pattern of equipment subject both to early failures and deterioration phenomena. However, to our knowledge, statistical inference procedures for the Lee model have not been developed yet.

Another NHPP with bathtub-type intensity, denoted the *exponential power process* (EPP)

$$\rho(t) = (\beta/\eta)(t/\eta)^{\beta-1}\exp[(t/\eta)^{\beta}] \qquad \eta > 0;\ 0 \le \beta \le 1 \qquad (18.38)$$

can be found in [54], where the EPP has been proposed in a wider context that includes minimal, perfect, and imperfect repairs. However, the EPP does not seem adequate to describe actual failure data because of its analytical simplicity.

More recently, an NHPP (called S-PLP) that arises from the superposition [5] (p.68–71) of two PLPs with different power-law parameters has been proposed by Pulcini [55]. The S-PLP intensity has the form

$$\rho(t) = \frac{\beta_1}{\alpha_1}\left(\frac{t}{\alpha_1}\right)^{\beta_1-1} + \frac{\beta_2}{\alpha_2}\left(\frac{t}{\alpha_2}\right)^{\beta_2-1} \qquad \alpha_1, \beta_1, \alpha_2, \beta_2 > 0 \qquad (18.39)$$

If, and only if, either β_1 or $\beta_2 < 1$, the intensity given by Equation 18.39 is non-monotonic and has a bathtub-type behavior. The expected number of failures up to a given time t is

$$M(t) = \left(\frac{t}{\alpha_1}\right)^{\beta_1} + \left(\frac{t}{\alpha_2}\right)^{\beta_2} \qquad (18.40)$$

and the likelihood relative to a time-truncated sample $t_1 < \cdots < t_n \le T$ is

$$L(\alpha_1, \beta_1, \alpha_2, \beta_2) = \prod_{i=1}^{n}\left[\frac{\beta_1}{\alpha_1}\left(\frac{t_i}{\alpha_1}\right)^{\beta_1-1} + \frac{\beta_2}{\alpha_2}\left(\frac{t_i}{\alpha_2}\right)^{\beta_2-1}\right] \times \exp\left[-\left(\frac{T}{\alpha_1}\right)^{\beta_1} - \left(\frac{T}{\alpha_2}\right)^{\beta_2}\right] \qquad (18.41)$$

A closed form solution for the ML estimators of the S-PLP parameters does not exist, so that the ML estimates can be obtained by maximization of the modified three-parameter log-likelihood:

$$\ell'(\alpha_1, \beta_1, \beta_2) = \sum_{i=1}^{n}\ln\left\{\frac{\beta_1}{\alpha_1}\left(\frac{t_i}{\alpha_1}\right)^{\beta_1-1} + \left[n - \left(\frac{T}{\alpha_1}\right)^{\beta_1}\right]\frac{\beta_2}{T}\left(\frac{t_i}{T}\right)^{\beta_2-1}\right\} - n \qquad (18.42)$$

subject to the nonlinear constraints $\hat{\alpha}_1 > T/n^{1/\hat{\beta}_1}$, $\hat{\beta}_1 > 0$, and $\hat{\beta}_2 > \hat{\beta}_1$, which assure $\hat{\alpha}_2 > 0$ and allow a unique solution to be obtained. For failure-truncated samples, $T \equiv t_n$.

A graphical approach, based on the transformations $x = \ln(t)$ and $y = \ln[M(t)]$, allows one to determine whether the S-PLP can adequately describe a given data set, and to obtain, if the sample size is sufficiently large, crude but easy estimates of process parameters. In fact, if a data set follows the S-PLP model, then the plot (on log–log paper) of the cumulative number of failures versus the observed failure times t_i should be reasonably close to a concave curve, with a slope that increases monotonically from the smaller β value to the larger one.

The S-PLP has been used to analyze the failure data of a 180 ton rear dump truck [56], and the failure data of the diesel-operated load–haul–dump LHD-A machine operating in a Swedish mine [57]. The application of the Laplace, Lewis and Robinson, and Vaurio trend tests has provided evidence, in both the failure data, of a bathtub-type intensity in a global context of increasing intensity.

18.4.3.1 Numerical Example

Let us consider the supergrid transmission failure data read from plots in Bendell and Walls [58]. The data consist of 134 failures observed until $t_{134} = 433.7$ (in thousands of unspecified time units) and are given in Table 18.1. Bendell and Walls [58] analyzed graphically the above data and concluded that there was a decreasing trend of the failure intensity in the initial period with a global tendency for the time between failures to become shorter, on average, over the entire observation period. To test the graphical conclusions of Bendell and Walls [58], the LA, Z, LR, and V_1 test statistics are evaluated

$$\text{LA} = 4.01 \quad Z = 243.0 \quad \text{LR} = 1.59 \quad V_1 = 1.62$$

and compared with the χ^2 and standard normal 0.10 and 0.90 quantiles. The Vaurio test provides evidence of a non-monotonic bathtub-type trend against the null hypothesis of no trend, and

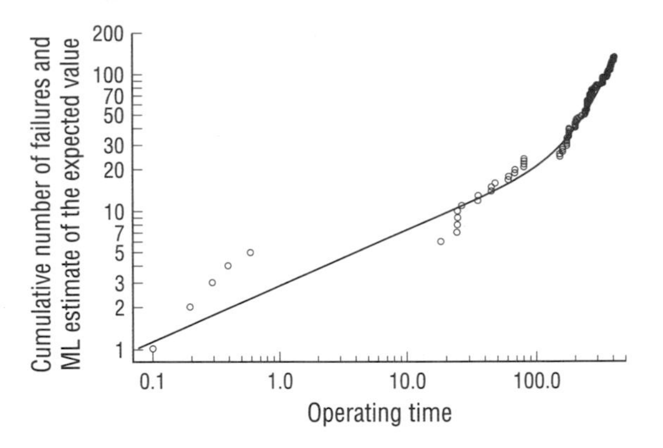

Figure 18.1. ML estimate of $M(t)$ and observed cumulative number of failures in a log–log plot for supergrid transmission data in numerical example

the LR test rejects the null hypothesis of a renewal process against the increasing trend alternative.

The LA and Z tests provide conflicting results. In fact, the Z test does not reject the null hypothesis of an HPP against the PLP alternative, whereas the LA test strongly rejects the HPP hypothesis against the hypothesis of an increasing log-linear intensity. The conflict between LA and Z results depends on the fact that, unlike the PLP, the LLP is able to describe the occurrence of a small number of early failures in a deteriorating process. Furthermore, the LA test can be effective in detecting deviation from the HPP hypothesis even in the case of bathtub intensity, provided that early failures constitute a small part of the observed failures.

Thus, the supergrid transmission failure data show statistically significant evidence of a bathtub intensity in a global context of increasing intensity. The ML estimates of S-PLP parameters are

$$\hat{\beta}_1 = 0.40 \quad \hat{\alpha}_1 = 0.074 \quad \hat{\beta}_2 = 2.53 \quad \hat{\alpha}_2 = 70$$

and, since $\hat{\beta}_1 < 1$ and $\hat{\beta}_2 > 1$, the bathtub behavior is confirmed. In Figure 18.1 the ML estimate of $M(t)$ is shown on a log–log plot and is compared with the observed cumulative number of failures.

Table 18.1. Supergrid transmission failure data in numerical example (read from plots in Bendell and Walls [58])

0.1	0.2	0.3	0.4	0.6	18.6	24.8	24.9	25.0	25.1	27.1	36.1
36.2	45.7	45.8	49.0	62.1	62.2	69.7	69.8	82.0	82.1	82.2	82.3
159.5	159.6	166.8	166.9	167.0	181.2	181.3	181.4	181.5	181.6	187.6	187.7
187.8	187.9	188.0	188.1	211.5	211.7	211.9	212.0	212.1	212.2	216.5	225.1
236.6	237.9	252.1	254.6	254.8	254.9	261.1	261.3	261.4	261.8	262.8	262.9
263.0	263.1	264.6	264.7	273.5	273.6	275.6	275.7	275.9	276.0	277.5	287.5
287.7	287.9	288.0	288.1	288.2	294.2	301.4	301.5	305.0	312.2	312.7	312.8
339.1	349.1	349.2	349.3	349.4	349.8	349.9	350.0	350.2	350.3	350.4	377.4
377.5	381.5	382.5	382.6	382.7	382.8	383.5	387.5	387.6	390.9	399.4	399.6
399.7	401.5	401.6	401.8	403.3	405.1	406.2	406.4	407.4	407.6	411.4	411.5
412.1	412.7	415.9	416.0	420.3	425.1	425.2	425.3	425.6	425.7	426.1	426.5
427.7	433.7										

18.4.4 Non-homogeneous Poisson Process Incorporating Covariate Information

The influence of the operating and environmental conditions on the intensity function is often significant and cannot be disregarded. One way is to relate the intensity function not only to the operating time but also to *explanatory variables*, or covariates, that provide a measure of the operating and environmental conditions. Several methods of regression analysis for repairable equipment have been discussed by Ascher and Feingold [1] (p.96–99).

More recently, Lawless [59] gave methods of regression analysis for minimally repaired equipment based on the so-called *proportional intensity Poisson process*, which is defined as follows. An equipment with $k \times 1$ covariate vector $\boldsymbol{x}$ has repeated failures that occur according to an NHPP with intensity:

$$\rho_{\boldsymbol{x}}(t) = \rho_0(t)\exp(\boldsymbol{x}'\boldsymbol{a}) \tag{18.43}$$

where $\rho_0(t)$ is a baseline intensity with parameter vector $\boldsymbol{\theta}$ and $\boldsymbol{a}$ is the vector of regression coefficients (regressors). Suppose that m units are observed over the time interval $(0, T_i)$ $(i = 1, \ldots, m)$ and that the ith equipment experiences n_i failures at times $t_{i,1} < t_{i,2} < \cdots < t_{i,n_i}$. If the ith equipment has covariate vector $\boldsymbol{x}_i$, the likelihood is

$$L(\boldsymbol{\theta}, \boldsymbol{a}) = \prod_{i=1}^{m}\left[\prod_{j=1}^{n_i}\rho_{\boldsymbol{x}_i}(t_{i,j})\right] \times \exp\left[-\left(\int_0^{T_i}\rho_0(t)\,\mathrm{d}t\right)\exp(\boldsymbol{x}'\boldsymbol{a})\right] \tag{18.44}$$

A semi-parametric analysis with $\rho_0(t)$ unspecified has been described, as well as a fully parametric analysis based on the assumption of a power-law form for the baseline intensity: $\rho_0(t) = \nu\beta t^{\beta-1}$. ML inferential procedures, model checks, and tests have also been discussed. Under the power-law intensity assumption, the parameter ν is included in the regression function by defining $\exp(a_0) = \nu$ and taking the covariate x_0 to be identically equal to one. When all T_i values are equal $(T_i = T)$, the ML estimate of the power-law parameter is:

$$\hat{\beta} = n \Big/ \sum_{i=1}^{m}\sum_{j=1}^{n_i}\ln(T/t_{i,j}) \tag{18.45}$$

and the ML estimate of regressors $\boldsymbol{a}$ can be obtained by solving iteratively:

$$\sum_{i=1}^{m} n_i x_{i,r} - \sum_{i=1}^{m} T^{\hat{\beta}} x_{i,r}\exp(\boldsymbol{x}_i'\boldsymbol{a}) = 0 \qquad r = 0, 1, \ldots, k \tag{18.46}$$

When the T_i values are unequal, the ML estimates can be obtained by maximization of the log-likelihood.

On the basis of the above model, Guida and Giorgio [60] analyzed failure data of repairable equipment in accelerating testing, where the covariate is the stress level at which the equipment is tested. They obtained an estimate of the baseline intensity $\rho_0(t)$ under use condition from failure data obtained under higher stresses. They assumed a power-law form for the baseline intensity, $\rho_0(t) = (\beta/\alpha_0)(t/\alpha_0)^{\beta-1}$, and a single acceleration variable (one-dimensional covariate vector) with two stress levels x_1 and $x_2 > x_1$, both greater than the use condition level. ML estimations of regression coefficient a and of parameters β and α_0 were also discussed, both when one unit and several units are tested at each stress level. In the case of one unit tested until T_1 under stress level x_1 and another unit tested until T_2 under x_2, the ML estimators are in closed-form:

$$\hat{\beta} = n \Bigg/ \sum_{i=1}^{2} \sum_{j=1}^{n_i} \ln(T_i/t_{i,j})$$

$$\hat{\alpha}_0 = (T_1/n_1^{1/\hat{\beta}})^{\xi} (T_2/n_2^{1/\hat{\beta}})^{1-\xi}$$

$$\hat{a} = \ln[(T_1/T_2)(n_2/n_1)^{1/\hat{\beta}}]/(x_2 - x_1) \quad (18.47)$$

where n_1 and n_2 are the number of failures observed under x_1 and x_2 respectively, and $\xi = x_2/(x_2 - x_1)$ is the extrapolation factor.

Proportional intensity models have found wide application, both in the case of minimal repair and when complex maintenance is carried out. A complete discussion on this topic has been presented by Kumar [61], where procedures for the estimation of regression coefficients and for the selection of a regression model are illustrated. Some proportional intensity models, specifically devoted to the analysis of complex maintenance policy, will be illustrated in Section 18.6.

18.5 Imperfect or Worse Repair

In many cases, the repair actions produce a noticeable improvement in the equipment condition, even without bringing the equipment to a new condition (*imperfect repair*). One explanation for this is that often a *minimum time* repair policy is performed rather than a *minimum repair* policy, in order to shorten the downtime of the equipment [62]. In other cases, repair actions can be incorrect or inject fresh faults into the equipment, so that the equipment after repair is in a worse condition than just before the failure occurrence (*worse repair*). This can happen when, for example, the repair actions are performed by inexperienced or unauthorized maintenance staff and/or under difficult conditions. Models that treat imperfect or worse maintenance situations have been widely proposed and studied. A complete discussion on this topic is given by Pham and Wang [2].

In the following, a number of families of models able to describe the effect of imperfect or worse repairs on equipment conditions are illustrated, with particular emphasis on the reliability modeling. The case of complex maintenance policies, such as those of minimal repairs interspersed with periodic preventive overhauls or perfect repairs, will be treated in Section 18.6.

18.5.1 Proportional Age Reduction Models

A general approach to model an imperfect maintenance action was proposed by Malik [63], who assumed that each maintenance reduces the age of the equipment by a quantity proportional to the operating time elapsed form the most recent maintenance (*proportional age reduction* (PAR) model). Hence, each maintenance action is assumed to reduce only the damage incurred since the previous maintenance epoch. This approach has since been largely used to describe the effect of corrective or preventive maintenance on the equipment condition, especially when the repairable equipment is subject to complex maintenance policy (see Sections 18.6.3 and 18.6.4).

In the case that, at each failure, the equipment is imperfectly repaired, the CIF for the PAR model is

$$\lambda(t;\, H_t) = h(t - \rho t_{i-1})$$
$$0 \le \rho \le 1,\ t_{i-1} < t \le t_i \quad (18.48)$$

where ρ is the *improvement parameter*. Then, the CIF (Equation 18.48) shows jumps downward at each failure time t_i ($i = 1, 2, \ldots$), except for the special case of $\rho = 0$, when the PAR model reduces to an NHPP with intensity $h(t)$ (minimal repair). The value of each jump depends on ρ and on the operating time since the most recent failure. When $\rho = 1$, the repair brings the equipment to a like-new condition (perfect repair) and the PAR model reduces to a renewal process.

The PAR model has been adopted by Shin *et al.* [64] in their Model A, where the equipment is assumed to be imperfectly repaired at each failure. Two different forms for the intensity function up to the first failure time have been considered, the power-law and the log-linear intensities, so that the CIF (Equation 18.48) results in

$$\lambda(t;\, H_t) = \frac{\beta}{\alpha}\left(\frac{t - \rho t_{i-1}}{\alpha}\right)^{\beta - 1}$$
$$0 \le \rho \le 1,\ \alpha, \beta > 0,\ t_{i-1} < t \le t_i \quad (18.49)$$

and

$$\lambda(t;\, H_t) = \exp[\alpha + \beta(t - \rho t_{i-1})]$$
$$0 \le \rho \le 1,\ -\infty < \alpha, \beta < \infty,\ t_{i-1} < t \le t_i \quad (18.50)$$

respectively. Depending on the value of parameters, both the models in Equations 18.49 and 18.50 are able to describe the failure pattern of equipment subject to imperfect repairs and experiencing reliability improvement or deterioration. Figure 18.2 shows the plot of the PAR model with power-law intensity (Equation 18.49) of deteriorating equipment ($\beta > 1$) for arbitrary failure times $t_1, \ldots, t_4$, and compares the PAR plot with the minimal ($\rho = 0$) and perfect ($\rho = 1$) repair plots.

ML estimates of model parameters, which require the maximization of the log-likelihood, can be found in [64] when data arise from several identical units. Since the effect of repair may be

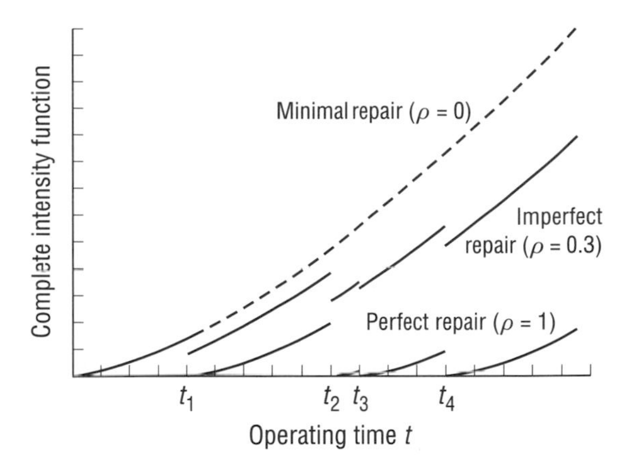

Figure 18.2. Plot of the complete intensity function for the PAR model (with $\rho = 0.3$ and power-law form) of deteriorating equipment. Comparison with the minimal ($\rho = 0$) and perfect ($\rho = 1$) repair plots

not uniform, the improvement factor ρ should be interpreted as the effect on average over the observation period.

Although the PAR model has been proposed to describe imperfect repair, it can adequately describe a worse maintenance by allowing the parameter ρ to be negative. In such a way, the effect of each maintenance action is that of increasing the age of the equipment by a quantity proportional to the operating time elapsed from the most recent maintenance.

18.5.2 Inhomogeneous Gamma Processes

The effect of imperfect or worse repairs on the equipment condition was modeled by Berman [65] by assuming that the equipment is subjected to shocks that occur according to an NHPP with intensity $\rho(t)$. The failure occurs not at every shock but at every kth shock. Then, if $k > 1$, the equipment is in a better condition just after a repair (and then the process describe an *imperfect repair*), whereas a value of $k < 1$ indicates that the system is in a worse condition just after a failure (*worse repair*).

The random variables $z_i = \int_{t_{i-1}}^{t_i} \rho(t)\,dt\,(i = 1, \ldots, n)$ (where $t_1 < t_2 < \cdots < t_n$ are the first n failure times) are independently and identically distributed according to the gamma distribution with unit scale parameter and shape parameter k. The CIF at any time t in the interval (t_{i-1}, t_i) depends on the history only through the time t_{i-1} of the most recent failure:

$$\begin{aligned}\lambda(t; H_t) = &\{\rho(t)[U(t_{i-1}, t)]^{k-1} \\ &\times \exp[-U(t_{i-1}, t)]/\Gamma(k)\} \\ &\times \left\{\int_t^\infty \rho(z)[U(t_{i-1}, z)]^{k-1}\right. \\ &\left.\times \exp[-U(t_{i-1}, z)]/\Gamma(k)\,dz\right\}^{-1} \\ &t > t_{i-1}\end{aligned} \quad (18.51)$$

where $U(t_{i-1}, t) = \int_{t_{i-1}}^{t} \rho(z)\,dz$ is the expected number of shocks from the most recent failure time up to t. A process with CIF given by Equation 18.51 is called an *inhomogeneous gamma process* and the likelihood relative to a failure-truncated sample $t_1 < t_2 < \cdots < t_n$ is

$$\begin{aligned}L(\boldsymbol{\theta}, k) = &\frac{1}{[\Gamma(k)]^n}\left\{\prod_{i=1}^{n} \rho(t_i)[U(t_{i-1}, t_i)]^{k-1}\right\} \\ &\times \exp[-U(0, t_n)]\end{aligned} \quad (18.52)$$

where $\boldsymbol{\theta}$ is the parameter's vector of $\rho(t)$. Although lacking a shock model interpretation, Equation 18.51 still defines the CIF of a point process when k is positive but not an integer. If $k = 1$, the inhomogeneous gamma process reduces to the NHPP (the repair is *minimal*), whereas if $\rho(t) = c$ (constant), then the process reduces to a renewal process with times between failures distributed as a gamma random variable with scale parameter c and shape parameter k.

A special form of an inhomogeneous gamma process, called the *modulated gamma process*, can be found in [65]. In a modulated gamma process the shock process is modeled by an NHPP with intensity $\rho(t) = \mu \exp\{\boldsymbol{\beta}'\boldsymbol{z}(t)\}$, where $\boldsymbol{\beta}' = (\beta_1, \ldots, \beta_p)$ and $\boldsymbol{z}(t)' = \{z_1(t), \ldots, z_p(t)\}$. Existence of time trend in the data can be modeled simply by including a single term $z_1(t) = t$ or, less simply, by including a further term $z_2(t) = t^2$.

Another special form of inhomogeneous gamma process, which has been studied greatly, is the *modulated PLP* (MPLP). In the MPLP the process of shocks is modeled by a PLP with intensity $\rho(t) = (\beta/\alpha)(t/\alpha)^{\beta-1}$, so that $U(t_{i-1}, t) = (t/\alpha)^\beta - (t_{i-1}/\alpha)^\beta$. When $k = 1$, the MPLP reduces to a PLP, whereas when $\beta = 1$ the process reduces to a gamma renewal process. Finally, when both $k = 1$ and $\beta = 1$, the MPLP becomes an HPP. The likelihood relative to a failure-truncated sample $t_1 < t_2 < \cdots < t_n$ results in

$$\begin{aligned}L(\alpha, \beta, k) = &\frac{(\beta/\alpha)^n}{[\Gamma(k)]^n}\left\{\prod_{i=1}^{n}\left(\frac{t_i}{\alpha}\right)^{\beta-1}\right. \\ &\left.\times\left[\left(\frac{t_i}{\alpha}\right)^\beta - \left(\frac{t_{i-1}}{\alpha}\right)^\beta\right]^{k-1}\right\} \\ &\times \exp\left[-\left(\frac{t_n}{\alpha}\right)^\beta\right]\end{aligned} \quad (18.53)$$

A complete discussion on the MPLP and ML procedures is given by Rigdon and coworkers [66, 67]. In particular, the ML estimates of parameters can be obtained by solving

$$\begin{aligned}\hat{\alpha} &= \frac{t_n}{(n\hat{k})^{1/\hat{\beta}}} \\ \hat{\beta} &= n\Bigg/\left[\left(\frac{t_n}{\hat{\alpha}}\right)^{\hat{\beta}} \ln\left(\frac{t_n}{\hat{\alpha}}\right) + n\hat{k}\ln\hat{\alpha} - \sum_{i=1}^{n}\ln t_i\right. \\ &\qquad \left. - (\hat{k}-1)\sum_{i=1}^{n}\frac{t_i^{\hat{\beta}}\ln t_i - t_{i-1}^{\hat{\beta}}\ln t_{i-1}}{t_i^{\hat{\beta}} - t_{i-1}^{\hat{\beta}}}\right] \\ \psi(\hat{k}) &= \frac{\sum_{i=1}^{n}\ln(t_i^{\hat{\beta}} - t_{i-1}^{\hat{\beta}})}{n} - \hat{\beta}\ln\hat{\alpha}\end{aligned} \quad (18.54)$$

where ψ denotes the di-gamma function: $\psi(x) = \Gamma'(x)/\Gamma(x)$. Approximate confidence intervals and hypothesis tests for the MPLP parameters can be obtained on the basis of the asymptotic distribution of ML estimators.

Bayesian inference, based on non-informative and vague prior densities, was proposed by Calabria and Pulcini [68] to obtain point estimates of MPLP parameters and credibility intervals that do not rely on asymptotic results. Bayesian prediction procedures for future failure times, which perform well even in the case of small samples, can be found in [68].

The MPLP expected number of failures $M(t)$ can be well approximated by $M(t) \cong (1/k)(t/\alpha)^{\beta}$, and hence the (unconditional) intensity function $\lambda(t)$ is approximately $\lambda(t) \cong (1/k)(\beta/\alpha)(t/\alpha)^{\beta-1}$ [69]. On the basis of such approximations, Bayesian inference on the MPLP parameters, as well as on $M(t)$ and $\lambda(t)$, can be found in [69] when prior information on β, k and/or $M(t)$ is available. Bayesian prediction on the failure times in a future sample can also be found. Finally, testing procedures for assessing the time trend and the effect of repair actions in one or two MPLP samples can be found in [70].

The MPLP model has been applied for analyzing the failure data of several aircraft air-conditioning equipment [18], an aircraft generator [71] and a photocopier [31].

18.5.3 Lawless–Thiagarajah Models

A very useful family of models that incorporates both time trend and the effect of past failures, such as renewal-type behavior, has been proposed by Lawless and Thiagarajah [3]. The CIF is of the form:

$$\lambda(t;\, H_t) = \exp\{\boldsymbol{\theta}' z(t)\} \qquad (18.55)$$

where $z(t) = \{z_1(t), \ldots, z_n(t)\}'$ is a vector of functions that may depend on both t and the history H_t, and $\boldsymbol{\theta} = (\theta_1, \ldots, \theta_p)'$ is a vector of unknown parameters. This model includes, as special cases, many commonly used processes, such as the PLP (when $z(t) = \{1, \ln(t)\}'$), the LLP (when $z(t) = \{1, t\}'$), and the renewal processes (when $z(t)$ is a function of the time $u(t)$ since the most recent repair: $u(t) = t - t_{N(t^-)}$). Models with CIF given by Equation 18.55 are then suitable for exploring failure data of repairable equipment when a renewal-type behavior may be expected, perhaps with a time trend superimposed.

The ML estimation procedures for the general form given by Equation 18.55, which in general need numerical integration, are given by Lawless and Thiagarajah [3]. A particular form of Equation 18.55 is discussed:

$$\lambda(t;\, H_t) = \exp\{\alpha + \beta t + \gamma u(t)\}$$
$$-\infty < \alpha,\, \beta,\, \gamma < \infty \qquad (18.56)$$

This shows jumps at each failure time t_i ($i = 1, 2, \ldots$), except for the special case ($\gamma = 0$), when it reduces to the NHPP with log-linear intensity (LLP). When $\beta = 0$, the process reduces to a renewal process with interarrival times distributed as a Gumbel random variable. The complete intensity (Equation 18.56) is then piecewise continuous: when the repair is effective and produces an improvement in the conditions of the equipment (imperfect repair), the jumps are downward, whereas when the repair is incorrect (worse repair) these jumps are upward. Test procedures for time trend can be found in [3], where adequacy of the large sample approximation to moderate samples has been assessed.

Other suitable forms of Equation 18.55, in particular the so-called *power-law–Weibull renewal* (PL–WR) and *log-linear–Weibull renewal* (LL–WR), can be found in [72]. The CIFs are

$$\lambda(t;\, H_t) = \gamma t^{\beta-1}[u(t)]^{\delta-1}$$
$$\gamma > 0,\ \beta + \delta > 1 \qquad (18.57)$$

and

$$\lambda(t;\, H_t) = \delta \exp(\theta + \beta t)[u(t)]^{\delta-1}$$
$$-\infty < \theta,\, \beta < \infty,\ \delta > 0 \qquad (18.58)$$

respectively. Both the CIFs show jumps at each failure time t_i ($i = 1, 2, \ldots$), except for the special case ($\delta = 1$), when they reduce to the PLP and the LLP respectively. The PL–WR and LL–WR models are suitable to describe the failure pattern of a wide range of repairable mechanical units that experience reliability improvement or deterioration with operating time and are subjected to imperfect or worse repairs. For example, when $\delta > 1$ ($\delta < 1$) the PL–WR model describes a repeated beneficial (harmful) effect of repair actions with respect to the minimal repair condition. The more δ differs from unity,

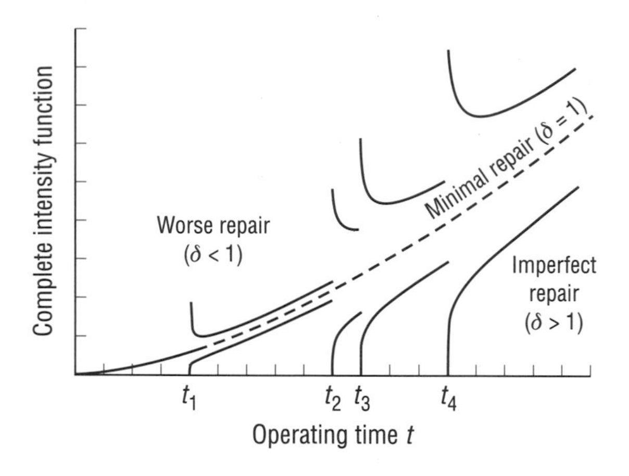

Figure 18.3. Plots of the complete intensity function of the PL–WR model for deteriorating equipment under imperfect ($\delta > 1$) or worse ($\delta < 1$) repairs. Comparison with the minimal repair case ($\delta = 1$)

the more the failure process departs from a minimal repair condition. When $\beta > 1$ ($1 - \delta < \beta < 1$) the PL–WR process describes the failure pattern of equipment experiencing reliability deterioration (improvement) with operating time.

Figure 18.3 shows the plots of the CIF for arbitrary failure times $t_1, \ldots, t_4$ of equipment that experiences reliability deterioration with operating time ($\beta > 1$) and is subject to imperfect ($\delta > 1$) or worse ($\delta < 1$) repairs. The plots are also compared with the minimal repair case ($\delta = 1$).

Point and interval ML estimates of parameters of PL–WR and LL–WR models can be found in [72], as well as test procedures, based on the likelihood ratio statistic, for assessing the departure from minimal or perfect repair assumption. In particular, under a time-truncated sampling, the ML estimators of the PL–WR parameters β and δ can be found by maximization of the modified two-parameter log-likelihood:

$$\ell'(\beta, \delta) = n \ln\left[\frac{n}{\sum_{i=1}^{n+1} I_i(\beta, \delta)}\right] + (\beta - 1)\ln\left(\prod_{i=1}^{n} t_i\right) + (\delta - 1)\ln\left[\prod_{i=1}^{n}(t_i - t_{i-1})\right] - n \tag{18.59}$$

where $I_i(\beta, \delta) = \int_{t_{i-1}}^{t_i} t^{\beta-1}(t - t_{i-1})^{\delta-1}\,\mathrm{d}t$ (with $t_0 = 0$ and $t_{n+1} = T$), subject to the constraint $\beta + \delta > 1$. The ML estimate of γ is in closed form:

$$\hat{\gamma} = \frac{n}{\sum_{i=1}^{n+1} I_i(\hat{\beta}, \hat{\delta})} \tag{18.60}$$

The Lawless–Thiagarajah family of models has been applied to several failure data, such as the air-conditioning equipment data [18] denoted as Planes 6 and 7 by Cox and Lewis [10], the aircraft generator data [71], and the failure data of automobile #3 [46].

18.5.4 Proportional Intensity Variation Model

An alternative way to describe a beneficial effect of preventive maintenance on the equipment reliability has been given by Chan and Shaw [73] by assuming that each maintenance reduces the (non-decreasing) intensity of a quantity that can be constant (*fixed reduction*) or proportional to the current intensity value (*proportional reduction*).

The latter assumption has been adopted by Calabria and Pulcini [74] to model the failure pattern of equipment subject to imperfect or worse repairs. Each repair alters the CIF of the equipment in such a way that the failure intensity immediately after the repair is proportional to the value just before the failure occurrence:

$$\lambda(t_i^+; H_{t_i^+}) = \delta\lambda(t_i^-; H_{t_i^-}) \quad \delta > 0 \tag{18.61}$$

Hence, the CIF shows jumps at each failure time t_i, except in the case of $\delta = 1$. Such jumps are proportional to the value of the complete intensity and, unlike the PAR model, do not depend explicitly on the operating time since the most recent failure. If $\rho(t)$ denotes the intensity function up to the first failure time, the CIF is

$$\lambda(t; H_t) = \delta^i \rho(t) \quad t_i < t \le t_{i+1} \tag{18.62}$$

(with $t_0 = 0$) and depends on the history of the process only through the number i of previous failures. The parameter δ is a measure of the effectiveness of repair actions. When $\delta = 1$, the model in Equation 18.62 reduces to an NHPP with intensity $\rho(t)$. When $\delta < 1$ ($\delta > 1$), the repair action produces an improvement (worsening) in the equipment condition, and then the model can describe the effect of imperfect or worse repair for equipment that, depending on the behavior of $\rho(t)$, experiences reliability improvement or deterioration.

Two different forms for $\rho(t)$ have been discussed, namely the power-law and the log-linear intensity, and the ML estimators of model parameters can be found in [74]. Procedures for testing the departure from the minimal repair assumption, based on the log-likelihood ratio and Wald statistics, can be found both for large and moderate samples. In particular, by assuming the power-law intensity $\rho(t) = (\beta/\alpha)(t/\alpha)^{\beta}$, the likelihood relative to a failure-truncated sample of size n is

$$L(\alpha, \beta, \delta) = \delta^{\nu} \frac{\beta^n}{\alpha^{n\beta}} \prod_{i=1}^{n} t_i^{\beta-1} \times \exp\left\{ -\delta^{i-1}\left[\left(\frac{t_i}{\alpha}\right)^{\beta} - \left(\frac{t_{i-1}}{\alpha}\right)^{\beta}\right]\right\} \quad (18.63)$$

where $\nu = (n^2 - n)/2$. The ML estimators of β and δ can be found by maximization of the modified two-parameter log-likelihood:

$$\ell'(\beta, \delta) = \nu \ln \delta - n \ln \left[\frac{\sum_{i=1}^{n} \delta^{i-1}(t_i^{\beta} - t_{i-1}^{\beta})}{n\beta}\right] + (\beta - 1)\ln\left(\prod_{i=1}^{n} t_i\right) - n \quad (18.64)$$

subject to the constraints $\beta > 0$ and $\delta > 0$, and the ML estimate of α:

$$\hat{\alpha} = \left[\frac{\sum_{i=1}^{n} \hat{\delta}^{i-1}(t_i^{\hat{\beta}} - t_{i-1}^{\hat{\beta}})}{n}\right]^{1/\hat{\beta}} \quad (18.65)$$

Figure 18.4 shows the plots of the CIF (Equation 18.62) for arbitrary failure times $t_1, \ldots, t_4$

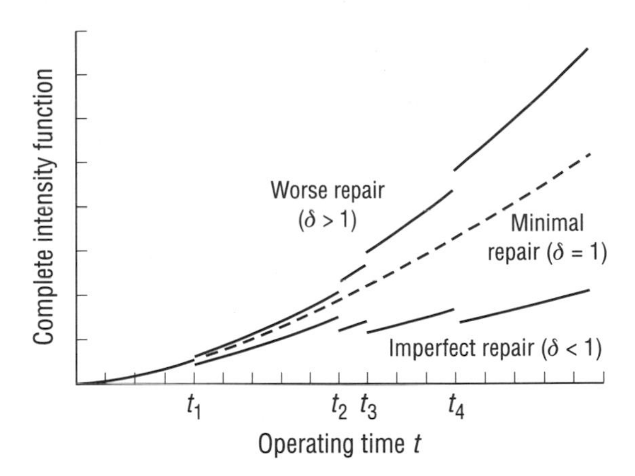

Figure 18.4. Plots of the complete intensity function for the proportional intensity variation model of deteriorating equipment subject to imperfect ($\delta < 1$) or worse ($\delta > 1$) repairs. Comparison with the minimal repair case ($\delta = 1$)

of equipment that experiences reliability deterioration with operating time ($\beta > 1$) and is subject to imperfect ($\delta < 1$) or worse ($\delta > 1$) repairs. The plots are also compared with the minimal repair case ($\delta = 1$). By comparing Figure 18.2 with Figure 18.4, one can see that the improvement caused by the third repair is very small under the PAR model, since the time interval ($t_3 - t_2$) is narrow. On the contrary, the improvement of the same third repair under the proportional intensity variation model is more noticeable since the complete intensity just before the third failure is large.

18.6 Complex Maintenance Policy

This section illustrates some reliability models that are able to describe the failure pattern of repairable equipment subject to a maintenance policy consisting in general of a sequence of corrective repairs interspersed with preventive maintenance.

Except for the model illustrated in Section 18.6.5, all models discussed in Sections 18.6.1–18.6.4 are based on specific assumptions regarding the effect of corrective and/or preventive maintenance on the equipment conditions. Such assumptions are summarized in Table 18.2. In most models, the preventive maintenance is assumed to improve the equipment reliability significantly, until the limit case of a complete restoration of the equipment (complete *overhaul*). However, as emphasized in Ascher and Feingold [1], not only do overhauls not necessarily restore the equipment to a same as new condition, but in some cases they can reduce reliability. This may happen when overhaul is performed by inexperienced men or/and under difficult working conditions: reassembling injects fresh faults into the equipment and burn-in techniques cannot be applied before putting the equipment in service again. Thus, the assumption that preventive maintenance improves equipment conditions should be assessed before using stochastic models based on such an assumption.

18.6.1 Sequence of Perfect and Minimal Repairs Without Preventive Maintenance

Let us consider repairable equipment that is repaired at failure and is not preventively maintained. By using the approach proposed by Nakagawa [75] to model preventive maintenance, Brown and Proschan [76] assumed that the equipment is perfectly repaired with probability p and it is minimally repaired with probability $1-p$. It is as though the equipment is subject to two types of failure: Type I failure (catastrophic failure) occurs with probability p and requires the replacement of the equipment, whereas Type II failure (minor failure) occurs with probability $1-p$ and is corrected with minimal repair. Note that both Nakagawa [75] and Brown and Proschan [76], as well as other authors, *e.g.* [77], use the term "imperfect repair" when they refer to "same as old repair".

If $p=0$, then the failure process reduces to a minimal repair process (NHPP), whereas if $p=1$ the process is a renewal one. When $0<p<1$, the process is a sequence of NHPPs interspersed with perfect repairs; each NHPP starts from the time epoch of the most recent perfect repair.

If $F_1(t)$ and $r_1(t)$ are respectively the cumulative distribution and the hazard rate of the occurrence time of the first failure, then the distribution of the time between successive perfect repairs is $F_p(t)=1-[1-F_1(t)]^p$ and the corresponding hazard rate is $r_p(t)=pr_1(t)$.

Let τ_j (with $j=1,\ldots,m$) denote the time epochs of m perfect repairs. For a generic time t in the interval (τ_{i-1},τ_i) (where $\tau_0=0$), the CIF depends on the operating time t and on the history H_t only through the difference $t-\tau_{i-1}$:

$$\lambda(t;H_t)=h(t-\tau_{i-1}) \qquad \tau_{i-1}<t\le\tau_i \tag{18.66}$$

It is numerically equal to the hazard rate $r_1(x)$ of the first failure time evaluated at $x=t-\tau_{i-1}$.

Non-parametric estimates of the cumulative distribution function $F_1(t)$ can be found in Whitaker and Samaniego [77] under the assumption that both failure times t_i $(i=1,2,\ldots)$ and the repair mode (minimal or perfect) following each failure are fully known (*complete data*). The likelihood relative to a complete data set observed until the occurrence of the nth failure is

$$L(p,F_1)=p^m(1-p)^{n-m-1}f_1(x_{(1)}) \times\prod_{i=1}^{n-1}\frac{f_1(x_{(i+1)})}{[1-F_1(x_{(i)})]^{1-z_{(i)}}} \tag{18.67}$$

where m is the number of perfect repairs, $x_{(i)}$ $(i=1,\ldots,n)$ are the ordered failure times measured from the most recent perfect repair epoch (the equipment installation corresponds to a perfect repair), and $z_{(i)}$ $(i=1,\ldots,n-1)$ is the (non-ordered) indicator variable for the mode of the ith ordered repair: $z_{(i)}=1$ for perfect repair, and $z_{(i)}=0$ for minimal repair. Of course: $\sum_{i=1}^{n-1}z_{(i)}=m$. If $z_{(n-1)}=1$, the non-parametric ML estimate of $F_1(t)$ is

$$\hat{F}_1(t)=\begin{cases}0 & t<x_{(1)}\\ 1-\prod_{j=1}^{i}\hat{\phi}_j & x_{(i)}\le t<x_{(i+1)}\\ 1 & t\ge x_{(n)}\end{cases} \qquad (i=1,\ldots,n-1) \tag{18.68}$$

Table 18.2. Assumptions regarding the effect of corrective and/or preventive maintenance for models in Sections 18.6.1–18.6.4

Maintenance	Section 18.6.1	Section 18.6.2	Section 18.6.3	Section 18.6.4
Corrective	Perfect or minimal	Minimal	Imperfect	Minimal
Preventive	No PM	Perfect	Perfect	Imperfect

where $\hat{\phi}_i = k_i/(k_i - 1)$ and k_i is the number of ordered failure times greater than $x_{(i)}$ for which the equipment was perfectly repaired. When $z_{(n-1)} = 0$, the non-parametric estimate does not exist. The estimate of $F_1(t)$ allows the CIF of Equation 18.66 between successive perfect repairs to be estimated. Whitaker and Samaniego [77] applied their estimation procedure to the failure data of the air-conditioning equipment of Boeing 720 airplane 7914 [18], by assigning arbitrary (but seemingly reasonable) values to the indicator variables $z_{(i)}$.

A Bayes procedure to estimate the unknown and random probability p of perfect repair and the waiting time between two successive perfect repairs when prior information on p is available can be found in [78].

A parametric form of the CIF (Equation 18.66) has been studied by Lim [79], who modeled the failure process between successive perfect repairs through a PLP. Then, the likelihood relative to a fully known data set, observed until the occurrence of the nth failure, is

$$L(p, \alpha, \beta) = p^m (1-p)^{n-m-1} \frac{\beta^n}{\alpha^{n\beta}} \prod_{i=1}^{n} x_i^{\beta-1} \times \exp\left[-\sum_{i=1}^{n} z_i \left(\frac{x_i}{\alpha}\right)^{\beta}\right] \quad (18.69)$$

where $m = \sum_{i=1}^{n-1} z_i$ is the number of perfect repairs and $z_n = 1$. Note that, in the parametric formulation, the failure times x_i $(i = 1, \ldots, n)$ from the most recent perfect repair need not be ordered. The ML estimates of model parameters for complete data are

$$\hat{p} = m/(n-1) \quad \hat{\alpha} = \left(\frac{1}{n}\sum_{i=1}^{n} z_i x_i^{\hat{\beta}}\right)^{1/\hat{\beta}}$$

$$\hat{\beta} = \left[\frac{\sum_{i=1}^{n} z_i x_i^{\hat{\beta}} \ln x_i}{\sum_{i=1}^{n} z_i x_i^{\hat{\beta}}} - \frac{1}{n}\sum_{i=1}^{n} \ln x_i\right]^{-1} \quad (18.70)$$

When the data set is incomplete, *i.e.* the repair modes are not recorded, the inferential procedures are much more complex. The ML estimate of the probability p of perfect repair, as well as that of the PLP parameters, can be obtained through an expectation–maximization algorithm based on an iterative performance of the expectation step and the maximization step. The proposed inferential procedure can be applied even when a different NHPP from the PLP is assumed to model the failure process between successive perfect repairs, by just modifying the maximization step.

When each repair mode can depend on the mode of previous repair, so that there is a first-order dependency between two consecutive repair modes, the procedure recently proposed by Lim and Lie [80] can be used. The failure process between successive perfect repairs is still modeled by a PLP and data can be partially masked if all failure times are available but some of the repair modes are not recorded. The ML estimates of model parameters and of transition probabilities $p_{j,k} = \Pr\{z_i = k | z_{i-1} = j\}$ $(j, k = 0, 1; i = 1, 2, \ldots)$ of repair process can be obtained by performing an expectation–maximization algorithm.

In some circumstances, the choice of repair mode does not depend only on external factors (such as the availability of a replacement) that are stable over time, but also on the equipment condition. For example, catastrophic failures, which require the replacement of the equipment, can occur with higher probability as the equipment age increases. This *age-dependent repair mode* situation has been modeled by Block *et al.* [81]. By extending the Brown–Proschan model, the repair is assumed to be perfect with probability $p(x)$ and minimal with probability $1 - p(x)$, where x is the age of the equipment measured from the last perfect repair epoch. The likelihood relative to a complete data set observed until the occurrence of the nth failure is given by

$$L(p(\cdot), F_1) = \prod_{i=1}^{n-1} p(x_i)^{z_i}[1 - p(x_i)]^{1-z_i} f_1(x_1) \times \prod_{i=1}^{n-1} \frac{f_1(x_{i+1})}{[1 - F_1(x_i)]^{1-z_{i-1}}} \qquad (18.71)$$

Non-parametric estimates of the cumulative distribution of the first failure time can be found in [77].

18.6.2 Minimal Repairs Interspersed with Perfect Preventive Maintenance

The failure pattern of repairable equipment subject to preventive maintenance that completely regenerates the intensity function (perfect maintenance) has been modeled by Love and Guo [82, 83]. Between successive preventive maintenance actions, the equipment is minimally repaired at failure, so that within each maintenance cycle the failure process can be regarded as a piecewise NHPP. In addition, the operating and environmental conditions of the equipment are not constant during each maintenance cycle but can change after each repair and then affect the next failure time. This operating and environmental influence is then modeled through suitable covariates [59].

Let $y_{i,j}$ denote the jth failure time in the ith maintenance cycle measured from the most recent maintenance epoch. Then, the CIF at any time y in the interval $(y_{i,j-1}, y_{i,j})$ is then given by

$$\rho_{i,j}(y) = \rho_0(y) \exp(z'_{i,j}\boldsymbol{a}) \quad y_{i,j-1} \le y < y_{i,j} \qquad (18.72)$$

where $y_{i,0} = 0$, $z_{i,j}$ is the $k \times 1$ vector of explanatory variables (covariates) that characterize the operating conditions of the equipment in the time interval $(y_{i,j-1}, y_{i,j})$, and $\boldsymbol{a}$ is the vector of regression coefficients.

A non-parametric form for the baseline intensity has been assumed in [82], whereas a power-law form for the baseline intensity $\rho_0(y) = (\beta/\alpha)(y/\alpha)^{\beta-1}$ has been chosen in [83]. In the latter case, the likelihood relative to m maintenance cycles is

$$L(\alpha, \beta, \boldsymbol{a}) = \left(\frac{\beta}{\alpha}\right)^n \left[\prod_{i=1}^{m}\prod_{j=1}^{n_i} \left(\frac{y_{i,j}}{\alpha}\right)^{\beta-1} \exp(z'_{i,j}\boldsymbol{a})\right] \times \exp\left\{-\sum_{i=1}^{m}\sum_{j=1}^{n_i+1}\left[\left(\frac{y_{i,j}}{\alpha}\right)^{\beta} - \left(\frac{y_{i,j-1}}{\alpha}\right)^{\beta}\right] \times \exp(z'_{i,j}\boldsymbol{a})\right\} \qquad (18.73)$$

where n_i is the number of failures occurred in the ith maintenance cycle, y_{i,n_i+1} is the ending time of the ith maintenance cycle, and $n = \sum_{i=1}^{m} n_i$ is the total number of observed failures. The ML estimate of α, β, and $\boldsymbol{a}$ can be found by solving the equations $\partial \ln(L)/\partial\beta = 0$, $\partial \ln(L)/\partial\alpha = 0$, and $\partial \ln(L)/\partial a_l = 0$ $(l = 1, \ldots, k)$ through the Newton–Raphson algorithm. Tests of the significance of the regression model can be performed on the basis of the log-likelihood ratio statistic under the null hypothesis of constant environmental and operating conditions (the regression coefficients are assumed to be zero). Significance tests for individual regression coefficients are based on the asymptotic distribution of the ML estimators.

The model in Equation 18.72 has also been applied to a large cement roller mill, whose data are the times between corrective or preventive maintenance actions. Three potential symptomatic covariates were considered: instantaneous change in pressure on the main bearing, recirculating load on the bucket elevator, and the power demanded by the main motor. The second and third covariates were found to be significant, and both the regression models that include one of the significant covariates fit the observed data rather better than the baseline model.

18.6.3 Imperfect Repairs Interspersed with Perfect Preventive Maintenance

The failure pattern of an equipment imperfectly repaired at failure and subject to perfect preventive maintenance can be described by the model given by Guo and Love [62]. The effect of imperfect repair is modeled through the PAR model of Section 18.5.1 and the covariate structure for the complete intensity given by Love and Guo [82, 83] can be adopted when different environmental and operating conditions have to be considered.

Let $y_{i,j}$ denote the jth failure time in the ith maintenance cycle measured from the most recent maintenance epoch. The CIF in any time interval $(y_{i,j-1}, y_{i,j})$ (where $y_{i,0}=0$) is then

$$\lambda_{i,j}(y; H_y) = h_0[y-(1-\delta)y_{i,j-1}]\exp(z'_{i,j}\boldsymbol{a}) \qquad 0\le\delta\le 1,\ y_{i,j-1}<y\le y_{i,j} \tag{18.74}$$

where $1-\delta$ is equal to the improvement factor ρ of Section 18.5.1. The effect of repair may be assumed to be constant (and unknown) over all failures (*constant repair impact*), or the effect of repair may vary from time to time (*repair-specific impact*). In the latter case, the value of the repair factor δ after each failure has to be assigned.

The baseline intensity up to the first failure has the power-law form and, under the assumption of constant repair effect (δ is then an unknown parameter), the likelihood relative to m maintenance cycles is

$$\begin{aligned} L(\alpha,\beta,\delta,\boldsymbol{a}) &= \left(\frac{\beta}{\alpha}\right)^n \left\{\prod_{i=1}^{m}\prod_{j=1}^{n_i}\left[\frac{y_{i,j}-(1-\delta)y_{i,j-1}}{\alpha}\right]^{\beta-1}\right\} \\ &\quad\times\exp(z'_{\mathrm{sumf}}\boldsymbol{a}) \\ &\quad\times\exp\left\{-\sum_{i=1}^{m}\sum_{j=1}^{n_i+1}\left\{\left[\frac{y_{i,j}-(1-\delta)y_{i,j-1}}{\alpha}\right]^{\beta} - \left(\frac{\delta y_{i,j-1}}{\alpha}\right)^{\beta}\right\}\exp(z'_{i,j}\boldsymbol{a})\right\} \end{aligned} \tag{18.75}$$

where n_i is the number of failures occurred during the ith maintenance cycle, y_{i,n_i+1} is the ending time of the ith maintenance cycle, and $z_{\mathrm{sumf}}=\sum_i\sum_j z_{i,j}$ is the summation vector of the covariates over the failure set. The ML estimate of α, β, δ, and $\boldsymbol{a}$ can be found in the traditional way by solving the equations $\partial\ln(L)/\partial\beta=0$, $\partial\ln(L)/\partial\alpha=0$, $\partial\ln(L)/\partial\delta=0$, and $\partial\ln(L)/\partial a_l=0$ $(l=1,\ldots,k)$.

When the effect of repair varies from time to time, the likelihood is conditioned on the vector $\boldsymbol{\Delta}=\{\delta_{1,1},\ldots,\delta_{1,n_1},\ldots,\delta_{m,1},\ldots,\delta_{m,n_m}\}'$ collecting the (assigned) value of the repair factors $\delta_{i,j}$. Then, the form of the likelihood $L(\alpha,\beta,\boldsymbol{a}|\boldsymbol{\Delta})$ can be obtained from Equation 18.75 by substituting δ with $\delta_{i,j-1}$.

Both the proposed models (the *constant repair impact* model and the *repair-specific impact* one) have been compared by Guo and Love [62] with the minimal repair model [83] and the *hybrid* model [77]. Data of the large cement roller mill [82] have been analyzed, and it is showed that the repair-specific impact model requires a good management judgement on the effect of each repair in order to be effective.

If the original intensity function (before accounting for repair discontinuities) shows a bathtub-type behavior in each maintenance cycle, the failure pattern of imperfectly repaired equipment can be described by the model given by Love and Guo [54]. The EPP intensity (Equation 18.38) describes the original intensity function, whereas the imperfect repair is described through the PAR model [63]. The conditional ML estimate of model parameters β and η, given the value of the repair factors $\delta_{i,j}$, can be obtained from the first- and second-order partial derivatives of the log-likelihood, by applying the Newton–Raphson algorithm. The mean square error of the ordered estimates of the generalized residuals with respect to the expected standard exponential order statistics can be used to compare the fit of different models to a given data set.

18.6.4 Minimal Repairs Interspersed with Imperfect Preventive Maintenance

A model suitable to describe the failure pattern of several units subject to pre-scheduled or random imperfect preventive maintenance has been proposed by both Shin *et al.* [64] (and there denoted as Model B) and Jack [84] (where it is denoted as Model I). The units are minimally repaired at failure and the effect of imperfect preventive maintenance is modeled according to the PAR criterion [63].

The failure pattern in each maintenance cycle is described by an NHPP, but the actual age of the equipment in the kth cycle is reduced by a fraction of the most recent maintenance epoch τ_{k-1}. The CIF at a time t in the kth maintenance cycle is

$$\lambda(t;\, H_t) = h(t - \rho\tau_{k-1}) \qquad \tau_{k-1} < t \leq \tau_k \quad (18.76)$$

Suppose that l units are observed until T_i $(i = 1, \ldots, l)$ and that each equipment is subject to m_i preventive maintenance actions at epochs $\tau_{i,1} < \cdots < \tau_{i,m_i} \leq T_i$. The ith equipment experiences $r_{i,k}$ failures during the kth maintenance cycle $(k = 1, \ldots, m_i + 1)$. Let $t_{i,k,j}$ $(j = 1, \ldots, r_{i,k})$ denote the jth failure experienced in the kth maintenance cycle by the ith equipment. Then, under the assumption that the improvement factor ρ is constant over all units and maintenance actions, the likelihood relative to the above data is

$$L_i = \prod_{i=1}^{l} \left\{ \prod_{k=1}^{m_i+1} \left[\prod_{j=1}^{r_{i,k}} h(t_{i,k,j} - \rho\tau_{i,k-1}) \right] \times \exp\left[-\sum_{k=1}^{m_i+1} \int_{\tau_{i,k-1}}^{\tau_{i,k}} h(x - \rho\tau_{i,k-1})\,\mathrm{d}x \right] \right\} \quad (18.77)$$

where $\tau_{i,0} = 0$ and $\tau_{i,m_i+1} \equiv T_i$. Since the effect of preventive maintenance may not be uniform over units and time, the improvement factor ρ can be considered as the averaged effect of preventive maintenance over the observation period.

Two different forms for the intensity function up to the first failure, namely the power-law and log-linear intensities, have been studied and related ML estimates can be found in [64] and [84]. In particular, under the power-law assumption, the likelihood results in

$$L(\alpha, \beta, \rho) = \left(\frac{\beta}{\alpha}\right)^n \left[\prod_{i=1}^{l} \prod_{k=1}^{m_i+1} \prod_{j=1}^{r_{i,k}} \left(\frac{t_{i,k,j} - \rho\tau_{i,k-1}}{\alpha} \right)^{\beta-1} \right] \times \exp\left\{ -\sum_{i=1}^{l} \sum_{k=1}^{m_i+1} \left[\left(\frac{\tau_{i,k} - \rho\tau_{i,k-1}}{\alpha} \right)^{\beta} - \left(\frac{\tau_{i,k-1} - \rho\tau_{i,k-1}}{\alpha} \right)^{\beta} \right] \right\} \quad (18.78)$$

where n is the total number of failures for l units during the whole observation periods. The ML estimate of the scale parameter α is in closed form, given the ML estimates of β and ρ:

$$\hat{\alpha} = \left\{ \sum_{i=1}^{l} \sum_{k=1}^{m_i+1} [(\tau_{i,k} - \hat{\rho}\tau_{i,k-1})^{\hat{\beta}} - (\tau_{i,k-1} - \hat{\rho}\tau_{i,k-1})^{\hat{\beta}}]/n \right\}^{1/\hat{\beta}} \quad (18.79)$$

and the ML estimates $\hat{\beta}$ and $\hat{\rho}$ can be found by maximizing the modified two-parameter log-likelihood:

$$\ell'(\beta, \rho) = n \ln \beta - n \ln \left\{ \sum_{i=1}^{l} \sum_{k=1}^{m_i+1} [(\tau_{i,k} - \rho\tau_{i,k-1})^{\beta} - (\tau_{i,k-1} - \rho\tau_{i,k-1})^{\beta}] \right\} + n \ln n + (\beta - 1) \times \left[\sum_{i=1}^{l} \sum_{k=1}^{m_i+1} \sum_{j=1}^{r_{i,k}} \ln(t_{i,k,j} - \rho\tau_{i,k-1}) \right] - n \quad (18.80)$$

Approximate confidence on model parameters can be obtained on the basis of the asymptotic distribution of ML estimators [64] or by using the likelihood ratio method [84]. Shin *et al.* [64] have analyzed the failure data of a central cooler system of a nuclear power plant subject to major overhauls, by assuming a power-law form for the CIF (Equation 18.76). This application will be discussed later on. Under the same power-law assumption, Jack [84] analyzed the failure data of medical equipment (syringe-driver infusion pumps) used in a large teaching hospital and estimated the model parameters and the expected number of failures in given future time intervals.

Bayes inferential procedures for the above model under the power-law intensity (denoted as PAR–PLP model) are given by Pulcini [85, 86] when data arise from a single piece of equipment observed until T. Point and interval estimates of model parameters and CIF can be obtained on the basis of a number of suitable prior informations that reflect different degrees of belief on the failure process and maintenance effect [85]. Unlike the ML procedure, the Bayesian credibility intervals do not rely on asymptotic results, and then, when they are based on non-informative or vague prior densities, they are a valid alternative to the confidence intervals for small or moderate samples. In addition, Bayes inference on the expected number of failures in a future time interval $(T, T + \Delta)$ can be made when preventive maintenance or no maintenance is assumed to be performed at T.

Tests on the effectiveness of preventive maintenance can be carried out on the basis of the Bayes factor that measures the posterior evidence provided by the observed data in favor of the null hypothesis of minimal or perfect preventive maintenance against the alternative of imperfect maintenance. Finally, Bayes prediction procedures, both of the future failure times and of the number of failures in a future time interval, can be found in [86] on the basis both of vague and informative prior densities on β and ρ.

The PAR criterion proposed by Malik [63] assumes that maintenance reduces only the damage incurred since the previous preventive maintenance, and thus prohibits a major improvement to the conditions of equipment that has a large age but has recently undergone preventive maintenance. A different age reduction criterion, where preventive maintenance reduces all previously incurred damage, has been proposed by Nakagawa [87]. The equipment is minimally repaired at failure and the CIF at any time t in the interval (τ_{k-1}, τ_k) between the $(k-1)$th and the kth maintenance epochs is given by

$$\lambda(t;\, H_t) = h\left\{t - \tau_{k-1} + \sum_{i=1}^{k-1}\left[(\tau_i - \tau_{i-1})\prod_{j=i}^{k-1} b_j\right]\right\} \qquad \tau_{k-1} < t \le \tau_k \tag{18.81}$$

where $\tau_0 = 0$ and $0 < b_j < 1 (j = 1, 2, \dots)$ is the improvement factor in age of the equipment reduced by the jth preventive maintenance. This criterion has been adopted in Model II of Jack [84], by assuming that all improvement factors are equal: $b_j = \delta (j = 1, 2, \dots)$. Thus, the CIF in Equation 18.81 at any time t in the interval (τ_{k-1}, τ_k) becomes

$$\lambda(t;\, H_t) = h\left[t + \sum_{i=1}^{k-1} \delta^i (\tau_{k-i} - \tau_{k-i-1}) - \tau_{k-1}\right] \qquad 0 \le \delta \le 1,\ \tau_{k-1} < t \le \tau_k \tag{18.82}$$

ML estimation procedures for the model in Equation 18.82 are similar to those for the model in Equation 18.76.

18.6.4.1 Numerical Example

Let us consider the failure data of a central cooler system of a nuclear power plant analyzed by Shin *et al.* [64]. Data consist of $n = 15$ failure times and $m = 3$ major overhaul epochs observed over 612 days and are given in Table 18.3. The failure pattern between successive overhauls is modeled through a PLP, and each overhaul reduces the age of the equipment according to the PAR model [63].

Figure 18.5 compares the ML estimate given by Shin *et al.* [64] of the CIF under the imperfect overhauls assumption with the alternative assumptions of (a) minimal overhauls ($\rho = 0$), and

Table 18.3. Failure data of a central cooler system in numerical example (given in Shin *et al.* [64])

116	151	154*	213	263*	386
387	395	407	463	492	494
501	512*	537	564	590	609

* denotes PM epochs.

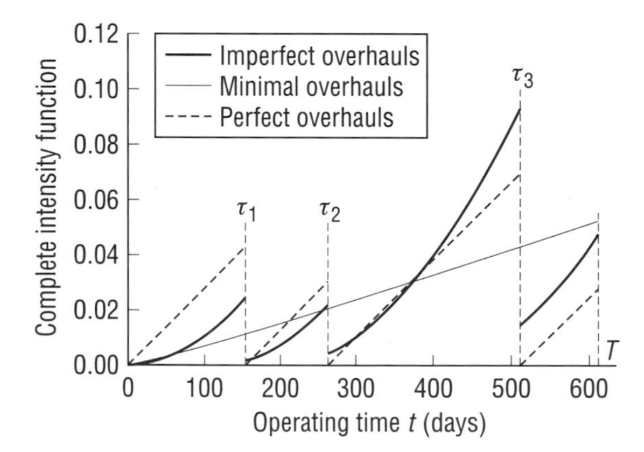

Figure 18.5. ML estimate of the complete intensity function under imperfect, minimal, and perfect overhaul assumptions for central cooler system data in numerical example

(b) perfect overhauls ($\rho = 1$). The assumption of minimal overhauls coincides with the assumption that the equipment is not preventively maintained but only minimally repaired at failure, and the whole failure process is then described by the PLP of Section 18.4.2.1. Under the assumption of perfect overhauls, the failure pattern is described by the model (deprived of covariates) given by Love and Guo [83] and discussed in Section 18.6.2.

In Table 18.4 the ML estimate of parameters β and ρ is compared with the Bayes estimate obtained by Pulcini [85] by assuming a uniform prior density for β over the interval (1, 5) and a gamma prior density for ρ with mean $\mu_\rho = 0.7$ and standard deviation $\sigma_\rho = 0.15$. Such densities formalize a prior ignorance on the value of β (except the prior conviction that the equipment experiences deterioration) and a vague belief that overhauls are quite effective. Table 18.5 gives the posterior mean and the equal-tails 0.80 credibility

Table 18.4. ML and Bayesian estimates of model parameters β and δ

	ML estimates		Bayes estimates	
	β	ρ	β	ρ
Point estimate	2.93	0.77	2.79	0.72
0.95 lower limit	1.82	0.50	1.73	0.50
0.95 upper limit	4.00	1.00	3.98	0.89

interval of the expected number of failures in the future time interval $(612, 612 + \Delta)$, for selected Δ values ranging from 30 to 100 days, under the different hypotheses that a major overhaul is or is not performed at $T = 612$ days.

18.6.5 Corrective Repairs Interspersed with Preventive Maintenance Without Restrictive Assumptions

The previously illustrated models are based on specific assumptions regarding the effect of corrective repairs and preventive maintenance on the equipment conditions. A model that does not require any restrictive assumption on the effect of maintenance actions on the equipment conditions is given by Kobbacy *et al.* [88]. The model assumes two different forms for the CIF after a corrective repair (CR) and a preventive maintenance (PM). In both the intensities, their dependence on the full history of the process is modeled through explanatory variables. In particular, the CIF following a CR or a PM is

$$\lambda_{\mathrm{CR}}(t; z) = \lambda_{\mathrm{CR},0}(t) \exp(z'\boldsymbol{a}) \qquad (18.83)$$

or

$$\lambda_{\mathrm{PM}}(t; \boldsymbol{y}) = \lambda_{\mathrm{PM},0}(t) \exp(\boldsymbol{y}'\boldsymbol{b}) \qquad (18.84)$$

respectively, where time t is measured from the time of applying this CR or PM, z and $\boldsymbol{y}$ are the $k \times 1$ and $l \times 1$ vectors of the explanatory variables, and $\boldsymbol{a}$ and $\boldsymbol{b}$ are the vectors of the

Table 18.5. Bayesian estimate of the expected number of failures in the future time interval $(\dot{T}, T + \Delta)$ when a major overhaul is or is not performed at $T = 612$ days

	No overhaul at T				Major overhaul at T			
	$\Delta = 30$	$\Delta = 50$	$\Delta = 80$	$\Delta = 100$	$\Delta = 30$	$\Delta = 50$	$\Delta = 80$	$\Delta = 100$
Point estimate	1.60	2.88	5.15	6.92	0.87	1.60	2.94	4.03
0.90 lower limit	0.97	1.72	2.98	3.91	0.40	0.79	1.56	2.21
0.90 upper limit	2.34	4.23	7.70	10.49	1.41	2.54	4.57	6.19

regression coefficients. Thus, the CIF following a failure or a preventive maintenance is given in terms of a baseline intensity ($\lambda_{CR,0}(t)$ or $\lambda_{PM,0}(t)$ respectively), which depends on the time t measured from this event, and of the value of the explanatory variables at a point in time just before the event occurrence. Potential explanatory variables are the age of the equipment (measured from the time the equipment was put into service), the total number of failures that have occurred, the time since the last CR, the time since the last PM, and so on. Hence, the explanatory variables measure the full history of the process until each CR and PM.

18.7 Reliability Growth

In many cases, the first prototype of mechanical equipment contains design and engineering flaws, and then prototypes are subjected to development testing programs before starting mass production. Reliability and performance problems are identified and design modifications are introduced to eliminate the observed failure modes. As a result, if the design changes are effective, the failure intensity decreases at each design change and a reliability growth is observed. Of course, in this context, the failure pattern of the equipment is strongly influenced by the effectiveness of design changes, as well as by failure mechanisms and maintenance actions. A complete discussion on reliability growth programs is given by Benton and Crow [89].

A large number of models that describe the reliability growth of repairable equipment have been proposed. Such models can be roughly classified into *continuous* and *discrete* models.

Continuous models are mathematical tools able to describe the *design process*, *i.e.* the decreasing trend in the ROCOF resulting from the repeated applications of design changes, which must not be confused with the behavior of the underlying failure process. Continuous models are commonly applied in test-analyze-and-fix (TAAF) programs when the equipment is tested until it fails; a design modification is then introduced in order to remove the observed failure cause, and the test goes on. This procedure is repeated until a desired level of reliability is achieved. The observed failure pattern is then affected by the failure mechanisms and design changes.

Discrete models incorporate a change in the failure intensity at each design change, which is not necessarily introduced at failure occurrence. Thus, more than one failure can occur during each development stage and the failure pattern depends on failure mechanisms, design changes and repair policy. In many cases, discrete models assume explicitly that the failure intensity is constant during each development stage, so that the resulting plot of the failure intensity as a function of test time is a non-increasing series of steps (*step-intensity* models). Thus, the failure process in each stage is modeled by an HPP (*e.g.* see the non-parametric model of Robinson and Dietrich [90] and the parametric model of Sen [91]). Under the step-intensity assumption, dealing with repairable units that, after repair and/or design modifications, continue to operate until the next failure is like dealing with non-repairable units (with constant hazard rate) which are replaced at each failure. Thus, models based on the explicit assumption of constant intensity in each development phase could be viewed as models for non-repairable units with exponentially distributed lifetime and will not be treated in this chapter, which deals with models specifically addressing repairable equipment.

18.7.1 Continuous Models

The most commonly used continuous model in the analysis of repairable equipment undergoing a TAAF program is the PLP model of Section 18.4.2.1, whose intensity function is

$$\rho(t) = \frac{\beta}{\alpha}\left(\frac{t}{\alpha}\right)^{\beta-1} \qquad \alpha > 0,\ 0 < \beta < 1 \quad (18.85)$$

In the TAAF context, the power-law parameter β is a measure of the management and engineering efforts in eliminating the failure modes during the development program. Small β values arise when an aggressive program is conducted by a knowledgeable staff.

The PLP was proposed and found its first application in a reliability growth context [12], specifically to explain the empirical results of Duane [71]. Duane analyzed the failure data of five types of mechanical unit, such as complex hydromechanical devices, aircraft generators and jet engines, undergoing development programs, and observed that a plot of the cumulative number of failures $N(t)$ observed until time t against the ratio $t/N(t)$ was nearly linear when a log–log plot was used.

In the application of the PLP in a reliability growth context, the inferential procedures are mainly addressed to estimate the reliability of the equipment at the end of the development program. In fact, if no further improvement is incorporated into the equipment design after the development program, the reliability level at the end of the program will characterize the reliability of the equipment as it goes into production. In addition, it is generally assumed that the *current* intensity function of the equipment as it goes into production remains constant during the so-called useful life period [92] and is equal to the intensity $\rho(T)$ at the termination time T of the development program. Thus, the failure process of the equipment as it goes into production can be modeled by an HPP with constant intensity $\rho(T)$, and the *current* reliability is given by

$$R_0(t_0) = \exp(-\rho(T)t_0) \qquad t_0 > 0 \qquad (18.86)$$

ML inferential procedures for current intensity and current reliability can be found in Lee and Lee [22], Bain and Engelhardt [23], Crow [93, 94], Calabria *et al.* [24], Tsokos and Rao [95], and Sen and Khattree [96]. These procedures, which refer both to failure- and time-truncated sampling, are based on pivotal properties of certain quantities and allow point estimates and exact confidence intervals to be obtained. In particular, if the development program terminates at the time t_n of the nth failure, confidence intervals on the current intensity $\rho(t_n)$ and the current reliability $R_0(t_0)$ can be obtained by exploiting the result that [22]

$$\frac{\rho(t_n)}{\hat{\rho}(t_n)} \sim \frac{1}{4n^2} ZS \qquad (18.87)$$

where Z and S are independent chi-squared random variables with $2(n-1)$ and $2n$ degrees of freedom respectively. For single or multiple copies undergoing a time-truncated development program, confidence intervals on $\rho(T)$ and $R_0(t_0)$ can be obtained by using tables given by Crow [93, 94].

A quasi-Bayes method to estimate the current intensity function has been proposed by Higgins and Tsokos [97], whereas a Bayes procedure for estimating both the current intensity and the current reliability on the basis of prior information on the parameter β can be found in Calabria *et al.* [98].

Some prediction problems have found a solution in Calabria and Pulcini [99] and Beiser and Rigdon [40]. In particular, Calabria and Dulcini [99] have found prediction procedures for the current lifetime t_0, both in the ML and Bayesian contexts. Exact confidence limits on t_0 can be computed by using a specific table, whereas, if prior information on the parameter β is available, prediction on t_0 can be made through a Bayesian procedure. Finally, prediction of the number of failures that new equipment will experience in a future time interval can be obtained by using the Bayesian procedure given by Beiser and Rigdon [40].

Another continuous model that has found application in reliability growth analysis is the NHPP with log-linear intensity of Section 18.4.2.2:

$$\rho(t) = \exp(\alpha + \beta t) \quad -\infty < \alpha < \infty,\ \beta < 0 \tag{18.88}$$

Unlike the PLP, whose intensity function tends to zero as t tends to infinity, the LLP intensity approaches more realistically to the positive value $\exp(\alpha)$ as t increases. In the reliability growth context, the LLP yields a defective distribution for the time to first failure [100] and then there is a positive probability that no failure will be observed during the development program.

When the TAAF program is carried out on several units, some of which are manufactured while the development program goes on, the initial reliability of all units put on test is not yet the same. In fact, the manufacturing process improves as the TAAF program goes on, since process-related defects, discovered during the TAAF program, are corrected. Thus, a unit manufactured a little while after the beginning of the production process will have fewer defects than one unit manufactured exactly at the beginning of the production process. A model that takes into account the dependence of the initial reliability of equipment on the age of the manufacturing process has been proposed by Heimann and Clark [101]. This *process-related reliability-growth* model describes the design process through a PLP (with $\beta < 1$) and assumes that the scale parameter α is not constant but depends on the age X of the manufacturing process when the unit is manufactured: $\alpha(X) = \alpha[1 - \exp(-bX)]$. Thus, the intensity function

$$\rho(t|X) = \frac{\beta t^{\beta-1}}{\{\alpha[1 - \exp(-bX)]\}^{\beta}} \tag{18.89}$$

depends through X on the maturity level of the manufacturing process. The ML estimates of model parameters, and then of the process-age-dependent intensity $\rho(t|X)$, can be obtained by maximization of the log-likelihood. Tests to assess the null hypothesis of constant initial reliability over all units (*i.e.* constant scale parameter over X) against the process-related model in Equation 18.89 can be performed through the log-likelihood ratio statistic.

18.7.2 Discrete Models

Discrete models find applications when one or more identical copies of the equipment are put on test and design modifications are not introduced a few at each failure but are implemented in block at the end of each development stage (*delayed fixes*) as a result of the experience gained during that stage. The next stage begins with new copies that incorporate the design changes and, if those modifications are effective, the reliability of the new design is higher than the reliability of the previous design.

One of the earliest models on this topic was proposed by Robinson and Dietrich [102]. It is based on the assumption that each time a failure occurs the equipment is perfectly repaired, so that the failure pattern in each development stage is described through a renewal process. The perfect repair assumption is not realistic when the development program involves complex equipment.

More recently, a discrete model based on the assumptions that the equipment is minimally repaired at failure and the failure process in each stage is modeled by a PLP has been proposed by Ebrahimi [103] and Calabria *et al.* [104]. The parameter β is constant over design modifications (changes do not alter the failure mechanism), whereas the scale parameter α increases from one stage to the next as a result of change effectiveness. Thus, the intensity function in the jth stage is

$$\rho_j(t) = \frac{\beta}{\alpha_j}\left(\frac{t}{\alpha_j}\right)^{\beta-1} \quad j = 1, 2, \ldots \tag{18.90}$$

where time t is measured from the beginning of the jth stage. ML estimates can be found in [103] where the intensity (Equation 18.90) has been reparameterized in terms of $\mu_j = \alpha_j^{-\beta} (j = 1, 2, \ldots)$. The parameters μ_j satisfy the conditions $\mu_1 \geq \mu_2 \geq \cdots$, since a reliability growth between stages is modeled. If $T_j (j = 1, 2, \ldots, m)$ denotes the test time of the jth stage and n_j failures occur in the jth stage at times $t_{1,j} < t_{2,j} < \cdots < t_{n_j,j}$, then the likelihood results in

$$L(\beta, \mu_1, \ldots, \mu_m) = \left[\prod_{j=1}^{m} \mu_j^{n_j}\right] \beta^n \left[\prod_{j=1}^{m} \prod_{i=1}^{n_j} t_{i,j}^{\beta-1}\right] \times \exp\left(-\sum_{j=1}^{m} \mu_j T_j^{\beta}\right) \tag{18.91}$$

where n is the total number of observed failures. The constrained ML estimate of model parameters can be found through a two-stage procedure that assures $\hat{\mu}_1 \geq \hat{\mu}_2 \geq \cdots \geq \hat{\mu}_m$, and approximate confidence intervals can be obtained through the asymptotic distribution of estimators.

The Bayesian approach proposed by Calabria *et al.* [104] is addressed to the solution of a decision-making problem rather than to an inferential one. Several identical copies of the equipment are put on test at each development stage. At the end of each stage, depending on the reliability level achieved, a decision between two alternative actions is made: (a) to accept the current design of the equipment for mass production, or (b) to continue the development program. The equipment reliability at stage j is measured by the expected number of failures in a given time interval $(0, \tau)$, say $\mathcal{N}_j(\tau)$, so that the decision process is based on the posterior density of $\mathcal{N}_j(\tau)$ and on specific loss functions that measure the economical consequences associated with each alternative action.

Acknowledgment

The author is indebted to Professor Maurizio Guida for valuable discussions and suggestions which helped the exposition of the article.

References

[1] Ascher H, Feingold H. Repairable systems reliability. modeling, inference, misconceptions and their causes. New York, Basel: Marcel Dekker; 1984.

[2] Pham H, Wang H. Imperfect maintenance. Eur J Oper Res 1996;94:425–38.

[3] Lawless JF, Thiagarajah K. A point-process model incorporating renewals and time trends, with application to repairable systems. Technometrics 1996;38:131–8.

[4] Cox DR, Isham V. Point processes. London, New York: Chapman and Hall; 1980.

[5] Thompson WA. Point process models with applications to safety and reliability. New York, London: Chapman and Hall; 1988.

[6] Thompson WA. On the foundations of reliability. Technometrics 1981;23:1–13.

[7] Berman M, Turner TR. Approximating point process likelihoods with GLIM. Appl Stat 1992;41:31–8.

[8] Ascher HE. Reliability models for repairable systems. In: Møltoft J, Jensen F, editors. Reliability technology—theory and applications. Oxford: Elsevier; 1986. p.177–85.

[9] Ascher H, Feingold H. "Bad-as-old" analysis of system failure data. In: Proceedings of 8th Reliability and Maintenance Conference, Denver, 1969; p.49–62.

[10] Cox DR, Lewis PAW. The statistical analysis of series of events. London: Chapman and Hall; 1978.

[11] Bain LJ, Engelhardt M, Wright FT. Tests for an increasing trend in the intensity of a Poisson process: a power study. J Am Stat Assoc 1985;80:419–22.

[12] Crow LH. Reliability analysis for complex, repairable systems. In: Proschan F, Serfling RJ, editors. Reliability and biometry. Philadelphia: SIAM; 1974. p.379–410.

[13] Lewis PAW, Robinson DW. Testing for a monotone trend in a modulated renewal process. In: Proschan F, Serfling RJ, editors. Reliability and biometry. Philadelphia: SIAM; 1974. p.163–82.

[14] Vaurio JK. Identification of process and distribution characteristics by testing monotonic and non-monotonic trends in failure intensities and hazard rates. Reliab Eng Syst Saf 1999;64:345–57.

[15] Lawless JF. Statistical models and methods for lifetime data. New York: John Wiley and Sons; 1982.

[16] Beiser JA, Rigdon SE. Bayesian prediction for the number of failures of a repairable system modeled by a homogeneous Poisson process. In: Vogt WG, Mickle MH, editors. Modeling and simulation, vol. 23, part 4. Control, signal processing, robotics, systems, power. Pittsburgh: Instrument Society of America; 1992. p.1783–9.

[17] Ascher H, Feingold H. Is there repair after failure? In: Proceedings of Annual Reliability and Maintenance Symposium, Los Angeles, 1978; p.190–7.

[18] Proschan F. Theoretical explanation of observed decreasing failure rate. Technometrics 1963;5:375–83.

[19] Kumar U, Klefsjø B. Reliability analysis of hydraulic systems of LHD machines using the power law process model. Reliab Eng Syst Saf 1992;35:217–24.

[20] Majumdar SK. Study on reliability modelling of a hydraulic excavator system. Qual Reliab Eng Int 1995;11:49–63.

[21] Finkelstein JM. Confidence bounds on the parameters of the Weibull process. Technometrics 1976;18:115–7.

[22] Lee L, Lee SK. Some results on inference for the Weibull process. Technometrics 1978;20:41–5.

[23] Bain LJ, Engelhardt M. Inferences on the parameters and current system reliability for a time truncated Weibull process. Technometrics 1980;22:421–6.

[24] Calabria R, Guida M, Pulcini G. Some modified maximum likelihood estimators for the Weibull process. Reliab Eng Syst Saf 1988;23:51–8.

[25] Lee L. Comparing rates of several independent Weibull processes. Technometrics 1980;22:427–30.

[26] Calabria R, Guida M, Pulcini G. Power bounds for a test of equality of trends in k independent power law processes. Commun Stat Theor Methods 1992;21:3275–90.

[27] Bain LJ, Engelhardt M. Sequential probability ratio tests for the shape parameter of a nonhomogeneous Poisson process. IEEE Trans Reliab 1982;R-31:79–83.

[28] Calabria R, Guida M, Pulcini G. Reliability analysis of repairable systems from in-service failure count data. Appl Stoch Models Data Anal 1994;10:141–51.

[29] Rigdon SE, Basu AP. The power law process: a model for the reliability of repairable systems. J Qual Technol 1989;21:251–60.

[30] Park WJ, Kim YG. Goodness-of-fit tests for the power-law process. IEEE Trans Reliab 1992;41:107–11.

[31] Baker RD. Some new tests of the power law process. Technometrics 1996;38:256–65.

[32] Engelhardt M, Bain LJ. Prediction intervals for the Weibull process. Technometrics 1978;20:167–9.

[33] Calabria R, Guida M, Pulcini G. Point estimation of future failure times of a repairable system. Reliab Eng Syst Saf 1990;28:23–34.

[34] Bain LJ, Engelhardt M. Statistical analysis of reliability and life-testing models. Theory and methods. (Second edition). New York, Basel, Hong Kong: Marcel Dekker; 1991.

[35] Rigdon SE, Ma X, Bodden KM. Statistical inference for repairable systems using the power law process. J Qual Technol 1998;30:395–400.

[36] Guida M, Calabria R, Pulcini G. Bayes inference for a non-homogeneous Poisson process with power intensity law. IEEE Trans Reliab 1989;38:603–9.

[37] Bar-Lev SK, Lavi I, Reiser B. Bayesian inference for the power law process. Ann Inst Stat Math 1992;44:623–39.

[38] Campodønico S, Singpurwalla ND. Inference and predictions from Poisson point processes incorporating expert knowledge. J Am Stat Assoc 1995;90:220–6.

[39] Calabria R, Guida M, Pulcini G. Bayes estimation of prediction intervals for a power law process. Commun Stat Theor Methods 1990;19:3023–35.

[40] Beiser JA, Rigdon SE. Bayes prediction for the number of failures of a repairable system. IEEE Trans Reliab 1997;46:291–5.

[41] Lilius WA. Graphical analysis of repairable systems. In: Proceedings of Annual Reliability and Maintenance Symposium, Los Angeles, USA, 1979; p.403–6.

[42] Härtler G. Graphical Weibull analysis of repairable systems. Qual Reliab Eng Int 1985;1:23–6.

[43] Nelson W. Graphical analysis of system repair data. J Qual Technol 1988;20:24–35.

[44] Martz HF. Pooling life test data by means of the empirical Bayes method. IEEE Trans Reliab 1975;R-24:27–30.

[45] Bassin WM. Increasing hazard functions and overhaul policy. In: Proceedings of Annual Symposium on Reliability, Chicago, USA, 1969; p.173–8.

[46] Ahn CW, Chae KC, Clark GM. Estimating parameters of the power law process with two measures of failure time. J Qual Technol 1998;30:127–32.

[47] Ascher HE, Hansen CK. Spurious exponentiality observed when incorrectly fitting a distribution to nonstationary data. IEEE Trans Reliab 1998;47:451–9.

[48] Leung FKN, Cheng ALM. Determining replacement policies for bus engines. Int J Qual Reliab Manage 2000;17:771–83.

[49] Aven T. Some tests for comparing reliability growth/deterioration rates of repairable systems. IEEE Trans Reliab 1989;38:440–3.

[50] Huang Y-S, Bier VM. A natural conjugate prior for the nonhomogeneous Poisson process with an exponential intensity function. Commun Stat Theor Methods 1999;28:1479–509.

[51] Engelhardt M, Bain LJ. On the mean time between failures for repairable systems. IEEE Trans Reliab 1986;R-35:419–22.

[52] Pulcini G. A bounded intensity process for the reliability of repairable equipment. J Qual Technol 2001;33:480–92.

[53] Lee L. Testing adequacy of the Weibull and log linear rate models for a Poisson process. Technometrics 1980;22:195–9.

[54] Love CE, Guo R. An application of a bathtub failure model to imperfectly repaired systems data. Qual Reliab Eng Int 1993;9:127–35.

[55] Pulcini G. Modeling the failure data of a repairable equipment with bathtub type failure intensity. Reliab Eng Syst Saf 2001;71:209–18.

[56] Coetzee JL. Reliability degradation and the equipment replacement problem. In: Proceedings of International Conference of the Maintenance Societies (ICOMS-96), Melbourne, 1996; paper 21.

[57] Kumar U, Klefsjö B, Granholm S. Reliability investigation for a fleet of load–haul–dump machines in a Swedish mine. Reliab Eng Syst Saf 1989;26:341–61.

[58] Bendell A, Walls LA. Exploring reliability data. Qual Reliab Eng Int 1985;1:37–51.

[59] Lawless JF. Regression methods for Poisson process data. J Am Stat Assoc 1987;82:808–15.

[60] Guida M, Giorgio M. Reliability analysis of accelerated life-test data from a repairable system. IEEE Trans Reliab 1995;44:337–46.

[61] Kumar D. Proportional hazards modelling of repairable systems. Qual Reliab Eng Int 1995;11:361–9.

[62] Guo R, Love CE. Statistical analysis of an age model for imperfectly repaired systems. Qual Reliab Eng Int 1992;8:133–46.

[63] Malik MAK. Reliable preventive maintenance scheduling. AIIE Trans 1979;11:221–8.

[64] Shin I, Lim TJ, Lie CH. Estimating parameters of intensity function and maintenance effect for repairable unit. Reliab Eng Syst Saf 1996;54:1–10.

[65] Berman M. Inhomogeneous and modulated gamma processes. Biometrika 1981;68:143–52.

[66] Lakey MJ, Rigdon SE. The modulated power law process. In: Proceedings of 45th Annual ASQC Quality Congress, Milwaukee, 1992; p.559–63.

[67] Black SE, Rigdon SE. Statistical inference for a modulated power law process. J Qual Technol 1996;28:81–90.

[68] Calabria R, Pulcini G. Bayes inference for the modulated power law process. Commun Stat Theor Methods 1997;26:2421–38.

[69] Calabria R, Pulcini G. Bayes inference for repairable mechanical units with imperfect or hazardous maintenance. Int J Reliab Qual Saf Eng 1998;5:65–83.

[70] Calabria R, Pulcini G. On testing for repair effect and time trend in repairable mechanical units. Commun Stat Theor Methods 1999;28:367–87.

[71] Duane JT. Learning curve approach to reliability. IEEE Trans Aerosp 1964;2:563–6.

[72] Calabria R, Pulcini G. Inference and test in modeling the failure/repair process of repairable mechanical equipments. Reliab Eng Syst Saf 2000;67:41–53.

[73] Chan J-K, Shaw L. Modeling repairable systems with failure rates that depend on age and maintenance. IEEE Trans Reliab 1993;42:566–71.

[74] Calabria R, Pulcini G. Discontinuous point processes for the analysis of repairable units. Int J Reliab Qual Saf Eng 1999;6:361–82.

[75] Nakagawa T. Optimum policies when preventive maintenance is imperfect. IEEE Trans Reliab 1979;R-28:331–2.

[76] Brown M, Proschan F. Imperfect repair. J Appl Prob 1983;20:851–9.

[77] Whitaker LR, Samaniego FJ. Estimating the reliability of systems subject to imperfect repair. J Am Stat Assoc 1989;84:301–9.

[78] Lim JH, Lu KL, Park DH. Bayesian imperfect repair model. Commun Statist Theor Meth 1998;27:965–84.

[79] Lim TJ. Estimating system reliability with fully masked data under Brown–Proschan imperfect repair model. Reliab Eng Syst Saf 1998;59:277–89.

[80] Lim TJ, Lie CH. Analysis of system reliability with dependent repair modes. IEEE Trans Reliab 2000;49:153–62.

[81] Block HW, Borges WS, Savits TH. Age-dependent minimal repair. J Appl Prob 1985;22:370–85.

[82] Love CE, Guo R. Using proportional hazard modelling in plant maintenance. Qual Reliab Eng Int 1991;7:7–17.

[83] Love CE, Guo R. Application of Weibull proportional hazards modelling to bad-as-old failure data. Qual Reliab Eng Int 1991;7:149–57.

[84] Jack N. Analysing event data from a repairable machine subject to imperfect preventive maintenance. Qual Reliab Eng Int 1997;13:183–6.

[85] Pulcini G. On the overhaul effect for repairable mechanical units: a Bayes approach. Reliab Eng Syst Saf 2000;70:85–94.

[86] Pulcini G. On the prediction of future failures for a repairable equipment subject to overhauls. Commun Stat Theor Methods 2001;30:691–706.

[87] Nakagawa T. Sequential imperfect preventive maintenance policies. IEEE Trans Reliab 1988;37:295–8.

[88] Kobbacy KAH, Fawzi BB, Percy DF, Ascher HE. A full history proportional hazards model for preventive maintenance scheduling. Qual Reliab Eng Int 1997;13:187–98.

[89] Benton AW, Crow LH. Integrated reliability growth testing. In: Proceedings of the Annual Reliability and Maintenance Symposium, Atlanta, USA, 1989; p.160–6.

[90] Robinson D, Dietrich D. A nonparametric-Bayes reliability-growth model. IEEE Trans Reliab 1989;38:591–8.

[91] Sen A. Estimation of current reliability in a Duane-based growth model. Technometrics 1996;40:334–44.

[92] Crow LH. Tracking reliability growth. In: Proceedings of 20th Conference on Design of Experiments. ARO Report 75-2, 1975; p.741–54.

[93] Crow LH. Confidence interval procedures for the Weibull process with application to reliability growth. Technometrics 1982;24:67–72.

[94] Crow LH. Confidence intervals on the reliability of repairable systems. In: Proceedings of the Annual Reliability and Maintenance Symposium, Atlanta, USA, 1993; p.126–33.

[95] Tsokos CP, Rao ANV. Estimation of failure intensity for the Weibull process. Reliab Eng Syst Saf 1994;45:271–5.

[96] Sen A, Khattree R. On estimating the current intensity of failure for the power-law process. J Stat Plan Infer 1998;74:253–72.

[97] Higgins JJ, Tsokos CP. A quasi-Bayes estimate of the failure intensity of a reliability-growth model. IEEE Trans Reliab 1981;R-30:471–5.

[98] Calabria R, Guida M, Pulcini G. A Bayes procedure for estimation of current system reliability. IEEE Trans Reliab 1992;41:616–20.

[99] Calabria R, Pulcini G. Maximum likelihood and Bayes prediction of current system lifetime. Commun Stat Theor Methods 1996;25:2297–309.

[100] Cozzolino JM. Probabilistic models of decreasing failure rate processes. Nav Res Logist Q 1968;15:361–74.

[101] Heimann DI, Clark WD. Process-related reliability-growth modeling. How and why. In: Proceedings of the Annual Reliability and Maintenance Symposium, Las Vegas, USA, 1992; 316–321.

[102] Robinson DG, Dietrich D. A new nonparametric growth model. IEEE Trans Reliab 1987;R-36:411–8.

[103] Ebrahimi N. How to model reliability-growth when times of design modifications are known. IEEE Trans Reliab 1996;45:54–8.

[104] Calabria R, Guida M, Pulcini G. A reliability-growth model in a Bayes-decision framework. IEEE Trans Reliab 1996;45:505–10.

Preventive Maintenance Models: Replacement, Repair, Ordering, and Inspection

Tadashi Dohi, Naoto Kaio and Shunji Osaki

19.1 Introduction

Many systems become more large-scale and more complicated and influence our society greatly, such as airplanes, computer networks, *etc.* Reliability/maintainability theory (engineering) plays a very important role in maintaining such systems. Mathematical maintenance policies, centering preventive maintenance, have been developed mainly in the research area of operations research/management science, to generate an effective preventive maintenance schedule. The most important problem in mathematical maintenance policies is to design a maintenance plan with two maintenance options: preventive replacement and corrective replacement. In preventive replacement, the system or unit is replaced by a new one before it fails. On the other hand, with corrective replacement it is the failed unit that is replaced. Hitherto, a huge number of replacement methods have been proposed in the literature. At the same time, some technical books on this problem have been published. For example, Arrow *et al.* [1], Barlow and Proschan [2, 3], Jorgenson *et al.* [4], Gnedenko *et al.* [5], Gertsbakh [6], Ascher and Feingold [7] and Osaki [8, 9] are the classical but very important texts to study regarding mathematical maintenance theory. Many authors have also published several monographs on specific topics. The reader is referred to Osaki and Hatoyama [10], Osaki and Cao [11], Ozekici [12], and Christer *et al.* [13]. Recently, Barlow [14] and Aven and Jensen [15] presented excellent textbooks reviewing mathematical maintenance theory. There are also some useful survey papers reviewing the history of this research context, such as those by McCall [16], Osaki and Nakagawa [17], Pierskalla and Voelker [18], Sherif and Smith [19], and Valdez-Flores and Feldman [20].

Since the mechanism that causes failure in almost all real complex systems may be considered to be uncertain, the mathematical technique to deal with maintenance problems should be based on the usual probability theory. If we are interested in the dynamic behavior of system failures depending on time, the problems are essentially reduced to studying the stochastic processes presenting phenomena on both failure and replacement. In fact, since the theory of stochastic processes depends strongly on mathematical maintenance theory, many textbooks on stochastic processes have treated several maintenance problems. See, for example, Feller [21], Karlin and Taylor [22, 23], Taylor and Karlin [24], and Ross [25]. In other words, in order to design a maintenance schedule effectively, both the underlying stochastic process governing the failure mechanism and the role of maintenance options carried out on the process have to be analyzed carefully. In that sense, mathematical maintenance theory is one of the most important parts in applied probability modeling.

In this tutorial article, we present the preventive maintenance policies arising in the context of maintenance theory. Simple but practically important preventive maintenance optimization models, that involve age replacement and block replacement, are reviewed in the framework of the well-known renewal reward argument. Some extensions to these basic models, as well as the corresponding discrete time models, are also introduced with the aim of the application of the theory to the practice. In almost all textbooks and technical papers, the discrete-time preventive maintenance models have been paid scant attention. The main reason is that discrete-time models are ordinarily considered as trivial analogies of continuous-time models. However, we often face some maintenance problems modeled in the discrete-time setting in practice. If one considers the situation where the number of take-offs from airports influences the damage to an airplane, the parts comprising the airplane should be replaced at a prespecified number of take-offs rather than after an elapsed time. Also, in the Japanese electric power company under our investigation, the failure time data of electric switching devices are recorded as group data (the number of failures per year) and it is not easy to carry out a preventive replacement schedule on a weekly or monthly basis, since the service team is engaged in other works, too. From our questionnaire, it is helpful for practitioners that a preventive replacement schedule should be carried out on a near-annual basis. This will motivate attention toward discrete-time maintenance models.

19.2 Block Replacement Models

For block replacement models, the preventive replacement is executed periodically at a prespecified time kt_0 $(t_0 \geq 0)$ or kN $(N = 0, 1, 2, \ldots)$, $(k = 1, 2, 3, \ldots)$. If the unit fails during the time interval $((k-1)t_0, kt_0]$ or $((k-1)N, kN]$, then the corrective maintenance is made at the failure time. The main property for the block replacement is that it is easier to administer in general, since the preventive replacement time is scheduled in advance and one need not observe the age of a unit. In this section, we develop the following three variations of block replacement model:

1. a failed unit is replaced instantaneously at failure (Model I);
2. a failed unit remains inoperable until the next scheduled replacement comes (Model II);
3. a failed unit undergoes minimal repair (Model III).

The cost components used in this section are as follows:

c_p (>0) unit preventive replacement cost at time kt_0 or kN, $(k = 1, 2, 3, \ldots)$

c_c (>0) unit corrective replacement cost at failure time

c_m (>0) unit minimal repair cost at failure time.

19.2.1 Model I

First, we consider the continuous-time block replacement model [2]. A failed unit is replaced by a new one during the replacement interval t_0, and the scheduled replacement for the non-failed unit is performed at kt_0 ($k = 1, 2, 3, \ldots$). Let $F(t)$ be the continuous lifetime distribution with finite mean $1/\lambda$ (>0). From the well-known renewal reward theorem, it is immediate to formulate the expected cost per unit time in the steady state for the block replacement model as follows:

$$B_c(t_0) = \frac{c_c M(t_0) + c_p}{t_0} \quad t_0 \geq 0 \tag{19.1}$$

where the function $M(t) = \sum_{k=1}^{\infty} F^{(k)}(t)$ denotes the mean number of failures during the time period $(0, t]$ (the renewal function) and $F^{(k)}(t)$ the k-fold convolution of the lifetime distribution. The problem is, of course, to derive the optimal block replacement time t_0^* that minimizes $B_c(t_0)$.

Define the numerator of the derivative of $B_c(t_0)$ as

$$j_c(t_0) = c_c[t_0 m(t_0) - M(t_0)] - c_p \tag{19.2}$$

where the function $m(t) = \mathrm{d}M(t)/\mathrm{d}t$ is the renewal density. Then, we have the optimal block replacement time t_0^* that minimizes the expected cost per unit time in the steady state $B_c(t_0)$.

Theorem 1.

(1) Suppose that the function $m(t)$ is strictly increasing with respect to t (>0).

(i) If $j_c(\infty) > 0$, then there exists one finite optimal block replacement time t_0^* ($0 < t_0^* < \infty$) that satisfies $j_c(t_0^*) = 0$. Then the corresponding minimum expected cost is

$$B_c(t_0^*) = c_c m(t_0^*) \tag{19.3}$$

(ii) If $j_c(\infty) \leq 0$, then $t_0^* \to \infty$, i.e. it is optimal to carry out only the corrective replacement, and the corresponding minimum expected cost is

$$B_c(\infty) = \lambda c_c \tag{19.4}$$

(2) Suppose that the function $m(t)$ is decreasing with respect to t (> 0). Then, the optimal block replacement time is $t_0^* \to \infty$.

Next, we formulate the discrete-time block replacement model [29]. In the discrete-time setting, the expected cost per unit time in the steady state is

$$B_d(N) = \frac{c_c M(N) + c_p}{N} \quad N = 0, 1, 2, \ldots \tag{19.5}$$

where the function $M(n) = \sum_{k=1}^{\infty} F^{(k)}(n)$ is the discrete renewal function for the discrete lifetime distribution $F(n)$ ($n = 0, 1, 2, \ldots$) and $F^{(k)}(n)$ is the k-fold convolution; for more details, see Munter [26] and Kaio and Osaki [27]. Define the numerator of the difference of $B_d(N)$ as

$$j_d(N) = c_c[Nm(N+1) - M(N)] - c_p \tag{19.6}$$

where the function $m(n) = M(n) - M(n-1)$ is the renewal probability mass function. Then, we have the optimal block replacement time N^* that minimizes the expected cost per unit time in the steady state $B_d(N)$.

Theorem 2.

(1) Suppose that the $m(n)$ is strictly increasing with respect to n (>0).

(i) If $j_d(\infty) > 0$, then there exists one finite optimal block replacement time N^* ($0 < N^* < \infty$) that satisfies $j_d(N-1) < 0$ and $j_d(N) \geq 0$. Then the corresponding minimum expected cost satisfies the inequality

$$c_c m(N^*) < B_d(N^*) \leq c_c m(N^* + 1) \tag{19.7}$$

(ii) If $j_d(\infty) \leq 0$, then $N^* \to \infty$, i.e. *it is optimal to carry out only the corrective replacement, and the corresponding minimum expected cost is*

$$B_d(\infty) = \lambda c_c \tag{19.8}$$

(2) Suppose that the function $m(n)$ is decreasing with respect to n (> 0). Then, the optimal block replacement time is $N^* \to \infty$.

Remarks. A large number of variations on the block replacement model have been studied in the literature. Though we assume in the model above that the cost component is constant, some modifications are possible. Tilquin and Cleroux [30] and Berg and Epstein [31] extended the original model in terms of cost structure.

19.2.2 Model II

For the first model, we have assumed that a failed unit is detected instantaneously just after failure. This implies that a sensing device monitors the operating unit. Since such a case is not always general, however, we assume that the failure is detected only at kt_0 ($t_0 \geq 0$) or kN ($N = 0, 1, 2, \dots$), ($k = 1, 2, 3, \dots$) (see Osaki [9]). Consequently, in Model II, a unit is always replaced at kt_0 or kN, but is not replaced at the time of failure, and the unit remains inoperable for the time duration from the occurrence of failure until its detection.

In the continuous-time model, since the expected duration from the occurrence of failure until its detection per cycle is given by $\int_0^{t_0}(t_0 - t)\,\mathrm{d}F(t) = \int_0^{t_0} F(t)\,\mathrm{d}t$, we have the expected cost per unit time in the steady state:

$$C_\mathrm{c}(t_0) = \frac{c_\mathrm{c}\int_0^{t_0} F(t)\,\mathrm{d}t + c_\mathrm{p}}{t_0} \tag{19.9}$$

where c_c is changed to the cost of failure per unit time, *i.e.* the cost occurs per unit time for system down. Define the numerator of the derivative of $C_\mathrm{c}(t_0)$ with respect to t_0 as $k_\mathrm{c}(t_0)$, *i.e.*

$$k_\mathrm{c}(t_0) = c_\mathrm{c}\left[F(t_0)t_0 - \int_0^{t_0} F(t)\,\mathrm{d}t\right] - c_\mathrm{p} \tag{19.10}$$

Theorem 3.

(i) If $k_\mathrm{c}(\infty) > 0$, then there exists one unique optimal block replacement time t_0^* ($0 < t_0^* < \infty$) that satisfies $k_\mathrm{c}(t_0^*) = 0$, and the corresponding minimum expected cost is

$$C_\mathrm{c}(t_0^*) = c_\mathrm{c}F(t_0^*) \tag{19.11}$$

(ii) If $k_\mathrm{c}(\infty) \leq 0$, then $t_0^* \to \infty$ and $C_\mathrm{c}(\infty) = c_\mathrm{c}$.

On the other hand, in the discrete-time setting, the expected cost per unit time in the steady state is

$$C_\mathrm{d}(N) = \frac{c_\mathrm{c}\sum_{k=1}^{N-1} F(k) + c_\mathrm{p}}{N} \qquad N = 0, 1, 2, \dots \tag{19.12}$$

where the function $F(n)$ is the lifetime distribution ($n = 0, 1, 2, \dots$). Define the numerator of the difference of $C_\mathrm{d}(N)$ as

$$i_\mathrm{d}(N) = c_\mathrm{c}\left[NF(N) - \sum_{k=1}^{N-1} F(k)\right] - c_\mathrm{p} \tag{19.13}$$

Then, we have the optimal block replacement time N^* that minimizes the expected cost per unit time in the steady state $C_\mathrm{d}(N)$.

Theorem 4.

(i) If $j_\mathrm{d}(\infty) > 0$, then there exists one finite optimal block replacement time N^* ($0 < N^* < \infty$) that satisfies $i_\mathrm{d}(N-1) < 0$ and $i_\mathrm{d}(N) \geq 0$. Then the corresponding minimum expected cost satisfies the inequality

$$c_\mathrm{c}F(N^* - 1) < C_\mathrm{d}(N^*) \leq c_\mathrm{c}F(N^*) \tag{19.14}$$

(ii) If $i_\mathrm{d}(\infty) \leq 0$, then $N^* \to \infty$.

Remarks. It is noted that Model II has not been widely studied in the literature, since this cannot detect the failure instantly and is not certainly superior to Model I in terms of cost minimization. However, as described previously, one can see that the continuous monitoring of the operating unit is not always possible for all practical applications.

19.2.3 Model III

In the final model, we assume that minimal repair is performed when a unit fails and the failure rate is not disturbed by each repair. If we consider a stochastic process $\{N(t), t \geq 0\}$ in that $N(t)$ represents the number of minimal repairs up to time t, the process $\{N(t), t \geq 0\}$ is governed by a non-homogeneous Poisson process with mean value function

$$\Lambda(t) = \int_0^t r(x)\,\mathrm{d}x \tag{19.15}$$

that is also called the hazard function, where the function $r(t) = f(t)/\bar{F}(t)$ is called the failure rate or the hazard rate; in general $\bar{\psi}(\cdot) = 1 - \psi(\cdot)$. Noting this fact, Barlow and Hunter [28] gave the expected cost per unit time in the steady state for the continuous-time model:

$$V_c(t_0) = \frac{c_m \Lambda(t_0) + c_p}{t_0} \tag{19.16}$$

Define the numerator of the derivative of $V_c(t_0)$ as

$$l_c(t_0) = c_m[t_0 r(t_0) - \Lambda(t_0)] - c_p \tag{19.17}$$

Then, we have the optimal block replacement time (with minimal repair) t_0^* that minimizes the expected cost per unit time in the steady state $V_c(t_0)$.

Theorem 5.

(1) *Suppose that $F(t)$ is strictly increasing failure rate (IFR), i.e. the failure rate is strictly increasing.*

  (i) *If $l_c(\infty) > 0$, then there exists one finite optimal block replacement time with minimal repair t_0^* $(0 < t_0^* < \infty)$ that satisfies $l_c(t_0^*) = 0$. Then the corresponding minimum expected cost is*

$$V_c(t_0^*) = c_m r(t_0^*) \tag{19.18}$$

  (ii) *If $l_c(\infty) \leq 0$, then $t_0^* \to \infty$ and the corresponding minimum expected cost is*

$$V_c(\infty) = c_m r(\infty) \tag{19.19}$$

(2) *Suppose that $F(t)$ is decreasing failure rate (DFR), i.e. the failure rate is decreasing. Then, the optimal block replacement time with minimal repair is $t_0^* \to \infty$.*

Next, we formulate the discrete-time block replacement model with minimal repair [29]. In the discrete-time setting, the expected cost per unit time in the steady state is

$$V_d(N) = \frac{c_m \Lambda(N) + c_p}{N} \qquad N = 0, 1, 2, \ldots \tag{19.20}$$

where the function $\Lambda(n)$ is the mean value function of the discrete non-homogeneous Poisson process. Define the numerator of the difference of $V_d(N)$ as

$$l_d(N) = c_m[N r(N+1) - \Lambda(N)] - c_p \tag{19.21}$$

where the function $r(n) = \Lambda(n) - \Lambda(n-1) = f(n)/\bar{F}(n-1)$ is the failure rate function. Then, we have the optimal block replacement time with minimal repair N^* that minimizes the expected cost per unit time in the steady state $V_d(N)$.

Theorem 6.

(1) *Suppose that $F(n)$ is strictly IFR.*

  (i) *If $l_d(\infty) > 0$, then there exists one finite optimal block replacement time with minimal repair N^* $(0 < N^* < \infty)$ that satisfies $l_d(N-1) < 0$ and $l_d(N) \geq 0$. Then the corresponding minimum expected cost satisfies the inequality*

$$c_m r(N^*) < V_d(N^*) \leq c_m r(N^* + 1) \tag{19.22}$$

  (ii) *If $l_d(\infty) \leq 0$, then $N^* \to \infty$.*

(2) *Suppose that $F(n)$ is DFR. Then, the optimal block replacement time with minimal repair is $N^* \to \infty$.*

Remarks. It is apparent that a great number of papers on minimal repair models have been published. Morimura [32], Tilquin and Cleroux [33] and Cleroux *et al.* [34] involved several interesting modifications. Later, Park [35], Nakagawa [36–40], Nakagawa and Kowada [41], Phelps [42], Berg and Cleroux [43], Boland [44], Boland and Proschan [45], Block *et al.* [46], and Beichelt [47] proposed extended minimal repair models from the standpoint of generalization. The most interesting of the models with minimal repair is the (t, T) policy. The (t, T) policy is a combined policy with three kinds of maintenance options: minimal repair, failure replacement, and preventive replacement. That is, the minimal repair is executed for failures during the first period $[0, t)$, but the failure replacement is made for $[t, T]$, where T is the preventive replacement time.

Tahara and Nishida [48] formulated this model first. It was then investigated by Phelps [49] and Segawa *et al.* [50]; recently, Ohnishi [51] proved the optimality of the (t, T) policy under average cost criterion, via a dynamic programming approach. This tells us that the (t, T) policy is optimal if we have only three kinds of maintenance option.

19.3 Age Replacement Models

As is well known, in the age replacement model, if the unit does not fail until a prespecified time t_0 $(t_0 \geq 0)$ or N $(N = 0, 1, 2, \dots)$, then it is replaced by a new one preventively, otherwise, it is replaced at the failure time. Denote the corrective and the preventive replacement costs by c_c and c_p respectively, where, without loss of generality, $c_c > c_p$. This model plays a central role in all replacement models, since the optimality of the age replacement model has been proved by Bergman [52] if the replacement by a new unit is the only maintenance option (*i.e.* if no repair is considered as an alternative option). In the remainder of this section, we introduce three kinds of age replacement model.

19.3.1 Basic Age Replacement Model

See Barlow and Proschan [2], Barlow and Hunter [53], and Osaki and Nakagawa [54]. From the renewal reward theorem, it can be seen that the expected cost per unit time in the steady state for the age replacement model is

$$A_c(t_0) = \frac{c_c F(t_0) + c_p \bar{F}(t_0)}{\int_0^{t_0} \bar{F}(t)\, dt} \qquad t_0 \geq 0 \qquad (19.23)$$

If one can assume that there exists the density $f(t)$ for the lifetime distribution $F(t)$ $(t \geq 0)$, the failure rate $r(t) = f(t)/\bar{F}(t)$ necessarily exists. Define the numerator of the derivative of $A_c(t_0)$ with respect to t_0, divided by $\bar{F}(t_0)$ as $h_c(t_0)$, *i.e.*

$$h_c(t_0) = r(t_0) \int_0^{t_0} \bar{F}(t)\, dt - F(t_0) - \frac{c_p}{c_c - c_p} \qquad (19.24)$$

Then, we have the optimal age replacement time t_0^* that minimizes the expected cost per unit time in the steady state $A_c(t_0)$.

Theorem 7.

(1) Suppose that the lifetime distribution $F(t)$ is strictly IFR.

(i) If $r(\infty) > K = \lambda c_c/(c_c - c_p)$, then there exists one finite optimal age replacement time t_0^* $(0 < t_0^* < \infty)$ that satisfies $h_c(t_0^*) = 0$. Then the corresponding minimum expected cost is

$$A_c(t_0^*) = (c_c - c_p) r(t_0^*) \qquad (19.25)$$

(ii) If $r(\infty) \leq K$, then $t_0^* \to \infty$ and $A_c(\infty) = B_c(\infty) = \lambda c_c$.

(2) If $F(t)$ is DFR, then $t_0^* \to \infty$.

Next, let us consider the case where the cost is discounted by the discount factor α $(\alpha > 0)$ (see Fox [55]). The present value of a unit cost after t $(t \geq 0)$ time is $\exp(-\alpha t)$. In the continuous-time age replacement problem, the expected total discounted cost over an infinite time horizon is

$$A_c(t_0; \alpha) = \frac{c_c \int_0^{t_0} e^{-\alpha t} f(t)\, dt + c_p e^{-\alpha t_0} \bar{F}(t_0)}{\alpha \int_0^{t_0} e^{-\alpha t} \bar{F}(t)\, dt} \qquad t_0 \geq 0 \qquad (19.26)$$

Define the numerator of the derivative of $A_c(t_0; \alpha)$ with respect to t_0, divided by $\bar{F}(t_0) \exp(-\alpha t_0)$ as $h_c(t_0; \alpha)$:

$$h_c(t_0; \alpha) = r(t_0) \int_0^{t_0} e^{-\alpha t} \bar{F}(t)\, dt - \int_0^{t_0} e^{-\alpha t} f(t)\, dt - \frac{c_p}{c_c - c_p} \qquad (19.27)$$

Then, we have the optimal age replacement time t_0^* that minimizes the expected total discounted cost over an infinite time horizon $A_c(t_0; \alpha)$.

Theorem 8.

(1) Suppose that the lifetime distribution $F(t)$ is strictly IFR.

(i) *If $r(\infty) > K(\alpha)$, then there exists one finite optimal age replacement time t_0^* $(0 < t_0^* < \infty)$ that satisfies $h_c(t_0^*; \alpha) = 0$, where*

$$K(\alpha) = \frac{c_c F^*(\alpha) + c_p \bar{F}^*(\alpha)}{(c_c - c_p)\bar{F}^*(\alpha)/\alpha} \quad (19.28)$$

$$F^*(\alpha) = \int_0^\infty e^{-\alpha t} f(t)\, dt \quad (19.29)$$

Then the corresponding minimum expected cost is

$$A_c(t_0^*; \alpha) = \frac{(c_c - c_p) r(t_0^*)}{\alpha} - c_p \quad (19.30)$$

(ii) *If $r(\infty) \le K(\alpha)$, then $t_0^* \to \infty$ and*

$$A_c(\infty; \alpha) = c_c F^*(\alpha)/\bar{F}^*(\alpha) \quad (19.31)$$

(2) *If $F(t)$ is DFR, then $t_0^* \to \infty$.*

Following Nakagawa and Osaki [56], let us consider the discrete-time age replacement model. Define the discrete-time lifetime distribution $F(n)$ $(n = 0, 1, 2, \ldots)$, the probability mass function $f(n)$, and the failure rate $r(n) = f(n)/\bar{F}(n-1)$. From the renewal reward theorem, it can be seen that the expected cost per unit time in the steady state for the age replacement model is

$$A_d(N) = \frac{c_c F(N) + c_p \bar{F}(N)}{\sum_{i=1}^{N} \bar{F}(i-1)} \quad N = 0, 1, 2, \ldots \quad (19.32)$$

Define the numerator of the difference of $A_d(N)$ as

$$h_d(N) = r(N+1) \sum_{i=1}^{N} \bar{F}(i-1) - F(N) - \frac{c_p}{c_c - c_p} \quad (19.33)$$

Then, we have the optimal age replacement time N^* that minimizes the expected cost per unit time in the steady state $A_d(N)$.

Theorem 9.

(1) *Suppose that the lifetime distribution $F(N)$ is strictly IFR.*

(i) *If $r(\infty) > K$, then there exists one finite optimal age replacement time N^* $(0 < N^* < \infty)$ that satisfies $h_d(N^* - 1) < 0$ and $h_d(N^*) \ge 0$. Then the corresponding minimum expected cost satisfies the following inequality*

$$(c_c - c_p) r(N^*) < A_d(N^*) \le (c_c - c_p) r(N^* + 1) \quad (19.34)$$

(ii) *If $r(\infty) \le K$, then $N^* \to \infty$ and $A_d(\infty) = B_d(\infty) = \lambda c_c$.*

(2) *If $F(N)$ is DFR, then $N^* \to \infty$.*

We introduce the discount factor β $(0 < \beta < 1)$ in the discrete-time age replacement problem. The present value of a unit cost after n $(n = 0, 1, 2, \ldots)$ period is β^n. In the discrete-time age replacement problem, the expected total discounted cost over an infinite time horizon is

$$A_d(N; \beta) = \left[c_c \sum_{j=0}^{N} \beta^j f(j) + c_p \beta^N \bar{F}(N) \right] \times \left[\frac{1-\beta}{\beta} \sum_{i=1}^{N} \beta^i \bar{F}(i-1) \right]^{-1} \quad N = 0, 1, 2, \ldots \quad (19.35)$$

Define the numerator of the difference of $A_d(N; \beta)$ as

$$h_d(N; \beta) = r(N+1) \sum_{i=1}^{N} \beta^i \bar{F}(i-1) - \sum_{j=0}^{N} \beta^j f(j) - \frac{c_p}{c_c - c_p} \quad (19.36)$$

Then, we have the optimal age replacement time N^* that minimizes the expected total discounted cost over an infinite time horizon $A_d(N; \beta)$.

Theorem 10.

(1) *Suppose that the lifetime distribution $F(N)$ is strictly IFR.*

(i) *If $r(\infty) > K(\beta)$, then there exists one finite optimal age replacement* *time* N^* $(0 < N^* < \infty)$ *that satisfies* $h_d(N^* - 1; \beta) < 0$ *and* $h_d(N^*; \beta) \geq 0$. *Then the corresponding minimum expected cost satisfies the following inequalities:*

$$\frac{(c_c - c_p)r(N^*)}{(1-\beta)/\beta} - c_p < A_d(N^*; \beta) \tag{19.37}$$

and

$$A_d(N^*; \beta) \leq \frac{(c_c - c_p)r(N^*+1)}{(1-\beta)/\beta} - c_p \tag{19.38}$$

where

$$K(\beta) = \left\{c_c \sum_{j=0}^{\infty} \beta^j f(j) + c_p \sum_{j=0}^{\infty} (1-\beta^j) f(j)\right\} \times \left\{(c_c - c_p) \sum_{i=1}^{\infty} \beta^i \bar{F}(i-1)\right\}^{-1} \tag{19.39}$$

(ii) *If* $r(\infty) \leq K(\beta)$, *then* $N^* \to \infty$ *and*

$$A_d(\infty; \beta) = \frac{c_c \sum_{j=0}^{\infty} \beta^j f(j)}{\sum_{j=0}^{\infty} (1-\beta^j) f(j)} \tag{19.40}$$

(2) *If* $F(N)$ *is DFR, then* $N^* \to \infty$.

Theorem 11.

(1) *For the continuous-time age replacement problems, the following relationships hold:*

$$A_c(t_0) = \lim_{\alpha \to 0} \alpha A_c(t_0; \alpha) \tag{19.41}$$

$$h_c(t_0) = \lim_{\alpha \to 0} h_c(t_0; \alpha) \tag{19.42}$$

$$K = \lim_{\alpha \to 0} K(\alpha) \tag{19.43}$$

(2) *For the discrete-time age replacement problems, the following relationships hold.*

$$A_d(N) = \lim_{\beta \to 1} (1-\beta) A_d(N; \beta) \tag{19.44}$$

$$h_d(N) = \lim_{\beta \to 1} h_d(N; \beta) \tag{19.45}$$

$$K = \lim_{\beta \to 1} K(\beta) \tag{19.46}$$

Remarks. Glasser [57], Scheaffer [58], Cleroux and Hanscom [59], Osaki and Yamada [60], and Nakagawa and Osaki [61] extended the basic age replacement model mentioned above. Here, we introduce an interesting topic on the age replacement policy under the different cost criterion from the expected cost per unit time in the steady state. Based on the seminal idea by Derman and Sacks [62], Ansell *et al.* [63] analyzed the age replacement model under an alternative cost criterion. In the continuous-time model with no discount, let Y_i and S_i denote the total cost and the time length for ith cycle ($i = 1, 2, \ldots$) respectively, where $Y_i = c_c I_{\{X_i < t_0\}} + c_p I_{\{X_i \geq t_0\}}$, $S_i = \min\{X_i, t_0\}$, X_i is the lifetime for the ith cycle, and $I_{\{\cdot\}}$ is the indicator function.

In Equation 19.23, we find that

$$\lim_{t \to \infty} \frac{E[\text{total cost on } (0, t]]}{t} = E[Y_i]/E[S_i] = A_c(t_0) \tag{19.47}$$

On the other hand, let $NU(t)$ denote the number of cycles up to time t. Then, we define

$$\eta(t) \equiv \frac{1}{NU(t)} \sum_{i=1}^{NU(t)} E[Y_i/S_i] \tag{19.48}$$

where $\eta(t)$ is the mean of the ratio $E[Y_i/S_i]$ during $NU(t)$ cycles. From the independence of each cycle, we have

$$A_c^*(t_0) = \lim_{t \to \infty} \eta(t) = E[Y_i/S_i] = \int_0^{t_0} (c_c/t)\, \mathrm{d}F(t) + \int_{t_0}^{\infty} (c_p/t_0)\, \mathrm{d}F(t) \tag{19.49}$$

This interesting cost criterion is called the expected cost ratio and is of course different from $E[Y_i]/E[S_i]$. Ansell *et al.* [63] compared this model with an approximated age replacement policy with finite time horizon by Christer [64,65] and Christer and Jack [66].

19.4 Ordering Models

In both block and age replacement problems, a spare unit is available whenever the original unit fails. However, it should be noted that this assumption is questionable in most practical cases. In fact, if a sufficiently large number of spare units is always kept on hand, a large inventory holding cost will be needed. Hence, if the system failure may be considered as a rare event for the operating system, then the spare unit can be ordered when it is required. There were seminal contributions by Wiggins [67], Allen and D'Esopo [68, 69], Simon and D'Esopo [70], Nakagawa and Osaki [71], and Osaki [72]. A large number of ordering/order replacement models have been analyzed by many authors. For instance, the reader should refer to Thomas and Osaki [73, 74], Kaio and Osaki [75–78] and Osaki *et al.* [79]. A comprehensive bibliography in this research area is listed in Dohi *et al.* [80].

19.4.1 Continuous-time Model

Let us consider a replacement problem for a one-unit system where each failed unit is scrapped and each spare is provided, after a lead time, by an order. The original unit begins operating at time $t = 0$, and the planning horizon is infinite. If the original unit does not fail up to a prespecified time $t_0 \in [0, \infty)$, the regular order for a spare is made at the time t_0 and after a lead time L_r (>0) the spare is delivered. Then if the original unit has already failed by $t = t_0 + L_r$, the delivered spare takes over its operation immediately. But even if the original unit is still operating, the unit is replaced by the spare preventively. On the other hand, if the original unit fails before the time t_0, the expedited order is made immediately at the failure time and the spare takes over its operation just after it is delivered after a lead time L_e (>0). In this situation, it should be noted that the regular order is not made. The same cycle repeats itself continually.

Under this model, define the interval from one replacement to the following replacement as one cycle. Let c_e (>0), c_r (>0), k_f (>0), w (>0), and s (<0) be the expedited ordering cost, the regular ordering cost, the system down (shortage) cost per unit time, the operation cost per unit time, and the salvage cost per unit time respectively. Then, the expected cost per unit time in the steady state is

$$O_c(t_0) = V_c(t_0)/T_c(t_0) \tag{19.50}$$

where

$$\begin{aligned}
V_c(t_0) &= c_e \int_0^{t_0} dF(t) + c_r \int_{t_0}^{\infty} dF(t) \\
&\quad + k_f \left[\int_0^{t_0} L_e \, dF(t) \right. \\
&\quad \left. + \int_{t_0}^{t_0+L_r} (t_0 + L_r - t) \, dF(t) \right] \\
&\quad + w \left[\int_0^{t_0+L_r} t \, dF(t) \right. \\
&\quad \left. + \int_{t_0+L_r}^{\infty} (t_0 + L_r) \, dF(t) \right] \\
&\quad + s \int_{t_0+L_r}^{\infty} (t - t_0 - L_r) \, dF(t) \\
&= c_e F(t_0) + c_r \bar{F}(t_0) \\
&\quad + k_f \left[(L_e - L_r) F(t_0) + \int_{t_0}^{t_0+L_r} F(t) \, dt \right] \\
&\quad + w \int_0^{t_0+L_r} \bar{F}(t) \, dt + s \int_{t_0+L_r}^{\infty} \bar{F}(t) \, dt \\
t_0 &\geq 0
\end{aligned} \tag{19.51}$$

and

$$\begin{aligned}
T_c(t_0) &= \int_0^{t_0} (t + L_e) \, dF(t) + \int_{t_0}^{\infty} (t_0 + L_r) \, dF(t) \\
&= (L_e - L_r) F(t_0) + L_r + \int_0^{t_0} \bar{F}(t) \, dt
\end{aligned} \tag{19.52}$$

Define the numerator of the derivative of $O_c(t_0)$ with respect to t_0, divided by $\bar{F}(t_0)$ as $q_c(t_0)$, *i.e.*

$$\begin{aligned}
q_c(t_0) &= \{(k_f - w + s) R(t_0) + (w - s) \\
&\quad + [k_f(L_e - L_r) + (c_e - c_r)] r(t_0)\} \\
&\quad \times \left[(L_e - L_r) F(t_0) + L_r + \int_0^{t_0} \bar{F}(t) \, dt \right] \\
&\quad - \left\{ w \int_0^{t_0+L_r} \bar{F}(t) \, dt + c_e F(t_0) + c_r \bar{F}(t_0) \right.
\end{aligned}$$

$$+ k_{\rm f}\left[(L_{\rm e} - L_{\rm r})F(t_0) + \int_{t_0}^{t_0+L_{\rm r}} F(t)\,{\rm d}t\right]$$
$$+ s\int_{t_0+L_{\rm r}}^{\infty} \bar{F}(t)\,{\rm d}t\Big\}[(L_{\rm e} - L_{\rm r})r(t_0) + 1] \quad (19.53)$$

where the function

$$R(t_0) = \frac{F(t_0 + L_{\rm r}) - F(t_0)}{\bar{F}(t_0)} \quad (19.54)$$

has the same monotone properties as the failure rate $r(t_0)$, *i.e.* $R(t)$ is increasing (decreasing) if and only if $r(t)$ is increasing (decreasing). Then, we have the optimal order replacement (ordering) time t_0^* that minimizes the expected cost per unit time in the steady state $O_{\rm c}(t_0)$.

Theorem 12.

(1) Suppose that the lifetime distribution $F(t)$ is strictly IFR.

(i) If $q_{\rm c}(0) < 0$ and $q_{\rm c}(\infty) > 0$, then there exists one finite optimal order replacement time t_0^* $(0 < t_0^* < \infty)$ that satisfies $q_{\rm c}(t_0^*) = 0$. Then the corresponding minimum expected cost is

$$O_{\rm c}(t_0^*) = \frac{(k_{\rm f} - w + s)R(t_0^*) + (w - s) + \mu(t_0^*)}{(L_{\rm e} - L_{\rm r})r(t_0^*) + 1} \quad (19.55)$$

where

$$\mu(t_0) = [k_{\rm f}(L_{\rm e} - L_{\rm r}) + (c_{\rm e} - c_{\rm r})]r(t_0) \quad (19.56)$$

(ii) If $q_{\rm c}(\infty) \le 0$, then $t_0^* \to \infty$ and

$$O_{\rm c}(\infty) = \frac{w/\lambda + c_{\rm e} + k_{\rm f}L_{\rm e}}{L_{\rm e} + 1/\lambda} \quad (19.57)$$

(iii) If $q_{\rm c}(0) \ge 0$, then $t_0^* = 0$ and

$$O_{\rm c}(0) = \frac{1}{L_{\rm r}}\left[w\int_0^{L_{\rm r}} \bar{F}(t)\,{\rm d}t + c_{\rm r} + k_{\rm f}\int_0^{L_{\rm r}} F(t)\,{\rm d}t + s\int_{L_{\rm r}}^{\infty} \bar{F}(t)\,{\rm d}t\right] \quad (19.58)$$

(2) Suppose that $F(t)$ is DFR. If the inequality

$$\left[w\int_0^{L_{\rm r}} \bar{F}(t)\,{\rm d}t + c_{\rm r} + k_{\rm f}\int_0^{L_{\rm r}} F(t)\,{\rm d}t + s\int_{L_{\rm r}}^{\infty} \bar{F}(t)\,{\rm d}t\right](L_{\rm e} + 1/\lambda) < (w/\lambda + c_{\rm e} + k_{\rm f}L_{\rm e})L_{\rm r} \quad (19.59)$$

holds, then $t_0^* = 0$, otherwise $t_0^* \to \infty$.

19.4.2 Discrete-time Model

In the discrete order-replacement model, the function in Equation 19.54 is given by

$$R(N) = \frac{\sum_{n=1}^{N+L_{\rm r}} f(n) - \sum_{n=1}^{N} f(n)}{\sum_{n=N+1}^{\infty} f(n)} \quad (19.60)$$

Then, the expected cost per unit time in the steady state is

$$O_{\rm d}(N) = V_{\rm d}(N)/T_{\rm d}(N) \quad (19.61)$$

where

$$V_{\rm d}(N) = w\sum_{i=1}^{N+L_{\rm r}}\sum_{n=i}^{\infty} f(n) + c_{\rm e}\sum_{n=1}^{N-1} f(n) + c_{\rm r}\sum_{n=N}^{\infty} f(n) + k_{\rm f}\left[(L_{\rm e} - L_{\rm r})\sum_{n=1}^{N-1} f(n) + \sum_{i=N+1}^{N+L_{\rm r}}\sum_{n=1}^{i-1} f(n) + s\sum_{i=N+L_{\rm r}+1}^{\infty}\sum_{n=i}^{\infty} f(n)\right] \quad (19.62)$$

and

$$T_{\rm d}(N) = (L_{\rm e} - L_{\rm r})\sum_{n=1}^{N-1} f(n) + L_{\rm r} + \sum_{i=1}^{N}\sum_{n=i}^{\infty} f(n) \quad (19.63)$$

Notice in the equations above that $L_{\rm e}$ and $L_{\rm r}$ are positive integers.

Similar to Equation 19.53, define the numerator of the difference of $O_{\rm d}(N)$ with respect to N, divided by $\bar{F}(N)$ as $q_{\mathrm{d}}(N)$, *i.e.*

$$\begin{aligned}
q_{\mathrm{d}}(N) &= \{(k_{\mathrm{f}} - w + s)R(N) + (w - s) \\
&\quad + [k_{\mathrm{f}}(L_{\mathrm{e}} - L_{\mathrm{r}}) + (c_{\mathrm{e}} - c_{\mathrm{r}})]r(N)\} \\
&\quad \times \left[(L_{\mathrm{e}} - L_{\mathrm{r}}) \sum_{n=1}^{N-1} f(n) + L_{\mathrm{r}} \right. \\
&\quad \left. + \sum_{i=1}^{N} \sum_{n=i}^{\infty} f(n) \right] \\
&\quad - \left\{ w \sum_{i=1}^{N+L_{\mathrm{r}}} \sum_{n=i}^{\infty} f(n) + c_{\mathrm{e}} \sum_{n=1}^{N-1} f(n) \right. \\
&\quad + c_{\mathrm{r}} \sum_{n=N}^{\infty} f(n) + k_{\mathrm{f}} \\
&\quad \times \left[(L_{\mathrm{e}} - L_{\mathrm{r}}) \sum_{n=1}^{N-1} f(n) + \sum_{i=N+1}^{N+L_{\mathrm{r}}} \sum_{n=1}^{i-1} f(n) \right] \\
&\quad \left. + s \sum_{i=N+L_{\mathrm{r}}+1}^{\infty} \sum_{n=i}^{\infty} f(n) \right\} \\
&\quad \times [(L_{\mathrm{e}} - L_{\mathrm{r}})r(N) + 1]
\end{aligned} \tag{19.64}$$

Then, we have the optimal order replacement time N^* that minimizes the expected cost per unit time in the steady state $O_{\mathrm{d}}(N)$.

Theorem 13.

(1) *Suppose that the lifetime distribution $F(N)$ is strictly IFR.*

(i) *If $q_{\mathrm{d}}(0) < 0$ and $q_{\mathrm{d}}(\infty) > 0$, then there exists one finite optimal order replacement time N^* $(0 < N^* < \infty)$ that satisfies $q_{\mathrm{d}}(N^* - 1) < 0$ and $q_{\mathrm{d}}(N^*) \geq 0$.*

(ii) *If $q_{\mathrm{d}}(\infty) \leq 0$, then $N^* \to \infty$.*

(iii) *If $q_{\mathrm{d}}(0) \geq 0$, then $N^* = 0$.*

(2) *Suppose that $F(N)$ is DFR. Then $N^* = 0$, otherwise $N^* \to \infty$.*

Remarks. This section has dealt with typical order-replacement models in both continuous- and discrete-time settings. These models can be extended from the various viewpoints. Thomas and Osaki [74] and Dohi *et al.* [80] presented continuous models with stochastic lead times. Recently, Dohi and coworkers [81–83] applied the order-replacement model to analyze the special continuous review cyclic inventory control problems.

19.4.3 Combined Model with Minimal Repairs

In the preceding subsections, if the original unit fails and there is no spare on hand, the system remains down until the spare is delivered. In this subsection, we modify the above-mentioned model to accommodate systems where, whenever the original unit fails, minimal repairs are made to the failed unit and it continues its operation until it can be replaced by the spare. We add the cost c_{m} suffered for each minimal repair instead of k_{f} and use the uniform cannibalization cost c_{s} instead of s. We discuss the combined model with minimal repairs for the continuous-time case.

Consider a one-unit system, where each spare is only provided after a lead time by an order. Each failure is detected instantaneously and for each failed unit a *minimal repair* is executed in a negligible time. The original unit starts operating at time zero. The planning horizon is infinite. If the original unit does not fail before a prespecified time $t_0 \in [0, \infty)$, the regular order for a spare is made at the time t_0 and the spare is delivered after a lead time L. Then, the original unit is exchanged by that spare immediately. If the original unit fails up to the delivery of the spare, then the minimal repair is executed for each failure. On the other hand, if the original unit fails before the time t_0, then at the failure time the minimal repair and the expedited order are executed simultaneously, and the original unit is exchanged by the spare as soon as it is delivered after a lead time L. If the original unit fails up to the delivery of the spare, then the minimal repair is executed for each failure. In this case, the regular order is not made. After each minimal repair, the unit continues its operation. Each exchange is made instantaneously, and after an exchange the system starts operating immediately. We define an interval from one exchange to the following exchange as one cycle. The cycle then repeats itself continually.

The lifetime for each unit obeys an identical and arbitrary distribution $F(t)$ with a finite mean $1/\lambda$ and a density $f(t)$. Let us introduce the following four costs: cost c_e is suffered for each expedited order made before the time t_0; cost c_r is suffered for each regular order made at the time t_0; the cost c_m is suffered for each minimal repair made at each failure time. Further, cost c_s is suffered as a salvage cost at each exchange time. We assume that $c_e > c_r$. Let us define the failure rate as follows:

$$r(t) = f(t)/\bar{F}(t) \tag{19.65}$$

where $\bar{F}(t) = 1 - F(t)$. This failure rate $r(t)$ is assumed to be differentiable. These assumptions seem to be reasonable.

Under these assumptions, we analyze this model and discuss the optimal ordering policies, that minimize the expected cost per unit time in the steady state.

The expected cost per one cycle $A(t_0)$ is given by

$$A(t_0) = c_e F(t_0) + c_r \bar{F}(t_0) + c_m \left[F(t_0+L) + \int_0^{t_0} \int_t^{t+L} r(\tau)\, d\tau\, dF(t) + \int_{t_0}^{t_0+L} \int_t^{t_0+L} r(\tau)\, d\tau\, dF(t) \right] + c_s \tag{19.66}$$

Also, the mean time of one cycle $M(t_0)$ is given by

$$M(t_0) = L + \int_0^{t_0} \bar{F}(t)\, dt \tag{19.67}$$

Thus, the expected cost per unit time in the steady state $C(t_0)$ is given by

$$C(t_0) = A(t_0)/M(t_0) \tag{19.68}$$

and

$$C(0) = \frac{c_r + c_m \int_0^L r(\tau)\, d\tau + c_s}{L} \tag{19.69}$$

$$C(\infty) = \frac{c_e + c_m[1 + \int_0^\infty \int_t^{t+L} r(\tau)\, d\tau\, dF(t)] + c_s}{L + 1/\lambda} \tag{19.70}$$

Define the following function from the numerator of the derivative of $C(t_0)$:

$$a(t_0) = [(c_e - c_r) r(t_0) + c_m r(t_0 + L)] M(t_0) - A(t_0) \tag{19.71}$$

The sufficient conditions for the existence of optimal ordering policies are presented as follows.

Theorem 14.

(i) If $a(0) < 0$, then there exists at least an optimal ordering time t_0^* $(0 < t_0^* \leq \infty)$ minimizing the expected cost per unit time in the steady state $C(t_0)$.

(ii) If $a(\infty) > 0$, then there exists at least an optimal ordering time t_0^* $(0 \leq t_0^* < \infty)$ minimizing the expected cost per unit time in the steady state $C(t_0)$.

Next, we have the following interesting results supposing the monotonic properties of the failure rate.

Theorem 15.

(1) Suppose that the failure rate is strictly increasing.

(i) If $a(0) < 0$ and $a(\infty) > 0$, then there exists a finite and unique optimal ordering time t_0^* $(0 < t_0^* < \infty)$, that minimizes the expected cost $C(t_0)$, satisfying $a(t_0^*) = 0$ and the corresponding expected cost is given by

$$C(t_0^*) = (c_e - c_r) r(t_0^*) + c_m r(t_0^* + L) \tag{19.72}$$

(ii) If $a(0) \geq 0$, then the optimal ordering time is $t_0^* = 0$, i.e. the order for a spare is made when the unit is put in service. The corresponding expected cost is given by Equation 19.69.

(iii) If $a(\infty) \leq 0$, then the optimal ordering time is $t_0^* \to \infty$, i.e. the order for a spare is made at the same time as the first failure of the original unit. The corresponding expected cost is given by Equation 19.70.

(2) Suppose that the failure rate is decreasing.

(i) If

$$\left[c_r + c_m \int_0^L r(\tau)\,d\tau + c_s\right](L + 1/\lambda) < \left\{c_e + c_m\left[1 + \int_0^\infty \int_t^{t+L} r(\tau)\,d\tau\,dF(t)\right] + c_s\right\} L \tag{19.73}$$

then $t_0^* = 0$.

(ii) If

$$\left[c_r + c_m \int_0^L r(\tau)\,d\tau + c_s\right](L + 1/\lambda) \geq \left\{c_e + c_m\left[1 + \int_0^\infty \int_t^{t+L} r(\tau)\,d\tau\,dF(t)\right] + c_s\right\} L \tag{19.74}$$

then $t_0^* \to \infty$.

So far we have treated non-negative ordering time t_0. However, if we choose an initial time zero carefully, it can be allowed that the ordering time t_0 is negative. That is, if the regular order for the spare is made at time $t_0 \in (-L, 0)$, the original unit begins operation after time interval $-t_0$, *i.e.* at time zero, and the spare is delivered at time $t_0 + L$ and the original unit is exchanged by that spare immediately as the above rule. In this case the regular order is only made for the order, since the order, is always made before the original unit fails.

Here, we treat the ordering policy not only with the non-negative ordering time t_0 but also with the negative one as mentioned above, *i.e.* the regular ordering time t_0 is over $(-L, \infty)$. Then the expected cost per unit time in the steady state $C(t_0)$ in Equation 19.68 is still valid for $t_0 \in (-L, \infty)$, since it is clear that $F(t_0) = 0$ and $\bar{F}(t_0) = 1$ for negative t_0. Furthermore, we assume that $c_r + c_s > 0$.

Thus, we can obtain the theorems corresponding to those above-mentioned, respectively.

Theorem 16. *If $a(\infty) > 0$, then there exists at least a finite optimal ordering time t_0^* $(-L < t_0^* < \infty)$ minimizing the expected cost per unit time in the steady state $C(t_0)$ in Equation 19.68.*

Theorem 17.

(1) Suppose that the failure rate is strictly increasing.

(i) If $a(\infty) > 0$, then there exists a finite and unique optimal ordering time t_0^* $(-L < t_0^* < \infty)$, that minimizes the expected cost $C(t_0)$ in Equation 19.68, satisfying $a(t_0^*) = 0$, and the corresponding expected cost is given by Equation 19.72.

(ii) If $a(\infty) \leq 0$, then the optimal ordering time is $t_0^* \to \infty$.

(2) Suppose that the failure rate is decreasing. Then $t_0^* \to \infty$.

Remarks. When we consider a negative ordering time, it is of no interest if we cannot choose the operational starting point of the system, *i.e.* an initial time zero definitely. For example, it is of no interest to consider a negative ordering time in the ordering model that when the regular ordered spare is delivered and the original unit does not fail once, that spare is put into inventory until the original one fails first, since we cannot choose the operational starting point of the system definitely.

19.5 Inspection Models

There are several systems where failures are not detected immediately they occur, usually those in that failure is not catastrophic and an inspection is needed to reveal the fault. If we execute too many inspections, then system failure is detected more rapidly, but we incur a high inspection cost. Conversely, if we execute few inspections, the interval between the failure and its detection increases and we incur a high cost of failure. The optimal inspection policy minimizes the total expected cost composed of costs for inspection and system failure. From this viewpoint, many authors have discussed optimal and/or nearly optimal inspection policies [2, 84–93].

Among those policies, the inspection policy discussed by Barlow and coworkers [2, 84] is the most famous. They have discussed the optimal inspection policy in the following model. A one-unit system is considered, that obeys an arbitrary lifetime distribution $F(t)$ with a probability density function (p.d.f.) $f(t)$. The system is inspected at prespecified times t_k $(k = 1, 2, 3, \dots)$, where each inspection is executed perfectly and instantaneously. The policy terminates with an inspection that can detect a system failure. The costs considered are c_i (>0), the cost of an inspection, and $k_f(>0)$, the cost of failure per unit of time. Then the total expected cost is

$$C_B = \sum_{k=0}^{\infty} \int_{t_k}^{t_{k+1}} [c_i(k+1) + k_f(t_{k+1} - t)]\, dF(t) \tag{19.75}$$

They obtained Algorithm 1 to seek the optimal inspection-time sequence that minimizes the total expected cost in Equation 19.75 by using the recurrence formula

$$t_{k+1} - t_k = \frac{F(t_k) - F(t_{k-1})}{f(t_k)} - \frac{c_i}{k_f} \qquad k = 1, 2, 3, \dots \tag{19.76}$$

where $f(t)$ is a PF_2 (Pólya frequency function of order two) with $f(t + \Delta)/f(t)$ strictly decreasing for $t \geq 0$, $\Delta > 0$, and with $f(t) > 0$ for $t > 0$, and $t_0 = 0$.

Algorithm 1.

begin:

choose t_1 to satisfy $c_i = k_f \int_0^{t_1} F(t)\, dt$;

repeat

compute $t_2, t_3, \dots,$ recursively using Equation 19.76;
if any $t_{k+1} - t_k > t_k - t_{k-1}$,
then reduce t_1;
if any $t_{k+1} - t_k < 0$,
then increase t_1;

until $t_1 < t_2 < \cdots$ are determined to the degree of accuracy required;
end.

However, Algorithm 1 by Barlow and coworkers is complicated to execute, because one must apply trial and error to decide the first inspection time t_1, and the assumption on $f(t)$ is restrictive. To overcome these difficulties, some improved procedures to obtain the nearly optimal inspection policy have been proposed [85, 87–93].

We review the nearly optimal inspection policies proposed by Kaio and Osaki [85, 92, 93], Munford and Shahani [87], and Nakagawa and Yasui [90]. We follow the inspection model and notation introduced by Barlow and coworkers. For details, see each of the contributed papers in the references.

19.5.1 Nearly Optimal Inspection Policy by Kaio and Osaki (K&O Policy)

Introduce the inspection density at time t, $n(t)$, that is a smooth function and denotes the approximate number of inspections per unit time at time t. Then the total expected cost up to the detection of the failure is approximately

$$C_n(n(t)) = c_i \int_0^{\infty} n(t)\bar{F}(t)\, dt + k_f \int_0^{\infty} \frac{1}{2n(t)}\, dF(t) \tag{19.77}$$

where $\bar{\psi} = 1 - \psi$, in general. The density $n(t)$ that minimizes the functional $C_n(n(t))$ in Equation 19.77 is

$$n(t) = [k_c r(t)]^{1/2} \tag{19.78}$$

where $k_c = k_f/(2c_i)$, and $r(t) = f(t)/\bar{F}(t)$, a failure rate. The inspection times t_k $(k = 1, 2, 3, \dots)$ satisfy

$$k = \int_0^{t_k} n(t)\, dt \qquad k = 1, 2, 3, \dots \tag{19.79}$$

Substituting $n(t)$ in Equation 19.78 into Equation 19.79 yields the nearly optimal inspection-time sequence.

Kaio and Osaki obtained this procedure by developing that of Keller [91]. For details, see Kaio and Osaki [85]. Note that the procedure does not depend on assumptions about the p.d.f. $f(t)$.

19.5.2 Nearly Optimal Inspection Policy by Munford and Shahani (M&S Policy)

Let

$$\frac{F(t_k) - F(t_{k-1})}{\bar{F}(t_{k-1})} = p$$
$$k = 1, 2, 3, \ldots ; \quad 0 < p < 1 \tag{19.80}$$

Then the inspection times t_k $(k = 1, 2, 3, \ldots)$ are

$$t_k = F^{-1}(1 - \bar{p}^k) \quad k = 1, 2, 3, \ldots \tag{19.81}$$

where the probability p is chosen such that the nearly total expected cost up to the detection of the failure, $C_{\mathrm{p}}(p)$, is minimized:

$$C_{\mathrm{p}}(p) = \frac{c_i}{p} + k_{\mathrm{f}}\left[\sum_{k=1}^{\infty} t_k \bar{p}^{k-1} p - \int_0^{\infty} t f(t)\,\mathrm{d}t\right] \tag{19.82}$$

This procedure does not depend on assumptions about the p.d.f. $f(t)$. For details, see Munford and Shahani [87]; additionally, see Munford and Shahani [88] for the case of the Weibull distribution and Tadikamalla [89] for the gamma distribution.

19.5.3 Nearly Optimal Inspection Policy by Nakagawa and Yasui (N&Y Policy)

This procedure is based on one by Barlow and coworkers [2, 84] (abbreviated to *B policy* below). If the p.d.f. $f(t)$ is a PF_2, then Algorithm 2 is obtained.

Algorithm 2.

begin:

choose d appropriately for $0 < d < \dfrac{c_i}{k_{\mathrm{f}}}$;
determine t_n after sufficient time has elapsed to give the degree of accuracy required;
compute t_{n-1} to satisfy

$$t_n - t_{n-1} - d = \frac{F(t_n) - F(t_{n-1})}{f(t_n)} - \frac{c_i}{k_{\mathrm{f}}}$$

repeat
compute $t_{n-2} > t_{n-3} > \cdots$ recursively using Equation 19.76;
until $t_i < 0$ or $t_{i+1} - t_i > t_i$;

end.

For details, see Nakagawa and Yasui [90].

Remarks. From numerical comparisons with Weibull and gamma distributions, we conclude that there are no significant differences between the optimal and nearly optimal inspection policies [93], and consequently we should adopt the simplest policy to compute, by Kaio and Osaki [85]. There are the following advantages when we use the *K&O* policy:

(i) We can obtain the nearly optimal inspection policy uniquely, immediately and easily from Equations 19.78 and 19.79 for any distributions, whereas the *B* and *N&Y* policies cannot treat non-PF_2 distributions.
(ii) We can analyze more complicated models and easily obtain their nearly optimal inspection policies, *e.g.* see Kaio and Osaki [85].

19.6 Concluding Remarks

This chapter has discussed the basic preventive maintenance policies and their extensions, in terms of both continuous- and discrete-time modeling. For further reading on the discrete models, see Nakagawa [29, 94] and Nakagawa and Osaki [95]. Since the mathematical maintenance models are applicable to a variety of real problems, such a modeling technique will be useful for practitioners and researchers. Though we have reviewed only the limited maintenance models in the limited space available here, a number of earlier models should be reformulated in a discrete-time setting, because, in most cases, the continuous-time models can be regarded as approximated models for actual maintenance problems and the maintenance schedule is often desired in discretized circumstances. These motivations for discrete-time setting will be evident from the recent development of computer technologies and their related computation abilities.

Acknowledgments

This work was partially supported by a Grant-in-Aid for Scientific Research from the Ministry of Education, Sports, Science and Culture of Japan under grant no. 09780411 and no. 09680426, and by the Research Program 1999 under the Institute for Advanced Studies of the Hiroshima Shudo University, Hiroshima, Japan.

References

[1] Arrow KJ, Karlin S, Scarf H, editors. Studies in applied probability and management science. Stanford: Stanford University Press; 1962.

[2] Barlow RE, Proschan F. Mathematical theory of reliability. New York: John Wiley & Sons; 1965.

[3] Barlow RE, Proschan F. Statistical theory of reliability and life testing: probability models. New York: Holt, Rinehart and Winston; 1975.

[4] Jorgenson DW, McCall JJ, Radner R. Optimal replacement policy. Amsterdam: North-Holland; 1967.

[5] Gnedenko BV, Belyayev YK, Solovyev AD. Mathematical methods of reliability theory. New York: Academic Press; 1969.

[6] Gertsbakh IB. Models of preventive maintenance. Amsterdam: North-Holland; 1977.

[7] Ascher H, Feingold H. Repairable systems reliability. New York: Marcel Dekker; 1984.

[8] Osaki S. Stochastic system reliability modeling. Singapore: World Scientific; 1985.

[9] Osaki S. Applied stochastic system modeling. Berlin: Springer-Verlag; 1992.

[10] Osaki S, Hatoyama Y, editors. Stochastic models in reliability theory. Lecture Notes in Economics and Mathematical Systems, vol. 235. Berlin: Springer-Verlag; 1984.

[11] Osaki S, Cao J, editors. Reliability theory and applications. Singapore: World Scientific; 1987.

[12] Ozekici S, editor. Reliability and maintenance of complex systems. NATO ASI Series. Berlin: Springer-Verlag; 1996.

[13] Christer AH, Osaki S, Thomas LC, editors. Stochastic modelling in innovative manufacturing. Lecture Notes in Economics and Mathematical Systems, vol. 445. Berlin: Springer-Verlag; 1997.

[14] Barlow RE. Engineering reliability. Philadelphia: SIAM; 1998.

[15] Aven T, Jensen U. Stochastic models in reliability. New York: Springer-Verlag; 1999.

[16] McCall JJ. Maintenance policies for stochastically failing equipment: a survey. Manage Sci. 1965;11:493–521.

[17] Osaki S, Nakagawa T. Bibliography for reliability and availability of stochastic systems. IEEE Trans Reliab 1976;R-25:284–7.

[18] Pierskalla WP, Voelker JA. A survey of maintenance models: the control and surveillance of deteriorating systems. Nav Res Logist Q 1976;23:353–88.

[19] Sherif YS, Smith ML. Optimal maintenance models for systems subject to failure—a review. Nav Res Logist Q 1981;28:47–74.

[20] Valdez-Flores C, Feldman RM. A survey of preventive maintenance models for stochastically deteriorating single-unit systems. Nav Res Logist 1989;36:419–46.

[21] Feller W. An introduction to probability theory and its applications. New York: John Wiley & Sons; 1957.

[22] Karlin S, Taylor HM. A first course in stochastic processes. New York: Academic Press; 1975.

[23] Karlin S, Taylor HM. A second course in stochastic processes. New York: Academic Press; 1981.

[24] Taylor HM, Karlin S. An introduction to stochastic modeling. New York: Academic Press; 1984.

[25] Ross SM. Applied probability models with optimization applications. San Francisco: Holden-Day; 1970.

[26] Munter M. Discrete renewal processes. IEEE Trans Reliab 1971;R-20:46–51.

[27] Kaio N, Osaki S. Review of discrete and continuous distributions in replacement models. Int J Sys Sci 1988;19:171–7.

[28] Barlow RE, Hunter LC. Optimum preventive maintenance policies. Oper Res 1960;8:90–100.

[29] Nakagawa T. A summary of discrete replacement policies. Eur J Oper Res 1979;17:382–92.

[30] Tilquin C, Cleroux R. Block replacement with general cost structure. Technometrics 1975;17:291–8.

[31] Berg M, Epstein B. A modified block replacement policy. Nav Res Logist Q 1976;23:15–24.

[32] Morimura H. On some preventive maintenance policies for IFR. J Oper Res Soc Jpn 1970;12:94–124.

[33] Tilquin C, Cleroux R. Periodic replacement with minimal repair at failure and adjustment costs. Nav Res Logist Q 1975;22:243–54.

[34] Cleroux R, Dubuc S, Tilquin C. The age replacement problem with minimal repair and random repair costs. Oper Res 1979;27:1158–67.

[35] Park KS. Optimal number of minimal repairs before replacement. IEEE Trans Reliab 1979;R-28:137–40.

[36] Nakagawa T. A summary of periodic replacement with minimal repair at failure. J Oper Res Soc Jpn 1979;24:213–28.

[37] Nakagawa T. Generalized models for determining optimal number of minimal repairs before replacement. J Oper Res Soc Jpn 1981;24:325–57.

[38] Nakagawa T. Modified periodic replacement with minimal repair at failure. IEEE Trans Reliab 1981;R-30:165–8.

[39] Nakagawa T. Optimal policy of continuous and discrete replacement with minimal repair at failure. Nav Res Logist Q 1984;31:543–50.

[40] Nakagawa T. Periodic and sequential preventive maintenance policies. J Appl Prob 1986;23:536–542.

[41] Nakagawa T, Kowada M. Analysis of a system with minimal repair and its application to replacement policy. Eur J Oper Res 1983;12:176–82.

[42] Phelps RI. Replacement policies under minimal repair. J Oper Res Soc 1981;32:549–54.

[43] Berg M, Cleroux R. The block replacement problem with minimal repair and random repair costs. J Stat Comput Sim 1982;15:1–7.

[44] Boland PJ. Periodic replacement when minimal repair costs vary with time. Nav Res Logist Q 1982;29:541–6.

[45] Boland PJ, Proschan F. Periodic replacement with increasing minimal repair costs at failure. Oper Res 1982;30:1183–9.

[46] Block HW, Borges WS, Savits TH. Age-dependent minimal repair. J Appl Prob 1985;22:370–85.

[47] Beichelt F. A unifying treatment of replacement policies with minimal repair. Nav Res Logist 1993;40:51–67.

[48] Tahara A, Nishida T. Optimal replacement policy for minimal repair model. J Oper Res Soc Jpn 1975;18:113–24.

[49] Phelps RI. Optimal policy for minimal repair. J Oper Res Soc 1983;34:425–7.

[50] Segawa Y, Ohnishi M, Ibaraki T. Optimal minimal-repair and replacement problem with average dependent cost structure. Comput Math Appl 1992;24:91–101.

[51] Ohnishi M. Optimal minimal-repair and replacement problem under average cost criterion: optimality of (t, T)-policy. J Oper Res Soc Jpn 1997;40:373–89.

[52] Bergman B. On the optimality of stationary replacement strategies. J Appl Prob 1980;17:178–86.

[53] Barlow RE, Hunter LC. Reliability analysis of a one-unit system. Oper Res 1961;9:200–8.

[54] Osaki S, Nakagawa T. A note on age replacement. IEEE Trans Reliab 1975;R-24:92–4.

[55] Fox BL. Age replacement with discounting. Oper Res 1966;14:533–7.

[56] Nakagawa T, Osaki S. Discrete time age replacement policies. Oper Res Q 1977;28:881–5.

[57] Glasser GJ. The age replacement problem. Technometrics 1967;9:83–91.

[58] Scheaffer RL. Optimum age replacement policies with an increasing cost factor. Technometrics 1971;13:139–44.

[59] Cleroux R, Hanscom M. Age replacement with adjustment and depreciation costs and interest charges. Technometrics 1974;16:235–9.

[60] Osaki S, Yamada S. Age replacement with lead time. IEEE Trans Reliab 1976;R-25:344–5.

[61] Nakagawa T, Osaki S. Reliability analysis of a one-unit system with unrepairable spare units and its optimization applications. Oper Res Q 1976;27:101–10.

[62] Derman C, Sacks J. Replacement of periodically inspected equipment. Nav Res Logist Q 1960;7:597–607.

[63] Ansell J, Bendell A, Humble S. Age replacement under alternative cost criteria. Manage Sci 1984;30:358–67.

[64] Christer AH. Refined asymptotic costs for renewal reward process. J Oper Res Soc 1978;29:577–83.

[65] Christer AH. Comments on finite-period applications of age-based replacement models. IMA J Math Appl Bus Ind 1987;1:111–24.

[66] Christer AH, Jack N. An integral-equation approach for replacement modelling over finite time horizons. IMA J Math Appl Bus Ind 1991;3:31–44.

[67] Wiggins AD. A minimum cost model of spare parts inventory control. Technometrics 1967;9:661–5.

[68] Allen SG, D'Esopo DA. An ordering policy for repairable stock items. Oper Res 1968;16:669–74.

[69] Allen SG, D'Esopo DA. An ordering policy for stock items when delivery can expedited. Oper Res 1968;16:880–3.

[70] Simon RM, D'Esopo DA. Comments on a paper by S.G. Allen and D.A. D'Esopo: An ordering policy for repairable stock items. Oper Res 1971;19:986–9.

[71] Nakagawa T, Osaki S. Optimum replacement policies with delay. J Appl Prob 1974;11:102–10.

[72] Osaki S. An ordering policy with lead time. Int J Sys Sci 1977;8:1091–5.

[73] Thomas LC, Osaki S. A note on ordering policy. IEEE Trans Reliab 1978;R-27:380–1.

[74] Thomas LC, Osaki S. An optimal ordering policy for a spare unit with lead time. Eur J Oper Res 1978;2:409–19.

[75] Kaio N, Osaki S. Optimum ordering policies with lead time for an operating unit in preventive maintenance. IEEE Trans Reliab 1978;R-27:270–1.

[76] Kaio N, Osaki S. Optimum planned maintenance with salvage costs. Int J Prod Res 1978;16:249–57.

[77] Kaio N, Osaki S. Discrete-time ordering policies. IEEE Trans Reliab 1979;R-28:405–6.

[78] Kaio N, Osaki S. Optimum planned maintenance with discounting. Int J Prod Res 1980;18:515–23.

[79] Osaki S, Kaio N, Yamada S. A summary of optimal ordering policies. IEEE Trans Reliab 1981;R-30:272–7.

[80] Dohi T, Kaio N, Osaki S. On the optimal ordering policies in maintenance theory—survey and applications. Appl Stoch Models Data Anal 1998;14:309–21.

[81] Dohi T, Kaio N, Osaki S. Continuous review cyclic inventory models with emergency order. J Oper Res Soc Jpn 1995;38:212–29.

[82] Dohi T, Shibuya T, Osaki S. Models for 1-out-of-Q systems with stochastic lead times and expedited ordering options for spares inventory. Eur J Oper Res 1997;103:255–72.

[83] Shibuya T, Dohi T, Osaki S. Spare part inventory models with stochastic lead times. Int J Prod Econ 1998;55:257–71.

[84] Barlow RE, Hunter LC, Proschan F. Optimum checking procedures. J Soc Indust Appl Math 1963;11:1078–95.

[85] Kaio N, Osaki S. Some remarks on optimum inspection policies. IEEE Trans Reliab 1984;R-33:277–9.

[86] Kaio N, Osaki S. Analytical considerations on inspection policies. In: Osaki S, Hatoyama Y, editors. Stochastic models in reliability theory. Heidelberg: Springer-Verlag; 1984. p.53–71.

[87] Munford AG, Shahani AK. A nearly optimal inspection policy. Oper Res Q 1972;23:373–9.

[88] Munford AG, Shahani AK. An inspection policy for the Weibull case. Oper Res Q 1973;24:453–8.

[89] Tadikamalla PR. An inspection policy for the gamma failure distributions. J Oper Res Soc 1979;30:77–80.

[90] Nakagawa T, Yasui K. Approximate calculation of optimal inspection times. J Oper Res Soc 1980;31:851–3.

[91] Keller JB. Optimum checking schedules for systems subject to random failure. Manage Sci 1974;21: 256–60.

[92] Kaio N, Osaki S. Optimal inspection policies: A review and comparison. J Math Anal Appl 1986;119:3–20.

[93] Kaio N, Osaki S. Comparison of inspection policies. J Oper Res Soc 1989;40:499–503.

[94] Nakagawa T. Modified discrete preventive maintenance policies. Nav Res Logist Q 1986;33:703–15.

[95] Nakagawa T, Osaki S. Analysis of a repairable system that operates at discrete times. IEEE Trans Reliab 1976;R-25:110–2.

Maintenance and Optimum Policy

Toshio Nakagawa

Chapter 20

20.1 Introduction

It has been well known that high system reliabilities can be achieved by the use of redundancy or maintenance. These techniques have actually been used in various real systems, such as computers, generators, radars, and airplanes, where failures during actual operation are costly or dangerous.

A failed system is replaced or repaired immediately. But whereas the maintenance of a system after failure may be costly, it may sometimes require a long time to undertake it. It is an important problem to determine when to maintain a system before its failure. However, it is not wise to maintain a system too often. From these viewpoints, the following three policies are well known.

1. Replacement policy: a system with no repair is replaced before failure with a new one.
2. Preventive maintenance (PM) policy: a system with repair is maintained preventively before failure.
3. Inspection policy: a system is checked to detect its failure.

Suppose that a system has to operate for an infinite time span. Then, it is appropriate to adopt, as an objective function, the expected cost per unit of time from the viewpoint of economics, and the availability from reliability. We summarize the theoretical results of optimum policies that minimize or maximize the above quantities. In Section 20.2 we discuss (1) age replacement, (2) block replacement, (3) periodic replacement, and (4) modified replacements with discounting, in discrete time, with two types of unit and of a shock model. In Section 20.3, PMs of (1) a one-unit system, (2) a two-unit system, (3) an imperfect PM and (4) a modified PM are discussed. Section 20.4 covers (1) standard inspection, (2) inspection with PM and (3) inspection of a storage system.

Barlow and Proschan [1] summarized the known results of maintenance and their optimization. Since then, many papers have been published and the subject surveyed extensively by Osaki and Nakagawa [2], Pierskalla and Voelker [3], Thomas [4], Valdez-Flores and Feldman [5], and Cho and Parlar [6]. The recent published books edited by Özekici [7], Christer *et al.* [8] and Ben-Daya *et al.* [9] and Osaki [10] collected many reliability and maintenance models, discussed their optimum policies, and applied them to actual systems. Most contents in this chapter are based on our original work [11–36].

The theories of renewal and Markov renewal processes are used in analyzing maintenance policies, and are discussed in the books by Cox [37], Çinlar [38], and Osaki [39].

20.2 Replacement Policies

In this section, we consider a one-unit system where a unit is replaced upon failure. It would be wise to replace an operating unit before failure at a small cost. We introduce a cost incurred by failure of an operating unit and a cost incurred by replacement before failure. Most units are replaced at failures and their ages, or at scheduled times according to their sizes, characters, circumstances, and so on.

A unit has to be operating for an infinite time span in the following replacement costs: a cost c_1 is suffered for each failed unit that is replaced; this includes all costs resulting from its failure and replacement. A cost $c_2(< c_1)$ is suffered for each non-failed unit that is exchanged. Let $N_1(t)$ denote the number of failures during $(0, t]$ and $N_2(t)$ denote the number of exchanges of non-failed units during $(0, t]$. Then, the expected cost during $(0, t]$ is

$$\hat{C}(t) \equiv c_1 E\{N_1(t)\} + c_2 E\{N_2(t)\} \qquad (20.1)$$

When the planning horizon is infinite, it is appropriate to adopt the expected cost per unit of time $\lim_{t\to\infty} \hat{C}(t)/t$ as an objective function.

We summarize three replacement policies of (1) age replacement, (2) block replacement, and (3) periodic replacement with minimal repair at failure, and moreover, consider (4) their modified models. We discuss optimum policies that minimize the expected costs per unit of time of each replacement, and summarize their derivation results.

20.2.1 Age Replacement

If the age of an operating unit is always known and its failure rate increases with age, it may be wise to replace it before failure on its age. A commonly considered age replacement policy for such a unit is made if it is replaced at time T $(0 < T \leq \infty)$ after its installation or at failure, whichever occurs first. We call the specified time T a planned replacement time. Berg [40] and Bergman [41] proved that an age replacement policy is optimal among all reasonable policies. It is assumed that failures are instantly detected and a failed unit is replaced by a new unit. A new installed unit also begins to operate instantly. Suppose that the failure time X_k $(k = 1, 2, \dots)$ of each operating unit is independent and has an identical distribution $F(t)$ with finite mean $1/\lambda$.

An age replacement policy has been treated by many authors. Barlow and Proschan [1] studied the optimum policy that minimizes the expected cost. Cox [37] gave a sufficient condition for a finite optimum solution to exist. Glasser [42] gave the replacement time for the cases of several failure time distributions. Scheaffer [43], Cléroux and coworkers [44,45], and

Subramanian and Wolff [46] gave more general cost structures of replacement. Fox [47] and Ran and Rosenland [48] proposed an age replacement with continuous discounting. Further, Berg and Epstein [49] compared age replacement with block replacement. Ingram and Scheaffer [50], Frees and Ruppert [51], and Léger and Cléroux [52] showed the confidence intervals of the optimum age replacement policy when a failure time distribution F is unknown. Christer [53] and Ansell *et al.* [54] gave the asymptotic costs of age replacement for a finite time span. Popova and Wu [55] applied fuzzy set theory to age replacement policies. Opportunistic replacement policies are important for the PM of complex systems. This area is omitted in this chapter; the reader is referred to Zheng [56], for example.

A new unit begins to operate at $t = 0$. Then, an age replacement procedure generates an ordinary renewal process as follows. Let $\{X_k\}_{k=1}^{\infty}$ be the failure times of successive operating units, and define a new random variable $Z_k \equiv \min\{X_k, T\}$. Then, $\{Z_k\}_{k=1}^{\infty}$ represent the intervals between replacements caused by either failure or planned replacement, and are independently, identically distributed random variables. Thus, a sequence of $\{Z_k\}_{k=1}^{\infty}$ forms a renewal process with

$$\Pr\{Z_k \le t\} = \begin{cases} F(t) & t < T \\ 1 & t \ge T \end{cases} \tag{20.2}$$

We term one cycle as being from the replacement to the next one. Then, the expected cost on one cycle is

$$E\{c_1 I_{(X_k<T)} + c_2 I_{(X_k \ge T)}\} = c_1 F(T) + c_2 \bar{F}(T) \tag{20.3}$$

and the mean time of one cycle is

$$E\{Z_k\} = \int_0^T t \, \mathrm{d}F(t) + T\bar{F}(T) = \int_0^T \bar{F}(t) \, \mathrm{d}t \tag{20.4}$$

where $\bar{F} \equiv 1 - F$ and I_A is an indicator.

Therefore, from an elementary renewal theorem (*e.g.* theorem 2.6 of Barlow and Proschan [1, p.55]), the expected cost per unit of time for an infinite time span is

$$C(T) \equiv \lim_{t\to\infty} \frac{\hat{C}(t)}{t} = \frac{\text{Expected cost on one cycle}}{\text{Mean time of one cycle}} = \frac{c_1 F(T) + c_2 \bar{F}(T)}{\int_0^T \bar{F}(t) \, \mathrm{d}t} \tag{20.5}$$

If $T = \infty$ then the policy corresponds to the replacement only at failure, and the expected cost is $C(\infty) = \lambda c_1$.

We derive an optimum replacement time T^* that minimizes the expected cost $C(T)$ in Equation 20.5. It is assumed that there exists a density $f(t)$ of the failure time distribution $F(t)$ with mean $1/\lambda$. Let $r(t) \equiv f(t)/\bar{F}(t)$ be the failure rate (or the hazard rate) and its limit is $r(\infty) \equiv \lim_{t\to\infty} r(t)$, which may possibly be infinite.

Theorem 1. *Suppose that $r(t)$ is continuous and strictly increasing, and $K \equiv \lambda c_1/(c_1 - c_2)$.*

(i) If $r(\infty) > K$ then there exists a finite and unique $T^*(0 < T^* < \infty)$ that satisfies

$$r(T) \int_0^T \bar{F}(t) \, \mathrm{d}t - F(T) = \frac{c_2}{c_1 - c_2} \tag{20.6}$$

and the resulting expected cost is

$$C(T^*) = (c_1 - c_2) r(T^*) \tag{20.7}$$

(ii) If $r(\infty) \le K$ then the optimum replacement time is $T^* = \infty$, i.e. *a unit should be replaced only at failure.*

Proof. Differentiating $C(T)$ (or $\log C(T)$) with respect to T and setting it equal to zero implies Equation 20.6. Denoting the left-hand side of Equation 20.6 by $Q(T)$, we easily have that $Q(T)$ is strictly increasing since $r(T)$ is strictly increasing, and $Q(0) = 0$:

$$Q(\infty) \equiv \lim_{T\to\infty} Q(T) = \frac{1}{\lambda} r(\infty) - 1$$

If $r(\infty) > K$ then $Q(\infty) > c_2/(c_1 - c_2)$. Hence, from the monotonicity and continuity of $Q(T)$, there exists a finite and unique $T^*(0 < T^* < \infty)$ that satisfies Equation 20.6, and it minimizes

$C(T)$. We easily have Equation 20.7 from Equation 20.6.

Conversely, if $r(\infty) \leq K$ then $Q(\infty) \leq c_2/(c_1 - c_2)$ for any finite T, which implies $\mathrm{d}C(T)/\mathrm{d}T < 0$ for any T. Hence, the optimum replacement time is $T^* = \infty$, *i.e.* a unit is replaced only at failure. □

20.2.2 Block Replacement

If a system consists of a block or group of components and only their failures are known, all components may be replaced periodically independent of their ages in use. The policy is commonly used with complex electronic systems and many electrical parts.

A new unit begins to operate at $t = 0$, and a failed unit is discovered instantly and replaced by a new one. Further, a unit is replaced at periodic times kT $(k = 1, 2, \dots)$ independent of its age. Suppose that each unit has an identical failure time distribution $F(t)$ with finite mean $1/\lambda$, and $F^{(n)}(t)$ $(n = 1, 2, \dots)$ is the n-fold Stieltjes convolution of $F(t)$ with itself, *i.e.* $F^{(n)}(t) \equiv \int_0^t F^{(n-1)}(t-u)\,\mathrm{d}F(u)$ $(n = 1, 2, \dots)$ and $F^{(0)}(t) \equiv 1$ for $t \geq 0$.

Barlow and Proschan [1] studied block replacement and compared it with age replacement. After that, Marathe and Nair [57] and Jain and Nair [58] defined the n-stage block replacement and compared it with other replacements. Schweitzer [59] compared block replacement with replacement only of individual failures for hyper-exponentially and uniformly distributed failure times. Savits [60] derived the cost relationship between age and block replacements. Tilquin and Cléroux [61] introduced the adjustment costs, which are increasing with the age of a unit. Archibald and Dekker [62] and Sheu [63–66] considered more general replacement policies and summarized their results.

Consider one cycle with a constant time T from the planned replacement to the next one. Then, since the expected number of failed units during one cycle is $M(T) \equiv \sum_{n=1}^{\infty} F^{(n)}(T)$, the expected cost on one cycle is

$$c_1 E\{N_1(T)\} + c_2 E\{N_2(T)\} = c_1 M(T) + c_2$$

Therefore, from Equation 20.5, the expected cost per unit of time for an infinite time span (Equation 20.6 of Barlow and Proschan [1, p.95]) is

$$C(T) \equiv \lim_{t\to\infty} \frac{\hat{C}(t)}{t} = \frac{c_1 M(T) + c_2}{T} \tag{20.8}$$

If a unit is replaced only at failures, *i.e.* $T = \infty$, then $\lim_{T\to\infty} M(T)/T = \lambda$ and the expected cost is $C(\infty) = \lambda c_1$.

We seek an optimum planned replacement time T^* that minimizes the expected cost $C(T)$ in Equation 20.8. It is assumed that $M(t)$ is differentiable and define $m(t) \equiv \mathrm{d}M(t)/\mathrm{d}t$, where $M(t)$ is called *renewal function* and $m(t)$ is called *renewal density* in a stochastic process. Then, differentiating $C(T)$ with respect to T and setting it equal to zero, we have

$$Tm(T) - M(T) = \frac{c_2}{c_1} \tag{20.9}$$

This equation is a necessary condition that there exists a finite T^*, and in this case the resulting expected cost is

$$C(T^*) = c_1 m(T^*) \tag{20.10}$$

Let σ^2 be the variance of $F(T)$. Then, from Cox [37, p.119], there exists a large T such that $C(T) < C(\infty) = \lambda c_1$ if $c_2/c_1 < (1 - \lambda^2\sigma^2)/2$.

It may be wasteful to replace a failed unit by a new one just before a planned replacement. Three modified models from this viewpoint are well known. When a failure occurs just before the planned replacement, it remains failed until the replacement time [37, 67, 68] or it is replaced by a used unit [69–72]. An operating unit of young age is not replaced at planned replacement and remains in operation [73,74].

We now discuss two typical modified models.

20.2.2.1 No Replacement at Failure

A unit is always replaced at times kT, but it is not replaced at failure, and hence it remains a failure for the time interval from the occurrence of failure to its detection. Let c_1 be the cost of the time elapsed between failure and its detection per unit of time, and c_2 be the cost of planned replacement. Then, the expected cost per unit of time is

$$C(T) = \frac{1}{T}\left[c_1 \int_0^T F(t)\,\mathrm{d}t + c_2\right] \qquad (20.11)$$

Differentiating $C(T)$ with respect to T and setting it equal to zero, we have

$$TF(T) - \int_0^T F(t)\,\mathrm{d}t = \frac{c_2}{c_1} \qquad (20.12)$$

Thus, if $1/\lambda > c_2/c_1$ then there exists an optimum replacement time T^* uniquely that satisfies Equation 20.12, and the resulting expected cost is

$$C(T^*) = c_1 F(T^*) \qquad (20.13)$$

20.2.2.2 Replacement with Two Variables

Cox [37] considered the modification of block replacement in which if a failure occurs just before a planned replacement, it may postpone replacing a failed unit until the next replacement. That is, if a failure occurs in an interval $(kT - T_\mathrm{d}, kT)$, the replacement is not made until time kT, and a unit is down for the time interval. Suppose that the cost suffered for unit failure is proportional to the downtime, *i.e.* let $c_3 t$ be the cost of time t elapsed between failure and its detection. Then, the expected cost is

$$C(T, T_\mathrm{d}) = \frac{1}{T}\left[c_1 M(T - T_\mathrm{d}) + c_2 + c_3 \int_{T-T_\mathrm{d}}^{T} (T-t)\bar{F}(T-t)\,\mathrm{d}M(t)\right] \qquad (20.14)$$

When $T_\mathrm{d} = 0$, this corresponds to the block replacement, and when $T_\mathrm{d} = T$, this corresponds to the modified model in Section 20.2.2.1. Nakagawa [28] discussed the optimum T^* and T_d^* that minimize $C(T, T_\mathrm{d})$ in Equation 20.14.

20.2.3 Periodic Replacement

When we consider large and complex systems that consist of many kinds of components, we should make only minimal repair at each failure, and make the planned replacement or PM at periodic times. Barlow and Proschan [1] considered the following replacement policy. A unit is replaced periodically at periodic times kT $(k = 1, 2, \ldots)$. After each failure, only minimal repair is made so that the failure rate remains undisturbed by any repair of failures between successive replacements. This policy is commonly used with computers and airplanes. Holland and McLean [75] provided a practical procedure for applying the policy to large motors and small electrical parts. Tilquin and Cléroux [76], Boland and coworkers [77, 78] and Chen and Feldman [79] gave more general cost structures, and Aven [80] and Bagai and Jain [81] considered the modification of a minimal repair model. Dekker and coworkers [83, 84] presented a general framework for three replacement policies.

A new unit begins to operate at $t = 0$ and, when it fails, only minimal repair is made. That is, the failure rate of a unit remains undisturbed by repair of failures. Further, a unit is replaced at periodic times kT $(k = 1, 2, \ldots)$ independent of its age, and any units are as good as new after replacement. It is assumed that the repair and replacement times are negligible. Suppose that the failure times of each unit are independent, and have a distribution $F(t)$ and the failure rate $r(t) \equiv f(t)/\bar{F}(t)$, where f is a density of F. Then, failures of a unit occur in a non-homogeneous Poisson process with a mean-value function $R(t)$, where $R(t) \equiv \int_0^t r(u)\,\mathrm{d}u$ and $\bar{F}(t) = \mathrm{e}^{-R(t)}$ (*e.g.* see Nakagawa and Kowada [29] and Murthy [82]).

Consider one cycle with a constant time T from the planned replacement to the next one. Then, since the expected number of failures during one cycle is $E\{N_1(T)\} = R(T)$, the expected cost on one cycle is

$$c_1 E\{N_1(T)\} + c_2 E\{N_2(T)\} = c_1 R(T) + c_2$$

where note that c_1 is the cost of minimal repair.

Therefore, from Equation 20.5, the expected cost per unit of time for an infinite time span (equation (2.9) of Barlow and Proschan [1, p.99]) is

$$C(T) \equiv \lim_{t\to\infty} \frac{\hat{C}(t)}{t} = \frac{1}{T}[c_1 R(T) + c_2] \qquad (20.15)$$

If a unit is never replaced, *i.e.* $T = \infty$, then $\lim_{T\to\infty} R(T)/T = r(\infty)$, which may possibly be infinite, and $C(\infty) = c_1 r(\infty)$.

We seek an optimum planned replacement time T^* that minimizes the expected cost $C(T)$ in Equation 20.15. Differentiating $C(T)$ with respect to T and setting it equal to zero, we have

$$Tr(T) - R(T) = \frac{c_2}{c_1} \tag{20.16}$$

Suppose that the failure rate $r(t)$ is continuous and strictly increasing. Then, if a solution T^* to Equation 20.16 exists, it is unique, and the resulting cost is

$$C(T^*) = c_1 r(T^*) \tag{20.17}$$

Further, Equation 20.16 can be rewritten as $\int_0^T t\, \mathrm{d}r(t) = c_2/c_1$. Thus, if $\int_0^\infty t\, \mathrm{d}r(t) > c_2/c_1$ then there exists a solution to Equation 20.16.

When $r(t)$ is strictly increasing, we easily have the inequality

$$Tr(T) - \int_0^T r(t)\, \mathrm{d}t \geq r(T) \int_0^T \bar{F}(t)\, \mathrm{d}t - F(T) \tag{20.18}$$

Thus, an optimum T^* is not greater than that of an age replacement in Section 20.2.1. Further, taking $T \to \infty$ in Equation 20.18, if $r(\infty) > \lambda(c_1 + c_2)/c_1$ then a solution to Equation 20.16 exists.

If the cost of minimal repair depends on the age x of a unit and is given by $c_1(x)$, the expected cost is [77]

$$C(T) = \frac{1}{T}\left[\int_0^T c_1(x) r(x)\, \mathrm{d}x + c_2\right] \tag{20.19}$$

Several modifications of this replacement policy have been proposed.

20.2.3.1 Modified Models with Two Variables

See Nakagawa [28]. (a) A used unit begins to operate at $t = 0$ and is replaced at times kT by the same used unit of age $x (0 \leq x < \infty)$. Then, the expected cost is [21]

$$C(T, x) = \frac{1}{T}\left[c_1 \int_x^{T+x} r(t)\, \mathrm{d}t + c_2(x)\right] \tag{20.20}$$

where $c_2(x)$ is an acquisition cost of a used unit of age x.

In this case, we consider the optimum problem of determining economically the age of a used unit. Suppose that x is a variable for a specified T and $c_2(x)$ is differentiable. Then, differentiating $C(T, x)$ with respect to x and setting it equal to zero implies

$$r(T + x) - r(x) = -\frac{c_2'(x)}{c_1} \tag{20.21}$$

which is a necessary condition that a finite x minimizes $C(T, x)$ for a fixed T.

(b) Mine and Kawai [85] considered the replacement of a unit with random and wearout failures, where an operating unit enters a wearout failure period at a fixed time T_0, after it has been operating in a random failure period. It is assumed that a unit is replaced at planned time $T + T_0$, where T_0 is constant and previously given, and it undergoes only minimal repair at failures between replacements.

Suppose that a unit has a constant failure rate λ in a random failure period and $\lambda + r(t)$ in a wearout failure period. Then, the expected cost is [26]

$$C(T, T_0) = \lambda c_1 + \frac{c_1 \int_0^T r(t)\, \mathrm{d}t + c_2}{T + T_0} \tag{20.22}$$

(c) If a failure occurs in an interval $(kT - T_\mathrm{d}, kT)(0 \leq T_\mathrm{d} \leq T)$, a unit is not repaired in this interval and is down for the interval from its failure to the replacement (see Section 20.2.2.2). Then, the expected cost is

$$C(T, T_\mathrm{d}) = \frac{1}{T}\Bigg\{c_1 \int_0^{T-T_\mathrm{d}} r(t)\, \mathrm{d}t + c_2 + c_3 \frac{1}{\bar{F}(T - T_\mathrm{d})} \int_{T-T_\mathrm{d}}^{T} [F(t) - F(T - T_\mathrm{d})]\, \mathrm{d}t\Bigg\} \tag{20.23}$$

In the above policy, we may not sometimes leave a failed unit as it is until the planned replacement time. To overcome this, we consider the following model. If a unit fails in $(T - T_\mathrm{d}, T)$ then it is replaced by a new one [86]. Tahara and

Nishida [87] termed this policy the (t, T) policy. The expected cost is

$$C(T, T_d) = \left\{c_1 \int_0^{T-T_d} r(t)\,dt + c_2 + c_4[F(T) - F(T-T_d)]/\bar{F}(T-T_d)\right\} \times \left\{T - T_d + \int_{T-T_d}^{T} \bar{F}(t)\,dt/\bar{F}(T-T_d)\right\}^{-1} \quad (20.24)$$

where c_4 is an additional replacement cost caused by failure. Note that $C(T, T)$ agrees with Equation 20.5. Phelps [88, 89] and Park and Yoo [90] proposed a modification of this model.

20.2.3.2 Replacement at N Variables

See Nakagawa [27]. Morimura [91] proposed a new replacement where a unit is replaced at the Nth failure. This is a modification of the periodic replacement, and would be useful in the case where the total operating time of a unit is not recorded or it is too time consuming and costly to replace it in operation. Therefore, this could be used in maintaining complex systems with a lot of equipment of the same type.

Recalling that failures occur by a non-homogeneous Poisson process with a mean-value function $R(t)$, we have

$$H_j(t) \equiv \Pr\{N_1(t) = j\} = \frac{[R(t)]^j}{j!} e^{-R(t)} \quad (j = 0, 1, 2, \ldots)$$

and hence the mean time to the Nth failure is, from Nakagawa and Kowada [29], $\sum_{j=0}^{N-1} \int_0^\infty H_j(t)\,dt$. Thus, the expected cost is

$$C(N) = \frac{(N-1)c_1 + c_2}{\sum_{j=0}^{N-1} \int_0^\infty H_j(t)\,dt} \quad (N = 1, 2, \ldots) \quad (20.25)$$

Suppose that $r(t)$ is continuous and strictly increasing. Then, if there exists a finite solution N^* that satisfies

$$\frac{1}{\int_0^\infty H_N(t)\,dt} \sum_{j=0}^{N-1} \int_0^\infty H_j(t)\,dt - (N-1) \geq \frac{c_2}{c_1} \quad (N = 1, 2, \ldots) \quad (20.26)$$

it is unique and minimizes $C(N)$. Further, if $r(\infty) > \lambda c_2/c_1$, then a finite solution to Equation 20.26 exists.

Further, suppose that a unit is replaced at time T or at the Nth failure, whichever occurs first. Then, the expected cost is

$$C(N, T) = \left\{c_1\left[N - 1 - \sum_{j=0}^{N-1}(N-1-j)H_j(T)\right] + c_2 \sum_{j=N}^{\infty} H_j(T) + c_3 \sum_{j=0}^{N-1} H_j(T)\right\} \times \left\{\sum_{j=0}^{N-1} \int_0^T H_j(t)\,dt\right\}^{-1}$$

where c_2 is the cost of planned replacement at the Nth failure and c_3 is the cost of planned replacement at time T. Similar replacement policies were discussed by Park [92], Nakagawa [30], Tapiero and Ritchken [93], Lam [94–96], Ritchken and Wilson [97], and Sheu [98].

20.2.4 Other Replacement Models

20.2.4.1 Replacements with Discounting

When we adopt the total expected cost as an appropriate objective function for an infinite time span, we have to evaluate the present values of all replacement costs. A continuous discounting with rate $\alpha (0 < \alpha < \infty)$ will be used for costs at the replacement time. Then, the cost of one cycle starting at time t is

$$c_1\, e^{-\alpha(t+X_k)} I_{(X_k<T)} + c_2\, e^{-\alpha(t+T)} I_{(X_k \geq T)}$$

The cost at each cycle is the same, except for a discount rate, and hence the total cost is equal to the sum of discounted costs incurred on the individual cycles. Thus, we have [13] that the expected cost of an age replacement is

$$C(T, \alpha) = \frac{c_1 \int_0^T e^{-\alpha t}\, dF(t) + c_2\, e^{-\alpha T} \bar{F}(T)}{\alpha \int_0^T e^{-\alpha t} \bar{F}(t)\, dt} \tag{20.27}$$

and Equations 20.6 and 20.7 are

$$r(T) \int_0^T e^{-\alpha t} \bar{F}(t)\, dt - \int_0^T e^{-\alpha t}\, dF(t) = \frac{c_2}{c_1 - c_2} \tag{20.28}$$

and

$$C(T^*; \alpha) = \frac{1}{\alpha}(c_1 - c_2) r(T^*) - c_2 \tag{20.29}$$

respectively.

Similarly, for a block replacement:

$$C(T; \alpha) = \frac{c_1 \int_0^T e^{-\alpha t}\, dM(t) + c_2\, e^{-\alpha T}}{1 - e^{-\alpha T}} \tag{20.30}$$

$$\frac{1 - e^{-\alpha T}}{\alpha} m(T) - \int_0^T e^{-\alpha t} m(t)\, dt = \frac{c_2}{c_1} \tag{20.31}$$

$$C(T^*; \alpha) = \frac{c_1}{\alpha} m(T^*) - c_2 \tag{20.32}$$

For a periodic replacement:

$$C(T; \alpha) = \frac{c_1 \int_0^T e^{-\alpha t} r(t)\, dt + c_2\, e^{-\alpha T}}{1 - e^{-\alpha T}} \tag{20.33}$$

$$\frac{1 - e^{-\alpha T}}{\alpha} r(T) - \int_0^T e^{-\alpha t} r(t)\, dt = \frac{c_2}{c_1} \tag{20.34}$$

$$C(T^*; \alpha) = \frac{c_1}{\alpha} r(T^*) - c_2 \tag{20.35}$$

Note that $\lim_{\alpha \to 0} \alpha C(T; \alpha) = C(T)$, which is the expected cost of three replacements with no discounting.

20.2.4.2 Discrete Replacement Models

See Nakagawa and coworkers [18, 33]. In failure studies, the time to unit failure is often measured by the number of cycles to failure. In actual situations, the tires on jet fighters are replaced preventively at 4–14 flights, which may depend on the kinds of use. In other cases, the life times are sometimes not recorded at the exact instant of failure and are collected statistically per day, per month, or per year. In any case, it would be interesting and possibly useful to consider discrete time processes [99].

Consider the time over an infinitely long cycle n $(n = 1, 2, \dots)$ that a unit has to be operating. Let $\{p_n\}_{n=1}^{\infty}$ be the discrete time failure distribution with finite mean $1/\lambda \equiv \sum_{n=1}^{\infty} n p_n$ that a unit fails at cycle n. Further, let $\{p_n^{(j)}\}_{n=1}^{\infty}$ be the jth convolution of p_n, *i.e.* $p_n^{(1)} \equiv p_n$ $(n = 1, 2, \dots)$, $p_n^{(j)} \equiv \sum_{i=1}^{n-1} p_i p_{n-i}^{(j-1)}$ $(j = 2, 3, \dots, n;\ n = 2, 3, \dots)$, $p_n^{(j)} \equiv 0$ $(j = n+1, n+2, \dots)$, let $r(n) \equiv p_n / \sum_{j=n}^{\infty} p_j$ $(n = 1, 2, \dots)$ be the failure rate, and let $m(n) \equiv \sum_{j=1}^{n} p_n^{(j)}$ $(n = 1, 2, \dots)$ be the renewal density of the discrete time distribution.

In an age replacement, a unit is replaced at cycle $N (1 \le N \le \infty)$ after its installation or at failure, whichever occurs first. Then, the expected cost for an infinite time span is

$$C(N) = \frac{c_1 \sum_{j=1}^{N} p_j + c_2 \sum_{j=N+1}^{\infty} p_j}{\sum_{i=1}^{N} \sum_{j=i}^{\infty} p_j} \quad (N = 1, 2, \dots) \tag{20.36}$$

Theorem 2 can be rewritten as follows.

Theorem 2. *Suppose that $r(n)$ is strictly increasing and $r(\infty) \equiv \lim_{n \to \infty} r(n) \le 1$.*

(i) If $r(\infty) > K$ then there exists a finite and unique minimum that satisfies

$$r(N+1) \sum_{i=1}^{N} \sum_{j=i}^{\infty} p_j - \sum_{j=1}^{N} p_j \ge \frac{c_2}{c_1 - c_2} \quad (N = 1, 2, \dots) \tag{20.37}$$

(ii) If $r(\infty) \le K$ then $N^* = \infty$, and $C(\infty) = \lambda c_1$.

In a block replacement, a unit is replaced at cycle kN $(k = 1, 2, \dots)$ independent of its age and a failed unit is replaced by a new one between planned replacements. Then, the expected cost is

$$C(N) = \frac{1}{N}\left[c_1 \sum_{j=1}^{N} m(j) + c_2\right] \quad (N = 1, 2, \dots) \tag{20.38}$$

and Equation 20.9 is

$$Nm(N+1) - \sum_{j=1}^{N} m(j) \geq \frac{c_2}{c_1} \quad (N = 1, 2, \ldots) \tag{20.39}$$

Example 1. Suppose that the failure time of a unit has a negative binomial distribution with a shape parameter 2, *i.e.* $p_n = np^2q^{n-1}$ $(n = 1, 2, \ldots; 0 < p < 1)$ where $q \equiv 1 - p$. Then, we have $1/\lambda = (1+q)/p$, $r(n) = np^2/(np+q)$, and $m(n) = p^2 \sum_{j=0}^{n-1} q^{2j}$.

In an age replacement, if $q/(1+q) \leq c_2/c_1$, then we should make no planned replacement; conversely, if $q/(1+q) > c_2/c_1$, then we should adopt an optimum replacement cycle N^* that is a unique minimum such that

$$\frac{2(N+1)p + q^{N+2}}{Np+q} \geq \frac{c_1}{c_1 - c_2} \tag{20.40}$$

Note that if $c_2 < c_1 \leq 2c_2$ then a unit should be replaced only at failure, and $C(\infty) = pc_1/(1+q)$.

In a block replacement, from Equation 20.39

$$p^2 \sum_{j=1}^{N} jq^{2j} \geq \frac{c_2}{c_1} \tag{20.41}$$

Thus, if $q/(1+q) > \sqrt{c_2/c_1}$ then there exists a finite and unique minimum N^* that satisfies Equation 20.41, since the left-hand side of Equation 20.41 is strictly increasing from $(p/q)^2$ to $[q/(1+q)]^2$. Conversely, if $q/(1+q) \leq \sqrt{c_2/c_1}$ then a unit should be replaced only at failure. Note that for $c_2/c_1 \geq 1/4$, no finite value of N satisfies Equation 20.41.

In a periodic replacement, a unit is replaced at cycle $kN (k = 1, 2, \ldots)$ and a failed unit between planned replacements undergoes only minimal repair. Then, the expected cost is

$$C(N) = \frac{1}{N}\left[c_1 \sum_{j=1}^{N} r(j) + c_2\right] \quad (N = 1, 2, \ldots) \tag{20.42}$$

and Equation 20.16 is

$$Nr(N+1) - \sum_{j=1}^{N} r(j) \geq \frac{c_2}{c_1} \quad (N = 1, 2, \ldots) \tag{20.43}$$

Example 2. Suppose that the failure time has a discrete Weibull distribution with a shape parameter 2, *i.e.* $p_n = q^{(n-1)^2} - q^{n^2}$ $(n = 1, 2, \ldots; \quad 0 < q < 1)$ [100]. Then, $r(n) = 1 - q^{2n-1}$ $(n = 1, 2, \ldots)$, which is strictly increasing from $1 - q$ to 1. Thus, from Equation 20.43, if $q/(1-q^2) > c_2/c_1$ then there exists a finite and unique minimum N^* that satisfies

$$\frac{q}{1-q^2}\{1 - [1 + N(1-q^2)]q^{2N}\} \geq \frac{c_2}{c_1} \tag{20.44}$$

and, otherwise, a unit undergoes only minimal repair at failures.

20.2.4.3 Replacements with Two Types of Unit

See Nakagawa [35]. Most systems consist of vital and non-vital parts or essential and non-essential components. If vital parts fail then a system becomes dangerous or suffers a high cost. It would be wise to make the planned replacement or overhaul at suitable times. We may classify failures into two types; partial and total failures, slight and serious failures, or simply faults and failures. Several authors [101–107] have studied the replacement policies for systems with two types of unit. Beichelt and coworkers [108, 109], Murthy and Maxwell [110], Berg [111], Block *et al.* [112, 113], and Sheu [66] proposed generalized replacement models with two types of failure.

We consider a system with unit 1 and unit 2 which operate independently, where unit 1 corresponds to non-vital parts and unit 2 to vital parts. It is assumed that unit 1 is always replaced together with unit 2. Unit i has a failure time distribution $F_i(t)$, failure rate $r_i(t)$, and cumulative hazard $R_i(t) (i = 1, 2)$, *i.e.* $\bar{F}_i(t) = \exp[-R_i(t)]$ and $R_i(t) = \int_0^t r_i(u)\,du$. Then, we consider the following four replacement policies, which combine age, block, and periodic replacements.

Case (a). Unit 2 is replaced at failure or time T, whichever occurs first, and when unit 1 fails between replacements, it is replaced by a new unit. Then, the expected cost is

$$C(T) = \frac{c_1 \int_0^T m_1(t)\bar{F}_2(t)\,\mathrm{d}t + c_2 F_2(T) + c_3}{\int_0^T \bar{F}_2(t)\,\mathrm{d}t} \tag{20.45}$$

where c_1 is the cost of replacement for a failed unit 1, c_2 is the additional replacement for a failed unit 2, and c_3 is the cost of replacement for units 1 and 2.

Case (b). In case (a), when unit 1 fails between replacements it undergoes only minimal repair. Then, the expected cost is

$$C(T) = \frac{c_1 \int_0^T r_1(t)\bar{F}_2(t)\,\mathrm{d}t + c_2 F_2(T) + c_3}{\int_0^T \bar{F}_2(t)\,\mathrm{d}t} \tag{20.46}$$

where c_1 is the cost of minimal repair for failed unit 1, and c_2 and c_3 are the same costs as case (a).

Case (c). Unit 2 is replaced at periodic times kT $(k = 1, 2, \dots)$ and undergoes only minimal repair at failures between planned replacements, and when unit 1 fails between replacements it is replaced by a new unit. Then, the expected cost is

$$C(T) = \frac{1}{T}[c_1 M_1(T) + c_2 R_2(T) + c_3] \tag{20.47}$$

where c_2 is the cost of minimal repair for a failed unit 2, and c_1 and c_3 are the same costs as case (a).

Case (d). In case (c), when unit 1 fails between replacements it also undergoes minimal repair. Then, the expected cost is

$$C(T) = \frac{1}{T}[c_1 R_1(T) + c_2 R_2(T) + c_3] \tag{20.48}$$

where c_1 is the cost of minimal repair for a failed unit 1, and c_2 and c_3 are the same costs as case (c).

We can similarly discuss optimum replacement policies that minimize the expected costs $C(T)$ in Equations 20.45–20.48. Further, we can easily extend these policies to the replacements of a system with one main unit and several sub-units.

20.2.4.4 Replacement of a Shock Model

We summarize briefly the replacement policies for a shock model. Shocks occur randomly in time at a stochastic process and give the amount of damage to a system. This damage accumulates and gradually weakens the system. A system fails when the total damage has exceeded a failure level.

The general concept of such processes was due to Smith [114]. Mercer [115] considered the model where shocks occur in a Poisson distribution and the amount of damage due to each shock has a gamma distribution. Cox [37], Esary *et al.* [116], and Nakagawa and Osaki [117] investigated the various properties of failure time distributions.

Suppose that a system is replaced at failure by a new one. It may be wise to exchange a system before failure at a smaller cost. Taylor [118] and Feldman [119–121] derived the optimal control-limit policies where a system is replaced before failure when the total damage has exceeded a threshold level. On the other hand, Zuckerman [122–125] proposed the replacement model where a system is replaced before failure at time T. Recently, Wortman *et al.* [126] and Sheu and coworkers [127–129] studied some replacement models of a system subject to shocks. This section is based on Nakagawa [15, 33], Satow *et al.* [130] and Qian *et al.* [131].

Consider a unit that has to operate for an infinite time span. Shocks occur by a non-homogeneous Poisson process with an intensity function $r(t)$ and a mean-value function $R(t)$, *i.e.* $R(t) \equiv \int_0^t r(u)\,\mathrm{d}u$. Thus, the probability that j shocks occur during $(0, t]$ is $H_j(t) \equiv \{[R(t)]^j/j!\}\,\mathrm{e}^{-R(t)}$ $(j = 0, 1, 2, \dots)$. Further, the amount W_j of damage to the jth shock has an identical distribution $G(x) \equiv \Pr\{W_j \le x\}$ $(j = 1, 2, \dots)$. A unit fails when the total damage has exceeded a failure level K.

A unit is replaced at time $T (0 < T \le \infty)$ or at failure, whichever occurs first. Let c_1 and $c_2 (< c_1)$ be the respective replacement costs at failure and at time T. Then, the expected cost per unit of time for an infinite span is

$$C(T) = \left\{c_1 \sum_{j=0}^{\infty}[G^{(j)}(K) - G^{(j+1)}(K)] \times \int_0^T H_j(t) r(t)\,\mathrm{d}t + c_2 \sum_{j=0}^{\infty} G^{(j)}(K) H_j(T)\right\} \times \left\{\sum_{j=0}^{\infty} G^{(j)}(K) \int_0^T H_j(t)\,\mathrm{d}t\right\}^{-1} \quad (20.49)$$

where $\Phi^{(j)}(x)$ $(j = 1, 2, \ldots)$ is the j-fold Stieltjes convolution of $\Phi(x)$ and $\Phi^{(0)}(x) \equiv 1$ for $x \geq 0$.

When successive times between shocks have an identical distribution $F(t)$, the expected cost in Equation 20.49 is rewritten as

$$C(T) = \left\{c_1 \sum_{j=0}^{\infty}[G^{(j)}(K) - G^{(j+1)}(K)] F^{(j+1)}(T) + c_2 \sum_{j=0}^{\infty} G^{(j)}(K)[F^{(j)}(T) - F^{(j+1)}(T)]\right\} \times \left\{\sum_{j=0}^{\infty} G^{(j)}(K) \times \int_0^T [F^{(j)}(t) - F^{(j+1)}(t)]\,\mathrm{d}t\right\}^{-1} \quad (20.50)$$

Next, a unit is replaced before failure at shock N $(N = 1, 2, \ldots)$. Then, the expected cost is

$$C(N) = \frac{c_1[1 - G^{(N)}(K)] + c_2 G^{(N)}(K)}{\sum_{j=0}^{N-1} G^{(j)}(K) \int_0^{\infty} H_j(t)\,\mathrm{d}t} \quad (N = 1, 2, \ldots) \quad (20.51)$$

Further, a unit is replaced preventively when the total damage has exceeded a threshold level $Z (0 \leq Z \leq K)$. Then, the expected cost is

$$C(Z) = \left[c_1 - (c_1 - c_2)\left\{G(K) - \int_0^Z [1 - G(K - x)]\,\mathrm{d}M(x)\right\}\right] \times \left[\sum_{j=0}^{\infty} G^{(j)}(Z) \int_0^{\infty} H_j(t)\,\mathrm{d}t\right]^{-1} \quad (20.52)$$

where $M(x) \equiv \sum_{j=1}^{\infty} G^{(j)}(x)$.

By the similar methods of obtaining optimum replacement policies, we can discuss the optimum T^*, N^*, and Z^* analytically that minimize $C(T)$, $C(N)$, and $C(Z)$.

20.2.5 Remarks

It has been assumed that the time required for replacement is negligible, and at any time there is an unlimited supply of units available for replacement. The results and methods in this section could be theoretically extended and modified in such cases, and be useful for actual replacements in practical fields.

In general, the results of block and periodic replacements are summarized as follows: The expected cost is

$$C(T) = \frac{1}{T}\left[c_1 \int_0^T \varphi(t)\,\mathrm{d}t + c_2\right] \quad (20.53)$$

where $\varphi(t)$ is $m(t)$, $F(t)$, and $r(t)$ respectively. Differentiating $C(T)$ with respect to T and setting it equal to zero:

$$T\varphi(T) - \int_0^T \varphi(t)\,\mathrm{d}t = \frac{c_2}{c_1} \quad (20.54)$$

and if a solution T^* to Equation 20.54 exists, then the expected cost is

$$C(T^*) = c_1 \varphi(T^*) \quad (20.55)$$

In a discount case:

$$C(T; \alpha) = \frac{c_1 \int_0^T \mathrm{e}^{-\alpha t} \varphi(t)\,\mathrm{d}t + c_2\,\mathrm{e}^{-\alpha T}}{1 - \mathrm{e}^{-\alpha T}} \quad (20.56)$$

$$\frac{1-e^{-\alpha T}}{\alpha}\varphi(T) - \int_0^T e^{-\alpha t}\varphi(t)\,dt = \frac{c_2}{c_1} \quad (20.57)$$

$$C(T^*;\alpha) = \frac{c_1}{\alpha}\varphi(T^*) - c_2 \quad (20.58)$$

20.3 Preventive Maintenance Policies

A unit is repaired upon failure. If a failed unit undergoes repair, it needs a repair time that may not be negligible. After the completion of repair, a unit begins to operate again. A unit represents the most fundamental unit which repeats up and down alternately. Such a unit forms an alternating renewal process [37] and a Markov renewal process with two states [132, 133].

When a unit is repaired after failure, it may require much time and high cost. In particular, the downtime of computers and radars should be made as short as possible by decreasing the number of unit failures. In this case, we need to maintain a unit to prevent failures, but not to do it too often from the viewpoint of reliability or cost.

A unit has to be operating for an infinite time span. We define that X_k and Y_k ($k = 1, 2, \dots$) denote the uptime and the downtime respectively, and $A(t)$ is the probability that a unit is operating at time t. Then, from Hosford [134] and Barlow and Proschan [1], $A(t)$ is called the *pointwise availability* and $\int_0^t A(u)\,du/t$ is the *interval availability*. Further, we have

$$A \equiv \lim_{t\to\infty}\frac{1}{t}\int_0^t A(u)\,du = \lim_{t\to\infty} A(t) = \frac{E\{X_k\}}{E\{X_k\}+E\{Y_k\}} \quad (20.59)$$

which is called the *steady-state availability* and is well known as the usual and standard definition of availability. For an infinite time span, an appropriate objective function is the availability A in Equation 20.59.

Morse [135] first derived the optimum PM policy that maximizes the steady-state availability. Barlow and Proschan [1] pointed out that this problem is reduced to an age replacement one if the mean time to repair is replaced by the replacement cost of a failed unit and the mean time to PM by the cost of exchanging a non-failed unit. Optimum PM policies for more general systems were discussed by Nakagawa [17], Sherif and Smith [136], Jardine and Buzacott [137], and Reineke *et al.* [138]. Liang [139] considered the PM policies for series systems by modifying the opportunistic replacement policies. Aven [140] and Smith and Dekker [141] treated the PM of a system with spare units. Silver and Fiechter [142] considered the PM model where the failure time distribution is uncertain. Further, Van der Duyn Schouten and Scarf [143] presented several maintenance models in Europe and gave a good survey of applied PM models. Chockie and Bjorkelo [144], Smith [145], and Susova and Petrov [146] gave the PM programs of plant and aircraft.

We consider the following PM policies: (1) PM of a one-unit system; (2) PM of a two-unit system; (3) imperfect PM; and (4) modified PM. We discuss optimum PM policies that maximize the availabilities or minimize the expected costs of each model, and summarize their derivation results.

20.3.1 One-unit System

When a unit fails, it undergoes repair immediately, and once repaired it is returned to the operating state. It is assumed that the failure time X is independent and has an identical distribution $F(t)$ with finite mean $1/\lambda$, and the repair time Y_1 is also independent and has an identical distribution $G_1(t)$ with finite mean $1/\mu_1$.

If the operating time of a unit is always known and its failure rate increases with time, it may be wise to maintain preventively it at time T before failure on its operating time. We call time T the planned PM time. The distribution of time Y_2 to PM completion is $G_2(t)$ with finite mean $1/\mu_2$, which may be smaller than the repair time Y_1.

A new unit begins to operate at $t = 0$. We define one cycle as being from the beginning of operation to the completion of PM or repair. Then, the mean time of downtime of one cycle is

$$E\{Y_1 I_{(X<T)} + Y_2 I_{(X\geq T)}\} = \frac{1}{\mu_1}F(T) + \frac{1}{\mu_2}\bar{F}(T) \tag{20.60}$$

and from Equation 20.4, the mean time of uptime of one cycle is $\int_0^T \bar{F}(t)\,\mathrm{d}t$, where $\bar{F} \equiv 1 - F$.

Therefore, from Equation 20.59, the steady-state availability is

$$A(T) \equiv \frac{\text{Mean time of uptime}}{\text{Mean time of one cycle}} = \frac{\int_0^T \bar{F}(t)\,\mathrm{d}t}{\int_0^T \bar{F}(t)\,\mathrm{d}t + \frac{1}{\mu_1}F(T) + \frac{1}{\mu_2}\bar{F}(T)} \tag{20.61}$$

Thus, the policy maximizing the availability is the same one as minimizing the expected cost $C(T)$ in Equation 20.5, as pointed out by Barlow and Proschan [1].

20.3.1.1 Interval Reliability

See Mine and Nakagawa [16]. Barlow and Proschan [1] defined that *interval reliability* $R(x, T_0)$ is the probability that, at a specified time T_0, a system is operating and will continue to operate for an interval of time x. The interval reliability is simply called reliability when $T_0 = 0$; furthermore, it becomes pointwise availability at time T_0 as $x \to 0$. Thus, the interval reliability is an important measure from the viewpoints of both reliability and availability.

Consider the PM of the above one-unit system, where a unit is repaired at failure or is maintained preventively at time T, whichever occurs first. However, the PM of an operating unit is not made during the interval $[T_0, T_0 + x]$ even if the time for PM comes. It is assumed that the distribution of the time to pm is the same as the repair time distribution $G(t)$ with finite mean $1/\mu$, for the simplicity of computations.

We set the PM time T to an operating unit and obtain the interval reliability $R(T; x, T_0)$ by a method similar to Barlow and Proschan [1, p.82]. Let $D(t)$ be the distribution of a degenerate random variable placing unit mass at T, *i.e.* $D(t) \equiv 0$ for $t < T$, and $D(t) \equiv 1$ for $t \geq T$. Then, we have

$$R(T; x, T_0) = \bar{F}(T_0 + x)\bar{D}(T_0) + \int_0^{T_0} \bar{F}(T_0 + x - u)\bar{D}(T_0 - u)\,\mathrm{d}M_{11}(u) \tag{20.62}$$

where $\bar{D} \equiv 1 - D$, $M_{11}(t)$ represents the expected number of occurrences of a completed repair during $(0, t]$, and from an elementary renewal theorem (*e.g.* theorem 2.9 of Barlow and Proschan [1, p.55]):

$$\lim_{t\to\infty} \frac{M_{11}(t)}{t} = \frac{1}{\int_0^T \bar{F}(t)\,\mathrm{d}t + \frac{1}{\mu}}$$

in which the denominator is equal to the mean time of one cycle.

Thus, the limiting interval reliability is

$$R(T; x) \equiv \lim_{T_0\to\infty} R(T; x, T_0) = \frac{\int_x^{T+x} \bar{F}(t)\,\mathrm{d}t}{\int_0^T \bar{F}(t)\,\mathrm{d}t + \frac{1}{\mu}} \tag{20.63}$$

We seek an optimum time T^* that maximizes $R(T; x)$ in Equation 20.63 for a fixed $x > 0$. Let $H(t) \equiv [F(t + x) - F(t)]/\bar{F}(t)$ for $t, x \geq 0$ and $F(t) < 1$. Then, both $H(t)$ and $r(t) \equiv f(t)/\bar{F}(t)$ are called the failure rates and have the same properties (Barlow and Proschan [1, p.23]). Then, we have the following similar theorem to Theorem 1.

Theorem 3. *Suppose that $H(t)$ is continuous and strictly increasing, where*

$$K(x) \equiv \left[\int_0^x \bar{F}(t)\,\mathrm{d}t + 1/\mu\right] \Big/ (1/\lambda + 1/\mu).$$

(i) *If $H(\infty) > K(x)$ then there exists a finite and unique $T^*(0 < T^* < \infty)$ that satisfies*

$$H(T)\left[\int_0^T \bar{F}(t)\,\mathrm{d}t + \frac{1}{\mu}\right] - \int_0^T [\bar{F}(t) - \bar{F}(t + x)]\,\mathrm{d}t = \frac{1}{\mu} \tag{20.64}$$

and the resulting interval reliability is

$$R(T^*; x) = 1 - H(T^*) \tag{20.65}$$

(ii) If $H(\infty) \leq K(x)$ then the optimum PM time is $T^* = \infty$, and

$$R(\infty; x) = \frac{\int_x^\infty \bar{F}(t)\,dt}{(1/\lambda + 1/\mu)}$$

If the time for PM has $G_2(t)$ with mean $1/\mu_2$ and the time for repair has $G_1(t)$ with mean $1/\mu_1$, then the limiting interval reliability is easily given by

$$R(T; x) = \frac{\int_x^{T+x} \bar{F}(t)\,dt}{\int_0^T \bar{F}(t)\,dt + \frac{1}{\mu_1} F(T) + \frac{1}{\mu_2} \bar{F}(T)} \tag{20.66}$$

which agrees with Equation 20.61 as $x \to 0$, and with Equation 20.63 when $1/\mu_1 = 1/\mu_2 = 1/\mu$.

Example 3. Suppose that the failure time of a unit has a gamma distribution with order 2, *i.e.* $f(t) = \beta^2 t\, e^{-\beta t}$. Then, we have

$$H(t) = 1 - e^{-\beta x} - \frac{\beta x}{1 + \beta t} e^{-\beta x}$$

$$K(x) = \frac{(2/\beta)(1 - e^{-\beta x}) - x\, e^{-\beta x} + 1/\mu}{2/\beta + 1/\mu}$$

The failure rate $H(t)$ is strictly increasing from $H(0) = F(x)$ to $H(\infty) = 1 - e^{-\beta x}$. Thus, from Theorem 3, if $x > 1/\mu$ then the optimum time T^* is a unique solution of

$$\beta T \left(x - \frac{1}{\mu}\right) - x(1 - e^{-\beta T}) = \frac{1 + \beta x}{\mu}$$

and the interval reliability is

$$R(T^*; x) = \frac{1 + \beta(T^* + x)}{1 + \beta T^*} e^{-\beta x}$$

20.3.2 Two-unit System

A two-unit standby redundant system with a single repairman is one of the most fundamental and important redundant systems in reliability theory. A system consists of two units where one unit is operating and the other is in standby as an initial condition. If an operating unit fails, then it undergoes repair immediately and the other standby unit takes over its operation. Either of two units is alternately operating. It can be said that a system failure occurs when two units are down simultaneously.

Gaver [147] obtained the distribution of time to system failure and its mean time for the model with exponential failure and general repair times. Gnedenko [148, 149] and Srinivasan [150] extended the results for the model with both general failure and repair times. Further, Osaki [151] obtained the same results by using a signal-flow graph method.

PM policies for a two-unit standby system have been studied by many authors. Rozhdestvenskiy and Fanarzhi [152] obtained one method of approximating the optimum PM policy that minimizes the mean time to system failure. Osaki and Asakura [153] showed that the mean time to system failure of a system with PM is greater than that with only repair maintenance under suitable conditions. Berg [154–156] considered the replacement policy for a two-unit system where, at the failure points of one unit, the other unit is replaced if its age exceeds a control limit.

Consider a two-unit standby system where two units are statistically identical. An operating unit has a failure time distribution $F(t)$ with finite mean $1/\lambda$ and a failed unit has a repair time distribution $G_1(t)$ with finite mean $1/\mu_1$. When an operating unit operates for a specified time T without failure, we stop its operation. The time for PM has a general distribution $G_2(t)$ with finite mean $1/\mu_2$. Further, we make the following three assumptions:

1. A unit is as good as new upon repair or PM completion.
2. PM of an operating unit is done only if the other unit is in standby.
3. An operating unit, which forfeited PM because of (1), undergoes PM just upon repair or PM completion of the other unit.

Under the above assumptions, the steady-state availability is [11]

$$A(T) = \left\{\left[1/\gamma_1 + \int_0^T \bar{F}(t)G_1(t)\,dt\right] \times \left[1 - \int_T^\infty G_2(t)\,dF(t)\right] + \left[1/\gamma_2 + \int_0^T \bar{F}(t)G_2(t)\,dt\right] \times \int_T^\infty G_1(t)\,dF(t)\right\} \times \left\{\left[1/\mu_1 + \int_0^T \bar{F}(t)G_1(t)\,dt\right] \times \left[1 - \int_T^\infty G_2(t)\,dF(t)\right] + \left[1/\mu_2 + \int_0^T \bar{F}(t)G_2(t)\,dt\right] \times \int_T^\infty G_1(t)\,dF(t)\right\}^{-1} \quad (20.67)$$

where $\bar{F} \equiv 1 - F$ and $1/\gamma_i \equiv \int_0^\infty \bar{F}(t)\bar{G}_i(t)\,dt$ $(i = 1, 2)$.

When an operating unit undergoes PM immediately upon repair or PM completion, *i.e.* $T = 0$, the availability is

$$A(0) = \frac{\theta_2/\gamma_1 + (1-\theta_1)/\gamma_2}{\theta_2/\mu_1 + (1-\theta_1)/\mu_2} \quad (20.68)$$

where $\theta_i \equiv \int_0^\infty \bar{G}_i(t)\,dF(t)$ $(i = 1, 2)$. When no PM is done, *i.e.* $T = \infty$, the availability is

$$A(\infty) = \frac{1/\lambda}{1/\lambda + 1/\mu_1 - 1/\gamma_1} \quad (20.69)$$

We find an optimum planned PM time T^* that maximizes the availability $A(T)$ in Equation 20.67. It is assumed that $G_1(t) < G_2(t)$ for $0 < t < \infty$. That is, the probability that the repair is completed up to time t is less than the probability that the PM is completed up to time t. Let $r(t) \equiv f(t)/\bar{F}(t)$, where $f(t)$ is a density of F.

Theorem 4. *Suppose that $G_1(t) < G_2(t)$ for $0 < t < \infty$, and $r(t)$ is continuous and strictly increasing.*

(i) *If $r(\infty) > K$, $\beta_1\mu_1 > \beta_2\mu_2$ and $r(0) < k$, or $r(\infty) > K$ and $\beta_1\mu_1 \le \beta_2\mu_2$, then there exists a finite and unique $T^*(0 < T^* < \infty)$ that satisfies*

$$r(T)\left[\int_0^T \bar{F}(t)G(t)\,dt + \int_0^\infty \bar{G}(t)\,dt\right] - \int_0^T G(t)\,dF(t) = \frac{\beta_1 \int_0^\infty \bar{G}(t)\,dF(t) + \beta_2 \int_0^\infty G(t)\,dF(t)}{\beta_1 - \beta_2} \quad (20.70)$$

and the resulting availability is

$$A(T^*) = 1 - (1/\mu_1 - 1/\gamma_1) \times \left\{1/\mu_1 - 1/\gamma_1 + 1/\lambda + \int_{T^*}^\infty G_1(t)\,dF(t)/r(T^*) - \int_{T^*}^\infty \bar{F}(t)G_1(t)\,dt\right\}^{-1} \quad (20.71)$$

(ii) *If $r(\infty) \le K$ then the optimum PM time is $T^* = \infty$,* i.e. *no PM is done, and the availability is given by Equation 20.69.*

(iii) *If $\beta_1\mu_1 > \beta_2\mu_2$ and $r(0) \ge k$ then the optimum PM time is $T^* = 0$, and the availability is given by Equation 20.68 where*

$$\beta_i \equiv \int_0^\infty F(t)\bar{G}_i(t)\,dt \qquad (i = 1, 2)$$

$$G(t) \equiv \frac{\beta_1 G_2(t) - \beta_2 G_1(t)}{\beta_1 - \beta_2}$$

$$k \equiv \frac{\beta_1\theta_2 + \beta_2(1-\theta_1)}{\beta_1/\mu_1 - \beta_2/\mu_1}$$

$$K \equiv \frac{\lambda\beta_1}{\beta_1 - \beta_2}$$

20.3.3 Imperfect Preventive Maintenance

Most PM models have assumed that a unit after PM is as good as new. Actually, this assumption may not be true. A unit after PM usually may be younger at PM, and occasionally it may be worse than before PM because of faulty procedures. Generally, the improvement of a unit by PM would depend on the resources spent for PM.

Weiss [157] first assumed that the inspection to detect failures may not be perfect. Chan and Downs [158], Nakagawa [20, 24], and Murthy and Nguyen [159] considered the imperfect PM where a unit after PM is not like new and discussed the optimum policies that maximize the availability or minimize the expected cost. Further, Lie and Chun [160] and Jayabalan and Chaudhuri [161] introduced an improvement factor in failure rate or age after maintenance, and Canfield [162] considered a system degradation with time where the PM restores the hazard function to the same shape.

Brown and Proschan [163], Fontenot and Proschan [164], and Bhattacharjee [165] assumed that a failed unit is as good as new with a certain probability and investigated some properties of a failure time distribution. Similar imperfect repair models were studied by Ebrahimi [166], Kijima *et al.* [167], Natvig [168], Stadje and Zuckerman [169], and Makis and coworkers [170, 171]. Recently, Wang and Pham [172, 173] considered extended PM models with imperfect repair and discussed the optimum policies that minimize the expected cost and maximize the availability.

A unit begins to operate at time zero and has to be operating for an infinite time span [36]. When a unit fails, it is repaired immediately and its mean repair time is $1/\mu_1$. To prevent failures, a unit undergoes PM at periodic times kT ($k = 1, 2, \ldots$), where the PM time is negligible. Then, one of the following three cases after PM results.

(a) A unit is not changed with probability p_1, *i.e.* PM is imperfect.
(b) A unit is as good as new with probability p_2, *i.e.* PM is perfect.
(c) A unit fails with probability p_3, *i.e.* PM becomes failure, where $p_1 + p_2 + p_3 \equiv 1$ and $p_2 > 0$. In this case, the mean repair time for PM failure is $1/\mu_2$.

The probability that a unit is renewed by repair upon actual failure is

$$\sum_{j=1}^{\infty} p_1^{j-1} \int_{(j-1)T}^{jT} \mathrm{d}F(t) = (1 - p_1) \sum_{j=1}^{\infty} p_1^{j-1} F(jT) \tag{20.72}$$

The probability that it is renewed by perfect PM is

$$p_2 \sum_{j=1}^{\infty} p_1^{j-1} \bar{F}(jT) \tag{20.73}$$

The probability that it is renewed by repair upon PM failure is

$$p_3 \sum_{j=1}^{\infty} p_1^{j-1} \bar{F}(jT) \tag{20.74}$$

where Equations 20.72 + 20.73 + 20.74 = 1.

The mean time until a unit is renewed by either repair or perfect PM is

$$\sum_{j=1}^{\infty} p_1^{j-1} \int_{(j-1)T}^{jT} t\,\mathrm{d}F(t) + (p_2 + p_3) \sum_{j=1}^{\infty} (jT) p_1^{j-1} \bar{F}(jT) = (1 - p_1) \sum_{j=1}^{\infty} p_1^{j-1} \int_0^{jT} \bar{F}(t)\,\mathrm{d}t \tag{20.75}$$

Thus, from Equations 20.59 and 20.61, the availability is

$$A(T) = \left\{(1 - p_1) \sum_{j=1}^{\infty} p_1^{j-1} \int_0^{jT} \bar{F}(t)\,\mathrm{d}t\right\} \times \left\{(1 - p_1) \sum_{j=1}^{\infty} p_1^{j-1} \int_0^{jT} \bar{F}(t)\,\mathrm{d}t + \frac{1 - p_1}{\mu_1} \sum_{j=1}^{\infty} p_1^{j-1} F(jT) + \frac{p_3}{\mu_2} \sum_{j=1}^{\infty} p_1^{j-1} \bar{F}(jT)\right\}^{-1} \tag{20.76}$$

which agrees with Equation 20.61 when $p_1 = 0$ and $p_3 = 1$.

Let

$$H(T;\, p_1) \equiv \frac{\sum_{j=1}^{\infty} p_1^{j-1} jf(jT)}{\sum_{j=1}^{\infty} p_1^{j-1} j\bar{F}(jT)}$$

Then, by a similar method to Theorem 1, we have the following theorem.

Theorem 5. *Suppose that $H(T;\, p_1)$ is continuous and strictly increasing, and $K \equiv \lambda(1-p_1)/p_2$.*

(i) If $H(\infty;\, p_1) > K$ and $(1-p_1)/\mu_1 > p_3/\mu_2$ then there exists a finite and unique T^* that satisfies

$$H(T;\, p_1)\sum_{j=1}^{\infty} p_1^{j-1}\int_0^{jT}\bar{F}(t)\,\mathrm{d}t - \sum_{j=1}^{\infty} p_1^{j-1}F(jT) = \frac{p_3}{1-p_1}\frac{\frac{1}{\mu_2}}{\frac{1-p_1}{\mu_1}-\frac{p_3}{\mu_2}} \tag{20.77}$$

and the resulting availability is

$$A(T^*) = \frac{1}{1+\left(\frac{1}{\mu_1}-\frac{p_3}{\mu_2(1-p_1)}\right)H(T^*)} \tag{20.78}$$

(ii) If $H(\infty;\, p_1) \le K$ or $(1-p_1)/\mu_1 \le p_3/\mu_2$ then the optimum time is $T^* = \infty$, and $A(\infty) = (1/\lambda)(1/\lambda + 1/\mu_1)$.

We adopt the expected cost per unit of time as an objective function and give the following two imperfect PM models.

20.3.3.1 Imperfect with Probability

An operating unit is repaired at failure or is maintained preventively at time T, whichever occurs first. Then, a unit after PM has the same age, *i.e.* the same failure rate, as it had before PM with probability p $(0 \le p < 1)$ and is as good as new with probability $q \equiv 1-p$. Then, the expected cost per unit of time is

$$C(T;\, p) \equiv \left\{c_1 q\sum_{j=1}^{\infty} p^{j-1}F(jT) + c_2\sum_{j=1}^{\infty} p^{j-1}\bar{F}(jT)\right\} \times \left\{\sum_{j=1}^{\infty} p^{j-1}\int_{(j-1)T}^{jT}\bar{F}(t)\,\mathrm{d}t\right\}^{-1} \tag{20.79}$$

where c_1 is the cost for repair and c_2 is the cost for PM.

Next, suppose that an operating unit is maintained preventively at times kT $(k = 1, 2, \dots)$ and undergoes only minimal repair at failures between PMs. Then the expected cost is

$$C(T;\, p) = \frac{1}{T}\left[c_1 q^2\sum_{j=1}^{\infty} p^{j-1}\int_0^{jT} r(t)\,\mathrm{d}t + c_2\right] \tag{20.80}$$

where c_1 is the cost for minimal repair and c_2 is the cost for PM.

20.3.3.2 Reduced Age

An operating unit is maintained preventively at times kT $(k = 1, 2, \dots)$ and undergoes only minimal repair at failures between PMs. Further, the age of a unit becomes x $(0 \le x \le T)$ units of younger at each PM, and it is replaced if it operates for the time interval NT $(N = 1, 2, \dots)$. Then, the expected cost per unit of time is

$$C(N;\, T, x) = \frac{1}{NT}\left[c_1\sum_{j=0}^{N-1}\int_{j(T-x)}^{T+j(T-x)} r(t)\,\mathrm{d}t + c_2 + (N-1)c_3\right] \tag{20.81}$$

where c_1 is the cost for minimal repair, c_2 is the cost for replacement at NT, and c_3 is the cost for PM.

In this model, if the age after PM reduces to at $(0 < a \le 1)$ when it was t before PM, the expected cost is

$$C(N;T,a)=\frac{1}{NT}\left[c_1\sum_{j=0}^{N-1}\int_{A_jT}^{(A_j+1)T} r(t)\,\mathrm{d}t + c_2+(N-1)c_3\right] \quad (20.82)$$

where $A_j \equiv a+a^2+\cdots+a^j$ $(j=1,2,\ldots)$ and $A_0 \equiv 0$.

The above model is extended to the following sequential PM policy. The PM is done at fixed intervals x_k $(k=1,2,\ldots,N-1)$ and is replaced at the Nth PM. Further, the age after the kth PM reduces to $a_k t$ when it was t before PM, where $0=a_0<a_1\le a_2\le\cdots\le a_N<1$. Then, the expected cost per unit of time is

$$C(x_1,x_2,\ldots,x_N) = \frac{c_1\sum_{k=1}^{N}\int_{a_{k-1}Y_{k-1}}^{Y_k} r(t)\,\mathrm{d}t + c_2+(N-1)c_3}{\sum_{k=1}^{N-1}(1-a_k)Y_k+Y_N} \quad (20.83)$$

where

$$Y_k \equiv x_k + a_{k-1}x_{k-1}+\cdots+a_{k-1}a_{k-2}\cdots a_2a_1x_1 \quad (k=1,2,\ldots)$$

20.3.4 Modified Preventive Maintenance

See Nakagawa [34]. We consider the following modified PM policy. Failures of a unit occur by a non-homogeneous Poisson process and the PM is made only at periodic times kT $(k=1,2,\ldots)$. If the total number of failures has exceeded a specified number N, then the PM should be made at the next planned time, otherwise no PM should be done. This policy was applied to the PM of hard disks by Sandoh *et al.* [174].

A unit has to operate for infinite time span and assume that failures occur by a non-homogeneous Poisson process with an intensity function $r(t)$ and a mean-value function $R(t)$, i.e. $R(t)\equiv\int_0^t r(u)\,\mathrm{d}u$. Then, the probability that j failures exactly occur during $(0,t]$ is $H_j(t)\equiv \{[R(t)]^j/j!\}\,\mathrm{e}^{-R(t)}$ $(j=0,1,2,\ldots)$. The PM is scheduled at times kT $(k=1,2,\ldots)$, and if the total number of failures has exceeded a specified number N, then the PM is made at the next PM time. Otherwise, a unit is left as it is. A unit undergoes minimal repair at each failure.

The probability that the PM is done at time $(k+1)T$ $(k=0,1,2,\ldots)$, because more than N failures have occurred during $(0,(k+1)T]$ when the number of failures was less than N until kT, is

$$\sum_{j=0}^{N-1} H_j[R(kT)]\sum_{i=N-j}^{\infty} H_i[R((k+1)T)-R(kT)] = \sum_{j=0}^{N-1}\{H_j[R(kT)]-H_j[R((k+1)T)]\}$$

Thus, the mean time to PM is

$$\sum_{k=0}^{\infty}[(k+1)T]\times\sum_{j=0}^{N-1}\{H_j[R(kT)]-H_j[R((k+1)T)]\} = T\sum_{k=0}^{\infty}\sum_{j=0}^{N-1}H_j[R(kT)] \quad (20.84)$$

Further, the expected number of failures until PM is

$$\sum_{k=0}^{\infty}\sum_{j=0}^{N-1}H_j[R(kT)]\times\sum_{i=N-j}^{\infty}(i+j)H_i[R((k+1)T)-R(kT)] = \sum_{k=0}^{\infty}[R((k+1)T)-R(kT)]\sum_{j=0}^{N-1}H_j[R(kT)] \quad (20.85)$$

Therefore, from Equations 20.84 and 20.85, the expected cost per unit of time is

$$C(N;\, T) = \left\{c_1 \sum_{k=0}^{\infty}[R((k+1)T) - R(kT)] \times \sum_{j=0}^{N-1} H_j[R(kT)] + c_2\right\} \times \left\{T \sum_{k=0}^{\infty}\sum_{j=0}^{N-1} H_j[R(kT)]\right\}^{-1} \tag{20.86}$$

where c_1 is the cost for minimal repair and c_2 is the cost for planned PM.

We seek an optimum number N^* that minimizes $C(N;\, T)$ in Equation 20.86 for a fixed $T > 0$. By a similar method to that of Section 20.2.3.2, we have the following results. From the inequality $C(N+1;\, T) > C(N;\, T)$:

$$q(N) \sum_{k=0}^{\infty}\sum_{j=0}^{N-1} H_j[R(kT)] - \sum_{k=0}^{\infty}[R((k+1)T) - R(kT)] \sum_{j=0}^{N-1} H_j[R(kT)] \geq \frac{c_2}{c_1} \quad (N = 1, 2, \ldots) \tag{20.87}$$

where

$$q(N) \equiv \left\{\sum_{k=0}^{\infty}[R((k+1)T) - R(kT)] H_N[R(kT)]\right\} \times \left\{\sum_{k=0}^{\infty} H_N[R(kT)]\right\}^{-1}$$

When $r(t)$ is strictly increasing, $q(N)$ is strictly increasing, and hence the left-hand side of Equation 20.87 is also strictly increasing. Thus, if there exists a finite and minimum solution N^* which satisfies Equation 20.87, it is unique and minimizes $C(N;\, T)$.

20.4 Inspection Policies

Suppose that failures are not detected immediately and can be done only through inspections. A typical example is standby electric generators in hospitals and other public facilities. It is extremely serious if a standby generator fails at the very moment of electric power supply stop. Similar examples can be found in defense systems, in which all weapons are in standby, and hence must be checked at periodic times. For example, missiles are in storage for a great part of their lifetimes after delivery. It is important to test the functions of missiles as to whether they can operate normally or not. Therefore, we need to check such systems at suitable times, and, if necessary, replace or repair them.

Barlow and Proschan [1] summarized the schedules of inspections that minimize two expected costs until detection of failure and per unit of time. Luss and Kander [175], Luss [176], and Wattanapanom and Shaw [177] considered the modified models where checking times are non-negligible and a system is inoperative during checking times. Platz [178] gave the availability of periodic checks. Zacks and Fenske [179], Luss and Kander [180], Anbar [181], Kander [182], Zuckerman [183, 184], and Qiu [185] treated much more complicated systems. Further, Weiss [157], Coleman and Abrams [186], Morey [187], and Apostolakis and Bansal [188] considered imperfect inspections where some failures might not be detected.

It would be especially important to check and maintain standby and protective units. Nakagawa [22], Thomas *et al.* [189], Sim [190], and Parmigiani [191] discussed the inspection policy for standby units, and Chay and Mazumdar [192], Inagaki *et al.* [193], and Shima and Nakagawa [194] did so for protective devices.

A unit has to be operating for an infinite time span. Let c_1 be the cost for one check and c_2 be the loss cost for the time elapsed between failure and its detection at the next checking time per unit of time. Then, the expected cost until detection of failure is, from Barlow and Proschan [1]:

$$C \equiv \int_0^{\infty} \{c_1[E\{N(t)\} + 1] + c_2 E\{\gamma_t\}\}\, \mathrm{d}F(t) \tag{20.88}$$

where $N(t)$ is the number of checks during $(0, t]$, γ_t is the interval from failure to its detection when a failure occurs at time t, and F is a failure time distribution of a unit.

We consider the following three inspection policies: (1) standard inspection and asymptotic checking time; (2) inspection with PM; and (3) inspection of a storage system. We summarize optimum policies that minimize the total expected cost until detection of failure.

20.4.1 Standard Inspection

A unit is checked at successive times x_k $(k = 1, 2, \dots)$ where $x_0 \equiv 0$. Any failures are detected at the next checking time and are replaced immediately. A unit has a failure time distribution $F(t)$ with finite mean $1/\lambda$ and its failure rate $r(t) \equiv f(t)/\bar{F}(t)$ is not changed by any checks, where f is a density of F and $\bar{F} \equiv 1 - F$. It is assumed that all times needed for checks and replacement are negligible.

Then, the total expected cost until detection of failure, from Equation 20.88, is

$$\begin{aligned} C(x_1, x_2, \dots) &= \sum_{k=0}^{\infty} \int_{x_k}^{x_{k+1}} [c_1(k+1) + c_2(x_{k+1} - t)]\, \mathrm{d}F(t) \\ &= \sum_{k=0}^{\infty} [c_1 + c_2(x_{k+1} - x_k)]\bar{F}(x_k) - \frac{c_2}{\lambda} \end{aligned} \tag{20.89}$$

Differentiating the total expected cost $C(x_1, x_2, \dots)$ with x_k and putting at zero:

$$x_{k+1} - x_k = \frac{F(x_k) - F(x_{k-1})}{f(x_k)} - \frac{c_1}{c_2} \tag{20.90}$$

Balrow and Proschan [1] proved that the optimum checking intervals are decreasing when f is PF_2 and gave the algorithm for computing the optimum schedule.

It is difficult to compute this algorithm numerically, because the computations are repeated until the procedures are determined to the required degree by changing the first checking time. We summarize three asymptotic calculations of optimum checking procedures.

Munford and Shahani [195] suggested that, when a unit was operating at time x_{k-1}, the probability that it fails during $(x_{k-1}, x_k]$ is constant for all k, *i.e.*

$$\frac{F(x_k) - F(x_{k-1})}{\bar{F}(x_k)} \equiv p \tag{20.91}$$

Then, the total expected cost in Equation 20.89 is rewritten as

$$C(p) = \frac{c_1}{p} + c_2 \sum_{k=1}^{\infty} x_k q^{k-1} p - \frac{c_2}{\lambda} \tag{20.92}$$

where $q \equiv 1 - p$. We may seek p that minimizes $C(p)$.

Keller [196] defined a smooth density $n(t)$, which denotes the number of checks per unit of time, expressed as

$$\int_0^{x_k} n(t)\, \mathrm{d}t = k \tag{20.93}$$

Then, the total expected cost is approximately

$$C(n(t)) \approx c_1 \int_0^{\infty} n(t)\bar{F}(t)\, \mathrm{d}t + \frac{c_2}{2} \int_0^{\infty} \frac{1}{n(t)}\, \mathrm{d}F(t) \tag{20.94}$$

and the approximate checking times are given by the following equation:

$$k = \sqrt{\frac{c_2}{2c_1}} \int_0^{x_k} \sqrt{r(t)}\, \mathrm{d}t \quad (k = 1, 2, \dots) \tag{20.95}$$

Finally, Nakagawa and Yasui [197] considered that if x_n is sufficiently large, we may assume approximately that

$$x_{n+1} - x_n + \varepsilon = x_n - x_{n-1} \tag{20.96}$$

where $0 < \varepsilon < c_1/c_2$. In this case, Equation 20.90 becomes

$$\frac{c_1}{c_2} - \varepsilon = \frac{\int_{x_{n-1}}^{x_n} [f(x) - f(x_n)]\, \mathrm{d}x}{f(x_n)} \tag{20.97}$$

We can determine the asymptotic schedule by computing $x_{n-1} > x_{n-2} > \cdots$ recursively from Equation 20.97, from starting at some time x_n of accuracy required.

Kaio and Osaki [198, 199] and Viscolani [200] discussed many modified inspection policies, using the above approximation methods, and compared them.

Munford [201] and Luss [202] proposed the following total expected cost for a continuous production system:

$$
\begin{aligned}
C(x_1, x_2, \dots) &= \sum_{k=0}^{\infty} \int_{x_k}^{x_{k+1}} [c_1(k+1) + c_2(x_{k+1} - x_k)]\,\mathrm{d}F(t) \\
&= c_1 \sum_{k=0}^{\infty} \bar{F}(x_k) \\
&\quad + c_2 \sum_{k=0}^{\infty} (x_{k+1} - x_k)[F(x_{k+1}) - F(x_k)]
\end{aligned} \tag{20.98}
$$

In this case, Equation 20.90 is rewritten as

$$
x_{k+1} - 2x_k + x_{k-1} = \frac{F(x_{k+1}) - 2F(x_k) + F(x_{k-1})}{f(x_k)} - \frac{c_1}{c_2} \tag{20.99}
$$

In particular, when a unit is checked at periodic times and the failure time is exponential, *i.e.* $x_k = kT$ $(k = 0, 1, 2, \dots)$ and $F(t) = 1 - \mathrm{e}^{-\lambda t}$, the total expected cost, from Equation 20.89, is

$$
C(T) = \frac{c_1 + c_2}{1 - \mathrm{e}^{-\lambda T}} - \frac{c_2}{\lambda} \tag{20.100}
$$

and an optimum checking time T^* is given by a finite and unique solution of

$$
\mathrm{e}^{\lambda T} - (1 + \lambda T) = \frac{\lambda c_1}{c_2} \tag{20.101}
$$

Similarly, from Equations 20.98 and 20.99, the total expected cost is

$$
C(T) = \frac{c_1}{1 - \mathrm{e}^{-\lambda T}} + c_2 T \tag{20.102}
$$

and an optimum time T^* is a finite and unique solution of

$$
\mathrm{e}^{\lambda T}(1 - \mathrm{e}^{-\lambda T})^2 = \frac{\lambda c_1}{c_2} \tag{20.103}
$$

Note that the solution to Equation 20.103 is less than that of Equation 20.101.

20.4.2 Inspection with Preventive Maintenance

See Nakagawa [23, 32]. We consider a modified inspection policy in which an operating unit is checked and maintained preventively at times kT $(k = 1, 2, \dots)$ and the age of a unit after PM reduces to at when it was t before PM (see Section 20.3.3.2). Then, we obtain the mean time to failure and the expected number of PMs before failure, and discuss an optimum inspection policy that minimizes the expected cost.

A unit begins to operate at time zero, and is checked and undergoes PM at periodic times kT $(k = 1, 2, \dots)$. It is assumed that the age of a unit after PM reduces to at $(0 \le a \le 1)$ when it was t before PM. A unit has a failure time distribution $F(t)$ with finite mean $1/\lambda$ and the process ends with its failure.

Let $H(t, x)$ be the failure rate of a unit, *i.e.* $H(t, x) \equiv [F(t + x) - F(t)]/\bar{F}(t)$ for $x > 0$, $t \ge 0$ and $F(t) < 1$, where $\bar{F} \equiv 1 - F$. Then, the reliability function, which is the probability that a unit is operating at time t, is

$$
\bar{S}(t; T, a) = \bar{H}(A_k T, t - kT) \prod_{j=0}^{k-1} \bar{H}(A_j T, T) \qquad kT \le t < (k+1)T \tag{20.104}
$$

where $A_k \equiv a + a^2 + \cdots + a^k$ $(k = 1, 2, \dots)$, $A_0 \equiv 0$, $\bar{H} \equiv 1 - H$, and $\Pi_{j=0}^{-1} \equiv 1$.

Using Equation 20.104, the mean time to failure is

$$
\begin{aligned}
\gamma(T; a) &= \sum_{k=0}^{\infty} \int_{kT}^{(k+1)T} \bar{S}(t; T, a)\,\mathrm{d}t \\
&= \sum_{k=0}^{\infty} \left[\prod_{j=0}^{k-1} \bar{H}(A_j T, T) \right] \frac{1}{\bar{F}(A_k T)} \\
&\quad \times \int_{A_k T}^{(A_k + 1)T} \bar{F}(t)\,\mathrm{d}t
\end{aligned} \tag{20.105}
$$

and the expected number of PMs before failure is

$$M(T;a) = \sum_{k=0}^{\infty} kH(A_kT, T) \prod_{j=0}^{k-1} \bar{H}(A_jT, T)$$
$$= \sum_{k=0}^{\infty} \left[\prod_{j=0}^{k} \bar{H}(A_jT, T) \right] \quad (20.106)$$

From Equations 20.105 and 20.106, we have the following: if the failure rate $H(T, x)$ is increasing in T, then both $\gamma(T; a)$ and $M(T; a)$ are decreasing in a for any $T > 0$, and

$$M(T;a) \leq \frac{\gamma(T;a)}{T} \leq 1 + M(T;a) \quad (20.107)$$

When a unit fails, its failure is detected at the next checking time. Then, the total expected cost until the detection of failure, from Equation 20.88, is

$$C(T;a) = \sum_{k=0}^{\infty} \int_{kT}^{(k+1)T} \{c_1(k+1) + c_2[(k+1)T - t]\} \, \mathrm{d}[-\bar{S}(t; T, a)]$$
$$= (c_1 + c_2T)[M(T;a) + 1] - c_2\gamma(T;a) \quad (20.108)$$

In particular, if $a = 1$, *i.e.* a unit undergoes only checks at periodic times, then

$$C(T;1) = (c_1 + c_2T) \sum_{k=0}^{\infty} \bar{F}(kT) - \frac{c_2}{\lambda} \quad (20.109)$$

If $a = 0$, *i.e.* the PM is perfect, then

$$C(T;0) = \frac{1}{F(T)} \left[c_1 + c_2 \int_0^T F(t) \, \mathrm{d}t \right] \quad (20.110)$$

Since $M(T; a)$ is a decreasing function of a, and from Equation 20.107, we have the inequalities

$$C(T;a) \geq c_1[M(T;a) + 1] \geq c_1 \sum_{k=0}^{\infty} \bar{F}(kT)$$

Thus, $\lim_{T \to 0} C(T; a) = \lim_{T \to \infty} C(T; a) = \infty$, and hence there exists a positive and finite T^* that minimizes the expected cost $C(T; a)$ in Equation 20.108.

Table 20.1. Optimum times T^* for various c_2 when $\bar{F}(t) = \mathrm{e}^{-(t/100)^2}$ and $c_1 = 1$

a	T^*					
	0.1	0.5	1.0	2.0	5.0	10.0
0	76	47	38	31	23	18
0.2	67	40	32	26	19	15
0.4	60	34	27	22	16	13
0.6	53	29	22	18	13	10
0.7	47	23	18	14	10	8
1.0	42	19	13	9	6	4

Example 4. Suppose that the failure time of a unit has a Weibull distribution with shape parameter 2, *i.e.* $\bar{F}(T) = \mathrm{e}^{-(t/100)^2}$, and $c_1 = 1$. Table 20.1 gives the optimum checking times T^* for several a and c_2. These times are small when a is large. The reason would be that the failure rate increases quickly with age when a is large and a unit fails easily.

20.4.3 Inspection of a Storage System

Systems like missiles and spare parts for aircraft are stored for long periods until required. However, their reliability deteriorates with time [203], and we cannot clarify whether a system can operate normally or not. The periodic tests and maintenances of a storage system are indispensable to obtaining a highly reliable condition. But, because frequent tests have a cost and may degrade a system [204], we should not test it very frequently.

Martinez [205] studied the periodic inspection of stored electronic equipment and showed how to compute its reliability after 10 years of storage. Itô and Nakagawa [206, 207] considered a storage system that is overhauled if its reliability becomes lower than a prespecified value, and obtained the number of tests and the time to overhaul. They further discussed the optimum inspection policy that minimizes the expected cost. Wattanapanom and Shaw [177] proposed an inspection policy for a system where tests may hasten failures.

It is well known that large electric currents occur in an electronic circuit containing induction

Table 20.2. Optimum times T^* and expected costs $C(T^*)$ when $\lambda_3 = 9.24 \times 10^{-7}, c_2 = 1$, and $a = 0.9$

m	λ	c_1	T^*	$C(T^*)$
1.0	29.24×10^{-6}	10	510	603
		15	670	764
		20	800	903
		25	920	1027
		30	1020	1140
1.0	58.48×10^{-6}	10	460	490
		15	590	613
		20	680	718
		25	790	811
		30	840	897
1.2	29.24×10^{-6}	10	430	377
		15	540	461
		20	630	531
		25	670	593
		30	760	648
1.2	58.48×10^{-6}	10	350	286
		15	390	347
		20	470	399
		25	510	445
		30	550	487

parts, such as coils and motors, at power on and off. Missiles are constituted of various kinds of electrical and electronic parts, and some of them are degraded by power on-off cycles during each test [204].

We consider the following inspection policy for a storage system that has to operate when it is used at any time.

1. A system is new at time zero, and is checked and maintained preventively if necessary at periodic times kT $(k = 1, 2, \ldots)$, where $T(> 0)$ is previously specified.
2. A system consists mainly of two independent units, where unit 1 becomes like new after every check; however, unit 2 does not become like new and is degraded with time and at each check.
3. Unit 1 has a failure rate r_1, which is given by $r_1(t - kT)$ for $kT < t \leq (k+1)T$, because it is like new at time kT.
4. Unit 2 has two failure rate functions r_2 and r_3, which describe the failure rates of system degradations with time and at each check respectively. The failure rate $r_2(t)$ remains undisturbed by any checks. Further, since unit 2 is degraded by power on-off cycles during the checking time, r_3 increases by constant rate λ_3 at each check, and is defined as $r_3(t) = k\lambda_3$ for $kT < t \leq (k+1)T$.

Thus, the failure rate function $r(t)$ of a system, from (3) and (4), is

$$r(t) \equiv r_1(t - kT) + r_2(t) + k\lambda_3 \qquad (20.111)$$

and the cumulative hazard function is

$$\begin{aligned} R(t) &\equiv \int_0^t r(u)\,\mathrm{d}u \\ &= kR_1(T) + R_1(t - kT) + R_2(t) \\ &\quad + \sum_{j=0}^{k-1} j\lambda_3 T + k\lambda_3(t - kT) \end{aligned} \qquad (20.112)$$

for $kT < t \leq (k+1)T$ $(k = 0, 1, 2, \ldots)$, where $R_i(t) \equiv \int_0^t r_i(u)\,\mathrm{d}u$ $(i = 1, 2)$.

Using Equation 20.112, the reliability of a system at time t is

$$\begin{aligned} \bar{F}(t) &= \mathrm{e}^{-R(t)} \\ &= \exp\left\{ - kR_1(T) - R_1(t - kT) - R_2(t) \right. \\ &\quad \left. - k\lambda_3\left[t - \frac{1}{2}(k+1)T\right]\right\} \end{aligned} \qquad (20.113)$$

for $kT < t \leq (k+1)T$ $(k = 0, 1, 2, \ldots)$, the mean time to system failure is

$$\begin{aligned} \gamma(T) &\equiv \int_0^\infty \bar{F}(t)\,\mathrm{d}t \\ &= \sum_{k=0}^{\infty} \exp\left[- kR_1(T) - \frac{k(k-1)}{2}\lambda_3 T\right] \\ &\quad \times \int_0^T \exp[-R_1(t) - R_2(t + kT) \\ &\quad - k\lambda_3 t]\,\mathrm{d}t \end{aligned} \qquad (20.114)$$

and the expected number of checks before failure is

$$M(T) \equiv \sum_{k=1}^{\infty} k[\bar{F}(kT) - \bar{F}((k+1)T)]$$
$$= \sum_{k=1}^{\infty} \exp\left[-kR_1(T) - R_2(kT) - \frac{k(k-1)}{2}\lambda_3 T\right] \quad (20.115)$$

Therefore, the total expected cost until detection of failure, from Equation 20.108, is

$$C(T) = (c_1 + c_2 T)[M(T) + 1] - c_2\gamma(T) \quad (20.116)$$

Suppose that units have a Weibull distribution, *i.e.* $R_i(t) = \lambda_i t^m$ $(i = 1, 2)$. Then, the total expected cost is

$$C(T) = (c_1 + c_2 T) \times \sum_{k=0}^{\infty} \exp\left[-k\lambda_1 T^m - \lambda_2 (kT)^m - \frac{N(N-1)}{2}\lambda_3 T\right]$$
$$- c_2 \sum_{k=0}^{\infty} \exp\left[-k\lambda_1 T^m - \frac{k(k-1)}{2}\lambda_3 T\right] \times \int_0^T \exp[-\lambda_1 t^m - \lambda_2 (t + kT)^m - k\lambda_3 t]\,\mathrm{d}t \quad (20.117)$$

Changing T, we can calculate numerically an optimum T^* that minimizes $C(T)$.

Bauer *et al.* [204] showed that the degradation failure rate λ_3 at each check is $\lambda_3 = N_c K \lambda_{SE}$ where N_c is the ratio of total cycles to checking time, K is the ratio of cyclic failure rate to storage failure rate, and λ_{SE} is the storage failure rate of electronic parts; their values are $N_c = 2.3 \times 10^{-14}$, $K = 270$, and $\lambda_{SE} = 14.88 \times 10^{-6}\ \mathrm{h}^{-1}$. Hence, $\lambda_{SE} = 9.24 \times 10^{-7}\ \mathrm{h}^{-1}$.

Table 20.2 gives the optimum times T^* and the resulting costs $C(T^*)$ for λ and c_1 when $c_2 = 1$, $m = 1, 1.2$ and $a = 0.9$, where $\lambda_1 = a\lambda$ and $\lambda_2 = (1 - a)\lambda$. This indicates that both T^* and $C(T^*)$ increase when c_1 and $1/\lambda$ increase, and that a system should be checked about once a month. It is of interest that T^* in the case of $m = 1.2$ is much shorter than that for $m = 1$, because a system deteriorates with time.

References

[1] Barlow RE, Proschan F. Mathematical theory of reliability. New York: John Wiley & Sons, 1965.

[2] Osaki S, Nakagawa T. Bibliography for reliability and availability. IEEE Trans Reliab 1976;R-25:284–7.

[3] Pierskalla WP, Voelker JA. A survey of maintenance models: the control and surveillance of deteriorating systems. Nav Res Logist Q 1976;23:353–88.

[4] Thomas LC. A survey of maintenance and replacement models for maintainability and reliability of multi-item systems. Reliab Eng 1986;16:297–309.

[5] Valdez-Flores C, Feldman RM. A survey of preventive maintenance models for stochastically deteriorating single-unit system. Nav Res Logist Q 1989;36:419–46.

[6] Cho DI, Parlar M. A survey of maintenance models for multi-unit systems. Eur J Oper Res 1991;51:1–23.

[7] Özekici S. (ed.) Reliability and maintenance of complex systems. Berlin: Springer-Verlag; 1996.

[8] Christer AH, Osaki S, Thomas LC. (eds.) Stochastic modelling in innovative manufacturing. Lecture Notes in Economics and Mathematical Systems, vol. 445. Berlin: Springer-Verlag; 1997.

[9] Ben-Daya M, Duffuaa SO, Raouf A. (eds.) Maintenance, modeling and optimization. Boston: Kluwer Academic Publishers; 2000.

[10] Osaki S. (ed.) Stochastic models in reliability maintainance. Berlin: Springer-Verlag; 2002.

[11] Nakagawa T, Osaki S. Optimum preventive maintenance policies for a 2-unit redundant system. IEEE Trans Reliab 1974;R-23:86–91.

[12] Nakagawa T, Osaki S. Optimum preventive maintenance policies maximizing the mean time to the first system failure for a two-unit standby redundant system. Optim Theor Appl 1974;14:115–29.

[13] Osaki S, Nakagawa T. A note on age replacement. IEEE Trans Reliab 1975;R-24:92–4.

[14] Nakagawa T, Osaki S. A summary of optimum preventive maintenance policies for a two-unit standby redundant system. Z Oper Res 1976;20:171–87.

[15] Nakagawa T. On a replacement problem of a cumulative damage model. Oper Res Q 1976;27:895–900.

[16] Mine H, Nakagawa T. Interval reliability and optimum preventive maintenance policy. IEEE Trans Reliab 1977;R-26:131–3.

[17] Nakagawa T. Optimum preventive maintenance policies for repairable systems. IEEE Trans Reliab 1977;R-26:168–73.

[18] Nakagawa T, Osaki S. Discrete time age replacement policies. Oper Res Q 1977;28:881–5.

[19] Mine H, Nakagawa T. A summary of optimum preventive maintenance policies maximizing interval reliability. J Oper Res Soc Jpn 1978;21:205–16.

[20] Nakagawa T. Optimum policies when preventive maintenance is imperfect. IEEE Trans Reliab 1979;R-28:331–2.

[21] Nakagawa T. A summary of block replacement policies. RAIRO Oper Res 1979;13:351–61.

[22] Nakagawa T. Optimum inspection policies for a standby unit. J Oper Res Soc Jpn 1980;23:13–26.

[23] Nakagawa T. Replacement models with inspection and preventive maintenance. Microelectron Reliab 1980;20:427–33.

[24] Nakagawa T. A summary of imperfect preventive maintenance policies with minimal repair. RAIRO Oper Res 1980;14:249–55.

[25] Nakagawa T. Modified periodic replacement with minimal repair at failure. IEEE Trans Reliab 1981;R-30:165–8.

[26] Nakagawa T. A summary of periodic replacement with minimal repair at failure. J Oper Res Soc Jpn 1981;24:213–27.

[27] Nakagawa T. Generalized models for determining optimal number of minimal repairs before replacement. J Oper Res Soc Jpn 1981;24:325–37.

[28] Nakagawa T. A modified block replacement with two variables. IEEE Trans Reliab 1982;R-31:398–400.

[29] Nakagawa T, Kowada M. Analysis of a system with minimal repair and its application to replacement policy. Eur J Oper Res 1983;12:176–82.

[30] Nakagawa T. Optimal number of failures before replacement time. IEEE Trans Reliab 1983;R-32:115–6.

[31] Nakagawa T. Combined replacement models. RAIRO Oper Res 1983;17:193–203.

[32] Nakagawa T. Periodic inspection policy with preventive maintenance. Nav Res Logist Q 1984;31:33–40.

[33] Nakagawa T. A summary of discrete replacement policies. Eur J Oper Res 1984;17:382–92.

[34] Nakagawa T. Modified discrete preventive maintenance policies. Nav Res Logist Q 1986;33:703–15.

[35] Nakagawa T. Optimum replacement policies for systems with two types of units. In: Osaki S, Cao JH, editors. Reliability Theory and Applications Proceedings of the China–Japan Reliability Symposium, Shanghai, China, 1987.

[36] Nakagawa T, Yasui K. Optimum policies for a system with imperfect maintenance. IEEE Trans Reliab 1987;R-36:631–3.

[37] Cox DR. Renewal theory. London: Methuen; 1962.

[38] Çinlar E. Introduction to stochastic processes. Englewood Cliffs (NJ): Prentice-Hall; 1975.

[39] Osaki S. Applied stochastic system modeling. Berlin: Springer-Verlag; 1992.

[40] Berg M. A proof of optimality for age replacement policies. J Appl Prob 1976;13:751–9.

[41] Bergman B. On the optimality of stationary replacement strategies. J Appl Prob 1980;17:178–86.

[42] Glasser GJ. The age replacement problem. Technometrics 1967;9:83–91.

[43] Scheaffer RL. Optimum age replacement policies with an increasing cost factor. Technometrics 1971;13:139–44.

[44] Cléroux R, Hanscom M. Age replacement with adjustment and depreciation costs and interest charges. Technometrics 1974;16:235–9.

[45] Cléroux R, Dubuc S, Tilquin C. The age replacement problem with minimal repair and random repair costs. Oper Res 1979;27:1158–67.

[46] Subramanian R, Wolff MR. Age replacement in simple systems with increasing loss functions. IEEE Trans Reliab 1976;R-25:32–4.

[47] Fox B. Age replacement with discounting. Oper Res 1966;14:533–7.

[48] Ran A, Rosenland SI. Age replacement with discounting for a continuous maintenance cost model. Technometrics 1976;18:459–65.

[49] Berg M, Epstein B. Comparison of age, block, and failure replacement policies. IEEE Trans Reliab 1978;R-27:25–9.

[50] Ingram CR, Scheaffer RL. On consistent estimation of age replacement intervals. Technometrics 1976;18:213–9.

[51] Frees EW, Ruppert D. Sequential non-parametric age replacement policies. Ann Stat 1985;13:650–62.

[52] Léger C, Cléroux R. Nonparametric age replacement: bootstrap confidence intervals for the optimal cost. Oper Res 1992;40:1062–73.

[53] Christer AH. Refined asymptotic costs for renewal reward processes. J Oper Res Soc 1978;29:577–83.

[54] Ansell J, Bendell A, Humble S. Age replacement under alternative cost criteria. Manage Sci 1984;30:358–67.

[55] Popova E, Wu HC. Renewal reward processes with fuzzy rewards and their applications to T-age replacement policies. Eur J Oper Res 1999;117:606–17.

[56] Zheng X. All opportunity-triggered replacement policy for multiple-unit systems. IEEE Trans Reliab 1995;44:648–52.

[57] Marathe VP, Nair KPK. Multistage planned replacement strategies. Oper Res 1966;14:874–87.

[58] Jain A, Nair KPK. Comparison of replacement strategies for items that fail. IEEE Trans Reliab 1974;R-23:247–51.

[59] Schweitzer PJ. Optimal replacement policies for hyperexponentially and uniform distributed lifetimes. Oper Res 1967;15:360–2.

[60] Savits TH. A cost relationship between age and block replacement policies. J Appl Prob 1988;25:789–96.

[61] Tilquin C, Cléroux R. Block replacement policies with general cost structures. Technometrics 1975;17:291–8.

[62] Archibald TW, Dekker R. Modified block-replacement for multi-component systems. IEEE Trans Reliab 1996;R-45:75–83.

[63] Sheu SH. A generalized block replacement policy with minimal repair and general random repair costs for a multi-unit system. J Oper Res Soc 1991;42:331–41.

[64] Sheu SH. Extended block replacement policy with used item and general random minimal repair cost. Eur J Oper Res 1994;79:405–16.

[65] Sheu SH. A modified block replacement policy with two variables and general random minimal repair cost. J Appl Prob 1996;33:557–72.

[66] Sheu SH. Extended optimal replacement model for deteriorating systems. Eur J Oper Res 1999;112:503–16.

[67] Crookes PCI. Replacement strategies. Oper Res Q 1963;14:167–84.

[68] Blanning RW. Replacement strategies. Oper Res Q 1965;16:253–4.

[69] Bhat BR. Used item replacement policy. J Appl Prob 1969;6:309–18.

[70] Tango T. A modified block replacement policy using less reliable items. IEEE Trans Reliab 1979;R-28:400–1.

[71] Murthy DNP, Nguyen DG. A note on extended block replacement policy with used items. J Appl Prob 1982;19:885–9.

[72] Ait Kadi D, Cléroux R. Optimal block replacement policies with multiple choice at failure. Nav Res Logist 1988;35:99–110.

[73] Berg M, Epstein B. A modified block replacement policy. Nav Res Logist Q 1976;23:15–24.

[74] Berg M, Epstein B. A note on a modified block replacement policy for units with increasing marginal running costs. Nav Res Logist Q 1979;26:157–60.

[75] Holland CW, McLean RA. Applications of replacement theory. AIIE Trans 1975;7:42–7.

[76] Tilquin C, Cléroux R. Periodic replacement with minimal repair at failure and adjustment costs. Nav Res Logist Q 1975;22:243–54.

[77] Boland PJ. Periodic replacement when minimal repair costs vary with time. Nav Res Logist Q 1982;29:541–6.

[78] Boland PJ, Proschan F. Periodic replacement with increasing minimal repair costs at failure. Oper Res 1982;30:1183–9.

[79] Chen M, Feldman RM. Optimal replacement policies with minimal repair and age-dependent costs. Eur J Oper Res 1997;98: 75–84.

[80] Aven T. Optimal replacement under a minimal repair strategy—a general failure model. Adv Appl Prob 1983;15:198–211.

[81] Bagai I, Jain K. Improvement, deterioration, and optimal replacement under age-replacement with minimal repair. IEEE Trans Reliab 1994;43:156–62.

[82] Murthy DNP. A note on minimal repair. IEEE Trans Reliab 1991;40:245–6.

[83] Dekker R. A general framework for optimization priority setting, planning and combining of maintenance activities. Eur J Oper Res 1995;82:225–40.

[84] Aven T, Dekker R. A useful framework for optimal replacement models. Reliab Eng Syst Saf 1997;58:61–7.

[85] Mine H, Kawai H. Preventive replacement of a 1-unit system with a wearout state. IEEE Trans Reliab 1974;R-23:24–9.

[86] Muth E. An optimal decision rule for repair vs replacement. IEEE Trans Reliab 1977;26:179–81.

[87] Tahara A, Nishida T. Optimal replacement policy for minimal repair model. J Oper Res Soc Jpn 1975;18:113–24.

[88] Phelps RI. Replacement policies under minimal repair. J Oper Res Soc 1981;32:549–54.

[89] Phelps RI. Optimal policy for minimal repair. J Oper Res Soc 1983;34:452–7.

[90] Park KS, Yoo YK. (τ, k) block replacement policy with idle count. IEEE Trans Reliab 1993;42:561–5.

[91] Morimura H. On some preventive maintenance policies for IFR. J Oper Res Soc Jpn 1970;12:94–124.

[92] Park KS. Optimal number of minimal repairs before replacement. IEEE Trans Reliab 1979;R-28:137–40.

[93] Tapiero CS, Ritchken PH. Note on the (N, T) replacement rule. IEEE Trans Reliab 1985;R-34:374–6.

[94] Lam Y. A note on the optimal replacement problem. Adv Appl Prob 1988;20,479–82.

[95] Lam Y. A repair replacement model. Adv Appl Prob 1990;22:494–97.

[96] Lam Y. An optimal repairable replacement model for deteriorating system. J Appl Prob 1991;28:843–51.

[97] Ritchken P, Wilson JG. (m, T) group maintenance policies. Manage Sci 1990;36: 632–9.

[98] Sheu SH. A generalized model for determining optimal number of minimal repairs before replacement. Eur J Oper Res 1993;69:38–49.

[99] Munter M. Discrete renewal processes. IEEE Trans Reliab 1971;R-20:46–51.

[100] Nakagawa T, Osaki S. The discrete Weibull distribution. IEEE Trans Reliab 1975;R-24:300–1.

[101] Scheaffer RL. Optimum age replacement in the bivariate exponential case. IEEE Trans Reliab 1975;R-24:214–5.

[102] Berg M. Optimal replacement policies for two-unit machines with increasing running costs I. Stoch Process Appl 1976;4:89–106.

[103] Berg M. General trigger-off replacement procedures for two-unit systems. Nav Res Logist Q 1978;25:15–29.

[104] Yamada S, Osaki S. Optimum replacement policies for a system composed of components. IEEE Trans Reliab 1981;R-30:278–83.

[105] Bai DS, Jang JS, Kwon YI. Generalized preventive maintenance policies for a system subject to deterioration. IEEE Trans Reliab 1983;R-32.512–4.

[106] Murthy DNP, Nguyen DG. Study of two-component system with failure interaction. Nav Res Logist Q 1985;32:239–47.

[107] Pullen KW, Thomas MU. Evaluation of an opportunistic replacement policy for a 2-unit system. IEEE Trans Reliab 1986;R-35:320–4.

[108] Beichelt F, Fisher K. General failure model applied to preventive maintenance policies. IEEE Trans Reliab 1980;R-29:39–41.

[109] Beichelt F. A generalized block-replacement policy. IEEE Trans Reliab 1981;R-30:171–2.

[110] Murthy DNP, Maxwell MR. Optimal age replacement policies for items from a mixture. IEEE Trans Reliab 1981;R-30:169–70.

[111] Berg MP. The marginal cost analysis and its application to repair and replacement policies. Eur J Oper Res 1995;82:214–24.

[112] Block HW, Borges WS, Savits TH. Age-dependent minimal repair. J Appl Prob 1985;22:370–85.

[113] Block HW, Borges WS, Savits TH. A general age replacement model with minimal repair. Nav Res Logist Q 1988;35:365–72

[114] Smith WL. Renewal theory and its ramifications. J R Stat Soc Ser B 1958;20:243–302.

[115] Mercer A. On wear-dependent renewal process. J R Stat Soc Ser B 1961;23:368–76.

[116] Esary JD, Marshall AW, Proschan F. Shock models and wear processes. Ann Prob 1973;1:627–49.

[117] Nakagawa T, Osaki S. Some aspects of damage models. Microelectron Reliab 1974;13:253–7.

[118] Taylor HM. Optimal replacement under additive damage and other failure models. Nav Res Logist Q 1975;22:1–18.

[119] Feldman RM. Optimal replacement with semi-Markov shock models. J Appl Prob 1976;13:108–17.

[120] Feldman RM. Optimal replacement with semi-Markov shock models using discounted costs. Math Oper Res 1977;2:78–90.

[121] Feldman RM. Optimal replacement for systems governed by Markov additive shock processes. Ann Prob 1977;5:413–29.

[122] Zuckerman Z. Replacement models under additive damage. Nav Res Logist Q 1977;24:549–58.

[123] Zuckerman Z. Optimal replacement policy for the case where the damage process is a one-sided Lévy process. Stoch Process Appl 1978;7:141–51.

[124] Zuckerman Z. Optimal stopping in a semi-Markov shock model. J Appl Prob 1978;15:629–34.

[125] Zuckerman D. A note on the optimal replacement time of damaged devices. Nav Res Logist Q 1980;27:521–4.

[126] Wortman MA, Klutke GA, Ayhan H. A maintenance strategy for systems subjected to deterioration governed by random shocks. IEEE Trans Reliab 1994;43:439–45.

[127] Sheu SH, Griffith WS. Optimal number of minimal repairs before replacement of a system subject to shocks. Nav Res Logist 1996;43:319–33.

[128] Sheu SH. Extended block replacement policy of a system subject to shocks. IEEE Trans Reliab 1997;46:375–82.

[129] Sheu SH. A generalized age block replacement of a system subject to shocks. Eur J Oper Res 1998;108:345–62.

[130] Satow T, Yasui K, Nakagwa T. Optimal garbage collection policies for a database in a computer system. RAIRO Oper Res 1996;30:359–72.

[131] Qian CH, Nakamura S, Nakagawa T. Cumulative damage model with two kinds of shocks and its application to the backup policy. J Oper Res Soc Jpn 1999;42:501–11.

[132] Pyke R. Markov renewal processes: definitions and preliminary properties. Ann Math Stat 1961;32:1231–42.

[133] Pyke R. Markov renewal processes with finitely many states. Ann Math Stat 1961;32:1243–59.

[134] Hosford JE. Measures of dependability. Oper Res 1960;8:53–64.

[135] Morse PM. Queues, inventories and maintenance. New York: John Wiley & Sons; 1958. Chapter 11.

[136] Sherif YS, Smith ML. Optimal maintenance models for systems subject to failure—a review. Nav Res Logist Q 1981;28:47–74.

[137] Jardine AKS, Buzacott JA. Equipment reliability and maintenance. Eur J Oper Res 1985;19:285–96.

[138] Reineke DM, Murdock Jr WP, Pohl EA, Rehmert I. Improving availability and cost performance for complex systems with preventive maintenance. In: Proceedings Annual Reliability and Maintainability Symposium, 1999; p.383–8.

[139] Liang TY. Optimum piggyback preventive maintenance policies. IEEE Trans Reliab 1985;R-34:529–38.

[140] Aven T. Availability formulae for standby systems of similar units that are preventively maintained. IEEE Trans Reliab 1990;R-39:603–6.

[141] Smith MAJ, Dekker R. Preventive maintenance in a 1 out of n system: the uptime, downtime and costs. Eur J Oper Res 1997;99:565–83.

[142] Silver EA, Fiechter CN. Preventive maintenance with limited historical data. Eur J Oper Res 1995;82:125–44.

[143] Van der Duyn Schouten, Scarf PA. Eleventh EURO summer institute: operational research models in maintenance. Eur J Oper Res 1997;99:493–506.

[144] Chockie A, Bjorkelo K. Effective maintenance practices to manage system aging. In: Proceedings Annual Reliability and Maintainability Symposium, 1992; p.166–70.

[145] Smith AM. Preventive-maintenance impact on plant availability. In: Proceedings Annual Reliability and Maintainability Symposium, 1992; p.177–80.

[146] Susova GM, Petrov AN. Markov model-based reliability and safety evaluation for aircraft maintenance-system optimization. In: Proceedings Annual Reliability and Maintainability Symposium, 1992; p.29–36.

[147] Gaver Jr DP. Failure time for a redundant repairable system of two dissimilar elements. IEEE Trans Reliab 1964;R-13:14–22.

[148] Gnedenko BV. Idle duplication. Eng Cybernet 1964;2:1–9.

[149] Gnedenko BV. Duplication with repair. Eng Cybernet 1964;2:102–8.

[150] Srinivasan VS. The effect of standby redundancy in system's failure with repair maintenance. Oper Res 1966;14:1024–36.

[151] Osaki S. System reliability analysis by Markov renewal processes. J Oper Res Soc Jpn 1970;12:127–88.

[152] Rozhdestvenskiy DV, Fanarzhi GN. Reliability of a duplicated system with renewal and preventive maintenance. Eng Cybernet 1970;8:475–9.

[153] Osaki S, Asakura T. A two-unit standby redundant system with preventive maintenance. J Appl Prob 1970;7:641–8.

[154] Berg M. Optimal replacement policies for two-unit machines with increasing running cost I. Stoch Process Appl 1976;4:89–106.

[155] Berg M. Optimal replacement policies for two-unit machines with running cost II. Stoch Process Appl 1977;5:315–22.

[156] Berg M. General trigger-off replacement procedures for two-unit system. Nav Res Logist Q 1978;25:15–29.

[157] Weiss H. A problem in equipment maintenance. Manage Sci 1962;8:266–77.

[158] Chan PKW, Downs T. Two criteria for preventive maintenance. IEEE Trans Reliab 1978;R-27:272–3.

[159] Murthy DNP, Nguyen DG. Optimal age-policy with imperfect preventive maintenance. IEEE Trans Reliab 1981;R-30:80–1.

[160] Lie CH, Chun YH. An algorithm for preventive maintenance policy. IEEE Trans Reliab 1986;R-35:71–5.

[161] Jayabalan V, Chaudhuri D. Cost optimization of maintenance scheduling for a system with assured reliability. IEEE Trans Reliab 1992;R-41:21–5.

[162] Canfield RV. Cost optimization of periodic preventive maintenance. IEEE Trans Reliab 1986;R-35:78–81.

[163] Brown M, Proschan F. Imperfect repair. J Appl Prob 1983;20:851–9.

[164] Fontnot RA, Proschan F. Some imperfect maintenance models. In: Abdel-Hameed MS, Çinlar E, Quinn J, editors. Reliability theory and models. Orlando (FL): Academic Press; 1984.

[165] Bhattacharjee MC. New results for the Brown–Proschan model of imperfect repair. J Stat Plan Infer 1987;16:305–16.

[166] Ebrahimi N. Mean time to achieve a failure-free requirement with imperfect repair. IEEE Trans Reliab 1985;R-34:34–7.

[167] Kijima M, Morimura H, Suzuki Y. Periodic replacement problem without assuming minimal repair. Eur J Oper Res 1988;37:194–203.

[168] Natvig B. On information based minimal repair and the reduction in remaining system lifetime due to the failure of a specific module. J Appl Prob 1990;27: 365–375.

[169] Stadje WG, Zuckerman D. Optimal maintenance strategies for repairable systems with general degree of repair. J Appl Prob 1991;28:384–96.

[170] Makis V, Jardine AKS. Optimal replacement policy for a general model with imperfect repair. J Oper Res Soc 1992;43:111–20.

[171] Lie XG, Makis V, Jardine AKS. A replacement model with overhauls and repairs. Nav Res Logist 1995;42: 1063–79.

[172] Wang H, Pham H. Optimal age-dependent preventive maintenance policies with imperfect maintenance. Int J Reliab Qual Saf Eng 1996;3:119–35.

[173] Pham H, Wang H. Imperfect maintenance. Eur J Oper Res 1996;94:425–38.

[174] Sandoh H, Hirakoshi H, Nakagawa T. A new modified discrete preventive maintenance policy and its application to hard disk management. J Qual Maint Eng 1998;4:284–90.

[175] Luss H, Kander Z. Inspection policies when duration of checkings is non-negligible. Oper Res Q 1974;25:299–309.

[176] Luss H. Inspection policies for a system which is inoperative during inspection periods. AIIE Trans 1976;9:189–94.

[177] Wattanapanom N, Shaw L. Optimal inspection schedules for failure detection in a model where tests hasten failures. Oper Res 1979;27:303–17.

[178] Platz O. Availability of a renewable, checked system. IEEE Trans Reliab 1976;R-25:56–8.

[179] Zacks S, Fenske WJ. Sequential determination of inspection epochs for reliability systems with general lifetime distributions. Nav Res Logist Q 1973;20: 377–86.

[180] Luss H, Kander Z. A preparedness model dealing with N systems operating simultaneously. Oper Res 1974;22:117–28.

[181] Anbar D. An aysmptotically optimal inspection policy. Nav Res Logist Q 1976;23:211–8.

[182] Kander Z. Inspection policies for deteriorating equipment characterized by N quality levels. Nav Res Logist Q 1978;25:243–55.

[183] Zuckerman D. Inspection and replacement policies. J Appl Prob 1980;17:168–77.

[184] Zuckerman D. Optimal inspection policy for a multi-unit machine. J Appl Prob 1989;26:543–51.

[185] Qiu YP. A note on optimal inspection policy for stochastically deteriorating series systems. J Appl Prob 1991;28:934–9.

[186] Coleman JJ, Abrams IJ. Mathematical model for operational readiness. Oper Res 1962;10:126–38.

[187] Morey RC. A criterion for the economic application of imperfect inspections. Oper Res 1967;15:695–8.

[188] Apostolakis GE, Bansal PP. Effect of human error on the availability of periodically inspected redundant systems. IEEE Trans Reliab 1977;R-26:220–5.

[189] Thomas LC, Jacobs PA, Gaver DP. Optimal inspection policies for standby systems. Commun Stat Stoch Model 1987;3:259–73.

[190] Sim SH. Reliability of standby equipment with periodic testing. IEEE Trans Reliab 1987;R-36:117–23.

[191] Parmigiani G. Inspection times for stand-by units. J Appl Prob 1994;31:1015–25.

[192] Chay SC, Mazumdar M. Determination of test intervals in certain repairable standby protective systems. IEEE Trans Reliab 1975;R-24:201–5.

[193] Inagaki T, Inoue K, Akashi H. Optimization of staggered inspection schedules for protective systems. IEEE Trans Reliab 1980;R-29:170–3.

[194] Shima E, Nakagawa T. Optimum inspection policy for a protective device. Reliab Eng 1984;7:123–32.

[195] Munford AG, Shahani AK. A nearly optimal inspection policy. Oper Res Q 1972;23:373–9.

[196] Keller JB. Optimum checking schedules for systems subject to random failure. Manage Sci 1974;21:256–60.

[197] Nakagawa T, Yasui K. Approximate calculation of optimal inspection times. J Oper Res Soc 1980;31:851–3.

[198] Kaio N, Osaki S. Inspection policies: Comparisons and modifications. RAIRO Oper Res 1988;22:387–400.

[199] Kaio N, Osaki S. Comparison of inspection policies. J Oper Res Soc 1989;40:499–503.

[200] Viscolani B. A note on checking schedules with finite horizon. RAIRO Oper Res 1991;25:203–8.

[201] Munford AG. Comparison among certain inspection policies. Manage Sci 1981;27:260–7.

[202] Luss H. An inspection policy model for production facilities. Manage Sci 1983;29:1102–9.

[203] Menke JT. Deterioration of electronics in storage. In: Proceedings of National SAMPE Symposium, 1983; p.966–72.

[204] Bauer J *et al.* Dormancy and power on–off cycling effects on electronic equipment and part reliability. RADAC-TR-73-248(AD-768619), 1973.

[205] Martinez EC. Storage reliability with periodic test. In: Proceedings of Annual Reliability and Maintainability Symposium, 1984; p.181–5.

[206] Itô K, Nakagawa T. Optimal inspection policies for a system in storage. Comput Math Appl 1992;24:87–90.

[207] Itô K, Nakagawa T. Extended optimal inspection policies for a system in storage. Math Comput Model 1995; 22:83–7.

Optimal Imperfect Maintenance Models

Hongzhou Wang and Hoang Pham

Chapter 21

21.1 Introduction

Maintenance involves preventive (planned) and corrective (unplanned) actions carried out to retain a system in or restore it to an acceptable operating condition. Optimal maintenance policies aim to provide optimum system reliability and safety performance at the lowest possible maintenance costs. Proper maintenance techniques have been emphasized in recent years due to increased safety and reliability requirements of systems, increased complexity, and rising costs of material and labor [1]. For some systems, such as airplanes, submarines, and nuclear systems, it is extremely important to avoid failure during operation because it is dangerous and disastrous. One important research area in reliability engineering is the study of various maintenance policies in order to prevent the occurrence of system failure in the field and improve system availability.

In the past decades, maintenance, replacement, and inspection problems have been extensively studied in the literature, and hundreds of models have been developed. McCall [2], Barlow and Proschan [3], Pieskalla and Voelker [4], Sherif and Smith [1], Jardine and Buzacott [5], Valdez-Flores and Feldman [6], Cho and Parlar [7], Jensen [8], Dekker [9], Pham and Wang [10], and Dekker *et al.* [11] survey and summarize the research done in this field. However, problems in this field are still not solved satisfactorily and some maintenance models are not realistic. Next, an important problem in maintenance and reliability theory is discussed.

Maintenance can be classified into two major categories: corrective and preventive. Corrective maintenance (CM) occurs when the system fails. Some researchers refer to CM as repair and we will use them interchangeably throughout this chapter. According to MIL-STD-721B, CM means all actions performed as a result of failure, to restore an item to a specified condition. Preventive maintenance (PM) occurs when the system is operating. According to MIL-STD-721B, PM means all actions performed in an attempt to retain an item in specified condition by providing systematic inspection, detection, and prevention of incipient failures. We believe that maintenance can also be categorized according to the *degree* to which the operating conditions of an item is restored by maintenance in the following way.

(a) Perfect repair or perfect maintenance: a maintenance action that restores the system operating condition to "as good as new", *i.e.* upon perfect maintenance, a system has the same lifetime distribution and failure rate function as a brand new one. Complete overhaul of an engine with a broken connecting rod is an example of perfect repair. Generally, replacement of a failed system by a new one is a perfect repair.

(b) Minimal repair or minimal maintenance: a maintenance action that restores the system to the failure rate it had when it just failed. Minimal repair was first studied by Barlow and Proschan [3]. After this the system operating state is often called "as bad as old" in the literature. Changing a flat tire on a car is an example of minimal repair, because the overall failure rate of the car is essentially unchanged.

(c) Imperfect repair or imperfect maintenance: a maintenance action may not make a system "as good as new" but younger. Usually, it is assumed that imperfect maintenance restores the system operating state to somewhere between "as good as new" and "as bad as old". Clearly, imperfect maintenance is general, which can include two extreme cases: minimal and perfect repair maintenance. Engine tune-up is an example of imperfect maintenance.

(d) Worse repair or maintenance: a maintenance action that undeliberately makes the system failure rate or actual age increase but the system does not break down. Thus, upon worse repair a system's operating condition becomes worse than that just prior to its failure.

(e) Worst repair or maintenance: a maintenance action that undeliberately makes the system fail or break down.

Because worse and worst maintenance is less met in practice than imperfect maintenance, the focus of this chapter is on imperfect maintenance.

According to the above classification scheme, we can say that a PM is a minimal, perfect, imperfect, worst or worse PM. Similarly, a CM may be a minimal, perfect, imperfect, worst or worse CM. We will refer to imperfect CM and PM as imperfect maintenance later.

The type and degree of maintenance that is used in practice depend on the types of system, their costs as well as system reliability and safety requirements. In maintenance literature, most studies assume that the system after CM or PM is "as good as new" (perfect repair) or "as bad as old" (minimal repair). In practice, the perfect maintenance assumption may be reasonable for systems with one component that is structurally simple. On the other hand, the minimal repair assumption seems plausible for the failure behavior of systems when one of its many, non-dominating components is replaced by a new one [12]. However, many maintenance activities may not result in such two extreme situations but in a complicated intermediate one. For example, an engine may not be "as good as new" or "as bad as old" after tune-up, a type of PM. It usually becomes "younger" than at the time just prior to PM. Therefore, perfect maintenance and minimal maintenance are not true in many actual instances and realistic imperfect maintenance should be modeled. Recently, imperfect maintenance (both corrective and preventive) has received increasing attention in the reliability and maintenance community. In fact, the advent of the concept of imperfect maintenance has great meaning in reliability and maintenance engineering. We think that imperfect maintenance study indicates a significant breakthrough in maintenance theory.

Brown and Proschan [13] propose some possible causes for imperfect, worse or worst maintenance due to the maintenance performer:

- repair of the wrong part;
- only partially repair the faulty part;
- repair (partially or completely) the faulty part but damage adjacent parts;
- incorrectly assess the condition of the unit inspected;
- perform the maintenance action not when called for but at his convenience (the timing for maintenance is off the schedule).

Nakagawa and Yasui [14] mention three possible reasons causing worse or worst maintenance:

- hidden faults and failures that are not detected during maintenance;
- human errors, such as wrong adjustments and further damage done during maintenance;
- replacement with faulty parts.

Helvik [15] suggests that imperfectness of maintenance is related to the skill of the maintenance personnel, the quality of the maintenance procedure, and the maintainability of the system.

21.2 Treatment Methods for Imperfect Maintenance

Impefect maintenance research started as early as in the late 1970s. Kay [16] and Chan and Downs [17] studied the worst PM. Ingle and Siewiorek [18] examined imperfect maintenance. Chaudhuri and Sahu [19] explored the concept of imperfect PM. Most early research on imperfect, worse, and worst maintenance is for a single-unit system. Some researchers have proposed various methods for modeling imperfect, worse, and worst maintenance. It is useful to summarize/classify these methods. These modeling methods can be utilized in various maintenance and inspection policies. Although these methods are summarized mainly from the work of a single-unit system they will also be useful for modeling multicomponent systems, because the study methods of imperfect, worse, and worst maintenance for single-unit systems will also be effective for modeling the imperfect maintenance of individual subsystems that are constituting parts of multicomponent systems. Methods for treating imperfect, worse, and worst maintenance can be classified into eight categories [10], which will be described next.

21.2.1 Treatment Method 1

Nakagawa [20] treats the imperfect PM in this way: the component is returned to the "as good as new" state (perfect PM) with probability p and it is returned to the "as bad as old" state (minimal PM) with probability $q = 1 - p$ upon imperfect PM. Obviously, if $p = 1$ the PM coincides with perfect PM and if $p = 0$ it corresponds to minimal PM. In this sense, minimal and perfect maintenances are special cases of imperfect maintenance, and imperfect maintenance is a general maintenance. Using such a method for modeling imperfect maintenance, Nakagawa succeeded in determining the optimum PM policies minimizing the s-expected maintenance cost rate for a one-unit system under both age-dependent [20] and periodic PM [21] policies, given that PM is imperfect.

Similar to Nakagawa's work [20], Helvic [15] states that, after PM, though the fault-tolerant system is usually renewed after maintenance with probability θ_2, its operating condition sometimes remains unchanged (as bad as old) with probability θ_1 where $\theta_1 + \theta_2 = 1$.

Brown and Proschan [13] contemplate the following model of the repair process. A unit is repaired each time it fails. The executed repair is either a perfect one with probability p or a minimal one with probability $1 - p$. Assuming that all repair actions take negligible time, Brown and Proschan [13] established aging preservation properties of this imperfect repair process and the monotonicity of various parameters and random variables associated with the failure process. They obtained an important, useful result: if the life distribution of a unit is F and its failure rate is r, then the distribution function of the time between successive perfect repairs $F_p = 1 - (1 - F)^p$ and the corresponding failure rate $r_p = pr$. Using this result, Fontenot and Proschan [22] and Wang and Pham [23] both obtained optimal imperfect maintenance policies for a one-component system. Later on, we will refer to this method of modeling imperfect maintenance as the (p, q) *rule*, *i.e.* after maintenance (corrective or preventive) a system becomes "as good as new" with probability p and "as bad as old" with probability $1 - p$. In fact, quite a few imperfect maintenance models have used this rule in recent years.

Bhattacharjee [24] obtained the same results as Brown and Proschan [13] and also some new results for the Brown–Proschan model of imperfect repair via a shock model representation of the sojourn time.

21.2.2 Treatment Method 2

Block *et al.* [25] extend the above Brown–Proschan imperfect repair model with the (p, q) rule to the age-dependent imperfect repair for a one-unit system: an item is repaired at failure (CM). With probability $p(t)$, the repair is a perfect repair; with probability $q(t) = 1 - p(t)$, the repair is a minimal one, where t is the age of the item in use at the failure time (the time since the last perfect repair). Block *et al.* [25] showed that if the item's life distribution F is a continuous function and its failure rate is r, then the successive perfect repair times is a renewal process with interarrival time distribution

$$F_p = 1 - \exp\left\{\int_0^t p(x)[1 - F(x)]^{-1} F(\mathrm{d}x)\right\}$$

and the corresponding failure rate $r_p(t) = p(t)r(t)$. In fact, similar results were found by Beichelt and Fischer [26]. Block *et al.* [25] proved that the aging preservation results of Brown and Proschan [13] hold under suitable hypotheses on $p(t)$. This imperfect maintenance modeling method is called the $(p(t), q(t))$ *rule* herein.

Using the $(p(t), q(t))$ rule, Block *et al.* [27] explored a general age-dependent PM policy, where an operating unit is replaced whenever it reaches age T; if it fails at age $t < T$, it is either replaced by a new unit, with probability $p(t)$, or it undergoes minimal repair, with probability $q(t) = 1 - p(t)$. The cost of the ith minimal repair is a function, $c_i(t)$, of age and number of repairs. After a perfect maintenance, planned (preventive) or unplanned (corrective), the above process is repeated.

Both the Brown and Proschan models and the Block *et al.* [25, 27] model assume that the repair time is negligible. It is worthwhile mentioning that Iyer [28] obtained availability results for imperfect repair using the $(p(t), q(t))$ rule given the repair time is not negligible. Obviously, his treatment method is more realistic.

In a related study, Sumita and Shanthikumar [29] (see Pham and Wang [10]) proposed and studied an age-dependent counting process generated from a renewal process, and applied that counting process to the age-dependent imperfect repair for a one-unit system.

Whitaker and Samaniego [30] established an estimator for the life distribution when the above maintenance process proposed by Block *et al.* [25] is observed until the time of the mth perfect repair. This estimator was motivated by a non-parametric maximum likelihood approach, and was shown to be a "neighborhood maximum likelihood estimator". They derived large-sample results for this estimator. Hollander *et al.* [31] took the more modern approach of using a counting process and martingale theory to analyze these models. Their methods yielded extensions of Whitaker and Samaniego's results to the whole line and provide a useful framework for further work on the minimal repair model.

The (p, q) rule and $(p(t), q(t))$ rule for modeling imperfect maintenance seem practical and realistic. They make imperfect maintenance be somewhere between perfect and minimal maintenance. The *degree* to which the operating condition of an item is restored by maintenance can be measured by probability p or $p(t)$. Especially in the $(p(t), q(t))$ rule, the *degree* to which the operating conditions of an item is restored by maintenance is related to its age t. Thus, the $(p(t), q(t))$ rule seems more realistic, but mathematical modeling of imperfect maintenance through using it may be more complicated. We think that the two rules can be expected to be powerful in future imperfect maintenance modeling. In fact, both rules have received much attention and have been used in some recent imperfect repair models.

Makis and Jardine [32] proposed a general treatment method for imperfect maintenance and modeled imperfect repair at failure in a way that repair returns a system to the "as good as new" state with probability $p(n, t)$ or to the "as bad as old" state with probability $q(n, t)$, or with probability $s(n, t) = 1 - p(n, t) - q(n, t)$ the repair is unsuccessful, the system is scrapped and replaced by a new one; here, t is the age of the system and n is the number of failures since the last replacement. This treatment method will be referred to as the $(p(n, t), q(n, t), s(n, t))$ *rule* later on.

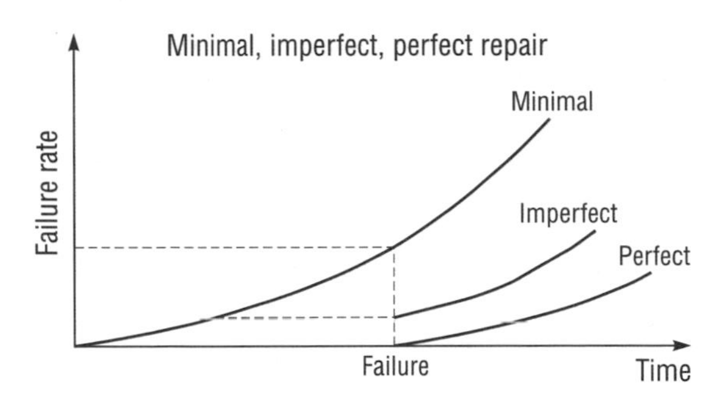

Figure 21.1. Minimal, perfect, imperfect repair versus failure rate change

21.2.3 Treatment Method 3

Malik [33] introduced the concept of improvement factor in the maintenance scheduling problem. He thinks that maintenance changes the system time of the failure rate curve to some newer time but not all the way to zero (not new), as shown in Figure 21.1. This treatment method for imperfect maintenance also makes the failure rate after PM lie between "as good as new" and "as bad as old". The degree of improvement in failure rate is called the improvement factor (note that PM is only justified by an increasing failure rate). Malik [33] contemplated that, since systems need more frequent maintenance with increased age, the successive PM intervals are decreasing in order to keep the system failure rate at or below a stated level (sequential PM policy), and proposed an algorithm to determine these successive PM intervals. Lie and Chun [34] presented a general expression to determine these PM intervals. Malik [33] relied on an expert judgment to estimate the improvement factor, whereas Lie and Chun [34] gave a set of curves as a function of maintenance cost and the age of the system for the improvement factor.

Using the improvement factor and assuming finite planning horizon, Jayabalan and Chaudhuri [35] introduced a branching algorithm to minimize the average total cost for a maintenance scheduling model with assured reliability, and they [36] discuss the optimal maintenance policy for a system with increased mean down time and assured failure rate. It is worthwhile noting that, using fuzzy set theory and an improvement factor, Suresh and Chaudhuri [37] established a PM scheduling procedure to assure an acceptable reliability level or tolerable failure rate given a finite planning horizon. They regard the starting condition, ending condition, operating condition, and type of maintenance of a system as fuzzy sets. The improvement factor is used to find out the starting condition of the system after maintenance.

Chan and Shaw [38] considered that the failure rate of an item is reduced after each PM and this reduction of failure rate depends on the item age and the number of PMs. They proposed two types of failure-rate reduction. (1) Failure rate with fixed reduction: after each PM, the failure rate is reduced such that all jump-downs of the failure rate are the same. (2) Failure rate with proportional reduction: after PM, the failure rate is reduced such that each jump-down is proportional to the current failure rate. Chan and Shaw [38] derived the cycle-availability for single-unit systems and discussed the design scheme to maximize the probability of achieving a specified stochastic-cycle availability with respect to the duration of the operating interval between PMs.

This kind of study method for imperfect maintenance is in terms of failure rate and seems useful and practical in engineering, which can be used as a general treatment method for imperfect maintenance or even *worse* maintenance. This treatment method will be called the *improvement factor method* herein.

In addition, Canfield [39] observed that PM at time t restores the failure rate function to its shape at $t - \tau$, whereas the level remains unchanged where τ is less than or equal to the PM intervention interval.

21.2.4 Treatment Method 4

Kijima *et al.* [40] proposed an imperfect repair model by using the idea of the virtual age process of a repairable system. In this imperfect repair model, if a system has the virtual age $V_{n-1} = y$ immediately after the $(n - 1)$th repair, the nth failure-time X_n is assumed to have the distribution function

$$\Pr\{X_n \leq x \mid V_{n-1} = y\} = \frac{F(x + y) - F(y)}{1 - F(y)}$$

where $F(x)$ is the distribution function of the time to failure of a new system. Let a be the degree of the nth repair, where $0 \leq a \leq 1$. Kijima *et al.* [40] constructed such a repair model: the nth repair cannot remove the damage incurred before the $(n - 1)$th repair. It reduces the additional age X_n to aX_n. Accordingly, the virtual age after the nth repair becomes

$$V_n = V_{n-1} + aX_n$$

Obviously, $a = 0$ corresponds to a perfect repair, whereas $a = 1$ is a minimal repair. In a later study Kijima [12] extended the above model to the case that a is a random variable taking a value between zero and one and proposed another imperfect repair model

$$V_n = A_n(V_{n-1} + X_n)$$

where A_n is a random variable taking a value between zero and one for $n = 1, 2, 3, \ldots$. For the extreme values zero and one, $A_n = 1$ means a minimal repair and $A_n = 0$ means a perfect repair. Comparing this treatment method with Brown and Proschan's [13], we can see that if A_n is independently and identically distributed (*i.i.d.*), then by taking the two extreme values zero and one they are the same. Therefore, the second treatment method by Kijima [12] is general. He also derived various monotonicity properties associated with the above two imperfect repair models.

This kind of treatment method is referred to as the *virtual age method* herein.

It is worth mentioning that Uematsu and Nishida [41] considered a more general model that includes the above two imperfect repair models by Kijima and coworkers [12, 40] as special cases, and obtained some elementary properties of the associated failure process. Let T_n denote the time interval between the $(n - 1)$th failure and the nth failure, and X_n denote the degree of repair. After performing the nth repair, the age of the system is $q(t_1, \ldots, t_n; x_1, \ldots, x_n)$ given that $T_i = t_i$ and $X_i = x_i$ $(i = 1, 2, \ldots, n)$ where T_i and X_i are random variables. On the other hand, $q(t_1, \ldots, t_n; x_1, \ldots, x_{n-1})$ represents the age of the system as just before the nth failure. The starting epoch of an interval is subject to the influence of all previous failure history, *i.e.* the nth interval is statistically dependent on $T_1 = t_1, \ldots, T_{n-1} = t_{n-1}$, $X_1 = x_1, \ldots, T_{n-1} = t_{n-1}$. For example, if

$$q(t_1, \ldots, t_n; x_1, \ldots, x_n) = \sum_{j=1}^{n} \sum_{i=j}^{n} x_i t_j,$$

then $X_i = 0$ $(X_i = 1)$ represents that a perfect repair (minimal repair) is performed at the ith failure.

21.2.5 Treatment Method 5

It is well known that the time to failure of a unit can be represented as a first passage time to a threshold for an appropriate stochastic process that describes the levels of damage. Consider a unit that is subject to shocks occurring randomly in time. At time $t = 0$, the damage level of the unit is assumed to be zero. Upon occurrence of a shock, the unit suffers from a non-negative random damage. Each damage, at the time of its occurrence, adds to the current damage level of the unit, and between shocks the damage level stays constant. The unit fails when its accumulated damage first exceeds a prespecified level. To keep the unit in an acceptable operating condition, PM is necessary [42].

Kijima and Nakagawa [42] proposed a cumulative damage shock model with imperfect periodic PM. The PM is imperfect in the sense that each PM reduces the damage level by $100(1-b)\%, 0 \leq b \leq 1$, of total damage. Note that if b=1 the PM is minimal and if b=0 the PM coincides with a perfect PM. Obviously, this modeling approach is similar to Treatment Method 1, the (p, q) *rule*. Kijima and Nakagawa [42] derived a sufficient condition for the time to failure to have an increasing failure rate (IFR) distribution and discussed the problem of finding the number of PMs that minimizes the expected maintenance cost rate.

Later, Kijima and Nakagawa [43] established a cumulative damage shock model with a sequential PM policy given that PM is imperfect. They modeled imperfect PM in a way that the amount of damage after the kth PM becomes $b_k Y_k$ when it was Y_k before PM, *i.e.* the kth PM reduces the amount Y_k of damage to $b_k Y_k$, where b_k is called the improvement factor. They assumed that a system is subject to shocks occurring according to a Poisson process and, upon occurrence of shocks, it suffers from a non-negative random damage, which is additive. Each shock causes a system failure with probability $p(z)$ when the total damage is z at the shock. In this model, PM is done at fixed intervals x_k for $k = 1, 2, 3, \ldots, N$, because more frequent maintenance is needed with age, and the Nth PM is perfect. If the system fails between PMs then it undergoes only minimal repair. They derive the expected maintenance cost rate until replacement assuming that $p(z)$ is an exponential function and damages are i.i.d., and discussed the optimal replacement policies.

This study approach for imperfect maintenance will be called *shock method* later in this chapter.

21.2.6 Treatment Method 6

Wang and Pham [23, 44] treated imperfect repair in a way that after repair the lifetime of a system will be reduced to a fraction α of its immediately previous one, where $0 < \alpha < 1$, *i.e.* the lifetime decreases with the number of repairs. The interarrival times between successive repairs constitute a quasi-renewal process, defined by Wang and Pham [23].

Wang and Pham [45] further considered that repair time is non-negligible, not as in most imperfect maintenance models, upon repair the next repair time becomes a multiple β of its current one, where $\beta > 1$, and the time to repair increases with the number of repairs. We refer to this treatment method as the (α, β) *rule*.

21.2.7 Treatment Method 7

Shaked and Shanthikumar [46] introduced the multivariate imperfect repair concept. They considered a system whose components have dependent lifetimes where each component is subject to an imperfect repair until it is replaced. For each component the repair is imperfect according to the (p, q) rule, *i.e.* at failure the repair is perfect with probability p and minimal with probability q. Assume that n components of the system start to function at the same time zero, and no more than one component can fail at a time. They established the joint distribution of the times to next failure of the functioning devices after a minimal repair or perfect repair, and derived the joint density of the resulting lifetimes of the components and other probabilistic quantities of interest, from which the distribution of the lifetime of the system can be obtained. Sheu and Griffith [47] extended this work further. This treatment method will be termed the *multiple* (p, q) *rule* herein.

21.2.8 Other Methods

Nakagawa [20] modeled imperfect PM in a way that, in the steady-state, PM reduces the failure rate of a unit to a fraction of its value just before PM and during operation the failure rate climbs back up. He proposed that the portion by which the failure rate is reduced is a function of some resource consumed in PM and a parameter. That is, after PM the failure rate of the unit becomes $\lambda(t) = g(c_1, \theta)\lambda(x + T)$, where the fraction reduction of failure rate $g(c_1, \theta)$ lies between zero and one, T is the time interval length between PM's, c_1 is the amount of resource consumed in PM, and θ is a parameter.

Nakagawa [48, 49] used two other methods to deal with imperfect PM for sequential PM policy. (a) The failure rate after PM k becomes $a_k h(t)$ given it was $h(t)$ in previous period, where $a_k \geq 1$; *i.e.* the failure rate increases with the number of PMs. (b) The age after PM k reduces to $b_k t$ when it was t before PM, where $0 \leq b_k < 1$. This modeling method is similar to the improvement factor method and will be called the *reduction method* herein. That is, PM reduces the age. In addition, in investigating periodic PM models, Nakagawa [21] treated imperfect PM in a way that the age of the unit becomes x units of time younger by each PM and further assumed that the x is in proportion to the PM cost, where x is less than or equal to the PM intervention interval.

Yak *et al.* [50] suggested that maintenance may result in system failure (worst maintenance), in modeling the mean time to failure and the availability of the system.

21.3 Some Results on Imperfect Maintenance

Next, we summarize some important results on imperfect maintenance from some recent work [10, 23, 51].

21.3.1 A Quasi-renewal Process and Imperfect Maintenance

Renewal theory had its origin in the study of strategies for replacement of technical components. In a renewal process, the times between successive events are supposed to be i.i.d. Most maintenance models using renewal theory are actually based on this assumption, *i.e.* "as good as new". However, in maintenance practice a system may not be "as good as new" after repair. Barlow and Hunter [52] (see Barlow and Proschan [3], p.89) introduced the notation of "minimal repair" and use non-homogeneous Poisson processes to model it. In fact, "as good as new" and "as bad as old" represent two extreme types of repair result. Most repair actions are somewhere between these extremes, which are imperfect repairs. Wang and Pham [23] explored a quasi-renewal process and its applications in modeling imperfect maintenance, which are summarized below.

Let $\{N(t),\ t > 0\}$ be a counting process and X_n denote the time between the $(n-1)$th and the nth event of this process, $n \geq 1$.

Definition 1. If the sequence of non-negative random variables $\{X_1, X_2, X_3, \ldots\}$ is independent and $X_i = \alpha X_{i-1}$ for $i \geq 2$, where $\alpha > 0$ is a constant, then the counting process $\{N(t),\ t \geq 0\}$ is said to be a quasi-renewal process with parameter α and the first interarrival time X_1.

When $\alpha = 1$, this process becomes the ordinary renewal process. Given that the probability distribution function (p.d.f.), cumulative distribution function (c.d.f.), survival function and failure rates of random variable X_1 are $f_1(x)$, $F_1(x)$, $s_1(x)$ and $r_1(x)$, the p.d.f., c.d.f., survival function, and failure rate of X_n for $n = 2, 3, 4, \ldots$ are

$$f_n(x) = \alpha^{1-n} f_1(\alpha^{1-n} x) \quad F_n(x) = F_1(\alpha^{1-n} x)$$
$$s_n(x) = s_1(\alpha^{1-n} x) \quad r_n(x) = \alpha^{1-n} r_1(\alpha^{1-n} x)$$

The following results are proved [23]:

Proposition 1. *If $f_1(x)$ belongs to IFR, DFR, IFRA, DFRA, NBU, NWU*[1]*, then $f_n(x)$ is in the same category for $n = 2, 3, \ldots$.*

Proposition 2. *The shape parameter of X_n are the same for $n = 1, 2, 3, 4, \ldots$ for a quasi-renewal process if X_1 follows the Gamma, Weibull, or lognormal distribution.*

Observe a quasi-renewal process with parameter α and the first interarrival time X_1. For the total number $N(t)$ of "renewals" that have occurred up to time t

$$\Pr\{N(t) = n\} = G_n(t) - G_{n+1}(t)$$

where $G_n(t)$ is the convolution of the interarrival times $F_1, F_2, \ldots, F_n$.

If the mean value of $N(t)$ is defined as the renewal function $M(t)$ then

$$M(t) = E[N(t)] = \sum_{n=1}^{\infty} G_n(t)$$

The derivative of $M(t)$ is known as renewal density:

$$m(t) = M'(t)$$

It is proved that the first interarrival distribution of a quasi-renewal process uniquely determines its renewal function.

Three applications of the quasi-renewal process to imperfect maintenance are described next.

21.3.1.1 Imperfect Maintenance Model A

Suppose that a unit is preventively maintained at times kT, $k = 1, 2, \ldots$, at a cost c_p, independently of the unit's failure history, where the constant $T > 0$ and the PM is perfect. The unit undergoes imperfect repair at failures between PM's at cost c_f in the sense that after repair the lifetime will reduce to a fraction α of the immediately previous one. Then, from the quasi-renewal theory, the following propositions follow.

Proposition 3. *The long-run expected maintenance cost per unit time, or maintenance cost rate, is*

$$L(T) = \frac{c_p + c_f M(T)}{T}$$

where $M(T)$ is the renewal function of a quasi-renewal process with parameter α.

Proposition 4. *There exists an optimum T^* that minimizes $L(T)$, where $0 < T^* \leq \infty$ and the resulting minimum value of $L(T)$ is $c_f m(T^*)$.*

21.3.1.2 Imperfect Maintenance Model B

This model is exactly like Model A, except that the PM at time kT is imperfect in the sense that, after PM, the unit is "as good as new" with probability p and "as bad as old" with probability $q = 1 - p$. From the quasi-renewal theory, the following propositions are proved.

Proposition 5. *The long-run expected maintenance cost per unit time is:*

$$L(T) = \frac{c_p + c_f p^2[M(T) + \sum_{i=2}^{\infty} q^{i-1} M(iT)]}{T}$$

where $M(iT)$ is the renewal function of a quasi-renewal process with parameter α.

Proposition 6. *There exists an optimum T^* that minimizes $L(T)$, where $0 < T^* \leq \infty$, and the resulting minimum value of $L(T)$ is $c_f p^2 \sum_{i=1}^{n} q^{i-1}[iT^* m(iT^*)]$.*

21.3.1.3 Imperfect Maintenance Model C

Assume that a unit is repaired at failure i at a cost $c_f + (i - 1)c_v$ if and only if $i \leq k - 1$, where $i = 1, 2, \ldots$, and the repair is imperfect in the sense that, after repair, the lifetime will reduce to a fraction α of the immediately previous one [23]. Notice that the repair cost increases by c_v for each next imperfect repair. Suppose that the first imperfect repair time is random variable Y_1 with mean η_1, the second imperfect repair time is βY_1 with mean $\beta\eta_1$, and the $(k-1)$th repair time is $\beta^{k-2} Y_1$ with mean $\beta^{k-2}\eta_1$, where the constant $\beta \geq 1$ means the repair times are increasing as the number of imperfect repairs increases. After the $(k - 1)$th imperfect repair at failure the unit is preventively maintained at times nT $(n = 1, 2, \ldots)$ at a cost c_p, where the constant

[1] See the Appendix 21.A.1 for definitions.

$T > 0$ and the PM is imperfect in the sense that, after PM, the unit is "as good as new" with probability p and restored to its condition just prior to failure (minimal repair) with probability $q = 1 - p$. Further suppose that the PM time is a random variable W with mean w. If there is a failure between nT values, an imperfect repair is performed at a cost c_{fr} with negligible repair time in the sense that, after repair, the lifetime of this unit will be reduced to a fraction λ of its immediately previous one where, $0 < \lambda < 1$.

One possible interpretation of this model is as follows: when a new unit is put into use, the first $k - 1$ times of failure repairs, because the unit is young at that time, will be performed at a low cost $c_f + (i - 1)c_v$ for $i = 0, 1, 2, \ldots, k - 2$, which result in imperfect repairs. After the $(k-1)$th imperfect repair at failure the unit will be in bad condition and then a better or perfect maintenance is necessary at a higher cost c_p or c_{fr}.

Proposition 7. *The long-run expected maintenance cost rate is*

$$L(T, k) = \left\{(k-1)c_f + \frac{(k-1)(k-2)}{2}c_v + c_p p^{-1} + p c_{fr} \sum_{i=1}^{\infty} q^{i-1} M(iT)\right\} \times \left\{\frac{\mu(1-\alpha^{k-1})}{1-\alpha} + \frac{\mu(1-\beta^{k-1})}{1-\beta} + \frac{T}{p} + w\right\}^{-1}$$

and the asymptotic average availability is

$$A(T, k) = \left\{\frac{\mu(1-\alpha^{k-1})}{1-\alpha} + \frac{T}{p}\right\} \times \left\{\frac{\mu(1-\alpha^{k-1})}{1-\alpha} + \frac{\eta(1-\beta^{k-1})}{1-\beta} + \frac{T}{p} + w\right\}^{-1}$$

where $M(t)$ is the renewal function of a quasi-renewal process with parameter λ and the first interarrival time distribution $F_1(\alpha^{1-k}t)$.

Sometimes the optimum maintenance policy to minimize the cost rate is desired while some availability requirements are satisfied. Thus, the following optimization problem can be formulated in terms of decision variables T and k:

Minimize

$$L(T, k) = \left\{(k-1)c_f + \frac{(k-1)(k-2)}{2}c_v + c_p p^{-1} + p c_{fr} \sum_{i=1}^{\infty} q^{i-1} M(iT)\right\} \times \left\{\frac{\mu(1-\alpha^{k-1})}{1-\alpha} + \frac{\eta(1-\beta^{k-1})}{1-\beta} + \frac{T}{p} + w\right\}^{-1}$$

Subject to

$$A(T, k) = \left\{\frac{\mu(1-\alpha^{k-1})}{1-\alpha} + \frac{T}{p}\right\} \times \left\{\frac{\mu(1-\alpha^{k-1})}{1-\alpha} + \frac{\eta(1-\beta^{k-1})}{1-\beta} + \frac{T}{p} + w\right\}^{-1} \geq A_0$$

$$k = 2, 3, \ldots$$

$$T > 0$$

where constant A_0 is the prespecified availability requirement.

Assume that the lifetime X_1 of a new system follows a normal distribution with mean μ and variance σ^2, and

$$\mu = 10 \quad \sigma = 1 \quad \eta = 0.9 \quad c_f = \$1 \quad c_v = \$0.06$$
$$c_p = \$3 \quad c_{fr} = \$4 \quad \alpha = 0.95 \quad \beta = 1.05$$
$$p = 0.95 \quad w = 0.2 \quad \lambda = 0.95 \quad A_0 = 0.94$$

From the above optimization model we can find the globally optimal solution (T^*, k^*) by using commercial optimization software that minimizes the maintenance cost rate given that the availability is at least 0.94:

$$T^* = 7.6530 \quad k^* = 3$$

and the corresponding minimum cost rate and availability are respectively

$$L(T^*, k^*) = \$0.2332 \quad A(T^*, k^*) = 0.9426$$

The results indicate that the optimal maintenance policy is to perform repair at the first two failures of the system at a cost of \$1 and \$1.06 respectively, and then do PM every 7.6530 time units at a cost of \$3 and repair the system upon failure between PMs at a cost \$4.

21.3.1.4 Imperfect Maintenance Model D

Model D is identical to Model C except that after the $(k-1)$th imperfect repair there are two types of failure [45], and that PM's at times $T, 2T, 3T, \ldots$ are perfect. Type 1 failure might be total breakdowns, whereas type 2 failure can be interpreted as a slight and easily fixed problem. Type 1 failures are subject to perfect repairs and type 2 failures are subject to minimal repairs. When a failure occurs it is a type 1 failure with probability $p(t)$ and a type 2 failure with probability $q(t) = 1 - p(t)$, where t is the age of the unit. Thus, the repair at failure can be modeled by the $(p(t), q(t))$ rule. Assume that the failure repair time is negligible, and PM time is a random variable V with mean v.

Consider T and k as decision variables. For this maintenance model, the times between consecutive perfect PM's constitute a renewal cycle. From the classical renewal reward theory, the long-run expected maintenance cost per system time, or cost rate, is

$$L(T, k) = \frac{C(T, k)}{D(T, k)}$$

where $C(T, k)$ is the expected maintenance cost per renewal cycle and $D(T, k)$ is the expected duration of a renewal cycle.

After the $(k-1)$th imperfect repair, let Y_p denote the time until the first perfect repair without PM since last perfect repair, *i.e.* the time between successive perfect repairs. As described in Section 21.2.2, the survival distribution of Y_p is given by

$$\begin{aligned}\bar{S}(t) &= \exp\left\{-\int_0^t p(x) r_k(x)\,\mathrm{d}x\right\} \\ &= \exp\left\{-\alpha^{1-k}\int_0^t p(x) r(\alpha^{1-k}x)\,\mathrm{d}x\right\}\end{aligned}$$

Assume that Z_t represents the number of minimal repairs during the time interval $(0, \min\{t, Y_p\})$ and $S(t) = 1 - \bar{S}(t)$. Wang and Pham [45] show:

$$\begin{aligned}E\{Z_t \mid Y_p < t\} &= \frac{1}{S(t)}\int_0^t\int_0^y q(x) r_k(x)\,\mathrm{d}x\,\mathrm{d}S(y) \\ &= \frac{\alpha^{1-k}}{S(t)} \\ &\quad\times \int_0^t\int_0^y q(x) r_1(\alpha^{1-k}x)\,\mathrm{d}x\,\mathrm{d}S(y) \\ E\{Z_t \mid Y_p \geq t\} &= \int_0^t q(x) r_k(x)\,\mathrm{d}x \\ &= \alpha^{1-k}\int_0^t q(x) r_1(\alpha^{1-k}x)\,\mathrm{d}x\end{aligned}$$

Let $N_1(t)$ and $N_2(t)$ denote the s-expected number of perfect repairs and minimal repairs in $(0, t)$ respectively; c_1, c_2, and c_p denote costs of perfect repair, minimal repair, and PM respectively. It is easy to obtain

$$\begin{aligned}D(T, k) &= \frac{\mu(1-\alpha^{k-1})}{1-\alpha} + \frac{\eta(1-\beta^{k-1})}{1-\beta} + T + v \\ C(T, k) &= (k-1)c_f + \frac{(k-1)(k-2)}{2}c_v \\ &\quad + c_1 N_1(T) + c_2 N_2(T) + c_p\end{aligned}$$

Obviously, $N_1(t)$ is the renewal function for the renewal process with the interarrival time distribution $S(t)$ and can be determined by the solution method to the renewal function in renewal theory. Wang and Pham [45] proved for $t \leq T$:

$$\begin{aligned}N_2(t) &= E\{Z_t \mid Y_p \geq t\}\bar{S}(t) \\ &\quad + \int_0^t [E\{Z_x \mid Y_p = x\} + N_2(t-x)]\,\mathrm{d}S(x)\end{aligned}$$

Note that

$$E\{Z_t \mid Y_p < t\}S(t) = \int_0^t E\{Z_x \mid Y_p = x\}\,\mathrm{d}S(x)$$

It follows that

$$\begin{aligned}N_2(t) &= E\{Z_t \mid Y_\text{p} \geq t\}\bar{S}(t) + E\{Z_t \mid Y_\text{p} < t\}S(t) \\ &\quad + \int_0^t N_2(t-x)\,\mathrm{d}S(t) \\ &= E(Z_t) + \int_0^t N_2(t-x)\,\mathrm{d}S(t) \\ &= \alpha^{1-k}\int_0^t \bar{S}(t)r_1(\alpha^{1-k}x)\,\mathrm{d}x - S(t) \\ &\quad + \int_0^t N_2(t-x)\,\mathrm{d}S(t)\end{aligned}$$

Therefore, $N_2(t)$ can be obtained by the Laplace transform or solving the above integral equation using numerical computation. The following proposition follows from the above equations.

Proposition 8. *The long-run expected maintenance cost rate is given by*

$$\begin{aligned}L(T,k) = \bigg\{&(k-1)c_\text{f} + \frac{(k-1)(k-2)}{2}c_\text{v} \\ &+ c_1N_1(T) + c_2N_2(T) + c_\text{p}\bigg\} \\ &\times \bigg\{\frac{\mu(1-\alpha^{k-1})}{1-\alpha} \\ &+ \frac{\eta(1-\beta^{k-1})}{1-\beta} + T + v\bigg\}^{-1}\end{aligned}$$

21.3.1.5 Imperfect Maintenance Model E

This model is the same as Model C except that at next failures, after the $(k-1)$th imperfect repair since time zero, repair cost is estimated by perfect inspection to determine whether to replace or imperfectly repair it [53]. Assume that the repair cost has a cumulative distribution function $C(x)$ that is independent of the age of the unit. If the estimated cost does not exceed a constant cost limit Q, then the unit is imperfectly repaired at an expected repair cost not exceeding Q. Otherwise, it is replaced by a new one at a higher fixed cost and the replacement time is W with mean w. Imperfect repair is modeled by the (p,q) rule. Assume that the repair time is V with mean v, and that W and V are independent of the previous failure history of the unit. Upon a perfect repair or replacement the process repeats. We consider k and Q as decision variables, and α, β, and p as parameters.

Proposition 9. *The long-run expected maintenance cost per unit time is*

$$\begin{aligned}&L(k,Q;\alpha,\beta,p) \\ &= \bigg\{(k-1)c_\text{f} + \frac{(k-1)(k-2)}{2}c_\text{v} \\ &\quad + \frac{c_2[1-C(Q)] + \bar{c}_1C(Q)}{1-pC(Q)}\bigg\} \\ &\quad \times \bigg\{\frac{\mu(1-\alpha^{k-1})}{1-\alpha} + \frac{\eta(1-\beta^{k-1})}{1-\beta} \\ &\quad + \frac{[1-C(Q)]w + pC(Q)v}{1-qC(Q)} \\ &\quad + \int_0^\infty \exp\{-H(\alpha^{1-k}t)[1-qC(Q)]\}\,\mathrm{d}t\bigg\}^{-1}\end{aligned}$$

and the asymptotic average availability is

$$\begin{aligned}&A(k,Q;\alpha,\beta,p) \\ &= \bigg\{\frac{\mu(1-\alpha^{k-1})}{1-\alpha} \\ &\quad + \int_0^\infty \exp\{-H(\alpha^{1-k}t)[1-qC(Q)]\}\,\mathrm{d}t\bigg\} \\ &\quad \times \bigg\{\frac{\mu(1-\alpha^{k-1})}{1-\alpha} + \frac{\eta(1-\beta^{k-1})}{1-\beta} \\ &\quad + \frac{[1-C(Q)]w + pC(Q)v}{1-qC(Q)} \\ &\quad + \int_0^\infty \exp\{-H(\alpha^{1-k}t)[1-qC(Q)]\}\,\mathrm{d}t\bigg\}^{-1}\end{aligned}$$

where $H(\alpha^{1-k}t) = \int_0^t \alpha^{1-k}r_1(\alpha^{1-k}x)\,\mathrm{d}x$ is the cumulative hazard of the unit right after the $(k-1)$th imperfect repair and $\bar{c}_1 = C^{-1}(Q)\int_0^L t\,\mathrm{d}C(t)$ is the mean of repair costs less than Q.

The optimum maintenance policy (k^*, Q^*) that minimizes $L(k,Q;\alpha,\beta,p)$ or maximizes $A(k,Q;\alpha,\beta,p)$ can be obtained using any nonlinear programming software.

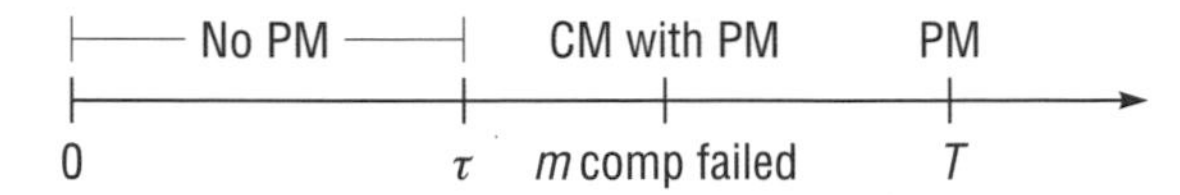

Figure 21.2. Imperfect maintenance policy for a k-out-of-n system

21.3.2 Optimal Imperfect Maintenance of k-out-of-n Systems

For a complex system, it may not be advisable to replace the entire system just because of the failure of one component. In fact, the system comes back into operation on repair or replacement of the failed component by a new one or by a used but operative one. Such maintenance actions do not renew the system completely but enable the system to continue to operate. However, the system is usually deteriorating with usage and time. At some point of time or usage, it may be in a bad operating condition and a perfect maintenance is necessary. Based on this situation, Pham and Wang [51] proposed the following imperfect maintenance policy for a k-out-of-n system.

A new system starts to operate at time zero. Each failure of a component of this system in the time interval $(0, \tau)$ is immediately removed by a minimal repair. Components that fail in the time interval (τ, T) can be lying idle (partial failure is allowed). Perform CM on the failed components together with PM on all unfailed but deteriorating ones at a cost of c_f once exactly m components are idle, or perform PM on the whole system at a cost of c_p once the total operating time reaches T, whichever occurs first. That is, if m components fail in the time interval (τ, T), then CM combined with PM is performed; if less than m components fail in the time interval (τ, T), then PM is carried out at time T. After a perfect maintenance, whether this is a CM combined with PM or a PM at T, the process repeats.

In the above maintenance policy, τ and T are decision variables. A k-out-of-n system is defined as a complex coherent system with n failure-independent components such that the system operates if and only if at least k of these components function successfully. This chapter assumes that m is a predetermined positive integer where $1 \leq m \leq n-k+1$, and real numbers $\tau < T$. This policy is demonstrated in Figure 21.2. In practice, m may take different values according to different reliability and cost requirements. Obviously, $m = 1$ means that the system is subject to maintenance whenever one component fails after τ. For a series system (n-out-of-n system) or a system with key applications, m may basically be required to be unity. If m is chosen as $n - k + 1$, then a k-out-of-n system is maintained once the system fails. In most applications, the whole system is subject to a perfect CM together with a PM upon a system failure ($m = n - k + 1$) or partial failure ($m < n - k + 1$). Here, partial failure means some components have failed but the system still functions. However, if inspection is not continuous and the system operating condition can be known only through inspection, m could be a number greater than $(n - k + 1)$. For simplicity, in this paper we assume that if CM together with PM is carried out, then both are perfect, given that CM combined with PM takes w_1 time units and PM at time T takes w_2 time units. Further suppose that every component has IFR, which is differentiable and remains undisturbed by minimal repair.

The justification of this policy is as follows: before τ, each component is "young" and no major repair is necessary. Therefore, only minimal repairs, which may not take much time and money, are performed. The component is deteriorating as the time passes. After τ, the component has a larger failure rate (due to IFR) and might be in a bad operating condition. Thus, a major or perfect repair may be needed. Because there exist economic dependence and availability requirements (less frequent shut-offs for maintenance), however, we may not replace failed components immediately but start CM until the number of failed components reaches some prespecified number m. In fact, when the number of failed components accumulates to m, the remaining $(n - m)$ operating components may degrade to a worse operating condition and need PM also. Note that, as long as m is less than $(n-k+1)$, the system will not fail and continue to operate.

Assume that for each component the cost of the ith minimal repair at age t consists of two parts: the deterministic part $c_1(t, i)$, which depends on the age t of this component and the number i of minimal repairs, and the age-dependent random part $c_2(t)$. The assumption that the failure rate of each component has IFR is essential. This is because the system may be subject to a PM at time T, which requires the system to be IFR after τ. The following proposition states the relationship between component and system IFRs.

Proposition 10. *If a k-out-of-n system is composed of independent, identical, IFR components, the system has an IFR also.*

Given that PM at time $T, 2T, 3T, \ldots$ is imperfect, the failure rate of each component is $q(t)$, the survival function of the time to failure of the k-out-of-n system is $\bar{F}_{n-k+1}(y)$, and the cost expectation at age t is $\mu(t)$, Pham and Wang [51] derived the following results using renewal theory:

Proposition 11. *If the PM is perfect with probability p and minimal with probability $q = 1 - p$, then the long-run expected system maintenance cost per unit time, or cost rate, for a k-out-of-n system:G is given by*

$$L(\tau, T \mid p) = \left\{ n \int_0^{\tau} \mu(y) q(y)\, \mathrm{d}y + c_{\mathrm{p}} \sum_{j=1}^{\infty} q^{j-1} \bar{F}_m(jT-\tau) + c_{\mathrm{f}} \left[1 - p \sum_{j=1}^{\infty} q^{j-1} \bar{F}_m(jT-\tau) \right] \right\} \times \left\{ \tau + \int_0^{T-\tau} \bar{F}_m(t)\, \mathrm{d}t + \sum_{j=2}^{\infty} q^{j-1} \int_{(j-1)T-\tau}^{jT-\tau} \bar{F}_m(t)\, \mathrm{d}t + w_1 \left[1 - p \sum_{j=1}^{\infty} q^{j-1} \bar{F}_m(jT-\tau) \right] + w_2 \left[\sum_{j=1}^{\infty} q^{j-1} \bar{F}_m(jT-\tau) \right] \right\}^{-1}$$

and the limiting average system availability is

$$A(\tau, T \mid p) = \left\{ \tau + \int_0^{T-\tau} \bar{F}_m(t)\, \mathrm{d}t + \sum_{j=2}^{\infty} q^{j-1} \int_{(j-1)T-\tau}^{jT-\tau} \bar{F}_m(t)\, \mathrm{d}t \right\} \times \left\{ \tau + \int_0^{T-\tau} \bar{F}_m(t)\, \mathrm{d}t + \sum_{j=2}^{\infty} q^{j-1} \int_{(j-1)T-\tau}^{jT-\tau} \bar{F}_m(t)\, \mathrm{d}t + w_1 \left[1 - p \sum_{j=1}^{\infty} q^{j-1} \bar{F}_m(jT-\tau) \right] + w_2 \left[\sum_{j=1}^{\infty} q^{j-1} \bar{F}_m(jT-\tau) \right] \right\}^{-1}$$

The above model includes at least 12 existing maintenance models as special cases. For example, when $k = 1$, and $n > 1$, then the k-out-of-n system is reduced to a parallel system. If we further let $k = 1$ and $m = n$, it follows that the long-run expected system maintenance cost rate for a parallel system with n components is

$$L(\tau, T) = \left\{ n \int_0^{\tau} \mu(y) q(y)\, \mathrm{d}y + (c_{\mathrm{f}} - c_{\mathrm{p}}) [\bar{F}(\tau) - \bar{F}(T)]^n \bar{F}^{-n}(\tau) + c_{\mathrm{p}} \right\} \times \left\{ \tau + \int_0^{T-\tau} \{1 - [\bar{F}(\tau) - \bar{F}(\tau+t)]^n \bar{F}^{-n}(\tau)\}\, \mathrm{d}t + (w_1 - w_2) [\bar{F}(\tau) - \bar{F}(T)]^n \bar{F}^{-n}(\tau) + w_2 \right\}^{-1}$$

If we set $\tau = 0$ and $w_1 = w_2 = 0$, then the above equation becomes

$$L(0, T) = \frac{(c_{\mathrm{f}} - c_{\mathrm{p}}) F^n(T) + c_{\mathrm{p}}}{\int_0^T [1 - F^n(t)]\, \mathrm{d}t}$$

which coincides with the result by Yasui *et al.* [54].

Sometimes, optimal maintenance policies may be required so that the system maintenance cost rate is minimized while some availability requirements are satisfied, or the system availability is maximized given that the maintenance cost rate is not larger than some predetermined value. For example, for the above maintenance model, the following optimization problem can be formulated in terms of decision variables T and τ:

Minimize

$$\begin{aligned}
L(\tau, T \mid p) &= \Bigg\{ n \int_0^{\tau} \mu(y) q(y)\,\mathrm{d}y \\
&\quad + c_{\mathrm{p}} \sum_{j=1}^{\infty} q^{j-1} \bar{F}_m(jT - \tau) \\
&\quad + c_{\mathrm{f}} \Bigg[1 - p \sum_{j=1}^{\infty} q^{j-1} \bar{F}_m(jT - \tau) \Bigg] \Bigg\} \\
&\quad \times \Bigg\{ \tau + \int_0^{T-\tau} \bar{F}_m(t)\,\mathrm{d}t \\
&\quad + \sum_{j=2}^{\infty} q^{j-1} \int_{(j-1)T-\tau}^{jT-\tau} \bar{F}_m(t)\,\mathrm{d}t \\
&\quad + w_1 \Bigg[1 - p \sum_{j=1}^{\infty} q^{j-1} \bar{F}_m(jT - \tau) \Bigg] \\
&\quad + w_2 \Bigg[\sum_{j=1}^{\infty} q^{j-1} \bar{F}_m(jT - \tau) \Bigg] \Bigg\}^{-1}
\end{aligned}$$

Subject to

$$\begin{aligned}
A(\tau, T \mid p) &\geq A_0 \\
\tau &\geq 0 \\
T &> 0
\end{aligned}$$

where constant A_0 is the pre-decided minimum availability requirement.

The optimal system maintenance policy (T^*, τ^*) can be determined from the above optimization model by using nonlinear programming software. Similarly, other optimization models can be formulated based on different requirements. Such optimization models should be truly optimal, since both availability and maintenance costs are addressed.

An application of the above model to a 2-out-of-3 aircraft engine system is discussed by Pham and Wang [51], given the time to failure of each engine follows a Weibull distribution with shape parameter $\beta = 2$ and scale parameter $\theta = 500$ (days). The optimal maintenance policy for this 2-out-of-3 aircraft engine system to minimize the system maintenance cost rate is determined from the above model: before $\tau^* = 335.32$ (days), only minimal repairs are performed; after $\tau^* = 335.32$, the failed engine will be subject to perfect repair together with PM on the remaining two once any engine fails; if no engine fails until time $T^* = 383.99$ (days), then PM is carried out at $T^* = 383.99$.

21.4 Future Research on Imperfect Maintenance

The following further work on imperfect maintenance is necessary:

- Investigate optimum PM policy and reliability measures for *multicomponent* systems because most previous work on imperfect maintenance has been focused on a *one-unit* system.
- Develop more methods for treating imperfect maintenance.
- Establish statistical estimation of parameters for imperfect maintenance models.
- Explore more realistic imperfect maintenance models, *e.g.* including non-negligible repair time, finite horizon. In most literature on maintenance theory, the maintenance time is assumed to be negligible. This assumption makes availability and mean-time-between-failures modeling impossible or unrealistic. Considering maintenance time will result in realistic system reliability measures.
- Use the reliability measures as the optimality criteria instead of cost rates, or combine both, as shown in Section 21.3.1: optimal maintenance policy to minimize cost rate under given reliability requirement. Most existing

optimal maintenance models use the optimization criterion of minimizing the system maintenance cost rate but ignoring the reliability performance. However, maintenance aims to improve system reliability performance. Therefore, the optimal maintenance policy must be based on not only cost rate but also on reliability measures. It is important to note that, for multicomponent systems, minimizing system maintenance cost rate may not imply maximizing the system reliability measures. Sometimes, when the maintenance cost rate is minimized the system reliability measures are so low that they are not acceptable in practice [54]. This is because various components in the system may have different maintenance costs and a different reliability importance in the system [53, 55, 56]. Therefore, to achieve the best operating performance, an optimal maintenance policy needs to consider both maintenance cost and reliability measures simultaneously.

- Consider the structure of a system to obtain optimal system reliability performance and optimal maintenance policy. For example, once a subsystem of a series system fails it is necessary to repair it immediately. Otherwise, the system will have longer downtime and worse reliability measures. However, when one subsystem of a parallel system fails, the system will still function even if this subsystem is not repaired right away. In fact, its repair can be delayed until it is time to do the PM on the system considering economic dependence; or repair can begin at such a time that only one subsystem operates and the other subsystems have failed and are awaiting repairs [51]; or at the time that all subsystems fail, and thus the system fails, if the system failure during actual operation is not critical.

21.A Appendix

21.A.1 Acronyms and Definitions

NBU	new better than used
NWU	new worse than used
NBUE	new better than used in expectation
NWUE	new worse than used in expectation
IFRA	increasing failure rate in average
DFRA	decreasing failure rate in average
IFR	increasing failure rate
DFR	decreasing failure rate.

Assume that lifetime has a distribution function $F(t)$ and survival function $s(t) = 1 - F(t)$ with mean μ. The following definitions are given (for details see Barlow and Proschan [3]).

- s is NBU if $s(x + y) \le s(x)s(y)$ for all $x, y \ge 0$.
  s is NWU if $s(x + y) \ge s(x)s(y)$ for all $x, y \ge 0$.
- s is NBUE if $\int_t^\infty s(x)\,\mathrm{d}x \le \mu s(t)$ for all $t \ge 0$.
  s is NWUE if $\int_t^\infty s(x)\,\mathrm{d}x \ge \mu s(t)$ for all $t \ge 0$.
- s is IFRA if $[s(x)]^{1/x}$ is decreasing in x for $x > 0$.
  s is DFRA if $[s(x)]^{1/x}$ is increasing in x for $x > 0$.
- s is IFR if and only if $[F(t + x) - F(t)]/s(t)$ is increasing in t for all $x > 0$.
  s is DFR if and only if $[F(t + x) - F(t)]/s(t)$ is decreasing in t for all $x > 0$.

21.A.2 Exercises

1. What is minimal repair? What is imperfect repair? Their relationships?
2. Prove Proposition 1.
3. Assume the imperfect maintenance Model C in Section 21.3.1.3 is modified, and the new model is exactly like it except that the imperfect PM is treated by the x rule, *i.e.* the age of the unit becomes x units of time younger after PM; that the unit undergoes minimal repair at failures between PMs at cost c_{fm} instead of imperfect repairs in terms of parameter λ in Model C. Assume further that the Nth PM since the last perfect PM is perfect, where N is a positive integer. A cost $c_{N\mathrm{p}}$ and an independent replacement time V with mean v is suffered for the perfect PM at time NT. Assume that imperfect PM at other times takes W time with mean w and

imperfect PM cost is c_p. Suppose that $c_{N\mathrm{p}} > c_\mathrm{p}$, $v \geq w$, and W and V are independent of the previous failure history of the unit. Derive the long-run expected maintenance cost per unit time and the asymptotic average availability.

4. A new model is identical to imperfect maintenance Model C in Section 21.3.1.3 except that after the $(k-1)$th repair at failure the unit will be either replaced at next failure at a cost of c_fr, or preventively replaced at age T at a cost c_p, whichever occurs first, *i.e.* after time zero a unit is imperfectly repaired at failure i at a cost $c_\mathrm{f} + (i-1)c_\mathrm{v}$ for $i \leq k-1$, where c_f and c_v are constants. The repair is imperfect in the sense that, after repair, the lifetime will be reduced to a fraction α of the immediately previous lifetime, that the repair times will be increased to a multiple β of the immediately previous one, and that the successive lifetimes and repair times are independent. Note that the lifetime of the unit after the $(k-2)$th repair and the $(k-1)$th repair time are respectively $\alpha^{k-2}Z_{k-1}$ and $\beta^{k-2}\zeta_{k-1}$ with means $\alpha^{k-2}\mu$ and $\beta^{k-2}\eta$, where Z_i and ζ_i are respectively i.i.d. random variable sequences with $Z_1 = X_1$ and $\zeta_1 = Y_1$. Derive the long-run expected maintenance cost per unit time, given T and k are decision variables, and α and β are parameters.
5. Show Proposition 6.

References

[1] Sherif YS, Smith ML. Optimal maintenance models for systems subject to failure—a review. Nav Res Logist Q 1981;28(1):47–74.

[2] McCall JJ. Maintenance policies for stochastically failing equipment: a survey. Manage Sci 1965;11(5):493–524.

[3] Barlow RE, Proschan F. Mathematical theory of reliability. New York: John Wiley & Sons; 1965.

[4] Pierskalla WP, Voelker JA. A survey of maintenance models: the control and surveillance of deteriorating systems. Nav Res Logist Q 1976;23:353–88.

[5] Jardine AKS, Buzacott JA. Equipment reliability and maintenance. Eur J Oper Res 1985;19:285–96.

[6] Valdez-Flores C, Feldman RM. A survey of preventive maintenance models for stochastically deteriorating single-unit systems. Nav Res Logist 1989;36:419–46.

[7] Cho ID, Parlar M. A survey of maintenance models for multi-unit systems. Eur J Oper Res 1991;51:1–23.

[8] Jensen U. Stochastic models of reliability and maintenance: an overview. In: Ozekici S, editor. Reliability and maintenance of complex systems, NATO ASI Series, vol. 154 (Proceedings of the NATO Advanced Study Institute on Current Issues and Challenges in the Reliability and Maintenance of Complex Systems, held in Kemer-Antalya, Turkey, June 12–22, 1995). Berlin: Springer-Verlag; 1995. p.3–36.

[9] Dekker R. Applications of maintenance optimization models: a review and analysis. Reliab Eng Syst Saf 1996;51(3):229–40.

[10] Pham H, Wang HZ. Imperfect maintenance. Eur J Oper Res 1996;94:425–38.

[11] Dekker R, Wilderman RE, van der Duyn Schouten FA. A review of multi-component maintenance models with economic dependence. Math Methods Oper Res 1997;45(3):411–35.

[12] Kijima M. Some results for repairable systems with general repair. J Appl Prob 1989;26:89–102.

[13] Brown M, Proschan F. Imperfect repair. J Appl Prob 1983;20:851–9.

[14] Nakagawa T, Yasui K. Optimum policies for a system with imperfect maintenance. IEEE Trans Reliab 1987;R-36:631–3.

[15] Helvic BE. Periodic maintenance, on the effect of imperfectness. In: 10th International Symposium on Fault-Tolerant Computing, 1980; p.204–6.

[16] Kay E. The effectiveness of preventive maintenance. Int J Prod Res 1976;14:329–44.

[17] Chan PKW, Downs T. Two criteria for preventive maintenance. IEEE Trans Reliab 1978;R-27:272–3.

[18] Ingle AD, Siewiorek DP. Reliability models for multiprocessor systems with and without periodic maintenance. In: 7th International Symposium on Fault-Tolerant Computing, 1977; p.3–9.

[19] Chaudhuri D, Sahu KC. Preventive maintenance intervals for optimal reliability of deteriorating system. IEEE Trans Reliab 1977;R-26:371–2.

[20] Nakagawa T. Imperfect preventive maintenance. IEEE Trans Reliab 1979;R-28(5):402.

[21] Nakagawa T. Summary of imperfect maintenance policies with minimal repair. RAIRO Rech Oper 1980;14;249–55.

[22] Fontenot RA, Proschan F. Some imperfect maintenance models. In: Abdel-Hameed MS, Cinlar E, Quinn J, editors. Reliability theory and models. Orlando (FL): Academic Press; 1984.

[23] Wang HZ, Pham H. A quasi renewal process and its application in the imperfect maintenance. Int J Syst Sci 1996;27(10):1055–62 and 1997;28(12):1329.

[24] Bhattacharjee MC. Results for the Brown–Proschan model of imperfect repair. J Stat Plan Infer 1987;16:305–16.

[25] Block HW, Borges WS, Savits TH. Age dependent minimal repair. J Appl Prob 1985;22:370–85.

[26] Beichelt F, Fischer K. General failure model applied to preventive maintenance policies. IEEE Trans Reliab 1980;R-29(1):39–41.

[27] Block HW, Borges WS, Savits TH. A general age replacement model with minimal repair. Nav Res Logist 1988;35(5):365–72.

[28] Iyer S. Availability results for imperfect repair. Sankhya: Indian J Stat 1992;54(2):249–59.

[29] Sumita U, Shanthikumar JG. An age-dependent counting process generated from a renewal process. Adv Appl Probab 1988;20(4):739–755

[30] Whitaker LR, Samaniego FJ. Estimating the reliability of systems subject to imperfect repair. J Am Stat Assoc 1989;84(405):301–9.

[31] Hollander M, Presnell B, Sethuraman J. Nonparametric methods for imperfect repair models. Ann Stat 1992;20(2):879–87.

[32] Makis V, Jardine AKS. Optimal replacement policy for a general model with imperfect repair. J Oper Res Soc 1992;43(2):111–20.

[33] Malik MAK. Reliable preventive maintenance policy. AIIE Trans 1979;11(3):221–8.

[34] Lie CH, Chun YH. An algorithm for preventive maintenance policy. IEEE Trans Reliab 1986;R-35(1):71–5.

[35] Jayabalan V, Chaudhuri D. Cost optimization of maintenance scheduling for a system with assured reliability. IEEE Trans Reliab 1992;R-41(1):21–6.

[36] Jayabalan V, Chaudhuri D. Sequential imperfect preventive maintenance policies: a case study. Microelectron Reliab 1992;32(9):1223–9.

[37] Suresh PV, Chaudhuri D. Preventive maintenance scheduling for a system with assured reliability using fuzzy set theory. Int J Reliab Qual Saf Eng 1994;1(4):497–513.

[38] Chan JK, Shaw L. Modeling repairable systems with failure rates that depend on age & maintenance. IEEE Trans Reliab 1993;42:566–70.

[39] Canfield RV. Cost optimization of periodic preventive maintenance. IEEE Trans Reliab 1986;R-35(1):78–81.

[40] Kijima M, Morimura H, Suzuki Y. Periodical replacement problem without assuming minimal repair. Eur J Oper Res 1988;37(2):194–203.

[41] Uematsu K, Nishida T. One-unit system with a failure rate depending upon the degree of repair. Math Jpn 1987;32(1):139–47.

[42] Kijima M, Nakagawa T. Accumulative damage shock model with imperfect preventive maintenance. Nav Res Logist 1991;38:145–56.

[43] Kijima M, Nakagawa T. Replacement policies of a shock model with imperfect preventive maintenance. Eur J Oper Res 1992;57:100–10.

[44] Wang HZ, Pham H. Optimal age-dependent preventive maintenance policies with imperfect maintenance. Int J Reliab Qual Saf Eng 1996;3(2):119–35.

[45] Wang HZ, Pham, H. Some maintenance models and availability with imperfect maintenance in production systems. Ann Oper Res 1999;91:305–18.

[46] Shaked M, Shanthikumar JG. Multivariate imperfect repair. Oper Res 1986;34:437–48.

[47] Sheu S, Griffith WS. Multivariate imperfect repair. J Appl Prob 1992;29(4):947–56.

[48] Nakagawa T. Periodic and sequential preventive maintenance policies. J Appl Prob 1986;23(2):536–42.

[49] Nakagawa T. Sequential imperfect preventive maintenance policies. IEEE Trans Reliab 1988;37(3):295–8.

[50] Yak YW, Dillon TS, Forward KE. The effect of imperfect periodic maintenance on fault tolerant computer systems. In: 14th International Symposium on Fault-Tolerant Computing, 1984; p.67–70.

[51] Pham H, Wang HZ. Optimal (τ, T) opportunistic maintenance of a k-out-of-n:G system with imperfect PM and partial failure. Nav Res Logist 2000;47:223–39.

[52] Barlow RE, Hunter LC. Optimum preventive maintenance policies. Oper Res 1960;8:90–100.

[53] Wang HZ, Pham H. Optimal maintenance policies for several imperfect maintenance models. Int J Syst Sci 1996;27(6):543–9.

[54] Yasui K, Nakagawa T, Osaki S. A summary of optimal replacement policies for a parallel redundant system. Microelectron Reliab 1988;28:635–41.

[55] Wang HZ, Pham H. Availability and optimal maintenance of series system subject to imperfect repair. Int J Plant Eng Manage 1997;2:75–92.

[56] Wang HZ, Pham H, Izundu AE. Optimal preparedness maintenance of multi-unit systems with imperfect maintenance and economic dependence. In: Pham H, editor. Recent advances in reliability and quality engineering. New Jersey: World Scientific; 2001.

Accelerated Life Testing

Elsayed A. Elsayed

Chapter 22

22.1 Introduction

The intensity of the global competition for the development of new products in a short time has motivated the development of new methods such as robust design, just-in-time manufacturing, and design for manufacturing and assembly. More importantly, both producers and customers expect that the product will perform the intended functions for extended periods of time. Hence, extended warranties and similar assurances of product reliability have become standard features of the product. These requirements have increased the need for providing more accurate estimates of reliability by performing testing of materials, components, and systems at different stages of product development. Testing under normal operating conditions requires a very long time (possibly years) and the use of an extensive number of units under test, so it is usually costly and impractical to perform reliability testing under normal conditions. This has led to the development of accelerated life testing (ALT), where units are subjected to a more severe environment (increased or decreased stress levels) than the normal operating environment so that failures can be induced in a short period of test time. Information obtained under accelerated conditions is then used in conjunction with a reliability prediction (inference) model to relate life to stress and to estimate the characteristics of life distributions at design conditions (normal operating conditions). Conducting an accelerated life test requires careful allocation of test units to different stress levels so that accurate estimation of reliability at normal conditions can be obtained using relatively small units and short test durations.

The accuracy of the inference procedure has a profound effect on the reliability estimates at normal operating conditions and the subsequent decisions regarding system configuration, warranties, and preventive maintenance schedules. Therefore, it is important that the inference procedure be robust and accurate. In this chapter we present ALT models and demonstrate their use in relating failure data at stress conditions to normal operating conditions. Before presenting such models, we briefly describe the design of ALT plans, including methods of applying stress and determination of the number of units for each stress level.

In this chapter, we introduce the subject of ALT and its importance in assessing the reliability of components and products under normal operating conditions. We describe briefly the methods of stress application and types of stress, and focus on the reliability prediction models that utilize the failure data under stress conditions to obtain reliability information under normal conditions. Several methods and examples that demonstrate its application are also presented. Before presenting such models, we briefly describe the design of ALT plans, including methods of applying stress and determination of the number of units for each stress level.

22.2 Design of Accelerated Life Testing Plans

A detailed test plan is usually designed before conducting an accelerated life test. The plan requires determination of the type of stress, methods of applying stress, stress levels, the number of units to be tested at each stress level, and an applicable accelerated life testing model that relates the failure times at accelerated conditions to those at normal conditions. Typical stress loadings are described in the following section.

22.2.1 Stress Loadings

Stress in ALT can be applied in various ways. Typical loadings include constant, cyclic, step, progressive, random stress loading, and combinations of such loadings. Figure 22.1 depicts different forms of stress loadings. The choice of a stress loading depends on how the product or unit is used in service and other practical and theoretical limitations [1]. For example, a typical automobile experiences extreme environmental temperature changes ranging from below freezing to more than 120 °F. Clearly, the corresponding accelerated stress test should be cyclic with extreme temperature changes.

The stress levels range from low to high. A low stress level should be higher than the operating conditions to ensure failures during the test, whereas the high stress should be the highest possible stress that can be applied without inducing failure mechanisms different from those that are likely to occur under normal operating conditions. Meeker and Hahn [2] provide extensive tables and practical guidelines for planning an ALT. They present a statistically optimum test plan and then suggest an alternative plan that meets practical constraints, and has desirable statistical properties. Their tables allow assessment of the effect of reducing the testing stress levels (thereby reducing the degree of extrapolation) on statistical precision.

Typical accelerated testing plans allocate equal units to the test stresses. However, units tested at stress levels close to the design or operating conditions may not experience enough failures that can be effectively used in the acceleration models. Therefore, it is preferred to allocate more test units to the low stress conditions than to the high stress conditions so as to obtain an equal expected number of failures at both conditions.

22.2.2 Types of Stress

The type of applied stress depends on the intended operating conditions of the product and the potential cause of failure. We classify the types of the stresses as follows.

1. *Mechanical stresses.* Fatigue stress is the most commonly used accelerated test for mechanical components. When the components are subject to elevated temperature, then creep testing (which combines both temperature and load) should be applied. Shock and vibration testing is suitable for components or products subject to such conditions as in the case of bearings, shock absorbers, tires and circuit boards in airplanes and automobiles.
2. *Electrical stresses.* These include power cycling, electric field, current density, and electromigration. Electric field is one of the most common electrical stresses, as it induces

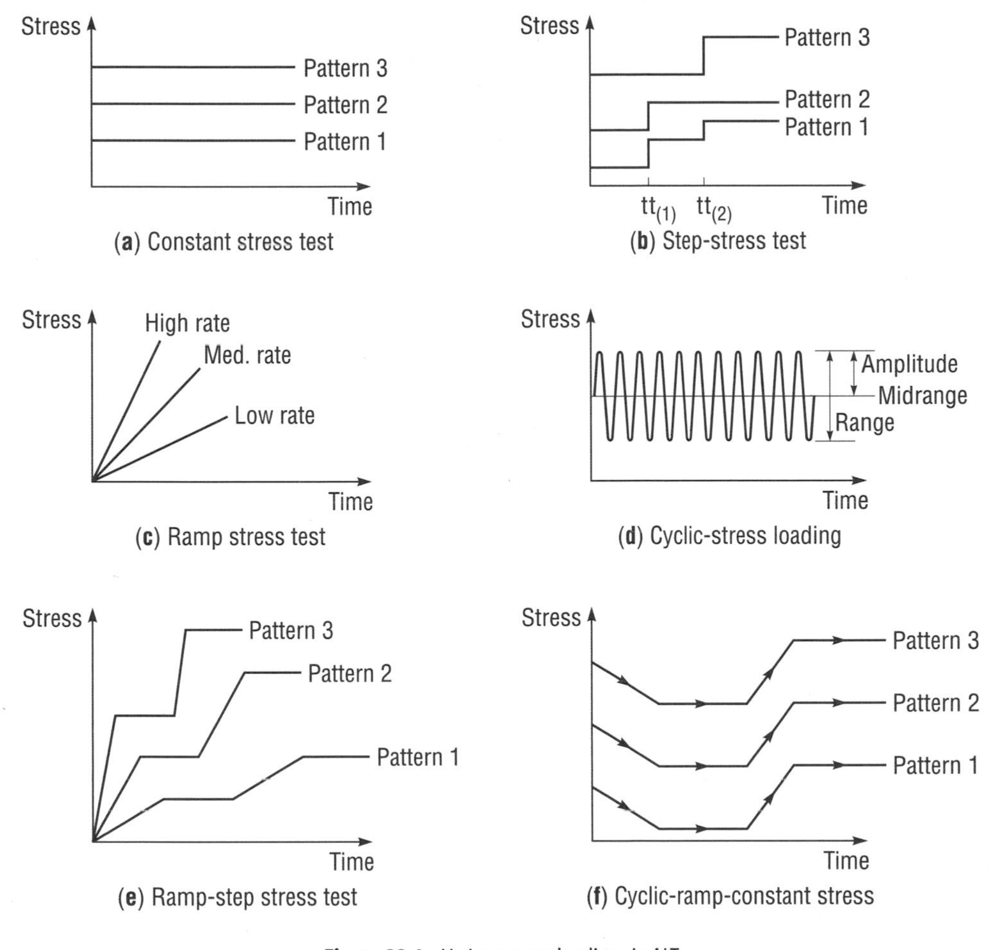

Figure 22.1. Various stress loadings in ALT

failures in relatively short times; its effect is also significantly higher than other types of stress.

3. *Environmental stresses.* Temperature and thermal cycling are commonly used for most products. Of course, it is important to use appropriate stress levels that do not induce different failure mechanisms than those under normal conditions. Humidity is as critical as temperature, but its application usually requires a very long time before its effect is noticed. Other environmental stresses include ultraviolet light, sulfur dioxide, salt and fine particles, and alpha and gamma rays.

22.3 Accelerated Life Testing Models

Elsayed [3] classified the inference procedures (or models) that relate life under stress conditions to life under normal or operating conditions into three types: *statistics-based models*, *physics-statistics-based models*, and *physics–experimental-based models* as shown in Figure 22.2.

The statistics-based models are further classified as parametric models and non-parametric models. The underlying assumption in relating the failure data, when using any of the models,

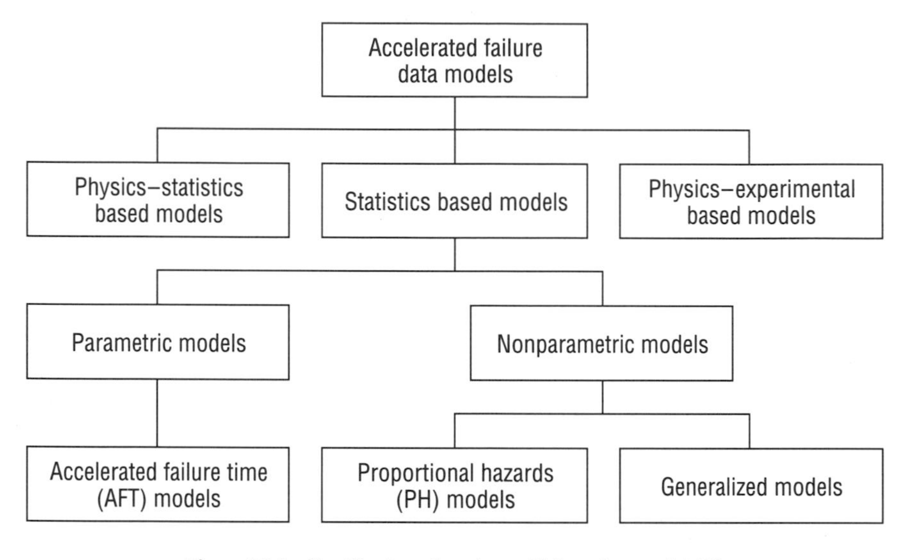

Figure 22.2. Classification of accelerated failure data models [3]

is that the components/products operating under normal conditions experience the same failure mechanism as those occurring at the accelerated conditions.

Statistics-based models are generally used when the exact relationship between the applied stresses and the failure time of the component or product is difficult to determine based on physics or chemistry principles. In this case, components are tested at different stress levels and the failure times are then used to determine the most appropriate failure time distribution and its parameters. The most commonly used failure time distributions are the exponential, Weibull, normal, lognormal, gamma, and the extreme value distributions. The failure times follow the same general distributions for all different stress levels, including the normal operating conditions. When the failure time data involve a complex lifetime distribution or when the number of observations is small, making it difficult to fit the failure time distribution accurately, semiparametric or non-parametric models appear to be a very attractive approach to predict reliability at different stress levels. The advantage of these models is that they are essentially "distribution free".

In the following sections, we present the details of some of these models and provide numerical examples that demonstrate the procedures for reliability prediction under normal operating conditions.

22.3.1 Parametric Statistics-based Models

As stated above, statistics-based models are generally used when the exact relationship between the applied stresses (temperature, humidity, voltage, *etc.*) and the failure time of the component (or product) is difficult to determine based on physics or chemistry principles. In this case, components are tested at different accelerated stress levels $s_1, s_2, \ldots, s_n$. The failure times at each stress level are then used to determine the most appropriate failure time probability distribution, along with its parameters. Under the parametric statistics-based model assumptions, the failure times at different stress levels are linearly related to each other.

Moreover, the failure time distribution at stress level s_1 is expected to be the same at different stress levels $s_2, s_3, \ldots$ as well as under the normal operating conditions. In other words, the shape parameters of the distributions are the same for all stress levels (including normal conditions) but the scale parameters may be different. Thus, the fundamental relationships between the operating conditions and stress conditions are summarized as follows [4]:

- failure times

$$t_o = A_F t_s \tag{22.1}$$

  where t_o is the failure time under operating conditions, t_s is the failure time under stress conditions, and A_F is the acceleration factor (the ratio between product life under normal conditions and life under accelerated conditions);
- cumulative distribution functions (CDFs)

$$F_o(t) = F_s\left(\frac{t}{A_F}\right) \tag{22.2}$$

- probability density functions

$$f_o(t) = \left(\frac{1}{A_F}\right) f_s\left(\frac{t}{A_F}\right) \tag{22.3}$$

- failure rates

$$h_o(t) = \left(\frac{1}{A_F}\right) h_s\left(\frac{t}{A_F}\right) \tag{22.4}$$

The most widely used parametric models are the exponential and Weibull models. Therefore, we derive the above equations for both models and demonstrate their use.

22.3.2 Acceleration Model for the Exponential Model

This is the case where the time to failure under stress conditions is exponentially distributed with a constant failure rate λ_s. The CDF at stress s is

$$F_s(t) = 1 - e^{-\lambda_s t} \tag{22.5}$$

and the CDF under normal conditions is

$$F_o(t) = F_s\left(\frac{t}{A_F}\right) = 1 - e^{-\lambda_s t / A_F} \tag{22.6}$$

Table 22.1. Failure times of the capacitors in hours

Temperature 145 °C	Temperature 240 °C	Temperature 305 °C
75	179	116
359	407	189
701	466	300
722	571	305
738	755	314
1015	768	403
1388	1006	433
2285	1094	440
3157	1104	468
3547	1493	609
3986	1494	634
4077	2877	640
5447	3001	644
5735	3160	699
5869	3283	781
6242	4654	813
7804	5259	860
8031	5925	1009
8292	6229	1176
8506	6462	1184
8584	6629	1245
11 512	6855	2071
12 370	6983	2189
16 062	7387	2288
17 790	7564	2637
19 767	7783	2841
20 145	10 067	2910
21 971	11 846	2954
30 438	13 285	3111
42 004	28 762	4617
mean = 9287	mean = 5244	mean = 1295

The failure rates are related as

$$\lambda_o = \frac{\lambda_s}{A_F} \tag{22.7}$$

Example 1. In recent years, silicon carbide (SiC) is used as an optional material for semiconductor devices, especially for those devices operating under high temperatures and high electric fields conditions. An extensive accelerated life experiment is conducted by subjecting 6H-SiC metal-oxide-silicon (MOS) capacitors to temperatures of 145, 240, and 305 °C. The failure times are recorded in Table 22.1.

Table 22.2. Temperatures and the 50th percentiles

Temperature (°C)	145	240	305
50th percentile	6437	3635	898

Determine the mean time to failure (MTTF) of the capacitors at 25 °C and plot the reliability function.

Solution. The data for every temperature are fitted using an exponential distribution and the means are shown in Table 22.1. In order to estimate the acceleration factor we chose some percentile of the failed population, which can be done non-parametrically using the rank distribution or a parametric model. In this example, the exponential distribution is used and the time at which 50% of the population fails is

$$t = \lambda(-\ln 0.5) \tag{22.8}$$

The 50th percentiles are given in Table 22.2.

We use the Arrhenius model to estimate the acceleration factor

$$t = k\,\mathrm{e}^{c/T}$$

where t is the time at which a specified portion of the population fails, k and c are constants and T is the absolute temperature (measured in degrees kelvin). Therefore

$$\ln t = \ln k + \frac{c}{T}$$

Using the values in Table 22.2 and least-squares regression we obtain $c = 2730.858$ and $k = 15.844\,32$. Therefore, the estimated 50th percentile at 25 °C is

$$t_{25\,^\circ\mathrm{C}} = 15.844\,32\,\mathrm{e}^{2730.858/(25+273)} = 151\,261\ \mathrm{h}$$

The acceleration factor at 25 °C is

$$A_\mathrm{F} = \frac{151\,261}{6437} = 23.49$$

and the failure rate under normal operating conditions is $1/(1295 \times 23.49) = 3.287 \times 10^{-5}$ failures/h, the mean time to failure is 30 419 h and the plot of the reliability function is shown in Figure 22.3.

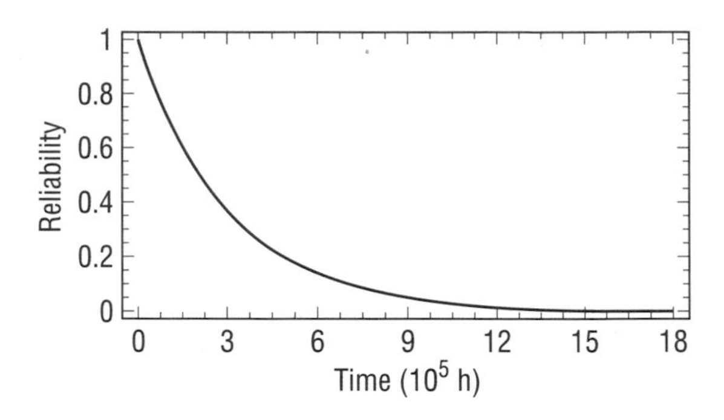

Figure 22.3. Reliability function for the capacitors

22.3.3 Acceleration Model for the Weibull Model

Again, we consider the true linear acceleration case. Therefore, the relationships between the failure time distributions at the accelerated and normal conditions can be derived using Equations 22.2–22.4. Thus

$$F_s(t) = 1 - \mathrm{e}^{-(t/\theta_s)^{\gamma_s}} \qquad t \geq 0,\ \gamma_s \geq 1,\ \theta_s > 0 \tag{22.9}$$

where γ_s is the shape parameter of the Weibull distribution under stress conditions and θ_s is the scale parameter under stress conditions. The CDF under normal operating conditions is

$$\begin{aligned} F_\mathrm{o}(t) &= F_s\left(\frac{t}{A_\mathrm{F}}\right) = 1 - \mathrm{e}^{-[t/A_\mathrm{F}\theta_s]^{\gamma_s}} \\ &= 1 - \mathrm{e}^{-[t/\theta_\mathrm{o}]^{\gamma_\mathrm{o}}} \end{aligned} \tag{22.10}$$

The underlying failure time distributions under both the accelerated stress and operating conditions have the same shape parameters, *i.e.* $\gamma_s = \gamma_\mathrm{o}$, and $\theta_\mathrm{o} = A_\mathrm{F}\theta_s$. If the shape parameters at different stress levels are significantly different, then either the assumption of true linear acceleration is invalid or the Weibull distribution is inappropriate to use for analysis of such data.

Let $\gamma_s = \gamma_\mathrm{o} = \gamma \geq 1$. Then the probability density function under normal operating conditions is

$$f_o(t) = \frac{\gamma}{A_F\theta_s}\left(\frac{t}{A_F\theta_s}\right)^{\gamma-1} e^{-[t/A_F\theta_s]^\gamma}$$
$$t \geq 0,\ \theta_s \geq 0 \qquad (22.11)$$

The MTTF under normal operating conditions is

$$\mathrm{MTTF_o} = \theta_o^{1/\gamma}\Gamma\left(1+\frac{1}{\gamma}\right) \qquad (22.12)$$

The failure rate under normal operating conditions is

$$h_o(t) = \frac{\gamma}{A_F\theta_s}\left(\frac{t}{A_F\theta_s}\right)^{\gamma-1} = \frac{h_s(t)}{A_F^\gamma} \qquad (22.13)$$

Example 2. A manufacturer of Bourdon tubes (used as a part of pressure sensors in avionics) wishes to determine its MTTF. The manufacturer defines the failure as a leak in the tube. The tubes are manufactured from 18 Ni (250) maraging steel and operate with dry 99.9% nitrogen or hydraulic fluid as the internal working agent. Tubes fail as a result of hydrogen embrittlement arising from the pitting corrosion attack. Because of the criticality of these tubes, the manufacturer decides to conduct ALT by subjecting them to different levels of pressures and determining the time for a leak to occur. The units are continuously examined using an ultrasound method for detecting leaks, indicating failure of the tube. Units are subjected to three stress levels of gas pressures and the times for tubes to show leak are recorded in Table 22.3.

Determine the mean lives and plot the reliability functions for design pressures of 80 and 90 psi.

Solution. We fit the failure times to Weibull distributions, which results in the following parameters for pressure levels of 100, 120, and 140 psi.

For 100 psi: $\gamma_1 = 2.87,\ \theta_1 = 10\,392$

For 120 psi: $\gamma_2 = 2.67,\ \theta_2 = 5375$

For 140 psi: $\gamma_3 = 2.52,\ \theta_3 = 943$

Since $\gamma_1 = \gamma_2 = \gamma_3 \cong 2.65$, then the Weibull model is appropriate to describe the relationship between failure times under accelerated conditions and normal operating conditions. Moreover, we have a true linear acceleration. Following Example 1, we determine the time at which 50% of the population fails as

$$t = \theta[-\ln(0.5)]^{1/\gamma}$$

The 50th percentiles are shown in Table 22.4.

The relationship between the failure time t and the applied pressure P can be assumed to be

Table 22.3. Time (hours) to detect leak

100 psi	120 psi	140 psi
1557	1378	215
4331	2055	426
5725	2092	431
5759	2127	435
6207	2656	451
6529	2801	451
6767	3362	496
6930	3377	528
7146	3393	565
7277	3433	613
7346	3477	651
7668	3947	670
7826	4101	708
7885	4333	710
8095	4545	743
8468	4932	836
8871	5030	865
9652	5264	894
9989	5355	927
10 471	5570	959
11 458	5760	966
11 728	5829	1067
12 102	5968	1124
12 256	6200	1139
12 512	6783	1158
13 429	6952	1198
13 536	7329	1293
14 160	7343	1376
14 997	8440	1385
17 606	9183	1780

Table 22.4. Percentiles at different pressures

Pressure (psi)	100	120	140
50th percentile	9050	4681	821

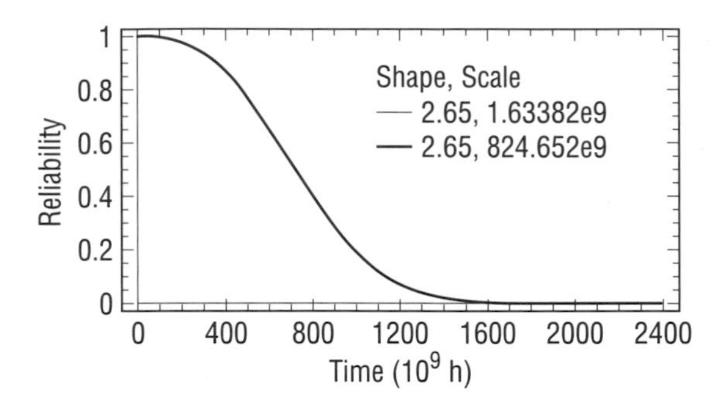

Figure 22.4. Reliability functions at 80 and 90 psi

similar to the Arrhenius model; thus

$$t = k\,\mathrm{e}^{c/P}$$

where k and c are constants. By making a logarithmic transformation, the above expression can be written as

$$\ln t = \ln k + \frac{c}{P}$$

Using a linear regression model, we obtain $k = 3.319$ and $c = 811.456$. The estimated 50th percentiles at 80 psi and 90 psi are 84 361 h and 27 332 h respectively. The corresponding acceleration factors are 9.32 and 3.02. The failure rates under normal operating conditions are

$$h_{\mathrm{o}}(t) = \frac{\gamma}{A_{\mathrm{F}}\theta_s}\left(\frac{t}{A_{\mathrm{F}}\theta_s}\right)^{\gamma-1} = \frac{h_s(t)}{A_{\mathrm{F}}^{\gamma}}$$

or

$$h_{80}(t) = \frac{2.65}{1.633\,82 \times 10^{13}} t^{1.65}$$

and

$$h_{90}(t) = \frac{2.65}{8.246\,52 \times 10^{13}} t^{1.65}$$

The reliability functions are shown in Figure 22.4.

The MTTFs for 80 and 90 psi are calculated as

$$\begin{aligned}\mathrm{MTTF}_{80} &= \theta^{1/\gamma}\Gamma\left(\frac{1}{\gamma}\right)\\ &= (1.633\,82 \times 10^{13})^{1/2.65}\Gamma\left(1 + \frac{1}{2.65}\right)\\ &= 96\,853.38 \times 0.885 = 85\,715\ \mathrm{h}\end{aligned}$$

and

$$\mathrm{MTTF}_{90} = 31\,383.829 \times 0.885 = 27\,775\ \mathrm{h}$$

22.3.4 The Arrhenius Model

Elevated temperature is the most commonly used environmental stress for accelerated life testing of microelectronic devices. The effect of temperature on the device is generally modeled using the Arrhenius reaction rate equation given by

$$r = A\,\mathrm{e}^{-(E_{\mathrm{a}}/kT)} \tag{22.14}$$

where r is the speed of reaction, A is an unknown non-thermal constant, E_{a} (eV) is the activation energy (*i.e.* energy that a molecule must have before it can take part in the reaction), k is the Boltzmann constant (8.623×10^{-5} eV K^{-1}), and T (K) is the temperature.

Activation energy E_{a} is a factor that determines the slope of the reaction rate curve with temperature, *i.e.* it describes the acceleration effect that temperature has on the rate of a reaction and is expressed in electron volts (eV). For most applications, E_{a} is treated as a slope of a curve rather than a specific energy level. A low value of E_{a} indicates a small slope or a reaction that has a small dependence on temperature. On the other hand, a large value of E_{a} indicates a high degree of temperature dependence [3].

Assuming that device life is proportional to the inverse reaction rate of the process, then Equation 22.14 can be rewritten as

$$\begin{aligned}L_{30} &= 719 \exp \frac{0.42}{4.2998 \times 10^{-5}}\\ &\quad \times \left(\frac{1}{30+273} - \frac{1}{180+273}\right)\\ &= 31.0918 \times 10^{6}\end{aligned}$$

The median lives of the units at normal operating temperature L_{o} and accelerated temperature L_s are related by

$$\frac{L_{\mathrm{o}}}{L_s} = \frac{A\,\mathrm{e}^{E_{\mathrm{a}}/kT_{\mathrm{o}}}}{A\,\mathrm{e}^{E_{\mathrm{a}}/kT_s}}$$

or

$$L_{\mathrm{o}} = L_s \exp \frac{E_{\mathrm{a}}}{k}\left(\frac{1}{T_{\mathrm{o}}} - \frac{1}{T_s}\right) \tag{22.15}$$

The thermal acceleration factor is

$$A_{\mathrm{F}} = \exp \frac{E_{\mathrm{a}}}{k}\left(\frac{1}{T_{\mathrm{o}}} - \frac{1}{T_s}\right)$$

Table 22.5. Failure time data (hours) for oxide breakdown

Temperature 180 °C	Temperature 150 °C
112	162
260	188
298	288
327	350
379	392
487	681
593	969
658	1303
701	1527
720	2526
734	3074
736	3652
775	3723
915	3781
974	4182
1123	4450
1157	4831
1227	4907
1293	6321
1335	6368
1472	7489
1529	8312
1545	13 778
2029	14 020
4568	18 640

The calculation of the median life (or percentile of failed units) is dependent on the failure time distribution. When the sample size is small it becomes difficult to obtain accurate results. In this case, it is advisable to use different percentiles of failures and obtain a weighted average of the median lives. One of the drawbacks of this model is the inability to obtain a reliability function that relates the failure times under stress conditions to failure times under normal condition. We can only obtain a point estimate of life. We now illustrate the use of the Arrhenius model in predicting median life under normal operating conditions.

Example 3. The gate oxide in MOS devices is often a source of device failure, especially for high-density device arrays that require thin gate oxides. The reliability of MOS devices on bulk silicon and the gate oxide integrity of these devices have been the subject of investigation over the

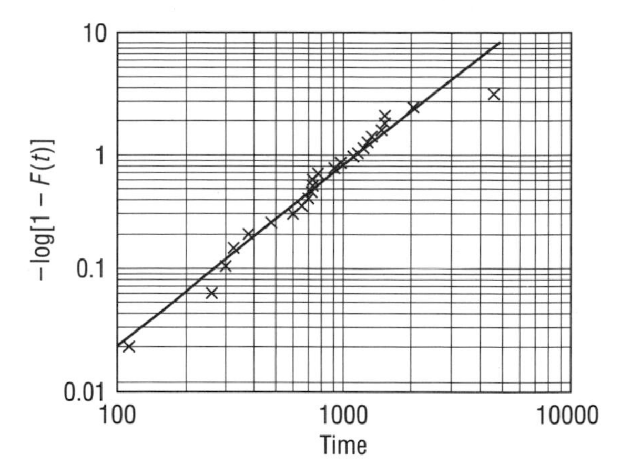

Figure 22.5. Weibull probability plot indicating appropriate fit of data

years. A producer of MOS devices conducts an accelerated test to determine the expected life at 30 °C. Two samples of 25 devices each are subjected to stress levels of 150 °C and 180 °C. The oxide breakdown is determined when the potential across the oxide reaches a threshold value. The times of breakdown are recorded in Table 22.5. The activation energy of the device is 0.42 eV. Obtain the reliability function of these devices.

Solution. Figure 22.5 shows that it is appropriate to fit the failure data using Weibull distributions with shape parameters approximately equal to unity. This means that they can be represented by exponential distributions with means of 1037 and 4787 h for the respective temperatures of 180 °C and 150 °C respectively. Therefore, we determine the 50th percentiles for these temperatures using Equation 22.8 as being 719 and 3318 respectively.

$$3318 = 719 \exp \frac{0.42}{k} \left(\frac{1}{150 + 273} - \frac{1}{180 + 273} \right)$$

which results in $k = 4.2998 \times 10^{-5}$.

The median life under normal conditions of 30 °C is

$$L_{30} = 719 \exp \frac{0.42}{4.2998 \times 10^{-5}} \times \left(\frac{1}{30+273} - \frac{1}{180+273} \right) = 31.0918 \times 10^{6}\ \text{h}$$

The mean life is 44.8561×10^6 and the reliability function is

$$R(t) = \mathrm{e}^{-44.8561 \times 10^6 t}$$

22.3.5 Non-parametric Accelerated Life Testing Models: Cox's Model

Non-parametric models relax the assumption of the failure time distribution, *i.e.* no predetermined failure time distribution is required. Cox's proportional hazards (PH) model [5, 6] is the most widely used among the non-parametric models. It has been successfully used in the medical area for survival analysis. In the past few years Cox's model and its extensions have been applied in the reliability analysis of ALT. Cox made two important contributions: the first is a model that is referred to as the PH model; the second is a new parameter estimation method, which is later referred to as the *maximum partial likelihood.*

The PH model is structured based on the traditional hazard function, $\lambda(t; \mathbf{z})$, the hazard function at time t for a vector of covariates (or stresses) $\mathbf{z}$. It is the product of an unspecified baseline hazard function $\lambda_0(t)$ and a relative risk ratio. The baseline hazard function $\lambda_0(t)$ is a function of time t and independent of $\mathbf{z}$, whereas the relative risk ratio is a function of $\mathbf{z}$ only. Thus, when a function $\exp(\cdot)$ is considered as the relative risk ratio, its PH model has the following form:

$$\lambda(t; \mathbf{z}) = \lambda_0(t) \exp(\boldsymbol{\beta}\mathbf{z}) = \lambda_0(t) \exp\left(\sum_{j=1}^{p} \beta_j z_j \right) \tag{22.16}$$

where

$\mathbf{z} = (z_1, z_2, \ldots, z_p)^{\mathrm{T}}$
: a vector of the covariates (or applied stresses), where "T" indicates transpose. For ALT it is the column vector of stresses used in the test and/or their interactions

$\boldsymbol{\beta} = (\beta_1, \beta_2, \ldots, \beta_p)^{\mathrm{T}}$
: a vector of the unknown regression coefficients

p number of the covariates.

The PH model implies that the ratio of the hazard rate functions for any two units associated with different vectors of covariates, $\mathbf{z}_1$ and $\mathbf{z}_2$, is constant with respect to time. In other words, $\lambda(t; \mathbf{z}_1)$ is directly proportional to $\lambda(t; \mathbf{z}_2)$.

If the baseline hazard function $\lambda_0(t)$ is specified, then the usual maximum likelihood method can be used to estimate $\boldsymbol{\beta}$. However, for the PH model, the regression coefficients can be estimated without the need for specifying the form of $\lambda_0(t)$. Therefore, the partial likelihood method is used to estimate $\boldsymbol{\beta}$.

Suppose that we have a random sample of n units; let $t_1 < t_2 \cdots < t_k$ represent k $(k \le n)$ distinct failure times with corresponding covariates $\mathbf{z}_1, \mathbf{z}_2, \ldots, \mathbf{z}_k$. If $k < n$, then the remaining $n - k$ units are assumed to be censored. The partial likelihood has the following form:

$$L(\boldsymbol{\beta}) = \prod_{i=1}^{k} \frac{\exp(\boldsymbol{\beta}\mathbf{z}_i)}{\sum_{l \in R(t_i)} \exp(\boldsymbol{\beta}\mathbf{z}_l)} \tag{22.17}$$

where $R(t_i)$, the risk set at t_i, denotes the number of units survived and uncensored just prior to t_i.

To accommodate tied survival times, Breslow [7] modified the above form to

$$L(\boldsymbol{\beta}) = \prod_{i=1}^{k} \left\{ \exp(\boldsymbol{\beta}\mathbf{S}_i) \Big/ \left[\sum_{l \in R(t_i)} \exp(\boldsymbol{\beta}\mathbf{z}_l) \right]^{d_i} \right\} \tag{22.18}$$

where d_i is the number of failure times equal to t_i, and $\mathbf{S}_i$ is the sum of the vectors $\mathbf{z}$ for these d_i units.

Though $\lambda_0(t)$ is not specified in the PH model, the only unknown parameter in the partial likelihood function is $\boldsymbol{\beta}$. Therefore, the vector $\boldsymbol{\beta}$ can be estimated by maximizing the log of partial likelihood function using numerical methods, such as the Newton–Raphson method. The estimate of $\boldsymbol{\beta}$ needs to be checked for significance using analytical or graphical methods. Those insignificant covariates can be discarded from the model, and a new estimate of $\boldsymbol{\beta}$ corresponding to the remaining covariates needs to be recalculated.

After estimating the unknown parameters $\beta_1, \beta_2, \ldots, \beta_p$ using the maximum partial likelihood method, the remaining unknown in the PH model is $\lambda_0(t)$. Some authors [8–10] suggest the use of a piecewise step function estimate where the function $\lambda_0(t)$ is assumed to be constant between subdivisions of the time scale. The time scale can be subdivided at arbitrary convenient points or at those points where failures occur. This idea of estimating $\lambda_0(t)$ is a variation of Kaplan–Meier's product limit estimate. The only difference is that, for the PH model, failure data are collected at different stresses, and they need to be weighted and transformed by the relative risk ratio to a comparable scale before using the Kaplan–Meier method. Another method to estimate $\lambda_0(t)$, proposed by Anderson and Senthilselvan [11], describes a piecewise smooth estimate. They assumed $\lambda_0(t)$ to be a quadratic spline instead of being constant between time intervals. They also suggest subtracting a roughness penalty function, $K_0 \int [\lambda_0'(t)]^2 \, dt$, from the log-likelihood and then maximizing the penalized log-likelihood in order to obtain a smooth estimate for $\lambda_0(t)$. To determine K_0, Anderson and Senthilselvan [11] plotted the estimates for several values of K_0 and chose the value that seemed to "best represent the data". K_0 may be adjusted if the hazard function is negative for some t.

After the unknown parameters $\beta_1, \beta_2, \ldots, \beta_p$ and the unknown baseline hazard rate function $\lambda_0(t)$ in the PH model are estimated, *i.e* all of the PH model's unknowns are estimated, we can use these estimates in the PH model to estimate the reliability at any stress level and time.

Example 4. The need for long-term reliability in electronic systems, which are subject to time-varying environmental conditions and experience various operating modes or duty cycles, introduces a complexity in the development of thermal control strategy. Coupling this with other applied stresses, such as electric field, results in significant deterioration of reliability. Therefore, the designer of a new electronic packaging system wishes to determine the expected life of the design at design conditions of 30 °C and 26.7 V. The designer conducted an accelerated test at several temperatures and voltages and recorded the failure times as shown in Table 22.6. Plot the reliability function and determine the MTTF under design conditions.

Solution. This ALT has two types of stresses: temperature and voltage. The sample sizes are relatively small and it is appropriate to use a non-parametric method to estimate the reliability and MTTF. We utilize the PH model and express it as

$$\lambda(t;\, T,\, V) = \lambda_0(t) \exp\left(\frac{\beta_1}{T} + \beta_2 V\right)$$

where T (K) is temperature V is voltage. The parameters β_1 and β_2 are determined using SAS® PROC PHREG [12] and their values are −24 538 and 30.332 739 respectively. The unspecified baseline hazard function can be estimated using any of the methods described by Kalbfleisch and Prentice [9] and Elsayed [3]. We can estimate the reliability of function under design conditions by using their values with PROC PHREG. The reliability values are then plotted and a regression model is fit accordingly, as shown in Figure 22.6.

We assume a Weibull baseline hazard function and fit the reliability values obtained under design conditions to the Weibull function, which results in

$$R(t;\, 30\,^\circ\text{C},\, 26.7) = \mathrm{e}^{-1.550\,65 \times 10^{-8} t^2}$$

The MTTF is obtained using the mean time to failure expression of the Weibull models as shown by Elsayed [3]:

$$\text{MTTF} = \theta^{1/\gamma} \Gamma\left(1 + \frac{1}{\gamma}\right) = 7166\ \text{h}$$

Table 22.6. Failure times (hours) at different temperatures and voltages

Failure time	Temperature (°C)	$Z_1 = [1/(\text{Temperature} + 273.16)]$	Z_2 (V)
1	25	0.003 353 9	27
1	25	0.003 353 9	27
1	25	0.003 353 9	27
73	25	0.003 353 9	27
101	25	0.003 353 9	27
103	25	0.003 353 9	27
148	25	0.003 353 9	27
149	25	0.003 353 9	27
153	25	0.003 353 9	27
159	25	0.003 353 9	27
167	25	0.003 353 9	27
182	25	0.003 353 9	27
185	25	0.003 353 9	27
186	25	0.003 353 9	27
214	25	0.003 353 9	27
214	25	0.003 353 9	27
233	25	0.003 353 9	27
252	25	0.003 353 9	27
279	25	0.003 353 9	27
307	25	0.003 353 9	27
1	225	0.002 007 4	26
14	225	0.002 007 4	26
20	225	0.002 007 4	26
26	225	0.002 007 4	26
32	225	0.002 007 4	26
42	225	0.002 007 4	26
42	225	0.002 007 4	26
43	225	0.002 007 4	26
44	225	0.002 007 4	26
45	225	0.002 007 4	26
46	225	0.002 007 4	26
47	225	0.002 007 4	26
53	225	0.002 007 4	26
53	225	0.002 007 4	26
55	225	0.002 007 4	26
56	225	0.002 007 4	26
59	225	0.002 007 4	26
60	225	0.002 007 4	26
60	225	0.002 007 4	26
61	225	0.002 007 4	26

22.4 Extensions of the Proportional Hazards Model

The PH model implies that the ratio of the hazard functions for any two units associated with different vectors of covariates, $\mathbf{z}_1$ and $\mathbf{z}_2$, is constant with respect to time. In other words, $\lambda(t;\, \mathbf{z}_1)$ is directly proportional to $\lambda(t;\, \mathbf{z}_2)$. This is the so-called PH model's hazard rate proportionality assumption. The PH model is a valid model to analyze ALT data only when the data satisfy the PH model's proportional hazard rate assumption. So, checking the validity of the PH model and the assumption of the covariates' multiplicative effect is critical for

Table 22.6. *Continued.*

Failure time	Temperature (°C)	$Z_1 = [1/(\text{Temperature} + 273.16)]$	Z_2 (V)
1365	125	0.002 511 6	25.7
1401	125	0.002 511 6	25.7
1469	125	0.002 511 6	25.7
1776	125	0.002 511 6	25.7
1789	125	0.002 511 6	25.7
1886	125	0.002 511 6	25.7
1930	125	0.002 511 6	25.7
2035	125	0.002 511 6	25.7
2068	125	0.002 511 6	25.7
2190	125	0.002 511 6	25.7
2307	125	0.002 511 6	25.7
2309	125	0.002 511 6	25.7
2334	125	0.002 511 6	25.7
2556	125	0.002 511 6	25.7
2925	125	0.002 511 6	25.7
2997	125	0.002 511 6	25.7
3076	125	0.002 511 6	25.7
3140	125	0.002 511 6	25.7
3148	125	0.002 511 6	25.7
3736	125	0.002 511 6	25.7

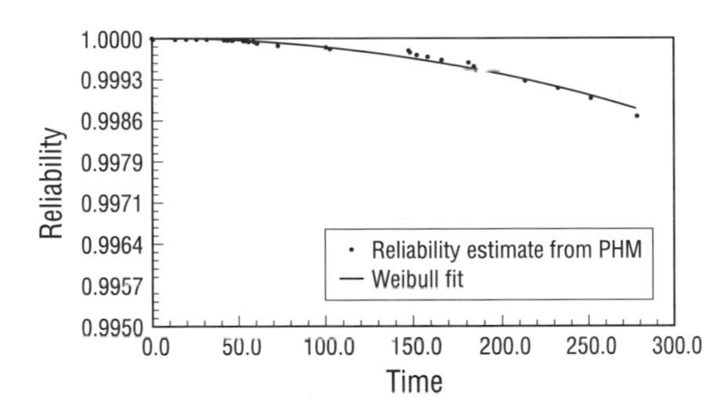

Figure 22.6. Reliability function under design conditions

applying the model to the failure data. When the proportionality assumption is violated, there are several extensions of the PH model that are proposed to handle the situation.

If the proportionality assumption is violated and there are one or more covariates that totally occur on q levels, a simple extension of the PH model is stratification [13], as given below:

$$\lambda_j(t; \mathbf{z}) = \lambda_{0j}(t) \exp(\boldsymbol{\beta}\mathbf{z}) \qquad j = 1, 2, \ldots, q$$

A partial likelihood function can be obtained for each of the q strata and $\boldsymbol{\beta}$ is estimated by maximizing the multiplication of the partial likelihood of the q strata. The baseline hazard rate $\lambda_{0j}(t)$, estimated as before, is completely unrelated among the strata. This model is most useful when the covariate is categorical and of no direct interest.

Another extension of the PH model includes the use of time-dependent covariates. Kalbfleisch and Prentice [13] classified the time-dependent covariates as internal and external. An internal covariate is the output of a stochastic process generated by the unit under study; it can be observed as long as the unit survives and is uncensored. An external covariate has a fixed value or is defined in advance for each unit under study.

Many other extensions exist in the literature [14]. However, one of the most generalized extensions is the extended linear hazard regression model proposed by Elsayed and Wang [15]. Both accelerated failure time models and PH models are indeed special cases of the generalized model, whose hazard function is

$$\lambda(t;\, \mathbf{z}) = \lambda_0(t\, e^{(\beta_0+\beta_1 t)\mathbf{z}})\, e^{(\alpha_0+\alpha_1 t)\mathbf{z}}$$

where $\lambda(t;\, \mathbf{z})$ is the hazard rate at time t and stress vector $\mathbf{z}$, $\lambda_0(\cdot)$ is the baseline hazard function, and β_0, β_1, α_0, and α_1 are constants. This model has been validated through extensive experimentation and simulation testing.

References

[1] Nelson WB. Accelerated life testing step-stress models and data analysis. IEEE Trans Reliab 1980;R-29:103–8.

[2] Meeker WQ, Hahn GJ. How to plan an accelerated life test-some practical guidelines. In: Statistical techniques, vol. 10, ASQC basic reference in QC. Milwaukee (WI): American Society for Quality Control; 1985.

[3] Elsayed EA. Reliability engineering. Massachusetts: Addison-Wesley; 1996.

[4] Tobias PA, Trindade D. Applied reliability. New York: Van Nostrand Reinhold; 1986.

[5] Cox DR. Regression models and life tables (with discussion). J R Stat Soc Ser B 1972;34:187–220.

[6] Cox DR. Partial likelihood. Biometrika 1975;62:267–76.

[7] Breslow NE. Covariance analysis of censored survival data. Biometrika 1974;66:188–90.

[8] Oakes D. Comment on "Regression models and life tables (with discussion)" by D. R. Cox. J R Stat Soc Ser B 1972;34:208.

[9] Kalbfleisch JD, Prentice RL. Marginal likelihood based on Cox's regression and life model. Biometrika 1973;60:267–78.

[10] Breslow NE. Comment on "Regression models and life tables (with discussion)" by D. R. Cox. J R Stat Soc Ser B 1972;34:187–202.

[11] Anderson JA, Senthilselvan A. Smooth estimates for the hazard function. J R Stat Soc Ser B 1980;42(3):322–7.

[12] Allison PD. Survival analysis using the SAS® system: a practical guide. Cary (NC): SAS Institute Inc.; 1995.

[13] Kalbfleisch JD, Prentice RL. Statistical analysis of failure time data. New York: Wiley; 1980.

[14] Wang X. An extended hazard regression model for accelerated life testing with time varying coefficients. PhD dissertation, Department of Industrial Engineering, Rutgers University, New Brunswick, NJ, 2001.

[15] Elsayed EA, Wang X. Confidence limits on reliability estimates using the extended linear hazards regression model. In: Ninth Industrial Engineering Research Conference, Cleveland, Ohio, May 21–23, 2000.

Accelerated Test Models with the Birnbaum–Saunders Distribution

Chapter 23

W. Jason Owen and William J. Padgett

23.1 Introduction

When modeling the time to failure T of a device or component whose lifetime is considered a random variable, the *reliability* measure $R(t)$, defined as

$$R(t) = \Pr(T > t)$$

describes the probability of the component of failing or exceeding a time t. In basic statistical inference procedures, various statistical cumulative distribution functions (CDFs), denoted by $F(t)$, can be used to model a lifetime characteristic for a device on test. In doing so, the relationship between the reliability and CDF for T is straightforward: $R(t) = 1 - F(t)$. Choosing a model for F is akin to assuming the behavior of observations of T, and frequent choices for CDFs in reliability modeling include the Weibull, lognormal, and gamma distributions. A description of these models is given by Mann *et al.* [1] and Meeker and Escobar [2], for example. This chapter will focus on the Birnbaum–Saunders distribution (B–S) and its applications in reliability and life testing, and this distribution will be described later in this section. The B–S fatigue life distribution has applications in the general area of accelerated life testing, and two approaches given here yield various three-parameter generalizations or extensions to the original form of their model. The first approach uses the (inverse) power law model to develop an accelerated form of the distribution using standard methods. The second approach uses different assumptions of cumulative damage arguments (with applications to strengths of complex systems) to yield various strength distributions that are similar in form. Estimation and asymptotic theory for the models are discussed with some applications to various data sets from engineering laboratories.

The following is a list of notation used throughout this chapter:

s, s_{ij}	strength variable, observed strength
t, t_{ij}	time to failure variable, observed failure time
F	CDF

f probability density function (PDF)
$E(T)$ expected value of random variable T
Φ standard normal (Gaussian) CDF
$\alpha, \beta, \delta, \gamma, \eta, \theta, \rho, \psi$ various CDF model parameters
V level of accelerated stress
L, L_i system size, for the ith system, $i = 1, \ldots, k$
$\mathcal{L}$ likelihood function
s_p distribution $p100$th percentile.

23.1.1 Accelerated Testing

Many high-performance items made today can have extremely large reliabilities (and thus tend not to wear out as easily) when they are operating as intended. For example, through advances in engineering science, a measurement such as *time to failure* for an electronic device could be quite large. For these situations, calculating the reliability is still of interest to the manufacturer and to the consumer, but a long period of time may be necessary to obtain sufficient data to estimate reliability. This is a major issue, since it is often the case with electronic components that the length of time required for the test may actually eclipse the "useful lifetime" of the component. In addition, performing a test over a long time period (possibly many years) could be very costly. One solution to this issue of obtaining meaningful life test data in a timely fashion is to perform an accelerated life test (ALT). This procedure involves observing the performance of the component at a higher than usual level of operating stress to obtain failures more quickly. Types of stress or acceleration variable that have been suggested in the literature include temperature, voltage, pressure, and vibration [1,3].

Obviously, it can be quite a challenge to use the failure data observed at the conditions of higher stress to predict the reliability (or even the average lifetime) under the "normal use" conditions. To extrapolate from the observed accelerated stresses to the normal-use stress, an acceleration model is used to describe the relationship between the parameters of the reliability model and the accelerated stresses. This functional relationship usually involves some unknown parameters that will need to be estimated in order to draw inference at the normal-use stress. An expository discussion of ALTs is given by Padgett [4].

Many different parametric acceleration models have been derived from the physical behavior of the material under the elevated stress level. Common acceleration models include the (inverse) power law, Arrhenius, and Eyring models [4]. The power law model has applications in the fatigue failure of metals, dielectric failure of capacitors under increased voltage, and the aging of multi-component systems. This model will be used in this chapter, and it is given by

$$h(V) = \gamma V^{-\eta} \tag{23.1}$$

where V represents the level of accelerated stress. Typically, $h(V)$ is substituted for a mean or scale parameter in a baseline lifetime distribution. Thus, by performing this substitution, it is assumed that the increased environmental stress has the effect of changing the mean or scale of the lifetime model, but the distribution "family" remains intact. The parameters γ and η in $h(V)$ are (usually) unknown and are to be estimated (along with any other parameters in the baseline failure model) using life-test data observed at elevated stress levels $V_1, V_2, \ldots, V_k$. ALT data are often represented in the following way: let t_{ij} characterize the observed failure (time) for an item under an accelerated stress level V_i, where $j = 1, 2, \ldots, n_i$ and $i = 1, 2, \ldots, k$. After parameter estimation using the ALT data, inferences can be drawn for other stress levels not observed, such as the "normal use" stress V_0. Padgett *et al.* [5] provide an example using the Weibull distribution with the power-law model (Equation 23.1) substituted for the Weibull scale parameter; this accelerated model was used for fitting tensile strength data from an experiment testing various gage lengths of carbon fibers and tows (micro-composites). In this application, "length" is considered an acceleration variable, since longer specimens of these materials tend to exhibit lower strengths due to increased occurrences of inherent flaws that ultimately weaken the material [5].

23.1.2 The Birnbaum–Saunders Distribution

While considering the physical properties of fatigue failure for materials subjected to cyclic stresses and strains, Birnbaum and Saunders [6] derived a new family of lifetime distributions. In their derivation, they assume (as is often the case with metals and concrete structures) that failure occurs by the progression of a dominant crack within the material to an eventual "critical length", which ultimately results in fatigue failure. Their derivation resulted in the CDF

$$F_{\text{B-S}}(t;\,\alpha,\,\beta) = \Phi\left[\frac{1}{\alpha}\left(\sqrt{\frac{t}{\beta}} - \sqrt{\frac{\beta}{t}}\right)\right] \quad t > 0;\,\alpha,\,\beta > 0 \tag{23.2}$$

where $\Phi[\cdot]$ represents the standard normal CDF. A random variable T following the distribution in Equation 23.2 will be denoted as $T \sim \text{B–S}(\alpha, \beta)$. The parameter α is a shape parameter and β is a scale parameter; also, β is the median for the distribution. This CDF has some interesting qualities, and many are listed by Birnbaum and Saunders [6, 7]. Of interest for this discussion are the first and first reciprocal moments

$$E(T) = \beta(1+\alpha^2/2) \quad E(T^{-1}) = (1+\alpha^2/2)/\beta \tag{23.3}$$

and the PDF for T given by

$$\begin{aligned} F'_{\text{B-S}}(t;\,\alpha,\,\beta) &= f(t;\,\alpha,\,\beta) \\ &= \frac{1}{2\alpha\beta}\sqrt{\frac{t}{\beta}}\left[1+\left(\frac{t}{\beta}\right)^{-1}\right]\frac{1}{\sqrt{2\pi}} \\ &\quad \times \exp\left[-\frac{1}{2\alpha^2}\left(\frac{t}{\beta}-2+\frac{\beta}{t}\right)\right] \\ &\quad t > 0;\,\alpha,\,\beta > 0 \end{aligned} \tag{23.4}$$

Desmond [8] showed that Equation 23.2 also describes the failure time that is observed when any amount of accumulating damage (not unlike crack propagation) exceeds a critical threshold. This research has been the foundation of the "cumulative damage" models that will be illustrated in Section 23.2.2.

Statistical inference for data following Equation 23.2 has been developed for single samples. To estimate the parameters in Equation 23.2 for a data set, the maximum likelihood estimators (MLEs) of α and β must be found using an iterative, root-finding technique. The equations for the MLEs are given in Birnbaum and Saunders [7]. Achcar [9] derived an approximation to the Fisher information matrix for the Birnbaum–Saunders distribution by using the Laplace approximation [10] for complicated integrals, and used the information matrix to provide some Bayesian approaches to estimation using Jeffery's (non-informative) priors for α and β. The Fisher information matrix for Equation 23.2 is also useful for other inferences (*e.g.* confidence intervals) using the asymptotic normality theory for MLEs (see Lehman and Casella [11] for a comprehensive discussion on this topic).

The remainder of this chapter is outlined as follows. In Section 23.2, the power-law accelerated Birnbaum–Saunders distribution will be presented, using the methods discussed in this section. In addition, a family of cumulative damage models developed for estimation of the strength of complex systems will be presented. These models assume various damage processes for material failure, and it will be shown that they possess a generalized Birnbaum–Saunders form—not unlike the ALT Birnbaum–Saunders form. Estimation techniques and large-sample theory will be discussed for all of these models in Section 23.3. Applications of the models will be discussed in Section 23.4, along with some conclusions.

23.2 Accelerated Birnbaum–Saunders Models

In this section, various accelerated life models will be presented that follow a generalized Birnbaum–Saunders form. The five models that will be presented are similar in that they represent three-parameter generalizations of the original two-parameter form given by Equation 23.2. A more detailed derivation for each model, along with other theoretical aspects, can be found in the accompanying references.

23.2.1 The Power-law Accelerated Birnbaum–Saunders Model

Owen and Padgett [12] describe an accelerated life model using the models in Equations 23.1 and 23.2. The power law model (Equation 23.1) was substituted for the scale parameter β in Equation 23.2 to yield

$$G(t;\, V) = F_{\text{B-S}}[t;\, \alpha,\, h(V)] = \Phi\left[\frac{1}{\alpha}\left(\sqrt{\frac{t}{\gamma V^{-\eta}}} - \sqrt{\frac{\gamma V^{-\eta}}{t}}\right)\right] \quad t > 0;\, \alpha,\, \gamma,\, \eta > 0 \qquad (23.5)$$

Note that Equation 23.5 is a distribution with *three* parameters; it would represent an overparameterization of Equation 23.2 unless at least two distinct levels of V are considered in an experiment (*i.e.* $k \geq 2$). The mean for the distribution in Equation 23.5 can easily be calculated from the equation given in Equation 23.3 and by substituting the power law model Equation 23.1 for β. Thus, it can be seen that as V (the level of accelerated stress) increases, the expected value for T decreases. Owen and Padgett [12] used the accelerated model Equation 23.5 to fit observed cycles-to-failure data of metal "coupons" (rectangular strips) that were stressed and strained at three different levels (see Birnbaum and Saunders [7]). A discussion of those results will be given in Section 23.4.

The ALT model (Equation 23.5) is related to the investigation by Rieck and Nedelman [13], where a log–linear form of the Birnbaum–Saunders model (Equation 23.2) was considered. If $T \sim$ B–S(α, β), then the random variable can be expressed as $T = \beta X$, where $X \sim$ B–S$(\alpha, 1)$ since β is a scale parameter. Therefore, if β is replaced by the acceleration model $\exp[a + bV]$, the random variable has the log–linear form since $\ln(T) = a + bV + \ln(X)$, where the error term $\ln(X)$ follows the sinh–normal distribution [13]. Thus, least-squares estimates (LSEs) of a and b can be found quite simply using the t_{ij} data defined previously, but the parameter α (confounded in the error term) can only be estimated using alternative methods. These are described by Rieck and Nedelman [13], along with the properties of the LSEs of a and b. Their approach and the acceleration model selected, although functionally related to Equation 23.1, make the log–linear model attractive in the general linear model (GLM) framework. Owen and Padgett [12] use the more common parameterization (Equation 23.1) of the power-law model and likelihood-based inference. The discussion of inference for Equation 23.5 is given in the next section.

23.2.2 Cumulative Damage Models

Many high-performance materials are actually complex systems whose strength is a function of many components. For example, the strength of a fibrous carbon composite (carbon fibers embedded in a polymer material) is essentially determined by the strength of the fibers. When a composite specimen is placed under tensile stress, the fibers themselves may break within the material, which ultimately weakens the material. Since industrially made carbon fibers are known to contain random flaws in the form of cracks, voids, *etc.*, and these flaws tend to weaken the fibers [14] and ultimately the composite, statistical distributions have often been proposed to model composite strength data. It is important to note that since carbon fibers are often on the order of 8 μm in diameter, it is nearly impossible to predict when a specimen is approaching failure based on "observable damage" (like cracks or abrasions, as is the case with metals or concrete structures). The models that have been used in the past are based on studies of "weakest link" [5, 15] or competing risk models [16]. Recent investigations on the tensile strengths for fibrous carbon composite materials [17, 18] have yielded new distributions that are based on "cumulative damage" arguments. This method models the (random) damage a material aggregates as stress is steadily increased; the damage accumulates to a point until it exceeds the strength of the material. The models also contain a length or "size" variable, given by L, which accounts for the scale of the system, and the derivation for these models will now be summarized. Four models using the cumulative damage approach will be given here, and for simplicity, the CDFs will be distinguished by $F_l(s; L)$, where s represents the strength of the specimen and the model type $l = 1, 2, 3, 4$. The simplicity of this notation will be evident later.

Suppose that a specimen of length or "size" L is placed under a tensile load that is steadily increased until failure. Assume that:

1. conceptually, the load is incremented by small, discrete amounts until system failure;
2. each increment of stress causes a random amount of damage $D > 0$, with D having CDF F_D;
3. the specimen has a fixed theoretical strength, denoted by ψ. The initial strength of the specimen is a random quantity, however, that represents the reduction of the specimen's theoretical strength by an amount of "initial damage" from the most severe flaw in the specimen. The random variable W represents the initial strength of the system.

As the tensile strength is incremented in this fashion, the cumulative damage after $n + 1$ increments of stress is given by [8, 17]

$$X_{n+1} = X_n + D_n h(X_n) \tag{23.6}$$

where $D_n > 0$ is the damage incurred at load increment $n + 1$, for $n = 0, 1, 2, \ldots$, and the D_n values are independent and distributed according to F_D. The function $h(u)$ is called the damage model function. If $h(u) = 1$ for all u, the incurred damage is said to be *additive*; if $h(u) = u$ for all u, the incurred damage to the specimen is said to be *multiplicative*.

Let N be the number of increments of tensile stress applied to the system until failure. Then, $N = \sup_n\{n : X_1 < \psi, \ldots, X_{n-1} < \psi\}$, and $N = 1$ if the set is empty. Let $F_n(w) = P(N > n \mid W = w)$. Now, since ψ is only the theoretical strength, we alternately define N according to W, the initial strength.

Let X_0 denote the "initial damage" due to flaws or other pre-existing weakening damage. With additive damage, $W = \psi - X_0$ so that $N = \sup_n\{n : X_1 - X_0 < \psi - X_0, \ldots, X_{n-1} - X_0 < \psi - X_0\}$. With multiplicative damage, $W = \psi/X_0$ with $X_0 > 1$, so that $N = \sup_n\{n : X_1/X_0 < \psi/X_0, \ldots, X_{n-1}/X_0 < \psi/X_0\}$. Regardless of how W is defined, the survival probability of the specimen after n increments of stress is simply [17]

$$P(N > n) = \int_0^\infty F_n(w)\, \mathrm{d}G_W(w) \tag{23.7}$$

with $G_W(w)$ representing the distribution for initial strength of the specimen. The distribution function $F_n(w)$ depends on the type of damage model $h(u)$. In either case, $F_n(w)$ can be approximated for large n for both additive and multiplicative damage.

23.2.2.1 Additive Damage Models

For an additive damage model, $h(u) = 1$ in Equation 23.6. Since $D_n = X_{n+1} - X_n$, the damage incurred to the specimen after n increments of stress is $\sum_{i=0}^{n-1} D_i = \sum_{i=0}^{n-1}(X_{i+1} - X_i) = X_n - X_0$. Using the idea that the increments of stress are small (so that the *number* of increments will often be very large), by the central limit theorem, $X_n - X_0$ has an approximate normal (Gaussian) distribution with mean $n\mu$ and variance $n\sigma^2$, where μ and σ^2 are the mean and variance of D. Therefore, for large n

$$\begin{aligned} F_n(w) &= P(X_n - X_0 \le w) \\ &\cong \Phi[(w - n\mu)/(n^{1/2}\sigma)] \end{aligned} \tag{23.8}$$

Substituting Equation 23.8 and supplying an appropriate distribution for the initial strength in Equation 23.7 gives an expression for the survival probability of the specimen after n increments of stress.

Durham and Padgett [17] proposed two distributions for $G_W(w)$, the initial strength of the specimen of length L. The first is a three-parameter Weibull distribution given by

$$G_W(w) = 1 - \exp\{-L[(w - w_0)/\delta]^\rho\} \quad (23.9)$$

Using Equations 23.9 and 23.8 in Equation 23.7 and letting S be the total tensile load after N increments, the "additive Gauss–Weibull" strength distribution for the specimen of length L is obtained and is given by

$$F_1(s;\, L) = \Phi\left[\frac{1}{\alpha}\left(\frac{\sqrt{s}}{\beta} - \frac{L^{-1/\rho}}{\sqrt{s}}\right)\right] \quad s > 0 \quad (23.10)$$

where α and β represent two new parameters obtained in the derivation of Equation 23.10 and are actually functions of the parameters δ, ρ, μ, and σ. See Padgett and coworkers [17, 19] for the specifics of the derivation.

The second initial strength model assumes a "flaw process" over the specimen of length L that accounts for the reduction of theoretical strength due to the most severe flaw. This process is described by a standard Brownian motion, which yields an initial strength distribution found by integrating the tail areas of a Gaussian PDF. The PDF for W is given by [17]

$$G'_W(w) = g_W(w) = \frac{2}{\sqrt{2\pi L}} \exp\left[-\frac{(w-\psi)^2}{2L}\right] \quad 0 < w < \psi \quad (23.11)$$

Substituting Equations 23.11 and 23.8 in Equation 23.7 and letting S be the total tensile load after N increments (again, see Durham and Padgett [17] for the mathematical details), the "additive Gauss–Gauss" strength distribution for the specimen of length L is given by

$$F_2(s;\, L) = \Phi\left[\frac{1}{\alpha}\left(\frac{\sqrt{s}}{\beta} - \frac{\psi - \sqrt{2L/\pi}}{\sqrt{s}}\right)\right] \quad s > 0 \quad (23.12)$$

23.2.2.2 Multiplicative Damage Models

For a multiplicative damage model, $W = \psi/X_0$ and $h(u) = u$ in Equation 23.6. Since $D_n = (X_{n+1} - X_n)/X_n$, the damage incurred to the specimen after n increments of stress is

$$\sum_{i=0}^{n-1} D_i = \sum_{i=0}^{n-1} (X_{i+1} - X_i)/X_i \approx \ln(X_n) - \ln(X_0)$$

since n will be large. Thus, for large n, X_n/X_0 has an approximate *log-normal* distribution (by the central limit theorem) with parameters $n\mu$ and $n\sigma^2$. Therefore, under multiplicative damage, we have

$$F_n(w) = P(X_n/X_0 \le w) \cong \Phi\{[\ln(w) - n\mu]/(n^{1/2}\sigma)\} \quad (23.13)$$

As with the additive damage case, two models have been proposed for the distribution of initial strength [18]. The first is the two-parameter Weibull distribution (this is Equation 23.9 with $w_0 = 0$) and the second is the "flaw process" model given by the PDF (Equation 23.11). Using each of these with Equation 23.13 in Equation 23.7, the "multiplicative Gauss–Weibull" model and the "multiplicative Gauss–Gauss" model [18] are

$$F_3(s;\, L) = \Phi\left[\frac{1}{\alpha}\left(\frac{\sqrt{s}}{\beta} - \frac{\gamma - \ln(L)}{\sqrt{s}}\right)\right] \quad s > 0;\ \alpha, \beta, \gamma > 0 \quad (23.14)$$

and

$$F_4(s;\, L) = \Phi\left[\frac{1}{\alpha}\left(\frac{\sqrt{s}}{\beta} - \frac{\ln(\psi) - (\sqrt{2L/\pi}/\psi) - (L/2\psi^2)}{\sqrt{s}}\right)\right] \quad s > 0;\ \alpha, \beta, \psi > 0 \quad (23.15)$$

respectively.

The cumulative damage models presented in this section have been used by Padgett and coworkers [17–19] to model strength data from an experiment with 1000-fiber tows (microcomposite specimens) tested at several gage lengths. A discussion of these will be given in Section 23.4. It is important to note that although the cumulative damage models presented in this section are not "true" ALT models, insofar that they were not derived from a baseline model, they do resemble them, since the length variable L acts as a level of accelerated stress (equivalent to the V variable from Section 23.1.2). Obviously, the longer the specimen, the weaker it becomes due to an increased occurrence of flaws and defects. Estimation and other inference procedures for these models are discussed by Padgett and coworkers [17–19]. These matters will be summarized in the next section.

23.3 Inference Procedures with Accelerated Life Models

Parameter estimation based on likelihood methods has been derived for the models given in Section 23.2. For the model in Equation 23.5 the MLEs for α, γ, and η are given by the values that jointly maximize the likelihood function

$$\mathcal{L}(\alpha, \gamma, \eta) = \prod_{i=1}^{k} \prod_{j=1}^{n_i} g(t_{ij}; V_i)$$

where $g(t; V)$ represents the PDF for the distribution (Equation 23.5) and the significance of the values i, j, k, n_i and t_{ij} are defined in Section 23.1.1. For simplicity, it is often the case that the natural logarithm of $\mathcal{L}$ is maximized, since this function is usually more tractable. Owen and Padgett [12] provide the likelihood equations for maximization of the likelihood function, along with other inferences (*e.g.* approximate variances of the MLEs), but since the methods are similar for the cumulative damage models, these will not be presented here.

Durham and Padgett [17] provide estimation procedures for the additive damage models (Equations 23.10 and 23.12). The authors use the fact that the Birnbaum–Saunders models can be approximated by the first term of an inverse Gaussian CDF [20, 21]. This type of approximation is very useful, since the inverse Gaussian distribution is an exponential form and parameter estimates are (generally) based on simple statistics of the data. Owen and Padgett [19] observed that the models in Equations 23.10, 23.12, 23.14, and 23.15 all have a similar structure, and they derived a *generalized* three-parameter Birnbaum–Saunders model of the form

$$F_l(s; L) = \Phi\left[\frac{1}{\alpha}\left(\frac{\sqrt{s}}{\beta} - \frac{\lambda_l(\theta; L)}{\sqrt{s}}\right)\right]$$
$$s > 0;\ \alpha, \beta, \theta > 0;\ l = 1, 2, 3, 4 \quad (23.16)$$

where the "acceleration function" $\lambda_l(\theta; L)$ corresponds to a known function of L, the length variable, and θ, a model parameter. For example, for the "additive Gauss–Weibull" model (Equation 23.10), the acceleration function $\lambda_1(\theta; L) = L^{-1/\theta}$. These functions are different for the four cumulative damage models, but the overall functional form for F is similar. This fact allows for a unified theory of inference using Equation 23.16, which we will now describe [19].

The PDF for Equation 23.16 is

$$f_l(s; L) = \frac{s + \beta\lambda_l(\theta; L)}{2\sqrt{2\pi}\alpha\beta s^{3/2}} \times \exp\left\{-\frac{[s - \beta\lambda_l(\theta; L)]^2}{2\alpha^2\beta^2 s}\right\} \quad s > 0 \quad (23.17)$$

Using Equation 23.17 to form the likelihood function $\mathcal{L}(\alpha, \beta, \theta)$ for strength data observed over the k gage lengths $L_1, L_2, \ldots, L_k$, the log-likelihood can be expressed as

$$\begin{aligned}\ell &= \ln[\mathcal{L}(\alpha, \beta, \theta)] \\ &= -m\ln(2\sqrt{2\pi}) - m\ln(\alpha) - m\ln(\beta) \\ &\quad + \sum_{i=1}^{k}\sum_{j=1}^{n_i} \ln\left[\frac{s_{ij} + \beta\lambda_l(\theta; L_i)}{s_{ij}^{3/2}}\right] \\ &\quad - \frac{1}{2\alpha^2\beta^2}\sum_{i=1}^{k}\sum_{j=1}^{n_i}\frac{[s_{ij} - \beta\lambda_l(\theta; L_i)]^2}{s_{ij}}\end{aligned}$$

where $m = \sum_{i=1}^{k} n_i$. The three partial derivatives of ℓ with respect to each of α, β, and θ are nonlinear likelihood equations that must be solved numerically. The equation for α can be solved in terms of β and θ, but each equation for β and θ must be solved numerically (*i.e.* the roots must be found, given values for the other two parameters), as follows:

$$\alpha = \sqrt{\frac{1}{m\beta^2}\sum_{i=1}^{k}\sum_{j=1}^{n_i}\frac{[s_{ij}-\beta\lambda_l(\theta; L_i)]^2}{s_{ij}}} \tag{23.18}$$

$$\begin{aligned} 0 = &-\frac{m}{\beta} + \sum_{i=1}^{k}\lambda_l(\theta; L_i)\sum_{j=1}^{n_i}[s_{ij}+\beta\lambda_l(\theta; L_i)]^{-1} \\ &+ \frac{1}{\alpha^2\beta^3}\sum_{i=1}^{k}\sum_{j=1}^{n_i}\frac{[s_{ij}-\beta\lambda_l(\theta; L_i)]^2}{s_{ij}} \\ &+ \frac{1}{\alpha^2\beta^2}\sum_{i=1}^{k}\lambda_l(\theta; L_i)\sum_{j=1}^{n_i}\frac{[s_{ij}-\beta\lambda_l(\theta; L_i)]}{s_{ij}} \end{aligned} \tag{23.19}$$

$$\begin{aligned} 0 = &\beta\sum_{i=1}^{k}\left[\frac{\partial}{\partial\theta}\lambda_l(\theta; L_i)\right]\sum_{j=1}^{n_i}[s_{ij}+\beta\lambda_l(\theta; L_i)]^{-1} \\ &+ \frac{1}{2\alpha^2\beta^2}\sum_{i=1}^{k}n_i\left[\frac{\partial}{\partial\theta}\lambda_l(\theta; L_i)\right] \\ &- \frac{1}{\alpha^2}\sum_{i=1}^{k}\lambda_l(\theta; L_i)\left[\frac{\partial}{\partial\theta}\lambda_l(\theta; L_i)\right]\sum_{j=1}^{n_i}\frac{1}{s_{ij}} \end{aligned} \tag{23.20}$$

Iteration over these three equations with the specific acceleration function $\lambda_l(\theta; L)$ can be performed using a nonlinear root-finding technique until convergence to an approximate solution for the MLEs $\hat{\alpha}$, $\hat{\beta}$, and $\hat{\theta}$. A procedure for obtaining the starting values for the parameters that are necessary for initiating the iterative root-finding procedure is detailed by Owen and Padgett [19]. From Equation 23.20, we see that an additional prerequisite for the acceleration function $\lambda_l(\theta; L)$ in the generalized distribution (Equation 23.17) is that the $\lambda_l(\theta; L)$ must be differentiable with respect to θ for MLEs to exist.

The $100p$th percentile for the generalized Birnbaum–Saunders distribution (Equation 23.17) for a given value of L and specified acceleration function $\lambda_l(\theta; L)$ can be expressed as [19]

$$s_p(L) = \frac{\beta^2}{4}\left[\alpha z_p + \sqrt{(\alpha z_p)^2 + 4\lambda_l(\theta; L_i)/\beta}\right]^2 \tag{23.21}$$

where z_p represents the $100p$th percentile of the standard normal distribution. The MLE for $s_p(L)$ can be obtained by substituting the MLEs $\hat{\alpha}$, $\hat{\beta}$, and $\hat{\theta}$ into Equation 23.21. Estimated values for the lower percentiles of a strength distribution are often of interest to engineers for design issues with composite materials. In addition, approximate lower confidence bounds on $s_p(L)$ can be calculated using its limiting distribution via the Cramér delta method, and this method is outlined by Owen and Padgett [19].

For a large value of m, approximate confidence intervals for the parameters α, β, and θ can be based on the asymptotic theory for MLEs (conditions for this are given by Lehmann and Casella [11]). As m increases without bound (and $n_i/m \to p_i > 0$ for all $i = 1, \ldots, k$), the standardized distribution of $\hat{\tau} = (\hat{\alpha}, \hat{\beta}, \hat{\theta})'$ approaches that of the trivariate normal distribution (TVN) with zero mean vector and identity variance–covariance matrix. Thus, the sampling distribution of $\hat{\boldsymbol{\tau}}$ can be approximately expressed as $\hat{\boldsymbol{\tau}} \mathrel{\dot\sim} \mathrm{TVN}(\boldsymbol{\tau}, \mathbf{I}_m^{-1}(\boldsymbol{\tau}))$, which is the trivariate normal distribution with mean vector $\boldsymbol{\tau}$ and variance–covariance matrix $\mathbf{I}_m^{-1}(\boldsymbol{\tau})$, the mathematical inverse of the Fisher information matrix for Equation 23.16 based on m independent observations. The negative expected values for the six second partial derivatives for the log of the likelihood function $\ell = \ln \mathcal{L}(\alpha, \beta, \theta)$ for Equation 23.16 are quite complicated but can be obtained. The quantities $E(-\partial^2\ell/\partial\alpha^2)$, $E(-\partial^2\ell/\partial\alpha\partial\beta)$ and $E(-\partial^2\ell/\partial\alpha\partial\theta)$ can be found explicitly, but the other three terms require an approximation for the intractable integral. The method used is explained in detail by Owen and Padgett [19], but here we simply present the six entries for $\mathbf{I}_m(\boldsymbol{\tau})$:

$$E(-\partial^2\ell/\partial\alpha^2) = 2m/\alpha^2 \tag{23.22a}$$

$$E(-\partial^2\ell/\partial\alpha\partial\beta) = m/(\alpha\beta) \tag{23.22b}$$

$$E(-\partial^2\ell/\partial\alpha\partial\theta) = -\frac{1}{\alpha}\sum_{i=1}^{k}n_i\frac{(\partial/\partial\theta)\lambda_l(\theta; L_i)}{\lambda_l(\theta; L_i)} \tag{23.22c}$$

$$E(-\partial^2\ell/\partial\beta^2) \cong 3m/(4\beta^2) + \frac{1}{\alpha^2\beta^3}\sum_{i=1}^{k} n_i\lambda_l(\theta;\, L_i) \tag{23.22d}$$

$$E(-\partial^2\ell/\partial\beta\partial\theta) \cong -\frac{1}{4\beta}\sum_{i=1}^{k} n_i\frac{(\partial/\partial\theta)\lambda_l(\theta;\, L_i)}{\lambda_l(\theta;\, L_i)} + \frac{1}{\alpha^2\beta^2}\sum_{i=1}^{k} n_i\frac{\partial}{\partial\theta}\lambda_l(\theta;\, L_i) \tag{23.22e}$$

$$E(-\partial^2\ell/\partial\theta^2) \cong \frac{3}{4}\sum_{i=1}^{k} n_i\left[\frac{(\partial/\partial\theta)\lambda_l(\theta;\, L_i)}{\lambda_l(\theta L_i)}\right]^2 + \frac{1}{\alpha^2\beta}\sum_{i=1}^{k} n_i\frac{[(\partial/\partial\theta)\lambda_l(\theta;\, L_i)]^2}{\lambda_l(\theta;\, L_i)} \tag{23.22f}$$

Using these expressions, the asymptotic variances of the three MLEs can be found from the diagonal elements of $\mathbf{I}_m^{-1}(\boldsymbol{\tau})$, and these variances can be estimated by substitution of the MLEs for the three unknown parameters involved. The variances can be used to calculate approximate confidence intervals for the parameters and lower confidence bounds (LCBs) for $s_p(L)$ from Equation 23.21.

In the next section, the power-law accelerated Birnbaum–Saunders model (Equation 23.16) and the cumulative damage models from Section 23.3 will be used to fit accelerated life data taken from two different industrial experiments for materials testing. A discussion of the results will follow.

23.4 Estimation from Experimental Data

Here, we will look at two applications of the three-parameter accelerated life models from Section 23.2, using the methods of inference given in Section 23.3.

23.4.1 Fatigue Failure Data

In the original collection of articles by Birnbaum and Saunders [6, 7], a data set was published on the fatigue failure of aluminum coupons (rectangular strips cut from 6061-T6 aluminum sheeting). The coupons were subjected to repeated cycles of alternating stresses and strains, and the recorded measurement for a specimen on test was the number of cycles until failure. In the experiment, three levels of *maximum stress* for the periodic loading scheme were investigated (*i.e.* $k = 3$), and these stress levels were $V_1 =$ 21 000, $V_2 = 26\,000$, and $V_3 = 31\,000$ psi with respective sample sizes $n_1 = 101$, $n_2 = 102$, and $n_3 = 101$. The data are given by Birnbaum and Saunders [7], and a scaled version of the data set is given by Owen and Padgett [12]. This data set was the motivation for the derivation of Birnbaum and Saunders' original model (Equation 23.2), and they used their distribution to fit each data set corresponding to one of the three maximum stress levels to attain *three* estimated lifetime distributions. Although the fitted distributions matched well with the empirical results, no direct inference could be drawn on how the level of maximum stress functionally affected the properties of the distribution. Owen and Padgett [12] used the power-law accelerated Birnbaum–Saunders model (Equation 23.16) to fit across all levels of maximum stress. For the specifics of the analyses, which include parameter estimates and approximate confidence intervals, see Owen and Padgett [12].

23.4.2 Micro-Composite Strength Data

Bader and Priest [22] obtained strength data (measured in giga-pascals, GPa) for 1000-fiber impregnated tows (micro-composite specimens) that were tested for tensile strength. In their experiment, several gage lengths were investigated: 20 mm, 50 mm, 150 mm, and 300 mm with 28, 30, 32, and 29 observed specimens tested at the respective gage lengths. These data are given explicitly by Smith [23]. The strength measurements range between 1.889 and 3.080 GPa over the entire data set, and, in general, it is seen that the specimens with longer gage lengths tend to have smaller strength measurements. All of the cumulative damage models presented in Section 23.3 have been used to model these data, and the details of the procedures are given by Padgett and coworkers [17–19]. Owen and Padgett [19] compared the fits of the four models by observing their respective overall mean squared error (MSE); this is calculated by averaging the squared distance of the estimated model (using the MLEs) to the empirical CDF for each of the gage lengths over all of the $m = 119$ observations. A comparison was also made with the "power-law Weibull model" that was used to fit these strength data by Padgett *et al.* [5]. Overall, the "multiplicative Gauss–Gauss" strength model (Equation 23.15) had the smallest MSE of 0.004 56, compared with the other models, and this casts some degree of doubt on the standard "weakest link" theory with the Weibull model for these types of material. From [19], the MLEs of the parameters in Equation 23.15 were calculated to be $\hat{\alpha} = 0.144\,36$, $\hat{\beta} = 0.832\,51$, and $\hat{\psi} = 34.254$ (recall that the parameter ψ represented the "theoretical strength" of the system). Using these MLEs in Equation 23.22a–f, the (approximate) quantities of the Fisher information matrix for Equation 23.15 can be estimated (for this case, the parameter θ is given by ψ). These estimates of Equations 23.22a–f are, respectively, 11 421.1037, 990.1978, −10.1562, 32 165.7520, 324.8901, and 3.4532. The diagonal elements of the inverse of the Fisher information matrix give the asymptotic variances of the MLEs, and the asymptotic standard deviations (ASDs) of the MLEs are estimated to be $\widehat{\mathrm{ASD}}(\hat{\alpha}) = 0.0105$, $\widehat{\mathrm{ASD}}(\hat{\beta}) = 0.0281$ and $\widehat{\mathrm{ASD}}(\hat{\psi}) = 2.7125$. Using the asymptotic normality result stated in the last Section 23.3, the individual 95% confidence intervals for the three parameters are: (0.1238, 0.1649) for α, (0.7774, 0.8876) for β, and (28.9377, 39.5703) for ψ. An estimation of $s_{0.1}(L)$, the 10th percentile for the strength distribution Equation 23.15 for the given values of L, and the 90% LCBs on $s_{0.1}(L)$ are presented by Owen and Padgett [19]. Thus, not only does the model in Equation 23.15 provide an excellent fit, it also gives insight into the physical nature of the failure process due to the scientific nature of the cumulative damage process at the microscopic level.

In summary, it is seen that the Birnbaum–Saunders distribution (Equation 23.2) has great utility in the general area of accelerated life testing. The power-law accelerated model (Equation 23.16) was derived using fairly standard techniques from the ALT literature, and is a three-parameter generalization of Equation 23.2. The cumulative damage models from Section 23.2 are also three-parameter generalizations of Equation 23.2 and are accelerated life models in the general sense. They are unique, however, since nowhere in the individual development of the models was a baseline Birnbaum–Saunders model assumed. In addition, given that the models were derived from physical arguments, the parameters involved have specific meanings and interpretations in the physical process of material failure.

Acknowledgments

The second author's work was partially supported by the National Science Foundation under grant number DMS-9877107.

References

[1] Mann NR, Schafer RE, Singpurwalla ND. Methods for statistical analysis of reliability and life data. New York: John Wiley and Sons; 1974.

[2] Meeker WQ, Escobar LA. Statistical methods for reliability data. New York: John Wiley and Sons; 1998.

[3] Nelson W. Accelerated testing: statistical models, test plans, and data analysis. New York: John Wiley and Sons; 1990.

[4] Padgett WJ. Inference from accelerated life tests. In: Abdel-Hameed MS, Cinlar E, Quinn J, editors. Reliability theory and models. New York: Academic Press; 1984. p.177–98.

[5] Padgett WJ, Durham SD, Mason AM. Weibull analysis of the strength of carbon fibers using linear and power law models for the length effect. J Compos Mater 1995;29:1873–84.

[6] Birnbaum ZW, Saunders SC. A new family of life distributions. J Appl Prob 1969;6:319–27.

[7] Birnbaum ZW, Saunders SC. Estimation for a family of life distributions with applications to fatigue. J Appl Prob 1969;6:328–47

[8] Desmond AF. Stochastic models of failure in random environments. Can J Stat 1985;13:171–83.

[9] Achcar JA. Inferences for the Birnbaum–Saunders fatigue life model using Bayesian methods. Comput Stat Data Anal 1993;15:367–80.

[10] Tierney L, Kass RE, Kadane JB. Fully exponential Laplace approximations to expectations and variances of nonpositive functions. J Am Stat Assoc 1989;84:710–6.

[11] Lehmann EL, Casella G. Theory of point estimation. New York: Springer; 1998.

[12] Owen WJ, Padgett WJ. A Birnbaum–Saunders accelerated life model. IEEE Trans Reliab 2000;49:224–9.

[13] Rieck JR, Nedelman JR. A log-linear model for the Birnbaum–Saunders distribution. Technometrics 1991;33:51–60.

[14] Stoner EG, Edie DD, Durham SD. An end-effect model for the single-filament tensile test. J Mater Sci 1994;29:6561–74.

[15] Wolstenholme LC. A non-parametric test of the weakest link property. Technometrics 1995;37:169–75.

[16] Goda K, Fukunaga H. The evaluation of strength distribution of silicon carbide and alumna fibres by a multi-modal Weibull distribution. J Mater Sci 1986;21: 4475–80.

[17] Durham SD, Padgett WJ. A cumulative damage model for system failure with application to carbon fibers and composites. Technometrics 1997;39:34–44.

[18] Owen WJ, Padgett WJ. Birnbaum–Saunders-type models for system strength assuming multiplicative damage. In: Basu AP, Basu SK, Mukhopadhyay S, editors. Frontiers in reliability. River Edge (NJ): World Scientific; 1998. p.283–94.

[19] Owen WJ, Padgett WJ. Acceleration models for system strength based on Birnbaum–Saunders distributions. Lifetime Data Anal 1999;5:133–47.

[20] Bhattacharyya GK, Fries A. Fatigue failure models—Birnbaum–Saunders vs. inverse Gaussian. IEEE Trans Reliab 1982;31:439–41.

[21] Chhikara RS, Folks JL. The inverse Gaussian distribution. New York: Marcel Dekker; 1989.

[22] Bader S, Priest A. Statistical aspects of fibre and bundle strength in hybrid composites. In: Hayashi T, Katawa K, Umekawa S, editors. Progress in Science and Engineering of Composites, ICCM-IV, Tokyo, 1982; p.1129–36.

[23] Smith RL. Weibull regression models for reliability data. Reliab Eng Syst Saf 1991;34:55–76.

Multiple-steps Step-stress Accelerated Life Test

Loon-Ching Tang

24.1 Introduction

The step-stress accelerated life test (SSALT) is a reliability test in which not only products are tested at higher than usual stress but the stress applied also changes, usually increases, at some predetermined time intervals during the course of the test. The stress levels and the time epochs at which the stress changes constitute a step-stress pattern (SSP) that resembles a staircase with uneven step and length. There are many advantages for this type of stress loading. Not only can the test time be reduced by increasing the stress during the course of a test, but experimenters need not start with a high stress that could be too harsh for the product, hence avoiding excessive extrapolation of test results. Moreover, only a single piece of test equipment, *e.g.* a temperature chamber, is required for each SSP; and a single SSP is sufficient to validate the assumed stress–life model (see Section 24.4). Having said that, the obvious drawback is that it requires stronger assumptions and more complex analysis compared with a constant-stress accelerated life test (ALT). In this chapter, we look at two related issues for an SSALT, *i.e.* how to plan/design a multiple-steps SSALT and how to analyze data obtained from an SSALT. To facilitate subsequent presentation and discussion, we shall adopt the following further acronyms:

AF	acceleration factor
AV	asymptotic variance
c.d.f.	cumulative distribution function
FFL	failure-free life
LCEM	linear cumulative exposure model
MLE	maximum likelihood estimator
MTTF	mean time to failure
N-M	Nelson cumulative exposure model
NLP	nonlinear programming
p.d.f.	probability density function.

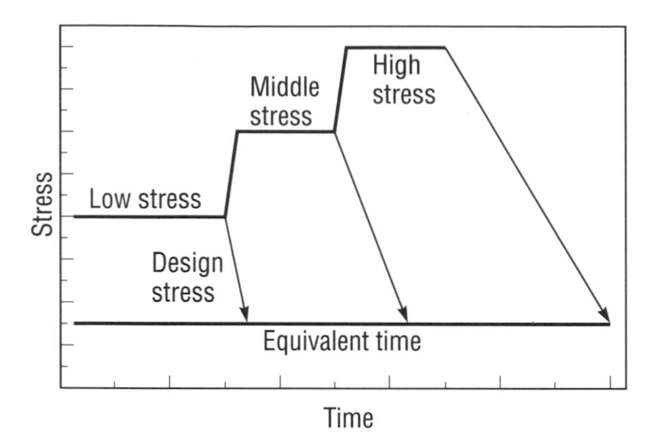

Figure 24.1. An example of an SSALT stress pattern and a possible translation of test time to a constant reference stress

The challenge presented in the SSALT can be illustrated as follows. Consider the situation where some specimens of a product are tested under an SSP, as depicted in Figure 24.1. Most specimens are expected to survive after the low stress regime, and a significant fraction may survive after the middle stress. When the test is finally terminated at high stress, we have failure data that cannot be described by a single distribution, as they come from three different stress levels. In particular, if there is indeed a stress–life model that adequately relates a percentile of the failure time to the stress, we would have to translate the failure times to equivalent failure times at a reference stress before the model could be applied. The translation of SSALT failure times must be done piecewise, as the acceleration factors are different at different stress levels. This means that, unlike a constant-stress ALT, failure data obtained under an SSALT may not be analyzed independently of the stress–life model without further assumptions. In particular, for specimens that fail at high stress, assumptions on how the "damage" due to exposure to lower stress levels is accumulated must be made. We shall defer the discussion on this to the next section.

The simplest from of SSALT is the partial ALT considered by Degroot and Goel [1], in which products are first tested under use condition for a period of time before the stress is increased and maintained at a level throughout the test. This is a special case of a simple SSALT, where only two stress levels are used. Much work has been done in this area since that by Nelson [2]. Literature appearing before 1989 has been covered by Nelson [3], in which a chapter has been devoted to step-stress and progressive-stress ALT assuming exponential failure time. Here, we give a brief review of the subsequent work.

We first begin with literature dealing with the planning problem in an SSALT. The commonly adopted optimality criterion is to minimize the AV of the log(MTTF) or some percentile of interest. Using the estimates for probabilities of failure at design and high stress levels by the end of the test as inputs, Bai and coworkers [4, 5] obtained the optimum plan for simple SSALTs under type I censoring assuming exponential [4] and Weibull [5] failure times. Khamis and Higgins [6] considered a quadratic stress–life relation and obtained the associated optimum plan for a three-step SSALT using parameters of the assumed quadratic function as inputs. They then proposed a compromised test plan for a three-step SSALT by combining the results of the optimum quadratic plan with that of the optimum simple SSALT obtained by Bai *et al.* [4]. Assuming complete knowledge of the stress–life relation with multiple stress variables, Khamis [7] proposed an optimal plan for a multiple-step SSALT. Park and Yum [8] developed statistically optimal modified ALT plans for exponential lifetimes, taking into consideration the rate of change in stress, under continuous inspection and type I censoring, with similar assumptions as in Bai *et al.* [4]. The above plans solve for optimal hold time and assumed that low stress levels are given. Yeo and Tang [9] presented an NLP for planning a two-step SSALT, with the target acceleration factor and the expected proportion of failure at high stress by the end of the test as inputs, in which both the low stress levels and the respective hold times are optimized. They proposed an algorithm for planning a

Table 24.1. A summary of the characteristics in the literature on optimal design of an SSALT

Reference	Problem addressed	Input	Output	Lifetime distribution
Bai *et al.* [4]	Planning 2-step SSALT	p_d, p_h	Optimal hold time	Exponential
Bai and Kim [5]	Planning 2-step SSALT	p_d, p_h, shape parameter	Optimal hold time	Weibull
Khamis and Higgins [6]	Planning 3-step SSALT with no censoring	All parameters of stress–life relation	Optimal hold time	Exponential
Khamis [7]	Planning M-step SSALT with no censoring	All parameters of stress–life relation	Optimal hold time	Exponential
Yeo and Tang [9]	Planning M-step SSALT	p_h and target acceleration factor	Optimal hold time and lower stress	Exponential
Park and Yum [8]	Planning 2-step SSALT with ramp rate	p_d, p_h, ramp rate	Optimal hold time	Exponential

multiple-step SSALT in which the number of steps was increased sequentially. Unfortunately, the algorithm inadvertently omitted a crucial step that would ensure optimality. A revised algorithm that addresses this will be presented in a later section. A summary of the above literature is presented in Table 24.1.

In the literature on modeling and analysis of an SSALT beyond Nelson [2], van Dorp *et al.* [10] developed a Bayes model and the related inferences of data from SSALT assuming exponential failure time. Tang *et al.* [11] generalized the Nelson cumulative exposure model (N-M), the basic model for data analysis of an SSALT. They presented an LCEM, which, unlike N-M, is capable of analyzing SSALT data with FFL. Their model also overcomes the mathematical tractability problem for N-M under a multiple-step SSALT with Weibull failure time. An attempt to overcome such difficulty is presented by Khamis and Higgins [12], who proposed a time transformation of the exponential cumulative exposure (CE) model for analyzing Weibull data arising from an SSALT. Recently, Chang *et al.* [13] used neural networks for modeling and analysis of step-stress data.

Interest in SSALTs appears to be gathering momentum, as evidenced by the recent literature, where a variety of other problems and case studies associated with SSALTs have been reported. Xiong and Milliken [14] considered an SSALT with random stress-change times for exponential data. Tang and Sun [15] addressed ranking and selection problems under an SSALT. Tseng and Wen [16] presented degradation analysis under step stress. McLinn [17] summarized some practical ground rules for conducting an SSALT. Sun and Chang [18] presented a reliability evaluation on an application-specific integrated circuit (ASIC) flash RAM under an SSALT using a Peck temperature–RH model and Weibull analysis. Gouno [19] presented an analysis of an SSALT using exponential failure time in conjunction with an Arrhenius model, where the activation energy and failure rate under operational conditions were estimated both graphically and using an MLE.

In the following, we first discuss the two basic cumulative exposure models, *i.e.* N-M and LCEM, that are pivotal in the analysis of SSALT data. This is followed by solving the optimal design problem of an SSALT with exponential lifetime. We then derive the likelihood functions using an LCEM for multiply censored data under continuous monitoring and read-out data with censoring. Finally, these methodologies are implemented on Microsoft Excel™ with a numerical example. The notation used in each section is defined at the beginning of each section.

24.2 Cumulative Exposure Models

The notation used in this section is as follows:

h	total number of stress levels in an SSP
S_i	stress level i, $i = 1, 2, \ldots, h$

N	index for stress level when failure occurs
t_i	stress change times, $i = 1, 2, \ldots, h-1$
Δt_i	sojourn time at S_i ($= t_i - t_{i-1}$, for survivors)
R	reliability
F, $F_i(\cdot)$	c.d.f., c.d.f. at S_i
$t(i)$	lifetime of a sample tested at S_i
$\tau_{e:i}$	equivalent start time of any sample at S_{i+1}, $\tau_{e:0} \equiv 0$
$t_e(i)$	equivalent total operating time of a sample at S_i
θ_i	MTTF at stress S_i.

As depicted in Figure 24.1, to facilitate the analysis of an SSALT we need to translate the test times over different stress levels to a common reference stress; say, the design stress. In other words, one needs to relate the life under an SSP to the life under constant stress. To put this into statistical perspective, Nelson [2] proposed a CE model with the assumption that "the remaining life of specimens depends only on the current cumulative fraction failed and current stress—regardless of how the fraction accumulated. Moreover, if held at the current stress, survivors will fail according to the c.d.f. for the stress but starting at the previously accumulated fraction failed". By this definition, we have

$$F(y) = \begin{cases} F_1(y) & 0 \le y < t_1 \\ F_i(\tau_{e;i-1} + y - t_{i-1}) & t_{i-1} \le y < t_i; \\ & \quad 1 < i < h \\ F_h(\tau_{e;h-1} + y - t_{h-1}) & t_{h-1} \le y < \infty \end{cases} \tag{24.1}$$

where the equivalent start time $\tau_{e:i}$ at S_i is the solution of the following:

$$F_{i+1}(\tau_{e:i}) = F_i(t_i - t_{i-1}) \quad \text{for } i = 1, \ldots, h-1$$

This N-M model has been widely accepted and commonly adopted by many authors [4–9, 14]. The problem, however, is that the model is not general enough to cater for life distributions with a location parameter representing FFL. This is because it is constructed by matching the probability of failure across various stresses. For the N-M model, when $t_i <$ FFL, $F(t_i) = 0$, implying that CE is zero. Consequently, no CE is registered. This does not seem reasonable, as there should be CE even if the test time is below FFL. Having an FFL is not uncommon among engineering products, particularly for products that fail through degradation and/or wear-out processes. For example, some semiconductors may take a few years before failure emerges; other examples include the life of batteries and the storage life of printer cartridges, photo-films, *etc.*

Another drawback of N-M is that it is defined purely on statistical grounds, and thus it is hard to give a physical interpretation. Here, we present an alternative CE model that is based on exposure times that a sample experienced under a test, call the LCEM, which has been shown to include N-M as its special case when failure time follows a Weibull distribution [11].

Suppose that the lifetime of a product can be described by a distribution and a sample is randomly selected from its population. The lifetime of the sample is an unknown constant at a fixed stress. Let the life of the sample operating at S_i be $t(i)$. The LCEM is defined as follows.

1. For a sample having tested for Δt_i at S_i, its fractional exposure accumulated at this S_i is $\Delta t_i / t(i)$.
2. For a sample test under an SSP, its fractional exposure is linearly accumulated. That is, if a sample has operated Δt_k, at S_k, $k = 1, 2, \ldots, i$, its accumulated fractional exposure is
$$\sum_{k=1}^{i} \Delta t_k / t(k).$$
3. For a survivor at any stress level, its equivalent start time at that stress depends only on the previously accumulated fractional exposure, regardless of how that exposure is accumulated:
$$[\Delta t_i + \tau_{e:i-1}]/t(i) = \tau_{e:i}/t(i+1)$$
$$\tau_{e:0} \equiv 0, \ i = 1, 2, \ldots, h-1$$

From the definition, it follows that the equivalent start time is given by:

$$\tau_{e:i} = t(i+1)\sum_{k=1}^{i} \Delta t_k / t(k) \quad i = 1, 2, \ldots, h-1 \tag{24.2}$$

Note that, compared with Equation 24.1, the equivalent time is independent of the distribution of the failure time. The LCEM allows the translation of test time to any reference stress quite effortlessly. From Equation 24.2, the equivalent test time of the sample at S_i after going through $\Delta t_k, k = 1, \ldots, i$, is:

$$\begin{aligned} t_e(i) &= \tau_{e:i-1} + \Delta t_i \\ &= t(i)\sum_{k=1}^{i-1} \Delta t_k / t(k) + \Delta t_i \\ &= t(i)\sum_{k=1}^{i} \Delta t_k / t(k) \end{aligned} \tag{24.3}$$

From Equation 24.2, Equation 24.3 can be written in the equivalent test time at any stress, say S_j, where $\Delta t_j = 0$:

$$t_e(j) = t(j)\sum_{k=1}^{i} \Delta t_k / t(k)$$

$$\Rightarrow \frac{t_e(j)}{t(j)} = \frac{t_e(i)}{t(i)} = \sum_{k=1}^{i} \Delta t_k / t(k) \tag{24.4}$$

Equation 24.4 is true as long as $t_e(j) \leq t(j)$, because failure occurs when $t_e(j) = t(j)$. The interpretation of Equation 24.4 is interesting: the equivalent test time at a stress is directly proportional to its failure time at the stress and, more importantly, the total fractional CE is invariant across stress levels. The failure condition, *i.e.*

$$\sum_{i=1}^{N} \Delta t_i / t(i) = 1 \tag{24.5}$$

is identical to the well-known Miner's rule. It can be interpreted that, once we have picked a sample, regardless of the step-stress spectrum it undergoes, the cumulative damage that triggers a failure is a deterministic unknown. Moreover, for all samples, the exposure is linearly accumulated until failure. In fact, this seems to be the original intent of Nelson [2], and it is only natural that N-M is a special case of LCEM (shown by Tang *et al.* [11]).

From Equations 24.2 to 24.4, it is noted that LCEMs need estimates of $t(i)$, $i = 1, 2, \ldots, N$. These must be estimated from the empirical c.d.f./reliability and the assumed stress–life model. For example, in the case of exponential failure time, suppose that the lifetime of a sample corresponds to the $(1 - R)$-quantile. Then

$$t(i, R) = -\theta_i \ln(R) \Rightarrow \hat{t}(i, R) = -\theta_i \ln(\hat{R}) \tag{24.6}$$

The MTTF at S_i, θ_i will depend on the stress–life model. We shall use the LCEM in the following sections.

24.3 Planning a Step-stress Accelerated Life Test

The notation adopted in this section is as follows:

S_0, S_i, S_h	stress (design, low, high) levels, $i = 1, 2, \ldots, h-1$
x_i	$\frac{S_i - S_0}{S_h - S_0}$: normalized low stress levels, $i = 1, 2, \ldots, h-1$
h	total number of stress levels
n	number of test units
p	expected failure proportion when items are tested only at S_h
p_i	expected failure proportion at S_i
n_i	number of failed units at S_i, $i = 1, 2, \ldots, h$
n_c	number of censored units at S_h (at end of test)
$y_{i,j}$	failure time j of test units at S_i, $i = 1, 2, \ldots, h$
$t_e(j, i)$	equivalent failure time for the jth failure at S_i
$R(j, i)$	reliability of the jth failure at S_i

θ_i	mean life at stress S_i, $i = 1, 2, \ldots, h$
τ_i	hold time at low S_i, $i = 1, 2, \ldots, h-1$
T	censoring time
$V(x, \tau)$	$n \cdot \mathrm{AV}[\log(\text{mean-life estimate})]$.

Table 24.1 shows a list of the work related to optimal design of an SSALT. The term "optimal" usually refers to minimizing the asymptotic variance of the log(MTTF) or that of a percentile under use condition.

As we can see from Table 24.1, with the exception of Bai and Kim [5], all work deals with exponential failure time. This is due to simplicity, as well as practicality, for it is hard to know the shape parameter of a Weibull distribution in advance.

The typical design problem for a two-step SSALT is to determine the optimal hold time, with a given low stress level. This solution is not satisfactory, because, in practice, the selection of the low stress is not obvious. Choosing a low stress close to the design stress might result in too few failures at low stress for meaningful statistical inference. Choosing a low stress close to a high stress could result in too much extrapolation of the stress–life model. In fact, it is well known that in the optimal design of a two-stress constant-stress ALT, both the sample allocation and the low stress are optimally determined. It is thus not an unreasonable demand for the same outputs in the design of an SSALT. This motivated the work of Yeo and Tang [9], who used p and an AF as inputs so that both the hold time and low stress can be optimally determined.

The use of an AF as input is more practical than using the probability of failure at design stress, as the latter is typically hard to estimate. A common interpretation of the AF is that it is the time compression factor. Given the time constraint that determines the maximum test duration and some guess of how much time the test would have taken if tested under use conditions, the target AF is simply the ratio of the two. In fact, as we shall see in Equation 24.9, the target AF can easily be derived using an LCEM once the test duration is given.

24.3.1 Planning a Simple Step-stress Accelerated Life Test

We consider a two-level SSALT in which n test units are initially placed on S_1. The stress is changed to S_2 at $\tau_1 = \tau$, after which the test is continued until all units fail or until a predetermined censoring time T. For each stress level, the life distribution of the test units is exponential with mean θ_i.

24.3.1.1 The Likelihood Function

From Equations 24.3 and 24.6, under exponential failure time assumption, the equivalent failure time of the jth failure at high stress is given by:

$$t_e(j, h) = \theta_2 \ln[R(j, h)] \times \left\{ \frac{\tau}{\theta_1 \ln[R(j, h)]} + \frac{y_{2,j} - \tau}{\theta_2 \ln[R(j, h)]} \right\}$$

As the term $\ln[R(j, h)]$ will be canceled by itself, we shall omit it whenever this happens. The contribution to the likelihood function for the jth failure at high stress is

$$\frac{1}{\theta_2} \exp\left[-\frac{t_e(j, h)}{\theta_2}\right] = \frac{1}{\theta_2} \exp\left(-\frac{y_{2,j} - \tau}{\theta_2} - \frac{\tau}{\theta_1}\right)$$

For the n_c survivors, the equivalent test time at high stress is

$$t_e(h) = \theta_2 \left(\frac{\tau}{\theta_1} + \frac{T - \tau}{\theta_2}\right)$$

so the contribution to the likelihood function is

$$\exp\left(-\frac{t_e(h)}{\theta_2}\right) = \exp\left(-\frac{T - \tau}{\theta_2} - \frac{\tau}{\theta_1}\right)$$

Putting these together, the likelihood function is

$$L(\theta_1, \theta_2) = \prod_{j=1}^{n_1} \left[\frac{1}{\theta_1} \exp\left(-\frac{y_{1,j}}{\theta_1}\right)\right] \times \prod_{j=1}^{n_2} \left[\frac{1}{\theta_2} \exp\left(-\frac{y_{2,j} - \tau}{\theta_2} - \frac{\tau}{\theta_1}\right)\right] \times \prod_{j=1}^{n_c} \exp\left(-\frac{\tau}{\theta_1} - \frac{T - \tau}{\theta_2}\right) \quad (24.7)$$

24.3.1.2 Setting a Target Accelerating Factor

The definition of an AF can be expressed as

$$\phi = \frac{\text{equivalent time to failure at design stress}}{\text{time to failure under test plan}} \tag{24.8}$$

From Equation 24.3, the equivalent operating time for the test duration T at the design stress is

$$t_e(0) = \theta_0 \left(\frac{\tau}{\theta_1} + \frac{T-\tau}{\theta_2} \right)$$

Substituting this into Equation 24.8 with the test time T as the denominator, the AF is

$$\phi = \frac{\tau(\theta_0/\theta_1) + (T-\tau)(\theta_0/\theta_2)}{T} \tag{24.9}$$

Suppose the mean life of a test unit is a log–linear function of stress:

$$\log(\theta_i) = \alpha + \beta S_i \tag{24.10}$$

where α and β ($\beta < 0$) are unknown parameters. (This is a common choice for the life–stress relationship because it includes both the power law and the Arrhenius relation as special cases.)

Without loss of generality, let $S_0 = 0$, $S_1 = x$, $S_2 = 1$, $T = 1$. Then, substituting Equation 24.10 into Equation 24.9 yields

$$\phi = (1-\tau)\exp(-\beta) + \tau\exp(-\beta x) \tag{24.11}$$

24.3.1.3 Maximum Likelihood Estimator and Asymptotic Variance

The MLE of $\log\theta_0$ can be obtained by differentiating the log-likelihood function in Equation 24.7:

$$\log(\hat{\theta}_0) = \frac{\log(U_1/n_1) - x\log(U_2/n_2)}{(1-x)}$$

where

$$U_1 = \sum_{j=1}^{n_1} y_{1,j} + (n-n_1)\tau$$

$$U_2 = \sum_{j=1}^{n_2} (y_{2,j} - \tau) + (n-n_c)(T-\tau)$$

From the Fisher information matrix, the AV of the MLE of the log(mean life) at the design stress $\text{AV}[\log(\hat{\theta}_0)]$ is

$$V(x,\tau) = \frac{\left(\frac{1}{1-x}\right)^2}{1-\exp\left(-\frac{\tau}{\theta_1}\right)} + \frac{\left(\frac{x}{1-x}\right)^2}{\exp\left(-\frac{\tau}{\theta_1}\right)\left[1-\exp\left(-\frac{1-\tau}{\theta_2}\right)\right]} \tag{24.12}$$

We need to express Equation 24.12 in terms of x, τ, p, β. From the log–linear relation of the mean in Equation 24.10, we have

$$\frac{\theta_2}{\theta_1} = \left(\frac{\theta_2}{\theta_0}\right)^{1-x} = \exp[\beta(1-x)]$$

And with

$$p = 1 - \exp\left(-\frac{1}{\theta_2}\right)$$

it follows that $V(x,\tau)$ is

$$V(x,\tau) = \frac{\left(\frac{1}{1-x}\right)^2}{1-(1-p)^{\omega}} + \frac{\left(\frac{x}{1-x}\right)^2}{(1-p)^{\omega}[1-(1-p)^{1-\tau}]} \tag{24.13}$$

where

$$\omega \equiv \tau\left(\frac{\theta_2}{\theta_0}\right)^{1-x} = \tau\{\exp[\beta(1-x)]\}$$

24.3.1.4 Nonlinear Programming for Joint Optimality in Hold Time and Low Stress

Given the numerical values of p and ϕ, the optimal (x,τ) can be obtained by solving the NLP:

$$\begin{array}{ll} \min & V(x,\tau) \\ \text{subject to} & (1-\tau)\exp(-\beta) + \tau\exp(-\beta x) = \phi \end{array} \tag{24.14}$$

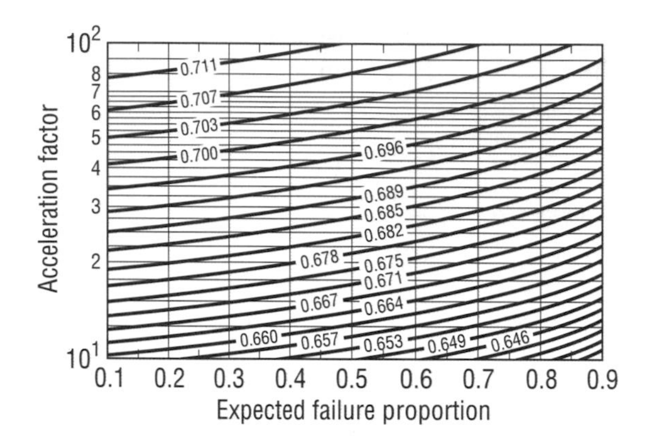

Figure 24.2. Contours of optimal hold time at low stress for a two-step SSALT. For a given (p, ϕ), the optimal hold time can be read off by interpolation between contours

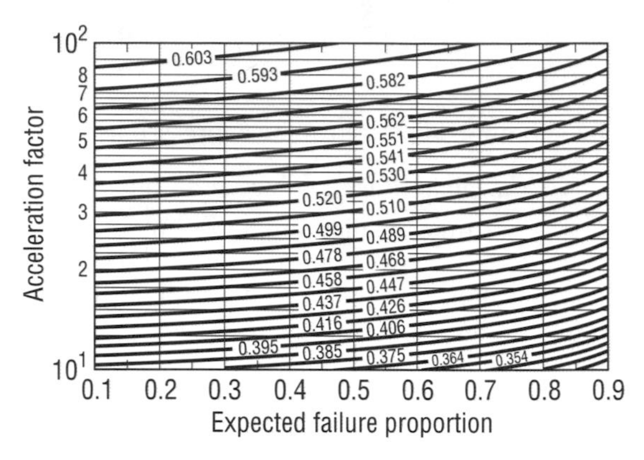

Figure 24.3. Contours of optimal low stress level for a two-step SSALT. For a given (p, ϕ), the optimal low stress can be read off by interpolation between contours

The results are given graphically in Figures 24.2 and 24.3, with (p, ϕ) on the x–y axis, for ϕ ranging from 10 to 100 and p ranging from 0.1 to 0.9. An upper limit for the range of ϕ is selected for; in practice, an AF beyond 100 may be too harsh. The range of p is chosen based on the consideration that $(1 - p)$ gives the upper bound for the fraction of censored data. For a well-planned experiment, the degree of censoring is usually less than 90% ($p = 0.1$) and it is not common to have more than 90% failure. Figure 24.2 shows the contours of the optimal normalized hold time τ and Figure 24.3 gives the contours of the optimal normalized low stress x. Given a pair of (p, ϕ), the simultaneous optimal low stress and hold time can be read from the graphs. An extensive grid search on the $[0, 1] \times [0, 1]$ solution space of (x, τ) has been conducted for selected combinations of input parameters (p, ϕ), from which the global optimality of the graphs has been verified.

Both sets of contours for the optimal hold time and the optimal low stress show an upward convex trend. The results can be interpreted as follows. For the same p, in situations where time constraints call for a higher AF, it is better to increase the low stress and extend the hold time at low stress. For a fixed AF, a longer test time (so that p increases) will result in a smaller optimal low stress and a shorter optimal hold time.

24.3.2 Multiple-steps Step-stress Accelerated Life Test Plans

From Figure 24.2, a higher AF gives higher optimal τ. This allows for further splitting of hold time if an intermediate stress is desirable. The idea is to treat testing at low stress as an independent test from that at high stress. In doing so, we need to ensure that, after splitting this low stress into a two-step SSALT with the optimal hold time as the censoring time, the AF achieved in the two-step SSALT is identical to that contributed by the low stress alone. This is analogous to dynamic programming, where the stage is the number of stress steps. At each stage, we solve the optimal design problem for a two-step SSALT by maintaining the AF contributed by the low stress test phase. For a three-step SSALT, this will give a middle stress that is slightly higher than the optimal low stress. Use this middle as the high stress and solve for the optimal hold time and low stress using Equation 24.14. Next, we also need to ensure that the solutions are consistent with the stress–life model in that the coefficients must be the same as before. For an m-step SSALT, there are $m-1$ cascading stages of SSALT. The resulting plan is "optimal", for a given number of steps and a target AF, whenever a test needs to be terminated prematurely at lower stress.

To make the idea illustrated above concrete, consider a three-step SSALT that has two cascading stages of a simple SSALT. The results of the stage #1 simple SSALT, which uses the maximum permissible stress as the high stress level, gives the optimal low stress level and its hold time. Stage #2 consists of splitting this low stress level into a simple SSALT that maintains the overall target AF. Since the AF is additive under an LCEM, the new target AF for stage #2 is the AF contributed by the stage #1 low stress.

From Equation 24.9 with $T=1$, and since $\theta_0/\theta_1 = \exp(-\beta x)$ from Equation 24.10, it follows that

$$\begin{aligned}\text{AF contributed by low stress} &= \text{New target AF}\\ &= \tau\exp(-\beta x)\end{aligned}\tag{24.15}$$

Note that to meet this target AF, x_m should be higher than the optimal low stress in stage #1. To solve the stage 2 NLP, however, this target AF needs to be normalized by the hold time τ, due to a change to the time scale. The other input needed is p_2, the expected proportion of failure at x_m at τ. Moreover, the solutions of the NLP must also satisfy the condition that the stress–life model at both stages #1 and #2 are consistent: having the same β. The resulting algorithm that iteratively solves for p_2, x_m, β, the new optimal low stress, and the hold time is summarized as follows.

Algorithm 1. This algorithm is for a multiple-step SSALT. With the values of (τ, x, β) from stage #1, do the following:

1. Compute the normalized AF ϕ_2 at stage #2 using Equation 24.15:

$$\phi_2 = (\text{New Target AF})/\tau = \exp(-\beta x)\tag{24.16}$$

2. Compute $p_2^{(0)}$ the expected failure proportion at the low stress of stage #1:

$$\begin{aligned}p_2^{(0)} &= 1-\exp\left(-\frac{\tau}{\theta_x}\right)\\ &= 1-\exp\left\{-\tau\log\left(\frac{1}{1-p}\right)\right.\\ &\qquad\left.\times\exp[\beta(1-x)]\right\}\end{aligned}\tag{24.17}$$

3. Solve the constrained NLP in Equation 24.14 to obtain $(\tau^*_{(1)}, x^*_{(1)}, \beta^*_{(1)})$.
4. Compute the new middle stress $x_m^{(1)}$

$$x_m^{(1)} = \beta^*_{(1)}\left\{\log\left[\frac{\log(1/(1-p_2^{(0)}))}{\tau\log(1/(1-p))}\right]+\beta^*_{(1)}\right\}^{-1}\tag{24.18}$$

which is the solution of

$$p_2^{(0)} = 1-\exp\left\{-\tau\log\left(\frac{1}{1-p}\right)\times\exp\left[\frac{\beta^*_{(1)}}{x_m^{(1)}}(1-x_m^{(1)})\right]\right\}$$

5. Update p_2 using $x_m^{(1)}$:

$$p_2^{(1)} = 1-\exp\left\{-\tau\log\left(\frac{1}{1-p}\right)\times\exp[\beta(1-x_m^{(1)})]\right\}\tag{24.19}$$

6. Repeat steps 3 to 5, with $(p_2^{(1)}, \phi_2)$, $(p_2^{(2)}, \phi_2)$, $(p_2^{(3)}, \phi_2)$, ... until $|\beta^*_{(k)} - x_m^{(k)}\beta| < \varepsilon$, for some pre-specified $\varepsilon > 0$.

Let $(\tau^*_{(k)}, x^*_{(k)}, \beta^*_{(k)})$ be the solutions from the scheme. By combining the stage #1 and #2 results, the plan for a three-step SSALT ($h=3$) is

$$x_1 = x_m^{(k)}x^*_{(k)} \quad x_2 = x_m^{(k)} \quad \tau_1 = \tau\tau^*_{(k)} \quad \tau_2 = \tau\tag{24.20}$$

Given the above parameters, the asymptotic variance for $\log(\hat{\theta}_0)$ is given by

$$n\text{AV}[\log(\hat{\theta}_0)] = \frac{\sum_{i=1}^{3} A_i x_i^2}{\sum_{(i,j)=\{(1,2),(1,3),(2,3)\}} A_i A_j (x_i - x_j)^2}$$

where

$$A_1 = 1 - \exp(-\tau_1/\theta_1)$$
$$A_2 = \exp(-\tau_1/\theta_1)\{1 - \exp[-(\tau_2 - \tau_1)/\theta_2]\}$$
$$A_3 = \exp(-\tau_1/\theta_1)\exp[-(\tau_2 - \tau_1)/\theta_2] \times \{1 - \exp(-(1 - \tau_2)/\theta_3)\}$$
$$\theta_i = \frac{\exp[-\beta(1 - x_i)]}{-\log(1 - p)}$$

From the above equations, the sample size can be determined given a desirable level of precision.

Algorithm 1 aims to adjust x_m upwards so as to maintain the AF computed in Equation 24.6 at each iteration. In doing so, p_2 will also increase. This adjustment is repeated until β of the stress–life model is identical to that obtained in stage #1. Note that the algorithm is a revised version of that of Yeo and Tang [9], which erroneously omitted step 1. As a result, the AF attained is lower than the target AF. The updating of x_m in Equation 24.18 is also different from [9].

For a four-step SSALT, the results from stage #2 are used to solve for stage #3 using the above algorithm. This updating scheme can be done recursively, if necessary, to generate the test plan for a multiple-step SSALT.

24.4 Data Analysis in the Step-stress Accelerated Life Test

The following notation is used in this section:

r_i	number of failures at S_i
r	number of failures at an SSP $(= \sum_i r_i)$
N_j	the stress level in which sample j failed or is being censored
S_i	stress level i, $i = 1, 2, \ldots, h$
t_i	stress change times, $i = 1, 2, \ldots, h - 1$
$\Delta t_{i,j}$	sojourn time of sample j at S_i
R_j	reliability of sample j
$t(i), t(i, R_j)$	life (of sample j) at S_i
$t_e(i), t_e(i, R_j)$	equivalent operating time (of sample j) at S_i; $t_e(0) = 0$.

24.4.1 Multiply Censored, Continuously Monitored Step-stress Accelerated Life Test

We consider a random sample of size n tested under an SSALT with an SSP. The test duration is fixed (time censored) and multiple censoring is permitted at other stress levels. Suppose that there are r failures and r_c survivors at the end of the test. All n samples have the same initial stress and stress increment for one SSP. Their hold times are the same at the same stress level, except at the stress level at which they fail or are being censored. The parameter estimation procedure using an LCEM is described as follows.

From Equation 24.4, the total test time corresponding to failed sample j at stress level N_j can be converted to the equivalent test time at S_{N_j} as:

$$t_e(N_j, R_j) = t(N_j, R_j)\left[\sum_{i=1}^{N_j-1} \Delta t_{i,j}/t(i, R_j)\right] + \Delta t_{N_j,j} \quad j = 1, 2, \ldots, r \tag{24.21}$$

Similarly, the total test time corresponding to survivor j at stress level N_j can be converted to the equivalent test time at S_{N_j} as

$$t_e(N_j, R_j) = t(N_j, R_j)\sum_{i=1}^{N_j} \Delta t_{i,j}/t(i, R_j) \quad j = r + 1, \ldots, n \tag{24.22}$$

From Equation 24.6, it can be seen that, given a stress-dependent c.d.f., *i.e.* $F(S, t, \theta)$, $t(i, R_j)$ in Equations 24.21 and 24.22, can be estimated from

$$F(S_i, t(i, R_j, \theta)) = 1 - \hat{R}_j \quad j = 1, 2, \ldots, n \tag{24.23}$$

θ is the set of parameters to be estimated.

The empirical c.d.f. can be estimated using the well-known Kaplan–Meier estimate or the cumulative hazard estimate for multiply censored data. For example, by sorting the failure times and censored times in ascending order, the cumulative hazard estimate for R_j is

$$\hat{R}_j = \exp\left[-\sum_{k=1}^{j}(1/m_k)\right] \quad (24.24)$$

where m_k is the number of samples remaining on test at the time before sample k fails.

The parameter set θ can be estimated by maximizing the likelihood function:

$$\left[\prod_{j=1}^{r} f(S_{N_j}, t_e(N_j, R_j), \theta)\right] \times \left[\prod_{j=r+1}^{n} R(S_{N_c}, t_e(N_c, R_j), \theta)\right] \quad (24.25)$$

For multiple SSPs, the parameters can be estimated from the joint likelihood function, which is simply the product of the likelihood functions of all patterns with the usual s-independence assumption.

24.4.1.1 Parameter Estimation for Weibull Distribution

The following additional notation is used:

η_i scale parameter of Weibull distribution at S_i
γ shape parameter of Weibull distribution.

For illustration purposes, we use a Weibull distribution with constant shape parameter throughout the SSP and stress-dependent η_i as an example. For an illustration of using three-parameter Weibull, refer to Tang *et al.* [11]. This model is widely used for describing failure data under mechanical stress, particularly for fatigue failure.

From Equation 24.23, the life corresponding to sample j and S_i, $t(i, R_j)$ is given by:

$$t(i, R_j) = [-\ln(\hat{R}_j)]^{1/\gamma}\eta_i \quad j = 1, 2, \ldots, n$$

Substituting into Equations 24.21 and 24.22 yields

$$t_e(N_j, R_j) = \eta_{N_j}\left(\sum_{i=1}^{N_j-1} \Delta t_{i,j}/\eta_i\right) + \Delta t_{N_j,j} = \eta_{N_j}\left(\sum_{i=1}^{N_j} \Delta t_{i,j}/\eta_i\right) \quad j = 1, 2, \ldots, n$$

This is the same expression given by Nelson [3] using the N-M model. We can convert these test times into the equivalent test time at the highest stress so that we can expect the most number of cancellations of η_i:

$$t_e(h, R_j) = \eta_h\left(\sum_{i=1}^{N_j} \Delta t_{i,j}/\eta_i\right) \quad j = 1, \ldots, n$$

Now we can analyze these failure/censored times as if they come from the same Weibull distribution under the design stress. The resulting likelihood function is given by

$$\left\{\prod_{j=1}^{r} \frac{\gamma}{\eta_h\left(\sum_{i=1}^{N_j} \Delta t_{i,j}/\eta_i\right)}\left(\sum_{i=1}^{N_j} \Delta t_{i,j}/\eta_i\right)^{\gamma} \times \exp\left[-\left(\sum_{i=1}^{N_j} \Delta t_{i,j}/\eta_i\right)^{\gamma}\right]\right\} \times \left\{\prod_{j=r+1}^{n} \exp\left[-\left(\sum_{i=1}^{N_j} \Delta t_{i,j}/\eta_i\right)^{\gamma}\right]\right\}$$

When the number of steps $h > 2$, but is small compared with the number of failures, η_i can be estimated individually so that the stress–life model can be identified. A set of constraints can be included to ensure that $\hat{\eta}_i > \hat{\eta}_j$ for $S_i < S_j$. Alternatively, a stress–life model, say, $\log(\eta_i) = \alpha + \beta S_i$, can be used to replace η_i in the above so that we have a likelihood function in terms of the unknown parameters α, β, and γ. The MLE of these parameters can be obtained via the usual means. This is left as an exercise.

24.4.2 Read-out Data

Suppose that the SSALT at read-out times coincides with the stress change time and there are r_i failures at S_i. The equivalent read-out time at the highest stress, after stress level i, can be expressed as

$$t_{e:i}(h) = t(h)\left[\sum_{k=1}^{i} \Delta t_k/t(k)\right] \quad i = 1, 2, \ldots, h$$

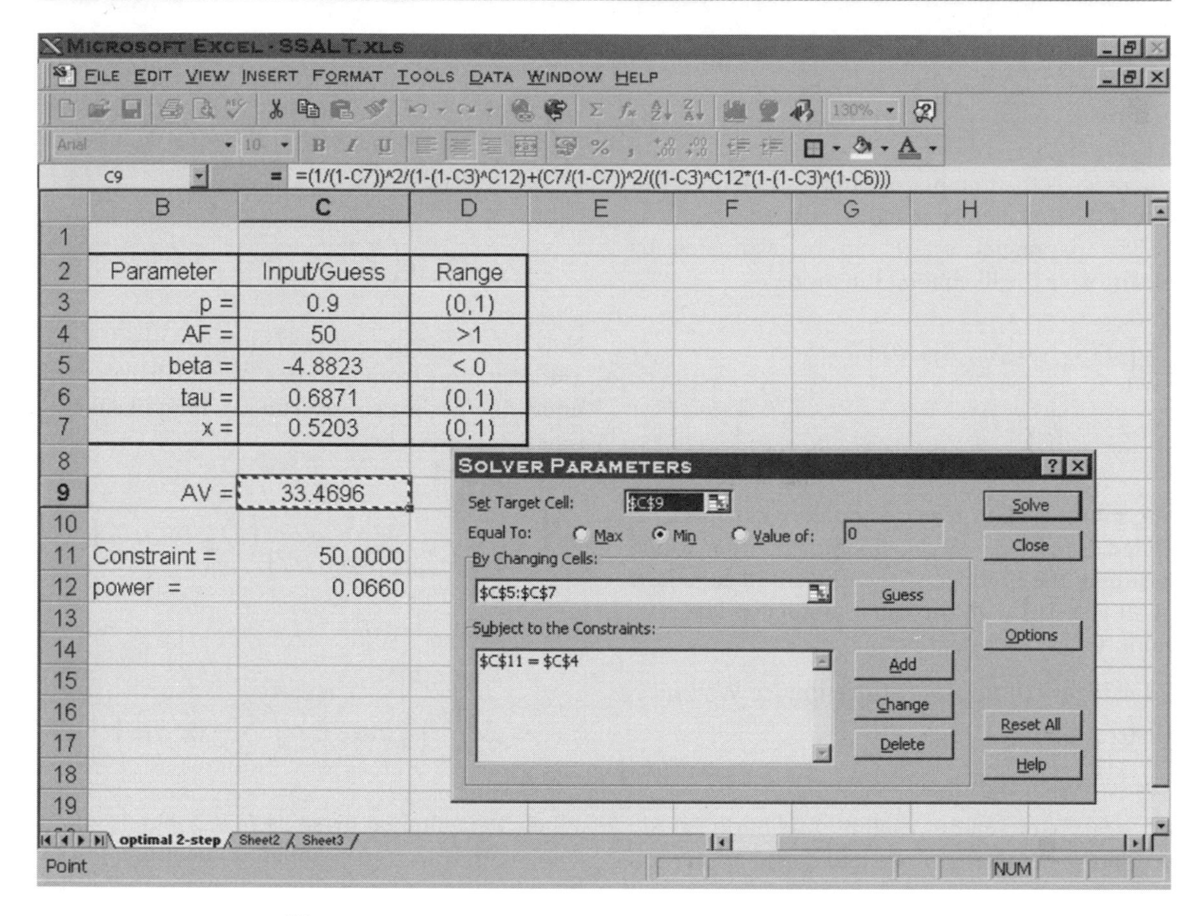

Figure 24.4. A Microsoft Excel[TM] template for designing an optimal two-step SSALT. The use of solver allows for optimization of a nonlinear function with nonlinear constraint

The likelihood function is given by

$$\left\{\prod_{i=1}^{h}[F_h(t_{e:i}(h)) - F_h(t_{e,i-1}(h))]^{r_i}\right\} \times [1 - F_h(t_{e:h}(h))]^{n-r}$$

where F_h is the c.d.f. of failure time at the highest stress.

In the case of Weibull failure time, we have

$$t_{e:i}(h) = \eta_h\left(\sum_{k=1}^{i}\Delta t_k/\eta_k\right) \quad i = 1, 2, \ldots, h$$

and the likelihood function is given by

$$\left(\prod_{i=1}^{h}\left\{\exp\left[-\left(\sum_{k=1}^{i-1}\Delta t_k/\eta_k\right)^{\gamma}\right] - \exp\left[-\left(\sum_{k=1}^{i}\Delta t_k/\eta_k\right)^{\gamma}\right]\right\}^{r_i}\right) \times \left\{\exp\left[-\left(\sum_{k=1}^{h}\Delta t_k/\eta_k\right)^{\gamma}\right]\right\}^{n-r} \tag{24.26}$$

Similarly, one could proceed with the estimation in the usual way.

In this section, we have confined our analysis to an SSP. For multiple SSPs, we simply multiply the marginal likelihood function by assuming

Table 24.2. A numerical illustration of the algorithm for planning a multiple-step SSALT. The equation numbers from which the results are computed are indicated

Iteration k	$p_2(k-1)$ Equations 24.17 and 24.19	$\beta(k)$ Equation 24.14	$\tau(k)$ Equation 24.14	$x(k)$ Equation 24.14	$x_m(k)$ Equation 24.18	$\beta(k) - \beta x_m(k)$
1	0.141 065	−3.395 32	0.667 67	0.419 06	0.591 777	−0.506 07
2	0.193 937	−3.395 20	0.666 83	0.416 88	0.630 109	−0.318 81
3	0.228 92	−3.393 81	0.665 67	0.415 32	0.652 689	0.207 18
4	0.251 933	−3.393 95	0.665 38	0.414 35	0.666 837	−0.138 25
5	0.2673	−3.394 05	0.665 17	0.413 66	0.676 017	−0.093 52
⋮	⋮	⋮	⋮	⋮	⋮	⋮
42	0.300 379	−3.394 19	0.664 67	0.412 08	0.695 201	$-2.023\,41 \times 10^{-8}$
43	0.300 379	−3.394 19	0.664 67	0.412 08	0.695 201	$-1.406\,68 \times 10^{-8}$

independence between data obtained under each SSP.

24.5 Implementation in Microsoft Excel™

The NLP in Equation 24.14 can easily be implemented using the solver tool in Microsoft Excel™. This is illustrated in Figure 24.4. For illustration purposes, the inputs are in cells C3 and C4, where $p = 0.9$ and $\phi = 50$. Equation 24.13 is stored in cell C9. Besides the two inputs, p and ϕ some initial guess values for β, x, and τ are required to compute the AV in Equation 24.13. The solver will minimize the cell containing the equation of the AV by changing cells C5–C7, which contain β, x, and τ. A constraint can be added to the solver by equating cell C11, which contains Equation 24.11 to cell C4, which contains the target AF ϕ. After running the solver, cells C5–C7 contain the optimal solutions for β, x, and τ.

Algorithm 1 for solving the design of a multiple-step SSALT can also be implemented in Microsoft Excel™. With the optimal solutions from above, Table 24.2 shows the equations used in generating results for planning a three-step SSALT and how the numerical results converge at each iteration. The inputs and results are summarized in Table 24.3.

Table 24.3. A summary of the inputs and results for the numerical example

Input	p	0.9
	ϕ	50
Final solution	β	−4.8823
	τ_2	0.6871
	x (from stage #1)	0.5203
	τ_1	0.4567
	x_1	0.2865
	x_2	0.6952

Suppose that the above test plan is for evaluating the breakdown time of insulating oil. The high and use stress levels are 50 kV and 20 kV respectively and the stress is in log(kV). The total test time is 20 h and recall that $p = 0.9$ and $\phi = 50$ From Table 24.3, we have

$$\begin{aligned} S_1 &= \exp[S_0 + x_1(S_h - S_0)] \\ &= \exp\{\log(20) + 0.2865 \times [\log(50) - \log(20)]\} \\ &= 26.0 \text{ kV} \end{aligned}$$

$$\begin{aligned} S_2 &= \exp[S_0 + x_2(S_h - S_0)] \\ &= \exp\{\log(20) + 0.6952 \times [\log(50) - \log(20)]\} \\ &= 37.8 \text{ kV} \end{aligned}$$

$$t_1 = \tau_1 \times 20 = 9.134 \text{ h} = 548.0 \text{ min}$$

$$t_2 = \tau_2 \times 20 = 13.74 \text{ h} = 824.5 \text{ min}$$

Suppose that 50 samples were tested, resulting in 2, 10, and 21 failures after t_1, t_2, and T respectively. The data are analyzed using the

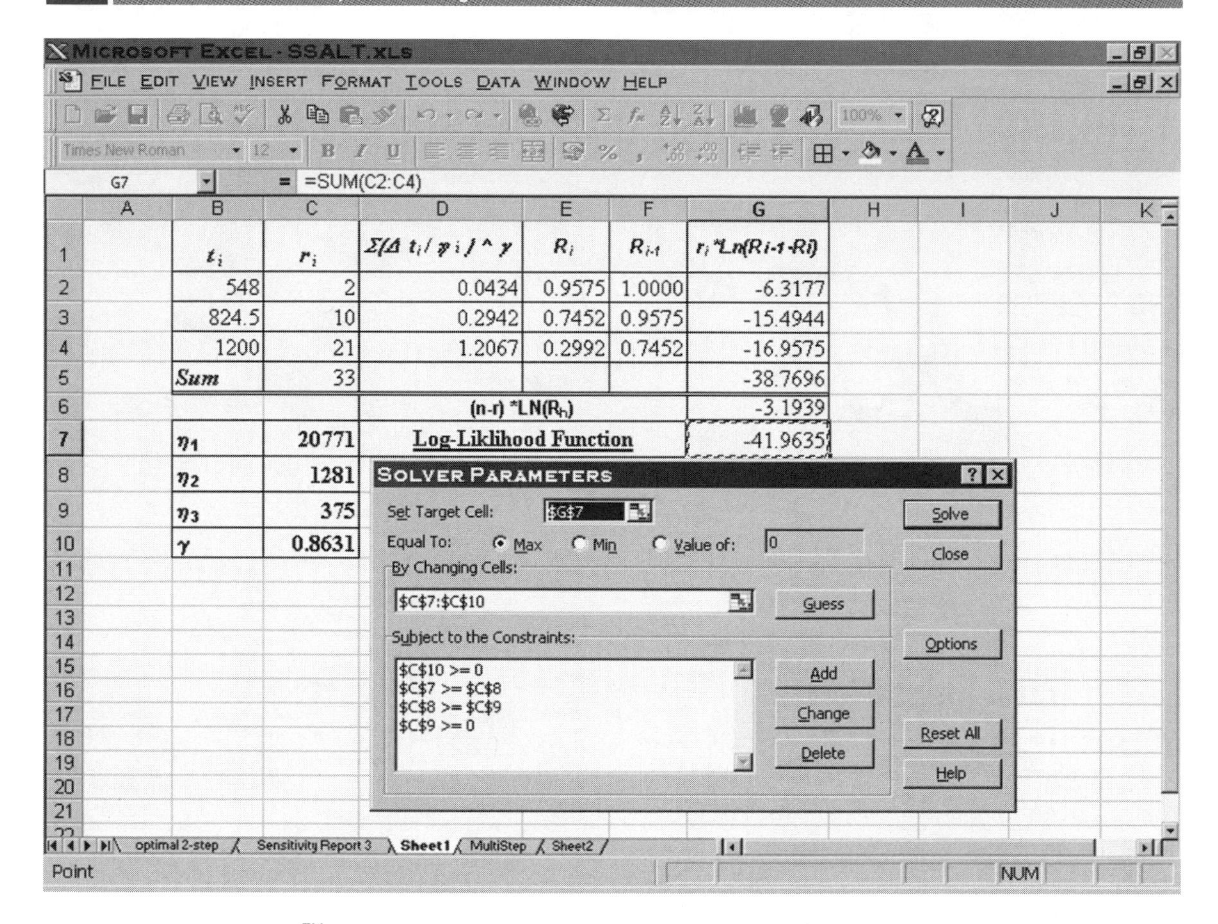

Figure 24.5. A Microsoft Excel™ template for the analysis of read-out data from an SSALT. A set of constraints is included to ensure that $\hat{\eta}_i > \hat{\eta}_j$ for $S_i < S_j$ and that all parameters are positive

solver in Microsoft Excel™ and presented in Figure 24.5. The equation in cell G7 is the log-likelihood function derived from Equation 24.26. One could proceed to fit a stress–life model for η_i. We shall leave this as an exercise.

24.6 Conclusion

In this chapter, some issues related to modeling, planning, and analysis of a multiple-step SSALT are addressed. In particular, a model that is additive in cumulative fractional exposure is presented and shown to be useful for solving optimal design and for data analysis of an SSALT. A new algorithm is provided for designing a multiple-step SSALT test plan, assuming that a target acceleration factor and a predetermined number of steps are given. Likelihood functions for two common testing situations under an SSALT are derived. The methods presented here are also implemented in Microsoft Excel™ with a numerical example.

References

[1] Degroot MH, Goel PK. Bayesian estimation and optimal designs in partially accelerated life testing. Nav Res Logist Q 1979;26:223–35.

[2] Nelson W. Accelerated life testing: step-stress models and data analysis. IEEE Trans Reliab 1980;29:103–8.

[3] Nelson W. Accelerated testing: statistical models, test plans and data analysis. New York: John Wiley & Sons; 1990.

[4] Bai DS, Kim MS, Lee SH. Optimum simple step-stress accelerated life tests with censoring. IEEE Trans Reliab 1989;38:528–32.

[5] Bai DS, Kim MS. Optimum simple step-stress accelerated life test for Weibull distribution and type I censoring. Nav Res Logist Q 1993;40:193–210.

[6] Khamis IH, Higgins JJ. Optimum 3-step step-stress tests. IEEE Trans Reliab 1996;45:341–5.

[7] Khamis IH. Optimum *M*-step, step-stress design with *k* stress variables. Commun Stat Comput Sim 1997;26:1301–13.

[8] Park SJ, Yum BJ. Optimal design of accelerated life tests under modified stress loading methods. J Appl Stat 1998;25:41–62.

[9] Yeo KP, Tang LC. Planning step-stress life-test with a target acceleration-factor. IEEE Trans Reliab 1999;48:61–7.

[10] VanDorp JR, Mazzuchi TA, Fornell GE, Pollock LR. A Bayes approach to step-stress accelerated life testing. IEEE Trans Reliab 1996;45:491–8.

[11] Tang LC, Sun YS, Goh TN, Ong HL. Analysis of step-stress accelerated-life-test data: a new approach. IEEE Trans Reliab 1996;45:69–74.

[12] Khamis IH, Higgins JJ. A new model for step-stress testing. IEEE Trans Reliab 1998;47:131–4.

[13] Chang DS, Chiu CC, Jiang ST. Modeling and reliability prediction for the step-stress degradation measurements using neural networks methodology. Int J Reliab Qual Saf Eng 1999;6:277–88.

[14] Xiong CJ, Milliken GA. Step-stress life-testing with random stress-change times for exponential data. IEEE Trans Reliab 1999;48:141–8.

[15] Tang LC, Sun Y. Selecting the most reliable population under step-stress accelerated life test. Int J Reliab Qual Saf Eng 1999;4:347–59.

[16] Tseng ST, Wen ZC. Step-stress accelerated degradation analysis for highly reliable products. J Qual Technol 2000;32:209–16.

[17] McLinn JA. Ways to improve the analysis of multi-level accelerated life testing. Qual Reliab Eng Int 1998;14:393–401.

[18] Sun FB, Chang WC. Reliability evaluation of a flash RAM using step-stress test results. In: Annual Reliability and Maintainability Symposium Proceedings, 2000; p.254–9.

[19] Gouno E. An inference method for temperature step-stress accelerated life testing. Qual Reliab Eng Int 2001;17:11–8.

Step-stress Accelerated Life Testing

Chengjie Xiong

25.1 Introduction

Accelerated life testing consists of a variety of test methods for shortening the life of test items or hastening the degradation of their performance. The aim of such testing is to obtain data quickly, which, when properly modeled and analyzed, yields the desired information on product life or performance under normal use conditions. Accelerated life testing can be carried out using constant stress, step-stress, or linearly increasing stress. The step-stress test has been widely used in electronics applications to reveal failure modes. The step-stress scheme applies stress to test units in such a way that the stress setting of test units is changed at certain specified times. Generally, a test unit starts at a low stress. If the unit does not fail at a specified time, stress on it is raised to a higher level and held a specified time. Stress is repeatedly increased and held, until the test unit fails. A simple step-stress accelerated life testing uses only two stress levels.

In this chapter, we plan to give a summary of a collection of results by the author, and some others, dealing with the statistical models and estimations based on data from a step-stress accelerated life testing. We consider two cases: when the stress-change times are either constants or random variables. More specifically, we use data from step-stress life testing to estimate unknown parameters in the stress–response relationship and the reliability function at the design stress. We also discuss the optimum test plan associated with the estimation of the reliability function under the assumption that the lifetime at a constant stress is exponential. The step-stress life testing with constant stress-change times is considered in Section 25.2. In Section 25.3, results are given when the stress-change times are random variables.

25.2 Step-stress Life Testing with Constant Stress-change Times

25.2.1 Cumulative Exposure Model

Since a test unit in a step-stress test is exposed to several different stress levels, its lifetime distribution combines lifetime distributions from all stress levels used in the test. The cumulative exposure model of the lifetime in a step-stress life testing continuously pieces these lifetime distributions in the order that the stress levels are applied. More specifically, the step-stress cumulative exposure model assumes that the remaining lifetime of a test unit depends on the current cumulative fraction failed and current stress, regardless of how the fraction is accumulated. In addition, if held at the current stress, survivors will fail according to the cumulative distribution for that stress but starting at the previously accumulated fraction failed. Moreover, the cumulative exposure model assumes that the change in stress has no effect on life—only the level of stress does. The cumulative exposure model has been used by many authors to model data from step-stress accelerated life testing [1–5]. Nelson [3], chapter 10, extensively studied cumulative exposure models when the stress is changed at prespecified constant times.

We denote the stress variable by x and assume that $x > 0$. In general, variable x represents a particular scale for the stress derived from the mechanism of the physics of failure. For example, x can be the voltage in a study of the voltage endurance of some transformer oil or the temperature in a study of the life of certain semiconductors. We assume that F is a strictly increasing and continuously differentiable probability distribution function on $(0, \infty)$. We also assume that the scale family $F(t/\theta(x))$ is the cumulative distribution function of the lifetime under constant stress x. Let $f(t) = \mathrm{d}F(t)/\mathrm{d}t$ for $t \geq 0$. Note that, at a constant stress x, the lifetime depends on the stress only through parameter $\theta(x)$. Since

$$\int_0^\infty t \,\mathrm{d}F\left(\frac{t}{\theta(x)}\right) = \theta(x) \int_0^\infty y \,\mathrm{d}F(y) \tag{25.1}$$

$\theta(x)$ is a multiple of the mean lifetime under constant stress x. We call $\theta(x)$ the stress–response relation. Further, we assume that $\ln \theta(x)$ is linearly and inversely related to stress x, *i.e.* $\ln \theta(x) = \alpha + \beta x$ with $\alpha > 0$ and $\beta < 0$. Parameters α and β are characteristics of the products and test methods. This simple stress–response relationship covers some of the most important models in engineering, such as the power law model, the Eyring model, and the Arrhenius model (*e.g.* see Mann *et al.* [6]).

We use x_0 to denote the design stress that is the stress level under normal use conditions. We use $x_1 < x_2 < \cdots < x_k$ to denote the k stress levels gradually applied in that order in a step-stress testing. Let c_i be the time that stress is changed from x_i to x_{i+1}, $1 \leq i \leq k-1$; $c_1 < c_2 < \cdots < c_{k-1}$. Let $c_0 = 0$ and $c_k = \infty$. Let T be the lifetime under the step-stress testing. Denote $\theta_i = \theta(x_i)$ and $F_i(t) = F(t/\theta_i)$, $i = 0, 1, 2, \ldots, k$. According to the cumulative exposure model, the cumulative distribution function of T is given by

$$G(t) = \begin{cases} F_1(t) & \text{for } 0 \leq t < c_1 \\ F_i(t - c_{i-1} + s_{i-1}) & \text{for } c_{i-1} \leq t < c_i \end{cases} \quad i = 2, 3, \ldots, k \tag{25.2}$$

where s_{i-1} is determined by $F_i(s_{i-1}) = F_{i-1}(c_{i-1} - c_{i-2} + s_{i-2})$, $i = 2, 3, \ldots, k$; $s_0 = 0$. Since $F_i(t) = F(t/\theta_i)$, it follows that $s_{i-1}/\theta_i = (c_{i-1} - c_{i-2} + s_{i-2})/\theta_{i-1}$, $i = 2, 3, \ldots, k$. Solving these equations inductively for s_i gives

$$s_{i-1} = \frac{\theta_i (c_{i-1} - c_{i-2} + s_{i-2})}{\theta_{i-1}} = \theta_i \sum_{j=1}^{i-1} \frac{c_j - c_{j-1}}{\theta_j}$$

$i = 2, 3, \ldots, k$. Therefore

$$G(t) = \begin{cases} F\left(\dfrac{t}{\theta_1}\right) & \text{for } 0 \leq t < c_1 \\ F\left(\dfrac{t - c_{i-1}}{\theta_i} + \displaystyle\sum_{j=1}^{i-1} \frac{c_j - c_{j-1}}{\theta_j}\right) & \text{for } c_{i-1} \leq t < c_i \end{cases} \quad i = 2, 3, \ldots, k \tag{25.3}$$

Thus, the probability density function of T is

$$g(t)=\begin{cases}\dfrac{1}{\theta_1}f\left(\dfrac{t}{\theta_1}\right) & \text{for } 0\le t<c_1\\ \dfrac{1}{\theta_i}f\left(\dfrac{t-c_{i-1}}{\theta_i}+\displaystyle\sum_{j=1}^{i-1}\dfrac{c_j-c_{j-1}}{\theta_j}\right) & \\ \qquad\qquad \text{for } c_{i-1}\le t<c_i & \end{cases}$$

$$i=2,3,\ldots,k \tag{25.4}$$

Notice that, at design stress x_0, the reliability function of the lifetime is

$$R_0(t)=1-F\left(\frac{t}{\theta(x_0)}\right) \tag{25.5}$$

One of the most important goals for carrying out a step-stress life testing is to obtain a statistical estimation on the reliability of lifetime under design stress x_0 by analyzing data from the test. When the functional form of F is given, the estimation of the reliability function becomes that of α and β.

25.2.2 Estimation with Exponential Data

When the lifetime under a constant stress is assumed exponential, *i.e.* $F(t)=1-\exp(-t)$, some exact estimations are available. We demonstrate this by using the simple step-stress testing. Many of these results can be generalized to general step-stress testing. Suppose that n test units are initially placed on low stress level x_1 and run until time c_1 when stress is changed to x_2 and the test is continued until the first $r (r>1)$ lifetimes are observed. We assume that n_1 test units fail before time c_1 and their failure times T_{1j}, $j=1,2,\ldots,n_1$, are observed under stress x_1. $n-n_1$ test units survive time c_1 and n_2 failure times T_{2j}, $j=1,2,\ldots,n_2$, are observed under stress x_2 after time c_1 ($n_2=r-n_1$). Let $T_{i\cdot}=\sum_{j=1}^{n_i}T_{ij}$ be the total lifetime under x_i, $i=1,2$. Let M_t and m_t be the observed maximum and minimum lifetime respectively. Note that n_1 may be zero or r with positive probabilities.

The assumptions of the cumulative exposure model and exponentially distributed life at any constant stress imply that the probability density function of a test unit under the simple step-stress test is

$$g(t)=\begin{cases}\dfrac{1}{\theta_1}\exp\left(-\dfrac{t}{\theta_1}\right) & \text{if } 0\le t<c_1\\ \dfrac{1}{\theta_2}\exp\left(-\dfrac{t-c_1}{\theta_2}-\dfrac{c_1}{\theta_1}\right) & \text{if } c_1\le t<\infty\end{cases} \tag{25.6}$$

The likelihood function from the first r lifetime observations T_{ij} is then

$$L(\alpha,\beta)=\frac{n!}{(n-r)!}\Pi_{ij}g(T_{ij})[1-G(M_t)]^{n-r}$$

Letting $\partial\log L/\partial\alpha=0$ and $\partial\log L/\partial\beta=0$ yields the maximum likelihood estimators for α and β when $r>n_1$ and $n_1U_2<n_2U_1$:

$$\hat{\alpha}=\frac{1}{x_2-x_1}\left(x_1\ln\frac{n_2}{U_2}-x_2\ln\frac{n_1}{U_1}\right) \tag{25.7}$$

$$\hat{\beta}=\frac{1}{x_2-x_1}\ln\left(\frac{n_1U_2}{n_2U_1}\right) \tag{25.8}$$

where

$$U_1=T_{1\cdot}+(n-n_1)c_1 \tag{25.9}$$

$$U_2=T_{2\cdot}-n_2c_1+(n-r)(M_t-c_1) \tag{25.10}$$

Parameter β describes the relationship between the mean of the lifetime of a test unit and the stress level. One may be interested in testing whether β equals zero, especially when the stress levels x_1 and x_2 are close to each other. To test $H_0:\beta=0$ versus $H_1:\beta<0$, we use the likelihood ratio test. When H_0 is true, the maximum likelihood estimator for α is

$$\bar{\alpha}=\ln\left(\frac{U_1+U_2}{r}\right)$$

The likelihood ratio is then

$$\Lambda=\frac{L(\bar{\alpha},0)}{L(\hat{\alpha},\hat{\beta})}=\left(\frac{r}{U_1+U_2}\right)^r\left(\frac{U_1}{n_1}\right)^{n_1}\left(\frac{U_2}{n_2}\right)^{n_2}$$

A size γ $(0<\gamma<1)$ test of H_0 rejects H_0 if $\Lambda<c$, where c is a constant such that

$$P(\Lambda\le c\mid H_0)=\gamma$$

Since the distribution of Λ is difficult to obtain, computer simulations can be used to approximate the distribution of Λ when H_0 is true.

Next, we construct confidence interval estimates to α, β, θ_0, and $R_0(t)$ at a given time t. We first observe that, if a random variable T has probability density function as in Equation 25.6, then it is easy to verify that the random variable

$$S = \begin{cases} \dfrac{T}{\theta_1} & \text{if } 0 \le T < c_1 \\ \dfrac{T - c_1}{\theta_2} + \dfrac{c_1}{\theta_1} & \text{if } c_1 \le T < \infty \end{cases} \tag{25.11}$$

is exponentially distributed with mean of unity (see Xiong [7]).

The following lemma is used in the derivation of confidence intervals for α, β, θ_0, and $R_0(t)$ at a given time t (see Lawless [8], theorem 3.5.1, p. 127).

Lemma 1. *Suppose that $S_i, i = 1, 2, \ldots, n$ are independent and identically distributed exponential random variables with mean of unity. Let $S_{(1)} \le S_{(2)} \le \cdots \le S_{(r)}$ be the r smallest ordered observations and $S_. = \sum_{i=1}^{r} S_{(i)}$. Then $2nS_{(1)}$ has a χ^2 distribution with two degrees of freedom and $2[S_. + (n - r)S_{(r)} - nS_{(1)}]$ has a χ^2 distribution with $2r - 2$ degrees of freedom. Further, $2nS_{(1)}$ and $2[S_. + (n - r)S_{(r)} - nS_{(1)}]$ are independent.*

Let $0 < \gamma < 1$. Now we transfer all random variables T_{ij} into S_{ij} through Equation 25.11. Denote

$$m_s = \min\{S_{ij}, i = 1, 2, j = 1, 2, \ldots, n_i\} = \begin{cases} \dfrac{m_t}{\theta_1} & \text{if } n_1 > 0 \\ \dfrac{m_t - c_1}{\theta_2} + \dfrac{c_1}{\theta_1} & \text{if } n_1 = 0 \end{cases}$$

and

$$M_s = \max\{S_{ij}, i = 1, 2, j = 1, 2, \ldots, n_i\} = \begin{cases} \dfrac{M_t}{\theta_1} & \text{if } r = n_1 \\ \dfrac{M_t - c_1}{\theta_2} + \dfrac{c_1}{\theta_1} & \text{if } r > n_1 \end{cases}$$

A direct application of Lemma 1 implies that

$$2nm_s \sim \chi^2(2)$$

and

$$D \sim \chi^2(2r - 2)$$

where

$$D = \begin{cases} \dfrac{2V_0}{\theta_1} & \text{if } r = n_1 > 0 \\ 2\left(\dfrac{V_1}{\theta_1} + \dfrac{V_2}{\theta_2}\right) & \text{if } r > n_1 > 0 \\ \dfrac{2[V_2 - n(m_t - c_1)]}{\theta_2} & \text{if } n_1 = 0 \end{cases}$$

and

$$V_0 = T_{1.} + (n - r)M_t - nm_t$$
$$V_1 = U_1 - nm_t$$
$$V_2 = U_2$$

U_1 and U_2 are given by Equations 25.9 and 25.10 respectively.

For any $0 < \gamma < 1$, let $F_\gamma(2, 2r - 2)$ be the upper $100\gamma\%$ point of the F distribution with degrees of freedom 2 and $2r - 2$. Denote $K_1 = F_{1-\gamma/2}(2, 2r - 2)$ and $K_2 = F_{\gamma/2}(2, 2r - 2)$. The independence between $2nm_s$ and D implies that

$$P\left[K_1 \le \frac{2n(r - 1)m_s}{D} \le K_2\right] = 1 - \gamma$$

Therefore

$$\begin{aligned} P\Bigg\{ & \frac{V_2 - n(m_t - c_1)}{n(r - 1)c_1}K_1 - \frac{m_t - c_1}{c_1} \\ & \le \exp[\beta(x_2 - x_1)] \le \frac{V_2 - n(m_t - c_1)}{n(r - 1)c_1}K_2 \\ & - \frac{m_t - c_1}{c_1}, n_1 = 0\Bigg\} \\ & + P\Bigg\{\frac{n(r - 1)m_t - V_1K_2}{V_2K_2} \\ & \le \exp[\beta(x_1 - x_2)] \\ & \le \frac{n(r - 1)m_t - V_1K_1}{V_2K_1}, r > n_1 > 0\Bigg\} \\ & + P\left\{K_1 \le \frac{n(r - 1)m_t}{V_0} \le K_2, r = n_1 > 0\right\} \\ = {} & 1 - \gamma \end{aligned}$$

Thus, if $r > n_1 > 0$, a $100(1-\gamma)\%$ confidence interval for β is $[l_1, l_2] \cap (-\infty, 0)$ when $[n(r-1)m_t - V_1K_2]/(V_2K_2) > 0$; and $[l_1, \infty) \cap (-\infty, 0)$ otherwise, where

$$l_i = \frac{1}{x_2 - x_1} \ln\left[\frac{V_2K_i}{n(r-1)m_t - V_1K_i}\right] \quad i = 1, 2$$

If $n_1 = 0$, a $100(1-\gamma)\%$ confidence interval for β is $[l_3, l_4] \cap (-\infty, 0)$ when $[V_2 - n(m_t - c_1)]\quad K_1/[n(r-1)c_1] - (m_t - c_1)/c_1 > 0$; and $(-\infty, l_4] \cap (-\infty, 0)$ otherwise, where

$$l_i = \frac{1}{x_2 - x_1} \ln\left[\frac{V_2 - n(m_t - c_1)}{n(r-1)c_1} K_{i-2} - \frac{m_t - c_1}{c_1}\right] \quad i = 3, 4$$

If $r = n_1 > 0$, a $100(1-\gamma)\%$ confidence interval for β is $(-\infty, 0)$ when $K_1 \le n(r-1)m_t/V_0 \le K_2$; and the empty set otherwise.

For $0 < \gamma < 1$, let $\chi^2_\gamma(k)$ be the upper $100\gamma\%$ point of the χ^2 distribution with degrees of freedom k. Let $0 < \gamma^* < 1$. Denote $B_1 = \chi^2_{1-\gamma/2}(2)$, $B_2 = \chi^2_{\gamma/2}(2)$, $B_3 = \chi^2_{1-\gamma^*/2}(2r-2)$ and $B_4 = \chi^2_{\gamma^*/2}(2r-2)$. From the independence of random variables $2nm_s$ and D, it follows that

$$\begin{aligned}
&(1-\gamma)(1-\gamma^*) \\
&\quad = P(B_1 \le 2nm_s \le B_2, B_3 \le D \le B_4) \\
&\quad = P\left(B_1 \le \frac{2nm_t}{\theta_1} \le B_2, B_3 \le \frac{2V_0}{\theta_1} \le B_4,\right. \\
&\qquad \left. r = n_1 > 0\right) \\
&\qquad + P\left(\frac{2nm_t}{B_2} \le \theta_1 \le \frac{2nm_t}{B_1}, \frac{B_3}{2} - \frac{V_1}{\theta_1}\right. \\
&\qquad \left. \le \frac{V_2}{\theta_2} \le \frac{B_4}{2} - \frac{V_1}{\theta_1}, r > n_1 > 0\right) \\
&\qquad + P\left(\frac{B_1}{2n} - \frac{m_t - c_1}{\theta_2} \le \frac{c_1}{\theta_1} \le \frac{B_2}{2n}\right. \\
&\qquad - \frac{m_t - c_1}{\theta_2}, \frac{V_2 - n(m_t - c_1)}{B_4} \le \frac{\theta_2}{2} \\
&\qquad \left. \le \frac{V_2 - n(m_t - c_1)}{B_3}, n_1 = 0\right) \\
&\quad \le P\left(\exp(\alpha) \ge \max\left(\frac{2nm_t}{B_2}, \frac{2V_0}{B_4}\right),\right. \\
&\qquad \left. r = n_1 > 0\right) \\
&\qquad + P\left(\frac{2nm_t}{B_2} \le \theta_1 \le \frac{2nm_t}{B_1}, \frac{B_3}{2} - \frac{V_1B_2}{2nm_t}\right. \\
&\qquad \left. \le \frac{V_2}{\theta_2} \le \frac{B_4}{2} - \frac{V_1B_1}{2nm_t}, r > n_1 > 0\right) \\
&\qquad + P\left(\frac{B_1}{n} - \frac{(m_t - c_1)B_4}{V_2 - n(m_t - c_1)} \le \frac{2c_1}{\theta_1}\right. \\
&\qquad \le \frac{B_2}{n} - \frac{(m_t - c_1)B_3}{V_2 - n(m_t - c_1)}, \\
&\qquad \frac{V_2 - n(m_t - c_1)}{B_4} \le \frac{\theta_2}{2} \le \frac{V_2 - n(m_t - c_1)}{B_3}, \\
&\qquad \left. n_1 = 0\right)
\end{aligned}$$

Thus, if $r > n_1 > 0$, a confidence interval for α of confidence level at least $100(1-\gamma)(1-\gamma^*)\%$ is $[l_5, l_6]$ when $B_3 - V_1B_2/(nm_t) > 0$; and $(-\infty, l_6]$ otherwise, where

$$l_i = \frac{1}{x_2 - x_1}\left[x_2 \ln\left(\frac{2nm_t}{B_{7-i}}\right) - x_1 \ln\left(\frac{2nm_tV_2}{nm_tB_{i-2} - V_1B_{7-i}}\right)\right] \quad i = 5, 6$$

If $n_1 = 0$, a confidence interval for α of confidence level at least $100(1-\gamma)(1-\gamma^*)\%$ is $[l_7, l_8]$ when $B_1/n - (m_t - c_1)B_4/[V_2 - n(m_t - c_1)] > 0$; and $[l_7, \infty)$ otherwise, where

$$\begin{aligned}
l_i = \frac{1}{x_2 - x_1}\Big(&x_2 \ln\{2nc_1[V_2 - n(m_t - c_1)] \\
&\times [(V_2 - n(m_t - c_1))B_{9-i} \\
&- n(m_t - c_1)B_{i-4}]^{-1}\} \\
&- x_1 \ln\left\{\frac{2[V_2 - n(m_t - c_1)]}{B_{i-4}}\right\}\Big) \quad i = 7, 8
\end{aligned}$$

If $r = n_1 > 0$, a confidence interval for α of confidence level at least $100(1-\gamma)(1-\gamma^*)\%$ is $[\max(\ln(2nm_t/B_2), \ln(2V_0/B_4)), \infty)$.

Note that $\theta_i = \exp(\alpha + \beta x_i) = \exp[(\alpha + \beta x_0) + \beta(x_i - x_0)]$, $i = 1, 2$. Using a very similar argument to the way that the confidence interval for α is derived, we can also set up confidence intervals for the mean of lifetime $\theta(x_0) = \exp(\alpha + \beta x_0)$ at design stress. If $r > n_1 > 0$, a confidence interval for $\theta(x_0)$ of confidence level at least $100(1-\gamma)(1-\gamma^*)\%$ is $[l_9, l_{10}]$ when $B_3 - V_1 B_2/(nm_t) > 0$; and $(-\infty, l_{10}]$ otherwise, where

$$l_i = \exp\left[\frac{x_2 - x_0}{x_2 - x_1} \ln\left(\frac{2nm_t}{B_{11-i}}\right) - \frac{x_1 - x_0}{x_2 - x_1} \ln\left(\frac{2nm_t V_2}{nm_t B_{i-6} - V_1 B_{11-i}}\right)\right]$$
$$i = 9, 10$$

If $n_1 = 0$, a confidence interval for $\theta(x_0)$ of confidence level at least $100(1-\gamma)(1-\gamma^*)\%$ is $[l_{11}, l_{12}]$ when $B_1/n - (m_t - c_1)B_4/[V_2 - n(m_t - c_1)] > 0$; and $[l_{11}, \infty)$ otherwise, where

$$l_i = \exp\left(\frac{x_2 - x_0}{x_2 - x_1} \times \ln\left\{\frac{2nc_1[V_2 - n(m_t - c_1)]}{[V_2 - n(m_t - c_1)]B_{13-i} - n(m_t - c_1)B_{i-8}}\right\} - \frac{x_1 - x_0}{x_2 - x_1} \ln\left\{\frac{2[V_2 - n(m_t - c_1)]}{B_{i-8}}\right\}\right)$$
$$i = 11, 12$$

If $r = n_1 > 0$, a confidence interval for $\theta(x_0)$ of confidence level at least $100(1-\gamma)(1-\gamma^*)\%$ is $[\exp(\max(\ln(2nm_t/B_2), \ln(2V_0/B_4))), \infty)$.

A confidence interval for the reliability function $R_0(t) = \exp[-t/\theta(x_0)]$ at any given survival time t can be obtained based on the confidence interval for $\theta(x_0)$. More specifically, if $[\underline{l}, \bar{l}]$ is a confidence interval for $\theta(x_0)$ of confidence level at least $100(1-\gamma)(1-\gamma^*)\%$, then $[\exp(-t/\underline{l}), \exp(-t/\bar{l})]$ is a confidence interval of $R_0(t)$ of the same confidence level.

25.2.3 Estimation with Other Distributions

For most other life distributions in step-stress life testing, there are usually no closed form estimations. Standard asymptotic theory can be used to set up asymptotic confidence intervals for model parameters. More specifically, the likelihood function from the first r observations of a simple step-stress life testing is

$$L(\alpha, \beta) = \frac{n!}{(n-r)!} \Pi_{ij} g(T_{ij})[1 - G(M_t)]^{n-r}$$

A standard routine, such as the Newton–Raphson method, can be applied to find the maximum likelihood estimates for α and β by solving the estimation equations $\partial \log L/\partial\alpha = 0$ and $\partial \log L/\partial\beta = 0$. We assume that

$$\lim_{n\to\infty} \frac{r}{n} > F\left(\frac{c_1}{\theta_1}\right)$$

Under certain regularity conditions on $f(t)$ (see Halperin [9]), the maximum likelihood estimate $(\hat{\alpha}, \hat{\beta})^{\mathrm{T}}$ is asymptotically normally distributed with mean $(\alpha, \beta)^{\mathrm{T}}$ and covariance matrix $\Sigma = (\sigma_{rs})$, $r, s = 1, 2$, where $\Sigma = I^{-1}$, $I = (i_{rs})$, and

$$i_{11} = -\left.\frac{\partial^2 \ln L}{\partial^2 \alpha}\right|_{(\alpha,\beta)=(\hat{\alpha},\hat{\beta})}$$

$$i_{22} = -\left.\frac{\partial^2 \ln L}{\partial^2 \beta}\right|_{(\alpha,\beta)=(\hat{\alpha},\hat{\beta})}$$

$$i_{12} = i_{21} = -\left.\frac{\partial^2 \ln L}{\partial\alpha\partial\beta}\right|_{(\alpha,\beta)=(\hat{\alpha},\hat{\beta})}$$

Therefore, a $100(1-\gamma)\%$ $(0 < \gamma < 1)$ asymptotic confidence interval for α is

$$\hat{\alpha} \pm z_{\gamma/2}\sigma_{11}$$

where $z_{\gamma/2}$ is the upper $\gamma/2$ point of the standard normal distribution. A $100(1-\gamma)\%$ asymptotic confidence interval for β is

$$\hat{\beta} \pm z_{\gamma/2}\sigma_{22}$$

A $100(1-\gamma)\%$ asymptotic confidence interval for $\theta_0 = \theta(x_0) = \exp(\alpha + \beta x_0)$ is

$$\hat{\theta}_0 \pm z_{\gamma/2}\hat{\theta}_0^2(1, x_0)\Sigma(1, x_0)^{\mathrm{T}}$$

where $\hat{\theta}_0 = \exp(\hat{\alpha} + \hat{\beta}x_0)$. A $100(1-\gamma)\%$ asymptotic confidence interval for $R_0(t) = 1 - F(t/\theta(x_0))$ at time t can be obtained by applying the transformation $R_0(t) = 1 - F(t/\theta(x_0))$ to the $100(1-\gamma)\%$ asymptotic confidence interval of $\theta(x_0) = \exp(\alpha + \beta x_0)$.

25.2.4 Optimum Test Plan

At the design stage of a step-stress life testing, the choices of stress levels and stress-change times are very important decisions. The optimum test plan chooses these parameters by minimizing the asymptotic variance of the maximum likelihood estimate for $\ln \theta_0 = \alpha + \beta x_0$. In a simple step-stress life testing, assuming that the lifetime under constant stress is exponential, Miller and Nelson [4] showed that the optimum test plan chooses x_1 as low as possible and x_2 as high as possible as long as these choices do not cause failure modes different from those at the design stress. When all units are run to failure, they also proved that the optimum choice for the stress-change time is

$$c_1 = \theta_1 \ln\left(\frac{2\xi + 1}{\xi}\right)$$

where $\xi = (x_1 - x_0)/(x_2 - x_1)$ is called the amount of stress extrapolation. When a constant censoring time c ($c > c_1$) is used in the simple step-stress life testing, Bai *et al.* [10] showed that the optimum stress-change time c_1 is the unique solution to the equation

$$\left[\frac{F_1(c_1)}{F_2(c - c_1)(1 - F_1(c_1))}\right]^2 \times \left\{F_2(c - c_1) + \frac{\theta_1}{\theta_2}[1 - F_2(c - c_1)]\right\} = \left(\frac{1 + \xi}{\xi}\right)^2$$

where $F_i(t) = 1 - \exp(-t/\theta_i)$, $i = 1, 2$. Xiong [11] used a different optimization criterion and also obtained results for the optimum choice of stress-change time c_1. All these optimum test plans are called locally optimum by Chernoff [12], since they depend on unknown parameters α and β which can only be guessed or estimated from a pilot study. Bai *et al.* [10] and Xiong [11] also discussed the variance increase in the maximum likelihood estimation of $\ln \theta_0$ when incorrect α and β were used.

25.3 Step-stress Life Testing with Random Stress-change Times

In this section we consider the situation that the stress-change times in a step-stress life testing are random. The random stress-change times happen frequently in real-life applications. For example, in a simple step-stress testing, instead of increasing stress at a prespecified constant time, the stress can be increased right after a certain number of test units fail. The stress-change time in this case is then an order statistic from the lifetime distribution under the first stress level. In general, when the stress-change times are random, there are two sources of variations in the lifetime of a step-stress testing. One is the variation due to the different stress levels used in the test; the other is due to the randomness in the stress-change times. We study the marginal lifetime distribution under a step-stress life testing with random stress-change times in Section 25.3.1. The estimation and optimum test plan are presented in Sections 25.3.2 and 25.3.3.

25.3.1 Marginal Distribution of Lifetime

We first consider the case of a simple step-stress life testing for which the stress-change time is a random variable. Let us assume that C_1 is the random stress-change time with cumulative distribution function $h_{C_1}(c_1)$ and that T is the lifetime under such a simple step-stress testing. We also assume that the conditional cumulative distribution function $G_{T|C_1}$ of T, given the stress-change time $C_1 = c_1$, is given by the classic cumulative exposure model (Equation 25.2):

$$G_{T|C_1}(t) = \begin{cases} F_1(t) & \text{for } 0 \le t < c_1 \\ F_2(t - c_1 + s_1) & \text{for } c_1 \le t < \infty \end{cases}$$

where $s_1 = c_1\theta_2/\theta_1$. The conditional probability density function $g(t \mid c_1)$ of T, given $C_1 = c_1$, is then

$$g(t \mid c_1) = \begin{cases} \dfrac{1}{\theta_1} f\left(\dfrac{t}{\theta_1}\right) & \text{for } 0 \le t < c_1 \\ \dfrac{1}{\theta_2} f\left(\dfrac{t - c_1}{\theta_2} + \dfrac{t}{\theta_1}\right) & \text{for } c_1 \le t < \infty \end{cases}$$

The joint probability density function $d(t, c_1)$ of T and C_1 is now

$$d(t, c_1) = g(t \mid c_1) h_{C_1}(c_1)$$

Thus, the marginal probability density function $q(t)$ of T is

$$\begin{aligned} q(t) &= \int_0^{\infty} d(t, c_1)\, \mathrm{d}c_1 \\ &= \frac{1}{\theta_2} \int_0^t f\left(\frac{t - c_1}{\theta_2} + \frac{t}{\theta_1}\right) h_{C_1}(c_1)\, \mathrm{d}c_1 \\ &\quad + \frac{1}{\theta_1} f\left(\frac{t}{\theta_1}\right) P(C_1 > t) \end{aligned} \tag{25.12}$$

The integration is replaced by a sum if the stress-change time C_1 is discrete.

We now assume that n test units are initially placed on low stress level x_1 and run until r test units fail, $1 \le r < n$. The stress is then changed to x_2 and is continued until all units fail. The stress-change time C_1 is the rth order statistic of a sample of size n from the lifetime distribution under stress x_1. Thus, the probability density function of C_1 (see Lawless [8], p. 518) is

$$\begin{aligned} h_{C_1}(c_1) = \binom{n-1}{r-1} \frac{n}{\theta_1} f\left(\frac{c_1}{\theta_1}\right) F^{r-1}\left(\frac{c_1}{\theta_1}\right) \\ \times R^{n-r}\left(\frac{c_1}{\theta_1}\right) \end{aligned}$$

where $R(t) = 1 - F(t), t > 0$. The marginal probability density function of the lifetime under such a test plan is

$$\begin{aligned} q(t) = n \binom{n-1}{r-1} \Bigg[\frac{1}{\theta_1 \theta_2} \\ \times \int_0^t f\left(\frac{t - c_1}{\theta_2} + \frac{c_1}{\theta_1}\right) f\left(\frac{c_1}{\theta_1}\right) F^{r-1}\left(\frac{c_1}{\theta_1}\right) \\ \times R^{n-r}\left(\frac{c_1}{\theta_1}\right) \mathrm{d}c_1 \\ + \frac{1}{\theta_1^2} f\left(\frac{t}{\theta_1}\right) \int_t^{\infty} f\left(\frac{c_1}{\theta_1}\right) F^{r-1}\left(\frac{c_1}{\theta_1}\right) \\ \times R^{n-r}\left(\frac{c_1}{\theta_1}\right) \mathrm{d}c_1 \Bigg] \end{aligned} \tag{25.13}$$

When the lifetime distribution at a constant stress is assumed exponential, *i.e.* $f(t) = \exp(-t)$ for $t > 0$, $q(t)$ can be further simplified as

$$\begin{aligned} q(t) = n \binom{n-1}{r-1} \sum_{i=0}^{r-1} \frac{(-1)^i \binom{r-1}{i}}{(\xi_r^{n,i} + 2)\theta_2 - \theta_1} \\ \times \left[\exp\left(-\frac{t}{\theta_2}\right) - \exp\left(-\frac{t(\xi_r^{n,i} + 2)}{\theta_1}\right) \right] \\ + n \binom{n-1}{r-1} \sum_{i=0}^{r-1} \frac{(-1)^i \binom{r-1}{i}}{\theta_1 (\xi_r^{n,i} + 1)} \\ \times \exp\left(-\frac{t(\xi_r^{n,i} + 2)}{\theta_1}\right) \end{aligned} \tag{25.14}$$

where $\xi_r^{n,i} = n + i - r$.

The first two moments of T based on the marginal distribution (Equation 25.14) are

$$\begin{aligned} \mathrm{ET} = n \binom{n-1}{r-1} \sum_{i=0}^{r-1} \Bigg\{ \frac{(-1)^i \binom{r-1}{i} \theta_2^2}{(\xi_r^{n,i} + 2)\theta_2 - \theta_1} \\ - \frac{(-1)^i \binom{r-1}{i} \theta_1^2}{(\xi_r^{n,i} + 2)^2 [(\xi_r^{n,i} + 2)\theta_2 - \theta_1]} \\ + \frac{(-1)^i \binom{r-1}{i} \theta_1}{(\xi_r^{n,i} + 2)^2 (\xi_r^{n,i} + 1)} \Bigg\} \end{aligned} \tag{25.15}$$

$$\begin{aligned} \mathrm{ET}^2 = 2n \binom{n-1}{r-1} \sum_{i=0}^{r-1} \Bigg\{ \frac{(-1)^i \binom{r-1}{i} \theta_2^3}{(\xi_r^{n,i} + 2)\theta_2 - \theta_1} \\ - \frac{(-1)^i \binom{r-1}{i} \theta_1^3}{(\xi_r^{n,i} + 2)^3 [(\xi_r^{n,i} + 2)\theta_2 - \theta_1]} \\ + \frac{(-1)^i \binom{r-1}{i} \theta_1^2}{(\xi_r^{n,i} + 2)^3 (\xi_r^{n,i} + 1)} \Bigg\} \end{aligned} \tag{25.16}$$

Next we consider the general step-stress life testing for which k stress levels, $x_1 < x_2 < \cdots < x_k$, are used. Test units are initially placed under stress x_1. The stress is increased gradually from x_1 to x_k at random times $C_1 < C_2 < \cdots < C_{k-1}$ (continuous or discrete). Let $C_0 = c_0 = 0$ and $C_k = c_k = \infty$. Let T be the lifetime of a test unit under such a step-stress life testing plan. We assume that the conditional cumulative distribution function $G_{T|C_1C_2\cdots C_{k-1}}(t)$ of T, given the stress-change times $C_1 = c_1, C_2 = c_2, \ldots, C_{k-1} = c_{k-1}$, follows the cumulative exposure model (Equation 25.3), *i.e.*

$$G_{T|C_1C_2\cdots C_{k-1}}(t) = \begin{cases} F\left(\dfrac{t}{\theta_1}\right) & \text{for } 0 \le t < c_1 \\ F\left(\dfrac{t - c_{i-1}}{\theta_i} + \displaystyle\sum_{j=1}^{i-1} \dfrac{c_j - c_{j-1}}{\theta_j}\right) & \text{for } c_{i-1} \le t < c_i \end{cases}$$

$$i = 2, 3, \ldots, k$$

The conditional probability density function $g(t \mid c_1, c_2, \ldots, c_{k-1})$ of T, given the stress-change times $C_1 = c_1, C_2 = c_2, \ldots, C_{k-1} = c_{k-1}$, is then

$$g(t \mid c_1, c_2, \ldots, c_{k-1}) = \begin{cases} \dfrac{1}{\theta_1} f\left(\dfrac{t}{\theta_1}\right) & \text{for } 0 \le t < c_1 \\ \dfrac{1}{\theta_i} f\left(\dfrac{t - c_{i-1}}{\theta_i} + \displaystyle\sum_{j=1}^{i-1} \dfrac{c_j - c_{j-1}}{\theta_j}\right) & \text{for } c_{i-1} \le t < c_i \end{cases}$$

$$i = 2, 3, \ldots, k$$

Let $h_{C_i|C_1C_2\cdots C_{i-1}}(c_i)$ be the conditional distribution of C_i, given $C_1, C_2, \ldots, C_{i-1}$, for $i = 2, 3, \ldots, k-1$. Let $h_{C_1|C_0}(c_1) = h_{C_1}(c_1)$, the cumulative distribution of C_1. The joint probability density function $d(t, c_1, c_2, \ldots, c_{k-1})$ of T and $C_1, C_2, \ldots, C_{k-1}$ is now

$$d(t, c_1, c_2, \ldots, c_{k-1}) = g(t \mid c_1, c_2, \ldots, c_{k-1}) \prod_{i=1}^{k-1} h_{C_i|C_1C_2\cdots C_{i-1}}(c_i)$$

$$= \begin{cases} \dfrac{1}{\theta_1} f\left(\dfrac{t}{\theta_1}\right) \displaystyle\prod_{i=1}^{k-1} h_{C_i|C_1C_2\cdots C_{i-1}}(c_i) \\ \quad \text{for } 0 \le t < c_1 \\ \dfrac{1}{\theta_i} f\left(\dfrac{t - c_{i-1}}{\theta_i} + \displaystyle\sum_{j=1}^{i-1} \dfrac{c_j - c_{j-1}}{\theta_j}\right) \\ \quad \times \displaystyle\prod_{i=1}^{k-1} h_{C_i|C_1C_2\cdots C_{i-1}}(c_i) \\ \quad \text{for } c_{i-1} \le t < c_i,\ i = 2, 3, \ldots, k \end{cases}$$

Thus, the marginal probability density function $q(t)$ of T is

$$\begin{aligned} q(t) &= \int_0^\infty \int_0^\infty \cdots \\ &\quad \cdots \int_0^\infty d(t, c_1, c_2, \ldots, c_{k-1})\, \mathrm{d}c_1\, \mathrm{d}c_2 \ldots \\ &\quad \ldots \mathrm{d}c_{k-1} \\ &= \sum_{i=2}^{k-1} \int_0^t \cdots \left[\int_{c_{i-3}}^t \left(\int_{c_{i-2}}^t \left\{ \int_t^\infty \cdots \right.\right.\right. \\ &\quad \cdots \left[\int_{c_{k-2}}^\infty \frac{1}{\theta_i} f\left(\frac{t-c_{i-1}}{\theta_i} + \sum_{j=1}^{i-1} \frac{c_j - c_{j-1}}{\theta_j}\right) \right. \\ &\quad \left. \times \prod_{i=1}^{k-1} h_{C_i|C_1C_2\cdots C_{i-1}}(c_i)\, \mathrm{d}c_{k-1} \right] \cdots \\ &\quad \left.\left.\left. \ldots \mathrm{d}c_i \right\} \mathrm{d}c_{i-1} \right) \mathrm{d}c_{i-2} \right] \cdots \\ &\quad \ldots \mathrm{d}c_1 + \int_t^\infty \left\{ \int_{c_1}^\infty \cdots \left[\int_{c_{k-2}}^\infty \frac{1}{\theta_1} f\left(\frac{t}{\theta_1}\right) \right.\right. \\ &\quad \left. \times \prod_{i=1}^{k-1} h_{C_i|C_1C_2\cdots C_{i-1}}(c_i)\, \mathrm{d}c_{k-1} \right] \cdots \\ &\quad \left. \ldots \mathrm{d}c_2 \right\} \mathrm{d}c_1 + \int_0^t \left\{ \int_{c_1}^t \cdots \right. \\ &\quad \cdots \left[\int_{c_{k-2}}^t \frac{1}{\theta_k} f\left(\frac{t-c_{k-1}}{\theta_k} + \sum_{j=1}^{k-1} \frac{c_j - c_{j-1}}{\theta_j}\right) \right. \\ &\quad \left.\left. \times \prod_{i=1}^{k-1} h_{C_i|C_1C_2\cdots C_{i-1}}(c_i)\, \mathrm{d}c_{k-1} \right] \ldots \mathrm{d}c_2 \right\} \mathrm{d}c_1 \end{aligned} \tag{25.17}$$

The integrations need to be replaced by sums if the stress-change times $C_1, C_2, \ldots, C_{k-1}$ are discrete. Even though the marginal probability density function $q(t)$ is always theoretically available, it becomes very complex and mathematically intractable even with simple lifetime distributions when k is large. We consider several particular cases in the following.

We first note that when the stress-change times are prespecified constants $c_1 < c_2 < \cdots < c_{k-1}$, *i.e.* $P(C_1 = c_1) = P(C_2 = c_2) = \cdots = P(C_{k-1} = c_{k-1}) = 1$, $q(t)$ becomes

$$q(t) = \begin{cases} \dfrac{1}{\theta_1} f\left(\dfrac{t}{\theta_1}\right) & \text{for } 0 \le t < c_1 \\ \dfrac{1}{\theta_i} f\left(\dfrac{t - c_{i-1}}{\theta_i} + \displaystyle\sum_{j=1}^{i-1} \dfrac{c_j - c_{j-1}}{\theta_j}\right) & \\ \qquad\qquad\qquad \text{for } c_{i-1} \le t < c_i & \end{cases}$$

$$i = 2, 3, \ldots, k$$

which is the original cumulative exposure model studied by Nelson [1,2].

Next we study the case $k = 3$. Two stress-change times C_1 and C_2 are used. Assume that n test units are initially placed on low stress level x_1 and run until r_1 test units fail. The stress is then changed to x_2 and run until another r_2 units fail. The stress is finally changed to x_3 and continued until all units fail. The stress-change time C_1 is the r_1th-order statistic of a sample of size n from the lifetime distribution under stress x_1. The probability density function of C_1 is

$$h_{C_1}(c_1) = \binom{n-1}{r_1-1} \frac{n}{\theta_1} f\left(\frac{c_1}{\theta_1}\right) F^{r_1-1}\left(\frac{c_1}{\theta_1}\right) \times R^{n-r_1}\left(\frac{c_1}{\theta_1}\right)$$

The stress-change time C_2, given $C_1 = c_1$, is the r_2th-order statistic of a sample of size $n - r_1$ from the conditional probability distribution

$$\frac{1}{\theta_2} f\left(\frac{t - c_1}{\theta_2} + \frac{c_1}{\theta_1}\right) R^{-1}\left(\frac{c_1}{\theta_1}\right)$$

for $t > c_1$. Thus, the conditional probability density function of C_2, given $C_1 = c_1$, is

$$h_{C_2|C_1}(c_2) = \frac{(n-r_1)\binom{n-r_1-1}{r_2-1}}{\theta_2 R^{n-r_1}\left(\frac{c_1}{\theta_1}\right)} f\left(\frac{c_2 - c_1}{\theta_2} + \frac{c_1}{\theta_1}\right) \times \left[F\left(\frac{c_2 - c_1}{\theta_2} + \frac{c_1}{\theta_1}\right) - F\left(\frac{c_1}{\theta_1}\right)\right]^{r_2-1} \times R^{n-r_1-r_2}\left(\frac{c_2 - c_1}{\theta_2} + \frac{c_1}{\theta_1}\right)$$

When the lifetime distribution at a constant stress is exponential, straightforward integrations yield the marginal probability density function $q(t)$ of lifetime T under this step-stress life testing as

$$\begin{aligned} q(t) &= \int_0^\infty \int_0^\infty g(t \mid c_1, c_2) h_{C_2|C_1}(c_2) \times h_{C_1}(c_1)\, \mathrm{d}c_2\, \mathrm{d}c_1 \\ &= n(n-r_1)\binom{n-1}{r_1-1}\binom{n-r_1-1}{r_2-1} \\ &\quad \times \sum_{i=0}^{r_1-1}\sum_{j=0}^{r_2-1}\left(\frac{(-1)^{i+j}\binom{r_1-1}{i}\binom{r_2-1}{j}}{(\eta_{r_1,r_2}^{n,j}+1)(\xi_{r_1}^{n,i}+1)\theta_1} \times \exp\left[-\frac{(\xi_{r_1}^{n,i}+2)t}{\theta_1}\right]\right. \\ &\quad + \{(-1)^{i+j}\tbinom{r_1-1}{i}\tbinom{r_2-1}{j}\theta_2\}\{[(\eta_{r_1,r_2}^{n,j}+2)\theta_1 \\ &\quad - (\xi_{r_1}^{n,i}+2)\theta_2][\theta_2 - (\eta_{r_1,r_2}^{n,j}+2)\theta_3]\}^{-1} \\ &\quad \times \left\{\exp\left[-\frac{(\xi_{r_1}^{n,i}+2)t}{\theta_1}\right] - \exp\left[-\frac{(\eta_{r_1,r_2}^{n,j}+2)t}{\theta_2}\right]\right\} \\ &\quad - \frac{(-1)^{i+j}\binom{r_1-1}{i}\binom{r_2-1}{j}\theta_3}{[\theta_2 - (\eta_{r_1,r_2}^{n,j}+2)\theta_3][\theta_1 - (\xi_{r_1}^{n,i}+2)\theta_3]} \\ &\quad \times \left\{\exp\left[-\frac{(\xi_{r_1}^{n,i}+2)t}{\theta_1}\right] - \exp\left[-\frac{t}{\theta_3}\right]\right\} \\ &\quad + \frac{(-1)^{i+j}\binom{r_1-1}{i}\binom{r_2-1}{j}}{[(\eta_{r_1,r_2}^{n,j}+2)\theta_1 - (\xi_{r_1}^{n,i}+2)\theta_2](\eta_{r_1,r_2}^{n,j}+1)} \\ &\quad \times \left.\left\{\exp\left[-\frac{(\xi_{r_1}^{n,i}+2)t}{\theta_1}\right] - \exp\left[-\frac{(\eta_{r_1,r_2}^{n,j}+2)t}{\theta_2}\right]\right\}\right) \end{aligned} \tag{25.18}$$

where $\eta_{r_1,r_2}^{n,j} = n + j - r_1 - r_2$.

25.3.2 Estimation

When the functional form of $f(t)$ is given and independent failure times, $T_1, T_2, \ldots, T_n$, are observed, the maximum likelihood estimates of α and β can be obtained by maximizing $\prod_{i=1}^{n} q(T_i)$ over α and β, where $q(t)$ is given in Equation 25.17. The maximization can be carried out by using numerical methods such as the Newton–Raphson method and net search methods (Bazaraa and Shetty [13], chapter 8).

In the case of a simple step-stress accelerated life testing where the stress is changed when the first r lifetimes are observed. The only stress-change time C_1 is the rth-order statistic from a lifetime sample of size n under stress x_1. When the lifetime at a constant stress is exponential, the method of moment estimates for α and β can be found by numerically solving the following system for α and β:

$$\begin{cases} \mathrm{ET} = \dfrac{\sum_{i=1}^{n} T_i}{n} \\ \mathrm{ET}^2 = \dfrac{\sum_{i=1}^{n} T_i^2}{n} \end{cases}$$

where ET and ET^2 are given in Equations 25.15 and 25.16 respectively.

If the lifetime distribution at any constant stress is exponential, all the confidence interval estimates for α, β, θ_0, and $R_0(t)$ given in Section 25.2.2 from a simple step-stress life testing are now conditional confidence intervals for the corresponding parameters, given $C_1 = c_1$. However, these conditional confidence intervals are also unconditional confidence intervals with the same confidence levels. For example, if $[a_1, a_2]$ is a conditional $100(1-\gamma)\%$ $(0 < \gamma < 1)$ confidence interval for β, *i.e.*

$$P(a_1 \le \beta \le a_2 \mid C_1) = 1 - \gamma$$

then

$$\begin{aligned} P(a_1 \le \beta \le a_2) &= E_{C_1}[P(a_1 \le \beta \le a_2 \mid C_1)] \\ &= 1 - \gamma \end{aligned}$$

where E_{C_1} is the expectation taken with respect to the distribution of C_1. Therefore, $[a_1, a_2]$ is a $100(1-\gamma)\%$ unconditional confidence interval for β. A similar argument will prove that if $[a_3, a_4]$ is an at least $100(1-\gamma)(1-\gamma^*)\%$ $(0 < \gamma < 1, 0 < \gamma^* < 1)$ conditional confidence interval for α, or θ_0, or $R_0(t)$ at a given t, then $[a_3, a_4]$ is also an at least $100(1-\gamma)(1-\gamma^*)\%$ unconditional confidence interval for the corresponding parameters. In the general case of $k > 2$, given $C_1, C_2, \ldots, C_{k-1}$, conditional confidence intervals for α, β, θ_0, and $R_0(t)$ can be constructed by using a similar transformation as in Equation 25.11 and Lemma 1. Again, these conditional confidence intervals are also unconditional for the corresponding parameters.

For most other lifetime distributions in a step-stress life testing with random stress-change times, the conditional asymptotic confidence intervals for α, β, θ_0, and $R_0(t)$ can be constructed using the results in Section 25.2.3. These conditional asymptotic confidence intervals are also unconditional asymptotic confidence intervals for the corresponding parameters.

25.3.3 Optimum Test Plan

We consider a simple step-stress life testing where the stress is changed when the first r lifetimes are observed. We assume that the two stress levels are already chosen and discuss the optimum test plan for choosing r when all the test units run to failure. Suppose that n test units are tested under a simple step-stress life testing which uses only two stress levels $x_1 < x_2$. The only stress-change time C_1 is the rth-order statistic from a lifetime sample of size n under stress x_1. Suppose that the lifetimes at constant stress x_1 and x_2 are exponential with means θ_1 and θ_2 respectively, where $\theta_i = \exp(\alpha + \beta x_i)$, $i = 1, 2$.

Conditional on $C_1 = c_1$, the maximum likelihood estimators $\hat{\alpha}$ and $\hat{\beta}$ are given by Equations 25.7 and 25.8. The Fisher information matrix $\mathbf{I}$, conditional on $C_1 = c_1$, is (see also

Bai *et al.* [10])

$$\mathbf{I} = n \begin{pmatrix} 1 & F_1(c_1)x_1 + [1 - F_1(c_1)]x_2 \\ F_1(c_1)x_1 + [1 - F_1(c_1)]x_2 & F_1(c_1)x_1^2 + [1 - F_1(c_1)]x_2^2 \end{pmatrix}$$

where

$$F_1(c_1) = 1 - \exp\left(-\frac{c_1}{\theta_1}\right)$$

Since the mean life at design stress x_0 is $\theta_0 = \exp(\alpha + \beta x_0)$, the conditional maximum likelihood estimator of θ_0 is $\hat{\theta}_0 = \exp(\hat{\alpha} + \hat{\beta} x_0)$. The conditional asymptotic variance Asvar $(\ln \hat{\theta}_0 \mid C_1)$ is then given by

$$n \text{ Asvar}(\ln \hat{\theta}_0 \mid C_1) = \frac{(1+\xi)^2}{F_1(c_1)} + \frac{\xi^2}{1 - F_1(c_1)}$$

where $\xi = (x_1 - x_0)/(x_2 - x_1)$. Thus, the unconditional asymptotic variance Asvar$(\ln \hat{\theta}_0)$ of $\ln \hat{\theta}_0$ can be found as

$$\text{Asvar}(\ln \hat{\theta}_0) = E_{C_1}[\text{Asvar}(\ln \hat{\theta}_0 \mid C_1)]$$

When the distribution of C_1 degenerates at c_1, Asvar$(\ln \hat{\theta}_0)$=Asvar$(\ln \hat{\theta}_0 \mid C_1 = c_1)$. Miller and Nelson [4] obtained the asymptotic optimum proportion of test units failed under stress x_1 as $(1+\xi)/(1+2\xi)$ by minimizing the asymptotic variance Asvar$(\ln \hat{\theta}_0 \mid C_1 = c_1)$. We now let C_1 be the rth smallest failure time at stress x_1. A simple integration yields

$$n \text{ Asvar}(\ln \hat{\theta}_0) = n E_{C_1}[\text{Asvar}(\ln \hat{\theta}_0 \mid C_1)] = \frac{n(1+\xi)^2}{r-1} + \frac{n\xi^2}{n-r}$$

Our optimum criterion is to find the optimum number of test units failed under stress x_1 such that Asvar$(\ln \hat{\theta}_0)$ is minimized. The minimization of Asvar$(\ln \hat{\theta}_0)$ over r yields

$$r^* = \frac{n(1+\xi) + \xi}{1 + 2\xi}$$

Since r must be an integer, the optimum r should take either $\lceil r^* \rceil - 1$, or $\lceil r^* \rceil$, or $\lceil r^* \rceil + 1$, whichever gives the smallest Asvar$(\ln \hat{\theta}_0)$. ($\lceil r^* \rceil$ is the greatest integer lower bound of r^*.) The asymptotic optimum proportion of test units failed under stress x_1 is then

$$\lim_{n \to \infty} \frac{r^*}{n} = \frac{1+\xi}{1+2\xi}$$

which is the same as given by Miller and Nelson [4] when the stress-change time is a prespecified constant.

As an example, we use the data studied by Nelson and coworkers [4, 14]. The accelerated life testing consists of 76 times (in minutes) to breakdown of an insulating fluid at constant voltage stresses (kilovolts). The two test stress levels are the log transformed voltages: $x_1 = \ln(26.0) = 3.2581$, $x_2 = \ln(38.00) = 3.6376$. The design stress is $x_0 = \ln(20.00) = 2.9957$. The amount of stress extrapolation is $\xi = 0.6914$. Miller and Nelson [4] gave the maximum likelihood estimates of model parameters from those data: $\hat{\alpha} = 64.912, \hat{\beta} = -17.704, \hat{\theta}_0 \approx 143\,793$ min. Now, a future simple step–step life testing is to be designed to estimate θ_0 using x_1, x_2, and the same number of test units used in the past study; the question is to find the optimum test plan. Treating the stress-change time as a prespecified constant, Miller and Nelson [4] found the optimum stress-change time as $c_1 = 1707$ min. Using the rth smallest lifetime under stress x_1 as the stress-change time, we found that the optimum number of test units failed under stress x_1 is $r^* = 54$.

25.4 Bibliographical Notes

The problem of statistical data analysis and optimum test plan in a step-stress life testing has been studied by many authors. Yurkowski *et al.* [15] gave a survey of the work on the early statistical methods of step-stress life testing data. Meeker and Escobar [16] surveyed more recent work on accelerated life testing. Nelson [1] first implemented the method of maximum likelihood to data from a step-stress testing. Shaked and Singpurwalla [17] proposed a model based on shock and wear processes and obtained non-parametric estimators for the life distribution at the use condition. Barbosa *et al.* [18] analyzed exponential data from a step-stress life testing using the approach of a generalized linear model. Tyoskin and Krivolapov [19] presented a non-parametric model for the analysis of step-stress test data. Schäbe [20] used an axiomatic approach to construct accelerated life testing models for non-homogeneous Poisson processes. DeGroot and Goel [21] and Dorp *et al.* [22] developed Bayes models for data from a step-stress testing and studied their inferences. Khamis and Higgins [23, 24] obtained results on optimum test plan for a three-step test and studied another variation of the cumulative exposure model. Xiong [11] gave inference results on type-II censored exponential step-stress data. Xiong and Milliken [25] investigated the statistical inference and optimum test plan when the stress-change times are random in a step-stress testing. Xiong [26] introduced a threshold parameter into the cumulative exposure models and studied the corresponding estimation problem.

References

[1] Nelson WB. Accelerated life testing—step-stress models and data analysis. IEEE Trans Reliab 1980;R-29:103–8.

[2] Nelson WB. Applied life data analysis. New York: John Wiley & Sons; 1982.

[3] Nelson WB. Accelerated life testing, statistical models, test plans, and data analysis. New York: John Wiley & Sons; 1990.

[4] Miller RW, Nelson WB. Optimum simple step-stress plans for accelerated life testing. IEEE Trans Reliab 1983; R-32:59–65.

[5] Yin XK, Sheng BZ. Some aspects of accelerated life testing by progressive stress. IEEE Trans Reliab 1987;R-36:150–5.

[6] Mann NR, Schafer RE, Singpurwalla ND. Methods for statistical analysis of reliability and life data. New York: John Wiley & Sons; 1974.

[7] Xiong C. Optimum design on step-stress life testing. In: Proceedings of 1998 Kansas State University Conference on Applied Statistics in Agriculture, 1998; p.214–25.

[8] Lawless JF. Statistical models and methods for lifetime data. New York: John Wiley & Sons; 1982.

[9] Halperin M. Maximum likelihood estimation in truncated samples. Ann Math Stat 1952;23:226–38.

[10] Bai DS, Kim MS, Lee SH. Optimum simple step-stress accelerated life tests with censoring. IEEE Trans Reliab 1989;38:528–32.

[11] Xiong C. Inferences on a simple step-stress model with type II censored exponential data. IEEE Trans Reliab 1998;47:142–6.

[12] Chernoff H. Optimal accelerated life designs for estimation. Technometrics 1962;4:381–408.

[13] Bazaraa MS, Shetty CM. Nonlinear programming: theory and algorithms. New York: John Wiley & Sons; 1979.

[14] Meeker WQ, Nelson WB. Optimum accelerated life tests for Weibull and extreme value distributions and censored data. IEEE Trans Reliab 1975;R-24:321–32.

[15] Yurkowski W, Schafer RE, Finkelstein JM. Accelerated testing technology. Rome Air Development Center Technical Report 67-420, Griffiss AFB, New York, 1967.

[16] Meeker WQ, Escobar LA. A review of recent research and current issues in accelerated testing. Int Stat Rev 1993;61:147–68.

[17] Shaked M, Singpurwalla ND. Inference for step-stress accelerated life tests. J Stat Plan Infer 1983;7:694–99.

[18] Barbosa EP, Colosimo EA, Louzada-Neto F. Accelerated life tests analyzed by a piecewise exponential distribution via generalized linear models. IEEE Trans Reliab 1996;45:619–23.

[19] Tyoskin OI, Krivolapov SY. Nonparametric model for step-stress accelerated life testing. IEEE Trans Reliab 1996;45:346–350.

[20] Schäbe H. Accelerated life testing models for nonhomogeneous Poisson processes. Stat Pap 1998; 39:291–312.

[21] DeGroot MH, Goel PK. Bayesian estimation and optimal designs in partially accelerated life testing. Nav Res Logist Q 1979;26:223–35.

[22] Dorp JR, Mazzuchi TA, Fornell GE, Pollock LR. A Bayes approach to step-stress accelerated life testing. IEEE Trans Reliab 1996;45:491–8.

[23] Khamis IH, Higgins JJ. Optimum 3-step step-stress tests. IEEE Trans Reliab 1996;45:341–5.

[24] Khamis IH, Higgins JJ. A new model for step-stress testing. IEEE Trans Reliab 1998;47:131–4.

[25] Xiong C, Milliken GA. Step-stress life-testing with random stress-change times for exponential data. IEEE Trans Reliab 1999;48:141–8.

[26] Xiong C. Step-stress model with threshold parameter. J Stat Comput Simul 1999;63:349–60.

PART V

Practices and Emerging Applications

Statistical Methods for Reliability Data Analysis

Michael J. Phillips

26.1 Introduction

The objective of this chapter is to describe statistical methods for reliability data analysis, in a manner which gives the flavor of modern approaches. The chapter commences with a description of five examples of different forms of data encountered in reliability studies. These examples are from recently published papers in reliability journals and include right censored data, accelerated failure data, and data from repairable systems. However, before commencing any discussion of statistical methods, a formal definition of *reliability* is required. The *reliability* of a system (or component) is defined as the probability that the system operates (performs a function under stated conditions) for a stated period of time. Usually the period of time is the initial interval of length t, which is denoted by $[0, t)$. In this case the reliability is a function of t, so that the *reliability function* $R(t)$ can be defined as:

$$R(t) = P(\text{System operates during } [0, t))$$

where $P(A)$ denotes the probability of an event A, say. This enables the various features of continuous distributions, which are used to model failure times, to be introduced. The relationships between the reliability function, the probability density function, and the hazard function are detailed. Then there is an account of statistical methods and how they may be used in achieving the objectives of a reliability study. This covers the use of parametric, semi-parametric, and non-parametric models.

26.2 Nature of Reliability Data

To achieve the objectives of a reliability study, it is necessary to obtain facts from which the other facts required for the objectives may be inferred. These facts, or *data*, obtained for reliability studies

Table 26.1. Times to failure (in hours) for pressure vessels

274	28.5	1.7	20.8	871	363	1311	1661	236	828
458	290	54.9	175	1787	970	0.75	1278	776	126

may come from testing in factory/laboratory conditions or from field studies. A number of examples of different kinds of reliability data obtained in reliability studies will be given, which are taken from published papers in the following well-known reliability journals: *IEEE Transactions in Reliability, Journal of Quality Technology, Reliability Engineering and Safety Systems, and Technometrics.*

Example 1. (Single sample failure data) The simplest example of failure time data is a single sample taken from a number of similar copies of a system. An example of this was given by Keating *et al.* [1] for 20 pressure vessels, constructed of fiber/epoxy composite materials wrapped around metal liners. The failure times (in hours) are given in Table 26.1 for 20 similarly constructed vessels subjected to a certain constant pressure.

This is an example of a complete sample obtained from a factory/laboratory test which was carried out on 20 pressure vessels and continued until failure was observed on all vessels. There is no special order for the failure times. These data were used by Keating *et al.* [1] to show that the failure distribution could have a decreasing hazard function and that the average residual life for a vessel having survived a wear-in period exceeds the expected life of a new vessel.

Example 2. (Right censored failure data) If one of the objectives of collecting data is to observe the time of failure of a system, it may not always be possible to achieve this due to a limit on the time or resources spent on the data collection process. This will result in incomplete data as only partial information will be collected, which will occur if a system is observed for a period and then observation ceases at time C, say. Though the time of failure T is not observed, it is known that T must exceed C. This form of incomplete data is known as *right censored* data.

An example of this kind of data was given by Kim and Proschan [2] for telecommunication systems installed for 125 customers by a telephone operating company. The failure times (in days) are given in Table 26.2 for 16 telecommunication systems installed in 1985 and the censored times (when the system was withdrawn from service without experiencing a failure or the closing date of the observation period) for the remaining 109 systems. The systems could be withdrawn from service at any time during the observation period on the customer's request. As well as failure to function, an unacceptable level of static, interference or noise in the transmission of the telecommunication system was considered as a system failure.

This situation, where a high proportion of the observations are censored, is often met in practice. These data were used by Kim and Proschan [2] to obtain a smooth non-parametric estimate of the reliability (survivor) function for the systems.

Example 3. (Right censored failure data with a covariate) Ansell and Ansell [3] analyzed the data given in Table 26.3 in a study of the performance of sodium sulfur batteries. The data consist of lifetimes (in cycles) of two batches of batteries. There were 15 batteries in the first batch and four of the lifetimes are right censored observations. These are the four largest observations and all have the same value. There were 20 batteries in the second batch and only one of the lifetimes is right censored. It is not the largest observation. The principal interest in this study is to investigate any difference in performance between batteries from the two different batches.

Example 4. (Accelerated failure data with covariates) Elsayed and Chan [4] presented failure times for time-dependent dielectric breakdown of metal–oxide–semiconductor integrated circuits for accelerated failure testing. The objective was

Table 26.2. Times to failure and censored times (in days) for telecommunication systems during five months in 1985

164+	2	45	147+	139+	135+	3	155+	150+	101+
139+	135+	164+	155+	150+	146+	139+	135+	164+	155+
150+	1	139+	7	163+	139+	149+	143+	138+	134+
163+	152+	149+	143+	40	13	163+	152+	149+	143+
138+	134+	163+	152+	149+	142+	138+	134+	163+	152+
149+	10	138+	134+	163+	94	149+	141+	138+	133+
77	151+	149+	141+	138+	133+	162+	151+	149+	141+
138+	133+	162+	151+	149+	34	138+	133+	73+	151+
115	140+	138+	133+	63	151+	138+	140+	137+	133+
161+	151+	148+	140+	137+	64+	160+	151+	147+	140+
137+	133+	160+	151+	147+	140+	137+	133+	67	90+
147+	140+	137+	133+	141+	151+	147+	140+	137+	133+
156+	151+	147+	54	137+					

Note: Figures with + are right censored observations not failures.

Table 26.3. Lifetimes (in cycles) of sodium sulfur batteries

Batch 1	164	164	218	230	263	467	538	639	669
	917	1148	1678+	1678+	1678+	1678+			
Batch 2	76	82	210	315	385	412	491	504	522
	646+	678	775	884	1131	1446	1824	1827	2248
	2385	3077							

Note: Lifetimes with + are right censored observations not failures.

Table 26.4. Times to failure and censored times (in hours) for three different temperatures (170 °C, 200 °C, 250 °C)

				Temperature (170 °C)					
0.2	5.0	27.0	51.0	103.5	192.0	192.0	429.0	429.0	954.0
2495.0	2495.0	2495.0	2495.0	2495.0	2495.0	2495.0	2495.0	2948.0	2948.0
				Temperature (200 °C)					
0.1	0.2	0.2	2.0	5.0	5.0	25.0	52.5	52.5	52.5
52.5	123.5	123.5	123.5	219.0	219.0	524.0	524.0	1145.0	1145.0
				Temperature (250 °C)					
0.1	0.2	0.2	0.2	0.5	2.0	5.0	5.0	5.0	10.0
10.0	23.0	23.0	23.0	23.0	23.0	50.0	50.0	120.0	

Note: Figure with + is right censored observation not failure.

to estimate the dielectric reliability and hazard functions at operating conditions from the data obtained from the tests. Tests were performed to induce failures at three different elevated temperatures, where temperature is used as the stress variable (or covariate). These data for silicon dioxide breakdown can be used to investigate failure time models which have parameters dependent on the value of the covariate (temperature), and are given in Table 26.4. The nominal electric field was the same for all three temperatures. Elsayed and Chan [4] used these data to estimate the parameters in a semi-parametric proportional hazards model.

Table 26.5. System failure times (in hours)

System A	452	752	967	1256	1432	1999	2383	(3000)
System B	233	302	510	911	1717	2107		(3000)
System C	783	1805						(2000)
System D	782							(2000)
System E								(2000)

Note: Figures in parentheses are current cumulative times.

Example 5. (Repairable system failure data) Examples 1 to 4 are of failure data that are records of the time to first failure of a system. Some of these systems, such as the pressure vessels described in Example 1 and the integrated circuits described in Example 4, may be irreparably damaged, though others, such as the telecommunication systems described in Example 2, will be repaired after failure and returned to service. This repair is often achieved by replacing part of the system to produce a *repairable* system.

Newton [5] presented failure times for a repairable system. Five copies of a system were put into operation at 500-hour intervals. When failures occurred the component which failed was instantaneously replaced by a new component. Failures were logged at cumulative system hours and are given in Table 26.5. The times of interest are the times between the times of system failure.

The objective for which Newton [5] used these data was to consider the question of whether the failure rate for the systems was increasing or decreasing with time as the systems were repaired.

26.3 Probability and Random Variables

Reliability has been defined as a probability, which is a function of time, used for the analysis of the problems encountered when studying the occurrence of events in time. Several aspects of time may be important: (i) age, (ii) calendar time, (iii) time since repair. However, for simplicity, problems are often specified in terms of one time scale, values of which will be denoted by t. In order to study the reliability of a component, define a random variable T denoting the time of occurrence of an event of interest (failure). Then the event "system operates during $[0, t)$" used in Section 26.1 to define the reliability function is equivalent to "$T \geq t$". Then the reliability function of the failure time distribution can be defined by:

$$R_T(t) = P(T \geq t)$$

and is a non-increasing function of t. The reliability of a component cannot improve with time! The distribution of T can be defined using the *probability density function* of the failure time distribution, and this is defined by:

$$f_T(t) = \lim_{h \to 0} \frac{P(t \leq T < t + h)}{h} = \frac{-\mathrm{d}R_T(t)}{\mathrm{d}t}$$

Another function which can be used to define the distribution of T is the *hazard function* (*age-specific failure rate* or *force of mortality*) of time to failure, which is defined by:

$$\lambda_T(t) = \lim_{h \to 0} \frac{P(t \leq T < t + h \mid T \geq t)}{h} = \frac{f_T(t)}{R_T(t)}$$

A related function is the *cumulative* (*integrated*) *hazard function* of the failure time distribution, which is closely connected with the reliability function and is defined by:

$$\Lambda_T(t) = \int_0^t \lambda_T(u)\,\mathrm{d}u = -\log(R_T(t))$$

So the reliability function can be obtained from the cumulative hazard function by:

$$R_T(t) = \exp(-\Lambda_T(t))$$

These results can be applied to various families of distributions, such as the exponential, Weibull, extreme value or lognormal. Details of these families of distributions can be found in such texts as Ansell and Phillips [6].

26.4 Principles of Statistical Methods

Statistical methods are mainly based on using parametric models, though non-parametric methods and semi-parametric models are also used.

When considering possible failure-time distributions to model the behavior of observed failure times, it is not usual to choose a fully specified function to give the reliability or probability density function of the distribution, but instead to use a class or "family" of functions. This family of functions will be a function of some variables known as *parameters*, whose values have to be specified to "index" the particular function of the family to be used to define the failure-time distribution. More than one parameter may be used and then the r parameters, say, will in general be denoted by the vector $\boldsymbol{\beta} = (\beta_1, \beta_2, \ldots, \beta_r)^{\mathrm{T}}$ and the reliability or probability density function of the parametric distribution for a continuous failure time will be denoted by $R_T(t; \boldsymbol{\beta})$ and $f_T(t; \boldsymbol{\beta})$, respectively.

One of the objectives of a reliability study may be to study the behavior of the failure time distribution of T through the variation of k other variables represented by the vector $\mathbf{z} = (z_1, z_2, z_3, \ldots, z_k)^{\mathrm{T}}$, say. These variables are usually referred to as *covariates*. In order to perform any parametric statistical analysis, it is necessary to know the joint distribution of the failure time T and the covariates $\mathbf{z}$ whose probability density function will be denoted by $f_{T,Z}(t, \mathbf{z}; \boldsymbol{\beta})$, for parameters $\boldsymbol{\beta}$.

Statistical methods based on using parametric models, with information from a random sample, are concerned with such questions as:

1. what is the best choice of the values of the parameters $\boldsymbol{\beta}$, or some function of $\boldsymbol{\beta}$, on the basis of the data observed? (*estimation*);
2. are the data observed consistent with the parameters $\boldsymbol{\beta}$ having the value $\boldsymbol{\beta}_0$, say? (*hypothesis testing*).

There have been a variety of ways of answering the first question over the years, though it is generally accepted that there are essentially three main approaches: *Classical* (due to Neyman and Pearson), *Likelihood* (due to Fisher), and *Bayesian* (developed by the followers of Bayes).

To achieve the objectives of the reliability study, it may be decided to use a parametric model with parameters $\boldsymbol{\beta}$, say. This may be to obtain the reliability of the system or the expected value of the failure time, and in order to do this it will be necessary to specify the values of the parameters of the model. This will be done by using the "best" value obtained from the reliability data, which are the observations collected (in a possibly censored or truncated form) from a random sample. *Estimation* is the process of choosing a function of these observations (known as a *statistic*) which will give a value (or values) close to the "true" (unknown) value of the parameters $\boldsymbol{\beta}$. Such a statistic, known as an *estimator*, will be a random variable denoted by $\boldsymbol{\beta}$. A *point* estimator gives one value for each of the parameters, whilst an *interval* estimator gives intervals for each of the parameters. For interval estimators a probability (or *confidence level*, usually expressed as a percentage by multiplying the probability by 100) is specified, which is the probability that the interval contains the "true" values of the parameter. One method for obtaining the interval is to use the *standard error* of an estimator, which is the standard deviation of the estimator with any parameters replaced by their estimates. Often the estimator can be assumed to have a normal distribution (at least for a large sample size, though this depends on the proportion of censored observations). Then an approximate 95% interval estimator for β using the point estimator $\hat{\beta}$ with a standard error $\mathrm{se}(\hat{\beta})$ would be given by the interval $(\hat{\beta} - 1.96\mathrm{se}(\hat{\beta}),\ \hat{\beta} + 1.96\mathrm{se}(\hat{\beta}))$.

As an alternative to choosing a parametric failure-time distribution, it is possible to make no specification of the reliability or probability density function of this distribution. This has advantages and disadvantages. While not being restricted to use specific statistical methods, which may not be valid for the model being considered, the price to be paid for this is that it may be difficult to make very powerful inferences from the data. This is because a *non-parametric* model can be effectively thought of as a parametric model with an infinity of parameters. Hence, with an infinity of parameters it is not surprising that statistical methods may be unable to produce powerful results. However, this approach to models and the resulting statistical methods which follow from it are often successful and hence have been very popular.

As a compromise to choosing a parametric or a non-parametric failure-time distribution it is possible to make only a partial specification of the reliability, probability density, or hazard function of this distribution. This has the advantage of introducing parameters which model the covariates of interest while leaving the often uninteresting remainder of the model to have a non-parametric form. Such models are referred to as *semi-parametric* models. Probably the most famous example of a semi-parametric model is the hazard model introduced by Cox [7], known as the *proportional hazards* model. The idea behind using such a model is to be able to use the standard powerful approach of parametric methods to make inferences about the covariate parameters while taking advantage of the generality of non-parametric methods to cope with any problems produced by the remainder of the model.

26.5 Censored Data

It is not always possible to collect data for lifetime distributions in as complete a form as required. Example 1 is an example of a complete sample where a failure time was observed for every system. This situation is unusual and only likely to occur in the case of laboratory testing. Many studies lead to the collection of incomplete data, which are either censored or truncated. One of the most frequent ways that this occurs is because of a limit of time (or resources) for the study, producing the case of right censored data. This case will be considered first.

Non-parametric estimation methods have mainly been developed to deal with the case of right censored data for lifetime distributions, as this is the case most widely encountered in all branches of applied statistics. Consider right censored failure-time data for n components (or systems). Such data can be represented by the pairs, (t_i, d_i) for the ith component, $i = 1, 2, \ldots, n$. For the most part the case in which the T_i are independent and identically distributed is considered, and where it is necessary to estimate some characterization of their distribution. Sometimes it is necessary to refer to the ith-ordered observed *event time* (the time of failure or censoring of the ith event) when the notation $t_{(i)}$ (rather than t_i) will be used. On the other hand, it is sometimes necessary to refer to the ith observed failure time (rather than the time of the ith event, which could be a failure or censoring). Then the notation $t_{[i]}$ (rather than t_i) will be used. Corresponding to each observed failure time is a *risk set*: the set comprising those components which were under observation at that time, *i.e.* the set of components which *could* have been the observed failure. The risk set corresponding to $t_{[i]}$ will be denoted by R_i. The number of components in the risk set R_i will be denoted by r_i. These definitions are illustrated in Example 8 (Table 26.6).

Non-parametric estimators have been proposed for the reliability (survivor) function and the cumulative hazard function of the failure-time distribution. The Kaplan–Meier (KM) estimator $\hat{R}_T(t)$ of the reliability function, $R_T(t)$, which was proposed by Kaplan and Meier [8], is a step function. In the case of ties of multiplicity m_i at the failure time $t_{[i]}$, the estimator is modified. If censoring times tie with failure times, censoring is assumed to occur *just following the failure*. The KM estimator gives a "point" estimator, or a single value for the reliability function at any time t. If it is desired to obtain a measure of the variation of this estimator over different samples, then an estimate of the variance of the KM estimator is needed. Greenwood's formula provides an estimate of the variance of $\log(\hat{R}_T(t))$, denoted by $\widehat{\mathrm{Var}}(\log(\hat{R}_T(t)))$. Though this gives a measure of the variation of the log of the KM estimator, it is possible to use it to estimate the variance of $\hat{R}_T(t)$ by $(\hat{R}_T(t))^2\ \widehat{\mathrm{Var}}(\log(\hat{R}_T(t)))$.

The Nelson–Altschuler (NA) estimator $\Lambda_T(t)$ of the cumulative hazard function, $\Lambda_T(t)$, which was proposed by Nelson [9], is also a step function. Using the basic relation between the reliability function and the cumulative hazard function given in Section 26.3, it is possible to use the NA estimator to estimate the reliability function by:

$$\hat{R}_T(t) = \exp(-\hat{\Lambda}_T(t))$$

A number of examples are now presented to illustrate the features of these non-parametric estimators of the reliability function.

Example 6. An example of a single sample from a laboratory test of 20 failure times for pressure vessels was presented in Example 1. In that example all the observations were distinct failure times, so that the sample was complete and there was no truncation or censoring. In this case the KM estimator is a step function with equal steps of 0.05 (= 1/20) at the 20 failure times. A plot of the KM estimator is given in Figure 26.1, with 95 confidence limits obtained using Greenwood's formula.

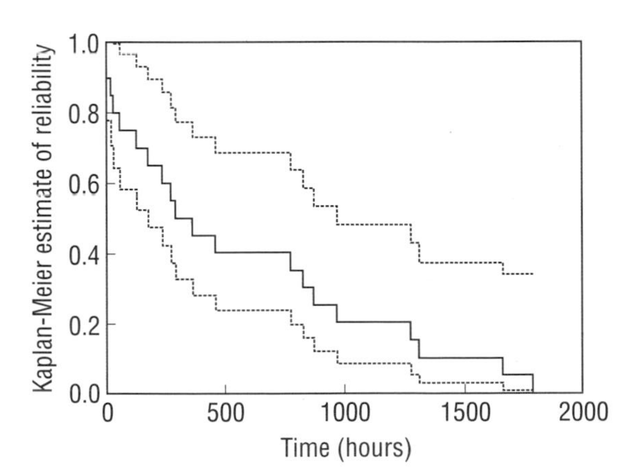

Figure 26.1. Plot of the Kaplan–Meier estimator of the reliability for the pressure vessels laboratory test data with 95% confidence limits

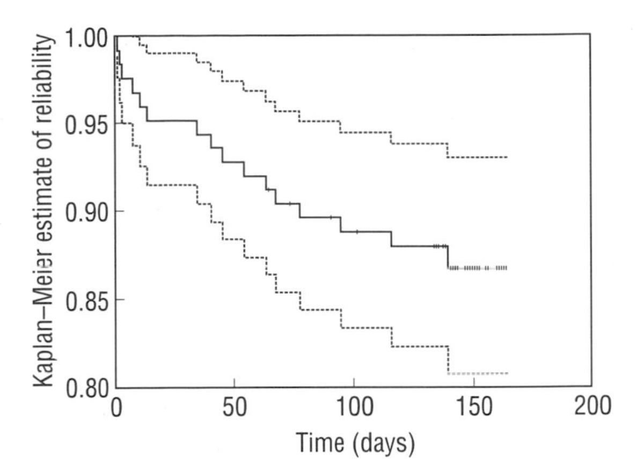

Figure 26.2. Plot of the Kaplan–Meier estimator of the reliability for the telecommunication systems data with 95% confidence limits

Example 7. An example of a sample with right censored failure times for 125 telecommunication systems was presented in Example 2. In that example the observations consisted of 16 failure times and 109 right censored observations. In this case the KM estimator is a step function with equal steps of 0.008 (= 1/125) for the first 11 failure times. After that time (63 days) right censored observations occur and the step size changes until the largest failure time (139 days) is reached. After that time the KM estimator remains constant until the largest censored time (164 days), after which the estimator is not defined. So in this example, unlike Example 6, the KM estimator is never zero. A plot of the KM estimator is given in Figure 26.2 with 95% confidence limits obtained using Greenwood's formula. (The vertical lines on the plot of the reliability represent the censored times.)

Table 26.6. Ordered event times (in hours)

Event number (i)	Censor indicator $d_{(i)}$	Event time $t_{(i)}$	Fail number $[i]$	Multiplicity $m_{[i]}$	Risk set $r_{[i]}$	$\hat{\Lambda}(t_{[i]})$	$\hat{R}(t_{[i]})$ NA	KM
1	1	69	1	1	21	0.048	0.953	0.952
2	1	176	2	1	20	0.098	0.907	0.905
3	0	196+	–	–	–	–	–	–
4	1	208	3	1	18	0.153	0.858	0.854
5	1	215	4	1	17	0.212	0.809	0.804
6	1	233	5	1	16	0.274	0.760	0.754
7	1	289	6	1	15	0.341	0.711	0.704
8	1	300	7	1	14	0.413	0.662	0.653
9	1	384	8	1	13	0.490	0.613	0.603
10	1	390	9	1	12	0.573	0.564	0.553
11	0	393+	–	–	–	–	–	–
12	1	401	10	1	10	0.673	0.510	0.498
13	1	452	11	1	9	0.784	0.457	0.442
14	1	567	12	1	8	0.909	0.403	0.387
15	0	617+	–	–	–	–	–	–
16	0	718+	–	–	–	–	–	–
17&18	1	783	13	1	5	0.309	0.270	0.232
19	1	806	14	1	3	0.642	0.194	0.155
20	0	1000+	–	–	–	–	–	–
21	1	1022	15	1	1	0.642	0.071	0.000

Note: Times with + are right censored times.

Example 8. This example is adapted from Newton [5], from the data which were described in Example 5. Five copies of a system were put into operation at 500-hour intervals. (For the purpose of this example the system will be considered to be equivalent to a single component, and will be referred to as a component.) When failures occurred the component was instantaneously replaced by a new component. Failures were logged at cumulative system hours and are given in Table 26.5, except that the failure time of system D has been changed from 782 to 783, for illustrative purposes, as this will produce two tied failure times and a multiplicity of 2. The details of the calculation of both the KM and NA estimators will now be illustrated.

For analysis of the distribution of the component failure times, the observed lifetimes (times between failures) are ranked as shown in Table 26.6. The events in parentheses in Table 26.6 are times from the last failure of a component to the current time. Such times are right censorings (times to non-failure) and are typical of failure data from field service. As most reliability data come from such sources, it is important not to just analyze the failure times and ignore the censoring times by treating the data as if they came from a complete sample as in a laboratory test. Ignoring censorings will result in pessimistic estimates for reliability. Data are often encountered, as in Example 7, where there are as many as four or five times more censorings than failures. The consequences of ignoring the censorings would result in wildly inaccurate conclusions about the lifetime distribution.

An estimate can be given for the reliability function at each observed component failure time. The estimate of the reliability function in the last column of Table 26.6 is that obtained by using the KM estimator. Alternatively, the NA estimator can be used. The estimate of the hazard function at each failure time is obtained as the number of failures (the multiplicity $m_{[i]}$) at each event time divided by the number of survivors immediately prior to that time (the number in the risk set, $r_{[i]}$) The cumulative sum of these values is the NA estimate, $\hat{\Lambda}(t)$, of the cumulative hazard function $\Lambda(t)$. Then $\exp(-\hat{\Lambda}(t))$ provides an estimate of $R(t)$, the reliability function, and is given in the penultimate column of Table 26.6. It can be seen that the NA estimate of the reliability function is consistently slightly higher than the KM estimate.

The improbability of tied lifetimes should lead to consideration of the possibility of the times not being independent, as would be the case in, for example, common cause failures. Both these estimates have been derived assuming that the lifetimes are independent and identically distributed (i.i.d.).

26.6 Weibull Regression Model

One of the most popular families of parametric models is that given by the Weibull distribution. It is usual for regression models to describe one or more of the distribution's parameters in terms of the covariates **z**. The relationship is usually linear, though this is not always the case. The Weibull distribution has a reliability function which can be given by:

$$R_T(t) = \exp(-\lambda t^{\kappa}) \quad \text{for } t > 0$$

where $\lambda = 1/\theta^{\kappa}$ and θ is the scale parameter, and κ is the shape parameter. Each of these parameters could be described in terms of the covariates **z**, though it is more usual to define either the scale or the shape parameter in terms of **z**. For example, if the scale parameter was chosen then a common model would be to have $\lambda(\mathbf{z};\boldsymbol{\beta}) = \exp(\boldsymbol{\beta}^{\mathrm{T}}\mathbf{z})$, where the number of covariates $k = r$, the number of parameters. Then the reliability function would be:

$$R_T(t \mid \mathbf{z};\boldsymbol{\beta}) = \exp(-\exp(\boldsymbol{\beta}^{\mathrm{T}}\mathbf{z})t^{\kappa}) \quad \text{for } t > 0$$

The probability density function is given by:

$$f_T(t \mid \mathbf{z};\boldsymbol{\beta}) = \kappa(\boldsymbol{\beta}^{\mathrm{T}}\mathbf{z})t^{\kappa-1} \times \exp(-\exp(\boldsymbol{\beta}^{\mathrm{T}}\mathbf{z})t^{\kappa}) \quad \text{for } t > 0$$

This model is commonly referred to as the Weibull regression model, but there are alternatives which have been studied, see Smith [10], where the shape parameter is dependent also on the covariate.

There are advantages to reparameterizing this model by taking logs, so that the model takes the form of Gumbel's extreme value distribution. A reason for this is to produce a model more akin to the normal regression model, but also it allows a more natural extension of the model and hence greater flexibility. Define $Y = \log T$ so that the reliability function is given by:

$$R_Y(y \mid \mathbf{z};\boldsymbol{\beta}) = \exp(-\exp(\kappa y + \boldsymbol{\beta}^{\mathrm{T}}\mathbf{z}))$$

so that

$$E(\log T) = -\frac{\gamma}{\kappa} - \frac{\boldsymbol{\beta}^{\mathrm{T}}\mathbf{z}}{\kappa}$$

where γ is Euler's constant. It is usual to estimate the parameters by using the maximum likelihood approach.

Standard errors may be obtained by the use of second derivatives to obtain the observed information matrix, $\mathbf{I}_{\mathrm{o}}$, and this matrix is usually calculated in the standard statistical software packages.

Example 9. Ansell and Ansell [3] analyzed the data given in Table 26.3 in a study of the performance of sodium sulfur batteries. The data consist of lifetimes (in cycles) of two batches of batteries. The covariate vector for the ith battery is given by $\mathbf{z}_i = (z_{i1}, z_{i2})^{\mathrm{T}}$, where $z_{i1} = 1$ and z_{i2} represents whether the battery comes from batch 1 or batch 2, so that:

$$z_{i2} = \begin{cases} 0 & \text{if battery } i \text{ is from batch 1} \\ 1 & \text{if battery } i \text{ is from batch 2} \end{cases}$$

Hence β_2 represents the difference in performance between batteries from batch 2 and batch 1.

Fitting the Weibull regression model results in $\hat{\beta}_2 = 0.0156$ and $\hat{\kappa} = 1.127$. Using the observed information matrix the standard error of $\hat{\beta}_2 = 0.386$. Hence a 95% confidence interval for β_2 is $(-0.740, 0.771)$.

An obvious test to perform is to see if β_2 is non-zero. If it is non-zero this would imply there is a difference between the two batches of batteries. The hypotheses are:

$$H_0: \quad \beta_2 = 0$$
$$H_1: \quad \beta_2 \neq 0$$

The log likelihood evaluated under H_0 is −49.7347 and under H_1 is −49.7339. Hence the likelihood ratio test statistic has a value of 0.0016. Under the null hypothesis the test statistic has a χ^2 distribution with one degree of freedom. Therefore the hypothesis that β_2 is zero, which is equivalent to no difference between the batches, is accepted. Using the estimate $\hat{\beta}_2$ of β_2 and its standard error gives a Wald statistic of 0.0016, almost the same value as the likelihood ratio test statistic, and hence leads to the same inference. Both these inferences are consistent with the confidence interval, as it includes zero.

Examination of the goodness of fit and the appropriateness of the assumptions made in fitting the regression model can be based on graphical approaches using residuals, see Smith [10]. The Cox and Snell generalized residuals, see Cox and Snell [11], are defined as:

$$e_i = -\log R_T(t_i \mid \mathbf{z}_i; \hat{\boldsymbol{\beta}})$$

where $R_T(t_i \mid \mathbf{z}_i; \hat{\boldsymbol{\beta}})$ is the reliability function evaluated at t_i and $\mathbf{z}_i$ with estimates $\hat{\boldsymbol{\beta}}$.

Obviously one problem that arises is with the residuals for censored observations and authors generally, see Lawless [12], suggest using:

$$e_i = -\log R_T(t_i \mid \mathbf{z}_i; \hat{\boldsymbol{\beta}}) + 1$$

Cox and Snell residuals should be i.i.d. random variables with a unit exponential distribution, *i.e.* with an expectation of one. Then the cumulative hazard function is a linear function with a slope of one. Also the cumulative hazard function is minus the log of the reliability function. Hence a plot of minus the log of the estimated reliability function for the residuals, $-\log \hat{R}(e_i)$, against e_i should be roughly linear with a slope of one when the model is adequate.

Example 9. (cont.) Using the data in Example 3 and fitting the Weibull regression model, the generalized residuals have been calculated and are presented in Figure 26.3. The plot is of minus the log of the reliability function of the generalized residuals against the generalized residuals. Since some of the points are far from the line with slope one, it would seem that the current model is not necessarily appropriate. Further investigation would be required to clarify if this was due to the choice of the distribution or the current model.

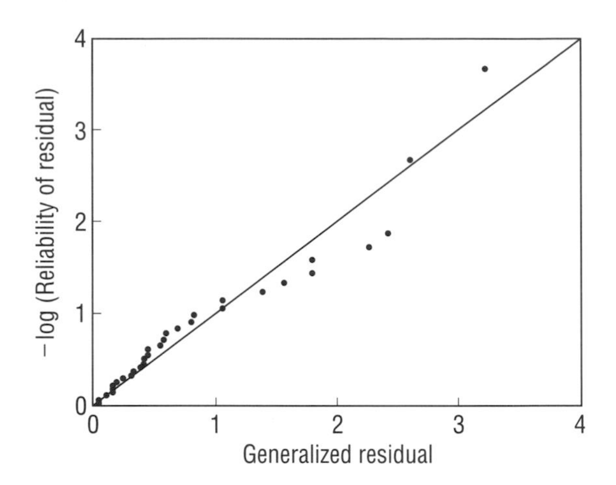

Figure 26.3. Plot of the generalized residuals of the Weibull regression model for the sodium sulfur battery data

The approach described above is applicable to many distributions. A special case of the Weibull regression model is the exponential regression model, when the shape parameter is taken to be 1 ($\kappa = 1$). This has been studied by Glasser [13], Cox and Snell [11] and Lawless [12]. Other lifetime regression models which can be used include: gamma, log-logistic, and lognormal, see Lawless [12]. Many of these models are covered by the general term location-scale models, or accelerated failure-time models.

Selection of an appropriate distribution model depends both on the physical context and on the actual fit achieved. There ought to be good physical justification for fitting a distribution using the context under study. Basing the decision about the distribution purely on fit can be very misleading, especially if there are a number of possible covariates to be chosen. It is always possible by fitting extra variables to achieve a better fit to a set of data, though such a fit will have little predictive power.

26.7 Accelerated Failure-time Model

The Weibull regression model discussed in the last section can be regarded as an example of an accelerated failure-time model. Accelerated failure-time models were originally devised to relate the performance of components put through a severe testing regime to a component's more usual lifetime. It was assumed that a variable or factor, such as temperature or number of cycles, could be used to describe the severity of the testing regime. This problem has been considered by a number of authors, see Nelson [14] for a comprehensive account of this area.

Suppose in a study that the covariate is z, which can take the values 0 and 1, and that it is assumed that the hazard functions are:

$$\lambda(t \mid z=0) = \lambda_0$$

and

$$\lambda(t \mid z=1) = \phi\lambda_0$$

so that ϕ is the *relative risk* for $z=1$ versus $z=0$. Then

$$R(t \mid z=1) = R(\phi t \mid z=0)$$

and, in particular,

$$E(T \mid z=1) = \frac{E(T \mid z=0)}{\phi}$$

So the time for components with $z=1$ is passing at a rate ϕ faster than for the components with $z=0$. Hence the name of the model.

The model can be extended as follows. Suppose ϕ is replaced by $\phi(z)$ with $\phi(0)=1$, then

$$R(t \mid z) = R(\phi(z)t \mid z=0)$$

and hence

$$\lambda(t \mid z) = \phi(z)\lambda(\phi(z)t \mid z=0)$$

and

$$E(T \mid z) = \frac{E(T \mid z=0)}{\phi(z)}$$

In using the model for analysis a parametric model is specified for $\phi(z)$ with β as the parameter, which will be denoted by $\phi(z;\beta)$. A typical choice would be

$$\phi(z;\beta) = \exp(\beta z)$$

This choice leads to a linear regression model for $\log T$ as $\exp(\beta z)T$ has a distribution which does not depend on z. Hence $\log T$ is given by:

$$\log T = \mu_0 - \beta z + \epsilon$$

where μ_0 is $E(\log T \mid z=0)$ and ϵ is a random variable whose distribution does not depend on the covariate z.

To estimate β there is the need to specify the distribution. If the distribution of T is lognormal then least squares estimation may be used as $\mu_0+\epsilon$ has a normal distribution. If the distribution of T is Weibull with a shape parameter κ, then $\kappa(\mu_0+\epsilon)$ has a standard extreme value distribution. Hence, as was stated at the beginning of this section, an example of the accelerated failure-time model is the Weibull regression model. Other such models have been widely applied in reliability, see Cox [15], Fiegl and Zelen [16], Nelson and Hahn [17], Kalbfliesch [18], Farewell and Prentice [19], and Nelson [14]. Plotting techniques, such as using Cox and Snell generalized residuals as defined for the Weibull regression model, may be used for assessing the appropriateness of the model.

Example 10. Elsayed and Chan [4] presented data collected from tests for time-dependent dielectric breakdown of metal-oxide-semiconductor integrated circuits, which was described in Example 4 with the data given in Table 26.4. The data consist of times to failure (in hours) for three different temperatures (170 °C, 200 °C and 250 °C). Elsayed and Chan [4] suggest a model where the covariate of interest is the inverse of the absolute temperature. So the covariate vector for the ith circuit is given by $\mathbf{z}_i = (z_{i1}, z_{i2})^{\mathrm{T}}$, where $z_{i1}=1$ and z_{i2} represents the inverse absolute temperature at

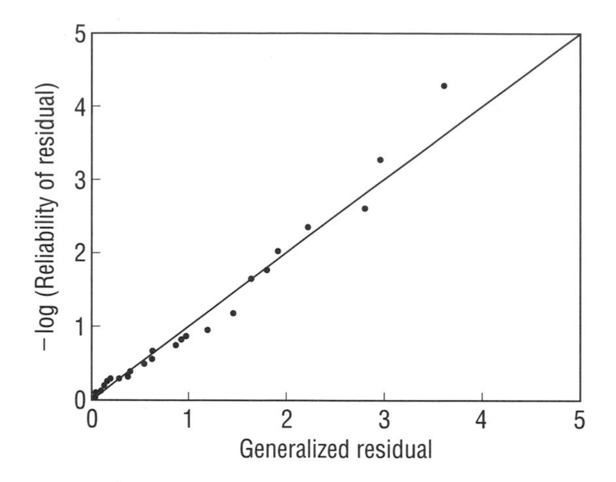

Figure 26.4. Plot of the generalized residuals of the Weibull regression model for the semiconductor integrated circuit data

which the test was performed, and takes the three values 0.001 911, 0.002 113, and 0.002 256. Hence β_2 represents the coefficient of the inverse absolute temperature covariate.

Fitting the Weibull regression model results in $\hat{\beta}_2 = -7132.4$ and $\hat{\kappa} = 0.551$. Using the observed information matrix the standard error of $\hat{\beta}_2 = 1222.0$. Hence a 95% confidence interval for β_2 is $(-9527.5, -4737.1)$. This interval indicates that β_2 is non-zero, and this would imply there is a difference in the performance of the circuits at the different temperatures. This can be confirmed by performing a hypothesis test to see if β_2 is non-zero. The hypotheses are:

$$H_0: \quad \beta_2 = 0$$
$$H_1: \quad \beta_2 \neq 0$$

The log likelihood evaluated under H_0 is -148.614 and under H_1 is -130.112. Hence the likelihood ratio test statistic has a value of 37.00. Under the null hypothesis the test statistic has a χ^2 distribution with one degree of freedom. Therefore the hypothesis that β_2 is zero is not accepted and this implies there is a difference between the circuits depending on the temperatures of the tests. Using the estimate $\hat{\beta}_2$ of β_2 and its standard error gives a Wald statistic of 34.06, almost the same value as the likelihood ratio test statistic, and hence leads to the same inference.

The generalized residuals are presented in Figure 26.4. The plot is of minus the log of the reliability function of the generalized residuals against the generalized residuals. The points lie closer to a line with slope one than was the case in Figure 26.3 for Example 9. So the current model may be appropriate.

26.8 Proportional Hazards Model

This model has been widely used in reliability studies by a number of authors: Bendell and Wightman [20], Ansell and Ansell [3] and Jardine and Anderson [21]. The model Cox [7] proposed assumed that the hazard function for a component could be decomposed into a baseline hazard function and a function dependent on the covariates. The hazard function at time t with covariates $\mathbf{z}$, $\lambda(t \mid \mathbf{z})$, would be expressed as:

$$\lambda(t \mid \mathbf{z}) = \psi[\lambda_0(t), \phi(\mathbf{z}; \boldsymbol{\beta})]$$

where ψ would be an arbitrary function, $\lambda_0(t)$ would be the baseline hazard function, ϕ would be another arbitrary function of the covariates, $\mathbf{z}$, and $\boldsymbol{\beta}$ the parameters of the function ϕ. It was suggested by Cox [7] that ψ might be a multiplicative function and that ϕ should be the exponential with a linear predictor for the argument. This *proportional hazards* model has the advantage of being well defined and is given by:

$$\lambda(t \mid \mathbf{z}) = \lambda_0(t) \exp(\boldsymbol{\beta}^{\mathrm{T}}\mathbf{z})$$

However, this is only one possible selection for ψ and ϕ. Using the multiplicative formulation it is usual to define $\phi(\mathbf{z}; \boldsymbol{\beta})$ so that $\phi(\mathbf{0}, \boldsymbol{\beta}) = 1$, so that $\phi(\mathbf{z}; \boldsymbol{\beta})$ is the relative risk for a component with covariate $\mathbf{z}$ compared to a component with covariate $\mathbf{z} = \mathbf{0}$. Thus the reliability function is given by:

$$R(t \mid \mathbf{z}) = R(t \mid \mathbf{z} = \mathbf{0})^{\phi(\mathbf{z}, \boldsymbol{\beta})}$$

where $R(t \mid \mathbf{z} = \mathbf{0})$, often denoted simply as $R_0(t)$, is the baseline reliability function. Etezardi-Amoli and Ciampi [22], amongst others, have considered an alternative additive model.

Cox [7] considered the case in which the hazard function is a semi-parametric model; $\lambda_0(t)$ is modeled non-parametrically (or more accurately using infinitely many parameters). It is possible to select a specific parametric form for $\lambda_0(t)$, which could be a hazard function from one of a number of families of distributions. In the case of the semi-parametric model, Cox [23] introduced the concept of *partial* likelihood to tackle the problems of statistical inference. This has been further supported by the work of Andersen and Gill [24]. In Cox's approach the partial likelihood function is formed by considering the components at risk at each of the n_0 failure times $t_{[1]}, t_{[2]}, t_{[3]}, \ldots, t_{[n_0]}$, as defined in Section 26.5. This produces a function which does not depend on the underlying distribution and can therefore be used to obtain estimates of $\boldsymbol{\beta}$. Ties often occur in practice in data, and adjustment should be made to the estimation procedure to account for ties, as suggested by Breslow [25] and Peto [26].

Example 11. Returning to the sodium sulfur batteries data used in Example 9, put $\mathbf{z} = z_2$, which was defined in Example 9 to indicate whether the battery comes from batch 1 or batch 2. (*Note:* The variable $z_1 = 1$ is not required as the arbitrary term $\lambda_0(t)$ will contain any arbitrary constant which was provided in the Weibull regression model in Example 9 by the parameter β_1.) Then $\hat{\beta} = -0.0888$, dropping the redundant suffix from $\hat{\beta}$. Using the information matrix, the standard error of $\hat{\beta}$ is 0.4034. Hence a 95% confidence interval for β is (−0.879, 0.702).

A test can be performed to see if β is non-zero. If it is non-zero this would imply there is a difference between the two batches of batteries. The hypotheses are:

$$H_0: \quad \beta = 0$$
$$H_1: \quad \beta \neq 0$$

It is possible to use the partial log likelihood, which when evaluated under H_0 is −81.262 and under H_1 is −81.238. Hence the "likelihood" ratio test statistic (twice the difference between these partial log likelihoods) has a value of 0.048. Under the null hypothesis this test statistic can be shown to have a χ^2 distribution with one degree of freedom, in the same way as for likelihood. Therefore the hypothesis that β is zero is accepted and this implies there is no difference between the batches.

An alternative non-parametric approach can be taken in the case when comparing two distributions to see if they are the same. In the case of two groups a test of whether $\beta = 0$ is equivalent to testing whether the two reliability functions, $R_1(t)$ for group 1 and $R_2(t)$ for group 2, are the same. The hypotheses are:

$$H_0: \quad R_2(t) = R_1(t)$$
$$H_1: \quad R_2(t) = R_1(t)^{\exp(\beta)} \quad \text{for } \beta \text{ not equal to } 0$$

A test statistic was proposed by Mantel [27], which under the null hypothesis can be shown to have a χ^2 distribution with one degree of freedom.

Example 12. Returning to Example 11, a test of whether $\beta = 0$ is equivalent to testing whether the reliability functions are the same for both batches of batteries. The statistic proposed by Mantel [27] is equal to 0.0485. This statistic, though not identical, is similar to that obtained from the likelihood ratio in Example 11, and hence leads to the same conclusion. Therefore the null hypothesis that β is zero is accepted.

These examples have used a categorical covariate which indicates different groups of observations. The next example uses a continuous covariate.

Example 13. Returning to the semiconductor integrated circuit data used in Example 10, put $\mathbf{z} = z_2$, which was defined in Example 10 as the inverse of the absolute temperature. Then $\hat{\beta} = -7315.0$. Using the information matrix, the standard error of $\hat{\beta}$ is 1345.0. Hence a 95% confidence interval for β is (−9951.2, −4678.8).

This interval indicates that β is non-zero, and this would imply there is a difference in the performance of the circuits at the different temperatures. This can be confirmed by performing a test to see if β is non-zero. The hypotheses are:

$$H_0: \quad \beta = 0$$

$$H_1: \quad \beta \neq 0$$

It is possible to use the partial log likelihood, which when evaluated under H_0 is -190.15 and under is -173.50. Hence the "likelihood" ratio test statistic (twice the difference between these partial log likelihoods) has a value of 33.31. Under the null hypothesis this test statistic can be shown to have a χ^2 distribution with one degree of freedom, in the same way as for likelihood. Therefore the hypothesis that β is zero is not accepted, and this implies there is a difference between the circuits depending on the temperatures of the tests.

Using the estimate $\hat{\beta}$ of β and its standard error gives a Wald statistic of 29.60, almost the same value as the likelihood ratio test statistic, and hence leads to the same inference.

The reliability function has to be estimated, and the usual approach is to first estimate the baseline reliability function, $R_0(t)$.

Example 14. The estimate of the baseline reliability function for the data on sodium sulfur batteries introduced in Example 9 with the estimate $\hat{\beta}$ of β as given in Example 11, is given in Figure 26.5 with 95% confidence limits.

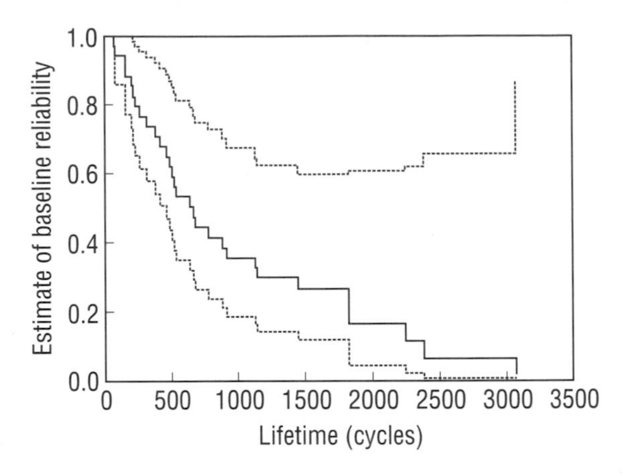

Figure 26.5. Plot of the baseline reliability function for the proportional hazards model for the sodium sulfur battery data with 95% confidence limits

Baseline has been defined as the case with covariate $z = 0$. However this is not always a sensible choice of the covariate. This is illustrated in the next example.

Example 15. For the semiconductor integrated circuit data used in Example 10, $\mathbf{z} = z$ was defined as the inverse of the absolute temperature. This will only be zero if the temperature is infinitely large. Hence it makes more sense to take the baseline value to be one of the temperatures used in the tests. The *smallest* temperature will be chosen, which corresponds to 0.002 256, the *largest* value of the inverse absolute temperature. Then the estimate of the baseline reliability function for the data, with the estimate of β as given in Example 13, is given in Figure 26.6 with 95% confidence limits.

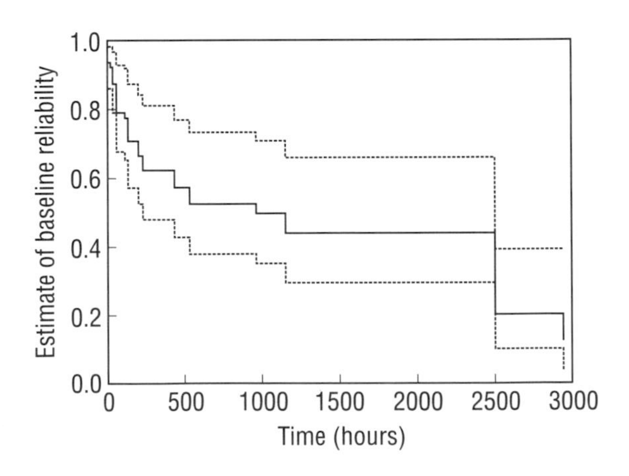

Figure 26.6. Plot of the baseline reliability function for the proportional hazards model for the semiconductor integrated circuit data with 95% confidence limits

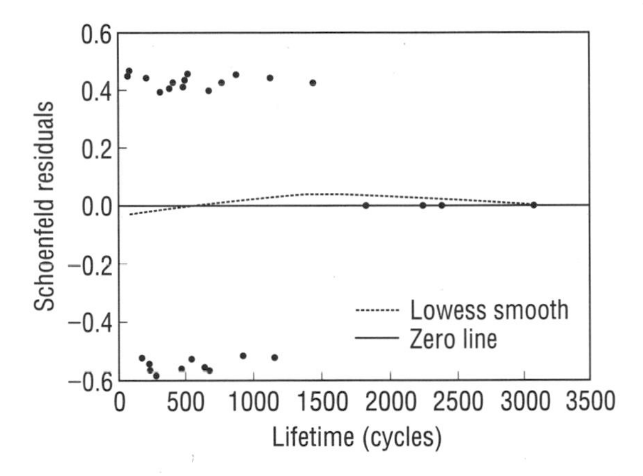

Figure 26.7. Plot of the Schoenfeld residuals for a batch of the proportional hazards model for the sodium sulfur battery data

26.9 Residual Plots for the Proportional Hazards Model

Given the estimates of the reliability function it is then possible to obtain the Cox and Snell generalized residuals. However, it has been found that other residuals are more appropriate. Schoenfeld [28] suggested partial residuals to examine the proportionality assumption made in the Cox model. These residuals can then be plotted against time, and if the proportional hazards assumption holds then the residuals should be randomly scattered about zero, with no time trend.

Example 16. Using the sodium sulfur battery data from Example 3 and fitting the proportional hazards model, the Schoenfeld residuals are calculated and presented in Figure 26.7 as a plot of the residuals against time. A non-parametric estimate of the regression line, the "lowess" line, see Cleveland [29], is included on the plot as well as the zero residual line. There seems to be no significant trend. A test for *linear* trend can be used. For this test the statistic is 0.0654, which would be from a χ^2 distribution with one degree of freedom if there was no linear trend. Hence there is no evidence of a linear trend, and it is probably safe to accept the proportional hazards assumption.

Example 17. Using the semiconductor integrated circuit data from Example 4 and fitting the proportional hazards model, the Schoenfeld residuals are calculated and presented in Figure 26.8 as a plot of the residuals against time. A non-parametric estimate of the regression line, the "lowess" line, is included on the plot as well as the zero residual line. For the test for *linear* trend the statistic is 1.60, which would be from a χ^2 distribution with one degree of freedom if there was no linear trend. Hence there is no evidence of a linear trend. However, it is not clear whether to accept the proportional hazards assumption.

A number of statistical packages facilitate proportional hazards modeling, including SAS and S-PLUS. Therneau *et al.* [30] suggest two alternative residuals, a martingale residual and a deviance residual. Except in the case of discrete covariates, these residuals are far from simple to calculate, however statistical software is available for their estimation. In the case of discrete covariates the martingale residuals are a transformation of the Cox and Snell generalized residuals. The deviance residuals are a transformation of

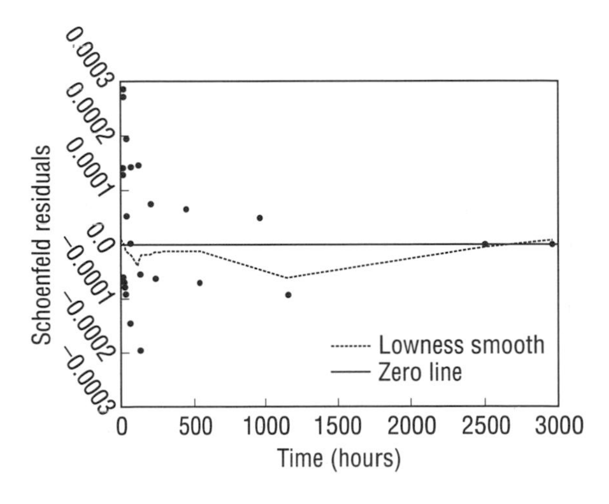

Figure 26.8. Plot of the Schoenfeld residuals for the inverse of absolute temperature of the proportional hazards model for the semiconductor integrated circuit data

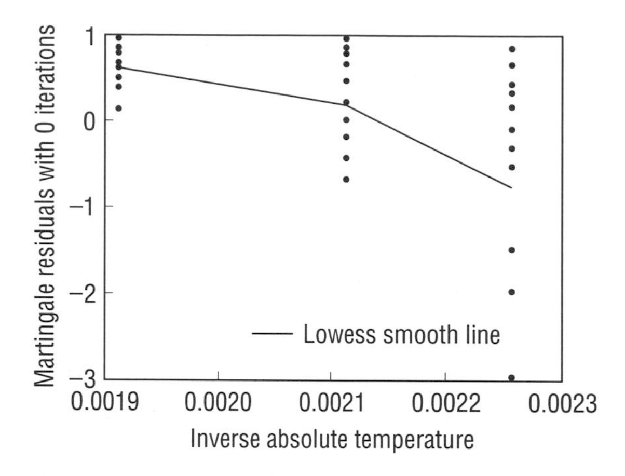

Figure 26.9. Plot of the martingale residuals versus the inverse of the absolute temperature of the null proportional hazards model for the semiconductor integrated circuit data

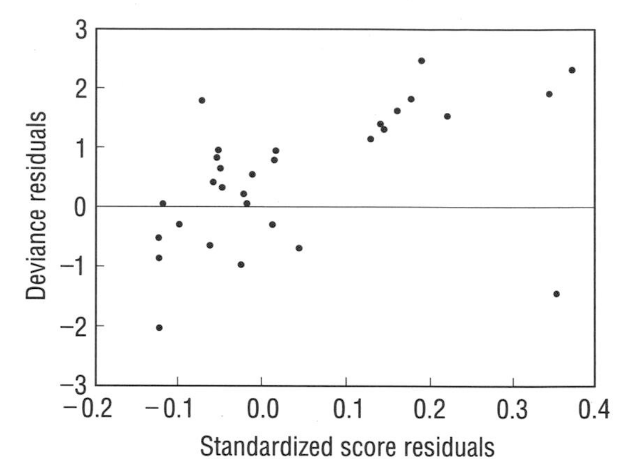

Figure 26.10. Plot of the deviance residuals versus the standardized score residuals for the inverse of the absolute temperature of the proportional hazards model for the semiconductor integrated circuit data

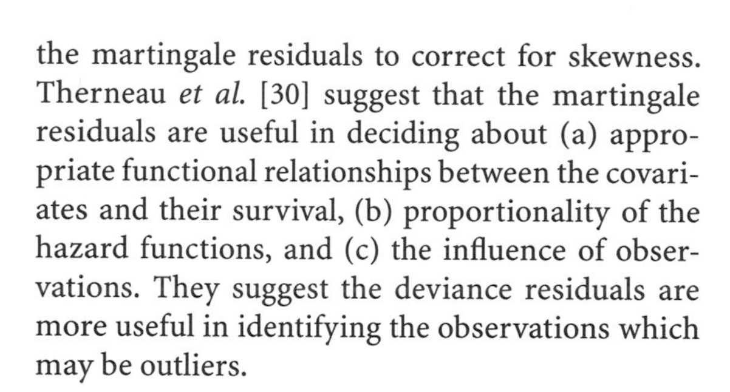

the martingale residuals to correct for skewness. Therneau *et al.* [30] suggest that the martingale residuals are useful in deciding about (a) appropriate functional relationships between the covariates and their survival, (b) proportionality of the hazard functions, and (c) the influence of observations. They suggest the deviance residuals are more useful in identifying the observations which may be outliers.

Example 18. Using the semiconductor integrated circuit data as in Example 4 and fitting the proportional hazards model, the martingale residuals are calculated and presented in Figure 26.9 as a plot of the residuals against the covariate, the inverse of the absolute temperature. A non-parametric estimate of the regression line, the "lowess" line, is included. There is some suggestion that a linear fit might not be best and a quadratic function might be an improvement, but this was found not to be the case.

The deviance residuals are calculated and can be used with the standardized score residuals to identify outliers. The plot of these residuals in Figure 26.10 suggests that there are some outliers with standardized score residuals which are larger than 0.3.

26.10 Non-proportional Hazards Models

An assumption of the Cox regression model was that the hazards are proportional. This can be interpreted as the distance between the log (−log) of the reliability functions not varying with time. Cox [7] suggested a test for this proportionality by adding an extra covariate of log time to the model with parameter β_2. In the case of two groups with z_i indicating membership of a group ($z_i = 0$ if the ith individual belongs to group 1, $z_i = 1$ if it belongs to group 2) with coefficient β_1, then the hazard functions become:

$$\begin{aligned} \text{for group 1:} \quad & \lambda_0(t) \\ \text{for group 2:} \quad & \lambda_0(t)\exp(\beta_1 + \beta_2 \log(t)) \\ & = \lambda_0(t) t^{\beta_2} \exp(\beta_1) \end{aligned}$$

If the coefficient of log time (β_2) is different from zero, the hazard functions are non-proportional as their ratio $t^{\beta_2}\exp(\beta_1)$ varies with t, otherwise they are proportional as the hazard function for group 2 reduces to $\lambda_0(t)\exp(\beta_1)$. The use of such models with time-dependent covariates was justified by Andersen and Gill [24].

Example 19. Using the sodium sulfur batteries data in Example 3 and fitting Cox's non-proportional hazards model, $\beta_2 = 0.0900$ with standard error 0.523. These give a Wald statistic of 0.0296, which is not significantly different from zero, for a χ^2 distribution with one degree of freedom. Hence the null hypothesis of proportionality of the hazards is accepted, which agrees with the conclusion made in Example 16, when using the Schoenfeld residuals.

Many authors have extended this use of time-dependent variables in studies of lifetimes, for example Gore *et al.* [31]. The other obvious problem is selection of an appropriate model. A plot of log (−log) of the reliability function or log hazard plot may reveal some suitable function, or the nature of the problem may suggest some particular structure. Alternatively one might use the smoothed martingale residuals suggested by Therneau *et al.* [30], plotted against the covariate to seek functional form. The danger is in producing too complex a model which does not increase insight. When the variables are quantitative rather than qualitative then the problem is exacerbated, and greater caution is necessary.

When predicting future performance there may be difficulties with time-dependent covariates and then it may be necessary to resort to simulation.

26.11 Selecting the Model and the Variables

The number of variables included in an analysis should be considerably less than the number of data points. Overfitting the data can be a major problem. Adding extra variables should improve the fit of the model to the data. However, this will not necessarily improve the ability of the model to predict future observations. Some authors have suggested that data in regression analyses should be split into two parts. With the first half one derives the model and with the second half judgment about the model should be made. Often there is too little data to allow such an approach. In the case of limited data one should be wary therefore of too good a fit to the data.

Most of the assessment of which variable to fit will be based on the analysis performed. In normal regression modeling considerable attention has been paid to variable selection, though no generally accepted methodology has been devised to obtain the best set. Two approaches taken are forward selection, in which variables are added to the model, and backward selection, in which variables are deleted until a "good" model is fitted. These approaches can be applied together, allowing for the inclusion and deletion of variables.

For the models considered in this chapter an equivalent to the F-value, used for models with normal errors, would be the change in *deviance*, which for the ith variable is

$$\mathrm{Dev}_i = -2(l(\beta_i = 0) - l(\beta_i \neq 0))$$

where $l(\beta_i = 0)$ is the log likelihood without the ith variable fitted and $l(\beta_i \neq 0)$ is with it fitted. This statistic is asymptotically distributed as χ^2 with one degree of freedom. Hence the addition, or deletion, of a variable given the current model can be decided on whether this value is significantly large or not. Whether the approach results in the "best" fit is always doubtful, but if there are a large number of variables it may be a sensible approach. Again if the above method is used then the model produced should be capable of physical explanation within the context. The model ought to seem plausible to others. If the model cannot be interpreted then there is a danger of treating the technique as a black box.

26.12 Discussion

Given the variety of models discussed it may seem crucial that the appropriate model is selected. Whilst in some specific cases this may be the case, generally the desire is to identify factors or variables which are having an effect on the component's or system's performance. In such cases most of the models will at least be able to detect effects of factors which are strongly associated with lifetimes, though they may overlook those with weak association provided the proportion of censored observations is small. Solomon [32] has also supported these comments when comparing accelerated failure-time and proportional hazards models. Further work is necessary in the case of highly censored data. Hence model selection may not be as crucial as initially suggested. Obviously if the desire is to model lifetimes of the component in detail then more care in choice should be taken.

This chapter has covered the analysis of lifetimes of components and systems. However, the data in Example 5 was for a repairable system. So it would be appropriate to estimate the rate of occurrence of failures (ROCOF). A non-parametric approach to this problem is given by Phillips [33]. This is beyond the scope of this chapter.

References

[1] Keating JP, Glaser RE, Ketchum NS. Testing hypotheses about the shape parameter of a gamma distribution. Technometrics 1990;32:67–82.

[2] Kim JS, Proschan F. Piecewise exponential estimator of the survivor function. IEEE Trans Reliab 1991;40:134–9.

[3] Ansell RO, Ansell JI. Modelling the reliability of sodium sulphur cells. Reliab Eng 1987;17:127–37.

[4] Elsayed EA, Chan CK. Estimation of thin-oxide reliability using proportional hazards models. IEEE Trans Reliab 1990;39:329–35.

[5] Newton DW. Some pitfalls in reliability data analysis. Reliab Eng 1991;34:7–21.

[6] Ansell JI, Phillips MJ. Practical methods for reliability data analysis. Oxford: Oxford University Press; 1994.

[7] Cox DR. Regression models and life tables. J R Stat Soc B 1972;34:187–220.

[8] Kaplan EL, Meier P. Non-parametric estimation from incomplete observations. J Am Stat Assoc 1958;53:457–81.

[9] Nelson W. Hazard plotting for incomplete failure data. J Qual Tech 1969;1:27–52.

[10] Smith RL. Weibull regression models for reliability analysis. Reliab Eng Safety Syst 1991;34:55–77.

[11] Cox DR, Snell EJ. A general definition of residuals (with discussion). J R Stat Soc B 1968;30:248–75.

[12] Lawless JF. Statistical models and methods for lifetime data. New York: Wiley; 1982.

[13] Glasser M. Exponential survival with covariance. J Am Stat Assoc 1967;62:561–8.

[14] Nelson W. Accelerated life testing. New York: Wiley; 1993.

[15] Cox DR. Some applications of exponential ordered scores. J R Stat Soc B 1964;26:103–10.

[16] Fiegl P, Zelen M. Estimation of exponential survival probabilities with concomitant information. Biometrics 1965;21:826–38.

[17] Nelson WB, Hahn GJ. Linear estimation of a regression relationship from censored data. Part 1. Simple methods and their applications. Technometrics 1972;14:247–76.

[18] Kalbfleisch JD. Some efficiency calculations for survival distributions. Biometrika 1974;61:31–8.

[19] Farewell VT, Prentice RL. A study of distributional shape in life testing. Technometrics 1977;19:69–76.

[20] Bendell A, Wightman DM. The practical application of proportional hazards modelling. Proc 5th Nat Reliab Conf, Birmingham, 1985.Culcheth: National Centre for Systems Reliability.

[21] Jardine AKS, Anderson M. Use of concomitant variables for reliability estimation and setting component replacement polices. Proc 8th Adv Reliab Tech Symp, Bradford, 1984. Culcheth: UKAEA.

[22] Etezardi-Amoli J, Ciampi A. Extended hazard regression for censored survival data with covariates: a spline approximation for the baseline hazard function. Biometrics 1987;43:181–92.

[23] Cox DR. Partial likelihood. Biometrika 1975;62:269–76.

[24] Andersen PK, Gill RD. Cox's regression model for counting processess: a large sample study. Ann Stat 1982;10:1100–20.

[25] Breslow NE. Covariate analysis of censored survival data. Biometrics 1974;30:89–99.

[26] Peto R. Contribution to the discussion of regression models and life tables. J R Stat Soc B 1972;34:205–7.

[27] Mantel N. Evaluation of survival data and two new rank order statistics arising from its consideration. Cancer Chemother Rep 1966;50:163–70.

[28] Schoenfeld D. Partial residuals for the proportional hazards regression model. Biometrika 1982;69:239–41.

[29] Cleveland WS. Robust locally weighted regression and smoothing scatterplots. J Am Stat Assoc 1979;74:829–36.

[30] Therneau TM, Grambsch PM, Fleming TR. Martingale based residuals for survival models. Biometrika 1990;77:147–60.

[31] Gore SM, Pocock SJ, Kerr GR. Regression models and non-proportional hazards in the analysis of breast cancer survival. Appl Stat 1984;33:176–95.

[32] Solomon PJ. Effect of misspecification of regression models in the analysis of survival data. Biometrika 1984;71:291–8.

[33] Phillips MJ. Bootstrap confidence regions for the expected ROCOF of a repairable system. IEEE Trans Reliab 2000;49:204–8.

The Application of Capture–Recapture Methods in Reliability Studies

Paul S. F. Yip, Yan Wang and Anne Chao

Chapter 27

27.1 Introduction

Suppose there are an unknown number ν of faults in a computer program. A fault is some defect in the code which (under at least one set of conditions) will produce an incorrect output, *i.e.* a failure. The program is executed for a total time τ under a range of conditions to simulate the operating environment. Failures occur on N_τ occasions at times $0 < t_1 < t_2 < \cdots < t_{N_\tau} < \tau$. The time until the detection of fault i is a positive random variable with hazard function $\lambda_i(t)$ $(i = 1, 2, \ldots, \nu)$. The first column of Table 27.1 gives a set of failure data for an information system (seconds CPU time). The data can be found in Moek [1]. The main purpose of this study is to estimate the total number of remaining faults in the system.

Estimation methods are available based on the observed number of faults. Information about ν comes from detecting new faults and the rate at which faults are being detected. However, it is shown that a large proportion of faults must be detected in order for the method to perform satisfactorily [2]. Alternatively, some other designs have been suggested to improve the estimation. For example:

- A recapture approach—the recapture idea is to assume a fault detected, corrected, and a counter inserted to record potential recurrences of the fault. The setting up of a counter where the fault is found, and the recording of the number of times the fault is re-detected, is assumed without causing system failure [3, 4]. The recapture faults assume that both the detection times and the fault type are recorded. Information about ν comes from (a) detecting new faults and (b) observing the proportion of revisit faults from subsequent testing. An estimator based on the optimal estimating equation is recommended as suggested by Lloyd *et al.* [4]. The estimator generated by the

Table 27.1. Real and simulated failure times of Moek's data [1]

Fault #	First	Subsequent simulated failure times					
1	880	64 653	329 685	520 016			
2	4310	305 937	428 364	432 134	576 243		
3	7170	563 910					
4	18 930	186 946	195 476	473 206			
5	23 680	259 496	469 180				
6	23 920	126 072	252 204	371 939			
7	26 220	251 385					
8	34 790	353 576					
9	39 410	53 878	147 409	515 884			
10	40 470	371 824	466 719				
11	44 290	83 996	327 296	352 035	395 324	494 037	
12	59 090	61 435	222 288	546 577			
13	60 860	75 630	576 386				
14	85 130						
15	89 930	205 224	292 321	294 935	342 811	536 573	553 312
16	90 400	228 283	334 152	360 218	368 811	377 529	547 048
17	90 440	511 836	511 967				
18	100 610	367 520	429 213				
19	101 730	162 480	534 444				
20	102 710	194 399	294 708	295 030	360 344	511 025	
21	127 010	555 065					
22	128 760						
23	133 210	167 108	370 739				
24	138 070	307 101	451 668				
25	138 710	232 743					
26	142 700	215 589					
27	169 540	299 094	428 902	520 533			
28	171 810	404 887					
29	172 010	288 750					
30	211 190						
31	226 100	378 185	446 070	449 665			
32	240 770	266 322	459 440				
33	257 080	374 384					
34	295 490	364 952					
35	296 610						
36	327 170	374 032	430 077				
37	333 380						
38	333 500	480 020					
39	353 710						
40	380 110	433 074					
41	417 910	422 153	479 514	511 308			
42	492 130						
43	576 570						

optimal estimating equation can be shown to be equivalent to the maximum likelihood estimator.

- A seeded fault approach without recapture—the seeded fault approach assumes a known number of faults are deliberately seeded in the system. Faults are detected and the number of real and seeded faults recorded. Information about ν comes from the proportion of real and seeded faults observed in subsequent testing. With sufficient seeded faults in the system, recaptures do not improve the performance significantly [4]. An estimator based on optimal estimating equation is also recommended in the present setting [5]. The estimator generated by optimal estimating equation can also be shown to be equivalent to the maximum likelihood estimator.

The two approaches suggested above make the homogeneity assumption for the failure intensities, *i.e.* $\lambda_i(t) = \alpha_t$ for all i. For heterogeneous models we consider the following three situations.

1. Non-parametric case (failure intensity $\lambda_i(t) = \gamma_i \alpha_t$) with recapture. The γ_i values are individual effects and can be unspecified.
2. Parametric case (failure intensity $\lambda_i(t) = \gamma_i$) without recapture, *i.e.* the Littlewood model [6] which also includes the Jelinski–Moranda model as a special case.
3. Parametric case (failure intensity $\lambda_i(t) = \gamma_i$) with recapture. The γ_i values are assumed from a gamma density.

For the homogeneous model we use the optimal estimating equation which is well established to be an alternative and efficient method to estimate population size [7, 8]. We shall introduce a sample coverage estimator and a Horvitz–Thompson type estimator [9] to estimate ν for heterogeneous models.

In this chapter we also discuss a sequential procedure to estimate the number of faults in a system. A procedure of this type is useful in the preliminary stages of testing. In practice it is expensive and generally infeasible to detect all the faults in the system in the initial stages of software development. The idea of the sequential procedure proposed here is not to find all the faults in the system. The goal is to detect designs of questionable quality earlier in the process by estimating the number of faults that go undetected after a design review. This can be valuable for deciding whether the design is of sufficient quality for the software to be developed further, as suggested by Van der Wiel and Votta [10]. In the early stages, we only need a rough idea how many faults there are before proceeding further. Usually, an estimate with some specified minimum level of accuracy would be sufficient [11]. Here we consider as accuracy criterion the ratio of the standard error of the estimate to the population size estimate, sometimes called the coefficient of variation.

The purpose of this chapter is to provide estimating methods to make inference on ν. This chapter is different from other reviews on the use of recapture in software reliability and epidemiology, which for example examine the case of using a number of inspectors to estimate the number of faults in a system or estimate the number of diseased persons based on a number of lists, respectively [12, 13]. Here, we concentrate on a continuous time framework. Section 27.2 gives a formulation of the problem which includes homogeneous and heterogeneous cases, parametrically and non-parametrically. A sequential procedure is suggested in Section 27.3 and a real example is given in Section 27.4 to illustrate the procedure. Section 27.5 gives a wide range of simulation results of the proposed estimators. A discussion can be found in Section 27.6.

27.2 Formulation of the Problem

Notation

ν	number of faults in a system
$N_i(t)$	counting process of the number of detections of fault i before time t

Δ_{dt}	small time interval increase after time t
$\mathcal{F}_t$	σ-algebra generated by $N_i(s)$, $s \in [0, t], i = 1, \ldots, \nu$
$\lambda_i(t)$	intensity process for $N_i(t)$
τ	stopping time of the experiment
γ_i	individual effect of fault i in the heterogeneous model
$\mathcal{M}_t, \mathcal{M}_t^*$	zero-mean martingale
M_t	number of distinct faults detected by time t
N_t	total number of detections by time t
K_t	number of re-detections of the already found faults by time t
$W_t, k(t)$	weight function
W_t^*	optimal weight function
$\hat{\nu}_W$	estimate of ν with weight function W
$\widehat{\text{se}}(\hat{\nu})$	estimated standard error of the estimated ν
D	number of seeded faults inserted at the beginning of the experiment
U_t	number of real faults detected in the system before time t
α_t	failure intensity of the real faults
β_t	failure intensity of the seeded faults
θ	proportionality between α_t and β_t
$\mathcal{R}_t, \mathcal{R}_t^*$	zero-mean martingale
C_t	proportion of total individual effects of detected faults
I	indicator function
$A_i(t)$	event that the fault i has not been detected up to t but detected at time $t + \Delta_{dt}$
Λ	cumulative hazard function
$f_i(t)$	number of individuals detected exactly i times by time t
R	remainder term in the extension of the expectation of νC_t
$Y_i(t)$	indicator function of the event whether fault i is at risk of being detected
$G(\phi, \beta)$	gamma distribution with shape, scale parameters ϕ, β
ϵ, ω	parameters after reparameterizing ϕ, β
n	number of distinct faults detected during $[0, \tau]$
δ_i	indicator function denoting at least one failure caused by fault i occurs in $[0, \tau]$
p_i	probability of δ_i taking value 1
$\Omega(\theta)$	limiting covariance matrix of parameter $\theta = (\epsilon, \omega)'$
d_0	given accuracy level
$\text{ave}(\hat{\nu})$	average estimated population size based on simulation
$\text{ave}(\widehat{\text{se}}(\hat{\nu}))$	average estimated standard error based on simulation.

In this chapter we consider various assumptions of the intensity process, $\lambda_i(t)$.

- Homogeneous model: assume all the faults are homogeneous with respect to failure intensity, *i.e.* $\lambda_i(t) = \alpha_t, t \geq 0$, for $i = 1, 2, \ldots, \nu$.
- Heterogeneous model: consider $\lambda_i(t) = \gamma_i \alpha_t$ (where α_t are specified).

The individual effects $\{\gamma_1, \gamma_2, \ldots, \gamma_\nu\}$ can be modeled in the following two ways.

1. Fixed-effect model: $\{\gamma_1, \gamma_2, \ldots, \gamma_\nu\}$ are regarded as fixed parameters.
2. Frailty model: $\{\gamma_1, \gamma_2, \ldots, \gamma_\nu\}$ are a random sample assumed from a gamma distribution whose parameters are unknown.

The optimal estimating function methods proposed here are based on results for continuous martingales and follow from the work of Aalen [14, 15]. For our purpose a *zero-mean martingale* (ZMM) is a stochastic process $\{\mathcal{M}_t \geq 0\}$ such that $E(\mathcal{M}_0) = 0$ and for all $t \geq 0$ we have $E|\mathcal{M}_t| < \infty$ as well as $E(\mathcal{M}_{t+h}|\mathcal{F}_t) = \mathcal{M}_t$ for every $h > 0$.

27.2.1 Homogeneous Model with Recapture

For a homogeneous model with intensity $\lambda_i(t) = \alpha_t$, $t \geq 0$, for $i = 1, 2, \ldots, \nu$, we make use of the estimating function to make inference about ν. The important feature is to identify an estimating function which involves the parameter ν. Let M_t denote the number of faults that are detected by time t and $N_t = \sum_{i=1}^{\nu} N_i(t)$ the total number of detections by time t. Then $K_t = N_t - M_t$ denotes the number of re-detections of the already found faults by time t. Define

$$d\mathcal{M}_u = M_u \, dM_u - (\nu - M_u) \, dK_u$$

then $d\mathcal{M}_u$ is a martingale difference with respect to $\mathcal{F}_u$, where $\mathcal{F}_u$ is the history generated by $\{N_1(s), N_2(s), \ldots, N_\nu(s); 0 \le s \le u\}$. Since we have

$$dM_u \mid \mathcal{F}_u = \mathrm{Bin}(\nu - M_u, \alpha_u)$$

and

$$dK_u \mid \mathcal{F}_u = \mathrm{Bin}(M_u, \alpha_u)$$

where Bin denotes "Binomially Distributed", and

$$E(dM_u \mid \mathcal{F}_u) = (\nu - M_u)\alpha_u$$

and

$$E(dK_u \mid \mathcal{F}_u) = M_u \alpha_u$$

then $E(d\mathcal{M}_u \mid \mathcal{F}_u) = 0$. Integrating $d\mathcal{M}_u$ from 0 to t, it follows that $\mathcal{M}_t$ is a ZMM [8]. For any predictable process W_u, we obtain the following ZMM:

$$\mathcal{M}_t^* = \int_0^t W_u \, d\mathcal{M}_u$$
$$= \int_0^t W_u\{M_u \, dM_u - (\nu - M_u) \, dK_u\} \quad (27.1)$$

A class of estimators of ν, evaluated at time τ, can be obtained by solving $\mathcal{M}_\tau^* = 0$, *i.e.*

$$\hat{\nu}_W(\tau) = \frac{\int_0^\tau W_u M_u \, dN_u}{\int_0^\tau W_u \, dK_u}$$

The variance of $\mathcal{M}_\tau^*$ is given by

$$\mathrm{Var}(\mathcal{M}_\tau^*) = E\left(\int_0^\tau W_u^2 \{M_u^2 \, dM_u + (\nu - M_u)^2 \, dK_u\} \right)$$

The term for variance follows from the standard results given by, for example, Andersen and Gill [16] or Yip [5]. The covariance is zero by virtue of the orthogonality of the result of the martingale, since dM_u and dK_u cannot jump simultaneously.

Using the result

$$\mathcal{M}_\tau^* / \sqrt{\mathrm{Var}(\mathcal{M}_\tau^*)} \longrightarrow N(0, 1)$$

so that $\mathcal{M}_\tau^* = (\hat{\nu}_W(\tau) - \nu) \int_0^\tau W_u \, dK_u$, a variance for the standard error of $\hat{\nu}_W(\tau)$ is given by

$$\widehat{\mathrm{se}}(\hat{\nu}_W(\tau)) = \frac{\left[\int_0^\tau W_u^2 \{M_u^2 \, dM_u + (\hat{\nu}_W(\tau) - M_u)^2 \, dK_u\} \right]^{1/2}}{\int_0^\tau W_u \, dK_u}$$

Here we consider two types of weight functions: $W_u = 1$ and the optimal weight suggested by Godambe [17], *i.e.*

$$W_u^* = \frac{E\left[\frac{\partial d\mathcal{M}_u}{\partial \nu} \,\middle|\, \mathcal{F}_u \right]}{E[d\mathcal{M}_u^2 \mid \mathcal{F}_u]} \quad (27.2)$$

The optimal property is in terms of giving the tightest asymptotic confidence intervals for the estimate. Note that ν is an integer value, so instead of taking the derivative w.r.t. ν, we compute the first difference of $d\mathcal{M}_t$.

If $W_u = 1$ we have

$$\hat{\nu}_1 = \frac{\int_0^\tau M_u \, dN_u}{K_\tau}$$

which is the Schnabel estimator [8, 18]. In the event that $K_\tau = 0$, we define $\hat{\nu}(\tau) = \infty$, $\widehat{\mathrm{se}}(\hat{\nu}(\tau)) = \infty$, and $\widehat{\mathrm{se}}(\hat{\nu}(\tau))/\hat{\nu}(\tau) = \infty$. The optimal weight suggested in Equation 27.2 can readily be shown to be

$$W_u^* = \frac{1}{(\nu - M_u)}$$

Hence, the optimal estimator $\hat{\nu}^*$ is the solution of the following equation:

$$\int_0^\tau \frac{1}{\nu - M_u} [M_u \, dM_u - (\nu - M_u) \, dK_u] = 0 \quad (27.3)$$

which is equivalent to the MLE [8,19]. An estimate of the standard error of the estimator $\hat{\nu}^*$ is given by

$$\widehat{\mathrm{se}}(\hat{\nu}^*) = \left(\hat{\nu}^* \Big/ \int_0^\tau \frac{M_u \, dM_u}{(\hat{\nu}^* - M_u)^2} \right)^{1/2}$$

The assumption of homogeneous intensity is not used in deriving the estimator from the martingale given in Equation 27.1. The intensity function may depend on time, t. It follows that the estimator is robust to time heterogeneity. For details, see Yip *et al.* [8].

27.2.2 A Seeded Fault Approach Without Recapture

A known number of faults D are seeded in the system at the beginning of the testing process [5]. The failure intensities of the seeded and real faults are allowed to be time dependent. The failure intensity of the seeded faults is assumed to differ from that of the real faults by a constant proportion, θ, assumed known. The same failure intensity is assumed to be applied to each type of fault in the system. Let U_t and M_t denote the number of real and seeded faults detected in the system in $[0, t]$, assuming once a fault is detected it is immediately removed from the system. No recapture information is available from detected faults (seeded or real). In the presence of the unknown parameter α_t, an identifiability problem occurs when we want to estimate ν by observing the failure times only, the information of ν and α_t is confounded. The extra effort of inserting a known number D of faults into the system at the beginning of the testing provides an extra equation to estimate ν. However, an identifiability problem can still occur with the extra unknown parameter β_t, which is the failure intensity of the seeded faults at time t. Here we assume a constant proportionality between the two intensities α_t and β_t, *i.e.* $\alpha_t = \theta\beta_t$ for all t and θ assumed known. Yip *et al.* [20] have examined the sensitivity of a misspecification of θ and the estimation performance with an unknown θ. If θ is unknown, similar estimating functions can be obtained for making inference on θ and ν.

Here, we define a "martingale difference" with the intent to estimate ν:

$$d\mathcal{R}_u = (D - M_u)\,dU_u - \theta(\nu - U_u)\,dM_u.$$

Note that $E[d\mathcal{R}_u \mid \mathcal{F}_u] = 0$. Let W_u be a measurable function w.r.t. $\mathcal{F}_u$. It follows that the process

$$\mathcal{R}^* = \{\mathcal{R}^*_t;\ t \geq 0\},$$

$$\mathcal{R}^*_t = \int_0^t W_u\{(D - M_u)\,dU_u - \theta(\nu - U_u)\,dM_u\} \tag{27.4}$$

is a ZMM. By equating Equation 27.4 to zero and evaluating at time τ, a class of estimators for ν is obtained, *i.e.*

$$\hat{\nu}_w = \left[\int_0^\tau W_u\{(D - M_u)\,dU_u + \theta U_u\,dM_u\}\right] \times \left[\theta \int_0^\tau W_u\,dM_u\right]^{-1} \tag{27.5}$$

which depends on the choice of W_u. The conditional variance of $\mathcal{R}^*_\tau$ in Equation 27.4 is given by

$$\mathrm{Var}(\mathcal{R}^*_\tau) = E\left[\int_0^\tau W_u^2\{(D - M_u)^2\,dU_u + \theta^2(\nu - U_u)^2\,dM_u\}\right]$$

Hence we can deduce a standard error of $\hat{\nu}_w$ for the seeded faults approach, given by

$$\left\{\left[\int_0^\tau W_u^2\{(D - M_u)^2\,dU_u + \theta^2(\hat{\nu}_w - U_u)^2\,dM_u\}\right]^{1/2}\right\} \times \left\{\theta\int_0^\tau W_u\,dM_u\right\}^{-1} \tag{27.6}$$

For the choice $W_u = 1$, we have explicit expressions for the estimate $\tilde{\nu}$ and its standard error, *i.e.*

$$\tilde{\nu} = \frac{\int_0^\tau\{(D - M_u)\,dU_u + \theta U_u\,dM_u\}}{\theta M_\tau} \tag{27.7}$$

and

$$\widehat{\mathrm{se}}(\tilde{\nu}) = \left\{\left[\int_0^\tau\{(D - M_u)^2\,dU_u + \theta^2(\hat{\nu}_1 - U_u)^2\,dM_u\}\right]^{1/2}\right\} \times \{\theta M_\tau\}^{-1} \tag{27.8}$$

Also, the optimal weight corresponding to Equation 27.4 is

$$W_u^* = \frac{1}{(\nu - U_u)[(D - M_u) + (\nu - U_u)\theta]}$$

Accordingly an optimal estimating equation in the sense of giving the tightest confidence for the estimator of ν is given by

$$\mathcal{R}_t^* = \int_0^t \frac{1}{(\nu - U_u)[(D - M_u) + (\nu - U_u)\theta]} \times \{(D - M_u)\,\mathrm{d}U_u - \theta(\nu - U_u)\,\mathrm{d}M_u\} \quad (27.9)$$

The optimal estimate $\tilde{\nu}^*$ is the solution of Equation 27.9. An explicit expression is not available and an iterative procedure is required. From Equation 27.6 we have an estimate of the standard error of $\tilde{\nu}^*$:

$$\widehat{\mathrm{se}}(\tilde{\nu}^*) = \left\{\int_0^\tau \frac{\{(D - M_u)^2\,\mathrm{d}U_u + \theta^2(\tilde{\nu}^* - U_u)^2\,\mathrm{d}M_u\}}{\{(\tilde{\nu}^* - U_u)[(D - M_u) + (\tilde{\nu}^* - U_u)\theta]\}^2}\right\} \times \left\{\theta\int_0^\tau \frac{\mathrm{d}M_u}{(\tilde{\nu}^* - U_u)[(D - M_u) + (\tilde{\nu}^* - U_u)\theta]}\right\}^{-1}$$

27.2.3 Heterogeneous Model

Here we introduce a sample coverage estimator and a two-step procedure to estimate ν for a heterogeneous model. The sample coverage method was initiated by Chao and Lee [21] and the two-step procedure was suggested by Huggins [22, 23] and Yip *et al.* [24]. The individual effects can be assumed to be specified or unspecified (parametrically or non-parametrically).

27.2.3.1 Non-parametric Case: $\lambda_i(t) = \gamma_i \alpha_t$

The individual effects $\{\gamma_1, \gamma_2, \ldots, \gamma_\nu\}$ can be modeled as either a fixed effect or a random effect. For a fixed-effect model, $\{\gamma_1, \gamma_2, \ldots, \gamma_\nu\}$ are regarded as fixed (nuisance) parameters. Instead of trying to estimate each of the failure intensities, the important parameters are the mean, $\bar{\gamma} = \sum_{i=1}^{\nu} \gamma_i/\nu$ and coefficient of variation (CV), $\gamma = \left[\sum(\gamma_i - \bar{\gamma})^2/\nu\right]^{1/2}/\bar{\gamma}$ in the sample coverage approach. The value of CV is a measure of the heterogeneity of the individual effects. CV = 0 is equivalent to all the γ_i values being equal. A larger CV implies the population is more heterogeneous in individual effects. For a random-effects model we assume $\{\gamma_1, \gamma_2, \ldots, \gamma_\nu\}$ are a random sample from an unspecified and unknown distribution F with mean $\bar{\gamma} = \int u\,\mathrm{d}F(u)$ and CV $\gamma = \int (u - \bar{\gamma})^2\,\mathrm{d}F(u)/\bar{\gamma}^2$. Both models lead to exactly the same estimator of population size using the sample coverage idea [25, 26]. The derivation procedure for the random-effects model is nearly parallel to that for the fixed-effects model. Thus in the following we only illustrate the approach by using the fixed-effects model.

Let us define the sample coverage, C_t, as the proportion of total individual fault effects of detected faults, given by

$$C_t = \left[\sum_{i=1}^{\nu} \gamma_i I(\text{the } i\text{th fault is detected at least once up to } t)\right]\left[\sum_{i=1}^{\nu} \gamma_i\right]^{-1}$$

where I is an indicator function. Let M_t and K_t be the numbers of marked faults in the system and the number of re-detections of those faults at time t, respectively, and let $A_i(t)$ be the event that the ith fault has not been detected up to t but detected at time $t + \Delta_{\mathrm{d}t}$. We have

$$E[\mathrm{d}M_t \mid \mathcal{F}_t] = E\left[\sum_{i=1}^{\nu} I\{A_i(t)\} \mid \mathcal{F}_t\right] = \sum_{i=1}^{\nu} \gamma_i \alpha_t\,\mathrm{d}t\,I(i\text{th fault is not detected up to } t)$$

It follows from the definition of C_t that

$$E[\mathrm{d}M_t \mid \mathcal{F}_t] = (1 - C_t)\left(\sum_{i=1}^{\nu} \gamma_i\right)\alpha_t\,\mathrm{d}t$$

and

$$E[\mathrm{d}K_t \mid \mathcal{F}_t] = (C_t)\left(\sum_{i=1}^{\nu} \gamma_i\right)\alpha_t\,\mathrm{d}t$$

for $t > 0$.

Consider the following ZMMs:

$$\begin{aligned}\mathcal{M}_t &= \int_0^t \Bigg\{(\nu C_u)\Bigg[\mathrm{d}M_u - (1 - C_u) \\ &\quad \times \left(\sum_{i=1}^{\nu} \gamma_i\right)\alpha_u \,\mathrm{d}u\Bigg] \\ &\quad - \nu(1 - C_u)\Bigg[\mathrm{d}K_u - (C_u) \\ &\quad \times \left(\sum_{i=1}^{\nu} \gamma_i\right)\alpha_u \,\mathrm{d}u\Bigg]\Bigg\} \\ &= \int_0^t \{(\nu C_u)\,\mathrm{d}M_u - (\nu - \nu C_u)\,\mathrm{d}K_u\}\end{aligned}$$

where $E[\mathcal{M}_t] = 0\ \forall t$. By eliminating the unknown parameters $\sum_{i=1}^{\nu} \gamma_i$ and α_u in $\mathcal{M}_t$ we can make inference about ν in the presence of the nuisance parameter νC_u which can be estimated accordingly. By integrating a predictable function W_u w.r.t. $\mathcal{M} = \{\mathcal{M}_t;\ t \geq 0\}$ we have another martingale, *i.e.*

$$\mathcal{M}_t^* = \int_0^t W_u\{(\nu C_u)\,\mathrm{d}M_u - (\nu - \nu C_u)\,\mathrm{d}K_u\} \tag{27.10}$$

where $\mathcal{M}_t^*$ involves the parameters ν and νC_u. Similarly, we consider two types of weight functions: $W_u = 1$ and the optimal weight suggested in Equation 27.2.

Suppose all the γ_i values are the same so we have $\nu C_u = M_u$ and $(\nu - \nu C_u) = \nu - M_u$ for the homogeneous case as in Section 27.2.1; the estimating function Equation 27.10 reduces to

$$\mathcal{M}_t^* = \int_0^t W_u[M_u\,\mathrm{d}M_u - (\nu - M_u)\,\mathrm{d}K_u] \tag{27.11}$$

which is equivalent to Equation 27.1. If we can find a "predictor" $\widehat{\nu C_u}$ for νC_u, we have the following estimator for ν:

$$\hat{\nu}_h = \frac{\int_0^t W_u(\widehat{\nu C_u})\,\mathrm{d}N_u}{\int_0^t W_u\,\mathrm{d}K_u} \tag{27.12}$$

The optimal weight can be shown to be

$$W_u^* = \frac{1}{1 - C_u}$$

Also, this weight reduces to the optimal weight used in the homogeneous case where all the γ values are the same. Here we suggest a procedure to find an estimator for $E[\nu C_u]$ and subsequently use it to replace νC_u in the estimating function. Note that

$$\begin{aligned}E[\nu C_u] &= \frac{1}{\bar{\gamma}} \sum_{i=1}^{\nu} \gamma_i E[\text{the } i\text{th fault is detected} \\ &\qquad\qquad \text{at least once up to } u] \\ &= \frac{1}{\bar{\gamma}} \sum_{i=1}^{\nu} \gamma_i[1 - \mathrm{e}^{-\gamma_i \Lambda}]\end{aligned}$$

where $\Lambda = \int_0^u \alpha_s\,\mathrm{d}s$ and we also have

$$E[M_u] = \sum_{i=1}^{\nu}[1 - \mathrm{e}^{-\gamma_i \Lambda}]$$

It then follows that

$$E[\nu C_u] = E(M_u) + G(\gamma_1, \ldots, \gamma_\nu)$$

where

$$G(\gamma_1, \ldots, \gamma_\nu) = \sum_{i=1}^{\nu}(1 - \gamma_i/\bar{\gamma})\exp(-\gamma_i \Lambda).$$

Define $\gamma_i^* = \bar{\gamma} + \theta(\gamma_i - \bar{\gamma})$, for some $0 < \theta < 1$. Expanding $G(\gamma_1, \ldots, \gamma_\nu)$ at $(\bar{\gamma}, \ldots, \bar{\gamma})$ we have

$$\begin{aligned}E[\nu C_u] &= E[M_u] + \gamma^2 \nu \Lambda \bar{\gamma}\,\mathrm{e}^{-\Lambda\bar{\gamma}} \\ &\quad - \frac{1}{2}\sum_{i=1}^{\nu} \Lambda^2\,\mathrm{e}^{-\gamma_i^* \Lambda}\frac{(\gamma_i - \bar{\gamma})^3}{\bar{\gamma}} \\ &= E[M_u] + \gamma^2 \sum_{i=1}^{\nu}(\Lambda\gamma_i)\,\mathrm{e}^{-\Lambda\gamma_i} \\ &\quad + \gamma^2 \Lambda \sum_{i=1}^{\nu}(\bar{\gamma}\,\mathrm{e}^{-\Lambda\bar{\gamma}} - \gamma_i\,\mathrm{e}^{-\Lambda\gamma_i}) \\ &\quad - \frac{1}{2}\sum_{i=1}^{\nu} \Lambda^2\,\mathrm{e}^{-\gamma_i^* \Lambda}\frac{(\gamma_i - \bar{\gamma})^3}{\bar{\gamma}} \\ &= E[M_u] + \gamma^2 E[f_1(u)] + R\end{aligned}$$

where $f_i(u)$ is the number of individuals that have been caught exactly i times in the sample by time u, and

$$R = \gamma^2 u \sum_{i=1}^{\nu} (\bar{\gamma}\, \mathrm{e}^{-\Lambda\bar{\gamma}} - \gamma_i\, \mathrm{e}^{-\Lambda\gamma_i}) - \frac{1}{2}\sum_{i=1}^{\nu} \Lambda^2\, \mathrm{e}^{-\gamma_i^{*}\Lambda} \frac{(\gamma_i - \bar{\gamma})^3}{\bar{\gamma}}$$

The remainder term, R, is usually negligible, as shown by Chao and Lee [25] when the variation of the γ_i values is not relatively large, say, $\gamma < 1$. One theoretical justification for the approximation is that when $\gamma_1, \gamma_2, \ldots, \gamma_\nu$ are a random sample from a gamma distribution, the remainder term R satisfies $R/\nu \to 0$ as $\nu \to \infty$. Further justifications can be found in Lin *et al.* [27]. The estimation of νC_u will then be based on the approximation ignoring R in Chao and Lee [21]. Thus we have the following estimator for νC_u:

$$\widehat{\nu C_u} = M_u + \hat{\gamma}^2 f_1(u)$$

where $\hat{\gamma}^2$ is an estimator of CV given by Chao and Lee [25]:

$$\hat{\gamma}^2 = \max\left\{\frac{M_t}{\hat{C}_t}\frac{\sum i(i-1) f_i(t)}{[\sum i f_i(t)]^2} - 1, 0\right\} \quad (27.13)$$

where

$$\hat{C}_t = 1 - \frac{f_1(t)}{\sum_i i f_i(t)}$$

An estimator of ν based on the sample coverage estimate in Chao and Lee [25] is

$$\hat{\nu}_2 = \frac{M_t}{\hat{C}_t} + \frac{f_1(t)}{\hat{C}_t}\hat{\gamma}^2 \quad (27.14)$$

Substituting $\widehat{\nu C_u}$ into Equation 27.12, we obtain the following two new estimators:

$$\hat{\nu}_3 = \frac{\int_0^t \{M_u + \hat{\gamma}^2 f_1(u)\}\, \mathrm{d}N_u}{K_t} \quad \text{for } W_u = 1$$

$$\hat{\nu}_4 = \left[\int_0^t \frac{1}{(1 - \hat{C}_u)} \{M_u + \hat{\gamma}^2 f_1(u)\}\, \mathrm{d}N_u\right] \times \left[\int_0^t \frac{1}{(1 - \hat{C}_u)}\, \mathrm{d}K_u\right]^{-1} \quad \text{for } W_u = \frac{1}{1 - \hat{C}_u}$$

The estimation of γ^2 is based on all data at the end of the experiment. However, we need to estimate C_u and νC_u sequentially for $0 < u \le \tau$. It is clear that the CV value plays an important role in the estimation procedure.

An estimate of the standard error for $\hat{\nu}_3$ and $\hat{\nu}_4$ is given by:

$$\widehat{\mathrm{se}}(\hat{\nu}_h) = \frac{\sqrt{\widehat{\mathrm{Var}}(\mathcal{M}_t^*)}}{\int_0^t W_u\, \mathrm{d}K_u} \quad (27.15)$$

where $\widehat{\mathrm{Var}}(\mathcal{M}_t^*) = \int_0^t (W_u)^2 \{(\widehat{\nu C_u})^2\, \mathrm{d}M_u + (\hat{\nu}_h - \widehat{\nu C_u})^2\, \mathrm{d}K_u\}$. Simulation results show that $\widehat{\mathrm{se}}(\hat{\nu}_3)$ and $\widehat{\mathrm{se}}(\hat{\nu}_4)$ severely underestimate the variability of $\hat{\nu}_3$ and $\hat{\nu}_4$. The expression in Equation 27.15 has not taken into account the additional variation introduced by substituting the estimates for the parameters such as νC_u and ν in the estimating function. The resulting expression for the estimate of $\widehat{\mathrm{se}}(\hat{\nu}_h)$ would be complicated if the variability of these estimates were taken into account. An alternative procedure by bootstrap is used to assess the variability instead [26].

27.2.3.2 Parametric Case: $\lambda_i(t) = \gamma_i$

Let $N_i(t)$ be the counting process for fault i and γ_i be its occurrence rate as a stochastic variable, *i.e.* the intensity process for the counting process depends on an unobservable random variable. Let $Y_i(t)$ indicate, by the value of 1 or 0, whether fault i has not been removed before t, thus $Y_i(t)$ is an observable non-negative predictable process. Let τ be the duration of the study and $\mathcal{F}_t$ denote the history generated by $\{N_i(s), Y_i(s), 0 \le s \le t\}$.

Suppose that γ_i has a gamma distribution with parameters ϕ and β, denoted by $G(\phi, \beta)$ with density function

$$f(\gamma_i \mid \phi, \beta) = \frac{\mathrm{e}^{-\beta\gamma_i} \gamma_i^{\phi-1} \beta^{\phi}}{\Gamma(\phi)} \quad (\gamma_i > 0)$$

Using the innovation theorem [15], the compensator of the counting process $N_i(t)$ is given by

$$\lambda_i(t) = Y_i(t)\left\{\frac{\phi + N_i(t)}{\beta + t}\right\} \quad (27.16)$$

The intensity function $\lambda_i(t)$ depends on time and the number of times the ith individual fault has been detected. The formulation in Equation 27.16 includes both the removal and recapture sampling methods. For the case of removal, in which $N_i(t)$ equals zero since the ith fault is removed from the system after being detected, the model becomes the *Littlewood model* [2, 6]. In addition, Equation 27.16 also includes the recapture experiment in which a counter is inserted at the location after the fault is detected and the counter registers the number of re-detections of a particular fault without causing the system failure [3]. The re-detection information has been shown to be important in determining the performance when estimating the number of faults in a system [4]. If we reparameterize the intensity in Equation 27.16 by letting

$$\epsilon = \frac{1}{\phi}, \qquad \omega = \frac{\beta}{\phi}$$

then we have

$$\lambda_i(t) = Y_i(t)\left\{\frac{1+\epsilon N_i(t-)}{\omega + \epsilon t}\right\} \tag{27.17}$$

This extension allows the case of $\epsilon \leq 0$ to have a meaningful interpretation [28]. Note that when $\epsilon = 0$ (*i.e.* $\phi = \infty$) for a removal experiment, the model reduces to the Jelinski–Moranda model in software reliability studies [29].

The full likelihood function is given by

$$L(\theta) = \frac{\nu!}{(\nu - n)!} \prod_{0\leq t\leq \tau} \left[\prod_{i=1}^{\nu} \lambda_i(t)^{\mathrm{d}N_i(t)} \times \{1 - \lambda_i(t)\,\mathrm{d}t\}^{1-\mathrm{d}N_i(t)}\right] \tag{27.18}$$

see Andersen *et al.* [2], p.402, which reduces to

$$L(\theta) = \frac{\nu!}{(\nu - n)!} \prod_{i=1}^{\nu} \left[\left\{\prod_{0\leq t\leq \tau} \lambda_i(t)^{\mathrm{d}N_i(t)}\right\} \times \exp\left\{-\int_0^{\tau} \lambda_i(t)\,\mathrm{d}t\right\}\right] \tag{27.19}$$

where n denotes the number of distinct faults detected in the experiment.

For the removal experiment, Littlewood [6] suggested using the maximum likelihood (ML) to estimate the parameters ω, ϵ (or ϕ, β) and ν. However, due to the complexity of the three highly nonlinear likelihood equations, it is difficult if not impossible to determine consistent estimates out of the possible multiple solutions of the likelihood equations. A similar problem exists for alternative estimation methods such as the M-estimator. An estimating function is used to estimate ϕ, β and ν which can be found by solving the equation

$$\sum_{i=1}^{\nu} \int_0^{\tau} k_j(t)\left[\mathrm{d}N_i(t) - Y_i(t)\left\{\frac{\phi + N_i(t)}{\beta + t}\right\}\mathrm{d}t\right] \tag{27.20}$$

for $j = 1, 2, 3$ with weight functions $k_j(t) = (\beta + t)t^{j-1}$, see Andersen *et al.* [2]. Accordingly, we obtained the linear system of equations for determining $\hat{\phi}$, $\hat{\beta}$ and $\hat{\nu}$. For the special case $N_i(t) = 0$ for all t, Andersen *et al.* [2] provided an estimate for ϕ, β and ν for the data in Table 27.1. However, the *ad hoc* approach does not provide stable estimates in the simulation. Also, for the recapture case, the parameter ν cannot be separated from the estimating Equation 27.20, and so we are unable to solve for $\hat{\nu}$. Furthermore, the optimal weights involve the unknown parameters in a nonlinear form, which makes Equation 27.20 intractable.

Here an alternative two-step estimation procedure is suggested. Firstly, with the specified form of intensity in Equation 27.16 or 27.17, we are able to compute the conditional likelihood of ω and ϵ (or ϕ and β) using the observed failure information. The likelihood equation in the first stage reduces to a lower dimension, so that it is easier to obtain the unique consistent solutions. The second stage employs a Horvitz–Thompson [9] type estimator which is the minimum variance unbiased estimator if the failure intensity is known. For the removal experiment, the conditional MLE are asymptotically equivalent to the unconditional MLE suggested by Littlewood [6].

Let $p_i = \Pr(\delta_i = 1)$ denote the probability of the ith fault being detected during the course of the experiment. These probabilities are the same for all faults under the model assumption, and with the application of the Laplace transform it can be shown that

$$p_i = p(\omega, \epsilon) = 1 - \left(\frac{\omega}{\omega + \epsilon\tau}\right)^{1/\epsilon}$$

The likelihood function for the given data can be rewritten as:

$$L(\theta) = L_1 \times L_2$$

where

$$L_1 = \prod_{i=1}^{\nu} \left\{ \frac{\prod_{0 \le t \le \tau} \lambda_i(t)^{dN_i(t)} \exp(-\int_0^\tau \lambda_i(t)\, dt)}{p_i} \right\}^{\delta_i} \tag{27.21}$$

$$L_2 = \frac{\nu!}{(\nu - n)!} \prod_{i=1}^{\nu} \left\{ \exp\left[-\int_0^\tau \lambda_i(t)\, dt\right]^{1-\delta_i} p_i^{\delta_i} \right\} \tag{27.22}$$

where δ_i indicates, by the value of 1 versus 0, whether or not the ith fault has ever been detected during the experiment. Since the marginal likelihood L_2 depends on the unknown parameter ν, we propose to make inference about $\theta = (\omega, \epsilon)'$ based on the conditional likelihood L_1 which does not depend on ν. The corresponding score function of the conditional likelihood L_1 is given by

$$\begin{aligned} U(\theta) &= \partial \log(L_1)/\partial\theta \\ &= \sum_{i=1}^{\nu} \delta_i \left\{ \int_0^\tau \frac{\partial \log \lambda_i(t)}{\partial\theta}\, dN_i(t) \right. \\ &\quad \left. - \int_0^\tau Y_i(t) \frac{\partial \lambda_i(t)}{\partial\theta}\, dt - \frac{\partial \log(p_i)}{\partial\theta} \right\} \end{aligned}$$

where

$$\lambda_i(t) = Y_i(t) \left\{ \frac{1 + \epsilon N_i(t-)}{\omega + \epsilon t} \right\}$$

Let $\hat{\theta} = (\hat{\omega}, \hat{\epsilon})'$ be the solution to $U(\theta) = 0$. The usual arguments ensure the consistency and $\nu^{1/2}(\hat{\theta} - \theta)$ converges to $N(0, \Omega^{-1}(\theta))$, where $\Omega(\theta) = \lim_{\nu\to\infty} \nu^{-1} I(\theta)$, $I(\theta) = -\partial U(\theta)/\partial\theta'$. The variance of the estimator $\hat{\theta}$ can thus be estimated by the negative of the first derivative of the score function, $I(\hat{\theta})$.

Since the probability of being detected is $p(\theta)$, it is natural to estimate the population size ν by the Horvitz–Thompson estimator [9], *i.e.*

$$\hat{\nu} = \sum_{i=1}^{\nu} \frac{\delta_i}{p(\hat{\theta})} = \frac{n}{p(\hat{\theta})} \tag{27.23}$$

where n denotes the number of distinct faults detected. To compute the variance of $\hat{\nu}$, we have

$$\hat{\nu}(\theta) - \nu = \sum_{i=1}^{\nu} \left[\frac{\delta_i}{p} - 1\right]$$

By a Taylor series expansion and some simple probabilistic arguments:

$$\nu^{-1/2}\{\hat{\nu}(\hat{\theta}) - \nu\} = \nu^{-1/2} \sum_{i=1}^{\nu} \left[\frac{\delta_i}{p} - 1\right] + H'(\theta)\nu^{1/2}(\hat{\theta} - \theta) + o_p(1) \tag{27.24}$$

where

$$H(\theta) = E\{\hat{H}(\theta)\}$$

$$\hat{H}(\theta) = \nu^{-1} \frac{\partial \hat{\nu}(\theta)}{\partial\theta} = \nu^{-1} \left\{ n \frac{\partial p/\partial\theta}{p^2} \right\}$$

It follows from the multivariate central limit theorem and the Cramér–Wold device that $\nu^{-1/2}\{\hat{\nu}(\hat{\theta}) - \nu\}$ converges in distribution to a zero-mean normal random variable. The variance for the first term on the right-hand side of Equation 27.24 is:

$$\nu^{-1} \sum_{i=1}^{\nu} \frac{\mathrm{Var}(\delta_i)}{p^2} = \nu^{-1} \sum_{i=1}^{\nu} \frac{p(1-p)}{p^2}$$

which can be consistently estimated by

$$\nu^{-1} \sum_{i=1}^{\nu} \frac{\delta_i(1-p)}{p^2} = \nu^{-1} \frac{n(1-p)}{p^2}$$

The variance for the second term is $H'(\theta)\Omega^{-1}(\theta)H(\theta)$. The covariance between the two terms is zero. A consistent variance estimator for $\nu^{-1/2}\{\hat{\nu}(\hat{\theta}) - \nu\}$ is then given by:

$$\hat{s}^2 = \hat{\nu}^{-1} \frac{n(1-\hat{p})}{\hat{p}^2} + \hat{H}'(\hat{\theta})\hat{\nu} I^{-1}(\hat{\theta})\hat{H}(\hat{\theta}) \tag{27.25}$$

where

$$\hat{H}(\theta) = \nu^{-1} \frac{\partial \hat{\nu}}{\partial\theta} \quad \text{and} \quad \hat{p} = \left(1 - \frac{\hat{\omega}}{\hat{\omega} + \hat{\epsilon}\tau}\right)^{1/\hat{\epsilon}}$$

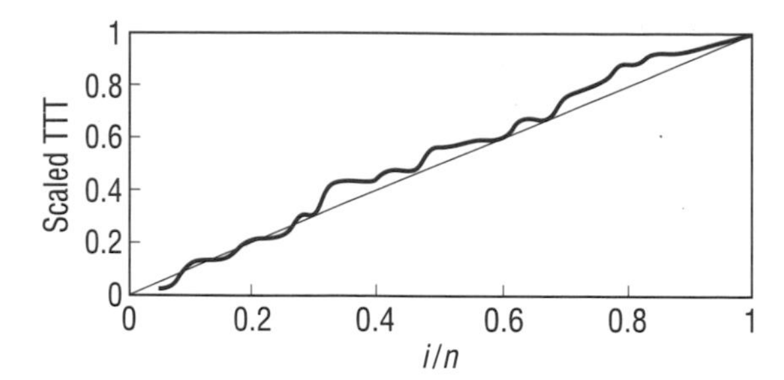

Figure 27.1. Scaled TTT plot of the failure data set from [1]

27.3 A Sequential Procedure

Here a sequential procedure is suggested to estimate the number of faults in a system and to examine the efficiency of such a procedure. A procedure of this type is useful in the preliminary stages of testing. The testing time needs to be long enough to allow sufficient information to be gathered to obtain an accurate estimate, but at the same time we do not want to waste testing effort as testing can be quite expensive.

In practice it is expensive and generally infeasible to detect all the faults in the system in the initial stages of software development. The idea of the sequential procedure proposed here is not to find all the faults in the system. The goal is to detect designs of questionable quality earlier in the process by estimating the number of faults that go undetected after a design review. Here we consider as accuracy criterion the ratio of the standard error of the estimate to the population size estimate, sometimes called the coefficient of variation. For a given accuracy level, d_0, the testing experiment stops at time τ_0 when the accuracy condition is met. Thus

$$\tau_0 = \inf\left[t : \frac{\widehat{\text{se}}\{\hat{\nu}(t)\}}{\hat{\nu}(t)} < d_0\right] \tag{27.26}$$

The criterion is applicable to any other estimators, so long as we can compute the ratio $\widehat{\text{se}}\{\hat{\nu}(t)\}/\hat{\nu}(t)$ at any time $t \in (0, \tau)$.

27.4 Real Examples

The proposed methods are applied to the aircraft movements data from Moek [1], as given in Table 27.1. The data comprises 43 occurrence times (in seconds CPU time) of first failures caused by 43 distinct faults of an information system for registering aircraft movements. Data are shown in the second column of Table 27.1. Figure 27.1 gives the total time on test (TTT) plot, which suggests that the failure times are identically exponentially distributed. The maximum likelihood estimate for the failure intensity $\hat{\lambda}$ is estimated to be 5.4×10^{-6}. Using the M-estimates suggested by Andersen *et al.* [2], we obtain $\hat{\omega} = 0.196(0.007)$, $\hat{\epsilon} = 0.089(0.102)$, and $\hat{\nu} = 46.4(10.14)$ based on the 43 failures reported. Here we illustrate how the recapture improves the estimation. For a recapture experiment, each fault is detected, corrected, and a counter inserted to record the number of redetections of a particular fault. In order to use the data set for a recapture experiment, we generated interfailure times for each of the detected faults based on the maximum likelihood estimate of the failure intensity $\hat{\gamma} = 5.4 \times 10^{6}$ [2] from an exponential distribution. The third column of Table 27.1 contains the simulated failure times after the first detection time. The time for the end of the experiment is 576 570 seconds in CPU time. The estimated number of total faults in the system using Equation 27.3 based on the recapture data is $\hat{\nu} = 47.4(2.5)$. The two-step procedure in Section 27.2.3.2 gives the estimate $\hat{\nu} = 45.2(1.6)$. Chao's coverage estimate in Equation 27.14 gives $\hat{\nu} = 45.9(2.4)$. The numbers inside the brackets represent standard errors. The recapture information improves the estimation. The two-step procedure and the sample coverage are almost the same.

Table 27.2 shows the results of applying the proposed sequential procedure in Section 27.3 when estimating ν for the data in Table 27.1. As τ increases the standard error of the estimate decreases with some fluctuation. The improvement of d_0 from 0.33 to 0.32 requires doubling the testing time. The distinct number of faults detected increased from 13 to 26. Also,

Table 27.2. Results of applying the sequential procedure for the Moek's data in Table 27.1

d_0	τ_0	N_{τ_0}	M_{τ_0}	$\hat{\nu}$	$\widehat{\text{se}}(\hat{\nu})$
0.05	576 570	126	43	47.40	2.54
0.06	520 016	114	42	47.56	2.82
0.07	459 440	101	41	48.05	3.33
0.08	428 902	93	41	49.13	3.91
0.09	374 032	83	39	48.95	4.38
0.10	368 811	79	39	49.95	4.89
0.11	353 576	73	38	50.40	5.48
0.12	329 685	67	36	49.58	5.80
0.13	295 030	59	33	48.27	6.21
0.14	294 708	57	33	49.54	6.85
0.15	288 750	55	33	51.05	7.62
0.16	266 322	54	33	51.90	8.07
0.17	252 204	51	32	52.16	8.68
0.18	232 743	48	31	52.65	9.45
0.19	228 283	47	31	54.00	10.21
0.20	222 288	45	30	53.47	10.47
0.21	215 589	44	30	55.14	11.44
0.22	205 224	42	29	54.77	11.87
0.23	205 224	42	29	54.77	11.87
0.24	195 476	41	29	56.92	13.16
0.25	194 399	40	29	59.45	14.74
0.26	194 399	40	29	59.45	14.74
0.27	186 946	39	29	62.50	16.71
0.28	167 108	35	26	56.89	15.81
0.29	167 108	35	26	56.89	15.81
0.30	167 108	35	26	56.89	15.81
0.31	162 480	34	26	60.75	18.49
0.32	162 480	34	26	60.75	18.49
0.33	83 996	18	13	30.80	9.96
0.34	83 996	18	13	30.80	9.96
0.35	83 996	18	13	30.80	9.96
0.36	83 996	18	13	30.80	9.96
0.37	83 996	18	13	30.80	9.96
0.38	83 996	18	13	30.80	9.96
0.39	75 630	17	13	35.25	13.44
0.40	75 630	17	13	35.25	13.44

Notes: N_{τ_0} = total number of detections by time τ_0.
M_{τ_0} = distinct number of faults detected by time τ_0.

the estimates during the period [83 996, 162 480] produce a larger estimate for ν with a larger standard error as well, and the ratio d_0 remains rather stable for the period. The improvement in standard error of the estimate is from 4.89 to 2.54 for the time 368 811 and 576 570, respectively. Also, the d_0 achieved at the end of the experiment is 0.054. If the desired accuracy level is 0.1, the testing experiment could stop at 368 811 with the estimate 49.9 and an estimated standard error 4.9. The information at that time is not much different from the estimate obtained at 576 570. We could save nearly one third of the testing time.

27.5 Simulation Studies

A number of Monte Carlo simulations were performed to evaluate the performance of the proposed estimators. Table 27.3 gives the simulation results for the seeded faults approach using the optimal estimator $\tilde{\nu}^*$ from Equation 27.9. Different values of the proportionality constant have been used. The capture proportions are assumed to be 0.5 and 0.9 respectively. Here we assume that the stopping time is determined by the removed proportion of the seeded faults. It is of interest to investigate the effects of θ, the stopping time, and the proportion of seeded faults placed in the system on the performance of the estimators. Table 27.3 lists all the different sets of parameter values used in the simulation study. One thousand repetitions were completed for each trial. The statistics computed were: average population size, ave($\hat{\nu}$); standard deviation of the 1000 observations, sd($\hat{\nu}$); average standard error, ave($\widehat{\text{se}}(\hat{\nu})$); and the coverage (cov), the proportion of the estimate falling between the 95% confidence intervals.

As the removed proportion of seeded faults increased, the standard error of the estimates reduced. The coverage is satisfactory. Also, a higher proportion of the seeded faults in the system also improves the estimation performance. As the value of θ increases from 0.5 to 1.5, the estimation performance also improves, since more data on real faults would be available for θ greater than 1.

Table 27.4 gives simulation results for $\hat{\nu}(t)$ using different stopping times and other related quantities for the case $\alpha_t = 1$ for the sequential procedure suggested in Section 27.3. The number of simulations used was 500 and the number of faults in the system was 400. The stopping time τ_0 increased from 0.64 for $d_0 = 0.1$ to 4.49 for $d_0 = 0.01$. Meanwhile, the average number

Table 27.3. Simulation results of the seeded faults approach using $p = 0.5$ and 0.9 and $\nu = 400$

D	θ	ave($\hat{\nu}$)	sd($\hat{\nu}$)	ave{$\widehat{\text{se}}(\hat{\nu})$}	cov
$p = 0.5$					
100	0.5	402.2	56.9	58.0	0.95
100	1.0	401.9	44.2	45.4	0.95
100	1.5	401.4	35.7	36.4	0.94
400	0.5	398.4	39.0	39.2	0.95
400	1.0	398.6	28.0	28.2	0.95
400	1.5	398.9	22.4	22.0	0.94
800	0.5	400.1	34.8	35.5	0.95
800	1.0	399.9	23.6	24.5	0.95
800	1.5	399.5	17.9	18.7	0.96
$p = 0.9$					
100	0.5	399.7	29.6	30.0	0.94
100	1.0	400.2	14.4	15.0	0.94
100	1.5	400.0	6.9	6.8	0.90
200	0.5	398.9	19.0	19.1	0.95
200	1.0	399.5	9.2	9.4	0.94
200	1.5	399.4	4.5	4.5	0.93
400	0.5	399.5	16.0	16.6	0.95
400	1.0	399.5	8.0	8.1	0.95
400	1.5	399.4	4.0	4.0	0.93

Notes: $\theta = \alpha_u/\beta_u$ where α_u and β_u are the failure intensities of the real and seeded faults respectively.
p = capture proportion for seeded faults.

of fault detections increased from 1022 to 1795 when d_0 decreased from 0.02 to 0.01. Also, the proportion of faults detected was 47% for $d_0 = 0.1$, whereas 70% were detected when using $d_0 = 0.05$. Figure 27.2 gives a plot for the stopping time versus the desired accuracy level d_0. The mean, $E\tau$, increased rapidly when d_0 decreased below 0.1.

Table 27.5 gives the simulation results for Chao's coverage estimator in Equation 27.14 and the two-step procedure proposed in Section 27.2.3.2. The failure intensities are generated from gamma distributions with different coverage values. The population size is 500 and the overall capture proportion is about 84%. It is shown from Table 27.5 that the sample coverage estimator underestimates the population size with a larger relative mean square error, while the two-step procedure performs satisfactorily with virtually no bias for $\hat{\nu}$ and $\widehat{\text{se}}(\hat{\nu})$. It should be noted here that

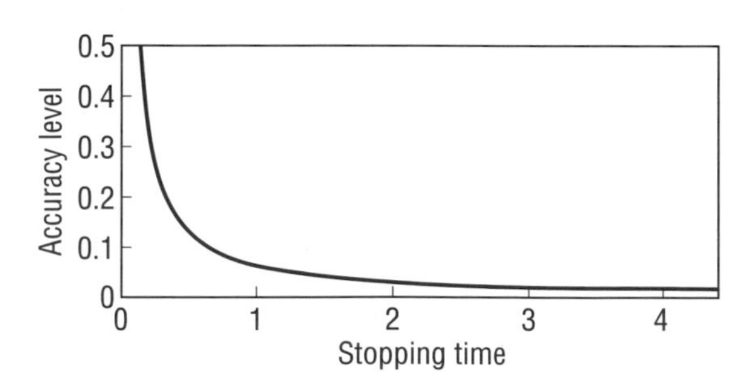

Figure 27.2. Average stopping time with different accuracy level d_0

Table 27.4. Simulation results for the sequential procedure with $\alpha_t = 1$ and $\nu = 400$

d_0	ave(τ_0)	P_{τ_0}	ave(N_{τ_0})	ave($\hat{\nu}$)	sd($\hat{\nu}$)	ave{se($\hat{\nu}$)/$\hat{\nu}$}	rel.MSE
0.01	4.49	0.99	1795.76	400.29	4.06	0.01	0.01
0.02	2.57	0.92	1022.99	400.88	8.23	0.02	0.02
0.03	1.84	0.84	733.15	401.08	11.90	0.03	0.03
0.04	1.44	0.76	575.99	401.39	15.77	0.04	0.04
0.05	1.19	0.70	475.26	401.16	20.01	0.05	0.05
0.06	1.01	0.64	404.58	400.61	24.33	0.06	0.06
0.07	0.89	0.59	352.91	400.64	27.87	0.07	0.07
0.08	0.79	0.54	313.21	400.71	32.15	0.08	0.08
0.09	0.71	0.51	281.32	400.38	35.62	0.09	0.09
0.10	0.64	0.47	255.72	400.75	40.14	0.10	0.10
0.15	0.44	0.36	176.26	401.98	61.35	0.15	0.15
0.20	0.33	0.28	133.93	399.97	81.03	0.20	0.20
0.25	0.27	0.24	108.36	399.65	99.52	0.24	0.25
0.30	0.23	0.20	90.34	394.00	116.27	0.29	0.29
0.35	0.19	0.18	77.67	390.47	132.18	0.34	0.33
0.40	0.17	0.16	68.47	393.15	152.93	0.38	0.38
0.45	0.15	0.14	61.84	393.91	174.55	0.42	0.44
0.50	0.14	0.13	54.98	394.65	192.51	0.47	0.48
0.60	0.12	0.11	46.82	395.62	223.85	0.54	0.56
0.70	0.09	0.09	37.53	400.17	278.94	0.67	0.70
0.80	0.09	0.09	36.96	395.08	283.20	0.67	0.71
0.90	0.09	0.08	34.94	376.65	293.51	0.69	0.74
1.00	0.06	0.06	25.52	422.32	411.85	0.94	1.03

Notes: $\text{ave}(\hat{\nu}) = \sum \hat{\nu}_i / R$; $\text{ave}(\tau) = \sum \tau_i / R$; $\text{ave}(N_\tau) = \sum N_{\tau_i} / R$;
$\widehat{\text{sd}}(\hat{\nu})^2 = \sum \{\hat{\nu}_i - \text{ave}(\hat{\nu})\}^2/(R-1)$; $\text{ave}\{\widehat{\text{se}}(\hat{\nu})/\hat{\nu}\} = \{\sum \widehat{\text{se}}(\hat{\nu}_i)/\hat{\nu}_i\}/R$;
$P_\tau = 1 - \exp\{\text{ave}(-\tau_0)\}$; $\text{rel.MSE}^2 = \sum\{(\hat{\nu}_i - \nu)/\nu\}^2/R$.
$R = 500$ is the number of simulations.

Table 27.5. Simulation results of the sample coverage estimator and the two-step procedure for heterogeneous models with gamma intensity for $\nu = 500$

CV	Estimator	ave($\hat{\nu}$)	sd($\hat{\nu}$)	ave($\widehat{\text{se}}(\hat{\nu})$)	RMSE
0.58	Two-step	500.3	18.6	17.9	17.8
	Chao & Lee	486.6	13.9	13.8	19.2
0.71	Two-step	500.3	18.8	18.6	18.6
	Chao & Lee	483.8	13.5	13.7	21.2
1.00	Two-step	500.9	21.5	20.8	20.9
	Chao & Lee	477.7	14.0	14.5	26.6
1.12	Two-step	500.5	22.9	23.6	23.6
	Chao & Lee	474.7	14.3	15.0	29.4

Note: Two-step = conditional likelihood with Horvitz–Thompson estimator.
Chao & Lee = sample coverage estimator.

the capture probability is fairly high in this simulation. Additional simulation results not shown here revealed that larger population sizes are required for the proposed procedure to perform well if the capture probabilities are low.

27.6 Discussion

Estimators based on the optimal estimating equation are recommended for the homogeneous models with or without recapture. The estimator is equivalent to the usual unconditional MLE derived by Becker and Heyde [19]. For the seeded fault model without recapture it can also be shown that the estimator using an optimal weight is identical to the unconditional MLE [30]. For the non-parametric case with γ_i values not specified, it seems that the coverage estimators of Chao and Lee [25] and Yip and Chao [26] are the best choices. However, for the parametric case the γ_i values depend on specified distributions and the two-step procedure should be highly reliable provided that the model holds. Also, the recommended Horvitz–Thompson estimator is equivalent to the conditional MLE suggested by Sanathanan [31,32] in the case of a homogeneous failure intensity.

Our approach adopted in this chapter is different from other recent reviews by Chao *et al.* [13] and Briand and Freimut [12], who examined the problem from a rather different perspective. They examined the problem when estimating the size of a target population based on several incomplete lists of individuals. Here, it is assumed we only have one tester equivalent, the data are the failure times of the faults (either seeded or re-detections). It has been shown that the recapture and seeded approach can provide substantial information for ν. Heterogeneity can be a cause for concern. However, it is interesting to note that most of the failure data set can be modeled by the Jelinski–Moranda model rather than the Littlewood model. Nevertheless, ignoring heterogeneity has been shown to provide estimates which are biased downward [20]. The present formulation allows removals during the testing process, and the other existing methods cannot handle that.

Sequential procedures are obtained for estimating the number of faults in a testing experiment using a fixed accuracy criterion. With a greater failure intensity and a greater number of faults in the system, a shorter testing time is needed to achieve the desired accuracy level for the estimates. The accuracy is specified by d_0, which prescribes an upper limit on the coefficient of variation of the estimator. Requiring greater accuracy results in an experiment of longer duration. Thus, a small value of d_0 results in a larger mean testing time, τ. Some accuracy levels may be unattainable. The mean testing time τ is found to increase exponentially when d_0 is smaller than 0.1. The value of $d_0 = 0.1$ (*i.e.* coefficient of variation 10%) seems to be an optimal choice in terms of accuracy and the cost of the experiment. The given results are consistent with the findings of Lloyd *et al.* [4] that when the proportion of faults found reaches a level of 70% of the total number of faults in the system, further testing adds little value to the experiment in terms of reducing the standard error of the estimate. Sequential estimation is an important tool in testing experiments. This chapter provides a practical way to use sequential estimation, which can be applied to other sampling designs and estimators.

Another potential application of the suggested procedure is in ecology [33]. In capture–recapture experiments in ecological studies, individuals are caught, marked, and released back into the population and subject to recapture again. The capture–recapture process is conducted for a period of time τ. Capturing all animals to determine the abundance is infeasible. The detection of total number of individuals is unwise, costly and expensive. The important issue here is not the estimator, but the use of the accuracy criterion to determine the stopping time. It is not our intention in this chapter to compare different estimators. Some comparison of estimators can be found in Yip *et al.* [34,35].

The conditional likelihood with a Horvitz–Thompson estimator in Section 27.2.3.2 can provide stable estimates. The proposed formulation includes both removal and recapture models. It also allows random removals during the recapturing process. The resulting conditional likelihood estimators have the consistency and asymptotic normality properties, which have also been confirmed by further simulation studies [36]. However, the asymptotic normality for the proposed frailty model requires a comparably large ν. For small values of ν and low capture probability, the distributions of $\hat{\nu}$ are skew. A similar situation occurred in removal studies with the use of auxiliary variables [37, 38] or in coverage estimators [30]. Ignoring frailty or heterogeneity in the population would seriously underestimate the population size, giving a misleading smaller standard error [35]. The likelihood or estimating function approach to estimate the unknown population size with heterogeneity has been shown to be unsatisfactory [39,40]. The proposed two-step estimating procedure is a promising alternative.

Acknowledgment

This work is supported by the Research Grant Council of Hong Kong (HKU 7267/2000M).

References

[1] Moek G. Comparison of software reliability models for simulated and real failure data. Int J Model Simul 1984;4:29–41.

[2] Andersen PK, Borgan Ø, Gill RD, Keiding N. Statistical models based on counting processes. New York: Springer-Verlag; 1993.

[3] Nayak TK. Estimating population sizes by recapture sampling. Biometrika 1988;75:113–20.

[4] Lloyd C, Yip PSF, Chan KS. A comparison of recapturing, removal and resighting design for population size estimation. J Stat Inf Plan 1998;71:363–73.

[5] Yip PSF. Estimating the number of errors in a system using a martingale approach. IEEE Trans Reliab 1995;44:322–6.

[6] Littlewood B. Theories of software reliability: how good are they and how can they be improved? IEEE Trans Software Eng 1980;6:489–500.

[7] Lloyd C, Yip PSF. A unification of inference from capture-recapture studies through martingale estimating functions. In: Godambe VP, editor. Estimating functions. Oxford: Oxford University Press; 1991. p.65–88.

[8] Yip PSF, Fong YT, Wilson K. Estimating population size by recapture via estimating function. Commun Stat-Stochast Models 1993;9:179–93.

[9] Horvitz DG, Thompson DJ. A generalization of sampling without replacement from a finite universe. J Am Stat Assoc 1952;47:663–85.

[10] Van der Wiel SA, Votta LG. Assessing software designs using capture-recapture methods. IEEE Trans Software Eng 1993;SE-19(11):1045–54.

[11] Huggins RM. Fixed accuracy estimation for chain binomial models. Stochast Proc Appl 1992;41:273–80.

[12] Briand LC, Freimut BG. A comprehensive evaluation of capture-recapture models for estimating software defect content. IEEE Trans Software Eng 2000;26: 518–38.

[13] Chao A, Tsay PK, Lin SH, Shau WY, Chao DY. The application of capture–recapture models to epidemiological data. Stat Med 2001;20:3123–57.

[14] Aalen OO. Statistical inference for a family of counting processes. Ph.D. thesis, University of California, Berkeley, 1975.

[15] Aalen OO. Non-parametric inference for a family of counting processes. Ann Stat 1978;6:701–26.

[16] Andersen PK, Gill RD. Cox's regression model for counting processes: a large sample study. Ann Stat 1982;10:1100–20.

[17] Godambe VP. The foundations of finite sample estimation in stochastic processes. Biometrika 1985;72: 419–28.

[18] Schnabel ZE. The estimation of the total fish population of a lake. Am Math Monthly 1938;39:348–52.

[19] Becker NG, Heyde CC. Estimating population size from multiple recapture experiments. Stochast Proc Appl 1990;36:77–83.

[20] Yip PSF, Xi L, Fong DYT, Hayakawa Y. Sensitivity analysis and estimating number of faults in removal debugging. IEEE Trans Reliab 1999;48:300–5.

[21] Chao A, Lee SM. Estimating the number of classes via sample coverage. J Am Stat Assoc 1992;87:210–7.

[22] Huggins RM. On the statistical analysis of capture experiments. Biometrika 1989;76:133–40.

[23] Huggins RM. Some practical aspects of a conditional likelihood approach to capture–recapture models. Biometrics 1991;47:725–32.

[24] Yip PSF, Wan E, Chan KS. A unified of capture–recapture methods in discrete time. J Agric, Biol Environ Stat 2001;6:183–94.

[25] Chao A, Lee SM. Estimating population size for continuous time capture–recapture models via sample coverage. Biometric J 1993;35:29–45.

[26] Yip PSF, Chao A. Estimating population size from capture–recapture studies via sample coverage and estimating functions. Commun Stat-Stochast Models 1996;12(1):17–36.

[27] Lin HS, Chao A, Lee SM. Consistency of an estimator of the number of species. J Chin Stat Assoc 1993;31:253–70.

[28] Nielsen GG, Gill RD, Andersen PK, Sørensen TIA. A counting process approach to maximum likelihood estimation in frailty models. Scand J Stat 1992;19:25–43.

[29] Jelinski Z, Moranda P. Software reliability research. Stat Comp Perform Eval 1972;465–84.

[30] Chao A, Yip PSF, Lee SM, Chu W. Population size estimation based on estimating functions for closed capture–recapture models. J Stat Plan Inf 2001;92:213–32.

[31] Sanathanan L. Estimating the size of a multinomial population. Ann Math Stat 1972;43:142–52.

[32] Sanathanan L. Estimating the size of a truncated sample. J Am Stat Assoc 1977;72:669–72.

[33] Pollock KH. Modeling capture, recapture and removal statistics for estimation of demographic parameters for fish and wildlife populations: past, present and future. J Am Stat Assoc 1991;86:225–38.

[34] Yip PSF, Huggins RM, Lin DY. Inference for capture–recapture experiments in continuous time with variable capture rates. Biometrika 1996;83:477–83.

[35] Yip PSF, Zhou Y, Lin DY, Fang XZ. Estimation of population size based on additive hazards models for continuous time recapture experiment. Biometrics 1999;55:904–8.

[36] Wang Y, Yip PSF, Hayakawa Y. A frailty model for detecting number of faults in a system. Department of Statistics and Actuarial Science Research Report 287, The University of Hong Kong,; 2001. p.1–15.

[37] Huggins R, Yip P. Statisitcal analysis of removal experiments with the use of auxillary variables. Stat Sin 1997;7:705–12.

[38] Lin DY, Yip PSF. Parametric regression models for continuous time recapture and removal studies. J R Stat Soc Ser B 1999;61:401–13.

[39] Burnham KP, Overton WS. Estimation of the size of a closed population when capture probabilities vary among animals. Biometrika 1978;65:625–33.

[40] Becker NG. Estimating population size from capture–recapture experiments in continuous time. Austr J Stat 1984;26:1–7.

Reliability of Electric Power Systems: An Overview

Roy Billinton and Ronald N. Allan

28.1 Introduction

The primary function of a modern electric power system is to supply its customers with electrical energy as economically as possible and with an acceptable degree of reliability. Modern society, because of its pattern of social and working habits, has come to expect the supply to be continuously available on demand. This degree of expectation requires electric power utilities to provide an uninterrupted power supply to their customers. It is not possible to design a power system with 100% reliability. Power system managers and engineers therefore strive to obtain the highest possible system reliability within their socio-economic constraints. Consideration of the two important aspects of continuity and quality of supply, together with other important elements in the planning, design, control, operation and maintenance of an electric power system network, is usually designated as reliability assessment. In the power system context, reliability can therefore be defined as concern regarding the system's ability to provide an adequate supply of electrical energy. Many utilities quantitatively assess the past performance of their systems and utilize the resultant indices in a wide range of managerial activities and decisions. All utilities attempt to recognize reliability implications in system planning, design and operation through a wide range of techniques [1–12].

Reliability evaluation methods can generally be classified into two categories: deterministic and probabilistic. Deterministic techniques are slowly being replaced by probabilistic methods [1–12]. System reliability can be divided into the two distinct categories of system security and system adequacy. The concept of security is associated

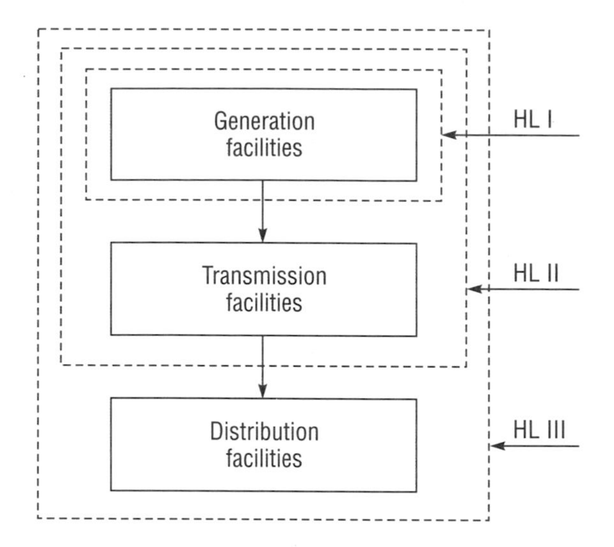

Figure 28.1. Hierarchical levels in an electric power system

with the dynamic response of the system to the perturbations to which it is subjected. Adequacy is related to static system conditions and the existence of sufficient facilities within the system to meet the system load demand. Overall adequacy evaluation of an electric power system involves a comprehensive analysis of its three principal functional zones of generation, transmission and distribution. The basic techniques for adequacy assessment are generally categorized in terms of their application to each of these zones.

The functional zones can be combined as shown in Figure 28.1 to obtain three distinct hierarchical levels (HL) and adequacy assessment can be conducted at each of these hierarchical levels [2]. At HLI, the adequacy of the generating system to meet the load requirements is examined. Both generation and the associated transmission facilities are considered in HLII adequacy assessment. This activity is sometimes referred to as composite or bulk system adequacy evaluation. HLIII adequacy assessment involves the consideration of all these functional zones in order to evaluate customer load point indices. HLIII studies are not normally conducted in a practical system because of the enormity of the problem.

Electric power utilities are experiencing considerable change with regard to their structure, operation and regulation. This is particularly true in those countries with highly developed systems. The traditional vertically integrated utility structure consisting of generation, transmission and distribution functional zones, as shown in Figure 28.1, has in many cases been decomposed into separate and distinct utilities in which each perform a single function in the overall electrical energy delivery system. Deregulation and competitive pricing will also make it possible for electricity consumers to select their supplier based on cost effectiveness. Requests for use of the transmission network by third parties are extending the traditional analysis of transmission capability far beyond institutional boundaries. There is also a growing utilization of local generation embedded in the distribution functional zone. The individual unit capacities may be quite small, but the combined total could represent a significant component of the total required generation capacity. Electric power system structures have changed considerably over the last decade and are still changing. The basic principles and concepts of reliability evaluation developed over many decades remain equally valid in the new operating environment. The objectives of reliability evaluation may, however, have to be reformulated and restructured to meet the new paradigms.

28.2 System Reliability Performance

Any attempt to perform quantitative reliability evaluation invariably leads to an examination of data availability and the data requirements to support such studies. Valid and useful data are expensive to collect, but it should be recognized that in the long run it would be more expensive not to collect them. It is sometimes argued as to which comes first: reliability data or reliability methodology. In reality, the data collection and reliability evaluation must evolve together and therefore the process is iterative. The principles of data collection are described in [13, 14] and various schemes have been implemented worldwide, for instance, by the Canadian Electricity Association (CEA) in Canada [15, 16], North American Electric Reliability Council (NERC) in the USA, and others which are well-established but less well-documented or less available publicly, *e.g.* the UK fault-reporting scheme (NAFIRS) [17] operated nationally and those of various individual North American Reliability Councils.

In conceptual terms [13, 14], data can be collected for one or both of two reasons: assessment of past performance and/or prediction of future system performance. The past performance assessment looks back at the past behavior of the system, whereas predictive assessment forecasts how the system is going to behave in the future. In order to perform predictive studies, however, it is essential to transform past experience into required future prediction. Consistent collection of data is therefore essential as it forms the input to relevant reliability models, techniques and equations. There are many reasons for collecting system and equipment performance data. Some of the more popular applications are as follows.

1. To furnish management with performance data regarding the quality of customer service on the electrical system as a whole and for each voltage level and operating area.
2. To provide data for an engineering comparison of electrical system performance among consenting companies.
3. To provide a basis for individual companies to establish service continuity criteria. Such criteria can be used to monitor system performance and to evaluate general policies, practices, standards and design.
4. To provide data for analysis to determine reliability of service in a given area (geographical, political, operating, *etc.*) to determine how factors such as design differences, environment or maintenance methods, and operating practices affect performance.
5. To provide reliability history of individual circuits for discussion with customers or prospective customers.
6. To identify substations and circuits with substandard performance and to ascertain the causes.
7. To obtain the optimum improvement in reliability per dollar expended for purchase, maintenance and operation of specific equipment/plant.
8. To provide equipment performance data necessary for a probabilistic approach to reliability studies. The purpose is to determine the design, operating and maintenance practices that provide optimum reliability per dollar expended and, in addition, to use this information to predict the performance of future generation, transmission and distribution system arrangements.
9. To provide performance data to regulatory bodies so that comparative performance records can be established between regulated monopolies and between planned and actual achievements within a regulated monopoly.

There are a wide range of data gathering systems in existence throughout the world. These systems vary from being very simple to exceedingly complex. However, for the purposes of identifying basic concepts, the protocols established by the CEA are used in this chapter to illustrate a possible framework and the data that can be determined. In the CEA system, component reliability data is collected under the Equipment Reliability Information System (ERIS). These protocols are structured using the functional zones of generation, transmission and distribution shown in Figure 28.1. The ERIS is described in more detail in a later section.

In addition to ERIS, the CEA has also created the Electric Power System Reliability Assessment (EPSRA) procedure, which is designed to provide data on past performance of the system. This procedure is in the process of evolution and at the present time contains systems for compiling information on bulk system disturbances, bulk system delivery point performance, and customer service continuity statistics. Customer satisfaction is a very important consideration for Canadian electric power utilities. The measured performance at HLIII and the ability to predict future performance are important factors in providing acceptable customer service and therefore customer satisfaction. The HLIII performance indices have been collected for many years by a number of Canadian utilities. The following text defines the basic customer performance indices and presents for illustration purposes some Canadian and UK indices for 1999.

Service performance indices can be calculated for the entire system, a specific region or voltage level, designated feeders or groups of customers. The most common indices [3] are as follows.

a. System Average Interruption Frequency Index (SAIFI)
This index is the average number of interruptions per customer served per year. It is determined by dividing the accumulated number of customer interruptions in a year by the number of customers served. A customer interruption is considered to be one interruption to one customer:

$$\text{SAIFI} = \frac{\text{total number of customer interruptions}}{\text{total number of customers}}$$

In the UK, this index is defined as security-supply interruptions per 100 connected customers [19].

b. Customer Average Interruption Frequency Index (CAIFI)
This index is the average number of interruptions per customer interrupted per year. It is determined by dividing the number of customer interruptions observed in a year by the number of customers affected. The customers affected should be counted only once regardless of the number of interruptions that they may have experienced during the year:

$$\text{CAIFI} = \frac{\text{total number of customer interruptions}}{\text{total number of customers affected}}$$

c. System Average Interruption Duration Index (SAIDI)
This index is the average interruption duration for customers served during a year. It is determined by dividing the sum of all customer interruption durations during a year by the number of customers served during the year:

$$\text{SAIDI} = \frac{\text{sum of customer interruption durations}}{\text{total number of customers}}$$

In the UK, this index is defined as availability-minutes lost per connected customer [19].

d. Customer Average Interruption Duration Index (CAIDI)
This index is the average interruption duration for customers interrupted during a year. It is determined by dividing the sum of all customer-sustained interruption durations by the number of sustained customer interruptions over a one-year period:

$$\text{CAIDI} = \frac{\text{sum of customer interruption durations}}{\text{total number of customer interruptions}} = \frac{\text{SAIDI}}{\text{SAIFI}}$$

This index is not explicitly published in the UK but can easily be calculated as availability/security.

e. Average Service Availability Index (ASAI)
This is the ratio of the total number of customer hours that service was available during a year to the total customer hours demanded. Customer hours demanded are determined as the 12-month average number of customers served times 8760 hours. This is sometimes known as the "Index of Reliability" (IOR):

$$\text{ASAI} = \frac{\text{customer hours of available service}}{\text{customer hours demanded}}$$

The complementary value to this index, *i.e.* the Average Service Unavailability Index (ASUI), may also be used. This is the ratio of the total number of customer hours that service was unavailable during a year to the total customer hours demanded.

Tables 28.1a and 28.1b show the overall Canadian [18] and UK [19] system performance

Table 28.1a. Overall Canadian service continuity statistics for 1999

SAIFI	2.59 int/y
SAIDI	4.31 hr/y
CAIDI	1.67 hr/int
IOR	0.999 508

Table 28.1b. Overall UK service continuity statistics for 1998/99

Security (SAIFI)	0.78 int/yr
Availability (SAIDI)	81 min/yr
Calculated CAIDI	104 min/int
Calculated ASAI/IOR	0.999 846

Table 28.2a. Differentiated Canadian service continuity statistics for 1999

	Urban utilities	Urban/rural utilities	Overall
SAIFI	2.28	2.88	2.59
SAIDI	2.68	5.06	4.31
CAIDI	1.17	1.76	1.67
IOR	0.999 694	0.999 422	0.999 508

indices for the 1999 period. The ASAI is known as the IOR in the EPSRA protocol. Very few utilities can provide the CAIFI parameter and therefore this is not part of EPSRA. This parameter is also not published in the UK performance statistics [19].

Tables 28.1a and 28.1b show the aggregate indices for the participating utilities. There are considerable differences between the system performance indices for urban utilities and those containing a significant rural component; this being true not only for Canada, but also the UK and elsewhere. This is shown in Table 28.2a for Canada [18] and Table 28.2b for the UK [19].

The indices shown in Tables 28.2a and 28.2b are overall aggregate values. Individual customer values will obviously vary quite widely depending on the individual utility, the system topology and the company operating philosophy. The indices shown in Tables 28.2a and 28.2b, however, do present a clear picture of the general level

Table 28.2b. Typical UK service continuity statistics for 1998/99

	Typical urban utility	Typical urban/rural utility
Security (SAIFI) (int/yr)	0.4	1.2
Availability (SAIDI) (min/yr)	50	100
Calculated CAIDI (min/int)	125	83.3
Calculated ASAI/IOR	0.999 905	0.999 810

of customer reliability. Similar statistics can be obtained at the bulk system delivery level, *i.e.* HLII. These indices [20, 21] are expressed as global values or in terms of delivery points rather than customers, as in the case of HLIII indices.

28.3 System Reliability Prediction

28.3.1 System Analysis

There are a wide range of techniques available and in use to assess the reliability of the functional zones and hierarchical levels shown in Figure 28.1. References [1–12] clearly illustrate the literature available on this subject. The following sections provide further discussion on the available techniques. The basic concepts associated with electric power system reliability evaluation are illustrated using a small hypothetical test system known as the RBTS [22–24]. This application illustrates the utilization of data such as that contained in the CEA-ERIS database to assess the adequacy of the system at all three hierarchical levels. The basic system data necessary for adequacy evaluation at HLI and HLII are provided in [22]. The capabilities of the RBTS have been extended to include failure data pertaining to station configurations in [25] and distribution networks that contain the main elements found in practical systems [24]. The RBTS is sufficiently small to permit a large number of reliability studies to be conducted with a reasonable solution time, yet sufficiently detailed to reflect the actual complexities involved in a practical reliability analysis. The RBTS described in [22] has two generation buses, five load buses (one of which is also a generation bus), nine transmission lines and 11 generating units. An additional line (Line 10) is connected between the 138-kV substations of Buses 2 and 4. A single line diagram of the RBTS at HLII is shown in Figure 28.2. A more complete diagram which shows the generation, transmission, station and distribution facilities is given in Figure 28.3. In order to reduce the complexity of the overall system, distribution networks are shown at Buses 2, 4 and 6. Bulk loads are used at Buses 3 and 5. The total installed capacity of the system is 240 MW with a system peak load of 185 MW. The transmission voltage level is 230 kV and the voltage limits are 1.05 and 0.97 pu. The studies reported in this chapter assume that there are no lines on a common right of way and/or a common tower. In order to obtain annual indices for the RBTS at HLII and HLIII, the load data described in [22] were arranged in descending order and divided into seven steps in increments of 10% of peak load as given in Table 28.3.

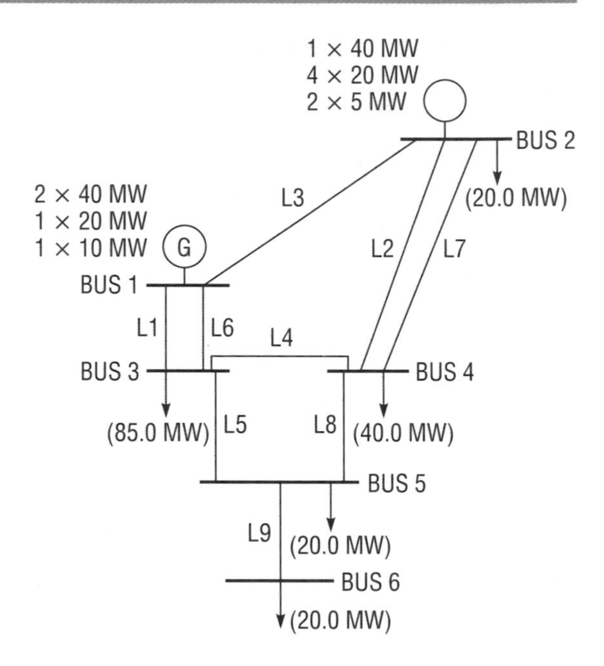

Figure 28.2. Single line diagram of the RBTS at HLII

28.3.2 Predictive Assessment at HLI

The basic reliability indices [3] used to assess the adequacy of generating systems are the loss of load expectation (LOLE) and the loss of energy expectation (LOEE). The main concern in an HLI study is to estimate the generating capacity required to satisfy the perceived system demand and to have sufficient capacity to perform corrective and preventive maintenance on the generation facilities. The generation and load models are convolved as shown in Figure 28.4 to create risk indices.

One possible casualty in the new deregulated environment is that of overall generation planning. There appears, however, to be a growing realization, particularly in view of the difficulties in California, that adequate generating capacity is an integral requirement in acceptable effective power supply.

The capacity model can take several forms and in analytical methods it is usually represented by a capacity outage probability table [3]. The LOLE is the most widely used probabilistic index at the present time. It gives the expected number of days (or hours) in a period during which the load will exceed the available generating capacity. It does not, however, indicate the severity of the shortage. The LOEE, also referred to as the expected energy not supplied (EENS), is the expected unsupplied energy due to conditions resulting in the load demand exceeding the available generating capacity. Normalized values of LOEE in the form of system-minutes (SM) or units per million (UPM) are used by some power utilities.

The basic adequacy indices of the RBTS resulting from an HLI study are given below:

$$\text{LOLE} = 1.09161 \text{ hr/yr}$$
$$\text{LOEE} = 9.83 \text{ MWh/yr}$$
$$\text{SM} = 3.188$$
$$\text{UPM} = 9.9$$

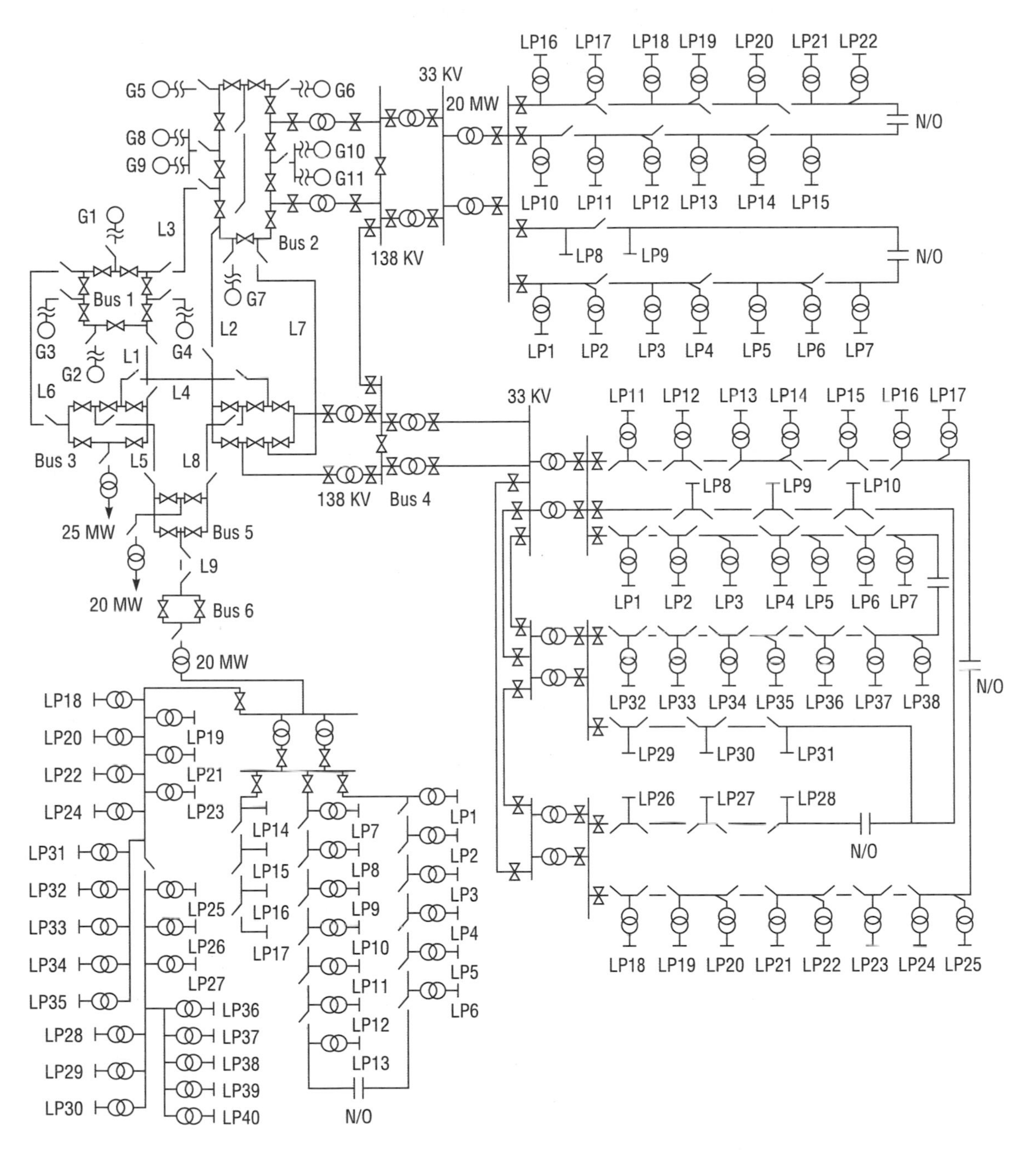

Figure 28.3. Single line diagram of the RBTS at HLIII

Table 28.3. Seven-step load data for the RBTS

Load (MW)	Probability	Duration (hr)
185.0	0.013 163 92	115.0
166.5	0.111 034 78	970.0
148.0	0.165 407 52	1445.0
129.5	0.232 028 37	2027.0
111.0	0.226 304 89	1883.0
92.5	0.226 304 89	1997.0
74.0	0.036 515 59	319.0

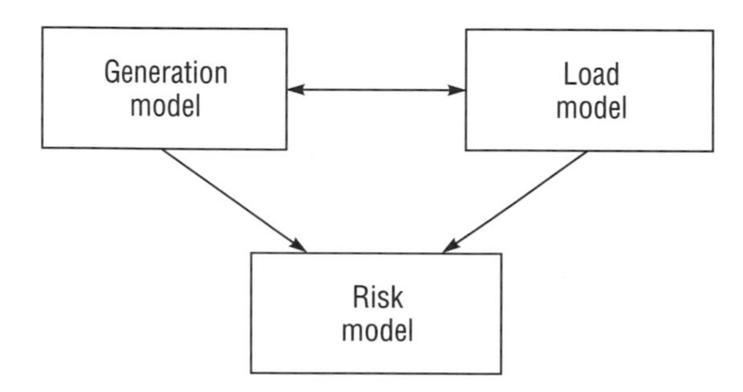

Figure 28.4. Conceptual model for HLI evaluation

The SM and UPM indices are obtained by dividing the LOEE by the system peak and the annual energy requirement respectively. Both indices are normalized LOEE indices and can be used to compare systems of different sizes or to use in a single system over time as the load changes. Both indices are in actual utility use.

A per-unitized value of LOLE gives LOLP, *i.e.* the probability of loss of load. This is the originally defined term which was soon converted into LOLE because of the latter's more physical interpretation. However, LOLP has been used in the UK [26] since privatization to determine the trading price of electricity, known as pool input price (PIP) and given by:

$$\text{PIP} = \text{SMP} + (\text{VOLL} - \text{SMP})\text{LOLP}$$

where SMP = system marginal price and VOLL = value of lost load (set at £2.389/kWh for 1994/95). The pool has now been replaced by bilateral trading and a balancing market known as NETA (the New Energy Trading Arrangement) in March 2001, when the use of LOLP in energy pricing ceased.

28.3.3 Predictive Assessment at HLII

HLII adequacy evaluation techniques are concerned with the composite problem of assessing the generation and transmission facilities in regard to their ability to supply adequate, dependable and suitable electric energy at the bulk load points [2–6]. A basic objective of HLII adequacy assessment is to provide quantitative input to the economic development of the generation and transmission facilities required to satisfy the customer load demands at acceptable levels of quality and availability. A number of digital computer programs have been developed to perform this type of analysis [6–12]. Extensive development work in the area of HLII adequacy assessment has been performed at the University of Saskatchewan and at UMIST. This has resulted in the development of several computer programs, including COMREL (analytical) and MECORE (simulation) at the University of Saskatchewan [27, 28] and RELACS (analytical) and COMPASS (sequential simulation) at UMIST [28,29]. The analytical programs are based on contingency enumeration algorithms and the simulation programs on Monte Carlo approaches. All include selection of disturbances, classification of these into specified failure events, and calculation of load point and system reliability indices. A range of other computer programs have been published [28], some of which are in-house assessment tools and others more widely available. These include MEXICO (EdF), SICRIT and SICIDR (ENEL), CREAM and TRELLS (EPRI, USA), ESCORT (NGC, UK), ZANZIBAR (EdP), TPLan (PTI), CONFTRA (Brazil), amongst others. These programs are not all equivalent; the modeling procedure, solution methods and calculated results vary quite considerably. Reliability evaluation of a bulk power system normally includes the independent outages of generating units, transformers and transmission lines. A major simplification that is normally used in HLII studies is that terminal stations are modeled as single busbars. This representation does not

Table 28.4. Annual HLII load point adequacy indices for the RBTS including station effects

Bus	a	b	c	d	e
2	0.000 292	0.1457	0.314	4.528	2.55
3	0.001 737	0.4179	12.823	683.34	15.17
4	0.000 323	0.1738	0.746	10.611	2.82
5	0.001 387	0.2530	2.762	115.26	12.12
6	0.001 817	1.6518	22.017	327.23	24.613

consider the actual station configurations as an integral part of the analysis. The HLII system used in this analysis is shown in Figure 28.2. Station-originated outages can have a significant impact on HLII indices. This chapter considers, in addition to independent overlapping outages of generators and lines, the effects on the reliability indices of station-originated outages. The HLIII system used in this analysis is shown in Figure 28.3.

HLII adequacy indices are usually expressed and calculated on an annual basis, but they can be determined for any period such as a season, a month, and/or a particular operating condition. The indices can also be calculated for a particular load level and expressed on an annual basis. Such indices are designated as annualized values. The calculated adequacy indices for HLII can be obtained for each load point or for the entire system. Both sets of indices are necessary to obtain a complete picture of HLII adequacy, *i.e.* these indices complement rather than substitute for each other. Individual load point indices can be used to identify weak points in the system, while overall indices provide an appreciation of global HLII adequacy and can be used by planners and managers to compare the relative adequacies of different systems. The load point indices assist in quantifying the benefits associated with individual reinforcement investments. The annual load point and system indices for the RBTS were calculated using COMREL and are presented in this section. The load point indices designated a, b, c, d and e in Table 28.4 are defined below:

a = Probability of failure

b = Frequency of failure (f/yr)

c = Total load shed (MW)

d = Total energy curtailed (MWh)

e = Total duration (hr)

It is also possible to produce overall system indices by aggregating the individual bus adequacy values. The system indices do not replace the bus values but complement them to provide a complete assessment at HLII and to provide a global picture of the overall system behavior. The system indices of the RBTS are as follows:

Bulk Power Supply Disturbances

$= 2.642(\text{occ/yr})$

Bulk Power Interruption Index

$= 0.209(\text{MW/MWyr})$

Bulk Power Energy Curtailment Index

$= 6.1674(\text{MWh/MWyr})$

Bulk Power Supply Average MW Curtailment

$= 14.633(\text{MW/disturbance})$

Modified Bulk Power Energy Curtailment Index

$= 0.000\,71$

$\text{SM} = 370.04(\text{system-minutes})$

The Modified Bulk Power Energy Curtailment Index of 0.000 71 can be multiplied by 10^6 to give 71 UPM. The SM and UPM indices at HLII are extensions of those obtained at HLI and illustrate the bulk system inadequacy.

28.3.4 Distribution System Reliability Assessment

The bulk of the outages seen by an individual customer occur in the distribution network and therefore practical evaluation techniques [3] have

Table 28.5. System indices for the distribution systems in the RBTS Buses 2, 4 and 6

Bus	SAIFI	SAIDI	CAIDI	ASAI
2	0.304 21	3.670 65	12.066 07	0.999 581
4	0.377 31	3.545 50	9.396 70	0.999 595
6	1.061 90	3.724 34	3.507 25	0.999 575

been developed for this functional zone. The basic load point indices at a customer load point are λ, r and U, which indicate the failure rate (frequency), average outage time and average annual outage time respectively. The individual load point indices can be combined with the number of customers at each load point to produce feeder, or area SAIFI, SAIDI, CAIDI and ASAI predictions. The distribution functional zones shown in Figure 28.3 have been analyzed and the results are shown in Table 28.5. Additional indices [3] can also be calculated.

28.3.5 Predictive Assessment at HLIII

HLIII adequacy evaluation includes all three segments of an electric power system in an overall assessment of actual consumer load point adequacy. The primary adequacy indices at HLIII are the expected failure rate, λ, the average duration of failure and the annual unavailability, U, of the customer load points [3]. The individual customer indices can then be aggregated with the average connected load of the customer and the number of customers at each load point to obtain the HLIII system adequacy indices. These indices are the SAIFI, the SAIDI, the CAIDI and the ASAI. Analysis of actual customer failure statistics indicates that the distribution functional zone makes the greatest individual contribution to the overall customer supply unavailability. Statistics collected by most, if not all, utilities indicate that, in general, the bulk power system contributes only a relatively small component to the overall HLIII customer indices. The conventional customer load point indices are performance parameters obtained from historical event reporting. The following illustrates how similar indices have been predicted. The HLIII adequacy assessment presented includes the independent outages of generating units, transmission lines, outages due to station-originated failures, sub-transmission element failures and distribution element failures. The method used is summarized in the three steps given below.

1. The COMREL computer program was used to obtain the probability, expected frequency and duration of each contingency at HLII that leads to load curtailment for each system bus. The contingencies considered include outages up to: four generation units, three transmission lines, two lines with one generator, and two generators with one line. All outages due to station-related failures and the isolation of load buses due to station failures are also considered. If the contingency results in load curtailment, a basic question is then how would the electric utility distribute this interrupted load among its customers? It is obvious that different power utilities will take different actions based on their experience, judgment and other criteria. The method used in this chapter assumes that load is curtailed proportionately across all the customers. For each contingency j that leads to load curtailment of L_{kj} at Bus k, the ratio of L_{kj} to bus peak load is determined. The failure probability and frequency due to an isolation case are not modified as the isolation affects all the customers.
2. At the sub-transmission system level, the impact of all outages is obtained in terms of average failure rate and average annual outage time at each distribution system supply point.

Table 28.6. Modified annual bus adequacy indices for the RBTS using Step 1

Bus	Failure probability	Failure frequency (f/yr)	Total duration (hr)
2	0.000 026	0.016 50	0.227
3	0.001 254	0.200 26	10.985
4	0.000 030	0.019 85	0.269
5	0.001 330	0.204 20	11.650
6	0.002 509	1.441 02	21.981

3. At the radial distribution level, the effects due to outages of system components such as primary main/laterals/low-voltage transformers, *etc*, are considered.

The modified annual indices of the RBTS were obtained using the load model given in Table 28.3 and the results are presented in Table 28.6. The indices in Table 28.6 are smaller than those shown in Table 28.4 as a significant number of the outage events included in Table 28.4 do not affect all the customers at a load point. The index modification approach described in Step 1 above was used to allocate these event contributions.

Table 28.7 presents a sample of the HLIII indices at Buses 2, 4 and 6. It can be seen from this table that the contribution of the distribution system to the HLIII indices is significant for Buses 2 and 4. Table 28.7 also shows that, for Bus 6, the contribution from HLII to the HLIII indices is very significant. The obvious reason is that customers at Bus 6 are supplied through a single circuit transmission line and through a single 230/33-kV transformer. Reinforcing the supply to Bus 6 would therefore obviously be an important consideration. Table 28.8 presents HLIII system indices at Buses 2, 4 and 6. The percentage contributions from the distribution functional zones to the total HLIII indices are given in Table 28.9.

The contribution of the distribution system to the HLIII indices is significant for Buses 2 and 4, while for Bus 6 the contribution from HLII to the HLIII indices is considerably larger. Table 28.8 shows the annual customer indices at the buses. These values can be aggregated to provide overall system indices, assuming that the customers at load points 2, 4 and 6 constitute all the system customers:

$$\text{SAIFI} = 1.226(\text{failures/system customer})$$
$$\text{SAIDI} = 12.588(\text{hours/system customer})$$
$$\text{CAIDI} = 10.296(\text{hours/customer interrupted})$$
$$\text{ASAI} = 0.998\,563$$

It can be seen from the above that the overall system indices, while important, mask the actual indices at any particular load bus. Reliability evaluation at each hierarchical level provides valuable input to the decision-making process associated with system planning, design, operation and management. Both past and predictive assessment are required to obtain a complete picture of system reliability.

28.4 System Reliability Data

It should be recognized that the data requirements of predictive methodologies should reflect the needs of these methods. This means that the data must be sufficiently comprehensive to ensure that the methods can be applied but restrictive enough to ensure that unnecessary data is not collected and irrelevant statistics are not evaluated. The data should therefore reflect and respond to the factors that affect the system reliability and enable it to be modeled and analyzed. This means that the data should relate to the two main processes involved in component behavior, namely the failure and restoration processes. It cannot be stressed too strongly that, in deciding which data is to be collected, a utility must make decisions on the basis of factors that have impact on its own planning and design considerations.

The quality of the data and the resulting indices depend on two important factors: confidence and relevance. The quality of the data and thus the confidence that can be placed in it is clearly dependent upon the accuracy and completeness of the information compiled by operating and maintenance personnel. It is therefore essential that they are made fully aware of the future use of the data and the importance it will play in later

Table 28.7. HLIII load point reliability indices for the RBTS Buses 2, 4 and 6

Load point	Failure rate (f/yr)	Outage time (hr)	Annual unavailability (hr/yr)	Contribution of distribution (%)
Bus 2				
1	0.3118	12.3832	3.8605	94.11
9	0.2123	3.7173	0.7890	71.20
11	0.3248	12.0867	3.9255	94.21
15	0.3150	12.2657	3.8638	94.12
20	0.3280	11.9778	3.9288	94.22
Bus 4				
1	0.3708	10.1410	3.7605	92.91
10	0.2713	2.6832	0.7280	63.39
19	0.4157	9.1589	3.8076	93.00
28	0.2935	2.3938	0.7025	61.75
Bus 6				
1	1.8248	12.5237	22.8530	3.81
16	1.7350	13.2443	22.9792	4.34
31	4.0580	7.8714	31.9422	31.18
40	3.9930	8.6799	34.6592	36.58

Table 28.8. HLIII system indices at Buses 2, 4 and 6 of the RBTS

Bus	SAIFI	SAIDI	CAIDI	ASAI
2	0.320 72	3.897 86	12.153 60	0.999 555
4	0.397 14	3.812 02	9.598 72	0.999 565
6	2.502 92	25.705 58	10.270 24	0.997 066

Table 28.9. Percentage contribution to HLIII inadequacy of the RBTS from the distribution system

Bus	SAIFI	Distribution (%)	SAIDI	Distribution (%)
2	0.320 72	94.85	3.897 86	94.17
4	0.397 14	95.01	3.812 02	93.01
6	2.502 92	92.43	25.7055	14.49

developments of the system. The quality of the statistical indices is also dependent upon how the data is processed, how much pooling is done, and the age of the data currently stored.

Many utilities throughout the world have established comprehensive procedures for assessing the performance of their systems. In Canada these procedures have been formulated and data is collected through the CEA. As this system is very much in the public domain, it is useful to use it as a means of illustrating general principles.

28.4.1 Canadian Electricity Association Database

All the major electric power utilities in Canada participate in a single data collection and analysis system of the CEA called ERIS. The CEA started collecting data on generation outages in 1977 and on transmission outages in 1978, and since then has published a number of reports in both areas [14–16, 18, 20]. The third stage, dealing with distribution equipment, is being implemented. This procedure is illustrated in Figure 28.5.

28.4.2 Canadian Electricity Association Equipment Reliability Information System Database for HLI Evaluation

The simplest representation of a generating unit is the two-state model shown in Figure 28.6, where λ and μ are the unit failure and repair rates respectively. Additional derated states can be added to the model to provide more accurate representation. This is shown in Figure 28.7 where a, b, c, d, e and f are the various state transition rates. The transition rates for these models can be calculated from the generation equipment status data collected by utilities. These data are available in CEA-ERIS, which provides a detailed reporting procedure for collecting generating unit operating and outage data. This database contains information regarding the various states in which a generating unit can reside and outage codes to explain why a unit is in a particular outage or derated state. The data provided by CEA-ERIS are quite adequate to perform reliability analysis at HLI. The basic parameters required in performing adequacy analysis are presented in Table 28.10 for typical generating unit classes using data for 1994/98 [15].

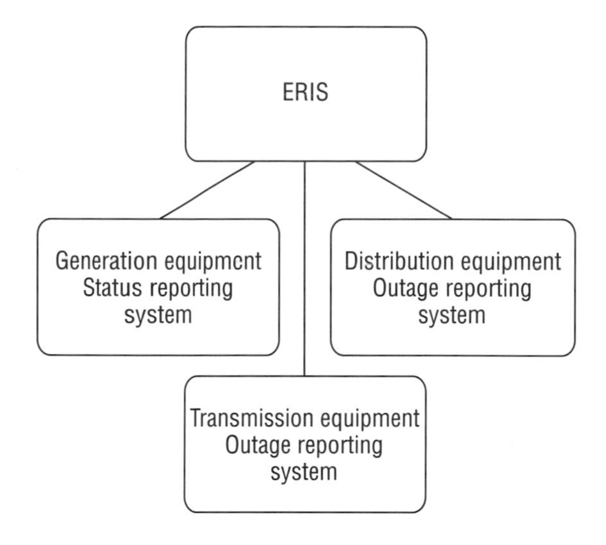

Figure 28.5. Basic components of ERIS

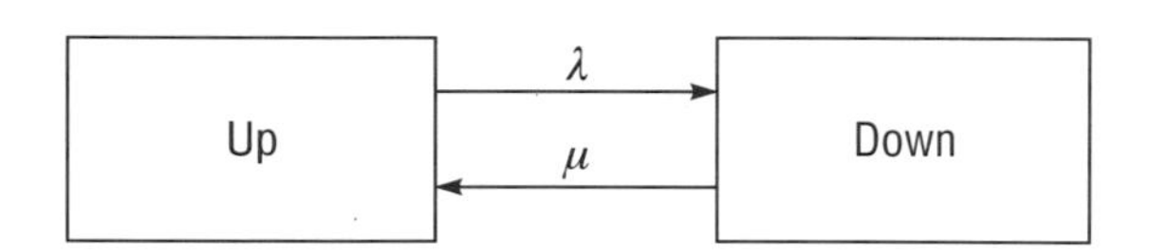

Figure 28.6. Basic two-state unit model

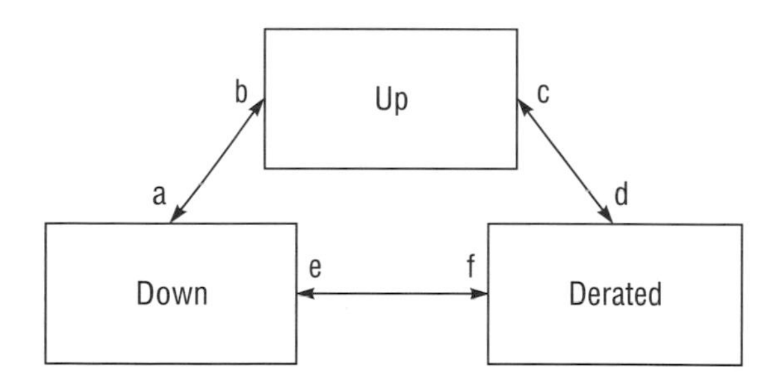

Figure 28.7. Basic three-state unit model

28.4.3 Canadian Electricity Association Equipment Reliability Information System Database for HLII Evaluation

Reliability evaluation at HLII requires outage statistics pertaining to both generation and transmission equipment. The generation outage statistics are available from [15] as discussed earlier. The basic equipment outage statistics required are failure rates and repair times of transmission lines and associated equipment such as circuit breakers and transformers. The transmission stage of CEA-ERIS was implemented in 1978 when Canadian utilities began supplying data on transmission equipment. The thirteenth CEA report on the performance of transmission equipment is based on data for the period from January 1, 1994 to December 31, 1998 [16]. This report covers transmission equipment in Canada with an operating

Table 28.10. Outage data for different generating unit types based on 1994/98 data

Unit type	FOR (%)	Failure rate (f/yr)
CTU	9.02*	8.72
Fossil	7.02	11.04
Hydraulic	2.00	2.81
Nuclear	11.46	3.33

CTU = combustion turbine unit.
* Indicates utilization forced outage probability (UFOP) used rather than forced outage rate. The UFOP of a generating unit is defined as the probability that the unit will not be available when required. The CEA-ERIS database contains unavailability data on units of different sizes and of different ages in addition to failure and repair rate information for frequency and duration reliability evaluation.

voltage of 110 kV and above. The transmission line performance statistics are given on a per-100-km-yr basis for line-related outages, and a summary of these statistics is provided in Table 28.11. The report also provides performance statistics of transformers and circuit breakers.

Tables 28.12 and 28.13 provide summaries of transformer and circuit breaker statistics by voltage classification for forced outages involving integral subcomponents and terminal equipment. The data presented in all three tables can be used to calculate HLII adequacy indices. However, it should be noted that, in order to include the effect of station-originated outages, additional data in the form of active failure rates, maintenance rates and switching times are required. Some of these parameters can also be obtained from the CEA database.

The reports produced by CEA on forced outage performance of transmission equipment [16] also provide summaries on a number of other statistics, including the following:

1. transmission line statistics for line-related transient forced outages;
2. cable statistics for cable-related and terminal-related forced outages;
3. synchronous compensator statistics;
4. static compensator statistics by voltage classification;
5. shunt reactor statistics by voltage classification;
6. shunt capacitator statistics by voltage classification;
7. series capacitor bank statistics by voltage classification.

28.4.4 Canadian Electricity Association Equipment Reliability Information System Database for HLIII Evaluation

HLIII adequacy assessment requires outage statistics of equipment in all parts of the system, *i.e.* generation, transmission and distribution. The outage statistics for the generation and transmission equipment are available in [15, 16] as discussed earlier. The distribution reliability data system recently commissioned by CEA completes the basic equipment reliability information system, containing generic performance and reliability information on all three major constituent areas of generation, transmission and distribution equipment. The required distribution component failure statistics are presently being collected. These data can be used for quantitative assessment of power system reliability and to assist in the planning, design, operation and maintenance of distribution systems. They will also support the quantitative appraisal of alternatives and the optimization of cost/reliability worth.

In the reporting system adopted for CEA-ERIS, distribution equipment has been divided into a number of major components and therefore the system can be referred to as component-oriented. Every event involves one major component (unless it is a common-mode failure) and the outage of a major component due to that of another component is not recorded. In this way it is possible to reconstruct the outage performance of any given configuration from the record of the various components.

The following devices have been selected as major components:

1. distribution line;
2. distribution cable;
3. distribution transformer;

Table 28.11. Summary of transmission line statistics

Voltage (kV)	Line-related sustained outages			Terminal-related sustained outages		
	Frequency (per yr)	Mean duration (hr)	Unavailability (%)	Frequency (per yr)	Mean duration (hr)	Unavailability (%)
110–149	2.4373	13.4	0.373	0.1074	5.8	0.007
150–199	0.8007	0.3	0.054	0.0263	21.8	0.007
200–299	1.0510	8.9	0.107	0.1572	9.7	0.017
300–399	0.2363	49.7	0.134	0.0593	15.8	0.011
500–599	1.7224	3.1	0.061	0.2032	16.0	0.037
600–799	0.2808	119.2	0.382	0.0143	13.9	0.023

Table 28.12. Summary of transformer bank statistics by voltage classification for forced outages

Voltage (kV)	Involving integral subcomponents		Involving terminal equipment	
	Frequency (per yr)	Mean duration (hr)	Frequency (per yr)	Mean duration (hr)
110–149	0.0504	163.8	0.0914	63.7
150–199	0.1673	113.9	0.0662	420.3
200–299	0.0433	169.4	0.0706	52.2
300–399	0.0505	119.4	0.0385	58.0
500–599	0.0389	277.3	0.0429	34.6
600–799	0.0320	176.9	0.0233	354.5

Table 28.13. Summary of circuit breaker statistics by voltage classification for forced outages

Voltage (kV)	Involving integral subcomponents		Involving terminal equipment	
	Frequency (per yr)	Mean duration (hr)	Frequency (per yr)	Mean duration (hr)
110–149	0.0373	136.4	0.0468	55.4
150–199	0.0313	74.9	0.0251	145.5
200–299	0.0476	164.2	0.0766	37.7
300–399	0.0651	134.6	0.0322	59.5
500–599	0.0795	125.1	0.0948	37.7
600–799	0.1036	198.8	0.0566	118.1

4. power transformer;
5. regulator;
6. capacitor.

There are other distribution equipment components which could also have been selected, but it was decided that they were either too difficult to define or that the added complexity was not justified. It was therefore decided to use a limited number of clearly defined major components.

28.5 System Reliability Worth

Adequacy studies of a system are only part of a required overall assessment. The economics of alternative facilities play a major role in the decision-making process. The simplest approach which can be used to relate economics with reliability is to consider the investment cost only. In this approach, the increase in reliability due to the various alternative reinforcement or expansion schemes is evaluated, together with the investment cost associated with each scheme. Dividing this cost by the increase in reliability gives the incremental cost of reliability, *i.e.* how much it will cost for a per unit increase in reliability. This approach is useful for comparing alternatives when it is known for certain that the reliability of a section of the power system must be increased, the lowest incremental cost of reliability being the most cost effective. This is a significant step forward compared with assessing alternatives and making major capital investment decisions using deterministic techniques.

The weakness of the approach is that it is not related to either the likely return on investment or the real benefit accruing to the consumer, utility and society. In order to make a consistent appraisal of economics and reliability, albeit only the adequacy, it is necessary to compare the adequacy cost (the investment cost needed to achieve a certain level of adequacy) with the adequacy worth (the benefit derived by the utility, consumer and society). A step in this direction is achieved by setting a level of incremental cost which is believed to be acceptable to consumers. Schemes costing less than this level would be considered viable, but schemes costing greater than this level would be rejected. A complete solution however requires a detailed knowledge of adequacy worth.

This type of economic appraisal is a fundamental and important area of engineering application, and it is possible to perform this kind of evaluation at the three hierarchical levels discussed. A goal for the future should be to extend this adequacy comparison within the same hierarchical structure to include security and therefore to arrive at reliability-cost and reliability-worth evaluation. The basic concept is relatively simple and can be illustrated using the cost/reliability curves of Figure 28.8.

The curves in Figure 28.8 show that the utility cost will generally increase as consumers are provided with higher reliability. On the other hand, the consumer costs associated with supply interruptions will decrease as the reliability increases. The total costs to society will therefore be the sum of these two individual costs. This total cost exhibits a minimum and so an "optimum" or target level of reliability is achieved. In this approach, the reliability level is not a fixed value and results from the cost minimization process.

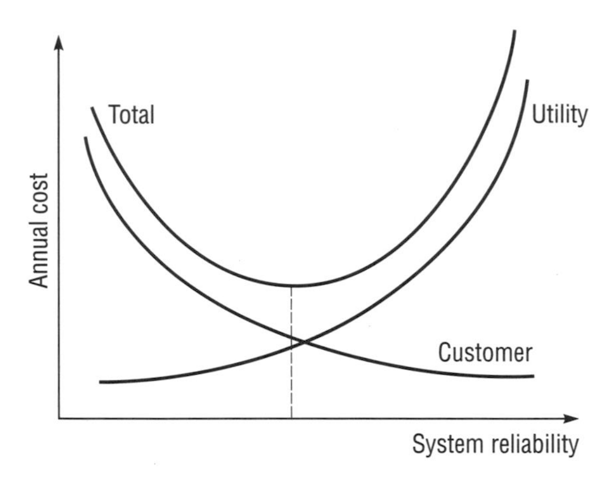

Figure 28.8. Consumer, utility and total cost as a function of system reliability

This concept is quite valid. Two difficulties arise in its assessment. First the calculated indices are assessments at the various hierarchical levels. Second, there are great problems in assessing consumer perceptions of outage costs.

There have been many studies concerning interruption and outage costs [7–12]. These studies show that, although trends are similar in virtually all cases, the costs vary over a wide range and depend on the country of origin and the type of consumer. It is apparent therefore that there is still considerable research needed on the subject of the cost of an interruption. This research should consider the direct and indirect costs associated with the loss of supply, both on a local and widespread basis. A recent CIGRE report [30] provides a wide range of international data on interruption costs and their utilization, together with a comprehensive bibliography of relevant material.

28.6 Guide to Further Study

As previously noted, there are a large number of excellent publications [7–12] on the subject of power system reliability evaluation. Reference [5] is a compendium of important publications selected from [7–11] and includes some seminal references. The basic techniques required in HLI, HLII and distribution system adequacy evaluation are described in detail in [3]. This reference contains many numerical examples and is used as a basic textbook on power system reliability evaluation. Reference [3] illustrates HLI adequacy evaluation in terms of the basic probability methods which provide LOLE and LOEE indices and in terms of frequency (F) duration (D) indices. The $F\&D$ approach provides additional information which can prove useful in capacity planning and evaluation. The recursive algorithms required when developing these indices for practical systems are illustrated by application to numerical examples. The text then extends the single system analysis by application to interconnected power systems.

Reliability evaluation at HLI includes recognition of operating or spinning reserve requirements. Reference [3] provides a chapter on this subject which includes both operating and response risk evaluation. The subject of HLII evaluation using contingency enumeration is presented by application to selected system configurations. The required calculations are shown in detail. Reference [3] also presents detailed examples of distribution system reliability evaluation in both radial and meshed networks. These applications include many practical system effects such as lateral distributor protection, disconnects and load transfer capability in radial networks. The meshed analysis includes the recognition of permanent and transient outages, maintenance considerations, adverse weather and common-mode outages. The meshed network analysis is also extended to include partial in addition to total loss of continuity. The concepts utilized in transmission and distribution system evaluation are then extended and applied to substation and switching stations. Reference [3] concludes by illustrating how basic reliability techniques can be used to assess general plants and equipment configurations. Reference [3] was extended in the second edition to include chapters dealing with reliability evaluation of power systems using Monte Carlo simulation and incorporating outage costs and reliability worth assessment. Reference [6] is focused entirely on the utilization of Monte Carlo simulation in power system reliability assessment. These two texts jointly provide the basic understanding required to read and appreciate the wide range of techniques described in [7–12]. The concepts associated with the inclusion of embedded generation in distribution system reliability evaluation are described in [31] and an extended review of the application of reliability concepts in the new competitive environment is detailed in [32]. A careful examination of [3] will provide the basic understanding required to read many of the papers detailed in References [7–12].

References

[1] Billinton R, Allan RN. Reliability evaluation of engineering systems, 2nd ed. New York: Plenum Press; 1992.

[2] Billinton R, Allan RN. Power system reliability in perspective. IEE Electron Power 1984;30(3):231–6.

[3] Billinton R, Allan RN. Reliability evaluation of power systems, 2nd ed. New York: Plenum Press; 1996.

[4] Billinton R, Allan RN. Reliability assessment of large electric power systems. Boston: Kluwer Academic; 1988.

[5] Billinton R, Allan RN, Salvaderi L. Applied reliability assessment in electric power systems. New York: IEEE Press; 1991.

[6] Billinton R, Li W. Reliability assessment of electric power systems using Monte Carlo methods. New York: Plenum Press; 1994.

[7] Billinton R. Bibliography on the application of probability methods in power system reliability evaluation. IEEE Trans Power App Syst 1972;PAS-91(2):649–60.

[8] IEEE Sub-Committee on Application of Probability Methods. Bibliography on the application of probability methods in power system reliability evaluation 1971–1977. IEEE Trans Power App Syst 1978;PAS-97(6):2235–42.

[9] Allan RM, Billinton R, Lee SH. Bibliography on the application of probability methods in power system reliability evaluation 1977–1982. IEEE Trans Power App Syst 1984;PAS-103(2):275–82.

[10] Allan RN, Billinton R, Shahidehpour SM, Singh C. Bibliography on the application of probability methods in power system reliability evaluation 1982–1987. IEEE Trans Power Syst 1988;3(4):1555–64.

[11] Allan RN, Billinton R, Breipohl AM, Grigg C. Bibliography on the application of probability methods in power system reliability evaluation 1987–1991. IEEE Trans Power Syst 1994;9(1):41–9.

[12] Allan RN, Billinton R, Breipohl AM, Grigg C. Bibliography on the application of probability methods in power system reliability evaluation 1992–1996. IEEE Trans Power Syst 1999;PWRS-14:51–7.

[13] CIGRE WG 38.03. Power system reliability analysis, vol. 1. Application guide. Paris: CIGRE Publication; 1987. Ch. 4, 5.

[14] Allan RN, Billinton R. Concepts of data for assessing the reliability of transmission and distribution equipment. 2nd IEE Conf on Reliability of Transmission and Distribution Equipment. Conference Publication 406; March 1995. p. 1–6.

[15] Canadian Electrical Association Equipment Reliability Information System Report. 1998 Annual Report—Generation equipment status. Montreal: CEA; September 1999.

[16] Canadian Electrical Association Equipment Reliability Information System Report. Forced outage performance of transmission equipment for the period January 1994 to December 1998. Montreal: CEA; February 2000.

[17] The Electricity Association: National Fault and Interruption Reporting Scheme (NAFIRS). London: Electricity Association.

[18] Canadian Electrical Association Equipment Reliability Information System Report. 1999 Annual Service Continuity Report on distribution system performance in Canadian Electrical Utilities. May 2000.

[19] Office of Gas and Electricity Markets (OFGEM). Report on distribution and transmission system performance 1998/99. London: OFGEM.

[20] Canadian Electrical Association Equipment Reliability Information System Report. Bulk electricity system delivery point interruptions 1998–1990 Report. October 1992.

[21] National Grid Company (UK). Performance of the Transmission System. Annual Reports to the Director General of Electricity Supply, OFFER.

[22] Billinton R, Kumar S, Chowdhury N, Chu K, Debnath K, Goel L, *et al.* A reliability test system for educational purposes—basic data. IEEE Trans PWRS 1989;4(3):1238–44.

[23] Billinton R, Kumar S, Chowdhury N, Chu K, Goel L, Khan E, *et al.* A reliability test system for educational purposes—basic results. IEEE Trans PWRS 1990;5(1):319–25.

[24] Allan RN, Billinton R, Sjarief I, Goel L. A reliability test system for educational purposes—basic distribution system data and results. IEEE Trans PWRS 1991;6(2):813–20.

[25] Billinton R, Vohra PK, Kumar S. Effect of station originated outages in composite system adequacy evaluation of the IEEE reliability test system. IEEE Trans PAS 1985;104:2649–56.

[26] NGC Settlements. An introduction to the initial pool rules. Nottingham; 1991.

[27] Medicherla TKP, Billinton R. Overall approach to the reliability evaluation of composite generation and transmission systems. IEE Proc C 1980;127(2):72–81.

[28] CIGRE WG 38.03. Power system reliability analysis, vol. 2. Composite power system reliability evaluation. Paris: CIGRE Publication; 1992.

[29] Roman Ubeda J, Allan RN. Sequential simulation applied to composite system reliability evaluation. Proc IEE C 1992;139:81–6.

[30] CIGRE TF 38.06.01. Methods to consider customer interruption costs in power system analysis. CIGRE; 2001.

[31] Jenkins N, Allan RN, Crossley P, Kirschen DS, Strbac G. Embedded generation. Stevenage: IEE Publishing; 2000.

[32] Allan RN, Billinton R. Probabilistic assessment of power systems. Proc IEEE 2000;88(2):140–62.

Chapter 29

Human and Medical Device Reliability

B. S. Dhillon

29.1 Introduction

The beginning of the human reliability field may be taken as the middle of the 1950s when the problem of human factors was seriously considered and the first human-engineering standards for the US Air Force were developed [1, 2]. In 1958, the need for human reliability was clearly identified [3] and some findings relating to human reliability were documented in two Convair-Astronautics reports [4, 5].

In 1962, a database (*i.e.* Data Store) containing time and human performance reliability estimates for human-engineering design features was established [6]. In 1986, the first book on human reliability was published [7]. All in all, over the years many other people have contributed to the field of human reliability and a comprehensive list of publications on the subject is given in [8].

The latter part of the 1960s may be regarded as the beginning of the medical device reliability field. During this period many publications on the subject appeared [9–13]. In 1980, an article listed most of the publications on the subject and in 1983, a book on reliability devoted an entire chapter to medical device/equipment reliability [14, 15]. A comprehensive list of publications on medical device reliability is given in [16].

This chapter presents various aspects of human and medical device reliability.

29.2 Human and Medical Device Reliability Terms and Definitions

Some of the common terms and definitions concerning human and medical device reliability are as follows [7, 15, 17–23].

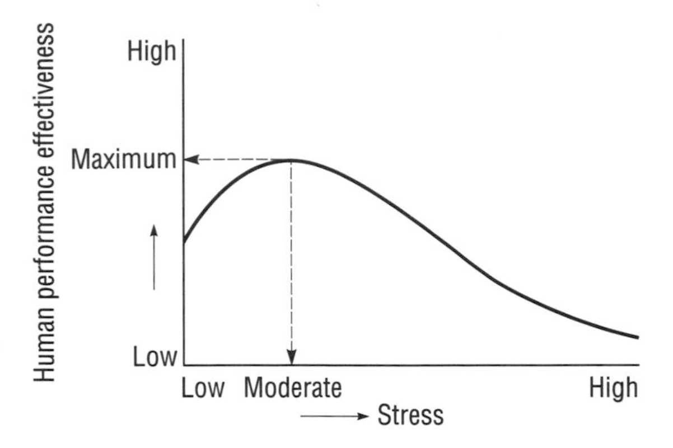

Figure 29.1. A hypothetical curve showing the human performance effectiveness versus stress relationship

- **Human reliability.** This is the probability of performing a given task successfully by the human at any required stage in system operations within a specified minimum time (if such time requirement is stated).
- **Human error.** This is the failure to perform a task (or the performance of a forbidden action) that could lead to the disruption of scheduled operations or damage to property and equipment.
- **Human factors.** This is the body of scientific facts concerning the characteristics of humans.
- **Human performance.** This is a measure of human functions and actions subject to specified conditions.
- **Medical device.** This is any machine, apparatus, instrument, implant *in vitro* reagent, implement contrivance, or other related or similar article, including any component, part, or accessory, that is intended for application in diagnosing diseases or other conditions or in the cure, treatment, mitigation, or prevention of disease or intended to affect the structure of any function of the body.
- **Reliability.** This is the probability that an item will perform its specified function satisfactorily for the stated time period when used according to the designed conditions.
- **Failure.** This is the inability of an item/equipment to function within the specified guidelines.
- **Safety.** This is conservation of human life and its effectiveness, and the prevention of damage to item/equipment according to operational requirements.
- **Quality.** There are many definitions used in defining quality and one of these is conformance to requirements.
- **Mean time to failure.** This is the sum of operating time of given items over the total number of failures (*i.e.* when times to failure are exponentially distributed).

29.3 Human Stress—Performance Effectiveness, Human Error Types, and Causes of Human Error

Human performance varies under different conditions and some of the factors that affect a person's performance are time at work, reaction to stress, social interaction, fatigue, social pressure, morale, supervisor's expectations, idle time, repetitive work, and group interaction and identification [24].

Stress is probably the most important factor that affects human performance, and in the past many researchers have studied the relationship between human performance effectiveness and stress. Figure 29.1 shows the resulting conclusion of their efforts [18,25]. This figure basically shows that the human performance for a specific task follows a curvilinear relation to the imposed stress. At a very low stress level, the task is dull and unchallenging, thus most humans' performance will not be at the optimal level. However, at a moderate stress level, the humans usually perform their task at optimal level. As the stress bypasses its moderate level, the human performance begins to decline and Figure 29.1 shows that in the highest stress region, human reliability is at its lowest.

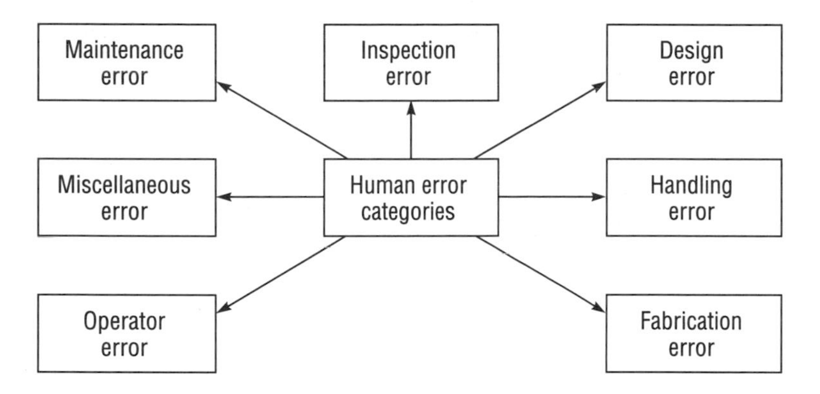

Figure 29.2. Types of human error

A human error may be categorized into many different classifications, as shown in Figure 29.2. These are design error, operator error, maintenance error, fabrication error, inspection error, handling error, and miscellaneous error [7, 26]. Although all these categories or types are self-explanatory, the miscellaneous error represents the situation where it is difficult to differentiate between human failure and equipment failure. Sometimes, the miscellaneous error is also referred to as the contributory error.

Human errors occur due to various reasons. Some of the important causes for the occurrence of human error are task complexity, poor equipment design, improper work tools, poorly written maintenance and operating procedures, inadequate training or skill, inadequate lighting in the work area, high noise level, crowded work space, high temperature in the work area, inadequate work layout, poor motivation, poor verbal communication, and inadequate handling of equipment [7,26].

29.4 Human Reliability Analysis Methods

There are many methods used in performing human reliability analysis [27]. These include the probability tree method, fault tree method, Markov method, throughput ratio method, block diagram method, personnel reliability index, technique for human error rate prediction (THERP), and Pontecorvo's method of predicting human reliability [7, 27]. The first three methods are described below.

29.4.1 Probability Tree Method

This approach is concerned with performing task analysis diagrammatically. More specifically, diagrammatic task analysis is represented by the branches of the probability tree. The tree branching limbs denote outcomes (*i.e.* success or failure) of each event, and in turn the occurrence probability is assigned to each branch of the tree.

There are many advantages of the probability tree method, including a visibility tool, lower probability of error due to computation because of computational simplification, the prediction of quantitative effects of errors, and the incorporation, with some modifications, of factors such as interaction stress, emotional stress, and interaction effects. The following example demonstrates this method.

Example 1. Assume that the task of an operator is composed of two subtasks, y and z. Each of these subtasks can either be performed successfully or unsuccessfully, and subtask y is accomplished before subtask z. The performance of subtasks unsuccessfully is the only error that can occur,

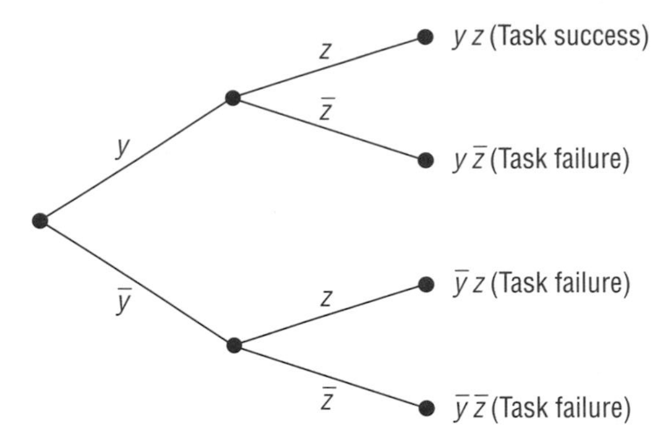

Figure 29.3. Probability tree diagram for Example 1

and the performance of both the subtasks is independent of each other.

Develop a probability tree and obtain an expression for the probability of performing the overall task incorrectly.

A probability tree for Example 1 is shown in Figure 29.3. The meanings of the symbols used in Figure 29.3 are defined below:

y = the subtask y is performed correctly

$\bar{\mathrm{y}}$ = the subtask y is performed incorrectly

z = the subtask z is performed correctly

$\bar{\mathrm{z}}$ = the subtask z is performed incorrectly

Using Figure 29.3, the probability of performing the overall task incorrectly is:

$$F_{\mathrm{t\,in}} = P_{\mathrm{y}} P_{\bar{\mathrm{z}}} + P_{\bar{\mathrm{y}}} P_{\mathrm{z}} + P_{\bar{\mathrm{y}}} P_{\bar{\mathrm{z}}} \tag{29.1}$$

where

$F_{\mathrm{t\,in}}$ is the probability of performing the overall task incorrectly

P_{y} is the probability of performing subtask y correctly

P_{z} is the probability of performing subtask z correctly

$P_{\bar{\mathrm{y}}}$ is the probability of performing subtask y incorrectly

$P_{\bar{\mathrm{z}}}$ is the probability of performing subtask z incorrectly.

29.4.2 Fault Tree Method

This is a widely used method in reliability analysis of engineering systems and it can also be used in performing human reliability analysis. This method is described in detail in [24]. Nonetheless, this approach makes use of the following symbols in developing a basic fault tree:

- **Rectangle.** This is used to denote a fault event that results from the combination of failure events through the input of a logic gate, e.g. AND or OR gate.
- **AND gate.** This gate denotes that an output fault event occurs if all the input fault events occur. Figure 29.4 shows a symbol for an AND gate.
- **OR gate.** This gate denotes that an output fault event occurs if any one or more of the input fault events occur. Figure 29.4 shows a symbol for an OR gate.
- **Circle.** This is used to denote basic fault events; more specifically, those events that need not be developed any further.

The following example demonstrates this method:

Example 2. A person is required to perform a job Z composed of four distinct and independent tasks A, B, C, and D. All of these tasks must be carried out correctly for the accomplishment of the job successfully. Tasks A and B are composed of two and three independent steps, respectively. More specifically, for the successful performance

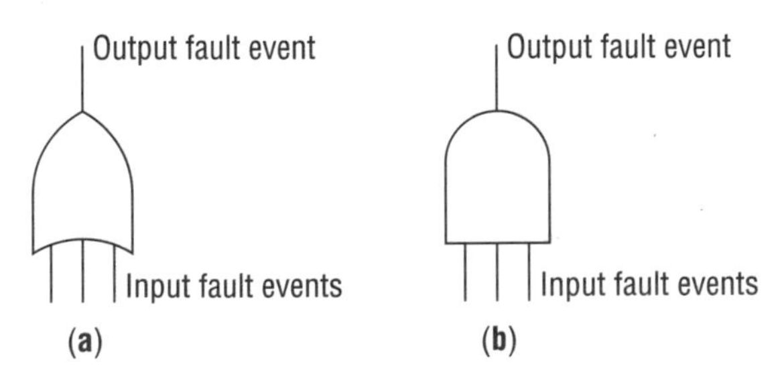

Figure 29.4. Symbols for basic logic gates: (a) OR gate; (b) AND gate

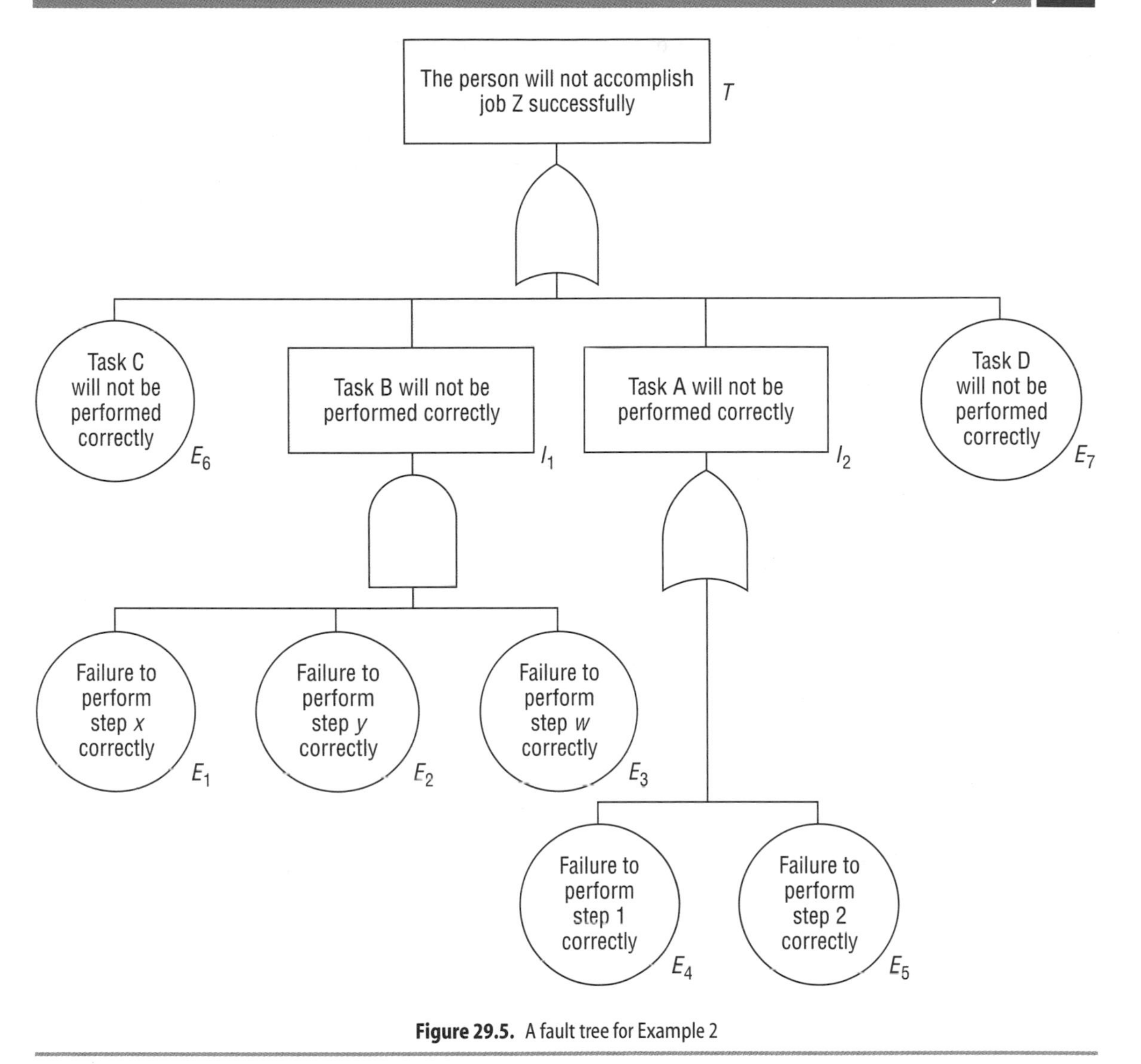

Figure 29.5. A fault tree for Example 2

of task A all the steps must be performed correctly, whereas for task B at least one of the steps must be accomplished correctly.

Develop a fault tree for this example by using the defined symbols above. Assume that the top fault event is "The person will not accomplish job Z successfully". Calculate the probability of occurrence of the top event, if the probabilities of failure to perform tasks C, D, and the steps involved in tasks A and B are 0.04, 0.05, and 0.1, respectively.

A fault tree for Example 2 is shown in Figure 29.5. The basic fault events are denoted by E_i, for $i = 1, 2, 3, 4, 5, 6$ and the intermediate fault events by I_i, for $i = 1, 2$. The letter T denotes the top fault event.

The probability of occurrence of intermediate event I_1 is given by [24]:

$$\begin{aligned} P(I_1) &= P(E_1)P(E_2)P(E_3) \\ &= (0.1)(0.1)(0.1) \\ &= 0.001 \end{aligned}$$

Similarly, the probability of occurrence of intermediate event I_2 is given by [24]:

$$\begin{aligned} P(I_2) &= P(E_4) + P(E_5) - P(E_4)P(E_5) \\ &= 0.1 + 0.1 - (0.1)(0.1) \\ &= 0.19 \end{aligned}$$

The probability of occurrence of the top fault event T is expressed by [24]:

$$\begin{aligned} P(T) &= 1 - [1 - P(E_6)][1 - P(E_7)] \\ &\quad \times [1 - P(I_1)][1 - P(I_2)] \\ &= 1 - [1 - 0.04][1 - 0.05] \\ &\quad \times [1 - 0.001][1 - 0.19] \\ &= 0.2620 \end{aligned}$$

It means the probability of occurrence of the top event T (*i.e.* the person will not accomplish job Z successfully) is 0.2620.

29.4.3 Markov Method

This is a powerful tool that can also be used in performing various types of human reliability analysis [24]. The following example demonstrates the application of this method in human reliability analysis [28].

Example 3. Assume that an operator is performing his/her tasks in fluctuating environments: normal, abnormal. Although human error can occur under both environments, it is assumed that the human error rate under the abnormal environment is greater than under the normal environment because of increased stress. The system state space diagram is shown in Figure 29.6. The numerals in the boxes denote system state. Obtain an expression for mean time to human error (MTTHE).

The following symbols were used to develop equations for the diagram in Figure 29.6:

- $P_i(t)$ is the probability of the human being in state i at time t; for $i = 0$ (means human performing task correctly under normal environment), $i = 2$ (means human committed error in normal environment), $i = 1$ (means human performing task correctly in abnormal environment), $i = 3$ (means human committed error in abnormal environment)
- λ_h is the constant human error rate from state 0
- λ_{ah} is the constant human error rate from state 1
- α_n is the constant transition rate from normal to abnormal environment
- α_a is the constant transition rate from abnormal to normal environment.

By applying the Markov method, we obtain the following system of differential equations for Figure 29.6:

$$\frac{dP_0(t)}{dt} + (\lambda_h + \alpha_n)P_0(t) = P_1(t)\alpha_a \tag{29.2}$$

$$\frac{dP_1(t)}{dt} + (\lambda_{ah} + \alpha_a)P_1(t) = P_0(t)\alpha_n \tag{29.3}$$

$$\frac{dP_2(t)}{dt} - P_0(t)\lambda_h = 0 \tag{29.4}$$

$$\frac{dP_3(t)}{dt} - P_1(t)\lambda_{ah} = 0 \tag{29.5}$$

At time $t = 0$, $P_0(0) = 1$ and $P_1(0) = P_2(0) = P_3(0) = 0$. Solving Equations 29.2–29.5, we get:

$$\begin{aligned} P_0(t) &= (s_2 - s_1)^{-1}[(s_2 + \lambda_{ah} + \alpha_n)\, e^{s_2 t} \\ &\quad - (s_1 + \lambda_{ah} + \alpha_n)\, e^{s_1 t}] \end{aligned} \tag{29.6}$$

where

$$\begin{aligned} s_1 &= [-a_1 + (a_1^2 - 4a_2)^{1/2}]/2 \\ s_2 &= [-a_1 - (a_1^2 - 4a_2)^{1/2}]/2 \\ a_1 &= \lambda_h + \lambda_{ah} + \alpha_n + \alpha_a \\ a_2 &= \lambda_h(\lambda_{ah} + \alpha_a) + \lambda_{ah}\alpha_n \end{aligned}$$

$$P_2(t) = a_4 + a_5\, e^{s_2 t} - a_6\, e^{s_1 t} \tag{29.7}$$

where

$$\begin{aligned} a_3 &= 1/(s_2 - s_1) \\ a_4 &= \lambda_h(\lambda_{ah} + \alpha_n)/s_1 s_2 \\ a_5 &= a_3(\lambda_h + a_4 s_1) \\ a_6 &= a_3(\lambda_h + a_4 s_2) \end{aligned}$$

$$P_1(t) = a_3\alpha_n(e^{s_2 t} - e^{s_1 t}) \tag{29.8}$$

$$P_3(t) = a_7[(1 + a_3)(s_1\, e^{s_2 t} - s_2\, e^{s_1 t}) \tag{29.9}$$

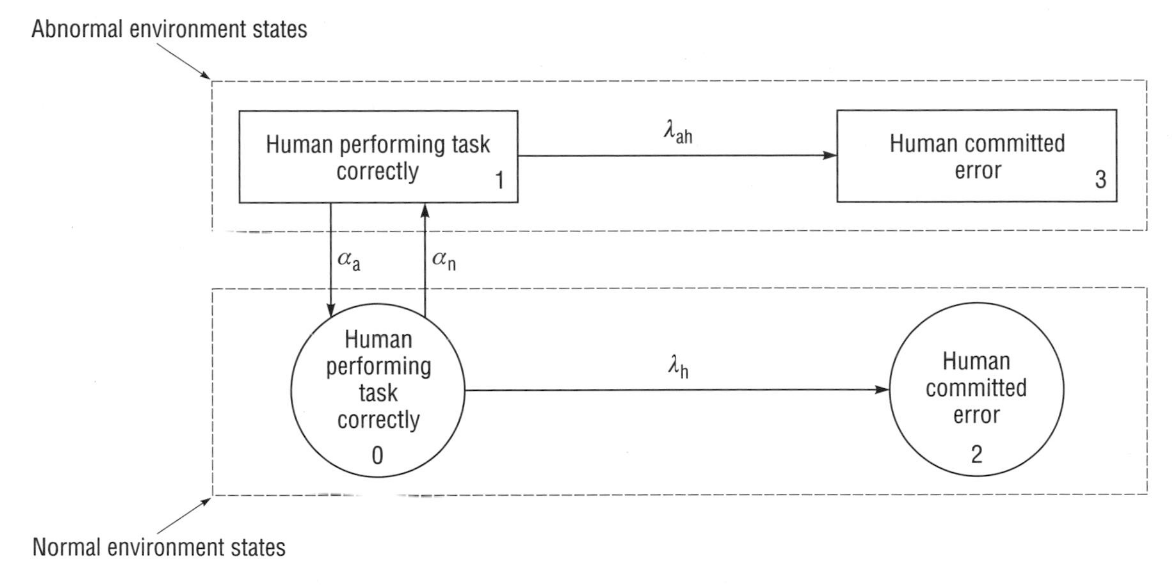

Figure 29.6. System state space diagram

where

$$a_7 = \lambda_{ah}\alpha_n / s_1 s_2$$

The reliability of the human is given by:

$$R_h(t) = P_0(t) + P_1(t) \quad (29.10)$$

Mean time to human error is given by:

$$\text{MTTHE} = \int_0^\infty R_h(t)\,dt = (\lambda_{ah} + \alpha_n + \alpha_a)/a_2 \quad (29.11)$$

29.5 Human Unreliability Data Sources

The United States Department of Defense and Atomic Energy Commission (AEC) were probably the first to provide an impetus to develop formal data collection and analysis methods for quantifying human performance reliability [29]. In 1962, the American Institute for Research (AIR) was probably the first organization to establish a human unreliability data bank called Data Store [6].

Over the years many other data banks have been established [7, 29]. Needless to say, today there are many sources for obtaining human error-related data and Table 29.1 presents some of these sources. References [7, 29] list many more such sources.

29.6 Medical Device Reliability Related Facts and Figures

Some of the facts and failures directly or indirectly related to medical device reliability are as follows.

- In 1983, the United States medical device industry employed around 200 000 persons and at least 3000 companies were involved with medical devices [38].
- In 1997, a total of 10 420 registered manufacturers were involved in the manufacture of medical devices in the United States alone [39].
- In 1969, 10 000 medical device-related injuries resulting in 731 deaths over a 10-year period were reported by the US Department of Health, Education, and Welfare special committee [40].

Table 29.1. Sources for obtaining human unreliability data

No.	Source
1	Data Store [6]
2	Book: Human reliability with human factors [7]
3	Book: Human reliability and safety analysis data handbook [30]
4	Operational Performance Recording and Evaluation Data System (OPREDS) [31]
5	Bunker–Ramo tables [32]
6	Aviation Safety Reporting System [33]
7	Technique for Establishing Personnel Performance Standards (TEPPS) [34]
8	Aerojet General Method [35]
9	Book: Mechanical reliability: theory, models, and applications [36]
10	Air Force Inspection and Safety Center Life Sciences Accident and Incident Reporting System [37]

- In 1997, the world market for medical devices was estimated to be around $120 billion [41].
- A study reported that more than 50% of all technical medical equipment problems were due to operator errors [42].
- In 1990, a study conducted by the Food and Drug Administration (FDA) over a period from October 1983 to September 1989 reported that approximately 44% of the quality-related problems led to the voluntary recall of medical devices [43]. The study also stated that effective design controls could have prevented such problems.
- In 1990, the Safe Medical Device Act (SMDA) was passed by the US Congress and in turn, it strengthened the FDA to implement the Preparation Quality Assurance Program [16].
- In 1969, a study reported that faulty instrumentation accounts for 1200 deaths per year in the United States [44, 45].

29.7 Medical Device Recalls and Equipment Classification

Defective medical devices are subject to recalls and repairs in the United States. The passage of the Medical Device Amendments of 1976 has helped to increase the public's awareness of these activities and their number. The FDA Enforcement Report publishes data on medical device recalls [46]. For example, this report for the period of 1980–1982 announced 230 medical device-related recalls and the following categories of problem areas:

- faulty product design;
- contamination;
- mislabeling;
- defects in material selection and manufacturing;
- defective components;
- no premarket approval and failure to comply with good manufacturing practices (GMPs);
- radiation (X-ray) violations;
- misassembly of parts;
- electrical problems.

The number of recalls associated with each of the above nine categories of problem areas were 40, 40, 27, 54, 13, 4, 25, 10, and 17, respectively. It is important to note that around 70% of the medical device recalls were due to faulty design, product contamination, mislabeling, and defects in material selection and manufacturing.

Six components of the faulty design were premature failure, potential for malfunction, electrical interference, alarm defects, potential for leakage of fields into electrical components, and failure to perform as required. There were four subcategories of product contamination: defective package seals, non-sterility, other package defects, and other faults. The mislabeling category included components such as incomplete labeling, misleading or incorrect labeling, disparity between label and product, and inadequate labeling. Five components of defects in material selection and manufacturing were manufacturing defects, material deterioration, inappropriate materials, separation of bounded components, and actual or potential breakage/cracking.

After the passage of the Medical Device Amendments of 1976, the FDA categorized medical devices prior to 1976 into three classifications [40].

- **Category I.** This included devices in which general controls such as GMPs were considered adequate in relation to efficacy and safety.
- **Category II.** This included devices in which general controls were deemed inadequate in relation to efficacy and safety but in which performance standards could be established.
- **Category III.** This included devices in which the manufacturer must submit evidence of efficacy and safety through well-designed studies.

Nonetheless, the health care system uses a large variety of electronic equipment and it can be classified into three groups [47].

- **Group X.** This contains equipment/devices that are considered responsible for the patient's life or may become so during emergency. More specifically, when such items fail, there is seldom sufficient time for repair, thus these items must have high reliability. Some examples of the equipment/devices belonging to this group are respirators, cardiac defibrillators, electro-cardiographic monitors, and cardiac pacemakers.
- **Group Y.** This includes the vast majority of equipment used for purposes such as routine or semi-emergency diagnostic or therapeutic. More specifically, the failure of equipment belonging to this group does not lead to the same emergency as in the case of group X equipment. Some examples of group Y equipment are gas analyzers, spectrophotometers, ultrasound equipment, diathermy equipment, electro-cardiograph and electroencephalograph recorders and monitors, and colorimeters.
- **Group Z.** This includes equipment not critical to a patient's life or welfare. Wheelchairs and bedside television sets are two typical examples of group Z equipment.

29.8 Human Error in Medical Devices

Past experience indicates that a large percentage of medical device-related problems are due to human error. For example, 60% of deaths or serious injuries associated with medical devices reported through the FDA Center for Devices and Radiological Health (CDRH) were due to user error [48]. Nonetheless, various studies conducted over the years have identified the most error-prone medical devices [49]. Some of these devices (in order of most error-prone to least error-prone) are as follows [49]:

- glucose meter;
- balloon catheter;
- orthodontic bracket aligner;
- administration kit for peritoneal dialysis;
- permanent pacemaker electrode;
- implantable spinal cord simulator;
- intra-vascular catheter;
- infusion pump;
- urological catheter;
- electrosurgical cutting and coagulation device;
- non-powered suction apparatus;
- mechanical/hydraulic impotence device;
- implantable pacemaker;
- peritoneal dialysate delivery system;
- catheter introducer;
- catheter guidewire.

29.9 Tools for Medical Device Reliability Assurance

There are many methods and techniques that can be used to improve the reliability of medical devices [16,24]. Some of the commonly used ones are described below.

29.9.1 General Method

This method is composed of a number of steps that can be used to improve medical equipment reliability. These steps are as follows [42]:

- specify required reliability parameters and their values in design specifications;
- allocate specified reliability parameter values;
- evaluate medical equipment reliability and compare the specified and theoretical values;
- modify design if necessary;
- collect field failure data and perform necessary analysis;
- make recommendations if necessary to improve weak areas of the medical equipment.

29.9.2 Failure Modes and Effect Analysis

The history of failure modes and effect analysis (FMEA) goes back to the early 1950s when the technique was used in the design and development of flight control systems. Since then the method has received widespread acceptance in the industry. FMEA is a tool to evaluate design at the initial stage from the reliability aspects. This criterion helps to identify the need for and the effects of design change. Furthermore, the procedures demand listing the potential failure modes of each and every component on paper and their effects on the listed subsystems. There are seven main steps involved in performing failure modes and effect analysis: (i) establishing system definition; (ii) establishing ground rules; (iii) describing system hardware; (iv) describing functional blocks; (v) identifying failure modes and their effects; (vi) compiling critical items; (vii) documenting.

This method is described in detail in [24] and a comprehensive list of publications on the technique is given in [50].

29.9.3 Fault Tree Method

As in the case of human reliability analysis, the fault tree method can also be used in medical device reliability analysis. This method was developed in the early 1960s at Bell Laboratories to perform analysis of the Minuteman Launch Control System. Fault tree analysis starts by identifying an undesirable event, called the top event, associated with a system. Events which could cause the top event are generated and connected by logic operators such as AND or OR. The fault tree construction proceeds by generation of events in a successive manner until the events (basic fault events) need not be developed further. The fault tree itself is the logic structure relating the top event to the basic events.

This method is described in detail in [16, 24].

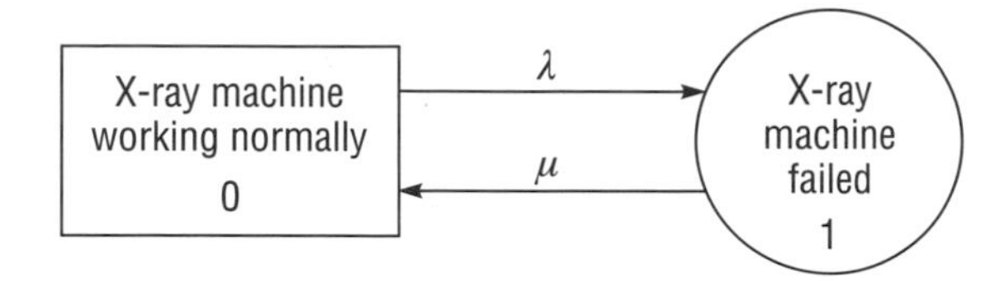

Figure 29.7. X-ray machine transition diagram

29.9.4 Markov Method

This is a very general approach and it can also be used in performing reliability analysis of medical devices. The method is quite appealing when the system failure and repair rates are constant. However, a problem may arise when solving a set of linear algebraic equations for systems with a large number of states. This method is described in detail in [24].

The application of the Markov method in reliability analysis of medical devices is demonstrated through the following example [16].

Example 4. An X-ray machine was observed for a long period and it was concluded that its failure and repair rates are constant. The state space diagram of the machine is given in Figure 29.7. The numerals in the boxes denote the X-ray machine states. Develop an expression for the X-ray machine unavailability by using the Markov method.

Table 29.2. Sources for obtaining data to conduct medical device reliability studies

No.	Source
1	MIL-HDBK-217: Reliability prediction of electronic equipment [51]
2	Medical Device Reporting System (MDRS) [52]
3	Universal Medical Device Registration and Regulatory Management System (UMDRMS) [53]
4	Hospital Equipment Control System (HECS^{TM}) [53]
5	Health Devices Alerts (HDA) [53]
6	Problem Reporting Program (PRP) [53]
7	Medical Device Manufacturers Database (MDMD) [53]
8	User Experience Network (UEN) [53]

The following symbols were used to develop equations for the state space diagram in Figure 29.7:

$P_i(t)$ is the probability that the X-ray machine is in state i at time t; for $i = 0$ (means the X-ray machine is operating normally), $i = 1$ (means the X-ray machine has failed)

λ is the X-ray machine constant failure rate

μ is the X-ray machine constant repair rate.

With the aid of the Markov method, we write the following equations:

$$\frac{\mathrm{d}P_0(t)}{\mathrm{d}t} + \lambda P_0(t) = \mu P_1(t) \quad (29.12)$$

$$\frac{\mathrm{d}P_1(t)}{\mathrm{d}t} + \lambda P_1(t) = \mu P_0(t) \quad (29.13)$$

At time $t = 0$, $P_0(0) = 1$ and $P_1(0) = 0$.

Solving Equations 29.12 and 29.13, we obtain:

$$P_0(t) - \frac{\mu}{\lambda + \mu} + \frac{\lambda}{\lambda + \mu} \mathrm{e}^{-(\lambda+\mu)t} \quad (29.14)$$

$$P_1(t) = \frac{\lambda}{\lambda + \mu} - \frac{\lambda}{\lambda + \mu} \mathrm{e}^{-(\lambda+\mu)t} \quad (29.15)$$

The X-ray machine unavailability is given by:

$$\mathrm{UA_x}(t) = P_1(t) = \frac{\lambda}{\lambda + \mu} - \frac{\lambda}{\lambda + \mu} \mathrm{e}^{-(\lambda+\mu)t} \quad (29.16)$$

where

$\mathrm{UA_x}(t)$ is the X-ray machine unavailability at time t.

29.10 Data Sources for Performing Medical Device Reliability Studies

As in the case of any other engineering product, the failure data are very important in medical device reliability studies. There are many data sources that can be quite useful in performing various types of reliability studies. Some of these sources are listed in Table 29.2.

29.11 Guidelines for Reliability Engineers with Respect to Medical Devices

Some of the guidelines for reliability engineers working in the area of medical devices are as follows [54].

- Remember that during the design, development, and manufacturing phases of a medical device, the responsibility for reliability rests with the manufacturer, but in the field with the user.
- Remember that not all device failures are equal in importance, thus direct your attention to critical failures for maximum returns.
- Avoid using sophisticated methods and techniques developed to improve aerospace systems reliability. For the time being, consider

yourself an expert on failure correction, simple fault tree analysis, failure mode and effects analysis, and so forth.

- Aim to become an effective cost-conscious reliability engineer, thus perform cost versus reliability trade-off analyses. Past experience indicates that some reliability improvement decisions require very little or no additional expenditure.
- Remember that even a simple failure reporting system in a health care organization will be more beneficial in terms of improving medical device reliability than having a large inventory of spares or standbys.
- Use design review, failure mode and effect analysis, qualitative fault tree analysis, and parts review for immediate results.

References

[1] Meister D. Human reliability. In: Muckler FA, editor. Human factors review. Santa Monica, CA: Human Factors Society; 1984. p.13–53.

[2] Swain AD. Overview and status of human factors reliability analysis. Proc 8th Annual Reliability and Maintainability Conf; 1969. p.251–4.

[3] Williams HL. Reliability evaluation of the human component in man-machine systems. Electr Manuf 1958; April: 78–82.

[4] Hettinger CW. Analysis of equipment failures caused by human errors. Report No. REL-R-005. Los Angles: Convair-Astronautics; April 15, 1958.

[5] Meister D. Human engineering survey of difficulties in reporting failure and consumption data. Report No. ZX-7-O35. Los Angles: Convair-Astronautics; September 1958.

[6] Munger SJ, Smith RW, Payne D. An index of electronic equipment operability: data store. Report No. AIR-C43-1/62 RP(29.1). Pittsburgh: American Institute for Research; 1962.

[7] Dhillon BS. Human reliability with human factors. New York: Pergamon Press; 1986.

[8] Dhillon BS. Reliability and quality control: bibliography on general and specialized areas. Gloucester, Canada: Beta Publishers; 1992.

[9] Johnson JP. Reliability of ECG instrumentation in a hospital. Proc Ann Symp Reliability; 1967. p.314–8.

[10] Crump JF. Safety and reliability in medical electronics. Proc Ann Symp Reliability; 1969. p.320–2.

[11] Taylor EF. The effect of medical test instrument reliability on patient risks. Proc Ann Symp Reliability; 1969. p.328–30.

[12] Meyer JL. Some instrument induced errors in the electrocardiogram. J Am Med Assoc 1967;201:351–8.

[13] Grechman R. Tiny flaws in medical design can kill. Hosp Top 1968;46:23–4.

[14] Dhillon BS. Bibliography of literature on medical equipment reliability. Microelectron. Reliab. 1980;20;737–42.

[15] Dhillon BS. Reliability engineering in systems design and operation. New York: Van Nostrand-Reinhold Company; 1983.

[16] Dhillon BS. Medical device reliability and associated areas. Boca Raton, FL: CRC Press; 2000.

[17] MIL-STD-721B. Definitions of effectiveness terms for reliability, maintainability, human factors and safety. Washington, DC: Department of Defense; 1966.

[18] Hagen EW. Human reliability analysis. Nucl Safety 1976;17:315–26.

[19] Meister D. Human factors in reliability. In: Ireson WG, editor. Reliability handbook. New York: McGraw-Hill; 1966. p.12.2–37.

[20] Federal Food, Drug, and Cosmetic Act (as Amended, Sec. 201 (h)). Washington, DC: US Government Printing Office; 1993.

[21] Naresky JJ. Reliability definitions. IEEE Trans Reliab 1970;19:198–200.

[22] Omdahl TP. Reliability, availability and maintainability (RAM) dictionary. Milwaukee, WI: ASQC Quality Press; 1988.

[23] McKenna T, Oliverson R. Glossary of reliability and maintenance terms. Houston, TX: Gulf Publishing Company; 1997.

[24] Dhillon BS. Design reliability: fundamentals and applications. Boca Raton, FL: CRC Press; 1999.

[25] Beech HR, Burns LE, Sheffield BF. A behavioural approach to the management of stress. New York: John Wiley and Sons; 1982.

[26] Meister D. The problem of human-initiated failures. Proc 8th Nat Symp Reliability and Quality Control; 1962, p.234–9.

[27] Meister D. Comparative analysis of human reliability models. Report No. AD 734-432; 1971. Available from the National Technical Information Service, Springfield, VA 22151.

[28] Dhillon BS. Stochastic models for predicting human reliability. Microelectron Reliab 1982;21:491–6.

[29] Dhillon BS. Human error data banks. Microelectron Reliab 1990;30:963–71.

[30] Gertman DI, Blackman HS, Human reliability and safety analysis data handbook. New York: John Wiley and Sons; 1994.

[31] Urmston R. Operational performance recording and evaluation data system (OPREDS). Descriptive Brochure, Code 3400. San Diego, CA: Navy Electronics Laboratory Center; November 1971.

[32] Hornyak SJ. Effectiveness of display subsystems measurement prediction technique. Report No. TR-67-292; September 1967. Available from the Rome Air Development Center (RADC), Griffis Air Force Base, Rome, NY.

[33] Aviation Safety Reporting Program. FAA Advisory Circular No. 00-4613; June 1, 1979. Washington, DC: Federal Aviation Administration (FAA).

[34] Topmiller DA, Eckel JS, Kozinsky EJ. Human reliability data bank for nuclear power plant operations: a review of existing human reliability data banks. Report No. NUREG/CR 2744/1; 1982. Available from the US Nuclear Regulatory Commission, Washington, DC.

[35] Irwin IA, Levitz JJ, Freed AM. Human reliability in the performance of maintenance. Report No. LRP 317/TDR-63-218; 1964. Available from the Aerojet-General Corporation, Sacramento, CA.

[36] Dhillon BS. Mechanical reliability: theory, models, and applications. Washington, DC: American Institute of Aeronatics and Astronautics; 1988.

[37] Life Sciences Accident and Incident Classification Elements and Factors. AFISC Operating Instruction No. AMISCM-127-6; December 1971. Available from the Department of Air Force, Washington, DC.

[38] Federal Policies and the Medical Devices Industry. Office of Technology Assessment. Washington, DC: US Government Printing Office; 1984.

[39] Allen D. California home to almost one-fifth of US medical device industry. Med Dev Diag Ind Mag 1997;19:64–7.

[40] Banta HD. The regulation of medical devices. Prevent Med 1990;19:693–9.

[41] Murray K. Canada's medical device industry faces cost pressures, regulatory reform. Med Dev Diag Ind Mag 1997;19:30–9.

[42] Dhillon BS. Reliability technology in health care systems. Proc IASTED Int Symp Computers and Advanced Technology in Medicine Health Care Bioengineering; 1990. p.84–7.

[43] Schwartz AP. A call for real added value. Med Ind Exec 1994;February/March:5–9.

[44] Walter CW. Instrumentation failure fatalities. Electron News, January 27, 1969.

[45] Micco LA. Motivation for the biomedical instrument manufacturer. Proc Ann Reliability and Maintainability Symp; 1972. p.242–4.

[46] Hyman WA. An evaluation of recent medical device recalls Med Dev Diag Ind Mag 1982;4:53–5.

[47] Crump JF. Safety and reliability in medical electronics. Proc Ann Symp Reliability; 1969. p.320–2.

[48] Bogner MS. Medical devices: a new frontier for human factors. CSERIAC Gateway 1993;IV:12–4.

[49] Wiklund ME. Medical device and equipment design. Buffalo Grove, IL: Interpharm Press Inc.; 1995.

[50] Dhillon BS. Failure mode and effects analysis: bibliography. Microelectron Reliab 1992;32:719–31.

[51] MIL-HDBK-217. Reliability prediction of electronic equipment. Washington, DC: Department of Defense.

[52] Medical Device Reporting System (MDRS). Center for Devices and Radiological Health, Food and Drug Administration (FDA), Rockville, MD.

[53] Emergency Care Research Institute (ECRI). 5200 Butler Parkway, Plymouth Meeting, PA.

[54] Taylor EF. The reliability engineer in the health care system. Proc Ann Reliability and Maintainability Symp; 1972. p.245–8.

Probabilistic Risk Assessment

Robert A. Bari

30.1 Introduction

Probabilistic risk assessment (PRA) is an analytical methodology for computing the likelihood of health, environmental, and economic consequences of complex technologies caused by equipment failure or operator error. It can also be used to compute the risks resulting from normal, intended operation of these technologies. PRAs are performed for various end uses. These include: understanding the safety characteristics of a particular engineering design, developing emergency plans for potential accidents, improving the regulatory process, and communicating the risks to interested parties. These are a few broadly stated examples of applications of the methodology.

The requirements of the end uses should determine the scope and depth of the particular assessment. A PRA often requires the collaboration of experts from several disciplines. For example, it could require the designers and operators of a facility to provide a characterization of the normal and potential off-normal modes of behavior of the facility; it could require analysts who develop and codify scenarios for the off-normal events; it could require statisticians to gather data and develop databases related to the likelihood of the occurrence of events; it could require physical scientists to compute the physical consequences of off-normal events; it could require experts in the transport of effluents from the facility to the environment; it could require experts in the understanding of health, environmental, or economic impacts of effluents. Experts in all phases of human behavior related to the normal and off-normal behavior of a facility and its environment could be required.

PRA is a methodology that developed and matured within the commercial nuclear power reactor industry. Therefore the main emphasis and examples in this chapter will be from that industry. Other industries also use disciplined methodologies to understand and manage the safe operation of their facilities and activities. The chemical process, aviation, space, railway, health, environmental, and financial industries have developed methodologies in parallel with the nuclear industry. Over the past decade there has been much activity aimed at understanding the commonalities, strengths, and weaknesses of methodological approaches across industries. Examples of these efforts are contained in the proceedings [1] of a series of conferences sponsored by the International Association for Probabilistic Safety Assessment and Management. In a series of six conferences from 1991 to 2002, risk assessment experts from over 30 countries have gathered to exchange information on accomplishments and challenges from within their respective fields and to forge new collaborations with practitioners across disciplines. Another useful reference in this regard is the proceedings [2] of a workshop by the European Commission on the identification of the need for further standardization and development of a top-level risk assessment standard across different technologies. This workshop included experts from the chemical process, nuclear power, civil structures, transport, food, and health care industries.

Even within the nuclear power industry there have been various approaches and refinements to the methodologies. One characteristic (but mostly symbolic) sign of the disparity in this community is the term PRA itself. Some practitioners prefer and use the term "PSA", which stands for probabilistic safety assessment. The methodologies are identical. Perhaps the emphasis on *safety* rather than *risk* reflects an interest in end use and communication. A closely associated nomenclature distinction exists for the field of risk assessment for high-level radioactive waste repositories. Here PRA is called "PA", which stands for performance assessment. Reference [3] provides a good discussion of the evolution of this terminology distinction and of how PRA is practiced in the areas of high-level radioactive waste, low-level radioactive waste, and decommissioning of nuclear facilities.

30.2 Historical Comments

The defining study for PRA is the Reactor Safety Study [4], which was performed by the Atomic Energy Commission (under the direction of Norman Rasmussen of the Massachusetts Institute of Technology) to provide a risk perspective to gauge insurance and liability costs for nuclear power plants. The US Congress was developing legislation, known as the Price–Anderson Act, which would specify the liabilities of owners and operators of nuclear facilities. Prior to the Reactor Safety Study, during the period 1957 to 1972, reactor safety and licensing was developed according to a deterministic approach. This approach is contained in the current version of the Federal Code of Regulations, Title 10.

Reactors were designed according to conservative engineering principles, with a strong containment around the plants, and built-in engineered safety features. Deterministic analyses were performed by calculating the temperatures and pressures in a reactor or in a containment building, or the forces on pipes that would result as a consequence of credible bounding accidents. This approach became the framework for licensing in the United States, and it is basically built into the regulations for nuclear power plant operation. During the same time period, there was informal usage of probabilistic ideas. In addressing events and how to deal with their consequences, one looked at events that were very likely to occur, events that might occur once a month, once a year, or once during the plant's lifetime. Systems were put into place to accommodate events on such a qualitative scale. Also, probabilistic ideas were introduced in an informal way to look at such questions as the probability of the reactor vessel rupturing. However, the studies were not yet integrated analyses of these probabilistic notions. The essential idea was to provide protection against the potential damage (or melting) of the reactor core and against the release of fission products to the environment.

On the consequence side, a study was made called WASH 740 [5], which was an analysis of a severe accident (which would melt the reactor core and lead to the release of fission products to the environment) that was postulated to occur at a nuclear power plant. It was performed to gauge insurance and liability costs for nuclear power plants, and to evaluate how law protects the public. The postulate was that a very large amount of radioactivity was released from the building, and the consequences were calculated on that basis. The results showed very large numbers of deaths and health effects from latent cancers. That study lacked a probabilistic perspective. It gave only the consequence side, not the probabilistic side. But by 1972, and probably before that, people were thinking about how to blend the probabilistic notions into a more formal and systematic approach. This culminated in the Reactor Safety Study [4], mentioned above.

In this study the Atomic Energy Commission tried to portray the risks from nuclear power plant operation. It was a landmark study in probabilistic risk assessment. To draw an analogy with the field of physics, this time period for probabilistic risk assessment is roughly analogous to the years 1925 to 1928 for the atomic theory. In a short period of time it laid the foundation for what we call the PRA technology, and the systematic integration of probabilistic and consequential ideas. It also shed light on where the significant safety issues were in nuclear power plants. For example, the conventional licensing approach advocated the deterministic analysis of a guillotine pipe rupture in the plant, where one of the large pipes feeding water to the vessel is severed, as if with a hatchet, and then the consequences are calculated in a deterministic way. Using probabilistic assessment instead, the Reactor Safety Study showed that small pipe breaks are the dominant ones in the risk profile.

Transient events, such as loss of power to part of the plant, were found to be important. The Reactor Safety Study challenged a notion that one always looks only at single failures in the licensing approach. There is a dictum that one looks at the response of the plant to a transient event in the presence of a single failure. Thus, the failure of an active component is postulated and it is demonstrated that the plant can safely withstand such an event. The Reactor Safety Study showed that is not the limiting (or bounding) case from the perspective of risk; multiple failures can occur, and, indeed, these have been studied since. The analysis also highlighted the role of the operator of the plant. In looking at failure rates in the plant, it was not the hardware failures for many systems that gave the lead terms in the failure rates, it was such terms as the incorrect performance of a maintenance act, or a failure on the part of an operator to turn a valve to the correct position. Another important piece of work in the Reactor Safety Study was the physical analysis of the core-melt sequence, coupling it with an evaluation of the response of the containment. Before that time, analysts, designers, and regulators did not seem to recognize the close coupling between how the core melts in the vessel and the responses of the containment. The Reactor Safety Study was the first publication to integrate these factors. Its overall message was that the risk was low for nuclear power plant operation.

The Reactor Safety Study was criticized, extensively analyzed, and reviewed. Congress commissioned a report now called the Lewis Report [6] (after Harold Lewis, University of California, Santa Barbara, who chaired the committee). This report made certain conclusions about how the Executive Summary of the Reactor Safety Study presented the data, how the uncertainties in the study were not stated correctly or, as they put it, were "understated". They commented on the scrutability of the report but, overall, they endorsed the methodology used in the study and advocated its further usage in the licensing, regulatory, and safety areas. Shortly after the Lewis Report, the accident at Three Mile Island occurred, that led some people to ask, Where have we gone wrong? An event occurred which was beyond what was expected to occur in the commercial reactor industry. The regulators reacted to Three Mile Island by imposing many new requirements on nuclear power plants.

A curious thing happened. The people in the probabilistic risk community went back, looked at the Reactor Safety Study, and asked, Where did we go wrong in our analysis? A major study of risk had been performed and it seemed to have missed Three Mile Island. But, upon closer inspection, it was, in principle, in the study. The Reactor Safety Study was done for two specific plants. One was a Westinghouse plant, the Surry Plant, and the other was Peach Bottom, a General Electric Plant. It was thought at the time that two power plants, one a boiling water reactor and one a pressurized water reactor, were fairly representative of the hundred or more nuclear plants that would be in place during the last part of the 20th century. However, each plant is unique. The risk must be assessed for each individual power plant, delineating the dominant contributors to risk at each plant. The Three Mile Island plant had certain design features and operational procedures for which the particular data and quantification in the Reactor Safety Study was not representative. Qualitatively however, the event, and many other sequences, were delineated in the study and by using failure data and plant modeling appropriate to TMI, the event could be explained within the framework of the study. This provided an impetus for further analysis in the area of probabilistic risk assessment by the regulators and also by the nuclear industry.

Industry itself took the big initiative in probabilistic risk assessment. They performed full plant-specific, probabilistic risk assessments. One of the first studies was made for the Big Rock Point Plant, located out in the Midwest. This small plant was at that time rather old, producing about 67 MW of electricity. It is now permanently shut down. Many requirements were put on it since the Three Mile Island accident, and, looked at from the risk perspective, it did not make sense to do the types of things that the regulators were promoting on a deterministic basis. Following that study, full-scale studies were done for three other plants, which had the special consideration that they were in areas of high population density: Indian Point, for example, is about 36 miles from New York City, Zion is near Chicago, and Limerick is close to Philadelphia. The basic conclusions, based on plant-specific features, were that the risks were low. Some features of each plant were identified as the risk outliers, and they were correctable. For example, one plant was found to be vulnerable to a potential seismic event, in which two buildings would move and damage each other. This was very simple to fix: a bumper was placed between the two buildings. The control room ceiling of another plant was predicted to be vulnerable to collapse, and it was reinforced appropriately.

30.3 Probabilistic Risk Assessment Methodology

The PRA methodology has several aspects. One is the logic model, which identifies events that lead to failures of subsystems and systems. There are two types of logic trees that are in widespread use. One is the event tree approach, which is inductive. It moves forward in time to delineate events through two-level logic-type trees; yes/no, fail/success. The other is a fault-tree approach, where one starts with a top event, which is the undesired event, and goes backward in time to find out what has led to this event. Current PRA combines both approaches, to go simultaneously backward and forward in time. The principal advantage is that it reduces the combination of events that would be present if one or the other type of approach were used alone.

Another very important feature of probabilistic risk assessment concerns the source of the data used to quantify risk assessments. The data is obtained in part from the nuclear power plant experience itself, and partly from other industries. For example, one could look at valve failures in fossil fuel plants and ask how the database from another part of the sample space applies to the events that we want to quantify. Another part of the database is judgment. Sometimes events are of very low probability or of very high uncertainty, so that judgment is used to supplement data. There is nothing wrong with this; in fact, the Bayesian approach in probabilistic theory easily accommodates judgment. The physical models describe how the core melts down, how the containment behaves, and how off-site consequences progress. Health effects and economic losses are then assessed on the basis of this information.

PRAs are sometimes categorized in terms of the end products or results that are reported in the evaluation. A Level 1 PRA determines the accident initiators and failure sequences that can lead to core damage. The Level 1 PRA reports risk results typically in terms of the (annualized) probabilities of accident sequences and the overall probability of core damage. Usually the uncertainty in the core damage probability is given as well as the sensitivity of this prediction to major assumptions of the analysis.

When the PRA analysis also includes an assessment of the physical processes in the reactor vessel, the containment structure, and in associated buildings at the plant, the study is referred to as a Level 2 PRA. Typical products of this analysis are the probability of release (both airborne and to liquid pathways) of radionuclides from the plant to the environment. The analysis includes the mode of failure of the reactor vessel, the resulting temperature and pressure loads to the containment boundaries, the timing of these loads, the quantity and character of radionuclides released to the containment and subsequently to the atmosphere, and the timing of the vessel failure relative to the containment failure. Again, uncertainty analyses are typically performed as well as sensitivity analyses related to major assumptions.

A Level 3 PRA extends the results of a Level 2 analysis to health effects and economic impacts outside the facility. Effects of weather, population conditions, evacuation options, terrain, and buildings are also considered. The results are reported in terms of probabilities of fatalities (acute and latent cancers), property damage, and land contamination.

Overall descriptions of the methodologies and tools for performing PRAs are given in procedures guides [7, 8] for this purpose. There are also several textbooks on this subject as well. A recent example is [9], which contains ample references to the specialized subtopics of PRA and to much of the recent literature. There are several technical reports that the reader might want to consult on special topics. Some of these are noted in [7–9], but are given here for convenience.

For data analysis, in particular for failure rates for many components and systems, of nuclear power plant risk studies that were performed during the 1980s and 1990s, the reader should see [10, 11]. These were a key data source for the NUREG-1150 risk study [12], which updated the early WASH-1400 study [4]. This data was also used in the Individual Plant Examination studies that were performed by the US nuclear industry over the last decade.

The use of logic models to delineate and describe accident sequences is given in [7, 8] and good examples of the implementation of these tools are given in [4, 12]. The reader should consult the Fault Tree Handbook [13] for an excellent discussion of how this methodology is used to decompose a system failure into its contributory elements. The SETS computer model [14] is an example of how large fault trees can be handled and how some dependent failures can be analyzed. Reference [9] provides a useful survey of the various tools that have been developed in this area. Event trees are typically used to delineate an accident sequence by moving forward in time and indicating, through binary branching, the success or failure of a specified system or action to mitigate the accident outcome. These are described and illustrated in [7, 8]. Reference [9] provides summary information on commercially available integrated computer code packages for accident sequence analysis, along with the particular features of the various tools.

The understanding and modeling of human behavior related to accident evolution has received much attention in the PRA area. One of the key insights from [4], which has been substantiated in many subsequent studies, is that the human is an important element of the overall risk profile of a nuclear power plant. Actions (or inactions) by the operations and maintenance crews, training programs, attitudes and cultures of management can all play a role in the aggravation or mitigation of an accident. Research continues to be done on various aspects of the role of the human in safe operation of power plants. An introduction to some of this work can be found in [15, 16]. More recent state-of-the-art developments and applications are contained in [1].

In order to obtain a complete as possible portrayal of the risks of a complex technological system, both the failures that are "internal" to the system and the failures that are "external" events to the system must be identified, and their likelihoods must be quantified. Examples of internal initiating events are stochastic failures of mechanical or electrical components that assure the safe operation of the facility. Examples of external initiating events are earthquakes, fires, floods, and hurricanes. In order to model an external event in a PRA, its likelihood and severity must be calculated, its impact on the plant systems must be described, and its consequences must be evaluated. This includes an assessment of the impact of the external event on emergency planning in the vicinity of the facility. There is much literature on the wide-ranging subject of risks due to external events. Some of the more recent work can be found in [1, 17].

The procedures described above are the essential elements of a Level 1 PRA for nuclear power reactors. They are sufficient to compute the frequency of core damage as well as less severe events, such as the unavailability of individual systems within the facility. This information is useful to a plant operator or owner because it provides a measure of overall facility performance and an indication of the degree of safety that is attained. The operator can also use this prediction of the core damage frequency to understand which systems, components, structures, and operational activities are important contributors to the core damage frequency. This will aid the operator in prioritizing resources for test, maintenance, upgrades, and other activities related to the safe operation of the facility. For the regulator, the computed core damage frequency provides a metric for the facility, which can be compared to a safety goal. This prediction of risk, especially if it is accompanied by uncertainty ranges for the calculated parameters, can be helpful ancillary information to the regulations when new decisions need to be made with regard to a safety issue.

There is additional information that can be derived from a PRA, which requires the calculation of the physical damage that would result from the predicted failures. This additional information would, in turn, be the input to the analysis of the health, environmental, and economic consequences of the predicted events. The physical damage assessment is usually referred to as the Level 2 part of the PRA, and the consequence assessment is termed the Level 3 part of the PRA. These elements of the PRA are discussed below.

The typical results of the Level 2 portion of the PRA may include: (1) the extent of damage to the reactor core, (2) the failure mode of the reactor vessel, (3) the disposition of fuel debris in the containment building, (4) the temperature and pressure histories of loads on the containment due to gas generation associated with the accident, (5) the containment failure mode and its timing relative to previous events during the accident, (6) the magnitude and timing of the release of fission products from the containment, (7) the radionuclide composition of the fission product "source term". This information is very useful to the operator and regulator in several ways. It provides the necessary characteristics of the accident progression that would be needed to formulate an accident management strategy [18] for the facility. The information derived from the Level 2 PRA could be useful to a regulator in the assessment of the containment performance, of a particular plant, in response to a severe accident.

The melt progression of core damage accidents is typically analyzed with complex computer codes. These, in turn, have been developed with the aid of experimental information from accident simulations with surrogate and/or prototypic materials and from the (thankfully) limited actual data in this area. There has been much international cooperation in this area, and the programs have been sponsored by both government and private funds. In the USA, the principal melt progression computer codes are MELCOR [19] and MAAP [20]. MELCOR was developed under the sponsorship of the USNRC and the current version of the code requires plant design and operational information and an initiating accident event. It then computes the accident progression and yields the time histories of key accident events, as indicated above. MAAP was developed under industry sponsorship and serves as an alternate choice of melt progression description. As with any computer codes, it is important for the user to understand the underlying assumptions of the code and to make sure that the application is commensurate with the physical situation that is being modeled. Specific accident sequences and containment failure modes have been given special attention in this regard. Two difficult areas and how they were analyzed are discussed in detail in [21, 22].

The third major element of the integrated PRA is the consequence analysis or the Level 3 portion of the assessment. The radiological source term from the damaged plant is the essential input to this analysis. The key factors are: the radionuclide composition and chemical forms of each radionuclide, the energy with which they are expelled from the containment building, alternative pathways for source term release (*e.g.* downward penetration through the basemat or through various compartments or pipes), and time of release (especially relative to initial awareness of the accident).

Weather conditions can play a crucial role in determining the disposition of airborne releases. Offsite population distributions and the effectiveness of emergency response (*e.g.* evacuation, sheltering) are factored into the consequence analysis. Large radiological dose exposures can lead to health effects in a very short term. Lower doses may lead to cancers after a period of many years. The accident consequence code that has been in widespread international use is the MACCS code [23]. This code has been the subject of many peer reviews, standard problem exercises, and currently receives the attention of an international users group.

Risk assessment tools can vary, depending upon the end user and styles and interests within a particular group. Assumptions can easily be built into models and codes and the user must be aware of this for a specific application. Many engineered systems can be analyzed on the basis of the general construct outlined above. There is a class of hazards that do not arise from a sudden or catastrophic failure of an engineered system. Rather, these hazards are continuously present in our environment and pose a risk in their own right. These risks are discussed below.

30.4 Engineering Risk Versus Environmental Risk

In order to bring risk insights to a particular assessment or decision, a method is needed to calculate the risk of the condition or activity in question. There are a wide range of activities and conditions for which risk tools have been used to calculate risk. The scope, depth of analysis, end products, and mode of inquiry are essential aspects of a risk assessment that define the particular methodology. For example, in the area of health science (or environmental) risk assessment, the top level approach is to address:

- hazard identification;
- dose response assessment;
- exposure assessment;
- risk characterization.

These four activities are well suited to a situation in which a hazard is continuously present (this could be regarded as a "chronic" risk) and the risk needs to be evaluated. For engineered systems, the favored top-level approach to risk assessment addresses the so-called risk triplet:

- What can go wrong?
- How likely is it?
- What are the consequences?

This approach is naturally suited to "episodic" risks, where something fails or breaks, a wrong action is taken, or a random act of nature (such as a tornado or a seismic event) occurs. This is the approach that has been taken, with much success, in probabilistic risk assessments of nuclear power reactors. A White Paper on Risk-Informed and Performance-Based Regulation [24] that was developed by the US Nuclear Regulatory Commission (NRC) has an evident orientation toward reactor risks that are episodic in origin.

The environmental risk assessment methodology and the engineered system methodology do have elements in common, particularly in the dose and exposure areas. Further, in the development of physical and biomedical models in the environmental area, uncertainties in the parameters of the models and the models themselves are expressed probabilistically. This is also sometimes the case in the engineered system area. Is it more appropriate to use an environmental risk approach or an engineered system risk approach for the risk assessments that are associated with waste management conditions? Is the distinction worth making?

To understand these questions a bit more, it is instructive to introduce some related concepts. In the waste and materials areas, the USNRC notes [25] that there are methodologies (not quite probabilistic risk assessments) that have been useful. Notably, PA is the method of choice for evaluation of the risk posed by a high-level waste repository. The USNRC paper defines PA to be a type of systematic safety assessment that characterizes the magnitude and likelihood of health, safety, and environmental effects of creating and using a nuclear waste facility. A review of some applications of PA indicates that PRA and PA appear to be very similar methodologies. In a recent paper, Eisenberg *et al.* [3] state the connection very succinctly and directly: "performance assessment is a probabilistic risk assessment method applied to waste management". Further, in a paper on this subject, Garrick in [2], notes that: "(PRA) is identified with the risk assessment of nuclear power plants, probabilistic performance assessment (PPA), or just PA, is its counterpart in the radioactive waste field". He adds that in the mid-1990s the USNRC began to equate PRA with PA.

30.5 Risk Measures and Public Impact

One benefit of performing a PRA is to have measures to compare with a risk standard or goal. This, of course, is not the only value of PRA. It can be especially beneficial in measuring changes in risk related to modifications in design or operational activities. Risk measures can be used as an aid to performing risk management and for structuring risk management programs that address unlikely, extreme consequence events and also less extreme, more likely events. In order to implement an effective risk management program, an appropriate set of risk measures (or indices) must be defined.

One approach to establishing risk criteria is to define and characterize a set of extreme consequence conditions (*e.g.* a large, uncontrolled release of a hazardous material). In this approach, if the operation is managed to prevent, control, or mitigate the extreme conditions, then less extreme conditions and events ought to be managed through this process as well. There are at least two potential flaws in this approach. One is that less extreme conditions or events may not be captured easily or at all by the approach. This is sometimes referred to as the "completeness" issue. Another problem is that failures and consequences that are necessary but not sufficient conditions for the extreme event could occur individually or in combination with other less extreme events. In both situations the less extreme events that might occur could have significant public impact or reaction in their own right.

An example of the first problem is the tritium leakage into the groundwater at Brookhaven National Laboratory from its High Flux Beam Reactor that was reported in 1997. Risks from the operation of this reactor were managed through several mechanisms: procedures, guidelines, orders, and oversight by the US Department of Energy, the Laboratory's own environment, safety and health program, and requirements and agreements with other federal (US Environmental Protection Agency), state (New York Department of Environmental Conservation), and local (Suffolk County Department of Health) organizations.

As part of the Laboratory's program, a PRA was performed approximately 10 years ago. The study focussed on events that could damage the reactor core and potentially lead to onsite and offsite radiological doses arising from the release of fission products from the damaged fuel. The hypothetical airborne doses were extremely low both onsite and offsite. Nevertheless, the risk assessment was used by the Laboratory to aid in decision-making regarding operation and potential safety upgrades.

The PRA did not focus on potential groundwater contamination because the downward penetration of damaged fuel was assessed to be very unlikely for the scenarios considered. The study, however, also did not address the routine contamination of the water in the spent fuel pool due to tritium (a byproduct of normal reactor operation). Nor did it address the potential leakage of water from the spent fuel pool. The leakage of tritiated water from the spent fuel pool resulted in the detection of concentrations of tritium in the groundwater, which exceeded the safe drinking water standard determined by the US EPA.

The contamination of the groundwater, in this case a sole source aquifer, led to a large public impact, which included the firing of the contractor for the Laboratory and to sustained public concern and scrutiny of the Laboratory. In principle, the probabilistic risk assessment performed 10 years ago could have identified this failure scenario if it: (1) included the EPA safe drinking water standard for tritium in its set of risk criteria, (2) identified the spent fuel pool as a source of contamination, (3) recognized the potential for leakage, and (4) identified the tritium source term in the pool water. However, the criteria and assessment tools were not focussed in this area and thus were not used to manage this risk and prevent the ensuing scenario.

Events occur that are not catastrophic but nevertheless capture the public attention. This can be due to the lack of public awareness of the significance (or insignificance) of the event. It could also be due to the perception that a less significant event undermines confidence in a technology and thereby portends a more severe, dreaded event. In either case, the occurrence of a less significant event undermines the case for the manager of the risks of the given technology. Thus the risk manager is faced with the task of managing both the risk of the less likely, higher consequence accidents and the risk of the more likely, lower consequence accidents.

One option would be to set stringent criteria for failure of systems, structures, components, and operations that define the performance and safety envelope for the technology in question. This would minimize the likelihood of the occurrence of small accidents (say below once in the lifetime of the facilities operation, or more stringently, once in the lifetime of the technology's operation). This option may be effective in eliminating the small accident, but it may be prohibitively costly.

Another option would be to develop a public education program that would sufficiently describe the technology and its potential off-normal behavior so that the small accidents would not result in a large public impact. Instilling a sense of trust in the public for the management of a complex technology would be key to assuring that reactions to off-normal events would not be disproportionate to their severity. The communication of risks is a challenging process and the acceptance of a small or insignificant risk by the public is difficult, especially when the benefits of a technology are not easily perceived.

These extreme options address the overall issue from either the cause (small accident) or the effect (large public impact). Let us further explore the issue from the perspective of the cause, *i.e.* consideration of risk measures that would aid in the management of the risks.

As the example from Brookhaven National Laboratory illustrates, it is important, when building models of the risk of a facility, to assure that they are able to address the relevant compliance criteria. This is an example in which the risk models must be as complete as practical to capture the risk envelope.

The case in which a subsystem failure leads to a small accident can perhaps be managed through an approach based on allocation of reliabilities. Systems, components, structures, and operational activities typically have a finite failure probability (or conversely, a finite reliability). A decision-theoretic approach to the allocation of reliability and risk was developed by Brookhaven National Laboratory [26, 27]. This was done in the context of the formulation of safety goals for nuclear power plants. In this approach the authors [26, 27] started with top-level safety goals for core damage and health risk and derived a subordinate set of goals for systems, components, structures, and operation. The methodology was based on optimization of costs (as a discriminator of options) for the allocation of subordinate goals that are consistent with the top-level goals. The method required that a probabilistic risk assessment model be available for the plant and that the costs of improving (or decreasing) the reliability of systems, components, structures, and operations can be specified. Because there are usually far more of these lower-tier elements than there are top-level goals, there will be a multiplicity of ways to be consistent with the top-level goals. The methodology searches for and finds those solutions that are non-inferior in the top-level goals. Thus, it will yield those sets of subordinate goals that are consistent with the top-level goals that do not yield less desirable results across the entire set of top-level goals. The decision-maker must perform a preference assessment among the non-inferior solutions.

Once these solutions are obtained, it is then possible to manage the lower-tier risks and hopefully minimize the likelihood of a "smaller" accident occurring. The methodology presents an interesting possibility for events that would have a large public impact. In addition to recognizing the need to minimize the economic penalties of changes in reliability of subtier elements, it may be possible to account for the psychological value of subtier element failures. In principle, one would give weightings to those failures that result in different public impact. The resulting solutions will still be consistent with the top-level goals of protecting public physical health.

An area in which risk-informed criteria setting is currently being considered [25] for less severe events is the regulation of nuclear materials uses and disposal. In this area, the risks to workers can be greater than the risk to the public, particularly for the use of nuclear byproduct materials [28]. Nevertheless, events that have occurred while causing no health impact to the public have raised concern. For example, in [28], the authors' objective was to consider both actual risks and risks perceived by the general public. The byproduct material systems considered in [28] included medical therapy systems, well logging, radiography, and waste disposal. The results of this study suggest that, because of the high variability between radiological doses for the highest-dose and lowest-dose systems analyzed, the approaches to management and regulation of these systems should be qualitatively different. The authors observed that risk for these systems was defined in terms of details of design, source strength, type of source, radiotoxicity, chemical form, and where and how the system is used. This, in turn, makes it difficult to characterize the risk of the system, to pose criteria based on this risk, and then to manage the risk.

Another area that has received attention recently is electrical energy system reliability. This was underscored during the summer of 1999 when power outages disrupted the lives of millions of people and thousands of businesses in regions of the USA. In response to public concerns, the US Department of Energy commissioned a study of these events. This is detailed in a report [29] that was authored by various experts on energy transmission and delivery systems. This report makes recommendations that are very relevant to the subject of this chapter. In particular, the authors recommend that there be mandatory standards for electric power systems. They note that our interconnected electric power system is being transformed from one that was designed to serve customers of full-service utilities (each integrated over generation, transmission, and distribution functions) to one that will support a competitive market. Because of this change, they observe that the current voluntary compliance with reliability standards is inadequate for ensuring reliability, and recommend mandatory standards for bulk-power systems to manage the emerging restructured electric power industry. The study also recommends that the US government consider the creation of a self-regulated reliability organization with federal oversight to develop and enforce reliability standards for bulk-power systems as part of a comprehensive plan for restructuring the electric industry.

Adverse public impact is always an undesirable outcome of a technological enterprise. This will be the case if there is a direct physical effect on the public or if there is only the perception of risk. In order to use risk management tools effectively to minimize adverse impact, it is important to understand the range of undesirable outcomes and then set criteria (or adopt them) against which performance can be measured. Clearly, resources would have to be allocated to a risk management program that accounts for a broad range of risks, from the high-consequence/low-likelihood to the low-consequence/high-likelihood end of the spectrum. Both pose risks to the successful operation of a facility or operation, and the prudent risk manager will assure adequate protection across the entire range.

30.6 Transition to Risk-informed Regulation

PRAs have been performed for many nuclear power plants. Individual Plant Examinations [30] were performed by the plant owners beginning in the late 1980s and continued through much of the 1990s. These studies were specifically scoped PRAs [31] designed to provide an increased understanding of the safety margins contained within the individual plants. These studies led to the identification of specific vulnerabilities at some of the plants and to greater insight, by the owners, to improved operation of their facilities. Over the same time frame, studies were undertaken of the risk posed by power reactors while in shutdown or low power conditions [32]. These studies resulted in and enhanced understanding of the overall risk envelope for reactor operation and led to improvements in activities associated with these other modes of operation.

More recently, the NRC has been sponsoring research on the development of approaches to risk-informing 10 CFR 50. This includes a framework document [33] and specific applications of an approach to risk informing 10 CFR 50.44 and 10 CFR 50.46. The former refers to combustible gas control in containment and the latter refers to the requirements on the emergency core cooling function. Progress on these activities will be reported in a future publication by the NRC and its contractors.

There has also been work performed recently [34] in the development of standards for the performance of probabilistic risk assessments. The need for standards has long been recognized. A fundamental dilemma has been how to develop a standard in an area that is still emerging and for which improved methods may yet be in the offing. Early attempts at standardization can be found in NUREG-2300 and NUREG/CR-2815. The advantage of a standard to the industry is that if they submit a request for a regulatory review to the NRC that is based on PRA methods, and if it is done within an accepted standard, then the review would not be encumbered by a case-specific review of the methodology itself.

30.7 Some Successful Probabilistic Risk Assessment Applications

There have been many accomplishments, which have in turn had lasting impact on improved safety. For example, the focus on the impact of human performance on nuclear power plant risk has been underscored in many PRAs. These have led the way to the formulation and execution of research programs by NRC, the industry, and by organizations in many countries on human performance. This has led to valuable insights that have affected emergency plans, accident management programs, and the day-to-day efficient operation of each unit.

The "Level 2" portion of the PRAs has been invaluable in providing great insight into the performance of containment under severe accident loads. This includes the timing of pressure and temperature loads as well as the behavior of combustible gases and core debris within containment. The relative benefits of each containment type are now far better understood because the phenomenology was evaluated and prioritized within the context of a Level 2 PRA.

Risk perspectives from performing PRAs for various modes of power operation were also of great benefit to the regulator and to the industry. These studies led to utilities having enhanced flexibility in managing outages and in extending allowed outage times. Further, the studies led to better decision-making by the NRC staff and to more effective incorporation of risk information in that process [35].

The Individual Plant Examination Program [30], which was essentially a PRA enterprise, led to many improvements, directly by the owners in the procedures and operations of their plants. The program also led to improvements in systems, structures, and components and to insights for improved decision-making by the regulator and the industry.

30.8 Comments on Uncertainty

Much has been said and written over the past two decades on the topic of uncertainty. Hopefully we are coming to closure on new insights on this subject. It has been a strength of PRA that it lends itself well to the expression of uncertainties. After all, uncertainty is at the very heart of risk. On the other hand, the elucidation of uncertainties by PRA has been taken by some to be a limitation of the PRA methodology (*i.e.* it deals in vagaries and cannot be trusted). It is also unfortunate that PRA practitioners and advocates have so often felt the need to dwell on the limitations of PRA. However, it should be recognized that deterministic methodologies are also fraught with uncertainties, but they are not as explicitly expressed as they are by the PRA methodologists. PRA allows decision-makers to be informed about *what they know about what they do not know*. And this is valuable information to have.

A good state-of-the-art presentation of papers on the subject of uncertainty in risk analysis can be found in a special issue of the journal *Reliability Engineering and Systems Safety* [36]. Among the 14 papers in this special volume are discussions of aleatory and epistemic uncertainties and applications to reactor technology and to waste management. In a recent paper, Chun *et al.* [37] consider various measures of uncertainty importance and given examples of their use to particular problems.

30.9 Deterministic, Probabilistic, Prescriptive, Performance-based

Four concepts that are central to the use of PRA in regulation are: deterministic, probabilistic, prescriptive, and performance-based. These concepts are sometimes used conjunctively, and sometimes interchanged. Figure 30.1 provides a two-dimensional view of how they should be correctly related. Prescriptive and deterministic are sometimes confused and used interchangeably. They tend to be within the comfort zone of some participants in regulatory matters, and perhaps that is the source of the interchangeability. Deterministic is really an approach to analysis and its opposite is probabilistic. Prescriptive is an approach to decision-making and its opposite is performance-based. Most analyses are not purely probabilistic or deterministic but an admixture of the two extremes. A classic "Chapter 15" safety analysis (required by the NRC for nuclear power plant applications) really starts by determining, however informally, what is likely and unlikely. Similarly, a probabilistic analysis as embodied in a PRA usually has some form of deterministic analysis, *e.g.* a heat transfer calculation. The important point is that this is a two-dimensional space with various regulatory activities falling in different parts of the plane. For example, if one were interested in compliance with a numerical safety criterion

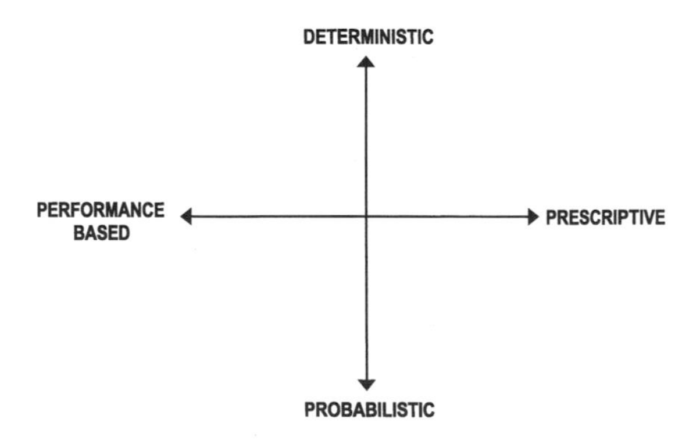

Figure 30.1. Two dimensional view of risk-related concepts

related to the likelihood of an event, it would fall in the lower right quadrant.

30.10 Outlook

In the early days of PRA, both regulators and the industry for the most part were not very comfortable with PRA and the use of risk management concepts to guide decisions with regard to the safe operation of plants. This has changed, in an evolutionary way, over the past quarter century. Now it is quite the custom to see, hear, and read of exchanges between the industry and the NRC (and among the organizations themselves) in which concepts like risk assessment and risk management form the currency for their exchange of "safety thought". There has been an ever-increasing attempt by the NRC and the industry to bring other stakeholders (states, public interest groups, and concerned citizens) into the communication loop on safety matters. However, the process, while moving forward, involves the exchange of different currency with regard to safety. This can be termed risk perception or more broadly, risk communication (see Figure 30.2). These are very important aspects of the safety enterprise and require much attention and development. It should be noted that the simple figure does not capture the fact that the indicated modes of currency are not exclusive to any of the interchanges. Each interchange is really an admixture of the two indicated types. The figure just indicates the predominant exchange mode.

Figure 30.2. Interactions among participants in the risk-related enterprise

Safety research, being an inductive discipline, will always have its frontiers. The following are three areas that might benefit from the now vast repertoire of knowledge and methodology that is contained in the PRA field.

Much of the work in PRA has focussed on the severe accident. This type of accident (core melt, large radiological release to the environment) tends to be unacceptable to all parties concerned. Fortunately, they are also very unlikely events—not expected to occur in the lifetime of a facility. There are events, however, that are much more likely, that do occur and, fortunately, have little or no health or environmental impact. Yet they do attract public attention and require much attention by all parties involved. A recent example of such an event is the steam generator tube rupture at a particular plant. The offsite radiological release was insignificant but the event drew much attention from the press and the political representatives (and therefore the NRC and the industry). From the PRA perspective, this was a very low risk event. While the PRA methodology can easily express the risk of such a small consequence event, can it be used to flag such events? Can these risks be managed? How should they be managed and by whom?

A related topic is risk perception. It is well known that an analytical quantification of health and environmental risks, an objective expression of reality, does not necessarily represent the risk that an individual or a group perceives. Further, it is sometimes said that perception is reality. The question is: can psychosocial measures of risk be defined and calculated in a way that is comparable to physical risks? If so, can risk management programs be developed that recognize these measures?

Finally, it should be recognized that the NRC and the civilian nuclear power industry have been at the vanguard of PRA methods development and applications. This has, no doubt, contributed to the safety and reliability of this technology. Other organizations and industries would benefit from the advances made in the uses of PRA and there should be a wider scale adoption (and adaptation) of these methods and applications.

Acknowledgment

I am pleased to thank J. Lehner for a review of this manuscript and for helpful comments and suggestions.

References

[1] See the website for this organization for information on past conference proceedings: www.iapsam.org

[2] Kirchsteiger C, Cojazzi G, editors. Proceedings of the Workshop on the Promotion of Technical Harmonization on Risk-Based Decision-Making. Organized by the European Commission, Stresa, Italy; May 22–24, 2000.

[3] Eisenberg NA, Lee MP, McCartin J, McConnell KI, Thaggard M, Campbell AC. Development of a performance assessment capability in waste management programs of the USNRC. Risk Anal 1999;19:847–75.

[4] WASH-1400, Reactor Safety Study: An assessment of accident risks in commercial nuclear power plants. US Nuclear Regulatory Commission. NUREG-75/014; October 1975.

[5] Theoretical possibilities and consequences of major nuclear accidents at large nuclear plants. US Atomic Energy Commission. WASH-740; March 1957.

[6] Lewis HW, Budnitz RJ, Kouts HJC, Loewenstein WB, Rowe WD, von Hippel F, *et al.* Risk assessment review group report to the US Nuclear Regulatory Commission. NUREG/CR-0400; September 1978.

[7] PRA Procedures Guide: A guide to the performance of probabilistic risk assessments for nuclear power plants. US Nuclear Regulatory Commission. NUREG/CR-2300; January 1983.

[8] Bari RA, Papazoglou IA, Buslik AJ, Hall RE, Ilberg D, Samanta PK. Probabilistic safety analysis procedures guide. US Nuclear Regulatory Commission. NUREG/CR-2815; January 1984.

[9] Fullwood RR. Probabilistic safety assessment in chemical and nuclear industries. Boston: Butterworth-Heinemann; 2000.

[10] Drouin MT, Harper FT, Camp AL. Analysis of core damage frequency from internal events: methodological guides. US Nuclear Regulatory Commission. NUREG/CR-4550, vol. 1; September 1987.

[11] Wheeler TA, *et al.* Analysis of core damage frequency from internal events: expert judgement elicitation. US Nuclear Regulatory Commission. NUREG/CR-4550, vol. 2; 1989.

[12] Reactor risk reference document. US Nuclear Regulatory Commission. NUREG-1150; February 1987.

[13] Vesely WE, Goldberg FF, Roberts NH, Haasl DF. Fault tree handbook. US Nuclear Regulatory Commission. NUREG-0492; January 1981.

[14] Worrell WB. SETS reference manual. Electric Power Research Institute. NP-4213; 1985.

[15] Swain AD, Gutmann HE. Handbook of human reliability analysis with emphasis on nuclear power plant applications. US Nuclear Regulatory Commission. NUREG/CR-1278; August 1983.

[16] Hannaman GW, Spurgin AJ. Systematic human action reliability procedure (SHARP). Electric Power Research Institute. NP-3583; June 1984.

[17] Bohn MP, Lambright JA. Recommended procedures for simplified external event risk analyses. US Nuclear Regulatory Commission. NUREG/CR-4840; February 1988.

[18] Bari RA, Pratt WT, Lehner J, Leonard M, DiSalvo R, Sheron B. Accident management for severe accidents. Proc Int ANS/ENS Conf on Thermal Reactor Safety. Avignon, France; October 1988. pp.269–76.

[19] MELCOR 1.8.5. Available by request from the US Nuclear Regulatory Commission at www.nrc.gov/RES/MELCOR/obtain.html

[20] Henry RE, *et al.* MAAP4—Modular accident analysis program for LWR power plant. Computer code manual 1-4. Electric Power Research Institute; May 1994.

[21] Theophanous TG, *et al.* The probability of Mark I containment failure by melt-attack of the liner. US Nuclear Regulatory Commission. NUREG/CR-6025; 1993.

[22] Pilch MM, *et al.* US Nuclear Regulatory Commission. NUREG/CR-6075; 1994.

[23] Chanin D, Young ML. Code manual for MACCS2. US Nuclear Regulatory Commission. NUREG/CR-6613; May 1998.

[24] Risk-informed and performance-based regulation. US Nuclear Regulatory Commission. www.nrc.gov/NRC/COMMISSION/POLICY/whiteppr.html

[25] Framework for risk-informed regulation in the Office of Nuclear Material Safety and Safeguards. US Nuclear Regulatory Commission. SECY-99-100; March 31,1999.

[26] Papazoglou IA, Cho NZ, Bari RA. Reliability and risk allocation in nuclear power plants: a decision-theoretic approach. Nucl Technol 1986;74:272–86.

[27] Cho NZ, Papazoglou IA, Bari RA. Multiobjective programming approach to reliability allocation in nuclear power plants. Nucl Sci Eng 1986;95:165–87.

[28] Schmidt ER *et al.* Risk analysis and evaluation of regulatory options for nuclear byproduct material systems. US Nuclear Regulatory Commission. NUREG/CR-6642; February 2000.

[29] Report of the US Department of Energy's Power Outage Study Team. Findings and recommendations to enhance reliability from the summer of 1999. US Department of Energy; March 2000. www.policy.energy.gov/electricity/postfinal.pdf

[30] Individual Plant Examinations Program. Perspectives on reactor safety and plant performance. US Nuclear Regulatory Commission. NUREG-1560; December 1997.

[31] Individual Plant Examination of severe accident vulnerabilities. 10 CFR 50.54(f). US Nuclear Regulatory Commission. Generic Letter GL-88-20; November 23, 1988.

[32] Perspectives report on low power and shutdown risk. PDF attachment to proposed staff plan for low power and shutdown risk analysis research to support risk-informed regulatory decision-making. US Nuclear Regulatory Commission. SECY 00-0007; January 12, 2000. http://www.nrc.gov/NRC/COMMISSION/SECYS/2000-0007scy.html

[33] Framework for risk-informed regulations. US Nuclear Regulatory Commission. Draft for Public Comment, Rev. 1.0; February 10, 2000. http://nrc-part50.sandia.gov/Document/framework_(4_21_2000).pdf

[34] Standard for probabilistic risk assessment for nuclear power plant applications. Rev. 12 of Proposed American National Standard. American Society of Mechanical Engineers; May 30, 2000.

[35] An approach for using probabilistic risk assessment in risk-informed decisions on plant-specific changes to the licensing basis. US Nuclear Regulatory Commission. Regulatory Guide 1.174; July 1998.

[36] Reliab Eng Syst Safety 1996;54:91–262.

[37] Chun M-H, Han S-J, Tak N-I. An uncertainty importance measure using a distance metric for the change in a cumulative distribution function. Reliab Eng Syst Safety 2000;70;313–21.

Total Dependability Management

Per Anders Akersten and Bengt Klefsjö

31.1 Introduction

Different methodologies and tools are available for the management and analysis of system dependability and safety. In the different phases of a system's lifecycle, some of these methodologies and tools are appropriate, and some are not. The choice is often made by the dependability or safety professional, according to personal experience and knowledge. However, this choice should be influenced by some management philosophy, based on a number of core values. In the literature, core values are sometimes referred to as principles, dimensions, elements, or cornerstones.

The core values constitute a basis for the culture of the organization. They have a great influence on the choice of strategies for accomplishing different kinds of goals, *e.g.* regarding dependability, safety, and other aspects of customer satisfaction. The strategies will in turn include the use of appropriate methodologies and tools.

In this chapter, a number of core values are discussed and their influence on the choice of methodologies and tools briefly analyzed. The aim is to present one approach to the application of a holistic view to the choice of methodologies and tools in dependability and safety management. Moreover, the aim is to give a conceptual base for the choice of appropriate methodologies and tools. The approach described is very general. The sets of core values, methodologies, and tools presented are by no means complete. They should be regarded as examples only.

The main objective is to make clear the usefulness of clearly expressed core values in the choice and design of strategies for the management of dependability and safety in all phases of a system's lifecycle. It is important that an implementation of a dependability and safety management system will focus on the totality. Just picking up a few methodologies or tools will not be sufficient.

A comparison is made to the structure of the IEC Dependability Standards, consisting of "Standards on Dependability Management", "Application Guides", and "Tools". However, the concept of "Tools", used in the IEC Dependability Standards structure, is wider than the concept of "Tool" used in this chapter.

31.2 Background

Dependability management has been an important issue for many years. Many articles and books, both in the academic and popular press, have been written on the subject of dependability;

Malcolm Baldrige National Quality Award	European Quality Award	The Swedish Quality Award
• Visionary leadership	• Results orientation	• Customer orientation
• Customer-driven excellence	• Customer focus	• Committed leadership
• Organizational and personal learning	• Leadership & constancy of purpose	• Participation by everyone
• Valuing employees and partners	• Management by processes & facts	• Competence development
• Agility	• People development & involvement	• Long-range perspective
• Focus on the future	• Continuous learning, innovation & improvement	• Public responsibility
• Management for innovation	• Partnership development	• Process orientation
• Management by fact	• Public responsibility	• Prevention
• Public responsibility and citizenship		• Continuous improvement
• Focus on results and creating value		• Learning from others
• Systems perspective		• Faster response
		• Management by facts
		• Partnership

Figure 31.1. The core values which are the basis for the Malcolm Baldrige National Quality Award, the European Quality Award, and the Swedish Quality Award (SQA, established in 1992 by the Swedish Institute for Quality and at that time strongly based on the Malcolm Baldrige Award). The values are taken from the versions of 2002.

not very many, however, on dependability management. A number of authors have addressed the subject of safety management. Still this number is small, compared to the number of articles and books on total quality management (TQM).

Interesting parallels can be drawn. Both dependability management and TQM are often identified as necessary to reach competitiveness, and there are also examples of companies that have failed in their implementation of dependability management or TQM. However, the concept of TQM is more generally known and it has been discussed on very many levels, sometimes clarifying the concept, sometimes not.

Still there exist different opinions about TQM. There are many vague descriptions and few definitions of what TQM really is. Regarding dependability management, it has not been described in too many different ways. According to the authors' view, the most comprehensive description is a result of work performed by the International Electrotechnical Commission (IEC) within their Technical Committee 56 "Dependability". See *e.g.* the IEC Standards [1–3]. (*Note*: These standards are currently under revision.)

Here we will discuss some of the problems with dependability management and describe and discuss our own view of dependability management as a management system consisting of values, methodologies, and tools.

31.3 Total Dependability Management

The concept of TQM is generally understood, and often also described, as some form of "management philosophy" based on a number of core values, such as customer focus, fact-based decisions, process orientation, everybody's commitment, fast response, result orientation, and learning from others; see Figure 31.1. What here are called core values are also in the literature named principles, dimensions, elements, or cornerstones. We prefer the term core value since it is a way to emphasize that these statements should work together to constitute the culture of the organization, and that they accordingly are basic concepts.

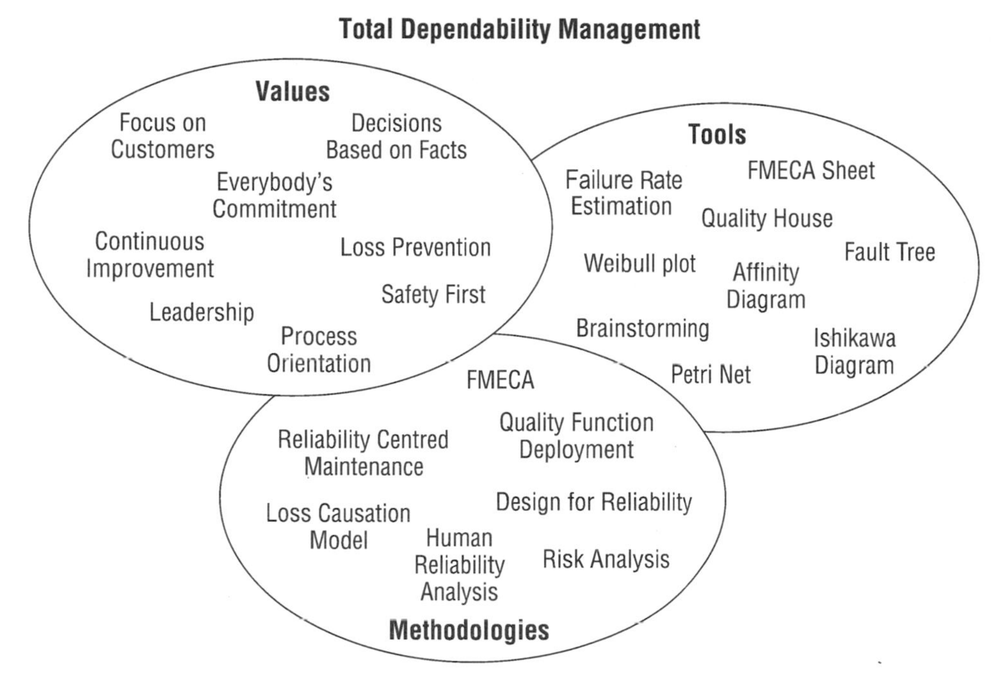

Figure 31.2. A dependability management system seen as a continuously evolving management system consisting of core values, methodologies, and tools. It is important to note that the methodologies and tools in the figure are just examples and not a complete list. In the same way the values may also vary a little between different organizations and over time.

A literature study [4] shows that a number of core values seem to be common in most descriptions of TQM, namely: "focus on customers", "management commitment", "everybody's commitment", "focus on processes", "continuous improvements", and "fact-based decisions" [5]. Further literature studies and informal contacts with a number of Swedish companies indicate that the same core values are common also in the context of dependability management.

31.4 Management System Components

Dependability management contains much more than core values. Like TQM it is a management system in the sense of Deming, *i.e.* as "a network of interdependent components that work together to try to accomplish the aim of the system" [6], p.50.

A very important component is constituted by the core values. These core values are the basis for the culture of the organization and also the basis for the goals set by the organization. Another component is a set of methodologies, *i.e.* ways to work within the organization to reach the goals. A methodology consists of a number of activities performed in a certain order. Sometimes it is appropriate to describe a methodology as a process. A third component is a set of tools, *i.e.* rather concrete and well-defined tools, which sometimes have a statistical basis, to support decision-making or facilitate analysis of data. These three components are interdependent and support each other, see Figure 31.2.

We believe that it is important to classify different terms related to dependability management according to any of the three components. For instance, FMECA (Failure Mode, Effects, and Criticality Analysis [7]) has often been considered

Methodologies	Customer focus	Fact-based decisions	Empowerment and participation of all staff	Top management commitment	Commitment to creativity	Continuous improvement and learning	Process orientation
Analysis of failure causes		s			s	w	
Analysis of failure consequences	w	s			s	w	
Design for reliability	s	s	w	w	s	s	w
Design for manufacturing			w		s		w
FMEA		s	s		s	s	w
FMECA		s	s	w	s	s	w
HAZOP		s	s		s		w
Human reliability analysis	w	s	s	w	s	s	s
Loss causation model		s			w	w	
Process management	s		w	w	w	w	s
QFD	s	s	w		s	s	
Reliability centred maintenance	s	s	s	w	w	s	w
Risk analysis		s	w	w	w		w

Legend:
s-strong relation
w-weak relation
(blank)-insignificant relation

Figure 31.3. The potential for mutual support between core values and methodologies. It must be emphasized that the lists of core values and methodologies are by no means complete, and the strengths of relations are based on the authors' present judgments. The table here is just an illustration of the idea.

as a tool. One reason for this is the very strong connection to the tabular FMECA sheet. Another example is QFD (Quality Function Deployment [8]), which often has been looked upon as a tool, mainly due to the confusion between QFD and the Quality House. However, QFD is "a system for translating consumer requirements into appropriate company requirements at each stage from research and product development to engineering and manufacturing to marketing/sales and distribution" [9], and is therefore, in our opinion, a methodology. The Quality House, on the other hand, is a tool to be used within that methodology. Another example is risk analysis [10], a methodology consisting of a number of steps, involving the use of different tools, e.g. hazard identification. By consistently using a terminology based on core values, methodologies, and tools, the "concepts" used within dependability will be clarified.

This management system is continuously evolving. Over time, some core values might change, and in particular the interpretation of some of them might be developed. New methodologies will also appear or be transferred from other management theories. New tools will be developed or taken from other disciplines.

One of the things that is important to notice is that the management system really should be looked upon as a system. It is often the case that a certain core value is necessary for the successful

Tools	QFD	HAZOP	FMEA	FMECA	Risk analysis	RCM	Process management	Loss causation model	Design for reliability	Design for manufacturing	Human reliability analysis
Affinity diagram	o	o	o	o	o	o	o	o	o	o	o
Analysis of trend						o					
Brainstorming		n	n	n	n	o		o	o	o	o
Cause and effect analysis diagram		o	o	o	o	o	o	o	o	o	
Checklists		o	o	o	o	o			o	o	o
Consequence modeling		o	o	o	n	o		o	o	o	o
Event tree		o		o	o	o			o		o
Failure rate estimation		o	o	n	o	o			o		o
Fault tree analysis		o	o	o	o	o			o		o
Field data collection				n	o	o		o	o		o
Guide word list		n	o	o	o	o			o		o
Hazard identification		n	n	n	n	n		o	n	o	n
Kano model	n									o	o
Markov analysis					o				o		
Multi-attribute prioritization		n		n	n	n		o	o	o	o
Petri nets					o				o		o
Power law process reliability growth model						o					
Process map							n	o	o	o	o
Quality house	n						o			o	
Relations diagram	o	o	o	o	o	o	o	o	o	o	o
Reliability block diagram			o	o	o	o			o	o	
Reliability prediction		o	o	n	o	o			n	o	o
Requirements identification	n	o	n	n	o	n	o		n	n	o
Simulation					o				o	o	o
Sneak circuit analysis					o	o			o		
Stress–strength test						o			o		
Weibull plot						o			o		

Legend:
n-necessary
o-optional

Figure 31.4. Scheme for the choice and implementation of various tools. It must be emphasized that the lists of methodologies and tools are by no means complete, and the usefulness of tools is based on the authors' present judgments. The table here is just an illustration of the idea.

implementation of a chosen methodology. At the same time, the systematic use of a methodology gives more or less support for the acceptance and establishing of different core values. For example, the core value of "everybody's commitment" cannot be implemented without the appropriate use of suitable methodologies. There is a potential for implementation of this core value, *e.g.* in the methodologies of FMECA, Reliability Centered Maintenance (RCM [11]) and Loss Control Management (LCM [12]). However, the methodologies chosen will not work efficiently without the skillful use of specific tools. As another example we can mention the value of "focus on customers". Here QFD is one useful methodology and the Quality House is then one useful tool for a systematic transformation. The potential for the support to core values of different methodologies is illustrated in Figure 31.3. In this figure we indicate a weak or strong relationship. Usually we can identify a mutual support in the sense that a certain core value is necessary for the successful implementation of a chosen technique. At the same time, the systematic use of the technique gives strong support for the acceptance and establishing of the core value.

There are several benefits with this system view of dependability management. One is that it emphasizes the role of top management. However, it is not obvious that the core value of top management commitment is automatically supported by the formal use of a single methodology. The way you perform work within this methodology is decisive. As a parallel, there is evidence that many of the organizations that have failed with TQM have not had sufficient top management commitment [13–15].

An important consequence of the use of a dependability management system, as described above, is that it focuses on the totality. Hopefully it will reduce the risk that an organization picks up just a few parts of the system. We have experienced companies failing to implement dependability work due to the use of only small parts of the system. They pick up one or a few tools or methodologies and believe that these will solve their problems. Illustrations of this are the non-systematic use of FMECA sheets and reliability prediction handbooks some decades ago.

We have to start with the core values and ask: Which core values should characterize our organization? When that is decided we have to identify methodologies that are suitable for our organization to use and which support our values. Finally, from that decision the suitable tools have to be identified and used in an efficient way to support the methodologies. Figure 31.4 gives an example of a scheme, useful for the choice and systematic implementation of appropriate tools. It is, of course, important to note that a particular methodology can support different core values and the same tool can be useful within many methodologies. If we can use such methodologies and tools we support several values, which of course is of benefit to the culture.

The idea of using this type of system approach is not quite new. A related approach is described in [16], giving a TQM definition based on a management system perspective.

31.5 Conclusions

The authors strongly believe that the system view of dependability management will facilitate organizations to work with dependability matters, since things are "put together to create a whole". Total dependability management builds upon the availability and use of all three management system components: a set of supporting core values, an appropriate methodology, and the skillful use of contextually adapted tools.

References

[1] IEC Standard 60300-1. Dependability management. Part 1. Dependability programme management. Geneva: International Electrotechnical Commission; 1993.

[2] IEC Standard 60300-2. Dependability management. Part 2. Dependability programme elements. Geneva: International Electrotechnical Commission; 1995.

[3] IEC Standard 60300-3-1. Dependability management. Part 3. Application guide. Section 1. Analysis techniques for dependability: guide on methodology. Geneva: International Electrotechnical Commission; 1991.

[4] Hellsten U. The Springboard—a TQM-based tool for self-assessment. Licentiate Thesis 1997: No. 42. Division of Quality Technology & Statistics, Luleå University of Technology; 1997.

[5] Bergman B, Klefsjö B. Quality from customer needs to customer satisfaction. London: McGraw-Hill and Lund, Studentlitteratur; 1994.

[6] Deming WE. The new economics for industry, government, education; 2nd Edition. Massachusetts Institute of Technology, Center for Advanced Studies; 1994.

[7] Stamatis DH. Failure mode and effects analysis. FMEA from theory to execution. Milwaukee: ASQ Press; 1995.

[8] Akao Y. Quality Function Deployment. Integrating customer requirements into product design. Cambridge: Productivity Press; 1990.

[9] Slabey WR. In: The Second Symposium on Quality Function Deployment/QPC, Automotive Division—ASQC and the American Supplier Institute, Novi, MI; 1990: 21–86.

[10] IEC Standard 60300-3-9. Dependability management. Part 3. Application guide. Section 9. Risk analysis of technological systems. Geneva: International Electrotechnical Commission; 1995.

[11] Anderson RT, Neri L. Reliability-centered maintenance. Management and engineering methods. London: Elsevier Applied Science; 1990.

[12] Bird FE, Germain GL. Practical loss control leadership; Revised Edition. Loganville: Det Norske Veritas (USA); 1996.

[13] Dahlgaard JJ, Kristensen K, Kanji GK. The quality journey: a journey without an end. Abingdon: Carfax Publishing Company; 1994.

[14] Oakland JS. Total quality management. The route to improving performance; 2nd Edition. Oxford: Butterworth-Heinemann; 1993.

[15] Tenner AR, DeToro IJ. Total Quality Management: three steps to continuous improvement. Reading, MA: Addison-Wesley; 1992.

[16] Hellsten U, Klefsjö B. TQM as a management system consisting of values, techniques and tools. TQM Magazine, 2000:12, 238–44.

Total Quality for Software Engineering Management

G. Albeanu and Fl. Popentiu Vladicescu

32.1 Introduction

32.1.1 The Meaning of Software Quality

Defining quality, in general, is a difficult task. Different people and organizations define quality in a number of different ways. Most people associate quality with a product or service in order to satisfy the customer's requirements. According to Uselac [1], "Quality is not only products and services but also includes processes, environment, and people". Also, Deming [2], in his famous work says "that quality has many different criteria and that these criteria change continually". This is the main reason "to measure consumer preferences and to remeasure them frequently", as Goetsch and Davis [3] observe. Finally, they conclude: "Quality is a dynamic state associated with products, services, people, processes and environments that meets or exceeds expectations".

What about software? Software is the component that allows a computer or digital device to perform specific operations and process data. Mainly, software consists of computer programs. However, databases, files, and operating procedures are closely related to software. Any user deals with two categories of software: the operating system controlling the basic operations of the digital system and the application software controlling the data processing for specific computer applications like computer-aided drafting, designing and manufacturing, text processing, and so on.

Like any other product, quality software is that which does what the customer expects it to do. The purpose for which a software system is intended is described in a document usually known as the user requirements specification. User requirements fall into two categories [4]:

1. capabilities needed by users to achieve an objective or to solve a problem;
2. constraints on how the objective is to be achieved or the problem solved.

Capability requirements describe functions and operations needed by customers. These include at least performance and accuracy attributes. Constraint requirements place restrictions on how software can be built and operated, and quality attributes of reliability, availability, portability, and security. All these user requirements are translated into software requirements. In order to build a high-quality software product, these requirements should be rigorously described. The software requirements can be classified into the following categories: functional, performance, interface, operational, resource, portability, security, documentation, verification and acceptance testing, quality, reliability, safety, and maintainability.

Functional requirements specify the purpose of the software and may include performance attributes. Performance requirements specify numerical values for measurable variables (in general as a range of values).

A user may specify interface requirements for hardware, software, and communications. Software interfaces consist of operating systems and software environments, database management systems and file formats. Hardware configuration is determined by the hardware interface requirements. The usage of some special network protocols and other constraints of the communication interface is registered as communication interface requirements.

Some users specify how the system will run and how it will communicate with human operators. Such a user interface, man–machine or human–computer interaction requirements (the help system, screen layout, content of error messages, *etc.*) are operational requirements.

As an end user, the specification of the upper limits on physical resources (processing power, internal memory, disk space, *etc.*) is necessary, especially when future extensions are very expensive. These constraints are classified as resource requirements. The degree to which computing resources are used by the software explains an important quality factor: the efficiency.

Following Ince [5], another quality factor is portability. "This is the effort required to transfer a system from one hardware platform to another". Possible computers and operating systems, other than those of the target platform, should be clearly outlined as portability requirements.

Confidentiality, integrity, and availability are fundamental security requirements. According to Ince, "Integrity is used to describe the extent to which the system and its data are immune to access by unauthorized users". Mathematically, integrity is the probability that a system operates without security penetration for a specified time when a rate of arrival of threats and a threat profile are specified. However, such a quality factor cannot be precisely measured. To increase the security, interlocking operator commands, inhibiting of commands, read-only access, a password system, and computer virus protection are important techniques. As Musa stated [6], the software availability is "the expected fraction of operating time during which a software component or system is functioning acceptably".

In order to have reliable software, the constraints on how the software is to be verified have to be stated as verification requirements. Simulation, emulation, tests with simulated inputs, tests with real inputs, and interfacing with the real environment are common verification requirements. It is also necessary to include the acceptance tests for software validation. However, the testability is very rarely specified by the customer.

Reliability requirements may have to be derived from the user's availability requirements. It is important to mention that reliability is user-oriented. For Ince, software reliability "describes the ability of a software system to carry on executing with little interruption to its functioning". Musa [6] says that "Software reliability is the most important aspect of quality to the user because it quantifies how well the software product will function with respect to his or her needs". We mention here that although hardware and software reliability have the same definition: probability of failure-free operation for a specified interval of time, there are differences between them (a failure being any behavior in operation that will cause user dissatisfaction). The main difference is that software reliability is changing with time, when faults are removed during reliability growth test or when new code (possibly containing additional faults) is added.

Safety requirements may identify critical functions in order to reduce the possibility of damage that can follow from software failure. As Musa [6] mentioned, software safety can be defined as "freedom from mishaps, where a mishap is an event that causes loss of human life, injury, or property damage".

The maintainability or modifiability "describes the ease with which a software system can be changed" according to Ince. He outlines three categories of modifications: corrective changes (due to error fixing), adaptive changes (due to the response to changes in requirements), and perfective changes (which improve a system). The ease of performing such tasks is quantified by the mean time to repair a fault. Maintainability requirements are derived from a user's availability and adaptability requirements.

An important quality factor is correctness—the software system has to conform to requirement specifications. Usability, as another quality factor, is "the effort required to learn, operate and interrupt a functioning system" as Ince says, because a system that satisfies all the functions in its requirement specification but has a poor interface is highly unusable.

For any software project a correct and clear documentation called the "software user manual" has to be provided. Two styles of user documentation are useful: the tutorial and the reference manual. The tutorial style is adequate for new users and the reference style is better for more experienced users searching for specific information.

There are some important reasons for establishing a software quality program. According to Breisford [7], these include: reduction of the risk of producing a low quality product, reduction of the cost of software development, increased customer satisfaction and limiting company responsibility. For some industrial software the emphasis is on reliability and lifecycle costs, while for others the safety is more important. An effective quality program is necessary when the customer requires it, like defense industry operations, space and nuclear industries. As mentioned in [8], "Safety management forms an integral part of project risk management and should be given equal consideration alongside Quality and Reliability management". This remark is also valid for computer software. It is enough to mention the report [9] on the Ariane 5 rocket failure in 1996, where the problems were caused by a few lines of Ada code containing three unprotected variables. According to Ince [5], a quality system "contains procedures, standards and guidelines". A procedure is "a set of instructions which describe how a particular software task should be carried out", and a standard is a set of compulsory tasks or items to be considered when implementing a procedure or a package of procedures. Guidelines only advise taking into consideration some procedures. The literature on quality uses the term standard with a modified meaning, but in the same manner.

32.1.2 Approaches in Software Quality Assurance

The IEEE defines software quality assurance as a "planned and systematic pattern of all actions necessary to provide adequate confidence that the item or product conforms to established technical requirements" [10]. A relevant standard family for the software industry is ISO 9000 developed by the International Organization for Standardization. More precisely, ISO 9001 [11] "Quality Systems—Model for Quality Assurance in Design, Development, Production, Installation and Servicing" describes the quality system used to support the development of a product which involves design; ISO 9003 [12] "Guidelines for the Application of ISO 9001 to the Development, Supply and Maintenance of Software" interprets ISO 9001 for the software developer; ISO 9004-2 [13] "Quality Management and Quality System

Elements—Part 2" provides guidelines for the servicing of software.

There are 20 items to describe the requirements: management responsibility; quality system; contract review; design control; document and data control; purchasing; purchaser supplied product; product identification and traceability; process control; inspection and testing; control of inspection, measuring, and test equipment; inspection and test status; control of non-conforming product; corrective and preventive action; handling, storage, packaging, preservation, and delivery; quality records; internal quality audits; training; servicing; statistical techniques.

Another approach is the "Capability Maturity Model for Software" (CMM) [14], developed by the Software Engineering Institute (SEI). The SEI has suggested that there are five levels of process maturity, ranging from initial stage (the least predictable and controllable) to repeatable, defined, managed, and optimizing (the most predictable and controllable).

The fundamental concepts underlying process maturity applied for software organization are: the software process; the capability; the performance; and the maturity. The software process is a "set of activities, methods, practices and transformations" used by people to develop and maintain the software and associated products (user manuals, architectural document, detailed design document, code and test plans). The capability of the software process run by an organization provides a basis to predict the expected results that can be achieved by following the software process. While the software process capability focusses on the results expected, the software process performance represents the actual results achieved. Maturity is a concept which implies the ability for growth in capability and gives the degree of maturation in defining, managing, measuring, controlling, and running. As a corollary, a software organization gains in software process maturity by running its software process via policies, standards, and organizational structures. A brief description of the CMM follows.

Except for Level 1, each of the four capability levels has a set of key process areas that an organization should focus on to improve its software process. The first level describes a software development process that is chaotic, an *ad hoc* approach based on individual efforts (a genius who has a bright idea), not on team accomplishments. The capability maturity model emphasizes quantitative control of the process, and the higher levels direct an organization to use measurement and quantitative analysis.

Each key process area comprises a set of key practices that show the degree (effective, repeatable, lasting) of the implementation and institutionalization of that area. The key process area of the second level contains the following key practices: requirements management; software project planning, software project tracking and oversight; software subcontract management; software quality assurance and software configuration management. At this level "basic project-management processes are established to track cost, schedule, and functionality". There is a discipline among team members, so that the team can repeat earlier successes on projects with similar applications. At the defined level (Level 3), "the software process for both management and engineering activities is documented, standardized, and integrated into a standard software process for the organization". Since some projects differ from others, the standard process, under management approvement, is tailored to the special needs. The key process, at this maturity level, is oriented to organizational aspects: organization process focus; organization process definition; training program; integrated software management; software product engineering; intergroup coordination and peer reviews. A managed process (Level 4) adds management oversight to a defined process: "detailed measures of the software process and product quality are collected". There is quantitative information about both the software process and products to understand and control them. Key process areas focus on quantitative process management and software quality management. An optimizing process is the ultimate level of process maturity, where quantitative feedback is incorporated into the process to produce continuous process improvements. Key process areas include: defect prevention; technology change management; and process change management.

Finally, the following items resume the CMM structure:

1. the maturity levels indicate the process capability and contain key process areas;
2. the key process areas achieve goals and are organized by common features;
3. the common features address implementation and institutionalization and contain key practices describing infrastructure or activities.

The CMM was the starting point in producing new process assessment methods. We note only the bootstrap approach [15] (an extension of the CMM developed by a European Community ESPRIT project) and the new standard called SPICE [16] (for Software Process Improvement and Capability dEtermination). Similar to CMM, SPICE can be used both for process improvement and capability determination. The SPICE model considers five activities: customer-supplied, engineering, project, support, and organization, while the generic practices, applicable to all processes, are arranged in six levels of capability: 0—"not performed", 1—"performed informally", 2—"planned and tracked", 3—"well-defined", 4—"quantitatively controlled", 5—"continuously improving". An assessment report, a profile, for each process area describes the capability level.

Ending this section, the reader has to observe that the CMM tends to address the requirement of continuous process improvement more explicitly than ISO 9001, and that CMM addresses software organizations while SPICE addresses processes. However, no matter which model is used, the assessment must be administered so as to minimize the subjectivity in the ratings.

32.2 The Practice of Software Engineering

32.2.1 Software Lifecycle

The increasing software complexity and the cost and time constraints against software reliability maximization for most software projects have made mandatory the usage of a standardized approach in producing such items.

The collection of techniques that apply an engineering approach to the construction and support of software is known as "software engineering". The theoretical foundations for designing and developing software are provided by computer science. However, software engineering provides the mechanism to implement the software in a controlled and scientific way. According to [17], software engineering is "the application of a systematic, disciplined, quantifiable approach to the development, operations, and maintenance of software, that is, the application of engineering to software". It is possible to say, as in [18, 19] that "Software engineering is evolving from an art to a practical engineering discipline".

This approach was also adopted by ESA [4], which establishes and maintains software engineering standards for the procurement, development, and maintenance of software. The ESA standard PSS-05-0 describes the processes involved in the complete lifecycle of the software. Organized in two parts, the standard defines a set of practices for making software products. Such practices can also be used to implement the requirements of ISO 9001. The ESA's Board for Software Standardization and Control says that [20]:

1. the ESA Software Engineering Standards are an excellent basis for a software quality management system;
2. the ESA standards cover virtually all the requirements for software developments—two thirds of the requirements in ISO 9001 are covered by ESA Software Engineering Standards, and the uncovered requirements are not related to software development;
3. the ESA standards do not contradict those of ISO 9001.

According to [4], a software lifecycle consists of the following six successive phases: (1) definition of the user requirements; (2) definition of the software requirements; (3) definition of the architectural design; (4) detailed design and production of code; (5) transfer of the software to operations; (6) operations and maintenance. These phases are mandatory whatever the size, application type, hardware configuration, operating system, or programming language used for coding. The software lifecycle begins with the delivery of the "user requirements document" to the developer for review. When this document is approved, three important phases have to be traversed before the software is transferred to users for operation. In order to start a new phase, the results of the previous phase are reviewed and approved. The software lifecycle ends after a period of operation, when the software is retired.

Pham [19], considering the software reliability realm, identifies five successive phases for the software lifecycle: analysis (requirements and functional specifications), design, coding, testing, and operating. From the reliability point of view, "a well-developed specification can reduce the incidence of faults in the software and minimize rework". It is well known (see [19]) that a well-done analysis will "generate significant rewards in terms of dependability, maintainability, productivity, and general software quality". The design phase consists of two stages of design: system architecture design and detailed design". The principal activities in the system architecture design stage are: (1) construction of the physical model; (2) specifying the architectural design; (3) selecting the programming language (or software development environments); and (4) reviewing the design.

Using implementation terminology, the developer will derive a physical model starting from the logical model developed in the analysis phase (software requirements stage). To build a physical model some activities are important: the decomposition of the software into components (a top-down approach, but care for the bottom-level components); the implementation of non-functional requirements; the design of quality criteria; and special attention paid to alternative designs. The last activity is needed because, in general, there is no unique design for a software system. The design has to be easy to modify and maintain, only make a minimal use of available resources, and must be understandable in order to be effectively built, operated and maintained. Other quality criteria related to design are: simplicity in form and function, and modularity.

The architectural design document contains diagrams showing the data flow and the control flow between components. For each component there are specified: data input, functions to be performed (derived from a software requirements document), and data output. Data input, data output, and temporary memory cells are clearly defined as data structures (with name, type, dimension, relationships between the elements, range of possible values of each element, and initial values of each element). The definition of the control flow between components describes sequential and parallel operations, including also synchronous and asynchronous behavior. About this phase, Pham says that "the system architecture document describes system components, subsystems and interfaces".

The selected data structures and algorithms, in the detailed design phase, will be implemented in a particular programming language on a particular platform (hardware and software system). The selected programming languages must support top-down decomposition, object-oriented programming, and concurrent production of the software. Also, the choice of a programming language depends on the non-functional requirements. Finally, the architectural design and the programming language should be compatible.

Detailed design is about designing the project and the algorithmic details. Lower-level components of the architectural design are decomposed until they can be presented as modules or functions in the selected programming language. For many recent projects, an object-oriented (OO) analysis and design approach has been considered. The goals of the OO analysis are to identify all major objects in the problem domain, including all data and major operations mandatory for establishing the software functions, and to produce a class diagram containing all the software project semantics in a set of concise but detailed definitions. A class specification and a data dictionary are also delivered. The OO design maps the analysis product development during the analysis phase to a structure that will allow the coding and execution. Various methodologies call this approach OOA/OOD because in a practical analysis/design activity, it is almost impossible to find the boundaries between OOA and OOD.

Producing software, after the detailed design phase, starts with coding. The code should be consistent (to reduce complexity) and structured (to reduce errors and enhance maintainability). From the structured programming point of view, each module has a single entry and an exit point, and the control flow proceeds from the beginning to the end. According to [4], "as the coding of a module proceeds, documentation of the design assumptions, function, structure, interface, internal data and resource utilisation should proceed concurrently". As Pham says, the coding process consists of the following activities: the identification of the "reusable modules", the code editing, the code "inspection", and the "final test planning". Existing modules of other systems or projects which are similar to the current system can be reused with modifications. This is an effective way to save time and effort. "Code inspection includes code reviews, quality, and maintainability". The aim of the code review process is to check module (software) logic and readability. "Quality verification ensures that all the modules perform the functionality as described in detailed design. Quality check focuses on reliability, performance, and security". Maintainability is checked to ensure the software project is easy to maintain. The final test plan provides the input to the testing phase: what needs to be tested, testing strategies and methods, testing schedules, *etc.*

Also, an important step in the production phase, even before testing, is integration. As ESA [4] mentions, "integration is the process of building a software system by combining components into a working entity" and shall proceed in an orderly function-by-function sequence.

Testing consists of the verification and validation of the software product. The goals for these activities are related to the software quality. The most important goal is to affirm the quality by detecting and removing faults in the software project. Next, all the software specified functionalities shall be present in the product. Finally, but as a considerable goal, is the estimation of the operational reliability of the software product.

The process of testing an integrated software system is called "system testing". Previous processes related to the testing phase are: unit testing and integration testing. Unit tests check the design and implementation of all components, "from the lowest level defined in the detailed design phase up to the lowest defined in the architectural design" according to ESA standards. Integration testing is directed at the interfacing part: to verify that major components interface correctly (both as data structures and control flow levels). The testing team shall include a large set of activities, like: end-to-end system tests (the input–execute–output pipe works properly); verification that user requirements will be entirely covered (acceptance tests); measurements of performance limits (stress tests); preliminary estimation of reliability and maintainability; and the verification of the "software user manual". According to Pham [19], the acceptance test consists of an internal test and a field test. "The internal test includes capability test and guest test, both performed in-house. The capability test tests the system in an environment configured similar to the customer environment. The guest test is conducted by the users in their software organization sites". The field test, also called the "beta test", allows the user to test the installed software, defining and developing the test cases. "Testing by an independent group provides assurance that the system satisfies the intent of the original requirements".

Considering the transfer phase, the developer will "install the software in the operational environment and demonstrate to the initiator and users that the software has all the capabilities" formulated in the user requirements phase. Acceptance tests will demonstrate the capabilities of the software in its operational environment and these are necessary for provisional acceptance. The final acceptance decision is made by the customer.

The final phase in the software lifecycle is operation, consisting of the following activities: training, support, and maintenance. Following Pham [19], maintenance is defined as "any change made to the software, either to correct a deficiency in performance, to compensate for environmental changes, or to enhance its operation". After a warranty period, the maintenance of the software is transferred from the developer to a dedicated maintenance organization.

32.2.2 Software Development Process

A software process model is different from a software methodology. The process model determines the tasks or phases involved in product development and all criteria for moving from one task to the next one. The software methodology defines the outputs of each task and describes how they should be achieved. The ESA approach, Part I, describes a software process model and for each phase describes the inputs, outputs, and activities. The second part of the ESA standard describes the procedures used to manage a software project. These aspects will be covered in Section 32.4.

It is difficult to clearly define which methods should be used to complete a phase of the product development and production. As Cederling remarks [21], "hardly any method is suitable to use in development of all kinds of applications and there are only a few methods that pretend to cover the whole life-cycle of the software". From a software quality point of view, the steps could be organized into three elements: software quality management, software quality engineering, and software quality control. Such aspects will be discussed in Section 32.3.

There are three favored software development processes: the waterfall model, the spiral model, and the generic integrated software development environment. ESA [20] also suggests three lifecycle models called: waterfall, incremental, and evolutionary. The waterfall model is the most used; it is a sequential approach and the implementation steps are the activities in the sequence in which they appear. This is the main reason that development schedules can readily be based on it. In the spiral model, each cycle involves a progression through the same series of steps, which are applied to everything from overall concept to detailed coding of software. Such an approach applies to development and maintenance. According to [20], the incremental model "splits the detailed design phase into manageable chunks". The generic integrated environment is used when an overall integration of the project's residual tools is necessary. The impact of object-oriented methods on software quality is quite important. Capper *et al.* [22], show how and when to use object orientation to improve software quality by taking advantage of reuse and code modularity. ESA [20] explains the advantages of an object-oriented approach: "the analysis more closely corresponds to the application domain", "the design (physical model) is an elaboration of the requirements analysis model (the logical model)", and "object-oriented methods use the concept of inheritance which, if properly used, permits the building of reusable software". Such object-oriented methods which recommend feedback loops between analysis, design and implementation phases when included in a software engineering standard require special treatment in order to maintain the PSS-05-0 concept of well-defined development phases.

Software development with prototypes is a new approach. As mentioned in [5], "prototyping is the process of developing a working model of a system early on in a project". The working model is shown to the customer, who suggest improvements. The improvements incorporation gives rise to an upgraded prototype which is shown again to the customer, and the process can continue. Producing a prototype is based on the following techniques: the relaxation of the quality assurance standards of the project; partial implementation; the table-driven processor approach; the usage of an application-oriented programming language, *etc.* Object-oriented programming languages are also an excellent medium for prototyping (see [23]). There are three popular prototyping models: throw-away prototyping, evolutionary prototyping, and incremental prototyping. By throw-away prototyping, effective for short projects, the developer builds a prototype and then starts the iterative process of showing–modifying already described. When the "game" is finished, the prototype is archived and conventional software development is started, based on a requirement specification written by examining the detailed functionality of the prototype. Evolutionary prototyping permits the developer "to keep the prototype alive" as Ince says. After the customer has decided that all requirements are included, the developer bases the remainder of the work on the prototype already generated. The incremental prototyping is effective for projects where the requirements can be partitioned into functionally separate areas, the system being developed as a series of small parts, each of them implementing a subset of functions.

A theory-based, team-oriented engineering process for developing high-quality software is the cleanroom approach. According to [24], the cleanroom process "combines formal methods of object-based box structure specification and design, function-theoretical correctness verification, and statistical usage testing for reliability certification to produce software approaching zero defects". The cleanroom software development method has three main attributes [25]: a set of attitudes; a series of carefully prescribed processes; and a rigorous mathematical basis. "Cleanroom processes are quite specific and lend themselves to process control and statistical measures". As Ince mentioned, "a formal method of software developments makes use of mathematics for specification and design of a system, together with mathematical proof as a validation method". Formal methods require the same conventional development models, unlike prototyping and object-oriented technology. A new useful mathematical framework important for software developers is the generalized analytic hierarchy process [26], which provides qualitative and quantitative information for resource allocation.

The late 1980s brought new graphically based software tools for requirements specification and design, called computer-aided software engineering (CASE) tools, with the following abilities: (1) creation of the diagrams describing the functions of the software system; (2) creation of the diagrams describing the system design of both the processes and the data in the system; (3) processing the graphical requirements specification and simulating its actions like a rudimentary prototype; (4) the identification of errors in the flow of data; (5) the generation of code from the design; and (6) the storage and manipulation of information related to the project management. These tools improve the quality management system of software development companies by enforcing their own requirements specification and system design standards, and doing a lot of bureaucratic checking activities.

32.2.3 Software Measurements

As Fenton and Pfleeger mention [27], we can neither predict nor control what we cannot measure. Measurement "helps us to understand what is happening during development and maintenance", "allows us to control", and "encourages us to improve our processes and products". Measurement is the first step towards software quality improvement. Due to the variety and complexity of software products, processes, and operating conditions, no single metric system can be identified as a measure of the software quality. Several attributes have to be measured, related to:

- the product itself: size, complexity, modularity, reuse, control flow, data flow, *etc.*;
- the development process: cost and effort estimation, productivity, schedule, *etc.*;
- quality and reliability: failures, corrections, times to failures, fault density, *etc.*;
- maintenance and upgrading: documentation, *etc.*

The quality of any measurement program is dependent on careful data collection. For instance, in order to perform reliability analysis, the following types of data should be collected: internal attributes (size, language, functions, verification and validation methods and tools, *etc.*) and external attributes (time of failure occurrence, nature of the failures, consequence, the current version of the software environment in which the fault has been activated, conditions under which the fault has been activated, type of faults, fault location, *etc.*). According to [27], "an internal attribute can be measured by examining the product, process or resource on its own, separate from its behavior", while external attributes are related to the behavior of the process, product, or resource and "can be measured only with respect to how the product, process or resource relates to its environment". Considering the process of constructing specification, internal attributes are: time, effort, number of requirements changes, *etc.* and quality, cost, stability are some external attributes. Any artefacts, deliverables, or documents that result from a process activity are identified as products: the specifications, designs, code, prototypes, documents, *etc.* For example, if we are interested in measuring the code of one software product, the internal attributes cover: size, reuse, modularity, functionality, coupling, algorithmic complexity, control-flow structures, hierarchical class structures (the number of classes, functions, and class interactions), *etc.* Some popular software-process metrics are: number of "problems" found by developer, number of "problems" found during beta testing, number of "problems" fixed that were found by developer in prior release, number of "problems" fixed that were found by beta testing in the prior release, number of "problems" fixed that were found by customer in the prior release, number of changes to the code due to new requirements, total number of changes to the code for any reason, number of distinct requirements that caused changes to the module, net increase in lines of code, number of team members making changes, number of updates to a module, *etc.*

Measuring the size of the software products is consistent with measurement theory principles [27]. The software size can be described with three attributes: length (the physical size of the product), functionality (the functions supplied by the product to the user), and complexity of the problem, algorithms, software structure, and cognitive effort.

The most used measure of source code program length is the LOC (number of lines of code), sometimes interpreted as NCLOC (only non-commented lines) or ELOC (effective lines of code). However, the dependence of the programming language and the impact of the visual programming windowing environments are important factors which have a great influence on the code. The length is also a measure for specifications and design, which consists of both text (the text length) and diagrams (the diagram length). Other statement metrics are: number of distinct included files, number of control statements, number of declarative statements, number of executable statements, number of global variables used, *etc.*

The functionality captures "the amount of function contained in a delivered product or in a description of how the product is supposed to be" [27]. There are three commonly used approaches: the Albrecht's effort estimation method (based on function points), COCOMO 2.0 (based on object points), and the DeMarco's specification weight.

Complexity is difficult to measure. The complexity of an approach is given in terms of the resources needed to implement such an approach (the computer running time or the computer memory used for processing). As mentioned by Pham [19], "Halstead's theory of software metric is probably the best-known technique to measure the complexity in a software program and amount of difficulty involved in testing and debugging the software". However, the focus of Halstead's measurements is the source code for an imperative language. Such an approach to software based on declarative programming is not suitable. The following metrics, used by Halstead, give rise to empirical models: the number of distinct operators and the number of distinct operands in a program to develop expressions for the overall program length, the volume and the number of remaining defects in a program. Another complexity metric measure of software is the cyclomatic number, proposed by McCabe (see [19, 27] and references cited therein), that provides a quantitative measure of the logical complexity of a program by counting the decision points. The above remark also applies for visual software environments, especially for object-oriented programming, which needs special complexity measures.

The software structure has at least three parts: control-flow structure (addressing the sequence in which instructions are executed), data-flow structure (showing the behavior of the data as it interacts with the software), and data structure (including data arrangements and algorithms for creating, modifying, or deleting them). The complexity of the control-flow structures is obtained using hierarchical measures, every computer program being built up in a unique way from the so-called prime structures according to the structured programming principles. Some popular control-flow graph metrics (for flowcharts) are: number of arcs that are not conditional arcs, number of non-loop conditional arcs (if–then constructs), number of loop constructs, number of internal nodes, number of entry nodes, number of exit nodes, *etc.*

External software product attributes will be considered in the next section. Without presenting equations, the above descriptive approach shows what the developers, the software managers have to measure in order to understand, control, and evaluate. Special attention will be given to data collection not only for different internal measurements, but for the investigation of relationships and future trends.

32.3 Software Quality Models

32.3.1 Measuring Aspects of Quality

In the first section we pointed out that quality is a composite of many characteristics. Early models of Boehm *et al.* [28] and McCall *et al.* [29] described quality by decomposition. Both models identify key attributes of quality from the customer's perspective, called quality factors, which are decomposed into lower-level attributes, called quality criteria and directly measurable. For example, McCall's model considers the following quality factors: usability, integrity, efficiency, correctness, reliability, maintainability, testability, flexibility, reusability, portability, and interoperability. We observe that most of these factors also describe the software requirements which were outlined in the first section.

A derivation of McCall's model, called "Software Product Evaluation: Quality Characteristics and Guidelines for their Use", is known as the ISO 9126 standard quality model, decomposing the quality only into six quality factors: functionality, reliability, efficiency, usability, maintainability, and portability [30]. Unfortunately, the definitions of these attributes differ from one standard to another.

Another approach extends Gilb's ideas [31] and provides an automated way to use them: the COQUAMO model of Kitchenham and Walker [32]. Other approaches view software quality as an equation based on defect-based measures, usability measures, and maintainability measures. Some popular measures are: the defect density (given as the division of the number of known defects by the product size), the system spoilage (the division of the time to fix post-release defects by the total system development time), the user performance measures defined by the MUSiC project [33], the total effort expended on maintenance [34], Gunning's readability measure [35] (for written documentation), *etc.* Other measures will appear in the next sections. Recently, sophisticated software tools to support software developers, with software-quality models, appeared. For example, [36] describes the decision support tool EMERALD (Enhanced Measurement for Early Risk Assessment of Latent Defects system). As [37] remarks, such tools "are the key to improve software-quality" and can "predict which modules are likely to be fault-prone" using the database of available measurements.

32.3.2 Software Reliability Engineering

Improving the reliability of a software product can be obtained by measuring different attributes during the software development phase. Measurement uses the outputs of a data collection process. Two types of data are related to software reliability: time-domain data and interval-domain data. The time-domain data approach requires that individual times at which failure occurred should be recorded. For the interval-domain approach it is necessary to count the number of failures occurring during a fixed period. From an accuracy point of view in the reliability estimation process, the time-domain approach is adequate but more data collection efforts are necessary. In the following, the reliability will be presented as an engineering concept, improving the framework of the total quality for software engineering management.

According to Musa [6], "engineering software reliability means developing a product in such a way that the product reaches the market at the right time, at an acceptable cost, and with satisfactory reliability". The reliability incorporates all those properties of the software that can be associated with the software running (correctness, safety, usability, and user friendliness). There are two approaches to measure the software reliability: a developer-oriented approach which counts the faults or defects found in a software product, and a user-oriented approach taking into account the frequency with which problems occur.

A basic approach of the software reliability engineering (SRE) process, according to [6], requires the following activities.

1. Identify the modules that should have a separate test.
2. Define the failure with severity classes (a failure severity class is a set of failures that have the same per-failure impact on users, based on a quality attribute).
3. Choose the natural or time unit (a natural unit is the one related to the output of a software-based product).
4. Set the system failure intensity objective for each module to be tested (failures per natural unit).
5. Determine the developed software failure intensity objective.
6. Apply, in an engineering manner, the reliability strategies in order to meet the failure intensity objective.
7. Determine the operational modes of each module to be tested.
8. For each module to be tested, develop the module and the operational profile (by identifying the operation initiators, creating the operations list, determining occurrence rates and occurrence probabilities).
9. Prepare test cases (involving the following steps: estimation of the number of new test cases needed for the current release, allocation of the test cases among the modules to be tested, for each module allocate new test cases among its new operations, specify the new test cases, and add these new test cases to those from previous releases).
10. Prepare test procedures (one test procedure for each operational mode).
11. Execute the test by allocation of the test time, test the systems (acquired components, product and variations, and supersystems), and carefully identify the failures (by analyzing the test output for deviations, determining which deviations are failures, establishing when the failures occurred and assigning failure severity classes).
12. Apply failure data to guide decisions.

Introduction of SRE into an organization it is a strong function of the software process maturity of that organization. The necessary period for introduction can range from six months to several years. An incremental approach is recommended. The process has to start with activities needed for establishing a baseline and learning about the product and about customer expectations.

An important set of SRE activities are concerned with measurement and prediction of software reliability and availability. This includes modeling of software failure behavior and modeling of the process that develops and removes faults. According to [6, 21, 38–40] and references cited therein, a large set of metrics and models are available for such a purpose.

There are many types of software reliability models. Some well-known models are: Halstead's software metric and McCabe's cyclomatic complexity metric (both deterministic models). Halstead's metric can be used to estimate the number of errors in the program, while McCabe's cyclomatic complexity metric can be used to determine an upper bound on the number of tests in a program. Models included in the second group, called "error seeding models", estimate the number of errors in a program by using a multistage sampling technique, when the errors are divided into indigenous errors and induced errors (seeded errors). Examples of such models are [19]: Mills' error seeding model, the hypergeometric distribution model, and Cai's model. Another group of models is used to study the software failure rate per fault at the failure intervals. Models included in this group are: Jelinski and Moranda (J-M), negative-binomial Poisson, Goel–Okumoto imperfect debugging, and a large number of variations of the J-M model. Other models, namely curve fitting models, use statistical regression analysis to study the relationship between software complexity and the number of faults in a program, the number of changes, or failure rate. When assuming that the future of the process is statistically independent of the past, NHPP (non-homogeneous Poisson process) models are proposed (see [19, ch.4]): Duane, Musa exponential, Goel–Okumoto, S-shaped growth models, hyperexponential growth models, generalized NHPP, *etc.* The last group, in Pham's classification, includes Markov models [19]. Unfortunately, because reliability is defined in terms of failures, it is impossible to measure before development is complete. Even carefully collecting data on interfailure times and using software reliability growth models, it is difficult to produce accurate predictions on all data sets in all environments; that means the above-mentioned techniques work effectively only if the software's future operational environment is similar to that in which the failure data was collected. The advice of Fenton and Pfleeger [27] is full of importance: "If we must predict reliability for a system that has not been released to a user, then we must simulate the target operational environment in our testing".

From an SRE point of view, it is essential that a person selecting models and making reliability predictions be appropriately trained in both software (reliability) engineering and statistics (see the cleanroom approach [25]). Data filtering and outliers identification are fundamental steps for data validation. Data partitioning is also important. Applying reliability growth models to the most severe failures allows, for example, evaluation of the software failure rate corresponding to the most critical behavior. Such a rate is more significant than the overall software failure rate which incorporates failures that do not have a major impact on system behavior.

Reliability improvement programs will help the companies to ameliorate the maturity of their software production, adopting the CMM approach, ESA standard, or other standards specially developed for producing safety-critical applications.

32.3.3 Effort and Cost Models

One of the major objections to applying programs for developing under reliability requirements is the total effort, which has been considered for a long time as increasing with the level required. This is why the effort estimation is crucial for the software development organization. Overestimated effort may convince general management to disapprove proposed systems that might significantly contribute to the quality of the development process. Underestimated effort may convince general management to approve, but the exceeding of their budgets and the final failing is the common way. A model which considers the testing cost, cost of removing errors detected during testing phase, cost of removing errors detected during the warranty period, and risk cost due to software failure is proposed by Pham and Zhang [41]. As the authors mention: "this model can be used to estimate realistic total software cost" for some software products, and "to determine the optimal testing release policies of the software system". Other software cost models based on the NHPP software reliability functions can be found in [19].

There are many proposed solutions, but, in general, it is very restrictive to apply them across a large set of software projects. Applying data analysis to empirical data indicates that effort trends depend on certain measurable parameters. There are two classes of models for the effort estimation: cost models and constraint models. COCOMO is an empirical cost model (a composite one) which provides direct estimates of effort or duration. Often, the cost models have one primary input parameter and a number of cost drivers (characteristics of the project, process, product, or resources having a significant influence on effort or duration). Constraint models describe a relationship over time between two or more parameters of effort, duration, or staffing level. Because these mathematical models are defined in terms of an algorithm, they are termed algorithmic models. Boehm [42] classifies algorithmic models used for software cost estimation as follows:

1. linear models that try and fit a simple line to the observed data;
2. multiplicative models that describe effort as a product of constants with various cost drivers as their exponents;
3. analytic models that usually describe effort as a function that is neither linear nor multiplicative;
4. tabular models that represent the relationship between cost drivers and development effort in a matrix form;
5. composite models using a combination of all or some of the above-mentioned approaches, which are generic enough to represent a large class of situations.

Composite models are mostly used in practice. Another composite model, widely used in industry, is Putnam's SLIM (software lifecycle management) model. Both the COCOMO and Putnam models use the Rayleigh distribution as an approximation to the smoothed labor distribution curve. SLIM uses separate Rayleigh curves for design and code, test and validation, maintenance, management. Boehm and his colleagues [43] have defined an updated COCOMO, useful for a wider collection of software development techniques and technologies, including re-engineering, applications generators, and object-oriented approaches.

However, practitioners also use informal considerations: an expert's subjective opinion, availability of resources, analogy, *etc.* To succeed, they use more than one technique simultaneously. For quality improving reasons, it is useful to assign the responsibilities of the cost estimation to a specific group of people [44]. According to [27], "the group members are familiar with the estimation and calibration techniques", "their estimating experience gives them a better sense of when a project is deviating from the average or standard one", "by monitoring the database, the group can identify trends and perform analyses that are impossible for single projects", and with the advantage of being separated from the project staff they can re-estimate periodically different external attributes.

32.4 Total Quality Management for Software Engineering

32.4.1 Deming's Theory

The US Department of Defense defines the total-quality approach (also called total quality management, or TQM) as follows (from ref. 11, ch.1, cited in [3]): "TQM consists of continuous improvement activities involving everyone in the organization—managers and workers—in a totally integrated effort toward improving performance at every level". A definition of TQM has two components: "the what and the how of total quality". The how component has 10 critical elements: customer focus, obsession with quality, scientific approach, long-term commitment, teamwork, continual improvement of systems, education and training, freedom throughout control, unity of purpose, and employee involvement and empowerment. Many people and organizations contribute to develop various concepts, collectively known as TQM. The best known quality pioneer is Dr W. Edwards Deming [2]. Deming's "fourteen points" philosophy is also applicable to software quality. According to Goetsch and Davis [3] and Zultner [45], following the plan-do-check-act-analyze Deming cycle, the 14 rules are (the numbers represent neither an order of progression nor relative priorities).

1. Create constancy of purpose for the improvement of development with the aim of becoming excellent (optimizing level in SEI approach).
2. Adopt a new philosophy: software projects fail because of a lack of management and control. Software engineering management must learn out of experience that quality is vital for survival.
3. Achieving quality has to be independent of inspection process. Develop for quality.
4. Stop awarding contracts based on the price tag alone.
5. Improve constantly and forever the system development process to improve quality and productivity, creating increasing value with lower cost.
6. Institute training on the job. Training is the best way to improve people on a continual basis.
7. Institute leadership (to help people and technology to work better).
8. Drive out fear so that everyone may work effectively.
9. Break down barriers between departments. Teamwork, especially for the front end of the project, improves the final product quality.
10. Eliminate slogans and targets that ask for new levels of productivity or competitive degrees. They create adversarial relationships.
11. Eliminate quotas.
12. Remove barriers that rob employees of their pride of workmanship.
13. Institute a vigorous program of education/training and self-improvement.
14. Put everyone to work to accomplish the transformation.

The above rules summarize Deming's views on what the software organization must do to effect a positive transition from "business-as-usual to world-class quality". To build quality software, the software house must take substantial changes in the way systems are developed and managed (see Zultner [45]). The factors that may inhibit such a transformation are termed "Deming's seven deadly diseases".

By adapting them to software quality, the following list is obtained.

1. Lack of constancy of purpose to plan software products that satisfy the user requirements and keep the company in business. Excessive maintenance costs.
2. Emphasis on short-term schedule.
3. Reviewing systems and managing by objectives without providing methods or resources to accomplish objectives.
4. Mobility of systems professionals and management.
5. Managing by "visible figures" alone with little consideration, or no consideration given to the figures that are unknown or cannot be known.
6. Excessive personnel cost.
7. Excessive costs of legal liabilities.

TQM can eliminate or reduce the impact of a lack of constancy, the reviewing system, the mobility and usage of visible figures. However, total quality will not free the organization from excessive personal costs or legal liabilities from pressure to produce short-term profits, these being "the diseases of the nation's financial, health care, and legal systems", as mentioned by Goetsch and Davis [3].

32.4.2 Continuous Improvement

Continuous improvement is the most fundamental element of TQM. It applies not only to products but also to processes and the people who operate them. The process under improvement has to be defined (all the activities to be performed are clearly described). Also, the process has to be used (applied) over many projects to create experience. To decide on process behavior, data and metrics have to be collected and analyzed. The above section on measurement and the cited literature explain the important role of measurement for TQM. Other models and references can be found in Popentiu and Burschy [46].

Process improvement can be addressed at an organizational level, at a project level, and at an individual level. Both CMM and ISO 9001 identify processes at organizational level. The CMM identifies key process areas mandatory for achieving a certain level of maturity. According to Paulk [14], in the CMM, process improvement is implied in the concept of maturity itself, being more specifically addressed in the key process areas of Level 5, while ISO 9001 just states the minimum criteria to achieve the ISO certification. At a project level, the SPICE model deals with process improvement, while the personal software process model, developed by Humphrey [47], addresses the process improvement at an individual level. In Humphrey's model, each individual goes through a series of four levels, in which new steps are added that make a more mature process. Let us observe that even though some quality models incorporate the concept of software improvement, they do not explain how it should be obtained. The ESA standard, and its updates to the new technologies, is more specific. It provides an algorithm both from a software engineering and a management point of view. A collaborative ESPRIT project involving nine European centers of excellence, called "ami" [48], is an adaptation of the goal–question–metric (GQM) method devised by Basili [49]. The "ami" approach had the support of the ESA. Another system engineering process which transforms the desires of the customer/user into the language required, at all project levels, is the quality function deployment (QFD). QFD, according to Goetsch and Davis [3], brings a number of benefits to organizations trying to enhance their competitiveness by continually improving quality and productivity. The process has the benefits of being: customer-focused, time-efficient, teamwork-oriented, and documentation-oriented.

The goal of the management activities is to build the software product within the budget, according to schedule, and with the required quality. To achieve this, according to ESA [4], an organization must establish plans for: software project management, software configuration management, software verification and validation, and software quality assurance.

The software project management plan (SPMP) contains "the technical and managerial project functions, activities and tasks necessary to satisfy the requirements of a software project". Following the continuous improvement paradigm, the SPMP has to be updated and refined, throughout the lifecycle, "as more accurate estimates of the effort involved become possible", and whenever changes of some attributes or goals occur.

Software configuration management, essential for proper software quality control, is both a managerial and a technical activity. Ince [5] identifies the following activities which make up the process of configuration management: configuration items identification, configuration control, status accounting, and configuration auditing.

All software verification and validation activities shall be documented in the corresponding plan. According to ESA [4], the verification activities include: reviewing, inspecting, testing, formal checking, and auditing. By validation, the organization will determine whether the final product meets the user requirements.

The software quality assurance plan defines how the standards adopted for project development will be monitored. Such a plan is a checklist for activities that have to be carried out to ensure the quality of the product. For each document, the ESA approach provides a clear layout, with strict format and content. Other software management practices exist, and Kenet and Baker [50] give a pragmatic view of the subject.

32.5 Conclusions

Strategies to obtain software quality are examined. The basic approaches in software quality assurance are covered; ISO 9000 and CMM standards being considered. The practice of software engineering is illustrated by ESA standards and modern software development techniques. Other methodologies are: software reliability engineering, software engineering economics, and software management. Putting them together, and using the adapted Deming's philosophy, the software organization will implement the total quality management paradigm.

References

[1] Uselac S. Zen leadership: the human side of total quality team management. Londonville, OH: Mohican Publishing Company; 1993.

[2] Deming WE. Out of the crisis. Cambridge, MA: Massachusetts Institute of Technology Center for Advanced Engineering Study; 1986.

[3] Goetsch DL, Davis S. Introduction to total quality. quality, productivity, competitiveness: Englewood Cliffs, NJ: Prentice Hall; 1994.

[4] ESA PSS-05-0: ESA Software Engineering Standards, Issue 2; February 1991.

[5] Ince D. An introduction to quality assurance and its implementation. London: McGraw-Hill; 1994.

[6] Musa JD. Software reliability engineering. New York: McGraw-Hill; 1999.

[7] Breisford JJ. Establishing a software quality program. Quality Progress, November 1988.

[8] CODERM: "Is it Safe?" The Newsletter 1998/99;15,(2):6–7.

[9] Lions JL. Ariane 5 Flight 501 Failure: Report of the Inquire Board. Paris; July 19, 1996.

[10] IEEE Standard for Software Quality Assurance Plans: IEEE Std. 730.1-1989.

[11] International Standards Organization. ISO 9001: Quality systems—model for quality assurance in design, development, production, installation and servicing. International Standards Organization; 1987.

[12] International Standards Organization. Quality management and quality assurance standards—Part 3. Guidelines for the application of ISO 9001 to the development, supply and maintenance of software. ISO/IS 9000-3. Geneva; 1990.

[13] International Standards Organization. Quality management and quality system elements guidelines—Part 2. ISO 9004-2. Geneva; 1991.

[14] Paulk MC. How ISO 9001 compares with the CMM. IEEE Software 1995;12(1):74–83.

[15] Kuvaja P, Bicego A. Bootstrap: a European assessment methodology. Software Qual J 1994;3(3):117–28.

[16] International Standards Organization. SPICE Baseline Practice Guide, Product Description, Issue 0.03 (Draft); 1993.

[17] IEEE. Glossary of software engineering terminology. IEEE Std. 610.12-1990.

[18] Lyu MR. Handbook of software reliability engineering. New York: McGraw Hill; 1996.

[19] Pham H. Software reliability. Singapore: Springer; 2000.

[20] Jones M, Mazza C, Mortensen UK, Scheffer A. 1977–1997: Twenty years of software engineering standardisation in ESA. ESA Bull 1997; 90.

[21] Cederling U. Industrial software development—a case study. In: Sommerville I, Paul M, editors. Software engineering—ESEC'93. Berlin: Springer-Verlag; 1993. p.226–37.

[22] Capper NP, Colgate RJ, Hunter JC, James. The impact of object-oriented technology on software quality: three case histories. Software Qual 1994;33(1):131–58.

[23] Blaschek G. Object-oriented programming with prototypes. Berlin: Springer-Verlag; 1994.

[24] Hausler PA. Linger RC, Trammell CJ. Adopting cleanroom software engineering with a phased approach. Software Qual 1994;33(1):89–110.

[25] Mills HD, Poore JH. Bringing software under statistical quality control. Qual Prog 1988; Nov.

[26] Lee M, Pham H, Zhang X. A methodology for priority setting with application to software development process. Eur J Operat Res 1999;118(2):375–89.

[27] Fenton NE, Pfleeger SL. Software metrics: a rigorous & practical approach, 2nd ed. International Thomson Computer Press; 1996.

[28] Boehm BW, Brown JR, Kaspar H, Lipow M, Macleod G, Merrit M. Characteristics of software quality. TRW series of software technology. Amsterdam: North Holland; 1978.

[29] McCall JA, Richards PK, Walters GF. Factors in software quality. RADC TR-77-369; 1977.

[30] International Standards Organization. Information technology—software product evaluation—quality characteristics and guide lines for their use. ISO/IEC IS 9126. Geneva; 1991.

[31] Gilb T. Software metrics. Cambridge, MA: Chartwell-Bratt; 1976.

[32] Kitchenham BA, Walker JG. A quantitative approach to monitoring software development. Software Eng J 1989;4(1):2–13.

[33] Bevan N. Measuring usability as quality of use. Software Qual J 1995,4(2):115–30.

[34] Belady LA, Lehman MM. A model of large program development. IBM Syst J 1976;15(3):225–52.

[35] Gunning R. The technique of clear writing. New York: McGraw-Hill; 1968.

[36] Hudepohl JP, Aud SJ, Khoshoftaar TM, *et al.* EMERALD: software metrics and models on the desktop. IEEE Software 1996;13(Sept):56–60.

[37] Khoshgoftaar TM, Allen EB, Jones WD, *et al.* Classification-tree models of software-quality over multiple releases. IEEE Trans Reliab 2000;49(1):4–11.

[38] Musa JD, Iannino A, Okumoto K. Software reliability: measurement, prediction, application. New York: McGraw Hill; 1987.

[39] Xie M. Software reliability modelling. Singapore: World Scientific; 1991.

[40] Burtschy B, Albeanu G, Boros DN, Popentiu FL, Nicola V. Improving software reliability forecasting. Microelectron Reliab 1997;37(6):901–7.

[41] Pham H, Zhang X. A software cost model with warranty and risk costs. IEEE Trans Comput 1999;48(1):71–5.

[42] Boehm BW. Software engineering economics. Englewood Cliffs, NJ: Prentice Hall; 1981.

[43] Boehm BW, Clark B, Horowitz E, Westland JC, Madachy RJ, Selby RW. Cost models for future life cycle processes: COCOMO 2.0. Ann Software Eng 1995;1(1):1–24.

[44] DeMarco T. Controlling software projects. New York: Yourdon Press; 1982.

[45] Zultner R. The Deming approach to software quality engineering. Qual Progr 1988; Nov.

[46] Popentiu Fl, Burschy B. Total quality for software engineering management. Final report for NATO—HTECH.LG 941434; September 1997.

[47] Humphrey WS. A discipline for software engineering. Reading, MA: Addison Wesley; 1995.

[48] Kuntzmann-Combelles A. Quantitative approach to software management: the "ami" method. In: Sommerville I, Paul M, editors. Software engineering—ESEC'93. Berlin: Springer-Verlag; 1993. p.238–50.

[49] Basili VR, Rombach D. The TAME project: towards improvement oriented software environments. IEEE Trans Software Eng 1988;14(6):758–73.

[50] Kenet RS, Baker ER. Software process quality: management and control. New York: Marcel Dekker; 1999.

Software Fault Tolerance

Xiaolin Teng and Hoang Pham

33.1 Introduction

Software fault tolerance is achieved through special programming techniques that enable the software to detect and recover from failures. This requires redundant software elements that provide alternative means of fulfilling the same specifications. The different versions must be developed independently such that they do not have common faults, if any, to avoid simultaneous failures in response to the same inputs.

The difference between software fault tolerance and hardware fault tolerance is that software faults are usually design faults, and hardware faults are usually physical faults. Hardware physical faults can be tolerated in redundant (spare) copies of a component that are identical to the original, since it is commonly assumed that hardware components fail independently. However, software design faults cannot generally be tolerated in this way because the error is likely to recur on the spare component if it is identical to the original [1].

Two of the best known fault-tolerant software schemes are N-version programming (NVP) and recovery block (RB). Both schemes are based on software component redundancy and the assumption that coincident failures of components are rare.

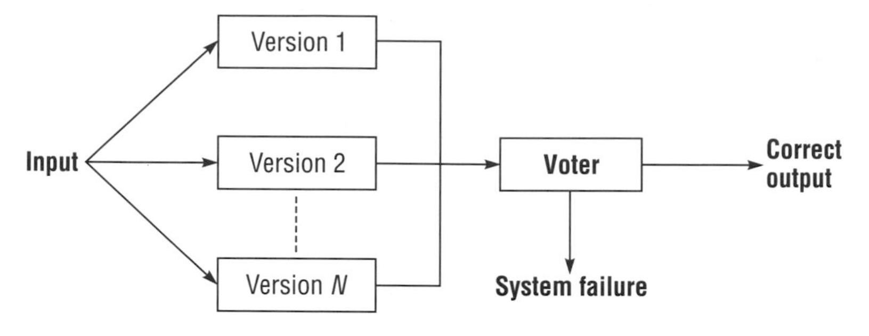

Figure 33.1. N-version programming scheme

33.2 Software Fault-tolerant Methodologies

33.2.1 N-version Programming

NVP was proposed by Chen and Avizienis. In this scheme, there are $N \geq 2$ functionally equivalent programs, called versions, which are developed independently from the same initial specification. An independent development means that each version will be developed by different groups of people. Those individuals and groups do not communicate with each other during the software development. Whenever possible, different algorithms, techniques, programming languages and tools are used in each version. NVP is actually a software generalization of the N-modular redundancy (NMR) approach used in hardware fault tolerance. All these N versions are usually executed simultaneously (*i.e.* in parallel), and send their outputs to a voter, which determines the "correct" or best output if one exists, by a voting mechanism, as shown in Figure 33.1.

Assume all N versions are statistically independent of each other and have the same reliability r, and if majority voting is used, then the reliability of the NVP scheme (R_{NVP}) can be expressed as:

$$R_{\text{NVP}} = \sum_{i=\lceil N/2 \rceil}^{N} \binom{N}{i} r^i (1-r)^{N-i} \qquad (33.1)$$

Several different voting techniques have been proposed. The simplest one is majority voting, which has been investigated by a number of researchers [3–6]. In majority voting, usually N is odd, and the voter needs at least $\lceil N/2 \rceil$ software versions to produce the same output to determine a "correct" result. Consensus voting is designed for multiversion software with small output space, in which case software versions can give identical but incorrect outputs [7]. The voter will select the output that the most software versions give. For example, suppose there are six software versions and three possible outputs. If three versions give output A, two versions give output B, and one version gives output C, then the voter will consider the correct result as output A. Leung [8] proposes a maximum likelihood voting method, which applies a maximum likelihood function to decide the most likely correct result. Just like consensus voting, this maximum likelihood voting method also needs small output space.

33.2.2 Recovery Block

The RB scheme was proposed by Randell [9]. This scheme consists of three elements: primary module, acceptance tests (AT), and alternate modules for a given task. The simplest scheme of the recovery block is as shown in Figure 33.2.

The process begins when the output of the primary module is tested for acceptability. If the acceptance test determines that the output of the primary module is not acceptable, it restores, recovers, or "rolls back" the state of the

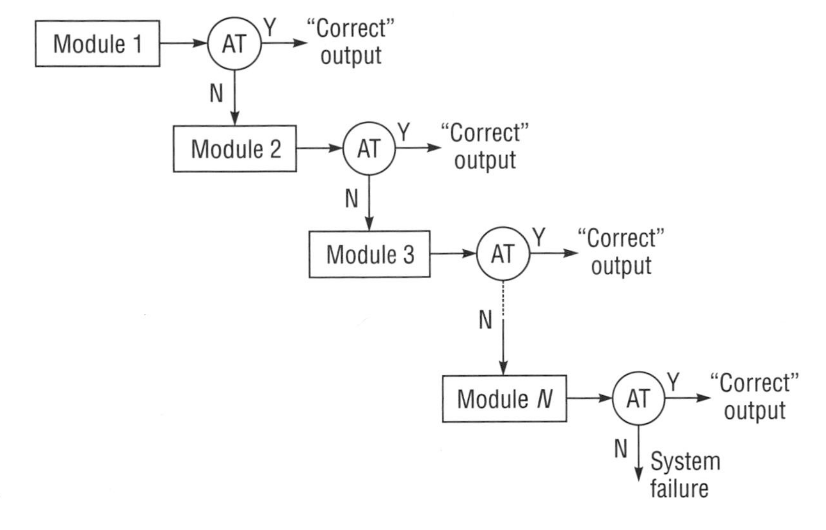

Figure 33.2. Recovery blocks scheme

system before the primary module was executed. It then allows the second module to execute, and evaluates its output, and if the output is not acceptable, it then starts the third module, and evaluates its output, and so on. If all modules execute and none gives the acceptable results, then the system fails.

One problem with this scheme is how to find a simple and highly reliable AT, because the AT is quite dependent on the application. There may also be a different acceptance test for each module. In some cases, the AT will be very simple. An interesting example is given in Fairley [10]. An implicit specification of the square root function SQRT can serve as an AT:

$$(\mathrm{ABS}((\mathrm{SQRT}(x)^{*}\mathrm{SQRT}(x)) - x) < \mathrm{Error})$$
$$\text{for } 0 \leq x \leq y$$

where x and y are real numbers. Error is the permissible error range.

In this case, it is much easier to do AT than SQRT itself, then we can easily set up a reliable AT for SQRT. Unfortunately, sometimes it is not easy to develop a reliable and simple AT, and an AT can be complex and as costly as the full problem solution. In some applications, it is enough for the AT to check only the output of the software, but in some other applications, this may not be sufficient. The software internal states also need to be checked to detect a failure. Compared with the voter in an NVP scheme, the AT is much more difficult to build.

One of the differences between RB and NVP is that the modules are executed sequentially in RB, and simultaneously in NVP. Also the AT in RB can determine whether a version gives the correct result, but no such kind of mechanism is built in NVP. NVP can only rely on the voter to decide the best output. Also, since the RB needs to switch to the backup modules when a primary software module fails, it takes much longer computation times to finish the computation in RB than in NVP if many modules fail in a task. Therefore, the recovery block generally is not applicable to critical systems where real-time response is of great concern.

33.2.3 Other Fault-tolerance Techniques

There are some studies that combine RB and NVP to create new hybrid techniques, such as consensus recovery block [5], and acceptance voting [11].

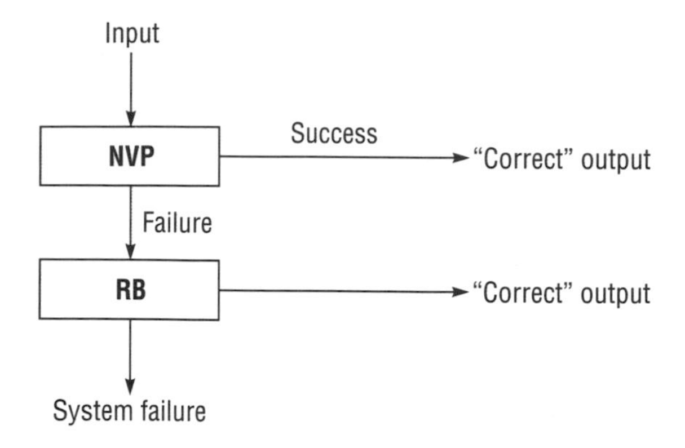

Figure 33.3. Consensus recovery block

Consensus recovery block is a hybrid system that combines NVP and RB in that order (Figure 33.3). If NVP fails, the system reverts to RB using the same modules (the same module results can be used). Only when both NVP and RB fail does the system fail.

Acceptance voting is also an NVP scheme, but it adds an AT to each version (Figure 33.4). The output of each version is first sent to the AT, if the AT accepts the output, then the output will be passed to the voter. The voter sees only those outputs that have been passed by the acceptance test. This implies that the voter may not process the same number of outputs at each invocation, and hence the voting algorithm must be dynamic. The system fails if no outputs are submitted to the voter. If only one output is submitted, the voter must assume it to be correct, then pass it to the next stage.

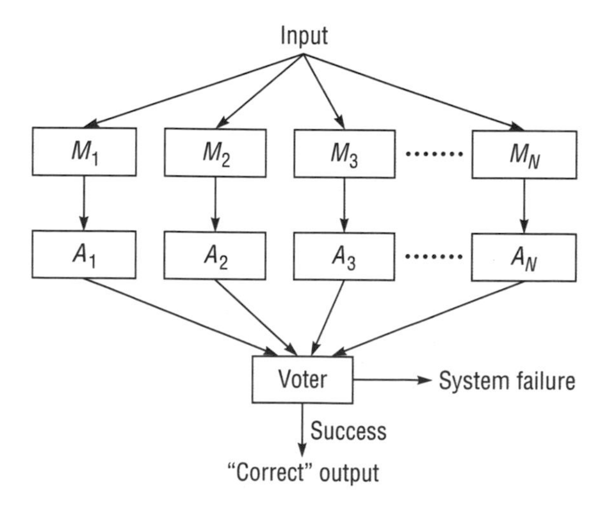

Figure 33.4. Acceptance voting

33.3 *N*-version Programming Modeling

A number of systematic experimental studies of fault-tolerant software issues have been conducted over the last 20 years by both academia and industry [6, 12, 13]. Modeling of an NVP scheme gives insight into its behavior and allows quantification of its merit. Some research has been conducted in the reliability modeling of fault-tolerant software systems. The reliability modeling for fault-tolerant software systems falls into two categories: data-domain modeling and time-domain modeling. Both analyses use the assumption that the failure events are independent between or among different versions.

33.3.1 Basic Analysis

33.3.1.1 Data-domain Modeling

Data-domain modeling uses the assumption that failure events among different software versions are independent of each other. This abstraction simplifies the modeling and provides insight into the behavior of NVP schemes.

If r_i represents the reliability of the ith software version, and R_V represents the reliability of the voter, which is also assumed independent of software version failures, then the system reliability of a three-version NVP fault-tolerant system is:

$$R_{NVP3}(r_1, r_2, r_3, R_V) = R_V(r_1 r_2 + r_1 r_3 + r_2 r_3 - 2 r_1 r_2 r_3) \quad (33.2)$$

If all versions have the same reliability r, then the probability that an NVP system will operate

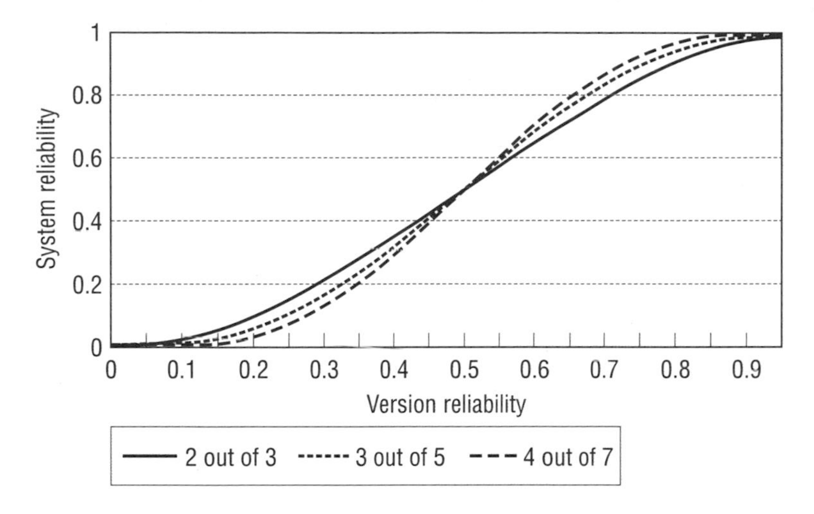

Figure 33.5. Comparison of system reliability and version reliability

successfully under majority voting strategy is given by the following expression:

$$R_{\mathrm{NVP}}(r, R_{\mathrm{V}}) = R_{\mathrm{V}} \sum_{i=m}^{N} \binom{N}{i} r^{i}(1-r)^{N-i} \tag{33.3}$$

where m $(m \leq N)$ is the lower bound on the required space number of versions that have the same output. Figure 33.5 shows various system reliability functions versus version reliability for $R_{\mathrm{V}} = 1$.

To achieve higher reliability, the reliability of a single version must be greater than 0.5, then the system reliability will increase with the number of versions involved. If the output space has cardinality ρ, then NVP will result in a system that is more reliable than a single component only if $r > 1/\rho$ [7].

33.3.1.2 Time-domain Modeling

Time-domain modeling is concerned with the behavior of system reliability over time. The simplest time-dependent failure model assumes that failures arrive randomly with interarrival times exponentially distributed with constant rate λ. The reliability of a single software version will be:

$$r(t) = \mathrm{e}^{-\lambda t} \tag{33.4}$$

If we assume that all versions have the same reliability level, then the system reliability of NVP is given by:

$$R_{\mathrm{NVP}}(r(t), R_{\mathrm{V}}) = R_{\mathrm{V}} \sum_{i=m}^{N} \binom{N}{i} r^{i}(t)(1-r(t))^{N-i} \tag{33.5}$$

If $N = 3$, then

$$R_{\mathrm{NVP3}}(t) = 3\,\mathrm{e}^{-2\lambda t} - 2\,\mathrm{e}^{-3\lambda t} \tag{33.6}$$

Given $\lambda = 0.05$ per hour, then we can plot $r(t)$ and $R_{\mathrm{NVP3}}(t)$ function curves, and compare the reliability improvement (Figure 33.6). It is easy to see that when $t \leq t^* = 14 \approx 0.7/\lambda$, the system is more reliable than a single version, but when $t > t^*$, the system is less reliable than a single version.

The above analysis is very simple, and the actual problem is far more complex. It has been shown that independently developed versions may not fail independently due to the failure correlations between software versions [12, 14]. The next section discusses several existing models that consider the failure correlation between software versions.

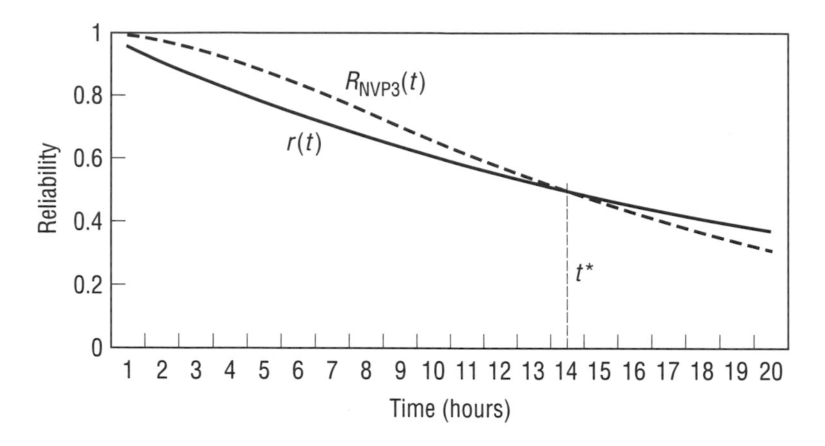

Figure 33.6. Single version reliability versus three-version system reliability

33.3.2 Reliability in the Presence of Failure Correlation

Experiments show that incidence of correlated failures of NVP system components may not be negligible in the context of current software development and testing techniques [5, 15, 16]. Although software versions are developed independently, many researchers have revealed that those independently developed software versions do not necessarily fail independently [12, 17]. According to Knight and Leveson [14], experiments have shown that the use of different languages and designs philosophy has little effect on the reliability in NVP because developers tend to make similar logical mistakes in a difficult-to-program part of the software.

In NVP systems, a coincident failure occurs when a related fault between versions is activated by some input or two unrelated faults in different versions are activated at the same input.

Laprie *et al.* [18] classify the software faults according to their independence into either related or independent. Related faults manifest themselves as similar errors and lead to common-mode failures, whereas independent faults usually cause distinct errors and separate failures.

Due to the difficulty of distinguishing the related faults from the independent faults, in this chapter we simplify this fault classification as follows. If two or more versions give identical but all wrong results, then the failures are caused by the related faults between versions; if two or more versions give dissimilar but wrong results, then the faults are caused by unrelated or *independent* software faults.

For the rest of the chapter, we refer to related faults as *common faults* for simplicity. Figure 33.7 illustrates the common faults and the independent faults in NVP systems.

Common faults are those which are located in the functionally equivalent modules among two or more software versions because their programmers are prone to making the same or similar mistakes, although they develop the versions independently. These faults will be activated by the same inputs to cause those versions to fail simultaneously, and these failures by common faults are called common failures. Independent faults are usually located in different or functionally unequivalent modules between or among different software versions. Since they are independent of each other and their resulting failures are typically distinguishable from the decision mechanism, they are often considered harmless to the fault-tolerant systems. However, there is still a probability, though very small compared with that of common failures (CF), that an unforeseeable input activates two independent faults in different software versions that will lead

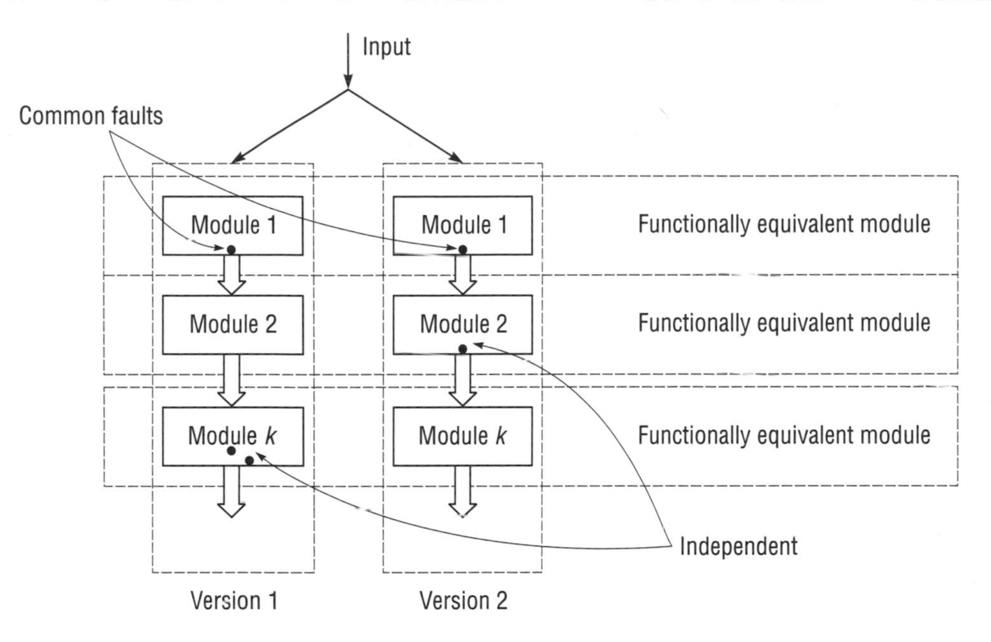

Figure 33.7. Common faults and independent faults

Table 33.1. Common failures and concurrent independent failures

	Common failures	Concurrent independent failures
Fault type	Common faults	Independent faults
Output	Usually the same	Usually different
Fault location (logically)	Same	Different
Voting result (majority voting)	Choose wrong solution	Unable to choose correct solution

these two versions to fail at the same time. These failures by independent faults are called concurrent independent failures (CIF). Table 33.1 shows the differences between the common failures and the concurrent independent failures.

33.3.3 Reliability Analysis and Modeling

The reliability model of NVP systems should consider not only the common failures, but also the concurrent independent failures among or between software versions. Some reliability models have been proposed to incorporate the interversion failure dependency.

Eckhardt and Lee [4] developed a theoretical model that provides a probabilistic framework for empirically evaluating the effectiveness of a general multiversion strategy when component versions are subject to coincident errors by assuming the statistical distributions of input choice and program choice. Their work is among the first that showed independently developed program versions to fail dependently.

Littlewood and Miller [6] further showed that there is a precise duality between input choice and program choice, and considered a generalization in which different versions may be developed using diverse methodologies. The use of diverse methodologies is shown to decrease the probability of simultaneous failure of several versions.

Nicola and Goyal [12] proposed to use a beta-binomial distribution to model correlated failures in multi-version software systems, and presented a combinatorial model to predict the reliability of a multi-version software configuration.

The above models focus on software diversity modeling, and are based on the detailed analysis of the dependencies in diversified software. Other researchers focus on modeling fault-tolerant software system behavior.

Dugan and Lyu [19] presented a quantitative reliability analysis for an N-version programming application. The software systems of this application were developed and programmed by 15 teams at the University of Iowa and the Rockwell/Collins Avionics Divisions. The overall model is a Markov reward model in which the states of the Markov chain represent the long-term evolution of the structure of the system.

Figure 33.8 illustrates the Markov model in Dugan and Lyu [19]. In the initial state, three independently developed software versions are running on three active processors. The processors have the same failure rate λ. After the first hardware fails, the TMR-system is reconfigured to a simplex system successfully with probability c. So the transition rate to the reconfiguration state is $3\lambda c$ and the transition rate to the failure state caused by an uncovered failure is $3\lambda(1-c)$. The system fails when the single remaining processor fails, thus the transition rate from the reconfiguration state to the failure state is λ.

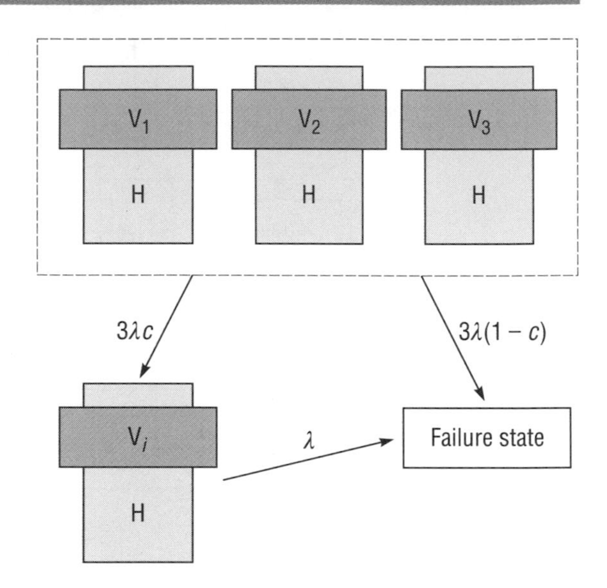

Figure 33.8. Markov model of system structure in Dugan and Lyu [19]

Their model considers independent software faults, related software faults, transient hardware faults, permanent hardware faults, and imperfect coverage. The experimental results from this application were used to estimate the probabilities associated with the activation of software faults. The overall NVP reliability is obtained by combining the hardware reliability and 3VP software reliability.

Tai *et al.* [20] proposed a performance and dependability model to assess and improve the effectiveness of NVP systems. In their research, the performance model of NVP is a renewal process, and the dependability is a fault-manifestation model. They propose a scheme to enhance the performance and dependability of NVP systems. This scheme is a 3VP scheme, in which the slowest version is utilized as a *tie-breaker* (TB) when the first two versions "tie" by disagreeing. An *acceptance test* (AT) is added in the NVP-AT scheme, where AT will be used to decide whether the result is correct after the decision function reaches a consensus decision. In this research, the failures by related faults and unrelated faults were well modeled, and their effects on the effectiveness of these NVP schemes are well evaluated.

While the modeling methods of Eckhardt and Lee [4], Littlewood and Miller [6], and Nicola and Goyal [12] are quite different from all other modeling methods [19, 20], Goseva-Popstojanova and Grnarov [21] incorporate their methodologies into a Markovian performance model and present a unified approach aimed at modeling the joint behavior of the N-version system and its operational environment. This new approach provides some insight into how the reliability is affected by version characteristics and the operational environment.

Lin and Chen [22] proposed two software-debugging models with linear dependence, which appears to be the first effort to apply a nonhomogeneous Poisson process (NHPP) to an NVP system. These two models define a correlating parameter α_i for the failure intensity of software version i, and the mean value function of version i is a logarithmic Poisson execution-time:

$$m_i(\theta_i, \lambda_i, t) = (1/\theta_i)\log(\lambda_i\theta_i t + 1) \qquad (33.7)$$

In model I, the effect of s-dependency among software versions is modeled by increasing the initial failure intensity of each version:

$$\lambda_i' = \alpha_1\lambda_1 + \cdots + \alpha_{i-1}\lambda_{i-1} + \lambda_i + \alpha_{i+1}\lambda_{i+1} + \cdots + \alpha_N\lambda_N \qquad (33.8)$$

In model II, the effect of s-dependency among software versions is modeled by increasing the nominal mean-value function of each version, to:

$$\begin{aligned} m_i'(\theta, \lambda, t) = {} & \alpha_1 m_1(\theta_1, \lambda_1, t) + \cdots \\ & \cdots + \alpha_{i-1} m_{i-1}(\theta_{i-1}, \lambda_{i-1}, t) \\ & + m_i(\theta_i, \lambda_i, t) \\ & + \alpha_{i+1} m_{i+1}(\theta_{i+1}, \lambda_{i+1}, t) + \cdots \\ & \cdots + \alpha_N m_N(\theta_N, \lambda_N, t) \end{aligned} \qquad (33.9)$$

Although these two models use the nonhomogeneous Poisson process method, they are not software reliability growth models (SRGMs). This chapter does not consider the software reliability growth due to continuous removal of faults from the components of the NVP systems. The modifications due to the progressive removal of residual design faults from software versions cause their reliability to grow, which in turn causes the reliability of NVP software systems to grow. Then the reliability of NVP systems can grow as a result of continuous removal of faults in the software versions. Kanoun *et al.* [23] proposed a reliability growth model for NVP systems by using the hyperexponential model. The failure intensity function is given by:

$$h(t) = \frac{\omega\xi_{\text{sup}}\exp(-\xi_{\text{sup}}t) + \bar{\omega}\xi_{\text{inf}}\exp(-\xi_{\text{inf}}t)}{\omega\exp(-\xi_{\text{sup}}t) + \bar{\omega}\exp(-\xi_{\text{inf}}t)} \qquad (33.10)$$

where $0 \le \omega \le 1$, $\bar{\omega} = 1 - \omega$, and ξ_{sup} and ξ_{inf} are, respectively, the hyperexponential model parameters characterizing reliability growth due to the removal of the faults. It should be noted that the function $h(t)$ in Equation 33.10 is nonincreasing with time t for $0 \le \omega \le 1$, from $h(0) = \omega\xi_{\text{sup}} + \bar{\omega}\xi_{\text{inf}}$ to $h(\infty) = \xi_{\text{inf}}$.

This model is the first that considers the impact of reliability growth as a result of progressive removal of faults from each software version on software reliability. By interpreting the hyperexponential model as a Markov model that can handle reliability growth, this model allows the reliability growth of an NVP system to be modeled from the reliability growth of its components.

Sha [24] investigated the relationship between complexity, reliability, and development resources within an N-version programming system, and presented an approach to building a system that can manage upgrades and repair itself when complex software components fail. His result counters the belief that diversity results in improved reliability under the limited development resources: the reliability of single-version programming with undivided effort is superior to three-version programming over a wide range of development effort. He also pointed out that single-version programming might not always be superior to its N-version counterpart, because some additional versions can be obtained inexpensively.

Little research has been performed in reliability modeling of NVP systems. Most of the NVP system research either focuses on modeling software diversity [4, 6, 12], or aims primarily to evaluate some dependability measures for specific types of software systems [13, 19]. Most of these proposed models assume stable reliability, *i.e.* they do not consider reliability growth due to corrective maintenance, thus they may not be applicable to the developing and testing phases in the software lifecycle, where the reliability of NVP systems grows as a result of progressive removal of residual faults from the software components.

During the developing and testing phases, some important questions are:

- How reliable is the software?
- How many remaining faults are in the software?

- How much testing does it still need to attain the required reliability level (*i.e.* when to stop testing)?

For traditional single-version software, the SRGM can be used to provide answers to these questions. Kanoun *et al.* [23] developed the first SRGM for NVP systems, which does not consider the impact of imperfect debugging on both independent failures and common failures. The role of faults in NVP systems may change due to the imperfect debugging, some (potential) common faults may reduce to low-level common faults or even to independent faults. Because Kanoun's model is not based on the characteristics of the software and the testing/debugging process, users cannot obtain much information about the NVP system and its components, such as the initial number of independent and common faults. Finally, this model is extremely complicated, which prevents it from being successfully applied to large N ($N > 3$) versions of programming applications.

Thus, there is a great need to develop a new reliability model for NVP systems that provides insight into the development process of NVP systems and is able to answer the above questions. In other words, the motive of this research is to develop an SRGM for NVP systems. Different from the research of Kanoun *et al.* [23], this model will be the first attempt to establish a software reliability model for NVP systems with considerations of error removal efficiency and error introduction rate during testing and debugging. In this chapter, we present a generalized NHPP model for a single software program, then apply this generalized NHPP model to NVP systems to develop an NVP software reliability growth model.

33.4 Generalized Non-homogeneous Poisson Process Model Formulation

As a general class of well-developed stochastic process models in reliability engineering, NHPP models have been successfully applied to software reliability modeling. NHPP models are especially useful to describe failure processes that possess certain trends such as reliability growth or deterioration. Zhang *et al.* [25] proposed a generalized NHPP model with the following assumptions.

1. A software program can fail during execution.
2. The occurrence of software failures follows NHPP with mean value function $m(t)$.
3. The software failure detection rate at any time is proportional to the number of faults remaining in the software at that time.
4. When a software failure occurs, a debugging effort occurs immediately. This effort removes the faults immediately with probability p, where $p \gg 1 - p$. This debugging is s-independent at each location of the software failures.
5. For each debugging effort, whether the fault is successfully removed or not, some new faults may be introduced into the software system with probability $\beta(t)$, $\beta(t) \ll p$.

From the above assumptions, we can formulate the following equations

$$\begin{aligned} m'(t) &= b(t)(a(t) - pm(t)) \\ a'(t) &= \beta(t)m'(t) \end{aligned} \tag{33.11}$$

where

$m(t)$ = expected number of software failures by time t, $m(t) = E[N(t)]$
$a(t)$ = expected number of initial software errors plus introduced errors by time t
$b(t)$ = failure detection rate per fault at time t
p = probability that a fault will be successfully removed from the software
$\beta(t)$ = fault introduction rate at time t.

If the marginal conditions are given as $m(0) = 0$, $a(0) = a$, the solutions to Equation 33.11 are shown as follows:

$$m(t) = a \int_0^t b(u)\, \mathrm{e}^{-\int_0^u (p-\beta(\tau))b(\tau)\,\mathrm{d}\tau} \mathrm{d}u \tag{33.12}$$

$$a(t) = a\left(1 + \int_0^t \beta(u)b(u)\, \mathrm{e}^{-\int_0^u (p-\beta(\tau))b(\tau)\,\mathrm{d}\tau}\, \mathrm{d}u\right) \tag{33.13}$$

This model can be used to derive most of the known NHPP models. If we change the assumptions on $b(t)$, $\beta(t)$, and p, we can obtain all those known NHPP models. Table 33.2, for example, shows some well-known NHPP software reliability growth models can be derived from this generalized software reliability model [26].

33.5 Non-homogeneous Poisson Process Reliability Model for N-version Programming Systems

NVP is designed to attain high system reliability by tolerating software faults. In this section we only present the modeling of NVP where $N = 3$, based on the results of Teng and Pham [30], but the methodology can be directly applied to the modeling of NVP where $N > 3$.

Let us assume that we have three independently developed software versions 1, 2, and 3, and use majority voting, and the reliability of the voter is 1. The following notation is used:

CF	common failure
CIF	concurrent independent failure
A	independent faults in version 1
B	independent faults in version 2
C	independent faults in version 3
AB	common faults between version 1 and version 2
AC	common faults between version 1 and version 3
BC	common faults between version 2 and version 3
ABC	common faults among version 1, version 2, and version 3
$N_x(t)$	counting process which counts the number of type x faults discovered up to time t, $x = A, B, C, AB, AC, BC, ABC$
$N_d(t)$	$N_d(t) = N_{AB}(t) + N_{AC}(t) + N_{BC}(t) + N_{ABC}(t)$; counting process which counts common faults discovered in the NVP system up to time t
$m_x(t)$	mean value function of counting process $N_x(t)$, $m_x(t) = E[N_x(t)]$, $x = A, B, C, AB, AC, BC, ABC, d$
$a_x(t)$	total number of type x faults in the system plus those type x faults already removed from the system at time t. $a_x(t)$ is a non-decreasing function, and $a_x(0)$ denotes the initial number of type x faults in the system, $x = A, B, C, AB, AC, BC, ABC$
$b(t)$	failure detection rate per fault at time t
β_1, β_2, and β_3	probability that a new fault is introduced into version 1, 2, and 3 during the debugging, respectively
p_1, p_2, and p_3	probability that a new fault is successfully removed from version 1, 2, and 3 during the debugging, respectively
$X_A(t)$, $X_B(t)$, and $X_C(t)$	number of type A, B, and C faults at time t remaining in the system respectively, *i.e.* $X_A(t) = a_A(t) - p_1 m_A(t)$ $X_B(t) = a_B(t) - p_2 m_B(t)$ $X_C(t) = a_C(t) - p_3 m_C(t)$
$R(x \mid t)$	software reliability function for given mission time x and time to stop testing t; $R(x \mid t) = \Pr\{$no failure during mission x $\mid$ stop testing at $t\}$
K_{AB}, K_{AC}, and K_{BC}	failure intensity per pair of faults for CIFs between version 1 and 2, between 1 and 3, and between 2 and 3 respectively

Table 33.2. Some well-known NHPP software reliability models

Model name	Model type	Mean value function $m(t)$	Comments
Goel–Okumoto (G-O) [27]	Concave	$m(t) = a(1 - e^{-bt})$ $a(t) = a$ $b(t) = b$	Also called exponential model
Delayed S-shaped	S-shaped	$m(t) = a(1 - (1 + bt)\,e^{-bt})$	Modification of G-O model to make it S-shaped
Inflection S-shaped SRGM	S-shaped	$m(t) = \frac{a(1 - e^{-bt})}{1 + \beta\, e^{-bt}}$ $a(t) = a$ $b(t) = \frac{b}{1 + \beta\, e^{-bt}}$	Solves a technical condition with the G-O model. Becomes the same as G-O if $\beta = 0$
Y exponential	S-shaped	$m(t) = a\left(1 - e^{-r\alpha(1 - e^{(-\beta t)})}\right)$ $a(t) = a$ $b(t) = r\alpha\beta\, e^{-\beta t}$	Attempt to account for testing effort
Y Rayleigh	S-shaped	$m(t) = a\left(1 - e^{-r\alpha(1 - e^{(-\beta t^2/2)})}\right)$ $a(t) = a$ $b(t) = r\alpha\beta t\, e^{-\beta t^2/2}$	Attempt to account for testing effort
Y imperfect debugging model 1	Concave	$m(t) = \frac{ab}{\alpha + b}(e^{\alpha t} - e^{-bt})$ $a(t) = a\, e^{\alpha t}$ $b(t) = b$	Assume exponential fault content function and constant fault detection rate
Y imperfect debugging model 1	Concave	$m(t) = a[1 - e^{-bt}]\left[1 - \frac{\alpha}{b}\right] + \alpha a t$ $a(t) = a(1 + \alpha t)$ $b(t) = b$	Assume constant introduction rate α and fault detection rate
P-N-Z model [28]	S-shaped and concave	$m(t) = \frac{a[1 - e^{-bt}]\left[1 - \frac{\alpha}{b}\right] + \alpha a t}{1 + \beta\, e^{-bt}}$ $a(t) = a(1 + \alpha t)$ $b(t) = \frac{b}{1 + \beta\, e^{-bt}}$	Assume introduction rate is a linear function of testing time, and the fault detection rate function is non-decreasing with an inflexion S-shaped model
P-Z model [29]	S-shaped and concave	$m(t) = \frac{1}{(1 + \beta\, e^{-bt})}[(c + a)(1 - e^{-bt})]$ $- \frac{a}{b - \alpha}(e^{-\alpha t} - e^{-bt})$ $a(t) = c + a(1 - e^{-\alpha t})$ $b(t) = \frac{b}{1 + \beta\, e^{-bt}}$	Assume introduction rate is exponential function of the testing time, and the fault detection rate is non-decreasing with an inflexion S-shaped model
Z-T-P model [25]	S-shaped	$m(t) = \frac{a}{p - \beta}\left[1 - \left(\frac{(1 + \alpha)\, e^{-bt}}{1 + \alpha\, e^{-bt}}\right)^{c/b(p - \beta)}\right]$ $a'(t) = \beta(t) m'(t)$ $b(t) = \frac{c}{1 + \alpha\, e^{-bt}}$ $\beta(t) = \beta$	Assume constant fault introduction rate, and the fault detection rate function is non-decreasing with an inflexion S-shaped model

$N_{\overline{AB}}(t)$, $N_{\overline{AC}}(t)$, and $N_{\overline{BC}}(t)$	counting processes that count the number of CIFs involving version 1 and 2, version 1 and 3, and version 2 and 3 up to time t respectively
$N_I(t)$	counting process that counts the total number of CIFs up to time t; $N_I(t) = N_{\overline{AB}}(t) + N_{\overline{AC}}(t) + N_{\overline{BC}}(t)$
$m_{\overline{AB}}(t)$, $m_{\overline{AC}}(t)$, $m_{\overline{BC}}(t)$, and $m_I(t)$	mean value functions of the corresponding counting processes $m(t) = E[N(t)]$. *E.g.*, $m_{\overline{AB}}(t) = E[N_{\overline{AB}}(t)]$
$h_{\overline{AB}}(t)$, $h_{\overline{AC}}(t)$, and $h_{\overline{BC}}(t)$	failure intensity functions of CIFs involving version 1 and 2, between 1 and 3, and between 2 and 3; $h_{\overline{AB}}(t) = \frac{\mathrm{d}}{\mathrm{d}t} m_{\overline{AB}}(t)$
MLE	maximum likelihood estimation
NVP-SRGM	software reliability growth model for NVP systems
$R_{\text{NVP-SRGM}}(x \mid t)$	NVP system reliability function for given mission time x and time to stop testing t with consideration of common failures in the NVP system
$R_{\text{Ind}}(x \mid t)$	NVP system reliability function for given mission time x and time to stop testing t, assuming no CF in the NVP system, *i.e.* both versions fail independently

Different types of faults and their relations are shown in Figure 33.9. Generally, this notation scheme uses numbers to refer to software versions, and letters to refer to software fault types. For example, process $N_1(t)$ counts the number of failures in software version 1 up to time t, therefore:

$$N_1(t) = N_A(t) + N_{AB}(t) + N_{AC}(t) + N_{ABC}(t)$$

Similarly:

$$N_2(t) = N_B(t) + N_{AB}(t) + N_{BC}(t) + N_{ABC}(t)$$
$$N_3(t) = N_C(t) + N_{AC}(t) + N_{BC}(t) + N_{ABC}(t)$$

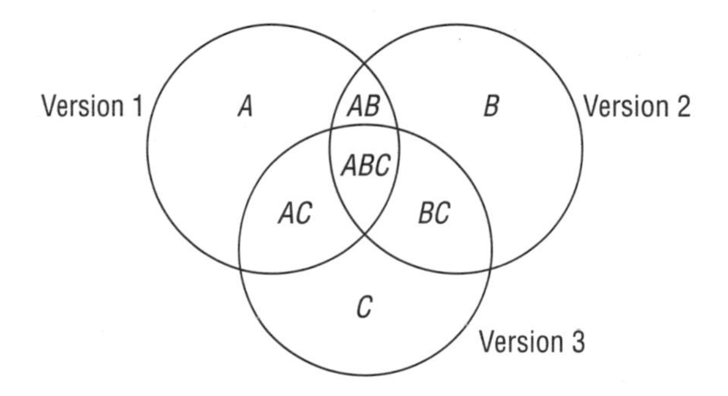

Figure 33.9. Different software faults in the three-version software system

There are two kinds of coincident failures in NVP systems: common failures (CFs) and concurrent independent failures (CIFs). CFs are caused by the common faults between or among versions, and CIFs are caused by the independent (unrelated) faults between versions. A CIF occurs when two or more versions fail at the same input on independent faults, *i.e.* on A, B, or C, not on AB, AC, *etc.*

$N_{AB}(t)$ and $N_{\overline{AB}}(t)$ both count the number of coincident failures between version 1 and version 2. The difference between them is that $N_{AB}(t)$ counts the number of CFs in version 1 and version 2, and $N_{\overline{AB}}(t)$ counts the number of CIFs in version 1 and version 2. Table 33.3 illustrates the different failure types between or among different software versions.

To establish the reliability growth model for an NVP fault-tolerant software system, we need to analyze the reliability of components as well as the correlation among software versions.

33.5.1 Model Assumptions

To develop an NHPP model for NVP systems, we make the following assumptions.

1. $N = 3$.
2. Voter is assumed perfect all the time, *i.e.* $R_{\text{voter}} = 1$.
3. Faster versions will have to wait for the slowest versions to finish (prior to voting).

Table 33.3. Failure types in the three-version programming software system

Failure time	Version 1	Version 2	Version 3	Failure type
t_1	✓			A
t_2			✓	C
t_3	✓			A
t_4		✓	✓	BC or $\overline{BC}$
t_5		✓		B
t_6	✓			A
t_7			✓	C
t_8	✓		✓	AC or $\overline{AC}$
t_9		✓		B
t_{10}	✓	✓		AB or $\overline{AB}$
...	...	...	...	...
t_i	✓	✓	✓	ABC
...	...	...	...	...

Note: 3VP system fails when more than two versions fail at the same time.

4. Each software version can fail during execution, caused by the faults in the software.
5. Two or more software versions may fail on the same input, which can be caused by either the common faults, or the independent faults between or among different versions.
6. The occurrence of software failures (by independent faults, two-version common faults or three-version common faults) follows an NHPP.
7. The software failure detection rate at any time is proportional to the number of remaining faults in the software at that time.
8. The unit error detection rates for all kinds of faults A, B, C, AB, AC, BC, and ABC are the same and constant, *i.e.* $b(t) = b$.
9. When a software failure occurs in any of the three versions, a debugging effort is executed immediately. That effort removes the corresponding fault(s) immediately with probability p_i, $p_i \gg 1 - p_i$ (i is the version number 1, 2, or 3).
10. For each debugging effort, whether the fault(s) are successfully removed or not, some new independent faults may be introduced into that version with probability β_i, $\beta_i \ll p_i$ (i is the version number 1, 2, or 3), but no new common fault will be introduced into the NVP system.
11. Some common faults may reduce to some low-level common faults or independent faults due to unsuccessful removal efforts.
12. The CIFs are caused by the activation of independent faults between different versions, and the probability that a CIF involves three versions is zero. Those failures only involve two versions.
13. Any pair of remaining independent faults between versions has the same probability to be activated.
14. The intensity for CIFs involving any two versions is proportional to the remaining pairs of independent faults in those two versions.

Following are some further explanations for assumptions 8–11,

All versions accept the same inputs and run simultaneously, the fastest version has to wait for the slowest version to finish. Therefore, the unit error detection rate, b, can be the same for all kinds of faults (assumption 8). N groups are assigned to perform testing and debugging on N versions independently, then the error removal efficiency p and the error introduction rate β are different for different software versions (assumption 9). When two groups are trying to remove an AB fault from version 1 and

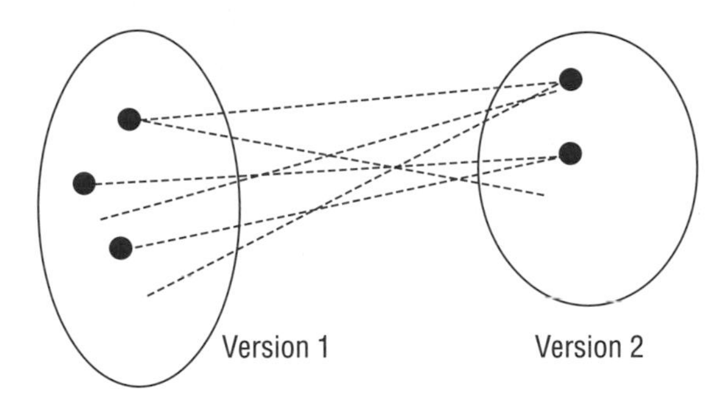

Figure 33.10. Independent fault pairs between version 1 and 2

version 2, each may introduce a new fault into its own version. But the probability that these two groups make the same mistake to introduce a new AB fault into both versions is zero (assumption 10), this means that only independent faults can be introduced into the system. Because an AB fault may be removed successfully from version 1, but not removed from version 2, then the previously common fault (AB) is no longer common to version 1 and version 2, it reduces to a B fault, which only exists in version 2 (assumption 11).

Figure 33.10 shows the pairs of independent faults between software version 1 and version 2. There are three independent faults (type A) in version 1, and there are two independent faults (type B) in version 2. Then there are $3 \times 2 = 6$ pairs of independent faults between version 1 and version 2. It is assumed that each of these six pairs has the same probability to be activated.

In this study, the voter is not expected to decide whether or not a software version fails during testing, but we assume that people have advanced methods or tools to determine exactly whether and when a software version fails or succeeds. Those methods or tools to detect a software failure are quite application-dependent, and they are either too awkward or too impractical to be built into an NVP system. Therefore, after the NVP software system is released, we do not have those tools to decide whether a version of the NVP system fails, we can only rely on the voter to decide the best or the correct output.

33.5.2 Model Formulations

Based on the assumptions given in the last section, we can establish the following NHPP equations for different types of software faults and failures.

33.5.2.1 Mean Value Functions

1. Type ABC

$$m'_{ABC}(t) = b(a_{ABC} - p_1 p_2 p_3 m_{ABC}(t)) \tag{33.14}$$

with marginal conditions $m_{ABC}(0) = 0$ and $a_{ABC}(0) = a_{ABC}$ where a_{ABC} is the initial number of type ABC faults in the 3VP software system. Equation 33.14 can be solved directly as:

$$m_{ABC}(t) = \frac{a_{ABC}}{p_1 p_2 p_3}(1 - e^{-bp_1 p_2 p_3 t}) \tag{33.15}$$

2. Type AB

$$m'_{AB}(t) = b(a_{AB} - p_1 p_2 m_{AB}(t))$$
$$a'_{AB}(t) = (1 - p_1)(1 - p_2) p_3 m'_{AB}(t) \tag{33.16}$$

Type AC

$$m'_{AC}(t) = b(a_{AC}(t) - p_1 p_3 m_{AC}(t))$$
$$a'_{AC}(t) = (1 - p_1)(1 - p_3) p_2 m'_{ABC}(t) \tag{33.17}$$

Type BC

$$m'_{BC}(t) = b(a_{BC}(t) - p_2 p_3 m_{BC}(t))$$
$$a'_{BC}(t) = (1 - p_2)(1 - p_3) p_1 m'_{ABC}(t) \tag{33.18}$$

with marginal conditions

$$m_{AB}(0) = 0, \quad a_{AB}(0) = a_{AB}$$
$$m_{AC}(0) = 0, \quad a_{AC}(0) = a_{AC}$$
$$m_{BC}(0) = 0, \quad a_{BC}(0) = a_{BC}$$

By applying Equation 33.15 to Equations 33.16–33.18, we can obtain the mean value functions for failure type AB,

AC, and BC respectively:

$$m_{AB}(t) = C_{AB1} - C_{AB2}\,\mathrm{e}^{-bp_1p_2t} + C_{AB3}\,\mathrm{e}^{-bp_1p_2p_3t} \quad (33.19)$$

$$m_{AC}(t) = C_{AC1} - C_{AC2}\,\mathrm{e}^{-bp_1p_3t} + C_{AC3}\,\mathrm{e}^{-bp_1p_2p_3t} \quad (33.20)$$

$$m_{BC}(t) = C_{BC1} - C_{BC2}\,\mathrm{e}^{-bp_2p_3t} + C_{BC3}\,\mathrm{e}^{-bp_1p_2p_3t} \quad (33.21)$$

where

$$C_{AB1} = \frac{a_{AB}}{p_1p_2} + \frac{a_{ABC}(1-p_1)(1-p_2)}{p_1^2p_2^2}$$

$$C_{AB2} = \frac{a_{AB}}{p_1p_2} + \frac{a_{ABC}(1-p_1)(1-p_2)}{p_1^2p_2^2} - \frac{a_{ABC}(1-p_1)(1-p_2)}{p_1^2p_2^2(1-p_3)}$$

$$C_{AB3} = -\frac{a_{ABC}(1-p_1)(1-p_2)}{p_1^2p_2^2(1-p_3)} \quad (33.22)$$

$$C_{AC1} = \frac{a_{AC}}{p_1p_3} + \frac{a_{ABC}(1-p_1)(1-p_3)}{p_1^2p_3^2}$$

$$C_{AC2} = \frac{a_{AC}}{p_1p_3} + \frac{a_{ABC}(1-p_1)(1-p_3)}{p_1^2p_3^2} - \frac{a_{ABC}(1-p_1)(1-p_3)}{p_1^2p_3^2(1-p_2)}$$

$$C_{AC3} = -\frac{a_{ABC}(1-p_1)(1-p_3)}{p_1^2p_3^2(1-p_3)} \quad (33.23)$$

$$C_{BC1} = \frac{a_{BC}}{p_2p_3} + \frac{a_{ABC}(1-p_2)(1-p_3)}{p_2^2p_3^2}$$

$$C_{BC2} = \frac{a_{BC}}{p_2p_3} + \frac{a_{ABC}(1-p_2)(1-p_3)}{p_2^2p_3^2} - \frac{a_{ABC}(1-p_2)(1-p_3)}{p_2^2p_3^2(1-p_1)}$$

$$C_{BC3} = -\frac{a_{ABC}(1-p_2)(1-p_3)}{p_2^2p_3^2(1-p_1)} \quad (33.24)$$

3. Type A

$$m'_A(t) = b(a_A(t) - p_1m_A(t))$$

$$a'_A(t) = \beta_1(m'_A(t) + m'_{AB}(t) + m'_{AC}(t) + m'_{ABC}(t)) + (1-p_1)p_2m'_{AB}(t) + (1-p_1)p_3m'_{AC}(t) + (1-p_1)p_2p_3m'_{ABC}(t) \quad (33.25)$$

Type B

$$m'_B(t) = b(a_B(t) - p_2m_B(t))$$

$$a'_B(t) = \beta_2(m'_B(t) + m'_{AB}(t) + m'_{BC}(t) + m'_{ABC}(t)) + (1-p_2)p_1m'_{AB}(t) + (1-p_2)p_3m'_{BC}(t) + (1-p_2)p_1p_3m'_{ABC}(t) \quad (33.26)$$

Type C

$$m'_C(t) = b(a_C(t) - p_3m_C(t))$$

$$a'_C(t) = \beta_3(m'_C(t) + m'_{AC}(t) + m'_{BC}(t) + m'_{ABC}(t)) + (1-p_3)p_1m'_{AC}(t) + (1-p_3)p_2m'_{BC}(t) + (1-p_3)p_1p_2m'_{ABC}(t) \quad (33.27)$$

with marginal conditions

$$m_A(0) = m_B(0) = m_C(0) = 0$$

$$a_A(0) = a_A, \quad a_B(0) = a_B, \quad a_C(0) = a_C$$

The solutions to these equations are very complicated and can be obtained from Teng and Pham [30].

33.5.2.2 Common Failures

After we obtain the common failure counting processes $N_{AB}(t)$, $N_{AC}(t)$, $N_{BC}(t)$, and $N_{ABC}(t)$, we can define a new counting process $N_d(t)$:

$$N_d(t) = N_{AB}(t) + N_{AC}(t) + N_{BC}(t) + N_{ABC}(t)$$

In a 3VP system that uses a majority voting mechanism, common failures in two or three versions lead to the system failures. Therefore, $N_d(t)$ counts the number of NVP software system failures due to the CFs among software versions.

Similarly, we have the mean value function as follows:

$$m_d(t) = m_{AB}(t) + m_{AC}(t) + m_{BC}(t) + m_{ABC}(t) \quad (33.28)$$

Given the release time t and mission time x, the probability that an NVP software system will survive the mission without a CF is:

$$\Pr\{\text{no CF during } x \mid T\} = e^{-(m_d(t+x)-m_d(t))} \tag{33.29}$$

The above equation is not the final system reliability function, since we need to consider the probability that the system fails at independent faults in the software system.

33.5.2.3 Concurrent Independent Failures

Usually the NVP software system fails by CFs involving multi-versions. However, there is still a small probability that two software versions fail on the same input because of independent faults. This kind of failure is concurrent independent failure.

From assumptions 12, 13, and 14, the failure intensity $h_{\overline{AB}}(t)$ for the CIFs between version 1 and version 2 is given by:

$$h_{\overline{AB}}(t) = K_{AB}X_A(t)X_B(t) \tag{33.30}$$

where $X_A(t)$ and $X_B(t)$ denote the number of remaining independent faults in version 1 and version 2 respectively, and K_{AB} is the failure intensity per pair of faults for CIFs between version 1 and version 2. Then we have another non-homogeneous Poisson process for CIFs, $N_{\overline{AB}}(t)$, with mean value function:

$$m_{\overline{AB}}(t) = \int_0^t h_{\overline{AB}}(\tau)\,d\tau \tag{33.31}$$

Given the release time t and mission time x, we can obtain the probability that there is no type $\overline{AB}$ CIF during $(t, t+x)$

$$\Pr\{\text{no type } \overline{AB} \text{ failure during } x\} = e^{-(m_{\overline{AB}}(t+x)-m_{\overline{AB}}(t))} \tag{33.32}$$

Similarly, we have another two NHPPs for CIFs between version 1 and version 3 and between version 2 and version 3, with mean value functions:

$$m_{\overline{AC}}(t) = \int_0^t h_{\overline{AC}}(\tau)\,d\tau \tag{33.33}$$

$$m_{\overline{BC}}(t) = \int_0^t h_{\overline{BC}}(\tau)\,d\tau \tag{33.34}$$

where

$$h_{\overline{AC}}(t) = K_{AC}X_A(t)X_C(t) \tag{33.35}$$

$$h_{\overline{BC}}(t) = K_{BC}X_B(t)X_C(t) \tag{33.36}$$

Also the conditional probabilities are

$$\Pr\{\text{no type } \overline{AC} \text{ failure during } x\} = e^{-(m_{\overline{AC}}(t+x)-m_{\overline{AC}}(t))}$$

$$\Pr\{\text{no type } \overline{BC} \text{ failure during } x\} = e^{-(m_{\overline{BC}}(t+x)-m_{\overline{BC}}(t))}$$

If we define a new counting process

$$N_i(t) = N_{\overline{AB}}(t) + N_{\overline{AC}}(t) + N_{\overline{BC}}(t) \tag{33.37}$$

with mean value function

$$m_i(t) = m_{\overline{AB}}(t) + m_{\overline{AC}}(t) + m_{\overline{BC}}(t) \tag{33.38}$$

then the probability that there is no CIF for an NVP system during the interval $(t, t+x)$ is:

$$\Pr\{\text{no CIF during } x\} = e^{-(m_1(t+x)-m_1(t))} \tag{33.39}$$

33.5.3 N-version Programming System Reliability

After we obtain the probability of common failures and concurrent independent failures, we can determine the reliability of an NVP ($N = 3$) fault-tolerant software system:

$$R_{\text{NVP-SRGM}}(x \mid t) = \Pr\{\text{no CF \& no CIF during } x \mid t\}$$

Because the CFs and CIFs are independent of each other, then:

$$R_{\text{NVP-SRGM}}(x \mid t) = \Pr\{\text{no CF during } x \mid t\} \times \Pr\{\text{no CIF during } x \mid t\}$$

From Equations 33.29 and 33.39, the reliability of an NVP system can be determined by:

$$R_{\text{NVP-SRGM}}(x \mid t) = e^{-(m_d(t+x)+m_i(t+x)-m_d(t)-m_i(t))} \tag{33.40}$$

Table 33.4. Available information and unknown parameters to be estimated

Available information			Parameters to be estimated
Failure type	Failure time	No. of cumulative failures	Unknown parameters in the mean value function
A	$t_i, i = 1, 2, \ldots$	$m_A(t_i)$	$a_{ABC}, a_{AB}, a_{AC}, a_A, b, \beta_1$
B	$t_i, i = 1, 2, \ldots$	$m_B(t_i)$	$a_{ABC}, a_{AB}, a_{BC}, a_B, b, \beta_2$
C	$t_i, i = 1, 2, \ldots$	$m_C(t_i)$	$a_{ABC}, a_{AC}, a_{BC}, a_C, b, \beta_3$
AB	$t_i, i = 1, 2, \ldots$	$m_{AB}(t_i)$	a_{ABC}, a_{AB}, b
AC	$t_i, i = 1, 2, \ldots$	$m_{AC}(t_i)$	a_{ABC}, a_{AC}, b
BC	$t_i, i = 1, 2, \ldots$	$m_{BC}(t_i)$	a_{BC}, a_{ABC}, b
ABC	$t_i, i = 1, 2, \ldots$	$m_{ABC}(t_i)$	a_{ABC}, b
$\overline{AB}$	$t_i^*, i = 1, 2, \ldots$	$m_{\overline{AB}}(t_i)$	$a_{ABC}, a_{AB}, a_{AC}, a_{BC}, a_A, a_B, b, \beta_1, \beta_2, K_{AB}$
$\overline{AC}$	$t_i^*, i = 1, 2, \ldots$	$m_{\overline{AC}}(t_i)$	$a_{ABC}, a_{AB}, a_{AC}, a_{BC}, a_A, a_C, b, \beta_1, \beta_3, K_{AC}$
$\overline{BC}$	$t_i^*, i = 1, 2, \ldots$	$m_{\overline{BC}}(t_i)$	$a_{ABC}, a_{AB}, a_{AC}, a_{BC}, a_B, a_C, b, \beta_2, \beta_3, K_{BC}$

* The failure time and the number of cumulative type $\overline{AB}$, $\overline{AC}$, and $\overline{BC}$ failures are also included in those of type A, B, and C failures.

Generally, common failures are the leading failures within NVP systems, and the testing groups should pay much more attention to common failures. After we obtain a final system reliability model for the NVP systems, we need to estimate the values of all unknown parameters. Parameter estimation for this NVP-SRGM will be discussed next.

33.5.4 Parameter Estimation

This software reliability growth model for NVP systems consists of many parameters. In this chapter, we use the maximum likelihood estimation (MLE) method to estimate all unknown parameters. To simplify the problem, we assume that the error removal efficiencies p_1, p_2, and p_3 are known. In fact, they can be obtained from several empirical testing data. Therefore, we need to estimate the following parameters: $b, a_{ABC}, a_{AB}, a_{AC}, a_{BC}, a_A, a_B, a_C, \beta_1, \beta_2, \beta_3, K_{AB}, K_{AC}$, and K_{BC}.

The unknown parameters in the NHPP reliability model can be estimated by using an MLE method based on either one of the following given data sets:

1. the cumulative number of type ABC, AB, AC, BC, A, B, and C failures up to a given time or the actual time that each failure occurs;
2. the cumulative number of concurrent independent failures up to a given time or the actual time that each failure occurs.

Table 33.4 shows all available information and those parameters to be estimated.

One can easily use the MLE method to obtain the estimates of all parameters in the model [30].

33.6 N-version Programming–Software Reliability Growth

33.6.1 Applications of N-version Programming–Software Reliability Growth Models

33.6.1.1 Testing Data

In this section, we illustrate the model results by analyzing a fault-tolerant software application of the water reservoir control system [31].

Consider a simplified software control logic for a water reservoir control (WRC) system. Water is supplied via a source pipe controlled by a source valve and removed via a drain pipe controlled by a drain valve. There are two level sensors, positioned at the high and low limits; the high sensor outputs above if the level is above it and

Table 33.5. Normalized failure data of WRC 2VT system

Fault no.	Failure time		Fault no.	Failure time	
	Version 1	Version 2		Version 1	Version 2
1	1.2	3.6	14	39.2	34.8
2	2.8	8.4	15	40	36.4
3	8.4	12.8	16	44	36.8
4	10	14.4	17	44.8	38
5	16.4	17.2	18	54	39.2
6	20	18	19	56	41.6
7	24.4	20	20	62.4	42
8	28	23.2	21	80	46.4
9	29.2	25.2	22	92	59.6
10	31.2	28	23	99.6	62.4
11	34	28.4	24		98.8
12	36	30.8	25		99.6
13	36.8	31.2	26		100

the low sensor outputs below if the level is below it. The control system should maintain the water level between these two limits, allowing for rainfall into and seepage from the reservoir. If, however, the water rises above the high level, an alarm should sound. The WRC system achieves fault tolerance and high reliability through the use of NVP software control logic with $N = 2$. The WRC NVP software system with $N = 2$ normalized test data is listed in Table 33.5.

Generally speaking, an NVP system consists of N software versions, where N should be greater than or equal to 3 so that a voting mechanism can be applied to choose a correct output. For this 2VP system application, we assume that the reliability of the voter is equal to 1, and the 2VP system fails only when both its components (software versions) fail at the same input data.

We will apply this set of data in two cases. One is that two software versions are assumed to fail s-independently, another is that both versions are not assumed to fail s-independently, and the proposed NVP–SRGM will be applied.

Case 1: Independent NVP–SRGM Assume that those two software versions are s-independent of each other, we can apply the generalized software reliability model in Section 33.4 to each version separately, and estimate the reliability of each version $R_1(x \mid t)$ and $R_2(x \mid t)$, and further obtain the reliability for the entire system by simply using the parallel system reliability model:

$$R_{\text{Ind}}(x \mid t) = 1 - (1 - R_1(x \mid t))(1 - R_2(x \mid t)) \tag{33.41}$$

where x is the mission time, and

$$R_1(x \mid t) = \mathrm{e}^{-(m_1(t+x)-m_1(t))}$$

$$R_2(x \mid t) = \mathrm{e}^{-(m_2(t+x)-m_2(t))}$$

Figures 33.11 and 33.12 show the software failure fitting curve of mean value function $m_1(t)$ (for version 1) and $m_2(t)$ for (version 2) respectively. Figures 33.13 and 33.14 show the independent reliability function curve of the 2VP system and the reliability function curve of each version at $x = 50$ and $x = 10$ respectively when two software versions fail independently.

The actual meaning of the system reliability $R_{\text{Ind}}(t)$ is the probability that at least one version does not fail during the mission time x given t (time to stop testing). If each version can fail at most once during the mission time x, then $R_{\text{Ind}}(t)$ is the 2VP system reliability while no common failures occur between versions.

Case 2: Dependent NVP SRGM Failure Types. From Table 33.5, we can observe that two versions fail simultaneously at some time, for example,

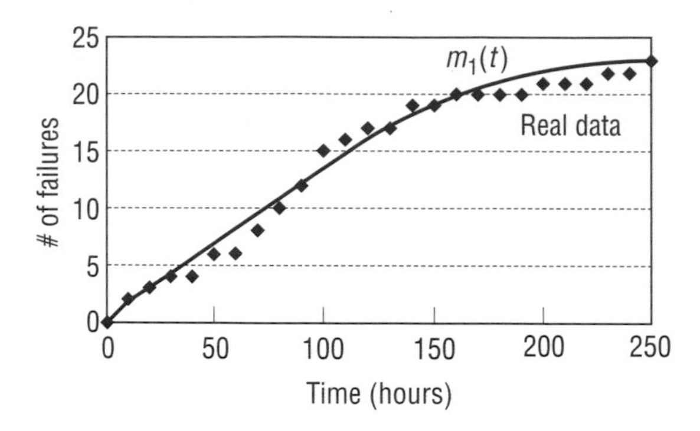

Figure 33.11. Mean value function $m_1(t)$ fitting curve for version 1

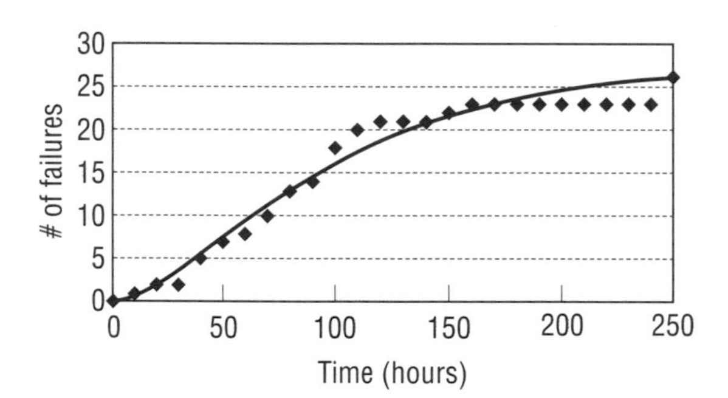

Figure 33.12. Mean value function $m_2(t)$ fitting curve for version 2

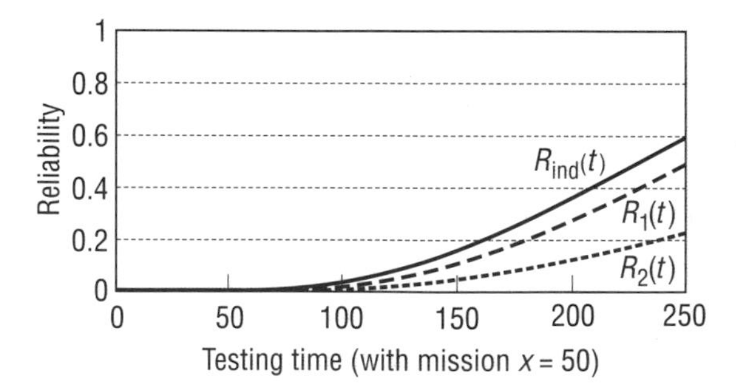

Figure 33.13. Independent system reliability with mission time $x = 50$

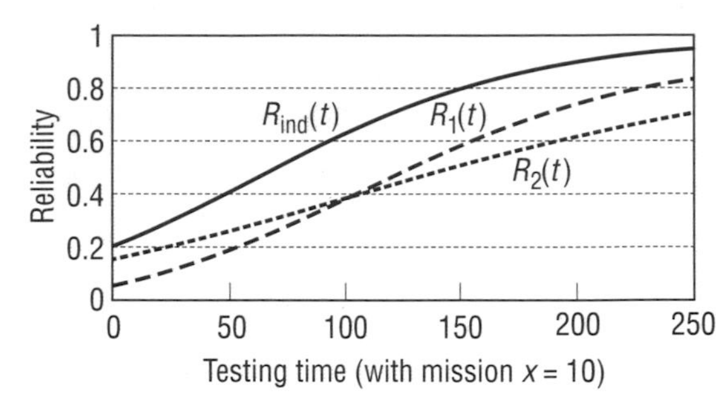

Figure 33.14. Independent system reliability curves with mission time $x = 10$

at $t = 8.4, 20, 28, \ldots, 99.6$. These failures are considered as coincident failures that are caused either by the common faults, or unrelated faults between two versions. Therefore, the assumption of independence is not valid for this testing data set.

In this example, we assume that all coincident failures are common failures. Then we can classify the software faults in this 2VP system according to the notation in Section 33.5. Table 33.6 is generated directly from Table 33.5 but it shows the different fault types in this 2VP system.

Reliability Modifications for NVP ($N = 2$) Systems. Since the system we presented in Section 33.5 is a 3VP system, and the system here is an NVP ($N = 2$) system, the reliability model can easily be modified. If we keep the same assumptions as those for NVP ($N = 3$) systems in Section 33.5, then we can obtain equations as follows:

1. Error type AB

$$m'_{AB}(t) = b(a_{AB} - p_1 p_2 m_{AB}(t)) \tag{33.42}$$

with marginal conditions $m_{AB}(0) = 0$ and $a_{AB}(0) = a_{AB}$. The solution to Equation 33.42 is:

$$m_{AB}(t) = \frac{a_{AB}}{p_1 p_2}(1 - e^{-bp_1 p_2 t}) \tag{33.43}$$

2. Fault type A

$$\begin{aligned} m'_A(t) &= b(a_A(t) - p_1 m_A(t)) \\ a'_A(t) &= (1 - p_1) p_2 m'_{AB}(t) \\ &\quad + \beta_1 (m'_A(t) + m'_{AB}(t)) \end{aligned} \tag{33.44}$$

Table 33.6. Fault type table for 2VP system

Fault no.	Failure time (hr)		
	Fault A	Fault B	Fault AB
1	1.2	3.6	8.4
2	2.8	12.8	20
3	10	14.4	28
4	16.4	17.2	31.2
5	24.4	18	36.8
6	29.2	23.2	39.2
7	34	25.2	62.4
8	36	28.4	99.6
9	40	30.8	
10	44	34.8	
11	44.8	36.4	
12	54	38	
13	56	41.6	
14	80	42	
15	92	46.4	
16		59.6	
17		98.8	
18		100	

with marginal conditions $m_A(0) = 0$ and $a_A(0) = a_A$, where a_A is the initial number of fault type A in the two-version programming software system.

Substitute Equation 33.43 into Equation 33.44 and solve, we then obtain the mean value function as:

$$m_A(t) = C_{A1} + C_{A2}\,\mathrm{e}^{-bp_1p_2t} + C_{A3}\,\mathrm{e}^{-b(p_1-\beta_1)t} \tag{33.45}$$

where

$$C_{A1} = \frac{a_A}{p_1-\beta_1} + \frac{((1-p_1)p_2+\beta_1)a_{AB}}{p_1p_2(p_1-\beta_1)}$$

$$C_{A2} = -\frac{((1-p_1)p_2+\beta_1)a_{AB}}{p_1p_2(p_1(1-p_2)-\beta_1)}$$

$$C_{A3} = \frac{((1-p_1)p_2+\beta_1)a_{AB}}{(p_1-\beta_1)(p_1(1-p_2)-\beta_1)} - \frac{a_A}{p_1-\beta_1}$$

3. Fault type B

$$\begin{aligned} m'_B(t) &= b(a_B(t) - p_2m_B(t)) \\ a'_B(t) &= (1-p_2)p_1m'_{AB}(t) \\ &\quad + \beta_2(m'_B(t) + m'_{AB}(t)) \end{aligned} \tag{33.46}$$

with marginal conditions $m_B(0) = 0$ and $a_B(0) = a_B$, where a_B is the initial number of fault type B in the two-version programming software system.

Substitute Equation 33.43 into Equation 33.46 and solve, we then obtain the mean value function as:

$$m_B(t) = C_{B1} + C_{B2}\,\mathrm{e}^{-bp_1p_2t} + C_{B3}\,\mathrm{e}^{-b(p_2-\beta_2)t} \tag{33.47}$$

where

$$C_{B1} = \frac{a_B}{p_2-\beta_2} + \frac{a_{AB}((1-p_2)p_1+\beta_2)}{p_1p_2(p_2-\beta_2)}$$

$$C_{B2} = -\frac{((1-p_2)p_1+\beta_2)a_{AB}}{p_1p_2(p_2(1-p_1)-\beta_2)}$$

$$C_{B3} = \frac{((1-p_2)p_1+\beta_2)a_{AB}}{(p_2-\beta_2)(p_2(1-p_1)-\beta_2)} - \frac{a_B}{p_2-\beta_2}$$

The likelihood functions are:

$$L_A = \prod_{i=1}^{n_A}\left\{\frac{[m_A(t_i)-m_A(t_{i-1})]^{y_{Ai}-y_{A(i-1)}}}{(y_{Ai}-y_{A(i-1)})!} \times \mathrm{e}^{-[m_A(t_i)-m_A(t_{i-1})]}\right\} \tag{33.48}$$

$$L_B = \prod_{i=1}^{n_B}\left\{\frac{[m_B(t_i)-m_B(t_{i-1})]^{y_{Bi}-y_{B(i-1)}}}{(y_{Bi}-y_{B(i-1)})!} \times \mathrm{e}^{-[m_B(t_i)-m_B(t_{i-1})]}\right\} \tag{33.49}$$

$$L_{AB} = \prod_{i=1}^{n_{AB}}\left\{\frac{[m_{AB}(t_i)-m_{AB}(t_{i-1})]^{y_{ABi}-y_{AB(i-1)}}}{(y_{ABi}-y_{AB(i-1)})!} \times \mathrm{e}^{-[m_{AB}(t_i)-m_{AB}(t_{i-1})]}\right\} \tag{33.50}$$

The united likelihood function is:

$$L = L_AL_BL_{AB} \tag{33.51}$$

and the log of the unified likelihood function is:

$$\ln(L) = \ln(L_A) + \ln(L_B) + \ln(L_{AB}) \tag{33.52}$$

Table 33.7. Different MLEs with respect to different p values

p_1	p_2	b	β_1	β_2	a_A	a_B	a_{AB}	MLES
0.8	0.8	0.01154	0.09191	0.00697	10.9847	15.0367	6.26062	−55.4742
0.8	0.85	0.0108	0.05699	0.0175	11.916	16.1536	6.43105	−55.4736
0.8	0.9	0.00979	0.0074	0.01101	13.2279	17.7376	7.00032	−55.4822
0.8	0.95	0.01151	0.0824	0.17392	10.9706	14.2399	6.86717	−55.4422
0.85	0.8	0.01317	0.1892	0.09393	9.5187	12.3296	6.10969	−55.4370
0.85	0.85	0.01043	0.09122	0.0069	12.2912	16.3465	7.01585	−55.4645
0.85	0.9	0.00965	0.05227	0.00823	13.2884	17.7446	7.40755	−55.4674
0.85	0.95	0.00901	0.01475	0.00893	14.3285	19.1661	7.7771	−55.4709
0.9	0.8	0.01105	0.15425	0.00372	11.6603	14.8114	7.10012	−55.4644
0.9	0.85	0.0114	0.18026	0.07047	11.0824	14.3879	6.91714	−55.4428
0.9	**0.9**	**0.00956**	**0.0983**	**0.00513**	**13.465**	**17.7861**	**7.77652**	−55.4583
0.9	0.95	0.00997	0.1196	0.09461	12.7394	16.5588	7.77522	−55.4468
0.95	0.8	0.01126	0.2196	0.01224	11.2847	14.5756	6.94388	−55.4478
0.95	0.85	0.01069	0.20046	0.0329	11.8325	15.4577	7.3401	−55.4449
0.95	0.9	0.00963	0.1448	0.02206	13.3545	17.2171	7.9219	−55.4485
0.95	0.95	0.00936	0.1384	0.05541	13.6155	17.7036	8.22915	−55.4484

or, equivalently,

$$\begin{aligned}\ln L = \sum_{i=1}^{n_A} \{ & (y_{Ai} - y_{A(i-1)}) \ln(m_A(t_i) - m_A(t_{i-1})) \\ & - (m_A(t_i) - m_A(t_{i-1})) \\ & - \ln((y_{Ai} - y_{A(i-1)})!)\} \\ + \sum_{i=1}^{n_B} \{ & (y_{Bi} - y_{B(i-1)}) \\ & \times \ln(m_B(t_i) - m_B(t_{i-1})) \\ & - (m_B(t_i) - m_B(t_{i-1})) \\ & - \ln((y_{Bi} - y_{B(i-1)})!)\} \\ + \sum_{i=1}^{n_{AB}} \{ & (y_{ABi} - y_{AB(i-1)}) \\ & \times \ln(m_{AB}(t_i) - m_{AB}(t_{i-1})) \\ & - (m_{AB}(t_i) - m_{AB}(t_{i-1})) \\ & - \ln((y_{ABi} - y_{AB(i-1)})!)\} \end{aligned} \tag{33.53}$$

Take derivatives on Equation 33.53 with respect to each unknown parameters and set them to be 0, then we get a set of equations. By solving those equations simultaneously, we can finally obtain the model parameters.

Maximum Likelihood Estimation. The data set in Table 33.5 does not provide sufficient information about what category (common failures or concurrent independent failures) those failures belong to. Since the common failures are dominant failures, then we just assume that all coincident failures are common failures to simplify this problem.

The error removal efficiency, p, is usually considered as a known parameter, mostly $0.8 \le p \le 0.95$. It can be determined from empirical data. In this study, we first choose the value of p arbitrarily, then compute the MLEs for all other unknown parameters.

Table 33.7 shows the computation results for MLEs with respect to various p_1 and p_2 values. Here we choose $p_1 = p_2 = 0.9$, so the MLEs for all parameters are:

$$\hat{a}_A = 15.47 \quad \hat{a}_B = 18.15 \quad \hat{a}_{AB} = 7.8$$

$$\hat{\beta}_1 = 0 \quad \hat{\beta}_2 = 0.002324 \quad \hat{b} = 0.009$$

Then we can obtain the software reliability of this NVP ($N = 2$) system as:

$$R_{\text{NVP-SRGM}}(t) = R(x \mid t) = \mathrm{e}^{-(m_{AB}(t+x) - m_{AB}(t))} \tag{33.54}$$

Table 33.8 shows the computation results for mean value functions. Figures 33.15–33.17 show the

Table 33.8. Mean value functions versus failure data

Time	Cumulative number of failures			Mean value functions		
	Fault A	Fault B	Fault AB	$m_A(t)$	$m_B(t)$	$m_{AB}(t)$
0	0	0	0	0.00	0.00	0.00
10	2	1	0	1.24	1.62	0.71
20	2	1	0	2.39	3.12	1.37
30	3	1	1	3.48	4.50	1.98
40	3	3	1	4.49	5.77	2.55
50	4	5	2	5.44	6.94	3.07
60	4	6	2	6.32	8.02	3.56
70	5	7	3	7.15	9.02	4.01
80	6	9	4	7.92	9.94	4.42
90	8	10	4	8.65	10.79	4.81
100	9	12	6	9.32	11.57	5.17
110	10	14	6	9.96	12.29	5.50
120	11	15	6	10.55	12.96	5.80
130	11	15	6	11.10	13.57	6.08
140	13	15	6	11.62	14.14	6.35
150	13	16	6	12.10	14.66	6.59
160	13	16	7	12.55	15.14	6.81
170	13	16	7	12.97	15.59	7.02
180	13	16	7	13.37	15.99	7.22
190	13	16	7	13.74	16.37	7.39
200	14	16	7	14.08	16.72	7.56
210	14	16	7	14.40	17.04	7.71
220	14	16	7	14.70	17.34	7.85
230	15	16	7	14.98	17.61	7.99
240	15	16	7	15.24	17.87	8.11
250	15	18	8	15.49	18.10	8.22

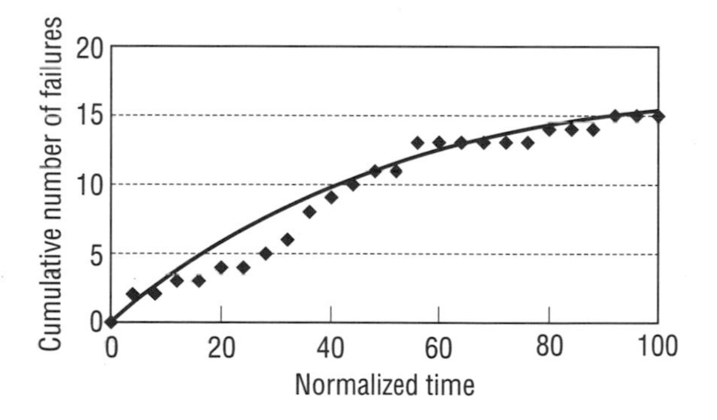

Figure 33.15. Mean value function $m_A(t)$ fitting curve

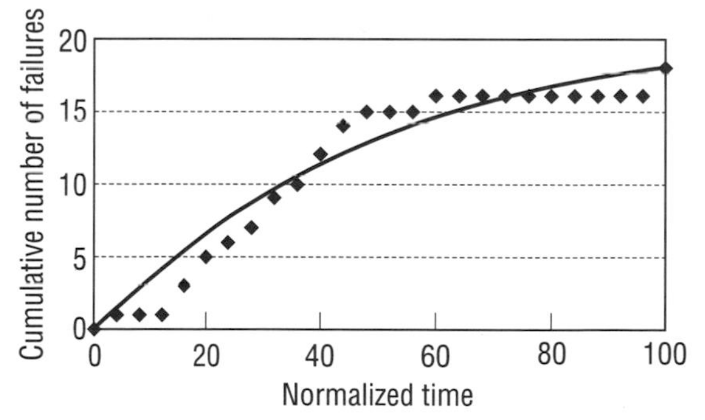

Figure 33.16. Mean value function $m_B(t)$ fitting curve

comparisons between the real failure data and the mean value functions.

Confidence Intervals. The confidence interval for all parameter estimates can also be obtained.

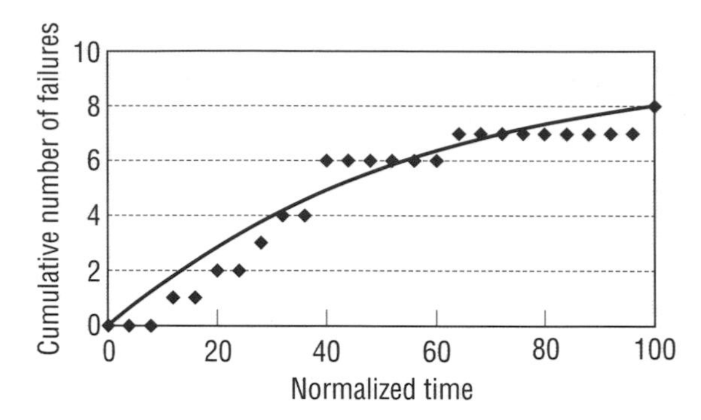

Figure 33.17. Mean value function $m_{AB}(t)$ fitting curve

Given the log likelihood function

$$L = \ln(L_A) + \ln(L_B) + \ln(L_{AB})$$

If we use $x_i, i = 1, 2, \ldots, 6$, to denote all parameters in the model to simplify the expression:

$$\begin{aligned} x_1 &\to a_A \\ x_2 &\to a_B \\ x_3 &\to a_{AB} \\ x_4 &\to \beta_1 \\ x_5 &\to \beta_2 \\ x_6 &\to b \end{aligned}$$

The actual numerical result for the Fisher information matrix is:

$$H = \begin{bmatrix} 0.0763 & 0 & 0.0042 & 1.028 & 0 & 39.48 \\ 0 & -0.051 & 0.0037 & 0 & 1.043 & 36.55 \\ 0.0042 & 0.0037 & 0.132 & 0.0825 & 0.133 & 40.11 \\ 1.028 & 0 & 0.0825 & 31.68 & 0 & -37.38 \\ 0 & 1.043 & 0.133 & 0 & 33.26 & -68.38 \\ 39.48 & 36.55 & 40.11 & -37.38 & -68.38 & 179746.12 \end{bmatrix} \tag{33.55}$$

and the variance matrix is:

$$V = H^{-1} = \begin{bmatrix} 41.473 & 40.4 & 5.625 & -1.384 & -1.33 & -0.0194 \\ 40.4 & 143.61 & 13.28 & -1.397 & -4.645 & -0.0431 \\ 5.625 & 13.28 & 9.511 & -0.215 & -0.467 & -0.0063 \\ -1.384 & -1.397 & -0.215 & 0.0778 & 0.046 & 0.00067 \\ -1.33 & -4.645 & -0.467 & 0.046 & 0.181 & 0.00142 \\ -0.0194 & -0.0431 & -0.0063 & 0.00067 & 0.00142 & 2.067 \times 10^{-5} \end{bmatrix}$$

Then the variances of estimations are:

$$\mathrm{Var}(\hat{a}_A) = 41.473 \quad \mathrm{Var}(\hat{a}_B) = 143.61$$
$$\mathrm{Var}(\hat{a}_{AB}) = 9.511$$
$$\mathrm{Var}(\hat{\beta}_1) = 0.0778 \quad \mathrm{Var}(\hat{\beta}_2) = 0.181$$
$$\mathrm{Var}(\hat{b}) = 2.067 \times 10^{-5}$$

Figures 33.18–33.20 show the mean value functions and their 95% confidence intervals as well as the number of cumulative failures.

Figure 33.21 shows the NVP system reliability and its 95% confidence interval given the fixed mission time $x = 10$ hours assuming the reliability estimation follows a normal distribution. One can see that the reliability confidence interval shrinks when people spend more time removing the faults from the NVP system. This means that after people

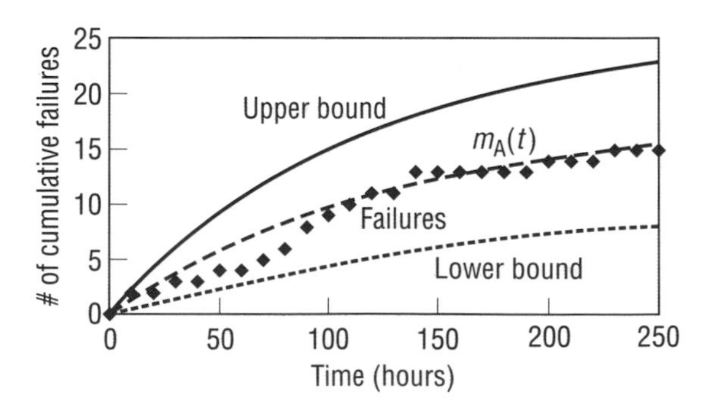

Figure 33.18. Confidence interval for mean value function $m_A(t)$

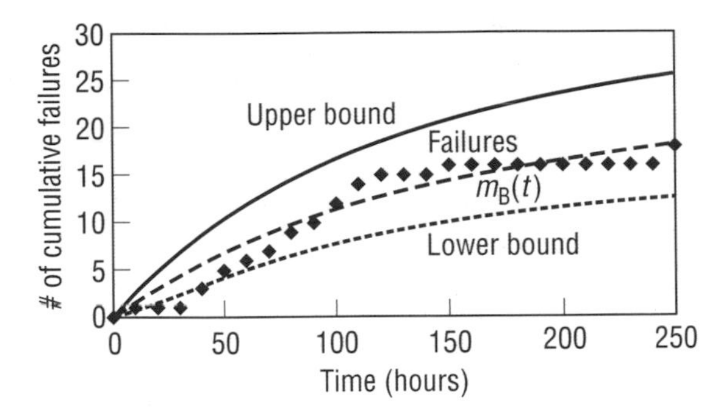

Figure 33.19. Confidence interval for mean value function $m_B(t)$

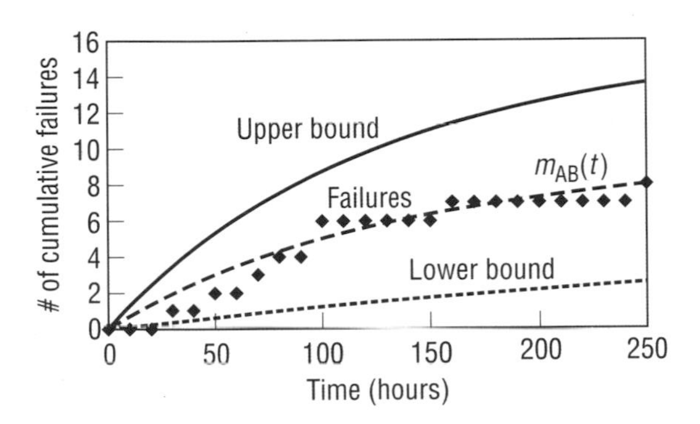

Figure 33.20. Confidence interval for mean value function $m_{AB}(t)$

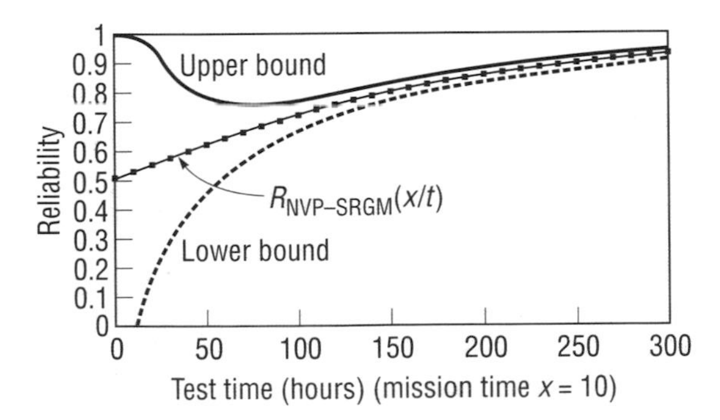

Figure 33.21. NVP system reliability and its 95% confidence interval

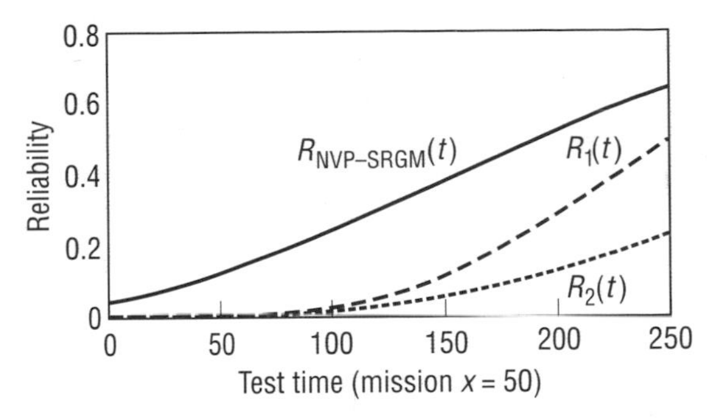

Figure 33.22. NVP system and single-version reliability curves with mission time $x = 50$

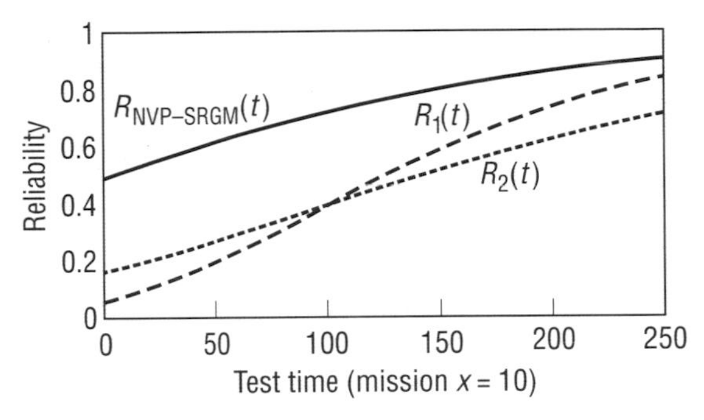

Figure 33.23. NVP system and single-version reliability curves with mission time $x = 10$

gain more knowledge about the NVP system, more accurate estimations can be made to assess the NVP system reliability.

The NVP system reliability function $R_{\text{NVP-SRGM}}(x \mid t)$ and single-version software reliability $R_1(x \mid t)$ and $R_2(x \mid t)$ are shown in Figures 33.22 and 33.23 with mission time $x = 50$ and $x = 10$, respectively.

From Figures 33.22 and 33.23, we can see that the 2VP scheme has higher reliability than any single component. This means that the N-version programming scheme is able to provide higher system reliability.

If we compare Figure 33.23 with Figure 33.14, then we can see clearly that although software versions are developed independently, common failures cannot be ignored in this NVP system.

When the component reliability is high, the independent assumption will lead to the overestimation of system reliability.

In this example, we assume that all concurrent failures are common failures. If the concurrent failure types are different, the reliability function will be a little different—the independent failures will be incorporated into the reliability function, though usually the common failures dominate the concurrent failures among versions.

As the first model of its kind in the area of NVP reliability modeling, the proposed NVP software reliability growth model can be used to overcome the shortcomings of the independent reliability model. It predicts the system reliability more accurately than the independent model and can be used to help determine when to stop testing, which will be one of the key questions in the testing and debugging phase of the NVP system lifecycle.

33.7 Conclusion

This chapter presents a non-homogeneous Poisson progress reliability model for N-version programming systems. We separate all faults within NVP systems into independent faults and common faults, and model each type of failure as NHPP. We further develop a reliability model for common failures in NVP systems and also present a model for concurrent independent failures in NVP systems. By combining the CF model and the CIF model together, we establish an NHPP reliability model for NVP systems. We also give an example to illustrate how to estimate all unknown parameters by using the maximum likelihood estimation method, and how to compute the variances for all parameter estimates in order to obtain the confidence intervals of NVP system reliability prediction.

References

[1] Voas J, Dugan JB, Hatton L, Kanoun K, Laprie J-C, Vouk MA. Fault tolerance roundtable. IEEE Software 2001;Jul/Aug:54–7.

[2] Chen L, Avizienis A. N-version programming: a fault tolerance approach to the reliable software. Proc 8th Int Symp Fault-Tolerant Computing, Toulouse, France; 1978. p.3–9.

[3] Avizienis A, Chen L. On the implementation of N-version programming for software fault-tolerance during program execution. Proc COMPASAC 77; 1977. p.149–55.

[4] Eckhardt D, Lee L. A theoretical basis for the analysis of multiversion software subject to coincident errors. IEEE Trans Software Eng 1985;SE-11(12):1511–7.

[5] Scott RK, Gault JW, McAllister DF. Fault-tolerant reliability modeling. IEEE Trans Software Eng 1987;SE-13(5):582–92.

[6] Littlewood B, Miller DR. Conceptual modeling of coincident failures in multiversion software. IEEE Trans Software Eng 1989;15(12):1596–614.

[7] McAllister DF, Sun CE, Vouk MA. Reliability of voting in fault-tolerant software systems for small output spaces. IEEE Trans Reliab 1990;39(5):524–34.

[8] Leung Y-W. Maximum likelihood voting for fault-tolerant software with finite output-space. IEEE Trans Reliab 1995;44(3):419–26.

[9] Randell B. System structure for software fault tolerance. IEEE Trans Software Eng 1975;SE-1(2):220–32.

[10] Fairley R. Software engineering concepts. McGraw-Hill: New York; 1985.

[11] Belli F, Jedrzejowicz P. Fault-tolerant programs and their reliability. IEEE Trans Reliab 1990;29(2):184–92.

[12] Nicola VF, Goyal A. Modeling of correlated failures and community error recovery in multiversion software. IEEE Trans Software Eng 1990;16(3):350–9.

[13] Lyu MR. Improving the N-version programming process through the evolution of a design paradigm. IEEE Trans Reliab 1993;42(2):179–89.

[14] Knight JC, Leveson NG. An experimental evaluation of the assumption of independence in multiversion programming. IEEE Trans Software Eng 1986;SE-12.

[15] Scott RK, Gault JW, McAllister DF, Wiggs J. Experimental validation of six fault tolerant software reliability models. Dig papers FTCS-14: 14th Ann Symp Fault-Tolerant Computing, Kissemmee, NY; 1984. p.102–7.

[16] Eckhardt DE, Caglayan AK, Knight JC, Lee LD, McAllister DF, Vouk MA, Kelly JPJ. An experimental evaluation of software redundancy as a strategy for improving reliability. IEEE Trans Software Eng 1991;17(7):692–702.

[17] Voas J, Ghosh A, Charron F, Kassab L. Reducing uncertainty about common-mode failures. In: Proc Int Symp on Software Reliability Engineering, ISSRE; 1997. p.308–19.

[18] Laprie J-C, Arlat J, Beounes C, Kanoun K. Definition and analysis of hardware and software fault-tolerant architectures. IEEE Comput 1990;23(7):39–51.

[19] Dugan JB, Lyu MR. System reliability analysis of an N-version programming application. IEEE Trans Reliab 1994;43(4):513–9.

[20] Tai AT, Meyer JF, Aviziems A. Performability enhancement of fault-tolerant software. IEEE Trans Reliab 1993;42(2):227–37.

[21] Goseva-Popstojanova K, Grnarov A. Performability modeling of N-version programming technique. In: Proc 6th IEEE Int Symp on Software Reliability Engineering (ISSRE'95), Toulouse, France.

[22] Lin H-H, Chen K-H. Nonhomogeneous Poisson process software-debugging models with linear dependence. IEEE Trans Reliab 1993;42(4):613–7.

[23] Kanoun K, Kaaniche M, Beounes C, Laprie J-C, Arlat J. Reliability growth of fault-tolerant software. IEEE Trans Reliab 1993;42(2):205–18.

[24] Sha L. Using simplicity to control complexity. IEEE Software 2001;Jul/Aug:20–8.

[25] Zhang X, Teng X, Pham H. A generalized software reliability model with error removal efficiency. IEEE Trans Syst, Man Cybernet 2002:submitted.

[26] Pham H. Software reliability. Springer-Verlag; 2000.

[27] Goel AL, Okumoto K. Time-dependent error-detection rate model for software and other performance measures. IEEE Trans Reliab1979;28:206–11.

[28] Pham H, Nordmann L, Zhang X. A general imperfect-software-debugging model with s-shaped fault-detection rate. IEEE Trans Reliab 1999;48(2):168–75.

[29] Pham H, Zhang X. An NHPP software reliability model and its comparison. Int J Reliab, Qual Safety Eng 1997;4(3):269–82.

[30] Teng X, Pham H. A software reliability growth model for N-version programming. IEEE Trans Reliab 2002;51(3):in press.

[31] Pham H, Pham M. Software reliability models for critical applications. INEL, EG&G-2663; 1991.

Markovian Dependability/Performability Modeling of Fault-tolerant Systems

Juan A. Carrasco

Chapter 34

34.1 Introduction

Increasing demand for system reliability (understood in its broad sense, *i.e.* as the capability of the system to perform properly) has motivated an increased interest in fault-tolerant systems. A fault-tolerant system is one that can continue correct operation with or without degraded performance in the presence of faults, *i.e.* physical defects, imperfections, external disturbances, or design flaws in hardware or software components. Fault tolerance can be achieved by fault masking, *i.e.* by preventing faults from producing errors without eliminating the faulty components from the operational system configuration, or by reconfiguration, *i.e.* by eliminating faulty components from the operational system configuration. The latter is, generally, more complex to implement and requires fault detection (recognizing that a fault has occurred), fault location (identifying the faulty component), fault containment (isolating the faulty component so that it does not produce errors which propagate throughout the system), and fault recovery (restoring the system to a correct state from which to continue operation if an incorrect state has been reached). All these fault-tolerance techniques require the addition of hardware redundancy (extra hardware components), information redundancy (redundant information), time redundancy (extra computations), or software redundancy (extra software components). Replication of hardware components is an example of hardware redundancy; error detecting and error correcting codes are examples of information redundancy; repeated computation and checkpointing are examples of time redundancy; consistency checks, N-version programming, and recovery blocks are examples of software redundancy. The addition of redundancy affects negatively some characteristics of the system such as cost, performance, size, weight, and power consumption, and, during the design of a fault-tolerant system, those impacts have to be balanced against the achieved increase in system reliability.

Fault tolerance is an attractive approach to design systems which, without fault tolerance, would have an unacceptable reliability level. This includes systems for critical-computation applications such as aircraft flight control and control of dangerous chemical processes, unrepairable systems with long mission times such as long-life spacecraft unmanned systems, systems requiring high availability of computational resources, data, or both such as call switching systems and airline reservation systems, and systems with large amounts of hardware/software such as large multiprocessors. Nanometric systems on chip is an area in which fault tolerance may become more and more attractive. This is because, as feature sizes scale down, nanoelectronic structures get more and more susceptible to manufacturing faults, degradation processes, and external perturbations, and it may happen in the near future that acceptable levels of yield/operational reliability for complex nanometric systems on chip can only be achieved through the use of fault tolerance.

Modeling plays an important role in the design and analysis of fault-tolerant systems. This is clearly true in the early design stages and when it has to be certified that an existing system achieves a very high reliability level. Modeling also allows us to study how changes in the design and operation of an existing system may affect its reliability without actually modifying the system. Component failures, fault recovery mechanisms, maintenance activities, and, often, performance-related activities have a stochastic behavior, and, thus, stochastic models have to be used. Typical parameters of those models are component failure rates, coverage probabilities (*i.e.* the probabilities that faults of certain classes are successfully recovered), characteristics of component repair time distributions, and characteristics of performance-related activities. Estimation of those parameters is the first step to construct a model. Such estimates can be obtained from system specifications, data collected from similar systems that are under operation, data provided by component manufacturers/developers (for instance, mean time to failures of hardware/software components), available standardized models (for instance, the MIL-HDBK-217E model for estimation of failure rates of hardware components), or experimentation on the real system, a prototype, or a more or less detailed simulation model of the system (fault injection experiments to estimate coverage parameters are an example; see Iyer and Tang [1] for a survey of those techniques).

The reliability of a fault-tolerant system can be quantified by several measures summarizing the behavior of the system as perceived by its users. Many systems can be regarded as either performing correctly (up) or performing incorrectly or not performing at all (down). For those systems, simple dependability measures such as the mean time to failure, the reliability measure (probability that the system has been continuously up), and the availability (probability that the system is up at a given time) are appropriate. Many other systems, however, are degradable, in the sense that their performance may degrade as components fail. Simple dependability measures can be generalized to evaluate the reliability of those systems by associating performance levels with system states and including in the up subset the states in which the system has a performance above or equal to each of those levels [2]. A more general approach is the performability concept introduced by Meyer [3]. In that approach, the user-perceived behavior of the system is quantified by a discrete or continuous set of accomplishment levels and the performability is defined as the probability of a measurable subset of accomplishment levels. An example of the performability concept would be the distribution of the accumulated performance (for instance, number of processed transactions in a transaction-oriented system) of a system over a time interval. Another example [4] would be the distribution of the fraction of time during a time interval in which a communication channel fails to provide a given quality of service to the admitted traffic sources.

Once a model and a measure have been selected, the model has to be specified and solved. The difficulty in solving the model depends on both the model characteristics and the measure. Combinatorial solution methods allow an efficient analysis of very complex systems, but have restricted scope of application. In combinatorial methods (see Abraham [5] and Aggarwal *et al.* [6]) the fault-tolerant system is conceptualized as made up of components which can be unfailed/failed and the system is up/down as determined from the unfailed/failed state of the components by a structure function, usually represented by a fault tree. Combinatorial methods compute the probability that the system is up/down from the probabilities that the components are unfailed/failed, assuming that component states are independent. This allows the computation of the availability of systems having components with independent behavior and the reliability of non-repairable systems with coherent structure functions and components with independent behavior, since for those systems the reliability at time t is equal to the availability at time t. Combinatorial methods allowing imperfect coverage have been developed recently [7–9]. When the fault-tolerant system can be decomposed into subsystems with independent behavior, some measures can be computed hierarchically. The SHARPE tool has been designed to accommodate such techniques, and the examples presented by Sahner and Trivedi [10] illustrate them very well. However, computation of more complex measures or computation of simple measures when the components in the system have interactions require direct analysis of a stochastic process representing the behavior of the whole system. In that context, homogeneous continuous-time Markov chains (CTMCs) are commonly used. CTMCs arise naturally when the activities modifying the state of the system (failure processes, repair processes, and performance-related activities) have durations with exponential distributions. Phase-type distributions [11], particularly acyclic phase-type distributions, for which efficient fitting algorithms exist [12], can be used to accommodate (approximately) non-exponential distributions in the CTMC framework, at the price of a, usually, significant increase in the size of the CTMC.

In this chapter we make a, necessarily incomplete, review of techniques for Markovian dependability/performability modeling of fault-tolerant systems. In the review, we will often make reference to the METFAC-2.1 tool, currently under development. The rest of the chapter is organized as follows. Section 34.2 defines and, in some cases, formalizes the computation of a set of generic measures defined over rewarded CTMCs encompassing many dependability measures, as well as particular instances of the general performability measure. Section 34.3 reviews model specification methodologies. Section 34.4 reviews numerical techniques for model solution. Section 34.5 reviews available techniques to deal with the largeness problem with emphasis on bounding methods. Section 34.6 illustrates some of the techniques reviewed in the previous sections with a case study. Finally, Section 34.7 presents some conclusions.

34.2 Measures

Rewarded CTMC models have emerged in the last years as a powerful modeling formalism. A rewarded CTMC is a CTMC with a reward structure imposed over it. The reward structure may include reward rates associated with states and impulse rewards associated with transitions. Reward rates specify the rate at which reward is earned while the CTMC is in particular states; impulse rewards are rewards earned each time a transition of the CTMC is followed. An appropriate reward structure may be used to quantify many aspects of system behavior: reliability, performance, cost of operation, energy consumption, *etc.* The probabilistic behavior of the resulting reward can be summarized using different reward measures. Many traditional dependability measures are obtained as particular instances of those generic measures by using particular reward structures. In this section, we will define and, in some cases, formalize the computation of eight reward measures. All those measures will be supported by the METFAC-2.1 tool and assume a reward structure including only reward rates.

To formalize the computation of some of those measures, it is necessary to use some basic concepts on CTMCs. The terminology on CTMCs is not well established and we will use our own. Also, we will use without either reference or proof both well-known results and results which can be obtained with some effort. Let $X = \{X(t);\ t \geq 0\}$ be a CTMC with finite state space Ω and initial probability distribution row vector $\boldsymbol{\alpha} = (\alpha_i)_{i \in \Omega}$, where $\alpha_i = P[X(0) = i]$. Let $\lambda_{i,j}$, $i, j \in \Omega$, $i \neq j$ denote the transition rate of X from state i to state j, let $\lambda_{i,B} = \sum_{j \in B} \lambda_{i,j}$, $i \in \Omega$, $B \subset \Omega - \{i\}$, let $\lambda_i = \sum_{j \in \Omega - \{i\}} \lambda_{i,j}$, $i \in \Omega$ denote the output rate of X from state i, and let $\mathbf{A} = (a_{i,j})_{i,j \in \Omega}$, $a_{i,i} = -\lambda_i$, $a_{i,j} = \lambda_{i,j}$, $i \neq j$ denote the transition rate matrix (also called infinitesimal generator) of X. The state transition diagram of X is a labeled digraph with set of nodes Ω and an arc labeled with $\lambda_{i,j}$ from each state i to each state j with $\lambda_{i,j} > 0$. The analysis of the state transition diagram together with the initial probability distribution row vector $\boldsymbol{\alpha}$ provides insight into the qualitative behavior of the transient probabilities $p_i(t) = P[X(t) = i]$, $i \in \Omega$ of the states of X.

Two states $i, j \in \Omega$ are said to be *strongly connected* if and only if there are paths in the state transition diagram both from i to j and from j to i. A state is strongly connected with itself. Strong state connectivity is an equivalence relation. The corresponding equivalence classes are called the *components* of the CTMC. A component is, then, a maximal subset of strongly connected states. A state i is said to be *unreachable* if $p_i(t) = 0$ for all t and *reachable* otherwise. It is easy to prove that either all states of a component are reachable or all states of a component are unreachable. Then, we can properly talk about reachable and unreachable components. To determine which components of a CTMC are unreachable, it is useful to define the *components digraph* of a CTMC. The *components digraph* of a CTMC is the acyclic digraph having a node for each component of the CTMC and an arc from component C to component C' if and only if the state transition diagram of the CTMC has an arc from some state in C to some state in C'. To illustrate the concepts defined so far, Figure 34.1 gives the state transition diagram of a small CTMC and the corresponding components digraph. Let $\alpha_B = \sum_{i \in B} \alpha_i$, $B \subset \Omega$. Then, a component C is unreachable if and only if $\alpha_C = 0$ and there is no path in the components digraph from a component C' with $\alpha_{C'} > 0$ to C. States in unreachable components can be discarded when analyzing a CTMC and, in the following, we will assume that they have been discarded and that all states of the CTMC are reachable.

A (reachable) state i is said to be *transient* if, starting at i, there is a non-null probability that X will leave i and never return to it. For a transient state i, $\lim_{t \to \infty} p_i(t) = 0$. For a (reachable) non-transient state i, $\lim_{t \to \infty} p_i(t) > 0$. A state i is said to be *absorbing* if and only if $\lambda_i = 0$. It can be proved that all states of a component are either transient or non-transient. Then, we can properly talk about transient components. A component which is not transient is said to be *trapping*. Classification of the (reachable) components of a CTMC into transient and trapping can be done easily by examining the components digraph. A component is trapping if and only if the component has no outgoing arc in the components digraph. For the example given in Figure 34.1, assuming that all components are reachable, components C_1, C_2, and C_3 would be transient and components C_4 and C_5 would be trapping. Trapping components are so called because, once X enters a trapping component, it never leaves it. It should be clear that an absorbing state constitutes itself a trapping component. A CTMC having a single (trapping) component is said to be *irreducible*.

We define in Sections 34.2.1–34.2.8 and, in some cases, formalize the computation of eight reward measures defined over rewarded CTMCs X

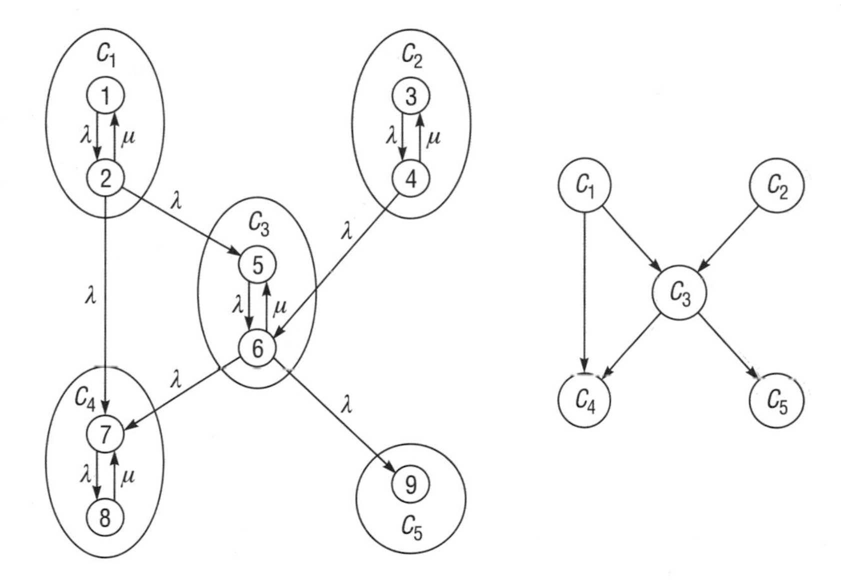

Figure 34.1. State transition diagram of a small CTMC (left) and the corresponding components digraph (right)

with finite state space Ω. The reward rate structure includes reward rates $r_i \geq 0$, with $i \in \Omega$ (measures ESSRR, ETRR(t), EARR(t), and CRD(t, s)) or $i \in B$, B being a subset of Ω (measures ECRTE, ECRDS, CRDTE(s), and CRDDS(s)). To be well defined some measures require X to have special properties. Later, in Section 34.2.9, we will show how more general reward structures can be accommodated for some measures.

34.2.1 Expected Steady-state Reward Rate

The measure is defined as ESSRR $= \lim_{t\to\infty} E[r_{X(t)}]$ and can be computed as:

$$\text{ESSRR} = \sum_{i\in\Omega} r_i p_i$$

where $p_i = \lim_{t\to\infty} p_i(t)$ is the steady-state probability of state i. Transient states have null steady-state probabilities. Non-transient states have steady-state probabilities > 0. Those probabilities can be computed as follows. Let S be the subset of transient states of X and let $C_1, C_2, \ldots, C_m$ be the trapping components of X. Let $\mathbf{p}^k = (p_i^k)_{i\in C_k}$ be the row vector having as components the conditional steady-state probabilities of the states in C_k (i.e. p_i^k is the steady-state probability that X is in state i conditioned to X being in C_k); formally, $p_i^k = \lim_{t\to\infty} P[X(t) = i \mid X(t) \in C_k]$. Let $\mathbf{0}$ and $\mathbf{1}$ be row vectors of appropriate dimensions with all elements equal to, respectively, 0 and 1; let $\mathbf{A}^{C_k,C_k}$ be the restriction of $\mathbf{A}$ to the states in C_k ($\mathbf{A}^{C_k,C_k} = (a_{i,j})_{i,j\in C_k}$); and let $\mathbf{x}^{\mathrm{T}}$ denote the transpose of a vector $\mathbf{x}$. Then, $\mathbf{p}^k$ is the normalized solution ($\mathbf{p}^k\mathbf{1}^{\mathrm{T}} = 1$) of the linear system $\mathbf{p}^k\mathbf{A}^{C_k,C_k} = \mathbf{0}$. Let $p_{C_k} = \lim_{t\to\infty} P[X(t) \in C_k] = \sum_{i\in C_k} p_i$. We have $p_i = p_{C_k} p_i^k$, $i \in C_k$. That result allows us to compute p_i, $i \in C_k$, $1 \leq k \leq m$ and, thus, ESSRR from the vectors $\mathbf{p}^k$ and p_{C_k}, $1 \leq k \leq m$. We discuss next the computation of p_{C_k}, $1 \leq k \leq m$.

If X has a single trapping component C_1, then X will be in C_1 with probability 1 for $t \to \infty$ and $p_{C_1} = 1$. If X has several trapping components and X does not have transient states, then $p_{C_k} = \alpha_{C_k}$. If X has transient states and more than one trapping component, $p_{C_k} = \alpha_{C_k} + \beta_k$, where β_k is the probability that X will leave S through C_k. The quantities β_k, $1 \leq k \leq m$, can be computed as follows. Let τ_i, $i \in S$ denote the expected value of the time spent by X in state i ($\tau_i = \int_0^\infty p_i(t)\,dt$, $< \infty$ since i is transient). Then, the row vector $\boldsymbol{\tau}^S = (\tau_i)_{i \in S}$ is the solution of the linear system $\boldsymbol{\tau}^S \mathbf{A}^{S,S} = -\boldsymbol{\alpha}^S$, where $\mathbf{A}^{S,S}$ and $\boldsymbol{\alpha}^S$ are the restrictions of, respectively, $\mathbf{A}$ and $\boldsymbol{\alpha}$ to S. Let the row vector $\boldsymbol{\Lambda}^k = (\lambda_{i,C_k})_{i \in S}$, then β_k can be obtained from $\boldsymbol{\tau}^S$ as $\beta_k = \boldsymbol{\tau}^S \boldsymbol{\Lambda}^{k\mathrm{T}}$.

As an example of the ESSRR measure, assume that X models a fault-tolerant system which can be either up or down. Then, if a reward rate 1 is assigned to the states in which the system is up and a reward rate 0 is assigned to the states in which the system is down, the ESSRR measure would be the steady-state availability of the fault-tolerant system.

34.2.2 Expected Cumulative Reward Till Exit of a Subset of States

In that measure, it is assumed that $\Omega = B \cup \{a\}$, where all states in B are transient and a is an absorbing state. The measure is defined as:

$$\mathrm{ECRTE} = E\left[\int_0^T r_{X(t)}\,dt\right]$$

where $T = \min\{t : X(t) = a\}$. In words, $\int_0^T r_{X(t)}\,dt$ is the random variable "reward earned by X until the time it leaves subset B" and ECRTE is its expectation. From a modeling point of view, the rewarded CTMC X can be seen as a CTMC keeping track of the behavior till exit from B of a larger rewarded CTMC Y actually modeling the system under study. Formally, X can be defined from Y as:

$$X(t) = \begin{cases} Y(t) & \text{if } Y(\tau) \in B, 0 \le \tau \le t \\ a & \text{otherwise} \end{cases}$$

The state transition diagram of X can be obtained from the state transition diagram of Y by eliminating the states not in B, adding the absorbing state a, and directing to a the transition rates from states in B to states outside B. The initial probability distribution of X is related to the initial probability distribution of Y by $\alpha_i = P[Y(0) = i], i \in B, \alpha_a = P[Y(0) \notin B]$ and X has associated with the states $i \in B$ the same reward rates as Y. Then, ECRTE is the expected value of the reward accumulated by Y till exit from B.

The ECRTE measure can be computed using $\mathrm{ECRTE} = \sum_{i \in B} r_i \tau_i$, where τ_i is the expected value of the time spent by X in i. Let the row vector $\boldsymbol{\tau}^B = (\tau_i)_{i \in B}$. Then, $\boldsymbol{\tau}^B$ is the solution of the linear system $\boldsymbol{\tau}^B \mathbf{A}^{B,B} = -\boldsymbol{\alpha}^B$, where $\mathbf{A}^{B,B}$ and $\boldsymbol{\alpha}^B$ are the restrictions of, respectively, $\mathbf{A}$ and $\boldsymbol{\alpha}$ to B. The condition that all states in B are transient is required for $\mathrm{ECRTE} < \infty$.

As an example of the ECRTE measure, assume that a CTMC Y models a fault-tolerant system which can be either up or down and that B is the subset of states of Y in which the system is up. Then, assigning a reward rate 1 to all states in B of the CTMC X keeping track of the behavior of Y until exit from B, ECRTE would be the mean time to failure (MTTF) of the system.

34.2.3 Expected Cumulative Reward During Stay in a Subset of States

Let B be a proper subset of Ω and assume that each trapping component of X has at least one state in B and one state outside B. Under those circumstances, X will switch indefinitely between the subsets B and $\Omega - B$. Let t_n be the time at which X makes its nth entry in B (by convention, if $X(0) \in B, t_1 = 0$) and let T_n be the time at which X makes its nth exit from B. The measure is defined as:

$$\mathrm{ECRDS} = \lim_{n \to \infty} E\left[\int_{t_n}^{T_n} r_{X(t)}\,dt\right]$$

In words, ECRDS is the expected value of the reward accumulated by X during its nth stay in the subset B as $n \to \infty$.

We discuss next how the ECRDS measure can be computed. Let E_n be the random variable "state through which X makes its nth entry in B" and let $q_i = \lim_{n \to \infty} P[E_n = i]$, $i \in B$. Let $C_1, C_2, \ldots, C_m$ be the trapping components of X and let $B_k = B \cap C_k$, $1 \le k \le m$. Let $p_{C_k} =$

$\lim_{t\to\infty} P[X(t) \in C_k]$, $1 \le k \le m$. Let q_i^k, $i \in B_k$, $1 \le k \le m$ be the limit for $n \to \infty$ of the probability that $E_n = i$ conditioned to $E_n \in B_k$; formally, $q_i^k = \lim_{n\to\infty} P[E_n = i \mid E_n \in B_k]$. Let $p_i^k = \lim_{t\to\infty} P[X(t) = i \mid X(t) \in C_k]$, $i \in C_k$, $1 \le k \le m$. The probabilities p_i^k, $i \in C_k$, $1 \le k \le m$ can be computed as described when dealing with the ESSRR measure. The probabilities q_i^k, $i \in B_k$, $1 \le k \le m$ can be computed from p_i^k, $i \in C_k$, $1 \le k \le m$ using:

$$q_i^k = \frac{\sum_{j\in C_k} p_j^k \lambda_{j,i}}{\sum_{l\in B_k} \sum_{j\in C_k} p_j^k \lambda_{j,l}}$$

Consider the rewarded CTMCs X_k, $1 \le k \le m$ keeping track of the behavior of X in B_k from entry in B_k with entry-state probability distribution q_i^k, $i \in B_k$. The rewarded CTMC X_k has state space $B_k \cup \{a\}$, the same transition rates among states in B_k as X, transition rates from $i \in B_k$ to a, $\lambda_{i,a}^k = \lambda_{i,C_k - B_k}$, the same reward rates in B_k as X, and initial probability distribution $P[X_k(0) = i] = q_i^k$, $i \in B_k$, $P[X_k(0) = a] = 0$. Let τ_i^k be the expected value of the time spent by X_k in $i \in B_k$. Note that all states in B_k of X_k are transient. Then, ECRDS can be computed from p_{C_k}, $1 \le k \le m$, and τ_i^k, $i \in B_k$, $1 \le k \le m$, using:

$$\mathrm{ECRDS} = \sum_{k=1}^{m} p_{C_k} \sum_{i\in B_k} r_i \tau_i^k$$

The row vectors $\boldsymbol{\tau}^k = (\tau_i^k)_{i\in B_k}$ are the solutions of the linear systems $\boldsymbol{\tau}^k \mathbf{A}^{B_k,B_k} = -\mathbf{q}^k$, where $\mathbf{A}^{B_k,B_k}$ is the restriction of the transition rate matrix of X_k to B_k and $\mathbf{q}^k$ is the row vector $(q_i^k)_{i\in B_k}$. The probabilities p_{C_k} and p_i^k, $i \in C_k$ can be computed as described when dealing with the ESSRR measure.

As an example of the ECRDS measure, assume that X models a repairable fault-tolerant system which can be either up or down, that B is the subset of up states, and that a reward rate 1 is assigned to the states in B. Then, the ECRDS measure would be the limiting expected duration of an up interval of the fault-tolerant system, *i.e.* the limit for $n \to \infty$ of the expected duration of the nth up interval.

34.2.4 Expected Transient Reward Rate

The measure is defined as:

$$\mathrm{ETRR}(t) = E[r_{X(t)}]$$

and can be computed in terms of the transient regime of X as:

$$\mathrm{ETRR}(t) = \sum_{i\in\Omega} r_i p_i(t)$$

As an example of the ETRR(t) measure, assume that X models a fault-tolerant system which can be either up or down, and that a reward rate 1 is assigned to the up states and a reward rate 0 is assigned to the down states. Then, ETRR(t) would be the availability of the system at time t.

34.2.5 Expected Averaged Reward Rate

The measure is defined as:

$$\mathrm{EARR}(t) = E\left[\frac{\int_0^t r_{X(\tau)}\,\mathrm{d}\tau}{t}\right]$$

and can be computed in terms of the transient regime of X as:

$$\mathrm{EARR}(t) = \frac{\sum_{i\in\Omega} r_i \int_0^t p_i(\tau)\,\mathrm{d}\tau}{t}$$

As an example of the EARR(t) measure, assume that X models a fault-tolerant system which can be either up or down, and that a reward rate 1 is assigned to the up states and a reward rate 0 is assigned to the down states. Then, EARR(t) would be the expected interval availability, *i.e.* the expected value of the fraction of time that the system is up in the time interval $[0, t]$.

34.2.6 Cumulative Reward Distribution Till Exit of a Subset of States

In that measure, it is assumed that $\Omega = B \cup \{a\}$, where a is an absorbing state and that each trapping component of X different from $\{a\}$ has some state i with $r_i > 0$. The measure is defined as:

$$\text{CRDTE}(s) = P\left[\int_0^T r_{X(t)}\,\text{d}t \le s\right]$$

where $T = \min\{t : X(t) = a\}$ or $T = \infty$ if $X(t) \in B$ for $t \in [0, \infty)$ and $s \ge 0$. From a modeling point of view, the rewarded CTMC X can be seen as a CTMC keeping track of the behavior till exit from B of a larger rewarded CTMC Y actually modeling the system under study. The relationships between X and Y are those discussed when dealing with the ECRTE measure.

Computation of the CRDTE(s) measure can be reduced to the transient analysis of a modified CTMC X^* [13]. The state transition diagram and initial probability distribution of the CTMC X^* are obtained from those of X by replacing the states $i \in B$ with $r_i = 0$ by instantaneous switches with jump probabilities to states j, $\gamma_{i,j} = \lambda_{i,j}/\lambda_i$, and by dividing the transition rates $\lambda_{i,j}$ from states i with $r_i > 0$ by r_i. The process of replacing states with null reward rate by instantaneous switches is analogous to eliminating vanishing markings in generalized stochastic Petri nets and several algorithms can be used [14]. Once X^* has been obtained, CRDTE(s) can be computed using CRDTE(s) $= P[X^*(s) = a]$.

Let $B^* \cup \{a\}$ be the state space of X^* and assume $B^* \neq \emptyset$. When not all the states in B^* are transient, CRDTE(s) can be computed by performing the transient analysis of a CTMC $X^{*\prime}$ of size smaller than X^*. The CTMC $X^{*\prime}$ can be constructed from X^*. Let $C_1, C_2, \ldots, C_m$ be the components of X^* different from $\{a\}$ not having a path in the components digraph of X^* to $\{a\}$. Let $C = \bigcup_{k=1}^{m} C_k$. Once X^* enters C, it will remain there forever and will never enter state a. Then, the CTMC $X^{*\prime}$ with state transition diagram and initial probability distribution obtained from those of X^* by deleting the states in C, adding an absorbing state b with initial probability equal to $P[X^*(0) \in C]$, and adding transition rates from states $i \in B^{*\prime} = B^* - C$ to b with rates $\lambda^{*\prime}_{i,b} = \lambda^*_{i,C}$, where $\lambda^*_{i,C}$ is the sum of the transition rates from i to the states in C in X^*, satisfies $P[X^{*\prime}(s) = a] = P[X^*(s) = a]$, and we can compute CRDTE(s) as $P[X^{*\prime}(s) = a]$.

As an example of the CRDTE(s) measure, assume that a CTMC Y models a fault-tolerant system which can be either up or down, and that B is the subset of states of Y in which the system is up. Then, assigning to the states in B of X a reward rate 1, CRDTE(t) would be the unreliability at time t (probability that the system has been failed in some time $\tau \in [0, t]$).

34.2.7 Cumulative Reward Distribution During Stay in a Subset of States

Let B be a proper subset of Ω and assume that each trapping component of X has at least one state in B and one state outside B. Under those circumstances, X will switch indefinitely between the subsets B and $\Omega - B$. Let t_n be the time at which X makes its nth entry in B (by convention, $t_1 = 0$ if $X(0) \in B$) and let T_n be the time at which X makes its nth exit from B. The measure is defined as:

$$\text{CRDDS}(s) = \lim_{n\to\infty} P\left[\int_{t_n}^{T_n} r_{X(t)}\,\text{d}t \le s\right]$$

with $s \ge 0$. In words, CRDDS(s) is the distribution of the reward accumulated by X during its nth stay in the subset B as $n \to \infty$.

We discuss next how CRDDS(s) can be computed. Let $C_1, C_2, \ldots, C_m$ be the trapping components of X and let $B_k = B \cap C_k$, $1 \le k \le m$. Let p_{C_k}, $1 \le k \le m$ and q_i^k, $i \in B_k$, $1 \le k \le m$ be defined as in the discussion of the ECRDS measure. Those quantities can be computed as described when dealing with the ECRDS measure. Consider the rewarded CTMCs X_k, $1 \le k \le m$, keeping track of the behavior of X in B_k from entry in B_k with entry-state probability distribution q_i^k, $i \in B_k$. The rewarded CTMC X_k has state space $B_k \cup \{a\}$, the same transition rates among states in B_k as X, transition rates from states $i \in B_k$ to a, $\lambda^k_{i,a} = \lambda_{i,C_k - B_k}$, the same reward rates in B_k as X, and initial probability distribution $P[X_k(0) = i] = q_i^k$,

$i \in B_k$, $P[X_k(0) = a] = 0$. Let $\text{CRDTE}_k(s)$ be the cumulative reward distribution of X_k till exit of B_k. Then, we have:

$$\text{CRDDS}(s) = \sum_{k=1}^{m} p_{C_k} \text{CRDTE}_k(s)$$

where $\text{CRDTE}_k(s)$ can be computed as described when dealing with the measure CRDTE(s), noting that, C_k being a trapping component of X and B_k a proper subset of C_k, all states of B_k will be transient in X_k.

34.2.8 Cumulative Reward Distribution

The measure is defined as:

$$\text{CRD}(t, s) = P\left[\int_0^t r_{X(\tau)}\, \mathrm{d}\tau \le s\right]$$

In words, CRD(t, s) is the probability that the reward earned by X up to time t is $\le s$. The measure can be seen as a particular instance of the generic performability measure where r_i has the meaning of "rate at which performance is accumulated", each possible value for the accumulated performance in the interval $[0, t]$ is an accomplishment level, and the measurable subset includes all accomplishment levels $\le s$.

The interval availability distribution IAVD(t, p), defined as the probability that the interval availability (*i.e.* the fraction of time that the system is up in the interval $[0, t]$) is $\le p$ can be seen as a particular instance of the CRD(t, s) measure when a reward rate 1 is assigned to the up states and a reward rate 0 is assigned to the down states. With that reward rate structure, CRD(t, s) is the probability that the up time during the interval $[0, t]$ is $\le s$ and, then, $\text{IAVD}(t, p) = \text{CRD}(t, pt)$.

34.2.9 Extended Reward Structures

With both reward rates $r_i \ge 0$ and impulse rewards $r_{i,j} \ge 0$ associated with transitions $i \to j$, the expected value of the reward accumulated during the time interval $[t, t + \Delta t]$, assuming $X(t) = i$, is $r_i \Delta t + \sum_{j \in \Omega - \{i\}} r_{i,j} \lambda_{i,j} \Delta t + O(\Delta t^2)$. This is because, assuming $X(t) = i$, the probability that X will remain in i during the whole interval is $1 - O(\Delta t)$, the probability that X will make a transition from i to j in the interval is $\lambda_{i,j} \Delta t + O(\Delta t^2)$, and the probability that X will make more than one transition in the interval is $O(\Delta t^2)$. This allows the computation of extended versions, including impulse rewards, of the measures ESSRR, ECRTE, ECRDS, ETRR(t), and EARR(t) as those measures with reward rates $r'_i = r_i + \sum_{j \in \Omega - \{i\}} r_{i,j} \lambda_{i,j}$. The extended measures would be formally defined as:

$$\text{ESSRR} = \lim_{t\to\infty} \lim_{\Delta t \to 0} \frac{E[\text{reward accumulated in } [t, t + \Delta t]]}{\Delta t}$$

$$\text{ECRTE} = E[\text{reward accumulated in } [0, T]]$$

$$\text{ECRDS} = \lim_{n\to\infty} E[\text{reward accumulated in } (t_n, T_n]]$$

$$\text{ETRR}(t) = \lim_{\Delta t \to 0} \frac{E[\text{reward accumulated in } [t, t + \Delta t]]}{\Delta t}$$

$$\text{EARR}(t) = E\left[\frac{\text{reward accumulated in } [0, t]}{t}\right]$$

where T, t_n, and T_n are defined as for the non-extended measures.

We have assumed $r_i \ge 0$, $i \in \Omega$ or $i \in B$, since this ensures the numerical stability of many solution methods. That restriction can, however, be circumvented for all eight reward measures defined in Sections 34.2.1–34.2.8 excepted the CRDTE(s) and CRDDS(s) measures by shifting, in case some r_i, $i \in \Omega$ or $i \in B$ is < 0, all reward rates by a positive amount d so that the new reward rates $r'_i = r_i + d$ be ≥ 0 for all $i \in \Omega$ and $i \in B$. The measures of the original rewarded CTMC are related to the measures of the rewarded CTMC with shifted reward rates, denoted by " $'$ ", by

$$\text{ESSRR} = \text{ESSRR}' - d,$$

$$\text{ECRTE} = \text{ECTRE}' - d\, \text{ECTTE},$$

$$\text{ECRDS} = \text{ECDRS}' - d\, \text{ECTDS},$$

$$\mathrm{ETRR}(t) = \mathrm{ETRR}'(t) - d,$$
$$\mathrm{EARR}(t) = \mathrm{EARR}'(t) - d,$$
$$\mathrm{CRD}(t, s) = \mathrm{CRD}'(t, s + dt),$$

where ECTTE is the ECRTE measure with reward rates $r_i'' = 1$, $i \in B$ and ECTDS is the ECRDS measure with reward rates $r_i'' = 1, i \in B$.

34.3 Model Specification

Direct specification of even medium-size CTMCs is both cumbersome and error-prone. To overcome that problem, most tools support more concise higher-level specifications from which CTMCs can be automatically derived. Some tools like SAVE [15] offer a modeling language specifically tailored to describe fault-tolerant systems. This provides maximum user-friendliness and makes explicit high-level knowledge about the model which can be exploited during model solution, but inevitably introduces restrictions making it difficult or even impossible to accommodate models other than those anticipated by the designers of the language. Formalisms from which arbitrary CTMCs can be derived include stochastic Petri nets, stochastic activity networks, stochastic process algebras, and production rule-based specifications. DSPNexpress [16], GreatSPN [17], SPNP [18], SURF-2 [19], and TimeNET [20] are examples of tools in which model specification is done using stochastic Petri nets. UltraSAN [21] supports model specification through stochastic activity networks. The PEPA Workbench [22] and TIPPtool [23] use stochastic process algebras as model specification formalism. METFAC-2.1 and TANGRAM [24] offer production rule-based modeling languages. The basic model specification formalism is, in some cases, too simple to accommodate easily complex model features and, for the sake of user-friendliness, some tools allow extensions such as enabling functions in generalized stochastic Petri nets. Another important feature which most tools incorporate is parametric model specification. Parameters may affect the "structure" of the state transition diagram, the values of the transition rates, the initial probability distribution, and the reward structure of the CTMC, and often allow a compact specification of models for a more or less wide "class" of fault-tolerant systems.

To illustrate model specification techniques, we will review the production rule-based modeling language which will be offered by METFAC-2.1 using a parametric model of a Level 5 RAID subsystem. The subsystem includes eight disks, two redundant disk controllers, and two redundant power supplies. The information stored in the disks is organized into groups of eight blocks, of which seven blocks contain data bits and one block contains parity bits. Each block of a group is stored in a different disk. Since parity blocks are accessed more often than data blocks, parity blocks are distributed evenly among the disks to achieve load balancing and maximum efficiency for the RAID subsystem. The use of parity blocks allows the system to continue operation without loss of data in the event of a disk failure. The failure of a second disk would take the system down and would involve loss of data. After a failed disk is repaired with the subsystem up, a reconstruction process generates the blocks which have to be stored in the repaired disk to have consistent block groups. Failure of a disk different from the disk under reconstruction takes the subsystem down. Disks fail with rate λ_{D} when the RAID has no disk under reconstruction and with rate λ_{DR} when the RAID has a disk under reconstruction. Since the load of the disks is higher when the RAID has a disk under reconstruction, one should expect $\lambda_{\mathrm{DR}} > \lambda_{\mathrm{D}}$. Controllers fail with rate λ_{C2} when both controllers are unfailed and with rate λ_{C1} when only one controller is unfailed. Typically, the access requests would be distributed between the controllers when both are unfailed, implying that they would have a lower activity than a single unfailed controller and, then, one should expect $\lambda_{\mathrm{C1}} > \lambda_{\mathrm{C2}}$. The power supplies work in cold standby redundancy, with the cold spare having null failure rate. The active power supply fails with rate λ_{P}. Controller and power supply failures are covered with probabilities C_{C} and C_{P}, respectively. An uncovered controller or power supply failure takes the subsystem down.

The reconstruction process is assumed to have a duration with an exponential distribution with parameter μ_{DR}. Components do not fail when the subsystem is down. When the subsystem is up, failed components are repaired by an unlimited number of repairmen with rate μ_U. A down system is brought to a fully operational state with no component failed and no disk under reconstruction with rate μ_D.

In METFAC-2.1, a model specification is given in a file called *name*.spec, where *name* is a string identifying the model. That model specification may invoke external model-specific C functions. In addition, other C functions with predefined names and prototypes may have to be provided to specify characteristics of the measure to be computed (the B subset in the ECRTE, ECRDS, CRDTE(s), and CRDDS(s) measures), or other uses (for instance, to check assertions on the state descriptions of the generated states, or to provide information required by some model solution methods). All these functions have to be included in an optional *name*.c file. Figure 34.2 shows the contents of a model specification file from which a rewarded CTMC appropriate to evaluate the steady-state unavailability of the RAID subsystem previously described using the generic measure ESSRR can be generated. The contents of that file have to follow a language with C-like syntax. The language is case-sensitive and, as in C, comments are enclosed by /* and */. The model specification starts with an optional declaration of the parameters of the model. Model parameters can be of two types: `double` and `int`. In the example, all parameters have type `double`. The next syntactic construction is a mandatory declaration of the state variables of the model. All state variables have type `int`. The set of state variables have to provide together a detailed enough description of the state of the model to allow an unambiguous specification of the initial probability distribution, transition rates, and reward rates of the CTMC. Five state variables have been used in the example. The state variable `DF` takes the value `yes` (implicitly defined as 1) when the RAID subsystem is up and one disk is failed and the value `no` (implicitly defined as 0) otherwise. The other state variables, `DR`, `CF`, `PF`, and `DOWN`, identify states in which, respectively, the subsystem is up and one disk is under reconstruction, the subsystem is up and one controller is failed, the subsystem is up and one power supply is failed, and the subsystem is down.

The use of external C functions enhances the flexibility of the modeling language. Those functions may be of type `double` or `int`, may only include `double` or `int` parameters, and have to be declared by an optional `external` construct which follows the declaration of the state variables of the model. The example uses a `double` function `rew_essrr()` with an `int` parameter which is called to compute the reward rates to be associated with the states of the CTMC.

The core of the model specification is a set of production rules that follow the declarative section and starts with the keyword `production_rules`. Those production rules determine how the state of the CTMC may change and with which rates. There are two types of production rules: *simple* and *with responses*. A simple production rule describes a simple action which may change the state of the CTMC and includes an optional condition, the keyword `action`, an optional name, a rate specification, and a state change description. The condition determines whether the action is active or not in a particular state (the action is active if the condition evaluates to a value different from 0 and is inactive otherwise). If the action description does not include any condition, the action is active in any state. For instance, the action described by the first production rule of the example models the failure of one disk when no disk is either failed or under reconstruction. That action is active when all `DOWN`, `DF`, and `DR` have value 0 (`no`), has name `DFAIL_NFDR`, occurs with rate $8\lambda_D$, and leads to a state which differs from the current one only in that the state variable `DF` has value 1 (`yes`). Production rules with responses describe actions with responses and include an optional action condition, the keyword `action`, an optional action name, a rate specification, and a set of responses, each with an optional response condition, the keyword `response`, an optional name,

```
parameters double
   LD,  /* disk failure rate when the RAID has no disk under reconstruction */
   LDR, /* disk failure rate when the RAID has one disk under reconstruction */
   LC2, /* controller failure rate when both controllers are unfailed */
   LC1, /* controller failure rate when only one controller is unfailed */
   LP,  /* power supply failure rate */
   CC,  /* coverage to controller failures */
   CP,  /* coverage to power supply failures */
   MDR, /* disk reconstruction rate */
   MU,  /* repair rate when RAID up */
   MD   /* repair rate when RAID down */

state_variables
   DF,   /* yes if RAID up and one disk is failed */
   DR,   /* yes if RAID up and one disk is under reconstruction */
   CF,   /* yes if RAID up and one controller is failed */
   PF,   /* yes if RAID up and one power supply is failed */
   DOWN  /* yes if RAID down */

external double rew_essrr(int)  /* 1 for down states, 0 for up states */

production_rules

/* Disk failure when no disk is either failed or under reconstruction */
if !DOWN && !DF && !DR action DFAIL_NFDR with_rate 8*LD new_state DF=yes

/* Disk failure when one disk is failed */
if DF action DFAIL_DF with_rate 7*LD new_state DF=no, DR=no, CF=no, PF=no, DOWN=yes

/* Disk failure when one disk is under reconstruction */
if DR action DFAIL_DR1 with_rate 7*LDR new_state DF=no, DR=no, CF=no, PF=no, DOWN=yes
if DR action DFAIL_DR2 with_rate LDR new_state DF=yes, DR=no

/* Controller failure when no controller is failed */
if !DOWN && !CF action CFAIL_NCF with_rate 2*LC2
   response COVERED with_prob CC new_state CF=yes
   response UNCOVERED with_prob 1-CC new_state DF=no, DR=no, CF=no, PF=no, DOWN=yes
end

/* Controller failure when one controller is failed */
if CF action CFAIL_CF with_rate LC1 new_state DF=no, DR=no, CF=no, PF=no, DOWN=yes

/* Power supply failure when no power supply is failed */
if !DOWN && !PF action PFAIL_NPF with_rate LP
   response COVERED with_prob CP new_state PF=yes
   response UNCOVERED with_prob 1-CP new_state DF=no, DR=no, CF=no, PF=no, DOWN=yes
end

/* Power supply failure when one power supply is failed */
if PF action PFAIL_PF with_rate LP new_state DF=no, DR=no, CF=no, PF=no, DOWN=yes

/* Disk repair */
if DF action DREP with_rate MU new_state DF=no, DR=yes

/* End of disk reconstruction */
if DR action DREC with_rate MDR new_state DR=no

/* Controller repair */
if CF action CREP with_rate MU new_state CF=no

/* Power supply repair */
if PF action PREP with_rate MU new_state PF=no

/* Global repair when system down */
if DOWN action GREP with_rate MD new_state DOWN=no

start_state DF=no, DR=no, CF=no, PF=no, DOWN=no

reward_rate rew_essrr(DOWN)
```

Figure 34.2. Model specification appropriate for the computation of the steady-state unavailability of the RAID subsystem using the ESSRR measure

an optional probability specification, and a state change description. Each response describes a way in which the state may change. Assuming that both the action and the response are active, such state change occurs at a rate given by the product of the action rate and the response probability (with a default value 1 if the response does not include a probability specification). The fifth production rule of the example illustrates the syntax of a production rule with responses. It models the failure of a controller when no controller is failed. That action is active only when the subsystem is up and no controller is failed and occurs at rate $2\lambda_{C2}$. With probability C_C, the failure of the controller is covered. This is modeled by the first response. With probability $1 - C_C$, the failure of the controller is not covered and takes the system down. This is modeled by the second response.

After the production rules, the model specification includes a `start_state` construct to define a "start" state. The CTMC is generated by applying all production rules to the "start" state and all generated states. The model specification ends with a `reward_rate` construct to specify the reward rate structure and an optional `initial_probability` construct to specify the initial probability distribution of the CTMC. The expressions given in those constructs are evaluated to obtain the reward rates and initial probabilities associated with the states of the CTMC. If the `initial_probability` construct is absent, a default initial probability distribution with probability 1 for the "start" state and probability 0 for the remaining states is used. In the example, the reward rate associated with a state is the value returned by the call `rew_cssrr(DOWN)`, which is 1.0 for the states with state variable DOWN equal to 1 (`yes`) and 0.0 for the states with state variable DOWN equal to 0 (`no`). This makes the generic measure ESSRR equal to the steady-state unavailability of the RAID subsystem. The specification of the model does not include any `initial_probability` construct and the default initial probability distribution is assigned to the CTMC. This is appropriate since the generated CTMC is irreducible, the ESSRR measure does not depend on the initial probability distribution of the CTMC, and any initial probability distribution is good enough.

Table 34.1. Descriptions of the states of the CTMC obtained from the model specification of Figure 34.2

State	DF	DR	CF	PF	DOWN
FOP	no	no	no	no	no
C	no	no	yes	no	no
D	yes	no	no	no	no
P	no	no	no	yes	no
R	no	yes	no	no	no
CD	yes	no	yes	no	no
CP	no	no	yes	yes	no
CR	no	yes	yes	no	no
DP	yes	no	no	yes	no
PR	no	yes	no	yes	no
CDP	yes	no	yes	yes	no
CPR	no	yes	yes	yes	no
DOWN	no	no	no	no	yes

Table 34.1 describes the states of the CTMC generated from the model specification of the RAID subsystem. Figure 34.3 gives the state transition diagram of the generated CTMC.

In METFAC-2.1, model specifications have to be "built" before CTMCs can be generated and solved. Building a model specification involves, first, preprocessing the model specification file to obtain a model-specific C source file and, after that, compiling that C source file and, if present, the file *name*.c, and linking the obtained object files with model-independent object files to obtain a model-specific executable file *name*.exe. All these steps are performed automatically by invoking an appropriate command. The CTMCs are generated and solved by running the executable *name*.exe. This approach makes model generation extremely efficient (for instance, the aggregated CTMC for the computation of the steady-state unavailability of the system with eight RAID subsystems described in Section 34.6, which has 125 970 states and 2 670 564 transitions was generated in a 167-MHz, 128-MB UltraSPARC1 workstation in 69 seconds).

34.4 Model Solution

Section 34.2 has reduced the computation of all measures defined there but the measure CRD(t, s)

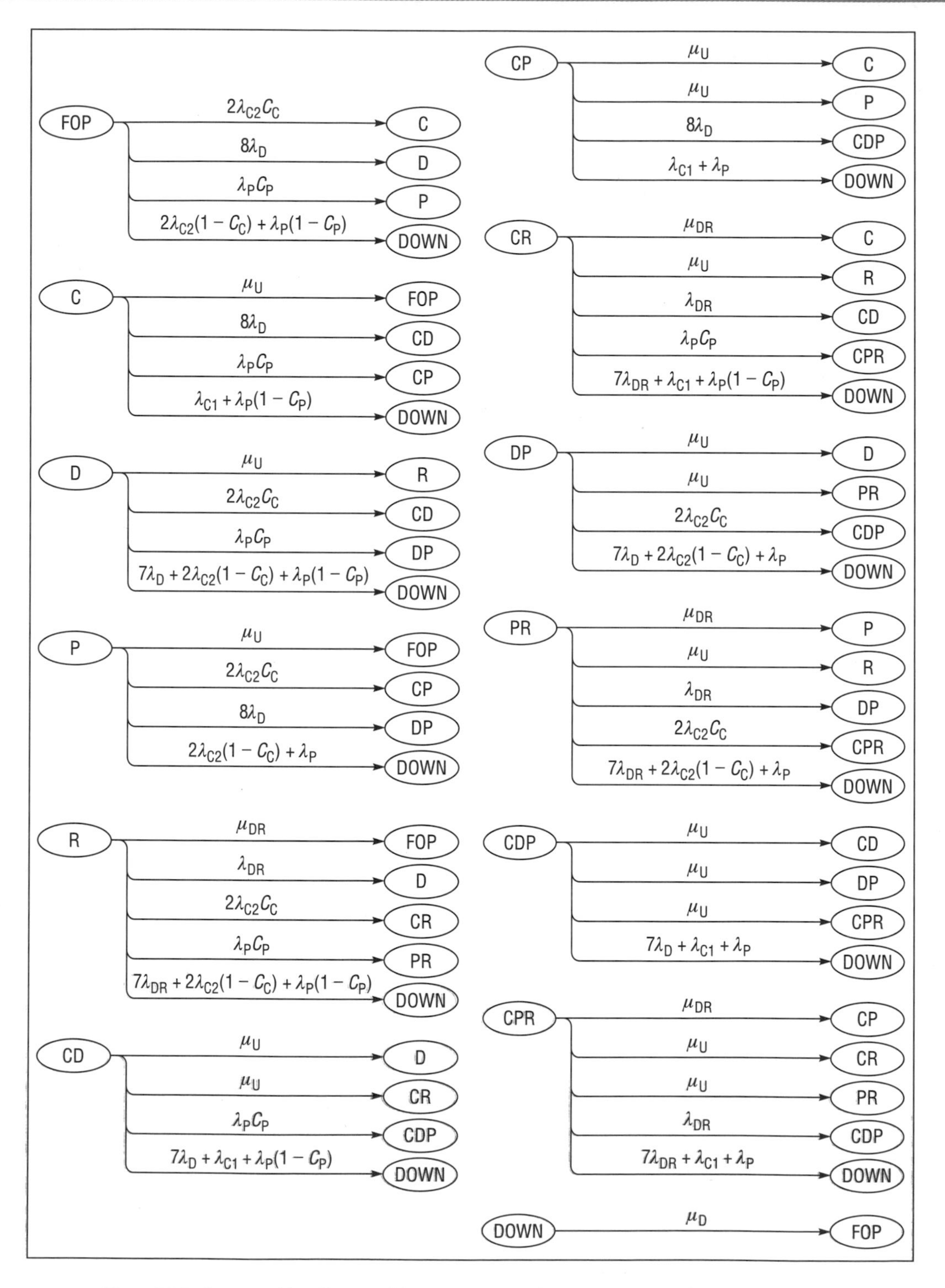

Figure 34.3. State transition diagram of the CTMC obtained from the model specification of Figure 34.2

to the solution of the following three numerical problems: (1) computation of the normalized solution ($\mathbf{p}\mathbf{1}^T = 1$) of the linear system

$$\mathbf{p}\mathbf{A}^{C,C} = \mathbf{0} \qquad (34.1)$$

where $\mathbf{A}^{C,C}$ is the restriction of the transition rate matrix of a CTMC to a trapping component C, $\mathbf{p}$ is a row vector and $\mathbf{0}$ and $\mathbf{1}$ are row vectors with all components equal to, respectively, 0 and 1; (2) computation of the solution of the linear system

$$\boldsymbol{\tau}\mathbf{A}^{S,S} = -\boldsymbol{\alpha} \qquad (34.2)$$

where $\mathbf{A}^{S,S}$ is the restriction of the transition rate matrix of a CTMC to a subset of transient states S, $\boldsymbol{\tau}$ is a row vector and $\boldsymbol{\alpha}$ is a row vector with components $\alpha_i \geq 0$, $i \in S$; and (3) computation of the transient probabilities $p_i(t) = P[X(t) = i]$, or their integrals, of a CTMC X. In this section we start by discussing available numerical methods to solve these problems. We will start by considering the first two numerical problems, then discuss available numerical methods to solve the third problem, and, finally, review briefly numerical methods to compute the CRD(t, s) measure, which are much more expensive.

The matrix $\mathbf{A}^{S,S}$ can be reducible. This occurs when the CTMC X^S having state transition diagram equal to the state transition diagram of the underlying CTMC restricted to S has several components. Let $C_1, C_2, \ldots, C_m$ be those components, and let $\mathbf{A}_{k,l}$, $1 \leq k, l \leq m$ be the block of $\mathbf{A}^{S,S} = (a_{i,j})_{i,j \in S}$ including the elements $a_{i,j}$, $i \in C_k$, $j \in C_l$. A topological sorting of the components in the components digraph of X^S puts $\mathbf{A}^{S,S}$ in upper block triangular form:

$$\mathbf{A}^{S,S} = \begin{pmatrix} \mathbf{A}_{1,1} & \mathbf{A}_{1,2} & \mathbf{A}_{1,3} & \cdots & \mathbf{A}_{1,m} \\ \mathbf{0} & \mathbf{A}_{2,2} & \mathbf{A}_{2,3} & \cdots & \mathbf{A}_{2,m} \\ \mathbf{0} & \mathbf{0} & \mathbf{A}_{3,3} & \cdots & \mathbf{A}_{3,m} \\ \vdots & \vdots & \vdots & \ddots & \vdots \\ \mathbf{0} & \mathbf{0} & \mathbf{0} & \cdots & \mathbf{A}_{m,m} \end{pmatrix}$$

where the components have been renamed so that the topological sorting of the components is $C_1, C_2, \ldots, C_m$. That upper block triangular form can be used to reduce the solution of the linear system (34.2) to the solution of similar linear systems with irreducible matrices $\mathbf{A}_{k,k}$, $1 \leq k \leq m$. Calling $\boldsymbol{\tau}_k$ and $\boldsymbol{\alpha}_k$ the restrictions of the vectors $\boldsymbol{\tau}$ and $\boldsymbol{\alpha}$ to the subset of states C_k, we have:

$$\begin{aligned} \boldsymbol{\tau}_1\mathbf{A}_{1,1} &= -\boldsymbol{\alpha}_1 \\ \boldsymbol{\tau}_2\mathbf{A}_{2,2} &= -\boldsymbol{\alpha}_2 - \boldsymbol{\tau}_1\mathbf{A}_{1,2} \\ \boldsymbol{\tau}_3\mathbf{A}_{3,3} &= -\boldsymbol{\alpha}_3 - \boldsymbol{\tau}_1\mathbf{A}_{1,3} - \boldsymbol{\tau}_2\mathbf{A}_{2,3} \\ &\vdots \\ \boldsymbol{\tau}_m\mathbf{A}_{m,m} &= -\boldsymbol{\alpha}_m - \boldsymbol{\tau}_1\mathbf{A}_{1,m} - \boldsymbol{\tau}_2\mathbf{A}_{2,m} \\ &\quad - \boldsymbol{\tau}_3\mathbf{A}_{3,m} - \cdots - \boldsymbol{\tau}_{m-1}\mathbf{A}_{m-1,m} \end{aligned}$$

This reduction is beneficial from a computational point of view and should be done whenever possible. Thus, in the following, we will assume $\mathbf{A}^{S,S}$ irreducible.

The linear systems 34.1 and 34.2 can be solved by either direct, *e.g.* Gaussian elimination, or iterative methods. Non-sparse implementations of direct methods are expensive when the matrix is large, since they require $O(n^3)$ time and $O(n^2)$ memory, n being the dimension of the matrix. Sparse implementations of direct methods can be significantly less costly, but are complex and very often result in excessive fill-in. In that case, available memory may rapidly be exhausted. Iterative methods, on the other hand, can be implemented without modifying the matrix and have minimum memory requirements, allowing the analysis of CTMCs with very large state spaces. For that reason, iterative methods have been traditionally preferred to direct methods. However, non-convergence or poor convergence rate are potential problems of iterative methods. Stewart [25] is an excellent recent source for both direct and iterative methods applied to the solution of CTMCs.

In order to avoid divisions, which tend to be expensive, it is convenient to transform the linear systems 34.1 and 34.2 into:

$$\mathbf{P}^{C,C}\mathbf{x} = \mathbf{0}^{\mathrm{T}} \qquad (34.3)$$

$$\mathbf{P}^{S,S}\mathbf{y} = -\boldsymbol{\alpha}^{\mathrm{T}} \qquad (34.4)$$

where $\mathbf{x}$ and $\mathbf{y}$ are column vectors, $\mathbf{P}^{C,C} = \mathbf{A}^{C,C\,\mathrm{T}}\mathrm{diag}(\mathbf{A}^{C,C})^{-1}$, and $\mathbf{P}^{S,S} =$

$\mathbf{A}^{S,S\,\mathrm{T}}\mathrm{diag}(\mathbf{A}^{S,S})^{-1}$, $\mathbf{B}^{\mathrm{T}}$ denoting the transpose of matrix $\mathbf{B}$ and $\mathrm{diag}(\mathbf{B})$ denoting the matrix with diagonal elements equal to the diagonal elements of $\mathbf{B}$ and null off-diagonal elements, and apply the numerical methods to those linear systems. The normalized solution of System 34.1 can be obtained from a non-null solution of System 34.3 using:

$$\mathbf{p}^{\mathrm{T}} = \mathrm{diag}(\mathbf{A}^{C,C})^{-1}\mathbf{x}/\|\mathrm{diag}(\mathbf{A}^{C,C})^{-1}\mathbf{x}\|_1$$

The solution of System 34.2 can be obtained from the solution of System 34.4 using:

$$\boldsymbol{\tau}^{\mathrm{T}} = \mathrm{diag}(\mathbf{A}^{S,S})^{-1}\mathbf{y}$$

Basic iterative methods include Jacobi, Gauss–Seidel, and successive overrelaxation (SOR). To describe those methods, consider a linear system $\mathbf{Bz} = \mathbf{b}$, where $\mathbf{z}$ and $\mathbf{b}$ are column vectors. Let $\mathbf{B}$ be partitioned as $\mathbf{D} - \mathbf{L} - \mathbf{U}$, where $\mathbf{D}$ is diagonal, $\mathbf{L}$ is strictly lower triangular, and $\mathbf{U}$ is strictly upper triangular. Basic iterative methods can be described as:

$$\mathbf{z}^{(k+1)} = \mathbf{H}\mathbf{z}^{(k)} + \mathbf{M}^{-1}\mathbf{b}$$

where the iterates $\mathbf{z}^{(k)}$ should converge to the solution of the linear system. The Jacobi method is defined by $\mathbf{H} = \mathbf{D}^{-1}(\mathbf{L} + \mathbf{U})$ and $\mathbf{M} = \mathbf{D}$. The Gauss–Seidel method is defined by $\mathbf{H} = (\mathbf{D} - \mathbf{L})^{-1}\mathbf{U}$ and $\mathbf{M} = \mathbf{D} - \mathbf{L}$. The SOR method is defined by $\mathbf{H} = (\mathbf{D} - \omega\mathbf{L})^{-1}[(1 - \omega)\mathbf{D} + \omega\mathbf{U}]$ and $\mathbf{M} = \mathbf{D}/\omega - \mathbf{L}$, where ω is the relaxation parameter. Note that for $\omega = 1$, SOR reduces to Gauss–Seidel. The following results [25–29] are known about the convergence of these methods when applied to the linear systems 34.3 and 34.4.

1. For the linear system 34.3: (a) Gauss–Seidel converges if $\mathbf{A}^{C,C}$ has a non-null subdiagonal element in all rows except the first one, and (b) SOR can only converge for $0 < \omega < 2$ and converges for $0 < \omega < 1$.
2. For the linear system 34.4: (a) Jacobi and Gauss–Seidel converge, (b) Gauss–Seidel converges asymptotically faster than Jacobi, (c) SOR can only converge for $0 < \omega < 2$ and converges for $0 < \omega \le 1$, and (d) for $0 < \omega \le 1$ SOR cannot converge asymptotically faster as ω decreases.

The condition which guarantees the convergence of Gauss–Seidel for the linear system 34.3 can easily be achieved by sorting the states in C as visited by a breadth-first traversal, following incoming arcs, of the restriction of the state transition diagram of the underlying CTMC to C, starting at any state. Usually, Gauss–Seidel will converge faster than Jacobi for the linear system 34.3. Also, SOR with a suitable selection of the value of the relaxation parameter can converge significantly faster than Gauss–Seidel. Except for matrices having special structures, there exists no theory supporting an optimal selection of ω. According to the available results, the search for the optimum ω has to be restricted to the interval $(0, 2)$ for the linear system 34.3 and to the interval $[1, 2)$ for the linear system 34.4. An apparently efficient and robust optimized SOR method is described by Suñé *et al.* [28]. That reference also compares the performance of basic iterative methods, some block iterative methods, and a projection method. Basic iterative methods may exhibit an extremely poor convergence rate for the linear system 34.4 for failure/repair CTMC models. However, an acceleration technique which works very well for those models has recently been developed [30].

Computation of the transient probabilities of a CTMC can be done by using either ODE (ordinary differential equation) solvers or randomization (also called uniformization). Good reviews of those methods with new results can be found in Malhotra *et al.* [31], Malhotra [32], and Reibman and Trivedi [33]. Letting the row vector $\mathbf{p}(t) = (p_i(t))_{i\in\Omega}$, $\mathbf{p}(t)$ satisfies the ordinary differential equation:

$$\frac{\mathrm{d}\mathbf{p}}{\mathrm{d}t} = \mathbf{p}(t)\mathbf{A}$$

where $\mathbf{A}$ is the transition rate matrix of the CTMC. This allows the use of ODE solvers to compute the transient probabilities $p_i(t)$, $i \in \Omega$. The performance of ODE solvers is, however, severely affected by the stiffness of the model. A practical measure of stiffness is $\max_{i\in\Omega} \lambda_i t$ [33]. Standard (non-stiff) ODE solvers require a very large number of steps when the model is stiff (has a large $\max_{i\in\Omega} \lambda_i t$ value). Stiff ODE solvers perform much better in those cases. Each step of a stiff ODE solver requires the solution of linear systems with matrices having the same non-null pattern as **A**. Typically, those linear systems are solved using an iterative method, *e.g.* Gauss–Seidel. For large $\max_{i\in\Omega} \lambda_i t$, the resulting number of matrix–vector multiplications is typically of the order of several thousands, and even stiff ODE solvers are expensive for large, stiff CTMCs.

The randomization method is attractive because it is numerically stable and the computation error is well-controlled and can be specified in advance. The randomization method is based on the following result [34, theorem 4.19]. Consider any $\Lambda \geq \max_{i\in\Omega} \lambda_i$ and define the randomized homogeneous discrete time Markov chain (DTMC) $\hat{X} = \{\hat{X}_k;\ k = 0, 1, 2, \ldots\}$ with the same state space and initial probability distribution as X and transition probabilities $P[\hat{X}_{k+1} = j \mid \hat{X}_k = i] = P_{i,j} = \lambda_{i,j}/\Lambda$, $i \neq j$, $P[\hat{X}_{k+1} = i \mid \hat{X}_k = i] = P_{i,i} = 1 - \lambda_i/\Lambda$. Let $Q = \{Q(t);\ t \geq 0\}$ be a Poisson process with arrival rate Λ independent of $\hat{X}$. Then, $X = \{X(t);\ t \geq 0\}$ is probabilistically identical to $\{\hat{X}_{Q(t)};\ t \geq 0\}$. We call this the "randomization result". The performance of the randomization method degrades as Λ increases and, for this reason, Λ is usually taken equal to $\max_{i\in\Omega} \lambda_i$. Let $\mathbf{q}(k) = (P[\hat{X}_k = i])_{i\in\Omega}$ be the probability row vector of $\hat{X}$ at step k. Then, using the randomization result, conditioning on the number of Poisson arrivals by time t, $Q(t)$, and using $P[Q(t) = k] = e^{-\Lambda t}(\Lambda t)^k/k!$, we can express $\mathbf{p}(t)$ as:

$$\mathbf{p}(t) = \sum_{k=0}^{\infty} \mathbf{q}(k)\, e^{-\Lambda t} \frac{(\Lambda t)^k}{k!} \tag{34.5}$$

The row vectors $\mathbf{q}(k)$ can be obtained using:

$$\mathbf{q}(k+1) = \mathbf{q}(k)\mathbf{P}$$

where $\mathbf{P} = (P_{i,j})_{i,j\in\Omega}$ is the transition probability matrix of $\hat{X}$. In a practical implementation of the randomization method, an approximated value for $\mathbf{p}(t)$ can be obtained by truncating the infinite series of Equation 34.5:

$$\mathbf{p}^a(t) = \sum_{k=0}^{N} \mathbf{q}(k)\, e^{-\Lambda t} \frac{(\Lambda t)^k}{k!}$$

The truncation point N can be chosen using a suitable error criterion, *e.g.* imposing $\|\mathbf{p}(t)^{\mathrm{T}} - \mathbf{p}^a(t)^{\mathrm{T}}\|_1 \leq \varepsilon$, where ε is a small enough quantity. Since $\|\mathbf{p}(t)^{\mathrm{T}} - \mathbf{p}^a(t)^{\mathrm{T}}\|_1 \leq \sum_{k=N+1}^{\infty} e^{-\Lambda t}(\Lambda t)^k/k!$, N can be chosen as:

$$N = \min\left\{m \geq 0 : \sum_{k=m+1}^{\infty} e^{-\Lambda t} \frac{(\Lambda t)^k}{k!} \leq \varepsilon\right\}$$

Stable and efficient computation of the Poisson probabilities $e^{-\Lambda t}(\Lambda t)^k/k!$ avoiding overflows and intermediate underflows is a delicate issue, and several alternatives have been proposed [35–38]. For large Λt, *i.e.* for stiff CTMCs, the required truncation point N is $\approx \Lambda t$ and, then, the randomization method will be expensive if the model is large. Although we have reviewed the randomization method for the computation of $\mathbf{p}(t)$, the method can easily be adapted to compute with well-controlled error the ETRR(t) and EARR(t) measures (see, for instance, Carrasco [39]).

Several variants of the (standard) randomization method have been proposed to improve its efficiency. Miller [40] has used selective randomization to solve reliability models with detailed representation of error handling activities. The idea behind selective randomization [41] is to randomize the model only in a subset of the state space. Reibman and Trivedi [33] have proposed an approach based on the multistep concept. The idea is to compute $\mathbf{P}^M$ explicitly, where M is the length of the multistep, and use the recurrence $\mathbf{q}(k+M) = \mathbf{q}(k)\mathbf{P}^M$ to advance $\hat{X}$ faster for steps which have negligible contributions to the transient solution of X. Since, for large Λt, the number of $\mathbf{q}(k)$s with significant contributions is of the order of $\sqrt{\Lambda t}$, the multistep concept allows a significant reduction in the required number of vector–matrix multiplications. However, when computing $\mathbf{P}^M$, significant fill-in can occur if $\mathbf{P}$ is sparse. Adaptive uniformization [42] is a recent method in which the randomization rate is adapted depending on the states in which the randomized DTMC can be at a given step. Numerical experiments have shown that adaptive uniformization can be faster than standard randomization for short to medium mission times. In addition, it can be used to solve models with infinite state spaces and not uniformly bounded output rates. Recently, the combination of adaptive and standard uniformization has been proposed to obtain a method which is faster for most models [38]. Another proposal to speed up the randomization method is steady-state detection [31]. Recently, a method based on steady-state detection which gives error bounds has been developed [43]. Steady-state detection is useful for models which reach their steady state before the largest time at which the measure has to be computed. Regenerative randomization [39, 44] is another randomization-like method recently proposed. The method requires the selection of a "regenerative" state and covers rewarded CTMC models with state space $\Omega = S \cup \{f_1, f_2, \ldots, f_A\}$, $|S| \geq 2$, $A \geq 0$, where f_i are absorbing states and either all states in S are transient or S has a single trapping component and the chosen regenerative state belongs to that component, having an initial probability distribution with $P[X(0) \in S] > 0$, and such that all states are readable. The regenerative state has to belong to S. A variant of that method, called bounding regenerative randomization [45], allows the computation of bounds for reliability-like measures. For a class of models, including typical failure/repair models, a natural selection for the regenerative state exists and, with that selection, the bounding regenerative randomization method is fast and seems to obtain tight bounds.

Available numerical methods [46–54] for the computation of the $\mathrm{CRD}(t, s)$ measure are expensive and, when the model is stiff, their applicability is limited to CTMCs of small size. Most of those methods [46, 48, 50, 51, 53, 54] are based on the randomization result. Some of them [50, 51, 54] allow both reward rates and impulse rewards. The methods described in Nabli and Sericola [48] and Qureshi and Sanders [51] have been proved to be numerically stable. As discussed in Section 34.2, the interval availability distribution $\mathrm{IAVD}(t, p)$ measure can be seen as a particular case of the $\mathrm{CRD}(t, s)$ measure and special numerical methods have been developed to compute it [55–58]. Many of those methods are based on the randomization result and are particularizations of methods for the computation of $\mathrm{CRD}(t, s)$.

34.5 The Largeness Problem

The modeling of fault-tolerant systems of even moderate complexity typically requires large CTMCs. Furthermore, the size of the CTMC tends to grow fast with the complexity of the system. That phenomenon is known as the largeness or state space explosion problem. Approaches to deal with the largeness problem include state aggregation, bounding methods, and "on-the-fly" solution techniques. In state aggregation, symmetries of the system are exploited to generate CTMC models of reduced size from a suitable model specification formalism making explicit those symmetries (see, for instance, Chiola *et al.* [59] and Sanders and Meyer [60]). Bounding methods compute bounds for the measure of interest using detailed knowledge of the CTMC model in a subset of states. For the bounds to be tight the subset should include the "more likely" states of the model. On-the-fly solution techniques (see Deavours and Sanders [61] and references cited therein) reduce memory requirements by avoiding the storage of the transition rate matrix of the CTMC. All three approaches can be combined. In the following, we will review bounding methods. X will denote the "exact" CTMC.

The measures $\mathrm{ETRR}(t)$, $\mathrm{EARR}(t)$, and $\mathrm{CRD}(t, s)$ are easy to bound. Let G be a proper subset of the state space Ω of the rewarded CTMC X. Let X_{lb} be the rewarded CTMC with state transition diagram obtained from that of X by deleting the states in $\Omega - G$, adding an absorbing state a and directing to a the transition rates of X from states $i \in G$ to $\Omega - G$; having an initial probability distribution in G equal to that of X and an initial probability in a equal to $P[X(0) \in \Omega - G]$; and having a reward rate structure with the same reward rates as X for the states in G and a reward rate $r_{\mathrm{lb}} \leq \min_{i \in \Omega} r_i$ for state a. Let X_{ub} be the rewarded CTMC differing from X_{lb} only in that a reward rate $r_{ub} \geq \max_{i\in\Omega} r_i$ is assigned to the absorbing state a. Then, the ETRR(t) and EARR(t) measures for X_{lb} lower bound the corresponding measures for X and the ETRR(t) and EARR(t) measures for X_{ub} upper bound the corresponding measures for X. Also, the CRD(t, s) measure for X_{ub} lower bounds the CRD(t, s) measure for X, and the CRD(t, s) measure for X_{lb} upper bounds the CRD(t, s) measure for X.

Bounding the CRDTE(s) measure is slightly more complex. Let G be a proper subset of B (X has state space $B \cup \{a\}$, where a is an absorbing state). Let X_{lb} be the CTMC with state transition diagram obtained from that of X by deleting the states in $B - G$, adding an absorbing state b and directing to b the transition rates of X from states $i \in G$ to $B - G$; having an initial probability distribution in $G \cup \{a\}$ identical to that of X and an initial probability in b equal to $P[X(0) \in B - G]$; and having a reward rate structure over $G \cup \{b\}$ with the same reward rates as X for the states in G and any reward rate > 0 for state b. Then, the CRDTE(s) measure for X_{lb} with subset "B" equal to $G \cup \{b\}$ lower bounds the CRDTE(s) measure for X. Let X_{ub} be the CTMC with state transition diagram obtained from that of X by deleting the states in $B - G$ and directing the transition rates from states $i \in G$ to a; having an initial probability distribution in G identical to that of X and an initial probability in a equal to $P[X(0) \in (B - G) \cup \{a\}]$; and having a reward rate structure over G with the same reward rates as X. Then, the CRDTE(s) measure for X_{ub} with subset "B" equal to G upper bounds the CRDTE(s) measure for X.

Bounding methods for the ESSRR measure are more sophisticated and have been developed recently [62–68]. We will review the method with state duplication described by Muntz *et al.* [66], with generalizations which use results obtained by Carrasco [62] and Mahévas and Rubino [65]. The method can be easily implemented using a general-purpose Markovian modeling tool such as METFAC-2.1. The method assumes X irreducible and requires to find a partition $\bigcup_{k=0}^{M} C_k$ of Ω with the following properties:

P1. $|C_0| = 1$.

P2. For each $i \in C_k$, $1 \leq k \leq M$, there exists a path within C_k going from i to the "left" (subset $\bigcup_{l=0}^{k-1} C_l$).

The method obtains the bounds using detailed knowledge of X in a subset $G = \bigcup_{k=0}^{K} C_k$, $K < M$ of Ω, and computes the bounds using two CTMCs X_{lb}, X_{ub} having the same state transition diagram and differing only in their reward rate structure. The state transition diagram of those CTMCs includes "bounding" transition rates $f_{k,l}^{+}$, $1 \leq k < M$, $1 \leq l \leq M - k$ and g_k^{-}, $1 \leq k \leq M$. The transition rate $f_{k,l}^{+}$ has to upper bound $\max_{i\in C_k} \lambda_{i,C_{k+l}}$. If all states $i \in C_k$ have a non-null transition rate to the left $\lambda_{i,\bigcup_{l=0}^{k-1} C_l}$, then any lower bound > 0 for $\min_{i\in C_k} \lambda_{i,\bigcup_{l=0}^{k-1} C_l}$ can be taken as g_k^{-}. Otherwise, it can be taken as g_k^{-} any lower bound > 0 for $\min_{i\in C_k} q_i / h_i$, where q_i is the probability that, starting at state i, X will exit C_k through the left, and h_i is the mean holding time of X in C_k, assuming entry in C_k through state i. Obtaining such a lower bound may be difficult, but it is theoretically possible because of property P2 of the partition. The CTMCs X_{lb} and X_{ub} have state space $G \cup \{c_1, c_2, \ldots, c_M\}$, transition rates between states in G as X, transition rates from states $i \in G$ to states c_k equal to λ_{i,C_k}, transition rate from each state c_k to each state c_l with $l > k$ equal to $f_{k,l-k}^{+}$, a transition rate from each state c_k, $k > 2$ to c_{k-1} equal to g_k^{-}, and a transition rate from c_1 to o, o being the state in C_0, equal to g_1^{-}. Figure 34.4 illustrates the state transition diagram of X_{lb} and X_{ub}. The states in G of CTMCs X_{lb} and X_{ub} have associated with them the same reward rates as X; the states c_k have associated with them a reward rate equal to r_{lb}, where $r_{lb} \leq \min_{i\in\Omega} r_i$, in X_{lb}, and a reward rate equal to r_{ub}, where $r_{ub} \geq \max_{i\in\Omega} r_i$, in X_{ub}. Then, the ESSRR measure for X_{lb} and X_{ub} lower and upper bounds, respectively, the ESSRR measure for X. A situation in which the bounding method can be efficient, *i.e.* can give tight bounds with moderate values of K, is when transitions to the right (*i.e.* from states $i \in C_k$, $0 \leq k < M$, to $\bigcup_{l=k+1}^{M} C_l$) are "slow" (have small rates) and increase the index k moderately, and there exists a "fast" path (made up of "fast" transitions)

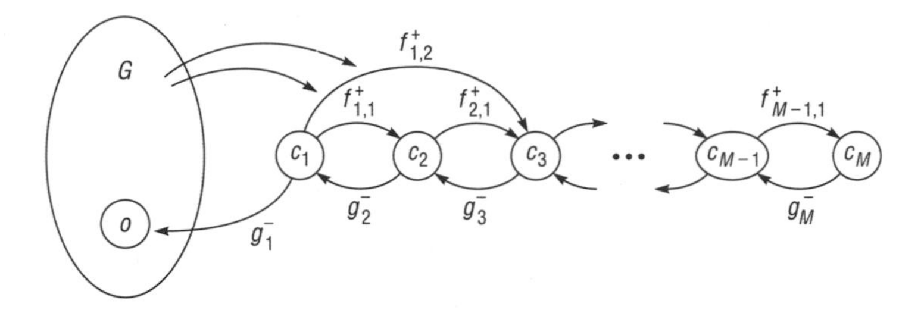

Figure 34.4. State transition diagram of the CTMCs bounding the ESSRR measure

within C_k to the left from any state $i \in C_k$, $1 \leq k \leq M$, which should allow the derivation of transition rates g^-_k significantly larger than the transition rates $f^+_{k,l}$, as it is desirable. This is typically the case for failure/repair models when C_k includes the states with exactly k failed components, since, for these models, failure rates are typically much smaller than repair rates.

Bounds for the ECRTE measure can be obtained using the previously described bounding method for the ESSRR measure. First, note that $\text{ECRTE} = \alpha_S \text{ECRTE}'$, where ECRTE' is the expected cumulative reward till exit of the subset B of the rewarded CTMC X' differing from X only in that its initial probability distribution restricted to B has been scaled so that the initial probabilities of the states in B add up to 1, *i.e.* $P[X'(0) = i] = \alpha_i/\alpha_S = \gamma_i$, $i \in B$, and the initial probability of the absorbing state a is 0. Let X'' be the irreducible rewarded CTMC whose state transition diagram is obtained from that of X' by adding a transition rate from state a to each state $i \in B$ with value $\gamma_i \Lambda$, where Λ is sufficiently large, for instance $\Lambda = \max_{i \in B} \lambda_i$, and has a reward rate structure with the same reward rates as X for the states in B and a reward rate 0 for state a. Let ECTTE' be the expected cumulative time till exit of B of X', let ESSRR'' be the expected steady-state reward rate of X'', and let p''_B be the steady-state probability of B in X''. Using regenerative process theory [69], taking as regeneration points the times at which X'' enters a, we have:

$$\text{ESSRR}'' = \frac{\text{ECRTE}'}{\text{ECTTE}' + 1/\Lambda}$$

$$p''_B = \frac{\text{ECTTE}'}{\text{ECTTE}' + 1/\Lambda}$$

from which we can obtain:

$$\text{ECRTE}' = \frac{\text{ESSRR}''}{(1 - p''_B)\Lambda}$$

Then, we can use the previously described bounding method for the ESSRR measure to obtain lower $([\text{ESSRR}'']_{\text{lb}}, [p''_B]_{\text{lb}})$ and upper $([\text{ESSRR}'']_{\text{ub}}, [p''_B]_{\text{ub}})$ bounds for ESSRR'' and p''_B, and, from them, obtain the following bounds for the ECRTE measure:

$$[\text{ECRTE}]_{\text{lb}} = \alpha_S \frac{[\text{ESSRR}'']_{\text{lb}}}{(1 - [p''_B]_{\text{lb}})\Lambda}$$

$$[\text{ECRTE}]_{\text{ub}} = \alpha_S \frac{[\text{ESSRR}'']_{\text{ub}}}{(1 - [p''_B]_{\text{ub}})\Lambda}$$

34.6 A Case Study

This section illustrates some of the techniques reviewed in the previous sections with a case study. The case study involves the computation of the steady-state unavailability and the unreliability of a fault-tolerant storage system made up of N RAID subsystems such as those described in Section 34.3. From a reliability point of view, the system is a series configuration of the RAID subsystems, *i.e.* the system is up if and only if every RAID subsystem is up. Failed components of up RAID subsystems are repaired by an unlimited number of repairmen with rate μ_U. However, a single repairman is available to bring at rate μ_D down RAID subsystems to a fully operational state. In case several RAID subsystems are down, the RAID subsystem which is brought up is selected at random among them by the repairman.

We will deal first with the computation of the steady-state unavailability. The model specification given in Section 34.3 can be trivially extended to the system under study by using an independent set of state variables to describe the state of each RAID subsystem. This has, however, two main drawbacks. The first one is that each value of N requires a different model specification. The second one is that the resulting CTMC would have 13^N states, a number which grows very fast with N and makes impossible the analysis of systems with even moderate values of N. The first problem can be solved and the second alleviated, by exploiting the fact that all RAID subsystems have exactly the same behavior and using a model specification yielding an aggregated CTMC with fewer states. Such an aggregated CTMC can be obtained by using state variables NFOP, NC, ND, NP, NR, NCD, NCP, NCR, NDP, NPR, NCDP, NCPR, and NDOWN, keeping track of the number of RAID subsystems in each of the states listed in Table 34.1. The production rules of such a model specification can easily be obtained from the state transition diagram of Figure 34.3, as illustrated in Figure 34.5, which gives the production rules corresponding to the transitions from the state FOP of the state transition diagram of Figure 34.3. The production rule corresponding to the transition from state DOWN to state FOP would have a rate μ_D, due to the fact that there is a single repairman to bring down RAID subsystems to the fully operational state. By defining N as model parameter and defining as start state the state in which NFOP has value N and all other state variables have value 0, a model specification independent of N can be obtained. Table 34.2 gives the size of the aggregated CTMCs obtained from such a model specification. The size of those CTMCs is substantially smaller than the size of the non-aggregated CTMCs (for instance, for $N = 8$, the non-aggregated CTMC would have 8.157×10^8 states while the aggregated CTMC has 125 970 states), but still grows fast with N, making impossible the analysis of systems with large values of N due to excessive storage requirements (the storage requirement for $N = 8$ was 91 MB).

Table 34.2. Size of the aggregated CTMC models for the computation of the steady-state unavailability of the storage system as a function of N

N	States	Transitions
1	13	53
2	91	689
3	455	4823
4	1820	24 115
5	6188	96 460
6	18 564	327 964
7	50 388	983 892
8	125 970	2 670 564

Bounding methods can be used to reduce further the storage requirements and allow the analysis of systems with larger values of N. Given the structure of the aggregated CTMC, it is possible to find a partition $\bigcup_{k=0}^{M} C_k$ with the properties required by the bounding method for the ESSRR measure described in Section 34.5, and for which the bounding method can be efficient. Let $N_C(s)$, $N_D(s)$, $N_P(s)$, $N_R(s)$, and $N_{DOWN}(s)$ be, respectively, the number of up RAID subsystems with one failed controller in state s, the number of up RAID subsystems with one failed disk in state s, the number of up RAID subsystems with one failed power supply in state s, the number of up RAID subsystems with one disk under reconstruction in state s, and the number of down RAID subsystems in state s. Let $N(s) = N_C(s) + N_D(s) + N_P(s) + N_R(s) + N_{DOWN}(s)$. Then, an appropriate partition is $C_k = \{s : N(s) = k\}$. With this partition, C_0 only includes the state in which all RAID subsystems are in their fully operational state, and, assuming failure rates much smaller than μ_{DR}, μ_U, and μ_D, transitions to the right

```
if NFOP>0 action FOP_1 with_rate NFOP*2*LC2*CC new_state NFOP=NFOP-1, NC=NC+1

if NFOP>0 action FOP_2 with_rate NFOP*8*LD new_state NFOP=NFOP-1, ND=ND+1

if NFOP>0 action FOP_3 with_rate NFOP*LP*CP new_state NFOP=NFOP-1, NP=NP+1

if NFOP>0 action FOP_4 with_rate NFOP*(2*LC2*(1-CC)+LP*(1-CP))
new_state NFOP=NFOP-1, NDOWN=NDOWN+1
```

Figure 34.5. Some production rules of a model specification from which the aggregated CTMC models for the computation of the steady-state unavailability of the storage system can be obtained

are slow and increase the k index moderately, and there exists a fast path within C_k to the left from any state $s \in C_k$, $k > 0$. To check that, Table 34.3 classifies the different types of events of the model into "slow" and "fast", depending on the values of the transition rates associated with them, and lists the values by which the event types may modify $N(s)$ (for each event type, the table gives the possible values for $\Delta N(s, s') = N(s') - N(s)$, where s and s' are states such that the event type causes a transition from s to s'). The possible values for $\Delta N(s, s')$ are obtained by analyzing how the event type may modify $N_C(s)$, $N_D(s)$, $N_P(s)$, $N_R(s)$, and $N_{DOWN}(s)$, taking into account that an up RAID subsystem cannot have both one disk failed and one disk under reconstruction. Thus, for instance, event type 2 will decrement $N_D(s)$ by 1, will increment $N_{DOWN}(s)$ by 1, and may decrement $N_C(s)$ and $N_P(s)$ by 1, since the subsystem including the failed disk could have had one controller and one power supply failed. From that table, it is clear that: (1) transitions to the right are slow and increase the k index by 1, (2) each state $s \in C_k$, $k > 0$ with $N_C(s) + N_P(s) + N_R(s) + N_{DOWN}(s) > 0$ has a fast transition to the left due to the end of a disk reconstruction in an up subsystem, the repair of a controller in an up subsystem, the repair of a power supply in an up subsystem, or a global repair of a down RAID subsystem, and (3) each state $s \in C_k$, $k > 0$ with $N_C(s) + N_P(s) + N_R(s) + N_{DOWN}(s) = 0$, which implies $N_D(s) > 0$, has a fast transition to a state $s' \in C_k$ with $N_C(s') + N_P(s') + N_R(s') + N_{DOWN}(s') > 0$ (because $N_R(s') = 1$) due to a disk repair in an up subsystem, implying with (2) the existence of a fast path within C_k to the left from any state $s \in C_k$, $k > 0$.

The maximum value of $N(s)$ is $M = 3N$. Transitions to the right (see Table 34.3) may only be due to disk failures in up subsystems with no disk either failed or under reconstruction, controller failures in up subsystems with no controller failed, and power supply failures in up subsystems with no power supply failed, and go from a subset C_k to a subset C_{k+1}. Then, as bounding transition rates to the right we can take:

$$f^{+}_{k,1} = N(8\lambda_D + 2\lambda_{C2} + \lambda_P) = f$$

$$f^{+}_{k,l} = 0 \quad l \neq 1$$

Obtaining appropriate transition rates g^{-}_{k} (*i.e.* significantly larger than f) is more difficult because some states in the subsets C_k, $1 \le k \le 3N$ do not have a fast transition to the left. Let $C^0_k = \{s\colon N(s) = k \wedge N_C(s) + N_P(s) + N_R(s) + N_{DOWN}(s) = 0\}$. Then, the states in C^0_k do not have a fast transition to the left. States $s \in C^1_k = C_k - C^0_k$, $k > 0$ have $N_C(s) + N_P(s) + N_R(s) + N_{DOWN}(s) > 0$ and, therefore, have a fast transition to the left.

For $N < k \le 3N$, $s \in C_k$, $N_D(s) \le N$ implies $N_C(s) + N_P(s) + N_R(s) + N_{DOWN}(s) > 0$ and $C^0_k = \emptyset$. Then, for $N < k \le 3N$, all states in C_k have a fast transition to the left and we can take a lower bound > 0 for the transition rate to the left from any state $s \in C_k$ as g^{-}_{k}. Since states s with

Table 34.3. Event types and properties in the CTMC model for the computation of the steady-state unavailability of the storage system

Event type	Description	Slow/fast	$\Delta N(s, s')$
1	Disk failure in up subsystem with no disk either failed or under reconstruction	slow	1
2	Disk failure in up subsystem with one disk failed	slow	$-2, -1, 0$
3	Failure of a disk which is not under reconstruction in up subsystem with one disk under reconstruction	slow	$-2, -1, 0$
4	Failure of the disk under reconstruction in up subsystem with one disk under reconstruction	slow	0
5	Covered controller failure in up subsystem with no controller failed	slow	1
6	Uncovered controller failure in up subsystem with no controller failed	slow	$-1, 0, 1$
7	Controller failure in up subsystem with one controller failed	slow	$-2, -1, 0$
8	Covered power supply failure in up subsystem with no power supply failed	slow	1
9	Uncovered power supply failure in up subsystem with no power supply failed	slow	$-1, 0, 1$
10	Power supply failure in up subsystem with one power supply failed	slow	$-2, -1, 0$
11	Disk repair in up subsystem	fast	0
12	End of disk reconstruction in up subsystem	fast	-1
13	Controller repair in up subsystem	fast	-1
14	Power supply repair in up subsystem	fast	-1
15	Global repair of down subsystem	fast	-1

$N_{\mathrm{C}}(s) > 0$ have a transition rate to the left $\geq \mu_{\mathrm{U}}$ (because in those states at least one controller of an up subsystem is being repaired), states s with $N_{\mathrm{P}}(s) > 0$ have a transition rate to the left $\geq \mu_{\mathrm{U}}$ (because in those states at least one power supply of an up subsystem is being repaired), states s with $N_{\mathrm{R}}(s) > 0$ have a transition rate to the left $\geq \mu_{\mathrm{DR}}$ (because in those states at least one disk of an up subsystem is under reconstruction), and states s with $N_{\mathrm{DOWN}}(s) > 0$ have a transition rate to the left $\geq \mu_{\mathrm{D}}$ (because in those states a down RAID subsystem is being brought up), we can take $g_k^- = \min\{\mu_{\mathrm{DR}}, \mu_{\mathrm{U}}, \mu_{\mathrm{D}}\} = g_1$.

For $1 \leq k \leq N$, $C_k^0 \neq \emptyset$ (C_k^0 includes just the state in which there are k up RAID subsystems with only one disk failed and the remaining $N - k$ RAID subsystems are in their fully operational state), and not all states in C_k have a fast transition to the left. However, the existence of a fast path within C_k to the left allows us to obtain a "large" lower bound >0 for $\min_{s \in C_k} q_s / h_s$, where q_s is the probability that, starting at s, X will exit C_k through the left, and h_s is the mean holding time of X in C_k, assuming entry in C_k through s, and that lower bound can be used as g_k^-. To derive that lower bound we will use the following two results, which can easily be proved.

Lemma 1. *Let Z be a transient CTMC with state space $B \cup \{a_{\mathrm{L}}, a_{\mathrm{R}}\}$, where a_{L} and a_{R} are absorbing states and all states in B are transient, and $P[Z(0) \in B] = 1$. Let $P_{\mathrm{L}} = \lim_{t\to\infty} P[Z(t) = a_{\mathrm{L}}]$ and $P_{\mathrm{R}} = \lim_{t\to\infty} P[Z(t) = a_{\mathrm{R}}]$ (P_{L} and P_{R} are, respectively, the probabilities that Z will be absorbed in a_{L} and a_{R}). Denote by g_i the transition rate from $i \in B$ to a_{L} and denote by f_i the* *transition rate from* $i \in B$ *to* a_R. *Assume* $g_i > 0$, $i \in B$, *let* $g^- \le \min_{i \in B} g_i$, $g^- > 0$, *and let* $f^+ \ge \max_{i \in B} f_i$. *Then,* $P_L \ge g^-/(g^- + f^+)$.

Lemma 2. *Let Z be a transient CTMC with state space* $B \cup \{a\}$, *where a is an absorbing state and all states in B are transient, and* $P[Z(0) \in B] = 1$. *Let* g_i, $i \in B$ *denote the transition rate from i to a, assume* $g_i > 0$, $i \in B$, *and let* $g^- \le \min_{i \in B} g_i$, $g^- > 0$. *Let* $h = \sum_{i \in B} \int_0^\infty P[Z(t) = i]\,dt$ *be the mean time to absorption of Z. Then,* $h \le 1/g^-$.

Let $s \in C_k^0$, $1 \le k \le N$. As noted before, s is the state in which there are k up RAID subsystems with only one disk failed and the remaining $N - k$ RAID subsystems are in their fully operational state. Let us call that state s_k. The repair of a disk in s_k will lead to a state s' with $N(s') = N(s_k)$ (see Table 34.3) and $N_R(s') > 0$ and, then, $s' \in C_k^1$. This implies that s_k has a transition rate $\ge k\mu_U$ to C_k^1. In addition, the transition rate from s_k to the right is $\le f$. Let $q_{s_k}^0$ be the probability that, starting at s_k, X will exit C_k^0 through C_k^1, and let $h_{s_k}^0$ be the mean holding time of X in s_k. Then, using Lemmas 1 and 2 we have:

$$q_{s_k}^0 \ge \frac{k\mu_U}{k\mu_U + f} \qquad 1 \le k \le N \tag{34.6}$$

$$h_{s_k}^0 \le \frac{1}{k\mu_U} \qquad 1 \le k \le N \tag{34.7}$$

Let now $s \in C_k^1$, $1 \le k \le N$. Such a state has a transition rate to the left $\ge g_1$ and a transition rate to the right $\le f$. Furthermore, s has a transition rate $\le N\lambda_{DR}$ to C_k^0. This can be proved by noting that the only event types which may lead from s to a state $s' \in C_k$ are the event types 2, 3, 4, 6, 7, 9, 10, and 11 listed in Table 34.3. Event types 2, 3, 6, 7, 9, and 10 lead to states s' with $N_{DOWN}(s') > 0$, which cannot belong to C_k^0. Event type 11 leads to a state s' with $N_R(s') > 0$, which cannot belong to C_k^0. Then, the only event type which may lead to a state $s' \in C_k^0$ is event type 4 and the rate of that event type is upper bounded by $N\lambda_{DR}$. Figure 34.6 clarifies the situation. Then, denoting by q_s^1 the probability that, starting at s, X will exit C_k^1 through the left, and by h_s^1 the mean holding time of X in C_k^1, starting at s, it follows from Lemmas 1 and 2 that:

$$q_s^1 \ge \frac{g_1}{g_1 + f + N\lambda_{DR}} \qquad s \in C_k^1,\ 1 \le k \le N \tag{34.8}$$

$$h_s^1 \le \frac{1}{g_1} \qquad s \in C_k^1,\ 1 \le k \le N \tag{34.9}$$

Consider now the state s_k. It is clear (see Figure 34.6) that q_{s_k} is lower bounded by $q_{s_k}^0 \min_{s' \in C_k^1} q_{s'}^1$. Then, using Relations 34.6 and 34.8 we have:

$$q_{s_k} \ge \left(\frac{k\mu_U}{k\mu_U + f}\right)\left(\frac{g_1}{g_1 + f + N\lambda_{DR}}\right) \quad 1 \le k \le N$$

Also, taking into account that $1 - q_s^1$, $s \in C_k^1$, $1 \le k \le N$, upper bounds the probability that, starting at s, X will exit C_k^1 through C_k^0, h_{s_k} is upper bounded (see Figure 34.6) by:

$$h_{s_k}^0 + \max_{s' \in C_k^1} h_{s'}^1 + \max_{s' \in C_k^1}(1 - q_{s'}^1)\Big[h_{s_k}^0 + \max_{s' \in C_k^1} h_{s'}^1 + \max_{s' \in C_k^1}(1 - q_{s'}^1)\Big[h_{s_k}^0 + \max_{s' \in C_k^1} h_{s'}^1 + \max_{s' \in C_k^1}(1 - q_{s'}^1)[\cdots]\Big]\Big]$$

and using Relations 34.7 and 34.9 and $1 - q_s^1 \le (f + N\lambda_{DR})/(g_1 + f + N\lambda_{DR})$, $s \in C_k^1$, $1 \le k \le n$, which follows from Relation 34.8, we have:

$$\begin{aligned}
h_{s_k} &\le \frac{1}{k\mu_U} + \frac{1}{g_1} + \left(\frac{f + N\lambda_{DR}}{g_1 + f + N\lambda_{DR}}\right) \times \left[\frac{1}{k\mu_U} + \frac{1}{g_1} + \left(\frac{f + N\lambda_{DR}}{g_1 + f + N\lambda_{DR}}\right)[\cdots]\right] \\
&= \left(\frac{1}{k\mu_U} + \frac{1}{g_1}\right)\sum_{n=0}^{\infty}\left(\frac{f + N\lambda_{DR}}{g_1 + f + N\lambda_{DR}}\right)^n \\
&= \left(\frac{1}{k\mu_U} + \frac{1}{g_1}\right)\left(\frac{1}{1 - \frac{f + N\lambda_{DR}}{g_1 + f + N\lambda_{DR}}}\right) \\
&= \left(\frac{1}{k\mu_U} + \frac{1}{g_1}\right)\left(\frac{g_1 + f + N\lambda_{DR}}{g_1}\right) \\
&= \left(\frac{k\mu_U + g_1}{k\mu_U g_1}\right) \times \left(\frac{g_1 + f + N\lambda_{DR}}{g_1}\right) \quad 1 \le k \le N
\end{aligned}$$

which gives:

$$\frac{q_{s_k}}{h_{s_k}} \geq \left(\frac{k\mu_U}{k\mu_U + f}\right)\left(\frac{g_1}{g_1 + f + N\lambda_{DR}}\right)^2 \times \left(\frac{k\mu_U g_1}{k\mu_U + g_1}\right) \quad 1 \leq k \leq N \qquad (34.10)$$

Let $s \in C_k^1$, $1 \leq k \leq N$. It is clear that q_s is lower bounded by q_s^1 and, then, using Relation 34.8:

$$q_s \geq \frac{g_1}{g_1 + f + N\lambda_{DR}} \quad s \in C_k^1,\ 1 \leq k \leq N \qquad (34.11)$$

On the other hand, $1 - q_s^1$ upper bounds the probability that, starting at s, X will exit C_k^1 through C_k^0, and, then, we have (see Figure 34.6):

$$h_s \leq h_s^1 + (1 - q_s^1)\Big[h_{s_k}^0 + \max_{s' \in C_k^1} h_{s'}^1 + \max_{s' \in C_k^1}(1 - q_{s'}^1) \times \Big[h_{s_k}^0 + \max_{s' \in C_k^1} h_{s'}^1 + \max_{s' \in C_k^1}(1 - q_{s'}^1)[\cdots]\Big]\Big] \quad s \in C_k^1,\ 1 \leq k \leq N$$

and, using Relations 34.7, 34.9 and

$$1 - q_s^1 \leq \left(\frac{f + N\lambda_{DR}}{g_1 + f + N\lambda_{DR}}\right) \quad s \in C_k^1,\ 1 \leq k \leq N$$

we get:

$$\begin{aligned}
h_s &\leq \frac{1}{g_1} + \left(\frac{f + N\lambda_{DR}}{g_1 + f + N\lambda_{DR}}\right)\left[\frac{1}{k\mu_U} + \frac{1}{g_1} + \left(\frac{f + N\lambda_{DR}}{g_1 + f + N\lambda_{DR}}\right)[\cdots]\right] \\
&= \frac{1}{g_1} + \left(\frac{1}{k\mu_U} + \frac{1}{g_1}\right) \times \sum_{n=1}^{\infty}\left(\frac{f + N\lambda_{DR}}{g_1 + f + N\lambda_{DR}}\right)^n \\
&= \frac{1}{g_1} + \left(\frac{1}{k\mu_U} + \frac{1}{g_1}\right)\left(\frac{\frac{f+N\lambda_{DR}}{g_1+f+N\lambda_{DR}}}{1 - \frac{f+N\lambda_{DR}}{g_1+f+N\lambda_{DR}}}\right) \\
&= \frac{1}{g_1} + \left(\frac{1}{k\mu_U} + \frac{1}{g_1}\right)\left(\frac{f + N\lambda_{DR}}{g_1}\right) \\
&= \frac{1}{g_1}\left[1 + \frac{(k\mu_U + g_1)(f + N\lambda_{DR})}{k\mu_U g_1}\right] \\
&\quad s \in C_k^1,\ 1 \leq k \leq N \qquad (34.12)
\end{aligned}$$

Combining Relations 34.11 and 34.12:

$$\begin{aligned}
\min_{s \in C_k^1} \frac{q_s}{h_s} &\geq \frac{\frac{g_1}{g_1+f+N\lambda_{DR}}}{\frac{1}{g_1}\left[1 + \frac{(k\mu_U+g_1)(f+N\lambda_{DR})}{k\mu_U g_1}\right]} \\
&= \left[\frac{k\mu_U g_1}{k\mu_U g_1 + (k\mu_U + g_1)(f + N\lambda_{DR})}\right] \times \left(\frac{g_1}{g_1 + f + N\lambda_{DR}}\right) g_1 \quad 1 \leq k \leq N
\end{aligned} \qquad (34.13)$$

Finally, combining Relations 34.10 and 34.13 we obtain, for $1 \leq k \leq N$:

$$\begin{aligned}
\min_{s \in C_k} \frac{q_s}{h_s} &= \min\left\{\frac{q_{s_k}}{h_{s_k}},\ \min_{s \in C_k^1} \frac{q_s}{h_s}\right\} \\
&\geq \min\left\{\left(\frac{k\mu_U}{k\mu_U + f}\right)\left(\frac{g_1}{g_1 + f + N\lambda_{DR}}\right)^2 \times \left(\frac{k\mu_U g_1}{k\mu_U + g_1}\right),\right. \\
&\qquad \left.\left[\frac{k\mu_U g_1}{k\mu_U g_1 + (k\mu_U + g_1)(f + N\lambda_{DR})}\right] \times \left(\frac{g_1}{g_1 + f + N\lambda_{DR}}\right) g_1\right\} \\
&= \left(\frac{g_1}{g_1 + f + N\lambda_{DR}}\right) \min\left\{\left(\frac{k\mu_U}{k\mu_U + f}\right) \times \left(\frac{g_1}{g_1 + f + N\lambda_{DR}}\right)\left(\frac{k\mu_U g_1}{k\mu_U + g_1}\right),\right. \\
&\qquad \left.\left[\frac{k\mu_U g_1}{k\mu_U g_1 + (k\mu_U + g_1)(f + N\lambda_{DR})}\right] g_1\right\} \\
&= g_2(k)
\end{aligned}$$

and, for $1 \leq k \leq N$, we can take $g_k^- = g_2(k)$.

It is possible to modify the model specification of the aggregated model to obtain a model specification implementing the bounding method with an `int` parameter specifying the value of the K parameter of the bounding method and another `int` parameter `UB` such that, when the `UB` parameter has the value `yes`, a reward rate 1 is assigned to the states of the bounding part (states $c_1, c_2, \ldots, c_M$) and the generic ESSRR

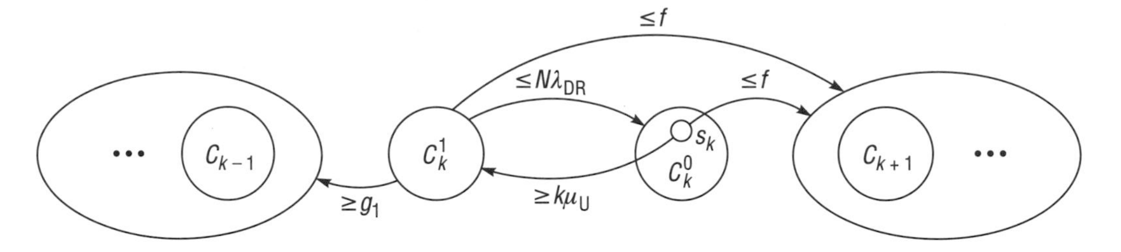

Figure 34.6. Known results about the transition rates from states in C_k, $1 \leq k \leq N$

Table 34.4. Bounds for the steady-state unavailability obtained using the bounding aggregated models for $N = 8$

K	Lower bound	Upper bound
1	$5.270\,7062 \times 10^{-5}$	$1.208\,9440 \times 10^{-4}$
2	$5.341\,5080 \times 10^{-5}$	$5.352\,3252 \times 10^{-5}$
3	$5.341\,6410 \times 10^{-5}$	$5.341\,6506 \times 10^{-5}$
4	$5.341\,6411 \times 10^{-5}$	$5.341\,6411 \times 10^{-5}$
5	$5.341\,6411 \times 10^{-5}$	$5.341\,6411 \times 10^{-5}$

Table 34.5. Size of the bounding aggregated models for the computation of the steady-state unavailability for $K = 4$ as a function of N

N	States	Transitions
1	13	53
2	84	575
3	197	1593
4	270	2158
5	273	2165
≥6	$258 + 3N$	$2135 + 6N$

measure becomes an upper bound for the steady-state unavailability, and, when that parameter has the value no, a reward rate 0 is assigned to the states of the bounding part and the generic ESSRR measure becomes a lower bound for the steady-state unavailability. A moderate value of K is enough to obtain very tight bounds. This is illustrated in Table 34.4, which gives the bounds obtained with increasing values of K for $N = 8$, $\lambda_D = 2 \times 10^{-6}$ h^{-1}, $\lambda_{DR} = 3 \times 10^{-6}$ h^{-1}, $\lambda_{C2} = 5 \times 10^{-6}$ h^{-1}, $\lambda_{C1} = 8 \times 10^{-6}$ h^{-1}, $\lambda_P = 6 \times 10^{-6}$ h^{-1}, $C_C = 0.99$, $C_P = 0.995$, $\mu_{DR} = 0.4$ h^{-1}, $\mu_U = 0.125$ h^{-1}, and $\mu_D = 0.02$ h^{-1}. Those values of failure rates, coverages, disk reconstruction rate, and repair rates will be used thereafter unless stated otherwise. Table 34.5 gives the size of the aggregated bounding models for $K = 4$ as a function of the number of RAID subsystems N. The bounding aggregated models are very small and their size grows very smoothly with N.

The bounding aggregated models allow the analysis of systems with very large N. Figure 34.7 shows the influence of the number of RAID subsystems N and the coverage to controller failures C_C on the steady-state unavailability. All results were obtained using $K = 4$, which yielded very tight bounds in all cases. As N increases, the steady-state unavailability increases. This is to be expected, since with more RAID subsystems the probability that any of them is down is larger. Improving the coverage to controller failures beyond the baseline value $C_C = 0.99$ has a significant impact on the steady-state unavailability only till C_C reaches a value ≈ 0.999. Beyond that point, the steady-state unavailability is little affected.

The unreliability of a RAID subsystem can be computed as the generic measure CRDTE(s) by modifying the model specification yielding the CTMC used for the computation of the steady-state unavailability of the RAID subsystem so that the state DOWN is absorbing and a reward rate 1 is assigned to the up states. That model specification can be trivially extended to generate

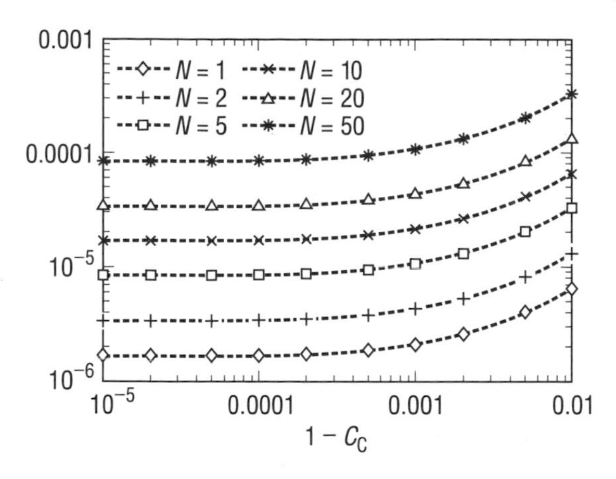

Figure 34.7. Influence of N and C_C on the steady-state unavailability of the storage system

a CTMC from which the unreliability of a storage system with N RAID subsystems can be computed by using an independent set of state variables DF, DR, CF, and PF for each RAID subsystem and a state variable DOWN to represent that the system has failed. The number of states of the resulting CTMC would be $12^N + 1$, which is unmanageable for large N. In addition, the model specification would be different for each value of N. A model specification independent of N yielding aggregated CTMCs of much smaller size can be developed using state variables counting the number of RAID subsystems in each of the up states listed in Table 34.1 and a state variable DOWN representing that the system has failed. Table 34.6 gives the size of the resulting aggregated CTMC as a function of N. Although smaller than the aggregated CTMCs which allow the computation of the steady-state unavailability of the storage system, the aggregated CTMCs are still too large to allow the analysis of systems with large N.

The bounding method for the measure CRDTE(s) described in Section 34.5 can be used to allow the analysis of systems with large N. To obtain tight bounds, G has to include the "more likely" up states. Given the structure of the aggregated CTMC, a reasonable selection is to include in G all states s in which the system is up and $N_C(s) + N_D(s) + N_P(s) + N_R(s) \leq K$.

Table 34.6. Size of the aggregated CTMC models for the computation of the unreliability of the storage system

N	States	Transitions
1	13	52
2	79	558
3	365	3484
4	1366	15 925
5	4369	58 968
6	12 377	187 096
7	31 825	526 864
8	75 583	1 384 542

Table 34.7. Bounds for the one-year unreliability obtained using the bounding aggregated models for $N = 8$

K	Lower bound	Upper bound
1	$9.290\,5804 \times 10^{-3}$	$1.423\,4973 \times 10^{-2}$
2	$9.314\,6951 \times 10^{-3}$	$9.319\,9494 \times 10^{-3}$
3	$9.314\,7213 \times 10^{-3}$	$9.314\,7248 \times 10^{-3}$
4	$9.314\,7213 \times 10^{-3}$	$9.314\,7213 \times 10^{-3}$
5	$9.314\,7213 \times 10^{-3}$	$9.314\,7213 \times 10^{-3}$

Table 34.8. Size of the bounding aggregated models for the computation of the unreliability of the storage system for $K = 4$ as a function of N

N	Lower bounding model		Upper bounding model	
	States	Transitions	States	Transitions
1	13	52	13	52
2	67	444	66	421
3	137	1004	136	931
≥ 4	172	1234	171	1126

It is possible to modify the model specification of the aggregated unreliability model to obtain a model specification implementing the bounding method with an int model parameter specifying the value of K and another int parameter UB such that, when that parameter has the value yes, the upper bounding model is generated and, when that parameter has the value no, the lower bounding model is generated. As for the steady-state unavailability, a moderate value of K is enough to obtain very tight bounds for the

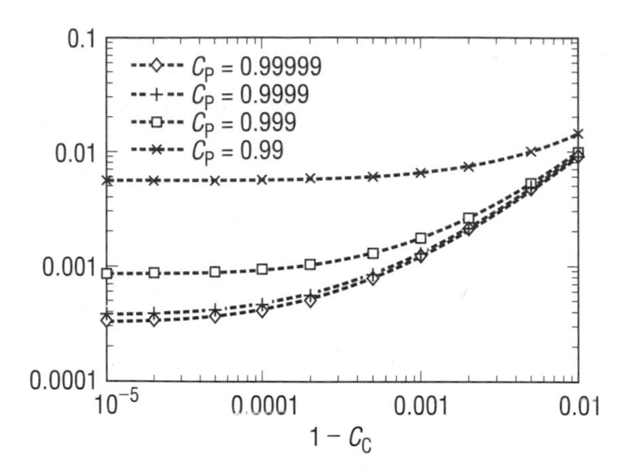

Figure 34.8. Influence of C_C and C_P on the one-year unreliability of a storage system with $N = 10$

unreliability. This is illustrated in Table 34.7, which gives the bounds obtained for the one-year unreliability, assuming that initially all RAID subsystems are in their fully operational state, for increasing values of K and $N = 8$. Table 34.8 gives the sizes of the lower and upper bounding aggregated CTMC models for $K = 4$ and several values of N. Both models are very small for any N.

As an illustration of the use of the bounding aggregated unreliability models, Figure 34.8 analyzes the impact of the coverages C_C and C_P on the one-year unreliability of a storage system with $N = 10$, assuming that initially all RAID subsystems are in their fully operational state. All results were obtained using $K = 4$, which gave very tight bounds in all cases.

34.7 Conclusions

Rewarded CTMC models are a flexible modeling formalism to evaluate dependability/performability measures for fault-tolerant systems. However, efficient use of that formalism requires the availability of powerful model specification and model solution methodologies. Although considerable advances have been performed in both directions, much work remains to be done. This is because increasing complexity of fault-tolerant systems results in larger and larger CTMC models, which are both difficult to specify and expensive to solve. The latter is particularly true for measures requiring the computation of transient probabilities and, even more, for measures with complex probabilistic structure such as the CRD(t, s) measure. More efficient numerical methods for the computation of those measures should be developed. Currently available approaches to deal with the largeness problem make feasible the numerical analysis of models of quite complex systems, as the case study has illustrated, but further work is required. Also, development of efficient methods for particular, but important model classes and measures, *e.g.* Suñé and Carrasco [70], should continue. Combinatorial methods should also be extended to deal efficiently with wider classes of models and measures.

References

[1] Iyer RK, Tang D. Experimental analysis of computer system dependability. In: Pradhan DK, editor. Fault-tolerant computer system design. Englewood Cliffs, NJ: Prentice-Hall; 1995. p.282–392.

[2] Laprie JC, Costes A. Dependability: a unifying concept for reliable computing. Proc 12th IEEE Int Symp on Fault-Tolerant Computing (FTCS-12); 1982. p.18–21.

[3] Meyer JF. On evaluating the performability of degradable computing systems. IEEE Trans Comput 1980;C-29(8):720–31.

[4] Meyer JF. Performability of an algorithm for connection admission control. IEEE Trans Comput 2001;50(7):724–33.

[5] Abraham JA. An improved algorithm for network reliability. IEEE Trans Reliab 1979;R-28:58–61.

[6] Aggarwal KB, Misra KB, Gupta JS. A fast algorithm for reliability evaluation. IEEE Trans Reliab 1975;R-24:83–5.

[7] Amari SV, Dugan JB, Misra RB. A separable method for incorporating imperfect fault-coverage into combinatorial models. IEEE Trans Reliab 1999;48(3):267–74.

[8] Doyle SA, Dugan JB, Patterson-Hine FA. A combinatorial approach to modeling imperfect coverage. IEEE Trans Reliab 1995;44(1):87–94.

[9] Dugan JB. Fault trees and imperfect coverage. IEEE Trans Reliab 1989;38(2):177–85.

[10] Sahner RA, Trivedi KS. Reliability modeling using SHARPE. IEEE Trans Reliab 1987;R-36(2):186–93.

[11] Neuts MF. Matrix-geometric solutions in stochastic models. An algorithmic approach. Dover Publications Inc; 1994.

[12] Bobbio A, Telek M. A benchmark for PH estimation algorithms: results for acyclic-PH. Commun Stat-Stochast Models 1994;10(3):661–77.

[13] Beaudry MD. Performance-related reliability measures for computing systems. IEEE Trans Comput 1978;C-27(6):540–7.

[14] Blakemore A. The cost of eliminating vanishing markings from generalized stochastic Petri nets. Proc 3rd IEEE Int Workshop on Petri Nets and Performance Models (PNPM89); 1989. p.85–92.

[15] Goyal A, Carter WC, de Souza e Silva E, Lavenberg SS. The system availability estimator. Proc 16th IEEE Int Symp on Fault-Tolerant Computing (FTCS-16); 1986. p.84–9.

[16] Lindemann C. DSPNexpress: a software package for the efficient solution of deterministic and stochastic Petri nets. Perform Eval 1995;22(1):3–21.

[17] Chiola G, Franceschinis G, Gaeta R, Ribaudo M. GreatSPN 1.7: graphical editor and analyzer for timed and stochastic Petri nets. Perform Eval 1995;24(1/2):47–68.

[18] Ciardo G, Muppala J, Trivedi K. SPNP: stochastic Petri net package. Proc 3rd IEEE Int Workshop on Petri Nets and Performance Models (PNPM89); 1989. p.142–51.

[19] Béounes C, Aguéra M, Arlat J, Bachmann S, Bourdeau C, Doucet J-E, Kanoun K, Laprie JC, Metze S, Moreira de Souza J, Powel D, Spiesser P. SURF-2: a program for dependability evaluation of complex hardware and software systems. Proc 23rd IEEE Int Symp on Fault-Tolerant Computing (FTCS-23); 1993. p.142–50.

[20] German R, Kelling C, Zimmerman A, Hommel G. TimeNET: a toolkit for evaluating non-Markovian stochastic Petri nets. Perform Eval 1995;24(1/2):69–87.

[21] Sanders WH, Obal II WD, Qureshi MA, Widjanarko FK. The UltraSAN modeling environment. Perform Eval 1995;24(1/2):89–115.

[22] Gilmore S, Hillston J. The PEPA Workbench: a tool to support a process algebra-based approach to performance modelling. Proc 7th Int Conf on Modelling Techniques and Tools for computer performance evaluation; 1994. Lecture Notes in Computer Science 794. Berlin: Springer-Verlag. p.353–68.

[23] Hermanns H, Herzog U, Klehmet U, Mertsiotakis V, Siegle M. Compositional performance modelling with the TIPtool. Perform Eval 2000;39(1–4):5–35.

[24] Berson S, de Souza e Silva E, Muntz RR. A methodology for the specification and generation of Markov models. In: Stewart WJ, editor. Numerical solution of Markov chains. New York: Marcel Dekker; 1991. p.11–36.

[25] Stewart WJ. Introduction to the numerical solution of Markov chains. Princeton, NJ: Princeton University Press; 1994.

[26] Barker GP, Plemmons RJ. Convergent iterations for computing stationary distributions of Markov chains. SIAM J Alg Discr Meth 1986;7(3):390–8.

[27] Berman A, Plemmons RJ. Nonnegative matrices in the mathematical sciences. SIAM; 1994.

[28] Suñé V, Domingo JL, Carrasco JA. Numerical iterative methods for Markovian dependability and performability models: new results and a comparison. Perform Eval 2000;39(1–4):99–125.

[29] Young DM. Iterative solution of large linear systems. New York: Academic Press; 1971.

[30] Heidelberger P, Muppala JK, Trivedi KS. Accelerating mean time to failure computations. Perform Eval 1996;27/28:627–45.

[31] Malhotra M, Muppala JK, Trivedi KS. Stiffness-tolerant methods for transient analysis of stiff Markov chains. Microelectron Reliab 1994;34(11):1825–41.

[32] Malhotra M. A computationally efficient technique for transient analysis of repairable Markovian systems. Perform Eval 1995;24(1/2):311–31.

[33] Reibman A, Trivedi KS. Numerical transient analysis of Markov models. Comput Operat Res 1988;15(1):19–36.

[34] Kijima M. Markov processes for stochastic modeling. London: Chapman and Hall; 1997.

[35] Bowerman PN, Nolty RG, Schener EM. Calculation of the Poisson cumulative distribution function. IEEE Trans Reliab 1990;39(2):158–61.

[36] Fox BL, Glynn PW. Computing Poisson probabilities. Commun ACM 1988;31(4):440–5.

[37] Knüsel L. Computation of the chi-square and Poisson distribution. SIAM J Sci Stat Comput 1986;7(3):1022–36.

[38] van Moorsel AP, Sanders WH. Transient solution of Markov models by combining adaptive & standard uniformization. IEEE Trans Reliab 1997;46(3):430–40.

[39] Carrasco JA. Computation of bounds for transient measures of large rewarded Markov models using regenerative randomization. Technical report DMSD_99_4. Universitat Politècnica de Catalunya; 1999. Available at ftp://ftp-eel.upc.es/techreports. To appear in Comput Operat Res.

[40] Miller DR. Reliability calculation using randomization for Markovian fault-tolerant computing systems. Proc 13th IEEE Int Symp on Fault-Tolerant Computing (FTCS-13); 1983. p.284–9.

[41] Melamed B, Yadin M. Randomization procedures in the computation of cumulative-time distributions over discrete state Markov processes. Operat Res 1984;32(4): 926–44.

[42] van Moorsel APA, Sanders WH. Adaptive uniformization. Commun Stat-Stochast Models 1994;10(3):619–48.

[43] Sericola B. Availability analysis of repairable computer systems and stationarity detection. IEEE Trans Comput 1999;48(11):1166–72.

[44] Carrasco JA. Transient analysis of large Markov models with absorbing states using regenerative randomization. Technical report DMSD_99_2. Universitat Politècnica de Catalunya; 1999. Available at ftp://ftp-eel.upc.es/techreports

[45] Carrasco JA. Computationally efficient and numerically stable bounds for repairable fault-tolerant systems. IEEE Trans Comput 2002;51(3):254–68.

[46] Donatiello L, Grassi V. On evaluating the cumulative performance distribution of fault-tolerant computer systems. IEEE Trans Comput 1991;40(11):1301–7.

[47] Islam SMR, Ammar HH. Performability of the hypercube. IEEE Trans Reliab 1989;38(5):518–26.

[48] Nabli H, Sericola B. Performability analysis: a new algorithm. IEEE Trans Comput 1996;45(4):491–4.

[49] Pattipati KR, Li Y, Blom HAP. A unified framework for the performability evaluation of fault-tolerant computer systems. IEEE Trans Comput 1993;42(3):312–26.

[50] Qureshi MA, Sanders WH. Reward model solution methods with impulse and rate rewards: an algorithm and numerical results. Perform Eval 1994;20:413–36.

[51] Qureshi MA, Sanders WH. A new methodology for calculating distributions of reward accumulated during a finite interval. Proc 26th IEEE Int Symp on Fault-Tolerant Computing; 1996. p.116–25.

[52] Smith RM, Trivedi KS, Ramesh AV. Performability analysis: measures, an algorithm, and a case study. IEEE Trans Comput 1988;37(4):406–17.

[53] de Souza e Silva E, Gail HR. Calculating availability and performability measures of repairable computer systems using randomization. J ACM 1989;36(1):171–93.

[54] de Souza e Silva E, Gail HR. An algorithm to calculate transient distributions of cumulative rate and impulse based reward. Commun Stat-Stochast Models 1998;14(3):509–36.

[55] Goyal A, Tantawi AN. A measure of guaranteed availability and its numerical evaluation. IEEE Trans Comput 1988;37(1):25–32.

[56] Rubino G, Sericola B. Interval availability distribution computation. Proc 23rd IEEE Int Symp on Fault-Tolerant Computing (FTCS-23); 1993. p.48–55.

[57] Rubino G, Sericola B. Interval availability analysis using denumerable Markov processes: application to multiprocessor subject to breakdowns and repair. IEEE Trans Comput 1995;44(2):286–91.

[58] de Souza e Silva E, Gail HR. Calculating cumulative operational time distributions of repairable computer systems. IEEE Trans Comput 1986;C-35(4):322–32.

[59] Chiola G, Dutheillet C, Franceschinis G, Haddad S. Stochastic well-formed colored nets and symmetric modeling applications. IEEE Trans Comput 1993;42(11): 1343–60.

[60] Sanders WH, Meyer JF. Reduced base model construction methods for stochastic activity networks. IEEE J Select Areas Commun 1991;9(1):25–36.

[61] Deavours DD, Sanders WH. "On-the-fly" solution techniques for stochastic Petri nets and extensions. IEEE Trans Software Eng 1998;24(10):889–902.

[62] Carrasco JA. Bounding steady-state availability models with group repair and phase type repair distributions. Perform Eval 1999;35(3/4):193–214.

[63] Lui JCS, Muntz R. Evaluating bounds on steady-state availability of repairable systems from Markov models. In: Stewart WJ, editor. Numerical solution of Markov chains. New York: Marcel Dekker; 1991. p.435–53.

[64] Lui JCS, Muntz RR. Computing bounds on steady state availability of repairable computer systems. J ACM 1994;41(4):676–707.

[65] Mahévas S, Rubino G. Bound computation of dependability and performance measures. IEEE Trans Comput 2001;50(5):399–413.

[66] Muntz RR, de Souza e Silva E, Goyal A. Bounding availability of repairable computer systems. IEEE Trans Comput 1989;38(12):1714–23.

[67] Semal P. Refinable bounds for large Markov chains. IEEE Trans Comput 1995;44(10):1216–22.

[68] de Souza e Silva E, Ochoa PM. State space exploration in Markov models. Perform Eval Rev 1992;20(1):152–66.

[69] Ross SM. Stochastic processes. New York: John Wiley & Sons; 1983.

[70] Suñé V, Carrasco JA. A failure-distance based method to bound the reliability of non-repairable fault-tolerant systems without the knowledge of minimal cuts. IEEE Trans Reliab 2001;50(1):60–74.

Random-request Availability

Kang W. Lee

35.1 Introduction

When allowing repair of a failed system, it is not meaningful to speak of the system reliability. The difficulty lies in the fact that system reliability does not allow consideration of system repairs. Consequently, since it should be to our advantage to repair failed systems as rapidly as possible, especially if their operation is critical to some desired objective, we need some additional measure of system performance that considers the effects of repair. For such a measure, "availability" has been proposed as a fundamental quantity of interest. It is a more appropriate measure than reliability for measuring the effectiveness of maintained systems because it includes reliability as well as maintainability. Lie *et al.* [1] surveyed and systematically classified the literature relevant to availability.

Three different widely used availability measures are: point availability, interval availability, and steady-state availability. Their definitions can be seen in various literature [2–4]. Those measures well represent a fraction of time that a system is in an operational state. In addition to these traditional availabilities, there are several other kinds of availabilities such as:

- mission availability, work mission availability, and joint availability [5];
- computation availability [6];
- equivalent availability [7].

These special measures have been proposed to suitably represent availabilities for specific application systems.

Another special availability measure, called random-request availability, has been proposed by Lee [8]. A repairable system is considered, which requires the performance of several tasks arriving randomly during the fixed mission duration. Examples include:

- a fault-tolerance mechanism which is required to handle errors resulting from transient errors (called single-event upsets) and permanent failures of system components;
- an information-measuring system which operates continuously and delivers information on user demand only at some random moments of time.

The stochastic model has already been developed [8], providing closed-form mathematical expressions for random-request availability. Since the joint availability gives the probability only at two distinctive times, the proposed measure of Lee can be considered as a useful extension of the joint availability. It is also general in the sense that the conventional interval availability can be represented as a special case.

The important characteristic of random-request availability is to introduce random task arrival into the availability measure. The task arrival rate might be constant, or a time-dependent function. Consider a packet switch network system, which has periods of operation and repair that form an alternating process. The arrival rate of incoming packets might not be constant. During the busy hour, the system can have an increased packet arrival rate. The random-request availability must have different values according to the average task arrival rate and arrival rate patterns. Therefore, the effect of "task arrival" element on the random-request availability value needs to be investigated. Due to the computational complexity, a simulation method using ARENA simulation language is used to get a solution. The simple approximation method based on interval availability is suggested. Its accuracy turns out to vary depending on the operational requirements of the system, the average task arrival rate, and the average number of on–off cycles during the mission time.

In Section 35.2, the three system elements for the stochastic model are described, and the random-request availability is defined. Sections 35.3 and 35.4 present mathematical expressions for random-request availability and numerical examples, respectively. In Section 35.5, several properties of random-request availability are derived from the simulation results. Section 35.6 presents the simple approximation method and discusses its accuracy. Section 35.7 contains concluding remarks.

35.2 System Description and Definition

The stochastic model of Lee [8] is interesting and potentially important because it provides closed-form mathematical expressions for random-request availability into which multiple system factors can be incorporated. The three elements of the system around which this model is developed are as follows.

1. Random Task Arrivals. A system is presented with a stream of tasks which arrive according to some random process. Examples include an air traffic control system which has to deal with arriving aircraft and a tank system which is confronted with an enemy which must be engaged in battle. Since the rate at which tasks arrive and the arrival rate pattern affect the overall system effectiveness, the task arrival process is included as one element of the model.

2. System State. At each task arrival time, the system can be in one of two states—on or off. If the system is on, it is operating; else the system is down (under repair). The system is in service for a random time T_{on}, until it fails. When it fails, it is then off for a random time T_{off}. It repeats the cycle of being on for a random time and being off for another random time. Successive times in the on state and in the off state are statistically independent.

3. Operational Requirements of the System. For mission success, it might not be necessary to complete all the tasks arriving randomly during the mission time, *i.e.* completion of parts of the arriving tasks might lead to mission success. For the operational requirements of the system, three systems are introduced: (a) perfect system, (b) $r(k)$-out-of-k system, and (c) scoring system.

The random-request availability is defined for each of the following three systems.

a. The Perfect System. The system needs to be in the on state at every task arrival time. The random-request availability is defined as the probability that the system is on at every task arrival time.

b. The $r(k)$-out-of-k System. The system needs to be in the on state at the times of at least some task arrivals. The minimum number, denoted by $r(k)$, depends on the number of task arrivals, k. The random-request availability is defined as the probability that the system is on at the times of at least "$r(k)$ out of k" task arrivals.

c. The Scoring System. If the system is in the on state at j ($j \le k$) out of k task arrival times, a score

$S_{j,k}$ is given to denote the probability of successful completion of the mission. The random-request availability is defined as the sum of the products from the probability of the system being on at j out of k task arrival times and the score $S_{j,k}$.

The scoring system includes the perfect and the $r(k)$-out-of-k systems as special cases. The perfect system is the case of $S_{j,k} = 1$ for $j = k$ and 0 otherwise. For the $r(k)$-out-of-k system we can set $S_{j,k} = 1$ for $j \geq r(k)$ and 0 otherwise. The conventional interval availability measure is a special case of the scoring system, where $S_{j,k} = j/k$.

35.3 Mathematical Expression for the Random-request Availability

35.3.1 Notation

$A(T)$ random-request availability for mission of length T

$A(t_1, t_2, \ldots, t_k)$ random-request availability, given task arrival times $t_1, t_2, \ldots, t_k$

M average number of on–off cycles during the mission time, $M = T/(1/\alpha + 1/\beta)$

$M(T)$ mean valuc function of a non-homogeneous Poisson process

$m(t)$ task arrival rate at time t, $m(t) = \mathrm{d}M(t)/\mathrm{d}t$

$N(T)$ number of tasks arriving during time T

$r(k)$ minimum number of tasks which must be in the on state, given k task arrivals

T mission duration time, a constant

$Z(t)$ indicator variable for the system state at time t, 0: off state, 1: on state

α rate parameter of the exponential distribution describing the on state of the system

β rate parameter of the exponential distribution describing the off state of the system.

35.3.2 Mathematical Assumptions

1. Task arrival is a non-homogeneous Poisson process with mean value function $M(t)$.
2. The random variable representing the system state (on or off) follows a time homogeneous Markov process. The sojourn time in each state is negatively exponentially distributed.
3. The mission time is of fixed duration (T).
4. A mission is defined to occur only when there are task arrivals.

35.3.3 Mathematical Expressions

The expression for random-request availability is given by Lee [8] as:

$$A(T) = \left\{ \sum_{k=1}^{\infty} \left[\iint \cdots \int_{(0 \leq t_1 < t_2 < \cdots < t_k \leq T)} A(t_1, t_2, \ldots, t_k) \times \prod_{i=1}^{k} m(t_i)\, \mathrm{d}t_1\, \mathrm{d}t_2 \cdots \mathrm{d}t_k \right] \exp[-M(T)] \right\} \times \{1 - \exp[-M(T)]\}^{-1} \quad (35.1)$$

where $A(t_1, t_2, \ldots, t_k)$ can be expressed as follows:

$$A(t_1, t_2, \ldots, t_k) = \begin{cases} \Pr\{z(t_1) = 1, z(t_2) = 1, \ldots, z(t_k) = 1\}, \\ \quad \text{for the perfect system} \\ \displaystyle\sum_{(\sum_{l=1}^{k} i_l \geq r(k), i_l = 0 \text{ or } 1)} \Pr\{z(t_1) = i_1, \\ \quad z(t_2) = i_2, \ldots, z(t_k) = i_k\}, \\ \quad \text{for the } r(k)\text{-out-of-}k \text{ system} \\ \displaystyle\sum_{j=0}^{k} S_{j,k} \left[\sum_{(\sum_{l=1}^{k} i_l = j)} \Pr\{z(t_1) = i_1, \right. \\ \quad \left. z(t_2) = i_2, \ldots, z(t_k) = i_k\} \right], \\ \quad \text{for the scoring system} \end{cases} \quad (35.2)$$

Using the Markov and time-homogeneous properties, $\Pr\{z(t_1) = i_1, z(t_2) = i_2, \ldots, z(t_k) = i_k\}$, $i_l = 0$ or 1, $l = 1, 2, \ldots, k$ can easily be derived.

Table 35.1. Mathematical expressions for $A(t_1, t_2)$

System	$A(t_1, t_2)$	Mathematical expression
Perfect	$\Pr\{z(t_1)=1, z(t_2)=1\}$	$\left(\frac{\beta}{\alpha+\beta}+\frac{\alpha}{\alpha+\beta}\exp[-(\alpha+\beta)(t_1)]\right) \times\left(\frac{\beta}{\alpha+\beta}+\frac{\alpha}{\alpha+\beta}\exp[-(\alpha+\beta)(t_2-t_1)]\right)$
$r(k)$-out-of-k	$\Pr\{z(t_1)=1, z(t_2)=1\}$ $+\Pr\{z(t_1)=1, z(t_2)=0\}$ $+\Pr\{z(t_1)=0, z(t_2)=1\}$	$\left(\frac{\beta}{\alpha+\beta}+\frac{\alpha}{\alpha+\beta}\exp[-(\alpha+\beta)(t_1)]\right) \times\left(\frac{\beta}{\alpha+\beta}+\frac{\alpha}{\alpha+\beta}\exp[-(\alpha+\beta)(t_2-t_1)]\right) +\left(\frac{\beta}{\alpha+\beta}+\frac{\alpha}{\alpha+\beta}\exp[-(\alpha+\beta)(t_1)]\right) \times\left(\frac{\alpha}{\alpha+\beta}-\frac{\alpha}{\alpha+\beta}\exp[-(\alpha+\beta)(t_2-t_1)]\right) +\left(\frac{\alpha}{\alpha+\beta}-\frac{\alpha}{\alpha+\beta}\exp[-(\alpha+\beta)(t_1)]\right) \times\left(\frac{\beta}{\alpha+\beta}-\frac{\beta}{\alpha+\beta}\exp[-(\alpha+\beta)(t_2-t_1)]\right)$
Scoring	$\frac{1}{2}[\Pr\{z(t_1)=1, z(t_2)=0\}$ $+\Pr\{z(t_1)=0, z(t_2)=1\}]$ $+\Pr\{z(t_1)=1, z(t_2)=1\}$	$\frac{1}{2}\left[\left(\frac{\beta}{\alpha+\beta}+\frac{\alpha}{\alpha+\beta}\exp[-(\alpha+\beta)(t_1)]\right) \times\left(\frac{\alpha}{\alpha+\beta}-\frac{\alpha}{\alpha+\beta}\exp[-(\alpha+\beta)(t_2-t_1)]\right) +\left(\frac{\alpha}{\alpha+\beta}-\frac{\alpha}{\alpha+\beta}\exp[-(\alpha+\beta)(t_1)]\right) \times\left(\frac{\beta}{\alpha+\beta}-\frac{\beta}{\alpha+\beta}\exp[-(\alpha+\beta)(t_2-t_1)]\right)\right] +\left(\frac{\beta}{\alpha+\beta}+\frac{\alpha}{\alpha+\beta}\exp[-(\alpha+\beta)(t_1)]\right) \times\left(\frac{\beta}{\alpha+\beta}+\frac{\alpha}{\alpha+\beta}\exp[-(\alpha+\beta)(t_2-t_1)]\right)$

Table 35.2. System parameters

• Task arrival (three types, see Figure 35.1)	$m_1(t)=0.016t$ $m_2(t)=0.08$ $m_3(t)=0.16-0.016t$
• System state	$\alpha=1$ $\beta=5$
• Operational requirements of the system	$r(k)=\begin{cases}[k/2], & k \text{ even}\\ [k/2]+1, & k \text{ odd}\end{cases}$ $S_{j,k}=j/k$
• Mission duration time	$T=10$

For instance, given $k = 2, r(2) = 1$ and $S_{j,2} = j/2$, $A(t_1, t_2)$ can be expressed as shown in Table 35.1.

35.4 Numerical Examples

Assume the system parameters are given as in Table 35.2.

Two examples are given. First one assumes the system is in the on state at time 0. In the second example, the system state is assumed to be in the off state at time 0. Tables 35.3 and 35.4 summarize the results. All the integrations are done using "Mathematica" software.

35.5 Simulation Results

In this section, we want to show some properties of the random-request availability. The important characteristic of the random-request availability is that the random task arrival is included in the model as one of the system elements. Therefore, the effect of the "task arrival" element on the random-request availability needs to be investigated. If the mean number of task arrivals is large, the computational complexity is high. Even if the mission has an average of two task arrivals under the Poisson arrival assumption, the seventh order of multiple integral is required to get a solution with error bound less than 10^{-2}. Therefore, a simulation method using ARENA simulation language is used to get a solution. The simulation results for the first example of Section 35.4 are compared with the analytical ones in Table 35.5. The simulation results are the average of 10^7 runs after some warm-up periods. We can see that the differences between two results are less than 10^{-3}.

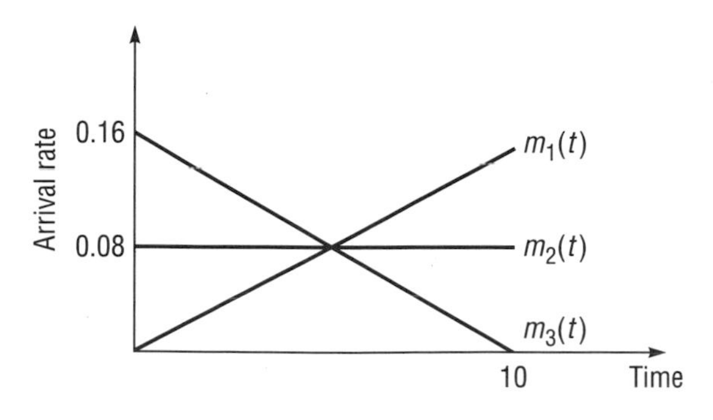

Figure 35.1. Task arrival rate

Varying the average task arrivals rate (1, 5, and 10), type of task arrival rate (constant, increasing, and decreasing), and the length of mean sojourn times in the on and off states ([0.1, 0.02], [1, 0.2], and [50/6, 10/6]), the random-request availability is obtained for each of the following three systems: perfect, $r(k)$-out-of-k, and scoring. Other system parameters are the same as those of Table 35.2. Table 35.6 summarizes the results, which show the following facts.

1. It can be seen that the random-request availability has different values depending on the average task arrival rate. We can also see that the effect of task arrival rate is different according to the operational requirements of the system. For the perfect system, a higher average task arrival rate gives a lower value of availability, and *vice versa*. As the average arrival rate increases, the values for the $r(k)$-out-of-k and the scoring systems can be increasing, decreasing, or even constant depending on the respective $r(k)$ value and score $S_{j,k}$ for a given k.

2. The random-request availability has different values according to the task arrival rate patterns, even if missions have the same average task arrivals during the mission time. For the system being on at time 0, $m_3(t)$ and $m_1(t)$ provide the highest and lowest random-request availabilities, respectively. If the system is off at time 0, $m_1(t)$ produces a higher random-request availability value than $m_2(t)$ and $m_3(t)$. These properties come directly from the facts that $m_3(t)$ gives a higher task arrival rate at the beginning part of the mission than at the latter part and $m_1(t)$ gives a higher task arrival rate at the latter part of the mission.

3. The above property 2 can be seen more clearly when the average number of on–off cycles (M) during the mission time has a small value, relative long sojourn times in the on and/or off states compared to the mission time. That is, for systems having small M the random-request availabilities have significantly different values according to the task arrival rate patterns, even

Table 35.3. Example 1: results

System	Task arrival rate		
	$m_1(t) = 0.016t$	$m_2(t) = 0.08$	$m_3(t) = 0.16 - 0.016t$
Perfect	0.776	0.779	0.782
$r(k)$-out-of-k	0.876	0.879	0.880
Scoring	0.833	0.836	0.838

Table 35.4. Example 2: results

System	Task arrival rate		
	$m_1(t) = 0.016t$	$m_2(t) = 0.08$	$m_3(t) = 0.16 - 0.016t$
Perfect	0.775	0.758	0.742
$r(k)$-out-of-k	0.875	0.865	0.857
Scoring	0.833	0.819	0.806

Table 35.5. Comparison of the simulation and the analytical results

System	Task arrival rate					
	$m_1(t) = 0.016t$		$m_2(t) = 0.08$		$m_3(t) = 0.16 - 0.016t$	
	Analytic	Simulation	Analytic	Simulation	Analytic	Simulation
Perfect	0.776	0.777	0.779	0.779	0.782	0.782
$r(k)$-out-of-k	0.876	0.876	0.879	0.879	0.880	0.880
Scoring	0.833	0.833	0.836	0.837	0.838	0.838

though there are the same average number of task arrivals during the mission time. All the while, the random-request availabilities for systems having large M have almost the same values irrespective of the arrival rate patterns if missions have the same average task arrivals during the mission time.

4. The simulation results show that the random-request availability has the lowest value under the perfect system and the highest value under the $r(k)$-out-of-k systems. From Table 35.6, we can also see that the effect of arrival rate pattern on the random-request availability value is more significant in the perfect system than in the $r(k)$-out-of-k and scoring systems.

5. As mentioned before, the random-request availability of the scoring system with $S_{j,k} = j/k$ is just a conventional interval availability. For example, in Table 35.6, the random-request availabilities of the scoring system with $\alpha = 6/50$ and $\beta = 6/10$ are 0.841, 0.857, and 0.874, respectively according to the task arrival rate patterns, $m_1(t)$, $m_2(t)$, and $m_3(t)$. These values are just conventional interval availabilities, which can be calculated as:

$$\int_0^{10} \Pr[z(t) = 1] \frac{m(t)}{M(T)}\, \mathrm{d}t = \begin{cases} \dfrac{1}{10} \displaystyle\int_0^{10} \left(\frac{5}{6} + \frac{1}{6} \exp\left(-\frac{36}{50} t \right) \right) \mathrm{d}t \\ \quad = 0.841, \quad \text{for } m(t) = m_1(t) \\ \dfrac{1}{50} \displaystyle\int_0^{10} \left(\frac{5}{6} + \frac{1}{6} \exp\left(-\frac{36}{50} t \right) \right) t\, \mathrm{d}t \\ \quad = 0.857, \quad \text{for } m(t) = m_2(t) \\ \dfrac{1}{50} \displaystyle\int_0^{10} \left(\frac{5}{6} + \frac{1}{6} \exp\left(-\frac{36}{50} t \right) \right) \\ \quad \times (10 - t)\, \mathrm{d}t \\ \quad = 0.874, \quad \text{for } m(t) = m_3(t) \end{cases}$$

Table 35.6. Simulation results for the random-request availability

		Average task arrival rate = 1		
		$m_1(t) = 0.02t$	$m_2(t) = 0.1$	$m_3(t) = 0.2 - 0.02t$
*				
$\alpha = 10$	Perfect	0.758	0.759	0.758
$\beta = 50$	$r(k)$-out-of-k	0.887	0.887	0.887
$(M = 83.33)$	Scoring	0.834	0.834	0.834
*				
$\alpha = 1$	Perfect	0.761	0.764	0.770
$\beta = 5$	$r(k)$-out-of-k	0.833	0.886	0.889
$(M = 8.33)$	Scoring	0.833	0.836	0.840
*				
$\alpha = 6/50$	Perfect	0.792	0.808	0.831
$\beta = 6/10$	$r(k)$-out-of-k	0.873	0.891	0.903
$(M = 1)$	Scoring	0.841	0.857	0.874
**				
$\alpha = 6/50$	Perfect	0.739	0.632	0.546
$\beta = 6/10$	$r(k)$-out-of-k	0.840	0.772	0.692
$(M = 1)$	Scoring	0.799	0.716	0.634

		Average task arrival rate = 5		
		$m_1(t) = 0.1t$	$m_2(t) = 0.5$	$m_3(t) = 1 - 0.1t$
*				
$\alpha = 10$	Perfect	0.434	0.436	0.435
$\beta = 50$	$r(k)$-out-of-k	0.967	0.968	0.968
$(M = 83.33)$	Scoring	0.833	0.834	0.834
*				
$\alpha = 1$	Perfect	0.463	0.467	0.474
$\beta = 5$	$r(k)$-out-of-k	0.958	0.961	0.961
$(M = 8.33)$	Scoring	0.834	0.836	0.839
*				
$\alpha = 6/50$	Perfect	0.619	0.623	0.671
$\beta = 6/10$	$r(k)$-out-of-k	0.904	0.929	0.936
$(M = 1)$	Scoring	0.841	0.857	0.874
**				
$\alpha = 6/50$	Perfect	0.529	0.353	0.27
$\beta = 6/10$	$r(k)$-out-of-k	0.529	0.822	0.723
$(M = 1)$	Scoring	0.800	0.718	0.633

*The system is on at time 0. **The system is off at time 0.

Table 35.6. *Continued.*

		Average task arrival rate = 10		
		$m_1(t) = 0.2t$	$m_2(t) = 1$	$m_3(t) = 2 - 0.2t$
*				
$\alpha = 10$	Perfect	0.195	0.195	0.196
$\beta = 50$	$r(k)$-out-of-k	0.991	0.992	0.992
$(M = 83.33)$	Scoring	0.833	0.833	0.834
*				
$\alpha = 1$	Perfect	0.245	0.247	0.255
$\beta = 5$	$r(k)$-out-of-k	0.982	0.984	0.984
$(M = 8.33)$	Scoring	0.834	0.836	0.839
*				
$\alpha = 6/50$	Perfect	0.512	0.537	0.564
$\beta = 6/10$	$r(k)$-out-of-k	0.910	0.938	0.943
$(M = 1)$	Scoring	0.841	0.857	0.874
**				
$\alpha = 6/50$	Perfect	0.394	0.198	0.138
$\beta = 6/10$	$r(k)$-out-of-k	0.884	0.837	0.741
$(M = 1)$	Scoring	0.800	0.717	0.633

*The system is on at time 0. **The system is off at time 0.

Table 35.7. Comparison of the approximation and the simulation results

System	Task arrival rate*	Mean sojourn times in the on and off states					
		$\alpha = 10, \beta = 50$ $(M = 83.33)$		$\alpha = 1, \beta = 5$ $(M = 8.33)$		$\alpha = 6/50, \beta = 6/10$ $(M = 1)$	
		Approx.	Simulation	Approx.	Simulation	Approx.	Simulation
Perfect	$m_2(t) = 0.1$ (1)	0.758	0.759	0.761	0.764	0.788	0.808
	$m_2(t) = 0.5$ (5)	0.431	0.436	0.437	0.461	0.484	0.623
	$m_2(t) = 1$ (10)	0.189	0.195	0.194	0.237	0.238	0.507
$r(k)$-out-of-k	$m_1(t) = 0.02t$ (1)	0.887	0.887	0.887	0.884	0.892	0.873
	$m_1(t) = 0.1t$ (5)	0.969	0.967	0.969	0.958	0.971	0.904
	$m_1(t) = 0.2t$ (10)	0.993	0.992	0.993	0.982	0.994	0.910
Scoring	$m_3(t) = 0.2 - 0.02t$ (1)	0.802	0.802	0.807	0.810	0.847	0.855
$(S_{j,k} = (j/k)^2)$	$m_3(t) = 1 - 0.1t$ (5)	0.731	0.732	0.738	0.743	0.791	0.814
	$m_3(t) = 2 - 0.2t$ (10)	0.711	0.712	0.719	0.724	0.775	0.802

*The number in parenthesis denotes the average.

Table 35.8. Maximum average task arrivals satisfying the accuracy requirement

System	Mean sojourn times in the on and off states	Maximum average task arrivals satisfying the accuracy requirement (|simulation result − approximation result| $\le 10^{-2}$)
Perfect	$\alpha = 10, \beta = 50\,(M = 83.33)$	∞
	$\alpha = 1, \beta = 5\,(M = 8.33)$	2
	$\alpha = 6/50, \beta = 6/10\,(M = 1)$	0.5
$r(k)$-out-of-k	$\alpha = 10, \beta = 50\,(M = 83.33)$	∞
	$\alpha = 1, \beta = 5\,(M = 8.33)$	10
	$\alpha = 6/50, \beta = 6/10\,(M = 1)$	1
Scoring ($S_{j,k} = (j/k)^2$)	$\alpha = 10, \beta = 50\,(M = 83.33)$	∞
	$\alpha = 1, \beta = 5\,(M = 8.33)$	∞
	$\alpha = 6/50, \beta = 6/10\,(M = 1)$	2

From Table 35.6, we can see that these values vary depending on the task arrival rate pattern, but not the average task arrival rate.

35.6 Approximation

It seems to be difficult to get robust approximation methods. Even if Finkelstein provided a heuristic approximation formula [9], it is good only for the perfect system. And its accuracy is guaranteed only within a limited range of system parameters. One simple intuitive method is to use the conventional interval availability. Taking the task arrival rate into consideration, it can be expressed as:

$$A(T) = \int_0^T \Pr[z(t) = 1]\frac{m(t)}{M(T)}\,\mathrm{d}t \qquad (35.3)$$

The approximate values for the perfect, $r(k)$-out-of-k, and scoring systems can be respectively expressed as follows:

$$\tilde{A}_{\mathrm{p}}(T) = \frac{\sum_{k=1}^{\infty}\{A(T)\}^k \Pr[N(T) = k]}{1 - \Pr[N(T) = 0]} \qquad (35.4)$$

$$\tilde{A}_{\mathrm{r}}(T) = \left(\sum_{k=1}^{\infty}\left\{\sum_{j=r(k)}^{k}\binom{k}{j}\{A(T)\}^j \times \{1 - A(T)\}^{k-j}\right\}\Pr[N(T) = k]\right) \times (1 - \Pr[N(T) = 0])^{-1} \qquad (35.5)$$

$$\tilde{A}_{\mathrm{s}}(T) = \left(\sum_{k=1}^{\infty}\left\{\sum_{j=0}^{k}\binom{k}{j}\{A(T)\}^j\{1-A(T)\}^{k-j} \times Sj,k\right\}\Pr[N(T) = k]\right) \times (1 - \Pr[N(T) = 0])^{-1} \qquad (35.6)$$

Using Equations 35.4–35.6, the approximation values for random-request availabilities are obtained under various system parameter values. The approximate results are compared with the simulation ones in Table 35.7. Unless specified otherwise, the system parameters are the same as those of Table 35.2.

For the perfect system the approximation inaccuracy grows quickly as M becomes small and the task arrival rate increases. The approximation results for the $r(k)$-out-of-k and the scoring systems are more accurate than those for the perfect system. They show better accuracy for larger M and lower task arrival rate. Given the value of M, we try to find the maximum average number of task arrivals which satisfies the accuracy requirement, |simulation result − approximation result| $\le 10^{-2}$. Table 35.8 shows the results. The task arrival rates for all the systems are assumed to be a decreasing function. For example, the task arrival rate rate corresponding to average one task arrival is $m_3(t) = 0.2 - 0.02t$.

From Table 35.8, we can see that the $r(k)$-out-of-k and the scoring systems can satisfy the approximation accuracy requirement at a higher average task arrival rate than the perfect system.

For the $r(k)$-out-of-k and the scoring systems, the approximation accuracy also depends on the respective $r(k)$ value and score $S_{j,k}$ for a given k. We can see low accuracy in the perfect system, which is the most strict case of the $r(k)$-out-of-k system ($r(k) = k$ for all k) or the scoring system ($S_{j,k} = 1$ for $j = k$, and 0 otherwise). Therefore, the more strict the operational requirement, the lower the approximation accuracy for the $r(k)$-out-of-k and the scoring system, and *vice versa*.

35.7 Concluding Remarks

For the system required to perform randomly arriving tasks during a fixed mission duration, an availability measure called random-request availability has been proposed. The stochastic model provides closed-form mathematical expressions, which incorporate three basic elements: the random task arrivals, the system state, and the operational requirements of the system.

The characteristic of the random-request availability is that the random task arrival is included as one of the system elements. Using a simulation method, the effect of the "task arrival" element on the random-request availability has been investigated.

If the mean number of task arrivals grows, the computational complexity for deriving the random-request availability becomes extremely high. A simple approximation method based on the conventional interval availability is suggested. Its accuracy varies according to the operational requirements of the system, the average task arrival rate, and the average number of on–off cycles (M) during the mission time.

Acknowledgment

This work was supported by the research fund of Seoul National University of Technology.

References

[1] Lie CH, Hwang CL, Tillman FA. Availability of maintained system: a state-of-the- art survey. AIIE Trans 1977;3:247–59.

[2] Lewis EE. Introduction to reliability engineering. New York: John Wiley & Sons; 1987.

[3] Kapur KC, Lamberson LR. Reliability in engineering design. New York: John Wiley & Sons; 1987.

[4] Rau JG. Optimization and probability in systems engineering. Amsterdam: Van Nostrand Reinhold; 1970.

[5] Birolini A. Quality and reliability of technical systems. Berlin: Springer; 1988.

[6] Reibman AL. Modeling the effect of reliability on performance. IEEE Trans Reliab 1990;3:314–20.

[7] Lazaroiu DF, Staicut E. Congestion–reliability–availability relationship in packet-switching computer networks. IEEE Trans Reliab 1983;4:354–7.

[8] Lee KW. Stochastic models for random-request availability. IEEE Trans Reliab 2000;1:80–4.

[9] Finkelstein MS. Multiple availability on stochastic demand. IEEE Trans Reliab 1999;1:19–24.

Index

D

E

F

G

H

I

J

K

L

M

N

O

P

Q

R

S

T

U

V

W

X

Z